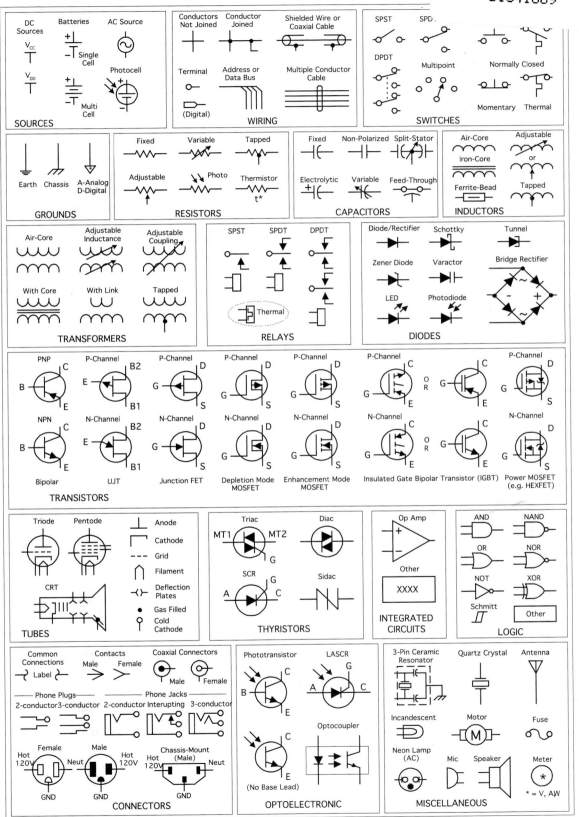

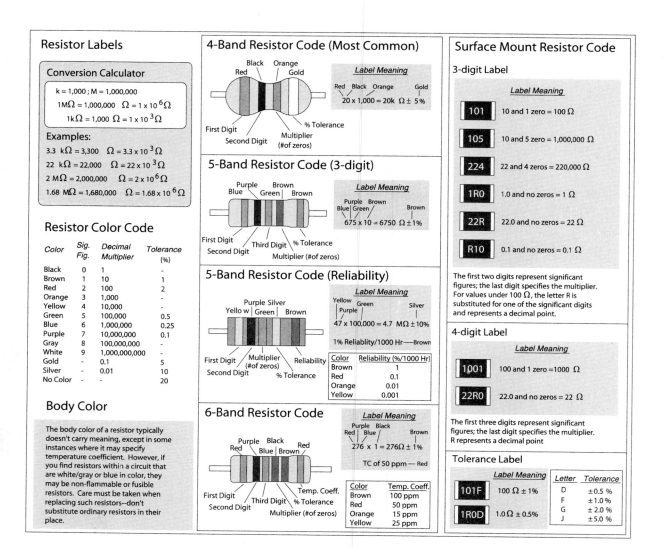

Resistor Labels

Conversion Calculator

k = 1,000 ; M = 1,000,000

$1M\Omega = 1,000,000\ \Omega = 1 \times 10^6 \Omega$

$1k\Omega = 1,000\ \Omega = 1 \times 10^3 \Omega$

Examples:

$3.3\ k\Omega = 3,300\ \Omega = 3.3 \times 10^3 \Omega$

$22\ k\Omega = 22,000\ \Omega = 22 \times 10^3 \Omega$

$2\ M\Omega = 2,000,000\ \Omega = 2 \times 10^6 \Omega$

$1.68\ M\Omega = 1,680,000\ \Omega = 1.68 \times 10^6 \Omega$

Resistor Color Code

Color	Sig. Fig.	Decimal Multiplier	Tolerance (%)
Black	0	1	-
Brown	1	10	1
Red	2	100	2
Orange	3	1,000	-
Yellow	4	10,000	-
Green	5	100,000	0.5
Blue	6	1,000,000	0.25
Purple	7	10,000,000	0.1
Gray	8	100,000,000	-
White	9	1,000,000,000	-
Gold	-	0.1	5
Silver	-	0.01	10
No Color	-	-	20

Body Color

The body color of a resistor typically doesn't carry meaning, except in some instances where it may specify temperature coefficient. However, if you find resistors within a circuit that are white/gray or blue in color, they may be non-flammable or fusible resistors. Care must be taken when replacing such resistors--don't substitute ordinary resistors in their place.

4-Band Resistor Code (Most Common)

Label Meaning: 20 x 1,000 = 20k Ω ± 5%

5-Band Resistor Code (3-digit)

Label Meaning: 675 x 10 = 6750 Ω ±1%

5-Band Resistor Code (Reliability)

Label Meaning: 47 x 100,000 = 4.7 MΩ ±10%

1% Reliablity/1000 Hr — Brown

Color	Reliability (%/1000 Hr)
Brown	1
Red	0.1
Orange	0.01
Yellow	0.001

6-Band Resistor Code

Label Meaning: 276 x 1 = 276Ω ± 1%

TC of 50 ppm — Red

Color	Temp. Coeff.
Brown	100 ppm
Red	50 ppm
Orange	15 ppm
Yellow	25 ppm

Surface Mount Resistor Code

3-digit Label

Label	Meaning
101	10 and 1 zero = 100 Ω
105	10 and 5 zero = 1,000,000 Ω
224	22 and 4 zeros = 220,000 Ω
1R0	1.0 and no zeros = 1 Ω
22R	22.0 and no zeros = 22 Ω
R10	0.1 and no zeros = 0.1 Ω

The first two digits represent significant figures; the last digit specifies the multiplier. For values under 100 Ω, the letter R is substituted for one of the significant digits and represents a decimal point.

4-digit Label

Label	Meaning
1001	100 and 1 zero = 1000 Ω
22R0	22.0 and no zeros = 22 Ω

The first three digits represent significant figures; the last digit specifies the multiplier. R represents a decimal point

Tolerance Label

Label	Meaning
101F	100 Ω ± 1%
1R0D	1.0 Ω ± 0.5%

Letter	Tolerance
D	±0.5 %
F	±1.0 %
G	± 2.0 %
J	±5.0 %

Standard Resistor Values (1%, 5% and 10% Tolerance)

1%								5%		10%
1.00	1.02	1.05	1.07	1.10	1.13	1.15	1.18	10	11	10
1.21	1.24	1.27	1.30	1.33	1.37	1.40	1.43	12	13	12
1.47	1.50	1.54	1.58	1.62	1.65	1.69	1.74	15	16	15
1.78	1.82	1.87	1.91	1.96	2.00	2.05	2.10	18	20	18
2.15	2.21	2.26	2.32	2.37	2.43	2.49	2.55	22	24	22
2.61	2.67	2.74	2.80	2.87	2.94	3.01	3.09	27	30	27
3.16	3.24	3.32	3.40	3.48	3.57	3.65	3.74	33	36	33
3.83	3.92	4.02	4.12	4.22	4.32	4.42	4.53	39	43	39
4.64	4.75	4.87	4.99	5.11	5.23	5.36	5.49	47	51	47
5.62	5.76	5.90	6.04	6.19	6.34	6.49	6.65	56	62	56
6.81	6.98	7.15	7.32	7.50	7.68	7.87	8.06	68	75	68
8.25	8.45	8.66	8.87	9.09	9.31	9.53	9.76	82	91	82

Standard resistance value is obtained from the above chart by multiply by powers of 10.

5% example resistors: 51Ω, 510Ω, 5.1kΩ, 51kΩ, 510kΩ, 5.1MΩ.

1% example resistors: 1.21Ω, 12.1Ω, 121Ω, 1.21kΩ, 12.1kΩ, 121kΩ, 1.21MΩ

Capacitor Markings

Capacitance Conversion Calculator

$1 F = 1 \times 10^6 \,\mu F = 1 \times 10^9 \, nF = 1 \times 10^{12} \, pF$
$1 \,\mu F = 1 \times 10^{-6} \, F = 1 \times 10^3 \, nF = 1 \times 10^6 \, pF$
$1 \, nF = 1 \times 10^{-9} \, F = 1 \times 10^{-3} \,\mu F = 1 \times 10^3 \, pF$
$1 \, pF = 1 \times 10^{-12} \, F = 1 \times 10^{-6} \,\mu F = 1 \times 10^{-3} \, nF$
$F = Farad, \ \mu = micro, \ n = nano, \ p = pico$

$1000 \,\mu F = 1{,}000{,}000 \, nF = 10 \times 10^8 \, pF$
$100 \,\mu F = 100{,}000 \, nF = 10 \times 10^7 \, pF$
$10 \,\mu F = 10{,}000 \, nF = 10 \times 10^6 \, pF$
$1 \,\mu F = 1{,}000 \, nF = 10 \times 10^5 \, pF$
$0.1 \,\mu F = 100 \, nF = 10 \times 10^4 \, pF$
$0.01 \,\mu F = 10 \, nF = 10 \times 10^3 \, pF$
$0.001 \,\mu F = 1 \, nF = 10 \times 10^2 \, pF$

Tantalum

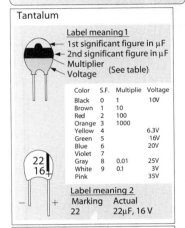

Label meaning 1
- 1st significant figure in μF
- 2nd significant figure in μF
- Multiplier
- Voltage (See table)

Color	S.F.	Multiplie	Voltage
Black	0	1	10V
Brown	1	10	
Red	2	100	
Orange	3	1000	
Yellow	4		6.3V
Green	5		16V
Blue	6		20V
Violet	7		
Gray	8	0.01	25V
White	9	0.1	3V
Pink			35V

Label meaning 2

Marking	Actual
22	22μF, 16 V

Mylar (Polyester Film) Polypropylene Dipped Mica

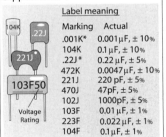

Label meaning

Marking	Actual
.001K*	0.001μF, ± 10%
104K	0.1μF, ± 10%
.22J*	0.22μF, ± 5%
472K	0.0047μF, ± 10%
221J	220 pF, ± 5%
470J	47pF, ± 5%
102J	1000pF, ± 5%
103F	0.01μF, ± 1%
223F	0.022μF, ± 1%
104F	0.1μF, ± 1%

Labels:
1st digit, 2nd digit, multiplier in pF (or μF if decimal before digits), and tolerance.

Metallized Polyester Film

Label meaning

Marking	Actual
2μ2	2.2μF
μ22	0.22μF
68n	68 nF
4n7	4.7 nF

Label:
"μ" place of decimal in microfarads
"n" place of decimal in nanofarads

Polyester Color Coded

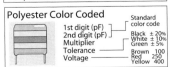

- 1st digit (pF)
- 2nd digit (pF)
- Multiplier
- Tolerance
- Voltage

Standard color code

Black	± 20%
White	± 10%
Green	± 5%
Brown	100
Red	250
Yellow	400

Ceramic Disc Capacitors

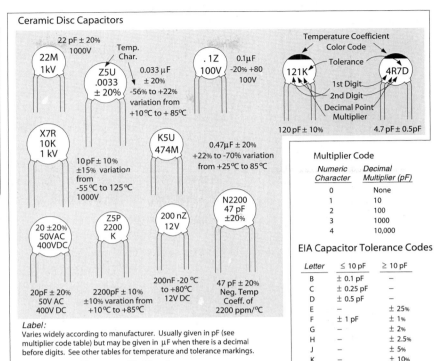

22 pF ± 20% 1000V — 22M 1kV

Temp. Char. — Z5U .0033 ± 20% — 0.033 μF ± 20% -56% to +22% variation from +10°C to + 85°C

.1Z 100V — 0.1μF 100V -20% +80

Temperature Coefficient Color Code — Tolerance — 121K — 1st Digit — 2nd Digit — Decimal Point — Multiplier — 120 pF ± 10%

4R7D — 4.7 pF ± 0.5pF

X7R 10K 1 kV — 10 pF ± 10% ±15% variation from -55°C to 125°C 1000V

K5U 474M — 0.47μF ± 20% +22% to -70% variation from +25°C to 85°C

20 ±20% 50VAC 400VDC — 20pF ± 20% 50V AC 400V DC

Z5P 2200 K — 2200pF ± 10% ±10% variation from +10°C to +85°C

200 nZ 12V — 200nF -20 °C to +80°C 12V DC

N2200 47 pF ±20% — 47 pF ± 20% Neg. Temp Coeff. of 2200 ppm/°C

Label:
Varies widely according to manufacturer. Usually given in pF (see multiplier code table) but may be given in μF when there is a decimal before digits. See other tables for temperature and tolerance markings.

Ceramic Disc (European Markings)

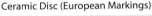

47p

Label Meaning

Marking	Actual	Marking	Actual
p68	0.68 pF	22p	22 pF
1p0	1.0 pF	n10	0.1 nF
4p7	4.7 pF	n27	0.27 nF

Label: p = picofarads, n = nanofarads; location of p or n signifies decimal point.

Fixed Ceramic Color Code

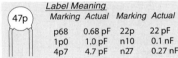

- 1st Digit
- 2nd Digit
- Multiplier
- Temp. Coeff.
- Tolerance

Color	S.F.	>10pF	<10pF	Temp. Coeff. ppm/°C
Black	0	1	± 20%	2.0 pF	-0
Brown	1	10	± 1%		-30
Red	2	100	± 2%		-80
Orange	3	1000			-150
Yellow	4				-220
Green	5		± 5%	0.5pF	-330
Blue	6				-470
Violet	7				-750
Gray	8	0.01		0.25pF	30
White	9	0.1	± 10%	1.0pF	500

Surface Mount Capacitors

SMD Ceramic

A1

Label meaning

Marking	Actual
N1	33 pF
A4	0.01 μF
S6	4.7 μF

SMD Electrolytic

10 6V — **Label meaning 1**

Marking	Actual
10 6V	10 μF, 6V

A475 — **Label meaning 2**

A475	4.7 μF, 10V

Significant Figure Code

Char.	S. F.	Char.	S. F.
A	1.0	T	5.1
B	1.1	U	5.6
C	1.2	V	6.2
D	1.3	W	6.8
E	1.5	X	7.5
F	1.6	Y	8.2
G	1.8	Z	9.1
H	2.0	a	2.5
J	2.2	b	3.5
K	2.4	d	4.0
L	2.7	e	4.5
M	3.0	f	5.0
N	3.3	m	6.0
P	3.6	n	7.0
Q	3.9	t	8.0
R	4.3	y	9.0
S	4.7		

Multiplier Code

Numeric Character	Decimal Multiplier (pF)
0	1
1	10
2	100
3	1,000
4	10,000
5	100,000
6	1,000,000
7	10,000,000
8	100,000,000
9	0.1

Label 2:
Voltage (see table below), 1st digit, 2nd digit, multiplier (pF).

Char.	Voltage
e	2.5
G	4
J	6.3
A	10
C	16
D	20
E	25
V	35
H	50

Multiplier Code

Numeric Character	Decimal Multiplier (pF)
0	None
1	10
2	100
3	1000
4	10,000

EIA Capacitor Tolerance Codes

Letter	≤ 10 pF	≥ 10 pF
B	± 0.1 pF	–
C	± 0.25 pF	–
D	± 0.5 pF	–
E	–	± 25%
F	± 1 pF	± 1%
G	–	± 2%
H	–	± 2.5%
J	–	± 5%
K	–	± 10%
M	–	± 20%
P	–	-0 + 100%
S	–	-20 + 50%
W	–	-0 + 200%
X	–	-20 + 40%
Z	–	-20 + 80%

EIA Temperature Characteristic Codes

Minimum temperature	Maximum temperature	Max cap. change over temp. range
X -55°C	2 +45°C	A ± 1.0%
Y -35°C	4 +65°C	B ± 1.5%
Z +10°C	5 +85°C	C ± 2.2%
	6 +105°C	D ± 3.3%
	7 +125°C	E ± 4.7%
		F ± 7.5%
		P ± 10%
		R ± 15%
		S ± 22%
		T -33%, + 22%
		U -56%, + 22%
		V -82%, + 22%

EIA Temperature Coefficient Color Codes

Color	Temp. Coeff. Industry	EIA
Black	NP0	C0G
Brown	N030/N033	S1G
Red	N075/N080	U1G
Orange	N 150	P2G
Yellow	N 220	R2G
Green	N 330	S2H
Blue	N 470	U2J
Violet	N 750	
Gray		
White	P 100	
Red/Violet	P 100	

Electrolytic Capacitors

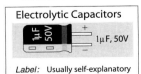

1μF 50V — 1μF, 50V

Label: Usually self-explanatory

Practical Electronics
for Inventors

ABOUT THE AUTHORS

Paul Scherz is a Systems Operation Manager who received his B.S. in physics from the University of Wisconsin. He is an inventor/hobbyist in electronics, an area he grew to appreciate through his experience at the University's Department of Nuclear Engineering and Engineering Physics and Department of Plasma Physics.

Dr. Simon Monk has a bachelor's degree in cybernetics and computer science and a Ph.D. in software engineering. He spent several years as an academic before he returned to industry, co-founding the mobile software company Momote Ltd. He has been an active electronics hobbyist since his early teens and is a full-time writer on hobby electronics and open-source hardware. Dr. Monk is author of numerous electronics books, including *Programming Arduino, Hacking Electronics,* and *Programming the Raspberry Pi.*

ABOUT THE TECHNICAL EDITORS

Michael Margolis has more than 40 years of experience developing and delivering hardware and software solutions. He has worked at senior levels with Sony, Lucent/Bell Labs, and a number of start-up companies. Michael is the author of two books, *Arduino Cookbook* and *Make an Arduino-Controlled Robot: Autonomous and Remote-Controlled Bots on Wheels.*

Chris Fitzer is a solutions architect and technical manager who received his Ph.D. in electrical and electronic engineering from the University of Manchester Institute of Science and Technology (UMIST) in 2003 and a first-class honors degree (B.Sc.) in 1999. He currently leads a global team developing and deploying Smart Grid technologies around the world. Previous positions have seen Chris drive the European interests of the ZigBee Smart Energy (ZSE) profile and lead the development of the world's first certified Smart Energy In Premise Display (IPD) and prototype smart meter. He has also authored or co-authored numerous technical journal papers within the field of Smarter Grids.

Practical Electronics for Inventors

Fourth Edition

Paul Scherz

Simon Monk

New York   Chicago   San Francisco
Athens   London   Madrid   Mexico City   Milan
New Delhi   Singapore   Sydney   Toronto

Library of Congress Control Number: 2016932853

McGraw-Hill Education books are available at special quantity discounts to use as premiums and sales promotions or for use in corporate training programs. To contact a representative, please visit the Contact Us page at www.mhprofessional.com.

Practical Electronics for Inventors, Fourth Edition

13 14 15 LOV 23 22 21

ISBN 978-1-25-958754-2
MHID   1-25-958754-1

This book is printed on acid-free paper.

Sponsoring Editor
Michael McCabe

Editorial Supervisor
Stephen M. Smith

Production Supervisor
Pamela A. Pelton

Acquisitions Coordinator
Lauren Rogers

Technical Editors
Michael Margolis and Chris Fitzer

Project Manager
Apoorva Goel,
Cenveo® Publisher Services

Copy Editor
Raghu Narayan Das,
Cenveo Publisher Services

Proofreader
Cenveo Publisher Services

Indexer
Cenveo Publisher Services

Art Director, Cover
Jeff Weeks

Composition
Cenveo Publisher Services

Illustration
Cenveo Publisher Services

CONTENTS

PREFACE

Inventors in the field of electronics are individuals who possess the knowledge, intuition, creativity, and technical know-how to turn their ideas into real-life electrical gadgets. We hope that this book will provide you with an intuitive understanding of the theoretical and practical aspects of electronics in a way that fuels your creativity.

This book is designed to help beginning inventors invent. It assumes little to no prior knowledge of electronics. Therefore, educators, students, and aspiring hobbyists will find this book a good initial text. At the same time, technicians and more advanced hobbyists may find this book a useful resource.

Notes about the Fourth Edition

The main addition to the fourth edition is a new chapter on programmable logic. This chapter focuses on the use of FPGAs (field-programmable gate arrays) and shows you how to program an FPGA evaluation board using both a schematic editor and the Verilog hardware definition language.

The book has also undergone numerous minor updates and fixes to errors discovered in the third edition. In addition, there has been some pruning of outdated material that is no longer relevant to modern electronics.

ACKNOWLEDGMENTS

We would like to thank the many people who have helped in the production of this book. Special thanks are due to the technical reviewers Michael Margolis, Chris Fitzer, and David Buckley.

We have been able to greatly improve the accuracy of the book thanks to the very detailed and helpful errata for the second edition that were collated by Martin Ligare at Bucknell University. Contributors to these errata were Steve Baker (Naval Postgraduate School), George Caplan (Wellesley College), Robert Drehmel, Earl Morris, Robert Strzelczyk (Motorola), Lloyd Lowe (Boise State University), John Kelty (University of Nebraska), Perry Spring (Cascadia Community College), Michael B. Allen, Jeffrey Audia, Ken Ballinger (EIT), Clement Jacob, Jamie Masters, and Marco Ariano. Thank you all for taking the time to make this a better book.

Many thanks to Michael McCabe, the ever-patient Apoorva Goel, and everyone from McGraw-Hill Education, for their support and skill in converting this manuscript into a great book.

—*Paul Scherz and Simon Monk*

Practical Electronics for Inventors

## CHAPTER 1

Introduction to Electronics

Perhaps the most common predicament newcomers face when learning electronics is figuring out exactly what it is they must learn. What topics are worth covering, and in which general order should they be covered? A good starting point for answering these questions is the flowchart presented in Fig. 1.1. This chart provides an overview of the basic elements that go into designing practical electrical gadgets and represents the information you will find in this book. This chapter introduces these basic elements.

At the top of the chart comes the theory. This involves learning about voltage, current, resistance, capacitance, inductance, and various laws and theorems that help predict the size and direction of voltages and currents within circuits. As you learn the basic theory, you will be introduced to basic passive components such as resistors, capacitors, inductors, and transformers.

Next down the line are discrete passive circuits. Discrete passive circuits include current-limiting networks, voltage dividers, filter circuits, attenuators, and so on. These simple circuits, by themselves, are not very interesting, but they are vital ingredients in more complex circuits.

After you have learned about passive components and circuits, you move on to discrete active devices, which are built from semiconductor materials. These devices consist mainly of diodes (one-way current-flow gates) and transistors (electrically controlled switches/amplifiers).

Once you have covered the discrete active devices, you get to discrete active/passive circuits. Some of these circuits include rectifiers (ac-to-dc converters), amplifiers, oscillators, modulators, mixers, and voltage regulators. This is where things start getting interesting.

Throughout your study of electronics, you will learn about various input/output (I/O) devices (transducers). Input devices (sensors) convert physical signals, such as sound, light, and pressure, into electrical signals that circuits can use. These devices include microphones, phototransistors, switches, keyboards, thermistors, strain gauges, generators, and antennas. Output devices convert electrical signals into physical signals. Output devices include lamps, LED and LCD displays, speakers, buzzers, motors (dc, servo, and stepper), solenoids, and antennas. These I/O devices allow humans and circuits to communicate with one another.

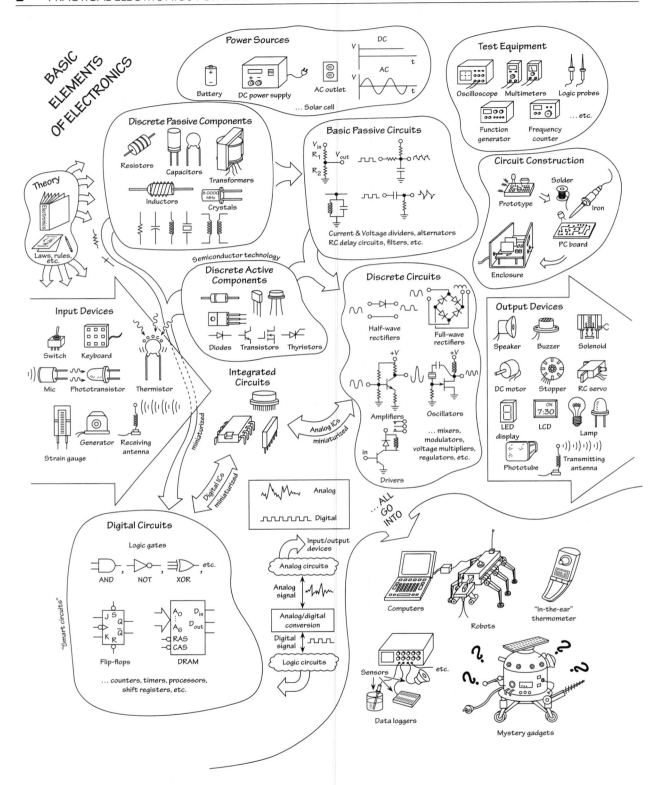

FIGURE 1.1

To make things easier on the circuit designer, manufacturers have created integrated circuits (ICs), which contain discrete circuits (like the ones mentioned in the previous paragraph) that are crammed onto a tiny chip of silicon. The chip is usually housed within a plastic package, where little internal wires link the chip to external metal terminals. ICs such as amplifiers and voltage regulators are referred to as *analog devices*, which means that they respond to and produce signals of varying degrees of voltage. (This is unlike *digital* ICs, which work with only two voltage levels.) Becoming familiar with ICs is a necessity for any practical circuit designer.

Digital electronics comes next. Digital circuits work with only two voltage states: *high* (such as 5 V) or *low* (such as 0 V). The reason for having only two voltage states has to do with the ease of processing data (numbers, symbols, and control information) and storage. The process of encoding information into signals that digital circuits can use involves combining bits (1s and 0s, equivalent to *high* and *low* voltages) into discrete-meaning "words." The designer dictates what these words will mean to a specific circuit. Unlike analog electronics, digital electronics uses a whole new set of components, which at the heart are all integrated in form.

A huge number of specialized ICs are used in digital electronics. Some of these ICs are designed to perform logical operations on input information; others are designed to count; while still others are designed to store information that can be retrieved later on. Digital ICs include logic gates, flip-flops, shift registers, counters, memories, processors, and so on. Digital circuits are what give electrical gadgets "brains." In order for digital circuits to interact with analog circuits, special analog-to-digital (A/D) conversion circuits are needed to convert analog signals into strings of 1s and 0s. Likewise, digital-to-analog conversion circuits are used to convert strings of 1s and 0s into analog signals.

With an understanding of the principals behind digital electronics, we are free to explore the world of microcontrollers. These are programmable digital electronics that can read values from sensors and control output devices using the I/O pins, all on a single IC controlled by a little program.

And mixed in among all this is the practical side of electronics. This involves learning to read schematic diagrams, constructing circuit prototypes using breadboards, testing prototypes (using multimeters, oscilloscopes, and logic probes), revising prototypes (if needed), and constructing final circuits using various tools and special circuit boards.

In the next chapter, we will start at the beginning by looking at the theory of electronics.

CHAPTER 2

Theory

2.1 Theory of Electronics

This chapter covers the basic concepts of electronics, such as current, voltage, resistance, electrical power, capacitance, and inductance. After going through these concepts, this chapter illustrates how to mathematically model currents and voltage through and across basic electrical elements such as resistors, capacitors, and inductors. By using some fundamental laws and theorems, such as Ohm's law, Kirchhoff's laws, and Thevenin's theorem, the chapter presents methods for analyzing complex networks containing resistors, capacitors, and inductors that are driven by a power source. The kinds of power sources used to drive these networks, as we will see, include direct current (dc) sources, alternating current (ac) sources (including sinusoidal and nonsinusoidal periodic sources), and nonsinusoidal nonperiodic sources. We will also discuss transient circuits, where sudden changes in state (such as flipping a switch within a circuit) are encountered. At the end of the chapter, the approach needed to analyze circuits that contain nonlinear elements (diodes, transistors, integrated circuits, etc.) is discussed.

We recommend using a circuit simulator program if you're just starting out in electronics. The web-based simulator CircuitLab (www.circuitlab.com) is extremely easy to use and has a nice graphical interface. There are also online calculators that can help you with many of the calculations in this chapter. Using a simulator program as you go through this chapter will help crystallize your knowledge, while providing an intuitive understanding of circuit behavior. Be careful—simulators can lie, or at least they can appear to lie when you don't understand all the necessary parameters the simulator needs to make a realistic simulation. It is always important to get your hands dirty—get out the breadboards, wires, resistors, power supplies, and so on, and construct. It is during this stage that you gain the greatest practical knowledge that is necessary for an inventor.

It is important to realize that components mentioned in this chapter are only "theoretically" explained. For example, in regard to capacitors, you'll learn how a capacitor works, what characteristic equations are used to describe a capacitor under certain conditions, and various other basic tricks related to predicting basic behavior. To get important practical insight into capacitors, however, such as real-life capacitor applications (filtering, snubbing, oscillator design, etc.), what type of real capacitors exist, how these real capacitors differ in terms of nonideal characteristics, which capacitors work best for a particular application, and, more important, how to read a capacitor label, requires that you jump to Chap. 3, Sec. 3.6, which is dedicated to these issues. This applies to other components mentioned in this theory portion of the book.

The theoretical and practical information regarding transformers and nonlinear devices, such as diodes, transistors, and analog and digital integrated circuits (ICs), is not treated within this chapter. Transformers are discussed in full in Chap. 3, Sec. 3.8, while the various nonlinear devices are treated separately in the remaining chapters of this book.

A word of advice: if the math in a particular section of this chapter starts looking scary, don't worry. As it turns out, most of the nasty math in this chapter is used to prove, say, a theorem or law or to give you an idea of how hard things can get if you do not use some mathematical tricks. The actual amount of math you will need to know to design most circuits is surprisingly small; in fact, basic algebra may be all you need to know. Therefore, when the math in a particular section in this chapter starts looking ugly, skim through the section until you locate the useful, nonugly formulas, rules, and so on, that do not have weird mathematical expressions in them. You don't have to be a mathematical whiz to be able to design decent circuits.

2.2 Electric Current

Electric current is the total charge that passes through some cross-sectional area A per unit time. This cross-sectional area could represent a disk placed in a gas, plasma, or liquid, but in electronics, this cross-sectional area is most frequently a slice through a solid material, such as a conductor.

If ΔQ is the amount of charge passing through an area in a time interval Δt, then the *average current* I_{ave} is defined as:

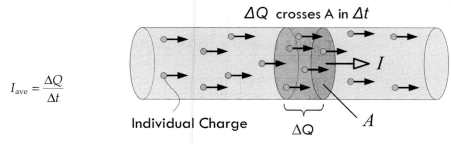

$$I_{ave} = \frac{\Delta Q}{\Delta t}$$

FIGURE 2.1

If the current changes with time, we define the *instantaneous current* I by taking the limit as $\Delta t \to 0$, so that the current is the instantaneous rate at which charge passes through an area:

$$I = \lim_{\Delta t \to 0} \frac{\Delta Q}{\Delta t} = \frac{dQ}{dt} \tag{2.1}$$

The unit of current is coulombs per second, but this unit is also called the *ampere* (A), named after Andre-Marie Ampere:

$$1\,A = 1\,C/s$$

To sound less nerdy, the term *amp* can be used in place of ampere. Because the ampere is a rather large unit, current is also expressed in *milliamps* (1 mA = 1 × 10⁻³ A), *micro-amps* (1 µA = 1 × 10⁻⁶ A), and *nanoamps* (1 nA = 1 × 10⁻⁹ A).

Within conductors such as copper, electrical current is made up of free electrons moving through a lattice of copper ions. Copper has one free electron per copper atom. The charge on a single electron is given by:

$$Q_{electron} = (-e) = -1.602 \times 10^{-19} \text{ C} \tag{2.2.a}$$

This is equal to, but opposite in sign of, the charge of a single copper ion. (The positive charge is a result of the atom donating one electron to the "sea" of free electrons randomly moving about the lattice. The loss of the electron means there is one more proton per atom than electrons.) The charge of a proton is:

$$Q_{proton} = (+e) = +1.602 \times 10^{-19} \text{ C} \tag{2.2.b}$$

The conductor, as a whole, is neutral, since there are equal numbers of electrons and protons. Using Eq. 2.2, we see that if a current of 1 A flows through a copper wire, the number of electrons flowing by a cross section of the wire in 1 s is equal to:

$$1 \text{ A} = \left(\frac{1 \text{ C}}{1 \text{ s}} \right) \left(\frac{\text{electron}}{-1.602 \times 10^{-19} \text{ C}} \right) = -6.24 \times 10^{18} \text{ electrons/s}$$

Now, there is a problem! How do we get a negative number of electrons flowing per second, as our result indicates? The only two possibilities for this would be to say that either electrons must be flowing in the opposite direction as the defined current, or positive charges must be moving in our wire instead of electrons to account for the sign. The last choice is an incorrect one, since experimental evidence exists to prove electrons are free to move, not positive charges, which are fixed in the lattice network of the conductor. (Note, however, there are media in which positive charge flow is possible, such as positive ion flow in liquids, gases, and plasmas.) It turns out that the first choice—namely, electrons flowing in the opposite direction as the defined current flow—is the correct answer.

Long ago, when Benjamin Franklin (often considered the father of electronics) was doing his pioneering work in early electronics, he had a convention of assigning positive charge signs to the mysterious (at that time) things that were moving and doing work. Sometime later, a physicist by the name of Joseph Thomson performed an experiment that isolated the mysterious moving charges. However, to measure and record his experiments, as well as to do his calculations, Thomson had to stick with using the only laws available to him—those formulated using Franklin's positive currents. But these moving charges that Thomson found (which he called electrons) were moving in the opposite direction of the conventional current *I* used in the equations, or moving against convention. See Fig. 2.2.

What does this mean to us, to those of us not so interested in the detailed physics and such? Well, not too much. We could pretend that there were positive charges moving in the wires and various electrical devices, and everything would work out fine: negative electrons going one way are equivalent to positive charges going in

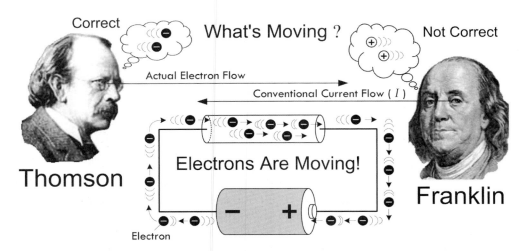

FIGURE 2.2   Thomson changed the notion that positive charges were what were moving in conductors, contrary to Franklin's notion. However, negative electrons going one way is equivalent to positive charges going the opposite direction, so the old formulas still work. Since you deal with the old formulas, it's practical to adopt Franklin's conventional current—though realize that what's actually moving in conductors is electrons.

the opposite direction. In fact, all the formulas used in electronics, such as Ohm's law ($V = IR$), "pretend" that the current I is made up of positive charge carriers. We will always be stuck with this convention. In a nutshell, it's convenient to pretend that positive charges are moving. So when you see the term *electron flow*, make sure you realize that the conventional current flow I is moving in the opposite direction. In a minute, we'll discuss the microscopic goings-on within a conductor that will clarify things a bit better.

Example 1: How many electrons pass a given point in 3 s if a conductor is carrying a 2-A current?

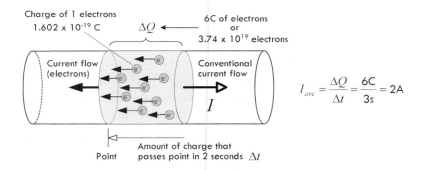

FIGURE 2.3

Answer: The charge that passes a given point in 3 s is:

$$\Delta Q = I \times \Delta t = (2 \text{ A})(3 \text{ s}) = 6 \text{ C}$$

One electron has a charge of 1.6×10^{-19} C, so 6 C worth of electrons is:

$$\# \text{ Electrons} = 6 \text{ C} / 1.602 \times 10^{-19} \text{ C} = 3.74 \times 10^{19}$$

Example 2: Charge is changing in a circuit with time according to $Q(t) = (0.001 \text{ C}) \sin [(1000/\text{s}) \, t]$. Calculate the instantaneous current flow.

$$I = \frac{dQ}{dt} = \frac{d}{dt}[(0.001 \, \text{C}) \sin(1000/\text{s} \cdot t)] = (0.001 \, \text{C})(1000/\text{s}) \cos(1000/\text{s} \cdot t)$$

$$= (1 \, \text{A}) \cos(1000/\text{s} \cdot t)$$

Answer: If we plug in a specific time within this equation, we get an instantaneous current for that time. For example, if $t = 1$, the current would be 0.174 A. At $t = 3$ s, the current would be -0.5 A, the negative sign indicating that the current is in the opposite direction—a result of the sinusoidal nature.

Note: The last example involved using calculus—you can read about the basics of calculus in App. C if you're unfamiliar with it. Fortunately, as we'll see, rarely do you actually need to work in units of charge when doing electronics. Usually you worry only about current, which can be directly measured using an ammeter, or calculated by applying formulas that usually require no calculus whatsoever.

2.2.1 Currents in Perspective

What's considered a lot or a little amount of current? It's a good idea to have a gauge of comparison when you start tinkering with electronic devices. Here are some examples: a 100-W lightbulb draws about 1 A; a microwave draws 8 to 13 A; a laptop computer, 2 to 3 A; an electric fan, 1 A; a television, 1 to 3 A; a toaster, 7 to 10 A; a fluorescent light, 1 to 2 A; a radio/stereo, 1 to 4 A; a typical LED, 20 mA; a mobile (smart) phone accessing the web uses around 200 mA; an advanced low-power microchip (individual), a few μA to perhaps even several pA; an automobile starter, around 200 A; a lightning strike, around 1000 A; a sufficient amount of current to induce cardiac/respiratory arrest, around 100 mA to 1 A.

2.3 Voltage

To get electrical current to flow from one point to another, a voltage must exist between the two points. A voltage placed across a conductor gives rise to an *electromotive force* (EMF) that is responsible for giving all *free electrons* within the conductor a push.

As a technical note, before we begin, voltage is also referred to as a *potential difference* or just *potential*—they all mean the same thing. We'll avoid using these terms, however, because it is easy to confuse them with the term *potential energy*, which is not the same thing.

A simple flashlight circuit, consisting of a battery connected to a lamp, through two conductors and a switch, is shown in Fig. 2.4. When the switch is open ("off"), no current will flow. The moment the switch is closed, however, the resistance of the switch falls to almost zero, and current will flow. This voltage then drives all free electrons, everywhere within the circuit, in a direction that points from negative to positive; conventional current flow, of course, points in the opposite direction (see Benjamin Franklin).

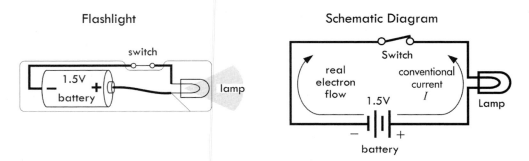

FIGURE 2.4

It is important to note that the battery needs the rest of the circuit, just as the rest of the circuit needs the battery. Without the linkage between its terminals, the chemical reactions within the battery cannot be carried out. These chemical reactions involve the transfer of electrons, which by intended design can only occur through a link between the battery's terminals (e.g., where the circuit goes). Figure 2.5 shows this process using an alkaline dry cell battery. Notice that the flow of current is conserved through the circuit, even though the nature of the current throughout the circuit varies—ionic current within sections of the battery, electron current elsewhere.

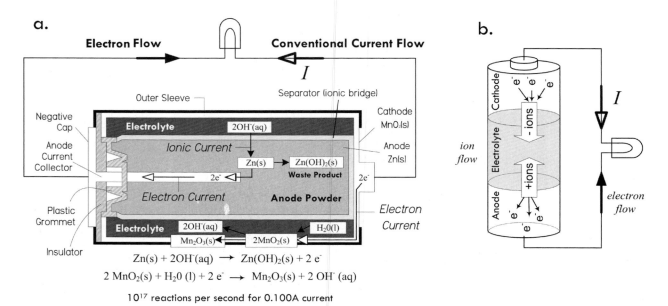

FIGURE 2.5

As free electrons within the lamp filament experience an EMF due to the applied voltage, the extra energy they gain is transferred to the filament lattice atoms, which result in heat (filament atomic vibrations) and emitted light (when a valence electron of a lattice atom is excited by a free electron and the bound electron returns to a lower energy configuration, thus releasing a photon).

A device that maintains a constant voltage across it terminals is called a *direct current voltage source* (or dc voltage source). A battery is an example of a dc voltage source. The schematic symbol for a battery is ⊣I⊢.

2.3.1 *The Mechanisms of Voltage*

To get a mental image of how a battery generates an EMF through a circuit, we envision that chemical reactions inside yield free electrons that quickly build in number within the negative terminal region (anode material), causing an electron concentration. This concentration is full of repulsive force (electrons repel) that can be viewed as a kind of "electrical pressure." With a load (e.g., our flashlight lamp, conductors, switch) placed between the battery's terminals, electrons from the battery's negative terminal attempt to alleviate this pressure by dispersing into the circuit. These electrons increase the concentration of free electrons within the end of the conductor attached to the negative terminal. Even a small percentage difference in free electron concentration in one region gives rise to great repulsive forces between free electrons. The repulsive force is expressed as a seemingly instantaneous (close to the speed of light) pulse that travels throughout the circuit. Those free electrons nearest to the pumped-in electrons are quickly repulsed in the opposite direction; the next neighboring electrons get shoved, and so on down the line, causing a chain reaction, or pulse. This pulse travels down the conductor near the speed of light. See Fig. 2.6.

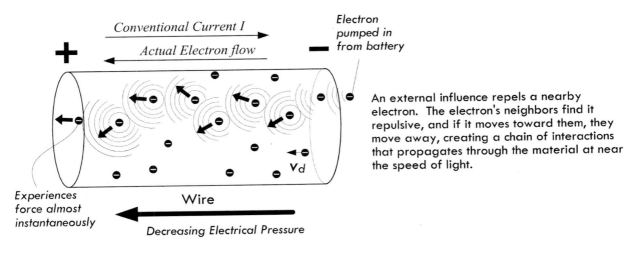

Conventional Current I

Actual Electron flow

Electron pumped in from battery

An external influence repels a nearby electron. The electron's neighbors find it repulsive, and if it moves toward them, they move away, creating a chain of interactions that propagates through the material at near the speed of light.

V_d

Experiences force almost instantaneously

Wire

Decreasing Electrical Pressure

FIGURE 2.6

The actual physical movement of electrons is, on average, much slower. In fact, the drift velocity (average net velocity of electrons toward the positive terminal) is usually fractions of a millimeter per second—say, 0.002 mm/s for a 0.1-A current through a 12-gauge wire. We associate this drift movement of free electrons with current flow or, more precisely, conventional current flow I moving in the opposite direction. (As it turns out, the actual motion of electrons is quite complex, involving thermal effects, too—we'll go over this in the next section.)

It is likely that those electrons farther "down in" the circuit will not feel the same level of repulsive force, since there may be quite a bit of material in the way which absorbs some of the repulsive energy flow emanating from the negative terminal (absorbing via electron-electron collisions, free electron–bond electron interactions, etc.). And, as you probably know, circuits can contain large numbers of components, some of which are buried deep within a complex network of pathways. It is possible to imagine that through some of these pathways the repulsive effects are reduced to a weak nudge. We associate these regions of "weak nudge" with regions of low "electrical pressure," or voltage. Electrons in these regions have little potential to do work—they have low potential energy relative to those closer to the source of pumped-in electrons.

Voltage represents the difference in potential energy. A unit charge has been at one location relative to another within a region of "electrical pressure"—the pressure attributed to new free electrons being pumped into the system. The relationship between the voltage and the difference in potential energy is expressed as:

$$V_{AB} = \frac{U_{AB}}{q} \text{ or } V_B - V_A = \frac{U_B - U_A}{q} \text{ or } \Delta V = \frac{\Delta U}{q}$$

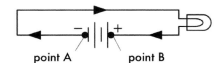

point A   point B

FIGURE 2.7

Implicit in the definition of *voltage* is the notion that voltage is always a measurement between two points, say point A and point B. That is the reason for the subscript "AB" in V_{AB}. The symbol ΔV means the same. Both infer that there is an absolute scale on which to measure and give individual points a specific voltage value. In electronics, we can create such a scale by picking a point, often the point where there is the lowest electrical pressure, and defining this point as the zero point, or 0-V reference. In many dc circuits, people choose the negative terminal of the battery as the 0-V reference, and let everyone know by inserting a ground symbol $\overset{\perp}{=}$ (more on this later). In practice, you rarely see voltages expressed using subscripts (V_{AB}) or deltas (ΔV), but instead you simply see the symbol V, or you may see a symbol like V_R. The "blank symbol" V, however, is always modified with a phrase stating the two points across which the voltage is present. In the second case, V_R, the subscript means that the voltage is measured across the component R—in this case, a resistor. In light of this, we can write a cleaner expression for the voltage/potential energy expression:

$$V = \frac{U}{q}$$

Just make sure you remember that the voltage and potential energy variables represent the difference in relation to two points. As we'll discover, all the big electronics laws usually assume that variables of voltage or energy are of this "clean form."

In our flashlight example, we can calculate the difference in potential energy between an electron emanating from the negative terminal of the 1.5-V battery and one entering the positive terminal.

$$\Delta U = \Delta Vq = (1.5 \text{ V})(1.602 \times 10^{-19} \text{ C}) = 2.4 \times 10^{-19} \text{ J}$$

Notice that this result gives us the potential energy difference between the two electrons, not the actual potential energy of either the electron emanating from the negative terminal (U_1) or the electron entering the positive terminal (U_0). However, if we make the assumption that the electron entering the positive terminal is at zero potential energy, we can figure that the electron emanating from the negative terminal has a relative potential energy of:

$$U_1 = \Delta U + U_0 = \Delta U + 0 = 2.4 \times 10^{-19} \text{ J}$$

Note: Increasing positive potential energy can be associated with similar charges getting closer together. Decreasing energy can be associated with similar charges getting farther apart. We avoided the use of a negative sign in front of the charge of the electron, because voltages are defined by a positive test charge. We are in a pickle similar to the one we saw with Benjamin Franklin's positive charges. As long as we treat the potential relative to the pumped-in electron concentration, things work out.

In a real circuit, where the number of electrons pumped out by the battery will be quite large—hundreds to thousands of trillions of electrons, depending on the resistance to electron flow—we must multiply our previous calculation by the total number of entering electrons. For example, if our flashlight draws 0.1 A, there will be 6.24×10^{17} electrons pumped into it by the battery per second, so you calculate the potential energy of all the new electrons together to be about 0.15 J/s.

What about the potential energies of free electrons at other locations throughout the circuit, such as those found in the lamp filament, those in the positive wire, those in the negative wire, and so on? We can say that somewhere in the filament of the lamp, there is an electron that has half the potential energy of a fresh pumped-in electron emanating from the negative terminal of the battery. We attribute this lower energy to the fact that other free electrons up the line have lost energy due to collision mechanisms, which in turn yields a weaker electrical repulsive pressure (shoving action) that our electron in question experiences. In fact, in our flashlight circuit, we attribute all loss in electrical pressure to be through the lamp filament as free-electron energy is converted into heat and light.

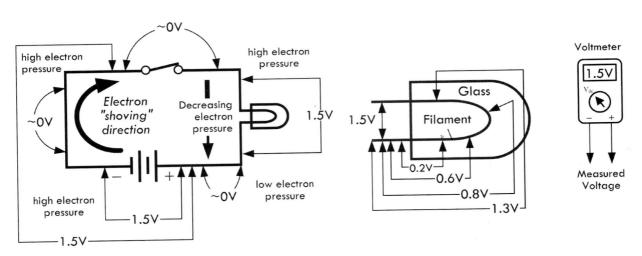

FIGURE 2.8

In regard to potential energies of free electrons within the conductors leading to and from the battery, we assume all electrons within the same conductor have the same potential energy. This assumes that there is no voltage difference between points in the same conductor. For example, if you take a voltmeter and place it between any two points of a single conductor, it will measure 0 V. (See Fig. 2.8.) For practical purposes, we accept this as true. However, in reality it isn't. There is a slight voltage drop through a conductor, and if we had a voltmeter that was extremely accurate we might measure a voltage drop of 0.00001 V or so, depending on the length of the conductor, current flow, and conductor material type. This is attributed to internal resistance within conductors—a topic we'll cover in a moment.

2.3.2 Definition of Volt and Generalized Power Law

We come now to a formal definition of the volt—the unit of measure of voltage. Using the relationship between voltage and potential energy difference $V = U/q$, we define a volt to be:

$$1 \text{ volt} = \frac{1 \text{ joule}}{1 \text{ coulomb}}, \ 1 \text{ V} \frac{1 \text{ J}}{1 \text{ C}} = \text{J/C (Energy definition)}$$

(Be aware that the use of "V" for both an algebraic quantity and a unit of voltage is a potential source of confusion in an expression like $V = 1.5$ V. The algebraic quantity is in italic.)

Two points with a voltage of 1 V between them have enough "pressure" to perform 1 J worth of work while moving 1 C worth of charge between the points. For example, an ideal 1.5-V battery is capable of moving 1 C of charge through a circuit while performing 1.5 J worth of work.

Another way to define a volt is in terms of power, which happens to be more useful in electronics. *Power* represents how much energy per second goes into powering a circuit. According to the conservation of energy, we can say the power used to drive a circuit must equal the power used by the circuit to do useful work plus the power wasted, as in the case of heat. Assuming that a single electron loses all its potential energy from going through a circuit from negative to positive terminal, we say, for the sake of argument, that all this energy must have been converted to work—useful and wasted (heat). By definition, power is mathematically expressed as dW/dt. If we substitute the potential energy expression $U = Vq$ for W, assuming the voltage is constant (e.g., battery voltage), we get the following:

$$P = \frac{dW}{dt} = \frac{dU}{dt} = V \frac{dq}{dt}$$

Since we know that current $I = dq/dt$, we can substitute this into the preceding expression to get:

$$P = VI \tag{2.3}$$

This is referred to as the *generalized power law*. This law is incredibly powerful, and it provides a general result, one that is independent of type of material and of the nature of the charge movement. The unit of this electrical power is watts (W), with 1 W = 1 J/s, or in terms of volts and amps, 1 W = 1 VA.

In terms of power, then, the volt is defined as:

$$1 \text{ volt} = \frac{1 \text{ watt}}{1 \text{ amp}}, \quad 1 \text{ V} = \frac{1 \text{ W}}{1 \text{ A}} = \text{W/A}$$

The generalized power law can be used to determine the power loss of any circuit, given only the voltage applied across it and the current drawn, both of which can easily be measured using a voltmeter and an ammeter. However, it doesn't tell you specifically how this power is used up—more on this when we get to resistance. See Fig. 2.9.

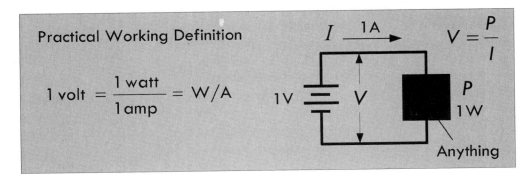

FIGURE 2.9

Example 1: Our 1.5-V flashlight circuit draws 0.1 A. How much power does the circuit consume?

Answer:

$$P = VI = (1.5 \text{ V})(0.1 \text{ A}) = 0.15 \text{ W}$$

Example 2: A 12-V electrical device is specified as consuming 100 W of power. How much current does it draw?

Answer:

$$I = \frac{P}{V} = \frac{100 \text{ W}}{12 \text{ V}} = 8.3 \text{ A}$$

2.3.3 Combining Batteries

To get a larger voltage capable of supplying more power, we can place two batteries in series (end to end), as shown in Fig. 2.10. The voltage across the combination is equal to the individual battery voltages added together. In essence, we have placed two charge pumps in series, increasing the effective electrical pressure. Chemically speaking, if the batteries are of the same voltage, we double the number of chemical reactions, doubling the number of electrons that can be pumped out into the circuit.

In Fig. 2.10, we use the notion of a *ground reference*, or 0-V reference, symbolized $\perp$. Though this symbol is used to represent an earth ground (which we define a bit later), it can also be used to indicate the point where all voltage measurements are to be

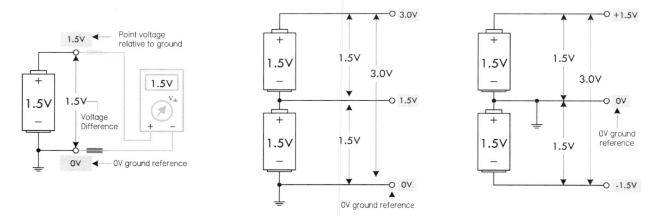

FIGURE 2.10

referenced within a circuit. Logically, whenever you create a scale of measure, you pick the lowest point in the scale to be zero—0 V here. For most dc circuits, the ground reference point is usually placed at the negative terminal of the voltage source. With the notion of ground reference point, we also get the notion of a *point voltage,* which is the voltage measured between the ground reference and a specific point of interest within the circuit. For example, the single battery shown in Fig. 2.10 has a voltage of 1.5 V. We place a ground reference at the negative terminal and give this a 0-V point voltage, and place a 1.5-V point voltage marker at the positive terminal.

In the center of Fig. 2.10, we have two 1.5-V batteries in series, giving a combined voltage of 3.0 V. A ground placed at the negative terminal of the lower battery gives us point voltages of 1.5 V between the batteries, and 3.0 V at the positive terminal of the top battery. A load placed between ground and 3.0 V will result in a load current that returns to the lower battery's negative terminal.

Finally, it is possible to create a split supply by simply repositioning the 0-V ground reference, placing it between the batteries. This creates +1.5 V and −1.5 V leads relative to the 0-V reference. Many circuits require both positive and negative voltage relative to a 0-V ground reference. In this case, the 0-V ground reference acts as a *common return.* This is often necessary, say, in an audio circuit, where signals are sinusoidal and alternate between positive and negative voltage relative to a 0-V reference.

2.3.4   *Other Voltage Sources*

There are other mechanisms besides the chemical reactions within batteries that give rise to an electromotive force that pushes electrons through circuits. Some examples include magnetic induction, photovoltaic action, thermoelectric effect, piezoelectric effect, and static electric effect. Magnetic induction (used in electrical generators) and photovoltaic action (used in photocells), along with chemical reactions, are, however, the only mechanisms of those listed that provide enough power to drive most circuits. The thermoelectric and piezoelectric effects are usually so small (mV range, typically) that they are limited to sensor-type applications. Static electric effect is based on giving objects, such as conductors and insulators, a surplus of charge. Though voltages can become very high between charged objects, if a circuit were connected between the objects, a dangerous discharge of current could flow, possibly damaging sensitive circuits. Also, once the discharge is complete—a matter of milliseconds—there is no more current to power the circuit. Static electricity is considered a nuisance in electronics, not a source of useful power. We'll discuss all these different mechanisms in more detail throughout the book.

2.3.5 Water Analogies

It is often helpful to use a water analogy to explain voltage. In Fig. 2.11, we treat a dc voltage source as a water pump, wires like pipes, Benjamin Franklin's positive charges as water, and conventional current flow like water flow. A load (resistor) is treated as a network of stationary force-absorbing particles that limit water flow. We'll leave it to you to compare the similarities and differences.

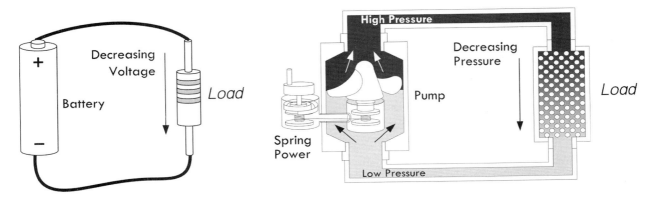

FIGURE 2.11

Here's another water analogy that relies on gravity to provide the pressure. Though this analogy falls short of being accurate in many regards, it at least demonstrates how a larger voltage (greater water pressure) can result in greater current flow.

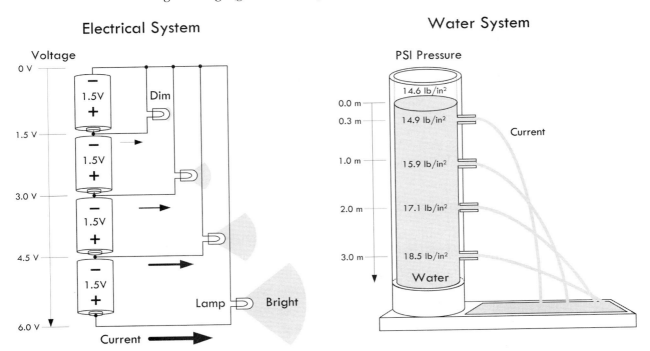

FIGURE 2.12

It's not wise to focus too much attention on these water analogies—they fall short of being truly analogous to electric circuits. Take them with a grain of salt. The next section will prove how true this is.

Example 1: Find the voltage between the various points indicated in the following figures. For example, the voltage between points A and B in Fig. 2.13a is 12 V.

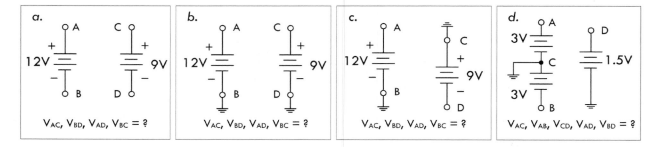

FIGURE 2.13

Answer: a. $V_{AC} = 0$, $V_{BD} = 0$, $V_{AD} = 0$, $V_{BC} = 0$. b. $V_{AC} = 3$ V, $V_{BD} = 0$ V, $V_{AD} = 12$ V, $V_{BC} = 9$ V. c. $V_{AC} = 12$ V, $V_{BD} = 9$V. $V_{AD} = 21$V, $V_{BC} = 0$ V. d. $V_{AC} = 3$ V, $V_{AB} = 6$ V, $V_{CD} = 1.5$ V, $V_{AD} = 1.5$ V, $V_{BD} = 4.5$ V.

Example 2: Find the point voltages (referenced to ground) at the various locations indicated in the following figures.

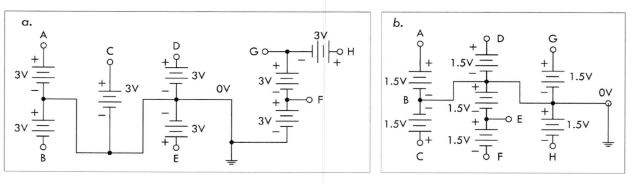

FIGURE 2.14

Answer: a. A = 3 V, B = –3 V, C = 3 V, D = 3 V, E = 3 V, F = 3 V, G = 6 V, H = 9 V. b. A = 1.5 V, B = 0 V, C = 1.5 V, D = 1.5 V, E = –1.5 V, F = –3.0 V, G = 1.5 V, H = –1.5 V.

2.4 A Microscopic View of Conduction (for Those Who Are Interested)

At a microscopic level, a copper conductor resembles a lattice of copper balls packed together in what's called a face-centered-cubic lattice structure, as shown in Fig. 2.15. For copper, as well as other metals, the bonding mechanism that holds everything together is referred to as *metallic bonding,* where outermost valence electrons from the metal atoms form a "cloud of free electrons" which fill the space between the metal ions (positively charged atoms missing an electron that became "free"—see the planetary model in Fig. 2.15b). This cloud of free electrons acts as a glue, holding the lattice metal ions together.

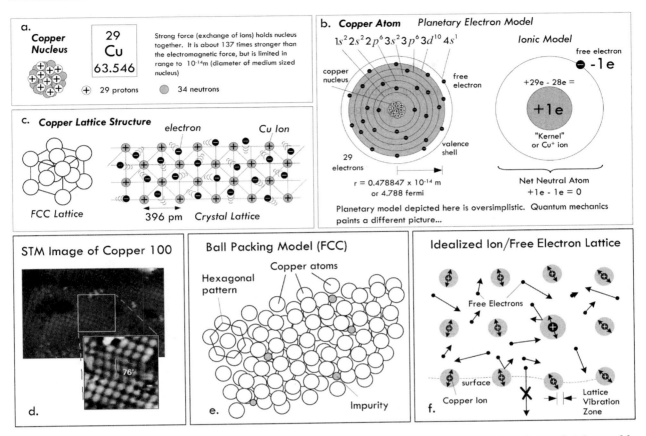

FIGURE 2.15 (a) Copper nucleus composed of protons and neutrons held together by nuclear forces that is roughly 137 times stronger than the electromagnetic force. (b) Copper atoms, as viewed by the classic planetary model, consisting of valance electrons held in orbit by electric forces. Quantum mechanics is required to explain why electrons exist in discrete energy levels, and why they don't fall into the nucleus or radiate electromagnetic energy as they orbit. (c) Copper lattice has a face-centered cubic packing arrangement. (d) Scanning tunneling electron microscope (STM) image of copper 100, courtesy of Institut für Allgemeine Physik, TU Wien. (e) Ball packing model of lattice, showing irregularities in lattice geometry, partly caused by impurities (other kinds of atoms). (f) Lattice view showing that lattice atoms vibrate due to external thermal interactions as well as interactions with free electrons. Free electrons move about randomly, at varying speeds and directions, colliding with other electrons and lattice ions. Under normal conditions, they do not leave the surface of the metal.

Each free electron within the cloud of free electrons moves about in random directions and speeds, colliding and rebounding "off" metal ions and other imperfections (impurities in lattice and grain boundary transitions, etc.). It is important to realize that this is occurring in a chunk of copper, at room temperature, without any applied voltage.

At room temperature, no free electrons ever leave the surface of the metal. A free electron cannot escape the coulomb (electric) attractive forces presented by the positive metal ions in the lattice. (We'll see later that under special conditions, using unique mechanisms, it is possible for electrons to escape.)

According to what's called the *free-electron model*—a classical model that treats free electrons as a gas of noninteracting charges—there is approximately one free electron per copper atom, giving a copper conductor a free electron concentration of $\rho_n = 8.5 \times 10^{28}$ electrons/m^3. This model predicts that, under normal conditions (a piece of copper just sitting there at room temperature), the *thermal velocity v* of electrons (or root-mean-square speed) within copper is about 120 km/s (1.2×10^5 m/s), but depends on temperature. The average distance an electron travels before it collides with something, called the *mean free path* λ, is about 0.000003 mm (2.9×10^{-9} m), with the average time between collisions τ of roughly 0.000000000000024 s (2.4×10^{-14} s). The free-electron model is qualitatively correct in many respects, but isn't as accurate as models based on quantum mechanics. (The speed, path, and time are related by $v = \lambda / \tau$.)

In quantum mechanics, electrons obey velocity-distribution laws based on quantum physics, and the movement of electrons depends on these quantum ideas. It requires that we treat electrons as though they were waves scattering from the lattice structure of the copper. The quantum view shows the thermal speed (now called Fermi velocity v_F) of a free electron to be faster than that predicted by the free-electron model, now around 1.57×10^6 m/s, and contrarily, it is essentially independent of temperature. In addition, the quantum model predicts a larger mean free path, now around 3.9×10^{-8} m, which is independent of temperature. The quantum view happens to be the accepted view, since it gives answers that match more precisely with experimental data. Table 2.1 shows the Fermi velocities of electrons for various metals.

TABLE 2.1   Condensed Matter Properties of Various Metals

MATERIAL	FERMI ENERGY E_F (eV)	FERMI TEMPERATURE ($\times 10^4$ K)	FERMI VELOCITY (M/S) $V_F = c\sqrt{\dfrac{2E_F}{m_e c^2}}$	FREE-ELECTRON DENSITY ρ_e (ELECTRONS/m³)	WORK FUNCTION W (eV)
Copper (Cu)	7.00	8.16	1.57×10^6	8.47×10^{28}	4.7
Silver (Ag)	5.49	6.38	1.39×10^6	5.86×10^{28}	4.73
Gold (Au)	5.53	6.42	1.40×10^6	5.90×10^{28}	5.1
Iron (Fe)	11.1	13.0	1.98×10^6	17.0×10^{28}	4.5
Tin (Sn)	8.15	9.46	1.69×10^6	14.8×10^{28}	4.42
Lead (Pb)	9.47	11.0	1.83×10^6	13.2×10^{28}	4.14
Aluminum (Al)	11.7	13.6	2.03×10^6	18.1×10^{28}	4.08

Note: 1 eV = 1.6022×10^{-19} J, $m_e = 9.11 \times 10^{-31}$ kg, $c = 3.0 \times 10^8$ m/s
Fermi energy and free electron density data from N.W. Ashcroft and N.D. Mermin, *Solid State Physics,* Saunders, 1976; work function data from Paul A. Tipler and Ralph A. Llewellyn, *Modern Physics,* 3rd ed., W.H. Freeman, 1999.

Also, the surface binding energy (caused by electrostatic attraction) that prevents electrons from exiting the surface of the metal, referred to as the *work function,* is about 4.7 eV for copper (1 eV = 1.6022×10^{-19} J). The only way to eject electrons is through special processes, such as thermionic emission, field emission, secondary emission, and photoelectric emission.

(*Thermionic emission:* increase in temperature provides free electrons enough energy to overcome work function of the material. The emitted electron is referred to as a thermoelectron. *Field emission:* additional energy from an electric field generated by a high-voltage conductor provides an attractive enough positive field to free electrons from the surface. This requires huge voltages [MV per cm between emitting surface and positive conductor]. *Secondary emission:* electrons are emitted from a metallic surface by the bombardment of high-speed electrons or other particles. *Photoelectric emission:* electron in material absorbs energy from incoming photon of particular frequency, giving it enough energy to overcome work function. A photon must be of the correct frequency, governed by $W = hf_0$, for this to occur [Planck's constant $h = 6.63 \times 10^{-34}$ J-s or 4.14×10^{-14} eV; f_0 is in hertz]).

2.4.1 Applying a Voltage

Next, we wish to see what happens when we apply a voltage across the conductor—say, by attaching a thick copper wire across a battery. When we do this, our randomly moving free electrons all experience a force pointing toward the positive end of the wire due to the electric field set up within the wire. (The field is due to the negative concentration of pumped-in electrons at one end relative to the neutral [positive relative to negative] concentration at the other end.) The actual influence this force has on the motion of the random free electrons is small—the thermal velocity is so large that it is difficult to change the momentum of the electrons. What you get is a slightly parabolic deviation in path, as shown in Fig. 2.16.

What's Going on Inside a Wire

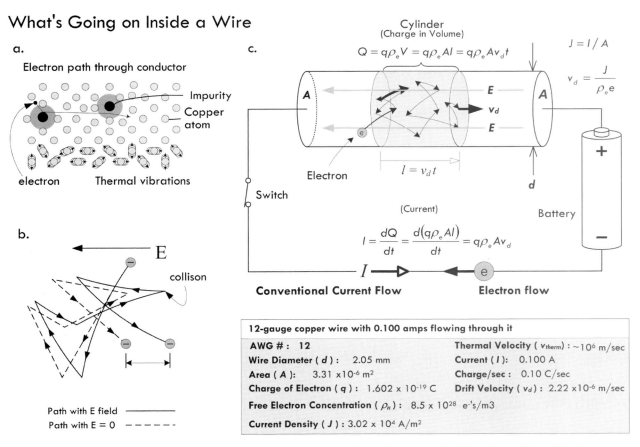

FIGURE 2.16   (a) Simplistic view of an electron randomly moving through a copper lattice, rebounding off lattice atoms and impurities. **(b)** An electron collides frequently with the ions and impurities in a metal and scatters randomly. In an electric field, the electron picks up a small component of velocity opposite the field. The differences in the paths are exaggerated. The electron's path in an electric field is slightly parabolic. **(c)** Model illustrating current density, drift velocity, charge density, thermal velocity, and current.

Normally, the field present in the wire would create a net acceleration component in the direction of the force; however, the constant collisions electrons experience create a drag force, similar to the drag experienced by a parachute. The net effect is an average group velocity referred to as the *drift velocity v_d*. Remarkably, this velocity is surprisingly small. For example, the voltage applied to a 12-gauge copper wire to yield a 0.100-A current will result in a drift velocity of about 0.002 mm per second! The drift velocity is related, determined by

$$v_d = J/(\rho_e e)$$

where J is the current density—the current flowing through an area ($J = I/A$), ρ_e is the free-electron density in the material, and e is the charge of an electron. Table 2.1 shows free-electron densities for various materials. As you can see, the drift velocity varies with current and diameter of the conductor.

The drift velocity is so slow, only fractions of a millimeter per second, that it is worth pondering how a measurable current can even flow. For example, what happens when you flip the switch on a flashlight? Of course, we don't have to wait hours for electrons to move down the conductors from the battery. When we throw the switch, the electric field of the incoming electron has a repulsive effect on its neighbor within the wire. This neighbor then moves away toward another neighbor, creating a chain of interactions that propagates through the material at near the speed of light. (See Fig. 2.17.) This reaction, however, really isn't the speed of light, but a fraction less, depending on the medium. The free electrons spreading throughout the conductor all start moving at once in response—those nearest the switch, as well as those nearest the light filament or LED. A similar effect occurs in fluid flow, as when you turn on a garden hose. Because if the hose is already full of water, the outflow starts immediately. The force of the water at the faucet end is quickly transmitted all along the hose, and the water at the open end of the hose flows almost at the moment the faucet is opened.

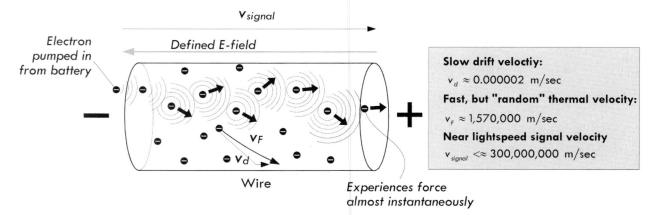

FIGURE 2.17   Illustration of how the electric field propagates down a wire as electrons are pumped into one end.

In the case of alternating current, the field reverses directions in a sinusoidal fashion, causing the drift velocity component of electrons to swish back and forth. If the alternating current has a frequency of 60 Hz, the velocity component would be vibrating back and forth 60 times a second. If our maximum drift velocity during an ac cycle is 0.002 mm/s, we could roughly determine that the distance between maximum swings in the drift distance would be about 0.00045 mm. Of course, this doesn't mean that electrons are fixed in an oscillatory position. It means only that the drift displacement component of electrons is—if there is such a notion. Recall that an electron's overall motion is quite random and its actual displacement quite large, due to the thermal effects.

2.5   Resistance, Resistivity, Conductivity

As was explained in the last section, free electrons in a copper wire at room temperature undergo frequent collisions with other electrons, lattice ions, and impurities within the lattice that limit their forward motion. We associate these microscopic mechanisms that impede electron flow with electrical resistance. In 1826, Georg Simon Ohm published experimental results regarding the resistance of different materials, using a qualitative approach that wasn't concerned with the hidden mechanisms, but rather considered only the net observable effects. He found a linear relationship between how much current flowed through a material when a given voltage was applied across it. He defined the *resistance* to be the ratio of the applied voltage divided by the resultant current flow, given by:

$$R \equiv \frac{V}{I} \tag{2.4}$$

This statement is called Ohm's law, where R is the resistance, given in volts per ampere or ohms (abbreviated with the Greek symbol omega, Ω). One ohm is the resistance through which a current of 1 A flows when a voltage of 1 V is applied:

$$1\,\Omega = 1\,V/1\,A$$

The symbol —⋀⋀— is used to designate a resistor.

Now, Ohm's law isn't really a law, but rather an empirical statement about the behavior of materials. In fact, there are some materials for which Ohm's law actually doesn't work.

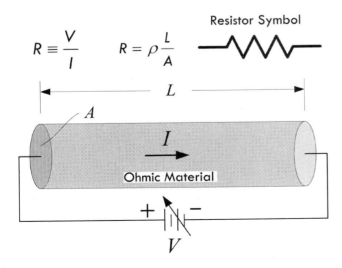

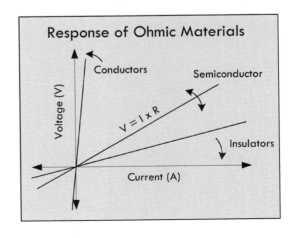

FIGURE 2.18

Ohm's law can be applied only to *ohmic materials*—materials whose resistance remains constant over a range of voltages. *Nonohmic materials,* on the other hand, do not follow this pattern; they do not obey Ohm's law. For example, a diode is a device that allows current to pass easily when the voltage is positive, but prevents current flow (creates a high resistance) when the voltage is negative.

CURIOUS NOTE ABOUT OHM'S LAW

Usually you see Ohm's law written in the following form:

$$V = I \times R$$

However, in this form it is tempting to define voltage in terms of resistance and current. It is important to realize that R is the resistance of an ohmic material and is independent of V in Ohm's law. In fact, Ohm's law does not say anything about voltage; rather, it defines resistance in terms of it and cannot be applied to other areas of physics such as static electricity, because there is no current flow. In other words, you don't define voltage in terms of current and resistance; you define resistance in terms of voltage and current. That's not to say that you can't apply Ohm's law to, say, predict what voltage must exist across a known resistance, given a measured current. In fact, this is done all the time in circuit analysis.

2.5.1 How the Shape of a Conductor Affects Resistance

The resistance of a conducting wire of a given material varies with its shape. Doubling the length of a wire doubles the resistance, allowing half the current to flow, assuming similar applied voltages. Conversely, doubling the cross-sectional area A has the opposite effect—the resistance is cut in half, and twice as much current will flow, again assuming similar applied voltages.

Increasing resistance with length can be explained by the fact that down the wire, there are more lattice ions and imperfections present for which an applied field (electric field instigated by added electrons pumped in by the source) must shove against. This field is less effective at moving electrons because as you go down the line, there are more electrons pushing back—there are more collisions occurring on average.

Decreasing resistance with cross-sectional area can be explained by the fact that a larger-volume conductor (greater cross-sectional area) can support a larger current flow. If you have a thin wire passing 0.100 A and a thick wire passing 0.100 A, the thinner wire must concentrate the 0.100 A through a small volume, while the thick wire can distribute this current over a greater volume. Electrons confined to a smaller volume tend to undergo a greater number of collisions with other electrons, lattice ions, and imperfections than a wire with a larger volume. The concentration of free-electron flow, according to Benjamin Franklin's convention, represents a concentration of conventional current flow in the opposite direction. This concentration of current flow is called the *current density J*—the rate of current flows per unit area. For a wire: $J = I/A$. Figure 2.19 demonstrates how the current density is greater in a thin 12-gauge wire than in a thicker 4-gauge wire. It also shows that the drift velocity in the thick wire is less than the drift velocity in the thin wire—a result of a "decrease in electron field pressure" lowering the average "push" in the current flow direction.

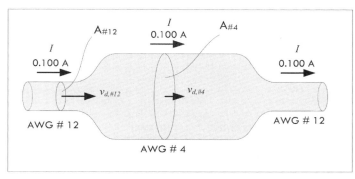

0.100A Current Flowing		
AWG #	**12**	**4**
Wire Diameter (d)	2.05 mm	5.19 mm
Area (A)	3.31×10^{-6} m²	2.11×10^{-5} m²
Drift Velocity (v_d)	2.22×10^{-6} m/sec	3.48×10^{-7} m/sec
Current Density (J)	30,211 A/m²	4,739 A/m²

Conductivity (ρ) : 1.7×10^{-8} $\Omega \cdot$m

Resistivity (σ) : 6.0×10^{7} $(\Omega \cdot$m$)^{-1}$

Properties of copper independent of diameter:

FIGURE 2.19   Effects of wire diameter on resistance. A thinner wire has more resistance per unit length than a thicker wire.

2.5.2 Resistivity and Conductivity

We have left out the most important aspect of resistance that has nothing to do with the physical length or diameter of the material. How does the "chemistry" of the material affect the resistance? For example, if you have a copper wire with the same dimensions as a brass wire, which metal has the greater overall resistance? To answer this question, as well as provide a way to categorize materials, we adopt the concept of resistivity. Unlike resistance, resistivity is entirely independent of the dimensions of the material. Resistivity is a property unique to the material. The *resistivity* ρ is defined as follows:

$$\rho \equiv R \frac{A}{L} \tag{2.5}$$

where A is the cross-sectional area, L is the length, and R is the overall resistance of the material, as measured across its length. The units of resistivity are ohm-meters (Ωm).

For some, resistivity is too negative a concept—it tells you how "bad" something is at passing current. Optimists prefer the concept of *conductivity*—how "good" something is at passing current. Conductivity, symbolized σ, is simply the mathematical inverse of resistivity:

$$\sigma \equiv \frac{1}{\rho} \tag{2.6}$$

The units of conductivity are siemens, $S = (\Omega m)^{-1}$. (*Mathematical note:* $[\Omega m]^{-1} = 1/[\Omega m]$). Both conductivity and resistivity contain the same important underlining information. Some prefer to play with equations that use the optimistic notion of conductivity ("glass half full"); others prefer the pessimistic notion of resistivity ("glass half empty").

In terms of resistivity and conductivity, we can rewrite Ohm's law as follows:

$$V = IR = \rho \frac{L}{A} I = \frac{I \times L}{\sigma \times A} \tag{2.7}$$

Table 2.2 shows the conductivities (resistivities, for the pessimists) of various materials. (Consult a technical handbook, such as *The Handbook of Chemistry and*

Physics, for a more complete list.) The conductivity of metals, such as copper and silver, is a factor of 10^{21} higher than that of a good insulator, such as Teflon. Though both copper and silver are great conductors, silver is simply too expensive for practical use. Though aluminum is a fairly good conductor and was actually used at one time in home circuitry, it quickly became apparent that it oxidized badly, inhibiting good electrical contacts and limiting current flow to channels of limited size. The result produced fire hazards.

TABLE 2.2   Conductivity of Various Materials

MATERIAL	RESISTIVITY ρ (ΩM)	CONDUCTIVITY σ (ΩM)$^{-1}$	TEMPERATURE COEFFICIENT α (°C^{-1})	THERMAL RESISTIVITY (W/cm °C)$^{-1}$	THERMAL CONDUCTIVITY K (W/cm °C)
Conductors					
Aluminum	2.82×10^{-8}	3.55×10^{7}	0.0039	0.462	2.165
Gold	2.44×10^{-8}	4.10×10^{7}		0.343	2.913
Silver	1.59×10^{-8}	6.29×10^{7}	0.0038	0.240	4.173
Copper	1.72×10^{-8}	5.81×10^{7}	0.0039	0.254	3.937
Iron	10.0×10^{-8}	1.0×10^{7}	0.0050	1.495	0.669
Tungsten	5.6×10^{-8}	1.8×10^{7}	0.0045	0.508	1.969
Platinum	10.6×10^{-8}	1.0×10^{7}	0.003927		
Lead	0.22×10^{-6}	4.54×10^{6}		2.915	0.343
Steel (stainless)	0.72×10^{-6}	1.39×10^{6}		6.757	0.148 (312)
Nichrome	100×10^{-8}	0.1×10^{7}	0.0004		
Manganin	44×10^{-8}	0.23×10^{7}	0.00001		
Brass	7×10^{-8}	1.4×10^{7}	0.002	0.820	1.22
Semiconductors					
Carbon (Graphite)	3.5×10^{-5}	2.9×10^{4}	-0.0005		
Germanium	0.46	2.2	-0.048		
Silicon	640	3.5×10^{-3}	-0.075	0.686	1.457 (pure)
Gallium arsenide				1.692	0.591
Insulators					
Glass	10^{10}–10^{14}	10^{-14}–10^{-10}			
Neoprene rubber	10^{9}	10^{-9}			
Quartz (fused)	75×10^{16}	10^{-16}			
Sulfur	10^{15}	10^{-15}			
Teflon	10^{14}	10^{-14}			

An important feature of resistivity (or conductivity) is its temperature dependency. Generally, within a certain temperature range, the resistivity for a large number of metals obeys the following equation:

$$\rho = \rho_0[1 + \alpha(T - T_0)] \tag{2.8}$$

where ρ is the calculated resistivity based on a set reference resistivity ρ_0 and temperature T_0. Alpha α is called the *temperature coefficient of resistivity,* given in units of 1/°C or (°C)$^{-1}$. The resistivity for most metals increases with temperature because lattice atom vibrations caused by thermal energy (increased temperature) impede the drift velocity of conducting electrons.

Are Water, Air, and Vacuums Considered Insulators or Conductors?

These media require special mention. See comments presented in Table 2.3.

TABLE 2.3   Resistivity of Special Materials

MATERIAL	RESISTIVITY, ρ (ΩM)	COMMENTS
Pure water	2.5×10^5	Distilled water is a very good insulator—it has a high resistivity. The rather weak mechanism of conduction is one based on ion flow, not electron flow, as we saw in metals. Water normally has an autoionization at room temperature that gives rise to an amount of H_3O^+ and OH^- ions, which number small in comparison to H_2O molecules (1:10^{-7}). If a battery of known voltage is connected to two copper electrodes placed apart within a bucket of distilled water, and a current is measured in the hardwire section of the circuit, using Ohm's law, we find that the water presents a resistance of around 20×10^6 Ω/cm. Between the electrodes, H_3O^+ flows toward the negative electrode, and OH^- flows toward the positive electrode. When each ion makes contact with the electrodes, it deposits an electron or accepts an electron, whereby the ion recombines in solution.
Saltwater	~0.2	Adding an ionic compound in the form of common salt (NaCl) to water increases the ion concentration within solution—NaCl ionizes into Na$^+$ and Cl$^-$. A gram of salt added introduces around 2×10^{22} ions. These ions act as charge carriers, which in turn, effectively lowers the solution's resistance to below an ohm per meter. If we use the solution as a conductor between a battery and a lamp—via electrodes placed in solution—there is ample current to light the lamp.
Human skin	~5.0×10^5	Varies with moisture and salt content.
Air		Considered an insulator since it is fairly void of free electrons. However, like liquids, air often contains concentrations of positive and negative airborne ions. The formation of an air ion starts out with an electron being knocked off a neutral air molecule, such as oxygen (O_2) or nitrogen (N_2), via naturally occurring bombardment with x-rays, gamma rays, or alpha particles emitted by decaying radon atoms. (There are about 5 to 10 ion pairs produced per second per centimeter at sea level; more around radon areas.) The positive molecule of oxygen or nitrogen rapidly attracts polar molecules (from 10 to 15 in number), mostly water. The resulting cluster is called a positive air ion. The liberated electron, on the other hand, will most likely attach itself to an oxygen molecule (nitrogen has no affinity for electrons). *Note:* air is considered neutral—ions are always formed in pairs, equal parts positive and negative.
		For current to flow through air, you must apply an electric field across the air—say, by using two parallel plates placed apart and set to different potentials (a voltage existing between them). At low voltage, the field strength is small and you may get movement of air ions between the plates, but since the concentration of free-floating ions is so small, there is practically no current. However, if you increase the field strength, it is possible to accelerate those free electrons (those liberated by natural processes) within the air to such high velocities (toward the positive plate) that these electrons themselves can have enough energy to collide with air molecules and generate more ion pairs (positive and negative). The breakdown field strength at which to instigate these ionizations is somewhere around 3 megavolts per meter between plane electrodes in air. Two metal plates 1 cm apart need about 30,000 V for ionization to take place between the plates. However, you can increase the strength of an electric field by replacing one of the plates with a sharp metal point or thin wire. The required voltage drops to a few kilovolts. What occurs can be classified into the following:

(Continued)

TABLE 2.3 **Resistivity of Special Materials (*Continued*)**

MATERIAL	RESISTIVITY, ρ (ΩM)	COMMENTS
Air		*Corona discharge:* Ionization is limited to a small region around the electrode, where the breakdown field strength is exceeded. In the rest of the field, there is just a current of slow-moving ions and even slower-moving charged particles finding their way to the counter electrode (which may be an oppositely charged plate, or simply the wall of the room or the ground). Such a discharge may be maintained as long as the breakdown field strength is exceeded in some region. *Spark discharge:* May take place between two well-rounded conductors at different potentials, one of them often grounded. Like corona discharge, the discharge starts at a point where the breakdown field strength is exceeded. But in contrast to corona discharge, in a spark the ionization takes place all the way between the electrodes. The discharge is very fast, and the energy of the discharge is confined to a narrow volume. The voltage necessary to cause breakdown is called the breakdown voltage. For example, when you become charged by walking across an insulated floor (e.g., carpet) with insulated shoes and acquire a charge due to contact and friction, then touch a grounded object, charge will flow between you, even before you make physical contact—the spark. Few people notice discharges at voltages lower than about 1000 V. Most people start to feel an unpleasant effect around 2000 V. Almost everyone will complain when exposed to discharges at voltages above 3000 V. *Brush discharge:* In between the corona discharge and the spark, which may take place, for example, between a charged material and a normally grounded electrode with a radius of curvature of some millimeters. If a brush discharge is maintained over longer periods, it may appear as irregular luminescent paths. Almost all discharges from insulators are brush discharges, like the crackle heard as you pick up a charged photocopy or that you feel when you pull a sweater over your head.
Vacuum		Considered a perfect insulator, since, by definition, it contains no free charges at all. However, this doesn't mean it isn't possible for charges to pass through it. Various mechanisms can eject electrons from a material into a so-called vacuum. (This applies for air, too.) Examples of such mechanisms include *thermionic emission*, where an increase in temperature provides free electrons enough energy to overcome surface potential barriers (the work function) of the material; *field emission*, whereby additional energy from an electric field generated by a high-voltage conductor provides an attractive enough positive field to free electrons from the surface, which requires huge voltages (MV per cm between emitting surface and positive conductor); *secondary emission*, where electrons are emitted from a metallic surface by the bombardment of high-speed electrons or other particles; *photoelectric emission*, where electrons are ejected from the surface of a material when a photon of a particular frequency strikes it (absorbed electron). Many of these mechanisms are employed in vacuum tube technology. However, without such mechanisms to source electrons, there will be no source of charge within the vacuum to support a current.

2.6 Insulators, Conductors, and Semiconductors

As we have seen, the electrical resistivities of materials vary greatly between conductors and insulators. A decent conductor has around 10^{-8} Ωm; a good insulator has around 10^{14} Ωm; a typical semiconductor from 10^{-5} to 10^3 Ωm—depending on temperature. What is the microscopic explanation for these differences?

The answer to this question rests upon the quantum nature of electrons. In classical physics the energy of an electron in a metal can take on any value—it is said the energy values form a continuum. (Here, electron energy is considered zero at infinite distance from the nucleus and becomes more negative in energy closer to the nucleus, relative to the zero reference state. Negative energy infers that there is electric attraction between the positive nucleus and the electron—it is electric potential energy.) However, a quantum description of electrons in metals shows that the energy values of electrons are quantized, taking on discrete values. This comes from the wavelike nature of electrons—analogous to standing waves on strings existing only at discrete frequencies. Figure 2.20 shows an energy diagram illustrating the possible energy levels of an electron (ignoring lattice influences). The diagram illustrates only the possible energy levels—electrons are not necessarily in each level.

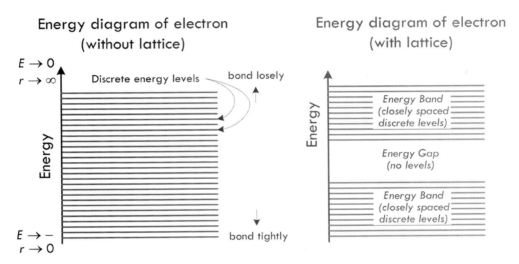

FIGURE 2.20   (Left) Energy diagram showing all possible energy levels of an electron in a solid; but it takes no account of the influence of an atomic crystalline lattice structure. **(Right)** Energy diagram that shows possible energy levels of an electron within a material made of a regular lattice of atoms. Electron energies are restricted to lie within allowed bands, and there is a large energy gap where no electrons are allowed. Even within the allowed bands, the possible electron energies are closely spaced discrete levels.

When a set of atoms forms a regular background lattice, the possible energy values of the electrons are altered even further. We still have discrete energy regions, called allowed bands, but we now get what are called energy gaps. Energy gaps are forbidden regions to the electron, and represent regions where no traveling wave (electron) can exist when placed in the periodic electric potential of the metal's positive lattice ions. These gaps are quite large in the scale of atomic physics—within electron-volts range. Again, energy levels presented in the band diagram specify only possible values of electron energies—they may or may not be occupied.

Now, quantum physics has an interesting property, called the *Pauli exclusion principle*, which has a critical role in determining the properties of materials. The Pauli exclusion principle says that no two electrons in an atom can be in the same quantum state. The lowest common divisor of quantum states is the spin quantum number m_s, which states that no more than two electrons with opposing spin (up or down) can be located in the same energy level. Now, if we consider a solid that has many free electrons, which is in an equilibrium state, electrons fill the lowest energy levels available in the allowed band, up to two in each level. Those electrons further down in energy are more tightly bound and are called innermost electrons. When all the electrons are placed in the lowest energy state, we are left with two possible outcomes. In the first case, the highest level to be filled is somewhere in the middle of a band. In the second case, the electrons just fill one or more bands completely. We assume that the material is at low enough temperature to prevent electrons from jumping to higher energy levels due to thermal effects.

Now if we add some energy to the free electrons by applying an electric field (attach a voltage source), for example, the electrons in the lower energy levels cannot accept that energy, because they cannot move into an already filled higher energy level. The only electrons that can accept energy are those that lie in the top levels, and then only if there are nearby empty levels into which they can move. Materials with electrons in only partly filled bands are *conductors.* When the top layer of their electrons moves freely into the empty energy levels immediately above, there is a current. The electrons that jump from a lower level to a higher level are said to be excited. The valance band is an allowed occupied band. The conduction band is an allowed empty band. The energy-band structure for conductors is shown in Fig. 2.21a and c.

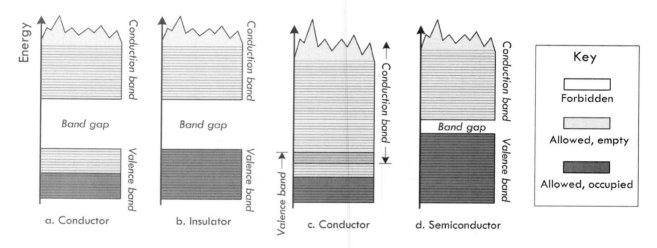

FIGURE 2.21 Four possible band structures for a solid: (a) Conductor—allowed band is only partially full, so electrons can be excited to nearby states. (b) Insulator—forbidden band with large energy gap between the filled band and the next allowed band. (c) Conductor—allowed bands overlap. (d) Semiconductor—energy gap between the filled band and the next allowed band is very small, so some electrons are excited to the conduction band at normal temperatures, leaving holes in the valence band.

If the highest-energy electrons of a material fill a band completely, then a small electric field will not give these electrons enough energy to jump the large energy gap to the bottom of the next (empty) band. We then have an *insulator* (see Fig. 2.21b). An example of a good insulator is diamond, whose energy gap is 6 eV.

In *semiconductors,* the highest-energy electrons fill a band (the valence band) at T = 0, as in insulators. However, unlike insulators, semiconductors have a small energy gap between that band and the next, the conduction band. Because the energy gap is so small, a modest electric field (or finite temperature) will allow electrons to jump the gap and thereby conduct electricity. Thus, there is a minimum electric field under the influence of which a material changes from insulator to conductor. Silicon and germanium have energy gaps of 1.1 eV and 0.7 eV, respectively, and are semiconductors. For semiconductors, an increase in temperature will give a fraction of the electrons enough thermal energy to jump the gap. For an ordinary conductor, a rise in temperature increases the resistivity, because the atoms, which are obstacles to electron flow, vibrate more vigorously. A temperature increase in a semiconductor allows more electrons into the empty band and thus lowers the resistivity.

When an electron in the valance band of a semiconductor crosses the energy gap and conducts electricity, it leaves behind what is known as a hole. Other electrons in the valence band near the top of the stack of energy levels can move into this hole, leaving behind their own holes, in which still other electrons can move, and so forth. The hole behaves like a positive charge that conducts electricity on its own as a positive charge carrier. An electron excited from the valence band to the conduction band is thus doubly effective at conducting electricity in semiconductors.

Besides the intrinsic elemental semiconductors, such as silicon and germanium, there are hybrid compounds—compounds such as gallium arsenide. Other semiconductors are made by introducing impurities into a silicon lattice. For example, an atom in the chemical group of phosphorous, arsenic, and antimony can replace one of the silicon atoms in a lattice without affecting the lattice itself too much. However, each of these impurities has one more electron in its valence level than the silicon atom has; this extra electron, for which there is no room in the valence band, takes a place in the conduction band and can conduct electricity. A semiconductor with impurities of this sort is called an *n-type semiconductor,* and the extra electrons are called donor electrons.

Atoms of elements in the same chemical group as boron, aluminum, and gallium have one less valence electron than silicon has. If an atom is added to a lattice of silicon as an impurity, there is one less electron than is needed to form a bond that holds the lattice together. This electron must be provided by the electrons of the valence band of the lattice material, and holes are created in this band. These holes act as positive charge carriers. The impurity atoms are called acceptors. A semiconductor with such impurities is called a *p-type semiconductor.*

We will see later on how *n*-type and *p*-type semiconductors are used for making one-way gates for current flow (diodes) and voltage-controlled current switches (transistors).

2.7   Heat and Power

In Sec. 2.3, we discovered the generalized power law, which tells us that if we can measure the current entering a device, as well as the voltage across the device, we know the power that is used by the device:

$$P = VI \tag{2.9}$$

The generalized power law tells us how much power is pumped into a circuit but doesn't say anything about how this power is used up. Let's consider a two-lead black box—an unknown circuit that may contain all sorts of devices, such as resistors, lamps, motors, or transistors. If all we can do is measure the current entering the black box and the voltage across it—say, using an ammeter and voltmeter (or singularly, a wattmeter)—we could apply the generalized power law, multiply the measured current and voltage readings together, and find the power pumped into the black box. For example, in Fig. 2.22, we measure 0.1 A entering when a voltage of 10 V is applied, giving us a total consumed power of 1 W.

Understanding $P = IV$ and $V = IR$

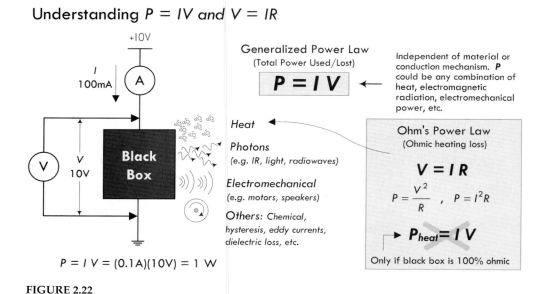

$$P = I\,V = (0.1A)(10V) = 1\ W$$

FIGURE 2.22

Knowing how much power is pumped into the black box is incredibly useful— it allows for quick power consumption measurements and often simplifies circuit analysis—as we'll see later. But let's say that we are interested in figuring out how much power is lost to heating (energy that goes into lattice vibrations, emissive radiation, etc.). We really can't say, assuming we aren't allowed to look inside the black box. There could be devices inside that take some of the initial energy and use it to do useful work, such as generating magnetic fields in the armature (rotor) and stator sections of a motor, causing the stator to rotate; or generating a magnetic field in a voice coil attached to a paper speaker cone that compresses air; or generating light energy, radio waves, and so on. There may be power converted into other weird forms not really coined as heat, such as driving chemical reactions, generating hysteresis effects, or eddy current in transformers.

The only time we can say for certain that power is totally converted into heat energy is if we assume our black box is a perfect resistor (100 percent ohmic in nature). Only then can we substitute Ohm's law into the generalized power equation:

$$P = VI = V(V/R) = V^2/R \tag{2.10}$$

or

$$P = VI = (IR)I = I^2R$$

In this form, the power lost due to heating is often called *Ohmic heating, Joule heating,* or *I^2R loss.* Be careful how you interpret this law. For example, let's consider our black box that drew 1 W of power. Given the power and the current—which we measure— it would be easy to assume that the resistance of the black box is:

$$R = \frac{P}{I^2} = \frac{1\ W}{(0.01\ A)} = 100\ \Omega$$

Accordingly, we would say the black box is a 100-Ω resistor generating 1 W of heat. As you can see, this is an erroneous assumption, since we have disregarded the internal workings of the black box—we didn't account for devices that perform useful work. You'll often see people treat any load (black box) as a resistor when doing circuit analysis and such. This will get you the right answer when solving for a particular variable, but it is an analysis trick and shouldn't be used to determine how much heat is being generated, unless, of course, the black box is actually a resistor.

The following example provides some insight into where power is being used and how much of it is being converted into heat.

EXAMPLE 1 (POWER LOSS IN CIRCUITS)

The total power pumped into this circuit gets converted into useful work and heat. The total "pump-in" power is:

$$P_{tot} = IV = (0.757 \text{ A})(12 \text{ V}) = 9.08 \text{ W}$$

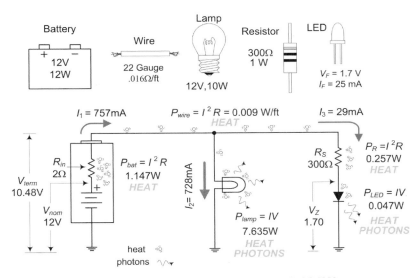

$$P_{tot} = P_{bat} + P_{wire} + P_{lamp} + P_R + P_{LED} = 9.086W$$

FIGURE 2.23

(This is based on an open terminal battery measurement, without the rest of the circuit attached.) From there, we notice that some of the total power is wasted within the internal resistance of the battery, within the internal resistance of the wires, and within the resistance of the current-limiting resistor used by the LED. The power used to create light from the lamp and LED can be considered useful power used. However, since we can't really separate the heat from the light power for these devices, we must apply the generalized power law to them and be content with that. According to the conservation of energy (or power), all individual powers within the circuit add up to the total power.

Example: With an ammeter and voltmeter, you measure the current drawn by a computer to be 1.5 A and the voltage entering to be 117 V. How much power does the computer consume? Can we say how much power is lost to heat?

Answer: $P = IV = (1.5 \text{ A})(117 \text{ V}) = 176 \text{ W}$. Knowing how much of the power is lost to heating is practically impossible to measure without taking the computer apart.

Example: Determine the resistance of the following four round rods of material, each 1 m long and 2 mm in diameter: copper, brass, stainless steel, and graphite. Also figure out how much power is lost to heating if a current of 0.2 A flows through each one.

Answer: Using Eq. (2.5),

$$R = \rho \frac{L}{A} = \rho \frac{L}{\pi r^2} = \rho \frac{1 \text{ m}}{\pi (0.001 \text{ m})^2} = \rho \frac{1 \text{ m}}{3.14 \times 10^{-6} \text{ m}^2} = \rho(3.18 \times 10^5 \text{ m}^{-1})$$

Using Table 2.2:

$$\rho_{copper} = 1.72 \times 10^{-8}\ \Omega m,\ \rho_{brass} = 7.0 \times 10^{-8}\ \Omega m,\ \rho_{steel} = 7.2 \times 10^{-7}\ \Omega m,\ \rho_{graphite} = 3.5 \times 10^{-5}\ \Omega m$$

Substituting this into the resistance expression, we get:

$$R_{copper} = 5.48 \times 10^{-3}\ \Omega,\ R_{brass} = 2.23 \times 10^{-2}\ \Omega,\ R_{steel} = 2.31 \times 10^{-1}\ \Omega,\ R_{graphite} = 11.1\ \Omega$$

The power loss we get by using Eq. 2.10: $P = I^2R = (0.2\ \text{A})^2R = (0.04\ \text{A}^2)R$:

$$P_{copper} = 2.2 \times 10^{-4}\ \text{W},\ P_{brass} = 8.9 \times 10^{-4}\ \text{W},\ P_{steel} = 9.2 \times 10^{-3}\ \text{W},\ P_{graphite} = 0.44\ \text{W}$$

2.8 Thermal Heat Conduction and Thermal Resistance

How is the energy transferred in heating? Within a *gas*, heat transfer represents the transfer of energy between colliding gas molecules. Gas molecules at a hotter temperature move around more quickly—they have a high kinetic energy. When they are introduced into another, colder-temperature gas, the "hotter," fast-moving molecules impart their energy to the slower-moving molecules. Gases tend to be the worst thermal conductors, due to a low density of molecules.

In *nonmetals*, heat transfer is a result of the transfer of energy due to lattice vibrations—energetically vibrating atoms in one region of a solid (e.g., the region near a flame) transfer their energy to other regions of a solid that have less energetically vibrating atoms. The transfer of heat can be enhanced by cooperative motion in the form of propagating lattice waves, which in the quantum limit are quantized as *phonons*. The thermal conductivity of nonmetals varies, depending greatly on the lattice structure.

In terms of *metals*, heat transfer is a result of both lattice vibration effects (as seen in nonmetals) as well as kinetic energy transfer due to mobile free electrons. Recall that free electrons within a metal are moving quite fast ($\sim 10^6$ m/s for most metals) at room temperature. Though quantum mechanics is required, it is possible to treat these electrons like a dense gas, capable of increasing its overall energy as heat is added, and likewise, capable of transporting this energy to regions of the metal that are lower in temperature. Notice, however, that an increase in metal temperature, as a whole, also increases the electrical resistance—the drift velocity component of the free electron goes down, due to both increased lattice vibrations and an increase in thermal velocity component. It becomes harder to influence the free electrons with an applied field. Metals are the best thermal conductors, due in part to the additional free electrons.

The energy an object has at temperature T is correlated to the internal energy—a result of its internal atomic/molecular/electron motions. It is not correct, however, to use the word *heat*, such as "object possesses heat." Heat is reserved to describe the process of energy transfer from a high-temperature object to a lower-temperature object. According to the first law of thermodynamics, which is a statement about the conservation of energy, the change in internal energy of a system ΔU is equal to the heat Q_H added to the system and the work W done by the system: $\Delta U = Q_H - W$. If we assume that no work is done (energy transferred to, say, move a piston, in the case of a gas), we say $\Delta U = Q_H$. With this assumption, we can take heat to be not a measure of internal energy of the system, but a change in internal energy. The main reason for this concept is that it is very difficult to determine the actual internal energy of a system; changes in internal energy are more meaningful and measurable.

In practice, what is most useful is the rate of heat transfer—the power loss due to heating. With the help of experimental data, the following formula can be used to determine how well certain materials transfer heat:

$$P_{\text{heat}} = \frac{dQ_H}{dt} = -k\nabla T \tag{2.11}$$

Here, k is called the *thermal conductivity* (measured in W/m °C) of the material in question, and ∇T is the temperature gradient:

$$\nabla T = \left(i\frac{\partial}{\partial t} + j\frac{\partial}{\partial t} + k\frac{\partial}{\partial t} \right)T$$

Now, the gradient is probably scary to a lot of you—it is simply a way to represent temperature distributions in 3-D, with time. To keep things simple, we will stick to 2-D, and represent the gradient as acting through an area A through a thickness L, and assume steady-state conditions:

$$P_{\text{heat}} = -k\frac{A\Delta T}{L} \tag{2.12}$$

where $\Delta T = T_{\text{hot}} - T_{\text{cold}}$, measured at points across the length L of the material. The material may be steel, silicon, copper, PC board material, and so on. Figure 2.24 shows a picture of the situation.

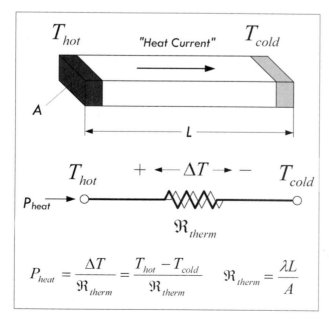

When one end of a block of material is placed at a hot temperature, heat will be conducted through the material to the colder end. The rate of heat transfer, or power due to heating, depends on the thermal resistance of the material, which in turn is dependent on the geometry of the material and the material's thermal resistivity. A weird-looking resistor-like symbol is used here to represent the thermal resistance. Table 2.4 shows the thermal resistivities of various materials.

FIGURE 2.24

TABLE 2.4   Typical Thermal Resistivities (λ) in Units of °C·in/W

Divide λ by 39 to get value in units of °C·m/W

MATERIAL	λ	MATERIAL	λ	MATERIAL	λ
Diamond	0.06	Lead	1.14	Quartz	27.6
Silver	0.10	Indium	2.1	Glass (774)	34.8
Copper	0.11	Boron nitride	1.24	Silicon grease	46
Gold	0.13	Alumina ceramic	2.13	Water	63
Aluminum	0.23	Kovar	2.34	Mica	80
Beryllia ceramic	0.24	Silicon carbide	2.3	Polyethylene	120
Molybdenum	0.27	Steel (300)	2.4	Nylon	190
Brass	0.34	Nichrome	3.00	Silicon rubber	190
Silicon	0.47	Carbon	5.7	Teflon	190
Platinum	0.54	Ferrite	6.3	P.P.O.	205
Tin	0.60	Pyroceram	11.7	Polystyrene	380
Nickel	0.61	Epoxy (high conductivity)	24	Mylar	1040
Tin solder	0.78			Air	2280

Thermal conductivity k, like electrical conductivity, has an inverse—namely, *thermal resistivity* λ. Again, one tells you "how good" a material is at transferring heat, the other tells you "how bad" it is at doing it. The two are related by $k = 1/\lambda$.

If we consider the geometry of the material, we can create a notion of *thermal resistance* $\Re_{\text{therm}}$ (analogous to electrical resistance), which depends on the cross-sectional area A, the length of the block of material L, and the thermal conductivity k, or resistivity λ:

$$\Re_{\text{therm}} = \frac{L}{kA} \quad \text{or} \quad \Re_{\text{therm}} = \frac{\lambda L}{A} \tag{2.13}$$

Thermal resistance has units of °C/W.

Thus, putting everything together, the power transfer of heat across a block of material, from one point at one temperature to another point at a different temperature, can be expressed as:

$$P_{\text{heat}} = \frac{dQ_{\text{heat}}}{dt} = k\left(\frac{A}{L}\right)\Delta T = \frac{1}{\lambda}\left(\frac{A}{L}\right)\Delta T = \frac{\Delta T}{\Re_{\text{therm}}} \tag{2.14}$$

(k = thermal conductance) (λ = thermal resistivity) ($\Re_{\text{therm}}$ = thermal resistance)

A very useful property of Eq. 2.14 is that it is exactly analogous to Ohm's law, and therefore the same principles and methods apply to heat flow problems as to circuit problems. For example, the following correspondences hold:

Thermal conductivity k [W/(m °C)]	Electrical conductivity σ [S/m or $(\Omega m)^{-1}$]
Thermal resistivity λ [m°C/W]	Electrical resistivity ρ [Ωm]
Thermal resistance $\mathfrak{R}_{therm}$ [°C/W]	Electrical resistance R [Ω]
Thermal current (heat flow) P_{heat} [W]	Electrical current I [A]
Thermal potential difference ΔT [°C]	Electrical potential difference or voltage V [V]
Heat source	Current source

Example: Calculate the temperature of a 4-in (0.1 m) piece of #12 copper wire at the end that is being heated by a 25-W (input power) soldering iron, and whose other end is clamped to a large metal vise (assume an infinite heat sink), if the ambient temperature is 25°C (77°F).

Answer: First, calculate the thermal resistance of the copper wire (diameter #12 wire is 2.053 mm, cross-sectional area is 3.31×10^{-6} m²):

$$\mathfrak{R}_{therm} = \frac{L}{kA} = \frac{(0.1 \text{ m})}{(390 \text{ W/(m°C)}(3.31 \times 10^{-6} \text{ m}^2))} = 77.4°C/W$$

Then, rearranging the heat flow equation and making a realistic assumption that only around 10 W of heat actually is transferred to the wire from the 25-watt iron, we get:

$$\Delta T = P_{heat}\mathfrak{R}_{therm} = (10 \text{ W})(77.4°C/W) = 774°C$$

So the wire temperature at the hot end is estimated at:

$$25°C + \Delta T = 799°C \text{ (or } 1470°F)$$

It's important to realize that in the preceding example, we assume steady-state conditions, where the soldering iron has been held in place for a long time. Also, the assumption that only 10 W were transferred is an important point, since there is a lot of heat being radiated off as heat to the air and the iron handle, and so on. In any case, things can get very hot, even when moderate power levels are in question.

2.8.1 *Importance of Heat Production*

Heat production has its place in electronics (toasters, hair dryers, water heaters, etc.), but most of the time heat represents power loss that is to be minimized whenever possible, or at least taken into consideration when selecting components. All real circuit components—not just resistors but things like capacitors, transformers, transistors, and motors—contain inherent internal resistances. Though these internal resistances can often be neglected, in some situations they cannot be ignored.

Major problems arise when unintended heat generation increases the temperature of a circuit component to a critical point, causing component failure by explosion, melting, or some other catastrophic event. Less severe problems may surface as a component becomes thermally damaged, resulting in a change in characteristic properties, such as a shift in resistance that may cause undesirable effects in circuit behavior.

To avoid problems associated with heat production, it's important to use components that are rated to handle two to three or more times the maximum power they are expected to dissipate. In cases where heat presents a shift in component parameter performance, selecting a component with a lower temperature coefficient (TC) will help.

Heat dissipation (more correctly, the efficient removal of generated heat) becomes very important in medium- to high-power circuits—power supplies, amplifier stages, transmitting circuits, and power-hungry circuits with power transistors. There are various techniques to remove heat from a circuit in order to lower the operating temperature of components to below critical levels. Passive methods include heat sinks, careful component layout, and ventilation. Heat sinks are special devices that are used to draw heat away from temperature-sensitive devices by increasing the radiating surface in air—which acts like a cooling fluid for conduction. Active methods include forced air (fans) or some sort of liquid cooling. We'll discuss these methods throughout the book.

Example: Figure 2.25 shows a thin-film resistor in an integrated circuit. How hot will it get with 2 W dissipated over its 0.1 × 0.2-in surface (100 W/in²)? Assume the ground plane is at 80°C.

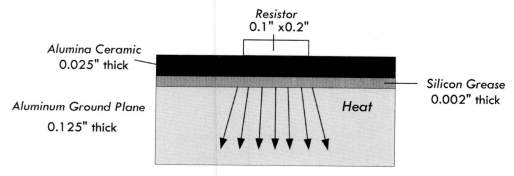

FIGURE 2.25

Answer: Since we have three different media through which this heat will be transferred, we must take into consideration thermal conduction within each. With the help of Eq. 2.14 and Table 2.4, we get the individual heat transfers through each region:

$$\Delta T_{1-2(\text{in ceramic})} = \lambda_{\text{ceramic}}\left(\frac{L}{A}\right)P_{\text{dis}} = \frac{(2.13)(0.025)(2)}{(0.1)(0.2)} = 5.3°C$$

$$\Delta T_{2-3(\text{in grease})} = \lambda_{\text{grease}}\left(\frac{L}{A}\right)P_{\text{dis}} = \frac{(46)(0.002)(2)}{(0.1)(0.2)} = 9.2°C$$

$$\Delta T_{3-4(\text{in aluminum})} = \lambda_{\text{aluminum}}\left(\frac{L}{A}\right)P_{\text{dis}} = \frac{(0.23)(0.125)(2)}{(0.1)(0.2)} = 2.9°C$$

Adding this together:

$$\Delta T_{1-4} = 5.3°C + 9.2°C + 2.9°C = 17.4°C$$

Adding this to the 80°C for the aluminum ground plane leads to an estimated 100°C for maximum resistor temperature. This is a conservative estimate, since it neglects transverse heat spreading.

2.9 Wire Gauges

In Sec. 2.5, we saw that the current density within a copper wire increased as the diameter of the wire decreased. As it turns out, a higher current density translates into a hotter wire; there are more collisions occurring between electrons and copper lattice ions. There is a point where the current density can become so large that the vibrating effect can overpower the copper-lattice binding energy, resulting in wire meltdown (also referred to as the fusing point). To prevent this from occurring, it is important to select the appropriate wire size for anticipated current levels. Wire size is expressed in gauge number—the common standard being the American Wire Gauge (AWG)—whereby a smaller gauge number corresponds to a larger-diameter wire (high current capacity). Table 2.5 shows a short list of common AWG wires. Section 3.1, on wires and cables, provides a more in-depth list.

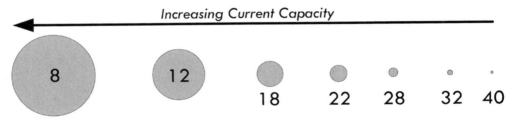

FIGURE 2.26

TABLE 2.5 **Copper Wire Specifications (Bare and Enamel-Coated Wire)**

WIRE SIZE (AWG)	DIAMETER (MILS)*	AREA (CM)†	FEET PER POUND BARE	OHMS PER 1000 FT, 25°C	CURRENT CAPACITY (AMPS)
4	204.3	41738.49	7.918	0.2485	59.626
8	128.5	16512.25	25.24	0.7925	18.696
10	101.9	10383.61	31.82	0.9987	14.834
12	80.8	6528.64	50.61	1.5880	9.327
14	64.1	4108.81	80.39	2.5240	5.870
18	40.3	1624.09	203.5	6.3860	2.320
20	32	1024.00	222.7	10.1280	1.463
22	25.3	640.09	516.3	16.2000	0.914
24	20.1	404.01	817.7	25.6700	0.577
28	12.6	158.76	2081	65.3100	0.227
32	8.0	64.00	5163	162.0000	0.091
40	3.1	9.61	34364	1079.0000	0.014

* 1 mil = 0.001 in or 0.0254 mm.
† A circular mil (CM) is a unit of area equal to that of a 1-mil-diameter circle. The CM area of a wire is the square of the mil diameter.
Diameters of wires in Fig. 2.26 are relative and not to scale.

Example: A load device that is known to vary in output power from 0.1 mW to 5 W is to be connected to a 12-V source that is 10 ft away from the load. Determine the minimum wire gauge, provided by Table 2.5, that can safely support any anticipated current drawn by the load.

Answer: We only care about the maximum power level, so using the generalized power law:

$$I = \frac{P}{V} = \frac{5 \text{ W}}{12 \text{ V}} = 0.42 \text{ A}$$

Given only the selection of wire gauges provided by Table 2.5, a 22-gauge wire with a 0.914-A rating would work, though we could be conservative and select an 18-gauge wire with a 2.32-A rating. Since the length is so short, there is no appreciable drop in voltage through the wire, so we can ignore the length.

Example: A 10-Ω heating device is powered by a 120-VAC source. How much current does it draw, and what size conductors should be used to connect to the device?

Answer: 120 VAC is an RMS value of a sinusoidal voltage—in this case, household line voltage. Though we'll discuss this later when we cover ac, it can be treated like a dc voltage in terms of power dissipated through a resistor. So,

$$P = \frac{V^2}{R} = \frac{(120 \text{ V}^2)}{10 \text{ }\Omega} = 1440 \text{ W}, \quad I\frac{P}{V} = \frac{1440 \text{ W}}{120 \text{ V}} = 12 \text{ A}$$

A 10-gauge wire would support this kind of current, though a larger 8-gauge wire would be safer.

Example: Why shouldn't you connect a wire across a voltage source? For example, if you connect a 12-gauge wire directly across a 120-V source (120-V mains outlet), what do you think will happen? What will happen when you do this to a 12-V dc supply, or to a 1.5-V battery?

Answer: In the 120-V mains case, you will likely cause a huge spark, possibly melting the wire and perhaps in the process receiving a nasty shock (if the wire isn't insulated). But more likely, your circuit breaker in the home will trip, since the wire will draw a huge current due to its low resistance—breakers trip when they sense a large level of current flowing into one of their runs. Some are rated at 10 A, others at 15 A, depending on setup. In a good dc supply, you will probably trip an internal breaker or blow a fuse, or in a bad supply, ruin the inner circuitry. In the case of a battery, there is internal resistance in the battery, which will result in heating of the battery. There will be less severe levels of current due to the internal resistance of the battery, but the battery will soon drain, possibly even destroying the battery, or in an extreme case causing the battery to rupture.

2.10   Grounds

As we left off in Sec. 2.3, we saw that understanding voltage is a relativity game. For example, to say that a point in a circuit has a voltage of 10 V is meaningless unless you have another point in the circuit with which to compare it. Often you define a point in the circuit to be a kind of 0-V reference point on which to base all other voltage measurements. This point is often called the *ground,* and is frequently represented by the symbol shown in Fig. 2.27:

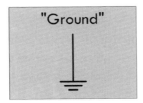

FIGURE 2.27

For example, Fig. 2.28 shows various ways in which to define voltages by selecting a ground—which in this case is simply a 0-V reference marker. The single battery provides a 1.5-V potential difference or voltage between its terminals. We can simply place a 0-V reference ground at the negative terminal and then state that the positive terminal sits at 1.5 V relative to the 0-V reference ground. The 0-V reference, or the negative terminal of the battery, is called the *return.* If a load, such as a lamp or resistor, is placed between the terminals, a load current will return to the negative terminal.

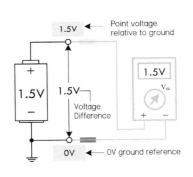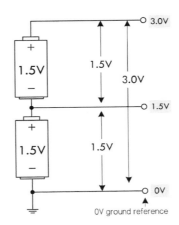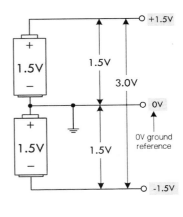

FIGURE 2.28

In the center diagram in Fig. 2.28, we have two 1.5-V batteries placed end to end. When batteries are linked this way, their voltages add, creating a combined total voltage of 3.0 V. With the 0-V reference ground at the bottom, we get 1.5-V and 3.0-V readings at the locations shown in the figure. A load placed across the two batteries (3.0-V difference) will result in a load current that returns to the lower battery's negative terminal. In this case, the return is through the 0-V reference—the lower battery's negative terminal.

Finally, it is possible to create a split supply by simply repositioning the 0-V ground reference, placing it between the batteries. This creates +1.5 V and −1.5 V leads relative to the 0-V reference. Many circuits require both positive and negative voltage relative to a 0-V ground reference. In this case, the 0-V ground reference acts as a *common return.* This is often necessary—say, in an audio circuit—where signals are sinusoidal and alternate between positive and negative voltage relative to a 0-V reference.

Now, the ground symbol shown in Fig. 2.27 as a 0-V reference, or as a return, is used all the time by various people. As it turns out, however, it really is supposed to represent a true earth ground—a physical connection to the earth through a conductive material buried in the earth. For whatever reasons, the symbol's dual meaning has survived, and this can often be a source of confusion to beginners.

2.10.1 Earth Ground

The correct definition of an *earth ground* is usually a connection terminated at a rod driven into the earth to a depth of 8 ft or more. This earth ground rod is wired directly to a mains breaker box's ground bar and sent to the various ac outlets in one's home via a green-coated or bare copper wire that is housed within the same mains cable as hot and neutral wires. The ground can then be accessed at the outlet at the ground socket. Metal piping buried in the earth is often considered an earth ground. See Fig. 2.29.

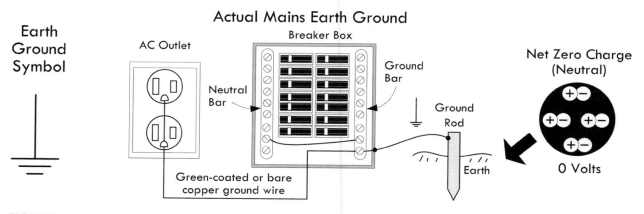

FIGURE 2.29   Earth ground.

A physical link to the earth is important because the earth provides an electrically neutral body; equal numbers of positive and negative charges are distributed through its entirety. Due to the earth's practically infinite charge neutrality, attempts at changing the earth potential, via electrical generators, batteries, static electrical mechanisms, or the like, will have essentially no measurable effect. Any introduction of new charge into the earth is quickly absorbed (the earth's moist soil is usually rather conductive). Such charge interactions occur constantly throughout the planet, and the exchanges average out to zero net charge.

For practical purposes, then, the earth is defined to be at a zero potential (relative to other things)—a potential that is practically immune to wavering. This makes the earth a convenient and useful potential on which to reference other signals. By connecting various pieces of electronics equipment to the earth ground, they can all share the earth's ground reference potential, and thus all devices share a common reference.

The actual physical connection to earth ground at a particular piece of equipment is usually through the power cord's ground wire that links to the mains ground wire network when the device is plugged in. The ground wire from the power cord is typically connected internally to the equipment chassis (frame) and, more important for our discussion, to the return portion of a channel that emanates from the interior circuitry. This is then brought out as a ground lead terminal. For example, in Fig. 2.30, an oscilloscope, function generator, and generic audiovisual device use

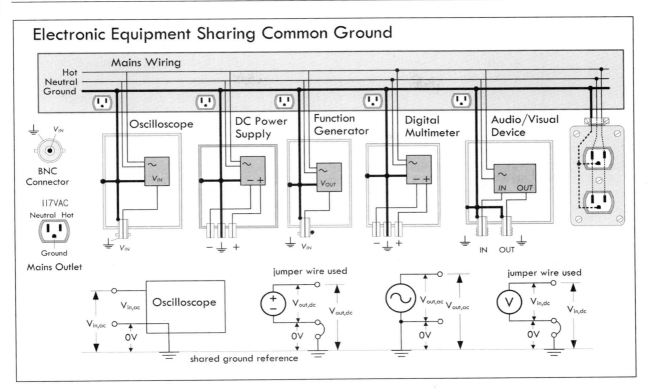

FIGURE 2.30 Illustration showing how various test instruments and an audiovisual device share a common ground connection through the mains ground wiring.

BNC and UHF connectors for input and output channels. Internally, the outer connector body of the BNC or UHF jack is wired to the return (or source) portion of the channel, while a central conductor wire (insulated from the outer body) is wired to the source (or return) portion of the channel. The important part, now, is that the return, or outer, connector is also internally wired to the mains ground wire through the power cord cable. This sets the return to an earth ground reference. In the case of the dc power supply, a separate earth ground terminal is presented at the face in the form of a banana jack terminal. In order to ground the dc supply, a jumper wire must be connected between the negative supply terminal and the ground terminal. If no jumper is used, the supply is said to be *floating*.

All the grounded pieces of equipment share a common ground. To prove this to yourself, try measuring the resistance between the ground terminals of any two separate pieces of test equipment in your lab. If the devices are properly grounded, you will get a measurement of 0 Ω (but a bit more for internal resistances).

Besides acting simply as a reference point, grounding reduces the possibility of electrical shock if something within a piece of equipment should fail and the chassis or cabinet becomes "hot." If the chassis is connected to a properly grounded outlet via a three-wire electrical system ground, the path of current flow from the hot chassis will be toward ground, not through your body (which is a more resistive path). A ground system to prevent shock is generally referred to as a dc ground. We'll discuss shock hazards and grounding later, when we cover ac.

Grounding also helps eliminate electrostatic discharge (ESD) when a statically charged body comes in contact with sensitive equipment. The charged body could be you, after a stroll you took across the carpet. Some ICs are highly vulnerable to damage from ESD. By providing a grounded work mat or using a grounded wrist strap while working with sensitive ICs, you can avoid destroying your chips by ensuring that charge is drained from your body before you touch anything.

Another big job the ground system does is provide a low-impedance path to ground for any stray RF current caused by stray radiofrequency-producing devices, such as electrical equipment, radio waves, and so on. Stray RF can cause equipment to malfunction and contributes to RFI problems. This low-impedance path is usually called RF ground. In most cases, the dc ground and the RF ground are provided by the same system.

Common Grounding Error

As previously mentioned, the ground symbol, in many cases, has been used as a generic symbol in circuit diagrams to represent the current return path, even though no physical earth ground is used. This can be confusing for beginners when they approach a three-terminal dc power supply that has a positive (+), negative (−), and ground terminal. As we have learned, the ground terminal of the supply is tied to the case of the instrument, which in turn is wired to the mains earth ground system. A common mistake for a novice to make is to attempt to power a load, such as a lamp, using the positive and ground terminals of the supply, as shown in Fig. 2.31a. This, however, doesn't complete a current return path to the energy source (supply), so no current will flow from the source; hence, the load current will be zero. The correct procedure, of course, is to either connect the load between the positive and negative terminals directly, thus creating a *floating load*, or, using a jumper wire between the ground and negative supply, create a *grounded load*. Obviously, many dc circuits don't need to be grounded—it will generally neither help nor hinder performance (e.g., battery-powered devices need no such connection).

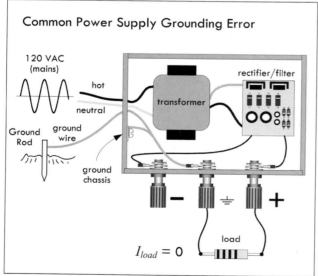

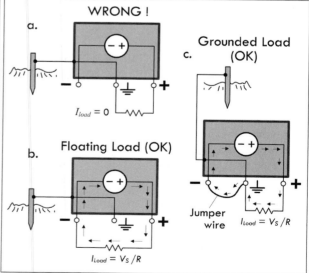

FIGURE 2.31

Circuits that require both positive and negative voltage require a power supply to provide each polarity. The supply for the positive voltage will have the negative terminal as a return, and the negative supply will have the positive terminal as the return. These two terminals are connected together, forming a common return path for load current, as shown in Fig. 2.32. The connection between the negative and the positive terminals of the supplies results in a *common* or *floating return*. The floating common may be connected to the earth ground terminal of a supply, if a particular circuit requests this. Generally, it will neither help nor hinder circuit performance.

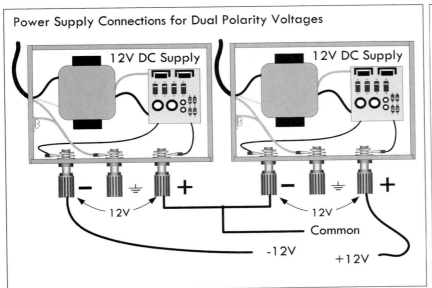
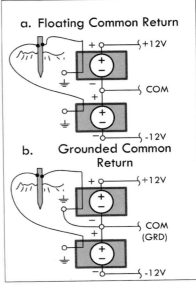

FIGURE 2.32

Unfortunately, the earth ground symbol is used a bit too loosely in electronics, often meaning different things to different people. It is used as a 0-V reference, even though no actual connection is made to Earth. Sometimes it actually means to connect a point in a circuit to earth ground. Sometimes it is used as a generic return—to eliminate the need to draw a return wire. It could be used as an actual earth ground return (using the mains ground copper wire), though this is unwise. See Fig. 2.33. To avoid complications, alternative symbols are used, which we discuss next.

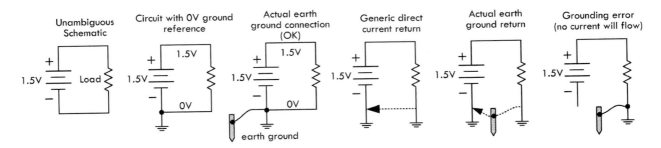

FIGURE 2.33

2.10.2 Different Types of Ground Symbols

To avoid misinterpretations regarding earth grounding, voltage references, and current return paths, less ambiguous symbols have been adopted. Figure 2.34 shows an earth ground (could be Earth or reference), a frame or chassis ground, and a digital and analog reference ground. Unfortunately, the common return for digital and analog is a bit ambiguous, too, but usually a circuit diagram will specify what the symbol is referring to.

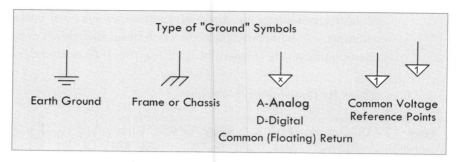

FIGURE 2.34

Table 2.6 provides a rundown on meanings behind these symbols.

TABLE 2.6   Types of Ground Symbols

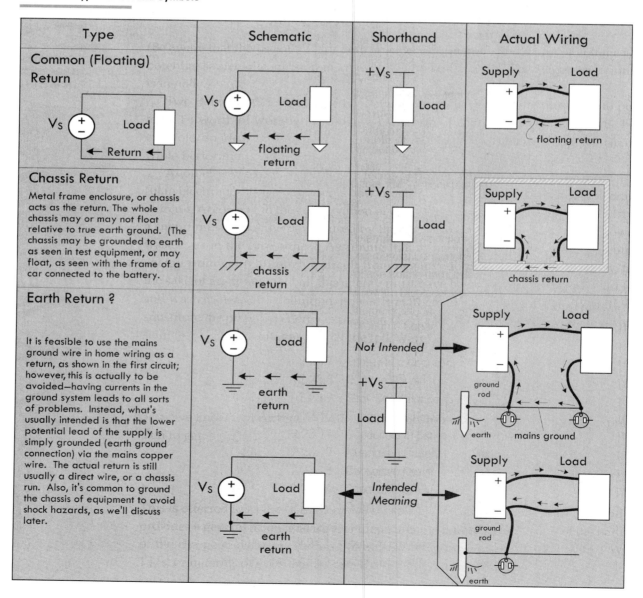

2.10.3 Loose Ends on Grounding

There are a few loose ends on grounding that need mentioning. They are discussed here, with reference to Fig. 2.35.

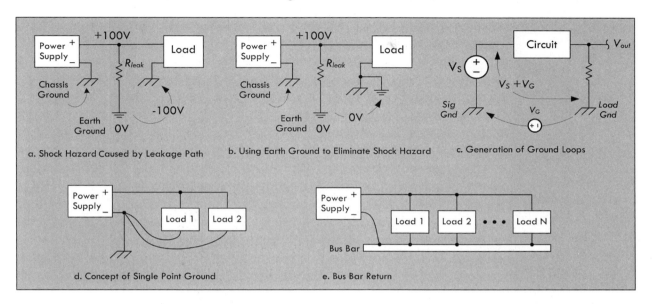

a. Shock Hazard Caused by Leakage Path   b. Using Earth Ground to Eliminate Shock Hazard   c. Generation of Ground Loops

d. Concept of Single Point Ground   e. Bus Bar Return

FIGURE 2.35

Shock Hazard

In instances where high voltages are required and chassis grounds or metal frames are used as return paths, shock hazardous conditions can be created if the earth grounds are neglected. For example, in Fig. 2.35a when a load circuit uses a metal enclosure as a chassis ground, resistive leakage paths (unwanted resistive paths) can exist, which result in high voltages between the enclosure and earth ground. If, inadvertently, an earth-grounded object, such as a grounded metal pipe, and the circuit chassis are simultaneously touched, a serious shock will result. To avoid this situation, the chassis is simply wired to an earth ground connection, as shown in Fig. 2.35b. This places the metal pipe and the enclosure at the same potential, eliminating the shock hazard. Similar hazardous conditions can develop in household appliances. Electrical codes require that appliance frames, such as washers and dryers, be connected to earth ground.

Grounding and Noise

The most common cause of noise in large-scale electronic systems is lack of good grounding practices. Grounding is a major issue for practicing design and system engineers. Though it is not within the scope of this book to get into the gory details, we'll mention some basic practices to avoid grounding problems in your circuits.

If several points are used for ground connections, differences in potential between points caused by inherent impedance in the ground line can cause troublesome *ground loops,* which will cause errors in voltage readings. This is illustrated in Fig. 2.35c, where two separated chassis grounds are used. V_G represents a voltage existing between signal ground and the load ground. If voltage measurements are made between the load ground and the input signal, V_S, an erroneous voltage, $(V_S + V_G)$ is measured. A way to circumvent this problem is to use a single-point ground, as shown in Fig. 2.35d.

The single-point ground concept ensures that no ground loops are created. As the name implies, all circuit grounds are returned to a common point. While this approach looks good on paper, it is usually not practical to implement. Even the simplest circuits can have 10 or more grounds. Connecting all of them to a single point becomes a nightmare. An alternative is to use a ground bus.

A *ground bus,* or bus bar found in breadboards and prototype boards, or which can be etched in a custom printed circuit board (PCB), serves as an adequate substitute for a single-point ground. A bus bar is simply a heavy copper wire or bar of low resistance that can carry the sum total of all load currents back to the power supply. This bus can be extended along the length of the circuitry so that convenient connections can be made to various components spaced about the board. Figure 2.35e shows a bus bar return. Most prototype boards come with two or three lines of connected terminals extending along the length of the board. One of these continuous strips should be dedicated as a circuit ground bus. All circuit grounds should be tied directly to this bus. Care must be taken to make sure that all lead and wire connections to the bus are secure. For prototype boards, this means a good solder joint; for wire-wrap board, a tight wire wrap; for breadboard, proper gauge wire leads to securely fit within sockets. Bad connections lead to intermittent contact that leads to noise.

Analog and Digital Grounds

Devices that combine analog and digital circuitry should, in general, have their analog and digital grounds kept separate, and eventually connected together at one single point. This is to prevent noise from being generated within the circuits due to a ground current. Digital circuits are notorious for generating spikes of current when signals change state. Analog circuits can generate current spikes when load currents change or during slewing. In either case, the changing currents are impressed across the ground-return impedance, causing voltage variations (use Ohm's law) at the local ground plane with respect to the system reference ground, often located near the power supply. The ground return impedance consists of resistive, capacitive, and inductive elements, though resistance and inductance are predominant. If a constant current is impressed across the ground return, resistance is primary, and a dc offset voltage will exist. If the current is alternating, resistance, inductance, and capacitance all play a role, and a resulting high-frequency ac voltage will exist. In either case, these voltage variations get injected into the local circuits and are considered noise—capable of screwing up sensitive signal levels used within the local circuits. There are a number of tricks to reduce noise (such as adding capacitors to counterbalance the inductance), but a good trick is to keep the digital and analog ground separated, then attach them together at one single point.

Example: What do the following symbols represent?

a. b. c. d. e.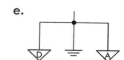

FIGURE 2.36

Answer: (a) Probably an analog circuit ground that is terminated at the supply to an actual earth ground connection. (b) A chassis ground that is connected to earth ground, probably to prevent shock hazards. (c) Appears like an analog ground return that is linked to both the chassis and the earth ground. (d) A floating chassis that is connected to circuit return ground—a potential shock hazard. (e) Separate analog and digital grounds that are linked to a common ground point near the supply, which in turn is grounded to earth.

2.11 Electric Circuits

Though we have already shown circuits, let's define circuits in basic terms. An electric circuit is any arrangement of resistors, wires, or other electrical components (capacitors, inductors, transistors, lamps, motors, etc.) connected together that has some level of current flowing through it. Typically, a circuit consists of a voltage source and a number of components connected together by means of wires or other conductive means. Electric circuits can be categorized as series circuits, parallel circuits, or a combination of series and parallel parts. See Fig. 2.37.

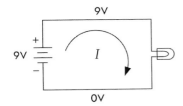

Basic Circuit

A simple lightbulb acts as a load (the part of the circuit on which work must be done to move current through it). Attaching the bulb to the battery's terminals, as shown, will initiate current flow from the positive terminal to the negative terminal. In the process, the current will power the filament of the bulb, and light will be emitted. (Remember that the term *current* here refers to conventional current—electrons are actually flowing in the opposite direction.)

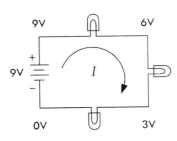

Series Circuit

Connecting load elements (lightbulbs) one after the other forms a series circuit. The current through all loads in series will be the same. In this series circuit, the voltage drops by a third each time current passes through one of the bulbs (all bulbs are exactly the same). With the same battery used in the basic circuit, each light will be one-third as bright as the bulb in the basic circuit. The effective resistance of this combination will be three times that of a single resistive element (one bulb).

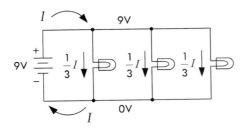

Parallel Circuit

A parallel circuit contains load elements that have their leads attached in such a way that the voltage across each element is the same. If all three bulbs have the same resistance values, current from the battery will be divided equally into each of the three branches. In this arrangement, lightbulbs will not have the dimming effect, as was seen in the series circuit, but three times the amount of current will flow from the battery, hence draining it three times as fast. The effective resistance of this combination will be one-third that of a single resistive element (one bulb).

FIGURE 2.37

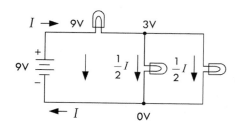

Combination of Series and Parallel

A circuit with load elements placed both in series and in parallel will have the effects of both lowering the voltage and dividing the current. The effective resistance of this combination will be three-halves that of a single resistive element (one bulb).

FIGURE 2.37   (*Continued*)

Circuit Analysis

Following are some important laws, theorems, and techniques used to help predict what the voltage and currents will be within a purely resistive circuit powered by a direct current (dc) source, such as a battery.

2.12   Ohm's Law and Resistors

Resistors are devices used in circuits to limit current flow or to set voltage levels within circuits. Figure 2.38 shows the schematic symbol for a resistor; two different forms are commonly used. Schematic symbols for variable resistors—resistors that have a manually adjustable resistance, as well as a model of a real-life resistor, are also shown. (The real-life model becomes important later on when we deal with high-frequency ac applications. For now, ignore the model.)

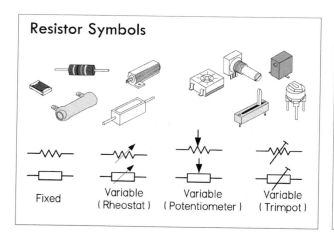

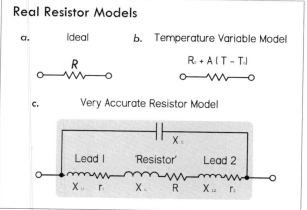

FIGURE 2.38

If a dc voltage is applied across a resistor, the amount of current that will flow through the resistor can be found using Ohm's law. To find the power dissipated as heat by the resistor, the generalized power (with Ohm's law substitution) can be used.

$$V = I \times R \qquad \text{(2.15) Ohm's law}$$

$$P = IV = V^2/R = I^2R \qquad \text{(2.16) Ohm's power law}$$

R is the resistance or the resistor expressed in ohms (Ω), P is the power loss in watts (W), V is the voltage in volts (V), and I is the current in amperes (A).

UNIT PREFIXES

Resistors typically come with resistance values from 1 Ω to 10,000,000 Ω. Most of the time, the resistance is large enough to adopt a unit prefix convention to simplify the bookkeeping. For example, a 100,000-Ω resistance can be simplified by writing 100 kΩ (or simply 100k, for short). Here k = ×1000. A 2,000,000-Ω resistance can be shortened to 2 MΩ (or 2M, for short). Here M = ×1,000,000.

Conversely, voltages, currents, and power levels are usually small fractions of a unit, in which case it is often easier to use unit prefixes such as m (milli or ×10^{-3}), µ (micro or ×10^{-6}), n (nano or ×10^{-9}), or even p (pico or ×10^{-12}). For example, a current of 0.0000594 A (5.94 × 10^{-5} A) can be written in unit prefix form as 59.4 µA. A voltage of 0.0035 (3.5 × 10^{-3} V) can be written in unit prefix form as 3.5 mV. A power of 0.166 W can be written in unit prefix form as 166 mW.

Example: In Fig. 2.39, a 100-Ω resistor is placed across a 12-V battery. How much current flows through the resistor? How much power does the resistor dissipate?

Answer: See Fig. 2.39.

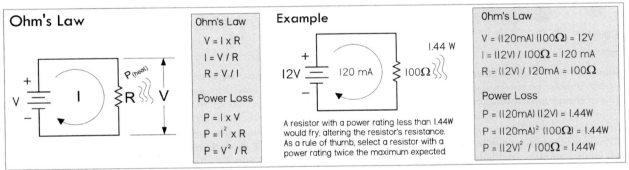

FIGURE 2.39

2.12.1 Resistor Power Ratings

Determining how much power a resistor dissipates is very important when designing circuits. All real resistors have maximum allowable *power ratings* that must not be exceeded. If you exceed the power rating, you'll probably end up frying your resistor, destroying the internal structure, and thus altering the resistance. Typical general-purpose resistors come in ⅛-, ¼-, ½-, and 1-W power ratings, while high-power resistors can range from 2 to several hundred watts.

So, in the preceding example, where our resistor was dissipating 1.44 W, we should have made sure that our resistor's power rating exceeded 1.44 W; otherwise, there could be smoke. As a *rule of thumb,* always select a resistor that has a power rating at least twice the maximum value anticipated. Though a 2-W resistor would work in our example, a 3-W resistor would be safer.

To illustrate how important power ratings are, we examine the circuit shown in Fig. 2.40. The resistance is variable, while the supply voltage is fixed at 5 V. As the resistance increases, the current decreases, and according to the power law, the power decreases, as shown in the graphs. As the resistance decreases, the current and power

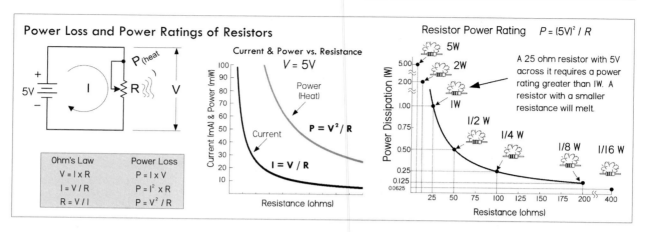

FIGURE 2.40

increase. The far right graph shows that as you decrease the resistance, the power rating of the resistor must increase; otherwise, you'll burn up the resistor.

Example 1: Using an ammeter, you measure a current of 1.0 mA through a 4.7-kΩ resistor. What voltage must exist across the resistor? How much power does the resistor dissipate?

Answer: $V = IR = (0.001\ A)(4700\ \Omega) = 4.7\ V$; $P = I^2 \times R = (0.001\ A)^2 \times (4700\ \Omega) = 0.0047\ W = 4.7\ mW$

Example 2: Using a voltmeter, you measure 24 V across an unmarked resistor. With an ammeter, you measure a current of 50 mA. Determine the resistance and power dissipated in the resistor.

Answer: $R = V/I = (24\ V)/(0.05\ A) = 480\ \Omega$; $P = I \times V = (0.05\ A) \times (24\ V) = 1.2\ W$

Example 3: You apply 3 V to a 1-MΩ resistor. Find the current through the resistor and the power dissipated in the process.

Answer: $I = V/R = (3\ V)/(1,000,000\ \Omega) = 0.000003\ A = 3\ \mu A$; $P = V^2/R = (3\ V)^2/(1,000,000\ \Omega) = 0.000009\ W = 9\ \mu W$

Example 4: You are given 2-Ω, 100-Ω, 3-kΩ, 68-kΩ, and 1-MΩ resistors, all with 1-W power ratings. What's the maximum voltage that can be applied across each of them without exceeding their power ratings?

Answer: $P = V^2/R \Rightarrow V = \sqrt{PR}$; voltages must not exceed 1.4 V (2 Ω), 10.0 V (100 Ω), 54.7 V (3 kΩ), 260.7 V (68 kΩ), 1000 V (1 MΩ)

2.12.2 Resistors in Parallel

Rarely do you see circuits that use a single resistor alone. Usually, resistors are found connected in a variety of ways. The two fundamental ways of connecting resistors are in series and in parallel.

When two or more resistors are placed in parallel, the voltage across each resistor is the same, but the current through each resistor will vary with resistance. Also, the total resistance of the combination will be lower than that of the lowest resistance value present. The formula for finding the total resistance of resistors in parallel is:

$$R_{\text{total}} = \frac{1}{\frac{1}{R1} + \frac{1}{R2} + \frac{1}{R3} + \frac{1}{R4} + \cdots} \tag{2.17}$$

$$R_{\text{total}} = \frac{R1 \times R2}{R1 + R2} \qquad \text{(2.18) Two resistors in parallel}$$

The dots in the equation indicate that any number of resistors can be combined. For only two resistances in parallel (a very common case), the formula reduces to Eq. 2.18. (You can derive the resistor-in-parallel formula by noting that the sum of the individual branch currents is equal to the total current: $I_{\text{total}} = I_1 + I_2 + I_3 + \cdots I_N$. This is referred to as Kirchhoff's current law. Then, applying Ohm's law, we get: $I_{\text{total}} = V_1/R_1 + V_2/R_2 + V_3/R_3 + \cdots V_N/R_N$. Because all resistor voltages are equal to V_{total} since they share the same voltage across them, we get: $I_{\text{total}} = V_{\text{total}}/R_1 + V_{\text{total}}/R_2 + V_{\text{total}}/R_3 + \cdots V_{\text{total}}/R_N$. Factoring out V_{total}, we get: $I_{\text{total}} = V_{\text{total}} (1/R_1 + 1/R_2 + 1/R_3 \cdots 1/R_4)$. We call the term in brackets R_{total}.

Note that there is a shorthand for saying that two resistors are in parallel. The shorthand is to use double bars $||$ to indicate resistors in parallel. So to say R_1 is in parallel with R_2, you would write $R_1 || R_2$. Thus, you can express two resistors in parallel in the following ways:

$$R_1 || R_2 = \frac{1}{1/R_1 + 1/R_2} = \frac{R_1 \times R_2}{R_1 + R_2}.$$

In terms of arithmetic order of operation, the $||$ can be treated similar to multiplication or division. For example, in the equation $Z_{in} = R_1 + R_2 || R_{load}$, you calculate R_2 and R_{load} in parallel first, and then you add R_1.

Example 1: If a 1000-Ω resistor is connected in parallel with a 3000-Ω resistor, what is the total or equivalent resistance? Also calculate total current and individual currents, as well as the total and individual dissipated powers.

$$R_{\text{total}} = \frac{R_1 \times R_2}{R_1 + R_2} = \frac{1000\ \Omega \times 3000\ \Omega}{1000\ \Omega + 3000\ \Omega} = \frac{3,000,000\ \Omega}{4000\ \Omega} = 750\ \Omega$$

To find how much current flows through each resistor, apply Ohm's law:

$$I_1 = \frac{V_1}{R_1} = \frac{12\ \text{V}}{1000\ \Omega} = 0.012\ \text{A} = 12\ \text{mA}$$

$$I_2 = \frac{V_2}{R_2} = \frac{12\ \text{V}}{3000\ \Omega} = 0.004\ \text{A} = 4\ \text{mA}$$

These individual currents add up to the total input current:

$$I_{\text{in}} = I_1 + I_2 = 12\ \text{mA} + 4\ \text{mA} = 16\ \text{mA}$$

This statement is referred to as Kirchhoff's current law. With this law, and Ohm's law, you come up with the *current divider*

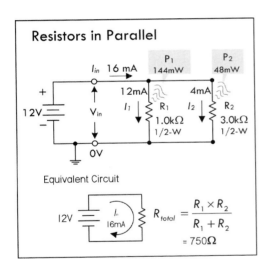

Resistors in Parallel

Equivalent Circuit

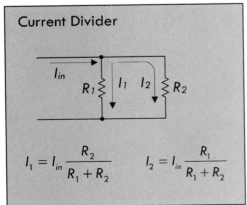

Current Divider

$$I_1 = I_{in}\frac{R_2}{R_1 + R_2} \qquad I_2 = I_{in}\frac{R_1}{R_1 + R_2}$$

FIGURE 2.41

equations, shown at the bottom of Fig. 2.41. These equations come in handy when you know the input current but not the input voltage.

We could have just as easily found the total current using:

$$I_{in} = V_{in}/R_{total} = (12 \text{ V})/(750 \text{ }\Omega) = 0.016 \text{ A} = 16 \text{ mA}$$

To find how much power resistors in parallel dissipate, apply the power law:

$$P_{tot} = I_{in}V_{in} = (0.0016 \text{ A})(12 \text{ V}) = 0.192 \text{ W} = 192 \text{ mW}$$

$$P_1 = I_1 V_{in} = (0.012 \text{ A})(12 \text{ V}) = 0.144 \text{ W} = 144 \text{ mW}$$

$$P_2 = I_2 V_{in} = (0.004 \text{ A})(12 \text{ V}) = 0.048 \text{ W} = 48 \text{ mW}$$

Example 2: Three resistors $R_1 = 1 \text{ k}\Omega$, $R_2 = 2 \text{ k}\Omega$, $R_3 = 4 \text{ k}\Omega$ are in parallel. Find the equivalent resistance. Also, if a 24-V battery is attached to the parallel circuit to complete a circuit, find the total current, individual currents through each of the resistors, total power loss, and individual resistor power losses.

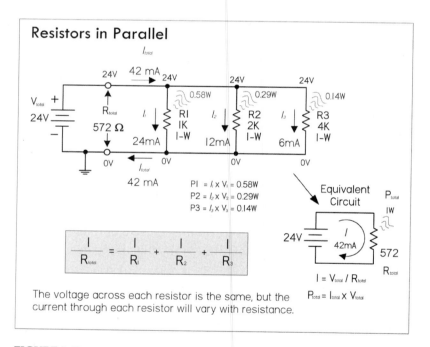

FIGURE 2.42

The total resistance for resistors in parallel:

$$\frac{1}{R_{total}} = \frac{1}{R_1} + \frac{1}{R_2} + \frac{1}{R_3}$$

$$= \frac{1}{1000 \text{ }\Omega} + \frac{1}{2000 \text{ }\Omega} + \frac{1}{4000 \text{ }\Omega} = 0.00175 \text{ }\Omega$$

$$R_{total} = \frac{1}{0.00175 \text{ }\Omega^{-1}} = 572 \text{ }\Omega$$

The current through each of the resistors:

$$I_1 = V_1/R_1 = 24\ V/1000\ \Omega = 0.024\ A = 24\ mA$$

$$I_2 = V_2/R_2 = 24\ V/2000\ \Omega = 0.012\ A = 12\ mA$$

$$I_3 = V_3/R_3 = 24\ V/4000\ \Omega = 0.006\ A = 6\ mA$$

The total current, according to Kirchhoff's current law:

$$I_{total} = I_1 + I_2 + I_3 = 24\ mA + 12\ mA + 6\ mA = 42\ mA$$

The total power dissipated by the parallel resistors:

$$P_{total} = I_{total} \times V_{total} = 0.042\ A \times 24\ V = 1.0\ W$$

The power dissipated by individual resistors is shown in Fig. 2.42.

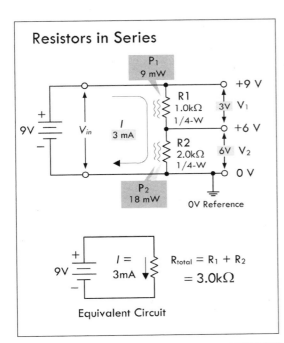

Resistors in Series

Equivalent Circuit

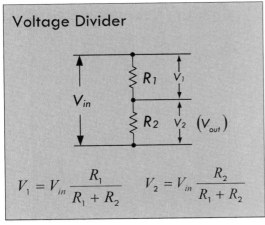

Voltage Divider

$$V_1 = V_{in}\frac{R_1}{R_1 + R_2} \qquad V_2 = V_{in}\frac{R_2}{R_1 + R_2}$$

2.12.3 *Resistors in Series*

When a circuit has a number of resistors connected in series, the total resistance of the circuit is the sum of the individual resistances. Also, the amount of current flowing through each resistor in series is the same, while the voltage across each resistor varies with resistance. The formula for finding the total resistance of resistors in series is:

$$R_{total} = R_1 + R_2 + R_3 + R_4 + \cdots \qquad (2.19)$$

The dots indicate that as many resistors as necessary may be added.

You can derive this formula by noting that the sum of all the voltage drops across each series resistor will equal the applied voltage across the combination $V_{total} = V_1 + V_2 + V_3 + \cdots + V_N$. This is referred to as Kirchhoff's voltage law. Applying Ohm's law, and noting that the same current I flows through each resistor, we get: $IR_{total} = IR_1 + IR_2 + IR_3 + \cdots + IR_N$. Canceling the I's you get: $R_{total} = R_1 + R_2 + R_3 + R_4 + \cdots$

Example 1: If a 1.0-kΩ resistor is placed in series with a 2.0-kΩ resistor, the total resistance becomes:

$$R_{tot} = R_1 + R_2 = 1000\ \Omega + 2000\ \Omega = 3000\ \Omega = 3\ k\Omega$$

When these series resistors are placed in series with a battery, as shown in Fig. 2.43, the total current flow I is simply equal to the applied voltage V_{in}, divided by the total resistance:

$$I = \frac{V_{in}}{R_{tot}} = \frac{9\ V}{3000\ \Omega} = 0.003\ A = 3\ mA$$

FIGURE 2.43

Since the circuit is a series circuit, the currents through each resistor are equal to the total current I:

$$I_1 = 3 \text{ mA}, I_2 = 3 \text{ mA}$$

To find the voltage drop across each resistor, apply Ohm's law:

$$V_1 = I_1 \times R_1 = 0.003 \text{ A} \times 1000 \text{ }\Omega = 3 \text{ V}$$

$$V_2 = I_2 \times R_2 = 0.003 \text{ A} \times 2000 \text{ }\Omega = 6 \text{ V}$$

Now, we didn't really have to calculate the current. We could have just plugged $I = V_{in}/R_{tot}$ into I_1 and I_2 in the preceding equations and got:

$$V_1 = IR_1 = \frac{V_{in}}{R_1 + R_2} \times R_1$$

$$V_2 = IR_2 = \frac{V_{in}}{R_1 + R_2} \times R_2$$

(Voltage divider equations)

These equations are called *voltage divider equations* and are so useful in electronics that it is worth memorizing them. (See Fig. 2.43.) Often V_2 is called the output voltage V_{out}.

The voltage drop across each resistor is directly proportional to the resistance. The voltage drop across the 2000-Ω resistor is twice as large as that of the 1000-Ω resistor. Adding both voltage drops together gives you the applied voltage of 9 V:

$$V_{in} = V_1 + V_2 \qquad 9 \text{ V} = 3 \text{ V} + 6 \text{ V}$$

The total power loss and individual resistor power losses are:

$$P_{tot} = IV_{in} = (0.003 \text{ A})(9 \text{ V}) = 0.027 \text{ W} = 27 \text{ mW}$$

$$(P_{tot} = I^2 R_{tot} = (0.003 \text{ A})^2(3000 \text{ }\Omega) = 0.027 \text{ W} = 27 \text{ mW})$$

$$P_1 = I^2 R_1 = (0.003 \text{ A})^2(1000 \text{ }\Omega) = 0.009 \text{ W} = 9 \text{ mW}$$

$$P_2 = I^2 R_2 = (0.003 \text{ A})^2(2000 \text{ }\Omega) = 0.018 \text{ W} = 18 \text{ mW}$$

The larger resistor dissipates twice as much power.

Example 2: The input of an IC requires a constant 5 V, but the supply voltage is 9 V. Use the voltage divider equations to create a voltage divider with an output of 5 V. Assume the IC has such a high input resistance (10 MΩ) that it practically draws no current from the divider.

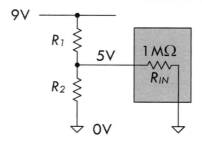

FIGURE 2.44

Answer: Since we assume the IC draws no current, we can apply the voltage divider directly:

$$V_{out} = V_{in} \frac{R_2}{R_1 + R_2}$$

We must choose voltage divider resistors, making sure our choice doesn't draw too much current, causing unnecessary power loss. To keep things simple for now, let's choose R_2 to be 10 kΩ. Rearranging the voltage divider and solving for R_1:

$$R_1 = R_2 \frac{(V_{\text{in}} - V_{\text{out}})}{V_{\text{out}}} = (10,000\ \Omega)\frac{(9-5)}{5} = 8000\ \Omega = 8\ \text{k}\Omega$$

Example 3: You have a 10-V supply, but a device that is to be connected to the supply is rated at 3 V and draws 9.1 mA. Create a voltage divider for the load device.

Answer: In this case, the load draws current and can be considered a resistor in parallel with R_2. Therefore, using the voltage divider relation without taking the load into consideration will not work. We must apply what is called the 10 percent rule.

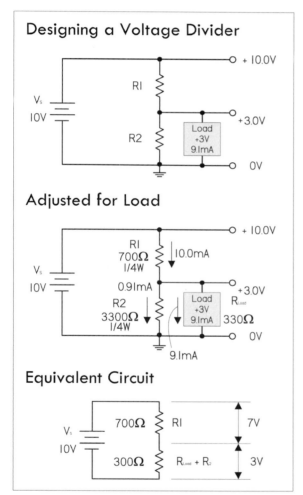

FIGURE 2.45

The 10 Percent Rule: This rule is a standard method for selecting R_1 and R_2 that takes into account the load and minimizes unnecessary power losses in the divider.

The first thing you do is select R_2 so that I_2 is 10 percent of the desired load current. This resistance and current are called the *bleeder resistance* and *bleeder current*. The bleeder current in our example is:

$$I_{\text{bleed}} = I_2 = (0.10)(9.1\ \text{mA}) = 0.91\ \text{mA}$$

Using Ohm's law, next we calculate the bleeder resistance:

$$R_{\text{bleed}} = R_2 = 3\ \text{V}/0.00091\ \text{A} = 3297\ \Omega$$

Considering resistor tolerances and standard resistance values, we select a resistor in close vicinity—3300 Ω.

Next, we need to select R_1, so that the output is maintained at 3 V.

To do this, simply calculate the total current through the resistor and use Ohm's law:

$$I_1 = I_2 + I_{\text{load}} = 0.91\ \text{mA} + 9.1\ \text{mA} = 10.0\ \text{mA} = 0.01\ \text{A}$$

$$R_1 \frac{10\ \text{V} - 3\ \text{V}}{0.01\ \text{A}} = 700\ \Omega$$

In terms of power ratings:

$$P_{R1} = V_1{}^2/R_1 = (7\ \text{V})^2/(700\ \Omega) = 0.07\ \text{W} = 70\ \text{mW}$$

$$P_{R2} = V_2{}^2/R_2 = (3\ \text{V})^2/(3300\ \Omega) = 0.003\ \text{W} = 3\ \text{mW}$$

Low-power ¼-W resistors will suffice.

In actual practice, the computed value of the bleeder resistor does not always come out to an even value. Since the rule of thumb for bleeder current is only an estimated value, the bleeder resistor can be of a value close to the computed value. (If the computed value of the resistance were 500Ω, a 5 percent 510Ω resistor could be used. See the standard resistor values table in the front matter of the book.) Once the actual value of the bleeder resistor is selected, the bleeder current must be recomputed. The voltage developed by the bleeder resistor must be equal to the voltage requirement of the load in parallel with the bleeder resistor. We'll discuss voltage dividers, as well as more complex divider arrangements, in the section on resistors in Chap. 3.

Example: Find the equivalent resistance of series resistors $R_1 = 3.3\text{k}$, $R_2 = 4.7\text{k}$, $R_3 = 10\text{k}$. If a 24-V battery is attached to the series resistors to complete a circuit, find the total current flow, the individual voltage drops across the resistors, the total power loss, and the individual power loss of each resistor.

The equivalent resistance of the three resistors:

$$R_\text{total} = R_1 + R_2 + R_3$$

$$R_\text{total} = 3.3\text{ k}\Omega + 4.7\text{ k}\Omega + 10\text{ k}\Omega = 18\text{ k}\Omega$$

The total current flow through the resistors is:

$$I_\text{total} = \frac{V_\text{total}}{R_\text{total}} = \frac{24\text{ V}}{18,000\text{ }\Omega}$$

$$= 0.00133\text{ A} = 1.33\text{ mA}$$

Using Ohm's law, the voltage drops across the resistors are:

$$V_1 = I_\text{total} \times R_1 = 1.33\text{ mA} \times 3.3\text{ k}\Omega = 4.39\text{ V}$$

$$V_2 = I_\text{total} \times R_2 = 1.33\text{ mA} \times 4.7\text{ k}\Omega = 6.25\text{ V}$$

$$V_3 = I_\text{total} \times R_3 = 1.33\text{ mA} \times 10\text{ k}\Omega = 13.30\text{ V}$$

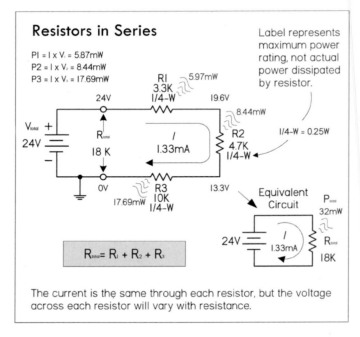

Resistors in Series

PI = I x V₁ = 5.87mW
P2 = I x V₂ = 8.44mW
P3 = I x V₃ = 17.69mW

Label represents maximum power rating, not actual power dissipated by resistor.

$$R_\text{total} = R_1 + R_2 + R_3$$

The current is the same through each resistor, but the voltage across each resistor will vary with resistance.

FIGURE 2.46

The total power dissipated is:

$$P_\text{total} = I_\text{total} \times V_\text{total} = 1.33\text{ mA} \times 24\text{ V} = 32\text{ mW}$$

Power dissipated by individual resistors is shown in Fig. 2.46.

2.12.4   Reducing a Complex Resistor Network

To find the equivalent resistance for a complex network of resistors, the network is broken down into series and parallel combinations. A single equivalent resistance for these combinations is then found, and a new and simpler network is formed. This new network is then broken down and simplified. The process continues over and over again until a single equivalent resistance is found. Here's an example of how reduction works.

Example 1: Find the equivalent resistance of the network attached to the battery by using circuit reduction. Afterward, find the total current flow supplied by the battery to the network, the voltage drops across all resistors, and the individual current through each resistor.

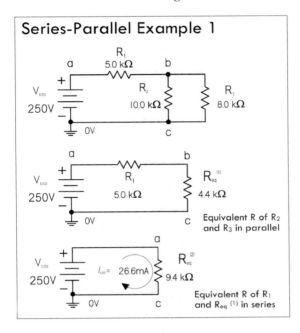

Series-Parallel Example 1

FIGURE 2.47

R_2 and R_3 form a parallel branch that can be reduced to:

$$R_{eq}^{(1)} = \frac{R_2 \times R_3}{R_2 + R_3} = \frac{10.0 \text{ k}\Omega \times 8.0 \text{ k}\Omega}{10.0 \text{ k}\Omega + 8.0 \text{ k}\Omega} = 4.4 \text{ k}\Omega$$

This equivalent resistance and R_1 are in series, so their combined resistance is:

$$R_{eq}^{(2)} = R_1 + R_{eq}^{(1)} = 5.0 \text{ k}\Omega + 4.4 \text{ k}\Omega = 9.4 \text{ k}\Omega$$

The total current flow through the circuit and through R_1 is:

$$I_{total} = \frac{V_{total}}{R_{eq}^{(2)}} = \frac{250 \text{ V}}{9.4 \text{ k}\Omega} = 26.6 \text{ mA} = I_1$$

The voltage across R_2 and R_3 is the same as that across $R_{eq}^{(1)}$:

$$V_{Req}^{(1)} = I_{total} \times R_{eq}^{(1)} = 26.6 \text{ mA} \times 4.4 \text{ k}\Omega = 117 \text{ V}$$

$$V_2 = V_3 = 117 \text{ V}$$

The current through R_2 and R_3 are found using Ohm's law:

$$I_2 = \frac{V_2}{R_2} = \frac{117 \text{ V}}{10 \text{ k}\Omega} = 11.7 \text{ mA}$$

$$I_3 = \frac{V_3}{R_3} = \frac{117 \text{ V}}{8.0 \text{ k}\Omega} = 14.6 \text{ mA}$$

The voltage across R_1 is, by Kirchhoff's voltage law:

$$V_1 = 250 \text{ V} - 117 \text{ V} = 133 \text{ V}$$

Alternatively we could have used Ohm's law:

$$V_1 = I_1 \times R_1 = (26.6 \text{ mA})(5.0 \text{ k}\Omega) = 133 \text{ V}$$

Example 2: Find the equivalent resistance of the following network, along with the total current flow, the individual current flows, and individual voltage drops across the resistors.

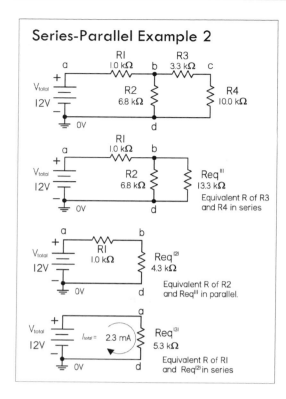

FIGURE 2.48

R_3 and R_4 can be reduced to resistors in series:

$$R_{eq}^{(1)} = R_3 + R_4 = 3.3 \text{ k}\Omega + 10.0 \text{ k}\Omega = 13.3 \text{ k}\Omega$$

This equivalent resistance is in parallel with R_2:

$$R_{eq}^{(2)} = \frac{R_2 \times R_{eq}^{(1)}}{R_2 + R_{eq}^{(1)}} = \frac{6.8 \text{ k}\Omega \times 13.3 \text{ k}\Omega}{6.8 \text{ k}\Omega + 13.3 \text{ k}\Omega} = 4.3 \text{ k}\Omega$$

This new equivalent resistance is in series with R_1:

$$R_{eq}^{(3)} = R_1 + R_{eq}^{(2)} = 1.0 \text{ k}\Omega + 4.3 \text{ k}\Omega = 5.3 \text{ k}\Omega$$

The total current flow is:

$$I_{total} = \frac{V_{total}}{R_{eq}^{(3)}} = \frac{12 \text{ V}}{5.3 \text{ k}\Omega} = 2.26 \text{ mA}$$

The voltage across $R_{eq}^{(2)}$ or point voltage at b is:

$$V_{Req(2)} = I_{total} \times R_{eq}^{(2)} = 2.26 \text{ mA} \times 4.3 \text{ k}\Omega = 9.7 \text{ V}$$

The voltage across R_1 is:

$$V_{R1} = I_{total} \times R_1 = 2.26 \text{ mA} \times 1.0 \text{ k}\Omega = 2.3 \text{ V}$$

You can confirm this using Kirchhoff's voltage law:

$$12 \text{ V} - 9.7 \text{ V} = 2.3 \text{ V}$$

The current through R_2:

$$I_2 = \frac{V_{Req^{(2)}}}{R_2} = \frac{9.7 \text{ V}}{6.8 \text{ k}\Omega} = 1.43 \text{ mA}$$

The current through $R_{eq}^{(1)}$, also through R_3 and R_4:

$$I_{Req^{(2)}} = I_3 = I_4 = \frac{V_{Req^{(2)}}}{R_{eq}^{(1)}} = \frac{9.7 \text{ V}}{13.3 \text{ k}\Omega} = 0.73 \text{ mA}$$

You can confirm this using Kirchhoff's current law:

$$2.26 \text{ mA} - 1.43 \text{ mA} = 0.73 \text{ mA}.$$

The voltage across R_3:

$$V_{R3} = I_3 \times R_3 = 0.73 \text{ mA} \times 3.3 \text{ k}\Omega = 2.4 \text{ V}$$

The voltage across R_4:

$$V_{R4} = I_4 \times R_4 = 0.73 \text{ mA} \times 10.0 \text{ k}\Omega = 7.3 \text{ V}$$

You can confirm this using Kirchhoff's voltage law:

$$9.7 \text{ V} - 2.4 \text{ V} = 7.3 \text{ V}$$

2.12.5 Multiple Voltage Dividers

Example 1: You wish to create a multiple voltage divider that powers three loads: load 1 (75 V, 30 mA), load 2 (50 V, 10 mA), and load 3 (25 V, 10 mA). Use the 10 percent rule and Fig. 2.49 to construct the voltage divider.

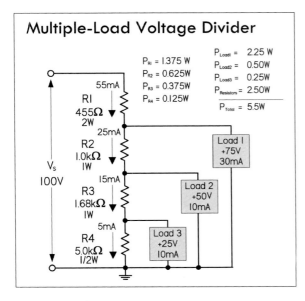

FIGURE 2.49

An important point in determining the resistor when applying the 10 percent rule of thumb is that to calculate the bleeder current, you must take 10 percent of the total load current. The steps are as follows:

Find the bleeder current, which is 10 percent (0.1) of the total current:

$$I_{R4} = 0.1 \times (10 \text{ mA} + 10 \text{ mA} + 30 \text{ mA}) = 5 \text{ mA}$$

To find R_4 (bleeder resistor), use Ohm's law:

$$R_4 = (25 \text{ V} - 0 \text{ V})/(0.005 \text{ A}) = 5000 \ \Omega$$

The current through R_3 is equal to the current through R_4 plus the current through load 3:

$$I_{R3} = I_{R4} + I_{\text{load3}} = 5 \text{ mA} + 10 \text{ mA} = 15 \text{ mA}$$

To find R_3, use Ohm's law, using the voltage difference between load 2 and load 3: $R_3 = (50 \text{ V} - 25 \text{ V})/(0.015 \text{ A}) = 1667 \ \Omega$ or 1.68 kΩ, considering tolerances and standard resistance values.

The current through R_2 is:

$$I_{R2} = I_{R3} + I_{\text{load2}} = 15 \text{ mA} + 10 \text{ mA} = 25 \text{ mA}$$

Using Ohm's law, $R_2 = (75 \text{ V} - 50 \text{ V})/(0.025 \text{ A}) = 1000 \ \Omega$
The current through R_1 is:

$$I_{R1} = I_{R2} + I_{\text{load1}} = 25 \text{ mA} + 30 \text{ mA} = 55 \text{ mA}$$

Using Ohm's law: $R_1 = (100 \text{ V} - 75 \text{ V})/(0.055 \text{ A}) = 455 \ \Omega$
To determine resistor power ratings and total power losses in loads use $P = IV$. See results in Fig. 2.49.

Example 2: In many cases, the load for a voltage divider requires both positive and negative voltages. Positive and negative voltages can be supplied from a single-source voltage by connecting the ground return between two of the divider resistors. The exact point in the circuit at which the ground is placed depends upon the voltages required by the loads.

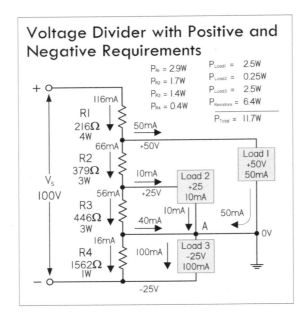

Voltage Divider with Positive and Negative Requirements

$P_{R1} = 2.9W$ $P_{Load1} = 2.5W$
$P_{R2} = 1.7W$ $P_{Load2} = 0.25W$
$P_{R3} = 1.4W$ $P_{Load3} = 2.5W$
$P_{R4} = 0.4W$ $P_{Resistors} = 6.4W$

$P_{Total} = 11.7W$

FIGURE 2.50

For example, the voltage divider in Fig. 2.50 is designed to provide the voltage and current to three loads from a given source voltage:

Load 1: +50 V @ 50 mA

Load 2: +25 V @ 10 mA

Load 3: –25 V @ 100 mA

The values of R_4, R_2, and R_1 are computed exactly as was done in the preceding example. I_{R4} is the bleeder current and can be calculated as follows:

$$I_{R4} = 10\% \times (I_{load1} + I_{load2} + I_{load3}) = 16 \text{ mA}$$

Calculating R_4:

$$R_4 = 25 \text{ V}/0.016 \text{ A} = 1562 \text{ }\Omega$$

To calculate current through R_3, use Kirchhoff's current law (at point A):

$$I_{R3} + I_{load2} + I_{load1} + I_{R4} + I_{load3} = 0$$

$$I_{R3} + 10 \text{ mA} + 50 \text{ mA} - 16 \text{ mA} - 100 \text{ mA} = 0$$

$$I_{R3} = 56 \text{ mA}$$

Calculating R_3: $R_3 = 25 \text{ V}/(0.056 \text{ A}) = 446 \text{ }\Omega$.
Calculating I_{R2}: $I_{R2} = I_{R3} + I_{load2} = 56 \text{ mA} + 10 \text{ mA} = 66 \text{ mA}$.
Then $R_2 = 25 \text{ V}/(0.066 \text{ A}) = 379 \text{ }\Omega$.
Calculating I_{R1}: $I_{R1} = I_{R2} + I_{load1} = 66 \text{ mA} + 50 \text{ mA} = 116 \text{ mA}$.
Then $R_1 = 25 \text{ V}/0.116 \text{ A} = 216 \text{ }\Omega$.

Though voltage dividers are simple to apply, they are not regulated in any way. If a load's resistance changes, or if there is variation in the supply voltage, all loads will experience a change in voltage. Therefore, voltage dividers should not be applied to circuits where the divider will be weighed down by changes in load. For applications that require steady voltage and that draw considerable current, it is best to use an active device, such as a voltage regulator IC—more on this later.

2.13 Voltage and Current Sources

An *ideal voltage* source is a two-terminal device that maintains a fixed voltage across its terminals. If a variable load is connected to an ideal voltage source, the source will maintain its terminal voltage regardless of changes in the load resistance. This means that an ideal voltage source will supply as much current as needed to the load in order to keep the terminal voltage fixed (in $I = V/R$, I changes with R, but V is fixed).

Now a fishy thing with an ideal voltage source is that if the resistance goes to zero, the current must go to infinity. Well, in the real world, there is no device that can supply an infinite amount of current. If you placed a real wire between the terminals of an ideal voltage source, the calculated current would be so large it would melt the wire. To avoid this theoretical dilemma, we must define a real voltage source (a battery, plug-in dc power supply, etc.) that can supply only a maximum finite amount of current. A *real voltage source* resembles an ideal voltage source with a small series internal resistance or source resistance r_s, which is a result of the imperfect conducting nature of the source (resistance in battery electrolyte and lead, etc.). This internal resistance tends to reduce the terminal voltage, the magnitude of which depends on its value and the amount of current that is drawn from the source (or the size of the load resistance).

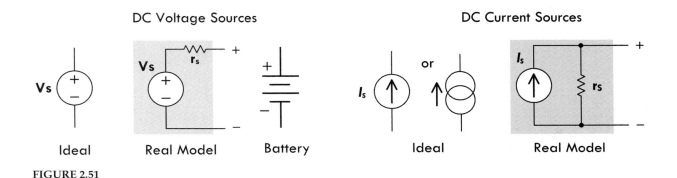

FIGURE 2.51

In Fig. 2.52, when a real voltage source is open-circuited (no load connected between its terminals), the terminal voltage (V_T) equals the ideal source voltage (V_S)—there is no voltage drop across the resistor, since current can't pass through it due to an incomplete circuit condition.

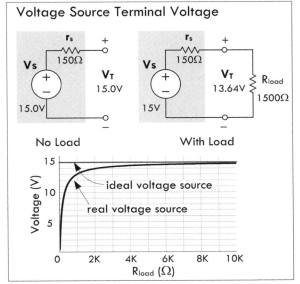

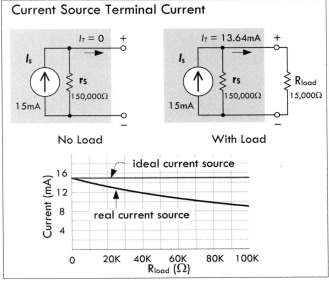

FIGURE 2.52

However, when a load R_{load} is attached across the source terminals, R_{load} and r_s are connected in series. The voltage at the terminal can be determined by using the voltage divider relation:

$$V_T = V_s \frac{R_{\text{load}}}{R_{\text{load}} + r_s}$$

From the equation, you can see that when R_{load} is very large compared to r_s, (1000 times greater or more), the effect of r_s is so small that it may be ignored. However, when R_{load} is small or closer to r_s in size, you must take r_s into account when doing your calculations and designing circuits. See the graph in Fig. 2.52.

In general, the source resistance for dc power supplies is usually small; however, it can be as high as 600 Ω in some cases. For this reason, it's important to always adjust the power supply voltage with the load connected. In addition, it is a good idea to recheck the power supply voltage as you add or remove components to or from a circuit.

Another symbol used in electronics pertains to dc current sources—see Fig. 2.52. An *ideal current source* provides the same amount of source current I_S at all times to a load, regardless of load resistance changes. This means that the terminal voltage will change as much as needed as the load resistance changes in order to keep the source current constant.

In the real world, current sources have a large internal shunt (parallel) resistance r_s, as shown in Fig. 2.52. This internal resistance, which is usually very large, tends to reduce the terminal current I_T, the magnitude of which depends on its value and the amount of current that is drawn from the source (or the size of load resistance).

When the source terminals are open-circuited, I_T must obviously be zero. However, when we connect a load resistance R_{load} across the source terminals, R_{load} and r_s form a parallel resistor circuit. Using the current divider expression, the terminal current becomes:

$$I_T = I_s \frac{r_s}{R_{\text{load}} + r_s}$$

From this equation, you can see that when R_{load} is small compared to r_s, the dip in current becomes so small it usually can be ignored. However, when R_{load} is large or closer to r_s in size, you must take r_s into account when doing your calculations.

A source may be represented either as a current source or a voltage source. In essence, they are duals of each other. To translate between the voltage source and current source representation, the resistance value is kept the same, while the voltage of the source is translated into the current of the source using Ohm's law. See Fig. 2.53 for details.

One way to look at an ideal current source is to say it has an internal resistance that is infinite, which enables it to support any kind of externally imposed potential difference across its terminals (e.g., a load's resistance changes). An approximation to an ideal current source is a voltage source of very high voltage V in series with a very large resistance R. This approximation would supply a current V/R into any load that has a resistance much smaller than R. For example, the simple resistive current source circuit shown in Fig. 2.54 uses a 1-kV voltage source in series with a 1-MΩ resistor. It

Source Transformation

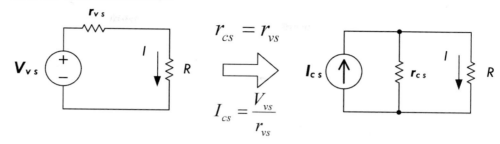

FIGURE 2.53

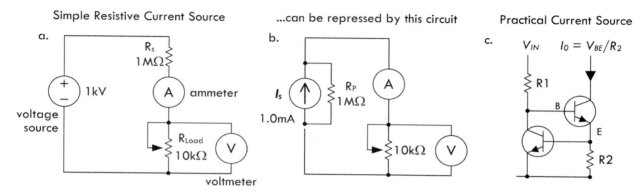

FIGURE 2.54

will maintain the set current of 1 mA to an accuracy within 1 percent over a 0- to 10-V range ($0 < R_{\text{load}} < 10$ kΩ). The current is practically constant even though the load resistance varies, since the source resistance is much greater than the load resistance, and thus the current remains practically constant. ($I = 1000$ V/(1,000,000 Ω + 10,0000 Ω. Since 1,000,000 Ω is so big, it overshadows R_{load}.)

A practical current source is usually an active circuit made with transistors, as shown in Fig. 2.54c. V_{in} drives current through R_1 into the base of the second transistor, so current flows into the transistor's collector, through it, and out its emitter. This current must flow through R_2. If the current gets too high, the first transistor turns on and robs the second transistor of base current, so its collector current can never exceed the value shown. This is an excellent way of either making a current source or limiting the available current to a defined maximum value.

2.14   Measuring Voltage, Current, and Resistance

Voltmeters, ammeters, and ohmmeters, used to measure voltage, current, and resistance, ideally should never introduce any effects within the circuit under test. In theory, an ideal voltmeter should draw no current as it measures a voltage between two points in a circuit; it has infinite input resistance R_{in}. Likewise, an ideal ammeter should introduce no voltage drop when it is placed in series within the circuit; it has zero input resistance. An ideal ohmmeter should provide no additional resistance when making a resistance measurement.

Real meters, on the other hand, have limitations that prevent them from making truly accurate measurements. This stems from the fact that the circuit of a meter requires a sample current from the circuit under test in order to make a display measurement. Figure 2.55 shows symbols for an ideal voltmeter, ammeter, and ohmmeter, along with the more accurate real-life equivalent circuits.

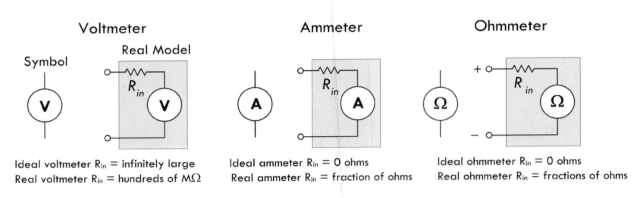

Voltmeter

Ammeter

Ohmmeter

Ideal voltmeter R_{in} = infinitely large
Real voltmeter R_{in} = hundreds of MΩ

Ideal ammeter R_{in} = 0 ohms
Real ammeter R_{in} = fraction of ohms

Ideal ohmmeter R_{in} = 0 ohms
Real ohmmeter R_{in} = fractions of ohms

FIGURE 2.55

An *ideal voltmeter* has infinite input resistance and draws no current; a real voltmeter's input resistance is several hundred MΩs. An *ideal ammeter* has zero input resistance and provides no voltage drop; a real ammeter's input resistance is fractions of ohms. An *ideal ohmmeter* has zero internal resistance; a real ohmmeter's internal resistance is fractions of ohms. It is important to read your instrument manuals to determine the internal resistances.

The effects of meter internal resistance are shown in Fig. 2.56. In each case, the internal resistance becomes part of the circuit. The percentages of error in measurements due to the internal resistances become more pronounced when the circuit resistances approach the meters' internal resistance.

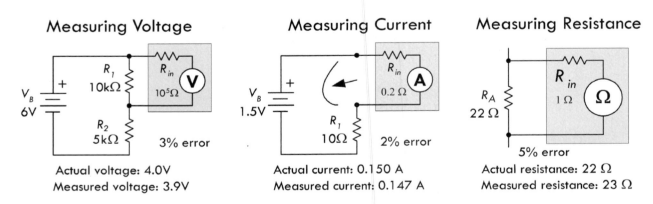

Measuring Voltage

Measuring Current

Measuring Resistance

Actual voltage: 4.0V
Measured voltage: 3.9V

Actual current: 0.150 A
Measured current: 0.147 A

Actual resistance: 22 Ω
Measured resistance: 23 Ω

FIGURE 2.56

Figure 2.57 shows the actual makeup of a basic analog multimeter. Note that this device is overly simplistic; real meters are much more sophisticated, with provisions for range selection, ac measurements, and so on. At any rate, it shows you that the meters are nonideal due to the sample current required to drive the galvanometer's needle.

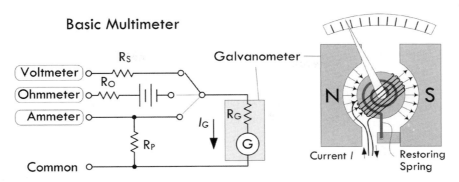

FIGURE 2.57   The heart of this simple analog multimeter is a galvanometer which measures current flow I_G. Current sent through the galvanometer's wire leads will generate a magnetic field from the coil windings in the center rotor. Since the coil is tilted relative to the N-S alignment of the permanent magnet, the rotor will rotate in accordance with the amount of current flow. Normally you get full deflection of the needle at extremely low current levels, so you can add a parallel resistor to divert current flow away from the galvanometer. The galvanometer can also be used as a voltmeter by placing its lead at the point voltages within the circuit. If there is a voltage difference at the leads, a current will flow into the galvanometer and move the needle in proportion to the voltage magnitude. A series resistor is used to limit current flow and needle deflection. A galvanometer can be used to make an ohmmeter, too, provided you place a battery in series with it. Again, to calibrate needle deflection, a series resistor R_O is used.

2.15   Combining Batteries

The two battery networks in Fig. 2.58 show how to increase the supply voltage and/or increase the supply current capacity. To increase the supply voltage, batteries are placed end to end or in series; the terminal voltages of each battery add together to give a final supply voltage equal to the sum from the batteries. To create a supply with added current capacity (increased operating time), batteries can be placed in parallel—positive terminals are joined together, as are negative terminals, as shown in Fig. 2.58. The power delivered to the load can be found using Ohm's power law: $P = V^2/R$, where V is the final supply voltage generated by all batteries within the network. Note that the ground symbol shown here acts simply as a 0-V reference, not as an actual physical connection to ground. Rarely would you ever connect a physical ground to a battery-operated device.

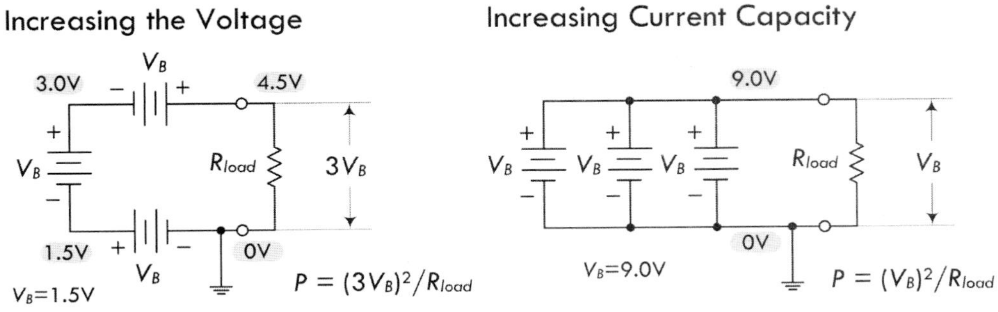

Increasing the Voltage

3.0V

V_B

4.5V

V_B R_{load} $3V_B$

1.5V

V_B 0V

$V_B = 1.5V$ $P = (3V_B)^2/R_{load}$

Increasing Current Capacity

9.0V

V_B V_B V_B R_{load} V_B

0V

$V_B = 9.0V$ $P = (V_B)^2/R_{load}$

FIGURE 2.58

Note: When placing batteries in parallel, it's important that the voltages and chemistry be the same—choose all similar batteries, all fresh. If the voltages are different, you can run into problems.

2.16   Open and Short Circuits

The most common problems (faults) in circuits are open circuits and shorts. A *short circuit* in all or part of a circuit causes excessive current flow. This may blow a fuse or burn out a component, which may result in an open-circuit condition. An *open circuit* represents a break in the circuit, preventing current from flowing. Short circuits may be caused by a number of things—from wire crossing, insulation failure, or solder splatter inadvertently linking two separate conductors within a circuit board. An open circuit may result from wire or component lead separation from the circuit, or from a component that has simply burned out, resulting in a huge resistance. Figure 2.59 shows cases of open- and short-circuit conditions. A fuse, symbolized ⌒⌒, is used in the circuit and will blow when the current through it exceeds its current rating, given in amps.

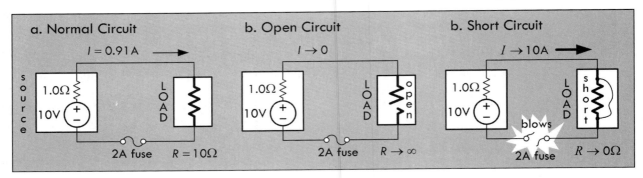

FIGURE 2.59

A weird thing about a short circuit, if you consider all components to be ideal in nature, is that an infinite current will flow if an ideal voltage source is short-circuited, while the voltage across the short goes to zero. Real voltage sources have internal resistance, as do the conductors, so the maximum current level is reduced. However, there is usually still plenty of current available to do damage.

You can often diagnose a short circuit by noting a burning smell, or by placing your hand nearby to sense any components overheating. To prevent short circuits from destroying circuits, various protection devices can be used, such as fuses, transient voltage suppressors, and circuit breakers. These devices sense when too much current is flowing, and they will blow or trip, thus creating an open-circuit condition to limit excessive current damage.

Example 1: In the series circuit in Fig. 2.60, determine how much current flows through the normal circuit (a), then determine how much current flows when there is

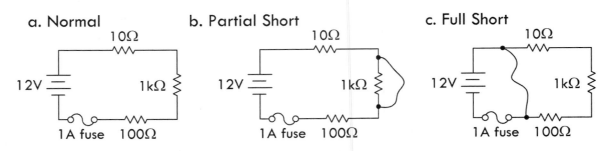

FIGURE 2.60

a partial short (b), and determine how much current would flow during a full short if the 1-A fuse weren't in place. Assume an internal 3-Ω circuit resistance at the moment of a full-short condition.

Answer: a. 11 mA, b. 109 mA, c. 4 A—fuse blows.

Example 2: In the parallel circuit in Fig. 2.61, determine the total current flow in the normal circuit (a), the open circuit (b), and the current flow in the short circuit (c). Assume the internal resistance of the battery is 0.2 Ω below 3 A, but goes to 2 Ω during a short-circuit condition.

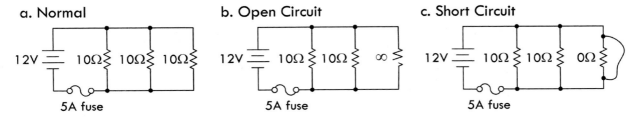

a. Normal   **b. Open Circuit**   **c. Short Circuit**

FIGURE 2.61

Answer: a. 3.4 A, b. 2.3 A, c. 6 A—fuse blows.

Example 3: In the parallel circuit in Fig. 2.62, loads B, C, and D receive no power when S2 is closed (assume all other switches are already closed); however, load A receives power. You notice the fuse is blown. After replacing the fuse, you close S2 with S3 and S4 open, and the fuse doesn't blow. Closing S3 power on loads B and C has no effect. Closing S4 causes B and C to turn off—load D receives no power, either. The fuse is blown again. What is the problem?

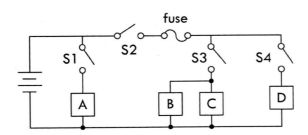

FIGURE 2.62

Answer: Load D has a short in it.

2.17 Kirchhoff's Laws

Often, you will run across a circuit that cannot be analyzed with simple resistor circuit reduction techniques alone. Even if you could find the equivalent resistance by using circuit reduction, it might not be possible for you to find the individual currents and voltage through and across the components within the network. Likewise, if there are multiple sources, or complex networks of resistors, using Ohm's law, as well as the current and voltage divider equations, may not cut it. For this reason, we turn to Kirchhoff's laws.

Kirchhoff's laws provide the most general method for analyzing circuits. These laws work for either linear (resistor, capacitors, and inductors) or nonlinear elements (diodes, transistors, etc.), no matter how complex the circuit gets. Kirchhoff's two laws are stated as follows.

Kirchhoff's Voltage Law (or Loop Rule): The algebraic sum of the voltages around any loop of a circuit is zero:

$$\sum_{closed\ path} \Delta V = V_1 + V_2 + \cdots + V_n = 0 \tag{2.20}$$

In essence, Kirchhoff's voltage law is a statement about the conservation of energy. If an electric charge starts anywhere in a circuit and is made to go through any loop in a circuit and back to its starting point, the net change in its potential is zero.

To show how Kirchhoff's voltage law works, we consider the circuit in Fig. 2.63. We start anywhere we like along the circuit path—say, at the negative terminal of the 5-V battery. Then we start making a loop trace, which in this case we choose to go clockwise, though it doesn't really matter which direction you choose. Each time we encounter a circuit element, we add it to what we'll call our ongoing loop equation. To determine the sign of the voltage difference, we apply the loop trace rules shown in the shaded section of the figure. We continue adding elements until we make it back to the start of the loop, at which point we terminate the loop equation with an "= 0."

Kirchhoff's Voltage Law

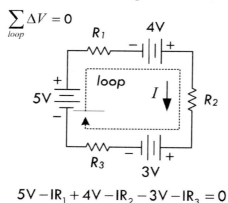

$$5V - IR_1 + 4V - IR_2 - 3V - IR_3 = 0$$

FIGURE 2.63

As noted, Kirchhoff's voltage law applies to any circuit elements, both linear and nonlinear. For example, the rather fictitious circuit shown here illustrates Kirchhoff's voltage law being applied to a circuit that has other elements besides resistors and dc sources—namely, a capacitor, an inductor, and a nonlinear diode, along with a sinusoidal voltage source. We can apply the loop trick, as we did in the preceding example, and come up with an equation. As you can see, the expressions used to describe the voltage changes across the capacitor, inductor, and diode are rather complex, not to mention the solution to the resulting differential equation. You don't do electronics this way (there are tricks), but it nevertheless demonstrates the universality of Kirchhoff's law.

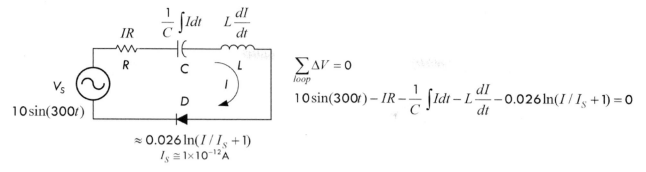

$$\sum_{loop} \Delta V = 0$$

$$10\sin(300t) - IR - \frac{1}{C}\int Idt - L\frac{dI}{dt} - 0.026\ln(I/I_S + 1) = 0$$

FIGURE 2.64

Kirchhoff's Current Law (or Junction Rule): The sum of the currents that enter a junction equals the sum of the currents that leave the junction:

$$\sum I_{in} = \sum I_{out} \tag{2.21}$$

Kirchhoff's current law is a statement about the conservation of charge flow through a circuit: at no time are charges created or destroyed.

The following example shows both Kirchhoff's current and voltage laws in action.

Example: By applying Kirchhoff's laws to the following circuit, find all the unknown currents, I_1, I_2, I_3, I_4, I_5, I_6, assuming that R_1, R_2, R_3, R_4, R_5, R_6, and V_0 are known. After that, the voltage drops across the resistors V_1, V_2, V_3, V_4, V_5, and V_6 can be found using Ohm's law: $V_n = I_n R_n$.

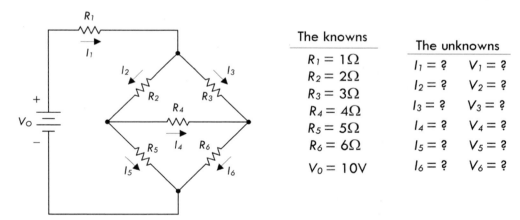

The knowns	The unknowns	
$R_1 = 1\,\Omega$	$I_1 = ?$	$V_1 = ?$
$R_2 = 2\,\Omega$	$I_2 = ?$	$V_2 = ?$
$R_3 = 3\,\Omega$	$I_3 = ?$	$V_3 = ?$
$R_4 = 4\,\Omega$	$I_4 = ?$	$V_4 = ?$
$R_5 = 5\,\Omega$	$I_5 = ?$	$V_5 = ?$
$R_6 = 6\,\Omega$	$I_6 = ?$	$V_6 = ?$
$V_0 = 10\text{V}$		

FIGURE 2.65

Answer: To solve this problem, you apply Kirchhoff's voltage law to enough closed loops and apply Kirchhoff's current law to enough junctions that you end up with enough equations to counterbalance the unknowns. After that, it is simply a matter of doing some algebra. Figure 2.66 illustrates how to apply the laws in order to set up the final equations.

In Fig. 2.66, there are six equations and six unknowns. According to the rules of algebra, as long as you have an equal number of equations and unknowns, you can usually figure out what the unknowns will be. There are three ways we can think of to solve for the unknowns in this case. First, you could apply the old "plug and chug"

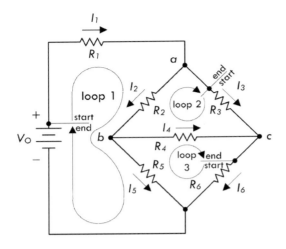

Equations resulting after applying Kirchhoff's current law:

$$I_1 = I_2 + I_3 \quad \text{(at junction a)}$$
$$I_2 = I_5 + I_4 \quad \text{(at junction b)}$$
$$I_6 = I_3 + I_4 \quad \text{(at junction c)}$$

Equations resulting after applying Kirchhoff's voltage law:

$$V_0 - I_1 R_1 - I_2 R_2 - I_5 R_5 = 0 \quad \text{(around loop 1)}$$
$$-I_3 R_3 + I_4 R_4 + I_2 R_2 = 0 \quad \text{(around loop 2)}$$
$$-I_6 R_6 + I_5 R_5 - I_4 R_4 = 0 \quad \text{(around loop 3)}$$

FIGURE 2.66

method, better known as the substitution method, where you combine all the equations together and try to find a single unknown, and then substitute it back into another equation, and so forth. A second method, which is a lot cleaner and perhaps easier, involves using matrices. A book on linear algebra will tell you all you need to know about using matrices to solve for the unknown.

A third method that we think is useful—practically speaking—involves using a trick with determinants and Cramer's rule. The neat thing about this trick is that you do not have to know any math—that is, if you have a mathematical computer program or calculator that can do determinants. The only requirement is that you be able to plug numbers into a grid (determinant) and press "equals." We do not want to spend too much time on this technique, so we will simply provide you with the equations and use the equations to find one of the solutions to the resistor circuit problem. See Fig. 2.67a.

A system of equations is represented by:

$$a_{11}x_1 + a_{21}x_2 + \ldots a_{1n}x_n = b_1$$
$$a_{12}x_1 + a_{12}x_2 + \ldots a_{2n}x_n = b_2$$
$$\vdots$$
$$a_{1n}x_1 + a_{n2}x_2 + \ldots a_{nn}x_n = b_n$$

a_{11} is the coefficient in front of variable x_1 of equation 1; a_{2n} is the coefficient in front of variable x_n of equation 2; b_2 is the constant to the right of the equal sign of equation 2.

The solution for variables in the system are:

$$x_1 = \frac{\Delta x_1}{\Delta}, x_2 = \frac{\Delta x_2}{\Delta}, \ldots, x_n = \frac{\Delta x_n}{\Delta}$$

where

The straight brackets represent a determinant

$$\Delta = \begin{vmatrix} a_{11} & a_{12} & \cdots & a_{1n} \\ a_{21} & a_{22} & \cdots & a_{2n} \\ \vdots & & & \vdots \\ a_{n1} & a_{n2} & \cdots & a_{nn} \end{vmatrix}, \Delta x_1 = \begin{vmatrix} b_1 & a_{12} & \cdots & a_{1n} \\ b_2 & a_{22} & \cdots & a_{2n} \\ \vdots & & & \vdots \\ b_n & a_{n2} & \cdots & a_{nn} \end{vmatrix}, \Delta x_2 = \begin{vmatrix} a_{11} & b_1 & \cdots & a_{1n} \\ a_{21} & b_2 & \cdots & a_{2n} \\ \vdots & & & \vdots \\ a_{n1} & b_n & \cdots & a_{nn} \end{vmatrix}, \ldots, \Delta x_n = \begin{vmatrix} a_{11} & a_{12} & \cdots & b_1 \\ a_{21} & a_{22} & \cdots & b_2 \\ \vdots & & & \vdots \\ a_{n1} & a_{n2} & \cdots & b_n \end{vmatrix}$$

FIGURE 2.67a

For example, you can find Δ for the system of equations from the resistor problem by plugging all the coefficients into the determinant and pressing the "evaluate" button on the calculator or computer. See Fig. 2.67b.

$$I_1 - I_2 - I_3 = 0$$
$$I_2 - I_5 - I_4 = 0$$
$$I_6 - I_4 - I_3 = 0$$
$$I_1 + 2I_2 + 5I_5 = 10$$
$$-3I_3 + 4I_4 + 2I_2 = 0$$
$$-6I_6 + 5I_5 - 4I_4 = 0$$

$$\Delta = \begin{vmatrix} 1 & -1 & -1 & 0 & 0 & 0 \\ 0 & 1 & 0 & -1 & -1 & 0 \\ 0 & 0 & -1 & -1 & 0 & 1 \\ 1 & 2 & 0 & 0 & 5 & 0 \\ 0 & 2 & -3 & 4 & 0 & 0 \\ 0 & 0 & 0 & -4 & 5 & -6 \end{vmatrix} = -587$$

FIGURE 2.67b

Now, to find, say, the current through R_5 and the voltage across it, you find ΔI_5, then use $I_5 = \Delta I_5 / \Delta$ to find the current. Then you use Ohm's law to find the voltage. Figure 2.67c shows how it is done.

$$\Delta I_5 = \begin{vmatrix} 1 & -1 & -1 & 0 & 0 & 0 \\ 0 & 1 & 0 & -1 & 0 & 0 \\ 0 & 0 & -1 & -1 & 0 & 1 \\ 1 & 2 & 0 & 0 & 10 & 0 \\ 0 & 2 & -3 & 4 & 0 & 0 \\ 0 & 0 & 0 & -4 & 0 & -6 \end{vmatrix} = -660$$

$$I_5 = \frac{\Delta I_5}{\Delta} = \frac{-660}{-587} = 1.124 A$$

$$V_5 = I_5 \times R_5 = (1.124 A)(5\Omega) = 5.62 V$$

FIGURE 2.67c

To solve for the other currents (and voltages), simply find the other ΔI's and divide by Δ.

The last approach, as you can see, requires a huge mathematical effort to get a single current value. For simplicity, you can find out everything that's going on in the circuit by running it through a circuit simulator program. For example, using MultiSim, we get the results in Fig. 2.68:

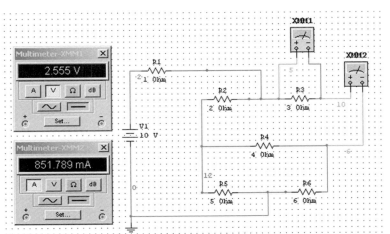

Doing long calculations is good theoretical exercise, but in practice it's a waste of time. A problem such as this only takes a few minutes to solve using a simulator. The results from simulation:

$V_1 = 2.027$ V	$I_1 = 2.027$ A
$V_2 = 2.351$ V	$I_2 = 1.175$ A
$V_3 = 2.555$ V	$I_3 = 0.852$ A
$V_4 = 0.204$ V	$I_4 = 0.051$ A
$V_5 = 5.622$ V	$I_5 = 1.124$ A
$V_6 = 5.417$ V	$I_6 = 0.903$ A

FIGURE 2.68

Plugging the results back into the diagram, you can see Kirchhoff's voltage and current laws in action, as shown in Fig. 2.69. Take any loop, sum the changes in voltage across components, and you get 0 (note that the voltages indicated in black shadow represent point voltages relative to 0-V reference ground). Also, the current that enters any junction will equal the sum of the currents exiting the junction, and vice versa—Kirchhoff's current law.

Kirchhoff's Voltage Law in Action

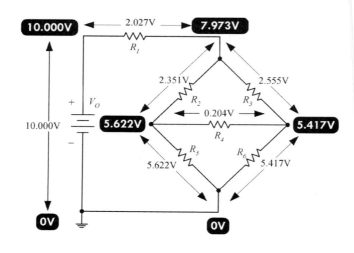

Kirchhoff's Current Law in Action

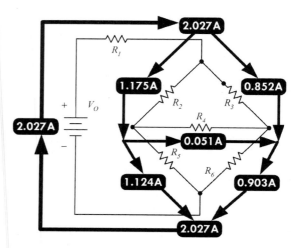

FIGURE 2.69

Now, before you get too gung ho about playing with equations or become lazy by grabbing a simulator, you should check out a special theorem known as Thevenin's theorem. This theorem uses some very interesting tricks to analyze circuits, and it may help you avoid dealing with systems of equations or having to resort to simulation. Thevenin's theorem utilizes something called the superposition theorem, which we must consider first.

2.18 Superposition Theorem

The superposition theorem is an important concept in electronics that is useful whenever a linear circuit contains more than one source. It can be stated as follows:

Superposition theorem: *The current in a branch of a linear circuit is equal to the sum of the currents produced by each source, with the other sources set equal to zero.*

The proof of the superposition theorem follows directly from the fact that Kirchhoff's laws applied to linear circuits always result in a set of linear equations, which can be reduced to a single linear equation in a single unknown. This means that an unknown branch current can thus be written as a linear superposition of each of the source terms with an appropriate coefficient. Be aware that the superposition should not be applied to nonlinear circuits.

It is important to make clear what it means to set sources equal to zero when interpreting the superposition theorem. A source may be a voltage source or a current source. If the source is a voltage source, to set it to zero means that the points in the circuit where its terminals were connected must be kept at the same potential. The only way to do this is to replace the voltage source with a conductor, thus creating a short circuit. If the source is a current source, to set it to zero means to simply remove it and leave the terminals open, thus creating an open circuit. A short circuit causes the voltage to be zero; an open circuit causes the current to be zero.

In Fig. 2.70, we will analyze the circuit using the superposition theorem. The circuit contains two resistors, a voltage source, and a current source.

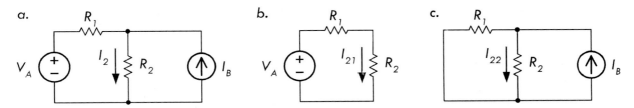

FIGURE 2.70   **The circuit in (a) can be analyzed using the superposition theorem by considering the simpler circuits in (b) and (c).**

First we remove the current source (create an open circuit at its terminals), as shown in Fig. 2.70b. The new current through R_2 due to the voltage source alone is just V_A divided by the equivalent resistance:

$$I_{21} = \frac{V_A}{R_1 + R_2}$$

This current is called the partial current in branch 2 due to source 1. Next, the voltage source is removed and set to zero by replacing it with a conductor (we short it); see Fig. 2.70c. The resulting circuit is a current divider, and the resulting partial current is given by:

$$I_{22} = \frac{I_B R_1}{R_1 + R_2}$$

Applying the superposition, we add the partial current to get the total current:

$$I_2 = I_{21} + I_{22} = \frac{V_A - I_B R_1}{R_1 + R_2}$$

The current through R_1 could have been determined in a similar manner, with the following result:

$$I_1 = \frac{V_A + I_B R_2}{R_1 + R_2}$$

The superposition theorem is an important tool whose theory makes possible complex impedance analysis in linear, sinusoidally driven circuits—a subject we will cover a bit later. The superposition is also an underlying mechanism that makes possible two important circuit theorems: Thevenin's theorem and Norton's theorem. These two theorems, which use some fairly ingenious tricks, are much more practical to use than the superposition. Though you will seldom use the superposition directly, it is important to know that it is the base upon which many other circuit analysis tools rest.

2.19   Thevenin's and Norton's Theorems

2.19.1   Thevenin's Theorem

Say that you are given a complex circuit such as that shown in Fig. 2.71. Pretend that you are only interested in figuring out what the voltage will be across terminals A and F (or any other set of terminals, for that matter) and what amount of current will flow through a load resistor attached between these terminals. If you were to apply Kirchhoff's laws to this problem, you would be in trouble—the amount of work required to set up the equations would be a nightmare, and then after that you would be left with a nasty system of equations to solve.

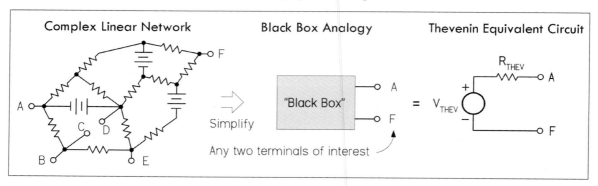

FIGURE 2.71   **The essence of Thevenin's theorem.**

Luckily, a man by the name of Thevenin came up with a theorem, or trick, to simplify the problem and produce an answer—one that does not involve "hairy" mathematics. Using what Thevenin discovered, if only two terminals are of interest, these two terminals can be extracted from the complex circuit, and the rest of the circuit can be considered a black box. Now the only things left to work with are these two terminals. By applying Thevenin's tricks (which you will see in a moment), you will discover that this black box, or any linear two-terminal dc network, can be represented by a voltage source in series with a resistor. (This statement is referred to as *Thevenin's theorem*.) The voltage source is called the *Thevenin voltage* V_{THEV}, and the resistance is called the *Thevenin resistance* R_{THEV}; together, the two form what is called the *Thevenin equivalent circuit*. From this simple equivalent circuit you can easily calculate the current flow through a load placed across its terminals by using Ohm's law: $I = V_{THEV}/(R_{THEV} + R_{LOAD})$.

Note that circuit terminals (black box terminals) might actually not be present in a circuit. For example, instead, you may want to find the current and voltage across a resistor (R_{LOAD}) that is already within a complex network. In this case, you must remove the resistor and create two terminals (making a black box) and then find the Thevenin equivalent circuit. After the Thevenin equivalent circuit is found, you simply replace the resistor (or place it across the terminals of the Thevenin equivalent circuit), calculate the voltage across it, and calculate the current through it by applying Ohm's law again: $I = V_{THEV}/(R_{THEV} + R_{LOAD})$. However, two important questions remain: What are the tricks? And what are V_{THEV} and R_{THEV}?

First, V_{THEV} is simply the voltage across the terminals of the black box, which can be either measured or calculated. R_{THEV} is the resistance across the terminals of the black box when all the dc sources (e.g., batteries) are replaced with shorts, and it, too, can be measured or calculated.

As for the tricks, we can generalize and say that the superposition theorem is involved. However, Thevenin figured out, using the tricks presented in the following example, that the labor involved with applying the superposition theorem (removing sources one at a time, calculating partial currents, and adding them, etc.) could be reduced by removing all sources at once, and finding the Thevenin resistance. An example is the best cure. See Fig. 2.72.

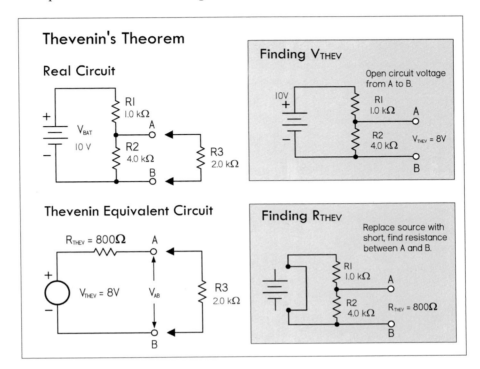

FIGURE 2.72   Here's an example of how Thevenin's theorem can be used in a voltage divider circuit to easily calculate the voltage across and the current flow through an attached load.

First remove the load R_3 and open up the terminals of interest (A and B). Then, determine the Thevenin voltage V_{THEV} using Ohm's law or the voltage divider equation—it's the open-circuit voltage across A and B.

Next, calculate the Thevenin resistance R_{THEV} across the terminals A and B by replacing the dc source (V_{BAT}) with a short and calculating or measuring the resistance across A and B. R_{THEV} is simply R_1 and R_2 in parallel.

The final Thevenin equivalent circuit is then simply V_{THEV} in series with R_{THEV}. The voltage across the load and the current through the load (R_3) are:

$$V_3 = \frac{R_3}{R_3 + R_{THEV}} \times V_{THEV} = \frac{2000\ \Omega}{2800\ \Omega} \times 8\ V = 5.7\ V$$

$$I_3 = \frac{V_{THEV}}{(R_{THEV} + R_3)} = \frac{8\ V}{2800\ \Omega} = 0.003\ A$$

2.19.2   Norton's Theorem

Norton's theorem is another tool for analyzing complex networks. Like Thevenin's theorem, it takes a complex two-terminal network and replaces it with a simple equivalent circuit. However, instead of a Thevenin voltage source in series with a Thevenin resistance, the Norton equivalent circuit consists of a current source in parallel with a resistance—which happens to be the same as the Thevenin resistance. The only new trick is finding the value of the current source, which is referred to as the Norton current I_{NORTON}. In essence, Norton's theorem is to current sources as Thevenin's theorem is to voltage sources. Its underlying mechanism, like that of Thevenin's, is the superposition theorem.

Figure 2.73 shows how the circuit just analyzed by means of Thevenin's theorem can be analyzed using Norton's theorem. The Norton current I_{NORTON} represents the short-circuit current through terminals A and B.

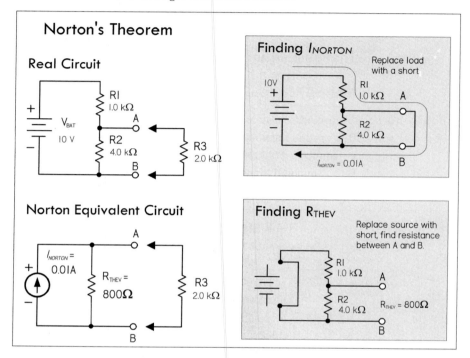

FIGURE 2.73   To find I_{NORTON}, first remove load R_3 and replace it with a short. Using Ohm's law, and noticing that no current will ideally flow through R_2 since it's shorted, you find the short-circuit current, or Norton current, to be:

$$I_{NORTON} = \frac{V_{BAT}}{R_1} = \frac{10 \text{ V}}{1000 \text{ } \Omega} = 0.01 \text{ A}$$

Next, find the Thevenin resistance—simply use the result from the previous example:

$$R_{THEV} = 800 \text{ } \Omega.$$

The final Norton equivalent circuit can then be constructed. Replacing R_3, you can now determine how much current will flow through R_3, using Ohm's law, or simply applying the current divider equation.

$$I_3 = \frac{R_{THEV}}{R_{THEV} + R_3} \times I_{NORTON} = \frac{800 \text{ } \Omega}{2800 \text{ } \Omega} \times 0.01 \text{ A} = 0.003$$

A Norton equivalent circuit can be transformed into a Thevenin equivalent circuit and vice versa. The equivalent resistor stays the same in both cases; it is placed in series with the voltage source in the case of a Thevenin equivalent circuit and in parallel with the current source in the case of the Norton equivalent circuit. The voltage for a Thevenin equivalent source is equal to the nonload voltage appearing across the resistor in the Norton equivalent circuit. The current for a Norton equivalent source is equal to the short-circuit current provided by the Thevenin source.

Examples: Find the Thevenin and Norton equivalent circuits for everything between points A and B for each of the four circuits in Fig. 2.74.

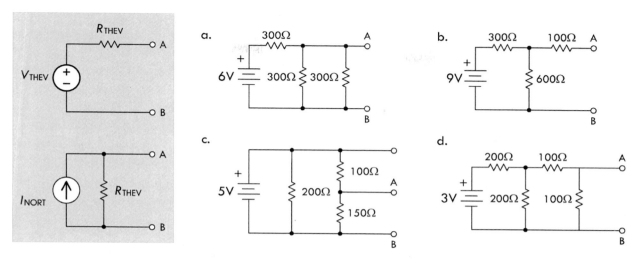

FIGURE 2.74

Answers: a: $V_{THEV} = 2$ V, $R_{THEV} = 100$ Ω, $I_{NORT} = 0.02$ A, b: $V_{THEV} = 6$ V, $R_{THEV} = 300$ Ω, $I_{NORT} = 0.02$ A, c: $V_{THEV} = 3$ V, $R_{THEV} = 60$ Ω, $I_{NORT} = 0.05$ A, d: $V_{THEV} = 0.5$ V, $R_{THEV} = 67$ Ω, $I_{NORT} = 0.007$ A

Example: Here's an example where Thevenin's theorem can be applied a number of times to simplify a complex circuit that has more than one source. In essence, you create Thevenin subcircuits and combine them. Often this is easier than trying to find the whole Thevenin equivalent circuit in one grand step. Refer to Fig. 2.75.

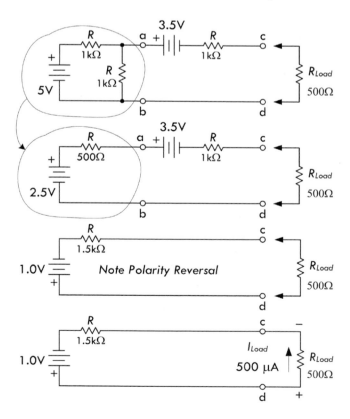

Here we're interested in finding the current that will flow through a load resistor R_{load} when attached to terminals c and d. To simplify matters, first find the Thevenin equivalent circuit for everything to the left of a and b. Using the voltage divider and resistors-in-parallel formulas:

$$V_{THEV}(a,b) = \frac{(1000\ \Omega)}{1000\ \Omega + 1000\ \Omega}(5\ \text{V}) = 2.5\ \text{V}$$

$$R_{THEV}(a,b) = \frac{(1000\ \Omega)(1000\ \Omega)}{1000\ \Omega + 1000\ \Omega} = 500\ \Omega$$

(Recall that we replaced the 5-V source with a short when finding R_{THEV}.)

Incorporating this equivalent circuit back into the main circuit, as shown in the second circuit down, we then determine the Thevenin equivalent circuit for everything to the left of c and d. Using Kirchhoff's voltage law and resistors-in-series formulas:

$$V_{THEV}(c,d) = 2.5\ \text{V} - 3.5\ \text{V} = -1.0\ \text{V}$$

(In terms of the diagram, this represents a polarity reversal, or simply flipping the battery.)

$$R_{THEV}(c,d) = 500\ \Omega + 1000\ \Omega = 1500\ \Omega$$

(Both sources were shorted to find R_{THEV}.)
Now we attach our load of 500 Ω and get the current:

$$I_{load} = \frac{1.0\ \text{V}}{1500\ \Omega + 500\ \Omega} = 5 \times 10^{-4}\ \text{A} = 0.5\ \text{mA}$$

FIGURE 2.75

Example: To increase the current capacity, batteries are placed in parallel. If the internal resistance is 0.2 Ω for each 1.5-V battery, find the equivalent Thevenin circuit. Refer to Fig. 2.76.

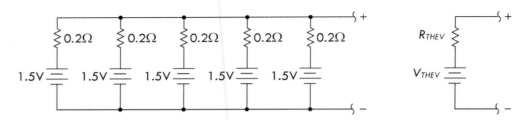

FIGURE 2.76

Answer: $R_{\text{THEV}} = 0.04\ \Omega$, $V_{\text{THEV}} = 1.5$ V. (Apply Thevenin's theorem in steps.) As you can see, the net internal resistance of the Thevenin circuit is much less—a result of placing batteries in parallel.

2.20 AC Circuits

A circuit is a complete conductive path through which electrons flow from source to load and back to source. As we've seen, if the source is dc, electrons will flow in only one direction, resulting in a direct current (dc). Another type of source that is frequently used in electronics is an alternating source that causes current to periodically change direction, resulting in an alternating current (ac). In an ac circuit, not only does the current change directions periodically, the voltage also periodically reverses.

Figure 2.77 shows a dc circuit and an ac circuit. The ac circuit is powered by a sinusoidal source, which generates a repetitive sine wave that may vary in frequency from a few cycles per second to billions of cycles per second, depending on the application.

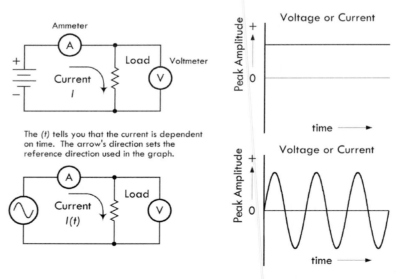

FIGURE 2.77

The positive and negative swings in voltage/current relative to a zero volt/amp reference line simply imply that the electromotive force of the source has switched directions, causing the polarity of the voltage source to flip, and forcing current to change directions. The actual voltage across the source terminals at a given instant in time is the voltage measured from the 0-V reference line to the point on the sinusoidal waveform at the specified time.

2.20.1 Generating AC

The most common way to generate sinusoidal waveforms is by electromagnetic induction, by means of an ac generator (or alternator). For example, the simple ac generator in Fig. 2.78 consists of a loop of wire that is mechanically rotated about an axis while positioned between the north and south poles of a magnet. As the loop rotates in the magnetic field, the magnetic flux through it changes, and charges are forced through the wire, giving rise to an effective voltage or induced voltage. According to Fig. 2.78, the magnetic flux through the loop is a function of the angle of the loop relative to the direction of the magnetic field. The resultant induced voltage is sinusoidal, with angular frequency ω (radians per second).

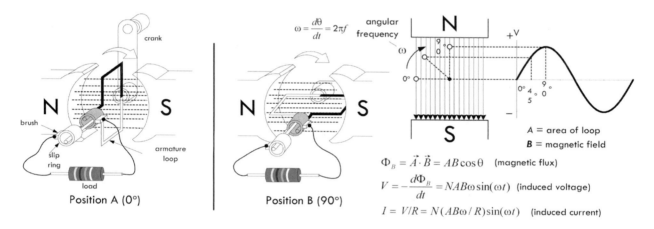

$$\omega = \frac{d\theta}{dt} = 2\pi f \quad \text{angular frequency}$$

A = area of loop
B = magnetic field

$$\Phi_B = \vec{A} \cdot \vec{B} = AB\cos\theta \quad \text{(magnetic flux)}$$

$$V = -\frac{d\Phi_B}{dt} = NAB\omega\sin(\omega t) \quad \text{(induced voltage)}$$

$$I = V/R = N(AB\omega/R)\sin(\omega t) \quad \text{(induced current)}$$

Position A (0°)

Position B (90°)

FIGURE 2.78 Simple ac generator.

Real ac generators are, of course, more complex than this, but they operate under the same laws of induction, nevertheless. Other ways of generating ac include using a transducer (e.g., a microphone) or even using a dc-powered oscillator circuit that uses special inductive and capacitive effects to cause current to resonate back and forth between an inductor and a capacitor.

Why Is AC Important?

There are several reasons why sinusoidal waveforms are important in electronics. The first obvious reason has to do with the ease of converting circular mechanical motion into induced current via an ac generator. However, another very important reason for using sinusoidal waveforms is that if you differentiate or integrate a sinusoid, you get a sinusoid. Applying sinusoidal voltage to capacitors and inductors leads to sinusoidal current. It also avoids problems on systems, a subject that we'll cover later. But one of the most important benefits of ac involves the ability to increase voltage or decrease voltage (at the expense of current) by using a transformer. In dc, a transformer is useless, and increasing or decreasing a voltage is a bit tricky, usually involving some resistive power losses. Transformers are very efficient, on the other hand, and little power is lost in the voltage conversion.

2.20.2 Water Analogy of AC

Figure 2.79 shows a water analogy of an ac source. The analogy uses an oscillating piston pump that moves up and down by means of a cam mechanism, driven by a hand crank.

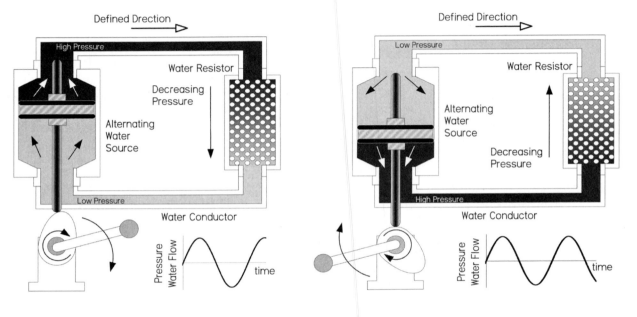

FIGURE 2.79

In the water analogy, water particles, on average, appear to simply swish back and forth as the crank is turned. In an ac electrical circuit, a similar effect occurs, though things are a bit more complex. One way to envision what's going on is that within a conductor, the drift velocity of the sea of electrons is being swished back and forth in a sinusoidal manner. The actual drift velocity and distance over which the average drift occurs are really quite small (fractions of millimeter-per-second range, depending on conductor and applied voltage). In theory, this means that there is no net change in position of an "average" electron over one complete cycle. (This is not to be confused with an individual electron's thermal velocity, which is mostly random, and at high velocity.) Also, things get even more complex when you start applying high frequencies, where the skin effect enters the picture—more on this later.

2.20.3 Pulsating DC

If current and voltage never change direction within a circuit, then from one perspective, we have a dc current, even if the level of the dc constantly changes. For example, in Fig. 2.80, the current is always positive with respect to 0, though it varies periodically in amplitude. Whatever the shape of the variations, the current can be referred to as "pulsating dc." If the current periodically reaches 0, it is referred to as "intermittent dc."

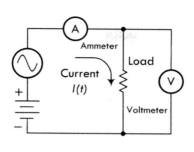

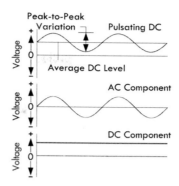

FIGURE 2.80

From another perspective, we may look at intermittent and pulsating dc as a combination of an ac and a dc current. Special circuits can separate the two currents into ac and dc components for separate analysis or use. There are also circuits that combine ac and dc currents and voltages.

2.20.4 Combining Sinusoidal Sources

Besides combining ac and dc voltages and currents, we can also combine separate ac voltages and currents. Such combinations will result in complex waveforms. Figure 2.81 shows two ac waveforms fairly close in frequency, and their resultant combination. The figure also shows two ac waveforms dissimilar in both frequency and wavelength, along with the resultant combined waveform.

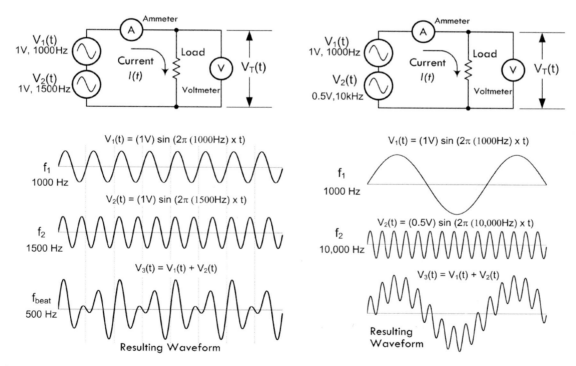

FIGURE 2.81 (Left) Two ac waveforms of similar magnitude and close in frequency form a composite wave. Note the points where the positive peaks of the two waves combine to create high composite peaks: this is the phenomenon of beats. The beat note frequency is $f_2 - f_1 = 500$ Hz. (Right) Two ac waveforms of widely different frequencies and amplitudes form a composite wave in which one wave appears to ride upon the other.

Later we will discover that by combining sinusoidal waveforms of the same frequency—even though their amplitudes and phases may be different—you always get a resultant sine wave. This fact becomes very important in ac circuit analysis.

2.20.5 AC Waveforms

Alternating current can take on many other useful wave shapes besides sinusoidal. Figure 2.82 shows a few common waveforms used in electronics. The squarewave is vital to digital electronics, where states are either true (on) or false (off). Triangular and ramp waveforms—sometimes called sawtooth waves—are especially useful in timing circuits. As we'll see later in the book, using Fourier analysis, you can create any desired shape of periodic waveform by adding a collection of sine waves together.

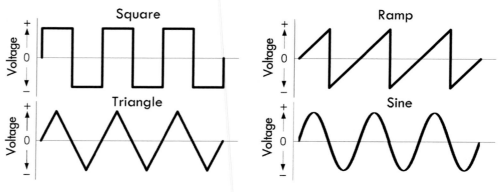

FIGURE 2.82

An ideal sinusoidal voltage source will maintain its voltage across its terminals regardless of load—it will supply as much current as necessary to keep the voltage the same. An ideal sinusoidal current source, on the other hand, will maintain its output current, regardless of the load resistance. It will supply as much voltage as necessary to keep the current the same. You can also create ideal sources of other waveforms. Figure 2.83 shows schematic symbols for an ac voltage source, an ac current source, and a clock source used to generate squarewaves.

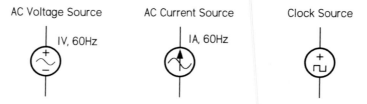

FIGURE 2.83

In the laboratory, a function generator is a handy device that can be used to generate a wide variety of waveforms with varying amplitudes and frequencies.

2.20.6 Describing an AC Waveform

A complete description of an ac voltage or current involves reference to three properties: amplitude (or magnitude), frequency, and phase.

Amplitude

Figure 2.84 shows the curve of a sinusoidal waveform, or sine wave. It demonstrates the relationship of the voltage (or current) to relative positions of a circular rotation through one complete revolution of 360°. The magnitude of the voltage (or current) varies with the sine of the angle made by the circular movement with respect to the zero point. The sine of 90° is 1, which is the point of maximum current (or voltage); the sine of 270° is –1, which is the point of maximum reverse current (or voltage); the sine of 45° is 0.707, and the value of current (or voltage) at the 45° point of rotation is 0.707 times the maximum current (or voltage).

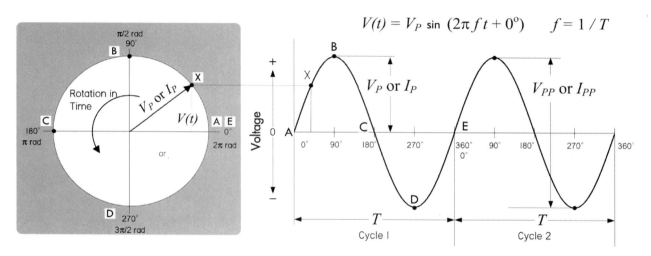

FIGURE 2.84

2.20.7 Frequency and Period

A sinusoidal waveform generated by a continuously rotating generator will generate alternating current (or voltage) that will pass through many cycles over time. You can choose an arbitrary point on any one cycle and use it as a marker—say, for example, the positive peak. The number of times per second that the current (or voltage) reaches this positive peak in any one second is called the frequency of the ac. In other words, frequency expresses the rate at which current (or voltage) cycles occur. The unit of frequency is cycles per second, or hertz—abbreviated Hz (after Heinrich Hertz).

The length of any cycle in units of time is the period of the cycle, as measured from two equivalent points on succeeding cycles. Mathematically, the period is simply the inverse of the frequency:

$$\text{Frequency in hertz} = \frac{1}{\text{Period in seconds}} \text{ or } f = \frac{1}{T} \qquad (2.22)$$

and

$$\text{Period in seconds} = \frac{1}{\text{Frequency in hertz}} \text{ or } T = \frac{1}{f} \qquad (2.23)$$

Example: What is the period of a 60-Hz ac current?

Answer:

$$T = \frac{1}{60 \text{ Hz}} = 0.0167 \text{ s}$$

Example: What is the frequency of an ac voltage that has period of 2 ns?

Answer:

$$f = \frac{1}{2 \times 10^{-9} \text{ s}} = 5.0 \times 10^8 \text{ Hz} = 500 \text{ MHz}$$

The frequency of alternating current (or voltage) in electronics varies over a wide range, from a few cycles per second to billions of cycles per second. To make life easier, prefixes are used to express large frequencies and small periods. For example: 1000 Hz = 1 kHz (kilohertz), 1 million hertz = 1 MHz (megahertz), 1 billion hertz = 1 GHz (gigahertz), 1 trillion hertz = 1 THz (terahertz). For units smaller than 1, as in the measurements of period, the basic unit of a second can become millisecond (1 thousandth of a second, or ms), microsecond (1 millionth of a second, or μs), nanosecond (1 billionth of a second, or ns), and picosecond (1 trillionth of a second, or ps).

2.20.8   Phase

When graphing a sine wave of voltage or current, the horizontal axis represents time. Events to the right on the graph take place later, while events to the left occur earlier. Although time can be measured in seconds, it actually becomes more convenient to treat each cycle of a waveform as a complete time unit, divisible by 360°. A conventional starting point for counting in degrees is 0°—the zero point as the voltage or current begins a positive half cycle. See Fig. 2.85a.

By measuring the ac cycle this way, it is possible to do calculations and record measurements in a way that is independent of frequency. The positive peak voltage or current occurs at 90° during a cycle. In other words, 90° represents the phase of the ac peak relative to the 0° starting point.

Phase relationships are also used to compare two ac voltage or current waveforms at the same frequency, as shown in Fig. 2.85b. Since waveform B crosses the zero point in the positive direction after A has already done so, there is a phase difference between the two waveforms. In this case, B lags A by 45°; alternatively, we can say that A leads B by 45°. If A and B occur in the same circuit, they add together, producing a composite sinusoidal waveform at an intermediate phase angle relative to the individual waveforms. Interestingly, adding any number of sine waves of the same frequency will always produce a sine wave of the same frequency—though the magnitude and phase may be unique.

In Fig. 2.85c we have a special case where B lags A by 90°. B's cycle begins exactly one-quarter cycle after A's. As one waveform passes through zero, the other just reaches its maximum value.

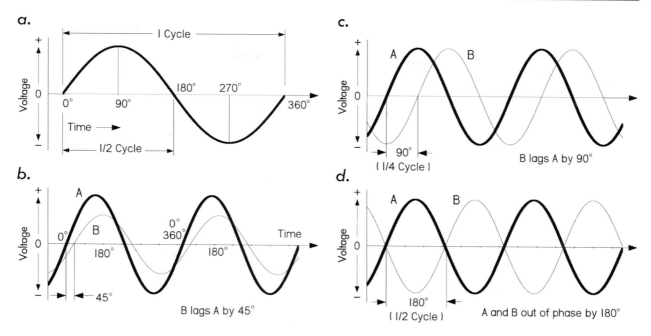

a.

b.

c.

d.

B lags A by 45°

B lags A by 90°

A and B out of phase by 180°

FIGURE 2.85 **(a) An ac cycle is divided into 360° that are used as a measure of time or phase. (b) When two waves of the same frequency start their cycles at slightly different times, the time difference or phase difference is measured in degrees. In this drawing, wave B starts 45° (one-eighth cycle) later than wave A, and so lags 45° behind A. (c and d) Two special cases of phase difference: In (c) the phase difference between A and B is 90°; in (d) the phase difference is 180°.**

Another special case occurs in Fig. 2.85d, where waveforms A and B are 180° out of phase. Here it doesn't matter which waveform is considered the leading or lagging waveform. Waveform B is always positive when waveform A is negative, and vice versa. If you combine these two equal voltage or current waveforms together within the same circuit, they completely cancel each other out.

2.21   AC and Resistors, RMS Voltage, and Current

Alternating voltages applied across a resistor will result in alternating current through the resistor that is in phase with the voltage, as seen in Fig. 2.86. Given the ac voltage and resistance, you can apply Ohm's law and find the ac current. For example, a sinusoidal waveform generated by a function generator can be mathematically described by:

$$V(t) = V_P \sin (2\pi \times f \times t) \tag{2.24}$$

where V_P is the peak amplitude of the sinusoidal voltage waveform, f is the frequency, and t is the time. Using Ohm's law and the power law, you get the following:

$$I(t) = \frac{V(t)}{R} = \frac{V_P}{R} \sin(2\pi \times f \times t) \tag{2.25}$$

If you graph both $V(t)$ and $I(t)$ together, as shown in Fig. 2.86, you notice that the current and the voltage are in phase with each other. As the voltage increases in one direction, the current also increases in the same direction. Thus, when an ac source is connected to a purely resistive load, the current and voltage are in phase. If the load isn't purely resistive (e.g., has capacitance and inductance), it's a whole different story—more on that later.

AC Current, Voltage, and Power Characteristics of Resistor (Ideal)

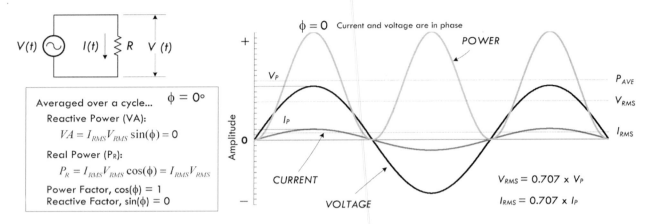

FIGURE 2.86

To find the power dissipated by the resistor under sinusoidal conditions, we can simply plug the sinusoidal voltage expression into Ohm's power law to get an instantaneous power expression:

$$P(t) = \frac{V(t)^2}{R} = \frac{V_P^2}{R} \sin^2(2\pi f t) \tag{2.26}$$

Expressing voltage, current, and power in an instantaneous fashion is fine, mathematically speaking; however, in order to get a useful result, you need to plug in a specific time—say, $t = 1.3$ s. But how often do you need to know that the voltage, current, or power are at exactly $t = 1.3$ s? Better yet, when do you start counting? These instantaneous values are typically inconvenient for any practical use. Instead, it is more important to come up with a kind of averaging scheme that can be used to calculate effective power dissipation without dealing with sinusoidal functions.

Now, you might be clever and consider averaging the sinusoidal voltage or current over a complete cycle to get some meaningful value. However, the average turns out to be zero—positive and negative sides of waveforms cancel. This may be a bit confusing, in terms of power, since the positive-going part of the wave still delivers energy, as does the negative-going part. If you've ever received a shock from 120-V line voltage, you'll be able to attest to that.

The measurement that is used instead of the average value is the RMS or *root mean square* value, which is found by squaring the instantaneous values of the ac voltage or current, then calculating their mean (i.e., their average), and finally taking the square root of this—which gives the effective value of the ac voltage or current. These effective, or RMS, values don't average out to zero and are essentially the ac equivalents of dc voltages and currents. The RMS values of ac voltage and current are based upon equating the values of ac and dc power required to heat a resistive element to exactly the same degree. The peak ac power required for this condition is twice the dc power needed. Therefore, the average ac power equivalent to a corresponding average dc power is half the peak ac power.

$$P_{ave} = \frac{P_{peak}}{2} \text{ (Average dc power equivalent of ac waveform)} \qquad (2.27)$$

Mathematically, we can determine the RMS voltage and current values for sinusoidal waveforms $V(t) = V_P \sin(2\pi \times f \times t)$, and $I(t) = I_P \sin(2\pi \times f \times t)$:

$$V_{RMS} = \sqrt{\frac{1}{T} \int_0^T V(t)^2 \, dt} = \frac{1}{\sqrt{2}} \times V_P = 0.707 \times V_P \qquad \text{(2.28) RMS voltage}$$

$$I_{RMS} = \sqrt{\frac{1}{T} \int_0^T I(t)^2 \, dt} = \frac{1}{\sqrt{2}} \times I_P = 0.707 \times I_P \qquad \text{(2.29) RMS current}$$

Notice that the RMS voltage and current depend only on the peak voltage or current—they are independent of time or frequency. Here are the important relations, without the scary calculus:

$$V_{RMS} = \frac{V_P}{\sqrt{2}} = \frac{V_P}{1.414} = 0.707 \times V_P \qquad V_P = V_{RMS} \times 1.414$$

$$I_{RMS} = \frac{I_P}{\sqrt{2}} = \frac{I_P}{1.414} = 0.707 \times I_P \qquad I_P = I_{RMS} \times 1.414$$

For example, a U.S. electric utility provides your home with 60 Hz, 120 VAC (in Europe and many other countries it's 50 Hz, 240V AC). The "VAC" unit tells you that the supplied voltage is given in RMS. If you were to attach an oscilloscope to the outlet, the displayed waveform on the screen would resemble the following function: $V(t) = 170 \text{ V} \sin(2\pi \times 60 \text{ Hz} \times t)$, where 170 V is the peak voltage.

With our new RMS values for voltage and current, we can now substitute them into Ohm's law to get what's called ac Ohm's law:

$$V_{RMS} = I_{RMS} \times R \qquad (2.30) \text{ ac Ohm's law}$$

Likewise, we can use the RMS voltage and current and substitute them into Ohm's power law to get what's called the ac power law, which gives the effective power dissipated (energy lost per second):

$$P = I_{RMS} \times V_{RMS} = \frac{V^2_{RMS}}{R} = I^2_{RMS} R \qquad (2.31) \text{ ac power law}$$

Again, these equations apply only to circuits that are purely resistive, meaning there is virtually no capacitance and/or inductance. Doing power calculation on circuits with inductance and capacitance is a bit more complicated, as we'll see a bit later.

Figure 2.87 shows the relationships between RMS, peak, peak-to-peak and half-wave average values of voltage and current. Being able to convert from one type to another is important, especially when dealing with component maximum voltage and current ratings—sometimes you'll be given the peak value, other times the RMS value. Understanding the differences also becomes crucial when making test measurements—see the note on pages 91–92 about making RMS test measurements. Most of the time, when dealing with ac voltage you can assume that voltage is expressed as an RMS value unless otherwise stated.

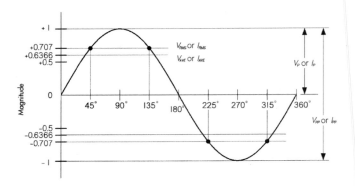

FIGURE 2.87

Conversion Factors for AC Voltage and Current

FROM	TO	MULTIPLY BY
Peak	Peak-to-peak	2
Peak-to-peak	Peak	0.5
Peak	RMS	$1/\sqrt{2}$ or 0.7071
RMS	Peak	$\sqrt{2}$ or 1.4142
Peak-to-peak	RMS	$1/(2\sqrt{2})$ or 0.35355
RMS	Peak-to-peak	$(2 \times \sqrt{2})$ or 2.828
Peak	Average*	$2/\pi$ or 0.6366
Average*	Peak	$\pi/2$ or 1.5708
RMS	Average*	$(2 \times \sqrt{2})/\pi$ or 0.9003
Average*	RMS	$\pi/(2 \times \sqrt{2})$ or 1.1107

* Represents average over half a cycle.

Example 1: How much current would flow through a 100-Ω resistor connected across the hot and neutral sockets of a 120-VAC outlet? How much power would be dissipated? What would the results be using 1000-Ω, 10,000-Ω, and 100,000-Ω resistors?

Answer:

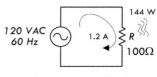

* VAC represents RMS voltage

AC Ohm's Law	AC Power Law
$V_{RMS} = I_{RMS} \times R = (1.2A)(100\Omega) = 120$ VAC	$P_{AVE} = I_{RMS} \times V_{RMS} = 120V \times 1.2A = 144$ W
$I_{RMS} = V_{RMS} / R = 120V / 100\Omega = 1.2$ A	$P_{AVE} = V_{RMS}^2 / R = (120V)^2 / 100\Omega = 144$ W
$R = V_{RMS} / I_{RMS} = 120V / 1.2A = 100\Omega$	$P_{AVE} = I_{RMS}^2 \times R = (1.2A)^2 \times 100\Omega = 144$ W

FIGURE 2.88

First, don't try this with any ordinary resistor; you'd need a power resistor or special heating element with a power rating of greater than 144 W! (Also, simply don't attempt attaching a resistor with the outlets powered.) A 1000-Ω resistor plugged into the same outlet would dissipate 14.4 W; a 10,000-Ω resistor would dissipate 1.44 W; a 100,000-Ω resistor would dissipate 0.14 W.

Example 2: What is the peak voltage on a capacitor if the RMS voltage of a sinusoidal waveform signal across it is 10.00 VAC?

Answer: VAC means RMS, so $V_P = \sqrt{2} \times V_{RMS} = 1.414 \times 10$ V $= 14.14$ V.

Example 3: A sinusoidal voltage displayed on an oscilloscope has a peak amplitude of 3.15 V. What is the RMS value of the waveform?

Answer:

$$V_{\text{RMS}} = \frac{V_P}{\sqrt{2}} = \frac{3.15 \text{ V}}{1.414} = 2.23 \text{ VAC}$$

Example 4: A 200-W resistive element in a heater is connected to a 120-VAC outlet. How much current is flowing through the resistive element? What's the resistance of the element, assuming it's an ideal resistor?

Answer: $I_{\text{RMS}} = P_{\text{AVE}}/V_{\text{RMS}} = 200 \text{ W}/120 \text{ VAC} = 1.7 \text{ A}$. $R = V_{\text{RMS}}/I_{\text{RMS}} = 120 \text{ V}/1.7 \text{ A} = 72 \text{ }\Omega$.

Example 5: A sinusoidal voltage supplied by a function generator is specified as 20 V peak to peak at 1000 Hz. What is the minimum resistance value of a ⅛-W resistor you can place across the generator's output before exceeding the resistor's power rating?

Answer: $V_P = 1/2 \text{ } V_{P-P} = 10 \text{ V}$; $V_{\text{RMS}} = 0.707 \times V_P = 7.1 \text{ VAC}$; $R = V_{\text{RMS}}^2/P_{\text{AVE}} = (7.1)^2/(1/8 \text{ W}) = 400 \text{ }\Omega$.

Example 6: The output of an oscillator circuit is specified as 680 mVAC. If you feed this into an op amp with an input resistance of 10 MΩ, how much current enters the IC?

Answer: $I_{\text{RMS}} = V_{\text{RMS}}/R = 0.68 \text{ V}/(10,000,000 \text{ }\Omega) = 0.000000068 = 68 \text{ nA}$.

MEASURING RMS VOLTAGES AND CURRENTS

There are many digital multimeters that do not measure the RMS value of an ac voltage directly. Often the meter will simply measure the peak value and calculate the equivalent RMS value—assuming the measured waveform is sinusoidal—and then display this value. Analog meters usually measure the half-wave average value, but are made to indicate the equivalent RMS.

True RMS multimeters, on the other hand, measure the true RMS value of voltages and current, and are especially handy for nonsinusoidal voltage and currents. Though relatively expensive, these meters are worth the price. It's important to note that true RMS meters will also include the contribution of any dc voltages or current components present along with the ac.

Fortunately, you can still get a fairly accurate idea of the RMS value of a sine waveform, knowing one of the other measurements such as the half-wave average, peak, or peak-to-peak value. This can be done by calculation, or using the table in Fig. 2.89. As you can see, it's also possible to work out the RMS value of a few other symmetrical and regular waveforms, such as square and triangular waves, knowing their peak, average, or peak-to-peak values.

An important thing to note when using the table is that you need to know the exact basis on which your meter's measurement is made. For example, if your meter measures the peak value, and then calculates and displays the equivalent sine wave RMS figure, this means you'll need to use the table dif-ferently when compared to the situation where the meter really measures the half-wave average and calculates the sine wave RMS figure from that. So take care, especially if you're not sure exactly how your meter works.

WAVEFORM	HALF-WAVE AVERAGE	RMS (EFFECTIVE)	PEAK	PEAK-TO-PEAK
Sine wave	1.00	1.11	1.567	3.14
	0.90	1.00	1.414	2.828
	0.637	0.707	1.00	2.00
	0.318	0.354	0.50	1.00
Squarewave	1.00	1.00	1.00	2.00
Triangle or Sawtooth	1.00	1.15	2.00	4.00
	0.87	1.00	1.73	3.46
	0.50	0.578	1.00	2.00
	0.25	0.289	0.50	1.00

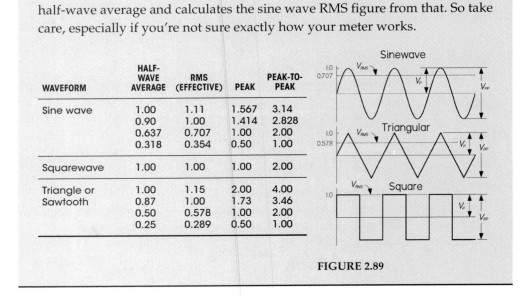

FIGURE 2.89

2.22 Mains Power

In the United States, three wires run from the pole transformers (or underground or surface enclosed transformer) to the main service panel at one's home. One wire is the A-phase wire (usually black in color), another is the B-phase wire (usually black in color), and the third is the neutral wire (white in color). Figure 2.90 shows where these three wires originate from the pole transformer. The voltage between the A-phase and the B-phase wires, or the hot-to-hot voltage, is 240 V, while the voltage between the neutral wire and either the A-phase or the B-phase wire, or the neutral-to-hot voltage, is 120 V. (These voltages are nominal and may vary from region to region, say 117 V instead of 120 V.)

At the home, the three wires from the pole/green box transformer are connected through a wattmeter and then enter a main service panel that is grounded to a long copper rod driven into the ground or to the steel in a home's foundation. The A-phase and B-phase wires that enter the main panel are connected through a main disconnect breaker, while the neutral wire is connected to a terminal referred to as the neutral bar or neutral bus. A ground bar also may be present within the main service panel. The ground bar is connected to the grounding rod or to the foundation's steel supports.

Within the main service panel, the neutral bar and the ground bar are connected together (they act as one). However, within subpanels (service panels that get their power from the main service panel but which are located some distance from the main service panel), the neutral and ground bars are not joined together. Instead, the subpanel's ground bar receives a ground wire from the main services panel. Often the metal conduit that is used to transport the wires from the main service panel to the subpanel is used as the ground wire. However, for certain critical applications (e.g., computer and life-support systems), the ground wire probably will be included within the conduit. Also, if a subpanel is not located in the same building as the main

Mains Power and Grounding

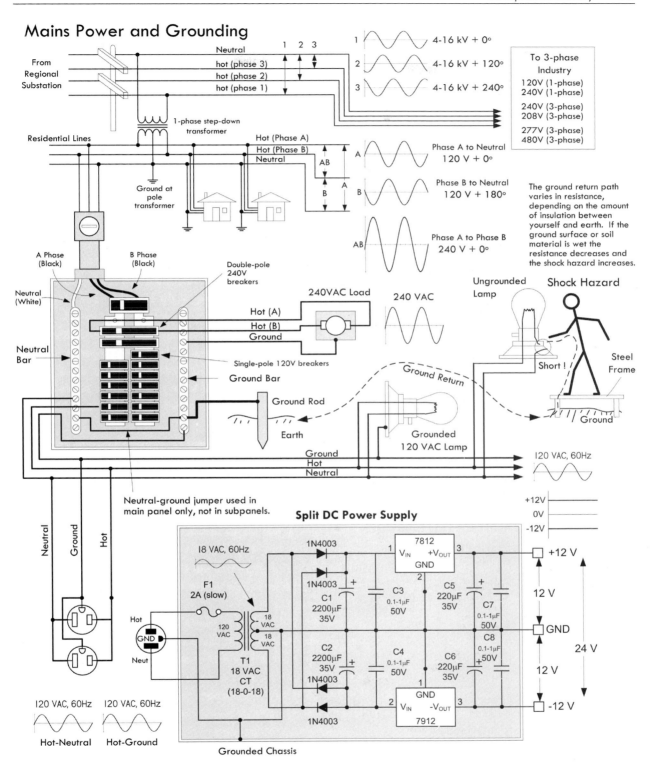

FIGURE 2.90

panel, a new ground rod typically is used to ground the subpanel. Note that different regions within the United States may use different wiring protocols. Therefore, do not assume that what we are telling you is standard practice where you live. Contact your local electrical inspector.

Within the main service panel, there are typically two bus bars into which circuit breaker modules are inserted. One of these bus bars is connected to the A-phase wire; the other bus bar is connected to the B-phase wire. To power a group of 120-V loads (e.g., upstairs lights and 120-V outlets), you throw the main breaker to the off position and then insert a single-pole breaker into one of the bus bars. (You can choose either the A-phase bus bar or the B-phase bus bar. The choice of which bus bar you use becomes important only when it comes to balancing the overall load—more on that in a moment.) Next, you take a 120-V three-wire cable and connect the cable's black (hot) wire to the breaker, connect the cable's white (neutral) wire to the neutral bar, and connect the cable's ground wire (green or bare) to the ground bar. You then run the cable to where the 120-V loads are located, connect the hot and neutral wires across the load, and fasten the ground wire to the case of the load (typically a ground screw is supplied on an outlet mounting or light figure for this purpose). To power other 120-V loads that use their own breakers, you basically do the same thing you did in the last setup. However, to maximize the capacity of the main panel (or subpanel) to supply as much current as possible without overloading the main circuit breaker in the process, it is important to balance the total load current connected to the A-phase breakers with the total load current connected to the B-phase breakers. This is referred to as "balancing the load."

Now, if you want to supply power to 240-V appliances (ovens, washers, etc.), you insert a double-pole breaker between the A-phase and B-phase bus bars in the main (or subpanel). Next, you take a 240-V three-wire cable and attach one of its hot wires to the A-phase terminal of the breaker and attach its other hot wire to the B-phase terminal of the breaker. The ground wire (green or bare) is connected to the ground bar. You then run the cable to where the 240-V loads are located and attach the wires to the corresponding terminals of the load (typically within a 240-V outlet). Also, 120-V/240-V appliances are wired in a similar manner, except you use a four-wire cable that contains an additional neutral (white) wire that is joined at the neutral bar within the main panel (or subpanel).

In addition, in many places, modifications to mains wiring must be carried out or checked by a certified electrician. As a note of caution, do not attempt home wiring unless you are sure of your abilities. If you feel that you are capable, just make sure to flip the main breaker off before you start work within the main service panel. When working on light fixtures, switches, and outlets that are connected to an individual breaker, tag the breaker with tape so that you do not mistakenly flip the wrong breaker when you go back to test your connections.

2.23 Capacitors

If you take two oppositely charged parallel conducting plates separated a small distance apart by an insulator—such as air or a dielectric such as ceramic—you have created what's called a capacitor. Now, if you apply a voltage across the plates of the capacitor using a battery, as shown in Fig. 2.91, an interesting thing occurs. Electrons are pumped out the negative battery terminal and collect on the lower plate, while electrons are drawn away from the upper plate into the positive battery terminal. The top plate becomes deficient in electrons, while the lower plate becomes abundant in electrons.

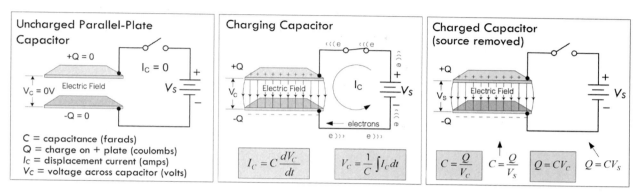

FIGURE 2.91

Very quickly, the top plate reaches a positive charge $+Q$ and the negative plate reaches a negative charge $-Q$. Accompanying the charge is a resultant electric field between the plates and a voltage equal to the battery voltage.

The important thing to notice with our capacitor is that when we remove the voltage source (battery), the charge, electric field, and corresponding voltage (presently equal to the battery voltage) remain. Ideally, this state of charge will be maintained indefinitely. Even attaching an earth ground connection to one of the plates—doesn't matter which one—will not discharge the system. For example, attaching an earth ground to the negative terminal doesn't cause the electrons within that plate to escape to the earth ground where neutral charge is assumed. (See Fig. 2.92.)

It might appear that the abundance of electrons would like to escape to the earth ground, since it is at a lower potential (neutral). However, the electric field that exists

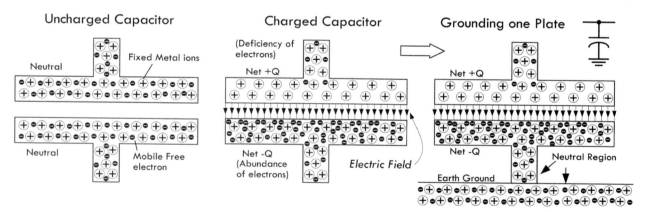

FIGURE 2.92

within the capacitor acts like a glue; the positive charge on the upper plate "holds" onto the abundance of electrons on the negative plate. In other words, the positive plate induces a negative charge in the grounded plate.

In reality, a real-life charged capacitor that is charged and removed from the voltage source would eventually lose its charge. The reason for this has to do with the imperfect insulating nature of the gas or dielectric that is placed between the plates. This is referred to as *leakage current* and, depending on the construction of the capacitor, can discharge a capacitor within as little as a few seconds to several hours, if the source voltage is removed.

To quickly discharge a capacitor you can join the two plates together with a wire, which creates a conductive path for electrons from the negative plate to flow to the positive plate, thus neutralizing the system. This form of discharge occurs almost instantaneously.

The ratio of charge on one of the plates of a capacitor to the voltage that exists between the plates is called *capacitance* (symbolized C):

$$C = \frac{Q}{V}$$

(2.32) Capacitance related to charge and voltage

C is always taken to be positive, and has units of farads (abbreviated F). One farad is equal to one coulomb per volt:

$$1\,F = 1\,C/1\,V$$

Devices that are specifically designed to hold charge (electrical energy in the form of an electric field) are called *capacitors*. Figure 2.93 shows various symbols used to represent capacitors, along with a real capacitor model that we'll discuss a bit later.

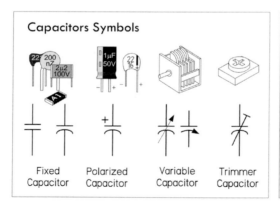

Capacitors Symbols

Fixed Capacitor · Polarized Capacitor · Variable Capacitor · Trimmer Capacitor

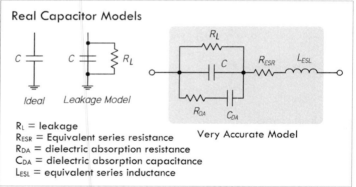

Real Capacitor Models

Ideal · Leakage Model · Very Accurate Model

R_L = leakage
R_{ESR} = Equivalent series resistance
R_{DA} = dielectric absorption resistance
C_{DA} = dielectric absorption capacitance
L_{ESL} = equivalent series inductance

FIGURE 2.93

The equation $C = Q/V$ is a general one; it really doesn't tell you why one capacitor has a larger or smaller capacitance than another. However, in practice, when you buy a capacitor all you'll be interested in is the capacitance value labeled on the device. (A voltage rating and other parameters are important, too, but we'll talk about them later.) Most commercial capacitors are limited to a range from 1 pF (1×10^{-12} F) to 4700 µF (1×10^{-6} F), with typical values for the first two digits of the capacitance of 10, 12, 15, 18, 22, 27, 33, 39, 47, 56, 68, 82, 100. (Examples: 27 pF, 100 pF, 0.01 µF, 4.7 µF, 680 µF).

Having a wide range of capacitances allows you to store different amounts of charge for a given potential difference, as well as maintaining different potential differences for a given charge. With the appropriate capacitor, you can therefore control the storage and delivery of charge, or control potential differences.

Example 1: Five volts are applied across a 1000-μF capacitor until the capacitor is fully charged. How much charge exists on the positive and negative plates?

Answer: $Q = CV = (1000 \times 10^{-6} \text{ F})(5 \text{ V}) = 5 \times 10^{-3}$ C. This is the charge on the positive plate. The charge on the negative plate is the same, but opposite in sign.

Example 2: A 1000-μF capacitor and a 470-μF capacitor are arranged in the circuit shown in Fig. 2.94, with a 10-V dc supply. Initially, the switch is at position B then thrown to position A, and then thrown to position B, and then to position A, and finally to position B. Assuming the capacitors have enough time to fully charge or discharge during the interval between switches, what is the final voltage across each capacitor after the last switch takes place?

Answer: When the switch is thrown from B to A the first time, C_1 charges to:

$$Q_1 = C_1V = (1000 \times 10^{-6} \text{ F})(10 \text{ V}) = 0.01 \text{ C}$$

When the switch is then thrown to B, the circuit becomes essentially one big capacitor equal to $C_1 + C_2$ or 1470 μF. Charge will flow from C_1 to C_2, since the system wants to go to the lowest energy configuration. The charge on each capacitor is the percentage of capacitance to the total capacitance for each capacitor multiplied by the initial charge on C_1 before the switch was thrown to position B:

$$Q_1 = \frac{1000 \text{ μF}}{1470 \text{ μF}}(0.01 \text{ C}) = 0.0068 \text{ C}$$

$$Q_2 = \frac{470 \text{ μF}}{1470 \text{ μF}}(0.01 \text{ C}) = 0.0032 \text{ C}$$

The voltage at the new equilibrium is:

$$V_1 = Q_1/C_1 = 0.0068/1000 \text{ μF} = 6.8 \text{ V}$$

$$V_2 = Q_2/C_2 = 0.0032/470 \text{ μF} = 6.8 \text{ V}$$

The rest of the results are obtained in using similar calculations—the final result yields 9.0 V, as shown in the graph to the right.

We could be content with this limited knowledge. However, if you want to build your own capacitors, as well as understand time-dependent behavior, such as displacement current and capacitive reactance, a deeper understanding of capacitance is needed.

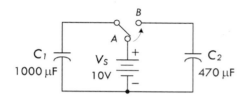

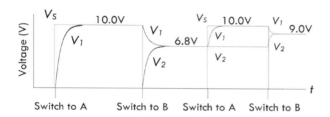

FIGURE 2.94

2.23.1 Determining Capacitance

The capacitance of a capacitor is determined by plate area A, plate separation d, and insulating material or dielectric. If a voltage V is applied between two parallel plates, an electric field equal to $E = V/d$ will be produced. From Gauss's law, each plate must contain an equal and opposite charge given by:

$$Q = \varepsilon AE = \frac{\varepsilon AV}{d} \tag{2.33}$$

where ε is the *permittivity* of the dielectric. Free space (vacuum) has a permittivity given by:

$$\varepsilon_0 = 8.85 \times 10^{-12} \, C^2/N \cdot m^2 \tag{2.34}$$

The constant $\varepsilon A/d$ term in the equation is the capacitance,

$$C = \frac{\varepsilon A}{d} \tag{2.35}$$

The relative permittivity of a material referenced to the permittivity in vacuum is referred to as the *dielectric constant*, which is given by:

$$k = \frac{\varepsilon}{\varepsilon_0}$$

Plugging this into the previous expression, we get the capacitance in terms of dielectric constant:

$$C = \frac{k\varepsilon_0 A}{d} = \frac{(8.85 \times 10^{-12} \, C^2/N \cdot m) \times k \times A}{d} \tag{2.36}$$

where C is in farads, A is in meters squared, and d is in meters.

The dielectric constant varies from 1.00059 for air (1 atm) to over 10^5 for some types of ceramic. Table 2.6 shows relative dielectric constants for various materials often used in constructing capacitors.

Capacitors often have more than two plates, the alternate plates being connected to form two sets, as shown in the lower drawing in Fig. 2.95. This makes it possible to obtain a fairly large capacitance in a small space. For a multiple-plate capacitor, we use the following expression to find the capacitance:

$$C = \frac{k\varepsilon_0 A}{d}(n-1) = \frac{(8.85 \times 10^{-12} \, C^2/N \cdot m) \times k \times A}{d}(n-1) \tag{2.37}$$

where the area A is in meters squared, the separation d is in meters, and the number of plates n is an integer.

Example: What is the capacitance of a multiple-plate capacitor containing two plates, each with an area of 4 cm^2, a separation of 0.15 mm, and a paper dielectric?

Answer:

$$C = \frac{k\varepsilon_0 A}{d}(n-1) = \frac{(8.85 \times 10^{-12} \, C^2/N \cdot m) \times 3.0 \times (4.0 \times 10^{-4} m^2)}{(1.5 \times 10^{-4} m)}(2-1) = 7.08 \times 10^{-11} F$$

$$= 70.8 \, pF$$

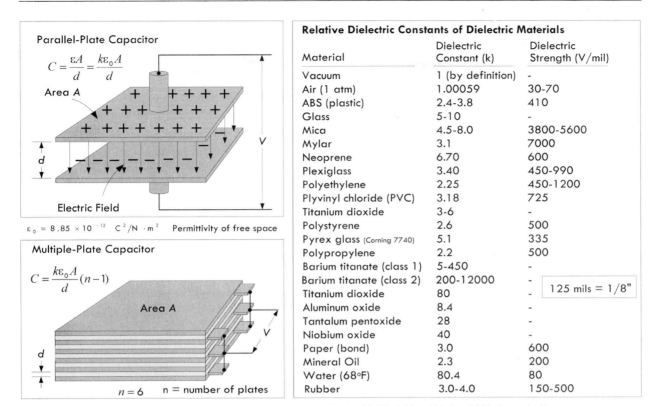

Parallel-Plate Capacitor

$$C = \frac{\varepsilon A}{d} = \frac{k\varepsilon_0 A}{d}$$

Area A

d

V

Electric Field

$\varepsilon_0 = 8.85 \times 10^{-12} \ C^2/N \cdot m^2$ Permittivity of free space

Multiple-Plate Capacitor

$$C = \frac{k\varepsilon_0 A}{d}(n-1)$$

Area A

d

V

$n = 6$ n = number of plates

Relative Dielectric Constants of Dielectric Materials

Material	Dielectric Constant (k)	Dielectric Strength (V/mil)
Vacuum	1 (by definition)	-
Air (1 atm)	1.00059	30-70
ABS (plastic)	2.4-3.8	410
Glass	5-10	-
Mica	4.5-8.0	3800-5600
Mylar	3.1	7000
Neoprene	6.70	600
Plexiglass	3.40	450-990
Polyethylene	2.25	450-1200
Plyvinyl chloride (PVC)	3.18	725
Titanium dioxide	3-6	-
Polystyrene	2.6	500
Pyrex glass (Corning 7740)	5.1	335
Polypropylene	2.2	500
Barium titanate (class 1)	5-450	-
Barium titanate (class 2)	200-12000	-
Titanium dioxide	80	-
Aluminum oxide	8.4	-
Tantalum pentoxide	28	-
Niobium oxide	40	-
Paper (bond)	3.0	600
Mineral Oil	2.3	200
Water (68°F)	80.4	80
Rubber	3.0-4.0	150-500

125 mils = 1/8"

Adapted from: Sears, F. W., Zemansky, M. W., Young, H. D., University Physics, 6th Ed., Addison-Wesley, 1982. Zemansky,Young Table 27-1; Charles A. Harper, Handbook of Components for Electronics, p 8-7

FIGURE 2.95

2.23.2 Commercial Capacitors

Commercial capacitors, like those shown in Fig. 2.95, are constructed from plates of foil with a thin solid or liquid dielectric sandwiched between, so relatively large capacitance can be obtained in a small unit. The solid dielectrics commonly used are mica, paper, polypropylene, and special ceramics.

Electrolytic capacitors use aluminum-foil plates with a semiliquid conducting chemical compound between them. The actual dielectric is a very thin film of insulating material that forms on one set of the plates through electrochemical action when a dc voltage is applied to the capacitor. The capacitance obtained with a given area in an electrolytic capacitor is very large compared to capacitors having other dielectrics, because the film is so thin—much less than any thickness practical with a solid dielectric. Electrolytic capacitors, due to the electrochemical action, require that one lead be placed at a lower potential than the other. The negative lead (–) is indicated on the package, and some surface mount electrolytics mark the positive end. This polarity adherence means that, with the exception of special nonpolarized electrolytics, electrolytic capacitors shouldn't be used in ac applications. It is okay to apply a superimposed ac signal riding upon a dc voltage, provided the peak voltage doesn't exceed the maximum dc voltage rating of the electrolytic capacitor.

2.23.3 Voltage Rating and Dielectric Breakdown

The dielectric within a capacitor acts as an insulator—its electrons do not become detached from atoms the way they do in conductors. However, if a high enough voltage is applied across the plates of the capacitor, the electric field can supply enough force on electrons and nuclei within the dielectric to detach them, resulting in a breakdown in the dielectric. Failed dielectrics often puncture and offer a low-resistance current path between the two plates.

The breakdown voltage of a dielectric depends on the dielectric's chemical composition and thickness. A gas dielectric capacitor breakdown is displayed as a spark or arc between the plates. Spark voltages are typically given in units of kilovolts per centimeter. For air, the spark voltage may range from 100 kV/cm for gaps as narrow as 0.005 cm to 30 kV/mm for gaps as wide as 10 cm. Other things that contribute to the exact breakdown voltage level are electrode shape, gap distance, air pressure or density, the voltage, impurities within the dielectric, and the nature of the external circuit (air humidity, temperature, etc.).

Dielectric breakdown can occur at a lower voltage between points or sharp-edged surfaces than between rounded and polished surfaces, since electric fields are more concentrated at sharp projections. This means that the breakdown voltage between metal plates can be increased by buffing the edges to remove any sharp irregularities. If a capacitor with a gas dielectric, such as air, experiences breakdown, once the arc is extinguished, the capacitor can be used again. However, if the plates become damaged due to the spark, they may require polishing, or the capacitor may need to be replaced. Capacitors with solid dielectrics are usually permanently damaged if there is dielectric breakdown, often resulting in a short or even an explosion.

Manufacturers provide what's called a dielectric withstanding voltage (dwv), expressed in voltage per mil (0.001 in) at a specified temperature. They also provide a dc working voltage (dcwv) that takes into account other factors such as temperature and safety margin, which gives you a guideline to the maximum safe limits of dc voltage that can be applied before dielectric breakdown. The dcwv rating is most useful in practice.

As a rule of thumb, it is not safe to connect a capacitor across an ac power line unless it is designed for it. Most capacitors with dc ratings may short the line. Special ac-rated capacitors are available for such tasks. When used with other ac signals, the peak value of ac voltage should not exceed the dc working voltage.

2.23.4 Maxwell's Displacement Current

An interesting thing to notice with our parallel-plate capacitor is that current appears to flow through the capacitor as it is charging and discharging, but doesn't flow under steady dc conditions. You may ask: How is it possible for current to flow through a capacitor, ever, if there is a gap between the plates of the capacitor? Do electrons jump the gap? As it turns out, no actual current (or electron flow) makes it across the gap, at least in an ideal capacitor.

As we calculated a moment ago using Gauss's law, the charge on an air-filled capacitor plate can be expressed in terms of the electric field, area, and permittivity of free space:

$$Q = \varepsilon_0 AE = \frac{\varepsilon_0 AV}{d} \tag{2.38}$$

Some time ago, Scottish physicist James Clerk Maxwell (1831–1879) noted that even if no real current passed from one capacitor plate to the other, there was nevertheless a changing electric flux through the gap of the capacitor that increased and decreased with the magnitude and direction of the electric flux. (Electric flux for a parallel-plate capacitor is approximated by $\Phi_E = EA$, while a changing electric flux is expressed as $d\Phi_E/dt$). Maxwell believed the electric flux permeated the empty space between the capacitor plates and induced a current in the other plate. Given the state of knowledge of electrodynamics at the time, he envisioned a displacement current (which he coined) crossing the empty gap, which he associated with a kind of stress within the ether (accepted at the time)—the "stress" being electric and magnetic fields. (The displacement current helped supply Maxwell with the missing component to complete a set of electromagnetic formulas known as Maxwell's equations.) He associated the displacement current with displacements of the ether. With a bit of theoretical reckoning, as well as some help from some experimental data, he came up with the following equation, known as the displacement current, to explain how current could appear to enter one end of a capacitor and come out the other end.

$$I_d = \frac{dQ}{dt} = \frac{d}{dt}(\varepsilon_0 A E) = \varepsilon_0 \frac{d\Phi_E}{dt} \tag{2.39}$$

Maxwell's displacement expression appears to provide the correct answer, even though his notion of the ether has lost favor in the realm of physics. Modern physics provides a different model for displacement current than that envisioned by Maxwell and his ether. Nevertheless, the results obtained using Maxwell's equation closely correlate with experiment.

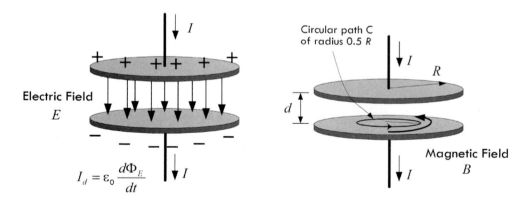

FIGURE 2.96

As a side note, there also exists a magnetic field due to the displacement current, as shown in the drawing on the right in Fig. 2.96. You can calculate the magnetic field using what's called Maxwell's generalized form of Ampere's law; however, the size of the magnetic field turns out to be so small that it essentially has no practical influence when compared to the electric field.

However deep you go when trying to explain the physical phenomena within a capacitor, such as using Maxwell's equations or even modern physics, the practical equations that are useful in electronics really don't require such detail. Instead, you can simply stick with using the following charge-based model.

2.23.5 Charge-Based Model of Current Through a Capacitor

Though Maxwell's displacement current provides a model to explain the apparent current flow through a capacitor in terms of changing electric fields, it really isn't needed to define capacitor performance. Instead, we can treat the capacitor as a black box with two leads, and define some rules relating the current entering and exiting the capacitor as the applied voltage across the capacitor changes. We don't need to worry about the complex physical behavior within.

Now, the question remains: how do we determine the rules if we don't understand the complex behavior within? Simple—we use the general definition of capacitance and the general definition of current, and combine the two. Though the mathematics for doing this is simple, understanding why this makes logical sense is not entirely obvious. The following parallel-plate example provides an explanation. Refer to Fig. 2.97.

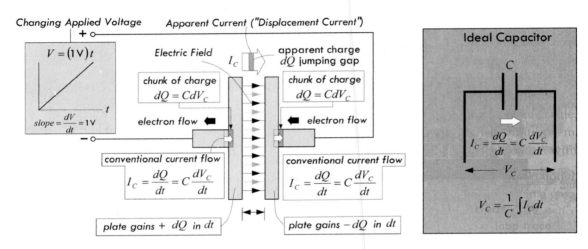

FIGURE 2.97

If we work in differentials (small changes), we can rewrite the general expression for capacitance as $dQ = CdV$, where C remains constant (with charge, voltage, or time). The general expression for current is $I = dQ/dt$, which, when combined with the last differential expression of capacitance, provides the expression:

$$I_C = \frac{dQ}{dt} = \frac{d(CV_C)}{dt} = C\frac{dV_C}{dt} \qquad \text{(2.40) Apparent current "through" capacitor}$$

Looking at Fig. 2.97, a "small" chunk dQ, which is equal to CdV_C, enters the right plate during dt, while an equal-sized chunk exits the left plate. So a current equal to $dQ/dt = CdV_C/dt$ enters the left plate while an equal-sized current exits the right plate. (Negative electrons flow in the opposite direction.) Even though no actual current (or electrons) passes across the gap, Eq. 2.40 makes it appear that it does. After our little exercise using differentials, however, we can see that there is really no need to assume that a current must flow across the gap to get an apparent current flow "through" the capacitor.

Moving on, we can take the capacitor current expression just derived, rearrange things, and solve for the voltage across the capacitor:

$$V_C = \frac{1}{C}\int I_C \, dt \qquad \text{(2.41) Voltage across capacitor}$$

It's important to note that these equations are representative of what's defined as an ideal capacitor. *Ideal capacitors,* as the equation suggests, have several curious properties that are misleading if you are dealing with real capacitors. First, if we apply a constant voltage across an ideal capacitor, the capacitor current would be zero, since the voltage isn't changing ($dV/dt = 0$). In a dc circuit, a capacitor thus acts like an open circuit. On the other hand, if we try to change the voltage abruptly, from 0 to 9 V, the quantity $dV/dt = 9\ V/0\ V = $ infinity and the capacitor current would have to be infinite (see Fig. 2.98). Real circuits cannot have infinite currents, due to resistivity, available free electrons, inductance, capacitance, and so on, so the voltage across the capacitor cannot change abruptly. A more accurate representation of a real capacitor, considering construction and materials, looks like the model shown in Fig. 2.93.

Charging and Discharging a Capacitor

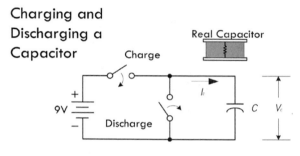

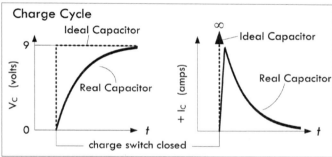

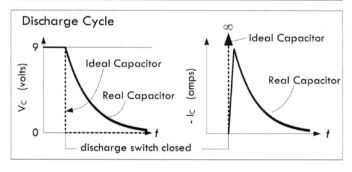

Under steady-state dc, a capacitor cannot pass current. It can only store or discharge charge (which it collects from current) when the voltage across it changes. Here, when the "charge switch" is closed (shorted), the 9-V battery voltage is applied instantly across the capacitor. In a real capacitor, the capacitor charges up to its maximum value practically instantly. But upon closer examination, the charge takes time to build up, due to internal resistance, meaning the displacement current cannot reach infinity. Instead, the current jumps to $V_{battery}/R_{internal}$ and quickly drops exponentially as the capacitor reaches full charge, during which the voltage rises exponentially until it levels off at the applied voltage. The graph to the left shows voltage and current curves as the capacitor charges up. Note the impossible behavior that an ideal treatment of a capacitor introduces.

When the discharge switch is closed, a conductive path from positive to negative plate is made, and charge electrons will flow to the plate with a deficiency of electrons. The current that results is in the opposite direction as the charging current, but resembles it in that it initially peaks to a value of $V_{battery}/R_{internal}$ and decays as the charge neutralizes. The voltage drops exponentially in the process.

FIGURE 2.98

If the equations for an ideal capacitor are screwy, what do we do for real-life calculations? For the most part, you don't have to worry, because the capacitor will be substituted within a circuit that has resistance, which eliminates the possibility of infinite currents. The circuit resistance is also usually much greater than the internal resistance of the capacitor, so that the internal resistance of the capacitor can typically be ignored. In a minute, we'll see a few resistor-capacitor circuits that will demonstrate this.

2.23.6 Capacitor Water Analogy

For those of you who are having problems with the previous explanations of apparent or displacement current, perhaps the following water analogy will help. Take it with a grain of salt, however, since what's going on in a real capacitor isn't analogous in all regards. Refer to Fig. 2.99.

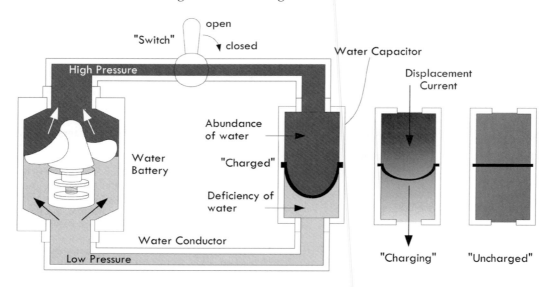

FIGURE 2.99

The water capacitor in Fig. 2.99 resembles a tube with a rubber membrane in the middle. The rubber membrane is somewhat analogous to the insulator or dielectric of a capacitor, while each separate compartment is analogous to each plate of a capacitor. If there is no pressure (analogous to voltage) across the water capacitor, each compartment contains the same amount of water (analogous to the number of free electrons). However, if pressure is suddenly applied across the water capacitor, the pressure within the top chamber increases, causing the membrane to expand downward, displacing water out from the lower chamber. Even though no water from the top makes it through the membrane, it appears as though current flows through the water capacitor, since the membrane is pushing the water within the lower chamber out. This is analogous to displacement current. Increasing the chamber size and altering the membrane strength are analogous to changing the capacitance and dielectric strength.

Example 1: A 10-µF capacitor is connected to a 50-mA constant current source. Determine the voltage across the capacitor after 10 µs, 10 ms, and 1 s.

Answer: Since IC is constant, it can be moved in front of the integral:

$$10 \text{ µs} : V_C = \frac{1}{C}\int I_C \, dt = \frac{I_C}{C}t = \frac{50 \times 10^{-3} \text{ A}}{10 \times 10^{-6} \text{ F}}(10 \times 10^{-6}\text{s}) = 0.05 \text{ V}$$

$$10 \text{ ms} : V_C = \frac{1}{C}\int I_C \, dt = \frac{I_C}{C}t = \frac{50 \times 10^{-3} \text{ A}}{10 \times 10^{-6} \text{ F}}(10 \times 10^{-3}\text{s}) = 50 \text{ V}$$

$$1 \text{ s} : V_C = \frac{1}{C}\int I_C \, dt = \frac{I_C}{C}t = \frac{50 \times 10^{-3} \text{ A}}{10 \times 10^{-6} \text{ F}}(1 \text{ s}) = 5000 \text{ V, obviously typical capacitor}$$

won't survive.

Example 2: A 47-μF capacitor is charged by the voltage sources having the following waveform. Determine the charging current. Refer to Fig. 2.100. Assume that the voltage source is ideal and has no resistance.

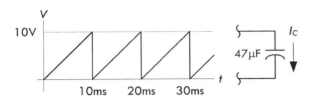

FIGURE 2.100

Answer: Since dV/dt represents the slope of the waveform, it is simply 10 V/10 ms, and the current becomes:

$$I_C = C\frac{dV_C}{dt} = (47 \times 10^{-6} \text{ F})\frac{10 \text{ V}}{10 \times 10^{-3} \text{ s}} = 0.047 \text{ A} = 47 \text{ mA}$$

Example 3: If the voltage across a 100-μF capacitor is 5 V e^{-t}, what is the capacitor current?

Answer:

$$I_C = C\frac{dV_C}{dt} = 100 \text{ μF}\frac{d}{dt}[5 \text{ V} e^{-t}] = -(100 \text{ μF})(5 \text{ V})e^{-t} = -(0.0005 \text{ A})e^{-t}$$

(Remember, these examples assume ideal capacitors. If you used real capacitors, the results would follow the same trends, but would be limited in current.)

2.23.7 Energy in a Capacitor

Finally, energy cannot be dissipated in an ideal capacitor. (This is not the case for real capacitors because of internal resistance, but since the internal resistance is so small, that energy lost to heating is often ignored.) Energy can only be stored in the electric field (or potential that exists between the plates) for later recovery. The energy stored in a capacitor is found by substituting the capacitor current into the generalized power law ($P = IV$), then inserting the resulting power into the definition of power ($P = dE/dt$), and solving for E by integration:

$$E_{\text{cap}} = \int VIdt = \int VC\frac{dV}{dt}dt = \int CVdV = \frac{1}{2}CV^2 \tag{2.42}$$

Example: How much energy is stored in a 1000-μF capacitor with an applied voltage of 5 V?

Answer:

$$E_{\text{cap}} = \frac{1}{2}CV^2 = \frac{1}{2}(1000 \times 10^{-6} \text{ F})(5 \text{ V})^2 = 0.0125 \text{ J}$$

2.23.8 RC Time Constant

When a capacitor is connected to a dc voltage source, it will charge up almost instantaneously. (In reality, a capacitor has internal resistance, as well as inductance, therefore the term "almost"—see Sec. 3.6, on capacitors, for real-life characteristics.) Likewise, a charged capacitor that is shorted with a wire will discharge almost instantaneously. However, with some resistance added, the rate of charge or discharge follows an exponential pattern, as shown in Fig. 2.101. There are numerous applications that use controlled charge and discharge rates, such as timing ICs, oscillators, waveform shapers, and low-discharge power backup circuits.

For a charging capacitor, the following equations are used.

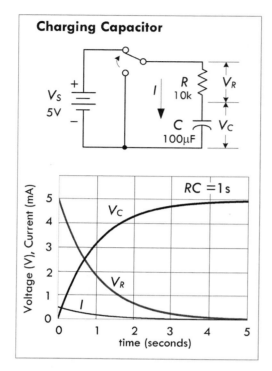

Charging Capacitor

$$RC = 1\,s$$

Current and voltage equations for RC charging circuit

$$I = \frac{V_s}{R} e^{-t/RC}$$

$$\frac{t}{RC} = -\ln\left(\frac{IR}{V_s}\right)$$

$$V_R = IR = V_s e^{-t/RC}$$

$$\frac{t}{RC} = -\ln\left(\frac{V_R}{V_s}\right)$$

$$V_C = \frac{1}{C}\int I\,dt = V_s(1 - e^{-t/RC})$$

$$\frac{t}{RC} = -\ln\left(\frac{V_s - V_C}{V_s}\right)$$

(2.43)

$$\tau = RC \text{ time constant}$$

where I is the current in amps, V_s is the source voltage in volts, R is the resistance in ohms, C is the capacitance in farads, t is the time in seconds after the source voltage is applied, $e = 2.718$, V_R is the resistor voltage in volts and V_C is the capacitor voltage in volts. Graph shown to the left is for circuit with $R = 10$ kΩ, and $C = 100$ μF. Decreasing the resistance means the capacitor charges up more quickly and the voltage across the capacitor rises more quickly.

FIGURE 2.101

(You can derive the preceding expressions using Kirchhoff's law, by summing the voltage around the closed loop: $V_s = RI + (1/C) \int I\,dt$. Differentiating and solving the differential equation, given an initial condition of current equal to V/R, the voltage across the resistor = V_s, and the voltage across the capacitor $V_C = 0$, you get the solution: $I = (V/R)e^{-t/RC}$. The voltages across the resistor and capacitor are simply found by plugging the current into the expression for the voltage across a resistor $V_R = IR$ and the voltage across a capacitor: $V_C = (1/C) \int I\,dt$. We'll discuss solving such circuits in detail in the section on transients in dc circuits.)

Theoretically, the charging process never really finishes, but eventually the charging current drops to an immeasurable value. A convention often used is to let $t = RC$, which makes $V(t) = 0.632\,V_S$. The RC term is called the *time constant* of the circuit and is the time in seconds required to charge the capacitor to 63.2 percent of the supply voltage. The lowercase tau τ is often used to represent RC: $\tau = RC$. After two time constants ($t = 2RC = 2\tau$) the capacitor charges another 63.2 percent of the difference between the capacitor voltage at one time constant and the supply voltage, for a total change of 86.5 percent. After three time constants, the capacitor reaches 95 percent of the supply voltage, and so on, as illustrated in the graph in Fig. 2.101. After five time constants, a capacitor is considered fully charged, having reached 99.24 percent of the source voltage.

Example: An IC uses an external RC charging network to control its timing. The IC requires 3.4 V at V_{IN} to trigger its output to switch from high to low, at which time an internal transistor (switch) is activated, allowing the capacitor to be discharged to ground. If a 5-s timing period before the trigger point is required, what value of R is required if $C = 10\ \mu F$?

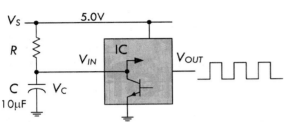

V_S — 5.0V

R

V_{IN} IC V_{OUT}

C — V_C

$10\mu F$

Transistor discharges capacitor when $V_C = 3.4V$

FIGURE 2.102

Answer: Use

$$\frac{t}{RC} = -\ln\left(\frac{V_s - V_C}{V_s}\right)$$

and solve for R:

$$R = \frac{t}{-\ln\left(\frac{V_s - V_C}{V_s}\right)C} = \frac{5.0\ \text{s}}{-\ln\left(\frac{5V - 3.4\ V}{5\ V}\right)(10 \times 10^{-6}\ \mu F)}$$

$$= 4.38 \times 10^5\ \Omega$$

For a discharging capacitor, the following equations are used.

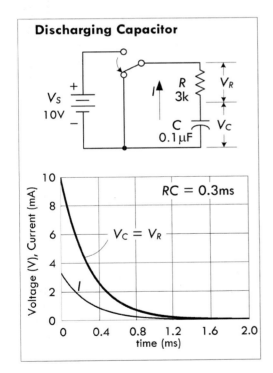

Discharging Capacitor

V_S 10V

R 3k V_R

C 0.1μF V_C

$RC = 0.3$ms

$V_C = V_R$

I

Voltage (V), Current (mA)

time (ms)

FIGURE 2.103

Current and voltage equations for discharging an RC circuit

$$I = \frac{V_s}{R}e^{-t/RC}, \qquad \frac{t}{RC} = -\ln\left(\frac{IR}{V_s}\right)$$

$$V_R = IR = V_s e^{-t/RC}, \qquad \frac{t}{RC} = -\ln\left(\frac{V_R}{V_s}\right) \qquad \text{(2.44)}$$

$$V_C = \frac{1}{C}\int_0^{} I dt = V_s e^{-t/RC}, \qquad \frac{t}{RC} = -\ln\left(\frac{V_C}{V_s}\right)$$

$$\tau = RC \text{ time constant}$$

where I is the current in amps, V_s is the source voltage in volts, R is the resistance in ohms, C is the capacitance in farads, t is the time in seconds after the source voltage is removed, $e = 2.718$, V_R is the resistor voltage in volts, and V_C is the capacitor voltage in volts. Graph shown to the left in Fig. 2.103 is for circuit with $R = 3\ k\Omega$ and $C = 0.1\ \mu F$.

(You can derive this expression using Kirchhoff's law, by summing the voltage around the closed loop: $0 = RI + (1/C)\int I dt$. Differentiating and solving the differential equation, given an initial condition of current equal to 0, the voltage across the resistor = 0, and the voltage across the capacitor $V_C = V_S$, you get the solution: $I = (V/R)e^{-t/RC}$. The voltage across the resistor and capacitor is found simply by plugging the current into the expression for the voltage across a resistor $V_R = IR$ and the voltage across a capacitor: $V_C = (1/C) \int I dt$. We'll discuss solving such circuits in detail in the section on transients in dc circuits.)

The expression for a discharging capacitor is essentially the inverse of that for a charging capacitor. After one time constant, the capacitor voltage will have dropped by 63.2 percent from the supply voltage, so it will have reached 37.8 percent of the supply voltage. After five time constants, the capacitor is considered fully discharged, it will have dropped 99.24 percent, or down to 0.76 percent of the supply voltage.

Example: If a 100-μF capacitor in a high-voltage power supply is shunted by a 100k resistor, what is the minimum time before the capacitor is considered fully discharged when power is turned off?

Answer: After five time constants, a capacitor is considered discharged:

$$t = 5\tau = 5RC = (5)(100 \times 10^3 \ \Omega)(100 \times 10^{-6} \ \text{F}) = 50 \ \text{s}$$

2.23.9 Stray Capacitance

Capacitance doesn't exist only within capacitors. In fact, any two surfaces at different electrical potential, and that are close enough together to generate an electric field, have capacitance, and thus act like a capacitor. Such effects are often present within circuits (e.g., between conductive runs or component leads), even though they are not intended. This unintended capacitance is referred to as *stray capacitance,* and it can result in a disruption of normal current flow within a circuit. Designers of electric circuits figure out ways to minimize stray capacitance as much as possible, such as keeping capacitor leads short and grouping components in such a way as to eliminate capacitive coupling. In high-impedance circuits, stray capacitance may have a greater influence, since capacitive reactance (which we will discuss shortly) may be a greater portion of the circuit impedance. In addition, since stray capacitance usually appears in parallel with a circuit, it may bypass more of the desired signal at higher frequencies. Stray capacitance usually affects sensitive circuits more profoundly.

2.23.10 Capacitors in Parallel

When capacitors are placed in *parallel,* their capacitances add, just like resistors in series:

$$C_{\text{tot}} = C_1 + C_2 + \ldots C_n \tag{2.45}$$

(You derive this formula by applying Kirchhoff's current law at the top junction, which gives $I_{\text{tot}} = I_1 + I_2 + I_3 + \ldots I_N$. Making use of the fact that the voltage V is the same across C_1 and C_2, you can substitute the displacement currents for each capacitor into Kirchhoff's current expression as follows:

$$I = C_1 \frac{dV}{dt} + C_2 \frac{dV}{dt} + C_3 \frac{dV}{dt} = (C_1 + C_2 + C_3) \frac{dV}{dt}$$

The term in brackets is the equivalent capacitance.)

Intuitively, you can think of capacitors in parallel representing one single capacitor with increased plate surface area. It's important to note that the largest voltage that can be applied safely to a group of capacitors in parallel is limited to the voltage rating of the capacitor with the lowest voltage rating. Both the capacitance and the voltage rating are usually included next to the capacitor symbol in schematics, but often the voltage rating is missing; you must figure out the rating based on the expected voltages present at that point in the circuit.

Capacitors In Parallel

Increases the total capacitance, but limits max voltage rating to that of smallest capacitor.

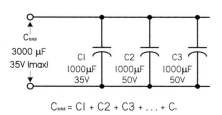

$$C_{total} = C1 + C2 + C3 + \ldots + C_n$$

Capacitors In Series

Increases max voltage rating, but decreases capacitance.

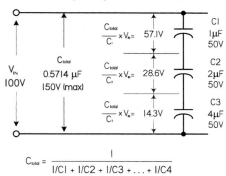

$$C_{total} = \frac{1}{1/C1 + 1/C2 + 1/C3 + \ldots + 1/C4}$$

FIGURE 2.104

2.23.11 Capacitors in Series

When two or more capacitors are connected in series, the total capacitance is less than that of the smallest capacitor in the group. The equivalent capacitance is similar to resistors in parallel:

$$\frac{1}{C_{tot}} = \frac{1}{C_1} + \frac{1}{C_2} + \cdots \frac{1}{C_N} \tag{2.46}$$

(You derive this by applying Kirchhoff's voltage law. Since the current I must flow through each capacitor, Kirchhoff's voltage expression becomes:

$$V = \frac{1}{C_1}\int Idt + \frac{1}{C_2}\int Idt + \frac{1}{C_3}\int Idt = \left(\frac{1}{C_1} + \frac{1}{C_2} + \frac{1}{C_3} + \cdots + \frac{1}{C_N}\right)\int Idt$$

The term in parentheses is called the equivalent capacitance for capacitors in series.)

Capacitors may be connected in series to enable the group to withstand a larger voltage than any individual capacitor is rated to withstand (the maximum voltage ratings add). The trade-off is a decrease in total capacitance—though that could be what you intend to do, if you can't find a capacitor or create a parallel arrangement that gives you the desired capacitance value. Notice in Fig. 2.104 that the voltage does not divide equally between capacitors. The voltage across a single capacitor—say, C_2—is a fraction of the total, expressed as $(C_{total}/C_2)V_{IN}$. There are various circuits that tap the voltage between series capacitors.

Use care to ensure that the voltage rating of any capacitor in the group is not exceeded. If you use capacitors in series to withstand larger voltages, it's a good idea to also connect an equalizing resistor across each capacitor. Use resistors with about

100 Ω per volt of supply voltage, and be sure they have sufficient power-handling capability. With real capacitors, the leakage resistance of the capacitor may have more effect on the voltage division than does the capacitance. A capacitor with a high parallel resistance will have the highest voltage across it. Adding equalizing resistors reduces this effect.

Example: (a) Find the total capacitance and maximum working voltage (WV) for the capacitor in the parallel network. (b) Find the total capacitance, WV, V_1, and V_2. (c) Find the total capacitance and WV for the network of capacitors. (d) Find the value of C that gives a total capacitance of 592 pF with a total WV of 200 V. Individual capacitor WV values are in parentheses. Refer to Fig. 2.105.

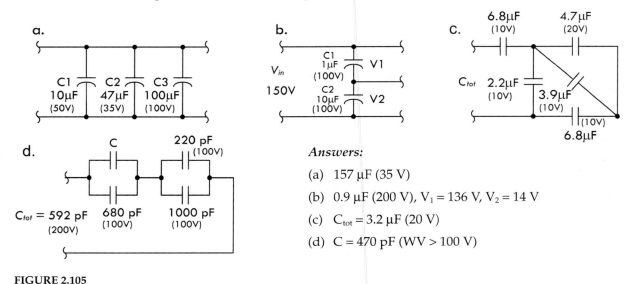

Answers:

(a) 157 μF (35 V)

(b) 0.9 μF (200 V), $V_1 = 136$ V, $V_2 = 14$ V

(c) $C_{tot} = 3.2$ μF (20 V)

(d) C = 470 pF (WV > 100 V)

FIGURE 2.105

2.23.12 *Alternating Current in a Capacitor*

Everything that was discussed about capacitors in dc circuits also applies in ac circuits, with one major exception. While a capacitor in a dc circuit will block current flow (except during brief periods of charging and discharging), a capacitor in an ac circuit will either pass or limit current flow, depending on frequency. Unlike a resistor that converts current energy into heat to reduce current flow, a capacitor stores electrical energy and returns it to the circuit.

The graph in Fig. 2.106 shows the relationship between current and voltage when an ac signal is applied to a capacitor. The ac sine wave voltage has a maximum value of 100. In interval 0 to A, the applied voltage increases from 0 to 38 and the capacitor charges up to that voltage. In interval AB, the voltage increases to 71, so the capacitor has gained 33 V more. During this interval, a smaller amount of charge has been added than 0A, because the voltage rise during AB is smaller than 0A. In interval BC, the voltage rises by 21, from 71 to 92. The increase in current in this interval is still smaller. In interval CD, the voltage increases only 8, and therefore the increase in current is also smaller.

If you were to divide the first quarter cycle into a very large number of intervals, you'd see that the current charging the capacitor has the shape of a sine wave, just like the applied voltage. The current is largest at the start of the cycle and goes to zero

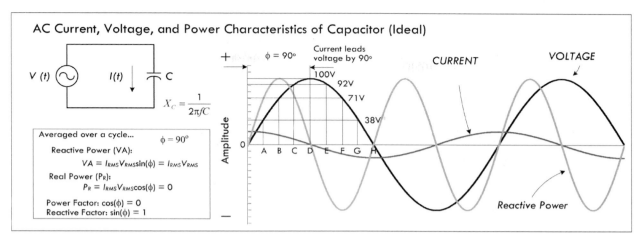

FIGURE 2.106

at the maximum value of voltage, so there is a phase difference of 90° between the voltage and the current.

In the second quarter cycle, from time D to H, the applied voltage decreases and the capacitor loses its charge. Using the same reasoning as before, it's apparent that the current is small during period DE and continues to increase during the other periods. The current is flowing against the applied voltage; however, since the capacitor is discharging into the circuit, the current flows in the negative direction during this quarter cycle.

The third and fourth quarter cycles repeat the events of the first and second, respectively, with one difference: the polarity of the applied voltage has reversed, and current changes to correspond. In other words, an alternating current flows in the circuit because of the alternate charging and discharging of the capacitor. As shown in the graph in Fig. 2.106, the current starts its cycle 90° before the voltage, so the current in a capacitor leads the applied voltage by 90°.

2.23.13 Capacitive Reactance

The amount of charge on a capacitor is equal to the voltage drop across the capacitor times the capacitance ($Q = CV$). Within an ac circuit, the amount of charge moves back and forth in the circuit every cycle, so the rate of movement of charge (current) is proportional to voltage, capacitance, and frequency. When the effect of capacitance and frequency are considered together, they form a quantity similar to resistance. However, since no actual heat is being generated, the effect is termed *capacitive reactance*. The unit for reactance is the ohm, just as for resistors, and the formula for calculating the reactance of a capacitor at a particular frequency is given by:

$$X_C = \frac{1}{2\pi f C} = \frac{1}{\omega C} \qquad \text{(2.47) Capacitive reactance}$$

where X_C is the capacitive reactance in ohms, f is the frequency in Hz, C is the capacitance in farads, and $\pi = 3.1416$. Often, omega ω is used in place of $2\pi f$, where ω is called the *angular frequency*.

(You can derive this by noting that a sinusoidal voltage source placed across a capacitor will allow displaced current to flow through it because the voltage across it is changing (recall that $I = CdV/dt$ for a capacitor). For example, if the voltage source is given by $V_0 \cos(\omega t)$, you plug this voltage into V in the expression for the displacement current for a capacitor, which gives:

$$I = C\frac{dV}{dt} = -\omega C V_0 \sin(\omega t)$$

Maximum current or peak current I_0 occurs when $\sin(\omega t) = -1$, at which point $I_0 = \omega C V_0$. The ratio of peak voltage to peak current V_0/I_0 resembles a resistance in light of Ohm's law, and is given in units of ohms. However, because the physical phenomenon for "resisting" is different from that of a traditional resistor (heating), the effect is given the name capacitive reactance.)

As the frequency goes to infinity, X_C goes to 0, and the capacitor acts like a short (wire) at high frequencies; capacitors like to pass current at high frequencies. As the frequency goes to 0, X_C goes to infinity, and the capacitor acts like an open circuit; capacitors do not like to pass low frequencies.

It's important to note that even though the unit of reactance is the ohm, there is no power dissipated in reactance. The energy stored in the capacitor during one portion of the cycle is simply returned to the circuit in the next. In other words, over a complete cycle, the average power is zero. See the graph in Fig. 2.106.

Example 1: Find the reactance of a 220-pF capacitor at an applied frequency of 10 MHz.

Answer:

$$X_C = \frac{1}{2\pi \times (10 \times 10^6 \ \text{Hz})(220 \times 10^{-12} \ \text{F})} = 72.3 \ \Omega$$

(*Note:* 1 MHz = 1×10^6 Hz, 1 μF = 1×10^{-6}, 1 nF = 1×10^{-9} F, 1 pF = 1×10^{-12} F.)

Example 2: What is the reactance of a 470-pF capacitor at 7.5 MHz and 15.0 MHz?

Answer: X_C @ 7.5 MHz = 45.2 Ω, X_C @ 15 MHz = 22.5 Ω

a. Ideal Capacitor Reactance vs. Frequency

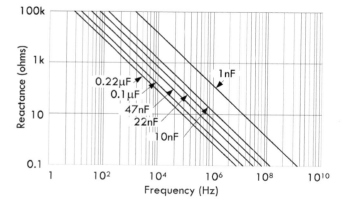

b. Real Capacitor Impedance vs. Frequency

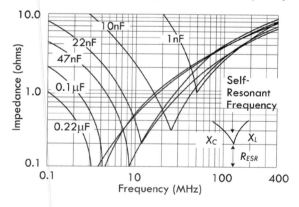

FIGURE 2.107   (a) Graph showing how reactance decreases with frequency for various sizes of capacitors—all capacitors are considered ideal in nature. (b) Graph showing the frequency response for real-life capacitors, which takes into account the parasitic resistances and inductances present within a real capacitor package. The pointy dips within the graph represent self-resonance, where the capacitive and inductive reactances cancel and only internal resistance within the capacitor package is left. The frequency at which point this occurs is called the self-resonant frequency.

As you can see, the reactance decreases with an increase in frequency and/or an increase in capacitance. The left graph in Fig. 2.107 shows the reactance versus frequency of a capacitor. Real capacitors don't follow the graph and equation so precisely, a result of parasitic effects—see "Real Capacitor Models" in Fig. 2.93.

2.23.14 Capacitive Divider

Capacitive dividers can be used with ac input signals or even dc, since the capacitors will rapidly reach a steady state. The formula for determining the ac output voltage of a capacitive divider is different from the resistive divider, because the series element, C_1 is in the numerator, not C_2, the shunt element. See Fig. 2.108.

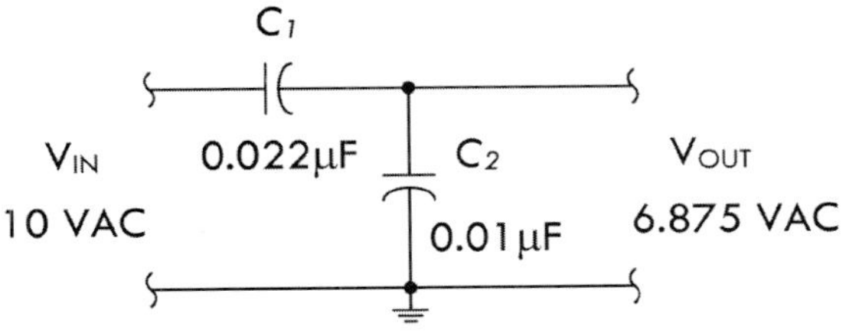

$$V_{out} = \frac{C_1}{C_1 + C_2} V_{in}$$

$$V_{out} = \frac{0.022\mu F}{0.032\mu F}(10\,VAC) = 6.875\,VAC$$

FIGURE 2.108

Note that the output voltage is independent of the input frequency. However, if the reactance of the capacitors is not large at the frequency of interest (i.e., the capacitance is not large enough), the output current capability will be very low.

2.23.15 Quality Factor

Components that store energy, like a capacitor (and as we'll see, an inductor), may be compared in terms of *quality factor Q*, also known as the merit. The Q of any such component is the ratio of its ability to store energy to the sum total of all energy losses within the component. Since reactance is associated with stored energy and resistance is associated with energy loss, we can express the quality factor as:

$$Q = \frac{\text{Reactance}}{\text{Resistance}} = \frac{X}{R} \tag{2.48}$$

Q has no units. When considering a capacitor, the reactance (in ohms) is simply the capacitive reactance $X = X_C$. (For an inductor, we'll see that $X = X_L$, where X_L is the inductive reactance.) R is the sum of all resistances associated with the energy losses in the component (in ohms). The Q of capacitors is ordinarily high. Good-quality ceramic capacitors and mica capacitors may have Q values of 1200 or more. Small ceramic trimmer capacitors may have Q values too small to ignore in some applications. Microwave capacitors can have poor Q values, 10 or less at 10 GHz or higher.

2.24 Inductors

In the preceding section we saw how a capacitor stored electrical energy in the form of an electric field. Another way to store electrical energy is in a magnetic field.

Circular radiating magnetic fields can be generated about a wire any time current passes through it. Increasing or decreasing current flow through the wire increases and decreases the magnetic field strength, respectively. During such changes in magnetic field strength, we encounter a phenomenon known as *inductance*. Inductance is a property of circuits somewhat analogous to resistance and capacitance; however, it is not attributed to heat production or charge storage (electric field), but rather it is associated with magnetic fields—more specifically, how changing magnetic fields influence the free electrons (current) within a circuit. Theoretically, any device capable of generating a magnetic field has inductance. Any device that has inductance is referred to as an inductor. To understand inductance requires a basic understanding of electromagnetic properties.

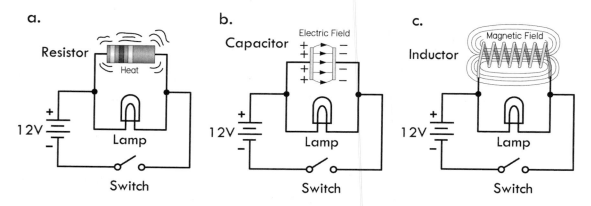

FIGURE 2.109 **The three cornerstones of electronics are resistance, capacitance, and inductance. Inductance, unlike the other two, involves alternations in a circuit's current and voltage characteristics as a result of forces acting upon free electrons resulting in the creation and collapse of magnetic fields, usually concentrated in a discrete inductor device. Like a capacitor, however, inductive effects occur only during times of change, when the applied voltage/current increases or decreases with time. Resistance doesn't have a time dependency. Can you guess what will happen to the brightness of the lamp in each of the circuits in the figure when the switch is closed? What do you think will occur when the switch is later opened? We'll discuss this a bit later.**

2.24.1 Electromagnetism

According to the laws of electromagnetism, the field of a charge at rest can be represented by a uniform, radial distribution of electric field lines or lines of force (see Fig. 2.110a). For a charge moving at a constant velocity, the field lines are still radial and straight, but they are no longer uniformly distributed (see Fig. 2.110b). At the same time, the electron generates a circular magnetic field (see Fig. 2.110c). If the charge accelerates, things get a bit more complex, and a "kink" is created in the electromagnetic field, giving rise to an electromagnetic wave that radiates out (see Fig. 2.110d and e).

As depicted in Fig. 2.110c, the electric field (denoted E) of a moving electron—or any charge for that matter—is, in effect, partially transformed into a magnetic field (denoted B). Hence, it is apparent that the electric and magnetic fields are part of the same phenomenon. In fact, physics today groups electric and magnetic fields together into one fundamental field theory, referred to as *electromagnetism*. (The work of Maxwell and Einstein helped prove that the two phenomena are linked. Today, certain fields in physics paint a unique picture of field interactions using virtual photons being emitted and absorbed by charges to explain electromagnetic forces. Fortunately, in electronics you don't need to get that detailed.)

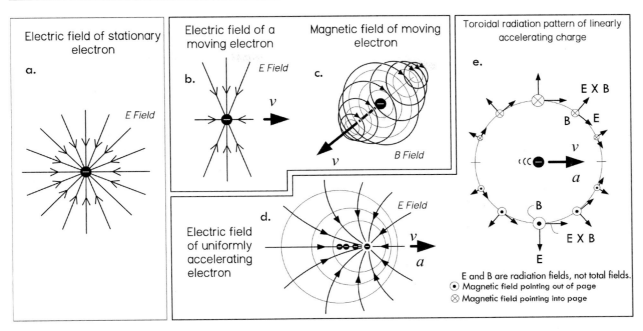

FIGURE 2.110   **The electric and magnetic fields are part of the same phenomenon called electromagnetism. Magnetic fields appear whenever a charge is in motion. Interestingly, if you move along with a moving charge, the observable magnetic field would disappear—thanks to Einstein's relativity.**

The simplest way to generate a magnetic field is to pass a current through a conductor. Microscopically, each electron within a wire should be generating a magnetic field perpendicular to its motion. However, without any potential applied across the wire, the sheer randomness of the electrons due to thermal effects, collision, and so on, cause the individual magnetic fields of all electrons to be pointing in random directions. Averaged over the whole, the magnetic field about the conductor is zero (see Fig. 2.111a.1). Now, when a voltage is applied across the conductor, free electrons gain a drift component pointing from negative to positive—conventional current in the opposite direction. In terms of electron speed, this influence is very slight, but it's enough to generate a net magnetic field (see Fig. 2.111a.2). The direction of this field is perpendicular to the direction of conventional current flow and curls in a direction described by the right-hand rule: your right thumb points in the direction of the conventional current flow; your finger curls in the direction of the magnetic field. See Fig. 2.111b. (When following electron flow instead of conventional current flow, you'd use your left hand.)

The magnetic field created by sending current through a conductor is similar in nature to the magnetic field of a permanent magnet. (In reality, the magnetic field pattern of a permanent bar magnet, in Fig. 2.111c, is more accurately mimicked when the wire is coiled into a tight solenoid, as shown in Fig. 2.111e.) The fact that both a current-carrying wire and a permanent magnet produce magnetic fields is no coincidence. Permanent magnets made from ferromagnetic materials exhibit magnetic closed-loop fields, mainly as a result of the motion of unpaired electrons orbiting about the nucleus of an atom, as shown in Fig. 2.112, generating a dipole magnetic field. The lattice structure of the ferromagnetic material has an important role of fixing a large portion of the atomic magnetic dipoles in a fixed direction, so as to set up a net magnetic dipole pointing from north to south. This microscopic motion of

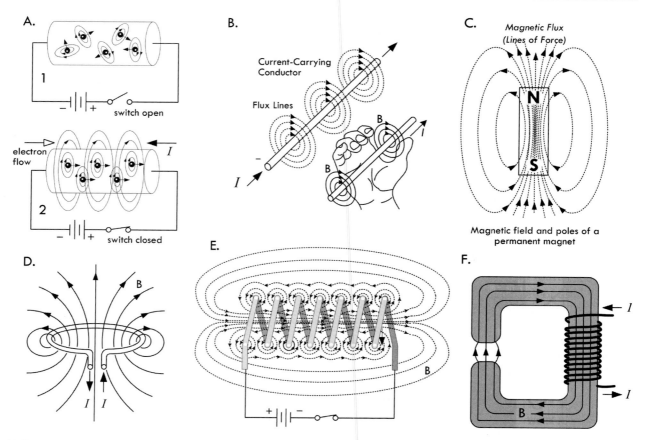

FIGURE 2.111 (a) Magnetic field generated free electrons moving in unison when voltage is applied. (b) Right-hand rule showing direction of magnetic field in relation to conventional current flow. (c) Permanent magnet. (d) Magnetic dipole radiation pattern created by current flowing through single loop of wire. (e) Energized solenoid with a magnetic field similar to permanent magnet. (f) Electromagnet, using ferromagnetic core for increased field strength.

Microscopic mechanism that generates magnetic fields in a permanent bar magnet

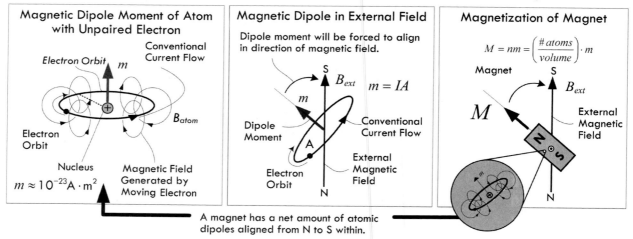

FIGURE 2.112 Microscopically, the magnetic field of a permanent magnet is a result of unpaired valence electrons fixed in a common direction, generating a magnetic dipole. The fixing in orientation is a result of atomic bonding in the crystalline lattice structure of the magnet.

the unpaired electron about the nucleus of the atom resembles the flow of current through a loop of wire, as depicted in Fig. 2.111d. (Electron spin is another source of magnetic fields, but is far weaker than that due to the electron orbital.)

By coiling a wire into a series of loops, a solenoid is formed, as shown in Fig. 2.111e. Every loop of wire contributes constructively to make the interior field strong. In other words, the fields inside the solenoid add together to form a large field component that points to the right along the axis, as depicted in Fig. 2.111e. By placing a ferromagnetic material (one that isn't initially magnetized) within a solenoid, as shown in Fig. 2.111f, a much stronger magnetic field than would be present with the solenoid alone is created. The reason for such an increase in field strength has to do with the solenoid's field rotating a large portion of the core's atomic magnetic dipoles in the direction of the field. Thus, the total magnetic field becomes the sum of the solenoid's magnetic field and the core's temporarily induced magnetic field. Depending on material and construction, a core can magnify the total field strength by a factor of 1000.

2.24.2 Magnetic Fields and Their Influence

Magnetic fields, unlike electric fields, only act upon charges that are moving in a direction that is perpendicular (or has a perpendicular component) to the direction of the applied field. A magnetic field has no influence on a stationary charge, unless the field itself is moving. Figure 2.113a shows the force exerted upon a moving charge placed within a magnetic field. When considering a positive charge, we use our right hand to determine the direction of force upon the moving charge—the back of the

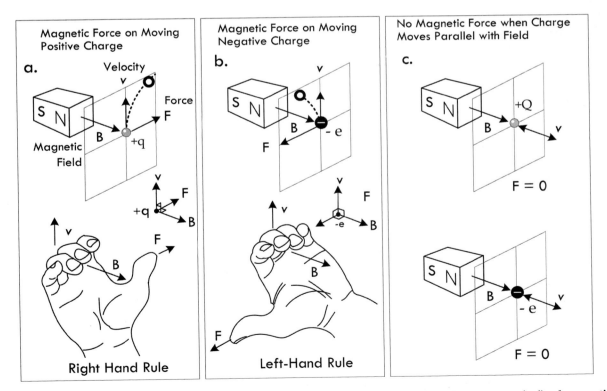

FIGURE 2.113   Illustration showing the direction of force upon a moving charge in the presence of a fixed magnetic field.

hand points in the direction of the initial charge velocity, the fingers curl in the direction of the external magnetic field, and the thumb points in the direction of the force that is exerted upon the moving charge. For a negative charge, like an electron, we can use the left hand, as shown in Fig. 2.113b. If a charge moves parallel to the applied field, it experiences no force due to the magnetic field—see Fig. 2.113c.

In terms of a group of moving charges, such as current through a wire, the net magnetic field of one wire will exert a force on the other wire and vice versa (provided the current is fairly large), as shown in Fig. 2.114. (The force on the wire is possible, since the electrostatic forces at the surface of the lattice structure of the wire prevent electrons from escaping from the surface.)

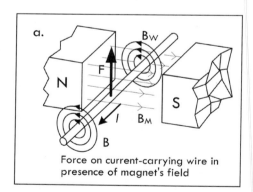
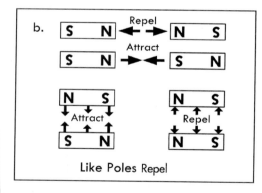

FIGURE 2.114   **Forces exerted between two current-carrying wires.**

Likewise, the magnetic field of a fixed magnet can exert a force on a current-carrying wire, as shown in Fig. 2.115.

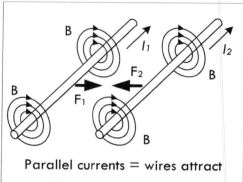
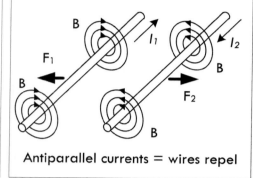

FIGURE 2.115   **(a) The force a current-carrying wire experiences in the presence of a magnet's field. (b) Illustration showing how bar magnets attract and repel.**

Externally, a magnet is given a north-seeking (or "north," for short) and south-seeking (or "south," for short) pole. The north pole of one magnet attracts the south pole of another, while like poles repel—see Fig. 2.115b. You may ask how two stationary magnets exert forces on each other. Isn't the requirement that a charge or field must be moving for a force to be observed? We associate the macroscopic (observed) force with the forces on the moving charges that comprise the microscopic internal magnetic dipoles which are, at the heart, electrons in motion around atoms. These orbitals tend to be fixed in a general direction called domains—a result of the lattice binding forces.

Another aspect of magnetic fields is their ability to force electrons within conductors to move in a certain direction, thus inducing current flow. The induced force is an electromotive force (EMF) being set up within the circuit. However, unlike, say, a battery's EMF, an induced EMF depends on time and also on geometry. According to Faraday's law, the EMF induced in a circuit is directly proportional to the time rate of change of the magnetic flux through the circuit:

$$\text{EMF} = -\frac{d\Phi_M}{dt}, \Phi_M = \int \vec{B} \cdot \vec{dA}, \text{EMF} = -N\frac{d\Phi_M}{dt} \text{ (for a coiled wire of } N \text{ loops)} \quad (2.49)$$

where Φ_M is the magnetic flux threading through a closed-loop circuit (which is equal to the magnetic field B dotted with the direction surface area—both of which are vectors, and summed over the entire surface area—as the integral indicates). According to the law, the EMF can be induced in the circuit in several ways: (1) the magnitude of B can vary with time; (2) the area of the circuit can change with time; (3) the angle between B and the normal of A can change with time; (4) any combination of these can occur. See Fig. 2.116.

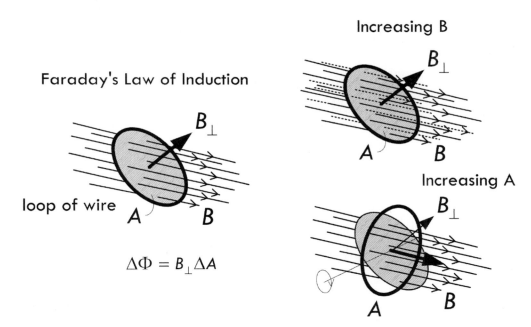

FIGURE 2.116   Illustration of Faraday's law of induction.

The simple ac generator in Fig. 2.117 shows Faraday's law in action. A simple rotating loop of wire in a constant magnetic field generates an EMF that can be used to power a circuit. As the loop rotates, the magnetic flux through it changes with time, inducing an EMF and a current in an external circuit. The ends of the loop are connected to slip rings that rotate with the loop, while the external circuit is linked to the generator by stationary brushes in contact with the slip rings.

A simple dc generator is essentially the same as the ac generator, except that the contacts to the rotating loop are made using a split ring or commutator. The result is a pulsating direct current, resembling the absolute value of a sine wave—there are no polarity reversals.

A motor is essentially a generator operating in reverse. Instead of generating a current by rotating a loop, a current is supplied to the loop by a battery, and the torque

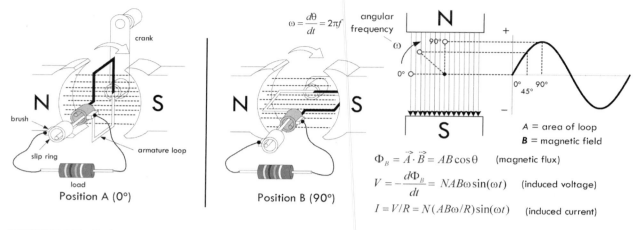

$$\omega = \frac{d\theta}{dt} = 2\pi f \quad \text{angular frequency}$$

$$\Phi_B = \vec{A} \cdot \vec{B} = AB\cos\theta \quad \text{(magnetic flux)}$$

$$V = -\frac{d\Phi_B}{dt} = NAB\omega\sin(\omega t) \quad \text{(induced voltage)}$$

$$I = V/R = N(AB\omega/R)\sin(\omega t) \quad \text{(induced current)}$$

A = area of loop
B = magnetic field

FIGURE 2.117   Basic ac generator.

acting on the current-carrying loop causes it to rotate. Real ac generators and motors are much more complex than the simple ones demonstrated here. However, they still operate under the same fundamental principles of electromagnetic induction.

The circuit in Fig. 2.118 shows how it is possible to induce current within a secondary coil of wire by suddenly changing the current flow through a primary coil of wire. As the magnetic field of the primary expands, an increasing magnetic flux permeates the secondary. This induces an EMF that causes current to flow in the secondary circuit. This is the basic principle behind how transformers work; however, a real transformer's primary and secondary coils are typically wound around a common ferromagnetic core to increase magnetic coupling.

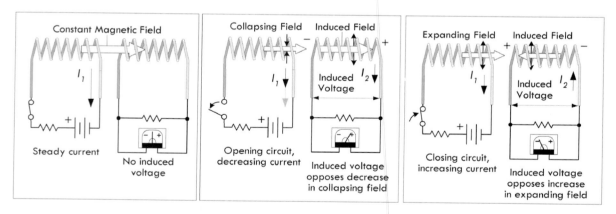

FIGURE 2.118   An induced voltage, or EMF, is generated in the secondary circuit whenever there is a sudden change in current in the primary.

2.24.3 Self-Inductance

In the previous section, we saw how an EMF could be induced in a closed-loop circuit whenever the magnetic flux through the circuit changed with time. This phenomenon of electromagnetic induction is used in a number of mechanisms, such as motors, generators, and transformers, as was pointed out. However, in each of these cases, the induced EMF was a result of an external magnetic field, such as the primary coil in relation to a secondary coil. Now, however, we will discuss a phenomenon called *self-induction*. As the name suggests, self-induction typically involves a looped wire inflicting itself with an induced EMF that is generated by the varying current that passes through it. According to Faraday's law of induction, the only time that our loop can self-inflict is when the magnetic field grows or shrinks in strength (as a result of an increase or decrease in current). Self-induction is the basis for the inductor, an important device used to store energy and release energy as current levels fluctuate in time-dependent circuits.

Consider an isolated circuit consisting of a switch, a resistor, and a voltage source, as shown in Fig. 2.119. If you close the switch, you might predict that the current flow through the circuit would jump immediately from zero to V/R, according to Ohm's law. However, according to Faraday's law of electromagnetic induction, this isn't entirely accurate. Instead, when the switch is initially closed, the current increases rapidly. As the current increases with time, the magnetic flux through the loop rises rapidly. This increasing magnetic flux then induces an EMF in the circuit that opposes the current flow, giving rise to an exponentially delayed rise in current. We call the induced EMF a *self-induced EMF*.

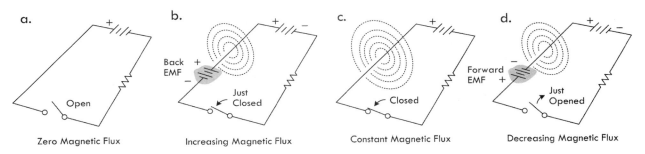

FIGURE 2.119 **(a) Circuit is open and thus no current or magnetic field is generated. (b) The moment the circuit is closed, current begins to flow, but at the same time, an increasing magnetic flux is generated through the circuit loop. This increasing flux induces a back EMF that opposes the applied or external EMF. After some time, the current levels off, the magnetic flux reaches a constant value, and the induced EMF disappears. (d) If the switch is suddenly opened, the current attempts to go to zero; however, during this transition, as the current goes to zero, the flux decreases through the loop, thus generating a forward induced voltage of the same polarity as the applied or external EMF. As we'll see later, when we incorporate large solenoid and toroidal inductors within a circuit, opening a switch such as this can yield a spark—current attempting to keep going due to a very large forward EMF.**

Self-induction within a simple circuit like that shown in Fig. 2.119 is usually so small that the induced voltage has no measurable effect. However, when we start incorporating special devices that concentrate magnetic fields—namely, discrete inductors—time-varying signals can generate significant induced EMFs. For the most part, unless otherwise noted, we shall assume the self-inductance of a circuit is negligible compared with that of a discrete inductor.

2.24.4 Inductors

Inductors are discrete devices especially designed to take full advantage of the effects of electromagnetic induction. They are capable of generating large concentrations of magnetic flux, and they are likewise capable of experiencing a large amount of self-induction during times of great change in current. (Note that self-induction also exists within straight wire, but it is usually so small that it is ignored, except in special cases, e.g., VHF and above, where inductive reactance can become significant.)

The common characteristic of inductors is a looplike geometry, such as a solenoid, toroid, or even a spiral shape, as shown in Fig. 2.120. A solenoid is easily constructed by wrapping a wire around a hollow plastic form a number of times in a tight-wound fashion.

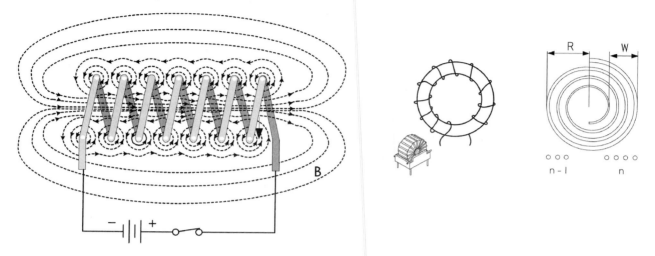

FIGURE 2.120   **Various coil configurations of an inductor—solenoid, toroid, and spiral.**

The basic schematic symbol of an air core inductor is given by ⎯⎯⟋⟍⟋⟍⟋⟍⎯⎯ . Magnetic core inductors (core is either iron, iron powder, or a ferrite-type ceramic), adjustable core inductors, and a ferrite bead, along with their respective schematic symbols are shown in Fig. 2.121.

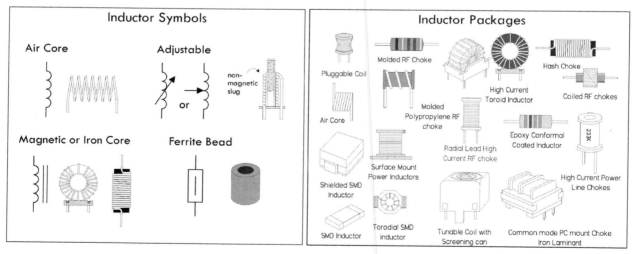

FIGURE 2.121

Magnetic core inductors are capable of generating much higher magnetic field densities than air core inductors as a result of the internal magnetization that occurs at the atomic level within the core material due to the surrounding wire coil's magnetic field. As a result, these inductors experience much greater levels of self-induction when compared to air core inductors. Likewise, it is possible to use fewer turns when using a magnetic core to achieve a desired inductance. The magnetic core material is often iron, iron powder, or a metallic oxide material (also called a ferrite, which is ceramic in nature). The choice of core material is a complex process that we will cover in a moment.

Air core inductors range from a single loop in length of wire (used at ultrahigh frequencies), through spirals in copper coating of an etched circuit board (used at very high frequencies), to large coils of insulated wire wound onto a nonmagnetic former. For radio use, inductors often have air cores to avoid losses caused by magnetic hysteresis and eddy current that occur within magnetic core–type inductors.

Adjustable inductors can be made by physically altering the effective coil length— say, by using a slider contact along the uncoated coil of wire, or more commonly by using a ferrite, powdered iron, or brass slug screwed into the center of the coil. The idea behind the slug-tuned inductor is that the inductance depends on the permeability of the material within the coil. Most materials have a relative permeability close to 1 (close to that of vacuum), while ferrite materials have a large relative permeability. Since the inductance depends on the average permeability of the volume inside the coil, the inductance will change as the slug is turned. Sometimes the slug of an adjustable inductor is made of a conducting material such as brass, which has a relative permeability near 1, in which case eddy currents flow on the outside of the slug and eliminate magnetic flux from the center of the coil, reducing its effective area.

A *ferrite bead*, also known as a *ferrite choke,* is a device akin to an inverted ferrite core inductor. Unlike a typical core inductor, the bead requires no coiling of wire (though it is possible to coil a wire around it for increased inductance, but then you've created a standard ferrite core inductor). Instead, a wire (or set of wires) is placed through the hollow bead. This effectively increases the inductance of the wire (or wires). However, unlike a standard inductor that can achieve practically any inductance based on the total number of coil turns, ferrite beads have a limited range over which they can influence inductance. Their range is typically limited to RF (radiofrequency). Ferrite beads are often slipped over cables that are known to be notorious radiators of RF (e.g., computers, dimmers, fluorescent lights, and motors). With the bead in place, the RF is no longer radiated but absorbed by the bead and converted into heat within the bead. (RF radiation can interfere with TV, radio, and audio equipment.) Ferrite beads can also be placed on cables entering receiving equipment so as to prevent external RF from entering and contaminating signals in the cable runs.

Inductor Basics

An inductor acts like a time-varying current-sensitive resistance. It only "resists" during changes in current; otherwise (under steady-state dc conditions), it passes current as if it were a wire. When the applied voltage increases, it acts like a time-dependent resistor whose resistance is greatest during times of rapid increase in current. On the other hand, when the applied voltage decreases, the inductor acts like a time-dependent voltage source (or negative resistance) attempting to keep current flowing. Maximal sourcing is greatest during times of rapid decreases in current.

In Fig. 2.122a, when an increasing voltage is applied across an inductor, resulting in an increasing current flow, the inductor acts to resist this increase by generating a reverse force on free electrons as a result of an increasing magnetic flux cutting across the coils of the solenoid (or crossing through the coil loops). This reverse force on free electrons can be viewed as an induced EMF that points in the opposite direction of the applied voltage. We deem this induced EMF a *reverse EMF*, also referred to as a *back EMF*. The result is that the inductor strongly resists during a sudden increase in current flow, but quickly loses its resistance once the current flow levels off to a constant value.

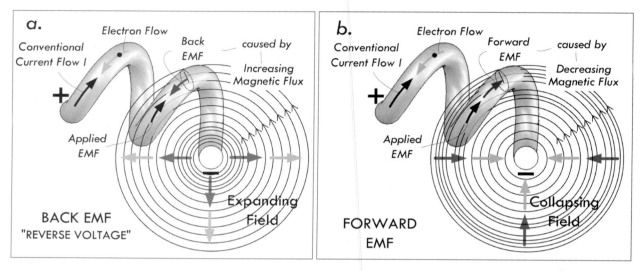

FIGURE 2.122

In Fig. 2.122b, when a decreasing voltage is applied across an inductor, resulting in a decreasing current flow, the inductor acts to resist this decrease by generating a forward force on free electrons as a result of a decreasing magnetic flux cutting across the coils of the solenoid (or crossing through the coil loops). This forward force on free electrons can be viewed as an induced EMF that points in the same direction as that of the applied voltage (the applied voltage that existed before the sudden change). We deem this induced EMF a *forward EMF*. The result is that the inductor acts like a voltage source during a sudden decrease in current flow, but quickly disappears once the current flow levels off to a constant value.

Another view of how inductors work is to consider energy transfer. The transfer of energy to the magnetic field of an inductor requires that work be performed by a voltage source connected across it. If we consider a perfect inductor where there is no resistance, the energy that goes into magnetic field energy is equal to the work performed by the voltage source. Or, in terms of power, the power is the rate at which energy is stored ($P = dW/dt$). Using the generalized power law $P = IV$, we can equate the powers and see that there must be a voltage drop across an inductor while energy is being stored in the magnetic field. This voltage drop, exclusive of any voltage drop due to resistance in a circuit, is the result of an opposing voltage induced in the circuit while the field is building up to its final value. Once the field becomes constant, the energy stored in the magnetic field is equal to the work performed by the voltage source.

Figure 2.123 illustrates what occurs when an inductor is energized suddenly as a switch is closed.

Energizing Inductor

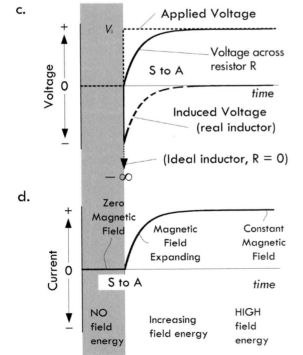

a.

Internal Resistance

Internal Resistance Model

b.

Conventional
Current Flow I

Electron Flow

Back
EMF

caused by

Increasing
Magnetic Flux

Applied
EMF

Expanding
Field

BACK EMF
"REVERSE VOLTAGE"

c.

Applied Voltage

V_s

Voltage across
resistor R

S to A

Voltage

0

time

Induced Voltage
(real inductor)

(Ideal inductor, R = 0)

$-\infty$

d.

Zero
Magnetic
Field

Magnetic
Field
Expanding

Constant
Magnetic
Field

Current

0

S to A

time

NO
field
energy

Increasing
field energy

HIGH
field
energy

FIGURE 2.123

When an inductor is being energized, we say that electrical energy from an externally applied source is transformed into a magnetic field about the inductor. Only during times of change—say, when the switch is thrown, and the magnetic field suddenly grows in size—do we see inductive behavior that affects circuit dynamics.

In the circuit to the left, when the switch is thrown from position B to position A, a sudden change in voltage is applied across the inductor, resulting in a sudden increase in current flow. Where there was no magnetic field, there now is a rapidly growing magnetic field about the inductor. The inductor is said to be energizing. The expanding magnetic field cuts across its own inductor coils, which, by Faraday's law, exerts a force on free electrons within the coil. The force on these electrons is such as to be pointing in the opposite direction as the applied voltage. This effective reverse force and the effect it has on the free electrons is deemed a reverse EMF—it's analogous to a little imaginary battery placed in the reverse direction of the applied voltage. (See Fig. 2.123b.) The result is that the inductor resists during increases in current. After a short while, the current flow through the circuit levels off, and the magnetic field stops growing and assumes a constant value. With no change in magnetic field strength, there is no increase in magnet flux through our fixed coils, so there is no more reverse EMF. The inductor acts as a simple conductor. Figure 2.123c shows the induced voltage and voltage across the resistor as a function of time.

Figure 2.123d shows the resultant current flow through the inductor as a result of the resultant voltage. Mathematically, the current flow is expressed by the following equation:

$$I = \frac{V_s}{R}(1 - e^{-t/(L/R)})$$

(You might question what happens if we assume an ideal inductor with zero internal resistance R. This is a very good question—one that we'll discuss when we get to defining inductance mathematically.)

In terms of energy, this can be viewed as the electrical energy being transformed into magnetic field energy. In terms of power, we see that a voltage drop occurs as energy is pumped into the magnetic field. (We associate the drop with the back EMF.) Once the current levels off, no more energy goes into the field; hence, no reverse voltage (or voltage drop) is present.

Figure 2.124 illustrates what occurs as an inductor is deenergized suddenly by opening a switch.

Deenergizing Inductor

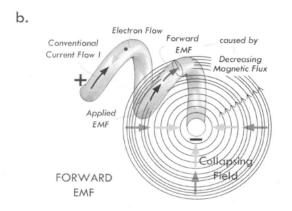

a.

Internal Resistance

Internal Resistance Model

b.

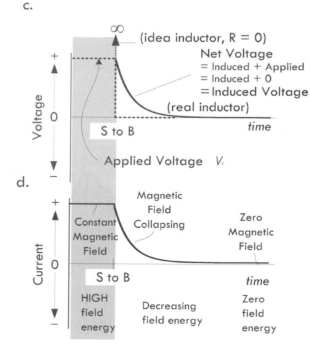

Electron Flow

Conventional Current Flow I

Forward EMF   caused by

Decreasing Magnetic Flux

Applied EMF

FORWARD EMF

Collapsing Field

c.

∞ (idea inductor, R = 0)

Net Voltage
= Induced + Applied
= Induced + 0
= Induced Voltage
(real inductor)

Voltage

0

time

S to B

Applied Voltage V_s

d.

Current

Magnetic Field Collapsing

Zero Magnetic Field

Constant Magnetic Field

0

S to B

time

HIGH field energy

Decreasing field energy

Zero field energy

Inductor momentarily sources current during sudden decrease in current

FIGURE 2.124

When an inductor is being deenergized, we say that the magnetic field energy of the inductor is "transformed" back into electrical energy. Again, only during times of change, which in our case involves a decreasing current flow, do we notice effects of induction.

In the circuit to the left, when the switch is thrown from position A to B, a sudden change voltage occurs across the inductor. The inductor initially opposes this decrease by generating a collapsing magnetic field that cuts across the inductor coils. According to Faraday's law, a decreasing magnetic flux passes through the loops of the inductor, thus imparting a force upon free electrons within the coil in the same direction the applied voltage was pointing right before the switch occurred. Since the force is in the same direction, we deem the effect a forward EMF. Hence, the inductor sources current when attempts are made to decrease current flow. The energy for it to do so comes from the magnetic field, whose energy drops in proportion to the electrical energy delivered to the circuit. See 2.124(b).

Part (c) shows the resultant voltage across an inductor adding the applied voltage to the inducted voltage.

Part (d) shows the resultant current flow through the inductor as a result of the resultant voltage.

Mathematically, the current flow is expressed by the following equation:

$$I = \frac{V_s}{R} e^{-t/(L/R)}$$

(Again, you might question what happens if we assume an ideal inductor with zero internal resistance R. This is a very good question—one that we'll discuss when we get to defining inductance mathematically.)

Note: We must assume that when the switch is thrown from A to B, the transition occurs instantaneously. We'll see in a moment that when a physical break occurs, cutting current flow through an inductive circuit, the collapsing magnetic field can be large enough to generate an EMF capable of causing a spark to jump between the break points (i.e., switch contacts).

2.24.5 Inductor Water Analogy

The property of inductance in electric circuits is closely analogous to mass inertia in mechanical systems. For example, the property of an inductor resisting any sudden changes in current flow (increasing or decreasing) is similar to a mass on a spinning wheel resisting any change in motion (increasing or decreasing in speed). In the following water analogy, we take this mass analogy to heart, incorporating a water turbine/flywheel device to represent a "water inductor."

To start, we consider a basic electrical inductor circuit, as shown to the left in Fig. 2.125. The field generated by a suddenly applied voltage creates a reverse induced voltage of opposite polarity that initially "resists" current flow. Quickly, depending on the inductance value, the reverse voltage disappears as the magnetic field becomes constant, at which point it has reached a maximum strength and energy. A collapsing field generated when the applied voltage is removed creates a forward induced voltage that attempts to keep current flowing. Quickly, depending on the inductance value, the forward voltage disappears and the magnetic field goes to zero.

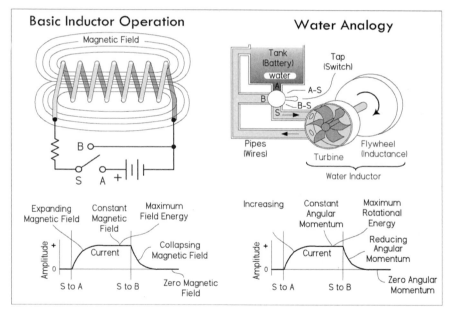

FIGURE 2.125

In the water analogy, the turbine with attached flywheel resists any sudden changes in current flow. If water pressure is suddenly applied, the turbine initially resists water flow due to its mass and that of the attached flywheel. However, the pressure exerted over the turbine blades quickly gives rise to mechanical motion. Depending on the mass of the flywheel, the time it takes for the flywheel to reach a steady angular velocity will vary—a heavier flywheel will require more time (analogous to a high-value inductor requiring more time to reach a constant current after a sudden increase in applied voltage). When the flywheel reaches this constant angular rotation, the water inductor has maximum rotational momentum and energy. This is analogous to the magnetic field strength reaching a constant maximum strength and energy when the reverse voltage disappears. If there is any sudden interruption in applied pressure—say, by turning the tap to position B-S, as shown in Fig. 2.125—the flywheel's angular momentum will attempt to keep current flowing. This is analogous to the collapsing magnetic field in an inductor inducing a forward voltage.

Example: What will happen when the switches are closed in the following three circuits? What happens when the switches are later opened? How does the size of the inductance and capacitance influence behavior? Refer to Fig. 2.126.

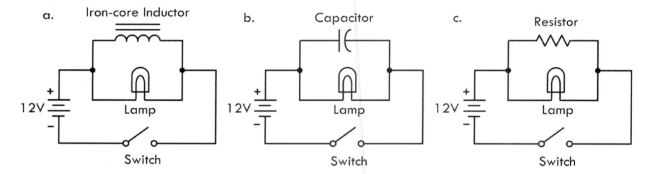

FIGURE 2.126

Answer: (a) When the switch is closed in the inductor circuit, the lamp will suddenly light up brightly, but then quickly dims out. This is because the moment the switch is closed, the inductor has very high impedance to current flow, but quickly loses its impedance as the current becomes constant (the magnetic field is no longer expanding). Once constant, the inductor acts like a short, and all current is diverted away from the lamp through the inductor. (We assume here that there is sufficient dc internal resistance in the inductor to prevent excessive current flow. Also, we assume the internal dc resistance of the inductor is much smaller than the internal resistance of the lamp.) A larger inductance value increases the time it takes for the lamp to dim out completely.

(b) When the switch is closed in the capacitor circuit, the opposite effect occurs; the lamp slowly builds up in brightness until it reaches maximum illumination. This is because when the switch is closed, the capacitor initially has very low impedance to current flow during rapid changes in applied voltage. However, as the capacitor charges up, the capacitor's impedance rises toward infinity, and consequently resembles an open circuit—hence, all current is diverted through the lamp. A larger capacitance value increases the time it takes for the lamp to reach full brightness.

(c) When the switch is closed in the resistor circuit, free-electron flow throughout the system is essentially instantaneous. Aside from small inherent inductance and capacitance built into the circuit, there are no time-dependent effects on current flow caused by discrete inductance or capacitance. Note that since the voltage source is ideal, no matter what value the parallel resistor has, there will always be 12 V across the lamp, and hence the brightness of the lamp does not change over time.

2.24.6 Inductor Equations

Conceptually, you should now understand that the amplitude of the induced voltage—be it reverse or forward induced—is proportional to the rate at which the current changes, or the rate at which the magnetic flux changes. Quantitatively, we can express this relationship by using the following equation:

$$V_L = L\frac{dI_L}{dt}$$

(2.50) Voltage across an inductor = induced EMF

If we integrate and solve for I_L, we get:

$$I_L = \frac{1}{L}\int V_L\, dt$$

(2.51) Current through an inductor

In Eq. 2.50, the proportionality constant L is called the inductance. This constant depends on a number of physical inductor parameters, such as coil shape, number of turns, and core makeup. A coil with many turns will have a higher L value than one with few turns, if both coils are otherwise physically similar. Furthermore, if an inductor is coiled around a magnetic core, such as iron or ferrite, its L value will increase in proportion to the permeability of that core (provided circuit current is below the point at which the core saturates).

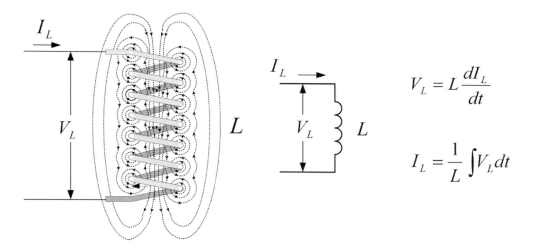

FIGURE 2.127 **The voltage measured across an ideal inductor is the induced voltage, or EMF, symbolized V_L. When there is a steady-state dc current flow, there is no induced voltage (V_L is zero) and the inductor resembles a short. As we'll see, the inductor equations can yield surprisingly unrealistic values, if we don't consider internal resistance and capacitance inherently present within a real inductor.**

The basic unit of inductance L is the henry, abbreviated H. One henry equals an induced voltage of 1 V when the current is varying at a rate of 1 A/s:

$$1\,\text{H} = \frac{1\,\text{V}}{1\,\text{A/s}} \quad \text{(Definition of a henry)}$$

Making inductors from scratch is common in electronics, unlike the construction of capacitors, which is left almost exclusively to the manufacturers. Though we'll examine how to make inductors in a moment, it's worth taking a look at some commercial inductors. Note the inductance range, core type, current, and frequency limits listed in Table 2.7.

TABLE 2.7 Typical Characteristics of Commercial Inductors

CORE TYPE	MINIMUM H	MAXIMUM H	ADJUSTABLE?	HIGH CURRENT?	FREQUENCY LIMIT
Air core, self-supporting	20 nH	1 mH	Yes	Yes	1 GHz
Air core, on former	20 nH	100 mH	No	Yes	500 MHz
Slug tuned open winding	100 nH	1 mH	Yes	No	500 MHz
Ferrite ring	10 mH	20 mH	No	No	500 MHz
RM Ferrite Core	20 mH	0.3 H	Yes	No	1 MHz
EC or ETD Ferrite Core	50 mH	1 H	No	Yes	1 MHz
Iron	1 H	50 H	No	Yes	10 kHz

Typical values for commercial inductors range from fractions of a nanohenry to about 50 H. Inductances are most commonly expressed in terms of the following unit prefixes:

nanohenry (nH): $1 \text{ nH} = 1 \times 10^{-9} \text{ H} = 0.000000001 \text{ H}$

microhenry (µH): $1 \text{ µH} = 1 \times 10^{-6} \text{ H} = 0.000001 \text{ H}$

millihenry (mH): $1 \text{ mH} = 1 \times 10^{-3} \text{ H} = 0.001 \text{ H}$

Example 1: Rewrite 0.000034 H, 1800 mH, 0.003 mH, 2000 µH, and 0.09 µH in a more suitable unit prefix format ($1 \leq$ numeric value < 1000).

Answer: 34 µH, 1.8 H, 3 µH, 2 mH, and 90 nH.

From another standpoint, inductance can be determined from basic physics principles. Theoretically, you can determine the inductance any time by stating that the inductance is always the ratio of the magnetic flux linkage ($N\Phi_M$) to the current:

$$L = \frac{N\Phi_M}{I} \tag{2.52}$$

For an air-filled solenoid, as shown in Fig. 1.128, if a current I flows through the coil, Ampere's law allows us to calculate the magnetic flux:

$$\Phi_M = BA = \left(\frac{\mu_0 NI}{\ell} \right) A = \mu_0 A n_{unit} I$$

where n_{unit} is the turns per unit length:

$$n_{unit} = N/\ell \tag{2.53}$$

where N is the total turns and ℓ is the length. The variable A is the cross-sectional area of the coil, and μ is the permeability of the material on which the coil is wound. For most materials (excluding iron and ferrite materials), the permeability is close to the permeability of free space:

$$\mu_0 = 4\pi \times 10^{-7} \, \text{T} \cdot \text{m}/\text{A}$$

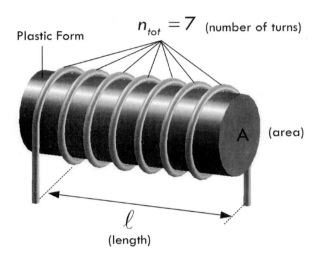

Plastic Form

$n_{tot} = 7$ (number of turns)

A (area)

ℓ (length)

FIGURE 2.128

According to Faraday's law, an induced voltage is in each loop of the solenoid, resulting in a net induced voltage across the solenoid equal to n times the change in magnetic flux:

$$V_L = N\frac{d\Phi_M}{dt} = \frac{\mu N^2 A}{\ell}\frac{dI}{dt}$$

The term in front of dI/dt in the equation we call the inductance of the solenoid:

$$L_{\text{sol}} = \frac{\mu N^2 A}{\ell} \tag{2.54}$$

Inductance varies as the square of the turns. If the number of turns is doubled, the inductance is quadrupled. This relationship is inherent in the equation but is often overlooked. For example, if you want to double the inductance of a coil, you don't double the turns; rather you make the number of turns $\sqrt{2}$ (or 1.41) times the original number of turns, or 40 percent more turns.

Example 2: Find the inductance of a cylindrical coil of length 10 cm, radius 0.5 cm, having 1000 turns of wire wrapped around a hollow plastic form.

Answer:

$$L = \mu_0 N^2 A/\ell = (4\pi \times 10^{-7}) \, 10^6 (\pi \times 0.005^2)/0.1$$
$$= 1 \times 10^{-3} \text{H} = 1 \text{ mH}$$

Luckily, there are some simple formulas, shown in Fig. 2.129, for air core inductors, as well as for a multiple-wound and a spiral-wound inductor. Note that the formulas' answers will not be in standard form—they assume the results are in units of microhenrys.

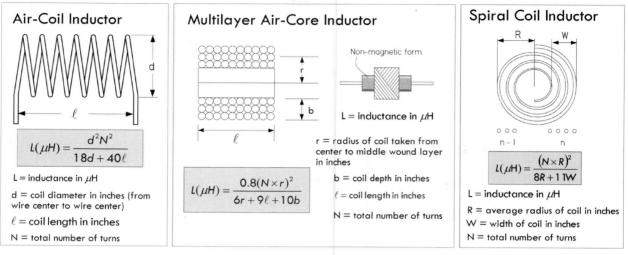

FIGURE 2.129   Practical air core inductor equations.

Example 3: What is the inductance of a coil wrapped around a 0.5-in-diameter plastic form if the coil has 38 turns wound at 22 turns per inch?

Answer: First the total length is determined:

$$\ell = \frac{N}{n_{\text{unit}}} = \frac{38 \text{ turns}}{22 \text{ turns/in}} = 1.73 \text{ in}$$

Next, using the equation for an air core inductor shown in Fig. 2.129, noting the result will be in units of microhenrys, we get:

$$L(\mu H) = \frac{d^2 N^2}{18d + 40\ell} = \frac{(0.50)^2 (38)^2}{18(0.50) + 40(1.73)} = \frac{361}{78} = 4.62 \ \mu H$$

Example 4: Design a solenoid inductor with an inductance of 8 μH if the form on which the coil is wound has a diameter of 1 in and a length of 0.75 in.

Answer: Rearranging the equation in the previous example:

$$N = \sqrt{\frac{L(18d + 40\ell)}{d^2}} = \sqrt{\frac{8(18(1) + 40(0.75))}{1^2}} = 19.6 \text{ turns}$$

A 20-turn coil would be close enough in practical work. Since the coil will be 0.75 in long, the number of turns per inch will be 19.6/0.75 = 26.1. A #17 enameled wire (or anything smaller) can be used. In practice, you wind the required number of turns on the form and then adjust the spacing between the turns to make a uniformly spaced coil of 0.75 in long.

On the Internet you'll find a number of free web-based inductor calculators. Some are quite advanced, allowing you to input inductance, diameter, and length values, and afterward providing you with the required number of turns, possible number of layers required, dc resistance of the wire, wire gauge to use, and so on. Check these tools out—you can save yourself the trouble of doing calculations and looking up wire dimensions and such.

2.24.7 Energy Within an Inductor

An ideal inductor, like an ideal capacitor, doesn't dissipate energy, but rather stores it in the magnetic field and later returns it to the circuit when the magnetic field collapses. The energy E_L stored in the inductor is found by using the generalized power law $P = IV$, along with the definition of power $P = dW/dt$, and the inductor equation $V = L\, dI/dt$. By equating the work W with E_L we get:

$$E_L = \int_{I=0}^{I=Ifinal} P\, dt = \int_{I=0}^{I=Ifinal} I\ V\, dt = \int_{I=0}^{I=Ifinal} I\, L\frac{dI}{dt}\, dt = \int_{I=0}^{I=Ifinal} L\ I\, dt = \frac{1}{2}LI^2 \qquad (2.55)$$

where E_L = energy in joules, I = current in amps, and L = inductance in henrys. Note that in a real inductor a small portion of energy is lost to resistive heating through the inductor's internal resistance.

2.24.8 Inductor Cores

To conserve space and material, inductors are often wound on a magnetic core material, such as laminated iron or a special molded mix made from iron powder or ferrite material (iron oxide mixed with manganese, zinc, nickel, and other ingredients). A magnetic core increases the magnetic flux density of a coil greatly, and thus increases the inductance. This is further intensified if the magnetic core is formed into a doughnut-shaped toroid. See Fig. 2.130.

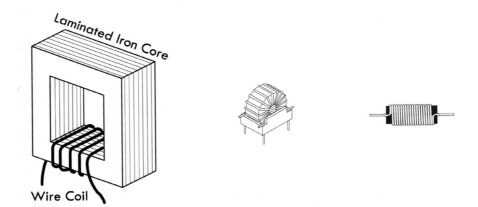

FIGURE 2.130 **Magnetic core inductors.**

The reason a magnetic core is so influential has to do with the magnetization that occurs within it as the outer coil passes current. When current is sent through the wire coil, a relatively weak magnetic field is set up at its center. This magnetic field, which we'll call the external magnetic field, causes the atomic rearrangement of magnetic dipoles (refer to Fig. 2.112) within the core material. This realignment is such as to rotate the dipole moments in a common direction. As more current is passed through the coil, more and more dipoles line up. The core itself is now generating a magnetic field as a result of the dipole alignment. The net magnetic field B_{total} generated by the whole inductor (coil and core) then becomes the sum of the external field (coil) and a term that is proportional to the magnetization M present in the core itself, as the external field is applied by the coil through the core. Mathematically, this is expressed:

$$B_{total} = B_{ext} + \mu_0 M \tag{1}$$

where μ_0 is the permeability of free space. The magnetic intensity H due to real current in the coil, as opposed to the magnetic intensity generated by atomic magnetization of the core, is expressed:

$$H = \frac{B_{ext}}{\mu_0} = \frac{B_{total}}{\mu_0} - M \tag{2}$$

This can be further reduced, using the notions of susceptibility and permeability (consult a physics book), to:

$$B_{total} = \mu H \tag{3}$$

where μ is the permeability of the core material.

The ratio of magnetic flux density produced by a given core material compared to the flux density produced by an air core is called the relative permeability of the material $\mu_R = \mu/\mu_0$. For example, an air core generating a flux density of 50 lines of force per square inch can be made to generate 40,000 lines of force per square inch with an iron core inserted. The ratio of these flux densities, iron core to air core, or the relative permeability is 40,000/50, or 800. Table 2.8 shows permeabilities of some popular high-permeability materials.

Problems with Magnetic Cores

When a magnetic core material is conductive (e.g., steel), there is a phenomenon known as eddy currents that arise within the core material itself when there is an applied magnetic field that is changing. For example, in Fig. 2.131a, when an increasing current is supplied through the outer coil, a changing magnetic flux passes through the core. This in turn induces a circular current flow within the core material. Eddy currents that are induced in the core represent loss in the form of resistive heating and can be a significant disadvantage in certain applications (e.g., power transformers). Eddy current losses are higher in materials with low resistivity.

TABLE 2.8 Permeability of Various Materials

MATERIAL	APPROXIMATE MAXIMUM PERMEABILITY (H/M)	APPROXIMATE MAXIMUM RELATIVE PERMEABILITY	APPLICATIONS
Air	1.257×10^{-6}	1	RF
Ferrite U60	1.00×10^{-5}	8	UHF chokes
Ferrite M33	9.42×10^{-4}	750	Resonant circuits
Ferrite N41	3.77×10^{-3}	3000	Power circuits
Iron (99.8% pure)	6.28×10^{-3}	5000	
Ferrite T38	1.26×10^{-2}	10,000	Broadband transformers
45 Permalloy	3.14×10^{-2}	25,000	
Silicon 60 steel	5.03×10^{-2}	40,000	Dynamos, transformers
78 Permalloy	0.126	100,000	
Supermalloy	1.26	1,000,000	Recording heads

To avoid eddy currents, the conductive core (steel, in this case) can be laminated together with insulating sheets of varnish or shellac. Current will still be induced in the sheets of the core, as shown in Fig. 2.131b, but because the area of the sheets is limited, so is the flux change, and therefore so are these currents.

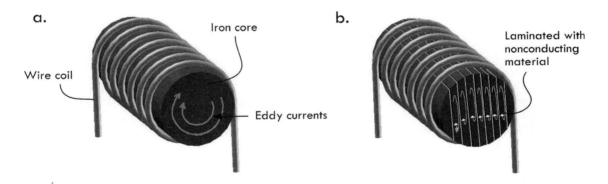

FIGURE 2.131 (a) Large eddy currents within core. (b) Eddy currents are reduced within laminated core.

Ferrite materials have quite high intrinsic resistivity compared to, say, steel (10 to 1,000 Ω-cm for Mn-Zn ferrites; 10^5 to 10^7 Ω-cm for Ni-Zn ferrites). Eddy current losses are therefore much less of a problem in ferrites, and this is the fundamental reason they are used in higher-frequency applications. Powdered iron cores, with their insulating compound present between iron particles, also reduce eddy currents, since the path for the eddy current is limited to the powder particle size.

A second difficulty with iron is that its permeability is not constant, but varies with the strength of the magnetic field and, hence, with the current in the windings. (It also varies with temperature.) In fact, at sufficiently high magnetic fields, the core will saturate and its relative permeability will drop to a value near unity. Not only that, but the magnetic field in the iron depends on the past history of the current in the winding. This property of remanence is essential in a permanent magnet, but in an inductor it gives rise to additional losses, called hysteresis losses. See Fig. 2.132.

Typical Hysteresis Curve for Magnetic Core

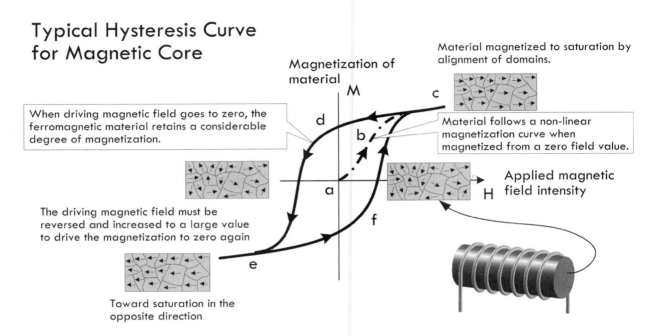

FIGURE 2.132 Hysteresis curve showing how the magnetization of the core material does not represent a reversible process. At point a, no current is applied through the coil. Along path a-b-c, coil current, and thus H (applied magnetic field or magnetizing force), increases, causing the core's magnetic dipole moments (in this case localized in domains) to rotate in proportion. As we approach point c, the core reaches saturation—increasing H causes no appreciable increase in M (magnetization or magnetic dipole density of core)—the domain's dipole moments are aligned as much as possible parallel to H. Saturation causes a rapid decrease in permeability. The saturation point of a magnetic core varies with material makeup; air and other nonmagnetic materials do not saturate—they have a permeability of 1. As H is decreased, the magnetization M does not follow the same path back down, but follows path c-d-e. Notice that when H goes to zero, the core remains magnetized. In essence, the core has become a permanent magnet. The term *retentivity* is used to describe this effect, and it presents another set of losses caused by hysteresis. In order to demagnetize the core, a reverse force is necessary to overcome the residual magnetism retained by the core. In other words, H must go negative in the opposite direction, driving the magnetic domains back into a random orientation. At point e, the core has again reached saturation; however, the magnetic dipoles (domains) are now pointing in the opposite direction. To reach saturation in the opposite direction again, H must be applied, as shown in path e-f-c. Air cores and other materials with a permeability of 1, such as brass and aluminum, are immune to hysteresis effects and losses.

To avoid loses associated with hysteresis, it is important to not run the core inductor into saturation. This can be accomplished by running the inductor at lower current, using a larger core, altering the number of turns, using a core with lower permeability, or using a core with an air gap.

It is possible for the eddy current and hysteresis losses to be so large that the inductor behaves more like a resistor. Furthermore, there is always some capacitance between the turns of the inductor, and under some circumstances an inductor may act like a capacitor. (We'll discuss this a bit later.)

Table 2.9 shows a comparison between the various core inductors.

Example 5: What is the inductance of a 100-turn coil on a powdered iron toroidal core with an index of inductance of 20?

TABLE 2.9   Comparison of Inductor Cores

Air core	Relative permeability is equal to 1. The inductance is independent of current-flow because air does not saturate. Limited to low inductance values, but can be operated at very high frequencies (e.g., RF applications ~1 GHz).
Iron core	Permeability is a factor of 1000 greater than for air core, but inductance is highly dependent on the current flowing through the coil because the core saturates. Chiefly used in power-supply equipment. Highly prone to eddy currents (power loss) due to the core's high conductivity. Losses caused by eddy currents can be reduced by laminating the core (cutting it into thin strips separated by insulation, such as varnish or shellac. Experience significant power losses due to hysteresis. Eddy currents and hysteresis losses in iron increase rapidly as the frequency of ac increases. Limited to power-line and audio frequencies up to about 15,000 Hz. Laminated iron cores are useless at radio frequencies.
Powdered iron	Ground-up iron (powder) mixed with a binder or insulating material reduces eddy currents significantly, since the iron particles are insulated from one another. Permeability is low compared to iron core due to insulator concentration. Slug-tuned inductors, where the slug is powered iron can be made adjustable and are useful in RF work, even up to the VHF range. Manufacturers offer a wide variety of core materials, or mixes, to provide units that will perform over a desired frequency range with a reasonable permeability. Toroidal Cores are considered self-shielding. Manufacturers provide an inductance index A_L for their toroidal cores. For powdered-iron toroids, A_L provides the inductance in µH per 100 turns of wire on the core, arranged in a single layer. Formulas found in Fig. 2.133, and example problems 5 and 6 illustrate how to calculate inductance of a powdered-iron toroidal inductor.
Ferrite	Composed of nickel-zinc ferrites for lower permeability ranges and of manganese-zinc ferrites for higher permeabilities, these cores span the permeability range from 20 to above 10,000. They are used for RF chokes and wideband transformers. Ferrites are often used, since they are nonconductors and are immune to eddy currents. Like powdered-iron toroids, ferrite toroids also have an A_L value provided by the manufacturers. However, unlike powdered-iron toroids, ferrite A_L values are given in mH per 1000 turns. Formulas found in Fig. 2.133 and example problems 7 and 8 illustrate how to calculate inductance of a ferrite toroidal inductor.

Answer: To solve this you'd have to consult the manufacturer's data sheets, but, here, use a T-12-2 from the table in Fig. 2.133. Using the equation for powdered iron toroids in Fig. 2.133, and inserting $A_L = 20$ for the T-12-2, and $N = 100$:

$$L(\mu H) = \frac{A_L \times N^2}{10,000} = \frac{20 \times 100^2}{10,000} = 20 \ \mu H$$

Example 6: Calculate the number of turns needed for a 19.0-µH coil if the index of inductance for the powdered iron toroid is 36.

Answer: Use:

$$N = 100 \sqrt{\frac{\text{desired } L \text{ in } \mu H}{A_L \text{ in } \mu H \text{ per 100 turns}}} = 100 \sqrt{\frac{19.0}{36}} = 72.6 \text{ turns}$$

Example 7: What is the inductance of a 50-turn coil on a ferrite toroid with an index of inductance of 68?

Answer: To solve this you'd have to consult the manufacturer's data sheets, but here we'll use a FT-50 with 61-Mix, as shown in the table in Fig. 2.133. Using the

Toroidal Inductor

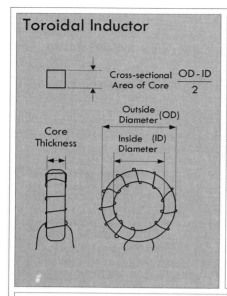

Cross-sectional Area of Core $\dfrac{\text{OD - ID}}{2}$

Outside Diameter (OD)

Core Thickness

Inside (ID) Diameter

Powdered-Iron Core:

$$L(\mu H) = \frac{A_L \times N^2}{10{,}000} \qquad N = 100\sqrt{\frac{\text{desired } L \text{ in } \mu H}{A_L \text{ in } \mu H \text{ per } 100 \text{ turns}}}$$

Inductance Index (A_L) for Powdered-Iron Toroids

Size	26	3	15	1	2	7	6	10	12	17	0
T-12	na	60	50	48	20	18	17	12	7.5	7.5	3.0
T-16	145	61	55	44	22	na	19	13	8.0	8.0	3.0
T-20	180	76	65	52	27	24	22	16	10	10	3.5
T-25	235	100	85	70	34	29	27	19	12	12	4.5
T-30	325	140	93	85	43	37	36	25	16	16	6
T-50	320	175	135	100	49	43	40	31	18	18	6.4
T-80	450	180	170	115	55	50	45	32	22	22	8.5
T-106	900	450	345	325	135	133	116	na	na	na	15
T-130	785	350	250	200	110	103	96	na	na	na	15
T-184	1640	720	na	500	240	na	195	na	na	na	na
T-200	895	425	na	250	120	105	100	na	na	na	na

* The units of A_L are in μH per 100 turns.

Ferrite Core:

$$L(\mu H) = \frac{A_L \times N^2}{1{,}000{,}000}$$

$$N = 1000\sqrt{\frac{(\text{desired } L \text{ in mH})}{(A_L \text{ in mH per } 1000 \text{ turns})}}$$

Inductance Index (A_L) for Ferrite Toroids

Size	63/67-Mix	61-Mix	43-Mix	77(72) Mix	J(75) Mix
FT-23	7.9	24.8	188.0	396	980
FT-37	19.7	55.3	420.0	884	2196
FT-50	22.0	68.0	523.0	1100	2715
FT-82	22.4	73.3	557.0	1170	NA
FT-114	25.4	79.3	603.0	1270	3179

* The units of A_L are in mH per 1000 turns.

FIGURE 2.133

equation for ferrite toroids in Fig. 2.133, and inserting $A_L = 68$ for the FT-50-61, and $N = 50$:

$$L(\mu H) = \frac{A_L \times N^2}{1{,}000{,}000} = \frac{68 \times 50^2}{1{,}000{,}000} = 0.17\ \mu H$$

Example 8: How many turns are needed for a 2.2-mH coil if the index of inductance for the ferrite toroid is 188?

Answer: Use

$$N = 1000\sqrt{\frac{(\text{desired } L \text{ in mH})}{(A_L \text{ in mH per } 1000 \text{ turns})}} = 1000\sqrt{\frac{2.2}{188}} = 108 \text{ turns}$$

2.24.9 Understanding the Inductor Equations

$$V_L = L\frac{dI_L}{dt}$$

The preceding inductor equation (which we derived earlier) has some curious properties that may leave you scratching your head. For starters, let's consider the dI_L/dt term. It represents the rate of change in current through the inductor with time.

If there is no change in current flow through the inductor, there is no measured voltage across the inductor. For example, if we assume that a constant dc current has been flowing through an inductor for some time, then dI_L/dt is zero, making V_L zero as well. Thus, under dc conditions the inductor acts like a short—a simple wire.

However, if the current I_L is changing with time (increasing or decreasing), dI_L/dt is no longer zero, and an induced voltage V_L is present across the inductor. For example, consider the current waveform shown in Fig. 2.134. From time interval 0 to 1s, the rate of current change dI_L/dt is 1 A/s—the slope of the line. If the inductance L is 0.1 H, the induced voltage is simply (1 A/s)(0.1 H) = 0.1 V during this time interval—see the lower waveform. During interval 1s to 2s, the current is constant, making the dI_L/dt zero and, hence, the induced voltage zero. During interval 2s to 3s, dI_L/dt is −1 A/s, making the induced voltage equal to (−1 A/s)(0.1 H) = −0.1 V. The induced voltage waveform paints the rest of the picture.

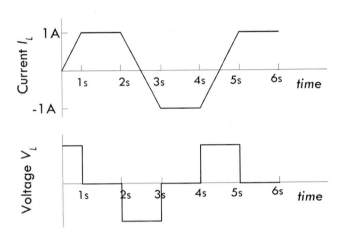

From 0 - 1 and 4 - 5 sec :

$$V = L\frac{dl}{dt} = (0.1\text{H})\left(\frac{1\,\text{A}}{1\,\text{s}}\right) = 0.1\,\text{V}$$

From 2 - 3 sec :

$$V = L\frac{dl}{dt} = (0.1\text{H})\left(-\frac{1\,\text{A}}{1\,\text{s}}\right) = -0.1\,\text{V}$$

From 1 - 2, 3 - 4, 5 - 6 sec

$$V = L\frac{dl}{dt} = (0.1\text{H})\left(\frac{0}{1\,\text{s}}\right) = 0\,\text{V}$$

FIGURE 2.134

Example 9: If the current through a 1-mH inductor is given by the function $2t$, what is the voltage across it?

Answer:

$$V_L = L\frac{dI_L}{dt} = (1\times10^{-3}\,\text{H})\frac{d}{dt}2tA = (1\times10^{-3}\,\text{H})\left(2\frac{A}{s}\right) = 2\times10^{-3}\,\text{V} = 2\,\text{mV}$$

Example 10: The current through a 4-mH inductor is given by $I_L = 3 - 2e^{-10t}$ A. What is the voltage across the inductor?

Answer:

$$V_L = L\frac{dI_L}{dt} = L\frac{d}{dt}[3 - 2e^{-10t}] = L(-2\times-10)e^{-10t} = (4\times10^{-3})(20)e^{-10t} = 0.08e^{-10t}\,\text{V}$$

Example 11: Suppose the current flow through a 1-H inductor decreases from 0.60 A to 0.20 A during a 1-s period. Find the average voltage across the inductor during this period. See how this compared to the average induced voltage if the period is 100 ms, 10 ms, and 1 ms.

Answer: Here we ignore changes between the 1-s intervals, and take averages:

$$V_{AVE} = L\frac{\Delta I}{\Delta t} = 1\,H\frac{0.20\,A - 0.60\,A}{1\,s} = -0.40\,V \qquad \text{(1 s)}$$

$$V_{AVE} = L\frac{\Delta I}{\Delta t} = 1\,H\frac{0.20\,A - 0.60\,A}{0.1\,s} = -4\,V \qquad \text{(100 ms)}$$

$$V_{AVE} = L\frac{\Delta I}{\Delta t} = 1\,H\frac{0.20\,A - 0.60\,A}{0.01\,s} = -40\,V \qquad \text{(10 ms)}$$

$$V_{AVE} = L\frac{\Delta I}{\Delta t} = 1\,H\frac{0.20\,A - 0.60\,A}{0.001\,s} = -400\,V \qquad \text{(1 ms)}$$

$$V_{AVE} = L\frac{\Delta I}{\Delta t} = 1\,H\frac{0.20\,A - 0.60\,A}{0\,s} = -\infty \qquad \text{(Instantaneous)}$$

From Example 11, notice how the induced voltage grows considerably larger as the change in current flow is more abrupt. When the change in current flow is instantaneous, the inductor equation predicts an infinite induced voltage. How can this be?

The answer to this dilemma is explained by the following example. Say you have an ideal inductor attached to a 10-V battery via a switch, as shown in Fig. 2.135. The moment the switch is closed, according to the inductor equation, dI/dt should be infinite (assuming ideal battery, wires, and coil wire). This means that the induced voltage should rise in proportion with the applied voltage, forever, and no current should flow. In other words, the reverse voltage should jump to infinity. (See Fig. 135b.) Likewise, if the switch is opened, as shown in Fig. 135c, an infinite forward voltage is predicted. The answer to this problem is rather subtle, but fundamentally important. As it turns

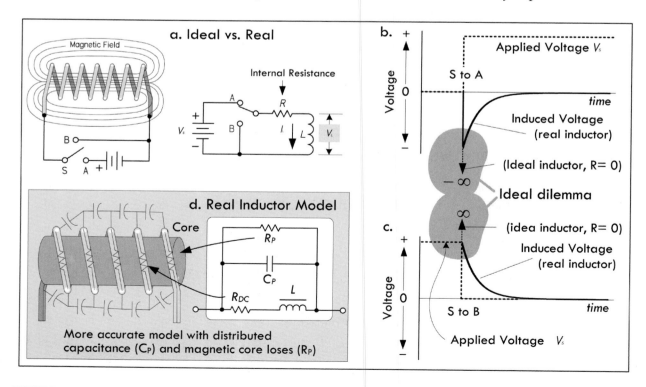

FIGURE 2.135   Real inductors never generate infinite induced voltages.

out, these infinities we predict are never observed in the real world. There is always internal resistance within a real inductor as well as internal capacitance (as well as internal resistance and capacitance in the rest of the circuit). A realistic model of an inductor is shown in Fig. 135d, incorporating internal resistance and capacitance. It is these "imperfections" that explain why the predictions are not observed.

As we can see from the last example, ignoring internal resistance can cause conceptual headaches. You may be wondering why someone didn't simply create an inductor equation that automatically included an internal resistance term. The fact of the matter is that we *should*, especially when considering a simple circuit like that shown in Fig. 2.135. However, it is important to define inductance as a unique quantity and associate its effects with changing magnetic field energy alone, not with resistive heating within the coil or magnetic core losses, or with distributed capacitance between the coil loops. As it turns out, when you start analyzing more complex circuits (say RL circuits and RLC circuits), the discrete resistance present within the circuit will keep the inductor equation from freaking out. In more precise circuits, knowing the internal resistance of the inductor is critical. A standard practice is to represent an inductor as an ideal inductor in series with a resistor R_{DC}, where R_{DC} is called the dc resistance of the inductor. More accurate models throw in parallel capacitance (interloop capacitance) and parallel resistance (representing magnetic core losses), which become important in high-frequency applications.

Before we move on, it should be noted that even though we must assume there is internal resistance within an inductor, it is still possible for induced voltages to reach surprisingly high values during transient conditions. For example, turning off inductive circuits can lead to some dangerously high voltages that can cause arcing and other problems requiring special handling.

Example 12: Suppose you apply a linearly increasing voltage across an ideal 1-H inductor. The initial current through the inductor is 0.5 A, and the voltage ramps from 5 to 10 V over a 10 ms interval. Calculate the current through the inductor as a function of time.

Answer: Kirchhoff's voltage law gives:

$$V_{\text{applied}} - L\frac{dI}{dt} = 0$$

Integrating this gives:

$$\int_{t'=0}^{t} \frac{dI}{dt'}\,dt' = \int_{t'=0}^{t} V_{\text{applied}}\,dt'$$

or:

$$I(t) - I(0) = \int_{t'=0}^{t} (mt' + b)\,dt'$$

Carrying out the integration gives:

$$I(t) = I(0) + \frac{1}{2}mt^2 + bt$$

At $t = 0.01$ s (the end of the ramp), the current is:

$$I(0.01) = 0.5 + \frac{1}{2} \times 500 \times 0.01^2 + 5 \times 0.01 = 0.575 \text{ A}$$

2.24.10 Energizing RL Circuit

When a resistor is placed in series with an inductor, the resistance controls the rate at which energy is pumped into the magnetic field of the inductor (or pumped back into the circuit when the field collapses). When we consider an RL circuit with a dc supply and a switch, as shown in Fig. 2.136, the energizing response that begins the moment the switch is closed ($t = 0$) is illustrated by the voltage and current response curves in Fig. 2.136, as well as the equations to the right.

You can derive the expressions for the energizing response of an RL circuit by applying Kirchhoff's law, summing the voltages around the closed loop:

Energizing Inductor

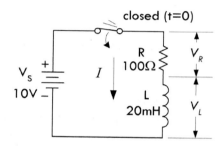

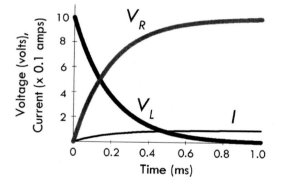

Current and voltage equations for RL energizing circuit:

$$I = \frac{V_s}{R}(1 - e^{-t/(L/R)}), \quad \frac{t}{(L/R)} = -\ln\left(\frac{1 - V_s/R}{V_s}\right)$$

$$V_R = IR = V_s(1 - e^{-t/(L/R)}), \quad \frac{t}{(L/R)} = -\ln\left(\frac{V_R - V_s}{V_s}\right)$$

$$V_L = L\frac{dI}{dt} = V_s e^{-t/(L/R)}, \quad \frac{t}{(L/R)} = -\ln\left(\frac{V_L}{V_s}\right)$$

$\tau = L/R$ time constant

where I is the current in amps, V_s is the source voltage in volts, R is the resistance in ohms, L is the inductance in henrys, t is the time in seconds after the source voltage is applied, $e = 2.718$, V_R is the resistor voltage in volts, and V_L is the inductor voltage in volts. The graph shown to the left is for a circuit with $R = 100\ \Omega$ and $L = 20$ mH.

FIGURE 2.136

$$V_s = IR + L\frac{dI}{dt}$$

Rewriting in the standard form gives:

$$\frac{dI}{dt} + \frac{R}{L}I = \frac{V}{L}$$

After solving this linear, first-order nonhomogeneous differential equation, using the initial condition that the current before the switch was closed was zero $I(0) = 0$, the solution for the current becomes:

$$I = \frac{V_s}{R}(1 - e^{-t/(L/R)})$$

We plug this into Ohm's law to find the voltage across the resistor:

$$V_R = IR = V_S \left(1 - e^{-t/(L/R)}\right)$$

and plug it into the expression for the inductor voltage:

$$V_L = L\frac{dI}{dt} = V_s e^{-t/(L/R)}$$

To understand what's going on within the energizing RL circuit, we first pretend the resistor value is zero. With no resistance, when the switch is closed, the current would increase forever (according to Ohm's law and assuming an ideal voltage source), always growing just fast enough to keep the self-induced voltage equal to the applied voltage.

But when there is resistance in the circuit, the current is limited—Ohm's law defines the value the current can reach. The reverse voltage generated in L must only equal the difference between the applied voltage and the drop across R. This difference becomes smaller as the current approaches the final Ohm's law value. Theoretically, the reverse voltage never quite disappears, and so the current never quite reaches the no-inductor value. In practical terms, the differences become immeasurable after a short duration.

The time in seconds required for the current to build up to 63.2 percent of the maximum value is called the time constant, and is equal to L/R. After each time interval equal to this constant, the circuit conducts an additional 63.2 percent of the remaining current. This behavior is graphed in Fig. 2.137. As is the case with capacitors, after five time constants the current is considered to have reached its maximum value.

How the Size of Inductance Affects Energizing Current Response

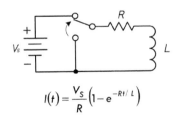

$$I(t) = \frac{V_S}{R}\left(1 - e^{-Rt/L}\right)$$

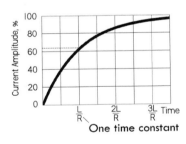

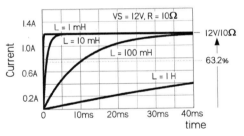

FIGURE 2.137

Example 13: If an RL circuit has an inductor of 10 mH and a series resistor of 10 Ω, how long will it take for the current in the circuit to reach full value after power is applied?

Answer: Since reaching maximum current takes approximately five time constants, $t = 5\tau = 5\,(L/R) = 5\,(10 \times 10^{-3}\text{ H})/10\text{ }\Omega = 5.0 \times 10^{-3}$ s or 5.0 ms.

Note that if the inductance is increased to 1.0 H, the required time increases to 0.5 s. Since the circuit resistance didn't change, the final current is the same for both cases in this example. Increasing inductance increases the time required to reach full current. Figure 2.137 shows current response curves for various inductors with the same resistance.

2.24.11 *Deenergizing RL Circuit*

Unlike a capacitor that can store energy in the form of an electric field when an applied voltage is cut (say, by means of a switch), an inductor does not remain "charged," or energized, since its magnetic field collapses as soon as current ceases. The energy stored in the magnetic field returns to the circuit. Now, when current flow is cut (say, by means of a switch), predicting the current flow and voltage drops within an RL circuit is a bit tricky. We can say that the instant the switch is opened, a rapid collapse of the magnetic field induces a voltage that is usually many times larger than the applied voltage, since the induced voltage is proportional to the rate at which the field changes. The common result of opening the switch in such a circuit is that a spark or arc forms at the switch contacts during the instant the switch opens—see Fig. 2.138a. When the inductance is large and the current in the circuit is high, large amounts of energy are released in a very short time. It is not unusual for the switch contacts to burn or melt under such circumstances. The spark or arc at the opened switch can be reduced or suppressed by connecting a suitable capacitor and resistor in series across the contacts. Such an RC combination is called a *snubber network.* Transistor switches connected to large inductive loads, such as relays and solenoids, also require protection. In most cases, a small power diode connected in reverse across the relay coil will prevent field-collapse currents from harming the transistor—more on this in a moment.

If the excitation is removed without breaking the circuit, as theoretically diagrammed in Fig. 2.138b, the current will decay according to the following waveforms and equations.

You can derive the expressions for a deenergizing RL circuit by applying Kirchhoff's law, summing the voltage around the closed loop:

$$V_s = IR + L\frac{dI}{dt} = 0$$

V_s is 0 because the battery is no longer in the circuit. Rewriting in the standard form gives:

$$\frac{dI}{dt} + \frac{R}{L}I = \frac{V}{L} = 0$$

After solving this linear, first-order nonhomogeneous differential equation, using the initial condition that the current before the switch was closed was simply:

$$I(0) = \frac{V_R}{R}$$

the solution for the current becomes:

$$I = \frac{V_s}{R}e^{-t/(L/R)}$$

We plug this into Ohm's law to find the voltage across the resistor:

$$V_R = IR = V_s e^{-t/(L/R)}$$

and plug it into the expression for the inductor voltage:

$$V_L = L\frac{dI}{dt} = -V_s e^{-t/(L/R)}$$

As with the energizing RL circuit, the deenergizing RL circuit current response can be modeled in terms of time constants. After five time constants, the inductor is considered fully deenergized. Increasing the inductance increases this time, as shown in Fig. 2.139.

A. Break in circuit

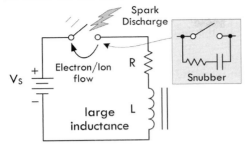

B. Assuming no break in circuit

In A, a break in current flow caused by opening a switch will cause a large inductive load's magnetic field to collapse, thus generating a large forward voltage. This voltage can be so big that the "electron pressure" between the switch contacts becomes so large that electrons escape the metal surface of one switch contact and jump toward the other contact. As the liberated electrons make the jump, they collide with airborne molecules, causing ionization reactions, which lead to a spark discharge across the switch contacts. The current and voltage response curves under such conditions are rather complex.

In B, if we remove the applied voltage but prevent a break in the circuit by moving the discharge to the ground (switch thrown to position B), we get predictable current and voltage expressions:

$$I = \frac{V_s}{R}e^{-t/(L/R)}), \quad \frac{t}{(L/R)} = -\ln\left(\frac{I \times R}{V_s}\right)$$

$$V_R = IR = V_s e^{-t/(L/R)}, \quad \frac{t}{(L/R)} = -\ln\left(\frac{V_R}{V_s}\right)$$

$$V_L = L\frac{dI}{dt} = V_s e^{-t/(L/R)}, \quad \frac{t}{(L/R)} = -\ln\left(-\frac{V_L}{V_s}\right)$$

$\tau = L/R$ **time constant**

where I is the current in amps, V_s is the source voltage in volts, R is the resistance in ohms, L is the inductance in henrys, t is the time in seconds after the source voltage is applied, $e = 2.718$, V_R is the resistor voltage in volts, and V_L is the inductor voltage in volts.

The graph shown in the figure is for a circuit with $R = 100\ \Omega$ and $L = 20$ mH.

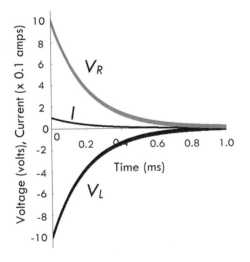

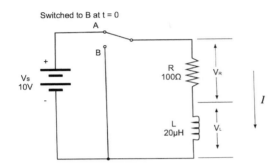

FIGURE 2.138

How the Size of Inductance Affects Deenergizing Current Response

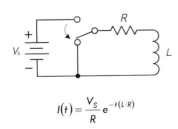

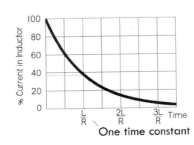

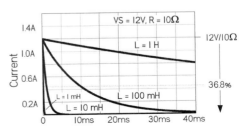

FIGURE 2.139

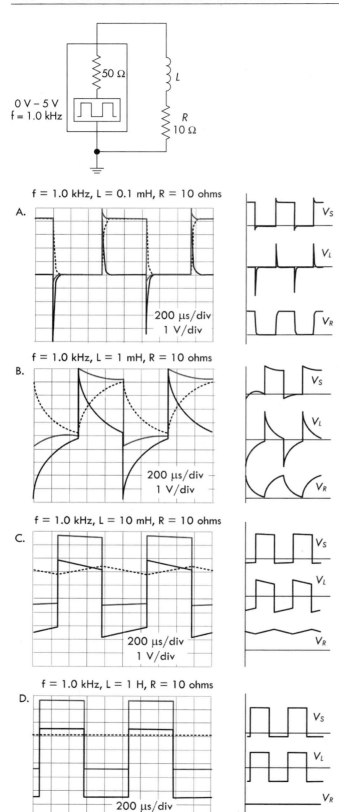

f = 1.0 kHz, L = 0.1 mH, R = 10 ohms

A.

200 μs/div
1 V/div

f = 1.0 kHz, L = 1 mH, R = 10 ohms

B.

200 μs/div
1 V/div

f = 1.0 kHz, L = 10 mH, R = 10 ohms

C.

200 μs/div
1 V/div

f = 1.0 kHz, L = 1 H, R = 10 ohms

D.

200 μs/div
1 V/div

FIGURE 2.140

Inductance, be it intended or not, can have a major influence on signals. For example, the output signals across the inductor and resistor in the RL circuit to the left become increasingly distorted as the inductance increases. With a constant 1.0-kHz, 0–5-V squarewave source, and fixed 10-Ω resistance, we increase the inductance and note the changes in the waveforms. First we note that the period of the squarewave is:

$$\text{Period: } T = \frac{1}{f} = \frac{1}{1000 \text{ Hz}} = 1 \text{ ms}$$

Now let's see how the waveforms change as we choose increasing inductance values.

Graph A: Inductance: $L = 0.1$ mH

Time constant: $\tau = 0.0001 \text{ H}/10 \text{ }\Omega = 0.01$ ms

Here the RL time constant is 1 percent of the period, so the induced voltage spikes are narrow during squarewave high-to-low and low-to-high transitions. Considering that an inductor is fully energized or deenergized after five time constants or, in this case, 0.05 ms, the inductor easily completely energizes and deenergizes during a half period of 0.5 ms. The voltage across the resistor is slightly rounded at the edges as a result.

Graph B: Inductance: $L = 1$ mH

Time constant: $\tau = 0.001 \text{ H}/10 \text{ }\Omega = 0.1$ ms

The RL time constant is 10 percent of the period, so induced voltage and the effects of exponential rise and fall during source voltage transitions are clearly visible. To fully energize or deenergize requires five time constants, or 0.5 ms, which is exactly equal to the half period of the signal. Hence, the magnetic field about the inductor is capable of absorbing and giving up all its magnetic field energy during each consecutive half cycle.

Graph C: Inductance: $L = 10$ mH

Time constant: $\tau = 0.01 \text{ H}/10 \text{ }\Omega = 1$ ms

The RL time constant is equal to the period of the squarewave. However, since it requires five time constants, or 5 ms, to be fully energized or deenergized, the resulting voltages appear linear—you get to see only a small portion of the exponential rise or fall. The magnetic field about the inductor isn't able to absorb or give up all its energy during a half cycle.

Graph D: Inductance: $L = 1$ H

Time constant: $\tau = 1 \text{ H}/10 \text{ }\Omega = 0.1$ s

The RL time constant is 100 times as large as the period of the squarewave. Again, to become fully energized or deenergized requires five time constants, or 500 times the period. This means that there is practically no time for the inductor to fully energize or deenergize. Though there is exponential rise and decay, practically, you see only the first ¹⁄₅₀₀ portion, which appears linear.

2.24.12 Voltage Spikes Due to Switching

Inductive voltage spikes are common in circuits where large inductive loads, such as relays, solenoids, and motors, are turned on and off by a mechanical or transistor-like switch. Spikes as high as a couple hundred volts are possible, even when the supply voltages are relatively small. Depending on the circuit design, these voltage spikes can cause arcing, leading to switch contact degradation, or damage to transistor or other integrated switching devices. Figure 2.141 shows a diode (a device that acts as a one-way gate to current flow) placed across the coil of a relay to provide a "pressure release" path for the inductive spike, should the supply voltage circuit be broken.

Transient Protection

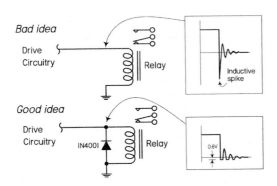

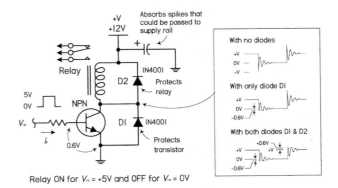

FIGURE 2.141

2.24.13 Straight-Wire Inductance

Every conductor passing current has a magnetic field associated with it and, therefore, inductance—the conductor need not be formed into a coil. For example, a straight wire has an inductance associated with it. This is attributed to the average alignment of magnetic fields of individual free electrons as they obtain a drift component under an applied EMF. The inductance of a straight nonmagnetic wire or rod in free space is given by:

$$L = 0.00508b \left[\ln\left(\frac{2b}{a}\right) - 0.75 \right] \tag{2.56}$$

where L is the inductance in μH, a is the wire radius in inches, b is the length in inches, and ln is the natural logarithm.

Example 14: Find the inductance of a #18 wire (diameter = 0.0403 in) that is 4 in long.

Answer: Here $a = 0.0201$ and $b = 4$, so:

$$L = 0.00508(4) \left[\ln\left(\frac{8}{0.0201}\right) - 0.75 \right] = 0.106 \ \mu\text{H}$$

Skin effects change Eq. 2.56 slightly at VHF (30–300 MHz) and above. As the frequency approaches infinity, the constant 0.75 in the preceding equation approaches 1.

Straight wire inductance is quite small, and is typically referred to as parasitic inductance. We can apply the concept of reactance to inductors, as we did with capacitive reactance earlier in this chapter. Parasitic inductive reactance at low frequencies (AF to LF) is practically zero. For instance, in our example, at 10 MHz the 0.106-μH inductance provides only 6.6 Ω of reactance. However, if we consider a frequency of 300 MHz, the inductive reactance goes to 200 Ω—a potential problem. For this reason, when designing circuits for VHF and above it is important to keep component leads as short as possible (capacitor leads, resistor leads, etc.). Parasitic inductance in a component is modeled by adding an inductor of appropriate value in series with the component (since wire lengths are in series with the component).

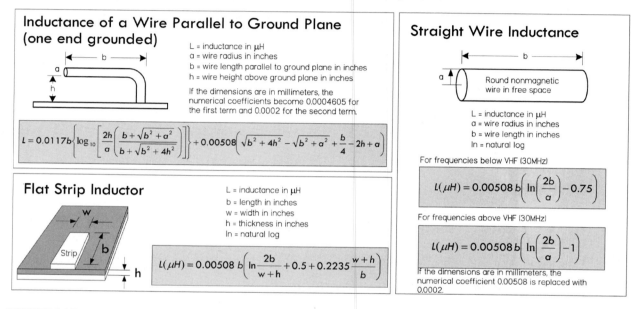

Inductance of a Wire Parallel to Ground Plane (one end grounded)

L = inductance in μH
a = wire radius in inches
b = wire length parallel to ground plane in inches
h = wire height above ground plane in inches

If the dimensions are in millimeters, the numerical coefficients become 0.0004605 for the first term and 0.0002 for the second term.

$$L = 0.0117b\left\{\log_{10}\left[\frac{2h}{a}\left(\frac{b+\sqrt{b^2+a^2}}{b+\sqrt{b^2+4h^2}}\right)\right]\right\} + 0.00508\left(\sqrt{b^2+4h^2} - \sqrt{b^2+a^2} + \frac{b}{4} - 2h + a\right)$$

Straight Wire Inductance

Round nonmagnetic wire in free space

L = inductance in μH
a = wire radius in inches
b = wire length in inches
ln = natural log

For frequencies below VHF (30MHz)

$$L(\mu H) = 0.00508\, b\left(\ln\left(\frac{2b}{a}\right) - 0.75\right)$$

For frequencies above VHF (30MHz)

$$L(\mu H) = 0.00508\, b\left(\ln\left(\frac{2b}{a}\right) - 1\right)$$

If the dimensions are in millimeters, the numerical coefficient 0.00508 is replaced with 0.0002.

Flat Strip Inductor

L = inductance in μH
b = length in inches
w = width in inches
h = thickness in inches
ln = natural log

$$L(\mu H) = 0.00508\, b\left(\ln\frac{2b}{w+h} + 0.5 + 0.2235\frac{w+h}{b}\right)$$

FIGURE 2.142

2.24.14   *Mutual Inductance and Magnetic Coupling*

When two inductor coils are placed near each other with their axes aligned, the current applied through coil 1 creates a magnetic field that propagates through coil 2. See Fig. 2.143.

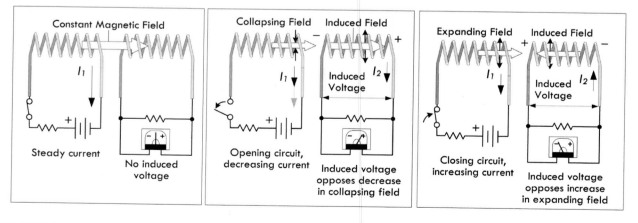

FIGURE 2.143

As a result, a voltage is induced in coil 2 whenever the field strength of coil 1 changes. The voltage induced in coil 2 is similar to the voltage of self-induction, but since it acts upon the external coil 2, it is called mutual induction—the two coils are said to be inductively coupled. The closer the coils are together, the greater the mutual inductance. If the coils are farther apart or are aligned off axis, the mutual inductance is relatively small—the coils are said to be loosely coupled. The ratio of actual mutual inductance to the maximum possible mutual inductance is called the coefficient of coupling, which is usually given as a percentage. The coefficient with air core coils may run as high as 0.6 to 0.7 if one coil is wound over the other, but much less if the two coils are separated. It is possible to achieve near 100 percent coupling only when the coils are wound on a closed magnetic core—a scenario used in transformer design. Mutual inductance also has an undesirable consequence within circuit design, where unwanted induced voltages are injected into circuits by neighboring components or by external magnetic field fluctuations generated by inductive loads or high-current alternating cables.

2.24.15 Unwanted Coupling: Spikes, Lightning, and Other Pulses

There are many phenomena, both natural and man-made, that generate sufficiently large magnetic fields capable of inducing voltages in wires leading into and out of electrical equipment. A mutual inductance exists between the external source and the affected circuit in question. Parallel-wire cables linking elements of electronic equipment consist of long wires in close proximity to each other. Signal pulses can couple both magnetically and capacitively from one wire to another. Since the magnetic field of a changing current decreases as the square of distance, separating the signal-carrying lines diminishes inductive coupling. Unless they are well shielded and filtered, however, the lines are still susceptible to the inductive coupling of pulses from other sources. This is often experienced when using a long ground lead for a scope probe—external magnetic interference can couple to the ground lead of the probe and appear within the displayed signal as unwanted noise. Such coupling between external sources and electrical equipment can be quite problematic—even more so when the source generates a "burst" in magnetic field strength. Sudden bursts have a tendency to induce high-level voltage spikes onto ac and dc power lines, which can migrate into the inner circuitry where sensitive components lay prone to damage. For example, lightning in the vicinity of the equipment can induce voltages on power lines and other conductive paths (even ground conductors) that lead to the equipment location. Lightning that may appear some distance away can still induce large spikes on power lines that ultimately lead to equipment. Heavy equipment with electrical motors can induce significant spikes into power lines within the equipment location. Even though the power lines are straight, the powerful magnetic fields of a spike source can induce damaging voltages on equipment left plugged in during electrical storms or during the operation of heavy equipment that inadequately filters its spikes.

2.24.16 Inductors in Series and Parallel

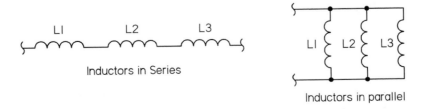

Inductors in Series

Inductors in parallel

FIGURE 2.144

When two or more inductors are connected in series, the total inductance is equal to the sum of the individual inductances, provided the coils are sufficiently separated, so that coils are not in the magnetic field of one another:

$$L_{tot} = L_1 + L_2 + L_3 + \ldots + L_N \qquad \text{(2.57) Inductors in series}$$

You can derive the inductors-in-series formula by applying Kirchhoff's voltage law. Taking the voltage drop across L_1 to be $L_1 dI/dt$ and the voltage drop across L_2 to be $L_2 dI/dt$, and across L_3 to be $L_3 dI/dt$, you get the following expression:

$$V = L_1 \frac{dI}{dt} + L_2 \frac{dI}{dt} + L_3 \frac{dI}{dt} = (L_1 + L_2 + L_3) \frac{dI}{dt}$$

$L_1 + L_2 + L_3$ is called the equivalent inductance for three inductors in series.

If inductors are connected in parallel, and if the coils are separated sufficiently, the total inductance is given by:

$$\frac{1}{L_{tot}} = \frac{1}{L_1} + \frac{1}{L_2} + \frac{1}{L_3} + \cdots + \frac{1}{L_N} \qquad \text{(2.58) Inductors in parallel}$$

When only two inductors are in parallel, the formula simplifies to $L_{tot} = (L_1 \times L_2)/(L_1 + L_2)$.

You derive this by applying Kirchhoff's current law to the junction, which gives $I = I_1 + I_2 + I_3$, making use of the fact that the voltage V is the same across L_1, L_2, and L_3. Thus I_1 becomes $1/L_1 \int V dt$, I_2 becomes $1/L_2 \int V dt$, and I_3 becomes $1/L_3 \int V dt$. The final expression for I is:

$$I = \frac{1}{L_1} \int V dt + \frac{1}{L_2} \int V dt + \frac{1}{L_3} \int V dt = \left(\frac{1}{L_1} + \frac{1}{L_2} + \frac{1}{L_3} \right) \int V dt$$

$1/L_1 + 1/L_2 + 1/L_3$ is called the reciprocal equivalent inductance of three inductors in parallel.

Example 15: What value of L_2 is required to make the total equivalent inductance of the circuit shown in Fig. 2.145 equal to 70 mH?

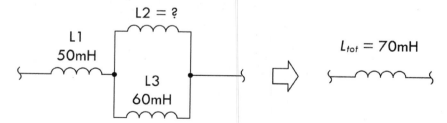

FIGURE 2.145

Answer: 30 mH; use $L_{tot} = L_1 + \dfrac{L_1 \times L_2}{L_1 + L_2}$

2.24.17 Alternating Current and Inductors

When an alternating voltage is applied to an ideal inductance, the current that flows through the inductor is said to lag the applied voltage by 90°. Or if you like, the applied voltage leads the current by 90°. (This is exactly the opposite of what we saw with capacitors under ac.) The primary cause for the current lag in an inductor is due to the reverse voltage generated in the inductance. The amplitude of the reverse voltage is proportional to the rate at which the current changes. This can be demonstrated by the graph in Fig. 2.146. If we start at time segment 0A, when the applied voltage is at its positive maximum, the reverse or induced voltage is also at a maximum, allowing the least current to flow. The rate at which the current is changing is the highest, a 38 percent change in the time period 0A. In the segment AB, the current changes by only 33 percent, yielding a reduced level of induced voltage, which is in step with the decrease in the applied voltage. The process continues in time segments BC and CD, the latter producing only an 8 percent rise in current as the applied and induced voltage approach zero.

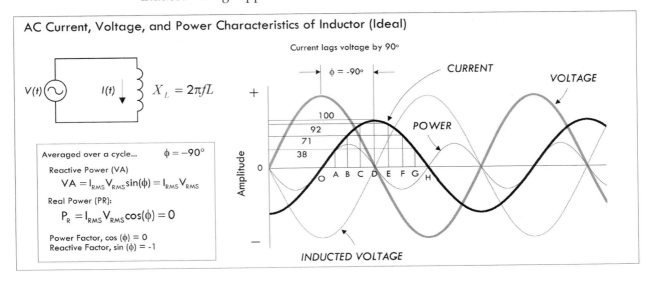

AC Current, Voltage, and Power Characteristics of Inductor (Ideal)

FIGURE 2.146

In interval DE, the applied voltage changes direction. The induced voltage also changes direction, returning current to the circuit as the magnetic field collapses. The direction of this current is now opposite to the applied voltage, which sustains the current in the positive direction. As the applied voltage continues to increase negatively, the current—although positive—decreases in value, reaching zero as the applied voltage reaches its negative maximum. The negative half cycle continues just as did the positive half cycle. Thus, we say that within a pure inductive ac circuit, the current lags the voltage by 90°.

2.24.18 Inductive Reactance

The amplitude of alternating current in an inductor is inversely proportional to the applied frequency. Since the reverse voltage is directly proportional to inductance for a given rate of current change, the current is inversely proportional to inductance for a given applied voltage and frequency. The combined effect of inductance and frequency is called inductive reactance. Like capacitive reactance, inductive reactance is expressed in ohms. The inductive reactance is given by the following formula:

$$X_L = 2\pi f L \qquad \text{(2.59) Inductive reactance}$$

where X_L is inductive reactance, $\pi = 3.1416$, f is frequency in Hz, and L is inductance in henrys. Inductive reactance, in angular form (where $\omega = 2\pi f$), is equal to:

$$X_L = \omega L.$$

You derive the expression for inductive reactance by connecting an inductor to a sinusoidal voltage source. To make the calculations more straightforward, we'll use a cosine function instead of a sine function—there is really no difference here. For example, if the source voltage is given by $V_0 \cos(\omega t)$, the current through the inductor becomes

$$I = \frac{1}{L}\int V dt = \frac{1}{L}\int V_0 \cos(\omega t)dt = \frac{V_0}{\omega L}\sin(\omega t)$$

the maximum current or peak current through an inductor occurs when $\sin(\omega t) = 1$, at which point it is equal to:

$$I_0 = \frac{V_0}{\omega L}$$

The ratio of peak voltage to peak current resembles a resistance and has units of ohms. However, because the physical phenomenon doing the "resisting" (e.g., reverse induced voltage working against forward voltage) is different from a resistor (heating), the effect is given a new name, inductive reactance:

$$X_L = \frac{V_0}{I_0} = \frac{V_0}{V_0/\omega L} = \omega L$$

As ω goes to infinity, X_L goes to infinity, and the inductor acts like an open circuit (inductors do not like to pass high-frequency signals). However, as ω goes to 0, the X_L goes to zero (inductors have an easier time passing low-frequency signals and ideally present no "resistance" to dc signals).

Figure 2.147 shows a graph of inductive reactance versus frequency for 1-μH, 10-μH, and 100-μH inductors. Notice the response is linear—increasing the frequency increases the reactance in proportion. However, in real inductors, the reactive response is a bit more complicated, since real inductors have parasitic resistance and capacitance built in. Figure 2.147 shows a real-life impedance versus frequency graph.

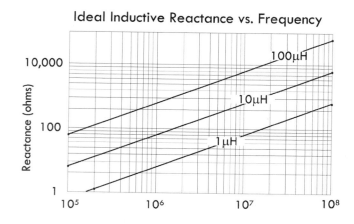

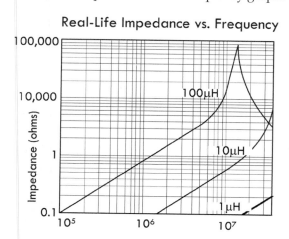

FIGURE 2.147

Notice that when the frequency approaches what is called the resonant frequency, the impedance is no longer linear in appearance, but peaks and falls. (This will make more sense when we cover resonant circuits a bit later.)

Example 16: What is the reactance of an ideal 100-μH coil with applied frequencies of 120 Hz and 15 MHz?

Answer:

120 Hz: $X_L = 2\pi f L = 2\pi(120 \text{ Hz})(100 \times 10^{-6} \text{ H}) = 0.075 \ \Omega$

15 MHz: $X_L = 2\pi f L = 2\pi(15 \times 10^6 \text{ Hz})(100 \times 10^{-6} \text{ H}) = 9425 \ \Omega$

Example 17: What inductance provides 100 Ω of reactance with an applied frequency of 100 MHz?

Answer:

$$L = \frac{X_L}{2\pi f} = \frac{100 \ \Omega}{2\pi(100 \times 10^6 \text{ Hz})} = 0.16 \ \mu\text{H}$$

Example 18: At what frequency will the reactance of a 1-μH inductor reach 2000 Ω?

Answer:

$$f = \frac{X_L}{2\pi L} = \frac{2000 \ \Omega}{2\pi(1 \times 10^{-6} \text{ H})} = 318.3 \text{ MHz}$$

By the way, inductive reactance has an inverse called *inductive susceptance*, given by the following expression:

$$B = \frac{1}{X_L} \tag{2.60}$$

Inductive susceptance uses units of siemens, abbreviated S (S = 1/Ω). It simply tells you "how well" an inductor passes current, as opposed to "how bad" it does—as reactance implies.

2.24.19 Nonideal Inductor Model

Though the ideal inductor model and associated ideal equations are fundamentally important in circuit analysis, using them blindly without consideration of real inductor imperfections, such as internal resistance and capacitance, will yield inaccurate results. When designing critical devices, such as high-frequency filters used in radio-frequency receivers, you must use a more accurate real-life inductor model.

In practice, a real inductor can be modeled by four passive ideal elements: a series inductor (L), a series resistor R_{DC}, a parallel capacitor C_P, and a parallel resistor R_P. R_{DC} represents the dc resistance, or the measured resistance drop when a dc current passes through the inductor. Manufacturers provide the dc resistance of their inductors on their specification sheets (e.g., 1900 series 100-μH inductor with an R_{DC} of 0.0065 Ω). R_P represents magnetic core losses and is derived from the self-resonant frequency f_0—the point at which the reactance of the inductor is zero (i.e., the impedance is purely resistive). It can be calculated from the quality factor Q, as we'll see in a moment. The parallel resistor limits the simulated self-resonance from rising to infinity. C_P represents the distributed capacitance that exists between coils and leads within the inductor, as shown in Fig. 2.148. When a voltage changes due to ac current passing through a coil, the effect is that of many small capacitors acting in parallel with the inductance of the coil. The graph in Fig. 2.148 shows how this distributed capacitance resonates with the inductance. Below resonance, the reactance is inductive, but it increases as the frequency increases. Above resonance, the reactance is capacitive and decreases with frequency.

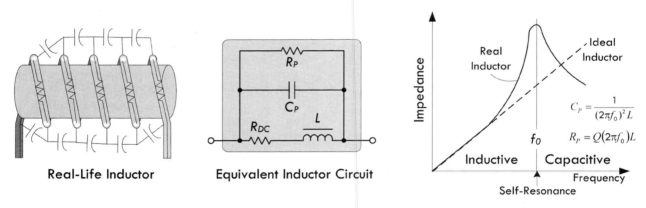

FIGURE 2.148   Inductors exhibit distributed capacitance, as explained in the text. The graph shows how this distributed capacitance resonates with the inductance. Below resonance, the reactance is inductive, but it decreases as frequency increases. Above resonance, the reactance is capacitive and increases with frequency.

Inductors are subject to many types of electrical energy losses, however—wire resistance, core losses, and skin effect. All electrical conductors have some resistance through which electrical energy is lost as heat. Moreover, inductor wire must be sized to handle the anticipated current through the coil. Wire conductors suffer additional ac losses because alternating current tends to flow on the conductors' surface. As the frequency increases, the current is confined to a thinner layer on the conductor surface. This property is called the *skin effect*. If the inductor's core is a conductive material, such as iron, ferrite, or brass, the core will introduce additional losses of energy.

2.24.20 *Quality Factor*

Components that store energy, such as capacitors and inductors, can be described in terms of a *quality factor Q*. The *Q* of such a component is a ratio of its ability to store energy to the total of all energy losses within the component. In essence, the ratio reduces to: $Q = X/R$, where *Q* is the quality factor (no units), *X* is the reactance (inductive or capacitive), and *R* is the sum of all resistance associated with the real energy losses within the component, given in ohms.

For a capacitor, *Q* is ordinarily high, with quality ceramic capacitors obtaining values of 1200 or more. Small ceramic trimmer capacitors may have *Q* values that are so small that they shouldn't be ignored in certain applications.

The quality factor for an inductor is given by $Q = 2\pi fL/R_{DC}$. The *Q* value for an inductor rarely if ever approaches capacitor *Q* in a circuit where both components work together. Although many circuits call for the highest *Q* inductor obtainable, other circuits call for a specific *Q*, which may, in fact, be very low.

Inductive Divider

Inductive dividers can be used with ac input signals. A dc input voltage would split according to the relative resistances of the two inductors by using the resistive voltage divider. The formula for determining the ac output voltage of an inductive divider (provided the inductors are separated—that is, not wound on the same core—and have no mutual inductance) is shown in Fig. 2.149.

$$V_{out} = \frac{L_2}{L_1 + L_2} V_{in}$$

$$V_{out} = \frac{50\text{mH}}{150\text{mH}}(10 \text{ VAC}) = 3.33 \text{ VAC}$$

FIGURE 2.149

Note that the output voltage is independent of the input frequency. However, if the reactance of the inductors is not high at the frequency of operation, (i.e., the inductance is not large enough), there will be a very large current drawn by the shunt element (L_2).

2.24.21 Inductor Applications

Inductors' basic function in electronics is to store electrical energy in a magnetic field. Inductors are used extensively in analog circuits and signal processing, including radio reception and broadcasting. Inductors, in conjunction with capacitors and other components, can form electric filters used to filter out specific signal frequencies. Two (or more) coupled inductors form a transformer that is used to step up or step down an ac voltage. An inductor can be used as the energy storage device in a switching regulator power supply—the inductor is charged for a specific fraction of the regulator's switching frequency and discharged for the remainder of the cycle. This charge/discharge ratio determines the output-to-input voltage ratio. Inductors are also employed in electrical transmission systems, where they are used to intentionally depress system voltages or limit fault current. In this field, they are more commonly referred to as reactors.

2.25 Modeling Complex Circuits

As a note, let us tell you that this section is designed to scare you. You may have a hard time understanding some parts of this section if you don't have a decent math background. However, this section is worth reading for theoretical footing and, more important, because it stresses the need to come up with alternative tricks to avoid the nasty math.

Theoretically, given enough parameters, any complex electric circuit can be modeled in terms of equations. In other words, Kirchhoff's laws always hold, whether our circuit consists of linear or nonlinear elements. *Linear devices* have responses that are proportional to the applied signal. For example, doubling the voltage across a resistor doubles the current through it. With a capacitor, doubling the frequency of the applied voltage across it doubles the current through it. With an inductor, doubling the frequency of voltage across it halves the current through it. We found that we could apply the following equations to model resistors, capacitors, and inductors:

$$V_R = IR, \; I_R = \frac{V_R}{R}$$

$$V_C = \frac{1}{C}\int I dt, \; I_C = C\frac{dV_C}{dt}$$

$$V_L = L\frac{dI}{dt}, \; I_L = \frac{1}{L}\int V_L dt$$

In terms of voltage and current sources, we have, up until now, mainly discussed dc and sinusoidal sources, which we can express mathematically as:

$$V_S = \text{Constant}, \; I_S = \text{Constant}, \; V_S = V_0 \sin(\omega t), \; I_S = I_0 \sin(\omega t)$$

If a circuit contains only resistors, capacitors, inductors, and one of these sources, we simply apply Kirchhoff's laws and come up with an equation or set of equations that accurately describes how the voltages and currents within our circuit will behave with time. *Linear dc circuits* are described by linear algebraic equations. On the other hand, *linear time-dependent circuits* are described by linear differential equations. The time dependence may be a result of a sinusoidal source, or it can simply be a dc source that is turned on or off abruptly—this is referred to as a transient.

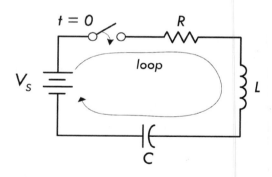

Kirchhoff's Voltage Law:

$$V_S - L\frac{dI}{dt} - RI - \frac{1}{C}\int I dt = 0$$

FIGURE 2.150

As an example, if we have a series RLC circuit (see Fig. 2.150) with a dc source V_S, we can write Kirchhoff's voltage loop equations as:

$$V_S - L\frac{dI}{dt} - RI - \frac{1}{C}\int I dt = 0$$

This equation isn't of any practical use at this stage. Mathematically, we need to simplify it in order to get rid of the integral. You'd start out by first differentiating everything with respect to time, then rearrange things to look like this:

$$L\frac{d^2 I}{dt^2} + R\frac{dI}{dt} + \frac{1}{C}I = 0$$

This is an example of a linear second-order homogeneous differential equation. To solve it requires some mathematical tricks and defining initial conditions—at which point the switch is turned on or off.

Now, let's take the same RLC circuit, but remove the switch and dc supply and insert a sinusoidal supply (see Fig. 2.151). The supply is expressed mathematically as $V_0 \cos(\omega t)$.

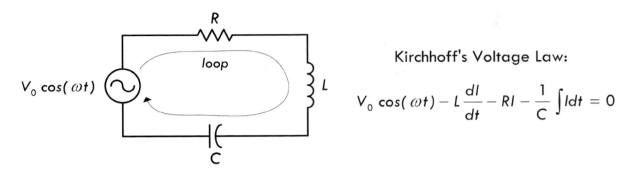

FIGURE 2.151

Applying Kirchhoff's voltage equation:

$$V_0 \cos(\omega t) - L\frac{dI}{dt} - RI - \frac{1}{C}\int I\,dt = 0$$

or,

$$L\frac{dI}{dt} + RI + \frac{1}{C}\int I\,dt = V_0 \cos(\omega t)$$

Again, we must simplify it to get rid of the integral:

$$L\frac{d^2I}{dt^2} + R\frac{dI}{dt} + \frac{1}{C}I = \omega V_0 \sin(\omega t)$$

This expression is a linear second-order nonhomogeneous differential equation. To find the solution to this equation, you could apply, say, the technique of variation of parameters or the method of undetermined coefficients. After the solution for the current is found, finding the voltages across the resistor, capacitor, and inductor is a simple matter of plugging the current into the characteristic voltage/current equation for that particular component. However, coming up with the solution for the current in this case is not easy because it requires advanced math.

As you can see, things don't look promising, mathematically speaking. Things get even worse when we start incorporating sources that are nonsinusoidal, such as a squarewave source or a triangle wave source. For example, how do we mathematically express a squarewave voltage source? As it turns out, the simplest way is to use the following Fourier series:

$$V(t) = \frac{4V_0}{\pi} \sum_{\substack{n=-\infty \\ n=\text{odd}}}^{\infty} \frac{\sin n\omega_0 t}{n}$$

where V_0 is the peak voltage of the squarewave. If we take our RLC circuit and attach this squarewave voltage source, then apply Kirchhoff's voltage law around the loop, we get:

$$V\frac{dI}{dt} + RI + \frac{I}{C}\int I\,dt = \frac{4V_0}{\pi} \sum_{\substack{n=-\infty \\ n=\text{odd}}}^{\infty} \frac{\sin n\omega_0 t}{n}$$

As you might guess, the solution to this equation isn't trivial.

There are other sources we haven't considered yet, such as nonsinusoidal nonrepetitive sources (impulse, etc.). And things, of course, get worse when you consider circuits with more than three linear elements. Let's not forget nonlinear devices, like diodes and transistors, which we haven't even discussed yet.

When circuits get complex and the voltage and current sources start looking weird, setting up Kirchhoff's equations and solving them can require fairly sophisticated mathematics. There are a number of tricks used in electronic analysis to prevent the math from getting out of hand, but there are situations where avoiding the nasty math is impossible. The comments in Fig. 2.152 should give you a feeling for the difficulties ahead.

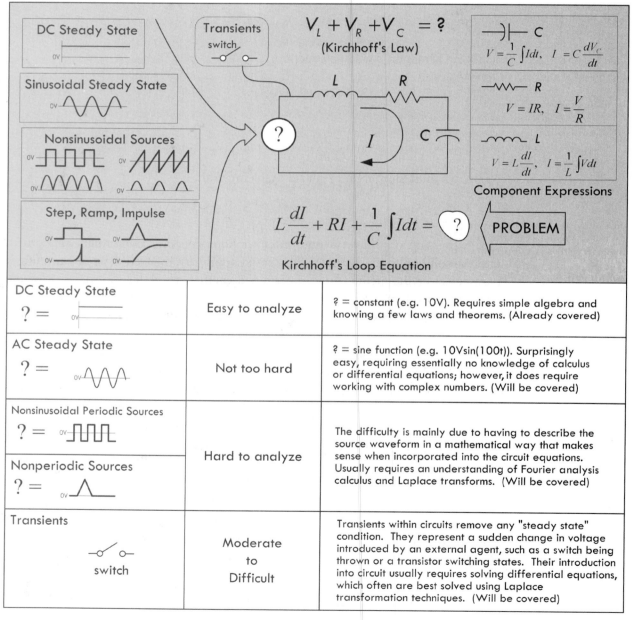

DC Steady State	Easy to analyze	? = constant (e.g. 10V). Requires simple algebra and knowing a few laws and theorems. (Already covered)
AC Steady State	Not too hard	? = sine function (e.g. 10Vsin(100t)). Surprisingly easy, requiring essentially no knowledge of calculus or differential equations; however, it does require working with complex numbers. (Will be covered)
Nonsinusoidal Periodic Sources / Nonperiodic Sources	Hard to analyze	The difficulty is mainly due to having to describe the source waveform in a mathematical way that makes sense when incorporated into the circuit equations. Usually requires an understanding of Fourier analysis calculus and Laplace transforms. (Will be covered)
Transients / switch	Moderate to Difficult	Transients within circuits remove any "steady state" condition. They represent a sudden change in voltage introduced by an external agent, such as a switch being thrown or a transistor switching states. Their introduction into circuit usually requires solving differential equations, which often are best solved using Laplace transformation techniques. (Will be covered)

FIGURE 2.152

In the next section, we'll discuss complex numbers. Complex numbers, along with a concept known as complex impedances, are tricks that we'll use to avoid setting up complex differential equations, at least under special circumstances.

2.26 Complex Numbers

Before we touch upon the techniques used to analyze sinusoidally driven circuits, a quick review of complex numbers is helpful. As you will see in a moment, a sinusoidal circuit shares a unique trait with a complex number. By applying some tricks, you will be able to model and solve sinusoidal circuit problems using complex numbers and the arithmetic that goes with it, and—this is the important part—you will be able to avoid differential equations in the process.

A complex number consists of two parts: a real part and an imaginary part. (See Fig. 2.153.)

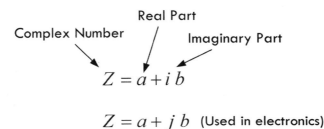

FIGURE 2.153

Both a and b are real numbers, whereas $i = \sqrt{-1}$ is an imaginary unit, thereby making the term ib an imaginary number or the imaginary part of a complex number. In practice, to avoid confusing i (imaginary unit) with the symbol i (current), the imaginary unit i is replaced with a j.

A complex number can be expressed graphically on a complex plane (argand or gaussian plane), with the horizontal axis representing the real axis and the vertical axis representing the imaginary axis. (See Fig. 2.154.)

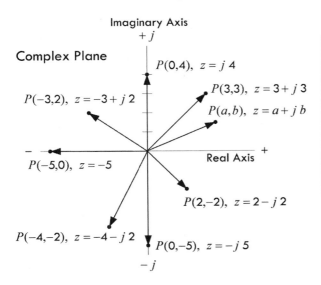

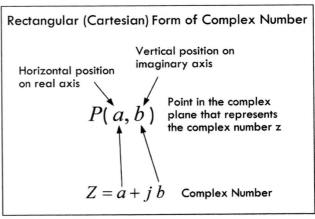

FIGURE 2.154

In terms of the drawing, a complex number can be interpreted as the vector from 0 to P having a magnitude of length of:

$$r = \sqrt{a^2 + b^2} \tag{2.61}$$

that makes an angle relative to the positive real axis of:

$$\theta = \arctan\left(\frac{b}{a}\right) = \tan^{-1}\left(\frac{b}{a}\right) \tag{2.62}$$

Now let's go a bit further—for the complex number to be useful in circuit analysis, it must be altered slightly. If you replace a with $r \cos \theta$ and replace b with $r \sin \theta$, the complex number takes on what is called the polar trigonometric form of a complex number. (See Fig. 2.155.)

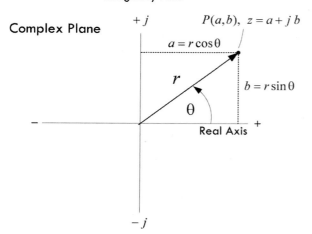
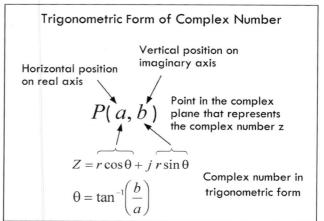

FIGURE 2.155

Okay, now you are getting there; just one more thing to cover. Long ago, a man by the name of Euler noticed that the $\cos \theta + j \sin \theta$ part of the trigonometric form of the complex number was related to $e^{j\theta}$ by the following expression:

$$e^{j\theta} = \cos \theta + j \sin \theta \tag{2.63}$$

You can prove this by taking the individual power series for $e^{j\theta}$, $\cos \theta$, $j \sin \theta$. When the power series for $\cos \theta$ and $j \sin \theta$ are added, the result equals the power series for $e^{j\theta}$. This means that the complex number can be expressed as follows:

$$z = re^{j\theta} \tag{2.64}$$

This represents the polar exponential form of a complex number. A shorthand version of this form can be written as:

$$z = r \angle \theta \tag{2.65}$$

Though this form goes by the name of polar coordinate form and has its basis in adding vectors and angles, and so on, it helps to think of it as a shorthand version of the polar exponential form, since it is really no different from the exponential form (same arithmetic rules apply), but it turns out to be a bit more intuitive, as well as easier to deal with in calculations.

Now we have basically four ways to express a complex number:

$$z = a + jb, \; z = r\cos\theta + jr\sin\theta, \; z = re^{j\theta}, \; z = r \angle \theta$$

Each form is useful in its own right. Sometimes it is easier to use $z = a + jb$, and sometimes it is easier to use $z = re^{j\theta}$ (or $z = r \angle \theta$)—it all depends on the situation, as we'll see in a moment.

Various Ways to Express a Complex Number

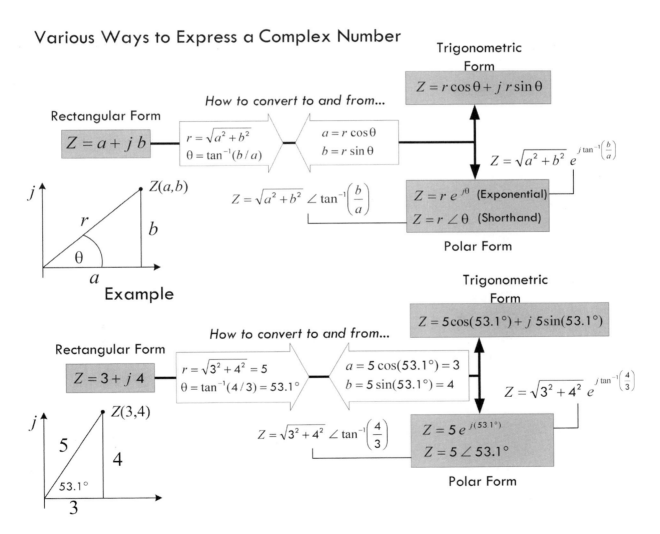

FIGURE 2.156

The model shown in Fig. 2.156 is designed to help you get a feeling for how the various forms of a complex number are related. For what follows, you will need

TABLE 2.10 Arithmetic Rules for Complex Numbers

FORM OF COMPLEX NUMBER	ADDITION/SUBTRACTION	MULTIPLICATION	DIVISION
Rectangular form $z_1 = a + jb$ $z_2 = c + jd$	$Z_1 \pm Z_2 = (a \pm c) + j(b \pm d)$ Example: $Z_1 = 3 + j4,\ Z_2 = 5 - j7$ $Z_1 + Z_2 = (3 + 5) + j(4 - 7) = 8 - j3$	$Z_1 \times Z_2 = (ac - bd) + j(ad + bc)$ Example: $Z_1 = 5 + j2,\ Z_2 = -4 + j3$ $Z_1 \times Z_2 = [5(-4) - 2(3)] + j[5(3) + 2(-4)]$ $= -26 + j7$	$\dfrac{Z_1}{Z_2} = \dfrac{ac + bd}{c^2 + d^2} + j\left(\dfrac{bc - ad}{c^2 + d^2}\right)$ Example: $Z_1 = 1 + j,\ Z_2 = 3 + j2$: $\dfrac{Z_1}{Z_2} = \dfrac{1(3) + 1(2)}{3^2 + 2^2} + j\left(\dfrac{1(3) - 1(2)}{3^2 + 2^2}\right) = \dfrac{5}{13} + j\dfrac{1}{13}$
Trigonometric form $Z_1 = r_1 \cos \theta_1 + jr_1 \sin \theta_1$ $Z_2 = r_2 \cos \theta_2 + jr_2 \sin \theta_2$	Can be done but involves using trigonometric identities; it is easier to convert this form into rectangular form and then add or subtract.	$z_1 \times z_2 = r_1 r_2 [\cos (\theta_1 + \theta_2) + j \sin (\theta_1 + \theta_2)]$	$\dfrac{z_1}{z_2} = \dfrac{r_1}{r_2}[\cos (\theta_1 - \theta_2) + j \sin (\theta_1 - \theta_2)]$
Exponential form $z_1 = r_1 e^{j\theta_1}$ $z_2 = r_2 e^{j\theta_2}$	Does not make much sense to add or subtract in this form because the result will not be in simplified form, except if r_1 and r_2 are equal; it is better to first convert to rectangular form and then add or subtract.	$Z_1 \times Z_2 = r_1 r_2 e^{j(\theta_1 + \theta_2)}$ Example: $Z_1 = 5e^{j(180°)},\ Z_2 = 2e^{j(90°)}$ $Z_1 \times Z_2 = 5\ (2)\ e^{j(180° + 90°)}$ $= 10\ e^{j(270°)}$	$\dfrac{Z_1}{Z_2} = \dfrac{r_1}{r_2}\ e^{j(\theta_1 - \theta_2)}$ Example: $Z_1 = 8e^{j(180°)},\ Z_2 = 2e^{j(60°)}$ $\dfrac{Z_1}{Z_2} = \dfrac{8}{2}\ e^{j(180° - 60°)} = 4e^{j(120°)}$
Polar coordinate form (shorthand) $z_1 = r_1 \angle\ \theta_1$ $z_2 = r_2 \angle\ \theta_2$	Does not make much sense to add or subtract in this form because the result will not be in simplified form, except if r_1 and r_2 are equal; it is better to first convert to rectangular form and then add or subtract.	$Z_1 \times Z_2 = r_1 r_2 \angle\ (\theta_1 + \theta_2)^*$ Example: $Z_1 = 5 \angle\ 180°,\ Z_2 = 2 \angle\ 90°$ $Z_1 \times Z_2$ $= 5(2) \angle\ (180° + 90°)$ $= 10 \angle\ 270°$	$\dfrac{Z_1}{Z_2} = \dfrac{r_1}{r_2} \angle\ (\theta_1 - \theta_2)^*$ Example: $Z_1 = 8 \angle\ 180°,\ Z_2 = 2 \angle\ 60°$ $\dfrac{Z_1}{Z_2} = \dfrac{8}{2}\ e^{j(180° - 60°)}$ $= 4 \angle\ 120°$

*Usually the most efficient form to use when doing your calculations—other forms are too difficult or provide unintuitive results.

to know the arithmetic rules for complex numbers, too, which are summarized in Table 2.10.

Here are some useful relationships to know when dealing with complex numbers:

$$X \text{ (in degrees)} = \frac{180°}{\pi} X \text{ (in radians)}, X \text{ (in radians)} = \frac{\pi}{180°} X \text{ (in degrees)}$$

$$j = \sqrt{-1}, \ j^2 = -1, \ \frac{1}{j} = -j, \ \frac{1}{A+jB} = \frac{A-jB}{A^2+B^2}$$

$$e^{j(0°)} = 1, \ e^{j(90°)} = j, \ e^{j(180°)} = -1, \ e^{j(270°)} = -j, \ e^{j(360°)} = 1$$

$$1 \angle 0° = 1, 1 \angle 90° = j, 1 \angle 180° = -1, 1 \angle 270° = -j, 1 \angle 360° = 1$$

$$Z^2 = (re^{j\theta})^2 = r^2 e^{j2\theta} \quad \text{or} \quad Z^2 = (r \angle \theta)^2 = r^2 \angle 2\theta$$

The following example shows a calculation that makes use of both rectangular and polar forms of a complex number to simplify a calculation involving addition, multiplication, and division of complex numbers:

$$\frac{(2+j5)+(3-j10)}{(3+j4)(2+j8)} = \frac{5-j5}{(3+j4)(2+j8)} = \frac{7.07 \angle 45.0°}{(5 \angle 53.1°)(8.25 \angle 76.0°)} = \frac{7.07 \angle 45.0°}{41.25 \angle 129.1°}$$

$$= 0.17 \angle -84.1°$$

The result can easily be converted into trigonometric or rectangular forms if the need exists:

$$0.17 \angle -84.1° = 0.17 \cos(-84.1°) + j0.17 \sin(-84.1°) = 0.017 - j0.17$$

Notice that when multiplication or division is involved, it is easier to first convert the complex terms into exponential form (shorthand version). In essence, for addition and subtraction you'll probably use rectangular form (though trigonometric form isn't difficult in this case), and use exponential (shorthand) form for multiplication and division. If you understand the preceding calculation, you should find the ac theory to come easy.

Note that sometimes the following notation is used to express a complex number:

$$|Z| = \sqrt{(\text{Re}\,Z)^2 + (\text{Im}\,Z)^2} \tag{2.66}$$

$$\arg(Z) = \arctan\left(\frac{\text{Im}\,Z}{\text{Re}\,Z}\right) = \tan^{-1}\left(\frac{\text{Im}\,Z}{\text{Re}\,Z}\right)$$

where $|Z|$ is the magnitude or modulus of a complex number (or r), Re Z is the real part of the complex number, and Im Z is the imaginary part of the number, while arg(Z) (or phase θ) represents the argument of Z or the phase angle (θ). For example, if $Z = 3 + j4$, then:

$$\text{Re } Z = 3 \qquad \text{Im } Z = 4 \qquad |Z| = \sqrt{(3)^2 + (4)^2} = 5 \qquad \arg(Z) = \arctan\left(\frac{4}{3}\right) = 53.1°$$

2.27 Circuit with Sinusoidal Sources

Suppose that you are given the two circuits shown in Fig. 2.157 containing linear elements (resistors, capacitors, inductors) driven by sinusoidal voltage sources.

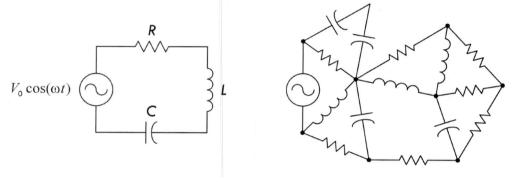

FIGURE 2.157

To analyze the simpler of the two circuits, you could apply Kirchhoff's voltage law to get the following:

$$V_0 \cos(\omega t) = IR + L\frac{dI}{dt} + \frac{1}{C}\int I \, dt$$

which reduces to:

$$L\frac{d^2 I}{dt^2} + R\frac{dI}{dt} + \frac{1}{C}I = -\omega V_0 \sin(\omega t)$$

This expression, as we discovered earlier, is a linear second-order nonhomogeneous differential equation. To find the solution to this equation, you could apply, say, the technique of variation of parameters or the method of undetermined coefficients. After the solution for the current is found, finding the voltages across the resistor, capacitor, and inductor is a simple matter of plugging the current into the characteristic voltage/current equation for that particular component. However, coming up with the solution for the current in this case is not easy because it requires advanced math (and the work is tedious).

Now, as if things were not bad enough, let's consider the more complex circuit in Fig. 2.157. To analyze this mess, you could again apply Kirchhoff's voltage and current laws to a number of loops and junctions within the circuit and then come up with a system of differential equations. The math becomes even more advanced, and finding the solution becomes ridiculously difficult.

Before we scare you too much with these differential equations, let us tell you about an alternative approach, one that does away with differential equations completely.

This alternative approach makes use of what are called complex impedances—something that will use the complex numbers we talked about in the last section.

2.27.1 Analyzing Sinusoidal Circuits with Complex Impedances

To make solving sinusoidal circuits easier, it's possible to use a technique that enables you to treat capacitors and inductors like special kinds of resistors. After that, you can analyze any circuit containing resistors, capacitors, and inductors as a "resistor" circuit. By doing so, you can apply all the dc circuit laws and theorems that were presented earlier. The theory behind how the technique works is a bit technical, even though the act of applying it is not hard at all. For this reason, if you do not have the time to learn the theory, we suggest simply breezing through this section and pulling out the important results. Here's a look at the theory behind complex impedances.

In a complex, linear, sinusoidally driven circuit, all currents and voltages within the circuit will be sinusoidal in nature after all transients have died out. These currents and voltages will be changing with the same frequency as the source voltage (the physics makes this so), and their magnitudes will be proportional to the magnitude of the source voltage at any particular moment in time. The phase of the current and voltage patterns throughout the circuit, however, most likely will be shifted relative to the source voltage pattern. This behavior is a result of the capacitive and inductive effects brought on by the capacitors and inductors.

As you can see, there is a pattern within the circuit. By using the fact that the voltages and currents will be sinusoidal everywhere, and considering that the frequencies of these voltages and currents will all be the same, you can come up with a mathematical trick to analyze the circuit—one that avoids differential equations. The trick involves using what is called the superposition theorem. The superposition theorem says that the current that exists in a branch of a linear circuit that contains several sinusoidal sources is equal to the sum of the currents produced by each source independently. The proof of the superposition theorem follows directly from the fact that Kirchhoff's laws applied to linear circuits always result in a set of linear equations that can be reduced to a single linear equation with a single unknown. The unknown branch current thus can be written as a linear superposition of each of the source terms with an appropriate coefficient. (Figure 2.158 shows the essence of superimposing of sine waves.)

What this all means is that you do not have to go to the trouble of calculating the time dependence of the unknown current or voltage within the circuit because you know that it will always be of the form $\cos(\omega t + \phi)$. Instead, all you need to do is calculate the peak value (or RMS value) and the phase, and apply the superposition theorem. To represent currents and voltages and apply the superposition theorem, it would seem obvious to use sine or cosine functions to account for magnitude, phase, and frequency. However, in the mathematical process of superimposing (adding, multiplying, etc.), you would get messy sinusoidal expressions in terms of sines and cosines that would require difficult trigonometric rules and identities to convert the answers into something you could understand. Instead, what you can do to represent amplitudes and phase of voltages and currents in a circuit is to use complex numbers.

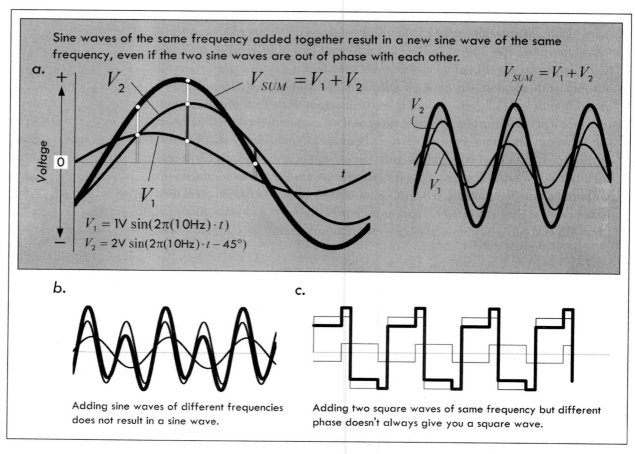

Sine waves of the same frequency added together result in a new sine wave of the same frequency, even if the two sine waves are out of phase with each other.

a.

V_2

$V_{SUM} = V_1 + V_2$

$V_{SUM} = V_1 + V_2$

V_2

V_1

Voltage

V_1

$V_1 = 1V \sin(2\pi(10Hz) \cdot t)$

$V_2 = 2V \sin(2\pi(10Hz) \cdot t - 45°)$

b.

Adding sine waves of different frequencies does not result in a sine wave.

c.

Adding two square waves of same frequency but different phase doesn't always give you a square wave.

FIGURE 2.158 **(a) Shows two sine waves and the resulting sum—another sine wave of the same frequency, but shifted in phase and amplitude. This is the key feature that makes it easy to deal with sinusoidally driven linear circuits containing resistors, capacitors, and inductors. Note that if you were to try this with sine waves of different frequencies, as shown in (b), the resultant waveform isn't sinusoidal. Superimposing nonsinusoidal waveforms of the same frequency, such as squarewaves, isn't guaranteed to result in a similar waveform, as shown in (c).**

Recall from the section on complex numbers that a complex number exhibits sinusoidal behavior—at least in the complex plane. For example, the trigonometric form of a complex number $z_1 = r_1 \cos \theta_1 + j r_1 \sin \theta_1$ will trace out a circular path in the complex plane when θ runs from 0 to 360°, or from 0 to 2π radians. If you graph the real part of z versus θ, you get a sinusoidal wave pattern. To change the amplitude of the wave pattern, you simply change the value of r. To set the frequency, you simply multiply θ by some number. To induce a phase shift relative to another wave pattern of the same frequency, you simply add some number (in degrees or radians) to θ. If you replace θ with ωt, where ($\omega = 2\pi f$), replace the r with V_0, and leave a place for a term to be added to ωt (a place for phase shifts), you come up with an expression for the voltage source in terms of complex numbers. You could do the same sort of thing for currents, too.

Now, the nice thing about complex numbers, as compared with sinusoidal functions, is that you can represent a complex number in various ways, within rectangular, polar-trigonometric, or polar-exponential forms (standard or shorthand versions). Having these different options makes the mathematics involved in the superimposing process easier. For example, by converting a number, say, into rectangular form, you can easily add or subtract terms. By converting the number into polar-exponential form (or shorthand form), you can easily multiply and divide terms (terms in the exponent will simply add or subtract).

It should be noted that, in reality, currents and voltages are always real; there is no such thing as an imaginary current or voltage. But then why are there imaginary parts? The answer is that when you start expressing currents and voltages with real and imaginary parts, you are simply introducing a mechanism for keeping track of the phase. (The complex part is like a hidden part within a machine; its function does not show up externally but does indeed affect the external output—the "real," or important, part, as it were.) What this means is that the final answer (the result of the superimposing) always must be converted back into a real quantity. This means that after all the calculations are done, you must convert the complex result into either polar-trigonometric or polar-exponential (shorthand form) and remove the imaginary part. For example, if you come across a resultant voltage expressed in the following complex form:

$$V(t) = 5 \text{ V} + j\, 10 \text{ V}$$

where the voltages are RMS, we get a meaningful real result by finding the magnitude, which we can do simply by converting the complex expression into polar exponential or shorthand form:

$$\sqrt{(5.0 \text{ V})^2 + (10.0 \text{ V})^2}\, e^{j(63.4°)} = (11.2 \text{ V})\, e^{j(63.5°)} = 11.2 \text{ V} \angle 63.5°$$

Whatever is going on, be it reactive or resistive effects, there is really 11.2 V RMS present. If the result is a final calculation, the phase often isn't important for practical purposes, so it is often ignored.

You may be scratching your head now and saying, "How do I really do the superimposing and such? This all seems too abstract or wishy-washy. How do I actually account for the resistors, capacitors, and inductors in the grand scheme of things?" Perhaps the best way to avoid this wishy-washiness is to begin by taking a sinusoidal voltage and converting it into a complex number representation. After that, you can apply it individually across a resistor, a capacitor, and then an inductor to see what you get. Important new ideas and concrete analysis techniques will surface in the process.

2.27.2 Sinusoidal Voltage Source in Complex Notation

Let's start by taking the following expression for a sinusoidal voltage:

$$V_0 \cos(\omega t) \qquad (\omega = 2\pi f)$$

and converting it into a polar-trigonometric expression:

$$V_0 \cos(\omega t) + j V_0 \sin(\omega t)$$

What about the $j V_0 \sin(\omega t)$ term? It is imaginary and does not have any physical meaning, so it does not affect the real voltage expression (you need it, however, for the superimposing process). To help with the calculations that follow, the polar-trigonometric form is converted into the polar-exponential form using Euler's relation $e^{j\theta} = r\cos(\theta) + jr\sin(\theta)$:

$$V_0 e^{j(\omega t)} \tag{2.67}$$

In polar-exponential shorthand form, this would be:

$$V_0 \angle (\omega t) \tag{2.68}$$

Graphically, you can represent this voltage as a vector rotating counterclockwise with angular frequency ω in the complex plane (recall that $\omega = d\theta/dt$, where $\omega = 2\pi f$). The length of the vector represents the maximum value of V—namely, V_0—while the projection of the vector onto the real axis represents the real part, or the instantaneous value of V, and the projection of the vector onto the imaginary axis represents the imaginary part of V.

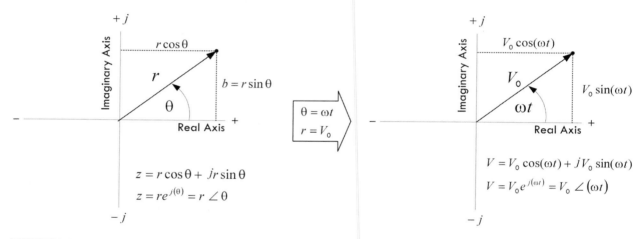

FIGURE 2.159

Now that you have an expression for the voltage in complex form, you can place, say, a resistor, a capacitor, or an inductor across the source and come up with a complex expression for the current through each component. To find the current through a resistor in complex form, you simply plug $V_0 e^{j(\omega t)}$ into V in $I = V/R$. To find the capacitor current, you plug $V_0 e^{j(\omega t)}$ into $I = C\, dV/dt$. Finally, to find the inductor current, you plug $V_0 e^{j(\omega t)}$ into $I = 1/L \int V dt$. The results are shown in Fig. 2.160.

Comparing the phase difference between the current and voltage through and across each component, notice the following:

Resistor: The current and voltage are in phase, $\phi = 0°$, as shown in the graph in Fig. 2.160. This behavior can also be modeled within the complex plane, where the voltage and current vectors are at the same angle with respect to each other, both of which rotate around counterclockwise at an angular frequency $\omega = 2\pi f$.

Capacitor: The current is out of phase with the applied voltage by $+90°$. In other words, the current leads the voltage by $90°$. By convention, unless otherwise stated, the

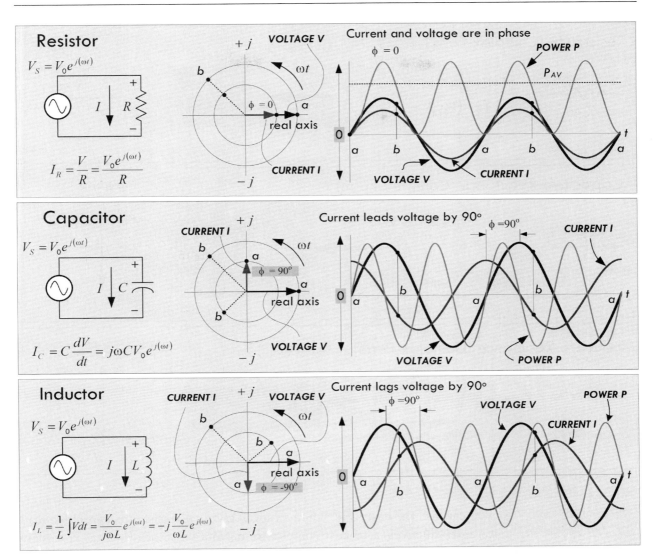

FIGURE 2.160

phase angle ϕ is referenced from the current vector to the voltage vector. If ϕ is positive, then current is leading (further counterclockwise in rotation); if ϕ is negative, current is lagging (further clockwise in rotation).

Inductor: The current is out of phase with the applied voltage by $-90°$. In other words, the current lags the voltage by $90°$.

We call the complex plane model, showing the magnitude and phase of the voltages and currents, a *phasor diagram*—where the term *phasor* implies phase comparison. Unlike a time-dependent mathematical function, a phasor provides only a snapshot of what's going on. In other words, it only tells you the phase and amplitude at a particular moment in time.

Now comes the important trick to making ac analysis easy to deal with. If we take the voltage across each component and divide it by the current, we get the following (see Fig. 2.161):

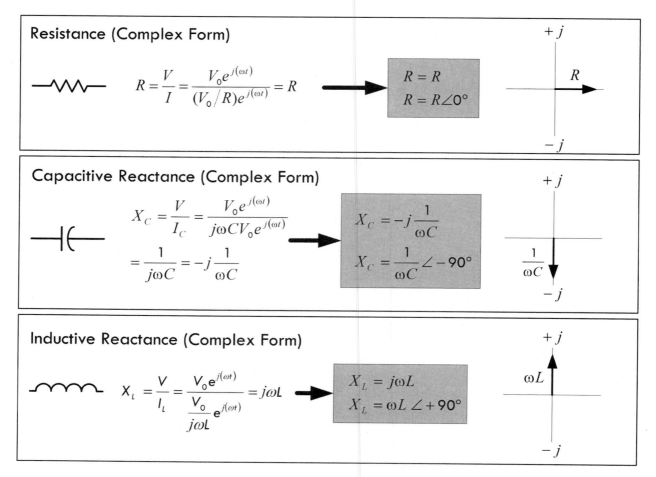

FIGURE 2.161

As you can see, the nasty $V_0 e^{j(\omega t)}$ terms cancel, giving us resistance, capacitive reactance, and inductive reactance in complex form. Notice that the resulting expressions are functions only of frequency, not of time. This is part of the trick to avoiding the nasty differential equations.

Now that we have a way of describing capacitive reactance and inductive reactance in terms of complex numbers, we can make an important assumption. We can now treat capacitors and inductors like frequency-sensitive resistors within sinusoidally driven circuits. These frequency-sensitive resistors take the place of normal resistors in dc circuit analysis. We must also replace dc sources with sinusoidal ones. If all voltages, currents, resistances, and reactances are expressed in complex form when we are analyzing a circuit, when we plug them into the old circuit theorems (Ohm's law, Kirchhoff's law, Thevinin's theorem, etc.) we will come up with equations whose solutions are taken care of through the mathematical operations of the complex numbers themselves (the superposition theorem is built in).

For example, *ac Ohm's law* looks like this:

$$V(\omega) = I(\omega) \times Z(\omega) \tag{2.69}$$

What does the Z stand for? It's referred to as complex impedance, which is a generic way of describing resistance to current flow, in complex form. The complex impedance may simply be resistive, it may be only capacitive, it may be inductive, or it could be a combination of resistive and reactive elements (e.g., an RLC circuit element). For example:

$$\text{Resistor: } V_R = I_R \times R$$

$$\text{Capacitor: } V_C = I_C \times X_C = I_C\left(-j\frac{1}{\omega C}\right) = -j\frac{I_C}{\omega C} = \frac{I_C}{\omega C}\angle -90°$$

$$\text{Inductor: } V_L = I_L \times X_L = I_L\,(j\omega L) = jI_L\omega L = I_L\omega L \angle +90°$$

Any complex impedance Z: $V_Z = I_Z \times Z$

Figure 2.162 shows what we just discussed, along with the phasor expression for a sinusoidal source, and a complex impedance Z composed of a resistor, a capacitor, and an inductor.

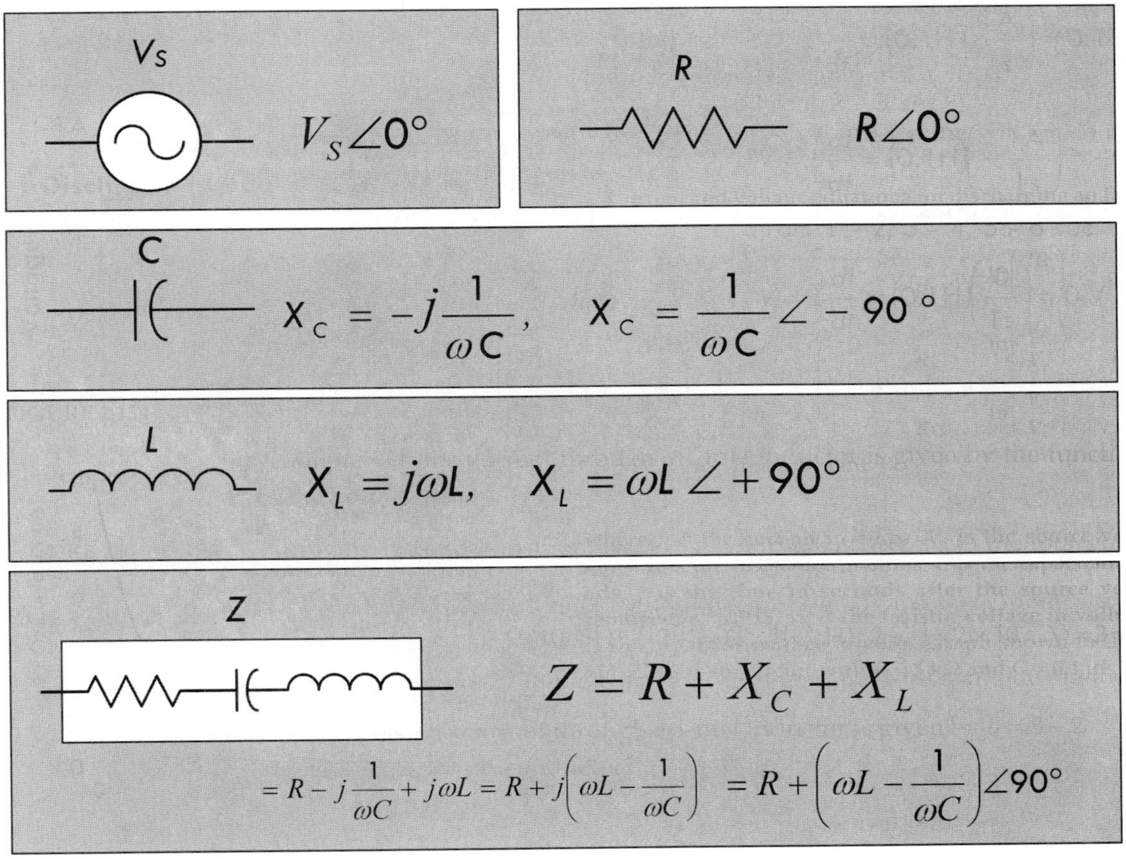

FIGURE 2.162

Continuing along the line of treating complex impedances as frequency-sensitive resistors, we can make use of the resistors in series equation, except now we take *impedances in series:*

$$Z_{\text{total}} = Z_1 + Z_2 + Z_3 + \ldots + Z_N \qquad (2.70) \ N \text{ in series}$$

Likewise, the old voltage divider now becomes the *ac voltage divider*:

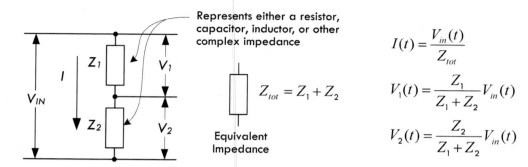

$$I(t) = \frac{V_{in}(t)}{Z_{tot}}$$

$$Z_{tot} = Z_1 + Z_2$$

$$V_1(t) = \frac{Z_1}{Z_1 + Z_2} V_{in}(t)$$

$$V_2(t) = \frac{Z_2}{Z_1 + Z_2} V_{in}(t)$$

FIGURE 2.163

To find the equivalent impedance for a larger number of impedances in parallel:

$$Z_{tot} = \frac{1}{1/Z_1 + 1/Z_2 + 1/Z_3 + \ldots 1/Z_N} \qquad \text{(2.71) } N \text{ in parallel}$$

$$Z_{total} = \frac{Z_1 Z_2}{Z_1 + Z_2} \qquad \text{(2.72) Two in parallel}$$

The *ac current divider* is consequently:

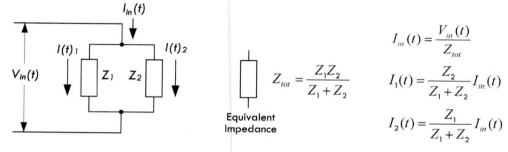

$$I_{in}(t) = \frac{V_{in}(t)}{Z_{tot}}$$

$$Z_{tot} = \frac{Z_1 Z_2}{Z_1 + Z_2}$$

$$I_1(t) = \frac{Z_2}{Z_1 + Z_2} I_{in}(t)$$

$$I_2(t) = \frac{Z_1}{Z_1 + Z_2} I_{in}(t)$$

FIGURE 2.164

And perhaps, most important, we can throw complex impedances in Kirchhoff's voltage and loop equations and solve complex circuits with many nodes:

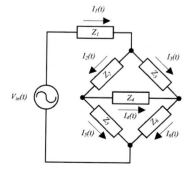

Applying Kirchhoff's current law, you get the following set of equations:

$$I_1(t) = I_2(t) + I_3(t)$$
$$I_2(t) = I_5(t) + I_4(t)$$
$$I_6(t) = I_4(t) + I_3(t)$$

Applying Kirchhoff's voltage law, you get the following set of equations:

$$V_{in}(t) - I_1(t)Z_1 - I_2(t)Z_2 - I_5(t)Z_5 = 0$$
$$- I_3(t)Z_3 + I_4(t)Z_4 + I_2(t)Z_2 = 0$$
$$- I_6(t)Z_6 + I_5(t)Z_5 - I_4(t)Z_4 = 0$$

FIGURE 2.165

Example 1: Find the complex impedances of the following networks.

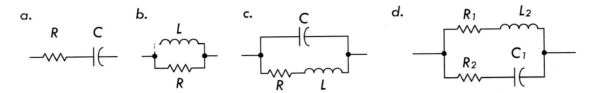

FIGURE 2.166

Answer:

(a) $R - j\dfrac{1}{\omega C}$ (b) $\dfrac{jR\omega L}{R + j\omega L}$ (c) $\dfrac{\dfrac{L}{C} - j\left(\dfrac{R}{\omega C}\right)}{R + j\left(\omega L - \dfrac{1}{\omega C}\right)}$ (d) $\dfrac{\left(R_1 R_2 + \dfrac{L_2}{C_1}\right) + j\left(R_2 \omega L_2 - \dfrac{R_1}{\omega C_1}\right)}{(R_1 + R_2) + j\left(\omega L_2 - \dfrac{1}{\omega C_1}\right)}$

You can simplify those results that have a complex number in the denominator by using:

$$\frac{1}{A + jB} = \frac{A - jB}{A^2 + B^2}$$

Example 2: Express networks (a) and (c) in the previous example in polar coordinate form.

Answer:

(a) $\sqrt{R^2 + \left(\dfrac{1}{\omega C}\right)^2} \, \angle \tan^{-1}\left(-\dfrac{1}{R\omega C}\right)$

(c) $Z_{TOT} = \dfrac{\sqrt{(L/C)^2 + [R/(\omega C)]^2}}{\sqrt{R^2 + [\omega L - 1/(\omega C)]^2}} \, \angle \left\{ -\tan^{-1}\left(\dfrac{R}{\omega C}\right) - \tan^{-1}\left[\dfrac{\omega L - 1/(\omega C)}{R}\right] \right\}$

What a mathematical nightmare, you may say, after doing the last example. As you can see, the expressions can get ugly if we don't start plugging values into the variables from the start. However, the act of finding a complex impedance and using it in ac analysis is vastly easier than using the characteristic equations of resistors, capacitors, and inductors; inserting them into Kirchhoff's laws; and solving the differential equations.

Example 3: The series RL circuit in Fig. 2.167 is driven by a 12-VAC (RMS), 60-Hz source. $L = 265$ mH, $R = 50$ Ω. Find I_S, I_R, I_L, V_R, and V_L and the apparent power, real power, reactive power, and power factor.

Series Impedance (RL Circuit)

Resistance and Inductive Reactance

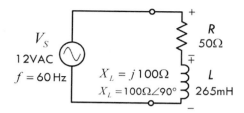

Equivalent Impedance and Current

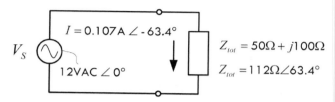

Voltage across R and L

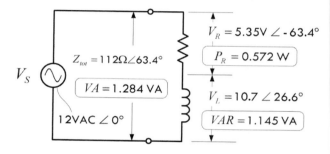

Sinusoidal Waveforms within Series RL Circuit

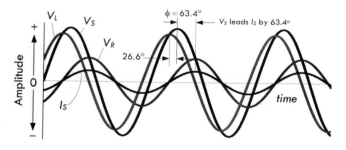

FIGURE 2.167

$V_S(t) = 17.0 \text{ V sin } (\omega t)$

$V_R(t) = 7.6 \text{ V sin } (\omega t - 63.4°)$

$V_L(t) = 15.1 \text{ V sin } (\omega t + 26.6°)$

$I_S(t) = 0.151 \text{ A sin } (\omega t - 63.4°)$

* **Peak voltages and currents used in these functions are the RMS equivalents multiplied by 1.414.**

First, calculate the reactance of the inductor:

$$X_L = j\omega L = j(2\pi \times 60 \text{ Hz} \times 265 \times 10^{-3} \text{ H}) = j\, 100 \ \Omega$$

Since the resistor and inductor are in series, the math is easy—simply add complex numbers in rectangular form:

$$Z = R + X_L = 50 \ \Omega + j100 \ \Omega$$

In polar form, the result is:

$$Z = \sqrt{50^2 + 100^2} \ \angle \ \tan^{-1}\left(\frac{100}{50}\right) = 112\Omega \ \angle \ 63.4°$$

Don't let the imaginary part or phase angle fool you. The impedance is real; it provides 112 ohms' worth, though only a portion of this is real resistance—the rest is inductive reactance.

The current can now be found using ac Ohm's law:

$$I_s = \frac{V_s}{Z_{tot}} = \frac{12 \text{ VAC} \ \angle \ 0°}{112 \ \Omega \ \angle \ 63.4°} = 0.107 \text{ A} \ \angle -63.4°$$

The −63.4° result means the current lags the applied voltage or total voltage across the network by 63.4°. Since this is a series circuit $I = I_R = I_L$. The voltage across the resistance and inductor can be found using ac Ohm's law or the ac voltage divider:

$$V_R = I \times R = (0.107\text{A} \ \angle -63.4°)(50 \ \Omega \ \angle \ 0°)$$

$$= 5.35 \text{ VAC} \ \angle -63.4°$$

$$V_L = I \times X_L = (0.107\text{A} \ \angle -63.4°)(100 \ \Omega \ \angle \ 90°)$$

$$= 10.7 \text{ VAC} \ \angle \ 26.6°$$

Note that the preceding calculations were for only an instant in time $t = 0$, as is the case by stating the starting condition: $V_S = 12$ VAC $\angle$ 0°. But that's all we need, since we know the phases and the voltages at those phases, which link amplitude. So to create an accurate picture of how the whole system behaves over time, we simply plug the more general ωt into the source voltage, and convert the RMS values into true values by multiplying by 1.414. We get $V_S = 17.0$ V $\angle$ ωt, which isn't a snapshot but a continuous mathematic description for all times. To make a graph, we must convert into trigonometric form, ignoring the imaginary part in the process: $V_S = 17.0$ V cos (ωt). Since we are only concerned with phase, we can express V_S in terms of a sine function, $V_S = 17.0$ V sin (ωt) and reference all other voltage and current waveforms from it by adding their phase terms and including their peak values, as shown in the following equations.

The apparent power due to the total impedance:

$$VA = I_{RMS} V_{RMS} = (0.107 A)(12 \text{ VAC}) = 1.284 \text{ VA}$$

Only resistance consumes power, however:

$$P_R = I_{RMS}{}^2 R = (0.107 A)^2 (50 \ \Omega) = 0.572 \text{ W}$$

The reactive power due to the inductor is:

$$VAR = I_{RMS}{}^2 X_L = (0.107 A)^2 (100 \ \Omega) = 1.145 \text{ VA}$$

Power factor (real power/apparent power) is:

$$PF = \frac{P_R}{VA} = \cos(\phi) = \cos(-63.4°) = 0.45 \text{ lagging}$$

where ϕ is the phase angle between V_S and I_S. Note that we haven't actually discussed apparent and reactive power yet—that's coming up.

2.27.3 Odd Phenomena in Reactive Circuits

In reactive circuits, circulation of energy accounts for seemingly odd phenomena. In our example in Fig. 2.167, it appears that Kirchhoff's law doesn't add up, since the arithmetical sum of the resistor and inductor voltages is:

$$5.35 \text{ VAC} + 10.70 \text{ VAC} = 16.05 \text{ VAC}$$

This is greater than the 12-VAC source voltage. The problem here is that we haven't taken into account the phase.

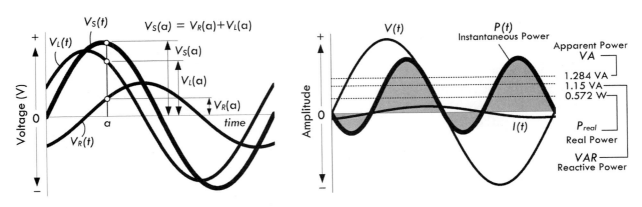

FIGURE 2.168

Their actual result, when phase is taken into account, is:

$$V_{total} = V_R + V_L = 5.35 \text{ VAC} \angle -63.4° + 10.70 \text{ VAC} \angle 26.6°$$

$$= 2.4 \text{ VAC} - j4.8 \text{ VAC} + 9.6 \text{ VAC} + j4.8 \text{ VAC} = 12 \text{ VAC}$$

Figure 2.168 graphically illustrates this point.

Note that the use of "VAC" to represent the fact that the voltages are expressed in RMS values is not always followed. Some people like to simply write "V" and assume that all voltages are sinusoidal and thus given in RMS form. The actual peak voltages are related to the RMS voltage value by $V_P = 1.414 \times V_{RMS}$.

In other cases, such as a series circuit with capacitance and inductance, the voltages across the components may exceed the supply voltage. This condition can exist because, while energy is being stored by the inductor, the capacitor is returning energy to the circuit from its previously charged state, and vice versa. In a parallel circuit with inductive and capacitive branches, the currents circulating through the components may exceed the current drawn from the source. Again, the phenomenon occurs because the inductor's collapsing field supplies current to the capacitor and the discharging capacitor provides current to the inductor. We will take a look at these cases in a moment.

2.28   Power in AC Circuits (Apparent Power, Real Power, Reactive Power)

In a complex circuit containing resistors, inductors, and capacitors, how do you determine what kind of power is being used? The best place to start is with the generalized power law $P = I_{RMS}V_{RMS}$. However, for now, let's replace P with VA, and call VA the *apparent power*:

$$VA = I_{RMS}V_{RMS} \qquad \text{(2.74) Apparent power}$$

In light of our RL series circuit in Fig. 2.167, we find the apparent power to be:

$$VA = I_{RMS}V_{RMS} = (0.107A)(12 \text{ V}) = 1.284 \text{ VA}$$

The apparent power VA is no different from the power we calculate using the generalized ac power expression. The reason for using VA instead of P is simply a convention used to help distinguish the fact that the calculated power isn't purely real and is not expressed in watts, as real power is. Instead, apparent power is expressed in volt-amperes, or VA, which happens to be the same letters used for the variable for apparent power. (This is analogous to the variable for voltage being similar to the unit volt, though the variable is distinguished from the unit by being italicized.) In essence, the apparent power takes into account both resistive power losses as well as reactive power. The reactive power, however, isn't associated with power loss, but is instead associated with energy storage in the form of magnetic fields within inductors and electric fields within capacitors. This energy is later returned to the circuit as the magnetic field in an inductor collapses, or the electric field vanishes as a capacitor discharges later in the ac cycle. Only if a circuit is purely resistive can we say that the apparent power is in watts.

So how do we distinguish what portions of the apparent power are real power and reactive power? Real power is associated with power loss due to heating through an ohmic material, so we can define *real power* using ac Ohm's law within the generalized power expression:

$$P_R = I_{RMS}{}^2 R \qquad \text{(2.75) Real power}$$

In our series RL circuit in Fig. 2.167, we find the real power to be:

$$P_R = I_{RMS}{}^2 R = (0.107 \text{ A})^2 \, (50 \, \Omega) = 0.572 \text{ W}$$

Notice that real power is always measured in watts.

Now, to determine the reactive power due to capacitance and inductance within a circuit, we create the notion of reactive power. The *reactive power* is given in volt-ampere reactive, or *VAR*. We can define the reactive power by using the Ohm's power law, replacing resistance (or impedance) with a generic reactance *X:*

$$VAR = I_{RMS}{}^2 X \qquad \text{(2.76) Reactive power}$$

At no time should watts be associated with reactive power. In fact, reactive power is called wattless power, and therefore is given in volt-amperes, or VA.

Considering the RL series circuit in Fig. 2.167, the reactive power is:

$$VAR = I_{RMS}{}^2 X_L = (0.107 \text{ A})^2 \, (100 \, \Omega) = 1.145 \text{ VA}$$

You may be saying, great, we can now add up the real power and reactive power, and this will give us the apparent power. Let's try it out for our RL circuit in Fig. 2.167:

$$0.572 + 1.145 = 1.717$$

But wait, the calculated apparent power was 1.284 VA, not 1.717 VA. What's wrong? The problem is that a simple arithmetic operation on variables that are reactive can't be done without considering phase (as we saw with adding voltages). Considering phase for our RL series circuit:

$$VAR = I^2{}_{RMS} X_L = (0.107 \text{ A} \angle -63.4°)^2 \, (100 \, \Omega \angle 90°) = 1.145 \text{ VA} \angle -36.8°$$

$$VAR = 0.917 \text{ VA} - j0.686 \text{ VA}$$

$$P_R = I^2{}_{RMS} R = (0.107 \text{ A} \angle -63.4°)^2 \, (50 \, \Omega \angle 0°) = 0.573 \angle -126.8°$$

$$P_R = -0.343 \text{ W} - j0.459 \text{ W}$$

Adding the reactive and real power together now gives us the correct apparent power:

$$VA = VAR + P_R = 0.574 \text{ VA} - j1.145 \text{ VA} = 1.281 \text{ VA} \angle -63.4°$$

To avoid doing such calculations, we note the following relationship:

$$VA = \sqrt{P_R^2 + VAR^2} \qquad (2.77)$$

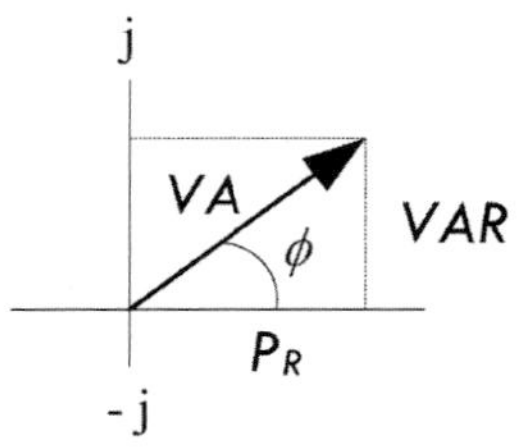

FIGURE 2.169

Here, we don't have to worry about phase angles—the Pythagorean theorem used in the complex plane, as shown in Fig. 2.169, takes care of that. Using the RL series example again and inserting our values into Eq. 2.77, we get an accurate expression relating real, apparent and reactive powers:

$$1.284 = \sqrt{(0.572)^2 + (1.145)^2}$$

2.28.1 Power Factor

Another way to represent the amount of apparent power to reactive power within a circuit is to use what's called the power factor. The *power factor* of a circuit is the ratio of consumed power to apparent power:

$$PF = \frac{P_{consumed}}{P_{apparent}} = \frac{P_R}{VA} \qquad (2.78)$$

In the example in Fig. 2.167:

$$PF = \frac{0.572 \text{ W}}{1.284 \text{ VA}} = 0.45$$

Power factor is frequently expressed as a percentage, in this case 45 percent.

An equivalent definition of power factor is:

$$PF = \cos \phi \qquad (2.79)$$

where ϕ is the phase angle (between voltage and current). In the example in Fig. 2.167, the phase angle is −63.4°, so:

$$PF = \cos(-63.4°) = 0.45$$

as the earlier calculation confirms.

The power factor of a purely resistive circuit is 100 percent, or 1, while the power factor of a purely reactive circuit is 0.

Since the power factor is always rendered as a positive number, the value must be followed by the words "leading" or "lagging" to identify the phase of the voltage with respect to the current. Specifying the numerical power factor is not always sufficient. For example, many dc-to-ac power inverters can safely operate loads having large net reactance of one sign but only a small reactance of the opposite sign.

The final calculation of the power factor in the example in Fig. 2.167 yields the value of 0.45 lagging.

In ac equipment, the ac components must handle reactive power as well as real power. For example, a transformer connected to a purely reactive load must still be capable of supplying the voltage and be able to handle the current required by the reactive load. The current in the transformer windings will heat the windings as a result of I^2R losses in the winding resistances.

As a final note, there is another term to describe the percentage of power used in reactance, which is called the reactive factor. The *reactive factor* is defined by:

$$RF = \frac{P_{\text{reactive}}}{P_{\text{apparent}}} = \frac{VAR}{VA} = \sin(\phi) \tag{2.80}$$

Using the example in Fig. 2.167:

$$RF = \frac{1.145 \text{ VA}}{1.284 \text{ VA}} = \sin(-63.4°) = -0.89$$

Example 4: The LC series circuit shown in Fig. 2.170 is driven by a 10-VAC (RMS), 127,323-Hz source. $L = 100 \text{ μH}$, $C = 62.5 \text{ nF}$. Find I_S, I_R, I_L, V_L, and V_C and the apparent power, real power, reactive power, and power factor.

Series Impedance (LC Circuit)

Inductive and Capacitive Reactances

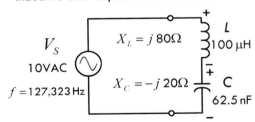

Equivalent Impedance and Current

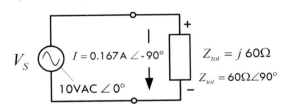

Voltage Across L and C

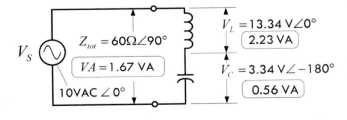

FIGURE 2.170

First we calculate the inductive and capacitive reactances:

$$X_L = j\omega L = j(2\pi \times 127{,}323 \text{ Hz} \times 100 \times 10^{-6} \text{ H})$$

$$= j \, 80 \, \Omega$$

$$X_C = -j\frac{1}{\omega C} = j\frac{1}{2\pi \times 127{,}323 \text{ Hz} \times 62.5 \times 10^{-9} \text{C}}$$

$$= -j20 \, \Omega$$

Since the inductor and capacitor are in series, the math is easy—simply add complex numbers in rectangular form:

$$Z = X_L + X_C = j80 \, \Omega + (-j20 \, \Omega) = j60 \, \Omega$$

In polar form, this is 60 Ω ∠ 90°. The fact that the phase is 90°, or the result is positive imaginary, means the impedance is net inductive. Don't let the imaginary part fool you—the impedance is actually felt—it provides 60 Ω of impedance, even though it's not resistive.

The current can now be found using ac Ohm's law:

$$I_s = \frac{V_s}{Z} = \frac{10 \text{ VAC} \angle 0°}{60 \, \Omega \angle 90°} = 0.167 \text{A} \angle -90°$$

Sinusoidal Waveforms within Series LC Circuit

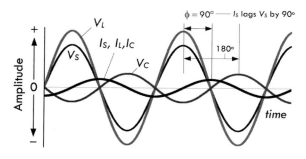

FIGURE 2.170 *(Continued)*

$V_S(t) = 14.1$ V sin (ωt)

$V_L(t) = 18.90$ V sin (ωt)

$V_C(t) = 4.72$ V sin $(\omega t - 180°)$

$I_S(t) = 0.236$ A sin $(\omega t - 90°)$

* **Peak voltages and currents used in these functions
are the RMS equivalents multiplied by 1.414.**

Note that we used 60 Ω ∠ 90° (polar) to make the division easy. The −90° result means the source current (total current) lags the source voltage by 90°. Since this is a series circuit: $I_S = I_L = I_C$.

The voltage across the inductor and capacitor can be found using ac Ohm's law (or the ac voltage divider):

$$V_L = I_S \times X_L = (0.167 \text{ A} \angle -90°)(80 \text{ Ω} \angle 90°)$$

$$= 13.36 \text{ VAC} \angle 0°$$

$$V_C = I_S \times X_C = (0.167 \text{ A} \angle -90°)(20 \text{ Ω} \angle -90°)$$

$$= 3.34 \text{ VAC} \angle -180°$$

Notice that the voltage across the inductor is greater than the supply voltage; the capacitor supplies current to the inductor as it discharges.

To convert these snapshots into a continuous set of functions, we convert all RMS values to true value (×1.414), add the ωt term to the phase angles, and convert to trigonometric form, then delete the imaginary part. Our results would all be in terms of cosines, but to make things pretty, we select sines. Doing this doesn't make any practical difference. See graphs and equations to left.

The apparent power due to the total impedance is:

$$VA = I_{\text{RMS}}V_{\text{RMS}} = (0.167 \text{ A})(10 \text{ VAC}) = 1.67 \text{ VA}$$

The real (true) power consumed by the circuit is:

$$P_R = I_{\text{RMS}}{}^2 R = (0.167 \text{ A})^2 (0 \text{ Ω}) = 0 \text{ W}$$

Only reactive power exists, and for the inductor and capacitor:

$$VAR_L = I_{\text{RMS}}{}^2 X_L = (0.167 \text{ A})^2 (80 \text{ Ω}) = 2.23 \text{ VA}$$

$$VAR_C = I_{\text{RMS}}{}^2 X_C = (0.167 \text{ A})^2 (20 \text{ Ω}) = 0.56 \text{ VA}$$

Power factor is:

$$PF = \frac{P_R}{VA} = \cos(\phi) = \cos(-90) = 0 \text{ (lagging)}$$

A power factor of 0 means the circuit is purely reactive.

Example 5: The LC parallel circuit in Fig. 2.171 is driven by a 10-VAC (RMS), 2893.7-Hz source. $L = 2.2$ mH, $C = 5.5$ μF. Find I_S, I_L, I_C, V_L, and V_C and the apparent power, real power, reactive power, and power factor.

Parallel Impedance (LC Circuit)

Inductive and Capacitive Reactances

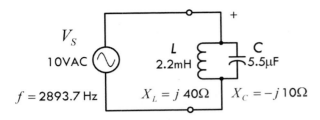

Equivalent Impedance and Current

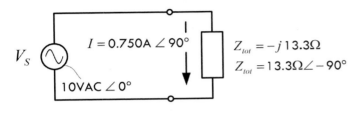

Current Through L and C

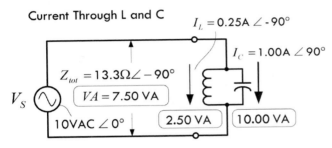

Sinusoidal Waveforms within Parallel LC Circuit

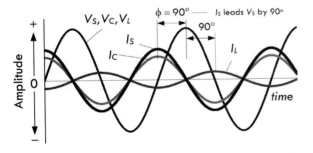

FIGURE 2.171

$V_S(t) = V_L(t) = V_C(t) = 14.1V \sin(\omega t)$

$I_S(t) = 1.061 \text{ A} \sin(\omega t + 90°)$

$I_L(t) = 0.354 \text{ A} \sin(\omega t - 90°)$

$I_C(t) = 1.414 \text{ A} \sin(\omega t + 90°)$

* Peak voltages and currents used in these functions are the RMS equivalents multiplied by 1.414.

First we find the inductive and capacitive reactances:

$$X_L = j\omega L = j(2\pi \times 2893.7 \text{ Hz} \times 2.2 \times 10^{-3} \text{ H})$$

$$= j40 \ \Omega$$

$$X_C = -j\frac{1}{\omega C} = -j\frac{1}{2\pi \times 2893.7 \text{ Hz} \times 5.5 \times 10^{-6} \text{ H}}$$

$$= -j10 \ \Omega$$

Since the inductor and capacitor are in parallel, the math is relatively easy—use two components in parallel formula, and multiply and add complex numbers:

$$Z = \frac{X_L X_C}{X_L + X_C} = \frac{(j40)(-j10)}{j40 + (-j10)} = \frac{400}{j30} = -j13.33 \ \Omega$$

(Tricks used to solve: $j \times j = -1$, $1/j = -j$)

In polar form, the result is 13.33 Ω $\angle -90°$. The fact that the phase is $-90°$, or negative imaginary, means the impedance is net capacitive. Don't let the imaginary part fool you—the impedance is actually felt—it provides 13.3 Ω of impedance, even though it's not resistive.

The total current can now be found using ac Ohm's law:

$$I_s = \frac{V_s}{Z} = \frac{10 \text{ VAC} \angle 0°}{13.33 \ \Omega \angle -90°} = 0.750 \text{ A} \angle 90°$$

Note that we used 13.3 Ω $\angle -90°$ (polar form) to make the division easy. The $+90°$ result means the total current leads the source voltage by $90°$. Since this is a parallel circuit: $V_S = V_L = V_C$.

The current through each component can be found using ac Ohm's law (or the current divider relation):

$$I_L = \frac{V_L}{X_L} = \frac{10 \text{ VAC} \angle 0°}{40 \ \Omega \angle 90°} = 0.25 \text{ A} \angle -90°$$

$$I_C = \frac{V_L}{X_C} = \frac{10 \text{ VAC} \angle 0°}{10 \ \Omega \angle -90°} = 1.0 \text{ A} \angle 90°$$

Notice that the capacitor current is larger than the supply current; the collapsing magnetic field of the inductor supplies current to the capacitor.

To convert these snapshots into a continuous set of functions, we convert all RMS values to true value (× 1.414), add the ωt term to the phase angles, and convert to trigonometric form, then delete the imaginary part. Our results would all be in terms of cosines, but to make things pretty, we select sines. Doing this doesn't make any practical difference. See graphs and equations in Fig. 2.171.

The apparent power is:

$$VA = I_{RMS}V_{RMS} = (0.750 \text{ A})(10 \text{ VAC}) = 7.50 \text{ VA}$$

Only reactive power exists, and for the inductor and capacitor:

$$VAR_L = I_L^2X_L = (0.25A)^2(40 \text{ }\Omega) = 2.50 \text{ VA}$$

$$VAR_C = I_C^2X_C = (1.00 \text{ A})^2(10 \text{ }\Omega) = 10.00 \text{ VA}$$

Power factor is:

$$PF = \frac{P_R}{VA} = \cos (\phi) = \cos (+90) = 0 \text{ (leading)}$$

Again, a power factor of 0 means the circuit is purely reactive.

Example 6: The LCR series circuit in Fig. 2.172 is driven by a 1.00-VAC (RMS), 1000-Hz source. $L = 25$ mH, $C = 1$ µF, and $R = 1.0$ Ω. Find the total impedance Z, V_L, V_C, V_R, and I_S, and the apparent power, real power, reactive power, and power factor.

Series Impedance (LC Circuit)

Resistance, Inductive and Capacitive Reactance

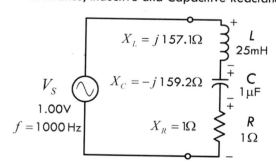

Equivalent Impedance and Current

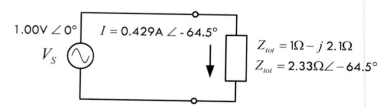

FIGURE 2.172

First let's find the reactances of the inductor and capacitor:

$$X_L = j\omega L = j(2\pi \times 1000 \text{ Hz} \times 25 \times 10^{-3} \text{ H})$$

$$= j157.1 \text{ }\Omega$$

$$X_C = -j\frac{1}{\omega C} = -j\frac{1}{2\pi \times 1000 \text{ Hz} \times 1 \times 10^{-6} C}$$

$$= -j159.2 \text{ }\Omega$$

To find the total impedance, take the L, C, and R in series:

$$Z = R + X_L + X_C = 1 \text{ }\Omega + j157.1 \text{ }\Omega - j159.2 \text{ }\Omega$$

$$= 1 \text{ }\Omega - j(2.1 \text{ }\Omega)$$

In polar form, the result is 2.33 Ω $\angle$ $-64.5°$. The fact that the phase is $-64.5°$, or the imaginary part is negative, means the

Voltage across R, L, and C

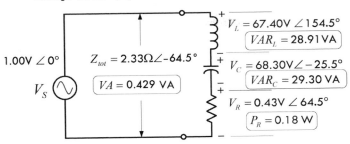

$V_L = 67.40V \angle 154.5°$
$VAR_L = 28.91VA$

$V_C = 68.30V \angle -25.5°$
$VAR_C = 29.30VA$

$V_R = 0.43V \angle 64.5°$
$P_R = 0.18W$

$1.00V \angle 0°$ $Z_{tot} = 2.33\Omega\angle -64.5°$ $VA = 0.429VA$

V_S

Sinusoidal Waveforms within Series RLC Circuit

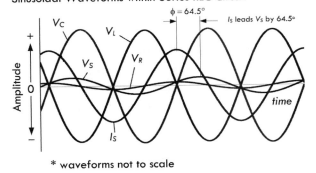

$\phi = 64.5°$ $I_S \text{ leads } V_S \text{ by } 64.5°$

V_C V_L V_R V_S I_S Amplitude   time

* waveforms not to scale

FIGURE 2.172   (*Continued*)

$V_S(t) = 1.41\,V \sin(\omega t)$

$V_L(t) = 95.32\,V \sin(\omega t + 154.5°)$

$V_C(t) = 96.60\,V \sin(\omega t - 25.5°)$

$V_R(t) = 0.61\,V \sin(\omega t + 64.5°)$

$I_S(t) = 0.607\,A \sin(\omega t + 64.5°)$

**   * Peak voltages and currents used in these functions are the RMS equivalents multiplied by 1.414.**

impedance is net capacitive. Don't let the imaginary part fool you—the impedance is actually felt—it provides 2.33 Ω of impedance, even though it's not entirely resistive.

The total current can now be found using ac Ohm's law:

$$I_s = \frac{V_s}{Z} = \frac{1.00\,\text{VAC} \angle 0°}{2.33\Omega \angle -64.5°} = 0.429A \angle 64.5°$$

Note that we used 2.33 Ω ∠ −64.5° (polar form) to make the division easy. The +64.5° result means the total current leads the source voltage by 64.5°. Since this is a series circuit: $I_S = I_L = I_C = I_R$.

The voltage across each component can be found using ac Ohm's law:

$$V_L = I_S X_L = (0.429\,A \angle 64.5°)(157.1\,\Omega \angle 90°)$$

$$= 67.40\,\text{VAC} \angle 154.5°$$

$$V_C = I_S X_C$$

$$= (0.429\,A \angle 64.5°)(159.2\,\Omega \angle -90°)$$

$$= 68.30\,\text{VAC} \angle -25.5°$$

$$V_R = I_S R = (0.429\,A \angle 64.5°)(1\,\Omega \angle 0°)$$

$$= 0.429\,\text{VAC} \angle 64.5°$$

Notice that the voltage across the inductor and capacitor are huge at this particular phase when compared to the supply voltage; the capacitor supplies current to the inductor as it discharges, while the inductor supplies current to the capacitor as its magnetic field collapses. To convert these snapshots into a continuous set of functions, we convert all RMS values to true value (× 1.414), add the ωt term to the phase angles, and convert to trigonometric form, then delete the imaginary part. Our results would all be in terms of cosines, but to make things pretty, we select sines. Doing this doesn't make any practical difference. See graphs and equations in Fig. 2.172.

The apparent power is:

$$VA = I_{\text{RMS}} V_{\text{RMS}} = (0.429\,A)(1.00\,\text{VAC}) = 0.429\,\text{VA}$$

The real (true) power, or power dissipated by resistor is:

$$P_R = I_S^2 R = (0.429\,A)^2(1\,\Omega) = 0.18\,W$$

The reactive powers for the inductor and capacitor:

$$VAR_L = I_L^2 X_L = (0.429 \text{ A})^2 (157.1 \text{ } \Omega) = 28.91 \text{ VA}$$

$$VAR_C = I_C^2 X_C = (0.429 \text{ A})^2 (159.2 \text{ } \Omega) = 29.30 \text{ VA}$$

Power factor is:

$$PF = \frac{P_R}{VA} = \cos(\phi) = \cos(64.5°) = 0.43 \text{ leading}$$

It should have become apparent from the last example that the VARs for the inductor and the capacitor became surprisingly large. When dealing with real-life components, the VAR values become important. Even though volt-amps do not contribute to the overall true power loss, this does not mean that the volt-amps aren't felt by the reactive components. Components like inductors and transformers (ideally reactive components) are usually given a volt-amp rating that provides the safety limit to prevent overheating the component. Again, it is the internal resistances within the inductor or transformer that must be considered.

Example 7: The LCR parallel circuit in Fig. 2.173 is driven by a 12.0-VAC (RMS), 600-Hz source. $L = 1.061$ mH, $C = 66.3$ μF, and $R = 10$ Ω. Find Z_{tot}, V_L, V_C, V_R, and I_S, and the apparent power, real power, reactive power, and power factor. First let's find the reactances of the inductor and capacitor:

Parallel Impedance (LC Circuit)

Inductive and Capacitive Reactances and Resistance

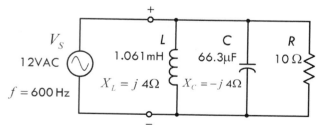

Equivalent Impedance and Current

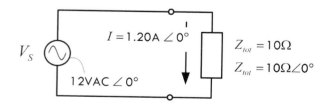

FIGURE 2.173

$$X_L = j\omega L = j(2\pi \times 600 \text{ Hz} \times 1.061 \times 10^{-3} \text{ H})$$

$$= j4.0 \text{ } \Omega$$

$$X_C = -j\frac{1}{\omega C} = -j\frac{1}{2\pi \times 600 \text{ Hz} \times 66.3 \times 10^{-6} \text{C}}$$

$$= -j4.0 \text{ } \Omega$$

Since the inductor and capacitor are in parallel, the math is relatively easy—use two components in general formula, and multiply and add complex numbers:

$$Z_{tot} = \frac{1}{\dfrac{1}{j4 \text{ } \Omega} + \dfrac{1}{-j4 \text{ } \Omega} + \dfrac{1}{10 \text{ } \Omega}} = 10 \text{ } \Omega$$

The fact that there is no reactive part to the total impedance is quite interesting and makes our life easy in terms of calculations. Before getting into the interesting matter, let's finish solving.

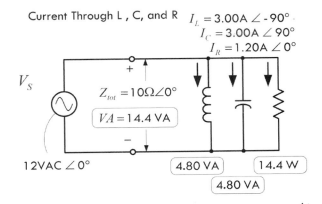

Current Through L , C, and R $I_L = 3.00A \angle -90°$
$I_C = 3.00A \angle 90°$
$I_R = 1.20A \angle 0°$

V_S
$Z_{tot} = 10\Omega \angle 0°$
$VA = 14.4\ VA$

12VAC $\angle 0°$ 4.80 VA   14.4 W
4.80 VA

Sinusoidal Waveforms within Parallel LC Circuit

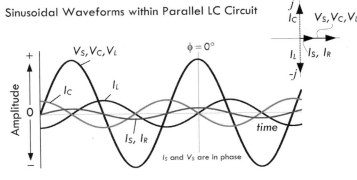

FIGURE 2.173   (*Continued*)

$V_S(t) = V_L(t) = V_C(t) = V_R(t) = 16.9\ \text{V} \sin(\omega t)$

$I_S(t) = 1.70\ \text{A} \sin(\omega t)$

$I_L(t) = 4.24\ \text{A} \sin(\omega t -90°)$

$I_C(t) = 4.24\ \text{A} \sin(\omega t + 90°)$

$I_R(t) = 1.70\ \text{A} \sin(\omega t)$

 * **Peak voltages and currents used in these functions are the RMS equivalents multiplied by 1.414.**

The total current can now be found using ac Ohm's law:

$$I_s = \frac{V_s}{Z} = \frac{12.0\ \text{VAC} \angle 0°}{10\ \Omega \angle 0°} = 1.20\ A \angle 0°$$

Since there is no phase angle, the source current and voltage are in phase. Since this is a parallel circuit, $V_S = V_L = V_C = V_R$.

The current through each component can be found using ac Ohm's law:

$$I_L = \frac{V_s}{X_L} = \frac{12.0\ \text{VAC} \angle 0°}{4\ \Omega \angle 90°} = 3.00\ \text{A} \angle -90°$$

$$I_C = \frac{V_s}{X_C} = \frac{12.0\ \text{VAC} \angle 0°}{4\ \Omega \angle -90°} = 3.00\ \text{A} \angle 90°$$

$$I_R = \frac{V_s}{R} = \frac{12.0\ \text{VAC} \angle 0°}{10\ \Omega \angle 0°} = 1.20\ \text{A} \angle 0°$$

You can convert the voltages and current back to true sinusoidal form to get the graph shown in Fig. 2.173.

The apparent power is:

$$VA = I_{RMS}V_{RMS} = (1.20\ \text{A})(12\ \text{VAC}) = 14.4\ \text{VA}$$

The real (true) power, or power dissipated by resistor, is:

$$P_R = I_s^2R = (1.20\ \text{A})^2(10\ \Omega) = 14.4\ \text{W}$$

The reactive powers for the inductor and capacitor:

$$VAR_L = I_L^2X_L = (1.20\ \text{A})^2(4\ \Omega) = 4.8\ \text{VA}$$

$$VAR_C = I_C^2X_C = (1.20\ \text{A})^2(4\ \Omega) = 4.8\ \text{VA}$$

Power factor is:

$$PF = \frac{P_R}{VA} = \cos(\phi) = \cos(0) = 1$$

A power factor of 1 indicates the circuit is purely resistive. How can this be? In this case we have a special condition where a circulating current is "trapped" within the LC section. This only occurs at a special frequency called the resonant frequency. We'll cover resonant circuits in a moment.

2.29 Thevenin's Theorem in AC Form

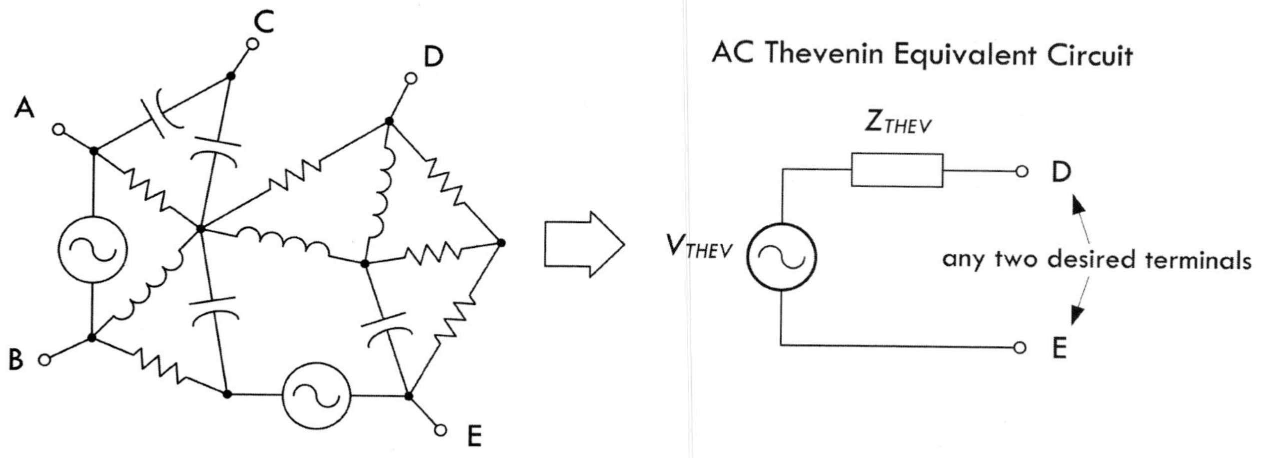

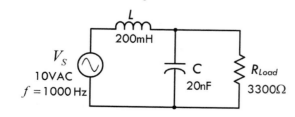

AC Thevenin Equivalent Circuit

FIGURE 2.174

Thevenin's theorem, like the other dc theorems, can be modified so that it can be used in ac linear circuits analysis. The revised ac form of Thevenin's theorem reads: Any complex network of resistors, capacitors, and inductors can be represented by a single sinusoidal power source connected to a single equivalent impedance. For example, if you want to find the voltage across two points in a complex, linear, sinusoidal circuit or find the current and voltage through and across a particular element within, you remove the element, find the Thevenin voltage $V_{\text{THEV}}(t)$, replace the sinusoidal sources with a short, find the Thevenin impedance $Z_{\text{THEV}}(t)$, and then make the Thevenin equivalent circuit. Figure 2.174 shows the Thevenin equivalent circuit for a complex circuit containing resistors, capacitors, and inductors. The following example will provide any missing details.

Example: Suppose that you are interested in finding the current through the resistor in the circuit in Fig. 2.175.

AC Thevenin Example

FIGURE 2.175

Answer: First, you remove the resistor in order to free up two terminals to make a black box. Next, you calculate the open circuit, or Thevenin voltage V_{THEV}, by using the ac voltage divider equation. First, however, let's determine the reactances of the capacitor and inductor:

$$X_L = j\omega L = j(2\pi \times 1000 \text{ Hz} \times 200 \times 10^{-3} \text{ H})$$

$$= j\,1257\ \Omega$$

$$X_C = -j\frac{1}{\omega C} = -j\frac{1}{2\pi \times 1000 \text{ Hz} \times 20 \times 10^{-9}\text{F}}$$

$$= -j7958\ \Omega$$

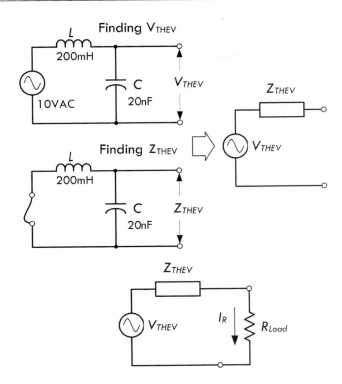

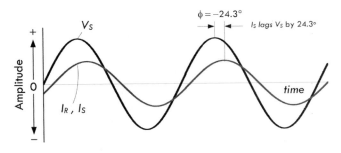

Sinusoidal Waveforms of Source Voltage and Resistor Current

FIGURE 2.175   *(Continued)*

$V_S(t) = V_C(t) = 14.1 \text{ V sin } (\omega t)$

$I_R(t) = I_S(t) = 4.64 \text{ mA sin } (\omega t - 24.3°)$

* Peak voltages and currents used in these functions are the RMS equivalents multiplied by 1.414.

Using the ac voltage divider:

$$V_{\text{THEV}} = V_C = \left(\frac{X_C}{X_C + X_L} \right) \times V_s$$

$$= \left(\frac{-j7958 \ \Omega}{-j7958 \ \Omega + j1257 \ \Omega} \right) \times 10 \text{ VAC}$$

$$= \left(\frac{7958 \ \angle -90°}{6701 \ \angle -90°} \right) \times 10 \text{ VAC}$$

$$= 11.88 \ VAC \ \angle \ 0°$$

To find Z_{THEV}, you short the source with a wire and take the impedance of the inductor and capacitor in parallel:

$$Z_{\text{THEV}} = \frac{X_C \times X_L}{X_C + X_L} = \frac{-j7958 \ \Omega \times j1257 \ \Omega}{-j7958 \ \Omega + -j1257 \ \Omega}$$

$$= \frac{(7958 \ \Omega \ \angle -90°)(1257 \ \Omega \ \angle \ 90°)}{6701 \ \Omega \ \angle -90°}$$

$$= \frac{10003206 \ \Omega^2 \ \angle \ 0°}{6702 \ \Omega \ \angle -90°}$$

$$= 1493 \ \Omega \ \angle \ 90° = j(1493 \ \Omega)$$

Next, you reattach the load resistor to the Thevenin equivalent circuit and find the current by combining V_{THEV} and R in series:

$$Z_{\text{TOTAL}} = R + Z_{\text{THEV}} = 3300 \ \Omega + j1493 \ \Omega$$

$$= 3622 \ \Omega \ \angle \ 24.3°$$

Using ac Ohm's law, we can find the current:

$$I_R = \frac{V_{\text{THEV}}}{Z_{\text{TOTAL}}} = \frac{11.88 \text{ VAC } \angle \ 0°}{3622 \ \Omega \ \angle \ 24.3°}$$

$$= 3.28 \text{ mA } \angle -24.3°$$

Don't let the complex expression fool you; the resistor current is indeed 3.28 mA, but lags the source voltage by 24.3°.

To turn our snapshots into real functions with respect to time, we add ωt to every phase angle expression, and convert from RMS to true values—see graphs and equations with Fig. 2.175.

Apparent power, real (true) power, reactive power, and power factor are:

$$VA = I_S^2 \times Z_{\text{TOTAL}} = (0.00328 \text{ A})^2(3622 \ \Omega) = 0.039 \text{ VA}$$

$$P_R = I_R^2 \times R = (0.00328 \text{ A})^2(3300 \ \Omega) = 0.035 \text{ W}$$

$$VAR = I_R^2 \times Z_{\text{THEV}} = (0.00328 \text{ A})^2(1493 \ \Omega) = 0.016 \text{ VA}$$

$$PF = \frac{P_R}{VA} = \cos(\phi) = \cos(-24.3)° = 0.91 \text{ lagging}$$

2.30 Resonant Circuits

When an LC circuit is driven by a sinusoidal voltage source at a special frequency called the resonant frequency, an interesting phenomenon occurs. For example, if you drive a series LC circuit (shown in Fig. 2.176) at its resonant angular frequency $\omega_0 = 1/\sqrt{LC}$, or equivalently its resonant frequency $f_0 = 1/(2\pi\sqrt{LC})$, the effective impedance across the LC network goes to zero. In effect, the LC network acts like a short. This means that the sourced current flow will be at a maximum. Ideally, it will go to infinity, but in reality it is limited by internal resistances of all the components in the circuit. To get an idea of how the series LC resonant circuit works, let's take a look at the following example.

LC Series Resonant Circuit

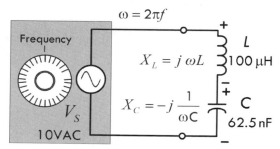

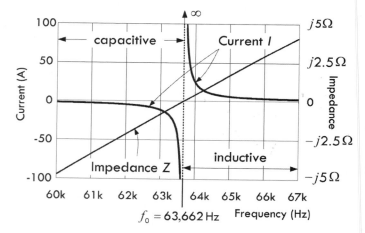

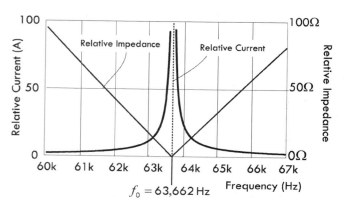

Example: To get an idea of how the LC series circuit works, we find the equivalent impedance of the circuit—taking L and C in series. Unlike the previous examples, the frequency is unknown, so it must be left as a variable:

$$Z_{TOT} = X_L + X_C$$

$$= j\omega L - j\frac{1}{\omega C} = j\left(\omega L - \frac{1}{\omega C}\right)$$

In polar form:

$$Z_{TOT} = \left(\omega L - \frac{1}{\omega C}\right) \angle 90°$$

Note that we got the phase angle by assuming that the arc tangent of anything divided by 0 is 90°.

The current through the series reactance is then:

$$I = \frac{V_S}{Z_{TOT}} = 10 \text{ VAC} \angle 0° / \left(\omega L - \frac{1}{\omega C}\right) \angle 90°$$

$$= \left[10 \text{ VAC} / \left(\omega L - \frac{1}{\omega C}\right)\right] \angle -90°$$

(Continued on next page.)

$$X_{L,0} = j2\pi f_0 L = j2\pi(62{,}663 \text{ s}^{-1})(100 \times 10^{-6}\text{H})$$

$$= j40\ \Omega = 40\ \Omega \angle 90°$$

$$X_{C,0} = -j\frac{1}{2\pi f_0 C} = -j\frac{1}{2\pi(62{,}663 \text{ s}^{-1})(62.5 \times 10^{-9}\text{F})}$$

$$= -j40\ \Omega = 40\ \Omega \angle -90°$$

FIGURE 2.176

If you plug in the $L = 100\ \mu H$ and $C = 62.5$ nF, and $\omega = 2\pi f$, the total impedance and current, ignoring phase angle, become:

$$|Z_{TOT}| = 6.28 \times 10^{-4} f - \frac{2,546,479}{f}\ \Omega$$

$$|I| = 10\ \text{VAC} / \left[6.28 \times 10^{-4} f - \frac{2,546,479}{f} \right] \Omega$$

Both the impedance and the current as a function of frequency are graphed in Fig. 2.176. Notice that as we approach the resonant frequency:

$$f_0 = \frac{1}{2\pi\sqrt{LC}} = \frac{1}{2\pi\sqrt{\left(100 \times 10^{-6}\text{H}\right)\left(62.5 \times 10^{-9}\text{F}\right)}} = 63,663\ \text{Hz}$$

the impedance goes to zero while the current goes to infinity. In other words, if you plug the resonant frequency into the impedance and current equations, the result gives you zero and infinity, respectively. In reality, internal resistance prevents infinite current.

Inductive and capacitive reactances at resonance are equal but opposite in phase, as depicted by the equations with Fig. 2.176.

Intuitively, you can imagine that the voltage across the capacitor and the voltage across the inductor are exactly equal but opposite in phase at resonance, within the LC series circuit. This means the effective voltage drop across the series pair is zero; therefore, the impedance across the pair must also be zero.

Resonance occurs in a parallel LC circuit as well. The angular resonant frequency is $\omega_0 = 1/\sqrt{LC}$ or equivalently its resonant frequency $f_0 = 1/(2\pi\sqrt{LC})$. This is the same resonant frequency expression for the series LC circuit; however, the circuit behavior is exactly opposite. Instead of the impedance going to zero and the current going to infinity at resonance, the impedance goes to infinity while the current goes to zero. In essence, the parallel LC network acts like an open circuit. Of course, in reality, there is always some internal resistance and parasitic capacitance and inductance within the circuit that prevent this from occurring. To get an idea of how the parallel LC resonant circuit works, let's take a look at the following example.

LC Parallel Resonant Circuit

Inductive and Capacitive Reactances

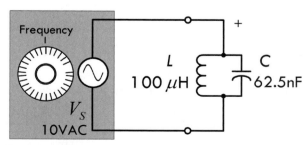

Frequency

V_S

10VAC

Frequency Generator

FIGURE 2.177

Example: For an LC parallel-resonant circuit, we take the inductor and capacitor in parallel (applying Eq. 2.72):

$$Z_{TOT} = \frac{X_L \times X_C}{X_L + X_C} = \frac{(j\omega L)(-j1/\omega C)}{j\omega L - j1/\omega C}$$

$$= \frac{L/C}{j(\omega L - 1/\omega C)} = -j\frac{L/C}{(\omega L - 1/\omega C)}$$

In polar form:

$$Z_{TOT} = \frac{L/C}{(\omega L - 1/\omega C)} \angle - 90°$$

Note that we got the phase angle by assuming that the negative arc tangent of anything over zero is −90°.

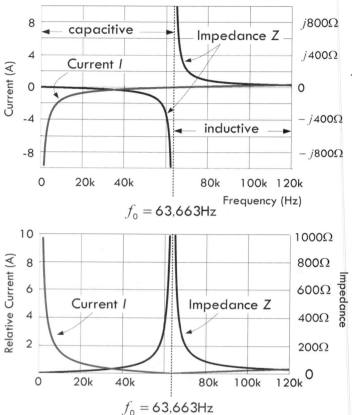

$$f_0 = 63{,}663\,\text{Hz}$$

$$f_0 = 63{,}663\,\text{Hz}$$

FIGURE 2.177   *(Continued)*

Inductive and capacitive reactances at resonance are equal but opposite in phase:

$$X_{L,0} = j2\pi f_0 L = j2\pi(62{,}663\ \text{s}^{-1})(100\times10^{-6}\,\text{H})$$

$$= j40\ \Omega = 40\ \Omega\ \angle\ 90°$$

$$X_{C,0} = -j\frac{1}{2\pi f_0 C} = -j\frac{1}{2\pi(62{,}663\ \text{s}^{-1})(62.5\times10^{-9}\,\text{F})}$$

$$= -j40\ \Omega = 40\ \Omega\ \angle\ -90°$$

The current through the parallel reactance is then:

$$I = \frac{V_s}{Z_{\text{TOT}}}$$

$$= 10\ \text{VAC}\ \angle\ 0°\Big/\frac{L/C}{(\omega L - 1/\omega C)}\ \angle\ -90°$$

$$= \left[10\ \text{VAC}\Big/\frac{L/C}{(\omega L - 1/\omega C)}\right]\ \angle\ 90°$$

If you plug in the $L = 100\ \mu\text{H}$ and $C = 62.5\ \text{nF}$, and $\omega = 2\pi f$, the total impedance and current, ignoring phase angle, are:

$$|Z_{\text{TOT}}| = 1600\Big/\left(6.28\times10^{-4}f - \frac{1}{3.92\times10^{-7}f}\right)\ \Omega$$

$$|I| = 0.00625\left(6.28\times10^{-4}f - \frac{1}{3.92\times10^{-7}f}\right)\ \text{A}$$

Both the impedance and the current as a function of frequency are graphed in Fig. 2.177. Notice that as we approach the resonant frequency:

$$f_0 = \frac{1}{2\pi\sqrt{LC}} = \frac{1}{2\pi\sqrt{(100\times10^{-6}\,\text{H})(62.5\times10^{-9}\,\text{F})}}$$

$$= 63{,}663\ \text{Hz}$$

the impedance goes to infinity while the current goes to zero. In other words, if you plug the resonant frequency into the impedance and current equations, you get infinity and zero, respectively. In reality, internal resistances and parasitic inductances and capacitors within the circuit prevent infinite current. Notice that as the frequency approaches zero, the current increases toward infinity, since the inductor acts like a short at dc. On the other hand, if the frequency increases toward infinity, the capacitor acts like a short, and the current goes toward infinity.

Intuitively, we can imagine that at resonance, the impedance and voltage across L are equal in magnitude but opposite in phase (direction) with respect to C. From this you can infer that an equal but opposite current will flow through L and C. In other words, at one moment a current is flowing upward through L and downward through C. The current through L runs into the top of C, while the current from C runs into the bottom of the inductor. At another moment the directions of the currents reverse (energy is "bounced" back in the other direction; L and C act as an oscillator pair that passes the same amount of energy back and forth, and the amount of energy is determined by the sizes of L and C). This internal current flow around the LC loop is referred to as a circulating current. Now, as this is going on, no more current will be supplied through the network by the source. Why? The power source doesn't "feel" a potential difference across it. Another way to put this would be to say that if an external current were to be supplied through the LC network, it would mean that one of the elements (L or C) would have to be passing more current than the other. However, at resonance, this does not happen because the L and C currents are equal and pointing in the opposite direction.

Example 1: What is the resonant frequency of a circuit containing an inductor of 5.0 μH and a capacitor of 35 pF?

Answer:

$$f_0 = 1/(2\pi\sqrt{LC}) = 1/[2\pi\sqrt{(5.0\times10^{-6}\text{ H})(35\times10^{-12}\text{ F})}] = 12\times10^{-6}\text{ Hz} = 12\text{ MHz}$$

Example 2: What is the value of capacitance needed to create a resonant circuit at 21.1 MHz, if the inductor is 2.00 μH?

Answer:

$$f_0 = 1/(2\pi\sqrt{LC}) \Rightarrow C = \frac{1}{L}\left(\frac{1}{2\pi f_0}\right)^2 = \frac{1}{2.0\times10^{-6}\text{ H}}\left[\frac{1}{2\pi(21.1\times10^6\text{ Hz})}\right]^2 = 2.85\times10^{-11}\text{ F}$$

$$= 28.5\text{ pF}$$

For most electronics work, these previous formulas will permit calculation of frequency and component values well within the limits of component tolerances. Resonant circuits have other properties of importance in addition to the resonant frequency, however. These include impedance, voltage drop across components in series-resonant circuits, circulating currents in parallel-resonant circuits, and bandwidth. These properties determine such factors as the selectivity of a tuned circuit and the component ratings for circuits handling considerable power. Although the basic determination of the tuned-circuit resonant frequency ignored any resistance in the circuit, that will play a vital role in the circuit's other characteristics.

2.30.1 Resonance in RLC Circuits

The previous LC series- and parallel-resonant circuits are ideal in nature. In reality, there is internal resistance or impedance within the components that leads to a deviation from the true resonant effects we observed. In most real LC resonant circuits, the most notable resistance is associated with losses in the inductor at high frequencies (HF range); resistive losses in a capacitor are low enough at those frequencies to be ignored. The following example shows how a series RLC circuit works.

RLC Series Resonant Circuit

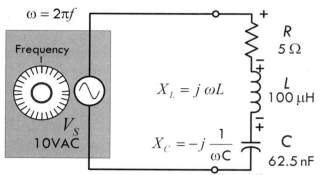

Frequency Generator

$\omega = 2\pi f$

Frequency

V_S
10VAC

$X_L = j\,\omega L$

$X_C = -j\dfrac{1}{\omega C}$

R
5 Ω

L
100 μH

C
62.5 nF

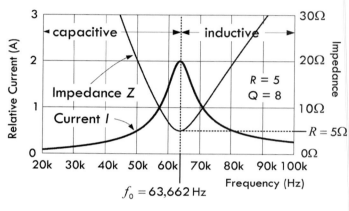

FIGURE 2.178

Inductive and capacitive reactances at resonance are equal but opposite in phase:

$$X_{L,0} = j2\pi f_0 L = j2\pi(62{,}663\ \text{s}^{-1})(100\times 10^{-6}\,\text{H})$$

$$= j40\ \Omega = 40\ \Omega\ \angle\ 90°$$

$$X_{C,0} = -j\frac{1}{2\pi f_0 C} = -j\frac{1}{2\pi(62{,}663\ \text{s}^{-1})(62.5\times 10^{-9}\,\text{F})}$$

$$= -j40\ \Omega = 40\ \Omega\ \angle -90°$$

Example: We start by finding the total impedance of the RLC circuit by taking R, L, and C in series:

$$Z_{\text{TOT}} = R + X_L + X_C = R + j\omega L - j\frac{1}{\omega C}$$

$$= R + j\left(\omega L - \frac{1}{\omega C}\right)$$

In polar form:

$$Z_{\text{TOT}} = \sqrt{R^2 + (\omega L - 1/\omega C)^2}\ \angle \tan^{-1}\left(\frac{\omega L - 1/\omega C}{R}\right)$$

The current through the total impedance is, ignoring phase:

$$I = \frac{V_s}{Z_{\text{TOT}}} = 10\ \text{VAC}/\sqrt{R^2 + (\omega L - 1/\omega C)^2}$$

If you plug in the $L = 100$ μH and $C = 62.5$ nF, and $\omega = 2\pi f$, the current expressed as a function of frequency is:

$$I = \frac{10\ \text{VAC}}{\sqrt{25 + (6.28\times 10^{-4} f - 2{,}546{,}479/f)^2}\ \Omega}$$

Now, unlike the ideal LC series-resonant circuit, when we plug in the resonant frequency:

$$f_0 = \frac{1}{2\pi\sqrt{LC}} = \frac{1}{2\pi\sqrt{(100\times 10^{-6}\,\text{H})(62.5\times 10^{-9}\,\text{F})}}$$

$$= 63{,}663\ \text{Hz}$$

the total current doesn't go to infinity; instead it goes to 2 A, which is just $V_S/R = 10$ VAC/ 5 Ω = 2 A. The resistance thus prevents a zero impedance condition that would otherwise be present when the inductor and capacitor reactances cancel at resonance.

The unloaded Q is the reactance at resonance divided by the resistance:

$$Q_U = \frac{1}{R}\sqrt{\frac{L}{C}} = \frac{X_{L,0}}{R} = \frac{\omega_0 L}{R} = \frac{2\pi f_0 L}{R} = \frac{40\ \Omega}{5\ \Omega} = 8$$

As pointed out in Fig. 2.178, at resonance, the reactance of the capacitor cancels the reactance of the inductor, and thus the impedance is determined solely by the resistance. We can therefore deduce that at resonance the current and voltage must be in phase—recall that in a sinusoidal circuit with a single resistor, the current and voltage are in phase. However, as we move away from the resonant frequency (keeping component values the same), the impedance goes up due to increases in reactance of the capacitor or the inductor. As you go lower in frequency from resonance, the capacitor's reactance becomes dominant—capacitors increasingly resist current as the frequency decreases. As you go higher in frequency from resonance, the inductor's reactance becomes dominant—inductors increasingly resist current as the frequency increases. Far from resonance in either direction, you can see that resistance has an insignificant effect on current amplitude.

Now if you take a look at the graph in Fig. 2.178, notice how the current curve looks like a pointy hilltop. In electronics, describing the pointiness of the current curve is an important characteristic of concern. When the reactance of the inductor or capacitor is of the same order of magnitude as the resistance, the current decreases rather slowly as you move away from the resonant frequency in either direction. Such a curve, or "hilltop," is said to be *broad*. Conversely, when the reactance of the inductor or capacitor is considerably larger than the resistance, the current decreases rapidly as you move away from the resonant frequency in either direction. Such a curve, or "hilltop," is said to be *sharp*. A sharp resonant circuit will respond a great deal more readily to the resonant frequency than to frequencies quite close to resonance. A broad resonant circuit will respond almost equally well to a group or band of frequencies centered about the resonant frequency.

As it turns out, both sharp and narrow circuits are useful. A sharp circuit gives good selectivity. This means that it has the ability to respond strongly (in terms of current amplitude) at one desired frequency and is able to discriminate against others. A broad circuit, on the other hand, is used in situations where a similar response over a band of frequencies is desired, as opposed to a strong response at a single frequency.

Next, we'll take a look at quality factor and bandwidth—two quantities that provide a measure of the sharpness of our RLC resonant circuit.

2.30.2   Q (Quality Factor) and Bandwidth

As mentioned earlier, the ratio of reactance or stored energy to resistance or consumed energy is by definition the quality factor Q. (Q is also referred to as the figure of merit, or multiplying factor.) As it turns out, within a series RLC circuit (where R is internal resistance of the components), the internal resistive losses within the inductor dominate energy consumption at high frequencies. This means the inductor Q largely determines the resonant circuit Q. Since the value of Q is independent of any external load to which the circuit might transfer power, we modify the resonant circuit Q, by calling it the unloaded Q or Q_U of the circuit. See Fig. 2.179.

In the example RLC series-resonant circuit from Fig. 2.178, we can determine the unloaded Q of the circuit by dividing the reactance of either the inductor or the capacitor (they have the same relative impedance at resonance) by the resistance:

$$Q_U = \frac{X_{L,0}}{R} = \frac{40\ \Omega}{5\ \Omega} = 8$$

$$Q_U = \frac{X_{C,0}}{R} = \frac{40\ \Omega}{5\ \Omega} = 8$$

Quality Factor (Q)

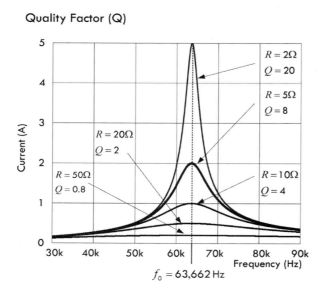

Bandwidth

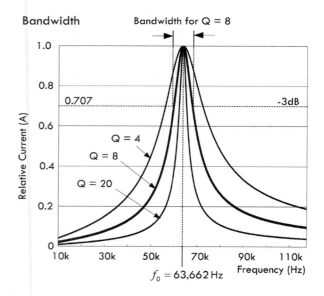

FIGURE 2.179

As you can see, if we increase the resistance, the unloaded Q decreases, giving rise to broad current response curves about resonance, as shown in the graph in Fig. 2.179. With resistances of 10, 20, and 50 Ω, the unloaded Q decreases to 4, 2, and 0.8, respectively. Conversely, if the resistance is made smaller, the unloaded Q increases, giving rise to a sharp current response curve about resonance. For example, when the resistance is lowered to 2 Ω, the unloaded Q becomes 20.

2.30.3 Bandwidth

An alternative way of expressing the sharpness of a series-resonant circuit is using what is called bandwidth. Basically, what we do is take the quality factor graph in Fig. 2.179 and convert it to the bandwidth graph in Fig. 2.179. This involves changing the current axis to a relative current axis and moving the family of curves for varying Q values up so that all have the same peak current. By assuming that the peak current of each curve is the same, the rate of change of current for various values of Q and the associated ratios of reactance to resistance are more easily compared. From the curves, you can see that lower Q circuits pass frequencies over a greater bandwidth of frequencies than circuits with a higher Q. For the purpose of comparing tuned circuits, bandwidth is often defined as the frequency spread between the two frequencies at which the current amplitude decreases to 0.707 or $1/\sqrt{2}$ times the maximum value. Since the power consumed by the resistance R is proportional to the square of the current, the power at these points is half the maximum power at resonance, assuming that R is constant for the calculations. The half-power, or -3-dB, points are marked on the drawing.

For Q values of 10 or greater, the curves shown in Fig. 2.179 are approximately symmetrical. On this assumption, bandwidth (BW) can be easily calculated:

$$\text{BW} = \frac{f_0}{Q_u} \tag{2.81}$$

where BW and f are in units of hertz.

Example: What is the bandwidth of the series-resonant circuit in Fig. 2.178 at 100 kHz and at 1 MHz?

Answer:

$$BW_1 = \frac{f_0}{Q_U} = \frac{100,000 \text{ Hz}}{8} = 12,500 \text{ Hz}$$

$$BW_2 = \frac{f_0}{Q_U} = \frac{1 \text{ MHz}}{8} = 125,000 \text{ Hz}$$

2.30.4   *Voltage Drop Across Components in RLC Resonant Circuit*

The voltage drop across a given inductor or a capacitor within an RLC resonant circuit can be determined by applying ac Ohm's law:

$$V_C = X_C I = \frac{1}{2\pi f_0 C} \times I \quad \text{and} \quad V_L = X_L I = 2\pi f_0 L \times I$$

As we discovered earlier, these voltages may become many times larger than the source voltage, due to the magnetic and electric stored energy returned by the inductor and capacitor. This is especially true for circuits with high Q values. For example, at resonance, the RLC circuit of Fig. 2.178 has the following voltage drops across the capacitor and inductor:

$$V_C = X_C I = 40 \ \Omega \ \angle -90° \times 2 \text{ A} \ \angle \ 0° = 80 \text{ VAC} \ \angle -90°$$

$$V_L = X_L I = 40 \ \Omega \ \angle +90° \times 2 \text{ A} \ \angle \ 0° = 80 \text{ VAC} \ \angle +90°$$

The actual amplitude of the voltage when we convert from the RMS values is a factor of 1.414 higher, or 113 V. High-Q circuits such as those found in antenna couplers, which handle significant power, may experience arcing from high reactive voltages, even though the source voltage to the circuit is well within component ratings. When Qs of greater than 10 are considered, the following equation gives a good approximation of the reactive voltage within a series-resonant RLC circuit at resonance:

$$V_X = Q_U V_S \tag{2.82}$$

2.30.5   *Capacitor Losses*

Note that although capacitor energy losses tend to be insignificant compared to inductor losses up to about 30 MHz within a series-resonant circuit, these losses may affect circuit Q in the VHF range (30 to 300 MHz). Leakage resistance, principally in the solid dielectric between the capacitor plates, is not exactly like the internal wire resistive losses in an inductor coil. Instead of forming a series resistance, resistance associated with capacitor leakage usually forms a parallel resistance with the capacitive reactance. If the leakage resistance of a capacitor is significant enough to affect the Q of a series-resonant circuit, the parallel resistance must be converted to an equivalent series resistance before adding it to the inductor's resistance. This equivalent series resistance is given by:

$$R_S = \frac{X^2_C}{R_P} = \frac{1}{R_P \times (2\pi f C)^2} \tag{2.83}$$

where R_P is the leakage resistance and X_C is the capacitive reactance. This value is then added to the inductor's internal resistance and the sum represents the R in an RLC resonant circuit.

Example: A 10.0-pF capacitor has a leakage resistance of 9,000 Ω at 40.0 MHz. What is the equivalent series resistance?

$$R_S = \frac{1}{R_P \times (2\pi f C)^2} = \frac{1}{9,000\ \Omega \times (6.283 \times 40.0 \times 10^6 \times 10.0 \times 10^{-12})^2} = 17.6\ \Omega$$

When calculating the impedance, current, and bandwidth of a series-resonant circuit, the series leakage resistance is added to the inductor's coil resistance. Since an inductor's resistance tends to increase with frequency due to what's called the skin effect (electrons forced to the surface of a wire), the combined losses in the capacitor and the inductor can seriously reduce circuit Q.

Example 1: What is the unloaded Q of a series-resonant circuit with a loss resistance of 4 Ω and inductive and capacitive components having a reactance of 200 Ω each? What is the unloaded Q if the reactances are 20 Ω each?

Answer:

$$Q_{u1} = \frac{X_1}{R} = \frac{200\ \Omega}{4\ \Omega} = 50$$

$$Q_{u2} = \frac{X_2}{R} = \frac{20\ \Omega}{4\ \Omega} = 5$$

Example 2: What is the unloaded Q of a series-resonant circuit operating at 7.75 MHz, if the bandwidth is 775 kHz?

$$Q_u = \frac{f}{BW} = \frac{7.75\ \text{MHz}}{0.775\ \text{MHz}} = 10.0$$

2.30.6 Parallel-Resonant Circuits

Although series-resonant circuits are common, the vast majority of resonant circuits are parallel-resonant circuits. Figure 2.180 shows a typical parallel-resonant circuit. As with series-resonant circuits, the inductor internal coil resistance is the main source of resistive losses, and therefore we put the series resistor in the same leg as the inductor. Unlike the series-resonant circuit whose impedance goes toward a minimum at resonance, the parallel-resonant circuit's impedance goes to a maximum. For this reason, RLC parallel-resonant circuits are often called *antiresonant* circuits or *rejector* circuits. (RLC series-resonant circuits go by the name *acceptor*). The following example will paint a picture of parallel resonance behavior.

RLC Parallel-Resonant Circuit

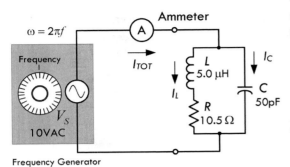

Frequency Generator

Behavior Near Resonance

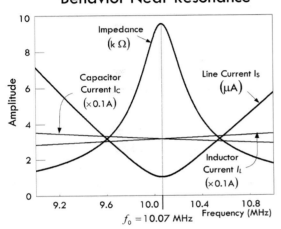

Quality Factor (Q)

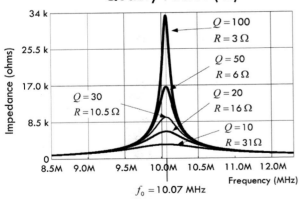

$f_0 = 10.07$ MHz

FIGURE 2.180

Example: The impedance of the RLC parallel circuit is found by combining the inductor and resistor in series, and then placing it in parallel with the capacitor (using the impedances in parallel formula):

$$Z_{TOT} = \frac{(R + X_L) \times X_C}{(R + X_L) + X_C} = \frac{(R + j\omega L)(-j1/\omega C)}{(R + j\omega L) + (-j1/\omega C)}$$

$$= \frac{L/C - j[R/(\omega C)]}{R + j[\omega L - 1/\omega C]}$$

In polar form:

$$Z_{TOT} = \frac{\sqrt{(L/C)^2 + [R/(\omega C)]^2} \angle \tan^{-1}\left[\dfrac{R/(\omega C)}{L/C}\right]}{\sqrt{R^2 + [\omega L - 1/(\omega C)]^2} \angle \tan^{-1}\left[\dfrac{\omega L - 1/(\omega C)}{R}\right]}$$

If you plug in $L = 5.0$ μH and $C = 50$ pF, $R = 10.5$ Ω and $\omega = 2\pi f$, the total impedance, ignoring phase, is:

$$Z_{TOT} = \frac{\sqrt{1.0 \times 10^{10} + \left(\dfrac{3.34 \times 10^{10}}{f}\right)^2}}{\sqrt{110.3 + \left(3.14 \times 10^{-5} f - \dfrac{3.18 \times 10^9}{f}\right)^2}} \; \Omega$$

The total current (line current), ignoring phase, is:

$$I_{TOT} = \frac{V_S}{Z_{TOT}}$$

$$= 10 \text{ V} \left/ \frac{\sqrt{1.0 \times 10^{10} + \left(\dfrac{3.34 \times 10^{10}}{f}\right)^2}}{\sqrt{110.3 + \left(3.14 \times 10^{-5} f - \dfrac{3.18 \times 10^9}{f}\right)^2}} \right. \; \Omega$$

Plugging these equations into a graphing program, you get the curves shown in Fig. 2.180. Note that at a particular frequency, the impedance goes to a maximum, while the total current goes to a minimum. This, however, is not at the point where $X_L = X_C$—the point referred to as the resonant frequency in the case of a simple LC parallel circuit or a series RLC circuit. As it turns out, the resonant frequency of a parallel RLC circuit is a bit more complex and can be expressed three possible ways. However, for now, we make an approximation that is expressed as before:

$$f_0 = \frac{1}{2\pi\sqrt{LC}} = 10,070,000 \text{ Hz} = 10.07 \text{ MHz}$$

The unloaded Q for this circuit, using the reactance of L, is:

$$Q_u = \frac{X_{L,0}}{R} = \frac{\omega_0 L}{R} = \frac{2\pi f_0 L}{R} = \frac{316.4 \text{ Ω}}{10.5 \text{ Ω}} = 30$$

The lower graph in Fig. 2.180 shows the quality factor and how it is influenced by the size of the parallel resistance in the inductor leg.

Unlike the ideal LC parallel resonant circuit we saw in Fig. 2.177, the addition of R changes the conditions of resonance. For example, when the inductive and capacitive reactances are the same, the impedances of the inductive and capacitive legs do not cancel—the resistance in the inductive leg screws things up. When $X_L = X_C$, the impedance of the inductive leg is greater than X_C and will not be 180° out of phase with X_C. The resultant current is not at a true minimum value and is not in phase with the voltage. See line (A) in Fig. 2.181.

Now we can alter the value of the inductor a bit (holding Q constant) and get a new frequency where the current reaches a true minimum—something we can accomplish with the help of a current meter. We associate the dip in current at this new frequency with what is termed the standard definition of RLC parallel resonance. The point where minimum current (or maximum impedance) is reached is called the *antiresonant point* and is not to be confused with the condition where $X_L = X_C$. Altering the inductance to achieve minimum current comes at a price; the current becomes somewhat out of phase with the voltage—see line (B) in Fig. 2.181.

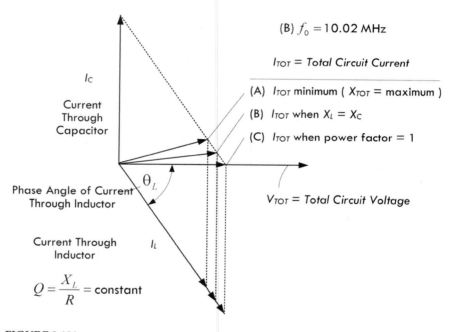

FIGURE 2.181

It's possible to alter the circuit design of our RLC parallel-resonant circuit to draw some of the resonant points shown in Fig. 2.181 together—for example, compensating for the resistance of the inductor by altering the capacitance (retuning the capacitor). The difference among the resonant points tends to get smaller and converge to within a percent of the frequency when the circuit Q rises above 10, in which case they can be ignored for practical calculations. Tuning for minimum current will not introduce a sufficiently large phase angle between voltage and current to create circuit difficulties.

As long as we assume Qs higher than 10, we can use a single set of formulas to predict circuit performance. As it turns out, what we end up doing is removing the series inductor resistance in the leg with the inductor and substituting a parallel-equivalent resistor of the actual inductor loss series resistor, as shown in Fig. 2.182.

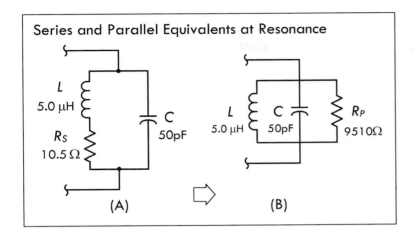

Series and parallel equivalents when both circuits are resonant. Series resistance R_S in (a) is replaced by the parallel resistance R_P in (b), and vice versa:

$$R_P = \frac{X_L^2}{R_S} = \frac{(2\pi fL)^2}{R_S} = Q_U X_L$$

FIGURE 2.182

This parallel-equivalent resistance is often called the dynamic resistance of the parallel-resonant circuit. This resistance is the inverse of the series resistance; as the series resistance value decreases, the parallel-equivalent resistance increases. Alternatively, this means that the parallel-equivalent resistance will increase with circuit Q. We use the following formula to calculate the approximate parallel-equivalent resistance:

$$R_P = \frac{X_L^2}{R_S} = \frac{(2\pi fL)^2}{R_S} = Q_U X_L \tag{2.84}$$

Example: What is the parallel-equivalent resistance for the inductor in Fig. 2.182b, taking the inductive reactance to be 316 Ω and a series resistance to be 10.5 Ω at resonance? Also determine the unloaded Q of the circuit.

Answer:

$$R_P = \frac{X_L^2}{R_S} = \frac{(316\ \Omega)^2}{10.5\ \Omega} = 9510\ \Omega$$

Since the coil Q_U remains the inductor's reactance divided by its series resistance, we get:

$$Q_U = \frac{X_L}{R_S} = \frac{316\ \Omega}{10.5\ \Omega} = 30$$

Multiplying Q_U by the reactance also provides the approximate parallel-equivalent resistance of the inductor's series resistance.

At resonance, assuming our parallel equivalent representation, $X_L = X_C$, and R_P now defines the impedance of the parallel-resonant circuit. The reactances just equal each other, leaving the voltage and current in phase with each other. In other words, at resonance, the circuit demonstrates only parallel resistance. Therefore, Eq. 2.84 can be rewritten as:

$$Z = \frac{X_L^2}{R_S} = \frac{(2\pi fL)^2}{R_S} = Q_U X_L \tag{2.85}$$

In the preceding example, the circuit impedance at resonance is 9510 Ω.

At frequencies below resonance, the reactance of the inductor is smaller than the reactance of the capacitor, and therefore the current through the inductor will be larger than that through the capacitor. This means that there is only partial cancellation of the two reactive currents, and therefore the line current is larger than the current with resistance alone. Above resonance, things are reversed; more current flows through the capacitor than through the inductor, and again the line current increases above a value larger than the current with resistance alone. At resonance, the current is determined entirely by R_P; it will be small if R_P is large and large if R_P is small.

Since the current rises off resonance, the parallel-resonant-circuit impedance must fall. It also becomes complex, resulting in an increasing phase difference between the voltage and the current. The rate at which the impedance falls is a function of Q_U. Figure 2.180 presents a family of curves showing the impedance drop from resonance for circuit Qs ranging from 10 to 100. The curve family for parallel-circuit impedance is essentially the same as the curve family for series-circuit current. As with series-tuned circuits, the higher the Q of a parallel-tuned circuit, the sharper the response peak. Likewise, the lower the Q, the wider the band of frequencies to which the circuit responds. Using the half-power (-3-dB) points as a comparative measure of circuit performance, we can apply the same equations for bandwidth for a series-resonant circuit to a parallel-resonant circuit, BW $= f/Q_U$. Table 2.11 summarizes performance of parallel-resonant circuits at high and low Qs, above and below resonance.

TABLE 2.11   Performance of Parallel-Resonant Circuits

A. High and Low Q Parallel-Resonant Circuits

	HIGH Q CIRCUIT	LOW Q CIRCUIT
Selectivity	High	Low
Bandwidth	Narrow	Wide
Impedance	High	Low
Line current	Low	High
Circulating current	High	Low

B. Off-Resonance Performance for Constant Values of Inductance and Capacitance

	ABOVE RESONANCE	BELOW RESONANCE
Inductive reactance	Increases	Decreases
Capacitive reactance	Decreases	Increases
Circuit resistance	Same*	Same*
Circuit impedance	Decreases	Decreases
Line current	Increases	Increases
Circulating current	Decreases	Decreases
Circuit behavior	Capacitive	Inductive

*True near resonance, but far from resonance skin effects within inductor alter the resistive losses.

Circulating Current at Resonance

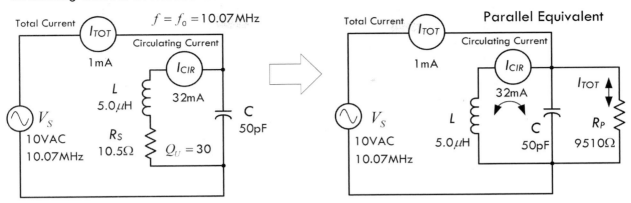

FIGURE 2.183

Note on Circulating Current

When we covered the ideal LC parallel-resonant circuit, we saw that quite a large circulating current could exist between the capacitor and the inductor at resonance, with no line current being drawn from the source. If we consider the more realistic RLC parallel-resonant circuit, we also notice a circulating current at resonance (which, too, can be quite large compared to the source voltage), but now there exists a small line current sourced by the load. This current is attributed to the fact that even though the impedance of the resonant network is high, it isn't infinite because there are resistive losses as the current circulates through the inductor and capacitor—most of which are attributed to the inductor's internal resistance.

Taking our example from Fig. 2.183 and using the parallel-equivalent circuit, as shown to the right in Fig. 2.183, we associate the total line current as flowing through the parallel inductor resistance R_P. Since the inductor, capacitor, and parallel resistor are all in parallel according to the parallel-equivalent circuit, we can determine the circulating current present between the inductor and capacitor, as well as the total line current now attributed to the parallel resistance:

$$I_R = \frac{V_s}{R_P} = \frac{10 \text{ VAC}}{9510 \ \Omega} = 1 \text{ mA}$$

$$I_L = \frac{V_S}{X_L} = \frac{V_S}{2\pi f L} = \frac{10 \text{ VAC}}{2\pi(10.07 \times 10^6 \text{ s}^{-1})(5.0 \times 10^{-6} \text{ H})} = \frac{10 \text{ VAC}}{316 \ \Omega} = 32 \text{ mA}$$

$$I_C = \frac{V_S}{X_C} = \frac{V_S}{1/(2\pi f C)} = \frac{10 \text{ VAC}}{1/[(2\pi)(10.07 \times 10^6 \text{ s}^{-1})(50.0 \times 10^{-12} \text{ F})]} = \frac{10 \text{ VAC}}{316 \ \Omega} = 32 \text{ mA}$$

The circuiting current is simply $I_{\text{CIR}} = I_C = I_L$ when the circuit is at the resonant frequency. For parallel-resonant circuits with an unloaded Q of 10 or greater, the circulating current is approximately equal to:

$$I_{\text{CIR}} = Q_U \times I_{\text{TOT}} \tag{2.86}$$

Using our example, if we measure the total line current to be 1 mA, and taking the Q of the circuit to be 30, the approximated circulating current is (30)(1 mA) = 30 mA.

Example: A parallel-resonant circuit permits an ac line current of 50 mA and has a Q of 100. What is the circulating current through the elements?

$$I_C = Q_U \times I_T = 100 \times 0.05 \text{ A} = 5 \text{ A}$$

Circulating current in high-Q parallel-tuned circuits can reach a level that causes component heating and power loss. Therefore, components should be rated for the anticipated circulating current, and not just for the line currents.

It is possible to use either series- or parallel-resonant circuits to do the same work in many circuits, thus providing flexibility. Figure 2.184 illustrates this by showing both a series-resonant circuit in the signal path and a parallel-resonant circuit shunted from the signal path to ground. Assuming both circuits are resonant at the same frequency f and have the same Q: the series-tuned circuit has its lowest impedance at the resonant frequency, permitting the maximum possible current to flow along the signal path; at all other frequencies, the impedance increases causing a decrease in current. The circuit passes the desired signal and tends to impede signals at undesired frequencies. The parallel circuit, on the other hand, provides the highest impedance at resonance, making the signal path lowest in impedance for the signal; at all frequencies off resonance, the parallel circuit presents a lower impedance, providing signals a route to ground away from the signal path. In theory, the effects displayed for both circuits are the same. However, in actual circuit design, there are many other things to consider, which ultimately determine which circuit is best for a particular application. We will discuss such circuits later, when we cover filter circuits.

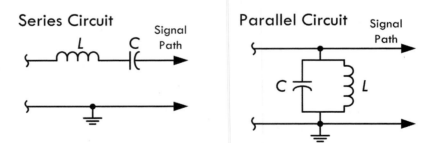

FIGURE 2.184

2.30.7 *The Q of Loaded Circuits*

In many resonant circuit applications, the only practical power lost is dissipated in the resonant circuit internal resistance. At frequencies below around 30 MHz, most of the internal resistance is within the inductor coil itself. Increasing the number of turns in an inductor coil increases the reactance at a rate that is faster than the accompanying internal resistance of the coil. Inductors used in circuits where high Q is necessary have large inductances.

When a resonant circuit is used to deliver energy to a load, the energy consumed within the resonant circuit is usually insignificant compared to that consumed by the load. For example, in Fig. 2.185, a parallel load resistor R_{LOAD} is attached to a resonant circuit, from which it receives power.

Loaded Circuit

Parallel Equivalent

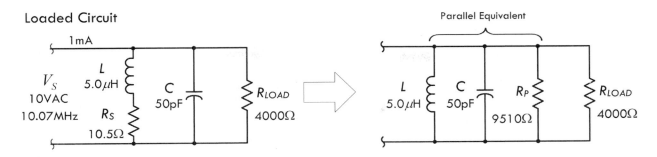

FIGURE 2.185

If the power dissipated by the load is at least 10 times as great as the power lost in the inductor and capacitor, the parallel impedance of the resonant circuit itself will be so high compared with the resistance of the load that for all practical purposes the impedance of the combined circuit is equal to the load impedance. In these circumstances, the load resistance replaces the circuit impedance in calculating Q. The Q of a parallel-resonant circuit loaded by a resistance is:

$$Q_{\text{LOAD}} = \frac{R_{\text{LOAD}}}{X} \qquad (2.87)$$

where Q_{LOAD} is the circuit-loaded Q, R_{LOAD} is the parallel load resistance in ohms, and X is the reactance in ohms of either the inductor or the capacitor.

Example 1: A resistive load of 4000 Ω is connected across the resonant circuit shown in Fig. 2.185, where the inductive and capacitive reactances at resonance are 316 Ω. What is the loaded Q for this circuit?

Answer:

$$Q_{\text{LOAD}} = \frac{R_{\text{LOAD}}}{X} = \frac{4000 \ \Omega}{316 \ \Omega} = 13$$

The loaded Q of a circuit increases when the reactances are decreased. When a circuit is loaded with a low resistance (a few kiloohms) it must have low-reactance elements (large capacitance and small inductance) to have reasonably high Q.

Sometimes parallel load resistors are added to parallel-resonant circuits to lower the Q and increase the circuit bandwidth, as the following example illustrates.

Example 2: A parallel-resonant circuit needs to be designed with a bandwidth of 400.0 kHz at 14.0 MHz. The current circuit has a Q_u of 70.0 and its components have reactances of 350 Ω each. What is the parallel load resistor that will increase the bandwidth to the specified value?

Answer: First, we determine the bandwidth of the existing circuit:

$$\text{BW} = \frac{f}{Q_u} = \frac{14.0 \ \text{MHz}}{70.0} = 0.200 \ \text{MHz} = 200 \ \text{kHz}$$

The desired bandwidth, 400 kHz, requires a loaded circuit Q of:

$$Q_{\text{LOAD}} = \frac{f}{\text{BW}} = \frac{140 \ \text{MHz}}{0.400 \ \text{MHz}} = 35.0$$

Since the desired Q is half the original value, halving the resonant impedance or parallel-resistance value of the circuit will do the trick. The present impedance of the circuit is:

$$Z = Q_U X_L = 70 \times 350 \ \Omega = 24{,}500 \ \Omega$$

The desired impedance is:

$$Z = Q_U X_L = 35.0 \times 350 \ \Omega = 12{,}250 \ \Omega$$

or half the present impedance.

A parallel resistor of 24,500 Ω will produce the required reduction in Q as bandwidth increases. In real design situations, things are a bit more complex—one must consider factors such as shape of the bandpass curve.

2.31 Lecture on Decibels

In electronics, often you will encounter situations where you will need to compare the relative amplitudes or the relative powers between two signals. For example, if an amplifier has an output voltage that is 10 times the input voltage, you can set up a ratio:

$$V_{\text{out}}/V_{\text{in}} = 10 \ \text{VAC}/1 \ \text{VAC} = 10$$

and give the ratio a special name—called *gain*. If you have a device whose output voltage is 10 times smaller than the input voltage, the gain ratio will be less than 1:

$$V_{\text{out}}/V_{\text{in}} = 1 \ \text{VAC}/10 \ \text{VAC} = 0.10$$

In this case, you call the ratio the *attenuation*.

Using ratios to make comparisons between two signals or powers is done all the time—not only in electronics. However, there are times when the range over which the ratio of amplitudes between two signals or the ratio of powers between two signals becomes inconveniently large. For example, if you consider the range over which the human ear can perceive different levels of sound intensity (average power per area of air wave), you would find that this range is very large, from about 10^{-12} to $1 \ \text{W/m}^2$. Attempting to plot a graph of sound intensity versus, say, the distance between your ear and the speaker, would be difficult, especially if you are plotting a number of points at different ends of the scale—the resolution gets nasty. You can use special log paper to automatically correct this problem, or you can stick to normal linear graph paper by first "shrinking" your values logarithmically. For this we use decibels.

To start off on the right foot, a *bel* is defined as the logarithm of a power ratio. It gives us a way to compare power levels with each other and with some reference power. The bel is defined as:

$$\text{bel} = \log\left(\frac{P_1}{P_0}\right) \tag{2.88}$$

where P_0 is the reference power and P_1 is the power you are comparing with the reference power.

In electronics, the bel is often used to compare electrical power levels; however, what's more common in electronics and elsewhere is to use decibels, abbreviated dB. A decibel is $\frac{1}{10}$ of a bel (similar to a millimeter being $\frac{1}{10}$ of a centimeter). It takes 10 decibels to make 1 bel. So in light of this, we can compare power levels in terms of decibels:

$$dB = 10 \, \log\left(\frac{P_1}{P_0}\right)$$

(2.89) Decibels in terms of power

Example: Express the gain of an amplifier (output power divided by input power) in terms of decibels, if the amplifier takes a 1-W signal and boosts it up to a 50-W signal.

Answer: Let P_0 represent the 1-W reference power, and let P_1 be the compared power:

$$dB = 10 \, \log\left(\frac{50 \text{ W}}{1 \text{ W}}\right) = 10 \, \log \, (50) = 17.00 \text{ dB}$$

The amplifier in this example has a gain of nearly 17.00 dB (17 decibels).

Sometimes when comparing signal levels in an electronic circuit, we know the voltage or current of the signal, but not the power. Though it's possible to calculate the power, given the circuit impedance, we take a shortcut by simply plugging ac Ohm's law into the powers in the decibel expression. Recall that $P = V^2/Z = I^2 Z$. Now, this holds true only as long as the impedance of the circuit doesn't change when the voltage or current changes. As long as the impedance remains the same, we get a comparison of voltage signals and current signals in terms of decibels:

$$dB = 10 \, \log\left(\frac{V_1^2}{V_0^2}\right) = 20 \, \log\left(\frac{V_1}{V_0}\right)$$

and

$$dB = 10 \, \log\left(\frac{I_1^2}{I_0^2}\right) = 20 \, \log\left(\frac{I_1}{I_0}\right)$$

(2.90) Decibels in terms of voltage and current levels

In the expressions above we used the laws of logarithms to remove square terms. For example:

$$10 \, \log \, (V_1^2/V_0^2) = 10(\log V_1^2 - \log V_0^2)$$

$$= 10(2 \, \log V_1 - 2 \, \log V_0) = 20(\log V_1 - \log V_2)$$

$$= 20 \, \log \, (V_1/V_0).$$

Notice that the impedance terms cancel and the final result is a factor of 2 greater—a result of the square terms in the log being pushed out (see law of logarithms). The power and voltage and current expressions are all fundamentally the same—they are all based on the ratio of powers.

There are several power ratios you should learn to recognize and be able to associate with the corresponding decibel representations.

For example, when *doubling power*, the final power is always 2 times the initial (or reference) power—it doesn't make a difference if you are going from 1 to 2 W, 40 to

80 W, or 500 to 1000 W, the ratio is always 2. In decibels, a power ratio of 2 is represented as:

$$dB = 10 \log (2) = 3.01 \text{ dB}$$

There is a 3.01-dB gain if the output power is twice the input power. Usually, people don't care about the .01 fraction and simply refer to the power doubling as a 3-dB increase in power.

When the *power is cut in half*, the ratio is always 0.5 or 1/2. Again, it doesn't matter if you're going from 1000 to 500 W, 80 to 40 W, or 2 to 1 W, the ratio is still 0.5. In decibels, a power ratio of 0.5 is represented as:

$$dB = 10 \log (0.5) = -3.01 \text{ dB}$$

A negative sign indicates a decrease in power. Again, people usually ignore the .01 fraction and simply refer to the power being cut in half as a −3-dB change in power or, more logically, a 3-dB decrease in power (the term "decrease" eliminates the need for the negative sign).

Now if you increase the power by 4, you can avoid using the decibels formula and simply associate such an increase with doubling two times: 3.01 dB + 3.01 dB = 6.02 dB or around 6 dB. Likewise, if you increase the power by 8, you, in effect, double four times, so the power ratio in decibels is 3.01 × 4 = 12.04, or around 12 dB.

The same relationship is true of power decreases. Each time you cut the power in half, there is a 3.01-dB, or around 3-dB decrease. Cutting the power by four times is akin to cutting in half twice, or 3.01 + 3.01 dB = 6.02 dB, or around a 6-dB decrease. Again, you can avoid stating "decrease" and simply say that there is a change of −6 dB.

Table 2.12 shows the relationship between several common decibel values and the power change associated with those values. The current and voltage changes are also included, but these are valid only if the impedance is the same for both values.

TABLE 2.12 Decibels and Power Ratios

COMMON DECIBEL VALUES AND POWER-RATIO EQUIVALENTS			COMMON DECIBEL VALUES AND POWER-RATIO EQUIVALENTS		
dB	P_2/P_1	V_2/V_1 or I_2/I_1	dB	P_2/P_1	V_2/V_1 or I_2/I_1
120	10^{12}	10^6	−120	10^{-12}	10^{-6}
60	10^6	10^3	−60	10^{-6}	10^{-3}
20	10^2	10.0	−20	10^{-2}	0.1000
10	10.00	3.162	−10	0.1000	0.3162
6.0206	4.0000	2.0000	−6.0206	0.2500	0.5000
3.0103	2.0000	1.4142	−3.0103	0.5000	0.7071
1	1.259	1.122	−1	0.7943	0.8913
0	1.000	1.000	0	1.000	1.000

* Voltage and current ratios hold only if the impedance remains the same.

2.31.1 Alternative Decibel Representations

It is often convenient to compare a certain power level with some standard reference. For example, suppose you measured the signal coming into a receiver from an antenna and found the power to be 2×10^{-13} mW. As this signal goes through the receiver, it increases in strength until it finally produces some sound in the receiver speaker or headphones. It is convenient to describe these signal levels in terms of decibels. A common reference power is 1 mW. The decibel value of a signal compared to 1 mW is specified as "dBm" to mean decibels compared to 1 mW. In our example, the signal strength at the receiver input is:

$$dB_m = 10 \log\left(\frac{2 \times 10^{-13} \text{ mW}}{1 \text{ mW}}\right) = -127 \text{ dBm}$$

There are many other reference powers used, depending upon the circuits and power levels. If you use 1 W as the reference power, then you would specify dBW. Antenna power gains are often specified in relation to a dipole (dBd) or an isotropic radiator (dBi). Anytime you see another letter following dB, you will know that some reference power is being specified. For example, to describe the magnitudes of a voltage relative to a 1-V reference, you indicate the level in decibels by placing a "V" at the end of dB, giving units of dBV (again, impedances must stay the same). In acoustics, dB, SPL is used to describe the pressure of one signal in terms of a reference pressure of 20 µPa. The term decibel is also used in the context of sound (see Sec. 15.1).

2.32 Input and Output Impedance

2.32.1 Input Impedance

Input impedance Z_{IN} is the impedance "seen" looking into the input of a circuit or device (see Fig. 2.186). Input impedance gives you an idea of how much current can be drawn into the input of a device. Because a complex circuit usually contains reactive components such as inductors and capacitors, the input impedance is frequency sensitive. Therefore, the input impedance may allow only a little current to enter at one frequency, while highly impeding current enters at another frequency. At low frequencies (less than 1 kHz) reactive components may have less influence, and the term *input resistance* may be used—only the real resistance is dominant. The effects of capacitance and inductance are generally more significant at high frequencies.

When the input impedance is small, a relatively large current can be drawn into the input of the device when a voltage of specific frequency is applied to the input.

Input Impedance

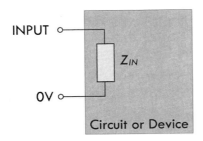

Output Impedance

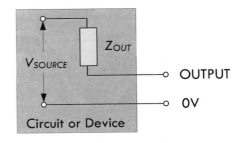

FIGURE 2.186

This typically has an effect of dropping the source voltage of the driver circuit feeding the device's input. (This is especially true if the driver circuit has a large output impedance.) A device with low input impedance is an audio speaker (typically 4 or 8 Ω), which draws lots of current to drive the voice coil.

When the input impedance is large, on the other hand, a small current is drawn into the device when a voltage of specific frequency is applied to the input, and, hence, doesn't cause a substantial drop in source voltage of the driver circuit feeding the input. An op amp is a device with a very large input impedance (1 to 10 MΩ); one of its inputs (there are two) will practically draw no current (in the nA range). In terms of audio, a hi-fi preamplifier that has a 1-MΩ radio input, a 500-kΩ CD input, and a 100-kΩ tap input all have high impedance input due to the preamp being a voltage amplifier not a current amplifier—something we will discuss later.

As a general rule of thumb, in terms of transmitting a signal, the input impedance of a device should be greater than the output impedance of the circuit supplying the signal to the input. Generally, the value should be 10 times as great to ensure that the input will not overload the source of the signal and reduce the strength by a substantial amount. In terms of calculations and such, the input impedance is by definition equal to:

$$Z_{in} = \frac{V_{in}}{I_{in}}$$

For example, Fig. 2.187 shows how to determine what the input impedance is for two resistor circuits. Notice that in example 2, when a load is attached to the output, the input impedance must be recalculated, taking R_2 and R_{load} in parallel.

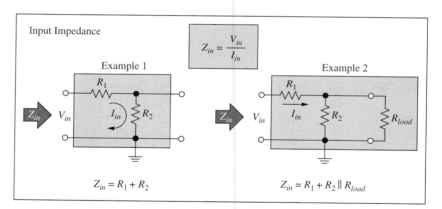

FIGURE 2.187

In Sec. 2.33.1, which covers filter circuits, you'll see how the input impedance becomes dependent on frequency.

2.32.2 Output Impedance

Output impedance Z_{OUT} refers to the impedance looking back into the output of a device. The output of any circuit or device is equivalent to an output impedance Z_{OUT} in series with an ideal voltage source V_{SOURCE}. Figure 2.186 shows the equivalent circuit; it represents the combined effect of all the voltage sources and the effective total impedance (resistances, capacitance, and inductance) connected to the output side of the circuit. You can think of the equivalent circuit as a Thevenin equivalent circuit, in which case it should be clear that the V_{SOURCE} present in Fig. 2.186 isn't necessarily the actual supply voltage of the circuit but the Thevenin equivalent voltage. As with input impedance, output impedance can be frequency dependent. The term *output resistance* is used in cases where there is little reactance within the circuit, or when the frequency of operation is low (say, less than 1 kHz) and reactive effects are of little consequence; the effects of capacitance and inductance are generally most significant at high frequencies.

When the output impedance is small, a relatively large output current can be drawn from the device's output without significant drop in output voltage. A source with an output impedance much lower than the input impedance of a load to which it is attached will suffer little voltage loss driving current through its output impedance. For example, a lab dc power supply can be viewed as an ideal voltage source in series with a small internal resistance. A decent supply will have an output impedance in the milliohm (mΩ) range, meaning it can supply considerable current to a load without much drop in supply voltage. A battery typically has a higher internal resistance and tends to suffer a more substantial drop in supply voltage as the current demands increase. In general, a small output impedance (or resistance) is considered a good thing, since it means that little power is lost to resistive heating in the impedance and a larger current can be sourced. Op amps, which, we saw, have large input impedance and tend to have low output impedance.

When the output impedance is large, on the other hand, a relatively small output current can be drawn from the output of a device before the voltage at the output drops substantially. If a source with a large output impedance attempts to drive a load that has a much smaller input impedance, only a small portion appears across the load; most is lost driving the output current through the output impedance.

Again, the rule of thumb for efficient signal transfer is to have an output impedance that is at least $\frac{1}{10}$ that of the load's input impedance to which it is attached.

In terms of calculations, the output impedance of a circuit is simply its Thevenin equivalent resistance R_{THEV}. The output impedance is sometimes called the source impedance. In terms of determining the output impedance in circuit analysis, it amounts to "killing" the source (shorting it) and finding the equivalent impedance between the output terminals—the Thevenin impedance.

For example, in Fig. 2.188, to determine the output impedance of this circuit, we effectively find the Thevinin resistance by "shorting the source" and removing the load, and determining the impedance between the output terminals. In this case, the output impedance is simply R_1 and R_2 in parallel.

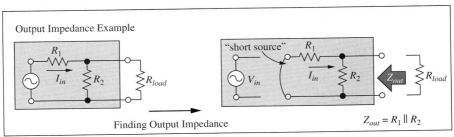

FIGURE 2.188

In Sec. 2.3.1, when we cover filter circuits, you'll see how the output impedance becomes dependent on frequency.

2.33   Two-Port Networks and Filters

2.33.1   Filters

By combining resistors, capacitors, and inductors in special ways, you can design networks that are capable of passing certain frequencies of signals while rejecting others. This section examines four basic kinds of filters: low-pass, high-pass, band-pass, and notch filters.

Low-Pass Filters

The simple RC filter shown in Fig. 2.189 acts as a low-pass filter—it passes low frequencies but rejects high frequencies.

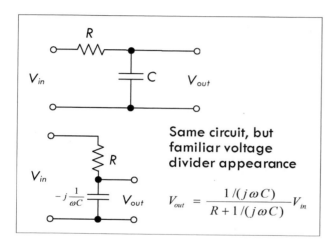

Example: To figure out how this network works, we find the transfer function. We begin by using the voltage divider to find V_{out} in terms of V_{in}, and consider there is no load (open output or $R_L = \infty$):

$$V_{out} = \frac{1/(j\omega C)}{R + 1/(j\omega C)} V_{in} = \frac{1}{1 + j\omega RC} V_{in}$$

The transfer function is then found by rearranging the equation:

$$H = \frac{V_{out}}{V_{in}} = \frac{1}{1 + j\omega RC}$$

$$H = \frac{V_{out}}{V_{in}} = \frac{1}{1 + j\tau\omega}, \quad \tau = RC$$

$$H = \frac{V_{out}}{V_{in}} = \frac{1}{1 + j\omega/\omega_c}, \quad \omega_c = \frac{1}{RC}$$

The magnitude and phase of H are:

$$|H| = \left| \frac{V_{out}}{V_{in}} \right| = \frac{1}{\sqrt{1 + \tau^2\omega^2}}$$

$$\arg(H) = \varphi = -\tan^{-1}\left(\frac{\omega}{\omega_c} \right)$$

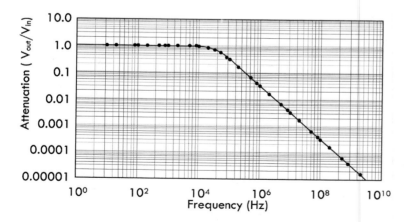

FIGURE 2.189

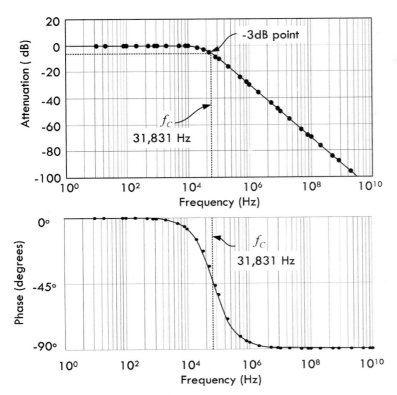

FIGURE 2.189   *(Continued)*

Here, τ is called the time constant, and ω_C is called the angular cutoff frequency of the circuit—related to the standard cutoff frequency by $\omega_C = 2\pi f_C$. The cutoff frequency represents the frequency at which the output voltage is attenuated by a factor of $1/\sqrt{2}$, the equivalent of half power. The cutoff frequency in this example is:

$$f_C = \frac{\omega_C}{2\pi} = \frac{1}{2\pi RC} = \frac{1}{2\pi(50\,\Omega)(0.1 \times 10^{-6}\,\text{F})}$$

$$= 31,831\,\text{Hz}$$

Intuitively, we imagine that when the input voltage is very low in frequency, the capacitor's reactance is high, so it draws little current, thus keeping the output amplitude near the input amplitude. However, as the frequency of the input signal increases, the capacitor's reactance decreases and the capacitor draws more current, which in turn causes the output voltage to drop. Figure 2.189 shows attenuation versus frequency graphs—one graph expresses the attenuation in decibels.

The capacitor produces a delay, as shown in the phase plot in Fig. 2.189. At very low frequency, the output voltage follows the input—they have similar phases. As the frequency rises, the output starts lagging the input. At the cutoff frequency, the output voltage lags by 45°. As the frequency goes to infinity, the phase lag approaches 90°.

Figure 2.190 shows an RL low-pass filter that uses inductive reactance as the frequency-sensitive element instead of capacitive reactance, as was the case in the RC filter.

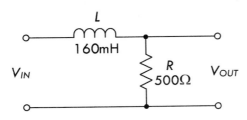

FIGURE 2.190

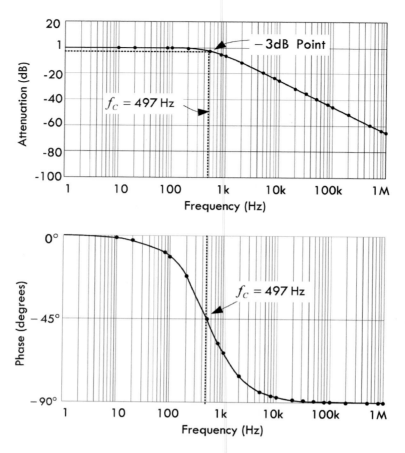

FIGURE 2.190   (*Continued*)

The input impedance can be found by definition $Z_{\text{in}} = \dfrac{V_{\text{in}}}{I_{\text{in}}}$, while the output impedance can be found by "killing the source" (see Fig. 2.191):

$$Z_{\text{in}} = R + \frac{1}{j\omega C} \quad \text{and} \quad Z_{\text{in}}\big|_{\min} = R$$

$$Z_{\text{out}} = R \,||\, \frac{1}{j\omega C} \quad \text{and} \quad Z_{\text{out}}\big|_{\max} = R$$

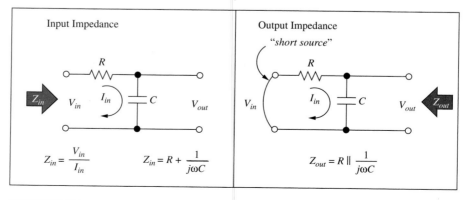

FIGURE 2.191

Now what happens when we put a finite load resistance R_L on the output? Doing the voltage divider stuff and preparing the voltage transfer function, we get:

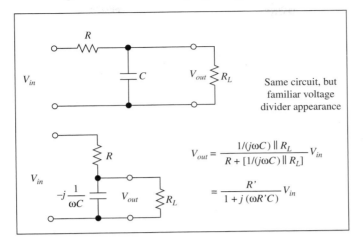

FIGURE 2.192

$$H = \frac{V_{\text{out}}}{V_{\text{in}}} = \frac{1/(j\omega C) \,||\, R_L}{R + \left[1/(j\omega C) \,||\, R_L\right]} = \frac{R'}{1 + j(\omega R'C)}$$

$$\text{where } R' = R \,||\, R_L$$

This is similar to the transfer function for the unterminated RC filter, but with resistance R being replaced by R'. Therefore:

$$\omega = \frac{1}{R'C} = \frac{1}{(R\,||\,R_L)C} \qquad \text{and} \qquad H = \frac{R'/R}{1 + j(\omega/\omega_C)}$$

As you can see, the load has the effect of reducing the filter gain ($K = R'/R < 1$) and shifting the cutoff frequency to a higher frequency as ($R' = R\,||\,R_L < R$).

The input and output impedance with load resistance become:

$$Z_{\text{in}} = R + \frac{1}{j\omega C}\,||\,R_L \qquad \text{and} \qquad Z_{\text{in}}\big|_{\text{min}} = R$$

$$Z_{\text{out}} = R\,||\,\frac{1}{j\omega C} \qquad \text{and} \qquad Z_{\text{out}}\big|_{\text{max}} = R$$

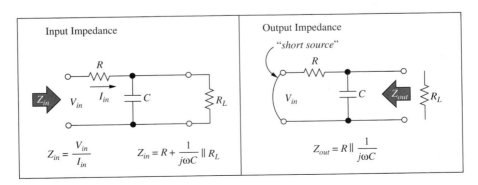

FIGURE 2.193

As long as $R_L \gg Z_{\text{out}}$ or $R \gg Z_{\text{out}}|_{\text{max}} = R$ (condition for good voltage coupling), $R' \approx R$ and the terminated RC filter will look exactly like an unterminated filter. The filter gain is one, the shift in cutoff frequency disappears, and the input and output resistance become the same as before.

Example: To find the transfer function or attenuation of the RL circuit with no load, we find V_{out} in terms of V_{in} using the voltage divider:

$$V_{\text{out}} = \frac{R}{R + j\omega L} V_{\text{in}} = \frac{1}{1 + j(\omega L/R)} V_{\text{in}}$$

$$H = \frac{V_{\text{out}}}{V_{\text{in}}} = \frac{1}{1 + j(\omega L/R)}$$

$$H = \frac{V_{\text{out}}}{V_{\text{in}}} = \frac{1}{1 + j(\omega/\omega_C)}, \quad \omega_C = \frac{R}{L}$$

The magnitude and phase become:

$$|H| = \left| \frac{V_{\text{out}}}{V_{\text{in}}} \right| = \frac{1}{\sqrt{1 + (\omega L/R)^2}}$$

$$\arg(H) = \varphi = \tan^{-1}\left(\frac{\omega L}{R} \right) = \tan^{-1}\left(\frac{\omega}{\omega_C} \right)$$

Here, ω_C is called the angular cutoff frequency of the circuit—related to the standard cutoff frequency by $\omega_C = 2\pi f_C$. The cutoff frequency represents the frequency at which the output voltage is attenuated by a factor of $1/\sqrt{2}$, the equivalent of half power. The cutoff frequency in this example is:

$$f_C = \frac{\omega_C}{2\pi} = \frac{R}{2\pi L} = \frac{500\,\Omega}{2\pi(160 \times 10^{-3}\ \text{H})} = 497\ \text{Hz}$$

Intuitively we imagine that when the input voltage is very low in frequency, the inductor doesn't have a hard time passing current to the output. However, as the frequency gets big, the inductor's reactance increases, and the signal becomes more attenuated at the output. Figure 2.190 shows attenuation versus frequency graphs—one graph expresses the attenuation in decibels.

The inductor produces a delay, as shown in the phase plot in Fig. 2.190. At very low frequency, the output voltage follows the input—they have similar phases. As the frequency rises, the output starts lagging the input. At the cutoff frequency, the output voltage lags by 45°. As the frequency goes to infinity, the phase lag approaches 90°.

The input impedance can be found using the definition of the input impedance:

$$Z_{\text{in}} = \frac{V_{\text{in}}}{I_{\text{in}}} = j\omega L + R$$

The value of the input impedance depends on the frequency ω. For good voltage coupling, the input impedance of this filter should be much larger than the output impedance of the previous stage. The minimum value of Z_{in} is an important number, and its value is minimum when the impedance of the inductor is zero ($\omega \to 0$):

$$Z_{in}\big|_{min} = R$$

The output impedance can be found by shorting the source and finding the equivalent impedance between output terminals:

$$Z_{out} = j\omega L \,||\, R$$

where the source resistance is ignored. The output impedance also depends on the frequency ω. For good voltage coupling, the output impedance of this filter should be much smaller than the input impedance of the next stage. The maximum value of Z_{out} is also an important number, and it is maximum when the impedance of the inductor is infinity ($\omega \to \infty$):

$$Z_{out}\big|_{max} = R$$

When the RL low-pass filter is terminated with a load resistance R_L, the voltage transfer function changes to:

$$H = \frac{V_{out}}{V_{in}} = \frac{1}{1 + j\omega/\omega_C} \quad \text{where } \omega_C = (R\,||\,R_L)/L$$

The input impedance becomes:

$$Z_{in} = j\omega L = R\,||\,R_L, \quad Z_{in}\big|_{min} = R\,||\,R_L$$

The output impedance becomes:

$$Z_{out} = (j\omega L)\,||\,R, \quad Z_{out}\big|_{max} = R$$

The effect of the load is to shift the cutoff frequency to a lower value. Filter gain is not affected. Again, for $R_L \gg Z_{out}$ or $R_L \gg Z_{out}\big|_{max} = R$ (condition for good voltage coupling), the shift in cutoff frequency disappears, and the filter will look exactly like an unterminated filter.

High-Pass Filters

Example: To figure out how this network works, we find the transfer function by using the voltage divider equation and solving in terms of V_{out} and V_{in}:

$$H = \frac{V_{out}}{V_{in}} = \frac{R}{R + 1/(j\omega C)} = \frac{1}{1 - j(1/\omega RC)} = \frac{j\omega\tau}{1 + j\omega\tau} \qquad \tau = RC$$

or

$$H = \frac{V_{out}}{V_{in}} = \frac{j(\omega/\omega_C)}{1 + j(\omega/\omega_C)} = \frac{1}{1 - j\omega_C/\omega} \qquad \omega_C = \frac{1}{RC}$$

RC High-Pass Filter Circuit

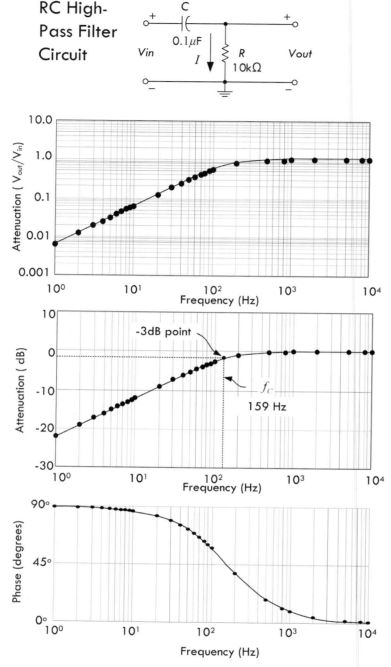

FIGURE 2.194

The magnitude and phase of H are:

$$|H| = \left|\frac{V_{\text{out}}}{V_{\text{in}}}\right| = \frac{\tau\omega}{\sqrt{1 + \tau^2\omega^2}}$$

$$\arg(H) = \varphi = \tan^{-1}\left(\frac{\omega_C}{\omega}\right)$$

Here τ is called the time constant and ω_C is called the angular cutoff frequency of the circuit—related to the standard cutoff frequency by $\omega_C = 2\pi f_C$. The cutoff frequency represents the frequency at which the output voltage is attenuated by a factor of $1/\sqrt{2}$, the equivalent of half power. The cutoff frequency in this example is:

$$f_C = \frac{\omega_C}{2\pi} = \frac{1}{2\pi RC}$$

$$= \frac{1}{2\pi(10,000\ \Omega)(0.1\times10^{-6}\ \text{F})} = 159\ \text{Hz}$$

Intuitively, we imagine that when the input voltage is very low in frequency, the capacitor's reactance is very high, and hardly any signal is passed to the output. However, as the frequency rises, the capacitor's reactance decreases, and there is little attenuation at the output. Figure 2.194 shows attenuation versus frequency graphs—one graph expresses the attenuation in decibels.

In terms of phase, at very low frequency the output leads the input in phase by 90°. As the frequency rises to the cutoff frequency, the output leads by 45°. When the frequency goes toward infinity, the phase approaches 0, the point where the capacitor acts like a short.

Input and output impedances of this filter can be found in a way similar to finding these impedances for low-pass filters:

$$Z_{\text{in}} = R + \frac{1}{j\omega C} \quad \text{and} \quad Z_{\text{in}}\big|_{\min} = R$$

$$Z_{\text{out}} = R\,||\,\frac{1}{j\omega C}, \quad Z_{\text{out}}\big|_{\max} = R$$

With a terminated load resistance, the voltage transfer function becomes:

$$H = \frac{V_{out}}{V_{in}} = \frac{R \,||\, R_L}{R \,||\, R_L + 1/(j\omega C)} = \frac{1}{1 - j(1/\omega R'C)} \text{ where } R' = R \,||\, R_L$$

This is similar to the transfer function for the unterminated RC filter, but with resistance R being replaced by R':

$$\omega_C = \frac{1}{R'C} = \frac{1}{(R \,||\, R_L)C} \text{ and } H = \frac{1}{1 - j\omega_C/\omega}$$

The load has the effect of shifting the cutoff frequency to a higher frequency $(R' = R \,||\, R_L < R)$.

The input and output impedances are:

$$Z_{in} = \frac{1}{j\omega C} + R \,||\, R_L, \ Z_{in}\big|_{min} = R \,||\, R_L$$

$$Z_{out} = R \,||\, \frac{1}{j\omega C}, \ Z_{out}\big|_{max} = R$$

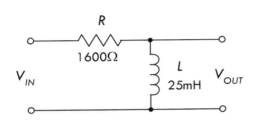

As long as $R_L \gg Z_{out}$ or $R_L \gg Z_{out}\big|_{max} = R$ (condition for good voltage coupling), $R' \approx R$ and the terminated RC filter will look like an unterminated filter. The shift in cutoff frequency disappears, and input and output resistance become the same as before.

RL High-Pass Filter

Figure 2.195 shows an RL high-pass filter that uses inductive reactance as the frequency-sensitive element instead of capacitive reactance, as was the case in the RC filter.

Example: To find the transfer function or attenuation of the RL circuit, we again use the voltage divider equation and solve for the transfer function or attenuation of the RL circuit in terms of V_{out} and V_{in}:

$$H = \frac{V_{out}}{V_{in}} = \frac{j\omega L}{R + j\omega L}$$

$$= \frac{\omega L \,\angle\, 90°}{\sqrt{R^2 + (\omega L)^2} \,\angle\, \tan^{-1}(\omega L/R)}$$

or

$$H = \frac{1}{1 - j\omega_C/\omega}, \qquad \omega_C = \frac{R}{L}$$

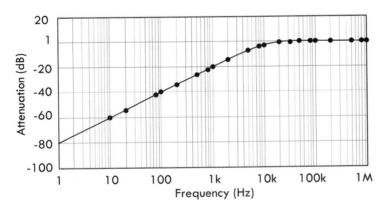

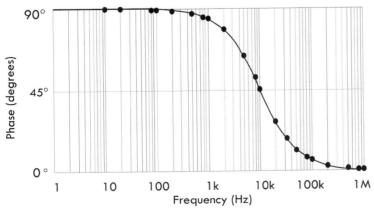

FIGURE 2.195

The magnitude and phase of H are:

$$|H| = \left| \frac{V_{out}}{V_{in}} \right| = \frac{\omega L}{\sqrt{R^2 + (\omega L)^2}} = \frac{\omega/\omega_C}{\sqrt{1 + (\omega/\omega_C)^2}}$$

$$arg(H) = \varphi = 90° - \tan^{-1} \left(\frac{\omega L}{R} \right)$$

Here ω_C is called the angular cutoff frequency of the circuit—related to the standard cutoff frequency by $\omega_C = 2\pi f_C$. The cutoff frequency represents the frequency at which the output voltage is attenuated by a factor of $1\sqrt{2}$, the equivalent of half power. The cutoff frequency in this example is:

$$f_C = \frac{\omega_0}{2\pi} = \frac{R}{2\pi L} = \frac{1600\ \Omega}{2\pi(25 \times 10^{-3}\ \text{H})} = 10,186\ \text{Hz}$$

Intuitively, we imagine that when the input voltage is very low in frequency, the inductor's reactance is very low, so most of the current is diverted to ground—the signal is greatly attenuated at the output. However, as the frequency rises, the inductor's reactance increases and less current is passed to ground—the attenuation decreases. Figure 2.195 shows attenuation versus frequency graphs—one graph expresses the attenuation in decibels.

In terms of phase, at very low frequency the output leads the input in phase by 90°. As the frequency rises to the cutoff frequency, the output leads by 45°. When the frequency goes toward infinity, the phase approaches 0, the point where the inductor acts like an open circuit.

The input and output impedances are:

$$Z_{in} = R + j\omega L,\ Z_{in}\big|_{min} = R$$

$$Z_{out} = R\,||\,j\omega L,\ Z_{out}\big|_{max} = R$$

For a terminated RL high-pass filter with load resistance, we do a similar calculation as we did with the RC high-pass filter, replacing the resistance with R':

$$H = \frac{V_{out}}{V_{in}} = \frac{R'/R}{1 - j\omega_C/\omega} \qquad \omega_C = \frac{R'}{L} \qquad R' = R\,||\,R_L$$

The input and output impedances are:

$$Z_{in} = R + j\omega L\,||\,R_L,\ Z_{in}\big|_{min} = R$$

$$Z_{out} = R\,||\,j\omega L,\ Z_{out}\big|_{max} = R$$

The load has the effect of lowering the gain, $K = R'/R < 1$, and it shifts the cutoff frequency to a lower value. As long as $R_L \gg Z_{out}$ or $R_L \gg Z_{out}\big|_{max} = R$ (condition for good voltage coupling), $R' \approx R$ and the terminated RC filter will look like an unterminated filter.

Bandpass Filter

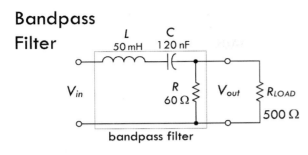

bandpass filter

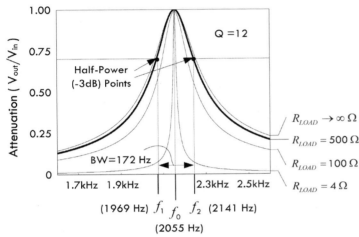

Parallel Bandpass Filter

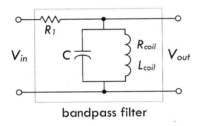

bandpass filter

FIGURE 2.196 The parallel bandpass filter shown here yields characteristics similar to those of the previous bandpass filter. However, unlike the previous filter, as you approach the resonant frequency of the tuned circuit, the LC (RL coil) section's impedance gets large, not allowing current to be diverted away from the load. On either side of resonance, the impedance goes down, diverting current away from load.

Bandpass Filter

The RLC bandpass filter in Fig. 2.196 acts to pass a narrow range of frequencies (band) while attenuating or rejecting all other frequencies.

Example: To find the transfer function or attenuation of the unloaded RLC circuit, we set up equations for V_{in} and V_{out}:

$$V_{in} = \left(j\omega L - j\frac{1}{\omega C} + R \right) \times I$$

$$V_{out} = R \times I$$

The transfer function becomes:

$$H = \frac{V_{out}}{V_{in}} = \frac{R}{R + j(\omega L - 1/\omega C)}$$

This is the transfer function for an unloaded output. However, now we get more realistic and have a load resistance attached to the output. In this case we must replace R with R_T, which is the parallel resistance of R and R_{LOAD}:

$$R_T = \frac{R \times R_{LOAD}}{R + R_{LOAD}} = \frac{500(60)}{500 + 60} = 54\ \Omega$$

Placing this in the unloaded transfer function and solving for the magnitude, we get:

$$|H| = \left| \frac{V_{out}}{V_{in}} \right| = \frac{R_T}{\sqrt{R_T^2 + [\omega L - 1/(\omega C)]^2}}$$

Plugging in all component values and setting $\omega = 2\pi f$, we get:

$$|H| = \left| \frac{V_{out}}{V_{in}} \right|$$

$$= \frac{54}{\sqrt{54^2 + [0.314 f - 1/(7.54 \times 10^{-7} f)]^2}}$$

The attenuation versus frequency graph based on this equation is shown in Fig. 2.196. (Three other families of curves are provided for loads of 4 Ω, 100 Ω, and an infinite resistance load.)

The resonant frequency, Q, bandwidth, and upper and lower cutoff frequencies are given by:

$$f_0 = \frac{1}{2\pi\sqrt{LC}} = \frac{1}{2\pi\sqrt{(50\times10^{-3})(120\times10^{-9})}} = 2055 \text{ Hz}$$

$$Q = \frac{X_{L,0}}{R_T} = \frac{2\pi f_0 L}{R_T} = \frac{2\pi(2055)(50\times10^{-3})}{54} = 12$$

$$\text{BW} = \frac{f_0}{Q} = \frac{2055}{12} = 172 \text{ Hz}$$

$$f_1 = f_0 - \text{BW}/2 = 2055 - 172/2 = 1969 \text{ Hz}$$

$$f_2 = f_0 + \text{BW}/2 = 2055 + 172/2 = 2141 \text{ Hz}$$

Notch Filter

The notch filter in Fig. 2.197 acts to pass a wide range of frequencies, while attenuating (rejecting) a narrow band of frequencies.

Notch Filter

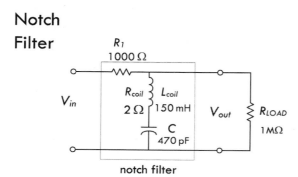

notch filter

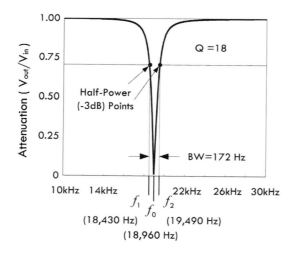

FIGURE 2.197 The notch filter shown here yields stopband characteristics similar to those of the previous notch filter. However, unlike the previous filter, as you approach the resonant frequency of the tuned circuit, the LC (RL coil) section's impedance gets large, not allowing current to be sent to the load. On either side of resonance, the impedance goes down, allowing current to reach load.

Example: To find the transfer function or attenuation of the unloaded RLC circuit, we set up equations for V_{in} and V_{out}:

$$V_{\text{in}} = \left(R_1 + R_{\text{coil}} + j\omega L - j\frac{1}{\omega C}\right) \times I$$

$$V_{\text{out}} = \left(R_{\text{coil}} + j\omega L - j\frac{1}{\omega C}\right) \times I$$

The transfer function becomes:

$$H = \frac{V_{\text{out}}}{V_{\text{in}}} = \frac{R_{\text{coil}} + j[\omega L - 1/\omega C]}{(R_1 + R_{\text{coil}}) + j[\omega L - 1/(\omega C)]}$$

This is the transfer function for an unloaded output. Now we get more realistic and have a load resistance attached to the output. However, in this case, the load resistance is so large that we can assume it draws inconsequential current, so we don't need to place it into the equation:

The magnitude of the transfer function is:

$$|H| = \left|\frac{V_{\text{out}}}{V_{\text{in}}}\right| = \frac{\sqrt{R_{\text{coil}}^2 + [\omega L - 1/\omega C]^2}}{\sqrt{(R_1 + R_{\text{coil}})^2 + j[\omega L - 1/(\omega C)]^2}}$$

Plugging in all component values and setting $\omega = 2\pi f$, we get:

$$|H| = \left|\frac{V_{\text{out}}}{V_{\text{in}}}\right| = \frac{\sqrt{4 + (0.94f - 3.38\times10^8/f)^2}}{\sqrt{1.00\times10^6 + (0.94f - 3.38\times10^8/f)^2}}$$

The attenuation versus frequency graph based on this equation is shown in Fig. 2.197.

Parallel Notch Filter

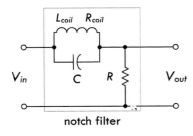

notch filter

FIGURE 2.197 *(Continued)*

The resonant frequency, Q, bandwidth, and upper and lower cutoff frequencies are given by:

$$f_0 = \frac{1}{2\pi\sqrt{LC}} = \frac{1}{2\pi\sqrt{(150\times10^{-3})(470\times10^{-12})}} = 18,960 \text{ Hz}$$

$$Q = \frac{X_{L,0}}{R_T} = \frac{2\pi f_0 L}{R_1} = \frac{2\pi(18,960)(150\times10^{-3})}{1000} = 18$$

$$BW = \frac{f_0}{Q} = \frac{18960}{18} = 1053 \text{ Hz}$$

$$f_1 = f_0 - BW/2 = 18,960 - 1053/2 = 18,430 \text{ Hz}$$

$$f_2 = f_0 + BW/2 = 18,960 + 1053/2 = 19,490 \text{ Hz}$$

2.33.2 *Attenuators*

Often it is desirable to attenuate a sinusoidal voltage by an amount that is independent of frequency. We can do this by using a voltage divider, since its output is independent of frequency. Figure 2.198 shows a simple voltage divider attenuator network inserted between a source and a load circuit to decrease the source signal's magnitude before it reaches the load. As we go through this example, you will pick up additional insight into input and output impedances.

Example of Input and Output Impedances

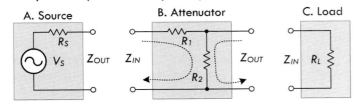

D. Input Impedance of Attenuator + Load

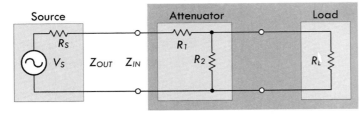

E. Output Impedance of Attenuator + Source

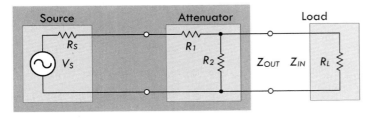

FIGURE 2.198

Example: In (a), the source has an output impedance equal to the internal resistance of the source:

$$Z_{\text{out}} = R_S$$

In (b), the attenuator network has input and output impedances of:

$$Z_{\text{in}} = R_1 + R_2$$

$$Z_{\text{out}} = R_2$$

We can come up with a transfer function for this, taking the same current to flow through R_2:

$$H = \frac{V_{\text{out}}}{V_{\text{in}}} = \frac{I\times Z_{\text{out}}}{I\times Z_{\text{in}}} = \frac{Z_{\text{out}}}{Z_{\text{in}}} = \frac{R_2}{R_1 + R_2}$$

If you rearrange terms, you can see this is simply a voltage divider.

In (c), the load has an input impedance of:

$$Z_{\text{in}} = R_L$$

F. Thevenin Equivalent Circuit

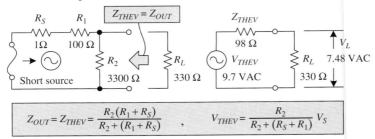

$$Z_{OUT} = Z_{THEV} = \frac{R_2(R_1 + R_S)}{R_2 + (R_1 + R_S)} \quad , \quad V_{THEV} = \frac{R_2}{R_2 + (R_S + R_1)} V_S$$

FIGURE 2.198 *(Continued)*

However, when we assemble the circuit, the useful input and output impedances change. Now, looking at things from the point of view of the source, as shown in (d), the source sees an input impedance of the attenuator combined with the load impedance, which is R_1 in series with the parallel combination of R_2 and R_L:

$$Z_{in} = R_1 + \frac{R_2 \times R_L}{(R_2 + R_L)} = 400\,\Omega$$

But now, from the point of view of the load, as shown in (e), the output impedance of the attenuator and source combined is R_2 in parallel with the series combination of R_1 and R_S:

$$Z_{OUT} = Z_{THEV} = \frac{R_2(R_1 + R_S)}{R_2 + (R_1 + R_S)} = \frac{3300\,\Omega\,(100\,\Omega + 1\,\Omega)}{3300\,\Omega + (100\,\Omega + 1\,\Omega)} = 98\,\Omega$$

(The input impedance to the load is still R_L.)

This output impedance is equivalent to the Thevinen equivalent impedance Z_{THEV}, as shown in (f). If we substitute $R_1 = 100\,\Omega$, $R_2 = 3300\,\Omega$, $R_S = 1\,\Omega$, and $V_S = 10$ VAC, we get the graph shown in (f). If we set the load $R_L = 330\,\Omega$, using the Thevenin circuit, we find:

$$V_L = \frac{R_L}{Z_{THEV} + R_L} V_{THEV} = \frac{330\,\Omega}{98\,\Omega + 330\,\Omega}(9.7\,\text{VAC}) = 7.48 \text{ VAC}$$

Compensated Attenuator

We just saw how a voltage divider could be used to attenuate a signal in a manner that is independent of frequency. However, in practice, there is always stray capacitance in a real circuit, and eventually a frequency is reached at which the voltage divider behaves like either a low- or a high-pass filter. This problem can be overcome by using a compensated attenuator circuit, as shown in Fig. 2.199.

Basic Attenuator

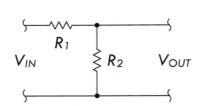

Compensated Attenuator

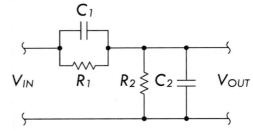

FIGURE 2.199

At low frequencies the circuit behaves like an ordinary resistive divider, but at high frequencies the capacitive reactance dominates, and the circuit behaves like a capacitive voltage divider. The attenuation is independent of frequency, provided that:

$$R_1 C_1 = R_2 C_2 \tag{2.91}$$

In practice, one of the capacitors is usually variable, so that the attenuator can be adjusted to compensate for any stray capacitance.

Such compensated attenuators are often used at the input of an oscilloscope to raise the input resistance and lower the input capacitance so as to make the oscilloscope into a more nearly ideal voltmeter. However, this results in a decrease in sensitivity of the scope to input voltage.

2.34 Transient Circuits

Transients within circuits remove any steady-state condition. They represent a sudden change in voltage introduced by an external agent, such as a switch being thrown or a transistor switching states. During a transient, the voltages and currents throughout the circuit readjust to a new dc value in a brief but nonnegligible time interval immediately following the transient event. The initial condition is a dc circuit; the final condition is a different dc circuit; but the interval in between—while the circuit is readjusting to the new conditions—may exhibit complex behavior. The introduction of a transient into a circuit containing reactive components usually requires solving differential equations, since the response is time dependent. The following simple example is an exception to the rule (no reactances—no differential equations), but is a good illustration of transient behavior, nevertheless. Refer to Fig. 2.200.

Example:

1. Initially the switch S is open. The instant it is closed ($t = 0$), V_S is applied across R and current flows immediately, according to Ohm's law. If the switch remains closed thereafter ($t > 0$), the current remains the same as it was at the instant S was closed:

$$I(t) = \begin{cases} 0 & t < 0 & \text{(before } S \text{ closed)} \\ V_S/R & t = 0 & \text{(}S \text{ just closed)} \\ V_S/R & t > 0 & \text{(time after } S \text{ closed)} \end{cases}$$

2. Initially S is closed. The instant it is open ($t = 0$), the voltage across and the current through the resistor go to zero. The voltage and current remain zero thereafter ($t > 0$), until the switch is opened again.

$$I(t) = \begin{cases} V_S/R & t < 0 & \text{(before } S \text{ opened)} \\ 0 & t = 0 & \text{(}S \text{ just opened)} \\ 0 & t > 0 & \text{(time after } S \text{ opened)} \end{cases}$$

3. Initially S is in position A ($t < 0$), and the voltage across R is V_1, thus making the current $I = V_1/R$. Neglecting the time delay for the switch element to move to position B, the instant S is thrown to position B ($t = 0$), the voltage immediately changes and the voltage across R is V_2, thus making $I = V_2/R$.

$$I(t) = \begin{cases} V_1/R & t < 0 & \text{(} S \text{ in position A)} \\ V_2/R & t = 0 & \text{(}S \text{ just switched to B)} \\ V_2/R & t > 0 & \text{(time after } S \text{ to B)} \end{cases}$$

1.

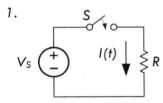

2.

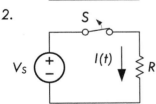

3.

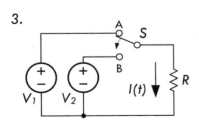

FIGURE 2.200

The preceding example may have seemed like child's play, but there is an important thing to notice from it. Just before the transient event of switching, the voltage across and the current through the resistor were both zero (or a constant value). Immediately after the switching event, the voltage and current jumped to new levels instantaneously. The resistor's natural response after the event was time independent as indicated by Ohm's law, $V = IR$. In other words, under a forced response (source is now in the picture), the resistor voltage can change instantly to a new steady state; under a forced response, the resistor current can change instantly to a new steady state.

When we consider capacitors and inductors, a forced response (applying or removing source influence) does not result in voltages and currents instantly jumping to a new steady state. Instead, there is a natural response after the forced response where the voltage and currents vary with time. With the help of Kirchhoff's laws, transient circuits can be modeled; the resulting differential equations take into account transient events by applying initial conditions. The following two examples, which were actually covered earlier in the capacitor and inductor sections, show what happens when voltage is suddenly applied to an RL and an RC circuit.

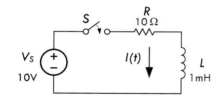

$$I(t) = \begin{cases} 0 & t < 0 \text{ (before S closed)} \\ 0 & t = 0 \text{ (S just closed)} \\ (V_S/R)(1 - e^{-(R/L)t}) & t > 0 \text{ (after S closed)} \\ V_S/R & t \to \infty \text{ (steady state)} \end{cases}$$

$$V_R(t) = \begin{cases} 0 & t < 0 \text{ (before S closed)} \\ 0 & t = 0 \text{ (S just closed)} \\ V_S(1 - e^{-(R/L)t}) & t > 0 \text{ (after S closed)} \\ V_S & t \to \infty \text{ (steady state)} \end{cases}$$

$$V_L(t) = \begin{cases} 0 & t < 0 \text{ (before S closed)} \\ V_S & t = 0 \text{ (S just closed)} \\ V_S e^{-(R/L)t} & t > 0 \text{ (after S closed)} \\ 0 & t \to \infty \text{ (steady state)} \end{cases}$$

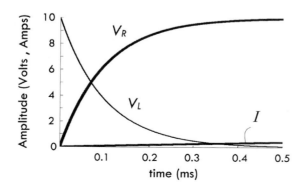

FIGURE 2.201

Example: The circuit in Fig. 2.201 can be modeled using Kirchhoff's voltage equation:

$$V_S - L\frac{dI}{dt} - RI = 0 \quad \text{or} \quad L\frac{dI}{dt} + RI = V_S$$

This is a first-order nonhomogeneous differential equation. To solve this equation, we separate the variables and integrate:

$$\int \frac{L}{V_S - RI} dI = \int dt$$

The solution of integration:

$$-\frac{L}{R}\ln(V_S - RI) = t + C$$

Using the initial conditions $t(0) = 0$, $I(0) = 0$, we can determine the constant of integration:

$$C = -\frac{L}{R}\ln V_S$$

Substituting this back into the solution, we get:

$$I = \frac{V_S}{R}(1 - e^{-Rt/L})$$

Using the component values shown in Fig. 2.201, this equation becomes:

$$I = \frac{10\text{ V}}{10\text{ }\Omega}(1 - e^{-10t/0.001}) = 1.0\ (1 - e^{-10,000t})\text{A}$$

Once the current is known, the voltage across the resistor and inductor can be easily determined:

$$V_R = IR = V_S(1 - e^{-Rt/L}) = 10\text{ V }(1 - e^{10,000t})$$

$$V_L = L\frac{dI_L}{dt} = V_S e^{-Rt/L} = 10\text{ V}e^{-10,000t}$$

The graph in Fig. 2.201 shows how these voltages change with time. The section on inductors explains some important details of the RL circuit not mentioned here and also explains how the RL deenergizing circuit works.

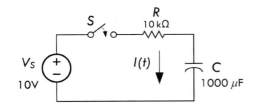

$$I(t) = \begin{cases} 0 & t < 0 \;\; \text{(before S closed)} \\ V_S/R & t = 0 \;\; \text{(S just closed)} \\ (V_S/R)e^{-t/RC} & t > 0 \;\; \text{(after S closed)} \\ 0 & t \to \infty \;\; \text{(steady state)} \end{cases}$$

$$V_R(t) = \begin{cases} 0 & t < 0 \;\; \text{(before S closed)} \\ V_S & t = 0 \;\; \text{(S just closed)} \\ V_S e^{-t/RC} & t > 0 \;\; \text{(after S closed)} \\ 0 & t \to \infty \;\; \text{(steady state)} \end{cases}$$

$$V_C(t) = \begin{cases} 0 & t < 0 \;\; \text{(before S closed)} \\ 0 & t = 0 \;\; \text{(S just closed)} \\ V_S(1 - e^{-t/RC}) & t > 0 \;\; \text{(after S closed)} \\ V_S & t \to \infty \;\; \text{(steady state)} \end{cases}$$

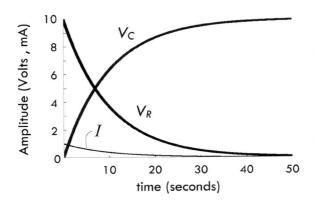

FIGURE 2.202

Example: The circuit in Fig. 2.202 can be modeled using Kirchhoff's voltage equation:

$$V_S = RI + \frac{1}{C}\int_0^t I(t)\,dt$$

The first step in solving such an equation is to eliminate the integral by differentiating each term:

$$0 = R\frac{dI}{dt} + \frac{1}{C}I$$

This is an example of a linear first-order homogeneous differential equation. It is linear because the unknown appears only once to the first power in each term. It is first-order because the highest derivative is the first, and it is homogeneous because the right-hand side, which would contain any terms not dependent on the unknown I, is zero. The solution to all linear first-order homogeneous differential equations is of the form:

$$I = I_0 e^{\alpha t}$$

where the constant α is determined by substituting the solution back into the differential equation and solving the resulting algebraic equation:

$$\alpha + \frac{1}{RC} = 0$$

In this case the solution is:

$$\alpha = -\frac{1}{RC}$$

The constant I_0 is determined from the initial condition at $t = 0$, at which time the voltage across the capacitor cannot change abruptly, and thus if the capacitor has zero voltage across it before the switch is closed, it will also have zero voltage immediately after the switch is closed. The capacitor initially behaves like a short circuit, and the initial current is:

$$I(0) = I_0 = \frac{V_S}{R}$$

Therefore, the complete solution for the transient series RC circuit for an initially discharged capacitor is:

$$I = \frac{V_S}{R} e^{-t/RC}$$

Using the component values shown in Fig. 2.202, this equation becomes:

$$I = 0.001 \text{ A } e^{-0.1t}$$

Once the current is known, the voltage across the resistor and capacitor can be easily determined:

$$V_R = IR = V_S e^{-t/RC} = 10 \text{ V } e^{-0.1t}$$

$$V_C = \frac{1}{C}\int_0^t I(t)\,dt = V_S(1 - e^{-t/RC}) = 10 \text{ V}(1 - e^{-0.1t})$$

The graph in Fig. 2.202 shows how these voltages change with time. The section on capacitors explains some important details of the RC circuit not mentioned here and also explains how the discharging RC circuit works.

Once you've memorized the response equations for a charging and discharging RC as well as the equations for an energizing and a deenergizing RL circuit (see following equations), it's often possible to incorporate these equations into transient circuits without starting from scratch and doing the differential equations.

RC charging	RC discharging	RL energizing	RL deenergizing
$I = \dfrac{V_S}{R} e^{-t/RC}$	$I = \dfrac{V_S}{R} e^{-t/RC}$	$I = \dfrac{V_S}{R}(1 - e^{-Rt/L})$	$I = \dfrac{V_S}{R} e^{-Rt/L}$
$V_R = V_S e^{-t/RC}$	$V_R = V_S e^{-t/RC}$	$V_R = V_S(1 - e^{-Rt/L})$	$V_R = V_S e^{-Rt/L}$
$V_C = V_S(1 - e^{-t/RC})$	$V_C = V_S e^{-t/RC}$	$V_L = V_S e^{-Rt/L}$	$V_L = -V_S e^{-Rt/L}$

The following example problems illustrate how this is done.

Example 1: The circuit in Fig. 2.203 was under steady state before the switch was opened. Determine the current flow I_2 the instant the switch is opened, and then determine both I_2 and the voltage across R_2 1 ms after that.

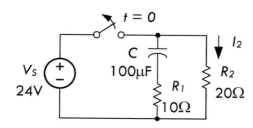

FIGURE 2.203

Answer: Unlike a resistor's voltage, a capacitor's voltage cannot change instantaneously (instead, it takes time). Because of this, the instant the switch is opened, the capacitor voltage remains as it was prior to the event:

$$V_C(0^+) = V_C(0^-) = 24 \text{ V}$$

where 0^+ means the instant after the switch is flipped, and 0^- means the instant before the switch is flipped.

The instant after the switch is opened ($t = 0^+$), the source is no longer part of the circuit, and we are left with the capacitor and two resistors all in series. Using Kirchhoff's law for this new circuit:

$$V_C + V_{R1} + V_{R2} = 0$$

$$24 \text{ V} + I(10 \text{ }\Omega + 20 \text{ }\Omega) = 0$$

$$I = -24 \text{ V}/30 \text{ }\Omega = 0.800 \text{ A}$$

This is the same current through R_2 the instant the switch is opened. To determine the current 1 ms ($t = 0.001$ s) after the switch is opened, simply treat the circuit as an RC discharge circuit (see Eq. 2.44), taking $R = R_1 + R_2$:

$$I = \frac{V_C}{R}e^{-t/RC} = \frac{24 \text{ V}}{(30 \text{ }\Omega)}e^{-0.001/(0.003)} = 0.573 \text{ A}$$

The voltage across R_2 at this time is:

$$V_{R2} = IR_2 = (0.573 \text{ A})(20 \text{ }\Omega) = 11.46 \text{ V}$$

Example 2: In the circuit in Fig. 2.204, determine the current I when the switch is opened ($t = 0$). Also determine the voltage across the resistor and inductor at $t = 0.1$ s.

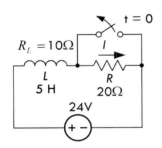

FIGURE 2.204

Answer: When the switch is opened, the inductor voltage cannot change instantaneously, so it resembles a short, which gives rise to the forced response:

$$I_f = \frac{24 \text{ V}}{(10 \text{ }\Omega + 20 \text{ }\Omega)} = 0.80 \text{ A}$$

where the inductor's resistance is in series with R.

For any time after $t = 0$, the natural, or source-free, response is simply a deenergizing RL circuit, where R is the combined resistance of the fixed resistor and the resistance of the inductor:

$$I_n = Ce^{-Rt/L} = Ce^{-30t/5} = Ce^{-6t}$$

The total current is the sum of the forced and natural responses:

$$I = I_f + I_n = 0.80 \text{ A} + Ce^{-6t}$$

The trick then is to find C by finding the current $I(0^+)$:

$$I(0^+) = V_S/R_L = 24 \text{ V}/10 \text{ }\Omega = 2.4 \text{ A}$$

$$C = \frac{2.4 \text{ A} - 0.80 \text{ A}}{e^{-6(0)}} = 1.60 \text{ A}$$

Thus:

$$I(t) = (0.8 + 1.60 \; e^{-6t}) \text{ A}$$

The voltage across the resistor and the inductor at $t = 0.1$ s can be found by first calculating the current at this time:

$$I(0.1) = (0.8 + 1.6e^{(-6 \times 0.1)})\text{A} = 1.68 \text{ A}$$

Then:

$$V_R = IR = 1.68 \text{ A} \times 20 \text{ }\Omega = 33.6 \text{ V}$$

$$V_L = 24 \text{ V} - V_R = 24 \text{ V} - 33.6 \text{ V} = -9.6 \text{ V}$$

Example 3: Calculate the current I_L in the circuit in Fig. 2.205 at $t = 0.3$ s.

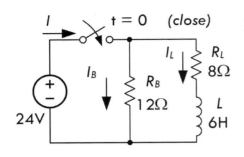

FIGURE 2.205

Answer: Notice that the 12-Ω resistor has no effect on the current I_L. Therefore, you get a simple RL energizing circuit with R_L in series with L:

$$I_L = \frac{V_S}{R_L}(1 - e^{-(R_L/L)t}) = \frac{24 \text{ V}}{8 \text{ Ω}}(1 - e^{-1.3t})$$

$$= (3 \text{ A})(1 - e^{-1.3t})$$

So at $t = 0.3$ s:

$$I_L = (3 \text{ A})(1 - e^{-1.3(0.3 \text{ s})}) = 0.99 \text{ A}$$

Here are some important things to notice during a forced response in regard to resistors, capacitors, and inductors:

Resistor: Under a forced response, a voltage is instantly placed across a resistor and a current immediately flows. There is no delay in voltage or current response (ideally).

Capacitor: Under a forced response, the voltage across a capacitor cannot change instantly, so at the instant a transition occurs it acts like an open circuit or constant voltage source. The voltage at instant $t = 0^-$ or $t = 0^+$ is a constant—the voltage that was present before the event. Also, at the instant $t = 0^-$ or $t = 0^+$ the current is zero, since no time transpires for charge to accumulate. However, after $t = 0^+$, the capacitor voltage and current have a natural response that is a function of time.

Inductor: Under a forced response, an inductor voltage cannot change instantaneously, so it acts like a short, meaning there is no voltage across it at $t = 0^-$ or $t = 0^+$. The current, however, at $t = 0^-$ or $t = 0^+$ will be a constant—the value of the current prior to the transient event. However, after $t = 0^+$, the inductor voltage and current have a natural response that is a function of time.

(Recall that $t = 0^-$ means the instant prior to the transient event, and $t = 0^+$ is the instant immediately after the transient event.)

Sometimes determining the voltage and currents within a transient circuit is a bit tricky, and requires a different approach than we encountered in the last three examples. The following few examples provide a good illustration of this.

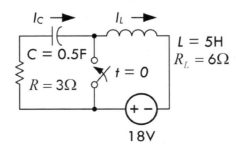

FIGURE 2.206

Example 4: In the circuit in Fig. 2.206, determine the current through the inductor and the voltage across the capacitor the instant before the switch is closed ($t = 0^-$) and the instant after the switch is closed ($t = 0^+$). Find I_C and I_L when the switch has been closed for $t = 0.5$ s.

Answer: Before the switch is closed, the capacitor acts as an open circuit, preventing current from flowing. The instant after the switch is closed, the capacitor cannot change in voltage instantaneously, so the current is still zero. The instant before and the instant after the switch is closed, the voltage across the capacitor is equal to the supply voltage. All this can be expressed by:

$$I_L(0^-) = I_L(0^+) = 0$$

$$V_C(0^-) = V_C(0^+) = 18 \text{ V}$$

When the switch is closed, we have a source-free RC circuit for which:

$$I_C = Be^{-t/R_1C} = Be^{-t/(1.5)}$$

At $t = 0^+$,

$$I_C(0^+)R_1 + V_C(0^+) = 18 \text{ V}$$

So,

$$I_C(0^+) = -\frac{V_C(0^+)}{R_1} = -\frac{18 \text{ V}}{3 \text{ }\Omega} = -6 \text{ A}$$

Plugging this back in to find B we get:

$$B = \frac{I_C(0^+)}{e^{-0/(1.5)}} = \frac{-6 \text{ A}}{1} = -6 \text{ A}$$

So the complete expression for I_C is:

$$I_C = (-6 \text{ A})e^{-t/(1.5)}$$

At $t = 0.5$ s,

$$I_C = (-6 \text{ A})e^{-0.5/(1.5)} = -4.3 \text{ A}$$

To find I_L at $t = 0.5$ s, we consider an RL circuit excited by an 18-V source (forced response) plus the natural response of an RL circuit:

$$I_L = I_f + I_n = \frac{V_S}{R_L} + Ae^{-6t/5}$$

Example 5: State all the initial conditions for the circuit in Fig. 2.207, which is under steady state for $t < 0$, and the switch is opened at $t = 0$. Determine the current in the circuit 0.6 s after the switch is opened. What is the induced voltage in the inductor at $t = 0.4$ s?

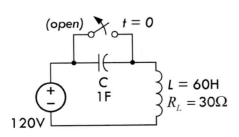

FIGURE 2.207

Answer: Since the switch is initially closed, the capacitor is shorted out, there is no voltage across it, and the current is simply equal to the supply voltage divided by the resistance of the inductor:

$$I_L(0^-) = I_L(0^+) = \frac{120 \text{ V}}{30 \text{ }\Omega} = 4 \text{ A}$$

$$V_C(0^-) = V_C(0^+) = 0 \text{ V}$$

By Kirchhoff's voltage law, we get the following equation for the circuit:

$$L\frac{dI}{dt} + R_L I + \frac{1}{C}\int I\,dt = 120 \text{ V}$$

$$(60 \text{ H})\frac{dI}{dt} + (30 \text{ }\Omega)I + \frac{1}{1 \text{ F}}\int I\,dt = 120 \text{ V}$$

from which the characteristic equation is:

$$60\,p^2 + 20\,p + 1 = 0$$

The characteristic roots are:

$$p = -0.46, -0.04$$

The complete current response is the sum of the forced and natural responses:

$$I = I_f + I_n = 0 + A_1 e^{-0.46t} + A_2 e^{-0.04t}$$

$I(0^+) = 4$ implies that $A_1 + A_2 = 4$.
   At $t = 0$,

$$(60 \text{ H})\frac{dI}{dt}(0^+) + (30 \text{ }\Omega)/(0^+) + V_C(0^+) = 120 \text{ V}$$

$$\frac{dI}{dt}(0^+) = 0 = 0.46A_1 + 0.04A_2$$

Solving for A_1 and A_2 gives $A_1 = -0.38$, $A_2 = 4.38$. Thus:

$$I = -0.38e^{-0.46t} + 4.38e^{-0.04t}$$

The induced voltage in the inductor at $t = 0.4$ s is simply found by plugging I into the definition of voltage for an inductor:

$$V_L = L\frac{dI}{dt} = 60[(-0.38)(-0.46)e^{-0.46t} + (4.38)(-0.04)e^{-0.04t}]$$

At $t = 0.4$ s,

$$V_L = 60(0.145 - 0.172) = -1.62 \text{ V}$$

Example 6: In the circuit in Fig. 2.208, the switch is moved from 1 to 2 at $t = 0$. Determine I as a function of time thereafter.

Answer: The complete current response when the switch is thrown is the sum of the forced response and the natural response:

$$I = I_f + I_n = \frac{24 \text{ V}}{R_2} + I_L e^{-Rt/L}$$

$$= \frac{24 \text{ V}}{2 \text{ }\Omega} + \left(\frac{12 \text{ V}}{5 \text{ }\Omega} - \frac{24 \text{ V}}{2 \text{ }\Omega}\right)e^{-2t/0.5}$$

$$= 12 \text{ A} - (9.6\text{A})e^{-4t}$$

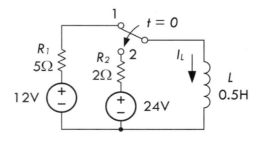

FIGURE 2.208

2.34.1   Series RLC Circuit

There's another transient example, which is a bit heavy in the math but is a classic analog of many other phenomena found in science and engineering. See Fig. 2.209.

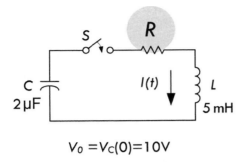

$V_0 = V_C(0) = 10V$

FIGURE 2.209

Assume the capacitor is charged to a voltage V_0, and then at $t = 0$, the switch is closed. Kirchhoff's voltage law for $t \geq 0$:

$$\frac{1}{C}\int I\,dt + IR + L\frac{dI}{dt} = 0$$

Rewriting in the standard form gives:

$$\frac{d^2I}{dt^2} + \frac{R}{L}\frac{dI}{dt} + \frac{1}{LC}I = 0$$

This is an example of a linear second-order homogeneous differential equation. It is reasonable to guess that the solution is of the same form as for the first-order homogeneous differential equation encountered earlier:

$$I = I_0 e^{\alpha t}$$

Substituting this into the differential equation gives:

$$\alpha^2 + \frac{R}{L}\alpha + \frac{1}{LC} = 0$$

Note that a solution of the form $e^{\alpha t}$ always reduces a linear homogeneous differential equation to an algebraic equation in which first derivatives are replaced by α and second derivates by α^2, and so forth. A linear second-order homogeneous differential equation then becomes a quadratic algebraic equation, and so on. This particular algebraic equation has the following solutions:

$$\alpha_1 = -\frac{R}{2L} + \sqrt{\frac{R^2}{4L^2} - \frac{1}{LC}}$$

$$\alpha_2 = -\frac{R}{2L} - \sqrt{\frac{R^2}{4L^2} - \frac{1}{LC}}$$

(2.92)

Since either value of α represents a solution to the original differential equation, the most general solution is one in which the two possible solutions are multiplied by arbitrary constants and added together:

$$I = I_1 e^{\alpha_1 t} + I_2 e^{\alpha_2 t}$$

The constants I_1 and I_2 must be determined from the initial conditions. An nth-order differential equation will generally have n constants that must be determined from initial conditions. In this case the constants can be evaluated from a knowledge of $I(0)$ and $dI/dt(0)$. Since the current in the inductor was zero for $t < 0$, and since it cannot change abruptly, we know that:

$$I(0) = 0$$

The initial voltage across the inductor is the same as across the capacitor, so that:

$$I_1 = -I_2 = \frac{V_0}{(\alpha_1 - \alpha_2)L}$$

The solution for the current in the series RLC circuit is thus:

$$I = \frac{V_0}{(\alpha_1 - \alpha_2)L}(e^{\alpha_1 t} - e^{\alpha_2 t}) \tag{2.93}$$

where α_1 and α_2 are given in Eq. 2.92.

The solution to the preceding equation has a unique character, depending on whether the quantity under the square root in Eq. 2.92 is positive, zero, or negative. We consider these three cases.

Case 1: Overdamped

For $R^2 > 4L/C$, the quantity under the square root is positive, and both values of α are negative with $|\alpha_2| > |\alpha_1|$. The solution is the sum of a slowly decaying positive term and a more rapidly decaying negative term of equal initial magnitude. The solution is sketched in Fig. 2.210. An important limiting case is the one in which $R^2 \gg 4L/C$. In that limit, the square root can be approximated as:

$$\sqrt{\frac{R^2}{4L^2} - \frac{1}{LC}} = \frac{R}{2L}\sqrt{1 - \frac{4L}{R^2 C}} \approx \frac{R}{2L} - \frac{1}{RC}$$

and the corresponding values of α are:

$$\alpha_1 = -\frac{1}{RC} \quad \text{and} \quad \alpha_2 = -\frac{R}{L}$$

Then the current in Eq. 2.93 is:

$$I \approx \frac{V_0}{R}(e^{-t/RC} - e^{-Rt/L}) \tag{2.94}$$

In this limit, the current rises very rapidly (in a time ~L/R) to a value near V_0/R and then decays very slowly back to zero. The overdamped response curve is shown in Fig. 2.210.

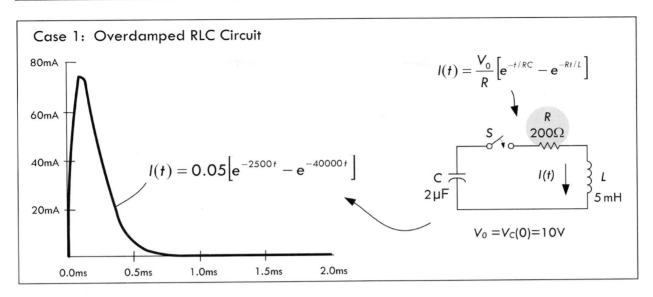

FIGURE 2.210

Case 2: Critically Damped

For $R^2 = 4L/C$, the quantity under the square root is zero, and $\alpha_1 = \alpha_2$. Equation 2.93 is then zero divided by zero, which is undefined. Therefore, the method of solution outlined here fails. A more productive approach is to let

$$\varepsilon = \sqrt{\frac{R^2}{4L^2} - \frac{1}{LC}}$$

and take the limit of Eq. 2.93 as $\varepsilon \to 0$. Then:

$$\alpha_1 = -\frac{R}{2L} + \varepsilon \quad \text{and} \quad \alpha_2 = -\frac{R}{2L} - \varepsilon$$

and Eq. 2.93 becomes:

$$I = \frac{V_0}{2\pi L} e^{-Rt/2L} (e^{\varepsilon t} - e^{-\varepsilon t})$$

Using the expansion $e^x \approx 1 + x$, for $|x| \ll 1$, the preceding equation becomes:

$$I = \frac{V_0}{L} e^{-Rt/2L} \tag{2.95}$$

The shape of the curve is not very different from the overdamped case, except that it approaches zero as fast as possible without overshooting the t axis and going negative. The critically damped response curve is shown in Fig. 2.211.

Critical damping is difficult to achieve: only a small change in R moves from this point. Small temperature change will cause that to occur. Energy transfer from C to L is now smaller than loss in R.

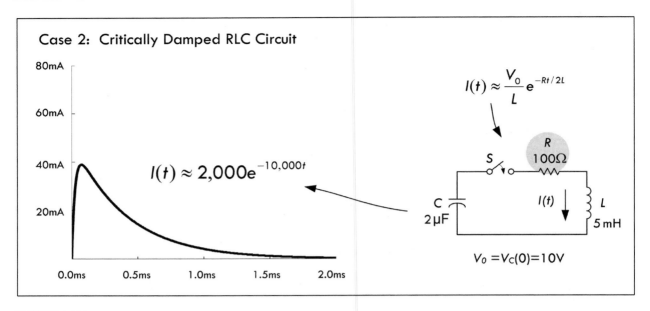

FIGURE 2.211

Case 3: Underdamped

For $R^2 < 4L/C$, the quantity under the square root is negative, and α can be written as:

$$\alpha = -\frac{R}{2L} \pm \frac{j}{\sqrt{LC}} \sqrt{1 - \frac{R^2C}{4L}}$$

where $j = \sqrt{-1}$

It is useful to define another symbol, ω, which we call the angular frequency:

$$\omega = \frac{1}{\sqrt{LC}} \sqrt{1 - \frac{R^2C}{4L}} \tag{2.96}$$

When $R^2 \ll 4L/C$, the angular frequency is:

$$\omega \approx \frac{1}{\sqrt{LC}}$$

and this approximation will usually suffice for most cases of interest. With these substitutions, Eq. 2.93 becomes:

$$I = \frac{V_0}{2j\omega L} e^{-Rt/2L} \left(e^{j\omega t} - e^{-j\omega t}\right)$$

We now make use of Euler's equation $e^{j\theta} = \cos\theta + j\sin\theta$ to express the current as follows:

$$I = \frac{V_0}{\omega L} e^{-Rt/2L} \sin(\omega t) \tag{2.97}$$

Notice that the solution is very different from the others, since it is oscillatory, with the oscillation amplitude decaying exponentially in time, as shown in Fig. 2.212.

Although ω is referred to as the angular frequency, note that it has units of radians per second, and it is related to the usual frequency f, which has units of cycles per second or hertz $\omega = 2\pi f$. The period of oscillation is $T = 1/f = 2\pi/\omega$.

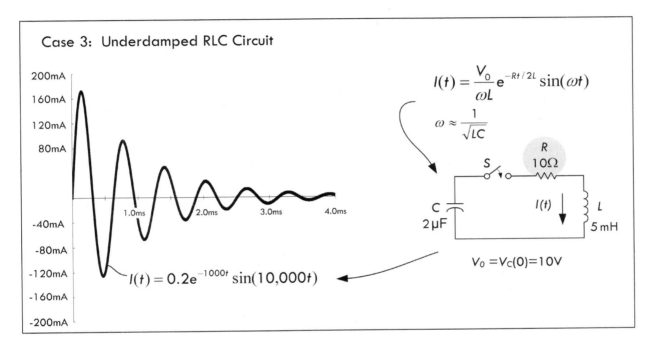

Case 3: Underdamped RLC Circuit

$$I(t) = \frac{V_0}{\omega L} e^{-Rt/2L} \sin(\omega t)$$

$$\omega \approx \frac{1}{\sqrt{LC}}$$

$$I(t) = 0.2e^{-1000t} \sin(10{,}000t)$$

$V_0 = V_C(0) = 10V$

FIGURE 2.212

The underdamped case is very interesting. At $t = 0$, all the energy is stored in the capacitor. As the current increases, energy is dissipated in the resistor and stored in the inductor until one-quarter of a cycle has passed, at which time there is no energy left in the capacitor. But as time goes on, the energy in the inductor decreases and the energy in the capacitor increases until one-half cycle has elapsed, at which time all the energy except that dissipated in the resistor is back in the capacitor. The energy continues to slosh back and forth, until it is eventually all dissipated by the resistor. A series LC circuit without any resistance would oscillate forever without damping.

This type of differential equation that we have seen for a series RLC circuit appears in many areas of science and engineering. It is referred to as a *damped harmonic oscillator*. Shock absorbers on an automobile, for example, are part of a mechanical harmonic oscillator designed to be nearly critically damped.

2.35   Circuits with Periodic Nonsinusoidal Sources

Suppose that you are given a periodic nonsinusoidal voltage (e.g., a squarewave, triangle wave, or ramp) that is used to drive a circuit containing resistors, capacitors, and inductors. How do you analyze the circuit? The circuit is not dc, so you cannot use dc theorems on it. The circuit is not sinusoidal, so you cannot directly apply complex impedances on it. What do you do?

If all else fails, you might assume that the only way out would be to apply Kirchhoff's laws on it. Well, before going any further, how do you mathematically represent the source voltage in the first place? That is, even if you could set up Kirchhoff's equations and such, you still have to plug in the source voltage term. For example, how do you mathematically represent a squarewave? In reality, coming up with an expression for a periodic nonsinusoidal source is not easy. However, for the sake of argument, let's pretend that you can come up with a mathematical representation of the waveform. If you plug this term into Kirchhoff's laws, you would again get differential equations (you could not use complex impedance then, because things are not sinusoidal).

To solve this dilemma most efficiently, it would be good to avoid differential equations entirely and at the same time be able to use the simplistic approach of complex impedances. The only way to satisfy both these conditions is to express a nonsinusoidal wave as a superposition of sine waves. In fact, a man by the name of Fourier discovered just such a trick. He figured out that a number of sinusoidal waves of different frequencies and amplitudes could be added together in a special manner to produce a superimposed pattern of any nonsinusoidal periodic wave pattern. More technically stated, a periodic nonsinusoidal waveform can be represented as a Fourier series of sines and cosines, where the waveform is a summation over a set of discrete, harmonically related frequencies.

2.35.1   Fourier Series

A time-dependent voltage or current is either periodic or nonperiodic. Figure 2.213 shows an example of a periodic waveform with period T.

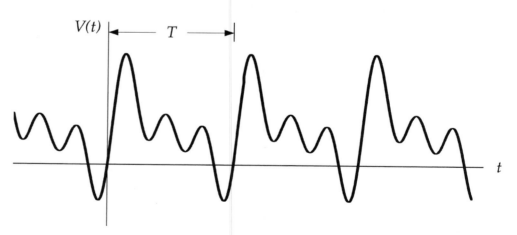

FIGURE 2.213

The wave is assumed to continue indefinitely in both the $+t$ and the $-t$ directions. A periodic function can be displaced by one period, and the resulting function is identical to the original function:

$$V(t \pm T) = V(t)$$

A periodic waveform can be represented as a Fourier series of sines and cosines:

$$V(t) = \frac{a_0}{2} + \sum_{n=1}^{\infty} (a_n \cos n\omega_0 t + b_n \sin n\omega_0 t) \tag{2.98}$$

where ω_0 is called the fundamental angular frequency,

$$\omega_0 = \frac{2\pi}{T} \tag{2.99}$$

$2\omega_0$ is called the second harmonic, and so on. The constants a_n and b_n are determined from:

$$a_n = \frac{2}{T} \int_{-T/2}^{T/2} V(t) \cos n\omega_0 t \, dt \tag{2.100}$$

$$b_n = \frac{2}{T} \int_{-T/2}^{T/2} V(t) \sin n\omega_0 t \, dt \tag{2.101}$$

The constant term $a_0/2$ is the average value of $V(t)$. The superposition theorem then allows you to analyze any linear circuit having periodic sources by considering the behavior of the circuit for each of the sinusoidal components of the Fourier series. Although most of the examples that we will use have voltage or current as the dependent variable and time as the independent variable, the Fourier methods are very general and apply to any sufficiently smooth function $f(t)$.

Using Euler's expression, $e^{j\theta} = \cos\theta + j\sin\theta$, we can convert Eq. 2.98 into a general expression for a periodic waveform as the sum of complex numbers:

$$V(t) = \sum_{n=-\infty}^{\infty} C_n e^{jn\omega_0 t} \tag{2.102}$$

By allowing both positive and negative frequencies ($n > 0$ and $n < 0$), it is possible to choose the C_n in such a way that the summation is always a real number. The value of C_n can be determined by multiplying both sides of Eq. 2.102 by $e^{-jm\omega_0 t}$, where m is an integer, and then integrating over a period. Only the term with $m = n$ survives, and the result is:

$$C_n = \frac{1}{T} \int_{-T/2}^{T/2} V(t) e^{-jn\omega_0 t} \, dt \tag{2.103}$$

Note that C_{-n} is the complex conjugate of C_n, and so the imaginary parts of Eq. 2.102 will always cancel, and the resulting $V(t)$ is real. The $n = 0$ term has a particularly simple interpretation; it is simply the average value of $V(t)$:

$$C_0 = \frac{1}{T} \int_{-T/2}^{T/2} V(t) \, dt \tag{2.104}$$

and corresponds to the dc component of the voltage. Whether the integrals in the preceding expression are over the interval $-T/2$ to $T/2$ or some other interval such as 0 to T is purely a matter of convenience, so long as the interval is continuous and has duration T.

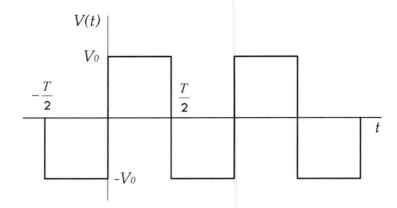

FIGURE 2.214

The following example illustrates a Fourier series of a squarewave, as depicted in Fig. 2.214.

To create a mathematical expression in terms of a series of complex numbers, as Eq. 2.102 requests, we first determine the constants from Eq. 2.103 by breaking the integral into two parts for which $V(t)$ is constant:

$$C_n = \frac{1}{T}\int_{-T/2}^{0}(-V_0)e^{-jn\omega_0 t}dt + \frac{1}{T}\int_{0}^{T/2}V_0 e^{-jn\omega_0 t}dt$$

$$= \frac{V_0}{jn\omega_0 T}(2 - e^{jn\omega_0 T/2} - e^{-jn\omega_0 T/2})$$

Since $\omega_0 T = 2\pi$, the preceding equation can be written as

$$C_n = \frac{V_0}{j2\pi n}(2 - e^{jn\pi} - e^{-jn\pi})$$

With the use of equation $e^{j\Theta} = \cos\Theta + j\sin\Theta$, the preceding equation becomes

$$C_n = \frac{V_0}{j\pi n}(1 - \cos n\pi)$$

Note that $\cos n\pi$ is +1 when n is even $(0,2,4,\ldots)$ and −1 when n is odd $(1,3,5,\ldots)$, so that all the even values of C_n are zero. Any periodic function, when displaced in time by half a period, is identical to the negative of the original function:

$$V\left(t \pm \frac{T}{2}\right) = -V(t)$$

In this case $V(t)$ is said to have half-wave symmetry, and its Fourier series will contain only odd harmonics. The squarewave is an example of such a function. If the wave remained at $+V_0$ and $-V_0$ for unequal times, the half-wave symmetry would be lost, and its Fourier series would then contain even as well as odd harmonics.

In addition to its half-wave symmetry, the squarewave shown in Fig. 2.214 is an odd function, because it satisfies the relation:

$$V(t) = -V(-t)$$

This property is not a fundamental property of the wave but arises purely out of the choice of where, with respect to the wave, the time origin $(t = 0)$ is assumed. For example, if the squarewave in Fig. 2.214 were displaced by a time of $T/4$, the resulting squarewave would be an even function, because it would then satisfy the relation:

$$V(t) = V(-t)$$

It's important to note that an odd function can have no dc component, since the negative parts exactly cancel the positive parts on opposite sides of the time axis. The cosine is an even function, and the sine is an odd function. Any even function can be written as a sum of cosines ($b_n = 0$ in Eq. 2.98), and any odd function can be written as a sum of sines ($a_n = 0$ in Eq. 2.98). Most periodic functions (such as the one in Fig. 2.213) are neither odd nor even.

The Fourier series calculation can often be simplified by adding or subtracting a constant to the value of the function, or by displacing the time origin so that the function is even or odd or so that it has half-wave symmetry.

The odd-numbered coefficients of the Fourier series representation of the squarewave are given by

$$C_n = \frac{2V_0}{n\pi j}$$

and the Fourier series is

$$V(t) = \frac{2V_0}{\pi j} \sum_{\substack{n=-\infty \\ n=\text{odd}}}^{\infty} \frac{1}{n} e^{jn\omega_0 t}$$

With the use of Euler's equation and the fact that $\sin\theta = -\sin(-\theta)$ and $\cos\theta = \cos(-\theta)$, the preceding equation becomes

$$V(t) = \frac{4V_0}{\pi} \sum_{\substack{n=-\infty \\ n=\text{odd}}}^{\infty} \frac{\sin n\omega_0 t}{n}$$

The first three terms of the preceding series ($n = 1,3,5$) along with their sum are plotted in Fig. 2.215. Note that the series, even with as few as three terms, is beginning to resemble the squarewave.

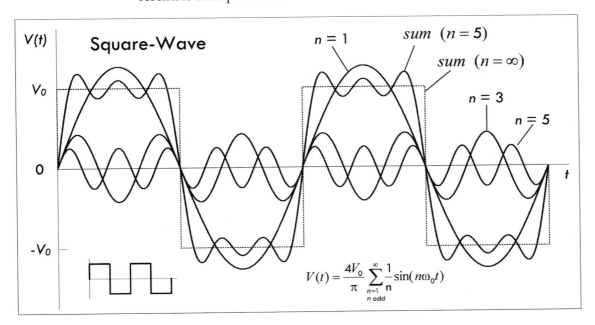

FIGURE 2.215

For waveforms more complicated than a squarewave, the integrals are more difficult to perform, but it is still usually easier to calculate a Fourier series for a periodic voltage than to solve a differential equation in which the same time-dependent voltage appears. Furthermore, tables of Fourier series for the most frequently encountered waveforms are available and provide a convenient shortcut for analyzing many circuits. Some common waveforms and their Fourier series are listed in Fig. 2.216.

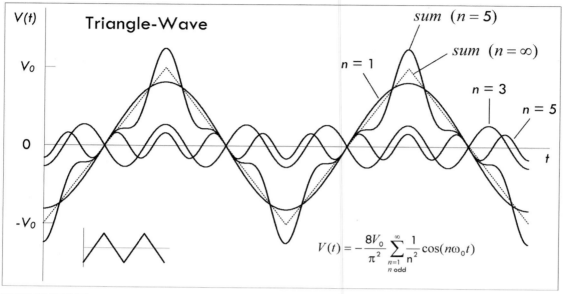

Triangle-Wave

$$V(t) = -\frac{8V_0}{\pi^2} \sum_{\substack{n=1 \\ n \text{ odd}}}^{\infty} \frac{1}{n^2} \cos(n\omega_0 t)$$

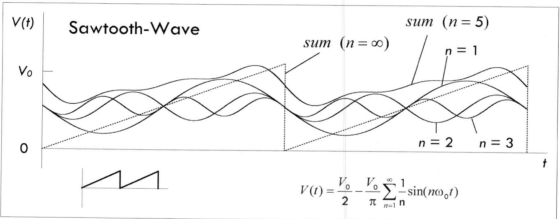

Sawtooth-Wave

$$V(t) = \frac{V_0}{2} - \frac{V_0}{\pi} \sum_{n=1}^{\infty} \frac{1}{n} \sin(n\omega_0 t)$$

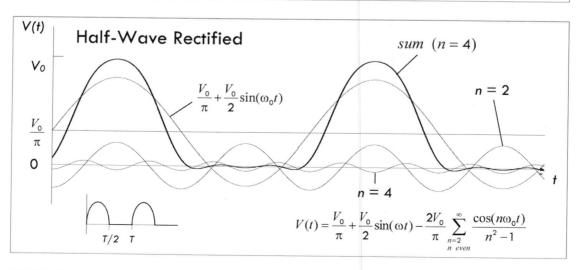

Half-Wave Rectified

$$V(t) = \frac{V_0}{\pi} + \frac{V_0}{2} \sin(\omega t) - \frac{2V_0}{\pi} \sum_{\substack{n=2 \\ n \text{ even}}}^{\infty} \frac{\cos(n\omega_0 t)}{n^2 - 1}$$

FIGURE 2.216

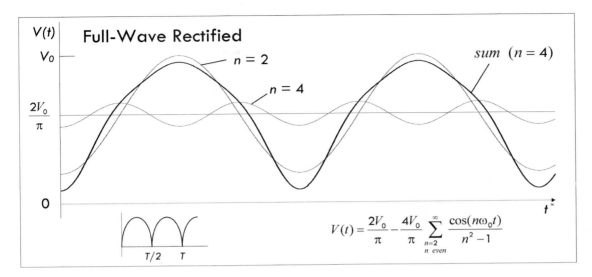

$$V(t) = \frac{2V_0}{\pi} - \frac{4V_0}{\pi} \sum_{\substack{n=2 \\ n \ even}}^{\infty} \frac{\cos(n\omega_0 t)}{n^2 - 1}$$

FIGURE 2.216 *(Continued)*

Example: Squarewave RC Circuit. The following example demonstrates the use of the Fourier series to analyze a circuit with a periodic source. Here a squarewave source is connected to a simple RC circuit. Refer to Fig. 2.217.

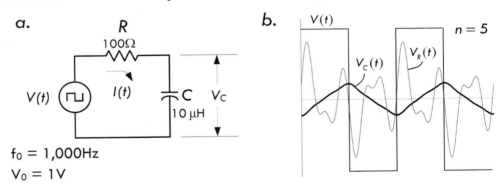

FIGURE 2.217

Since the source is periodic, the current $I(t)$ is also periodic with the same period, and it can be written as a Fourier series:

$$I(t) = \sum_{n=-\infty}^{\infty} C'_n e^{jn\omega_0 t}$$

Each C'_n is a phasor current representing one frequency component of the total current in the same way that each C_n represents a component of the phasor voltage in the previous section. The relationship between the two phasors is determined by dividing by the circuit impedance:

$$C'_n = \frac{C_n}{R + 1/j\omega C} = \frac{C_n}{R + 1/jn\omega_0 C}$$

Substituting the value of C_n derived earlier for the squarewave gives:

$$C'_n = \frac{2V_0}{n\pi(jR + 1/n\omega_0 C)} = \frac{2\omega_0 C(1 - jn\omega_0 RC)V_0}{\pi(n^2 \omega_0^2 R^2 C^2 + 1)}$$

for n odd. For n even, C'_n is zero, since C_n is zero for even n. The corresponding current is then

$$I(t) = \frac{2\omega_0 C V_0}{\pi} \sum_{\substack{n=-\infty \\ n \text{ odd}}}^{\infty} \frac{1 - jn\omega_0 RC}{\pi(n^2\omega_0^2 R^2 C^2 + 1)} e^{jn\omega_0 t}$$

With the use of Euler's relation, the preceding current can be written as:

$$I(t) = \frac{4\omega_0 C V_0}{\pi} \sum_{\substack{n=1 \\ n \text{ odd}}}^{\infty} \frac{\cos n\omega_0 t + n\omega_0 RC \sin n\omega_0 t}{n^2\omega_0^2 R^2 C^2 + 1}$$

The voltage across the resistor and capacitor can be determined from the definitions of an ideal resistor and an ideal capacitor:

$$V_R(t) = I(t)R = \frac{4\omega_0 CRV_0}{\pi} \sum_{\substack{n=1 \\ n \text{ odd}}}^{\infty} \frac{\cos n\omega_0 t + n\omega_0 RC \sin n\omega_0 t}{n^2\omega_0^2 R^2 C^2 + 1}$$

$$V_C(t) = \frac{1}{C} \int I(t)\,dt = \frac{4V_0}{\pi} \sum_{\substack{n=1 \\ n \text{ odd}}}^{\infty} \frac{\frac{1}{n}\sin n\omega_0 t + \omega_0 RC \cos n\omega_0 t}{n^2\omega_0^2 R^2 C^2 + 1}$$

The sum of the first three terms ($n = 1$, $n = 3$, $n = 5$) of the Fourier series for $V_C(t)$ and $V_R(t)$ are shown in Fig. 2.217b. As n approaches infinity, the waveforms approach the real deal.

Note that circuits with squarewave sources can also be analyzed as transient circuits. In the circuit in Fig. 2.217a, during a half period (such as $0 < t < T/2$) when the source voltage is constant, the voltage across the capacitor is expressed as:

$$V_C(t) = A + Be^{-t/RC}$$

The constants A and B can be determined from

$$V_C(\infty) = A = V_0$$

$$V_C(T/2) = A + Be^{-T/2RC} = -V_C(0) = -A - B$$

The first equation results from knowing that if the source remains at $+V_0$ forever, the capacitor would charge to voltage V_0. The second equation is required to ensure that the function has half-wave symmetry. Hence,

$$A = V_0$$

$$B = -\frac{2V_0}{1 + e^{-T/2RC}}$$

The capacitor voltage is then:

$$V_C(t) = V_0 - \frac{2V_0 e^{-t/RC}}{1 + e^{-T/2RC}}$$

for $0 < t < T/2$. The waveform repeats itself for $t > T/2$ with each half cycle alternating in sign.

2.36 Nonperiodic Sources

Nonperiodic voltages and currents can also be represented as a superposition of sine waves as with the Fourier series. However, instead of a summation over a set of discrete, harmonically related frequencies, the waveforms have a continuous spectrum of frequencies. It is possible to think of a nonperiodic function as a periodic function with an infinite period. The fundamental angular frequency, which was $\omega_0 = 2\pi/T$ for the Fourier series, approaches zero as the period approaches infinity. In this case, to remind us that we're now dealing with an infinitesimal quantity, we represent the angular frequency as $\Delta\omega$. The various harmonics within the waveform are separated by the infinitesimal $\Delta\omega$, so that all frequencies are present. The waveform as a summation, as was done with Fourier series, is:

$$V(t) = \sum_{n=-\infty}^{\infty} C_n e^{jn\omega_0 t} = \sum_{n=-\infty}^{\infty} C_n e^{j\omega t} \frac{T\Delta\omega}{2\pi} = \frac{1}{2\pi}\sum_{n=-\infty}^{\infty} C_n T e^{j\omega t} \Delta\omega$$

where we have used the fact that $\omega = n\omega_0$ and $T\Delta\omega = 2\pi$. Since $\Delta\omega$ is infinitesimal, the preceding summation can be replaced with an integral ($d\omega = \Delta\omega$):

$$V(t) = \frac{1}{2\pi}\int_{-\infty}^{\infty} C_n T e^{j\omega t}\, d\omega$$

As before, C_n is given by

$$C_n = \frac{1}{T}\int_{-T/2}^{T/2} V(t) e^{-j\omega t}\, dt$$

However, since T is infinite, we can write

$$C_n T = \int_{-\infty}^{\infty} V(t) e^{-j\omega t}\, dt$$

Although T is infinite, the term $C_n T$ is usually finite in value. The $C_n T$ term is referred to as the Fourier transform of $V(t)$ and is rewritten as $\overline{V}(\omega)$. After integration, it is only a function of angular frequency ω. The following two equations are called a Fourier transform pair:

$$V(t) = \frac{1}{2\pi}\int_{-\infty}^{\infty} \overline{V}(\omega) e^{j\omega t}\, d\omega \tag{2.105}$$

$$\overline{V}(\omega) = \int_{-\infty}^{\infty} V(t) e^{-j\omega t}\, dt \tag{2.106}$$

These two equations are symmetric. ($\overline{V}(\omega)$ is sometimes defined as $C_n T/\sqrt{2\pi}$ to make the symmetry even more perfect.)

Like the coefficients of the Fourier series, the Fourier transform $\overline{V}(\omega)$ is generally complex, unless $V(t)$ is an even function of time. When $V(t)$ is an odd function of time, the Fourier transform $\overline{V}(\omega)$ is entirely imaginary. For this reason, when plotting a Fourier transform, it is customary to plot either the magnitude $|\overline{V}(\omega)|$ or the square of the magnitude $|\overline{V}(\omega)|^2$—referred to as the *power spectrum*—as a function of ω.

As an example, we will calculate the Fourier transform of the square pulse shown in Fig. 2.218a and given by:

$$V(t) = \begin{cases} 0 & t < -\tau/2 \quad \text{and} \quad t > \tau/2 \\ V_0 & -\tau/2 \le t \le \tau/2 \end{cases}$$

From Eq. 2.106, the Fourier transform is:

$$\overline{V}(\omega) = V_0 \int_{-\tau/2}^{\tau/2} e^{-j\omega t}\, dt = \frac{2V_0}{\omega} \sin\frac{\omega\tau}{2}$$

The magnitude $|\overline{V}(\omega)|$ is plotted as a function of ω in Fig. 2.218b. As before, most of the Fourier spectrum is a band of frequencies within about $1/\tau$ of zero.

a.

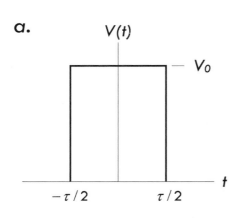

b.

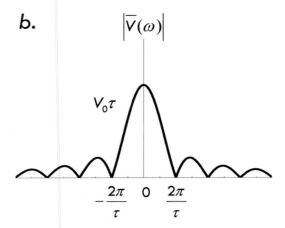

FIGURE 2.218

In practice, if we were to attach a nonperiodic pulse, such as our square pulse, to some complex circuit with total impedance $Z(\omega)$, we could determine the current as a function of time (and thus individual voltages and currents within the circuit) by first using the Fourier transform voltage for the square pulse,

$$\overline{V}(\omega) = V_0 \int_{-\tau/2}^{\tau/2} e^{-j\omega t}\, dt = \frac{2V_0}{\omega} \sin\frac{\omega\tau}{2}$$

and using this to find the Fourier transform of the current by dividing by the impedance:

$$\bar{I}(\omega) = \frac{\overline{V}(\omega)}{Z(\omega)} \tag{2.107}$$

Once this is found, the current as a function of time could be determined using the inverse Fourier transform:

$$I(t) = \frac{1}{2\pi} \int_{-\infty}^{\infty} \bar{I}(\omega)e^{j\omega t}\, d\omega$$

Using the Fourier transforms for a problem like this seems incredibly difficult—try placing a simple RCL network into the impedance and solving the integrals. However, the Fourier transform, as nasty as it can get, provides the easiest method of solution for the nonperiodic problems.

In summary, analyzing a circuit by this method involves first converting to the frequency domain by calculating the Fourier transform of the sources from Eq. 2.106, then using the impedance to determine the Fourier transform of the unknown Eq. 2.107, and finally converting back to the time domain by calculating the inverse Fourier transform of the unknown from Eq. 2.105. Solving things this way, using the difficult integrals, is actually usually much easier than solving the corresponding differential equation with a time-dependent source.

Note that special devices called *spectrum analyzers* can display the Fourier transform $|\overline{V}(\omega)|$ of a voltage as a function of frequency.

Enough of the difficult stuff; let's let a simulator do the thinking for us.

2.37 SPICE

SPICE is a computer program that simulates analog circuits. It was originally designed for the development of integrated circuits, from which it derives its name: Simulation Program with Integrated Circuit Emphasis, or SPICE for short.

The origin of SPICE is traced back to another circuit simulation program called CANCER (Computer Analysis of Non-Linear Circuits Excluding Radiation), developed by Ronald Rohrer, of University of California–Berkeley, along with some of his students. CANCER was able to perform dc, ac, and transient analysis, and included special linear-companion models for basic active devices like diodes (Shockley equation) and bipolar transistors (Ebers-Moll equations).

When Rohrer left Berkeley, CANCER was rewritten and renamed SPICE, released as version 1 to the public in 1972. SPICE 1 was based on nodal analysis and included revised models for bipolar transistors (using Gummel-Poon equations), as well as new models for JFET and MOSFET devices.

In 1975, SPICE 2 was introduced, with modified nodal analysis (MNA) that replaced the old nodal analysis, and now supported voltage sources and inductors. Many things were added and many alterations were made to SPICE 2. The last version of SPICE 2 to be written in FORTRAN, version SPICE 2G.6, came out in 1983.

In 1985, SPICE 3 appeared on the scene, written in the C programming language rather than FORTRAN. It included a graphical interface for viewing results and also polynomial capacitors and inductors and voltage-controlled sources, as well as models for MESFETs, lossy transmission lines, and nonideal switches. SPICE 3 also had improved semiconductor models and was designed to eliminate many convergence problems found in previous versions. From this time on, commercial versions of SPICE have appeared: HSPICE, IS_SPICE, and MICROCAP and PSPICE (MicroSim's PC version of SPICE).

Today, there are many user-friendly simulator programs out there that use SPICE as the brains behind the analysis. These high-level simulator programs allow you to click, drag, and drop components onto a page and draw wire connections. Test instruments, such as voltmeters, power meters, oscilloscopes, and spectrum analysis, can be dragged and connected to the circuit. Almost any type of source is available as well as any type of device (passive, active, digital, etc.)—MultiSim from Electronics Workbench contains a component library of over 13,000 models. Figure 2.219 shows an example screen shot depicting the various elements that come with a simulator. Three popular commercial simulators include MicroSim, TINAPro, and CircuitMaker. There are also online simulation programs such as CircuitLab (www.circuitlab.com) that are quick and easy to use.

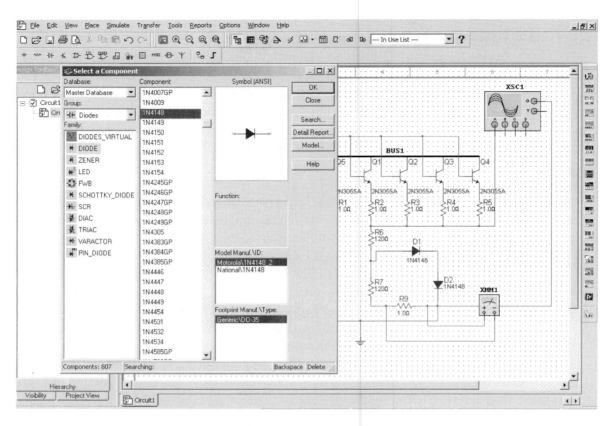

FIGURE 2.219 Screen shot of a simulator user interface, showing the various devices, sources, test equipment, and analysis capabilities. The simulator shown is MultiSim from Electronics Workbench.

Basic devices: Passive devices, diodes, LEDs, thyristors, transistors, analog amplifiers and comparators, TTL logic devices, CMOS logic devices, miscellaneous digital (e.g., TIL, VHDL, VERILOG HDL), mixed signal devices, indicators, RF devices, electromechanical devices, and so on.

Sources: dc and ac voltage and current sources; clock source; AM source and FM voltage and current sources; FSK source; voltage-controlled sine, square, and triangle sources as well as current-controlled ones; pulse voltage and current sources; exponential voltage and current sources; piecewise linear voltage and current sources; controlled one-shot, polynomial, and nonlinear dependent sources.

Analysis techniques: dc operating point, ac analysis, transient analysis, Fourier analysis, noise analysis, distortion analysis, dc sweep, sensitivity, parameter sweep, temperature sweep, pole zero, transfer function, worst case, Monte Carlo, trace width analysis, batched analysis, user-defined analysis, noise figure analysis, and so on.

Test equipment: Multimeter, function generator, wattmeter, oscilloscope, bode plotter, word generator, logic analyzer, logic converter, distortion analyzer, spectrum analyzer, network analyzer, and so on.

2.37.1 How SPICE Works

Here we take a look at the heart of simulators—SPICE. What are the mathematical tricks the code in SPICE uses to simulate electrical circuits described by nonlinear differential equations? At the core of the SPICE engine is a basic technique called nodal analysis. It calculates the voltage at any node, given all of the circuit resistances (or, inversely, their conductances) and the current sources. Whether the program is performing dc, ac, or transient analysis, SPICE ultimately casts its components (linear, nonlinear, and energy-storage elements) into a form where the innermost calculation is nodal analysis.

Kirchhoff discovered that the total current entering a node is equal to the total current leaving a node. Stated another way, the sum of currents in and out of a node is zero. These currents can be described by equations in terms of voltages and conductances. If you have more than one node, then you get more than one equation describing the same system (simultaneous equations). The trick then is finding the voltage at each node that satisfies all the equations simultaneously.

For example, let's consider the simple circuit in Fig. 2.220. In this circuit there are three nodes: node 0 (which is always ground), node 1, and node 2.

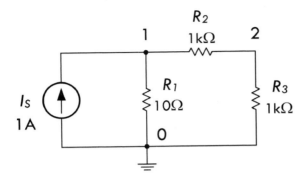

FIGURE 2.220

Using Kirchhoff's current law in the form of "the sum of current in and out of a node is zero," we can create two equations for the two nodes 1 and 2 (0 is by default ground):

$$-I_S + \frac{V_1}{R_1} + \frac{V_1 - V_2}{R_2} = 0$$

$$\frac{V_2 - V_1}{R_2} + \frac{V_2}{R_3} = 0$$

Since the mission here is to calculate the node voltages, we rewrite these equations in order to separate V_1 and V_2:

$$\left(\frac{1}{R_1} + \frac{1}{R_2}\right)V_1 + \left(-\frac{1}{R_2}\right)V_2 = I_S$$

$$\left(-\frac{1}{R_2}\right)V_1 + \left(\frac{1}{R_2} + \frac{1}{R_3}\right)V_2 = 0$$

The trick now becomes finding values of V_1 and V_2 that satisfy both equations. Although we could solve one variable in terms of the other from one equation and plug it into the second equation to find the other variable, things get messy (so many R variables to deal with). Instead, we'll take a cleaner approach—one that involves conductance G, where $G = 1/R$. This will make the bookkeeping easier, and it becomes especially important when the nodal number increases with complex circuits.

Writing resistors in terms of total conductance:

$$G_{11} = 1/R_1 + 1/R_2, \; G_{21} = -1/R_2, \; G_{12} = -1/R_2, \; G_{22} = 1/R_2 + 1/R_3.$$

Thus, the system of equations is transformed into:

$$G_{11}V_1 + G_{12}V_2 = I_S$$

$$G_{21}V_1 + G_{22}V_2 = 0$$

Solving the second equation for V_1:

$$V_1 = \frac{-G_{22}V_2}{G_{21}}$$

we then stick this into the first equation and solve for V_2:

$$V_2 = \frac{I_S}{\left[G_{12} - \dfrac{G_{11}G_{22}}{G_{21}}\right]}$$

See how much cleaner the manipulations were using conductances? V_2 is described by circuit conductances and I_S alone. We still must find the numerical value of V_2 and stick it back into the V_1 equation while, in the process, calculating the numerical values of the conductances. However, we have found circuit voltages V_1 and V_2 that satisfy both system equations.

Although the last approach wasn't that hard, or messy, there are times when the circuits become big and the node and device counts so large that bookkeeping terms become a nightmare. We must go a step further in coming up with a more efficient and elegant approach. What we do is use matrices.

In matrix form, the set of nodal equations is written:

$$\begin{bmatrix} \dfrac{1}{R_1} + \dfrac{1}{R_2} & \dfrac{-1}{R_2} \\ \dfrac{-1}{R_2} & \dfrac{1}{R_2} + \dfrac{1}{R_3} \end{bmatrix} \times \begin{bmatrix} V_1 \\ V_2 \end{bmatrix} = \begin{bmatrix} I_S \\ 0 \end{bmatrix}$$

Or in terms of total conductances and source currents:

$$\begin{bmatrix} G_{11} & G_{12} \\ G_{21} & G_{22} \end{bmatrix} \times \begin{bmatrix} V_1 \\ V_2 \end{bmatrix} = \begin{bmatrix} I_1 \\ I_2 \end{bmatrix} \tag{2.108}$$

Treating each matrix as a variable, we can rewrite the preceding equation in compact form:

$$\mathbf{G} \cdot \mathbf{v} = \mathbf{i} \tag{2.109}$$

In matrix mathematics, you can solve for a variable (almost) as you would in any other algebraic equation. Solving for $\mathbf{v}$ you get:

$$\mathbf{v} = \mathbf{G}^{-1} \cdot \mathbf{i} \tag{2.110}$$

where $\mathbf{G}^{-1}$ is the matrix inverse of $\mathbf{G}$. ($1/\mathbf{G}$ does not exist in the matrix world.) This equation is the central mechanism of the SPICE algorithm. Regardless of the analysis—ac, dc, or transient—all components or their effects are cast into the conductance matrix **G** and the node voltages are calculated by $\mathbf{v} = \mathbf{G}^{-1} \cdot \mathbf{i}$, or some equivalent method.

Substituting the component values present in the circuit in Fig. 2.218 into the conductance matrix and current matrix (we could use an Excel spreadsheet and apply formulas to keep things clean), we get:

$$\mathbf{G} = \begin{bmatrix} G_{11} & G_{12} \\ G_{21} & G_{22} \end{bmatrix} = \begin{bmatrix} \dfrac{1}{R_1} + \dfrac{1}{R_2} & \dfrac{-1}{R_2} \\ -\dfrac{1}{R_2} & \dfrac{1}{R_2} + \dfrac{1}{R_3} \end{bmatrix} = \begin{bmatrix} 0.101 & -0.001 \\ -0.001 & 0.002 \end{bmatrix}$$

$$\mathbf{i} = \begin{bmatrix} 1 \\ 0 \end{bmatrix}$$

Hence, the voltage is:

$$\begin{array}{ccccc} \mathbf{G}^{-1} & \times & \mathbf{i} & = & \mathbf{v} \\ \begin{bmatrix} 9.95 & 4.98 \\ 4.98 & 502.49 \end{bmatrix} & \times & \begin{bmatrix} 1 \\ 0 \end{bmatrix} & = & \begin{bmatrix} 9.9502 \\ 4.9751 \end{bmatrix} \end{array}$$

$V_1 = 9.9502$ and $V_2 = 4.9751$

2.37.2 Limitations of SPICE and Other Simulators

Simulation of a circuit is only as accurate as the behavior models in the SPICE devices created for it. Many simulations are based on simplified models. For more complex circuits or subtle behaviors, the simulation can be misleading or incorrect. This can mean disaster if you're relying entirely on SPICE (or an advanced simulator based on SPICE) when developing circuits. SPICE can also be deceiving, since simulations are free of noise, crosstalk, interference, and so on—unless you incorporate these behaviors into the circuit. Also, SPICE is not the best predictor of component failures. You must know what dangers to look for and which behaviors are not modeled in your SPICE circuit. In short, SPICE is not a prototype substitute—the performance of the actual breadboard provides the final answer.

2.37.3 A Simple Simulation Example

As an example, we will look at simulating a simple RLC crossover network (which we will meet again in Chap. 15) with the free and easy-to-use online simulator CircuitLab (www.circuitlab.com).

The first step is to draw the schematic using the editing tool (see Fig. 2.221). The LRC network is responsible for separating an audio signal into two parts: a low-pass filter to drive the low-frequency woofer speaker and a high-pass filter to drive the tweeter.

As well as adding the appropriate components to the design and connecting them, we have added an ac voltage source to represent the input signal and attached two labels (Tweeter and Woofer), so that we can see the results of the simulation (there will be labels next to our graph plots).

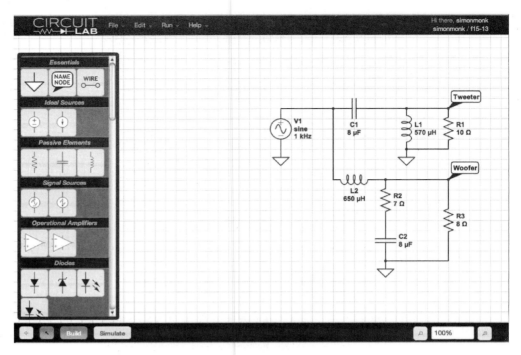

FIGURE 2.221   CircuitLab schematic diagram.

When we are ready, we can run a simulation. In this case, we are going to carry out a frequency-domain simulation so that we can determine the crossover frequency of the network. To do this, we specify V_1 as being the input as well as start and end frequencies. We also specify the outputs that we wish to see plotted (see Fig. 2.222).

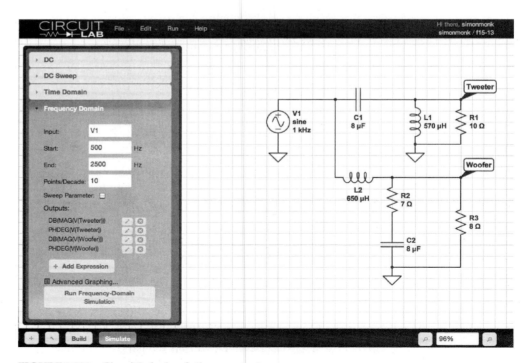

FIGURE 2.222   CircuitLab simulation parameters.

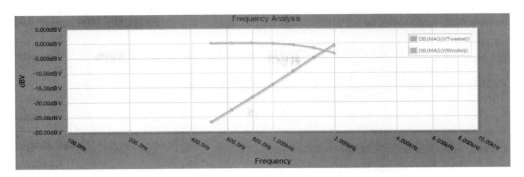

FIGURE 2.223   **CircuitLab simulation results.**

When we click the Simulate button, CircuitLab will run the simulation and produce an output plot, as shown in Fig. 2.223.

From Fig. 2.223, we can see that the crossover frequency where both speakers are receiving the same magnitude of signal is at around 1.6 kHz.

CircuitLab is a good starting point for understanding simulators. We strongly encourage you to try some of the sample schematics on the website and experiment with a few simulations.

CHAPTER 3

Basic Electronic Circuit Components

3.1 Wires, Cables, and Connectors

Wires and cables provide low-resistance pathways for electric currents. Most electrical wires are made from copper or silver and typically are protected by an insulating coating of plastic, rubber, or lacquer. Cables consist of a number of individually insulated wires bound together to form a multiconductor transmission line. Connectors, such as plugs, jacks, and adapters, are used as mating fasteners to join wires and cable with other electrical devices.

3.1.1 Wires

A wire's diameter is expressed in terms of a *gauge number*. The gauge system, as it turns out, goes against common sense. In the gauge system, as a wire's diameter increases, the gauge number decreases. At the same time, the resistance of the wire decreases. When currents are expected to be large, smaller-gauge wires (large-diameter wires) should be used. If too much current is sent through a large-gauge wire (small-diameter wire), the wire could become hot enough to melt. Table 3.1 shows various characteristics for B&S-gauged copper wire at 20°C. For rubber-insulated wire, the allowable current should be reduced by 30 percent.

Wire comes in solid core, stranded, or braided forms.

Solid Core

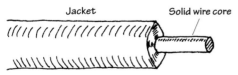

Jacket   Solid wire core

This wire is useful for wiring breadboards; the solid-core ends slip easily into breadboard sockets and will not fray in the process. These wires have the tendency to snap after a number of flexes.

FIGURE 3.1

Stranded Wire

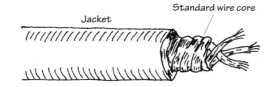

The main conductor is comprised of a number of individual strands of copper. Stranded wire tends to be a better conductor than solid-core wire because the individual wires together comprise a greater surface area. Stranded wire will not break easily when flexed.

Braided Wire

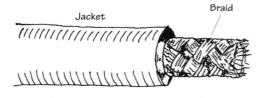

A braided wire is made up of a number of individual strands of wire braided together. Like stranded wires, these wires are better conductors than solid-core wires, and they will not break easily when flexed. Braided wires are frequently used as an electromagnetic shield in noise-reduction cables and also may act as a wire conductor within the cable (e.g., coaxial cable).

FIGURE 3.1 (*Continued*)

TABLE 3.1 Copper Wire Specifications (Bare and Enamel-Coated Wire)

WIRE SIZE (AWG)	DIAMETER (MILS)*	DIAMETER (MM)	OHMS PER 1000 FT	OHMS PER KM	CURRENT CARRYING CAPACITY (A)	NEAREST BRITISH SWG NO.
1	289.3	7.35	0.1239	0.41	119.564	1
2	257.6	6.54	0.1563	0.51	94.797	2
3	229.4	5.83	0.1971	0.65	75.178	4
4	204.3	5.19	0.2485	0.82	59.626	5
5	181.9	4.62	0.3134	1.03	47.268	6
6	162.0	4.12	0.3952	1.30	37.491	7
7	144.3	3.67	0.4981	1.63	29.746	8
8	128.5	3.26	0.6281	2.06	23.589	9
9	114.4	2.91	0.7925	2.60	18.696	11
10	101.9	2.59	0.9987	3.28	14.834	12
11	90.7	2.31	1.2610	4.13	11.752	13
12	80.8	2.05	1.5880	5.21	9.327	13
13	72.0	1.83	2.0010	6.57	7.406	15
14	64.1	1.63	2.5240	8.29	5.870	15
15	57.1	1.45	3.1810	10.45	4.658	16
16	50.8	1.29	4.0180	13.17	3.687	17
17	45.3	1.15	5.0540	16.61	2.932	18
18	40.3	1.02	6.3860	20.95	2.320	19
19	35.9	0.91	8.0460	26.42	1.841	20
20	32.0	0.81	10.1280	33.31	1.463	21

TABLE 3.1   Copper Wire Specifications (Bare and Enamel-Coated Wire) (*Continued*)

WIRE SIZE (AWG)	DIAMETER (MILS)*	DIAMETER (MM)	OHMS PER 1000 FT	OHMS PER KM	CURRENT CARRYING CAPACITY (A)	NEAREST BRITISH SWG NO.
21	28.5	0.72	12.7700	42.00	1.160	22
22	25.3	0.64	16.2000	52.96	0.914	22
23	22.6	0.57	20.3000	66.79	0.730	24
24	20.1	0.51	25.6700	84.22	0.577	24
25	17.9	0.46	32.3700	106.20	0.458	26
26	15.9	0.41	41.0200	133.90	0.361	27
27	14.2	0.36	51.4400	168.90	0.288	28
28	12.6	0.32	65.3100	212.90	0.227	29
29	11.3	0.29	81.2100	268.50	0.182	31
30	10.0	0.26	103.7100	338.60	0.143	33
31	8.9	0.23	130.9000	426.90	0.113	34
32	8.0	0.20	162.0000	538.30	0.091	35
33	7.1	0.18	205.7000	678.80	0.072	36
34	6.3	0.16	261.3000	856.00	0.057	37
35	5.6	0.14	330.7000	1079.00	0.045	38
36	5.0	0.13	414.8000	1361.00	0.036	39
37	4.5	0.11	512.1000	1716.00	0.029	40

* 1 Mil = 2.54×10^{-5} m

Kinds of Wires

Pretinned Solid Bus Wire

This wire is often referred to as hookup wire. It includes a tin-lead alloy to enhance solderability and is usually insulated with polyvinyl chloride (PVC), polyethylene, or Teflon. Used for hobby projects, preparing printed circuit boards, and other applications where small bare-ended wires are needed.

Speaker Wire

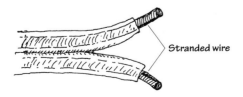

Stranded wire

This wire is stranded to increase surface area for current flow. It has a high copper content for better conduction.

FIGURE 3.2

Magnet Wire

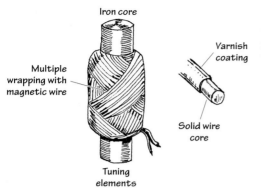

Iron core

Varnish coating

Multiple wrapping with magnetic wire

Solid wire core

Tuning elements

This wire is used for building coils and electromagnets or anything that requires a large number of loops, say, a tuning element in a radio receiver. It is built of solid-core wire and insulated by a varnish coating. Typical wire sizes run from 22 to 30 gauge.

FIGURE 3.2 *(Continued)*

3.1.2 Cables

A cable consists of a multiple number of independent conductive wires. The wires within cables may be solid core, stranded, braided, or some combination in between. Typical wire configurations within cables include the following:

FIGURE 3.3

Twin Lead

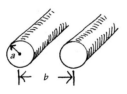

Coaxial

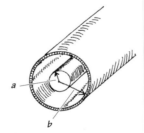

Ribbon and Plane

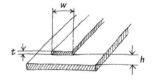

Twisted Pair

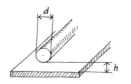

Wire and Plane

Strip Line

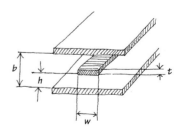

3.1.3 Connectors

The following is a list of common plug and jack combinations used to fasten wires and cables to electrical devices. Connectors consist of plugs (male-ended) and jacks (female-ended). To join dissimilar connectors together, an adapter can be used.

Kinds of Cables

Paired Cable

This cable is made from two individually insulated conductors. Often it is used in dc or low-frequency ac applications.

Twisted Pair

This cable is composed of two interwound insulated wires. It is similar to a paired cable, but the wires are held together by a twist.

The common CAT5 cable used for Ethernet, among other things, is based on a set of four twisted pairs.

Twin Lead

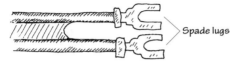

Spade lugs

This cable is a flat two-wire line, often referred to as 300-Ω line. The line maintains an impedance of 300 Ω. It is used primarily as a transmission line between an antenna and a receiver (e.g., TV, radio). Each wire within the cable is stranded to reduce skin effects.

Shielded Twin Lead

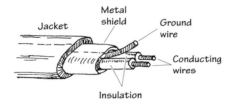

Jacket Metal shield Ground wire Conducting wires Insulation

This cable is similar to paired cable, but the inner wires are surrounded by a metal-foil wrapping that's connected to a ground wire. The metal foil is designed to shield the inner wires from external magnetic fields—potential forces that can create noisy signals within the inner wires.

Unbalanced Coaxial

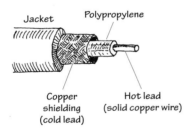

Jacket Polypropylene Copper shielding (cold lead) Hot lead (solid copper wire)

This cable typically is used to transport high-frequency signals (e.g., radio frequencies). The cable's geometry limits inductive and capacitive effects and also limits external magnetic interference. The center wire is made of solid-core copper or aluminium wire and acts as the hot lead. An insulative material, such as polyethylene, surrounds the center wire and acts to separate the center wire from a surrounding braided wire. The braided wire, or copper shielding, acts as the cold lead or ground lead. Characteristic impedances range from about 50 to 100 Ω.

Dual Coaxial

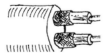

This cable consists of two unbalanced coaxial cables in one. It is used when two signals must be transferred independently.

Balanced Coaxial

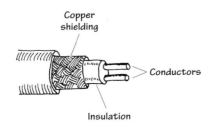

Copper shielding Conductors Insulation

This cable consists of two solid wires insulated from one another by a plastic insulator. Like unbalanced coaxial cable, it too has a copper shielding to reduce noise pickup. Unlike unbalanced coaxial cable, the shielding does not act as one of the conductive paths; it only acts as a shield against external magnetic interference.

FIGURE 3.4

Ribbon

This type of cable is used in applications where many wires are needed. It tends to flex easily. It is designed to handle low-level voltages and often is found in digital systems, such as computers, to transmit parallel bits of information from one digital device to another.

Multiple Conductor

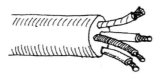

This type of cable consists of a number of individually wrapped, color-coded wires. It is used when a number of signals must be sent through one cable.

Fiberoptic

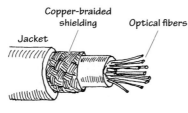

Fiberoptic cable is used in the transport of electromagnetic signals, such as light. The conducting-core medium is made from a glass material surrounded by a fiberoptic cladding (a glass material with a higher index of refraction than the core). An electromagnetic signal propagates down the cable by multiple total internal reflections. It is used in direct transmission of images and illumination and as waveguides for modulated signals used in telecommunications. One cable typically consists of a number of individual fibers.

FIGURE 3.4   (*Continued*)

117-Volt

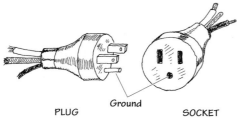

This is a typical home appliance connector. It comes in unpolarized and polarized forms. Both forms may come with or without a ground wire.

Banana

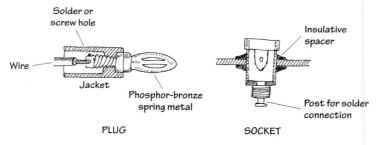

This is used for connecting single wires to electrical equipment. It is frequently used with testing equipment. The plug is made from a four-leafed spring tip that snaps into the jack.

Spade Lug/Barrier Strip

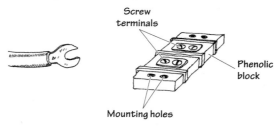

This is a simple connector that uses a screw to fasten a metal spade to a terminal. A barrier strip often acts as the receiver of the spade lugs.

FIGURE 3.5

Crimp

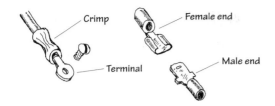

Crimp connectors are color-coded according to the wire size they can accommodate. They are useful as quick, friction-type connections in dc applications where connections are broken repeatedly. A crimping tool is used to fasten the wire to the connector.

Alligator

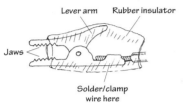

Alligator connectors are used primarily as temporary test leads.

Phone

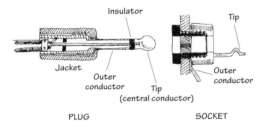

These connectors accept shielded braid, but they are larger in size. They come in two- or three-element types and have a barrel that is 1¼ in (31.8 mm) long. They are used as connectors in microphone cables and for other low-voltage, low-current applications.

3.5 mm and even 2.5 mm versions of these connectors are also commonly used.

Phono

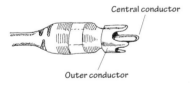

Phono connectors are often referred to as RCA plugs or pin plugs. They are used primarily in audio connections.

F-Type

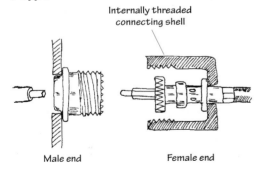

F-type connectors are used with a variety of unbalanced coaxial cables. They are commonly used to interconnect video components. F-type connectors are either threaded or friction-fit together.

Tip

These connectors are commonly used to supply low voltage dc between 3 and 15V.

FIGURE 3.5 (*Continued*)

Mini

PL-259

BNC

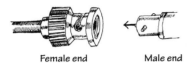

Female end   Male end

T-Connector

DIN Connector

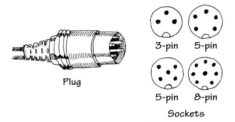

Plug

3-pin   5-pin

5-pin   8-pin

Sockets

Meat Hook

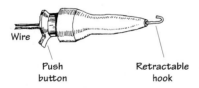

Wire

Push   Retractable
button   hook

D-Connector

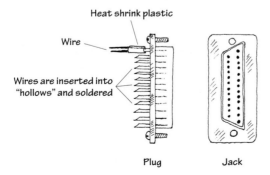

Heat shrink plastic

Wire

Wires are inserted into
"hollows" and soldered

Plug   Jack

IDC connectors are often found in computers. The plug is attached to ribbon cable using v-shaped teeth that are squeezed into the cable insulation to make a solderless contact.

These are often referred to as UHF plugs. They are used with RG-59/U coaxial cable. Such connectors may be threaded or friction-fit together.

BNC connectors are used with coaxial cables. Unlike the F-type plug, BNC connectors use a twist-on bayonet-like locking mechanism. This feature allows for quick connections

T-connectors consist of two plug ends and one central jack end. They are used when a connection must be made somewhere along a coaxial cable.

These connectors are used with multiple conductor wires. They are often used for interconnecting audio and computer accessories.

Smaller versions of these connectors (mini-DIN) are also widely used.

These connectors are used as test probes. The spring-loaded hook opens and closes with the push of a button. The hook can be clamped onto wires and component leads.

D-connectors are used with ribbon cable. Each connector may have as many as 50 contacts. The connection of each individual wire to each individual plug pin or jack socket is made by sliding the wire in a hollow metal collar at the backside of each connector. The wire is then soldered into place.

FIGURE 3.5   *(Continued)*

3.1.4 Wiring and Connector Symbols

Wiring

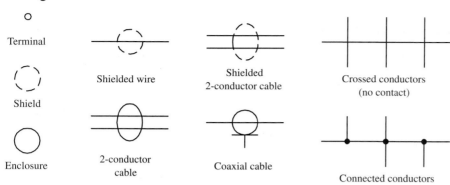

Terminal

Shield

Enclosure

Shielded wire

2-conductor cable

Shielded 2-conductor cable

Coaxial cable

Crossed conductors (no contact)

Connected conductors

Connectors

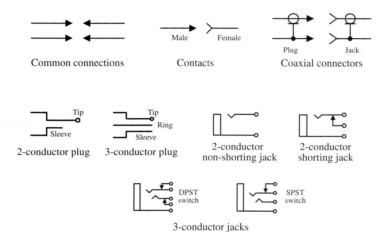

Common connections

Contacts

Male Female

Plug Jack

Coaxial connectors

2-conductor plug

3-conductor plug

2-conductor non-shorting jack

2-conductor shorting jack

DPST switch

SPST switch

3-conductor jacks

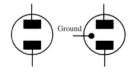

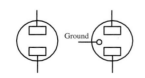

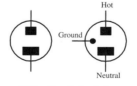

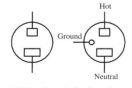

117-volt nonpolarized plug

117-volt nonpolarized socket

117-volt polarized plug

117-volt polarized socket

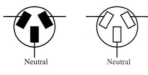

234-volt plug

234-volt socket

FIGURE 3.6

3.1.5 High-Frequency Effects Within Wires and Cables

Weird Behavior in Wires (Skin Effect)

When dealing with low-current dc hobby projects, wires and cables are straight-forward—they act as simple conductors with essentially zero resistance. However, when you replace dc currents with very high-frequency ac currents, weird things begin to take place within wires. As you will see, these "weird things" will not allow you to treat wires as perfect conductors.

First, let's take a look at what is going on in a wire when a dc current is flowing through it.

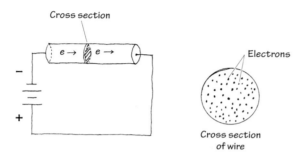

A wire that is connected to a dc source will cause electrons to flow through the wire in a manner similar to the way water flows through a pipe. This means that the path of any one electron essentially can be anywhere within the volume of the wire (e.g., center, middle radius, surface).

FIGURE 3.7

Now, let's take a look at what happens when a high-frequency ac current is sent through a wire.

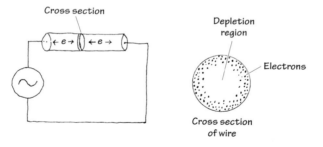

An ac voltage applied across a wire will cause electrons to vibrate back and forth. In the vibrating process, the electrons will generate magnetic fields. By applying some physical principles (finding the forces on every electron that result from summing up the individual magnetic forces produced by each electron), you find that electrons are pushed toward the surface of the wire. As the frequency of the applied signal increases, the electrons are pushed further away from the center and toward the surface. In the process, the center region of the wire becomes devoid of conducting electrons.

FIGURE 3.8

The movement of electrons toward the surface of a wire under high-frequency conditions is called the *skin effect*. At low frequencies, the skin effect does not have a large effect on the conductivity (or resistance) of the wire. However, as the frequency increases, the resistance of the wire may become an influential factor. Table 3.2 shows just how influential skin effect can be as the frequency of the signal increases (the table uses the ratio of ac resistance to dc resistance as a function of frequency).

One thing that can be done to reduce the resistance caused by skin effects is to use stranded wire—the combined surface area of all the individual wires within the conductor is greater than the surface area for a solid-core wire of the same diameter.

TABLE 3.2   AC/DC Resistance Ratio as a Function of Frequency

WIRE GAUGE	R_{AC}/R_{DC}			
	10^6 HZ	10^7 HZ	10^8 HZ	10^9 HZ
22	6.9	21.7	68.6	217
18	10.9	34.5	109	345
14	17.6	55.7	176	557
10	27.6	87.3	276	873

Weird Behavior in Cables (Lecture on Transmission Lines)

Like wires, cables also exhibit skin effects. In addition, cables exhibit inductive and capacitive effects that result from the existence of magnetic and electrical fields within the cable. A magnetic field produced by the current through one wire will induce a current in another. Likewise, if two wires within a cable have a net difference in charge between them, an electrical field will exist, thus giving rise to a capacitive effect.

Coaxial Cable

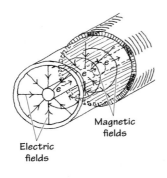

Paired Cable

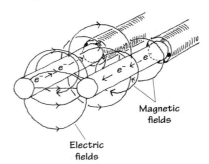

FIGURE 3.9   Illustration of the electrical and magnetic fields within a coaxial and paired cable.

Taking note of both inductive and capacitive effects, it is possible to treat a cable as if it were made from a number of small inductors and capacitors connected together. An equivalent inductor-capacitor network used to model a cable is shown in Fig. 3.10.

The impedance of a cable can be modeled by treating it as a network of inductors and capacitors.

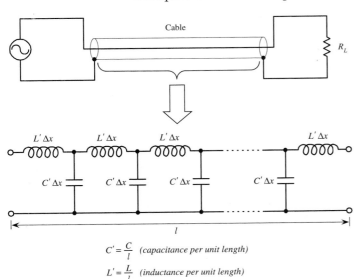

$C' = \dfrac{C}{l}$ *(capacitance per unit length)*

$L' = \dfrac{L}{l}$ *(inductance per unit length)*

FIGURE 3.10

To simplify this circuit, we apply a reduction trick; we treat the line as an infinite ladder and then assume that adding one "rung" to the ladder (one inductor-capacitor section to the system) will not change the overall impedance Z of the cable. What this means—mathematically speaking—is we can set up an equation such that $Z = Z +$ (LC section). This equation can then be solved for Z. After that, we find the limit as Δx goes to zero. The mathematical trick and the simplified circuit are shown below.

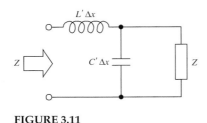

$$Z = j\omega L'\Delta x + \frac{Z/j\omega C'\Delta x}{Z + 1/j\omega C'\Delta x} = j\omega L'\Delta x + \frac{Z}{1 + j\omega C'Z\Delta x}$$

When $\Delta x \rightarrow$ small,

$$Z = \sqrt{L'/C'} = \sqrt{\frac{L/l}{C/l}} = \sqrt{L/C}$$

FIGURE 3.11

By convention, the impedance of a cable is called the *characteristic impedance* (symbolized Z_0). Notice that the characteristic impedance Z_0 is a real number. This means that the line behaves like a resistor despite the fact that we assumed the cable had only inductance and capacitance built in.

The question remains, however, what are L and C? Well, figuring out what L and C should be depends on the particular geometry of the wires within a cable and on the type of dielectrics used to insulate the wires. You could find L and C by applying some physics principles, but instead, let's cheat and look at the answers. The following are the expressions for L and C and Z_0 for both a coaxial and parallel-wire cable:

Coaxial

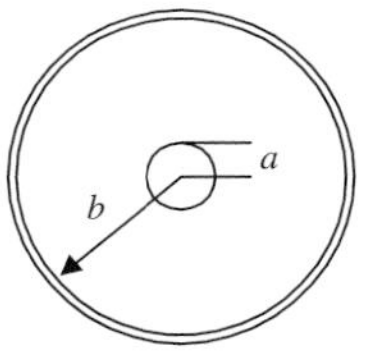

	L (H/m)	C (F/m)	$Z_0 = \sqrt{L/C}$ (Ω)
	$\dfrac{\mu_0 \ln(b/a)}{2\pi}$	$\dfrac{2\pi\varepsilon_0 k}{\ln(b/a)}$	$\dfrac{138}{\sqrt{k}} \log \dfrac{b}{a}$

Parallel Wire

	$\dfrac{\mu_0 \ln(D/a)}{\pi}$	$\dfrac{\pi\varepsilon_0 k}{\ln(D/a)}$	$\dfrac{276}{\sqrt{k}} \log \dfrac{D}{a}$

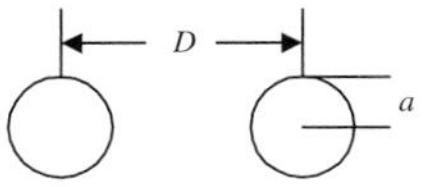

FIGURE 3.12   Inductance, capacitance, and characteristic impedance formulas for coaxial and parallel wires.

Here, k is the dielectric constant of the insulator, $\mu_0 = 1.256 \times 10^{-6}$ H/m is the permeability of free space, and $\varepsilon_0 = 8.85 \times 10^{-12}$ F/m is the permitivity of free space. Table 3.3 provides some common dielectric materials with their corresponding constants.

Often, cable manufacturers supply capacitance per foot and inductance per foot values for their cables. In this case, you can simply plug the given manufacturer's values into $Z_0 = \sqrt{L/C}$ to find the characteristic impedance of the cable. Table 3.4 shows capacitance per foot and inductance per foot values for some common cable types.

TABLE 3.3 **Common Dielectrics and Their Constants**

MATERIAL	DIELECTRIC CONSTANT (k)
Air	1.0
Pyrex glass	4.8
Mica	5.4
Paper	3.0
Polyethylene	2.3
Polystyrene	5.1–5.9
Quartz	3.8
Teflon	2.1

TABLE 3.4 **Capacitance Inductance and Impedance per Foot for Some Common Transmission-Line Types**

CABLE TYPE	CAPACITANCE/ FT (pF)	INDUCTANCE/ FT (μH)	CHARACTERISTIC IMPEDANCE (Ω)
RG-8A/U	29.5	0.083	53
RG-11A/U	20.5	0.115	75
RG-59A/U	21.0	0.112	73
214-023	20.0	0.107	73
214-076	3.9	0.351	300

Sample Problems (Finding the Characteristic Impedance of a Cable)

EXAMPLE 1

An RG-11AU cable has a capacitance of 21.0 pF/ft and an inductance of 0.112 μH/ft. What is the characteristic impedance of the cable?

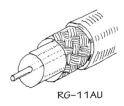

RG–11AU

FIGURE 3.13

You are given the capacitance and inductance per unit length: $C' = C$/ft, $L' = L$/ft. Using $Z_0 = \sqrt{L/C}$ and substituting C and L into it, you get

$$Z_0 = \sqrt{L/C}$$

$$Z_0 = \sqrt{\frac{0.112 \times 10^{-6}}{21.0 \times 10^{-12}}} = 73\ \Omega$$

EXAMPLE 2

What is the characteristic impedance of the RG-58/U coaxial cable with polyethylene dielectric ($k = 2.3$) shown below?

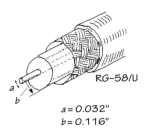

RG–58/U

$a = 0.032"$
$b = 0.116"$

FIGURE 3.14

$$Z_0 = \frac{138}{\sqrt{k}} \log \frac{b}{a}$$

$$Z_0 = \frac{138}{\sqrt{2.3}} \log \left(\frac{0.116}{0.032} \right) = 91 \times 0.056 = 51\Omega$$

EXAMPLE 3

Find the characteristic impedance of the parallel-wire cable insulated with polyethylene ($k = 2.3$) shown below.

$$Z_0 = \frac{276}{\sqrt{k}} \log \frac{D}{a}$$

$$Z_0 = \frac{276}{\sqrt{2.3}} \log \frac{0.270}{0.0127} = 242\Omega$$

FIGURE 3.15

Impedance Matching

Since a transmission line has impedance built in, the natural question to ask is, How does the impedance affect signals that are relayed through a transmission line from one device to another? The answer to this question ultimately depends on the impedances of the devices to which the transmission line is attached. If the impedance of the transmission line is not the same as the impedance of, say, a load connected to it, the signals propagating through the line will only be partially absorbed by the load. The rest of the signal will be reflected back in the direction it came from. Reflected signals are generally bad things in electronics. They represent an inefficient power transfer between two electrical devices. How do you get rid of the reflections? You apply a technique called *impedance matching*. The goal of impedance matching is to make the impedances of two devices—that are to be joined—equal. The impedance-matching techniques make use of special matching networks that are inserted between the devices.

Before looking at the specific methods used to match impedances, let's first take a look at an analogy that should shed some light on why unmatched impedances result in reflected signals and inefficient power transfers. In this analogy, pretend that the transmission line is a rope that has a density that is analogous to the transmission line's characteristic impedance Z_0. Pretend also that the load is a rope that has a density that is analogous to the load's impedance Z_L. The rest of the analogy is carried out below.

Unmatched Impedances ($Z_0 < Z_L$)

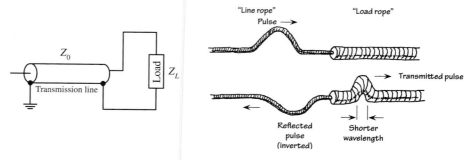

FIGURE 3.16a

A low-impedance transmission line that is connected to a high-impedance load is analogous to a low-density rope connected to a high-density rope. In the rope analogy, if you impart a pulse at the left end of the low-density rope (analogous to sending an electrical signal through a line to a load), the pulse will travel along without problems until it reaches the high-density rope (load). According to the laws of physics, when the wave reaches the high-density rope, it will do two things. First, it will induce a smaller-wavelength pulse within the high-density rope, and second, it will induce a similar-wavelength but inverted and diminished pulse that rebounds back toward the left end of the low-density rope. From the analogy, notice that only part of the signal energy from the low-density rope is transmitted to the high-density rope. From this analogy, you can infer that in the electrical case similar effects will occur—only now you are dealing with voltage and currents and transmission lines and loads.

Unmatched Impedances ($Z_0 > Z_L$)

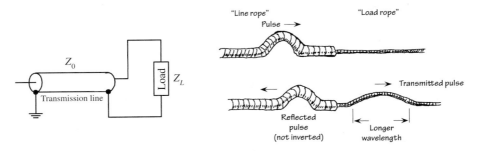

FIGURE 3.16b

A high-impedance transmission line that is connected to a low-impedance load is analogous to a high-density rope connected to a low-density rope. If you impart a pulse at the left end of the high-density rope (analogous to sending an electrical signal through a line to a load), the pulse will travel along the rope without problems until it reaches the low-density rope (load). At that time, the pulse will induce a longer-wavelength pulse within the low-density rope and will induce a similar-wavelength and diminished pulse that rebounds back toward the left end of the high-density rope. From this analogy, again you can see that only part of the signal energy from the high-density rope is transmitted to the low-density rope.

Matched Impedances ($Z_0 = Z_L$)

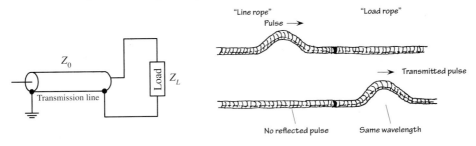

FIGURE 3.16c

Connecting a transmission line and load of equal impedances together is analogous to connecting two ropes of similar densities together. When you impart a pulse in the "transmission line" rope, the pulse will travel along without problems. However, unlike the first two analogies, when the pulse meets the load rope, it will continue on through the load rope. In the process, there will be no reflection, wavelength change, or amplitude change. From this analogy, you can infer that if the impedance of a transmission line matches the impedance of the load, power transfer will be smooth and efficient.

Standing Waves

Let's now consider what happens to an improperly matched line and load when the signal source is producing a continuous series of sine waves. You can, of course, expect reflections as before, but you also will notice that a superimposed standing-wave pattern is created within the line. The standing-wave pattern results from the interaction of forward-going and reflected signals. Figure 3.17 shows a typical resulting standing-wave pattern for an improperly matched transmission line attached between a sinusoidal transmitter and a load. The standing-wave pattern is graphed in terms of amplitude (expressed in terms of V_{rms}) versus position along the transmission line.

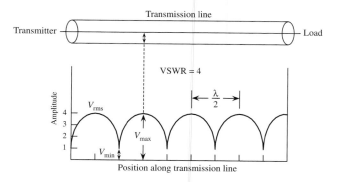

FIGURE 3.17   Standing waves on an improperly terminated transmission line. The VSWR is equal to V_{max}/V_{min}.

A term used to describe the standing-wave pattern is the *voltage standing-wave ratio* (VSWR). The VSWR is the ratio between the maximum and minimum rms voltages along a transmission line and is expressed as

$$VSWR = \frac{V_{rms,max}}{V_{rms,min}}$$

The standing-wave pattern shown in Fig. 3.17 has VSWR of 4/1, or 4.

Assuming that the standing waves are due entirely to a mismatch between load impedance and characteristic impedance of the line, the VSWR is simply given by either

$$VSWR = \frac{Z_0}{R_L} \qquad \text{or} \qquad VSWR = \frac{R_L}{Z_0}$$

whichever produces a result that is greater than 1.

A VSWR equal to 1 means that the line is properly terminated, and there will be no reflected waves. However, if the VSWR is large, this means that the line is not properly terminated (e.g., a line with little or no impedance attached to either a short or open circuit), and hence there will be major reflections.

The VSWR also can be expressed in terms of forward and reflected waves by the following formula:

$$VSWR = \frac{V_F + V_R}{V_F - V_R}$$

To make this expression meaningful, you can convert it into an expression in terms of forward and reflected power. In the conversion, you use $P = IV = V^2/R$. Taking P to be proportional to V^2, you can rewrite the VSWR in terms of forward and reflected power as follows:

$$VSWR = \frac{\sqrt{P_F} + \sqrt{P_R}}{\sqrt{P_F} - \sqrt{P_R}}$$

Rearranging this equation, you get the percentage of reflected power and percentage of absorbed power in terms of VSWR:

$$\% \text{ reflected power} = \left[\frac{VSWR - 1}{VSWR + 1}\right]^2 \times 100\%$$

$$\% \text{ absorbed power} = 100\% - \% \text{ reflected power}$$

EXAMPLE (VSWR)

Find the standing-wave ratio (VSWR) of a 50-Ω line used to feed a 200-Ω load. Also find the percentage of power that is reflected at the load and the percentage of power absorbed by the load.

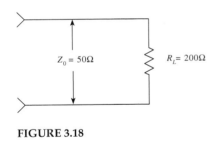

FIGURE 3.18

$$\mathbf{VSWR} = \frac{Z_0}{R_L} = \frac{200}{50} = 4$$

VSWR is 4:1

$$\% \text{ reflected power} = \left[\frac{VSWR - 1}{VSWR + 1}\right]^2 \times 100\%$$

$$= \frac{4 - 1}{4 + 1}^2 \times 100\% = 36\%$$

% absorbed power = 100% − 36% reflected power = 64%

Techniques for Matching Impedances

This section looks at a few impedance-matching techniques. As a rule of thumb, with most low-frequency applications where the signal's wavelength is much larger than the cable length, there is no need to match line impedances. Matching impedances is usually reserved for high-frequency applications. Moreover, most electrical equipment, such as oscilloscopes, video equipment, etc., has input and output impedances that match the characteristic impedances of coaxial cables (typically 50 Ω). Other devices, such as television antenna inputs, have characteristic input impedances that match the characteristic impedance of twin-lead cables (300 Ω). In such cases, the impedance matching is already taken care of.

IMPEDANCE-MATCHING NETWORK

IMPEDANCE-MATCHING NETWORK

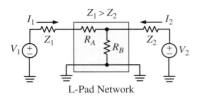

L-Pad Network

FIGURE 3.19

A general method used to match impedances when both source and load impedances are practically purely resistive (very small amounts of reactance) is to use a resistive L pad network, as shown to the left. Either side of the L pad can be the source or load, but the Z_1 side must be the side with higher impedance. The following equations are used to determine the L pad resistor values:

$$R_B = \frac{Z_2}{\sqrt{1 - Z_2 / Z_1}}, R_A = \frac{Z_1 Z_2}{R_B}$$

There is inherent insertion loss when an L pad is used:

$$Loss = 10 \log \frac{P_{pad} + P_{load}}{P_{load}} = -20 \log \frac{2\sqrt{Z_1 / Z_2}}{1 + (R_A + Z_1)(R_B + Z_2) / (R_B Z_2)}$$

where P_{load} is the power dissipated by load and P_{pad} is the power dissipated by the pad resistors. L pads are used in both audio and RF applications. For example, speaker L pads are designed to match the impedance of the speaker, and they are commonly available with 4, 8, 16 Ω impedances. A special configuration of variable resistors can we used to control volume while maintaining constant load impedance on the output of an audio amplifier. Other resistor networks, such as T pads and Pi pads, are also used for matching impedances: Pi networks are employed in amplifiers and T networks used in transmatches. To get a more in- depth overview of all these resistive networks, try an internet search, using L, T, and Pi pads as your keywords.

IMPEDANCE TRANSFORMER

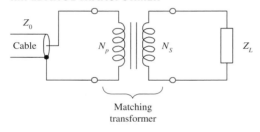

Matching transformer

FIGURE 3.20

Here, a transformer is used to match the characteristic impedance of a cable with the impedance of a load. By using the formula

$$N_p/N_s = \sqrt{Z_0/Z_L}$$

you can match impedances by choosing appropriate values for N_P and N_S so that the ratio N_P/N_S is equal to $\sqrt{Z_0/Z_L}$.

For example, if you wish to match an 800-Ω impedance line with an 8-Ω load, you first calculate

$$\sqrt{Z_0/Z_L} = \sqrt{800/8} = 10$$

To match impedances, you select N_P (number of coils in the primary) and N_S (number of coils in the secondary) in such a way that $N_P/N_S = 10$. One way of doing this would be to set N_P equal to 10 and N_S equal to 1. You also could choose N_P equal to 20 and N_S equal to 2 and you would get the same result.

BROADBAND TRANSMISSION-LINE TRANSFORMER

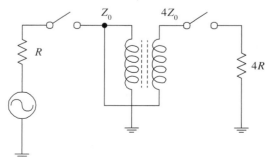

FIGURE 3.21

A broadband transmission-line transformer is a simple device that consists of a few turns of miniature coaxial cable or twisted-pair cable wound about a ferrite core. Unlike conventional transformers, this device can more readily handle high-frequency matching (its geometry eliminates capacitive and inductive resonance behavior). These devices can handle various impedance transformations and can do so with incredibly good broadband performance (less than 1 dB loss from 0.1 to 500 MHz).

QUARTER-WAVE SECTION

QUARTER-WAVE SECTION

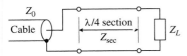

FIGURE 3.22

A transmission line with characteristic impedance Z_0 can be matched with a load with impedance Z_L by inserting a wire segment that has a length equal to one quarter of the transmitted signal's wavelength ($\lambda/4$) and which has an impedance equal to

$$Z_{sec} = \sqrt{Z_0 Z_L}$$

To calculate the segment's length, you must use the formula $\lambda = v/f$, where v is the velocity of propagation of a signal along the cable and f is the frequency of the signal. To find v, use

$$v = c/\sqrt{k}$$

where $c = 3.0 \times 10^8$ m/s, and k is the dielectric constant of the cable's insulation.

For example, say you wish to match a 50-Ω cable that has a dielectric constant of 1 with a 200-Ω load. If you assume the signal's frequency is 100 MHz, the wavelength then becomes

$$\lambda = \frac{v}{f} = \frac{c/\sqrt{k}}{f} = \frac{3 \times 10^8 / 1}{100 \times 10^6} = 3 \text{ m}$$

To find the segment length, you plug λ into $\lambda/4$. Hence the segment should be 0.75 m long. The wire segment also must have an impedance equal to

$$Z_{sec} = \sqrt{(50)(200)} = 100 \ \Omega$$

STUBS

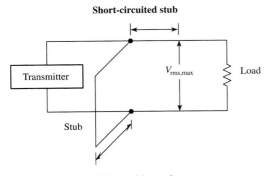

Pair of open-ended matching stubs

A short length of transmission line that is open ended or short-circuit terminated possesses the property of having an impedance that is reactive. By properly choosing a segment of open-circuit or short-circuit line and placing it in shunt with the original transmission line at an appropriate position along the line, standing waves can be eliminated. The short segment of wire is referred to as a *stub*. Stubs are made from the same type of cable found in the transmission line. Figuring out the length of a stub and where it should be placed is fairly tricky. In practice, graphs and a few formulas are required. A detailed handbook on electronics is the best place to learn more about using stubs.

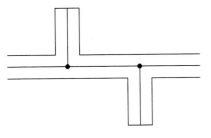

FIGURE 3.23

3.2 Batteries

A battery is made up of a number of *cells*. Each cell contains a positive terminal, or *cathode*, and a negative terminal, or *anode*. (Note that most other devices treat *anodes* as positive terminals and *cathodes* as negative terminals.)

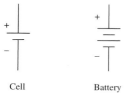

Cell   Battery

FIGURE 3.24

When a load is placed across a cell's terminals, a conductive bridge is formed that initiates chemical reactions within the cell. These reactions produce electrons in the anode material and remove electrons from the cathode material. As a result, a potential is created across the terminals of the cell, and electrons from the anode flow through the load (doing work in the process) and into the cathode.

A typical cell maintains about 1.5 V across its terminals and is capable of delivering a specific amount of current that depends on the size and chemical makeup of the cell. If more voltage or power is needed, a number of cells can be added together in either series or parallel configurations. By adding cells in series, a larger-voltage battery can be made, whereas adding cells in parallel results in a battery with a higher current-output capacity. Figure 3.25 shows a few cell arrangements.

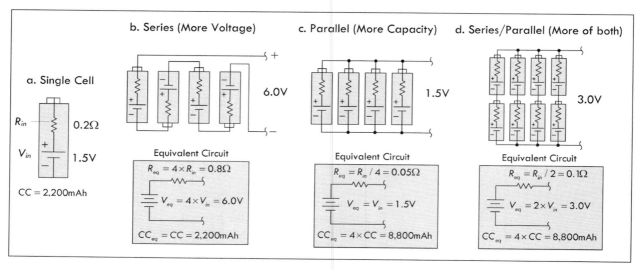

FIGURE 3.25

Battery cells are made from a number of different chemical ingredients. The use of a particular set of ingredients has practical consequences on the battery's overall performance. For example, some cells are designed to provide high open-circuit voltages, whereas others are designed to provide large current capacities. Certain kinds of cells are designed for light-current, intermittent applications, whereas others are designed for heavy-current, continuous-use applications. Some cells are designed for pulsing applications, where a large burst of current is needed for a short period of time. Some cells have good shelf lives; others have poor shelf lives. Batteries that are designed for one-time use, such as carbon-zinc and alkaline batteries, are called *primary batteries*. Batteries that can be recharged a number of times, such as nickel metal hydride and lead-acid batteries, are referred to as *secondary batteries*.

3.2.1 How a Cell Works

A cell converts chemical energy into electrical energy by going through what are called *oxidation-reduction reactions* (reactions that involve the exchange of electrons).

The three fundamental ingredients of a cell used to initiate these reactions include two chemically dissimilar metals (positive and negative electrodes) and an electrolyte (typically a liquid or pastelike material that contains freely floating ions). The following is a little lecture on how a simple lead-acid battery works.

For a lead-acid cell, one of the electrodes is made from pure lead (Pb); the other electrode is made from lead oxide (PbO_2); and the electrolyte is made from a sulfuric acid solution ($H_2O + H_2SO_4 \rightarrow 3H^+ + SO_4^{2-} + OH^-$).

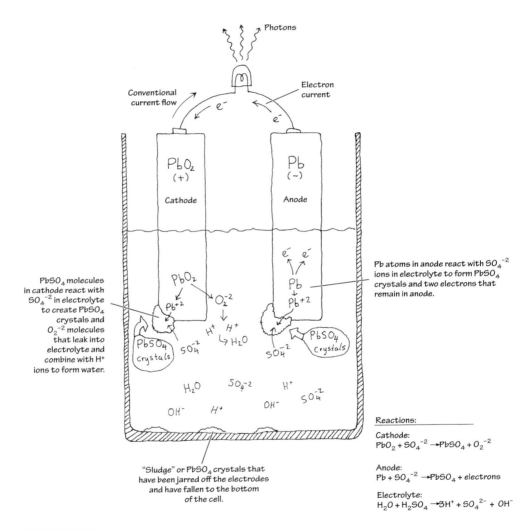

FIGURE 3.26

When the two chemically dissimilar electrodes are placed in the sulfuric acid solution, the electrodes react with the acid (SO_4^{-2}, H^+ ions), causing the pure lead electrode to slowly transform into $PbSO_4$ crystals. During this transformation reaction, two electrons are liberated within the lead electrode. Now, if you examine the lead oxide electrode, you also will see that it too is converted into $PbSO_4$ crystals. However, instead of releasing electrons during its transformation, it releases O_2^{-2} ions. These ions leak out into the electrolyte solution and combine with the hydrogen ions to form H_2O (water). By placing a load element, say, a lightbulb, across the electrodes, electrons will flow from the electron-abundant lead electrode, through the bulb's filament, and into the electron-deficient lead oxide electrode.

As time passes, the ingredients for the chemical reactions run out (the battery is drained). To get energy back into the cell, a reverse voltage can be applied across the cell's terminals, thus forcing the reactions backward. In theory, a lead-acid battery can be drained and recharged indefinitely. However, over time, chunks of crystals will break off from the electrodes and fall to the bottom of the container, where they are not recoverable. Other problems arise from loss of electrolyte due to gasing during electrolysis (a result of overcharging) and due to evaporation.

3.2.2 Primary Batteries

Primary batteries are one-shot deals—once they are drained, it is all over. Common primary batteries include carbon-zinc batteries, alkaline batteries, mercury batteries, silver oxide batteries, zinc air batteries, and silver-zinc batteries. Here are some common battery packages:

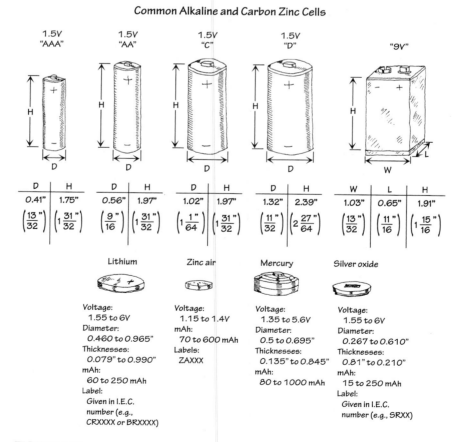

FIGURE 3.27

3.2.3 Comparing Primary Batteries

Carbon-Zinc Batteries

Carbon-zinc batteries ("standard-duty") are general-purpose primary-type batteries that were popular back in the 1970s, but have become obsolete with the advent of alkaline batteries. These batteries are not suitable for continuous use (only for intermittent use) and are susceptible to leakage. The nominal voltage of a carbon-zinc cell is about 1.5 V, but this value gradually drops during service. Shelf life also tends to be poor, especially at elevated temperatures. The only really positive aspect of these batteries is their low cost and wide size range. They are best suited to low-power applications with intermittent use, such as in radios, toys, and general-purpose low-cost devices. Don't use these cells in expensive equipment or leave them in equipment for long periods of time—there is a good chance they will leak. Standby applications and applications that require a wide temperature range should also be avoided. In all, these batteries are to be avoided, if you can even find them.

Zinc-Chloride Batteries

Zinc-chloride batteries ("heavy-duty") are a heavy-duty version of the carbon-zinc battery, designed to deliver more current and provide about 50 percent more capacity. Like the carbon-zinc battery, zinc-chloride batteries are essentially obsolete as compared to alkaline batteries. The terminal voltage of a zinc-chloride cell is initially about 1.5 V, but drops as chemicals are consumed. Unlike carbon-zinc batteries, zinc-chloride batteries perform better at low temperatures, and slightly better at higher temperatures, too. The shelf life is also longer. They tend to have lower internal resistance and higher capacities than carbon-zinc, allowing higher currents to be drawn for longer periods. These batteries are suited to moderate, intermittent use. However, an alkaline battery will provide better performance in similar applications.

Alkaline Batteries

Alkaline batteries are the most common type of household battery you can buy—they have practically replaced the carbon-zinc and zinc-chloride batteries. They are relatively powerful and inexpensive. The nominal voltage of an alkaline cell is again 1.5 V, but doesn't drop as much during discharge as the previous two battery types. The internal resistance is also considerably lower, and remains so until near the end of the battery's life cycle. They have very long shelf lives and better high- and low-temperature performance, too. General-purpose alkaline batteries don't work particularly well on high-drain devices like digital cameras, since the internal resistance limits output current flow. They will still work in your device, but the battery life will be greatly reduced. They are well suited to most general-purpose applications, such as toys, flashlights, portable audio equipment, digital cameras, and so on. Note that there is a rechargeable version of an alkaline battery, as well.

Lithium Batteries

Lithium batteries use a lithium anode, one of a number of different kinds of cathodes, and an organic electrolyte. They have a nominal voltage of 3.0 V—twice that of most other primary cells—that remains almost flat during the discharge cycle. They also have a very low self-discharge rate, giving them an excellent shelf life—as much as 10 years. The internal resistance is also quite low, and remains so during discharge. It performs well in both low and high temperatures, and advanced versions of this battery are used on satellites, on space vehicles, and in military applications. They are ideal for low-drain applications such as smoke detectors, data-retention devices, pacemakers, watches, and calculators.

Lithium-Iron Disulfide Batteries

Unlike other lithium cells that have chemistries geared to obtaining the greatest capacity in a given package, lithium-iron disulfide cells are a compromise. To match existing equipment and circuits, their chemistry has been tailored to the standard nominal 1.5-V output (whereas other lithium technologies produce double that). These cells are consequently sometimes termed *voltage-compatible lithium* batteries. Unlike other lithium technologies, lithium-iron disulfide cells are not rechargeable. Internally, the lithium-iron disulfide cell is a sandwich of a lithium anode, a separator, and an iron disulfide cathode with an aluminum cathode collector. The cells are sealed but vented. Compared to the alkaline cells—with which they are meant to compete—lithium-iron disulfide cells are lighter (weighing about 66 percent of same-size alkaline cells) and higher in capacity, and they have a much longer shelf life—even after 10 years on the shelf, lithium disulfide cells still retain most of their capacity. Lithium iron-disulfide cells operate best under heavier loads. In high-current applications, they can supply power for a duration exceeding 260 percent of the time that a similar-sized alkaline cell can supply. This advantage diminishes at lower loads, however, and at very light loads may disappear or even reverse. For example, under a 20-mA load, a certain manufacturer rates its AA-size lithium-disulfide cells to provide power for about 122 hours while its alkaline cells will last for 135 hours. However, under a heavy load of 1 A, the lithium disulfide cell overshadows the alkaline counterpart by lasting 2.1 hours versus only 0.8 hours for the alkaline battery.

Mercury Cells

Zinc-mercuric oxide, or "mercury," cells take advantage of the high electrode potential of mercury to offer a very high energy density combined with a very flat discharge curve. Mercuric oxide forms the positive electrode, sometimes mixed with manganese dioxide. The nominal terminal voltage of a mercury cell is 1.35 V, and this remains almost constant over the life of the cell. They have an internal resistance that is fairly constant. Although made only in small button sizes, mercury cells are capable of reasonably high-pulsed discharge current and are thus suitable for applications such as quartz analog watches and hearing aids as well as voltage references in instruments, and the like.

Silver Oxide Batteries

The silver oxide battery is the predominate miniature battery found on the market today. Silver oxide cells are made only in small button sizes of modest capacity but have good pulsed discharge capability. They are typically used in watches, calculators, hearing aids, and electronic instruments. This battery's general characteristics include higher voltage than comparable mercury batteries, flatter discharge curve than alkaline batteries, good low-temperature performance, good resistance to shock and vibration, essentially constant internal resistance, excellent service maintenance, and long shelf life—exceeding 90 percent charge after storage for five years. The nominal terminal voltage of a silver oxide cell is slightly over 1.5 V and remains almost flat over the life of the cell. Batteries built from cells range from 1.5 V to 6.0 V and come in a variety of sizes. Silver oxide hearing aid batteries are designed to produce greater volumetric energy density at higher discharge rates than silver oxide watch or photographic batteries. Silver oxide photo batteries are designed to provide constant voltage or periodic high-drain pulses with or without a low drain background current. Silver oxide watch batteries, using a sodium hydroxide (NaOH) electrolyte system are designed primarily for low-drain continuous use over long periods of time—typically five years. Silver oxide watch batteries using potassium hydroxide (KOH) electrolyte systems are designed primarily for continuous low drains with periodic high-drain pulse demands, over a period of about two years.

Zinc Air Batteries

Zinc air cells offer very high energy density and a flat discharge curve, but have relatively short working lives. The negative electrode is formed of powdered zinc, mixed with the potassium hydroxide electrolyte to form a paste. This is retained inside a small metal can by a separator membrane that is porous to ions, and on the other side of the membrane is simply air to provide the oxygen (which acts as the positive electrode). The air/oxygen is inside an outer can of nickel-plated steel that also forms the cell's positive connection, lined with another membrane to distribute the oxygen over the largest area. Actually there is no oxygen or air in the zinc-oxygen cell when it's made. Instead, the outer can has a small entry hole with a covering seal, which is removed to admit air and activate the cell. The zinc is consumed as the cell supplies energy, which is typically for around 60 days. The nominal terminal voltage of a zinc-oxygen cell is 1.45 V, and the discharge curve is relatively flat. The internal resistance is only moderately low, and they are not suitable for heavy or pulsed discharging. They are found mainly in button and pill packages, and are commonly used in hearing aids and pagers. Miniature zinc air batteries are designed primarily to provide power to hearing aids. In most hearing aid applications, zinc air batteries can be directly substituted for silver oxide or mercuric oxide batteries and will typically give the longest hearing aid service of any common battery system. Notable characteristics include high capacity-to-volume ratio for a miniature battery, more stable voltage at high currents when compared to mercury or silver oxide batteries, and essentially constant internal resistance. They are activated by removing the covering (adhesive tab) from the air access hole, and they are most effective in applications that consume battery capacity in a few weeks.

TABLE 3.5 Primary Battery Comparison Chart

TYPE (CHEMISTRY)	COMMON NAME(S)	NOMINAL CELL VOLTS	INTERNAL RESISTANCE	MAXIMUM DISCHARGE RATE	COST	PROS AND CONS	TYPICAL APPLICATIONS
Carbon-zinc	Standard-duty	1.5	Medium	Medium	Low	Low cost, various sizes, but terminal voltage drops steadily during cell life	Radios, toys, and general-purpose electrical equipment
Zinc-chloride	Heavy-duty	1.5	Low	Medium to high	Low to medium	Low cost at higher discharge rates and at lower temperature; terminal voltage still drops	Motor-driven portable devices, clocks, remote controls
Alkaline zinc-manganese dioxide	Alkaline	1.5	Very low	High	Medium to high	Better for high continuous or pulsed loads and at low temperatures, but terminal voltage drops	Photoflash units, battery shavers, digital cameras, handheld transceivers, portable CD players, etc.
Lithium-manganese dioxide	Lithium	3.0	Low	Medium to high	High	High energy density, very low self-discharge rate (excellent shelf-life), good temperature tolerance	Watches, calculators, cameras (digital and film), DMMs, and other test instruments
Zinc-mercuric oxide	Mercury cell	1.35	Low	Low	High	High energy density (compact), very flat discharge curve, good at higher temperatures	Calculators, pagers hearing aids, watches, test intruments
Zinc-silver oxide	Silver oxide cell	1.5	Low	Low	High	Very high energy density (very compact), very flat discharge curve, reasonable at lower temperatures	Calculators, pagers, hearing aids, watches test instruments
Zinc-oxygen	Zinc air cell	1.45	Medium	Low	Medium	High energy density, very lightweight, flat discharge curve, but must have access to air	Hearing aids and pagers

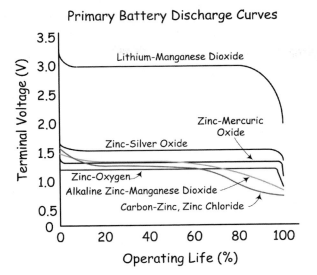

FIGURE 3.28

3.2.4 Secondary Batteries

Secondary batteries, unlike primary batteries, are rechargeable by nature. The actual discharge characteristics for secondary batteries are similar to those of primary batteries, but in terms of design, secondary batteries are made for long-term, high-power-level discharges, whereas primary batteries are designed for short discharges at low power levels. Most secondary batteries come in packages similar to those of primary batteries, with the exception of, say, lead-acid batteries and special-purpose batteries. Secondary batteries are used to power such devices as laptop computers, portable power tools, electric vehicles, emergency lighting systems, and engine starting systems.

Here are some common packages for secondary batteries:

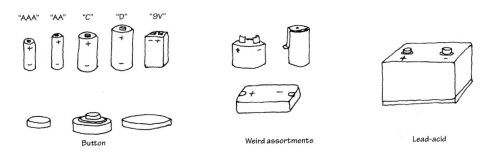

FIGURE 3.29

Comparing Secondary (Rechargeable) Batteries

LEAD-ACID BATTERIES

Lead-acid batteries are typically used for high-power applications, such as motorized vehicle power and battery backup applications. There are basically three types of lead-acid batteries: flooded lead-acid, valve-regulated lead-acid (VRLA), and sealed lead-acid (SLA). The flooded types must be stood upright and tend to lose electrolytes while producing gas over time. The SLA and VRLA are designed for a low overvoltage potential to prohibit the battery from reaching its gas-generating potential during discharge. However, SLA and VRLA can never be charged to their full potential. VRLA is generally used for stationary applications, while the SLA can be used in various positions. Lead-acid batteries typically come in 2-V, 4-V, 6-V, 8-V, and 12-V versions, with capacities ranging from 1 to several thousand amp-hours. The flooded lead-acid battery is used in automobiles, forklifts, wheelchairs, and UPS devices.

An SLA battery uses a gel-type electrolyte rather than a liquid to allow it to be used in any position. However, to prevent gas generation, it must be operated at a lower potential—meaning it's never fully charged. This means that it has a relatively poor energy density—the lowest for all sealed secondary batteries. However, they're the cheapest secondary, making them best suited for applications where low-cost, stationary power storage is the main concern. SLA batteries have the lowest self-discharge rate of any of the secondary batteries (about 5 percent per month). They do not suffer from memory effect (as displayed in NiCad batteries), and they perform well with shallow cycling; in fact, they tend to prefer it to deep cycling, although they perform well with intermittent heavy current demands, too. SLA batteries aren't designed for fast charging—typically 8 to 16 hours for full recharge. They must also always be stored in a charged state. Leaving them in a discharged state can lead to sulfation, a condition that makes the batteries difficult, if not impossible, to recharge. Also, SLA batteries have an environmentally unfriendly electrolyte.

The basic technique for recharging lead-acid batteries, be they flooded, sealed, or valve-regulated, is to read the technical directions that come with them. If you don't know what you're doing—say, trying to make your own battery recharger—you may run into a serious problem, such as blowing up batteries with too much pressure, melting them, or destroying the chemistry. (The procedure for charging lead-acid batteries is different from that for NiCad and NiMH batteries in that voltage limiting is used instead of current limiting.)

NICKEL-CADMIUM (NiCAD) BATTERIES

Nickel-cadmium batteries are made using nickel hydroxide as the positive electrode and cadmium hydroxide as the negative electrode, with potassium hydroxide as the electrolyte. Nickel-cadmium batteries have been a very popular rechargeable battery over the years; however, with the introduction of NiMH batteries, they have seen a decline in use. In practical terms, NiCad batteries don't last very long before needing a recharge. They put out less voltage (per cell) than a standard alkaline (1.2 V versus 1.5 V for alkaline). This means that applications that require four or more alkaline batteries might not work at all with comparable-sized NiCad batteries. During discharge, the average voltage of a sealed NiCad cell is about 1.2 V per cell. At nominal discharge rates, the characteristic is very nearly flat until the cell approaches full discharge. The battery provides most of its energy above 1.0 V per cell. The self-discharge rate of a NiCad is not great, either—around two to three months. However, like SLAs, sealed NiCads can be used in virtually any position. NiCads have a higher energy density than SLAs (about twice as much), and with a relatively low cost, they are popular for powering compact portable equipment: cordless power tools, model boats and cars, and appliances such as flashlights and vacuum cleaners. NiCads suffer from memory effect and are therefore not really suitable for applications that involve shallow cycling or spending most of their time on a float charger. They perform best in situations where they're deeply cycled. They have a high number of charge/discharge cycles—around 1000.

Use a recommended charger—a constant current–type charger with due regard for heat dissipation and wattage rating. Improper charging can cause heat damage or even high-pressure rupture. Observe proper charging polarity. The safe charge rate for sealed NiCad cells for extended charge periods is 10 hours, or C/10 rate.

NICKEL METAL HYDRIDE (NiMH) BATTERIES

NiMH batteries are very popular secondary batteries, replacing NiCad batteries in many applications. NiMH batteries use a nickel/nickel hydroxide positive electrode, a hydrogen-storage alloy (such as lanthanum-nickel or zirconium-nickel) as a negative electrode, and potassium hydroxide as the electrolyte. They have a higher energy density than standard NiCad batteries (about 30 to 40 percent higher) and don't require special disposal requirements, either. The nominal voltage of a NiMH battery is 1.2 V per cell, which must be taken into consideration when substituting them into devices that use standard 1.5-V cells such as alkaline cells. They self-discharge in about two to three months and do display some memory effect, but not as bad as NiCad batteries. They are not as happy with a deep discharge cycle as a NiCad battery, and they tend to have a shorter work life. Best results are achieved with load currents of 0.2-C to 0.5-C (one-fifth to one-half of the rated capacity). Typical applications include remote-control vehicles and power tools (although NiMH batteries are rapidly being superceded by Li-ion and LiPo batteries).

Recharging NiMH batteries is a bit complex due to significant heat generation; the charge uses a special algorithm that requires trickle charging and temperature sensing. Unlike NiCad batteries, NiMH batteries have little memory effect. The batteries require regular full discharge to prevent crystalline formation.

Li-ION BATTERIES

Lithium is the lightest of all metals and has the highest electrochemical potential, giving it the possibility of an extremely high energy density. However, the metal itself is highly reactive. While this isn't a problem with primary cells, it poses an explosion risk with rechargeable batteries. For these to be made safe, lithium-ion technology had to be developed; the technology uses lithium ions from chemicals such as lithium-cobalt dioxide, instead of the metal itself. Typical Li-ion batteries have a negative electrode of aluminum coated with a lithium compound such as lithium-cobalt dioxide, lithium-nickel dioxide, or lithium-manganese dioxide. The positive electrode is generally of copper, coated with carbon (generally either graphite or coke), while the electrolyte is a lithium salt such as lithium-phosphorous hexafluoride, dissolved in an organic solvent such as a mixture of ethylene carbonate and dimethyl carbonate. Li-ion batteries have roughly twice the energy density of NiCads, making them the most compact rechargeable yet in terms of energy storage. Unlike NiCad or NiMH batteries, they are not subject to memory effect, and have a relatively low self-discharge rate—about 6 percent per month, less than half that of NiCads. They are also capable of moderately deep discharging, although not as deep as NiCads, as they have a higher internal resistance. On the other hand, Li-ion batteries cannot be charged as rapidly as NiCads, and they cannot be trickle or float charged, either. They also are significantly more costly than either NiCads or NiMH batteries, making them the most expensive rechargeables of all. Part of this is that they must be provided with built-in protection against both excessive discharging and overcharging (both of which pose a safety risk). Most Li-ion batteries are therefore supplied in self-contained battery packs, complete with "smart" protective circuitry. They are subject to aging, even if not used, and have moderate discharge currents. The main applications for Li-ion batteries are in places where as much energy as possible needs to be stored in the smallest possible space, and with as little weight as possible. They are found in laptop computers, PDAs, camcorders, and cell phones.

Li-ion batteries require special voltage-limiting recharging devices. Commercial Li-ion battery packs contain a protection circuit that prevents the cell voltage from going too high while charging. The typical safety threshold is set to 4.30 V/cell. In addition, temperature sensing disconnects the charging device if the internal temperature approaches 90°C (194°F). Most cells feature a mechanical pressure switch that permanently interrupts the current path if a safe pressure threshold is exceeded. The charge time of all Li-ion batteries, when charged at a 1-C initial current, is about three hours. The battery remains cool during charge. Full charge is attained after the voltage has reached the upper voltage threshold and the current has dropped and leveled off at about 3 percent of the nominal charge current. Increasing the charge current on a Li-ion charger does not shorten the charge time by much. Although the voltage peak is reached more quickly with higher current, the topping charge will take longer.

LITHIUM POLYMER (Li-POLYMER) BATTERIES

The lithium polymer batteries are a lower-cost version of the Li-ion batteries. Their chemistry is similar to that of the Li-ion in terms of energy density, but uses a dry solid polymer electrolyte only. This electrolyte resembles a plastic-like film that does not conduct electricity but allows an exchange of ions (electrically charged atoms or groups of atoms). The dry polymer is more cost effective during fabrication, and the overall design is rugged, safe, and thin. With a cell thickness measuring as little as 1 mm, it is possible to use this battery in thin compact devices where space is an issue. It is possible to create designs which form part of a protective housing, are in the shape of a mat that can be rolled up, or are even embedded into a carrying case or piece of clothing. Such innovative batteries are still a few years away, especially for the commercial market.

Unfortunately, the dry Li-polymer suffers from poor ion conductivity, due to high internal resistance; it cannot deliver the current bursts needed for modern communication devices. However, it tends to increase in conductivity as the temperature rises, a characteristic suitable for hot climates. To make a small Li-polymer battery more conductive, some gelled electrolyte may be added. Most of the commercial Li-polymer batteries used today for mobile phones are hybrids and contain gelled electrolytes.

The charge process of a Li-polymer battery is similar to that of the Li-ion battery. The typical charge time is around one to three hours. Li-polymer batteries with gelled electrolyte, on the other hand, are almost identical to Li-ion batteries. In fact, the same charge algorithm can be applied.

NICKEL-ZINC (NiZn) BATTERIES

Nickel-zinc batteries are commonly used in light electric vehicles. They are considered the next generation of batteries used for high-drain applications, and are expected to replace sealed lead-acid batteries due to their higher energy densities (up to 70 percent lighter for the same power). They are also relatively cheap compared to NiCad batteries.

NiZn batteries are chemically very similar to NiCad batteries; both use an alkaline electrolyte and a nickel electrode, but they differ significantly in their voltage. The NiZn cell delivers more than 0.4 V of additional voltage both at open circuit and under load. With the additional 0.4 V per cell, multicell batteries can be constructed in smaller packages. For example, a 19.2-V pack can replace a 14.4-V NiCad pack, representing a 25 percent lower cell space and delivering higher power and a 45 percent lower impedance. They are also less expensive than most rechargeables. They are safe (abuse-tolerant). The life cycle is a bit better than for NiCad batteries for typical applications. They have superior shelf life when compared to lead-acid. Also, they are considered environmentally green—both nickel and zinc are nontoxic and easily recycled.

In terms of recharge times, it takes less than two hours to achieve full recharge; there is an 80 percent charge in one hour. This feature makes them useful in cordless power tools. Their high energy density and high discharge rate make them suitable for applications that demand large amounts of power in small, lightweight packages. They are found in cordless power tools, UPS systems, electric scooters, high-intensity dc lighting and the like.

NICKEL-IRON (NiFe) BATTERIES

Nickel-iron batteries, also called nickel alkaline or NiFe batteries, were introduced in 1900 by Thomas Edison. These are very robust batteries that are tolerant of abuse and can have very long life spans (30 years or more). The open-circuit voltage of these cells is 1.4 V, and the discharge voltage is about 1.2 V. They withstand overcharge and over-discharge. They accept high depth of discharge (deep cycling) and can remain discharged for long periods without damage, unlike lead-acid batteries that need to be stored in a charged state. They are, however, very heavy and bulky. Also, the low reactivity of the active components limits high-discharge performance. The cells take a charge slowly, give it up slowly, and have a steep voltage dropoff with state of charge. Furthermore, they have a low energy density compared to other secondary batteries, and a high self-discharge rate. NiFe batteries are used in applications similar to those for lead-acid batteries, but oriented toward a necessity of longevity. (A typical lead-acid battery will last around five years, compared to around 30 to 80 years for a NiFe battery.)

TABLE 3.6 Rechargeable Battery Comparison Chart

TYPE (CHEMISTRY)	NOMINAL CELL VOLTS (APPROX.)	ENERGY DENSITY (Wh/Kg)	CYCLE LIFE	CHARGE TIME	MAX. DISCHARGE RATE	COST	PROS AND CONS	TYPICAL APPLICATIONS
Sealed lead-acid	2.0	Low (30)	Long (shallow cycles)	8–16 h	Medium (0.2 C)	Low	Low cost, low self-discharge, happy float charging, but prefers shallow charging	Emergency lighting, alarm systems solar power systems, wheel-chairs, etc.
Recharge-able alkaline-manganese	1.5	High (75 initial)	Short to medium	2–6 h (pulsed)	Medium (0.3 C)	Low	Low cost, low self-discharge, prefer shallow cycling, no memory effect but short cycle life	Portable emergency lighting, toys, portable radios, CD players, test instruments, etc.
NiCad	1.2	Medium (40–60)	Long (deep cycles)	14–16 h (0.1 C) or <2 h with care (1 C)	High (>2 C)	Medium	Prefer deep cycling, good pulse capac-ity, but have memory effect, fairly high self-discharge rate, envi-ronmentally unfriendly	Portable tools and appliances, model cars and boats, data loggers, camcorders, portable transceivers, and test equipment
NiMH	1.2	High (60–80)	Medium	2–4 h	Medium (0.2–0.5 C)	Medium	Very compact en-ergy source, but have some memory effect, high self-discharge rate	Remote control vehicles, cordless mobile phones, personal DVD and CD players, power tools
NiZn	1.65	High (>170)	Medium to high	1–2 h	—	Medium	Low cost, environ-mentally green, twice energy density of NiCad	Exceptional perfor-mance, no memory, long shelf-life
NiFe	1.4	High (>200)	Extremely long	Long	—	Low	High cycle life, incred-ibly long life up to 80 years, environmentally friendly	Forklifts and other, simi-lar SLA-like applications, but where longevity is important
Li-ion/LiPo	3.6	Very high (>100)	Medium	3–4 h (1 C–0.03 C)	Med/high (<1 C)	High	Very compact, low maintenance, low self-discharge, but needs great care with charging	Compact cell phones and notebook PCs, digital cameras, and similar very small portable device

RECHARGEABLE ALKALINE-MANGANESE (RAM) BATTERIES

Rechargeable alkaline-manganese, or RAM, batteries are the rechargeable version of primary alkaline batteries. Like the primary technology, they use a manganese dioxide positive electrolyte and potassium hydroxide electrode, but the negative electrode is now a special porous zinc gel designed to absorb hydrogen during the charging process. The separator is also laminated to prevent it being pierced by zinc dendrites. These are often considered a poor substitute for a rechargeable, as compared to a NiCad or NiMH battery. RAM batteries have a tendency to plummet in capacity over few recharge cycles. It is feasible for a RAM battery to lose 50 percent of its capacity after only eight cycles. On the positive side, they are inexpensive and readily available. They can usually be used as a direct replacement for non-rechargeable batteries, but they usually have a lower nominal voltage, making them unsuitable for some devices, but not in high-drain devices like digital cameras. They have a low self-discharge rate and can be stored on standby for up to 10 years. Also, they are environmentally friendly (no toxic metals are used) and maintenance-free; there is no need for cycling or worrying about memory effect. On the short side, they have limited current-handling capability and are limited to light-duty applications such as flashlights and other low-cost portable electronic devices that require shallow cycling. Recharging a RAM battery requires a special recharger; if you charge them in a standard charger, they may explode.

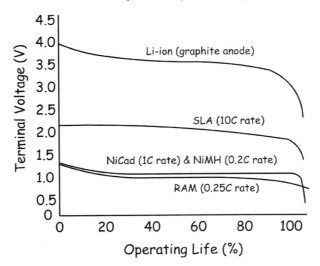

FIGURE 3.30

THE SUPERCAPACITOR

The supercapacitor isn't really a battery but a cross between a capacitor and a battery. It resembles a regular capacitor, but uses special electrodes and some electrolytes. There are three kinds of electrode material found in a supercapacitor: high-surface-area-activated carbons, metal oxide, and conducting polymers. The one using high-surface-area-activated carbons is the most economical to manufacture. This system is also called double layer capacitor (DLC) because the energy is stored in the double layer formed near the carbon electrode surface. The electrolyte may be aqueous or organic. The aqueous electrolyte offers low internal resistance but limits the voltage to 1 V. In contrast, the organic electrolyte allows 2 and 3 V of charge, but the internal resistance is higher.

To make the supercapacitor practical for use in electronic circuits, higher voltages are needed. Connecting the cells in series accomplishes this task. If more than three or four capacitors are connected in series, voltage balancing must be used to prevent any cell from reaching overvoltage.

Supercapacitors have values from 0.22 F upwards to several F. They have higher energy storage capacity than electrolytic capacitors, but a lower capacity than a battery (approximately ¹⁄₁₀ that of a NiMH battery). Unlike electrochemical batteries that deliver a fairly steady voltage, the voltage of a supercapacitor drops from full voltage to zero volts without the customary flat voltage curve characteristic of most batteries. For this reason, supercapacitors are unable to deliver the full charge. The percentage of charge that is available depends on the voltage requirements of the applications. For example, a 6-V battery is allowed to discharge to 4.5 V before the equipment cuts off; the supercapacitor reaches that threshold with the first quarter of the discharge. The remaining energy slips into an unusable voltage range.

The self-discharge of the supercapacitor is substantially higher than that of the electrochemical battery. Typically, the voltage of the supercapacitor with an organic electrolyte drops from full charge to the 30 percent level in as little as 10 hours. Other supercapacitors can retain the charged energy longer. With these designs, the capacity drops from full charge to 85 percent in 10 days. In 30 days, the voltage drops to roughly 65 percent, and to 40 percent after 60 days.

The most common supercapacitor applications are memory backup and standby power for real-time clock ICs. Only in special applications can the supercapacitor be used as a direct replacement for a chemical battery. Often the supercapacitor is used in tandem with a battery (placed across its terminals, with a provision in place to limit high influx of current when equipment is turned on) to improve the current handling of the battery: during low load current the battery charges the supercapacitor; the stored energy of the supercapacitor kicks in when a high load current is requested. In this way the supercapacitor acts to filter and smooth pulsed load currents. This enhances the battery's performance, prolongs the runtime, and even extends the longevity of the battery.

Limitations include an inability to use the full energy spectrum—depending on the application, not all energy is available. A supercapacitor has low energy density, typically holding ¹⁄₅ to ¹⁄₁₀ the energy of an electrochemical battery. Cells have low voltages—serial connections are needed to obtain higher voltages. Voltage balancing is required if more than three capacitors are connected in series. Furthermore, the self-discharge is considerably higher than that of an electrochemical battery.

Advantages include a virtually unlimited life cycle—supercapacitors are not subject to the wear and aging experienced by electrochemical batteries. Also, low impedance can enhance pulsed current demands on a battery when placed in parallel with the battery. Supercapacitors experience rapid charging—with low-impedance versions reaching full charge within seconds. The charge method is simple—the voltage-limiting circuit compensates for self-discharge.

Selecting the Right Battery (Comparison Chart)

Legend: ● Yes ○ Borderline (blank) No

	Carbon Zinc	Zinc Chloride	Alkaline	Lithium	Zinc air	Silver Oxide	Mercury	RAM	Lead Acid (SLA)	NiCad	NiMH	Lithium Ion	Lithium Polymer	NiZn	Supercapacitor
Characteristics of single cell															
Obsolete (not recommended)	●	●					●								
Rechargeable								●	●	●	●	●	●	●	●
Stable voltage				●	●	●	●				●	●	●	●	
High energy density (Wh/kg)		○	●	●	●	●	●	●			○	●	●	●	
High capacity rating (mAh)														●	
High peak load-current rating	○	●	●	●				●	●	●	●	○	○	●	
High pulsed discharge current		●	●	●		●	●		●	●		●	●	●	●
Low self-discharge rate			●	●				●	●	○		●	●	●	
Good at high temperatures		○	●	●			●								
Good at low temperatures		●	●	●		●									
High cycle life									●	●	●	●	●	●	●
Miniature				●	●	●	●							●	
Memory effect										●	○				
Expensive (cost)				●	●	●	●		●	●		●	●	○	
Environmentally unfriendly									●	●					
Applications															
Very small portable devices				●	●	●	●					●	●		
Pagers, hearing aids (h), watches (w)				●	h	w	●								
Radios, toys, general purpose	●	●	●					●							
Small motorized items		●	●					○			●				
Camcorders, digital cameras, test equipment			●							●	●				
Remotes, clocks, calculators		●	●	●				●							
Cellphones, mobile phones, laptops				○						○	●	●	●		
Low-self discharge (smoke detectors, data loggers)				●			●								
Power tools, model cars, electric toothbrush, etc.										●	●			●	
Motorized vehicles (electric bike, scooters, trolling motors, power tools, mower)									●	●	●			●	
High-power battery backup (emergency lighting, solar power storage, UPS)									●						
Power backup for short-term primary power outages (e.g. CMOS RAM, motor start, etc.)			●	●											●
Standard cell and battery packages															
AAA	●	●	●	●				●		●	●				
AA	●	●	●	●				●		●	●				
C	●	●	●	●				●		●	●				
D	●	●	●	●				●		●	●				
9V	●	●	●	●						●	●				
6V Lantern	●	●	●												
Button (coin)				●	●	●	●								●
Special battery pack			●	●					●	●	●	●	●	●	
Plastic Box									●					●	
PCB mount				●											●

FIGURE 3.31

3.2.5 Battery Capacity

Batteries are given a capacity rating that indicates how much electrical energy they are capable of delivering over a period of time. The capacity rating is specified in terms of ampere-hours (Ah) and millampere-hours (mAh). Knowing the battery capacity, it is possible to estimate how long the battery will last before being considered dead. The following example illustrates this.

Example: A battery with a capacity of 1800 mAh is to be used in a device that draws 120 mA continuously. Ignoring possible loss in capacity as a result of load current magnitude, how long should the battery be able to deliver power?

Answer: Ideally, this would be:

$$t = \frac{1800 \text{ mAh}}{120 \text{ mA}} = 15 \text{ h}$$

Note: In reality, you must consult the battery manufacturer's data sheets and analyze their discharge graphs (voltage as a function of time and of load current) to get an accurate determination of actual discharge time. As the load current increases, there is an apparent loss in battery capacity caused by internal resistance.

Typical capacity ratings for AAA, AA, C, D, and 9-V NiMH batteries are 1000 mAh (AAA), 2300 mAh (AA), 5000 mAh (C), 8500 mAh (D), 250 mAh (9).

C Rating

The charge and discharge currents of a battery are measured in capacity rating or C rating. The capacity represents the efficiency of a battery to store energy and its ability to transfer this energy to a load. Most portable batteries, with the exception of lead-acid, are rated at 1 C. A discharge rate of 1 C draws a current equal to the rated capacity that takes one hour (h). For example, a battery rated at 1000 mAh provides 1000 mA for 1 hour if discharged at 1 C rate. The same battery at 0.5 C provides 500 mA for 2 hours. At 2 C, the same battery delivers 2000 mA for 30 minutes. 1 C is often referred to as a 1-hour discharge; 0.5 C would be 2 hours, and 0.1C would be a 10-hour discharge. The discrepancy in C rates between different batteries is largely dependent on the internal resistance.

Example: Determine the discharge time and average current output of a battery with a capacity rating of 1000 mAh if it is discharged at 1 C. How long would it take to discharge at 5 C, 2 C, 0.5 C, 0.2 C, and 0.05 C?

Answer: At 1 C, the battery is attached to a load drawing 1000 mA (rated capacity/hour), so the discharge time is:

$$t = 1 \text{ hC}/C \text{ rating} = 1 \text{ hC}/1 \text{ C} = 1 \text{ h}$$

At 5 C, the battery is attached to a load drawing 5000 mA (five times rated capacity/hour), so the discharge time is:

$$t = 1 \text{ hC}/C \text{ rating} = 1 \text{ hC}/5 \text{ C} = 0.2 \text{ h}$$

At 2 C, the battery is attached to a load drawing 2000 mA (two times rated capacity/hour), so the discharge time is:

$$t = 1 \text{ hC}/C \text{ rating} = 1 \text{ hC}/2 \text{ C} = 0.5 \text{ h}$$

At 0.5 C, the battery is attached to a load drawing 500 mA (half the rated capacity/hour), so the discharge time is:

$$t = 1 \text{ hC}/C \text{ rating} = 1 \text{ hC}/0.5 \text{ C} = 2 \text{ h}$$

At 0.2 C, the battery is attached to a load drawing 200 mA (20 percent rated capacity/hour) so the discharge time is:

$$t = 1\,\text{hC}/\text{C rating} = 1\,\text{hC}/0.2\,\text{C} = 5\,\text{h}$$

At 0.05 C, the battery is attached to a load drawing 50 mA (5 percent rated capacity/hour) so the discharge time is:

$$t = 1\,\text{hC}/\text{C rating} = 1\,\text{hC}/0.05\,\text{C} = 20\,\text{h}$$

Again, note that these values are estimates. When load currents increase (especially when C values get large), the capacity level drops below nominal values—due to nonideal internal characteristics such as internal resistance—and must be determined using manufacturer's discharge curves and Peurkert's equation. Do a search of the Internet, using "Peurkert's equation" as a keyword, to learn more.

3.2.6   Note on Internal Voltage Drop of a Battery

Batteries have an internal resistance that is a result of the imperfect conducting elements that make up the battery (resistance in electrodes and electrolytes). Though the internal resistance may appear low (around 0.1 Ω for an AA alkaline battery, or 1 to 2 Ω for a 9-V alkaline battery), it can cause a noticeable drop in output voltage if a low-resistance (high-current) load is attached to it. Without a load, we can measure the open-circuit voltage of a battery, as shown in Fig. 3.32a. This voltage is essentially equal to the battery's rated nominal voltage—the voltmeter has such a high input resistance that it draws practically no current, so there is no appreciable voltage drop. However, if we attach a load to the battery, as shown in Fig. 3.32, the output terminal voltage of the battery drops. By treating the internal resistance R_{in} and the load resistance R_{load} as a voltage divider, you can calculate the true output voltage present across the load—see the equation in Fig. 3.32b.

A. Open Circuit Voltage

Voltmeter

R_{in} V_{out} 10MΩ

1.0Ω   1.5V

V_{in}

1.5V

B. Closed Circuit Voltage

R_{in} V_{out} R_{LOAD}

1.0Ω   1.36V   10Ω

V_{in}

1.5V

$$V_{Out} = V_{in}\,\frac{R_{LOAD}}{R_{LOAD} + R_{in}}$$

FIGURE 3.32

Batteries with large internal resistances show poor performance in supplying high current pulses. (Consult the battery comparison section and tables to determine which batteries are best suited for high-current, high-pulse applications.) Internal resistance also increases as the battery discharges. For example, a typical alkaline AA battery may start out with an internal resistance of 0.15 Ω when fresh, but may increase to 0.75 Ω when 90 percent discharged. The following list shows typical internal resistance for various batteries found in catalogs. The values listed should not be assumed to be universal—you must check the specs for your particular batteries.

9-V zinc carbon	35 Ω
9-V lithium	16 to 18 Ω
9-V alkaline	1 to 2 Ω
AA alkaline	0.15 Ω (0.30 Ω at 50 percent discharge)
AA NiMH	0.02 Ω (0.04 Ω at 50 percent discharge)
D Alkaline	0.1 Ω
D NiCad	0.009 Ω
D SLA	0.006 Ω
AC13 zinc air	5 Ω
76 silver	10 Ω
675 mercury	10 Ω

Battery OK / LOW Indicator

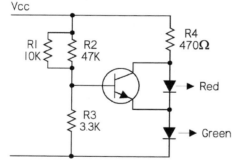

FIGURE 3.33

Here a green LED is used to indicate that the battery is okay. This stays on all the time to indicate that the battery is live, and the red LED comes on when the battery voltage falls below the set threshold. A green LED has around 2.0 V across it when it is illuminated. This value varies a bit with different manufacturers, but is pretty well matched within any batch. Add the base emitter voltage, and you need 2.6 V on the base of the right transistor (i.e., across the R_3) to turn on the transistor. 2.6 V across R_3 needs 9.1 across the supply rail. Below this threshold voltage, the transistor is off and the red LED is on. Above this voltage, the red LED is off. By adjusting the values of the three resistors, you can alter the threshold level. We'll discuss transistors and LEDs later on in this book.

3.3 Switches

A *switch* is a mechanical device that interrupts or diverts electric current flow within a circuit.

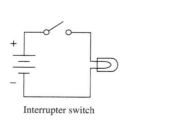

Interrupter switch

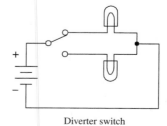

Diverter switch

FIGURE 3.34

3.3.1 How a Switch Works

Two slider-type switches are shown in Fig. 3.35. The switch in Fig. 3.35a acts as an interrupter, whereas the switch in Fig. 3.35b acts as a diverter.

Other kinds of switches, such as push-button switches, rocker switches, magnetic-reed switches, etc., work a bit differently than slider switches. For example, a magnetic-reed switch uses two thin pieces of leaflike metal contacts that can be forced together by a magnetic field. This switch, as well as a number of other unique switches, will be discussed later on in this section.

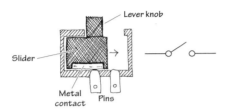

FIGURE 3.35

(*a*) **When the lever is pushed to the right, the metal strip bridges the gap between the two contacts of the switch, thus allowing current to flow. When the lever is pushed to the left, the bridge is broken, and current will not flow.**

(*b*) **When the lever is pushed upward, a conductive bridge is made between contacts *a* and *b*. When the lever is pushed downward, the conductive bridge is relocated to a position where current can flow between contact *a* and *c*.**

3.3.2 Describing a Switch

A switch is characterized by its number of *poles* and by its number of *throws*. A pole represents, say, contact a in Fig. 3.35b. A throw, on the other hand, represents the particular contact-to-contact connection, say, the connection between contact a and contact *b* or the connection between contact *a* and contact *c* in Fig. 3.35b. In terms of describing a switch, the following format is used: (number of poles) "*P*" and (number of throws) "*T*." The letter *P* symbolizes "pole," and the letter *T* symbolizes "throw." When specifying the number of poles and the number of throws, a convention must be followed: When the number of poles or number of throws equals 1, the letter S, which stands for "single," is used. When the number of poles or number of throws equals 2, the letter *D*, which stands for "double," is used. When the number of poles or number of throws exceeds 2, integers such as 3, 4, or 5 are used. Here are a few examples: SPST, SPDT, DPST, DPDT, DP3T, and 3P6T. The switch shown in Fig. 3.35a represents a single-pole single-throw switch (SPST), whereas the switch in Fig. 3.35b represents a single-pole double-throw switch (SPDT).

Two important features to note about switches include whether a switch has momentary contact action and whether the switch has a center-off position. Momentary-contact switches, which include mainly pushbutton switches, are used when it is necessary to only briefly open or close a connection. Momentary-contact switches come in either normally closed (NC) or normally open (NO) forms. A normally closed pushbutton switch acts as a closed circuit (passes current) when left untouched. A normally open pushbutton switch acts as an open circuit (broken circuit) when left untouched. Center-off position switches, which are seen in diverter switches, have an additional "off" position located between the two "on" positions. It is important to note that not all switches have center-off or momentary-contact features—these features must be specified.

Symbols for Switches

SPST SWITCHES

Throw switch

Normally open
push-button

Normally closed
push-button

SPDT SWITCHES

Throw switch

Normally open/closed
push-button

DPST SWITCHES

Throw switch

Normally open
push-button

DPDT SWITCHES

Throw switch

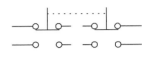

Normally open/closed
push-button

SP(n)T SWITCHES

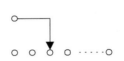

Multiple contact slider
switch

Multiple contact rotary
switch
(SP8T)

(n)P(m)T SWITCHES

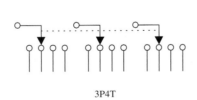

3P4T

2-deck rotary
(DP8T)

FIGURE 3.36

3.3.3 Kinds of Switches

Toggle Switch

SPST

SPDT

DPDT

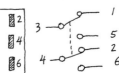

FIGURE 3.37

Pushbutton Switch

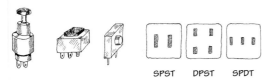

SPST   DPST   SPDT

Snap Switch (Microswitch)

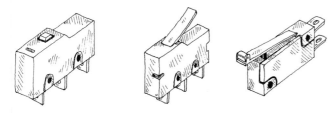

Rotary Switch

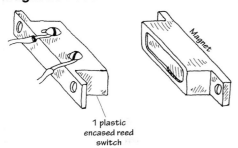

Magnetic Reed Switch

1 plastic
encased reed
switch

A reed switch consists of two closely spaced leaflike contacts that are enclosed in an air-tight container. When a magnetic field is brought nearby, the two contacts will come together (if it is a normally open reed switch) or will push apart (if it is a normally closed reed switch).

Binary-Coded Switches

Rotatable
dial

Decimal

Bottom views

1 2 4 8

Hexadecimal

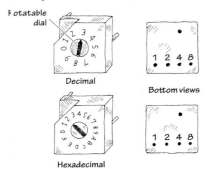

These switches are used to encode digital information. A mechanism inside the switch will "make" or "break" connections between the switch pairs according to the position of the dial on the face of the switch. These sitches come in either true binary/hexadecimal and complementary binary/hexa-decimal forms. The charts below show how these switches work:

True Binary/Hexadecimal

Type	Position	Code			
		1	2	4	8
Hexadecimal / Decimal	0				
	1	•			
	2		•		
	3	•	•		
	4			•	
	5	•		•	
	6		•	•	
	7	•	•	•	
	8				•
	9	•			•
	A		•		•
	B	•	•		•
	C			•	•
	D	•		•	•
	E		•	•	•
	F	•	•	•	•

Complementary Binary/Hexadecimal

Type	Position	Code			
		1	2	4	8
Hexadecimal / Decimal	0	•	•	•	•
	1		•	•	•
	2	•		•	•
	3			•	•
	4	•	•		•
	5		•		•
	6	•			•
	7				•
	8	•	•	•	
	9		•	•	
	A	•		•	
	B			•	
	C	•	•		
	D		•		
	E	•			
	F				

FIGURE 3.37   (*Continued*)

DIP Switch

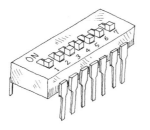

DIP stands for "dual-inline package." The geometry of this switch's pin-outs allows the switch to be placed in IC sockets that can be wired directly into a circuit board.

Mercury Tilt-Over

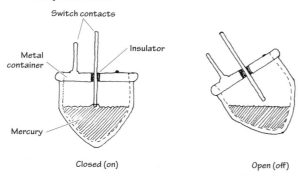

Closed (on)   Open (off)

This type of switch is used as a level-sensing switch. In a normally closed mercury tilt-over switch, the switch is "on" when oriented vertically (the liquid mercury will make contact with both switch contacts). However, when the switch is tilted, the mercury will be displaced, hence breaking the conductive path.

These days, a metal ball and pair of contacts are more common than the toxic and expensive mercury.

FIGURE 3.37   (*Continued*)

3.3.4 Simple Switch Applications

Simple Security Alarm

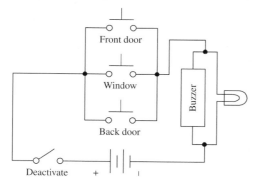

Here's a simple home security alarm that's triggered into action (buzzer and light go on) when one of the normally open switches is closed. Magnetic reed switches work particularly well in such applications.

FIGURE 3.38

Dual-Location On/Off Switching Network

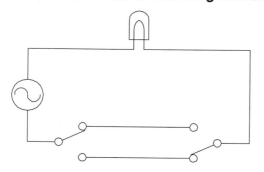

Here's a switch network that allows an individual to turn a light on or off from either of two locations. This setup is frequently used in household wiring applications.

FIGURE 3.39

Current-Flow Reversal

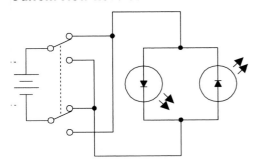

A DPDT switch, shown here, can be used to reverse the direction of current flow. When the switch is thrown up, current will flow throw the left light-emitting diode (LED). When the switch is thrown down, current will flow throw the right LED. (LEDs only allow current to flow in one direction.)

FIGURE 3.40

Multiple Selection Control of a Voltage-Sensitive Device via a Two-Wire Line

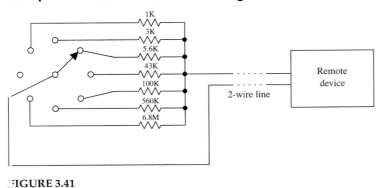

FIGURE 3.41

Say you want to control a remote device by means of a two-wire line. Let's also assume that the remote device has seven different operational settings. One way of controlling the device would be to design the device in such a way that if an individual resistor within the device circuit were to be altered, a new function would be enacted. The resistor may be part of a voltage divider, may be attached in some way to a series of window comparators (see op amps), or may have an analog-to-digital converter interface. After figuring out what valued resistor enacts each new function, choose the appropriate valued resistors and place them together with a rotary switch. Controlling the remote device becomes a simple matter of turning the rotary switch to select the appropriate resistor.

3.4 Relays

Relays are electrically actuated switches. The three basic kinds of relays include mechanical relays, reed relays, and solid-state relays. For a typical mechanical relay, a current sent through a coil magnet acts to pull a flexible, spring-loaded conductive plate from one switch contact to another. Reed relays consist of a pair of reeds (thin, flexible metal strips) that spring together whenever a current is sent through an encapsulating wire coil. A solid-state relay is a device that can be made to switch states by applying external voltages across *n*-type and *p*-type semiconductive junctions (see Chap. 4). In general, mechanical relays are designed for high currents (typically 2 to 15 A) and relatively slow switching (typically 10 to 100 ms). Reed relays are designed for moderate currents (typically 500 mA to 1 A) and moderately fast switching (0.2 to 2 ms). Solid-state relays, on the other hand, come with a wide range of current ratings (a few microamps for low-powered packages up to 100 A for high-power packages) and have extremely fast switching speeds (typically 1 to 100 ns). Some limitations of both reed relays and solid-state relays include limited switching arrangements (type of switch section) and a tendency to become damaged by surges in power.

Mechanical Relay   Reed Relay

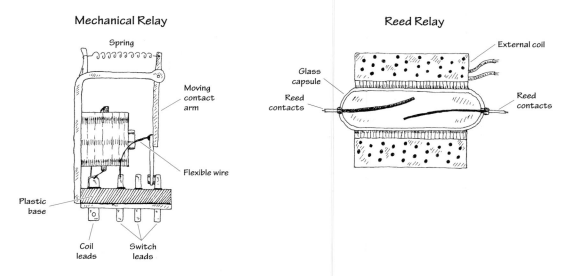

FIGURE 3.42

A mechanical relay's switch section comes in many of the standard manual switch arrangements (e.g., SPST, SPDT, DPDT, etc.). Reed relays and solid-state relays, unlike mechanical relays, typically are limited to SPST switching. Some of the common symbols used to represent relays are shown below.

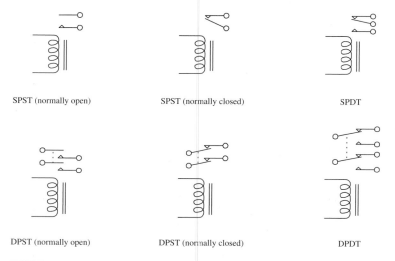

FIGURE 3.43

The voltage used to activate a given relay may be either dc or ac. For, example, when an ac current is fed through a mechanical relay with an ac coil, the flexible-metal conductive plate is pulled toward one switch contact and is held in place as long as the current is applied, regardless of the alternating current. If a dc coil is supplied by an alternating current, its metal plate will flip back and forth as the polarity of the applied current changes.

Mechanical relays also come with a latching feature that gives them a kind of memory. When one control pulse is applied to a *latching relay*, its switch closes. Even when the control pulse is removed, the switch remains in the closed state. To open the switch, a separate control pulse must be applied.

3.4.1 Specific Kinds of Relays

Subminiature Relays

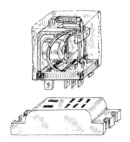

Typical mechanical relays are designed for switching relatively large currents. They come with either dc or ac coils. Dc-actuated relays typically come with excitation-voltage ratings of 6, 12, and 24 V dc, with coil resistances (coil ohms) of about 40, 160, and 650 Ω, respectively. AC-actuated relays typically come with excitation-voltage ratings of 110 and 240 V ac, with coil resistances of about 3400 and 13600 Ω, respectively. Switching speeds range from about 10 to 100 ms, and current ratings range from about 2 to 15 A.

Miniature Relays

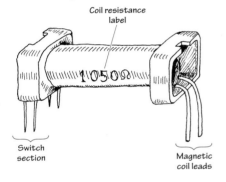

Miniature relays are similar to subminiature relays, but they are designed for greater sensitivity and lower-level currents. They are almost exclusively actuated by dc voltages but may be designed to switch ac currents. They come with excitation voltages of 5, 6, 9, and 12, and 24 V dc, with coil resistances from 50 to 3000 Ω.

Reed Relays

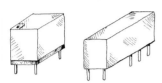

Coil resistance label

1050Ω

Switch section

Magnetic coil leads

Two thin metal strips, or reeds, act as movable contacts. The reeds are placed in a glass-encapsulated container that is surrounded by a coil magnet. When current is sent through the outer coil, the reeds are forced together, thus closing the switch. The low mass of the reeds allows for quick switching, typically around 0.2 to 2 ms. These relays come with dry or sometimes mercury-wetted contacts. They are dc-actuated and are designed to switch lower-level currents, and come with excitation voltages of 5, 6, 12, and 24 V dc, with coil resistances around 250 to 2000 Ω. Leads are made for PCB mounting.

Solid-State Relays

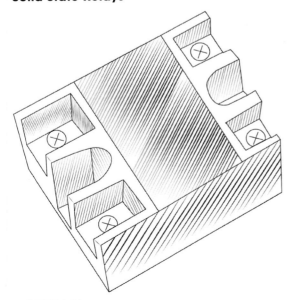

Solid-State Relays or SSRs are sealed modules designed to be used in the same way as electromechanical relays, but switching using opto-isolators and power transistors or Triacs. As such they are not really basic components, but modules.

They are usually divided in two varieties, AC and DC. An AC device usually uses an opto-isolator with zero-switching detector and a Triac to switch the load as the voltage is close to 0 V in the cycle, but a DC device uses a MOSFET or IGBT transistor (see Chap. 4) to switch the load.

Using an opto-isolator has the dual advantage of only requiring a couple of mA to switch the relay on, but also isolates the control side of the relay from the switching side.

FIGURE 3.44

3.4.2 A Few Notes about Relays

To make a relay change states, the voltage across the leads of its magnetic coil should be at least within ±25 percent of the relay's specified control-voltage rating. Too much voltage may damage or destroy the magnetic coil, whereas too little voltage may not be enough to "trip" the relay or may cause the relay to act erratically (flip back and forth).

The coil of a relay acts as an inductor. Now, inductors do not like sudden changes in current. If the flow of current through a coil is suddenly interrupted, say, a switch is opened, the coil will respond by producing a sudden, very large voltage across its leads, causing a large surge of current through it. Physically speaking, this phenomenon is a result of a collapsing magnetic field within the coil as the current is terminated abruptly. [Mathematically, this can be understood by noticing how a large change in current (dI/dt) affects the voltage across a coil ($V = L\,dI/dt$).] Surges in current that result from inductive behavior can create menacing voltage spikes (as high as 1000 V) that can have some nasty effects on neighboring devices within the circuit (e.g., switches may get zapped, transistors may get zapped, individuals touching switches may get zapped, etc.). Not only are these spikes damaging to neighboring devices, they are also damaging to the relay's switch contacts (contacts will suffer a "hard hit" from the flexible-metal conductive plate when a spike occurs in the coil).

The trick to getting rid of spikes is to use what are called *transient suppressors*. You can buy these devices in prepackaged form, or you can make them yourself. The following are a few simple, homemade transient suppressors that can be used with relay coils or any other kind of coil (e.g., transformer coils). Notably, the switch incorporated within the networks below is only one of a number of devices that may interrupt the current flow through a coil. In fact, a circuit may not contain a switch at all but may contain other devices (e.g., transistors, thyristors, etc.) that may have the same current-interrupting effect.

DC-Driven Coil

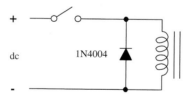

Placing a diode in reverse bias across a relay's coil eliminates voltage spikes by going into conduction before a large voltage can form across the coil. The diode must have a peak current capability able to handle currents equivalent to the maximum current that would have been flowing through the coil before the current supply was interrupted. A good general-purpose diode that works well for just such applications is the 1N4004 diode.

AC-Driven Coil

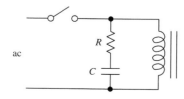

FIGURE 3.45

When dealing with ac-actuated relays, using a diode to eliminate voltage spikes will not work—the diode will conduct on alternate half-cycles. Using two diodes in reverse parallel will not work either—the current will never make it to the coil. Instead, an *RC* series network placed across the coil can be used. The capacitor absorbs excessive charge, and the resistor helps control the discharge. For small loads driven from the power line, setting $R = 100\ \Omega$ and $C = 0.05\ \mu F$ works fine for most cases. (Note: There are special devices, such as bidirectional TVs, MOVs, and MTLVs, that are designed for dc transients. See Sec. 4.5.)

3.4.3 *Some Simple Relay Circuits*

DC-Actuated Switch

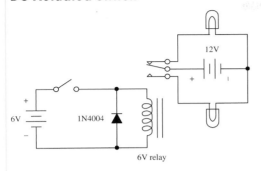

Here, a dc-powered SPDT relay is used to relay current to one of two light bulbs. When the switch in the control circuit is opened, the relay coil receives no current; hence the relay is relaxed, and current is routed to the upper bulb. When the switch in the control circuit is closed, the relay coil receives current and pulls the flexible-metal conductive plate downward, thus routing current to the lower bulb. The diode acts as a transient suppressor. Note that all components must be selected according to current and voltage ratings.

AC-Actuated Switch

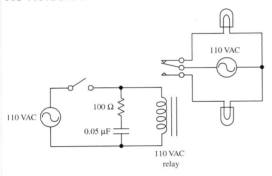

Here, an ac-actuated relay is used to switch ac current to one of two ac-rated light bulbs. The behavior in this circuit is essentially the same as in the preceding circuit. However, currents and voltages are all ac, and an *RC* network is used as a transient suppressor. Make sure that resistor and capacitor are rated for a potential transient current that is as large as the typical coil current. The capacitor must be rated for ac line voltage. A discrete transient suppressor (e.g., bipolar TVs or MOV) can take the place of an RC network.

Relay Driver

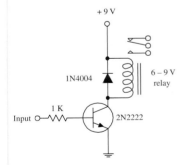

FIGURE 3.46

If a relay is to be driven by an arbitrary control voltage, this circuit can be used. The *npn* bipolar transistor acts as a current-flow control valve. With no voltage or input current applied to the transistor's base lead, the transistor's collector-to-emitter channel is closed, hence blocking current flow through the relay's coil. However, if a sufficiently large voltage and input current are applied to transistor's base lead, the transistor's collector-to-emitter channel opens, allowing current to flow through the relay's coil.

3.5 Resistors

There are various kinds of resistors available today. There are fixed resistors, variable resistors, digitally adjustable resistors, fusible resistors, photoresistors, and various resistor arrays (networks). Figure 3.47 shows schematic symbols and pictures of some of the most common types.

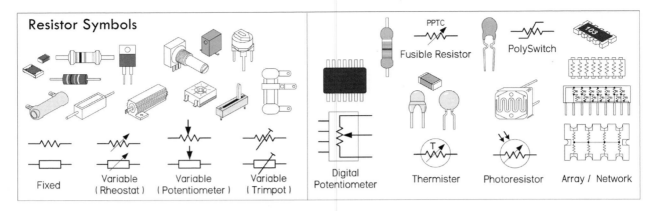

FIGURE 3.47

Resistors perform two basic functions in electronics: to limit current flow and to set voltage levels within a circuit. Figure 3.48a shows a resistor being used to reduce current flow to an LED. Without the resistor, the LED would receive excess current capable of melting its sensitive *p-n* junction. A variation of the LED circuit is shown with a variable resistor in series with the current-limiting resistor. The variable resistor (or potentiometer, or pot) provides additional current limiting with the desired effect of controlling the brightness of the LED.

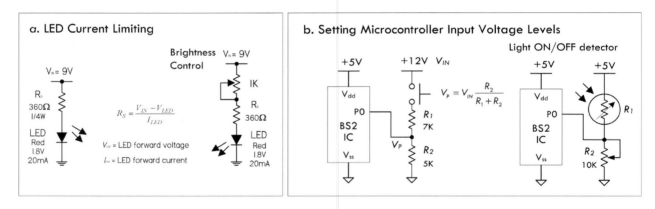

FIGURE 3.48

Figure 3.48b demonstrates how two resistors are used to create a voltage divider capable of providing a dc voltage that is a fraction of the input voltage. In this example, a voltage of 12 V is reduced to a voltage of 5 V—a usable logic HIGH level for a microcontroller's input. A photoresistor replacing one of the voltage divider resistors acts as a variable resistor whose resistance decreases with light intensity. When this resistance decreases, the voltage present at the microcontroller's input increases, eventually reaching a logic HIGH level. Once a logic HIGH is set up, it is then up to the microcontroller's program to determine what course of action to take next.

The key features of current limiting and voltage setting are implemented in various ways in electronics. Resistors are used to set operating current and signal levels in circuits, provide voltage reduction, set precise gain values in precision circuits, act as shunts in ammeters and voltage meters, behave like damping agents in oscillators and timer circuits, act as bus and line terminators in digital circuits, provide feedback networks for amplifiers, and act as pullup and pulldown elements in digital circuits. They are also used in attenuators and bridge circuits. Special kinds of resistors are even used as fuses.

3.5.1 Resistance and Ohm's Law

From Chap. 2, we learned that if a dc voltage is applied across a resistor, the amount of current that will flow through the resistor can be found using *Ohm's law*—simply rearrange the equation ($I = V/R$). To find the power dissipated as heat by the resistor, apply the second equation below. By plugging Ohm's law into the power equation, you also get $P = I^2 \times R$ and $P = V^2/R$, which come in handy, too.

$$V = I \times R \text{ (Ohm's law)}$$

$$P = I \times V \text{ (Power law)}$$

R is the *resistance*, or the resistor, expressed in ohms (Ω), P is the *power loss* in watts (W), V has the voltage in volts (V), and I is the current in amperes (A).

Resistance values are given in kiloohms (kΩ) or megaohms (MΩ), where k represents 1000 and M represents 1,000,000. So a 3.3-kΩ resistor is equal to 3300 Ω, while a 2M resistor is equal to 2,000,000 Ω. Voltage, current, and power are often expressed in millivolts (mV), milliamps (mA), and milliwatts (mW), where m is equal to 0.001. 1 mV = 0.001 V; 200 mA = 0.2 A, 33 mW = 0.033 W.

As an example, in Fig. 3.49, the amount of current through a 100-Ω resistor attached to a 12-V battery is $I = 12 \text{ V}/100 \ \Omega = 0.120$ A or 120 mA. The power loss due to heating becomes $P = 0.120 \text{ A} \times 12 \text{ V} = 1.44$ W.

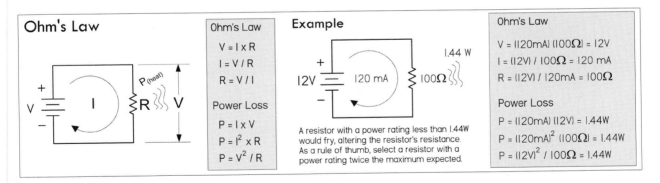

FIGURE 3.49

Determining the power loss is very important when designing circuits. All real resistors have maximum allowable *power ratings* that must not be exceeded. If you exceed the power rating you'll probably end up frying your resistor, destroying the internal structure, and thus altering the resistance. Typical general-purpose resistors come in ⅛-, ¼-, ½- and 1-W power ratings, while high-power resistors can range from 2 to several hundred watts.

So, going back to our example circuit in Fig. 3.49, the power rating of the resistor must be greater than the calculated dissipated power of 1.44 W. In reality, the power rating should be greater than this for safety. As a *rule of thumb*, always select a resistor that has a power rating at least twice the maximum value anticipated. Though a 2-W resistor would work in our example, a 3-W resistor would be safer. Other factors, such as ambient temperature, enclosures, resistor groupings, pulsed operation, and additional air cooling will increase or decrease the required power rating of a resistor—see the section on real resistor characteristics for more information.

3.5.2 Resistors in Series and Parallel

Rarely do you see circuits that use a single resistor alone. Usually, resistors are found connected in a variety of ways. The two fundamental ways of connecting resistors are in series and in parallel.

Resistors in Parallel

When two or more resistors are placed in parallel, the voltage across each resistor is the same, but the current through each resistor will vary with resistance. Also, the total resistance of the combination will be lower than that of the lowest resistance value present. The formula for finding the total resistance of resistors in parallel is:

$$R_{total} = \frac{1}{\dfrac{1}{R_1} + \dfrac{1}{R_2} + \dfrac{1}{R_3} + \dfrac{1}{R_4} + \cdots} \quad \text{(Resistors in parallel)}$$

$$R_{total} = \frac{R_1 \times R_2}{R_1 + R_2} \quad \text{(Two resistors in parallel)}$$

The dots in the equation indicate that any number of resistors can be combined. For only two resistances in parallel (a very common case), the formula reduces to the equation above.

Example: If a 1000-Ω resistor is connected in parallel with one of 3000 Ω, what is the total resistance or equivalent resistance?

$$R_{total} = \frac{R_1 \times R_2}{R_1 + R_2} = \frac{1000\ \Omega \times 3000\ \Omega}{1000\ \Omega \times 3000\ \Omega} = \frac{3,000,000\ \Omega^2}{4000\ \Omega} = 750\ \Omega$$

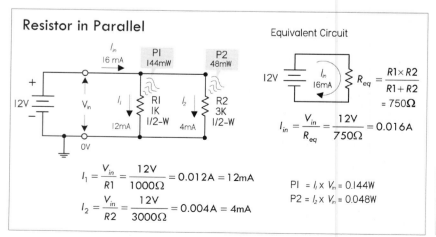

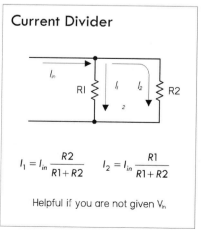

FIGURE 3.50

An important thing to note after applying these equations is that the current entering the top junction of the resistors in parallel equals the sum of the current entering the resistors ($I_{in} = I_1 + I_2$). This statement is referred to as Kirchhoff's current law. With this law, and Ohm's law, you come up with the current divider equation, shown to the right in Fig. 3.50. This equation comes in handy when you know the input current but not the input voltage.

To find how much power resistors in parallel dissipate, apply the power law, as shown in Fig. 3.49.

Resistors in Series

When a circuit has a number of resistors connected in series, the total resistance of the circuit is the sum of the individual resistances. Also, the amount of current flowing through each resistor in series is the same, while the voltage across each resistor varies with resistance. The formula for finding the total resistance of resistors in series is:

$$R_{total} = R_1 + R_2 + R_3 + R_4 + \cdots \text{ (Resistors in series)}$$

The dots indicate that as many resistors as necessary may be added.

Example: If a 1.0-kΩ resistor is placed in series with a 2.0-kΩ resistor, what is the total resistance?

$$R_{total} = R_1 + R_2 = 1000 \ \Omega + 2000 \ \Omega = 3000 \ \Omega = 3 \ k\Omega$$

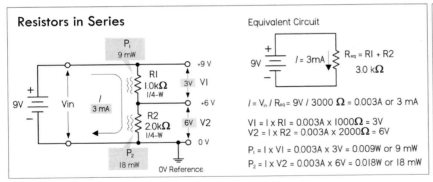

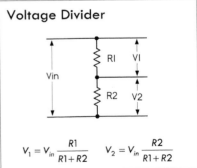

FIGURE 3.51

If we attach the two resistors to an input voltage $V_{in} = 9$ V, the total current flow, which will be the same as the individual current flow through each resistor, is:

$$I = \frac{V_{in}}{R_{tot}} = \frac{9 \ V}{3000 \ \Omega} = 0.003 \ A = 3 \ mA, \ I_1 = 3 \ mA, \ I_2 = 3 \ mA$$

To find the voltage drop across each resistor, apply Ohm's law:

$$V_1 = I_1 \times R_1 = 0.003 \ A \times 1000 \ \Omega = 3 \ V$$

$$V_2 = I_2 \times R_2 = 0.003 \ A \times 2000 \ \Omega = 6 \ V$$

You can use this approach to find the voltage drop across any number of resistors in series. Also, notice that the voltage drop across each resistor is directly proportional to the resistance. The 2000-Ω resistor value is twice as large as the 1000-Ω resistor and the voltage drop across the 2000-Ω resistor is twice as large.

An important thing to notice with the voltage drops is that they all add up to the supply voltage V_{in}. If you start at the positive terminal of the battery (+9 V) and subtract the 3-V drop across R_1 and then subtract the 6-V drop across R_2, you end up with zero: +9 V − 3 V − 6 V = 0 V. Another way to put it is to say that the sum of the voltage changes around a closed path is zero. The resistors are power sinks, while the battery is a power source. It is common to assign a + sign to power sources and a − sign to power sinks. This means the voltages across the resistors have the opposite sign from the battery voltage. Adding all the voltages yields zero. This is called *Kirchhoff's voltage law*.

Now, if there are only two resistors in series, you can avoid calculating currents and simply apply the following handy equations, referred to as the *voltage divider equations*:

$$V_1 = \frac{R_1}{R_1 + R_2} V_{in} = \frac{1000\,\Omega}{1000\,\Omega + 2000\,\Omega} \times 9\,V = 3\,V, V_2 = \frac{R_2}{R_1 + R_2} V_{in}$$

$$= \frac{2000\,\Omega}{1000\,\Omega + 2000\,\Omega} \times 9\,V = 6\,V$$

A voltmeter reading from ground to the middle of the two resistors, which we'll call the output voltage V_{out}, will read 6 V, since you're really just measuring the voltage drop across R_2.

3.5.3 Reading Resistor Labels

Axial lead resistors, such as carbon composition, carbon film, and metal film, use color bands to indicate resistance values. The most common labeling scheme uses four bands: the first band represents the first digit, the second band the second digit, the third band the multiplier (as an exponent of 10), and the fourth band the tolerance (if there is no fourth band, the tolerance is 20 percent). The table in Fig. 3.52 indicates the meaning of each color in regard to number, multiplier, and tolerance.

On precision resistors, you'll find five bands: the first three bands are used as significant figures, the fourth band is the multiplier, while a space between the fourth and fifth band that is wider than the others is used to identify the fifth tolerance band.

Another five-band labeling scheme that is typically reserved for military-spec resistors has a fifth band reserved for reliability level. The reliability band tells you the percentage change in resistance over a time interval (e.g., 1000 hours, brown = 1 percent, red = 0.1 percent, orange = 0.01 percent, yellow = 0.001 percent).

Surface-mount resistors use either a three-digit or a four-digit label. In the three-digit scheme, the first two digits represent significant figures, and the last digit is the multiplier. For values less than 100 Ω, the letter "R" is substituted for one of the significant digits and represents a decimal point (e.g., 1R0 = 1.0 Ω).

When tolerance levels become important (e.g., narrower than around ± 2 percent), an extra digit letter is placed at the end of the previous three-digit code to indicate tolerance (e.g., F = ± 1). See examples in Fig. 3.52.

Precision surface-mount resistors with a four-digit code use the first three digits as significant figures, while using the last digit as the multiplier. Again, the letter "R" is used as a decimal point—see the examples in Fig. 3.52.

Resistor Labels

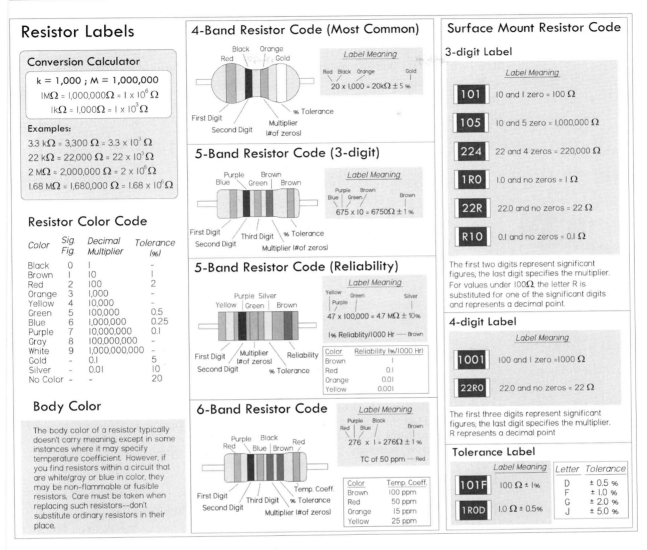

Conversion Calculator

k = 1,000 ; M = 1,000,000

$1M\Omega = 1,000,000\Omega = 1 \times 10^6 \Omega$

$1k\Omega = 1,000\Omega = 1 \times 10^3 \Omega$

Examples:

$3.3 \text{ k}\Omega = 3,300 \Omega = 3.3 \times 10^3 \Omega$

$22 \text{ k}\Omega = 22,000 \Omega = 22 \times 10^3 \Omega$

$2 \text{ M}\Omega = 2,000,000 \Omega = 2 \times 10^6 \Omega$

$1.68 \text{ M}\Omega = 1,680,000 \Omega = 1.68 \times 10^6 \Omega$

Resistor Color Code

Color	Sig. Fig.	Decimal Multiplier	Tolerance (%)
Black	0	1	–
Brown	1	10	1
Red	2	100	2
Orange	3	1,000	–
Yellow	4	10,000	–
Green	5	100,000	0.5
Blue	6	1,000,000	0.25
Purple	7	10,000,000	0.1
Gray	8	100,000,000	–
White	9	1,000,000,000	–
Gold	–	0.1	5
Silver	–	0.01	10
No Color	–	–	20

Body Color

The body color of a resistor typically doesn't carry meaning, except in some instances where it may specify temperature coefficient. However, if you find resistors within a circuit that are white/gray or blue in color, they may be non-flammable or fusible resistors. Care must be taken when replacing such resistors--don't substitute ordinary resistors in their place.

4-Band Resistor Code (Most Common)

Black / Orange
Red / Gold

First Digit
Second Digit
Multiplier (#of zeros)
% Tolerance

Label Meaning

Red Black Orange / Gold

$20 \times 1,000 = 20k\Omega \pm 5\%$

5-Band Resistor Code (3-digit)

Purple / Brown
Blue / Green / Brown

First Digit
Second Digit
Third Digit
Multiplier (#of zeros)
% Tolerance

Label Meaning

Purple Brown
Blue / Green / Brown

$675 \times 10 = 6750\Omega \pm 1\%$

5-Band Resistor Code (Reliability)

Purple Silver
Yellow / Green / Brown

First Digit
Second Digit
Multiplier (#of zeros)
% Tolerance
Reliability

Label Meaning

Yellow Green / Silver
Purple

$47 \times 100,000 = 4.7 \text{ M}\Omega \pm 10\%$

1% Reliablity/1000 Hr — Brown

Color	Reliability (%/1000 Hr)
Brown	1
Red	0.1
Orange	0.01
Yellow	0.001

6-Band Resistor Code

Purple Black Red
Red / Blue / Brown

First Digit
Second Digit
Third Digit
Multiplier (#of zeros)
% Tolerance
Temp. Coeff.

Label Meaning

Purple Black
Red / Blue / Brown

$276 \times 1 = 276\Omega \pm 1\%$

TC of 50 ppm — Red

Color	Temp. Coeff.
Brown	100 ppm
Red	50 ppm
Orange	15 ppm
Yellow	25 ppm

Surface Mount Resistor Code

3-digit Label

Label Meaning

101	10 and 1 zero = 100 Ω
105	10 and 5 zero = 1,000,000 Ω
224	22 and 4 zeros = 220,000 Ω
1R0	1.0 and no zeros = 1 Ω
22R	22.0 and no zeros = 22 Ω
R10	0.1 and no zeros = 0.1 Ω

The first two digits represent significant figures; the last digit specifies the multiplier. For values under 100Ω, the letter R is substituted for one of the significant digits and represents a decimal point.

4-digit Label

Label Meaning

1001	100 and 1 zero =1000 Ω
22R0	22.0 and no zeros = 22 Ω

The first three digits represent significant figures; the last digit specifies the multiplier. R represents a decimal point

Tolerance Label

Label Meaning

		Letter	Tolerance
101F	100 Ω ± 1%	D	± 0.5 %
1R0D	1.0 Ω ± 0.5%	F	± 1.0 %
		G	± 2.0 %
		J	± 5.0 %

FIGURE 3.52

NOTE OF CAUTION (RESISTOR BODY COLOR)

Resistor body color usually doesn't carry significant meaning. It sometimes represents the resistor's temperature coefficient, but this is seldom of great importance in most hobby-type work. However, a note of caution: there are two resistor body colors that you should be aware of if you go tinkering around inside consumer electronics equipment. Resistor body colors white and blue are used to mark nonflammable resistors and fusible resistors. If you encounter this type of resistor in a circuit, do not replace it with a normal resistor. Doing so may cause a fire hazard if something goes wrong in the circuit. Nonflammable resistors and fusible resistors are designed so that they don't catch fire when they overheat. When fusible resistors overheat, they cut the current flowing like a fuse. We'll discuss these resistors in greater detail later in this chapter.

3.5.4 Real Resistor Characteristics

There are a number of things to consider when selecting a resistor for a given application. Two primary considerations include selecting the appropriate nominal resistance and power rating. The next step is to develop an acceptable tolerance for the resistor that ensures it will function properly in all extremes of the application. This task can be a bit difficult because it requires understanding a variety of nonideal characteristics that vary from one resistor family (or even between resistors in the same family) to another.

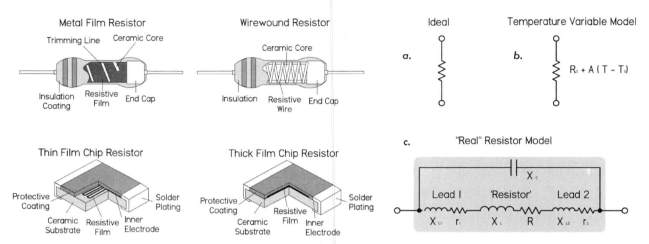

FIGURE 3.53   Some examples of real resistor construction, as well as various resistor models that are used to aid in predicting how a real resistor will behave. Model (a) represents an ideal resistor, while model (b) represents a temperature variable model for noninductive resistors. Model (c) takes into consideration the inherent inductive and capacitive elements within a resistor's construction. For UHF and microwave designs, model (c) could be used with *L* representing lead inductances.

There are many different kinds of resistors out there, each with its specific set of limitations and suitable applications. A resistor that is good for one application can be disastrous in another. Resistors designated as "precision" resistors (such as precision metal film) are designed for applications where tight resistance tolerance and stability are primary considerations. They generally have restricted operating temperature limits and power dissipation ratings. "Power" resistors (such as power wirewounds) tend to be designed to optimize power dissipation at the expense of precision, and generally have extended operating temperature limits. "General-purpose" resistors (such as carbon film) tend to be somewhere in between, and are suitable for most general applications.

The following is a rundown of the important specifications used when selecting resistors. You can find detailed specifications for real resistors by checking out the manufacturers' data sheets (e.g., www.vishay.com).

Voltage Rating

This is the maximum value of dc or RMS voltage that can be imposed across a resistor at specified ambient temperatures. The voltage rating is related to the power rating by $V = \sqrt{P \times R}$, where V is the voltage rating (in volts), P is the power rating (in watts), and R is the resistance (in ohms). For a given value of voltage and power rating, a critical value of resistance can be calculated. For values of resistance below the critical value, the maximum voltage is never reached; for values of resistance above the critical value, the power dissipated is lower than the rated power. One-half-watt and some 1-W resistors usually are rated only to 250 to 350 V. For high-voltage applications (e.g., a high-voltage amplifier), you may have to resort to, say, 1 W (continuous, 1000-V surge) or 2 W, 750-V-rated resistors.

Tolerance

This is expressed as the deviation (in percent) in resistance from the nominal value, measured at 25°C with no load applied. Typical resistor tolerances are 1 percent 2 percent, 5 percent, 10 percent, and 20 percent. Precision resistors, such as precision wirewounds, are made with tolerances as tight as ±0.005 percent. To understand what tolerance means, consider a 100-Ω resistor with 10 percent tolerance. The specified tolerance means that the resistor's resistance could actually be anywhere between 90 and 110 Ω. On the other hand, a 100-Ω resistor with 1 percent tolerance has a possible resistance range from 99 to 101 Ω.

Carbon-composition resistors, as a whole, have the worst tolerance, around 5 to 20 percent. Carbon-film resistors are about 1 to 5 percent, metal-film about 1 percent, and precision metal-film resistors as low as 0.1 percent. Most wirewound resistors are from 1 to 5 percent, while precision wirewounds can achieve ±0.005 percent tolerance. Foil resistors, a relatively new technology, can achieve 0.0005 percent. For most general-purpose applications, a resistor with a 5 percent tolerance is adequate.

Example Derating Curves

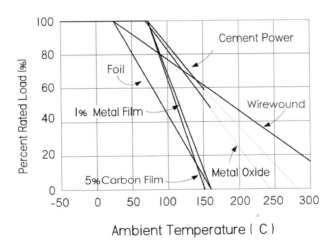

FIGURE 3.54

Power Rating

Resistors must be operated within specified temperature limits to avoid permanent damage to the materials. The temperature limit is defined in terms of the maximum power, called the *power rating*, and a derating curve that is provided by the resistor manufacturers (see Fig. 3.54). The power rating of a resistor is the maximum power in watts that the resistor can safely dissipate as heat, usually specified at +25°C. Beyond +25°C, the derating curve, which plots the maximum allowable power versus the ambient temperature, is used. The derating curve is usually linearly drawn from the full-rated load temperature to the maximum allowable no-load temperature. A resistor may be operated at ambient temperatures above the maximum full-load ambient temperature if operating at lower than full-rated power capacity. The maximum allowable no-load temperature is also the maximum storage temperature for the resistor.

In regard to resistor life, any change in temperature of 30 to 40°C is tolerated—the resistor will return to its normal resistance when the temperature returns to its nominal value. However, if the resistor gets too hot to touch, you may end up permanently damaging it. For this reason, it's important to be conservative when specifying the power rating of a resistor.

Standard power ratings for resistors include: $\frac{1}{16}$, $\frac{1}{10}$, $\frac{1}{8}$, $\frac{1}{4}$, $\frac{1}{2}$, 1, 2, 5, 10, 15, 25, 50, 100, 200, 250, and 300 W. To determine the power rating for a particular application, use $P = IV$ ($P = I^2R$ or $P = V^2/R$), then, as a rule of thumb, select a resistor that has a power rating two to four times greater than the calculated value. Note, however, that there are many factors that go into selecting the power rating of a resistor. For more accurate design applications, you may have to weigh in other factors, such as whether resistors are grouped or enclosed within a box, whether they are fan cooled, or whether they are pulsed. In that case, you can use the chart in Fig. 3.55 to calculate an approximate power rating.

A kit of 1/4-, 1/2- and 1-W carbon-film and metal-film resistors is usually sufficient for most applications. A selection of 1- to 5-W wirewound resistors often comes in handy, and in rare instances you may have to resort to 10- to 600-W aluminum-housed wirewound or thick-film power resistors.

Note that in some applications (e.g., input stages to amplifiers), an effect called *contact noise* can be reduced by increasing the power rating (size) of the resistor—see the section on noise later in this chapter.

Temperature Coefficient of Resistance (TCR or TC)

This tells you the amount of resistance change that occurs when the temperature of a resistor changes. TC values are typically expressed as parts per million (ppm) for each degree centigrade change from some nominal temperature—usually room temperature (25°C).

So a resistor with a TC of 100 ppm will change 0.1 percent in resistance over a 10°C change, and will change 1 percent over a 100°C change (provided the temperature change is within the resistor's rated temperature range, e.g., –55 to +145°C, measured at 25°C room temperature). A *positive* TC means an increase in resistance with increasing temperature, while a *negative TC* indicates a decreasing resistance with an increase in temperature.

Here's another example: A 1000-Ω resistor with a TC of +200 ppm/°C is heated from 27°C to 50°C. The change in resistance in parts per million is:

$$(200 \text{ ppm}/°C) \times (50°C - 27°C) = 4600 \text{ ppm}$$

meaning the new resistance is:

$$1000 \, \Omega \times (1 + 4600 \text{ ppm}/1{,}000{,}000 \text{ ppm}) = 1004.6 \, \Omega$$

There are a wide range of TC values available from ± 1 ppm/°C to ± 6700 ppm/°C.

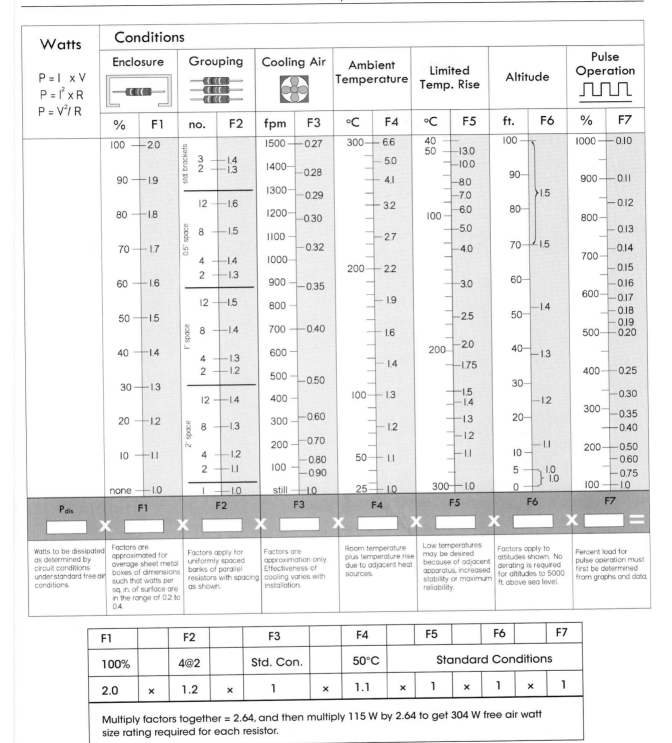

FIGURE 3.55 Calculator used for determining approximate power rating of a resistor, based on enclosure, grouping, cooling, ambient temperature, limited temperature rise, altitude, and pulsed load conditions. Example: Four resistors, each dissipating 115 W are to be mounted in a group, with spacing of two inches surface-to-surface, ambient to 50°C. The resistors are totally enclosed. Other factors are standard.

Specifying TC is important in applications where the change in resistance with temperature changes must be small. Equally important may be applications where a specific TC is required (e.g., temperature compensation circuits). Typically, there are two contributors to temperature-related resistance changes: increased temperature due to dissipated power and ambient temperature variations. The exact correspondence between dissipation, ambient temperature, and the TC is not a linear function, but rather resembles a bell-shaped or S-shaped curve. Therefore, for serious high-precision applications, you'll have to check out the manufacturer's resistor data sheets.

When designing circuits, you have to ask yourself which instances require high stability and low TC. For example, using a carbon-composition resistor in a stable, high-precision circuit is asking for trouble due to its high TC and weak tolerance. Using a more stable metal-film resistor with a TC of 50 to 100 ppm/°C will improve accuracy and stability considerably. Precision-film resistors have greatly improved accuracy and TCs—20, 10, 5, or 2 ppm/°C and accuracies as good as 0.01 percent. Carbon-film resistors have much higher TCs than metal-film resistors—around 500 to 800 ppm/°C. It's easy to mistakenly insert a drifty carbon-film resistor for the intended metal-film type. Carbon-film resistors are also unique among the major resistor families, since they alone have a negative temperature coefficient. They are often used to offset the thermal effects of the other components.

As a note: often, matching TCs for pairs or sets of resistors is more important than the actual TC itself. In these cases, matched sets are available that ensure that the resistance values of the set track in the same magnitude and direction as operating temperature changes.

Frequency Response

Resistors are not perfect—they have inherent inductive and capacitive features that can alter the device's impedance, especially as the applied ac voltage frequency increases. (Figure 3.56 shows a frequency model of a resistor.) For this reason, it's possible for a resistor to perform like an R/C circuit, a filter, or an inductor. The primary cause of these inductive and capacitive effects results from imperfect interior layout of the resistive element as well as the resistor's leads. In spiraled and wirewound resistors, these inductive and capacitive reactances are created by the loops and spaces formed by the spirals or turns of wire. In pulse applications, these reactive distortions result in a poor replication of the input. It is probable that a 20-ns pulse will be completely missed by a wirewound resistor, while a foil resistor, due to its superior design, achieves full replication in the time allotted. As the frequency increases, inductive reactance becomes more prevalent.

Even though the definition of the useful frequency range of a resistor is application dependent, typically the useful range of the resistor is the highest frequency at which the impedance differs from the resistance by more than the tolerance of the resistor. Typical reactive values for these special designs are less than 1 μH for a 500-Ω resistor, and less than 0.8-pF capacitance for a 1-MΩ resistor. A typical fast-rise-time resistor has a rise time of 20 ns or less. (*Rise time* is an associated parameter relating the resistor's response to a step or pulse input.)

Wirewound resistors are notorious for their poor frequency response due to their internal coil windings. In composition resistors, frequency response also suffers from the capacitances that are formed by the many conducting particles that are held in contact by the dielectric binder. The most stable resistors for high-frequency operations are film resistors. The impedance for film resistors remains constant until around 100 MHz and then decreases at higher frequencies. In general, a resistor with a smaller diameter will have better frequency response. Most high-frequency resistors have a length-to-diameter ratio between 4:1 and 10:1. Manufacturers often supply data sheets that show the frequency response of their resistors. Impedance analyzers can also aid in modeling a resistor's frequency response.

Frequency Response of Foil Resistor

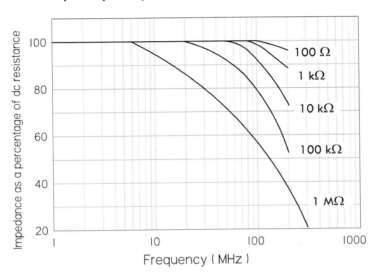

FIGURE 3.56

Noise

Resistors exhibit electrical noise in the form of small ac voltage fluctuations when dc voltage is applied. Noise is extremely difficult to measure accurately and doesn't affect the value of the resistor but can have a devastating effect on low-level signals, digital amplifiers, high-gain amplifiers, and other applications. Noise in a resistor is a function of the applied voltage, physical dimensions, and materials. The total noise is a sum of Johnson noise, current-flow noise, noise due to cracked bodies, loose end caps, and leads. For variable resistors, noise can also be caused by the brief jumping motion of the wiper as it moves along the resistive element (e.g., wire turns).

There are three main types of resistor noise: thermal, contact, and shot noise. Thermal noise is dependent mainly on temperature, bandwidth, and resistance, while shot noise is dependent on bandwidth and average dc current. Contact noise is dependent upon average dc current, bandwidth, material geometry, and type. The following is a brief summary of the various kinds of resistor noise.

Johnson noise is temperature-dependent thermal noise. Thermal noise is also called "white noise" because the noise level is the same at all frequencies. The magnitude of thermal noise, V_{rms}, is dependent on the resistance value and the temperature of the resistance due to thermal agitation:

$$V_{RMS} = \sqrt{4kRT\Delta f}$$

where V_{rms} is the root-mean-square value of the noise voltage (in volts), K is Boltzmann's constant (1.38×10^{-23} J/K), R is the resistance (in ohms), T is the temperature (in Kelvin), and Δf is the bandwidth (in hertz) over which the noise energy is measured. For resistors, the thermal noise is equivalent for equal valued resistors, regardless of material type (carbon, metal film, etc.). The only way to reduce thermal noise is to reduce the resistance value. This is why you try to avoid using a 10-MΩ resistor on the input stages of amplifiers.

Current noise varies inversely with frequency and is a function of the current flowing through the resistor and the value of the resistor—see Fig. 3.56. The magnitude of the current noise is directly proportional to the square root of the current. The current noise magnitude is usually expressed by a noise index given as the ratio of the root-mean-square current noise voltage (V_{rms}) over one decade bandwidth to the average voltage caused by a specified constant current passed through the resistor at a specified hot-spot temperature.

$$NI = 20 \log_{10} \left(\frac{\text{Noise voltage}}{\text{dc voltage}} \right)$$

$$V_{RMS} = V_{dc} \times 10^{NI/20} \sqrt{\log (f_2/f_1)}$$

where NI is the noise index, V_{dc} is the dc voltage drop across the resistor, and f_1 and f_2 represent the frequency range over which the noise is being computed. Units of noise index are μV/V. At higher frequencies the current noise becomes less dominant compared to Johnson noise.

Contact noise is directly proportional to a constant that depends on resistor material/size and upon the average dc current. In amplifiers, for example, using a larger 2-W carbon-composition resistor will improve performance over that of a ½-W equivalent under the same conditions. The predominant noise in carbon composition, carbon film, metal oxide and metal film is composed of contact noise, which can be very large at low frequencies due to a $1/f$ frequency characteristic. Wirewound resistors do not have this noise—only resistors made of carbon particles. If no current (ac or dc) flows in the resistor, the noise is equal to the thermal noise. The contact noise increases as the current is increased. This means that for low-noise operation, the dc or ac current should be kept low. The material and geometry of a resistor greatly affects the contact noise, and therefore doubles the power rating of the resistor, which increases the size and area, and will reduce the contact noise generated by the resistor.

Shot noise is dependent upon current—the more average dc current through a resistor, the more noise you get. To reduce this type of noise, dc current levels must be kept low. This is often put into action in the first amplifier stage or in low-level stages such as reverb-recover amps in audio, where it is the most critical. It's best to use a wirewound or metal-film resistor in these applications, unless you are making a high-frequency amp where inductance of the wirewound resistor comes into play.

The best resistors for low-noise applications are precision resistors. Precision wirewounds tend to be the quietest, having only thermal noise (unless terminations are faulty), but they aren't readily available in large resistance values and are usually inductive. Precision film resistors also have extremely low noise—you'll have to consult the manufacturers' websites to see how noise-free modern film resistors have become. Next down the line comes metal oxide, followed by carbon film, and, last, carbon composition. Composition resistors show some degree of noise due to internal electrical contacts between conducting particles held together with the binder. And remember that, when designing critical circuits, you'll cut down on contact noise (not applicable to wirewounds) by using the resistor with the largest practical wattage.

Don't forget potentiometers, which are most commonly carbon composition. They generally have large values (e.g., 1 MΩ for volume control), which make them major sources of noise in an amplifier. For absolute lowest noise, conductive plastic element pots should be used, employing the lowest practical value and largest practical power rating.

Voltage Coefficient of Resistance

Resistance is not always independent of the applied voltage. The voltage coefficient of resistance is the change in resistance per unit change in voltage, expressed as a percentage of the resistance at 10 percent of rated voltage. The voltage coefficient is given by the relationship:

$$\text{Voltage coefficient} = \frac{100 \times (R_1 - R_2)}{R_2(V_1 - V_2)}$$

where R_1 is the resistance at the rated voltage V_1 and R_2 is the resistance at 10 percent of rated voltage V_2. Voltage coefficient is associated with carbon-composition and carbon-film resistors, and is a function of the resistor's value and its composition.

Stability

This is defined as the repeatability of resistance of a resistor when measured at a reference temperature and subjected to a variety of operating and environmental conditions over time. Stability is difficult to specify and measure, since it is application dependent. Generally, wirewound and bulk-metal resistor designs are best, while designs using composition resistors are least stable. For highest resistance stability, it's best to operate critical resistors with limited temperature rise and limited load level. Changes in temperature alternately apply and relieve stresses on the resistive element, thus causing change in resistance. The wider these temperature variations and the more rapid these changes are, the greater the change in resistance. If severe enough, this can literally destroy a resistor. Humidity can also alter the resistance by causing the insulation of the resistor to swell, thus applying pressure to the resistive element.

Reliability

This is the degree of probability that a resistor will perform its desired function. It is typically rated as mean time between failures (MTBF) or failure rate per 1,000 hours of operation. Reliability isn't usually a critical specification for most general-purpose applications. It often does pop up in critical applications, such as those used by the military.

Temperature Rating

This is the maximum allowable temperature at which the resistor may be used. Often, two temperatures are used—one for full loads up to, say, +85°C derated to no load at +145°C. Temperature range may be specified, for example, from −55°C to +275°C.

3.5.5 Types of Resistors

There are numerous resistor technologies out there, with new technologies springing up all the time. The major technologies include carbon film, metal film, thick film, thin film, carbon composition, wirewound, and metal oxide. When selecting resistors for an application, you generally specify whether the resistor is to be a precision resistor, semiprecision resistor, general-purpose resistor, or power resistor.

Precision resistors have low voltage and power coefficients and excellent temperature and time stabilities, along with low noise and very low reactance. These resistors are available in metal-film or wire constructions and are typically designed for circuits having very close resistance tolerances on values.

Semiprecision resistors are smaller than precision resistors and are used primarily for current-limiting or voltage-dropping functions. They have long-term temperature stability.

General-purpose resistors are used in circuits that do not require tight resistance tolerances or long-term stability. For general-purpose resistors, initial resistance variations may be in the neighborhood of 5 percent, and the variation in resistance under full-rated power may approach 20 percent. Typically, general-purpose resistors have a high coefficient of resistance and high noise levels. However, good-quality metal film resistors are low cost and often used as general-purpose resistors.

Power resistors are used for power supplies, control circuits, and voltage dividers where operational stability of 5 percent is acceptable. Power resistors are available in wirewound and film construction. Film-type power resistors have the advantage of stability at high frequencies and have higher resistance values than wirewound resistors for a given size.

The following provides finer details highlighting the differences between the various available resistors.

Precision Wirewound

Precision wirewound resistors are very stable resistors manufactured with high tolerances. They are made by winding wire of nickel-chromium alloy onto a ceramic tube covered with a vitreous coating. They are designed to have a very low temperature coefficient of resistance (as low as 3 ppm/°C) and can achieve accuracies up to 0.005 percent. They are usually expected to operate in a temperature range from –55 to 200°C, with a maximum operating temperature of 145°C. Life is generally rated at 10,000 hours at rated temperature and load, though this can increase if operated below rated temperature. The allowable change in resistance under these conditions is about 0.10 percent. In terms of noise, there is little—only contact noise. The power-handling capability is generally low, but high-power versions are available with heat sinks.

Because of the wire-winding nature, these resistors have a component of inductances as well as capacitance associated with them. They tend to be inductive at lower frequencies and somewhat capacitive at higher frequencies, regardless of resistance value. They also have a resonant frequency—with a very low Q value. For this reason, they are unsuitable for operation above 50 kHz—forget about RF applications. Precision wirewounds are not to be used for general-purpose work, but are reserved for high-accuracy dc applications such as high-precision dc measuring equipment and as reference resistors for voltage regulators and decoding networks. (*Note:* There are certain precision wirewounds listed in manufacturers' catalogs as "type HS" wirewounds. These resistors have a special winding pattern that can greatly cut down on the inductance of the winds. There are two different types of HS wirewounds: one type has almost zero inductance but greatly increased interwinding capacitance; the other type has low inductance and low capacitance and is well suited for fast-settling amplifiers.)

Once considered the best and most stable resistors, precision wirewounds now have a competitor—precision film resistors, which can match them in most every regard.

Power Wirewound

Power wirewound resistors are similar to their precision counterparts but are designed to handle a lot more power. They will handle more power per unit volume than any other resistor. Some of the most powerful are wound similar to heater elements and require some form of cooling (e.g., fans or immersion in liquids such as mineral oil or high-density silicone liquids). These resistors are wound on a winding form, such as a ceramic tube, rod, heavily anodized aluminum, or fiberglass mandrel. The cores on which the windings are made have high heat conductivity (Steatite, Alumina, beryllium oxide, etc.). They come in various shapes—oval, flat, cylindrical—most shapes designed for heat dissipation. Chassis-mount wirewounds are generally cylindrical power resistors wound on a ceramic core molded and pressed into an aluminum heat sink and usually with heat-radiating fins. These are designed to be mounted to metal plates or a chassis to further conduct heat, which results in a rating about five times the normal power rating. Power wirewounds come in a variety of different accuracy and TCR ratings.

Metal Film

In applications that involve fast rise times (microseconds) or high frequencies (megahertz), metal-film resistors are usually the best. They are also quite cheap and come in small sizes (e.g., surface-mount). Metal-film resistors are often considered the best compromise of all resistors. Once considered less accurate and stable than wirewounds, the technology has greatly improved, with special precision metal-film resistors reaching TC values as low as 20, 10, 5, and even 2 ppm/°C, with accuracies as good as 0.01 percent. They also have much less inductance than wirewounds and are smaller in size and less expensive. When compared to carbon-film resistors, they have lower TCs, lower noise, linearity, and better frequency characteristics and accuracy. They also surpass carbon-film resistors in terms of high-frequency characteristics. Carbon-film resistors do, however, come with higher maximum resistance values.

A metal-film resistor is made from a base metal that is vaporized in a vacuum and deposited on a ceramic rod or wafer. The resistance value is then controlled by careful adjustment of the width, length, and depth of the film. The process is very exacting, resulting in resistors with very tight tolerance values. Metal-film resistors are used extensively in surface-mount technology.

Carbon Film

Carbon-film resistors are the most common resistor around. They are made by coating (dipping, rolling, printing, or spraying) a ceramic substrate with a special carbon-film mixture. The thickness and percentage of the carbon mixture roughly determine the resistance. To tailor resistances to precise values the ceramic pieces can be cut to a specific length. Further refinement is accomplished by cutting a spiral trimming groove—see Fig. 3.53. An alternative method of producing carbon film is to mechanically apply carbon dust dispersed in a curable polymeric binder. The material is painted on the substrate in a spiral pattern and cured at a moderately elevated temperature.

Carbon-film resistors, with 1 percent tolerances, are normally manufactured with spiral cuts and have the same kind of voltage-overload limitations as metal-film types. Though these resistors are very popular, they are drifty (TC values around 500 to 800 ppm/°C) and should not be used in circuits where metal-film resistors are intended. In other words, don't confuse the two when building or replacing blown components. Carbon-film resistors have many of the same characteristics as carbon-composition resistors—such as being noisy and having a voltage coefficient; they outperform carbon-composition resistors in terms of lower TCR ratings and tighter tolerances. Resistor types include general-purpose, through-hole, and surface-mount devices. They also come in specialty types, such as high-power, high-voltage, and fusible. Tolerances of 1 percent or even better can be achieved; however, caution must be used in getting tight tolerances for this type of resistor because the TC, voltage coefficient, and stability may mean that it is good only for that tolerance at the time it was installed. The TC of carbon-film resistors is in the neighborhood of 100 to 200 ppm and is generally negative. Frequency response of carbon-film resistors is among the best—far better than wirewounds, and much better than carbon composition.

Carbon Composition

Carbon-composition resistors, though not as popular as they once were, still find use in noncritical applications. They are composed of carbon particles mixed with a binder. The resistance value is varied by controlling the carbon concentration. This mixture is molded into a cylindrical shape and hardened by baking. Leads are attached axially to each end, and the assembly is encapsulated in a protective coating. Composition resistors are economical and exhibit low noise levels for resistances of about 1 MΩ. Composition resistors are usually rated for temperatures in the neighborhood of 70°C for power ranging from ⅛ to 2 W. They have end-to-end shunted capacitance that may be noticed at frequencies in the neighborhood of 100 kHz, especially for resistance values above 0.3 MΩ.

However, due to poor tolerances—from 5 to 20 percent—carbon-composition resistors should not be used in critical applications. Due to their construction, they generate considerable noise that varies depending on resistance value and the package size (though above 1 MΩ they exhibit low noise).

Though composition resistors have many poor characteristics, they do quite well in overvoltage conditions. Where a metal-film resistor's spiral gap would be zapped (breakdown causing it to short and destroy itself) during a severe overvoltage condition, a carbon-composition resistor wouldn't be so wimpy. A carbon-composition resistor uses a large chunk of resistive material that can handle large overloads for a short time without any flashover effects (shorting). So if you're planning to discharge a high-voltage capacitor through a series resistor where tolerances and such aren't important, carbon composition isn't a bad choice. Power-handling capability in relation to physical size is greater than with precision wirewounds but less than with power wirewounds.

Bulk-Metal Foil Resistors

Foil resistors are similar in characteristics to metal films; they have better stability and lower TCRs (approaching those of precision wirewounds), and accuracy about that of metal-film resistors. High-precision versions can achieve tolerances as low as 0.005 percent and TCR values of 0.2 ppm/°C. Their main shining point is their excellent frequency response. Foil resistors are manufactured by rolling the same wire materials as used in precision wirewound resistors to make thin strips of foil. This foil is then bonded to a ceramic substrate and etched to produce the value required. Their main disadvantage is limited high-value resistance values—less than those of metal films.

Filament Resistors

The filament resistors are similar to what's called "bathtub boat resistors," except they are not packaged in a ceramic shell (boat). The individual resistive element with the leads already crimped is coated with an insulating material, generally a high-temperature varnish. These are used in applications where tolerance, TCR, and stability are not important but the cost is. The cost of this type is slightly higher than for carbon composition and the electrical characteristics are better.

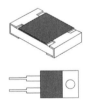

Thin- and Thick-Film Resistors

Thin-film resistors are made by depositing an extremely thin layer of NiCr resistive film (less than 1 μm) on an aluminum oxide substrate, while using NiCu materials as conducting electrodes. Thin-film technology offers extreme precision and stability (tight tolerances and low TCR values). However, these resistors have relatively limited surge capabilities due to the low mass of the resistive material. Thin-film resistors are designed as small surface-mount devices used in PCB designs and are frequently used as microwave passive and active power components such as microwave power resistors, microwave power terminations, microwave resistive power dividers, and microwave attenuators.

Thick-film resistors, in contrast to thin-film resistors, use a thicker film of RuO_2, and have PdAg electrodes. These materials are also mixed with glass-based material to form a paste for printing on the substrate. The thickness of the printing material is usually 12 μm. Thick-film resistors also exhibit decent precision and stability, perhaps approaching those of thin-film resistors; however, they far exceed thin films in terms of maximum surge capacity—one to two orders of magnitude difference. Thick-film resistors come in two-lead packages and in surface-mount form. Some thick-film resistors are designed as power resistors.

Both thin-film and thick-film technologies are constantly improving, and it is difficult to specify all characteristics. Your best bet is to consult manufacturers' data sheets for more details.

Power-Film Resistors

Power-film resistors are similar in manufacture to their respective metal-film or carbon-film resistors. They are manufactured and rated as power resistors, with power rating being the most important characteristic. Power-film resistors are available in higher maximum values than power wirewound resistors and have a very good frequency response. They are generally used in applications requiring good frequency response and/or higher maximum values. Generally, they are used for power applications, where tolerances are wider, and the temperature ratings are changed so that under full load the resistor will not exceed the maximum design temperature. Also, the physical size of the resistor is larger, and in some cases the core is made from a heat-conductive material attached to a heat sink to dissipate heat more efficiently.

Metal Oxide (Power-Metal-Oxide Film, Flameproof)

Metal-oxide resistors contain a resistance element formed by the oxidation reaction of a vapor or spray of tin-chloride solution on the heated surface of a glass or ceramic rod. The resulting tin-oxide film is adjusted to value by cutting a helix path through the film. These resistors can sustain high temperatures and electrical overloads, and have moderate-to-precision characteristics. Resistor types in this class include high-power and flameproof axial through-hole and surface-mount types. Axial versions are either blue in color or white. The outer shell of these resistors, which is flameproof like the interior, is also resistant to external heat and humidity. Metal-oxide resistors can be used to replace carbon-composition components in some applications. They are ideal for pulse-power applications. Small-sized power-type metal-oxide resistors come in a 0.5- to 5-W range, with standard tolerances of ±1 to ±5 percent and TCRs around ±300 ppm/°C. Metal-oxide resistors are used in general-purpose voltage dividers, RC timing circuits, and as pullup and pulldown resistor surge applications (e.g., RC snubber circuits, current-limiting circuits, and overload ground lines). They also come with maximum resistance values exceeding those of wirewound resistors. In general, they have decent electrical and mechanical stability and high reliability.

Fuse Resistors (Carbon Composition, Power Oxide, Metal Film)

A fuse resistor acts as both a resistor and a fuse. Fuse resistors are designed to open-circuit (fuse) when subjected to a large surge current or fault condition. They are specially spiraled to provide the fusible function with flame-retardant coating. The fusing current is calculated based on the amount of energy required to melt the resistive material (the melt temperature plus the amount of energy required to vaporize the resistive material). These resistors will typically run hotter than a normal precision or power resistor, so that a momentary surge will bring the resistive element up to fusing temperature. Some designs create a hot spot inside the resistor to assist in this fusing. The major unknown when using fuse resistors is the heat transfer of the materials, which can be quite significant for pulses of long duration and is very difficult to calculate. Mounting fuse resistors is critical, since this will affect the fusing current. Many fuse resistors are made to mount in fuse clips for more accurate fusing characteristics. They come in a variety of types, including carbon film, metal film, thin film, and wire-wound fusible. Fuse resistors are widely used in constant voltage and overload protection circuits found in battery chargers, TV sets, cordless phones, PC/CPU coolers, and so on. Like traditional fuses, they come in fast- and slow-burn types.

Chip Resistor Arrays

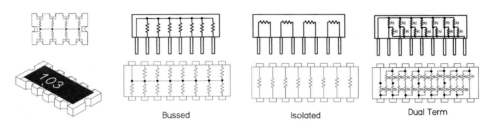

Bussed   Isolated   Dual Term

Resistor arrays contain any combination of two or more resistive elements produced on a single substrate. The resistive elements can be constructed using thick-film or thin-film technologies. These arrays come in SIP and DIP packages as well as leadless surface-mountable packages with solderable terminations. Various circuit schematics are available, including isolated resistors, single common and dual common bused resistors. Resistor arrays are used for a wide range of applications where economy of space and weight and placement costs are at a premium. Tolerances are 1 percent and 5 percent, temperature coefficients range from 50 ppm to 200 ppm, and power capabilities compare to individual resistors of similar size.

Cement Resistors

These resistors are designed as power resistors with the added provision of being heat and flame resistant. Typical power ratings range from 1 W to 20 W or more. Tolerances are around 5 percent, with TCR ratings of around 300 ppm/°C.

Zero-Ohm Resistors (Zero-Ohm Jumpers)

These are really nothing more than a piece of wire used for crossovers, permanent jumpers, or program jumpers (manual switch circuitry) in PCB design. They look like a signal diode with a single black stripe in the center—not to be confused with a diode that has its stripe nearer to one end. The single black stripe is meant to signify a 0-Ω value. Advantages to using these zero-ohm resistors over simple wires include ease of handling for mechanized PCB placement machinery, very low jumper-to-jumper capacitance (suitable for high-speed data lines), small footprint, and overall improvement of PC board performance.

3.5.6 *Variable Resistors (Rheostats, Potentiometers, Trimmers)*

Variable resistors are often called *potentiometers*, or "pots" for short, because one very common use for them is as an adjustable voltage divider. For many years they were called "volume controls," because another very common use was for adjusting the audio volume produced by amplifiers and radio and TV receivers. Another early name for essentially the same component (when it was used simply as a variable resistance) was "rheostat," meaning a device to set the flow (of current). See Fig. 3.57.

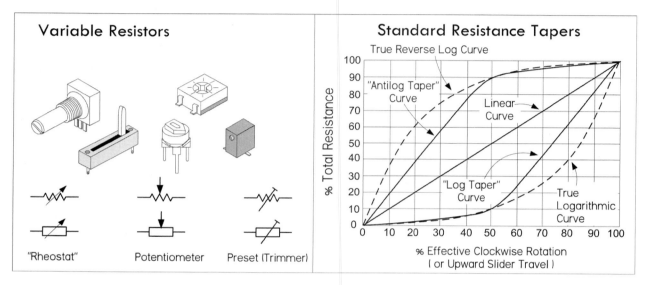

FIGURE 3.57

Pots are made in a variety of physical forms, with different kinds of resistance elements. Some pots are designed to handle frequent manual adjustment via a control knob, while others are designed to be adjusted only occasionally with a screwdriver (or similar tool) for fine-tuning a circuit. The latter type is usually referred to as a *preset pot*, or simply a *trimpot*. Most rotary pots made for manual control offer a total rotation of about 270°—¾ of a single turn. Such a limited range, however, can make accurate settings difficult, so multirevolution pots are also available. These pots have a resistance element arranged in a spiral or helix, and the wiper moves along as the control spindle is turned through multiple revolutions (typically, 10 or 20). It is possible to really home in on a resistance value with one of these pots. Multirevolution pots are rather expensive when compared to single-turn pots. For applications that require a logarithmic response (such as audio applications), you can get a pot with a logarithmic taper, as opposed to a linear taper. (Actually most logarithmic pots don't have a true logarithmic response, as shown in Fig. 3.57. It is difficult to design such a device cheaply; making one with a near-logarithmic response isn't so expensive.) Reverse logarithmic and antilog pots are also available—these are explained in the next section, on pot characteristics.

Trimpots can be made in circular, multiturn circular, linear-slider, and multiturn linear-slider form. Low-cost varieties generally use an open construction where the resistance element and slider are fully exposed and therefore are prone to contamination by dust and moisture. Higher-quality trimpots are generally housed inside a small plastic case, which is often hermetically sealed. Some multirevolution trimpots use a worm drive system with a circular element, while others use a linear element with a slider driven via a lead-screw. In both cases, the reduction drive is designed to have very low backlash, providing smooth and accurate adjustment.

Pots can be combined together and actuated by using a common spindle. Such pots are referred to as *ganged pots*. Generally, only two pots are combined this way, but it's possible to have more. Dual ganged pots with logarithmic taper are often used in amplifier design when dealing with stereo, where you have two distinct signals.

There are various materials used in the construction of pots, such as carbon, Cermet, conductive plastic, and simple wire. Many of the characteristics that apply to fixed resistors also apply to variable resistors of the same nature—see the section on fixed resistor characteristics. Characteristics unique to pots—such as resolution, resistance taper, hop-on/hop-off resistance, and contact resistance—will differ, depending on material makeup. We will discuss these characteristics in a moment.

There are basically two ways a pot can be used in electronics. It can be used to adjust current levels or it can be used to adjust voltage levels. Figure 3.58 shows the basic setups for each scenario.

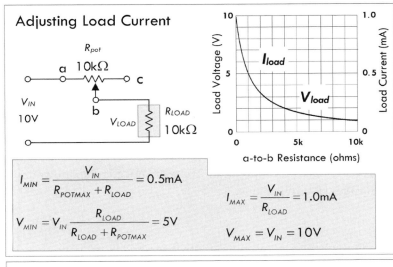

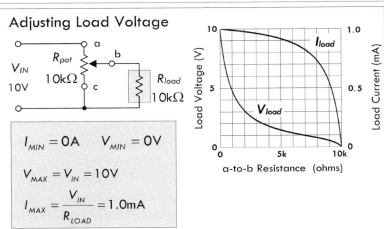

Adjusting load current: Here, a pot's variable resistance element (a-to-b) is in series with a load. Adjusting manual control of the pot alters the current to the load. The current as a function of pot resistance is:

$$I = \frac{V_{IN}}{R_{POT} + R_{LOAD}}$$

Notice in the graph that the load current follows a curve similar to that of the voltage. The equations in the figure display the maximum and minimum voltages and currents that can be applied across and through the load, where R_{POTMAX} represents the maximal pot resistance—which in this example is 10 kΩ.

Adjusting load voltage: The second configuration is essentially a variable voltage divider, which is used for adjusting the voltage applied to the load. Notice that in the graph, the load current doesn't fall as quickly as in the previous configuration. In fact, from 0 to 5 kΩ the current falls about only 1/10 its maximum value. However, from 5 kΩ on, the drop grows significantly.

FIGURE 3.58

3.5.7   Potentiometer Characteristics

Resistance Taper

Pots have either linear or log tapers. Pots with linear tapers have a linear relationship between the wiper position and the resistance; for example, moving the wiper by 10 percent down the resistive element changes the resistance by 10 percent. (See Fig. 3.57.) On the other hand, a potentiometer with a log taper follows a logarithmic resistance change with wiper position, as shown in Fig. 3.57.

Naturally, the main use of common log taper pots is for controls that need to adjust a quantity in an approximate logarithmic fashion—such as for audio. Linear pots are used for most other applications. As an example, if you use a linear pot for a volume control, you'd run into problems; its adjustment grows far too rapidly as the pot is turned up from zero, while the rest of the pot's rotation doesn't have much control—the pot's useful range as a volume control is squeezed into the first 60° or so of rotation, making it hard to set the right level. The log taper pot, on the other hand, is perfect as a volume control, since its logarithmic change with position matches that of the human ear's logarithmic response to sound levels.

Now, in reality, most modern log taper pots don't have a true logarithmic characteristic, but instead follow a rough approximation of a logarithmic curve, as shown in Fig. 3.57. The cost of manufacturing a true logarithmic taper is too high; it's cheaper to coat a strip of material with two resistive elements of different resistive composition. This is quite acceptable as a volume control and is close to a two-slope resistance element, with the transition at about 50 percent rotation. There are indeed pots with true logarithmic response, which are made using a wirewound element on a tapered form or are made with a bulk-metal element with a carefully granulated tapered pattern.

Also shown in Fig. 3.57 are the curves for rotary reverse logarithmic and antilog taper pots, which are virtually the same as true logarithmic and log taper pots but made for the opposite, or anticlockwise, operation. These are not common nowadays, but are still available for special applications.

Resolution

This represents the smallest change in tapping ratio that can be made by moving the pot's wiper. The resolution of wirewound pots tends to be fairly poor because the element is wound from discrete turns of resistance wire and the wiper contact can usually only slide from one turn to the next. The output of the pot, therefore, tends to vary in small regular steps, each corresponding to the voltage drop in one turn of the elements.

Pots using an etched bulk-metal element tend to have similar problems. However, pots that use a carbon-composition, hot-molded-carbon, or Cermet element tend to have somewhat better resolution because the resistance of their element is more finely graduated. Where high resolution is needed, multiturn pots tend to be more widely used than single-turn pots. It is argued that multiturn pots don't have superior settability. The next time you need a pot with superior settability, evaluate a multiturn pot and a single-turn pot. Set each one to the desired value, tap the pots with a pencil, and tell which one stays put. Normally, it is expected that a multiturn pot, whether it has a linear or a circular layout, is better, but in reality this is not true—it can be two to four times worse than a single-turn pot because the mechanical layout of a single-turn pot is more stable and balanced.

Contact Resistance

This represents the contact resistance between the pot's wiper and the resistance element, which has an effect on the resolution. Contact resistance also affects the noise generated by the pot, both when it's being adjusted and when it's simply in a fixed setting. The type of contact style and material used in pots affects contact resistance. For example, many carbon-composition pots use a simple wiper stamped from nickel-plated spring steel, with multiple fingers to give parallel wiping contacts in order to reduce contact resistance. Higher-power wirewound pots may use a wiper with a brush made from a single block of carbon, but the block is made with a loading of copper powder to keep the contact resistance as low as possible. High-grade pots with Cermet or bulk-metal elements generally have a multifinger spring metal wiper of gold-plated phosphor bronze or steel. Many small trimpots of the cheaper variety have a simple spring metal wiper with a pressed dimple contact that touches the element. This is okay when the trimpot doesn't get a lot of adjustment, but with frequent adjustment, the contact resistance tends to rise.

Hop-on and Hop-off Resistance

Most pots, whether rotary or linear-slider in design and regardless of type of resistance element, tend to have metal contact strips at the ends of the fixed element. The wiper contact generally touches and rests on these metal strips when it's at either extreme of its travel. However, when the wiper is moved away from these extremes, it soon hops on or off the actual resistance element. Ideally, the change in resistance that occurs at the hop-on and hop-off should be zero, so that there are no sudden changes when the pot is used as a volume control, for example. However, it isn't easy to manufacture a pot without this flaw; the best generally keep the hop-on and hop-off resistance below about 1 percent of the total value of the resistance element. This is so low that it usually can't be detected in most audio and similar applications.

Pot Markings

As well as being marked with a character string indicating the total value of the resistance element—for example, 100 K or 1 M—the case of a pot generally also carried a code letter showing its resistance taper curve. Today most pots are marked according to the simplified taper coding system adopted by Asian component manufacturers:

A = log taper

B = linear taper

However, in older equipment you may come across pots marked according to an earlier taper coding system:

A = linear taper

C = log or audio taper

F = antilog taper

Notice the possible source of confusion!

Notes on Pots

Don't exceed your pot's *I* and *V* ratings. If you put a constant voltage between the wiper and one end and turn the resistance way down, you will exceed the maximum wiper current rating and soon damage or destroy the wiper contact. Note that the power rating of most variable resistors is based on the assumption that the power dissipation is uniformly distributed over the entire element. If half of the element is required to dissipate the device's rated power, the pot may last for a short while. However, if a quarter of the element is required to dissipate the same amount of power, the pot will fail quickly. Also, some trimming pots are not rated to carry any significant dc current through the wiper. This dc current, even a milliamp, could cause electromigration, leading to an open circuit or noisy, unreliable wiper action. Carbon pots are not likely to be degraded by such a failure.

Digital Pots

Digital pots are basically analog variable resistors whose value is set by applying a digital code to their digital input leads. The actual inner circuitry is an array of digitally controlled switches that can add or subtract segments of an integrated array of single polycrystalline resistor elements. Depending on the digital code applied, a few or all the internal integrated resistors can add in series to a desired resultant resistance. Some digital pots are quite advanced, allowing you to store wiper positions in memory. Other control signals may enable the chip, increment or decrement the resistance a given amount, and so on. Digital pots can be used as three-terminal devices or as two-terminal devices. The most common is a three-terminal device, where the digital pot acts as a voltage divider. The two-terminal method is used to make the digital pot a variable current-control device—say, to step the amount of current flow through a diode. These devices are very handy in many types of analog circuits that interface with digital, such as amplifiers where the gain is determined by the digital pot's resistance in the feedback circuit. Another example is in filter design, where the digital pot replaces the resistance used to set the cutoff frequency. It is worth purchasing a few of these devices once you start playing with microcontrollers.

3.6 Capacitors

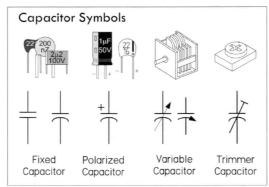

Capacitor Symbols

Fixed Capacitor | Polarized Capacitor | Variable Capacitor | Trimmer Capacitor

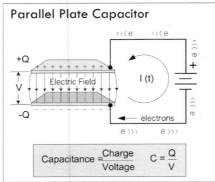

Parallel Plate Capacitor

$$\text{Capacitance} = \frac{\text{Charge}}{\text{Voltage}} \qquad C = \frac{Q}{V}$$

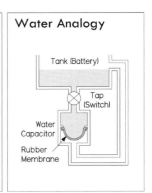

Water Analogy

FIGURE 3.59

Capacitors perform a number of functions in electronics. One major function is simple *energy storage*, where charge from an applied current is stored within the capacitor and later released back into the circuit as useful current. The rate of charging and discharging can be controlled by placing a resistor in series with the capacitor. This effect is often used in high-current discharge circuits (photoflashes, actuators, etc.), as well as small energy backup supplies for low-power memory ICs. It is also used to smooth out power supply ripple, control timing in ICs, and alter the shape of waveforms.

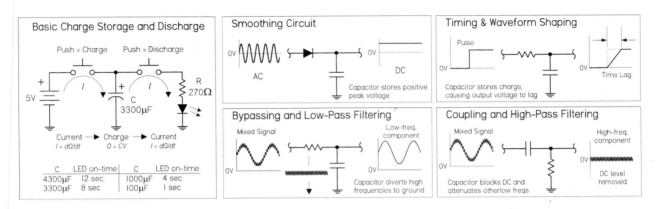

FIGURE 3.60

The second major function of a capacitor when placed in series with a signal path is to block dc while allowing ac signal components to pass. A capacitor used in this way is referred to as a *dc-blocking or ac-coupling capacitor*. At dc, a capacitor's impedance is ideally infinite—no current flows, no dc levels within a mixed signal are passed. However, if an ac signal is applied, the capacitor's impedance becomes a finite value, dependent on the frequency of the signal. The higher the frequency, the lower the impedance—ideally. So, in essence, a series capacitor can be used to couple two circuits together without introducing unwanted dc shifts into the next stage, and can control how much of a given frequency signal gets through—it controls the attenuation.

Now, a capacitor placed in parallel with a signal path (i.e., to ground) has an effect opposite that of the coupling capacitor. Instead, it acts as a *decoupling capacitor*, allowing dc to continue along the path, while diverting high-frequency signal components to ground—the capacitor acts as a low-impedance path to ground. A similar effect, known as *bypassing*, is used when a capacitor is placed across a particular circuit element to divert unwanted frequencies around it. Decoupling and bypassing become fundamental when removing unwanted random high-frequency ripple and other undesired alterations within a supply voltage (or voltage-critical location) caused by random noise, or sudden current demands generated by accompanying circuit elements. Without decoupling and bypassing, many sensitive circuits, especially those incorporating digital logic ICs, have a tendency to misbehave.

Capacitors are also in passive and active filter networks, LC resonant circuits, RC snubber circuits, and so on. In these applications, it's the reactive response to change in applied frequency that makes the capacitor useful. We'll take a closer look at capacitor applications later in this chapter.

3.6.1 Capacitance

When a dc voltage V is applied across the leads of a capacitor, one of its "plates" charges up to a value of $Q = CV$, while the other plate will charge up to $-Q$. Here Q is the charge in coulombs (C) and C is the capacitance, or simply the proportional constant that relates Q with V. Capacitance is measured in farads, F (1 F = 1 C/1 V). Once the capacitor is charged, obtaining a voltage nearly equal to the source voltage, it will not pass any dc current—the physical separation between plates prevents this.

Capacitors come with various capacitance values, typically from 1 pF (1×10^{-12}F) to 68,000 μF (0.068 F), and with various maximum voltage ratings, from a few volts to thousands of volts, depending on the type of capacitor.

In practical terms, the capacitance simply tells you how much charge can be stored within the capacitor. For example, the circuit in Fig. 3.60 shows how a 4300-μF capacitor holds more charge than a 100-μF capacitor, and, hence, will supply more current to keep the LED lit longer.

3.6.2 Capacitors in Parallel

When capacitors are placed in parallel, their capacitances add, just as they do for resistors in series:

$$C_{\text{tot}} = C_1 + C_2 + \cdots C_n \text{ (Parallel capacitors)}$$

Intuitively, you can think of capacitors in parallel representing one single capacitor with increased plate surface area. It's important to note that the largest voltage that can be applied safely to a group of capacitors in parallel is limited to the voltage rating of the capacitor with the lowest voltage rating. Both the capacitance and voltage rating are usually included next to the capacitor symbol in schematics, but often the voltage rating is missing; you must figure out the rating based on the expected voltages present at that point in the circuit.

Capacitors In Parallel

Increases the total capacitance, but limits max. voltage rating to that of smallest rated capacitor.

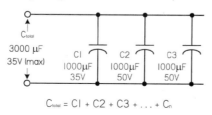

$$C_{\text{total}} = C1 + C2 + C3 + \ldots + C_n$$

Capacitors In Series

Increases max voltage rating, but decreases capcitance.

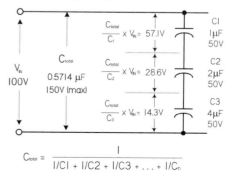

$$C_{\text{total}} = \frac{1}{1/C1 + 1/C2 + 1/C3 + \ldots + 1/C_n}$$

FIGURE 3.61

3.6.3 Capacitors in Series

When two or more capacitors are connected in series, the total capacitance is less than that of the smallest capacitor in the group. The equivalent capacitance is calculated similarly to the resistance calculation for resistors in parallel:

$$\frac{1}{C_{\text{tot}}} = \frac{1}{C_1} + \frac{1}{C_2} + \cdots \frac{1}{C_n} \text{ (Capacitors in series)}$$

Capacitors may be connected in series to enable the group to withstand a larger voltage than any individual capacitor is rated to withstand (the maximum voltage ratings add). The trade-off is a decrease in total capacitance—though that could be what you intend to do, if you can't find a capacitor or create a parallel arrangement that gives you the desired capacitance value. Notice in Fig. 3.61 that the voltage does not divide equally among capacitors. The voltage across a single capacitor—say, C_2—is a fraction of the total, expressed as $(C_{\text{tot}}/C_2)V_{\text{in}}$. There are circuits that tap the voltage between series capacitors.

Use care to ensure that the voltage rating of any capacitor in the group is not exceeded. If you use capacitors in series to withstand larger voltages, it's a good idea to also connect an equalizing resistor across each capacitor. Use resistors with about 100 Ω per volt of supply voltage, and be sure they have sufficient power-handling capability. With real capacitors, the leakage resistance of the capacitor may have more effect on the voltage division than does the capacitance. A capacitor with a high parallel resistance will have the highest voltage across it. Adding equalizing resistors reduces this effect.

3.6.4 RC Time Constant

When a capacitor is connected to a dc voltage source, it will charge up almost instantaneously. Likewise, a charged capacitor that is shorted with a wire will discharge almost instantaneously. However, with some resistance added, the rate of charge or discharge follows an exponential pattern, as shown in Fig. 3.62. There are numerous applications that use controlled charge and discharge rates, such as timing ICs, oscillators, waveform shapers, and low-discharge power backup circuits.

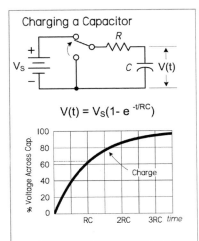

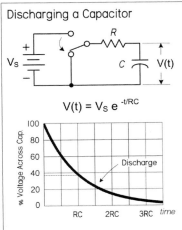

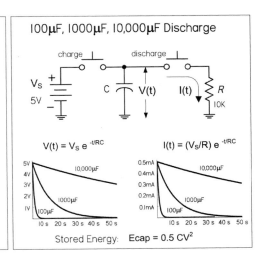

FIGURE 3.62

When charging a capacitor through a resistor, the voltage across the capacitor, with respect to time, is given as:

$$V(t) = V_S (1 - e^{-(t/RC)}) \text{ (Charging RC)}$$

where $V(t)$ is the capacitor voltage in volts at time t, V_S is the source voltage, t is the time in seconds after the source voltage is applied, $e = 2.718$, R is the circuit resistance in ohms, and C is the capacitance in farads. Theoretically, the charging process never really finishes, but eventually the charging current drops to an unmeasurable value. A convention often used is to let $t = RC$, which makes $V(t) = 0.632 V_s$. The RC term is called the time constant of the circuit and is the time in seconds required to charge the capacitor to 63.2 percent of the supply voltage. The lowercase tau (τ) is often used to represent RC: $\tau = RC$. After two time constants ($t = 2RC = 2\tau$), the capacitor charges another 63.2 percent of the difference between the capacitor voltage at one time constant and the supply voltage, for a total change of 86.5 percent. After three time constants, the capacitor reaches 95 percent of the supply voltage, and so on, as illustrated in the graph in Fig. 3.62. After five time constants, a capacitor is considered fully charged, having reached 99.24 percent of the source voltage.

For a discharging capacitor, the following equation is used:

$$V(t) = V_s e^{-(t/RC)} \text{ (Discharging RC)}$$

This expression is essentially the inverse of the previous expression for a charging capacitor. After one time constant, the capacitor voltage will have dropped by 63.2 percent from the supply voltage, so it will have reached 37.8 percent of the supply voltage. After five time constants, the capacitor is considered fully discharged; it will have dropped 99.24 percent, or down to 0.76 percent of the supply voltage.

3.6.5 Capacitive Reactance

The amount of charge that can be placed on a capacitor is proportional to the applied voltage and the capacitance ($Q = CV$). Within an ac circuit, the amount of charge moves back and forth in the circuit every cycle, so the rate of movement of charge (current) is proportional to voltage, capacitance, and frequency. When the effect of capacitance and frequency are considered together, they form a quantity similar to resistance. However, since no actual heat is being generated, the effect is termed capacitive reactance. The unit for reactance is the ohm, just as for resistors, and the formula for calculating the reactance of a capacitor at a particular frequency is given by:

$$X_C = \frac{1}{2\pi f C} \text{ (Capacitive reactance)}$$

where X_C is the capacitive reactance in ohms, f is the frequency in hertz, C is the capacitance in farads, and $\pi = 3.1416$. Often, omega (ω) is used in place of $2\pi f$.

It's important to note that even though the unit of reactance is the ohm, there is no power dissipated in reactance. The energy stored in the capacitor during one portion of the cycle is simply returned to the circuit in the next. In other words, over a complete cycle, the average power is zero. See the graph in Fig. 3.63.

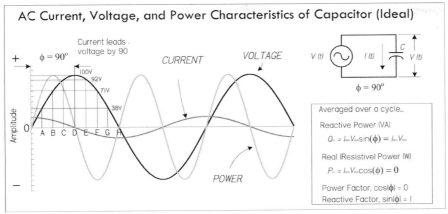

AC Current, Voltage, and Power Characteristics of Capacitor (Ideal)

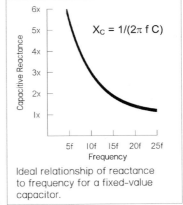

Ideal relationship of reactance to frequency for a fixed-value capacitor.

FIGURE 3.63

As an example, the reactance of a 220-pF capacitor at an applied frequency of 10 MHz is:

$$X_C = \frac{1}{2\pi \times (10 \times 10^6 \, \text{Hz})(220 \times 10^{-12} \, \text{F})} = 72.3 \, \Omega$$

As you can see, the reactance decreases with an increase in frequency and/or an increase in capacitance. The right graph in Fig. 3.63 shows the reactance versus the frequency of a capacitor. Real capacitors don't follow the graph and equation so precisely, a result of parasitic effects.

3.6.6   Real Capacitors

There are many different types of capacitors, used for various applications. Selecting the right one can be confusing. The main reason for the variety of different capacitors has to do with nonideal characteristics of real capacitors. They contain salient imperfections or parasitic effects that can mess up a particular circuit's performance. Some capacitors, due to their construction, may have larger resistive or inductive components inherent in their design. Others may act in a nonlinear fashion or may contain dielectric memory. Understanding the effects these parasitics have in each application largely determines which capacitor you select.

The four major nonideal capacitor parameters are *leakage* (parallel resistance), *equivalent series resistance* (ESR), *equivalent series inductance* (ESL), and *dielectric absorption* (memory). Figure 3.64 shows a schematic model of a real-life capacitor. These parameters, as well as a number of other specifications are explained in the following section.

3.6.7   Capacitor Specifications

DC Working Voltage (DCWV)

This is the maximum safe limit of dc voltage across a capacitor to prevent dielectric breakdown—a condition that usually results in a puncture in the dielectric offering a low-resistance current path between the two plates. It is not safe to connect capacitors

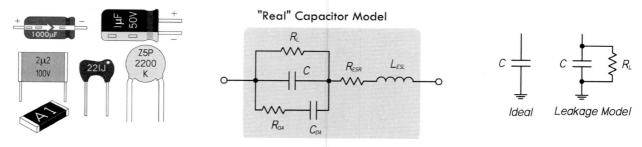

FIGURE 3.64

across an ac power line unless they are rated for it. Capacitors with dc ratings may short the line. Several manufacturers make capacitors specifically rated for use across the ac power line. For use with other ac signals, the peak value of ac voltage should not exceed the dc working voltage, unless otherwise specified in component ratings. In other words, the RMS value of ac should be 0.707 times the DCWV value or lower. With many types of capacitors, further derating is required as the operating frequency increases.

Capacitor Leakage (RL)

This is the internal leakage, as shown in the real capacitor model, that trickles off at a rate determined by the $R_L C$ time constant. Leakage becomes an important parameter when dealing with ac coupling applications and storage applications, and when capacitors are used in high-impedance circuits. Capacitors notorious for high leakage are electrolytic-type capacitors, on the order of 5 to 20 nA per μF. These capacitors are not suited for storage or high-frequency coupling applications. A better choice would be a film-type capacitor such as polypropylene or polystyrene, which has extremely low leakage current—insulation resistance typically greater than 10^6 MΩ.

Equivalent Series Resistance (ESR)

This is a mathematical construct, expressed in ohms, that allows all capacitor losses (resistance in capacitor leads, electrodes, dielectric losses, and leakage) at a single specific frequency to be expressed as a single series resistance with the capacitance. A high ESR causes a capacitor to dissipate more power (loss) when high-ac currents are flowing. This can degrade the capacitor. This can also have serious performance consequences at RF and in supply decoupling carrying high ripple currents. However, it is unlikely to have a significant effect in precision high-impedance, low-level analog circuitry. ESR can be calculated using the following equation:

$$\text{ESR} = X_C/Q = X_C \times DF$$

where X_C is the reactance, Q the quality factor, and DF the dissipation factor of the capacitor. Once you know the ESR, you can calculate how much power is lost due to internal heating, assuming a sine wave RMS current is known: $P = I_{\text{RMS}}^2 \times \text{ESR}$. So a capacitor that's considered lossy presents a large X_C and is highly resistive to signal power. Using capacitors with low ESR is important in high-current, high-performance applications, such as power supplies and high-current filter networks.

The lower the ESR, the higher the current-carrying ability. A few capacitors with low ESR include both mica and film types.

Equivalent Series Inductance (ESL)

The ESL of a capacitor models the inductance of the capacitor's leads in series with the equivalent capacitance of the capacitor plates. Like ESR, ESL can also be a serious problem at high frequencies (e.g., RF), even though the precision circuitry itself may be operating at dc or low frequencies. The reason is that the transistors used in the analog circuits may have gain extending up to transition frequencies of hundreds of megahertz, or even several gigahertz, and can amplify resonances involving low values of inductance. This makes it essential that the power supply terminals of such circuits be decoupled properly at high frequency. Electrolytic, paper, or plastic-film capacitors are a poor choice for decoupling at high frequencies; they basically consist of two sheets of foil separated by sheets of plastic or paper dielectric and formed into a roll. This kind of structure has considerable self-inductance and acts more like an inductor than a capacitor at frequencies exceeding just a few megahertz. An appropriate choice for HF decoupling is a monolithic, ceramic-type capacitor, which has very low series inductance. It consists of a multilayer sandwich of metal films and ceramic dielectric, and the films are joined in parallel to bus bars rather than rolled in series. A minor trade-off is that monolithic ceramic capacitors can be microphonic (i.e., sensitive to vibration), and some types may even be self-resonant, with comparatively high Q, because of the low series resistance accompanying their low inductance. Disc ceramic capacitors are often used, but they are often quite inductive, although less expensive. Lead length of a capacitor and its construction determine the capacitor's self-inductance and, thus, its resonant frequency.

Dissipation Factor (DF) or Tangent Delta (tan δ)

This is the ratio of all loss phenomena (dielectric and resistive) to capacitive reactance, usually expressed as a percent. It can be thought of as the ratio of energy dissipated per cycle to energy stored per cycle. It is also the ratio of the current in phase with the applied voltage to the reactive current. Dissipation factor also turns out to be equivalent to the reciprocal of the capacitor's figure of merit, or Q, which is also often included in the manufacturer's data sheet. DF must be given at a specific frequency to be meaningful. A lower DF indicates less power dissipated under otherwise identical conditions.

Dielectric Absorption (DA)

Monolithic ceramic capacitors are excellent for HF decoupling, but they have considerable dielectric absorption, which makes them unsuitable for use as the hold capacitor of a sample-hold amplifier. Dielectric absorption is a hysteresis-like internal charge distribution within the dielectric that causes a capacitor that is quickly discharged and then open-circuited to appear to recover some of its charge. Since the amount of charge recovered is a function of its previous charge, this is, in effect, a charge memory and will cause errors in any sample-hold amplifier where such a capacitor is used as the hold capacitor. DA is given as a percent of charge stored in a capacitor's dielectric, as opposed to the foil surfaces. It can be approximated by the ratio of the equilibrium value "self-recharge" voltage to the voltage before discharge.

Capacitors with low dielectric absorption, typically less than 0.01 percent, suitable for sample-hold applications, include polyester, polypropylene, and Teflon.

Temperature Coefficient (TC)

This represents the change in capacitance with temperature, expressed linearly as parts per million per degree centigrade (ppm/°C), or as a percent change over a specified temperature range. Most film capacitors are not linear; therefore, their TC is usually expressed as a percentage. TC should always be factored in to designs operating at temperature above or below 25°C.

Insulation Resistance (IR)

This is a measure of the resistance to a dc current flow through the capacitor under steady-state conditions. Values for film and ceramic capacitors are usually expressed in megaohm-microfarads for a given design and dielectric. The actual resistance of the capacitor is obtained by dividing the megohm-microfarads by the capacitance.

Quality Factor (Q)

This is the ratio of the energy stored to that dissipated per cycle. It is defined as $Q = X_C / R_{ESR}$. In one respect, Q is a figure of merit that defines a circuit component's ability to store energy compared to the energy it wastes. The rate of heat conversion is generally in proportion to the power and frequency of the applied energy. Energy entering the dielectric, however, is attenuated at a rate proportional to the frequency of the electric field and the loss tangent of the material. Thus, if a capacitor stores 1000 J of energy and dissipates only 2 J in the process, it has a Q of 500.

I_{RMS}

This is the maximum RMS ripple current in amps at a given frequency.

I_{peak}

This is the maximum peak current in amps at 25°C for nonrepetitive pulses or where the pulse time off is sufficient to allow cooling so overheating will not result.

Graphs Representing Real Capacitor Characteristics

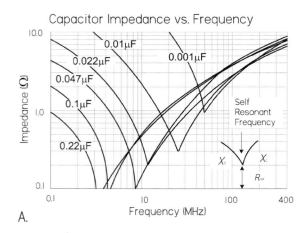

A.

Graph A: **Real capacitors aren't entirely capacitively reactive. They also have inductive and resistive elements (ESL, ESR) that influence their overall impedance. As shown in the graph, as the frequency increases toward the capacitor's self-resonant frequency, the impedance reaches a minimum, equal to the ESR. Past this frequency, the nonideal inductive reactance characteristic kicks in, leading to an increase in overall impedance. For many applications, the capacitor's series-resonant frequency sets the upper frequency limit, especially where the phase angle of the capacitor is expected to maintain near 90° voltage/current relationship. The type of construction of the capacitor and lead length affect the self-inductance and thus can shift the resonant frequency. Graphs like these are given by manufacturers and are useful when selecting decoupling capacitors.**

FIGURE 3.65

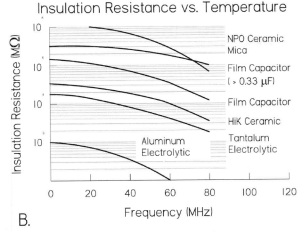

B.

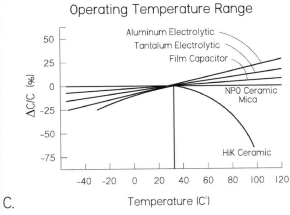

C.

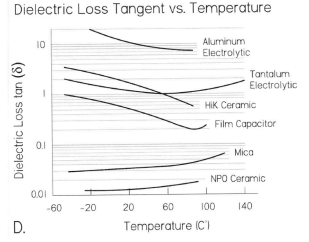

D.

Graph B: **Shows how the insulation resistance (IR) of various-capacitors changes with frequency.**

Graph C: **Shows the temperature characteristics for various capacitors. Note the extraordinary curve of the HiK ceramic capacitor. Utmost attention must be used when using this type of capacitor when considering temperature-sensitive applications. NPO ceramics have the best temperature characteristic, followed by film, mica, and tantalum.**

Graph D: **Shows the dielectric loss tangent as a function of temperature for various capacitors.**

FIGURE 3.65 *(Continued)*

3.6.8 Types of Capacitors

Here's an overview of the various capacitors available. More in-depth coverage is provided in Table 3.7.

Trimmer (Variable)

These are usually designated as tuning or trimmer capacitors. Trimmer capacitors use ceramic or plastic dielectrics and are typically within the picofarad range. Trimmer capacitors often have their capacitance range printed on them but also may use the following color code: yellow (1 to 5 pF), beige (2 to 10 pF), brown (6 to 20 pF), red (10 to 40 pF), purple (10 to 60 pF), black (12 to 100 pF), and so on. They are often used for fine-tuning sensitive circuits or for compensating as a circuit ages.

Air Core (Variable)

This is an air-spaced capacitor dielectric that is the best approximation to the ideal picture. These capacitors are large when compared to those of the same value using other dielectrics. Their capacitance is very stable over a wide temperature range. Leakage losses are low, and therefore a very high Q can be obtained. To vary the capacitance, the effective surface area of an array of parallel plates is altered via a mechanical turn-knob. Tuning capacitors are used mainly in radio-tuning applications.

Vacuum

These come in both fixed and variable types. They are rated by their maximum working voltages (3 to 60 kV), capacitance (1 to 5000 pF), and currents. Losses are specified as negligible for most applications. There is excellent leakage control. These are used in high-voltage applications such as RF transmitters.

Aluminum Electrolytic

These capacitors have space between their foil plates filled with a chemical paste. When voltage is applied, a chemical reaction forms a layer of insulating material on the foil. Electrolytic capacitors are popular because they provide high capacitance values in small packages at a reasonable cost. They leak badly, have bad tolerances, drift, and have high internal inductance, limiting them to low-frequency applications. Their range is from about 0.1 to 500,000 µF. Aluminum electrolytic capacitors are very popular and are heavily used in almost every kind of circuit, since they are cheap, readily available, and good for filters and storage of large amounts of charge. They will explode, however, if the rated working voltage is exceeded or polarity is reversed. This also means you shouldn't apply an ac voltage across an electrolytic capacitor. If an ac voltage is superimposed on a dc voltage, make sure that the peak value does not exceed the voltage rating. Aluminum electrolytic capacitors are not suited for high-frequency coupling applications due to poor isolation resistance and internal inductance. Electrolytic capacitors should not be used if the dc potential is well below the capacitor working voltage. Applications include power supply ripple filters, audio coupling, and bypassing. There are also nonpolarized electrolytic capacitors, but these are more expensive and larger than their polarized relatives.

Tantalum Electrolytic

These capacitors are made of tantalum pentoxide. Like electrolytic capacitors, they are polarized, so watch the "+" and "−." They are smaller, lighter, and more stable; they leak less and have less inductance when compared to aluminum electrolytics, but they tend to be more expensive, have a lower maximum working voltage and capacitance, and are easily prone to damage due to current spikes. For the last reason, tantalum electrolytics are used mostly in analog signal systems that lack high current-spike noise.

Tantalum capacitors are not suited for storage or high-frequency coupling applications due to poor isolation resistance and internal inductance. They should not be used if the dc potential is well below the capacitor working voltage. Applications include blocking, bypassing, decoupling, and filtering.

Polyester Film

These capacitors use a thin polyester film as their dielectric. They do not have as high a tolerance as polypropylene capacitors, but they have good temperature stability, and they are popular and cheap. Tolerances are from 5 to 10 percent. These are a good choice for coupling and/or storage applications due to high isolation resistance. They are typically used in moderately high-frequency circuits and audio and oscillator circuits.

Polypropylene Film

This type uses a polypropylene film dielectric and is used mainly when a higher tolerance is needed than what a polyester film capacitor can provide. Tolerance is about 1 percent. This is a good choice for coupling and/or storage applications due to high isolation resistance. This type exhibits stable capacitance for frequencies below 100 kHz. These capacitors are used for noise suppression, blocking, bypassing, coupling, filtering, and timing.

Silver Mica

These are made from depositing a thin layer of silver on a mica dielectric. They are very stable with respect to time (tolerances of 1 percent or less), and have a good temperature coefficient and excellent endurance, but they don't come in high capacitance values and can be expensive. They are used in resonance circuits and high-frequency filters, due to good stability with temperature. They are also used in high-voltage circuits, because of their good insulation. Their temperature coefficient, in regard to oscillators, isn't as low as can be achieved by other types, and some silver micas have been known to behave erratically.

Ceramic (Single-Layer)

These capacitors are constructed with materials such as titanium acid barium for the dielectric. Internally, these capacitors are not constructed as a coil, so they have low inductance and are well suited for use in higher-frequency applications. Together with electrolytics, they are the most widely used capacitor around. Ceramic capacitors come in three basic varieties:

Ultrastable or temperature compensating: This type is one of the most highly stable capacitors, made from a mixture of titanates. It has very predictable temperature coefficients (TCs) and, in general, does not have an aging characteristic. The most popular ultrastable ceramic capacitor is the NPO (negative-positive 0 ppm/°C) or COG (EIA designation); others include N030 (SIG) and N150 (P2G). The TC for these capacitors is specified in the capacitance change in parts per million (ppm) per degrees centigrade. To calculate the maximum capacitance change with temperature, the following equation is used. Here we used a 1000-pF NPO capacitor that has reached 35°C, or a change in temperature (ΔT) of 10°C above the standard reference temperature of 25°C. The TC of an NPO capacitor is 0 ± 30 ppm/°C.

$$\text{Capacitance change (pF)} = \frac{C \times TC \times \Delta T}{1,000,000} = \frac{1000 \text{ pF} \times \pm 30 \text{ (ppm)} \times 10}{1,000,000} = \pm 0.3 \text{ pF}$$

Therefore, a 1000-pF capacitor subjected to a 10°C change in temperature may result in a value as high as 1000.3 pF or as low as 999.7 pF. Ultrastable capacitors are best suited for applications where stability over a wide variation of temperatures and high Q are required. Filter networks and most circuits associated with tuning and timing, as well as various types of resonant circuits, generally require ultrastable capacitors. They are particularly suitable for oscillator construction in order to compensate for frequency drift with temperature. See Table 3.7 for more details.

Semistable: These are not nearly as temperature stable as the ultrastable capacitors; however, they have a higher electrostatic capacity. All semistable capacitors vary in capacitance value under the influence of temperature, operating voltage (both ac and dc), and frequency. These capacitors are best suited for applications where high capacitance values are important, while Q stability over temperature is not a major concern. The TC for semistable capacitors is expressed as a percentage. So a 1000-pF X7R capacitor, which has a TC of ± 15 percent, might be as high as 1150 pF or as low as 850 pF at temperatures above or below 25°C. EIA TC designations are as follows: the first character defines the low-temperature limit ($X = -55$°C, $Y = -30$°C, $Z = +10$°C); the second character defines the high-temperature limit ($5 = +85$°C, $7 = +125$°C); the third character defines the maximum capacitance change in percentage ($V = +22, -82$ percent, $U = +22, -56$ percent, $T = +22, -33$ percent, $S = \pm 22$ percent, $R \pm 15$ percent, $P = \pm 10$ percent, $F = \pm 7.5$ percent, $E = \pm 4.7$ percent). See Table 3.7 and Fig. 3.66 for more details and applications.

HiK: This variety has a high dielectric constant, or electrostatic capacity, but it has poor stability, poor DA, and a high voltage coefficient, and is sensitive to vibration— some types may be resonant, with comparatively high Q. There is very poor temperature drift, high voltage coefficient of capacitance, high voltage coefficient of dissipation, high frequency coefficient of capacitance, and significant aging rate. These also exhibit low inductance, a wide range of values, small size, and higher density than dipped ceramic. They are best suited for coupling (dc blocking) and power supply bypassing. They should be used only in linear applications where performance and stability are of no great concern.

Multilayer Ceramic

These capacitors were developed to meet the demand for high-density ceramic capacitors. They incorporate multiple printed layers of electrode plates made of thin ceramic sheets. These capacitors are more compact and have, in general, better temperature characteristics than single-layer ceramic capacitors. They are, however, more expensive. Like single-layer ceramic capacitors, they come in ultrastable, stable, and HiK types. See Table 3.7 for more details and applications.

TABLE 3.7 Capacitor Comparison

TYPE		1. WVDC 2. CAPACITANCE 3. DIELECTRIC ABSORPTION 4. STANDARD TOLERANCE	IR 1.<1μF 2.>1μF (MΩ-μF)	FREQUENCY RESPONSE 1. (1=POOR, 10=BEST) 2. MAX. FREQUENCY	TEMPERATURE RANGE	DF @ 1 KHZ, % (MAX.)	STABILITY 1000 HOURS %ΔC	ADVANTAGES/DISADVANTAGES	APPLICATIONS
Multilayer Ceramics	NPO	25–200 V 1pF–0.01 μF 0.6% [±1(F), ±2% (G), ±5% (J), ±10% (K)]	10^5 NA	9 100 MHz	−55°C, +125°C	0.1%	0.1%	Good stability, low inductance, low DA, good frequency response. Very low temperature drift, very low aging, voltage coefficient, frequency coefficient, leakage, and dissipation factor. More expensive than the other types of ceramics.	Excellent in HF decoupling (into the GHz range) due to low series inductance. High-frequency switch-mode power supplies. Used in many analog applications, such as HF switch-mode power supplies, but avoided in sample-hold and integrators, where DA may be a problem.
	Stable	25–200 V 220 pF–0.47 μF 2.5% [±5% (J), ±10% (K), ±20% (M)]	10^5 2500	8 10 MHz	−55°C, +125°C	2.5%	10%	Low inductance, wide range of values, small, higher density than dipped ceramic. Poor stability, poor DA, high voltage coefficient, and significant aging rate. Sensitive to vibration—some types may be resonant with comparatively high Q.	Best suited for coupling/dc blocking and power supply bypassing. They should be used only in linear applications where performance and stability are of no great concern.
	(High-K) HiK	25–100 V 0.25 pF–22 μF NA [±20% (M), ±80%–20% (Z)]	10^4 10^3	8 10 MHz	+10°C, +85°C and −55°C, +85°C	4.0%	20%	Very poor stability, especially with temperature variations. Poor DA and high voltage coefficient. Not suited for high-temperature environment. Short longevity.	Limited mainly to dc blocking and power supply bypassing. Even then, change in capacitance due to aging, temperature, and voltage coefficients must be taken into consideration. Use lowest-K material you can get.
Ceramic Disc (NPO, Stable, HiK)		50–10,000 V 1pF–0.1 μF Same as multilayers	Same as multilayers	8 Same as multilayers	−55°C, +85°C	0.1% – 4.0%	Same as multilayers	Inexpensive, wide range of values, and popular. Same features as multilayers.	Used in coupling and bypassing, but can be quite inductive if leads are long. Internal structure not coiled, so can be used in high-frequency applications. See applications of multilayers.
Polystyrene		30–600 V 100 pF–0.027 μF 0.05% ±65%	10^6 NA	6 NA	−55°C, +70°C	0.1%	2%	Inexpensive, low DA available, wide range of values, good stability. High isolation resistance. Damaged by temperatures >+70°C. Large case size, high inductance.	Not used in high-frequency applications—inside acts like an inductor coil. Works well in filter circuits or timing circuits that run at several hundred kHz or less. Good choice for coupling and/or storage applications due to high isolation resistance.

TABLE 3.7 Capacitor Comparison (*Continued*)

TYPE	1. WVDC / 2. CAPACITANCE / 3. DIELECTRIC ABSORPTION / 4. STANDARD TOLERANCE	IR 1.<1μF 2.>1μF (MΩ-μF)	FREQUENCY RESPONSE 1. (1= POOR, 10 = BEST) 2. MAX. FREQUENCY	TEMPERATURE RANGE	DF @ 1 KHZ, % (MAX.)	STABILITY 1000 HOURS %ΔC	ADVANTAGES/DISADVANTAGES	APPLICATIONS
Polypropylene Film	100–600 V / 0.001 μF to 0.47 μF / 0.05% / ±5%	10^5 / NA	6 / NA	−55°C, +85°C	0.35%	3%	Inexpensive, low DA available, wide range of values, high isolation resistance, damaged by temperatures >105°C, large case size.	Good choice for coupling and/or storage applications due to high isolation resistance. Most stable capacitance for frequencies below 100 kHz, but often used at higher frequencies. Used for noise suppression, blocking, bypassing, coupling, filtering, snubbing, and timing. Good general-purpose capacitor.
Metallized Polypropylene	100–1250 V / 47 pF–10 μF / 0.05% / ±20% (M), ±10%(K), ±5% (J)	10^5 / NA	6 / NA	−55°C, +105°C	0.05%	2%	More compact than film/foil types, but higher DF, lower IR, lower maximum current, lower ac—unique self-healing feature, unlike film/foil, voltage-frequency capability.	Used in moderately high-frequency, high-voltage circuits, and for noise suppression, timing, and snubbing. Used in switching power supplies, audio equipment (provide musically clean dynamic), and many other general-purpose applications.
Polyester Film (Mylar)	50–600 V / 0.001 μF–10 μF / 0.5% / ±10	10^4 / 10^3	6 / NA	−55°C, +125°C	2%	10%	Moderate stability, inexpensive, low DA available, wide range of values, high isolation resistance, large case size.	Good choice for coupling and/or storage applications due to high isolation resistance. Moderately high-frequency circuits, audio sound quality, oscillator circuits.
Metallized Polyester	63–1250 V / 470 pF–22 μF / 0.5% / ±20% (M), ±10%(K), ±5% (J)	10^4 / 10^3	6 / NA	−55°C, +125°C	0.8%	NA	More compact than film/foil types, but higher DF, lower IR, lower maximum current, lower ac-voltage-frequency capability. Does have a unique self-healing feature, unlike film/foil, which prevents dielectric breakdown from resulting in catastrophic permanent failure.	General-purpose applications, audio equipment, moderately high-frequency, high-voltage applications. Switching power supplies, blocking, bypassing, filtering, timing, coupling, decoupling, and interference suppression.

Type							Characteristics	Applications
Mica	50–500 V 1 pF–0.09 μF 0.3%–0.7% ±1%, ±5%	10^2 NA	7 100	−55°C, +125°C	0.1%	0.1%	Low loss at HF, low inductance, very stable, available in 1% values or better Large, low values (<10 nF), expensive.	Excellent capacitor, good at RF. Used in resonance circuits and high-frequency filters, due to good stability with temperature. Also used in high-voltage circuits due to their good insulation.
Multilayer Glass	50–2000 V 0.5 pF–0.01 μF 0.05% ±1%, ±5%	10^5 NA	9	−75°C, +200°C	0.2%	0.5%	Extremely low stable Q factor at high frequencies, low dielectric absorption, large RF current capability, high operating temperature range, high shock/vibration capability. Excellent stability and long-term stability.	Use in military applications and high-performance commercial sectors. Wide applications: high-temperature circuitry, modulators, RF amplifier output filters, variable-frequency oscillators, amplifier coupling, sample-hold, transistor biasing, ramp integrators, voltage snubbers, etc.
Aluminum Electrolytic	4 V–450 V 0.1 μF–1 F High +100%, −10%	NA 100	2 NA	−40°C, +85°C	8% at 120 Hz	10%	High currents, high voltages, small size. Very poor stability, poor accuracy, inductive. Usually polar, meaning they can be damaged if placed in reverse polarity.	Not suited for storage or HF coupling applications due to poor isolation resistance and internal inductance. Usually used as a ripple filter in power supplies or as a filter to bypass low-frequency signals. Used in audio bypassing and power supply filtering—at higher frequencies there is too much loss.
Tantalum Electrolytic	6.3–50 V 0.01–1000 μF High ±20%	10^2 10	5 0.002 MHz	−55°C, +125°C	8%– 24%	10%	Small size, large values, medium inductance. Better capacitance stability than aluminum with temperature. Quite high leakage, usually polarized, expensive, poor stability, poor accuracy.	Not suited for storage or HF coupling applications due to poor isolation resistance and internal inductance. Acts more like an inductor than a capacitor above a few MHz. Used in dc blocking, bypassing, decoupling, filtering, and timing. Usually used as a ripple filter in power supplies or as a filter to bypass low-frequency signals.
Double-Layer Supercapacitor Ultracapacitor	2.3 V, 5.5 V, 11 V, etc. 0.022–50 F High	NA	NA	−40°C, +70°C	NA	NA	Huge capacitance values, high power output. Exhibit relatively high ESR, and therefore are not recommended for ripple absorption in dc power supply applications. Low leakage, but poor temperature stability.	Actuator applications (relay-solenoid starters), primary power supply for LED displays, electric buzzers, etc. Power backup for CMOS microcomputers. Also used in many interesting low-powered circuits, such as solar-powered robots, where they store energy and act as the primary power source. Many other creative uses.

Polystyrene

These use a polystyrene dielectric. They are constructed like a coil inside, so they are not suitable for high-frequency applications. They are used extensively within filter circuits and timing applications, and also in coupling and storage applications due to a high isolation resistance. A warning worth noting: polystyrene capacitors exhibit a permanent change in value should they ever be exposed to temperatures much over 70°C; they do not return to their old value upon cooling.

Metallized Film (Polyester and Polypropylene)

These utilize thin conductors or plates that have a distinct size advantage over other types. They are formed by a vacuum-deposition process that laminates a film substrate with a thin aluminum coating measured in angstroms. These capacitors are used where small signal levels (low current/high impedance) and small physical size are primary factors.

Metallized-film capacitors are generally not appropriate for large-signal ac applications. Film and foil types are better suited for this purpose, since they have much thicker plates (foil) that help carry away heat buildups, thus lowering losses, extending life, and reducing effects on DF.

One major advantage of metallized-film capacitors is their self-healing characteristic—a feature resulting from the extreme thinness of the metallized electrode material. Whenever a flaw or weak spot in the dielectric results in a short condition, the stored electrons in the capacitor and the accompanying circuitry will immediately avalanche across the shorted point. This can vaporize the thin metallic electrode. The vaporized electrode forms a reasonably concentric pattern away from the point of the short. As a result of the vaporization, the short condition is removed and the capacitor is again operational. This effect is known as *clearing*, which is the self-healing process. In nonmetallized capacitors, the short condition results in a catastrophic permanent failure.

Some disadvantages over nonmetallized film capacitors include a slightly higher dissipation factor, slightly lower insulation resistance, lower maximum current, and lower maximum ac-voltage-frequency capability. In general, this type is a good-quality, low-drift, temperature-stable capacitor. Applications include moderately high-frequency, low-current circuits, noise suppression, timing, snubbing, switching power supplies, bypassing, and audio applications.

Supercapacitors (Double-Layer or Ultracapacitors)

These devices store extremely large amounts of charge (from 0.022 to 50 F)—much more than a typical capacitor. This level of energy storage approaches around 1/10 that of a low-density battery. However, unlike a battery, the power output can be 10 times greater—a useful feature in high-current pulse applications.

Supercapacitors consist of two nonreactive porous plates suspended within an electrolyte. A voltage applied to the positive plate attracts the negative ions in the electrolyte, while the voltage on the negative plate attracts the positive ions. This effectively creates two layers of capacitive storage, one where the charges are separated at the positive plate and another where the charges are separated at the negative plate. Conductive rubber membranes contain the electrode and electrolyte material and make contact with a cell. Several cells are stacked in series to achieve the desired voltage ratings. Typical voltage ratings are 3.5 and 5.5, in keeping with their common role as backup capacitors for 3.3 V or 5 V devices.

Battery/Supercapacitor Comparison

A supercapacitor can be charged to any voltage within its voltage rating extremely quickly, and it can be stored totally discharged, while many batteries are damaged by quick charging. The state of charge of a supercapacitor is simply a function of its voltage, while the state of charge of a battery is complex and often unreliable. Though a battery will store more energy than a supercapacitor, a supercapacitor is able to deliver frequent high-power pulses without any detrimental effect, while many batteries experience reduced life under similar conditions.

Supercapacitors can be used as an intermediate power source or a bridge between batteries and conventional capacitors. Many applications benefit from the use of supercapacitors, from those requiring short power pulses to those requiring low-power, long-duration support of critical memory systems—they are capable of maintaining contents of low-dissipation CMOS memory for several months. They are excellent solutions in a number of systems when used alone or combined with other energy sources. Examples include quick-charge applications that can be charged in seconds and then discharge over a few minutes (power tools and toys), short-term support for uninterruptible power systems, where the supercapacitor provides the power for short outages, or as a bridge to a generator or other continuous backup power supply. They can provide load-leveling to an energy-rich, power-poor energy source such as a solar array.

When supercapacitors are strategically placed within battery-powered systems, they can prevent peak power–induced battery stress by supplying the peak power demands for the battery. This often allows a smaller-capacity battery to be used, and it can even extend the overall life span of the battery.

Supercapacitors exhibit high ESR, so they are not recommended for ripple absorption in dc power supply applications. See the "Capacitor Applications" section for example supercapacitor circuits.

Oil-Filled Capacitor

This type is used in high-voltage, high-current applications that generate a lot of heat—the oil cools the capacitor. Applications include induction heating, high-energy pulsing, commutation, equipment bypassing, ignitions, frequency conversion, high-voltage ripple filtering, snubbing, coupling, and spark generation. Voltages range from 1 to 300 kV, and capacitance from around 100 pF to 5000 μF. Typically, they come in large packages.

Reading Capacitor Labels

There are many different schemes used for labeling capacitors. Some use color bands, and some use combinations of numbers and letters. Capacitors may be labeled with their capacitance value, tolerance, temperature coefficient, voltage rating, or some combination of these specifications. Figure 3.66 shows several popular labeling systems.

3.6.9 Capacitor Applications

Coupling and DC Blocking

Coupling capacitors act to pass a range of ac signals from one circuit to another, while preventing any dc components from passing. This is due to the capacitor's

Capacitor Markings

Capacitance Conversion Calculator

$1 \text{ F} = 1 \times 10^{6} \, \mu\text{F} = 1 \times 10^{9} \, \text{nF} = 1 \times 10^{12} \text{pF}$

$1 \, \mu\text{F} = 1 \times 10^{-6} \, \text{F} = 1 \times 10^{3} \text{nF} = 1 \times 10^{6} \text{pF}$

$1 \, \text{nF} = 1 \times 10^{-9} \, \text{F} = 1 \times 10^{-3} \mu\text{F} = 1 \times 10^{3} \text{pF}$

$1 \, \text{pF} = 1 \times 10^{-12} \, \text{F} = 1 \times 10^{-6} \mu\text{F} = 1 \times 10^{-3} \text{nF}$

F = Farad, μ = micro, n = nano, p = pico

$1000 \, \mu\text{F} = 1,000,000 \, \text{nF} = 10 \times 10^{8} \text{pF}$
$100 \, \mu\text{F} = 100,000 \, \text{nF} = 10 \times 10^{7} \text{pF}$
$10 \, \mu\text{F} = 10,000 \, \text{nF} = 10 \times 10^{6} \text{pF}$
$1 \, \mu\text{F} = 1,000 \, \text{nF} = 10 \times 10^{5} \text{pF}$
$0.1 \, \mu\text{F} = 100 \, \text{nF} = 10 \times 10^{4} \text{pF}$
$0.01 \, \mu\text{F} = 10 \, \text{nF} = 10 \times 10^{3} \text{pF}$
$0.001 \, \mu\text{F} = 1 \, \text{nF} = 10 \times 10^{2} \text{pF}$

Tantalum

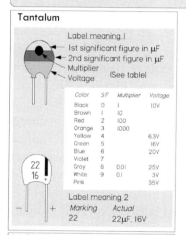

Label meaning 1

- 1st significant figure in μF
- 2nd significant figure in μF
- Multiplier
- Voltage (See table)

Color	SF	Multiplier	Voltage
Black	0	1	10V
Brown	1	10	
Red	2	100	
Orange	3	1000	
Yellow	4		6.3V
Green	5		16V
Blue	6		20V
Violet	7		
Gray	8	0.01	25V
White	9	0.1	3V
Pink			35V

Label meaning 2

Marking	Actual
22	22μF, 16V

Mylar (Polyester Film)
Polypropylene
Dipped Mica

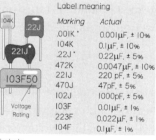

Label meaning

Marking	Actual
.001K	0.001μF, ± 10%
104K	0.1μF, ± 10%
.22J	0.22μF, ± 5%
472K	0.0047μF, ± 10%
221J	220 pF, ± 5%
470J	47pF, ± 5%
102J	1000pF, ± 5%
103F	0.01μF, ± 1%
223F	0.022μF, ± 1%
104F	0.1μF, ± 1%

Labels:

1st digit, 2nd digit, multiplier in pF (or μF if decimal before digits), and tolerance.

Metallized Polyester Film

Label meaning

Marking	Actual
2μ2	2.2 μF
μ22	0.22 μF
68n	68 nF
4n7	4.7 nF

Label:

"μ" place of decimal in microfarads
"n" place of decimal in nanofarads

Polyester Color Coded

- 1st digit (pF)
- 2nd digit (pF)
- Multiplier
- Tolerance
- Voltage

Standard color code

Black	± 20%
White	± 10%
Green	± 5%
Brown	100
Red	250
Yellow	400

Ceramic Disc Capacitors

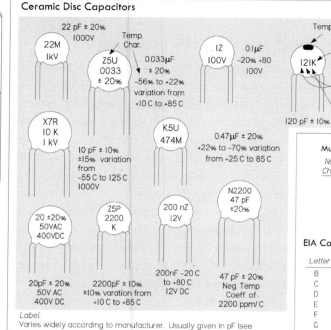

22 pF ± 20%
1000V
22M
1kV

Temp. Char.
Z5U
.0033
± 20%
0.033μF
± 20%
-56% to +22%
variation from
+10 C to +85 C

.IZ
100V
0.1μF
-20% +80
100V

Temperature Coefficient
Color Code
Tolerance
121K
1st Digit
2nd Digit
Decimal Point
Multiplier
120 pF ± 10%

4R7D
4.7 pF ± 0.5pF

X7R
10 K
1 kV
10 pF ± 10%
±15% variation
from
-55 C to 125 C
1000V

K5U
474M
0.47μF ± 20%
+22% to -70% variation
from +25 C to 85 C

N2200
47 pF
±20%

20 ±20%
50VAC
400VDC

Z5P
2200
K

200 nZ
12V

20pF ± 20%
50V AC
400V DC

2200pF ± 10%
±10% variation from
+10 C to +85 C

200nF -20 C
to +80 C
12V DC

47 pF ± 20%
Neg. Temp
Coeff. of
2200 ppm/C

Label:
Varies widely according to manufacturer. Usually given in pF (see multiplier code table) but may be given in μF when there is a decimal before digits. See other tables for temperature and tolerance markings.

Ceramic Disc (European Markings)

47p

Label Meaning

Marking	Actual	Marking	Actual
p68	0.68 pF	22p	22 pF
1p0	1.0 pF	n10	0.1 nF
4p7	4.7 pF	n27	0.27 nF

Label: p = picofarads, n = nanofarads; location of p or n signifies decimal point.

Fixed Ceramic Color Code

- 1st Digit
- 2nd Digit
- Multiplier
- Temp. Coeff.
- Tolerance

Color	S.F.	Tolerance >10pF	Tolerance <10pF	Temp. Coeff. ppm/C
Black	0	± 20%	2.0 pF	0
Brown	1	± 1%		-30
Red	2	± 2%		-80
Orange	3			-150
Yellow	4			-220
Green	5	± 5%	0.5pF	-330
Blue	6			-470
Violet	7			-750
Gray	8	0.01		0.25pF	30
White	9	0.1	± 10%	1.0pF	500

Surface Mount Capacitors

SMD Ceramic

A1

Label meaning

Marking	Actual
N1	33 pF
A4	0.01 μF
S6	4.7 μF

Significant Figure Code

Char.	S.F.		Char.	S.F.
A	1.0		T	5.1
B	1.1		U	5.6
C	1.2		V	6.2
D	1.3		W	6.8
E	1.5		X	7.5
F	1.6		Y	8.2
G	1.8		Z	9.1
H	2.0		a	2.5
J	2.2		b	3.5
K	2.4		d	4.0
L	2.7		e	4.5
M	3.0		f	5.0
N	3.3		m	6.0
P	3.6		n	7.0
Q	3.9		t	8.0
R	4.3		y	9.0
S	4.7			

SMD Electrolytic

10 6V

Label meaning 1

Marking	Actual
10 6V	10 μF, 6V

A475

Label meaning 2

Marking	Actual
A475	4.7 μF, 10V

Multiplier Code

Numeric Character	Decimal Multiplier (pF)
0	1
1	10
2	100
3	1,000
4	10,000
5	100,000
6	1,000,000
7	10,000,000
8	100,000,000
9	0.1

Label 2:
Voltage (see table below), 1st digit, 2nd digit, multiplier (pF).

Char.	Voltage
e	2.5
G	4
J	6.3
A	10
C	16
D	20
E	25
V	35
H	50

Multiplier Code

Numeric Character	Decimal Multiplier (pF)
0	None
1	10
2	100
3	1000
4	10,000

EIA Capacitor Tolerance Codes

Letter	≤ 10 pF	≥ 10 pF
B	± 0.1 pF	–
C	± 0.25 pF	–
D	± 0.5 pF	–
E	–	± 25%
F	± 1 pF	± 1%
G	–	± 2%
H	–	± 2.5%
J	–	± 5%
K	–	± 10%
M	–	± 20%
P	–	-0 + 100%
S	–	-20 + 50%
W	–	-0 + 200%
X	–	-20 + 40%
Z	–	-20 + 80%

EIA Temperature Characteristic Codes

Minimum temperature		Maximum temperature		Max cap. change over temp. range	
X	-55.C	2	+45.C	A	± 1.0%
Y	-35 C	4	+65.C	B	± 1.5%
Z	+10 C	5	+85.C	C	± 2.2%
		6	+105.C	D	± 3.3%
		7	+125.C	E	± 4.7%
				F	± 7.5%
				P	± 10%
				R	± 15%
				S	± 22%
				T	-33%, + 22%
				U	-56%, + 22%
				V	-82%, + 22%

EIA Temperature Coefficient Color Codes

Color	Temp. Coeff. Industry	EIA
Black	NP0	C0G
Brown	N030/N033	SIG
Red	N075/N080	UIG
Orange	N 150	P2G
Yellow	N 220	R2G
Green	N 330	S2H
Blue	N 470	U2J
Violet	N 750	
Gray		
White	P 100	
Red/Violet	P 100	

Electrolytic Capacitors

μF 50V
1μF, 50V

Label: Usually self-explanatory

FIGURE 3.66

Blocking and Coupling

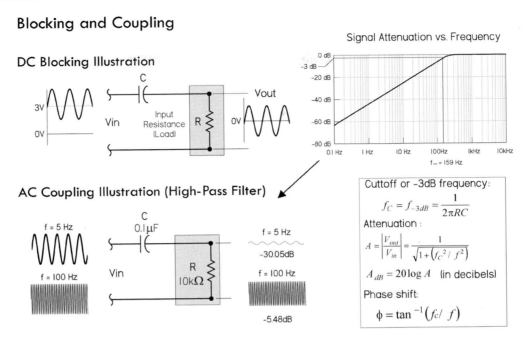

DC Blocking Illustration

AC Coupling Illustration (High-Pass Filter)

Signal Attenuation vs. Frequency

Cuttoff or –3dB frequency:

$$f_C = f_{-3dB} = \frac{1}{2\pi RC}$$

Attenuation :

$$A = \left|\frac{V_{out}}{V_{in}}\right| = \frac{1}{\sqrt{1 + \left(f_c^2 / f^2\right)}}$$

$$A_{dB} = 20\log A \quad \text{(in decibels)}$$

Phase shift:

$$\phi = \tan^{-1}\left(f_c / f\right)$$

FIGURE 3.67 *Blocking:* **A capacitor is used to block or prevent dc voltages from passing from one circuit to another. To block the dc voltage, the capacitor is placed in series with the circuit element.**

Coupling: **A coupling capacitor is used to couple or link together only the ac signal from one circuit element to another. The capacitor is connected in series between the input and the coupled load. Considering a purely resistive load, the attenuation and cutoff frequency (or -3-dB frequency) can be estimated by means of the formulas to the right. These formulas are based on an ideal capacitor.**

reactance—at dc the reactance is theoretically infinite, but at higher ac frequencies the reactance decreases. Effective coupling requires that the capacitor's impedance be as low as possible at the frequency range of interest. If not, certain frequencies may become more attenuated than others. The formulas in Fig. 3.67 can be used to find the cutoff frequency, or –3-dB frequency (½ power point), attenuation, and phase shift, assuming ideal conditions and purely resistive load.

Note that in many situations, the load (or coupled stage) to which the capacitor is coupled may be frequency sensitive—its impedance may change with frequency due to inductive and capacitive elements. For example, Fig. 3.70 shows how the input impedance of a transistor amplifier changes with frequency. Since the selection of the coupling capacitor determines which frequencies get attenuated, understanding the changing impedance of the coupled stage is important.

The major characteristics to look at when selecting a coupling capacitor are insulation resistance (IR), ESR, voltage rating, and overall frequency response. Consult Table 3.7 for suggested coupling capacitors. For example, for many audio applications, polypropylene and polyester film and even an electrolytic capacitor may be good choices, but for a high-frequency, high-stability decoupling application, up into the megahertz range, an NPO multilayer ceramic capacitor may be required.

Bypassing

Bypass capacitors are often used to bypass undesired alternating signals (supply ripple, noise, etc.) around a component or group of components to ground. Often the ac is removed (or greatly attenuated) from the ac/dc mixture, leaving the dc free to feed the bypassed component. Figure 3.68 shows the basics behind bypassing.

As a general rule of thumb, the impedance of the bypass capacitor should be 10 percent of the input impedance of the circuit element.

There are many types of capacitors used for bypassing, from electrolytic to ceramic NPO. The one you choose will depend on the kind of frequency response and stability you require. Major characteristics to consider include insulation resistance (IR), ESL, and ESR.

Power Supply Decoupling (Bypassing)

Decoupling becomes very important in both digital and analog dc circuits. In these dc circuits, any slight variations in voltage within the circuit may cause improper operation. For example, in Fig. 3.69, noise (random fluctuations in supply voltage) present on the V_{CC} line can cause problems by presenting improper voltage levels to an IC's sensitive supply lead. (Some ICs will act erratically if this happens.) However, by placing a bypass capacitor in parallel to the IC's input, the capacitor will bypass the high-frequency noise around the IC to ground, thus maintaining a steady dc voltage. The bypass capacitor acts to decouple the IC from the supply.

It's important to note that variations within the supply voltage line aren't caused just by random low-level fluctuations. They are also caused by sudden fluctuations in voltage caused by high-current switching action that draws sudden, large amounts of current from the supply line. The more current these devices draw, the bigger the ripple in the supply line. Relay and motor switching is notorious in this regard. (Usually these devices incorporate a snubber diode or some type of local transient suppressor to limit the magnitude of the transient. However, low-level, high-frequency ringing that occurs after switching will often sneak into the line.) Even TTL and CMOS ICs can generate current spikes in the power lines, due to a

Bypassing (Low-Pass Filter)

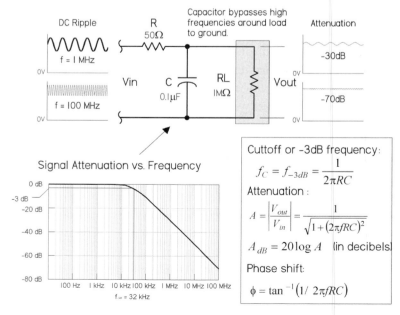

In this circuit, the RC section acts like a low-pass filter, which attenuates high frequencies from reaching the load or circuit element.

As $X_C < R_L$, signals bypass R_L through C.
As $X_C > R_L$, signals pass through R_L.

where $X_C = 1/(2\pi f C)$, the capacitor's reactance. In other words, at high frequencies, X_C gets small, so signals tend to be diverted around R_L through C.

The graph in the figure shows the attenuation versus frequency response, and the equations tell you how to calculate the cutoff frequency, attenuation, and phase shift.

Note that the R in the circuit in this figure isn't often physically present as a discrete component. It may represent, say, the inherent resistance present in the power supply line (which is usually much smaller than what's shown). Though R helps set the frequency response, it can reduce the clamping efficiency.

Cuttoff or −3dB frequency:
$$f_C = f_{-3dB} = \frac{1}{2\pi RC}$$

Attenuation :
$$A = \left|\frac{V_{out}}{V_{in}}\right| = \frac{1}{\sqrt{1 + (2\pi f R C)^2}}$$

$$A_{dB} = 20\log A \quad \text{(in decibels)}$$

Phase shift:
$$\phi = \tan^{-1}\left(1/\, 2\pi f R C\right)$$

FIGURE 3.68

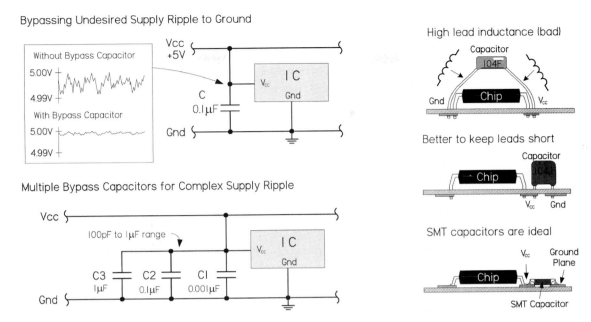

FIGURE 3.69   The top circuit uses a 0.1-μF decoupling capacitor to keep the dc supply line free of high-frequency transients. The lower circuit uses three capacitors of different capacitance levels to handle a wide range of transient frequencies.

transient state in which both output transistors are simultaneously on. The resistance between the 5-V supply terminals limits the supply current, and as speed increases this resistance gets smaller, and the transient currents increase to as high as 100 mA. These transient currents usually contain high frequencies due to the fast switching of the logic device. When the current spikes propagate down a power distribution system, they can develop 10- to 100-mV voltage spikes. Even worse, if an entire bus changes states, the effects are additive, resulting in transients as high as 500 mV propagating down the power lines. Such large transients wreak havoc within logic circuits.

It's important to regard the power supply and the distribution systems (wires, PCB bus, etc.) as nonideal. The power supply contains internal resistance, and the supply distribution system (wires, PCB traces, etc.) contains small amounts of resistance, inductance, and capacitance. Any sudden demands in current from a device attached to the distribution system will thus result in a voltage dip in the supply—use Ohm's law.

SELECTING AND PLACING DECOUPLING CAPACITORS

What Needs Bypassing: High clock-rate logic circuits and other sensitive analog circuits all require decoupling of the power supplies. As a general rule of thumb, use one 0.1-μF ceramic per digital chip, two 0.1-μF ceramics per analog chip, (one on each supply where positive and negative supplies are used) and one 1-μF tantalum per every eight ICs or per IC row, though you can often do with less. Also, a good place for bypass capacitors is on power connectors. Anytime power lines are leading off to another board or long wire, it's a good idea to throw in a bypass capacitor; long wires act like inductive antennas, picking up electrical noise from any magnetic field. A capacitor at both ends of the wire is a good idea—a 0.01-μF or 0.001-μF capacitor connected across the line will often do the trick.

Placement: Capacitor placement is crucial for good high-frequency decoupling. Place capacitors as close as possible to the IC, between power pin and ground pin, and ensure that leads consist of wide PC tracks. Run traces from device to capacitor, then to power planes. Capacitor lead lengths must be kept short (less than 1.5 mm); even a small amount of wire has considerable inductance, which can resonate with the capacitor. Surface-mount capacitors are excellent in this regard, since you can place them almost on top of the power leads, thus eliminating lead inductance.

Size of Capacitor: The frequency of the ripple has a role in choosing the capacitor value. A rule of thumb: the higher the frequency ripple, the smaller the bypass capacitor. In Fig. 3.69, a 0.01- to 0.1-µF capacitor with a self-resonant frequency from around 10 to 100 MHz is used to handle high-frequency transients. If you have very high-frequency components in your circuit, it might be worth using a pair of capacitors in parallel—one large value (say, 0.01 µF) and one small value (say, 100 pF). If there is a complex ripple, several bypass capacitors in parallel may be used, each one targeting a slightly different frequency. For example, in the lower circuit in Fig. 3.69, C_1 (1 µF) catches the lower voltage dips that are relatively low in frequency (associated with bus transients), C_2 (0.1 µF) the midrange frequencies, and C_3 (0.001 µF) the higher frequencies. In general, local decoupling values range from 100 pF to 1 µF. It's generally not a good idea to place a large 1-µF capacitor on each individual IC, except in critical cases; if there is less than 10 cm of reasonably wide PC track between each IC and the capacitor, it's possible to share it among several ICs.

Type of Capacitor: The type of capacitor used in decoupling is very important. Avoid capacitors with low ESR, high inductance, and high dissipation factor. For example, aluminum electrolytic capacitors are not a good choice for high-frequency decoupling. However, a 1-µF tantalum electrolytic, as mentioned before, is useful when decoupling at lower frequencies. Monolithic ceramic capacitors, especially

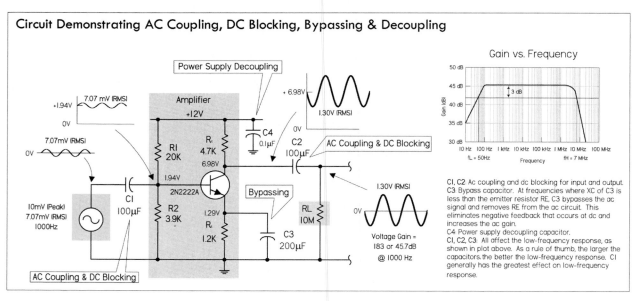

Circuit Demonstrating AC Coupling, DC Blocking, Bypassing & Decoupling

CI, C2: Ac coupling and dc blocking for input and output. C3: Bypass capacitor. At frequencies where XC of C3 is less than the emitter resistor RE, C3 bypasses the ac signal and removes RE from the ac circuit. This eliminates negative feedback that occurs at dc and increases the ac gain. C4: Power supply decoupling capacitor. CI, C2, C3: All affect the low-frequency response, as shown in plot above. As a rule of thumb, the larger the capacitors, the better the low-frequency response. CI generally has the greatest effect on low-frequency response.

FIGURE 3.70

surface-mount types, are an excellent choice for high-frequency decoupling due to their low ESL and good frequency response. Polyester and polypropylene capacitors are also good choices, provided you keep lead lengths short. The capacitor you eventually choose will depend on the frequency range you're trying to eliminate. See Table 3.7 for suggested decoupling capacitors.

3.6.10   Timing and Sample and Hold

To change the voltage across a capacitor, it takes a finite amount of time—a characteristic dependent on the capacitance value and any added series resistance. The charging time, as well as the acquired voltage level, can easily be predicted, for example, by using the RC time constant. This phenomenon is put to use in timing circuits, such as oscillators, signal generators, and latch timers. An explanation of how a simple relaxation oscillator works is shown in Fig. 3.71.

In sample-and-hold circuits, a capacitor will acquire an analog voltage level equal to that which is taken during a sample (switch closed). The stored charge within the capacitor holds onto this voltage level until a new sample is taken. See the sample-and-hold circuits in Fig. 3.72.

In both timing and sample-and-hold circuits, important capacitor characteristics to look for include high insulation resistance (IR), relatively low ESR, low dielectric absorption (DA), and good capacitance stability. Polystyrene capacitors with very high IR ratings are good in many timing and storage applications, except within timing circuits that exceed a few hundred kilohertz, where ESL becomes a factor. Monolithic ceramic capacitors, though good at high frequencies, have considerably high dielectric absorption, making them unsuitable for sample-and-hold

Simple Relaxation Oscillator

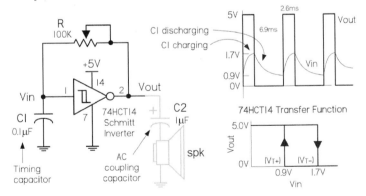

FIGURE 3.71

When a capacitor is charged through a resistor, its voltage follows a very predictable curve. By applying the capacitor's voltage to a level-sensitive input of an IC, the IC will trigger an output response, once the voltage across the capacitor reaches the set trigger level. For example, in the circuit in the figure, a 74HCT14 Schmitt inverter, along with an RC network, comprise a simple squarewave oscillator. Assuming the capacitor is discharged, the inverter's output is HIGH (+5 V). Current then passes through the resistor and starts collecting as charge within the capacitor. Once the capacitor reaches 1.7 V, the inverter interprets a HIGH input level, and sets the output LOW. C1 then begins discharging, and once its voltage reaches 0.9 V, the inverter interprets a LOW input level, and it again sets the output HIGH. The process continues. The frequency of oscillation is determined by the RC time constant. A smaller capacitance yields a higher frequency—it takes less time to charge up. Many different kinds of capacitors could be used, such as polypropylene, polyester, polystyrene (less than a few hundred kilohertz), and even electrolytic (at lower frequencies).

Sample-and-Hold Circuits

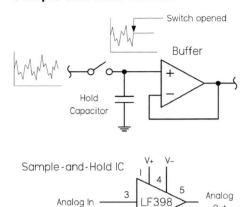

FIGURE 3.72

Sample-and-hold circuits are used to sample an analog signal and hold it so that it can be analyzed or converted into, say, a digital signal at one's leisure. In the first circuit, a switch acts as a sample/hold control. Sampling begins when the switch is closed and ends when the switch is opened. When the switch is opened, the input voltage present at that exact moment will be stored in C. The op amp acts as a buffer, relaying the capacitor's voltage to the output, but preventing the capacitor from discharging (ideally, no current enters the input of an op amp). The length of the sample voltage that can be held varies, depending on the capacitor's leakage. Use capacitors with high ESR, low DA, low ESL, and high stability. See text.

The LF398 is a special sample-and-hold IC whose control of sample-and-hold functions is enacted by a digital input signal. An external capacitor is still needed, however, to hold the sample voltage.

applications. Other capacitors often used in both cases include polypropylene and polyester, both of which have good IR ratings. In low-frequency relaxation-type oscillators and latch timers, electrolytic capacitors are often used, but they should not be used for sample-and-hold circuits; their leakage levels are too high to hold a constant sample voltage.

3.6.11   RC Ripple Filter

The full-wave rectifier dc supply, shown in Fig. 3.73, uses a capacitor to smooth out the rectified dc pulses in order to create a near constant dc output. The capacitor stores charge and delivers it to the load when the rectified or pulsing dc voltage decreases below its peak value. However, as it does this, the voltage drops according to the $R_L C$ time constant, until the capacitor recharges during another positive cycle. The resultant output voltage therefore takes on a small ac ripple. Since ripple is usually bad when driving sensitive circuits, it's important to keep it as low as possible by using a large capacitance value. Figure 3.74 shows how to derive and calculate the ripple voltage, average dc voltage, and ripple factor, given the input frequency, average dc rectified voltage, capacitance, and load resistance.

The ripple factor represents the maximum variation in output voltage (in rms) per the average dc output voltage. A typical and practical ripple factor to shoot for is usually around 0.05. Large capacitance values are desired to keep RF low—typically 1000 µF or more. Increasing the load resistance (decreasing load current) will also keep it down. To achieve higher capacitance levels, place filter capacitors in parallel.

Note that many times you'll probably include a voltage regulator IC between the filter capacitor and load to combat ac-line surges and maintain a constant output voltage with changing load-current demands. These regulators typically

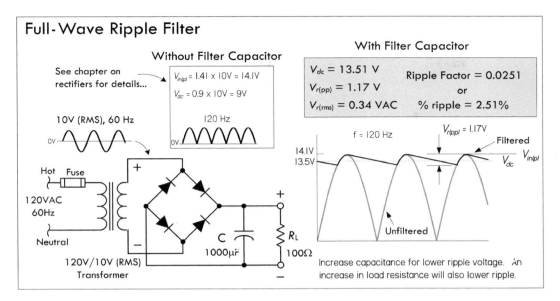

FIGURE 3.73   Here, a 1000-μF capacitor supply filter is used to reduce the output ripple voltage. With a 100-Ω load, the ripple voltage in this circuit is 0.34 VAC, or 1.17 V peak to peak. The ripple factor (rms ripple voltage divided by average dc voltage) is 0.00251. If we replaced the full-wave bridge with a half-wave rectifier, the ripple voltage would be twice as large. See equations in Fig. 3.74 and Problem 7 at the end of this chapter.

have additional built-in ripple rejection. See the chapter on voltage regulators and power supplies.

Important capacitor characteristics to consider in power supply filtering include ESR, voltage, and ripple current ratings. In most power supplies, the filtering capacitors are electrolytic, since most other types can't supply the necessary microfarads. In some high-voltage, high-current supplies, oil-filled capacitors may be used.

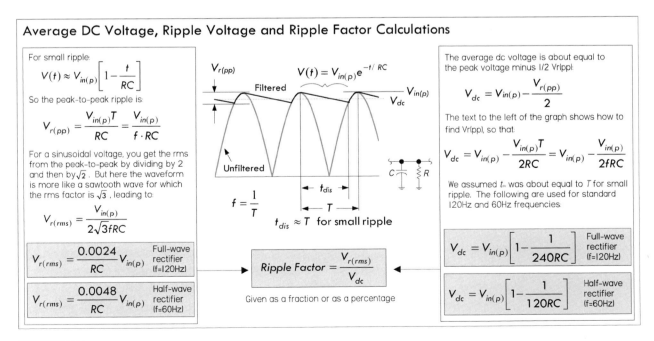

FIGURE 3.74

Speaker Crossover Network

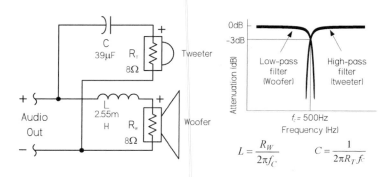

$$L = \frac{R_W}{2\pi f_C} \qquad C = \frac{1}{2\pi R_T f_C}$$

FIGURE 3.75 Crossover networks are used to provide each speaker in a system a range of frequencies most responsive to a given speaker's dynamic range. For example, the circuit in the figure is a simple first-order crossover for a system including a tweeter (high-frequency dynamic) and woofer (low-frequency dynamic). To divert only high frequencies to the tweeter, a capacitor is placed in series with the tweeter, where the internal resistance of the tweeter and the capacitance form a high-pass RC circuit. A low-pass filter is used to pass low frequencies to the woofer. It uses an inductor in series with the woofer's resistance. The response curves and formulas for calculating the cutoff frequency for each filter are provided to the right in the figure. Some popular crossover capacitors include polypropylene, polyester, and metallized-film capacitors.

3.6.12 Arc Suppression

There are two types of discharge that can damage switching contacts and generate noise in the process. The first is *glow discharges*, caused by the ignition of gases between contacts. These develop at around 320 V with about a 0.0003-in gap, and can be sustained at a much wider gap range. The other type of discharge is arc discharge, occurring at much smaller voltages, about 0.5 MV/cm. Minimum voltages and current are required for arc discharges to be sustained. Contact material also plays a role in sustained arc discharges (see table in Fig. 3.77).

The main cause of discharge or arcing is the sudden switching off of an inductive load, such as a relay, motor, solenoid, or transformer. These inductive loads do not instantly let go of a current. At the instant a contact opens, the current through the contact does not want to change. When the switch is closed, the resistance is essentially zero, making the voltage zero, too. As the contact opens, the resistance increases, which leads to a high voltage that leads to the arcing or discharge effect across the contacts, usually destructive to contact life—it promotes contact erosion. See Fig. 3.77.

The role of the *RC network*, or *snubber*, is to keep the voltage across the contacts below 300 V, while keeping the rate of voltage change below 1 V/µs, as well as keeping the current below the minimum current levels specified for a particular switch contact material. See the table in Fig. 3.77. The capacitor acts to absorb the instantaneous voltage change. When the contacts are closed, the capacitor doesn't play a role. As the contacts open, any change in voltage is restricted by the capacitor if its value is large enough to restrict the voltage change to less than 1 V/µs. However, a capacitor alone is not a good idea. When the contacts are opened, the capacitor charges up to the supply

Active Filters

Low-Pass RC Active Audio Filter

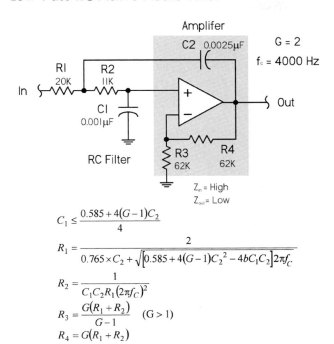

$$C_1 \le \frac{0.585 + 4(G-1)C_2}{4}$$

$$R_1 = \frac{2}{0.765 \times C_2 + \sqrt{\left[0.585 + 4(G-1)C_2{}^2 - 4bC_1C_2\right]}2\pi f_C}$$

$$R_2 = \frac{1}{C_1C_2R_1(2\pi f_C)^2}$$

$$R_3 = \frac{G(R_1+R_2)}{G-1} \quad (G>1)$$

$$R_4 = G(R_1+R_2)$$

G = Gain, f_c = -3dB cutoff point, C2 = a standard value near $10/f_c\mu$F. Note: For G = I, short R4 and omit R3.

High-Pass RC Active Audio Filter

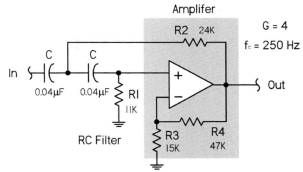

$$F_1 \le \frac{4}{\left[0.765 + \sqrt{0.585 + 8(G-1)2\pi f_C C}\right]}$$

$$R_2 = \frac{1}{C^2R_1(2\pi f_C)^2}$$

$$R_3 = \frac{GR_1}{G-1} \quad (G>1)$$

$$R_4 = GR_1$$

G = Gain, f_c = -3dB cutoff point, C = a standard value near $10/f_c\mu$F. Note: For G = I, short R4 and omit R3.

FIGURE 3.76   With the help of op amps, low-pass, high-pass, bandpass, and notch filters can be created with gain, if desired. Such filters are called _active_ filters, since they have an active element (op amp), as opposed to being passive (no additional power source). The circuits in the figure are low-pass and high-pass active filters suitable for audio applications. The gain and desired cutoff frequency (-3 dB) can be obtained by using the provided formulas to calculate the component values. Polyester, polypropylene film, and metallized capacitors are good in filters at audio ranges.

Contact Arcing Protection and Noise Suppression

Across the contacts

Across the load

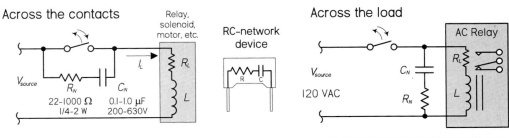

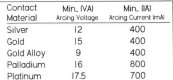

The circuit that is used is dependent on the particular situation. Either circuit may provide enough protection for a given application. For extreme cases, both load and contact protection may be required.

Functions of RC network include:

1. Keep voltage across contacts below 300V; $C \le (I_L/300)^2 L$

2. Keep the rate of voltage change below IV/μs; $C \ge I_L \times 10^{-6}$

3. Keep the current below that described in the table.

Contact Material	Min. (VA) Arcing Voltage	Min. (IA) Arcing Current (mA)
Silver	12	400
Gold	15	400
Gold Alloy	9	400
Palladium	16	800
Platinum	17.5	700

FIGURE 3.77

voltage. As the contacts close, an inrush of current results, limited only by residual resistance, and damage may still result. For this reason, a resistor is connected in series with the capacitor. The voltage across the contacts when opened is equal to the load current × resistance ($V = IR$). It is recommended that $V \leq$ supply voltage. In this case, the maximum resistance of the network will be equal to the load resistance.

Usually a 0.1- to 1-μF polypropylene film/foil or metallized-film capacitor with a voltage rating from 200 to 630 V will work for most applications. A carbon 22- to 1000-Ω, ¼- to 2-W capacitor is usually used for the resistance—the size and wattage depend on the load conditions. (In high-voltage applications, oil-filled capacitors with voltage ratings in the kilovolts are often used.) You can avoid building your own RC networks by simply buying prefab versions from the manufacturer.

Note that often shunt resistors, diodes, and gas discharge valves, along with various transient suppressor devices, can also be used to suppress arcing. However, some advantages RC networks have over these devices include bipolar operation suitable for ac application, relay operating time not very much affected, and no current consumption; electromagnetic interference suppression is also achieved (high-frequency EMI is present within the chaotic discharging sparks, but becomes dampened by the RC network).

In some ac circuits it may be necessary to connect the RC network across the load. For example, with ac relays, the relay contacts can also suffer from mechanical and electrical erosion due to arcing when subject to inductive spikes caused by sudden switching. Though diodes are used to suppress transients generated by dc relays, they cannot be used with ac relays due to the ac source's dual polarity. The rightmost circuit in Fig. 3.77 shows an RC network connected across the ac relay to reduce arcing effects.

3.6.13 Supercapacitor Applications

The extremely high capacitance levels and low leakage rates of supercapacitors make them nifty devices in many low-energy supplies that require recharging in mere seconds. Many interesting solar-power circuits can be made, such as light-seeking robots that use solar energy stored in a supercapacitor to drive small motors. Usually in these circuits special energy-saving modulating circuits are used to pulse the motor in order to conserve energy. Figure 3.78 shows a few simple supercapacitor circuits to get you started.

3.6.14 Problems

Problem 1: What do each of the following capacitor labels mean (see Fig. 3.79)?

Answer: (a) 0.022 μF ± 1%; (b) 0.1 μF ± 10%; (c) 47 pF ± 5%; (d) 0.47 μF, 100 V; (e) 0.047 μF ± 20%, −56% to +22% variation from +10°C to +85°C; (f) 0.68 nF; (g) 4.7 μF, 20 V; (h) 0.043 μF.

Problem 2: Find the equivalent capacitance and maximum voltage ratings for the capacitor networks shown in Fig. 3.80. Also, find V_1 and V_2 within circuit b.

Answer: (a) 157 μF, 35 V; (b) 0.9 μF, 200 V, $V_1 = 136$ V, $V_2 = 14$ V.

Solar-Energy Storage Without a Battery

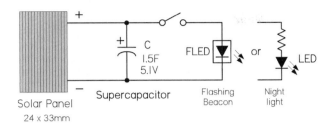

Similar to Above but with Photosensor

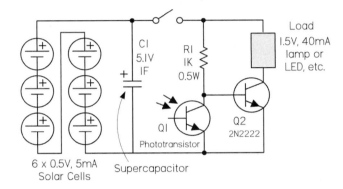

Power-Backup with Current-Limiting Circuit

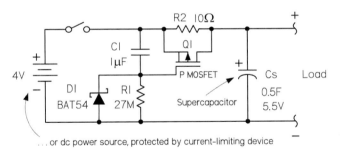

... or dc power source, protected by current-limiting device

FIGURE 3.78

The circuit at the top in the figure uses a supercapacitor to store charge from a solar panel over time. The larger the capacitance, the more charge that can be stored and the more current that can be delivered to a load. Here, a simple flashing LED (FLED) is used as a nighttime beacon, getting its power from the supercapacitor charged during the day. Assuming the solar panel has fully charged C if C = 0.047 F, then FLED flashes for around 10 min when the light is removed or the switch is opened. If C = 1.5 F, then FLED flashes for hours. There are a number of other cool solar-powered circuits you can make, such as solar-powered engines and backup supplies. Though the circuitry is a bit more complex, they all basically use the same principle of energy storage shown here.

The second circuit is similar to the first, except now there is a phototransistor Q_1 that keeps transistor Q_2 off if exterior light exists. This prevents the supercapacitor C_1 from discharging unless it's dark. When it's dark, Q_1 turns off and Q_2 turns on, providing approximately 10 s of illumination for one 40-mA, 1.5-V lamp. This can be used for short-term nightlights, for mailboxes, or near home entrances. This is a good circuit to avoid the hassle of dealing with dead batteries.

The circuit at the bottom uses a supercapacitor as a power backup for a battery (or a dc power source with current-limiting protection). As with the previous circuit, energy is stored in the supercapacitor; however, now a current-limiting regulator section is added to limit the current delivered to the supercapacitor when the power is first applied to the circuit. Thus, the supercapacitor can charge up without imposing a high current load on the battery. The MOSFET must be rated to dissipate the heat generated during the charging phase. This is a one-off event each time the device is turned on, which lasts only a few seconds, so the MOSFET doesn't have to be rated to dissipate that amount of power continuously.

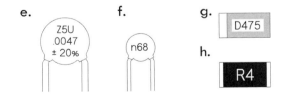

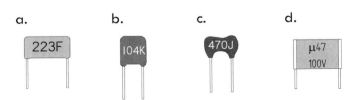

FIGURE 3.79

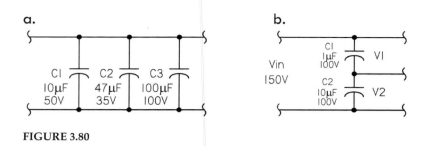

FIGURE 3.80

Problem 3: If a 100-μF capacitor in a high-voltage power supply is shunted by a 100-K resistor, what is the minimum time before the capacitor is considered fully discharged when power is turned off?

Answer: 50 s. Hint: After five time constants, a capacitor is considered discharged.

Problem 4: An IC uses an external RC charging network to control its timing. To match its internal circuitry, the IC requires 0.667 of the supply voltage (5 V), or 3.335 V. If the capacitor's value is 10 μF, what value resistor is required in the RC network to obtain a 5.0-s timing period?

Answer: 500 kΩ. Hint: Use

$$V(t) = V(1 - e^{-(t/RC)}) = 0.667 \text{ V}$$

Problem 5: What is the reactance of a 470-pF capacitor at 7.5 MHz and 15.0 MHz?

Answer: X_C(7.5 MHz) = 45.2 Ω, X_C(15 MHz) = 22.5 Ω

Problem 6: Find the f_{-3dB}, the attenuation in decibels A_{dB}, and phase shift ϕ for the given frequencies f_1, f_2, f_3, and f_4 of the filter circuits shown in Fig. 3.81.

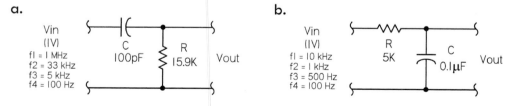

FIGURE 3.81

Answer: (a) f_{-3dB} = 100,097 Hz; f_1 = 1 MHz, A_{dB} = −0.043 dB, ϕ = 5.71°; f_2 = 33 kHz, A_{dB} = −10.08 dB, ϕ = 71.7°; f_3 = 5 kHz, A_{dB} = −26.04 dB, ϕ = 87.1°; f_4 = 100 Hz, A_{dB} = −60.00 dB, ϕ = 89.9°.

(b) f_{-3dB} = 318 Hz; f_1 = 10 kHz, A_{dB} = −29.95 dB, ϕ = −88.2°; f_2 = 1 kHz, A_{dB} = −10.36 dB, ϕ = −72.3°; f_3 = 500 Hz, A_{dB} = −5.40 dB, ϕ = −57.5°; f_4 = 100 Hz, A_{dB} = −0.41 dB, ϕ = −17.4°.

Problem 7: For the half-wave rectifier supply shown in Fig. 3.82, find the average dc output voltage, the peak-to-peak ripple voltage, the rms ripple voltage, the ripple factor, and percent ripple.

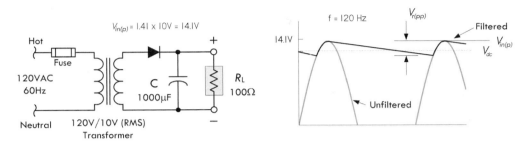

FIGURE 3.82

Answer: $V_{dc} = 12.92$ V, $V_{r(pp)} = 2.35$ V, $V_{r(rms)} = 0.68$ VAC, ripple factor = 0.0526, percent ripple = 5.26%. See Fig. 3.74.

3.7 Inductors

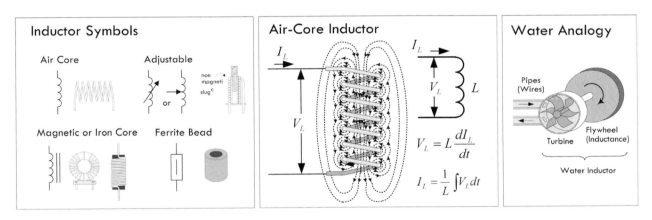

FIGURE 3.83

The basic role of an inductor is to prevent any sudden changes in current from flowing through it. (Refer to Sec. 2.24 for the details of how this works.) Under ac conditions, an inductor's impedance (reactance) increases with frequency; an inductor acts to block high-frequency signals while allowing low-frequency signals to pass through it. By selecting the proper inductance value, it is possible to create high-frequency chokes (e.g., RF/EMI chokes) that, when placed in series with power or signal paths, will prevent RF (radiofrequency) or EMI (electromagnetic interference) from entering into the main circuit, where they could introduce undesired hum and false triggering effects.

Filter networks can be created when inductors are used alongside other components such as resistors and capacitors. For example, Fig. 3.84a and b show

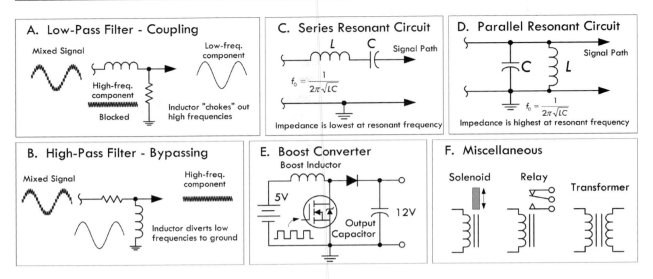

FIGURE 3.84

low-pass and high-pass filters using an inductor as the reactive element. In the low-pass filter, the inductor "chokes" out the high-frequency components, while in the high-pass filter the inductor passes the low-frequency components to ground—high-frequency components are prevented from taking the same path and follow the signal path. Series-resonant (bandpass) and parallel-resonant (notch) filters are shown in Figs. 3.84c and d. Parallel-resonant filters are used in oscillator circuits to eliminate, from an amplifier, any input frequencies significantly different from the resonant frequency of the LC filter. Such circuits are often used to generate radio-frequency carrier signals for transmitters. Resonant filters also act as tuned circuits used in radio reception.

The energy storage nature of an inductor can be utilized in switching power supplies. For example, in Fig. 3.84e, a step-up switching regulator or "boost converter" is used to increase a 5-V input voltage to a 12-V output voltage. When the control element (transistor) is on, energy is stored in the inductor. The load, isolated by the diode, is supplied by the charge stored in the capacitor. When the control element is off, the energy stored in the inductor is added to the input voltage. At this time, the inductor supplies load current as well as restoring charge to the capacitor. Other converter configurations include step-down and inverting switching regulators.

A heavily wound, iron core inductor can be used as an electromagnet, capable of attracting steel and other ferromagnetic materials. A solenoid is an electromagnet that has a mechanical mechanism that is pulled when the solenoid is energized by current. The movement mechanism may involve latching or unlatching a door, or opening or closing a valve (i.e., a solenoid valve), or making or breaking switch contacts (as in an electrical relay).

Coupled inductors that share magnetic flux linkage are used to create transformers—devices that utilize mutual inductance to step up or step down ac voltages and currents.

3.7.1 Inductance

The voltage across an inductor is proportional to the rate at which the current changes. The relationship between voltage and current are described by the following equations:

$$V_L = L\frac{dI_L}{dt} \quad \text{(Voltage across inductor = induced voltage)}$$

$$I_L = \frac{1}{L}\int V_L dt \quad \text{(Current through inductor)}$$

The proportionality constant L is the inductance, which depends on a number of physical parameters, such as coil shape, number of turns, and core makeup. The basic unit of inductance is the henry, abbreviated H. One henry equals an induced voltage of 1 V when the current is varying at a rate of 1 A/s:

$$1\,H = \frac{1\,V}{1\,A/s}$$

Typical values of inductance for commercial inductors vary from a few nanohenrys for small air core inductors to 50 H for large iron core inductors.

3.7.2 Constructing Inductors

Although constructing a capacitor from scratch is rare and usually unnecessary, constructing an inductor from scratch is quite common. Often it is necessary, since specific inductance values may be difficult to find or overly expensive in commercial units. Even stock inductors often require trimming to match an accurately desired value. Figure 3.85 shows some formulas used to construct inductors. Refer to Sec. 2.24 to see example calculations.

3.7.3 Inductors in Series and Parallel

When two or more inductors are connected in series, the total inductance is equal to the sum of the individual inductances, provided the coils are sufficiently separated so that coils are not in the magnetic field of one another:

$$L_{\text{TOT}} = L_1 + L_2 + L_3 + \cdots + L_N \quad \text{(Inductor in series)}$$

Air-Coil Inductor

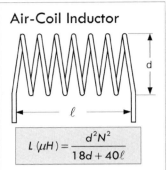

$$L\ (\mu H) = \frac{d^2 N^2}{18d + 40\ell}$$

L = inductance in μH

d = coil diameter in inches (from wire center to wire center)

ℓ = coil length in inches

N = total number of turns

Multilayer Air-Core Inductor

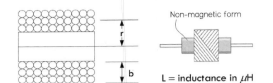

Non-magnetic form

L = inductance in μH

r = radius of coil taken from center to middle wound layer in inches

$$L\ (\mu H) = \frac{0.8\ (N \times r)^2}{6r + 9\ell + 10b}$$

b = coil length in inches

ℓ = coil length in inches

N = total number of turns

Spiral Coil Inductor

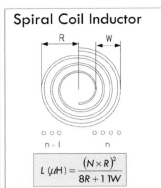

$$L\ (\mu H) = \frac{(N \times R)^2}{8R + 11W}$$

L = inductance in μH

R = average radius of coil in inches

W = width of coil in inches

N = total number of turns

Toroidal Inductor

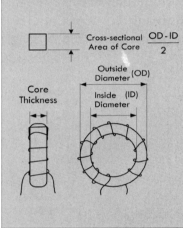

Cross-sectional Area of Core $\dfrac{OD - ID}{2}$

Outside Diameter (OD)

Inside (ID) Diameter

Core Thickness

Powdered-Iron Core:

$$L\ (\mu H) = \frac{A_L \times n^2}{10{,}000} \qquad N = 100 \sqrt{\frac{\text{desired } L \text{ in } \mu H}{A_L \text{ in } \mu H \text{ per 100 turns}}}$$

Inductance Index (A_L) for Powdered-Iron Toroids

Size	26	3	15	1	2	7	6	10	12	17	0
T-12	na	60	50	48	20	18	17	12	7.5	7.5	3.0
T-16	145	61	55	44	22	na	19	13	8.0	8.0	3.0
T-20	180	76	65	52	27	24	22	16	10	10	3.5
T-25	235	100	85	70	34	29	27	19	12	12	4.5
T-30	325	140	93	85	43	37	36	25	16	16	6
T-50	320	175	135	100	49	43	40	31	18	18	6.4
T-80	450	180	170	115	55	50	45	32	22	22	8.5
T-106	900	450	345	325	135	133	116	na	na	na	15
T-130	785	350	250	200	110	103	96	na	na	na	15
T-184	1640	720	na	500	240	na	195	na	na	na	na
T-200	895	425	na	250	120	105	100	na	na	na	na

* The units of A$_L$ are in μH per 100 turns.

Ferrite Core:

$$L\ (\mu H) = \frac{A_L \times N^2}{1{,}000{,}000}$$

$$N = 1000 \sqrt{\frac{(\text{desired } L \text{ in mH})}{(A_L \text{ in mH per 1000 turns})}}$$

Inductance Index (A_L) for Ferrite Toroids

Size	63/67-Mix	61-Mix	43-Mix	77(72) Mix	J(75) Mix
FT-23	7.9	24.8	188.0	396	980
FT-37	19.7	55.3	420.0	884	2196
FT-50	22.0	68.0	523.0	1100	2715
FT-82	22.4	73.3	557.0	1170	NA
FT-114	25.4	79.3	603.0	1270	3179

* The units of A$_L$ are in mH per 1000 turns.

FIGURE 3.85

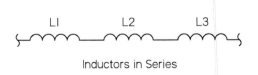

Inductors in Series

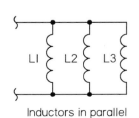

Inductors in parallel

FIGURE 3.86

If inductors are connected in parallel, and if the coils are separated sufficiently, the total inductance is given by:

$$\frac{1}{L_{\text{TOT}}} = \frac{1}{L_1} + \frac{1}{L_2} + \frac{1}{L_3} + \cdots \frac{1}{L_N} \quad \text{(Inductors in parallel)}$$

When only two inductors are in parallel, the formula simplifies to $L_{\text{TOT}} = (L_1 \times L_2)/(L_1 + L_2)$. Refer to Sec. 2.24 to see how these formulas were derived.

3.7.4 RL Time Constant

When a resistor is placed in series with an inductor, the resistance controls the rate at which energy is pumped into the magnetic field of an inductor (or pumped back into the circuit when the field collapses). Figure 3.87 shows energizing and deenergizing RL circuits with corresponding current-versus-time response curves.

In the energizing RL circuit, the inductor current and voltage as a function of time after the switch is thrown to the up position is given by the following equation:

$$I(t) = \frac{V_S}{R}\left(1 - e^{-Rt/L}\right) \quad \text{(Current through energizing inductor)}$$

$$V_L(t) = V_S e^{-Rt/L} \quad \text{(Voltage across energizing inductor)}$$

$$V_R(t) = V_S(1 - e^{-t(L/R)}) \quad \text{(Voltage across resistor)}$$

RL Current vs. Time Characteristics

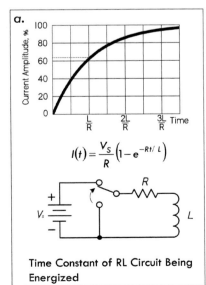

$$I(t) = \frac{V_S}{R}\left(1 - e^{-Rt/L}\right)$$

Time Constant of RL Circuit Being Energized

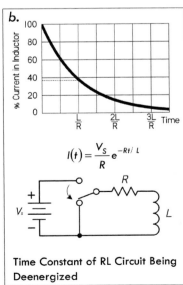

$$I(t) = \frac{V_S}{R} e^{-Rt/L}$$

Time Constant of RL Circuit Being Deenergized

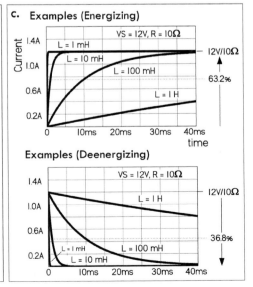

FIGURE 3.87

The time in seconds required for current to build up to 63.2 percent of the maximum value is called the time constant τ, which is equal to L/R. After five time constants, the current is considered to have reached its maximum value.

For a deenergizing RL circuit, things are a bit tricky. However, if we assume when we remove the applied voltage that there is no break in the circuit by moving the discharge to the ground, we can apply the following deenergizing equations:

$$I(t) = \frac{V_S}{R}\left(1 - e^{-Rt/L}\right) \quad \text{(Current through deenergizing inductor)}$$

$$V_L(t) = -V_s e^{-Rt/L} \quad \text{(Voltage across energizing inductor)}$$

$$V_R(t) = V_s e^{-Rt/L} \quad \text{(Voltage across resistor)}$$

It takes one time constant $\tau = L/R$ for the inductor to lose 63.2 percent of its initial current value. After five time constants, the current is considered to have reached its minimum value.

The deenergizing RL circuit requires special attention. In real life, when the current flow through an inductor is suddenly interrupted—say, by flipping a switch—there can be a very large induced voltage generated as the inductor's magnetic field rapidly collapses. The magnitude of the induced voltage can be so great that "electron pressure" at the switch contacts overcomes the work function of the metal, and a spark may jump between the contacts. Section 2.24 describes inductive spikes in greater detail.

3.7.5 Inductive Reactance

An inductor has a frequency-sensitive impedance, or inductive reactance, that increases with applied frequency. Unlike resistance, the inductive reactance doesn't dissipate energy in the form of heat, but temporarily stores energy in a magnetic field and later returns it (or reflects it back) to the source. Inductive reactance has units of ohms, and it increases with frequency according to the following equation:

$$X_L = 2\pi f L \quad \text{(Inductive reactance)}$$

For example, a 1-H coil with an applied frequency of 60 Hz will provide 377 Ω of reactance, while a 10-μH inductor at 20 MHz will provide 1257 Ω of reactance.

Note that the preceding equation is for an ideal inductor. Real inductors have imperfections, such as internal resistance and capacitance that tend to cause the reactance to deviate from the equation, especially at higher frequencies—see Sec. 2.24.

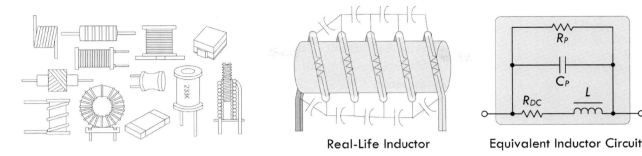

Real-Life Inductor   Equivalent Inductor Circuit

FIGURE 3.88

3.7.6 Real Inductors

There are many different kinds of inductors used for various applications. Selecting the right inductor requires understanding the nonideal characteristics of real inductors. Real inductors have imperfections, such as internal resistance and capacitance, that make them operate slightly differently than the ideal equations predict. Some inductors, due to their construction, may have larger resistive or capacitive components inherent in their design that can cause the inductor to act in a nonlinear fashion when approaching certain frequencies. (See Sec. 2.24 for details.) Other important differences include current-handling capacity, tolerance, maximum inductance and size, quality factor (Q), saturation characteristics, adjustability, radiated electromagnetic interference (EMI), and environmental endurance.

The following provides important inductor specifications that you'll find listed in manufacturer data sheets.

3.7.7 Inductor Specifications

Inductance (L): The property of an element that tends to oppose any change in the current flowing through it. The inductance depends on core material, core shape, size, and turn count of the coil. The unit of inductance is the henry (H). Most often, inductances are expressed in microhenries (μH).

1 henry (H) = 10^6 μF

1 millihenry (mH) = 10^3 μH

1 nanohenry (nH) = 10^{-3} μH

Inductive tolerance: Inductive tolerance is the allowed amount of variation from the nominal value specified by the manufacturer. Standard inductance tolerances are typically designated by a tolerance letter: F = $\pm$1%, G = $\pm$2%, H = $\pm$3%, J = $\pm$5%, K = $\pm$10%, L = $\pm$15% (some military products L = $\pm$20%), M = $\pm$20%.

Direct current resistance (*DCR*): The resistance of the inductor winding measured using dc current. The DCR is most often minimized in the design of an inductor and specified as a maximum rating.

Incremental current: The dc bias current flowing through an inductor that causes an inductance drop of 5 percent from the initial zero dc bias inductance value. This current level indicates where the inductance can be expected to drop significantly if the dc bias current is increased further. This applies mostly to ferrite cores rather than to powdered iron. Powdered iron cores exhibit soft saturation characteristics, which means their inductance drop from higher dc levels is much more gradual than that of ferrite cores. The rate at which the inductance will drop is also a function of the core shape.

Maximum dc current (IDC): The level of continuous direct current that can be passed through an inductor with no damage. The dc current level is based on a maximum temperature rise at the maximum rated ambient temperature. For low-frequency currents, the RMS current can be substituted for the dc-rated current.

Saturation current: The dc bias current flowing through an inductor that causes the inductance to drop by a specified amount from the initial zero dc bias inductance value. Common specified inductance drop percentages are 10 percent and 20 percent. It is useful to use the 10 percent inductance drop value for ferrite cores and 20 percent for powdered iron cores in energy storage applications. The cause of the inductance drop due to the dc bias current is related to the magnetic properties of the core. The core, and some of the space around the core, can store only a given amount of magnetic flux density. Beyond the maximum flux density point, the permeability of the core is reduced; thus, the inductance drops. Core saturation does not apply to air core inductors.

Self-resonant frequency (SRF or f_0): The frequency at which the inductor's distributed capacitance resonates with the inductance. At this frequency, the inductance is equal to the capacitance and they cancel each other. As a consequence, at SRF, the inductor acts as a purely resistive high-impedance element. Also at this frequency, the Q value of the inductor is zero. Distributed capacitance is caused by the turns of wire layered on top of each other and around the core. This capacitance is in parallel to the inductance. At frequencies above SRF, the capacitive reactance of the parallel combination will become the dominant component.

Quality factor (Q): The measure of the relative losses in the inductor. Q is defined as the ratio of inductive reactance to the effective resistance, or X_L/R_E. Since both X_L and R_E are functions of frequency, the test frequency must be given when specifying Q. As the self-resonant frequency, Q is zero since the inductance is zero at that point. Ideally, $Q = X_L/R_{DC} = 2\pi f L/R_{DC}$.

Inductance temperature coefficient: The change in inductance per unit temperature change. Measured under zero bias condition and expressed in parts per million (ppm).

Resistance temperature coefficient: The change in dc wire resistance per unit temperature change. Measured low dc bias (<1 V_{DC}) and expressed in parts per million (ppm).

Curie temperature (TC): The temperature beyond which the core material loses its magnetic properties.

Magnetic saturation flux density (B_{SAT}): A core parameter that indicates the maximum flux the material can be induced to hold. At this value of flux density, all magnetic domains within the core are magnetized and aligned.

Electromagnetic interference (EMI): For inductors, this refers to the amount of magnetic field radiated away from the inductor into space. The field may cause interference with other magnetically sensitive components and may require consideration in circuit design and layout.

3.7.8 Types of Inductors

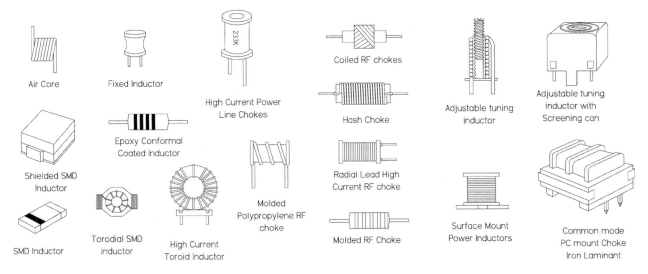

FIGURE 3.89

There are many different inductors out there. Some inductors are designed for general-purpose filtering, others for RF/EMI filtering, others for high-current choking, and still others for energy storage (in switching power supplies). Typically, manufacturers break down their inductor stock into the following categories: common-mode, general-purpose, high-current, high-frequency, power, and RF chokes. Fundamental specifications for selection include inductance value, dc-rated current, dc resistance, tolerance, and quality factor (Q). Table 3.8 provides a general overview matching inductance characteristics to specified application.

TABLE 3.8 Overview of Typical Applications for Inductors

APPLICATION	INDUCTANCE (L)	MAXIMUM DC CURRENT (IDC)	SELF-RESONANT FREQUENCY (SRF)	QUALITY FACTOR (Q)	DC RESISTANCE (RDC)
High-frequency (RF), resonance circuits	Low	Low	Very high	Very high	Low
EM coupling	High	☆	High	Low	Very low
Filter circuits	High	High	High	Low	Very low
Switch-mode power supplies, dc/dc converters	☆	High	Medium	Low	Low

Table 3.9 provides a selection guide to help determine what kind of inductor to use for a particular application. For in-depth details about specific inductors, check out manufacturers' websites and read through their data sheets. (Some noteworthy manufacturers: API Delevan, Bourns, C&D Technologies, Fastron, KOA, JW Miller Magnetics, muRata, Pulse, TRIAD, TDK, and VISHAY.)

TABLE 3.9   Inductor Selection Guide

INDUCTOR TYPE	DESCRIPTION	APPLICATION
Multilayer chip inductors	Used predominantly in high-density PCB surface-mount circuits where size, interboard magnetic coupling, and vibration are major concerns. They have fewer parasitic characteristics and less resistive loss when compared to most other leaded inductors. Come with excellent Q values, and high-frequency performance, while generating very little noise. Smaller boards and shorter traces also reduce EMI emissions and signal cross-coupling. Come in various current ratings.	Used for EMI/RFI attenuation and suppression, as well as reactive elements in LC resonant oscillators and impedance-matching networks, and other choke-coil type applications. They are found in A/D converters, bandpass filters, pulse generators, RF amplifiers, signal generators, switching power supplies, and telecom.
Molded inductors	Small with axial leads for PC board mounting. Outer coating protects coil from the environment. Comes in a shielded version as well. Frequency range is typically greater than 50 kHz.	Used in filters, AD converters, AM/FM radio, pulse generators, signal generators, switching power supplies, and telecom.
Shielded inductors	The magnetic shield is designed to prevent magnetic coupling and RF/EMI interference issues—especially important in densely packed boards where signal corruption is a major concern. Come in surface-mount, axial lead, and other configurations and a variety of current-handling ratings.	Used in high-reliability applications where magnetic coupling is to be avoided. Used for dc/dc converters, computers, telecom equipment, filters, LDC displays, etc.
Conformal coated (dipped) inductors	An inexpensive inductor that comes with axial or radial leads—similar to molded inductors. Outer coating protects inductor from environments	Typically used in less harsh environments than molded inductors. Usually used in less critical RFI/EMI applications. Higher-Q versions can be used in many of the same applications as molded inductors.

TABLE 3.9   **Inductor Selection Guide** (*Continued*)

INDUCTOR TYPE	DESCRIPTION	APPLICATION
High-current chokes Hash choke RF choke	These devices utilize ferrite or powdered iron cores to achieve large inductance values for low coil count and small volume. Fewer turns translates into lower dc resistance—an important feature for high-current applications.	High-current hash chokes are used for home appliances, communications systems, computer add-ons, dc-fc, switching power supplies, transmitters, and uninterrupted power supplies. Power line chokes are used for filters, power supplies, RFI suppression, power amps, switching regulators, SCR and Triac controls, and speaker crossover networks.
Wideband chokes	Wideband chokes attenuate unwanted signals in circuits without contributing to power losses at lower frequencies or dc. Impedances range from 20 to 500 Ω and are obtained over frequencies from 1.0 to 400 MHz. Currents are limited only by #22 or #24 AWG tinned copper wire used in construction.	Used mainly in PC boards to filter out EMI and RFI. Also used in RF circuits to suppress parasitic oscillation at VHF and UHF. Used in A/D converters, communications systems, computer add-ons, dc-fc, I/O boards, RF power amplifiers, signal generators, switching power supplies, telecom equipment, and uninterrupted power supplies.
Toroids	Exhibit high inductance due to powdered iron or ferrite core, and also display good self-shielding due to core shape (good low EMI source). Low coil count reduces dc resistance compared to other solenoid-wound inductors. Come in a variety of types: small surface-mount devices, larger general-purpose devices, and high-power toroids that can handle very large currents. Toroids are also less susceptible to induced noise from other components, as the applied magnetic field induces equal and opposite currents inside the toroid, thus canceling interference	Toroids are used in a variety of different applications. They are used as chokes in ac power lines and to reduce EMI. Also used in home appliances, audio generators, auto electronics, bandpass filters, audiovisual equipment, dc-fc, I/O boards, impedance match transformers, oscillators, pulse generators, switching power supplies, telecom equipment, transmitters, tuned amplifiers, uninterrupted power supplies, VHF/UHF repeaters, ripple filters, etc.
Pot cores	Provide ultra-high inductance values and high dc current ratings while maintaining a stable inductance due to high saturation currents and self-shielding. Excellent Q values in a small size.	Commonly used in telecom, audio, and automotive applications. Used as dc chokes, differential mode chokes, filters, and in switching circuits.

TABLE 3.9 Inductor Selection Guide (*Continued*)

INDUCTOR TYPE	DESCRIPTION	APPLICATION
Balun chokes	Typically used in impedance-matching applications. "Balun" refers to a balanced-unbalanced transformation of impedance levels.	Used in AM/FM radio, television and communications systems, I/O boards, impedance matching, pulse generators, transmitters, and walkie-talkies.
Air coils	Range from a single loop of wire (for ultra-high frequencies) to larger coils wrapped around nonmagnetic form. Air core inductors are used in high-frequency applications requiring high-Q characteristics, where losses and distortion (common in magnetic core inductors due to hysteresis and eddy currents) cannot be tolerated. Without a magnetic core, however, these inductors have low inductance values. Used primarily for RF applications. Surface-mount air coil are also available.	Used in RF resonant tank circuits for TV tuners, FM stereo receivers, garage door openers, pulse generators, RF power amplifiers, switching power supplies, RC toys, uninterrupted power supplies, and walkie-talkies.
Adjustable inductors (variable or tunable)	May use a slider contact along coils or, more commonly, a ferrite, powered iron, or brass slug screwed into the center of the coil. The inductance changes as the slug is screwed in and out of the coil. Magnetic cores increase the permeability or inductance, while brass slugs decrease inductance due to eddy current induction in a slug that reduces internal magnetic flux in the core. These devices are used to adjust resonant circuits that require high Q values. Used in circuits requiring very narrow bandwidths.	Typically used at frequencies above 50 kHz. Used in LC resonant tank circuits in AM/FM radio, TV, and other communication systems. Used in garage door openers, I/O boards, oscillators, pulse generators, RF power amplifiers, signal generators, switching power supplies, toys, transmitters, uninterrupted power supplies, VHF/UHF repeaters, and walkie-talkies.
Common-mode chokes	Common- and differential-mode chokes are used to eliminate noise on a pair of conductors. Common-mode noise is defined as noise that is present in or common to both conductors, and can be the result of induced noise caused by the antenna effect on a conductor or PC trace. Used to prevent EMI and RF interference from power supply lines and for prevention of malfunctioning of electronic equipment. Common-mode chokes usually have ferrite core material well suited to attenuate common-mode current.	Common-mode chokes are very useful devices found in many radio circuits. They may help nearly any interference problem, from cable TV to telephones to audio interference caused by RF picked up on speaker leads. Particularly well-suited for applications such as line filters in switch-mode power supplies, and are also commonly used in desktop computers, industrial electronics, office equipment, and consumer electronics such as TVs and audio equipment.

TABLE 3.9 Inductor Selection Guide (*Continued*)

INDUCTOR TYPE	DESCRIPTION	APPLICATION
Ferrite beads (ferrite chokes)	Unlike a typical core inductor, the bead requires no coiling of wire—though wire/cable from a circuit may be coiled around it or simply pass through it. This effectively increases the inductance of the wire (or wires). The inductance range is rather limited to RF. Used for removing RF energy that exists within a transmission line structure (PCB trace). To remove unwanted RF energy, chip beads are used as high-frequency resistors (attenuators) that allow dc to pass while absorbing RF energy and dissipating it as heat. Come in hollow and leaded core versions.	Ferrite beads are often slipped over cables that radiate RF (computers, dimmers, fluorescent lights, motors, etc.). Ferrite beads are also placed on cables entering receiving equipment to prevent external RF from entering and contaminating signals in the cable runs. The "bumps" on computer and appliance cables (mouse, keyboard, monitor, etc.) are usually a ferrite bead coated by the plastic cord coating.
Ceramic core inductor	These inductors use a special ceramic core that surpasses many ferrite-core-type inductors in terms of high-frequency operation, low IDC, high SRF, high Q, and tight tolerances.	Used in LC resonant circuits such as oscillators and signal generators. Used in impedance matching, circuit isolation, and RF filtration. Found in mobile phones, Bluetooth devices, wireless instruments, as well as audio, TV, and telecom devices.
Antenna rods ferrite core Phenolic core	Commonly used in antenna applications where narrow bandwidths are required. The core material, or "rod," on which the coil is wound is either ferrite, powered iron, or, alternatively, phenolic (essentially an air core). The magnetic core versions are far more popular, but the phenolic cores provide higher operating frequencies.	As antennas, ferrite cores with 800 permeability rods are suitable in 100-kHz to 1-MHz range. The 125 permeability rods are suitable in the 550-kHz to 1.6-MHz range. The 40 permeability rods are suitable in the 30-MHz range. The 20 permeability rods are suitable in the 150-MHz range. As the core's permeability decreases, the operating frequency increases.
Current sense inductor	Used to sense current passing through conductors. Specified by frequency range. Often come with a center tap.	Used in high-current, low-profile point of load (POL) converters, dc/dc converters in distributed power systems, and dc/dc converters for field-programmable gate arrays. Found in PDAs, notebooks, desktops, servers, and battery-powered devices.

3.7.9 Reading Inductor Labels

If you are lucky, the inductance value and tolerance will be printed on your inductor in an easy-to-read format (e.g., 82 μH ± 10%). However, many molded and dipped inductors come with a band color code. For smaller surface-mount inductors, a special shorthand printed code provides the inductance value and tolerance—see Fig. 3.90.

5-Band Inductor Code

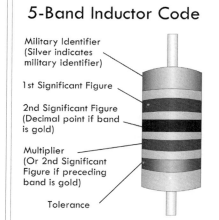

Military Identifier
(Silver indicates
military identifier)

1st Significant Figure

2nd Significant Figure
(Decimal point if band
is gold)

Multiplier
(Or 2nd Significant
Figure if preceding
band is gold)

Tolerance

Axial-leaded inductors are typically color banded. The military color code consists of 5 bands. A silver band that is double the width of the other four bands is located near one end. When present, the band identifies military radio frequency coils. The next three bands indicate the inductance value in microhenries, while the fourth band indicates tolerance.

For inductances less than 10, when either of the first two inductance bands is gold, the gold band represents a decimal point, while the other two bands represent significant figures. When the inductance is 10 or more, the first two bands represent significant figures, and the third band represents the multiplier.

Example

SILVER (Military)

GREEN (5)

GOLD (".")

BLUF (6)

BROWN (±1%)

5.6μH ± 1%

Color	1st S.F.	2nd S.F.	Multiplier	Tolerance
Black	0	0	1	
Brown	1	1	10	±1%
Red	2	2	100	±2%
Orange	3	3	1,000	±3%
Yellow	4	4	10,000	±4%
Green	5	5		
Blue	6	6		
Violet	7	7		
Grey	8	8		
White	9	9		
None				±20%
Silver				±10%
Gold		Dec. Point		±5%

4-Band Inductor Code

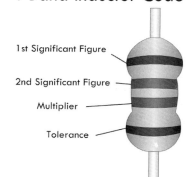

1st Significant Figure

2nd Significant Figure

Multiplier

Tolerance

The 4-band color code does not contain a military band. The first and second bands represent first and second significant figures. The third band represents the multiplier, where silver = x 0.01, and gold = x 0.1. The fourth band indicates the tolerance, where black is ±20%, silver is ±10%, and gold is ±5%. All values are given in microhenries.

Example

RED (2)

VIOLET (7)

RED (x100)

SILVER (±10%)

2700μH ± 10%

Color	1st S.F.	2nd S.F.	Multiplier	Tolerance
Black	0	0	1	±20%
Brown	1	1	10	
Red	2	2	100	
Orange	3	3	1,000	
Yellow	4	4		
Green	5	5		
Blue	6	6		
Violet	7	7		
Grey	8	8		
White	9	9		
None				
Silver			0.01	±10%
Gold			0.1	±5%

SMD Inductance Code

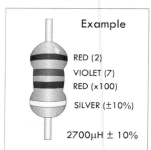

472K =4700μH ± 10%   221K =220μH ± 10%

22N =22nH ± 20%   5N6F =5.6nH ± 1%

391K = 390μH ± 10%   68M =0.068μH ± 20%

LETTER	TOLERANCE
F	± 1%
G	± 2%
H	± 3%
J	± 5%
K	± 10%
L	± 15%*
M	± 20%

* L = ± 20% for some Military devices

Value Code	Inductance	Value Code	Inductance	Value Code	Inductance	Value Code	Inductance	Value Code	Inductance	Value Code	Inductance
47	0.047μH	1R7	1.7μH	182	1800μH	260	26μH	390	39μH	6R8	6.8μH
68	0.068μH	1N8	1.8nH	R18	180nH	R27	270nH	4N7	4.7nH	601	600μH
82	0.082μH	1R8	1.8μH	181	180μH	271	270μH	4R7	4.7μH	600	60μH
R12	0.12μH	102	1000μH	18N	18nH	27N	27nH	401	400μH	650	65μH
R15	0.15μH	R10	0.1μH	180	18μH	270	27μH	400	40μH	682	6800μH
R18	0.18μH	101	100μH	190	19μH	3N3	3.3nH	421	420μH	681	680μH
R10	0.1μH	10N	10nH	1N0	1nH	3R3	3.3μH	450	45μH	68N	68nH
R22	0.22μH	100	10μH	1R0	1μH	3N9	3.9nH	472	4700μH	680	68μH
R27	0.27μH	110	11μH	2N2	2.2nH	3R9	3.9μH	471	470μH	700	70μH
R33	0.33μH	121	120μH	2R2	2.2μH	301	300μH	47N	47nH	751	750μH
R39	0.39μH	12N	12nH	2N7	2.7nH	300	30μH	470	47μH	750	75μH
R47	0.47μH	120	12μH	2R7	2.7μH	310	31μH	5N6	5.6nH	070	7μH
R56	0.56μH	141	140μH	202	2000μH	332	3300μH	5R6	5.6μH	8N2	8.2nH
R68	0.68μH	152	1500μH	201	200μH	R33	330nH	500	50μH	8R2	8.2μH
R82	0.82μH	R15	150μH	222	2200μH	331	330μH	561	560μH	800	80μH
1N2	1.2nH	151	150μH	R22	220nH	33N	33nH	56N	56nH	821	820μH
1R2	1.2μH	15N	15nH	221	220μH	330	33μH	560	56μH	82N	82nH
1N5	1.5nH	150	15μH	22N	22nH	391	390μH	050	5μH	820	82μH
1R5	1.5μH	170	17μH	220	22μH	39N	39nH	6N8	6.8nH	900	90μH

FIGURE 3.90

3.7.10 Inductor Applications

Filter Circuits

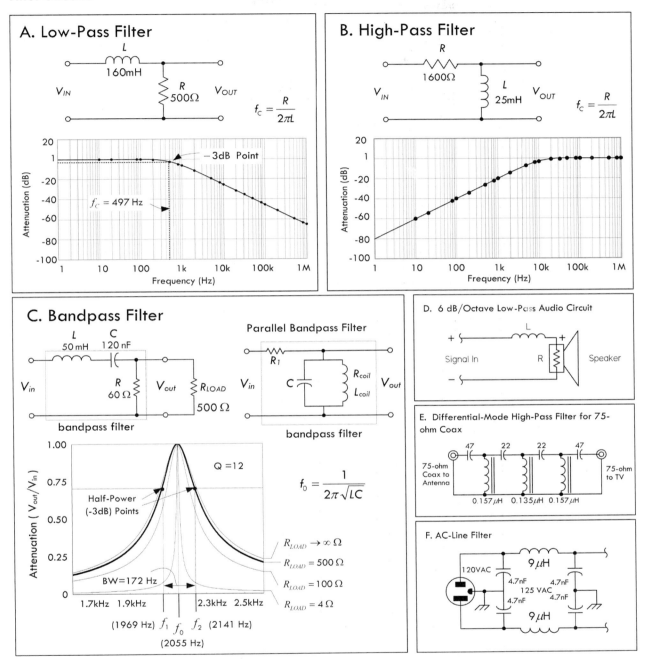

FIGURE 3.91   Filters offer little opposition to certain frequencies while blocking others. In (a), a low-pass filter is constructed using a resistor and inductor. The inductor's impedance increases with frequency, thus preventing high-frequency signals from passing. In (b), a high-pass filter blocks low frequencies—the inductor acts as a low-impedance path to ground for low-frequency signals. (c) shows a bandpass filter, allowing only a very narrow band of frequencies through. See Chap. 2 for the theory. (d) shows a low-pass filter used with a speaker. (e) shows a differential-mode high-pass filter for 75-Ω co-ax. It rejects high-frequency signals picked up by a TV antenna or that leak into a cable TV system. It is ineffective against common-mode signals, however. (f) shows an ac-line filter used to filter RF energy from power lines.

Switching Regulators

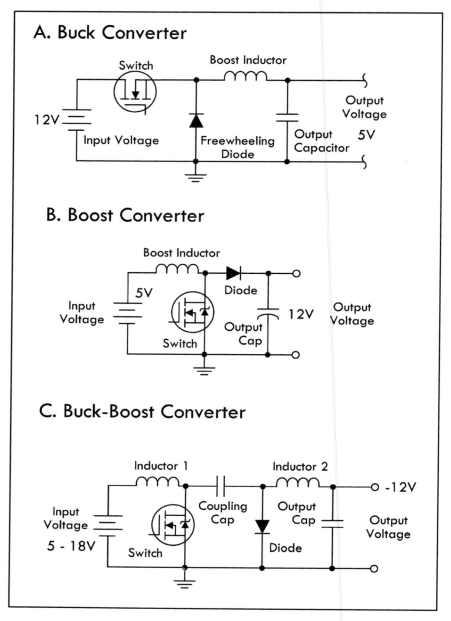

FIGURE 3.92 In switching regulator applications, an inductor is used as an energy storage device, when the semiconductor switch is on, the current in the inductor ramps up and energy is stored. When the switch is turned off, the stored energy is released into the load. Output voltages have a ripple that must be minimized by selecting appropriate inductance and output capacitor values. Figure 3.92a, b, and c shows various switching regulator configurations: *buck* (lower output voltage), *boost* (higher output voltage), and *buck-boost* (opposite polarity). Note that in the boost converter, the boost inductor current does not continuously flow to the load. During the switch "on" period, the inductor current flows to ground and the load current is supplied from the output capacitor.

Oscillators

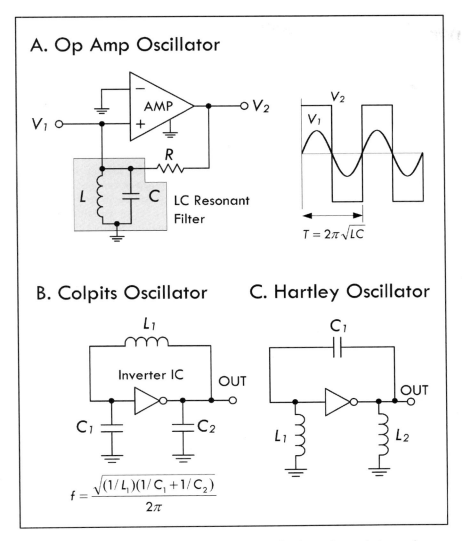

A. Op Amp Oscillator

LC Resonant Filter

$T = 2\pi\sqrt{LC}$

B. Colpits Oscillator

Inverter IC

$$f = \frac{\sqrt{(1/L_1)(1/C_1 + 1/C_2)}}{2\pi}$$

C. Hartley Oscillator

FIGURE 3.93 (*a*) An amplifier with positive feedback can be made to produce an output even in the absence of any input. Such circuits are called oscillators. In (a), an operational amplifier uses an LC resonant filter that eliminates from the amplifier input any frequencies significantly different from the LC resonant filter's resonant frequency: $f_0 = 1/(2\pi\sqrt{LC})$. The amplifier is alternately driven to saturation in the positive and negative direction, so it produces a squarewave at V_2. This squarewave has a Fourier component that is fed back to the noninverting input through resistor R in order to keep the oscillation from damping out. A sine wave is generated at V_1.

(b) There are two basic types of LC oscillators, Colpits and Hartley. The Colpits uses two capacitors, as shown here. It is generally favored over the Hartley, shown in (c), because of the simplicity of requiring only one inductor, which is usually more expensive and difficult to obtain than capacitors C_1 and C_2. The frequency of the Colpits oscillator is given by the formula in (b).

See Chap. 10 on oscillators for more details.

Radio Circuits

A. Short-Wave Receiver

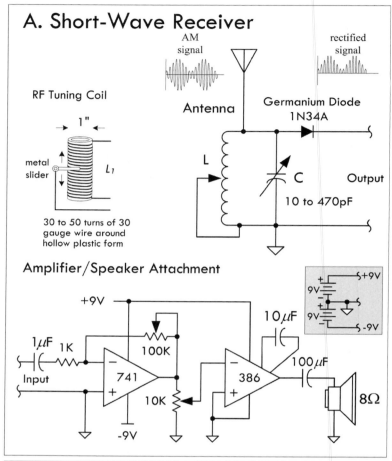

B. RF Oscillator/Transmitter

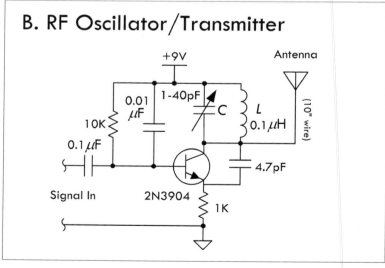

FIGURE 3.94

Here are a few simple radiofrequency circuits that make use of LC resonant filters to tune the circuits.

(a) The simplest radio receiver consists of nothing more than an antenna, a diode (germanium), and a pair of headphones. Such a receiver has no frequency selectivity, however, and will receive several of the strongest AM broadcast stations simultaneously. By adding a variable inductor that forms a resonant LC circuit with the antenna capacitance, a usable radio receiver can be constructed that is capable of tuning in to a number of different stations. (The variable capacitor provides additional tunability.) The encoded audio signal within the AM carrier is demodulated by the diode (the diode generates low-frequency [audio] Fourier components that are absent in the transmitted wave). After passing through the diode, only the positive half of the wave remains. This wave contains low-frequency components in addition to components at the frequency of the carrier wave. With the addition of a low-pass filter, only the low-frequency components remain. The frequency response of headphones and of a human ear will effectively act as a low-pass filter. The demodulated signal can be input into an amplifier to drive a speaker—see amplifier circuit. Real radio AM receivers are much more sophisticated than this, using a superheterodyne design scheme.

(b) A radio transmitter consists of an RF oscillator, one or more amplifier stages, and a modulator. In the simple FM transmitter shown here, the LC resonant filter sets the amplifier oscillatory frequency—the variable capacitor allows for adjustability. Audio signals that enter will be frequency-modulated into the carrier and radiated as radio waves. An FM radio receiver should be able to pick up the circuit signal. A homemade inductor can be made by tightly winding 10 turns of 22-gauge wire around a ¼-in form.

3.7.11   EMI/EMC Design Tips

The following are some tips to avoid EMI and EMC problems. Tips include proper design techniques for PCB layout, proper power supply considerations, and effective use of filtering components. (Referenced letters that follow refer to Fig. 3.95.)

PCB Design Tips

Avoid slit apertures in PCB layout, such as a ground plane that is divided into two parts. Regions of high impedance are sources of high EMI, so use wide tracks for power lines on the trace sides for increased conduction. Make signal tracks stripline (ground planes above and below) and include ground plane and power plane whenever possible. Keep HF and RF tracks short and lay out the HF tracks first—see (a). In Fig. 3.95. Avoid track stubs, since they cause reflections and harmonics—see (b). Use a surrounding guard ring and ground fill where possible on sensitive components and terminations. A guard ring around trace layers reduces emission out of board. Likewise, connect to ground only at a single point—see (e). Keep separate power planes over a common ground to reduce system noise and power coupling—see (d). Keep track runs as orthogonal as possible between adjacent layers—see (c). Also avoid loop track that acts as a receiving or radiating antenna. Avoid floating conductor areas, since they act as EMI radiators. Connect them to a ground plane otherwise.

Power Supply

Avoid loops in supply lines, as shown in (h). Also decouple supply lines at local boundaries—see (i). Position high-speed circuits closest to the power supply, while placing the slowest section farthest away to reduce power plane transients—see (g). When possible, isolate individual systems on both power supply and signal lines—see (j).

Filtering Components

Position biasing and pull-up/down components close to driver/bias points. Make use of common-mode chokes between current-carrying and signal lines to increase

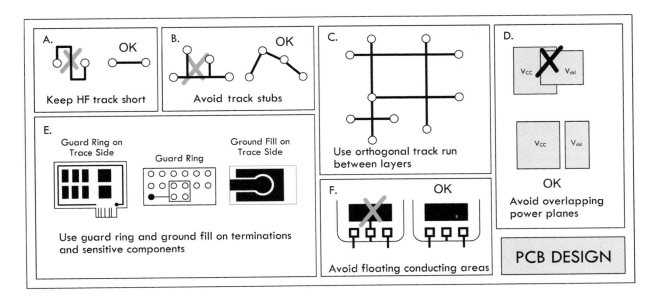

FIGURE 3.95

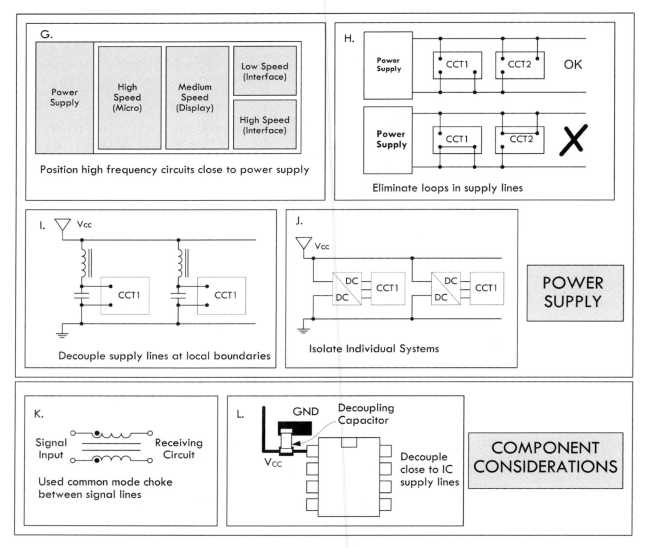

FIGURE 3.95 **(***Continued***)**

coupling and eliminate stray fields—see (k). Place decoupling capacitors close to chip supply lines to reduce component noise and power line transients—see (l).

These tips were adapted from an Engineering Note, "Electro-Magnetic Interference and Electro-Magnetic Compatibility (EMI/EMC)" written by David B. Fancher, Inductive Products Division, Vishay Dale.

3.8 Transformers

3.8.1 Basic Operations

A basic transformer is a two-port (four-terminal) device capable of transforming an ac input voltage into a higher or lower ac output voltage. Transformers are not designed to raise or lower dc voltages, however, since the conversion mechanism relies on a changing magnetic field generated by a changing current. A typical transformer consists of two or more insulated wire coils that share a common laminated iron core. One of the coils is called the primary (containing N_P turns), while the other coil is called the secondary (containing N_S turns). A simplistic representation of a transformer is shown in Fig. 3.96, along with its schematic symbol.

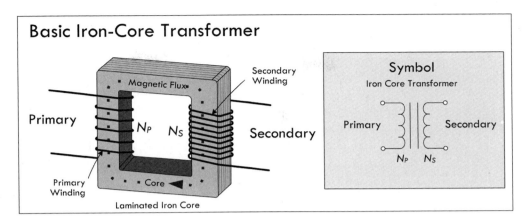

FIGURE 3.96

When an ac voltage is applied across the primary coil of the transformer, an alternating magnetic flux $\Phi_M = \int(V_{IN}/N_p)dt$ emanates from the primary, propagates through the iron-laminated core, and passes through the secondary coil. (The iron core increases the inductance, and the laminations decrease power-consuming eddy currents.) According to Faraday's law of induction, the changing magnetic flux induces a voltage of $V_S = N_S d\Phi_M/dt$, assuming there is perfect magnetic flux coupling (coefficient of coupling $k = 1$). Combining the primary flux equation with the secondary induced voltage equation results in the following useful expression:

$$V_S = V_P\left(\frac{N_S}{N_P}\right)$$ (3.1) Transformer voltage ratio

This equation says that if the number of turns in the primary coil is larger than the number of turns in the secondary coil, the secondary voltage will be smaller than the primary voltage. Conversely, if the number of turns in the primary coil is less than the number of turns in the secondary, the secondary voltage will be larger than the primary.

When a source voltage is applied across a transformer's primary terminals while the secondary terminals are open-circuited (see Fig. 3.97), the source treats the transformer as if it were a simple inductor with an impedance of $Z_P = j\omega L_P = \omega L \angle 90°$, where L_P represents the inductance of the primary coil. This means that the primary current will lag the voltage (source voltage) by 90°, and the primary current will be

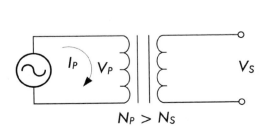

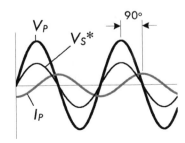

* In phase with V_P or 180° out of phase, depending on winding arrangement and ground reference.

FIGURE 3.97

equal to V_P/Z_P, according to Ohm's law. At the same time, a voltage of $(N_S/N_P)\,V_P$ will be present across the secondary and will be in phase with the primary voltage or 180° out of phase, depending on the secondary coil winding direction or depending on which secondary coil end you choose as a reference (more on this in a moment).

When there is no load attached to the secondary of a transformer, the current within the primary is called the magnetizing current of the transformer. An ideal transformer, with no internal losses, would consume no power, since the current through the primary inductor would be 90° out of phase with the voltage (in $P = IV$, I is imaginary and the "power" is imaginary or reactive). With no load in the secondary, the only losses in the transformer are associated with those losses in the iron core and losses within the primary coil wire itself.

Example 1: A transformer has a primary of 200 turns and a secondary of 1200 turns. If a 120 VAC is applied to the primary, what voltage appears across the secondary?

Answer: Rearranging Eq. 3.1,

$$V_S = V_P\left(\frac{N_S}{N_P}\right) = 120 \text{ VAC}\left(\frac{1200 \text{ turns}}{200 \text{ turns}}\right) = 720 \text{ VAC}$$

This is an example of a step-up transformer, since the secondary voltage is higher than the primary voltage.

Example 2: Using the same transformer from Example 1, flip it around so the secondary now acts as the primary. What will be the new secondary voltage?

Answer:

$$V_S = V_P\left(\frac{N_S}{N_P}\right) = 120 \text{ VAC}\left(\frac{200 \text{ turns}}{1200 \text{ turns}}\right) = 20 \text{ VAC}$$

This is an example of a step-down transformer, since the secondary voltage is lower than the primary voltage.

As you can see from the previous example, either winding of a transformer can be used as the primary, provided the windings have enough turns (enough inductance) to induce a voltage equal to the applied voltage without requiring an excessive current. The windings must also have insulation with a voltage rating sufficient for the voltage present.

Now let's take a look at what happens when you attach a load to the secondary, as shown in Fig. 3.98.

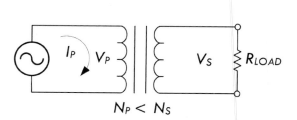

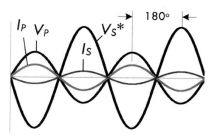

* In phase with V_P or 180° out of phase, depending on winding arrangement and ground reference.

FIGURE 3.98

When a load is attached to the secondary, the secondary current sets up a magnetic field that opposes the field set up by the primary current. For the induced voltage in the primary to equal the applied voltage, the original field must be maintained. The primary must draw enough additional current to set up a field exactly equal and opposite to the field set up by the secondary current. At this point, for practical purposes, we assume that the entire primary current is a result of the secondary load. (This is close to true, since the magnetizing current will be very small in comparison with the primary load current at rated power output.)

Current Ratio

To figure out the relationship between the primary and secondary currents, consider that an ideal transformer is 100 percent efficient (real transformers are around 65 to 99 percent efficient, depending on make), and then infer that all the power dissipated by the load in the secondary will be equal to the power supplied by the primary source. With the help of the generalized power law, we get:

$$P_P = P_S$$
$$I_P V_P = I_S V_S$$

Plugging our transformer voltage equation (3.1) into the V_S term, we get:

$$I_P V_P = I_S \left(V_P \frac{N_S}{N_P} \right)$$

Eliminating the V_P from both sides, we get the following useful current relation:

$$I_P = I_S \left(\frac{N_S}{N_P} \right) \qquad \text{(3.2) Ideal transformer current ratio}$$

Example 3: A transformer with a primary of 180 turns and a secondary with 1260 turns is delivering 0.10 A to a load. What is the primary current?

Answer: Rearranging Eq. 3.2 and solving for the primary current:

$$I_P = I_S \left(\frac{N_S}{N_P} \right) = 0.10\text{A} \left(\frac{1260 \text{ turns}}{180 \text{ turns}} \right) = 0.7 \text{ A}$$

Notice from the previous example that even though the secondary voltage is larger than the primary voltage, the secondary current is smaller than the primary current. The secondary current in an ideal transformer is 180° out of phase with the primary current, since the field in the secondary just offsets the field in the primary. The phase relationship between the currents in the windings holds true no matter what the phase difference between the current and the voltage of the secondary. In fact, the phase difference, if any, between voltage and current in the secondary will be reflected back to the primary as an identical phase difference. Note that phase, however, can be selected according to how you pull the secondary out—see the following note.

NOTE ABOUT PHASE

By now you may be a bit annoyed with the notion of phase. For example, to say that the primary voltage is out of phase with the secondary by 180° is a smack in the face of relativity. Couldn't you simply wind the secondary winding in a different direction or, more easily, simply reverse the secondary leads to get an output that is within phase? The answer is yes. It is a relativity game with the transformer pins. Figure 3.99 shows two transformers that are identical in very way except for the winding direction of the secondary. The winding A arrangement, when tested with a common ground and oscilloscope, yields in-phase voltage, while the winding in B yields voltage and currents that are 180° out of phase with the primary. To avoid confusion, phase dots are used to tell you the relative orientation of the windings. **The way to read the phase dots is that current goes into the primary dot and causes current to exit out the secondary dot.** The following drawing should help give you a sense of the direction of current flow and polarities that are defined by the position of the primary and secondary phase dots.

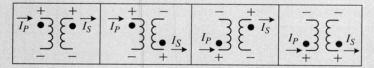

The transformer symbol actually isn't fully defined until phase dots are included. If you don't see them, either phase isn't important in the circuit, or the author forgot to include them. In many applications, keeping track of phase between the primary and secondary is not important since the secondary stage doesn't care about what the primary is doing. One the other hand, there are circuits whose secondary stages do indeed care and they won't work properly if phase isn't strictly adhered to. Usually these circuits have an additional linkage between the primary and secondary stages (e.g., transistor, capacitor additional transformer, etc.) and the fundamental operation of the circuit depends of interplay between the primary and secondary phase voltages. An example of a circuit that doesn't care about the secondary phase is a dc power supply that uses a step-down transformer's secondary to feed a rectifier stage. The rectifier stage doesn't care about phase. An example of a circuit that does care about phase is a Joule Thief circuit that links the secondary to primary through a bipolar transistor stage. Google "Joule Thief" to check out this interesting circuit. You'll also find phase dots frequently within radio circuitry where phase again is important.

Power Ratio

A moment ago, when deriving the transformer current equation, we assumed that the transfer of power from primary to secondary was 100 percent efficient. However, it is important to realize that there is always some power loss in the resistance of the

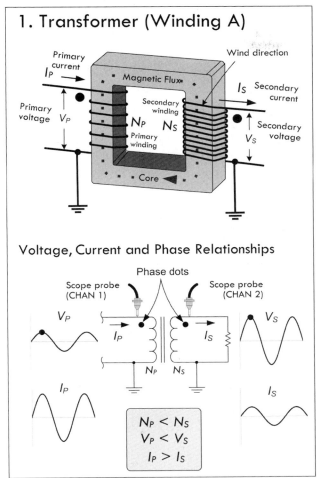

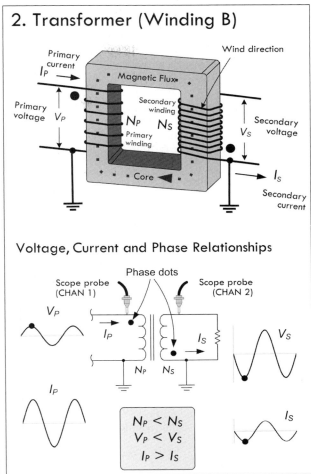

FIGURE 3.99

coils and in the iron core of the transformer. This means that the power taken from the source is greater than the power used in the secondary. This can be stated by the following expression:

$$P_S = n \times P_P$$ (3.3) Efficiency factor

where P_S is the power output from the secondary, P_P is the power input to primary and n is the efficiency factor. The efficiency n is always less than 1. It is usually expressed as a percentage—for example, 0.75 represents an efficiency of 75 percent.

Example 4: What is the power input to the primary if a transformer has an efficiency of 75 percent and its full load output at the secondary is 100 W?

Answer: Rearranging Eq. 3.3,

$$P_P = \frac{P_S}{n} = \frac{100 \text{ W}}{0.75} = 133 \text{ W}$$

Transformers are typically designed to have highest efficiency at the manufacturer's rated outputs. Above or below the rated output, the efficiency drops. The amount of power a transformer can handle depends on its own losses (heating of wire and core, etc.). Exceeding the rated power of a transformer can lead to wire meltdown or insulation breakdown. Even when the load is purely reactive, the transformer will still be generating heat loss due to internal resistance of the coils and losses in the core. For this reason, manufacturers also specify a maximum volt-amp rating, or VA-rating, that should not be exceeded.

Impedance Ratio

Using ac Ohm's law, $I_P = V_P/Z_P$, and assuming an ideal transformer, where power from the primary is 100 percent transferred to secondary, we can come up with an equation relating the primary and secondary impedances:

$$P_P = P_S$$

$$I_P V_P = I_S V_S$$

$$\frac{V^2{}_P}{Z_P} = \frac{V^2{}_S}{Z_S} \rightarrow (\text{plug in } V_S = V_P(N_S/N_P)) \rightarrow \frac{V^2{}_P}{Z_P} = \frac{V^2{}_P(N_S/N_P)^2}{Z_S}$$

Canceling the primary voltage terms, you get the following useful expression:

$$Z_P = Z_S \left(\frac{N_P}{N_S}\right)^2 \qquad \text{(3.4) Transformer impedance ratio}$$

where Z_P is the impedance looking into the primary terminal from the power source, and Z_S is the impedance of the load connected to the secondary. Figure 3.100 shows an equivalent circuit.

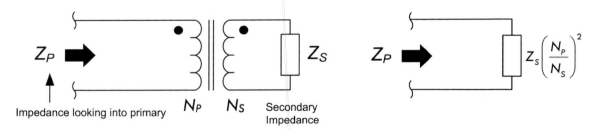

Impedance looking into primary N_P N_S Secondary Impedance

FIGURE 3.100

If the load impedance in the secondary increases, the impedance looking into the primary (from the source's point of view) will also increase in a manner that is proportional to the ratio of the turns squared.

Example 5: A transformer has a primary with 500 turns and a secondary with 1000 turns. What is the primary impedance if a 2000-Ω load impedance is attached to the secondary?

Answer: Using Eq. 3.4:

$$Z_P = 2000 \ \Omega \left(\frac{500 \ \text{turns}}{1000 \ \text{turns}}\right)^2 = 2000 \ \Omega (0.5)^2 = 500 \ \Omega$$

As you can see, by selecting the proper turns ratio, the impedance of a fixed load can be transformed to any desired value (ideally). If transformer losses can be neglected, the transformed (reflected) impedance has the same phase angle as the actual load impedance. Hence, if the load is purely resistive, the load presented by the primary to the power source will also be pure resistance. If the load impedance is complex (e.g., inductance and capacitance are thrown in so that load current and voltage are out of phase with each other), then the primary voltage and current will show the same phase angle.

In electronics, there are many instances where circuits require a specific load resistance (or impedance) for optimum performance. The impedance of the actual load dissipating power may differ widely from the impedance of the source. In this case, a transformer can be used to change the actual load into an impedance of desired value. This is referred to as impedance matching. We can rearrange Eq. 3.4 and get:

$$\frac{N_P}{N_S} = \sqrt{\frac{Z_P}{Z_S}} \tag{3.5}$$

where N_P/N_S is the required turns ratio—primary to secondary, Z_P is the primary impedance required, and Z_S is the impedance of the load connected to the secondary.

Example 6: An amplifier circuit requires a 500-Ω load for optimum performance, but is to be connected to an 8.0-Ω speaker. What turns ratio, primary to secondary, is required in the coupling transformer?

Answer:

$$\frac{N_P}{N_S} = \sqrt{\frac{Z_P}{Z_S}} = \sqrt{\frac{500\ \Omega}{8\ \Omega}} = 8$$

Hence, the primary must have eight times as many turns as the secondary.

Knowing what to set the primary count at depends on low internal losses and leakage current and making sure that the primary has enough inductance to operate with low magnetizing current at the voltage applied to the primary.

Example 7: What are the load impedances "seen" by the voltage sources in Fig. 3.101?

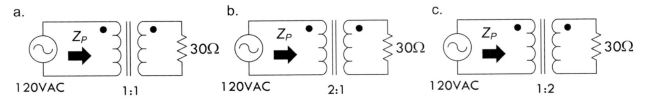

FIGURE 3.101

Answer: (a) 30 Ω, (b) 120 Ω, (c) 8 Ω.

Example 8: If a step-up transformer has a turns ratio of 1:3, what are the voltage ratio, current ratio, and impedance ratio? Assume ratios are given in the form "primary: secondary."

Answer: Voltage ratio is 1:3, current ratio is 3:1, impedance ratio is 1:9.

Transformer Gear Analogy

It is often helpful to think of transformers as gearboxes. For example, in the gearbox analogy in Fig. 3.102, the primary winding is analogous to the input shaft (where the motor is attached) and the secondary winding is analogous to the output shaft. Current is equivalent to shaft speed (rmp) and voltage is equivalent to torque. In a gearbox, mechanical power (speed multiplied by torque) is constant (neglecting losses) and is equivalent to electrical power (voltage times current), which is also constant. The gear ratio is equivalent to the transformer step-up or step-down ratio. A step-up transformer acts like a reduction gear (in which mechanical power is transferred from a small, rapidly rotating gear to a large, slowly rotating gear): it trades current (speed) for voltage (torque), by transferring power from a primary coil to a secondary coil having more turns. A step-down transformer acts similarly to a multiplier gear (in which mechanical power is transferred from a large gear to a small gear): it trades voltage (torque) for current (speed), by transferring power from a primary coil to a secondary coil having fewer turns.

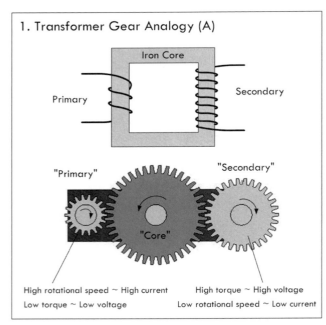

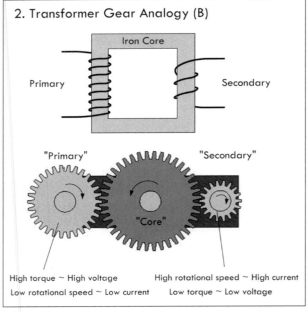

FIGURE 3.102

Center-Tap Transformers

Rarely do you see transformers in the real world with just four leads—two for the primary and two for the secondary. Many commercial transformers employ center taps. A center tap is simply an electrical connection that is made somewhere between the two ends of a transformer winding. By using a center tap, it is possible to utilize only a fraction of the winding voltage. For example, in Fig. 3.103, a transformer's

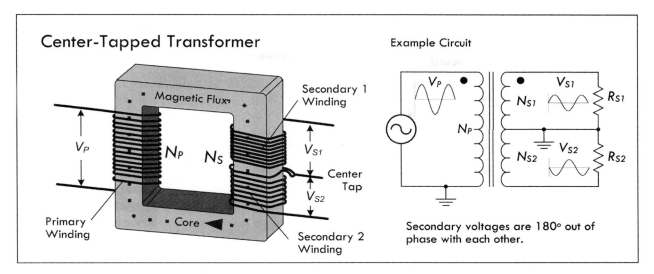

FIGURE 3.103

secondary is center-tapped midway between its winding, yielding two output voltages V_{S1} and V_{S2}. If we place a ground reference on the center tap (it is treated now as a common), we see the voltages in terms of phase, as shown in the example circuit in Fig. 3.103. In this case, the two secondary voltages are equal because we assumed that the number of turns on either end of the center tap were the same. In general, the secondary voltages are determined by the turns ratio.

Center taps can be placed on both the primary side and the secondary side, with multiple taps on either side. For example, a typical power transformer has several secondary windings, each providing a different voltage. Figure 3.104 shows a schematic of a typical power supply transformer. It is possible to join pins with a jumper to get the desired voltage ratios across other pins. Manufacturers will provide you with the voltages between the various tap points, usually specifying CT as the center-tap voltage. Center taps provide flexibility in design and allow varying outputs, which you implement by incorporating switches, for example. We'll see how

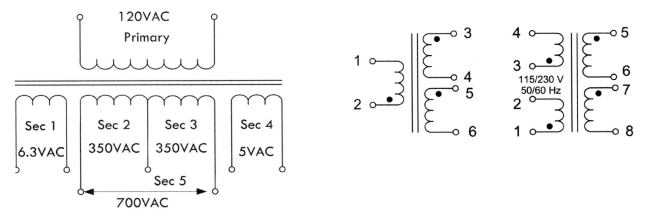

FIGURE 3.104

a center-tap transformer is used to split incoming 240 VAC for the main into two 120-VAC legs within the circuit breaker of your house, and we'll also discover how full-wave center-tap rectifier circuits are used in building dc power supplies.

Real Transformer Characteristics

A perfect or ideal transformer has a primary-to-secondary coupling coefficient of 1. This means that both coils link with all the magnetic flux lines, so that the voltage induced per turn is the same for both coils. This also means that the induced voltage per turn is the same for both primary and secondary coils. Iron core transformers operating at low frequencies come close to being ideal. However, due to various imperfections, such as eddy current, hysteresis losses, internal coil resistance, and skin effects at higher frequencies, this isn't quite true.

In real transformers, not all of the magnetic flux is common to both windings. Flux not associated with linkage is referred to as *leakage flux* and is responsible for a voltage of self-induction. There are small amounts of leakage inductance associated with both windings of a transformer. Leakage inductance acts in exactly the same manner as an equivalent amount of ordinary inductance inserted in series with the circuit. The reactance associated with leakage inductance is referred to as *leakage reactance*, which varies with transformer build and frequency. Figure 3.105 shows a real-life model of a transformer including leakage reactances for both primary and secondary coils, namely, X_{L1} and X_{L2}. When current flows through a leakage reactance, there is an associated voltage drop. The voltage drop becomes greater with increasing current and increases as more power is taken from the secondary.

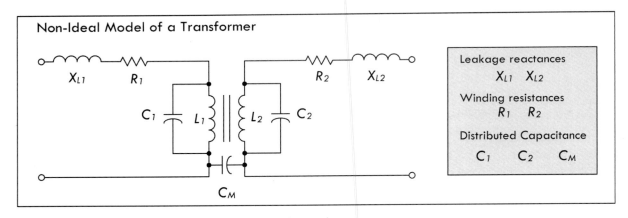

FIGURE 3.105

The internal resistances of a transformer's windings R_1 and R_2 also result in voltage drop when there is current flow. Although these voltage drops are not in phase with those caused by leakage reactance, together they result in a lower secondary voltage under load than is indicated by the transformer turns ratio formula.

Another nonideal characteristic of transformers is stray capacitance. An electric field exists between any two points having a different voltage. When current flows through a coil, each turn has a slightly different voltage than its adjacent turns. This results in capacitance between turns and is modeled by C_1 and C_2 in Fig. 3.105. A mutual capacitance C_M also exists between the primary and secondary windings for the same reason. It is also possible for transformer windings to exhibit capacitance relative to nearby metal, such as a chassis, shield, or even the core itself.

Stray capacitance tends to have little influence in power and audio transformers, but becomes influential as the frequencies increase. In RF applications where transformers are used, the stray capacitance can resonate with either the leakage reactance or, at lower frequencies, the winding reactances, especially under very light or zero-ohm loads. In the frequency region around resonance, transformers do not exhibit behavior as described by the previous transformer equations.

Iron core transformers also experience losses with hysteresis and eddy current, as was discussed in Sec. 2.24. These losses, which add to the required magnetizing current, are equivalent to adding a resistance in parallel to R_1 in Fig. 3.105.

TRANSFORMER PRECAUTIONS

There are three basic rules to observe when using a transformer. First, never apply a voltage that is greatly in excess of the transformer winding ratings. Second, never allow a significant direct current to flow through any winding not designed to handle it. Third, don't operate the transformer at a frequency outside the range specified by the manufacturer. Applying a voltage of, say, 120 VAC to a secondary in hopes of achieving 1200 VAC at the primary is a bad idea—expect smoke and combustion, accompanied by insulation failure. Similar results can be expected with excessive dc current through the primary. In terms of frequency, a 60-Hz transformer driven at 20 Hz will draw too much magnetizing current and will run dangerously hot.

3.8.2 Transformer Construction

Cores

Transformers used for power and audio frequencies have cores made of many thin laminations of silicon steel. The laminations reduce eddy currents, as discussed in Sec. 2.24. A typical laminated core is made from E-shaped and I-shaped pieces, sandwiched together, as shown in Fig. 3.106. Transformers made from these cores are therefore often referred to as EI transformers.

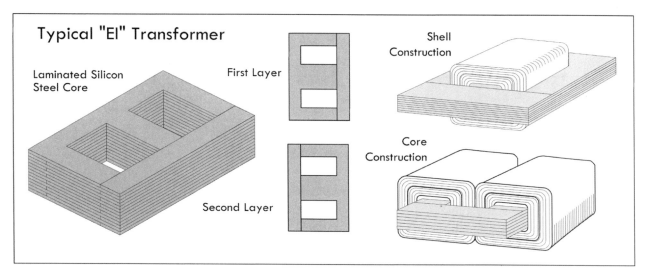

FIGURE 3.106

Two common core shapes in use are shown in Fig. 3.106. In the shell construction, both the primary and the secondary windings are wound around the same inner leg, while in the core construction, primary and secondary windings are wound on separate legs. The core construction is often implemented to minimize capacitive effects between primary and secondary windings, or when one winding is to be operated at very high voltage. The size, shape, and type of core material, as well as the frequency range, influence the required number of turns in each winding. In most transformers, the coils are wound in layers, with a sheet of special paper insulation placed between each layer. A thicker insulation is used between adjacent coils and between the core and the first coil.

Powdered iron cores, with their low eddy current characteristics, are used in transformers that operate above mains frequencies (60 Hz) up to several kilohertz. These cores have a very high permeability and thus provide decent stepping capability for their size. Transformers that are used in even higher-frequency applications, such as RF, often contain cores made from nonconductive magnetic ceramic materials or ferrites.

A common core shape for powdered iron and ferrite core transformers is the toroid, as shown in Fig. 3.107a. The closed ring shape of the toroid eliminates air gaps inherent in the construction of an EI core. The primary and secondary coils are often wound concentrically to cover the entire surface of the core. Ferrite cores are used at higher frequencies, typically between a few tens of kilohertz to a megahertz. In general, toroidal transformers are more efficient (around 95 percent) than cheaper laminated EI transformers; they are more compact (about half the size), weigh less (about half), have less mechanical hum (making them superior in audio applications), and have lower off-load losses (making them more efficient in standby circuits).

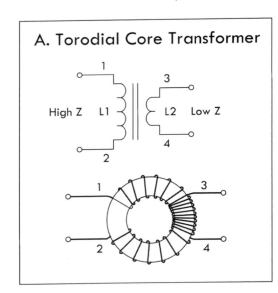

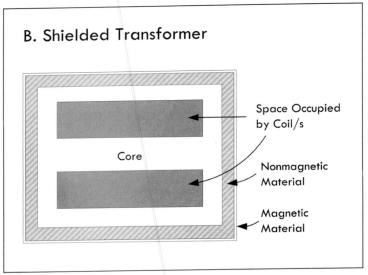

FIGURE 3.107

Shielding

To eliminate mutual capacitance between windings within a transformer, an electrostatic shield is often placed between the windings. Some transformers may incorporate a magnetic shield, as shown in Fig. 3.107b. The magnetic shield helps prevent outside magnetic fields (interference) from inducing currents within the inner windings. The shield also helps prevent the transformer from becoming an interference radiator itself.

Windings

For small-power and signal transformers, the windings are made from solid wire copper, insulated typically with enamel; sometimes additional insulation is used for safety. Larger-power transformers may be wound with copper or aluminum wire, or even strip conductors for very heavy current; in some cases multistrand conductors are used to reduce skin effect losses. High-frequency transformers operating in the kilohertz range often have windings made of Litz wire to minimize skin effects. For signal transformers, the windings may be arranged in a way to minimize leakage inductance and stray capacitance in order to improve high-frequency response.

3.8.3 *Autotransformers and Variable Transformers*

An autotransformer is similar to a standard transformer; however, it uses only one single coil and a center tap (or taps) to make primary and secondary connections. See Fig. 3.108. As with standard transformers, autotransformers can be used to step up or step down voltages, as well as match impedances; however, they will not provide electrical isolation like a standard transformer, since their primary and secondary are on the same coil—there is no electrical isolation between the two coils.

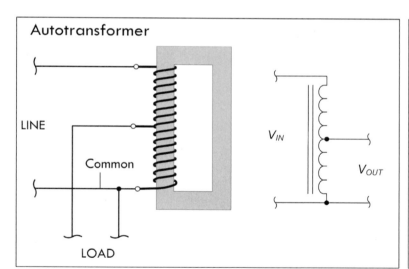
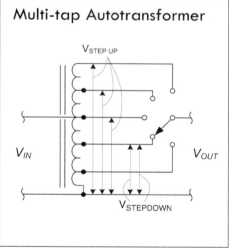

FIGURE 3.108

Although an autotransformer has only one winding, the laws of induction that were used with a standard transformer to step up and step down voltage can be applied just as well. This also applies to principles of current and impedance as a function of the number of winding turns. In Fig. 3.108, the current in the common winding is the difference between the line current (primary current) and the load current (secondary current), since these currents are out of phase. Hence, if the line and load currents are nearly equal, the common section of the winding may be wound with comparatively small wire. The line and load currents will be equal only when the primary (line) and secondary (load) voltages are close in magnitude.

Autotransformers are often used in impedance-matching applications. They are also frequently used for boosting or reducing the power-line voltage by relatively

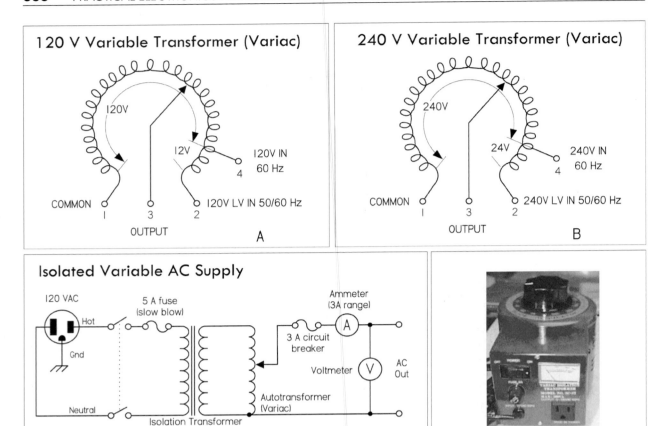

FIGURE 3.109 (a) Nonisolated 120-V Variac whose output voltage is varied by rotating a wiper. (b) Nonisolated 240-V Variac. (c) A homemade variable ac supply with isolation protection provided by means of an isolation transformer. (d) ac power supply that houses an isolation transformer, Variac, switch, fuse, ac outlet, and meter.

small amounts. Figure 3.108 shows a switch-stepped autotransformer whose output voltage can be set to any number of values determined by the switch contact position.

A variable transformer or Variac (commercial name) is similar to the switch-stepped autotransformer in Fig. 3.108; however, it has a continuous wiper action along a circular coil, as shown in Fig. 3.109. A Variac acts like an adjustable ac voltage source. Its primary is connected to the hot and neutral of the 120-V line voltage, while the secondary leads consist of the neutral and an adjustable wiper that moves along the single core winding.

Being able to adjust the line voltage is a very useful trick when troubleshooting line-power equipment, where the fuse instantly blows at normal line voltage. Even without a fuse blowing, troubleshooting at around 85 V may reduce the fault current.

It is important to note that a Variac by itself does not provide isolation protection like a standard transformer, since the primary and secondary shared a common winding. It is therefore important, if you plan to work on ungrounded, "hot chassis" equipment, that you place an isolation transformer before the Variac—never after it. If you don't, shock hazards await. Figure 3.109c shows a schematic of such an arrangement. It includes a switch and fuse protection, as well as current and voltage meters, all of which create an adjustable, fully isolated ac power source.

To avoid the hassle of cascading a Variac and isolation transformer together, simply get an ac power supply that houses both elements together in one package—see Fig. 3.109d.

Boosting and Bucking

We just saw how autotransformers are used in applications requiring a slight boost or reduction in voltage to a load. It is possible to accomplish the same effect by using a normal (isolated) transformer with just the right primary/secondary turns ratio. There is still another alternative—use a step-down configuration with secondary winding connected in a series-aiding ("boosting") or series-opposing ("bucking") configuration, as shown in Fig. 3.110.

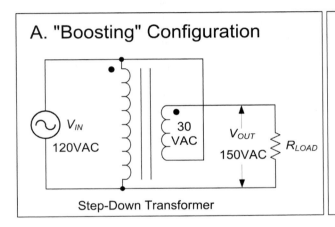 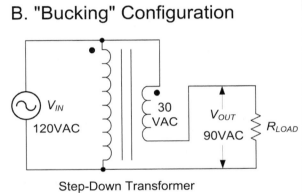

FIGURE 3.110

In the boosting configuration, the secondary coil's polarity is oriented so that its voltage directly adds to the primary voltage. In the bucking configuration, the secondary coil's polarity is oriented so that its voltage directly subtracts from the primary voltage. An autotransformer does the same job as the boosting and bucking functions displayed here, but using only a single winding, making it cheaper and lighter to manufacture.

3.8.4 Circuit Isolation and the Isolation Transformer

Transformers perform an important role in isolating one circuit from another. Figure 3.111 shows an example application that uses a transformer to isolate a load connected to an ac outlet. In this application, there is no need to step up or step down the voltage, so the transformer has a 1:1 winding ratio. Such a transformer is referred to as an *isolation transformer*. In Fig. 3.111, a mains isolation transformer is used to isolate a load from the source, as well as provide ground fault protection. An isolation transformer should be used whenever you work on nongrounded equipment, with no input isolation such as switch-mode power supplies.

In your home wiring, the neutral (white) and the ground (green) connections are tied together at the main junction box, so they are basically at the same potential— 0 V, or earth ground. If you accidentally touch the hot wire while being in contact

AC Line Isolation Transformer

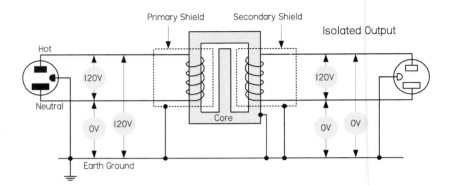

FIGURE 3.111

with a grounded object, current will pass through your body and give you a potentially fatal shock. With an isolation transformer, the secondary winding leads act as a 120-V source and return, similar to the mains' hot and neutral, but with an important difference. Neither the secondary source nor return runs are tied to earth ground! This means that if you touch the secondary source or return while being in contact with a grounded object, no current will flow through your body. Current only wants to pass between the secondary source and return runs. (Note that all transformers provide isolation, not just line isolation transformers. Therefore, equipment with input power transformers already have basic isolation protection built in. Isolation transformers used for laboratory work are explained in greater detail in Sec. 7.5.12.)

Isolation transformers are also typically constructed with two isolated Faraday shields between the primary and secondary windings. The use of the two shields diverts high-frequency noise, which would normally be coupled across the transformer to ground. Increasing the separation between the two Faraday shields minimizes the capacitance between the two and, hence, the coupling of noise between the two. Therefore, the isolation transformer acts to clean up line power noise before being delivered to a circuit.

3.8.5 Various Standard and Specialized Transformers

Power Transformers

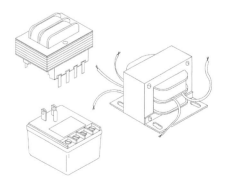

These transformers are used primarily to reduce line voltage. They come in a variety of different shapes, sizes, and primary and secondary winding ratios. They often come with taps and multiple secondary windings. Color-coded wires are frequently used to indicate primary and secondary leads (e.g., black wires for primary, green for secondary, and yellow for tap lead is one possibility). Other transformers use pins for primary, secondary, and tapped leads, allowing them to be mounted on a PC board. You can also find transformers in wall-mount packages that plug directly into an ac outlet, with screw-in terminals as secondary and tapped leads.

FIGURE 3.112

Audio Transformers

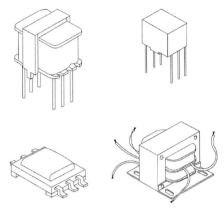

FIGURE 3.113

Audio transformers are used primarily to match impedances between audio devices (e.g., between microphone and amplifier or amplifier and speaker), though they can be implemented in other ways as well. They work best at audio frequencies from 20 Hz to 20 kHz. Outside this range they will reduce or block signals. They come in a variety of shapes and sizes and typically contain a center tap in both the primary and secondary windings. Some come with color-coded wires to specify leads, while other audio transformers have pinlike terminals that can be mounted on PC boards. Spec tables provide dc resistance values for primary and secondary windings to help you select the appropriate transformer for the particular matching application. Besides performing simple impedance matching, audio transformers can be used to step up or step down a signal voltage, convert a circuit from unbalanced to balanced and vice versa, block dc current in a circuit while allowing ac current to flow, and provide basic isolation between one device and another. Note that audio transformers have a maximum input level that cannot be exceeded without causing distortion. Also, audio transformers cannot step up a signal by more than about 25 dB when used in typical audio circuits. Because of this, an audio transformer cannot be substituted for a microphone preamp. If more than 25 dB of gain is required, an active preamp must be used instead of a transformer.

Air Core RF Transformers

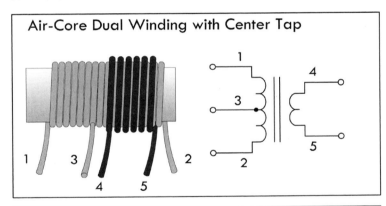

Air-Core Dual Winding with Center Tap

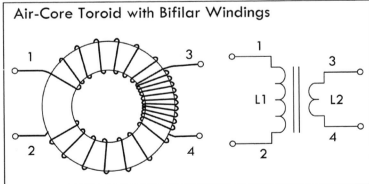

Air-Core Toroid with Bifilar Windings

FIGURE 3.114

Air core transformers are special devices used in radio-frequency circuits. (They are used for RF coupling, such as antenna tuning and impedance matching.) Unlike steel or ferrite core transformers, the core is made from a nonmagnetic form, usually a hollow tube of plastic. The degree of coupling between windings in an air core transformer is much less than that of a steel core transformer; however, there are no losses associated with eddy currents, hysteresis, saturation, and so on, as is the case with magnetic cores. This becomes critically important in RF applications—at high frequencies, steel core transformers experience significant losses. Toroidal air core transformers aren't common nowadays, except in VHF (very high frequency) work. Today, special coupling networks and RF powdered-iron and ferrite toroids have generally replaced air cores, except in situations where the circuit handles very high power or the coils must be very temperature stable.

Ferrite and Powdered-Iron Toroidal Transformers

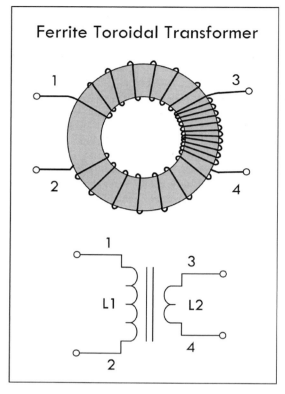

Ferrite Toroidal Transformer

FIGURE 3.115

Toroidal ferrite and powdered-iron transformers are used from a few hundred hertz well into the UHF spectrum. The principal advantage of this type of core is self-shielding and low losses due to eddy currents. The permeability/size ratio is also very large, making them compact devices requiring fewer coil turns than traditional transformers. The most common ferrite toroid transformer is the conventional broadband transformer. Broadband transformers provide dc isolation between the primary and secondary circuits. The primary of a step-down impedance transformer is wound to occupy the entire core, with the secondary wound over the primary, as shown in Fig. 3.115. This style of transformer is frequently used in impedance matching. In standard broadcast radio receivers, these transformers operate in a frequency range from 530 to 1550 kHz. In shortwave receivers, RF transformers are subjected to frequencies up to about 20 MHz; in radar, it approaches upward of 200 MHz.

Pulse and Small Signal Transformers

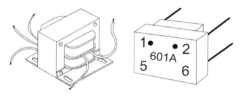

FIGURE 3.116

Pulse transformers are special transformers optimized for transmitting rectangular electrical pulses—ones with fast rise and fall times and constant amplitude. A small signal transformer is a small version of a pulse transformer. These devices are used in digital logic and telecom circuits, often for matching logic drivers to transmission lines. Medium-size power versions are used in power-control circuits such as camera flash controllers, while larger-power versions are used in electrical power distribution to interface low-voltage control circuitry with high-voltage power semiconductive gates, such as TRIACs, IGBTs, thyristors, and MOSFETs. Special high-voltage pulse transformers are used to generate high-power pulses for radar, particle accelerators, or other pulsed power applications.

To minimize pulse shape distortion, a pulse transformer requires very low leakage inductance and distributed capacitance, and a high open-circuit inductance. Low coupling capacitance is also important in power-pulse transformer applications to protect circuitry on the primary side from high-power transients created by load.

3.8.6 Transformer Applications

There are three principal uses for transformers: to transform voltages and currents from one level to another, to physically isolate the primary circuit from the secondary, and to transform circuit impedances from one level to another. Here are some examples in action.

Current Transformers

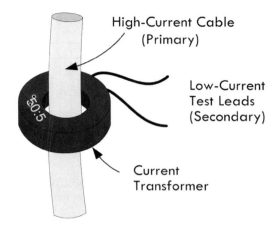

High-Current Cable (Primary)

Low-Current Test Leads (Secondary)

Current Transformer

FIGURE 3.117

Current transformers are special devices used primarily to measure larger currents that would be too dangerous to measure with an ammeter. They are designed to provide a current in their secondary that is proportional to the current flowing in the primary. A typical current transformer resembles a toroidal core inductor with many secondary windings. The primary coil consists of simply passing a single cable-to-be-measured (insulated) through the center of the toroid. The output current through the secondary is many times smaller than the actual current through the cable (primary). These transformers are specified by their input and output current ratio (400:5, 2000:5, etc.). Current transformers designed for electrical supply applications are designed to drive 5-A (full-scale) meters. There are also wideband current transformers used to measure high-frequency waveforms and pulsed currents.

Center-Tap Pole Transformer

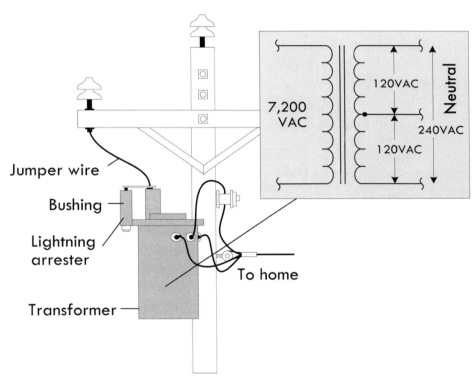

FIGURE 3.118   In the United States, main power lines carry ac voltages upward of several thousand volts. A center-tap pole transformer is used to step down the line voltage to 240 V. The tap then acts to break this voltage up into 120-V portions. Small appliances, such as TVs, lights, and hair dryers can use either the top line and the neutral line or the bottom line and the neutral line. (The neutral is grounded to a ground rod through a link between neutral and ground buses in the breaker box.) Larger appliances, such as stoves, refrigerators, and clothes dryers often make use of the 240-V terminals and often use the neutral terminal as well. See App. A for more on power distribution and home wiring.

Transformer Used for Landscape Lighting

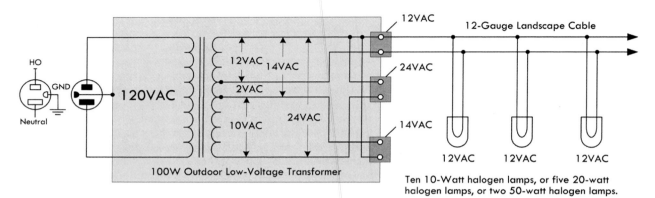

FIGURE 3.119 Here a step-down low-voltage transformer is used to drive quartz halogen landscape lamps. The lamps in this case don't care if the voltage is ac since the frequency (60 Hz) is too fast for there to be any noticeable variation in luminous output. Most commercial transformers used for landscape wiring, or for driving solenoid-powered sprinkler systems, will come with multiple outputs. This transformer provides 12-V, 24-V, and 14-V outputs. The 14-V output may be used to drive 12-V lamps if there is an anticipated voltage drop along long cable runs; the 24-V output may be used to drive 24-V devices. Note that the total load should not consume more power than the transformer's rated output capacity. For example, this 100-W transformer should not be used to drive more than, say, ten 10-watt lamps or five 20-watt lamps. Exceeding this will result in lamp dimming.

Step-Down Transformer for DC Power Supply

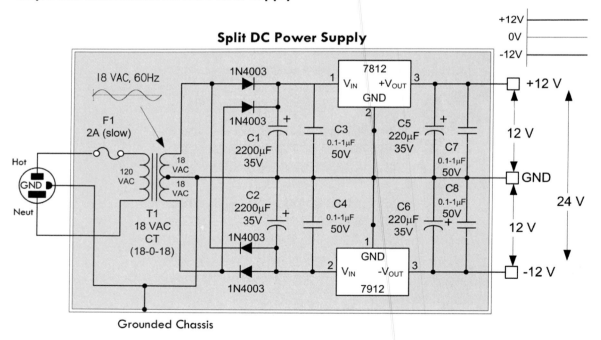

FIGURE 3.120 Transformers are essential ingredients in power supply design. Here a 120-V to 18V-0-18V center-tap transformer is used to create a split ±12-V dc power supply. The transformer acts to reduce the voltage to 18 VAC across each coil end and the center tap. The rectifier section built from diodes acts to eliminate negative swings in the upper positive section and eliminate positive swings in the lower section. Capacitors are thrown in to remove the pulsating dc and make the voltages appear dc. The regulators are used to set the dc voltages to exactly +12 V and −12 V. See Chap. 11 on power supplies for more details.

Various Transformer/Rectifier Arrangements

A. Dual Complementary Rectifier

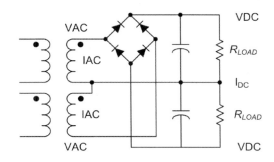

There are various ways in which to create dc power supplies. Figure 3.121 shows the four basic schemes used. Each scheme has its pros and cons, which are briefly described here and in greater detail in the sections on diodes and power supply in chapters to come.

(a) *Dual complementary rectifier:* Very efficient and best choice for two balanced outputs with a common return. The output windings are bifilar wound for precisely matched series resistances, coupling, and capacitance.

$$V_{AC} = 0.8 \times (V_{DC} + 2)$$
$$I_{AC} = 1.8 \times I_{DC}$$

B. Full-Wave Bridge

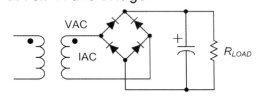

(b) *Full-wave bridge:* Most efficient use of transformer's secondary winding. Best for high-voltage outputs.

$$V_{AC} = 0.8 \times (V_{DC} + 2)$$
$$I_{AC} = 1.8 \times I_{DC}$$

C. Half-Wave Rectifier

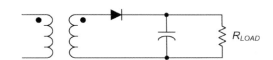

(c) *Half-wave rectifier:* This design should be avoided for power supply design, as it is an inefficient use of the transformer. This arrangement causes the core to become polarized and to saturate in one direction.

D. Full-Wave Center Tap

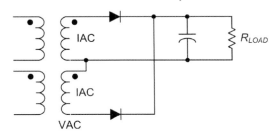

(d) *Full-wave center tap:* While more efficient than the half-wave rectifier circuit, the full wave does not make full use of secondaries, but is good for high-current, low-voltage applications, as there is only one diode drop per positive half cycle.

$$V_{AC} = 1.7 \times (V_{DC} + 1)$$
$$I_{AC} = 1.2 \times I_{DC}$$

FIGURE 3.121

Audio Impedance Matching

A. Need for Matching Impedances

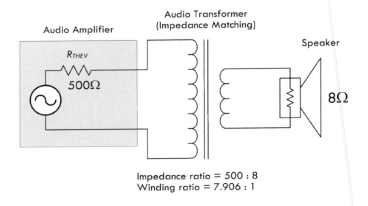

Impedance ratio = 500 : 8
Winding ratio = 7.906 : 1

(a) Maximum power is transferred to a load if the load impedance is equal to the Thevenin impedance of the network supplying power. To supply maximum power transfer from the audio amplifier with an output impedance of 500 Ω to an 8-Ω speaker, we must properly match the load impedance with that of the output impedance (or Thevenin impedance) of the source. If we were not to match impedance and attempt to drive the 8-Ω speaker directly, the impedance mismatch would result in very poor (low peak power) performance. Also, the amplifier would dissipate considerable power in the form of heat as it tries to drive the low-impedance speaker.

When going from a high-impedance (high-voltage, low-current) source to a low-impedance (low-voltage, high-current) load, we need to use a step-down transformer. To determine the turns ratio required, we refer back to Eq. 5:

$$\frac{N_P}{N_S} = \sqrt{\frac{Z_P}{Z_S}} = \sqrt{\frac{500 \ \Omega}{8 \ \Omega}} = 7.906$$

In other words, the required winding ratio is 7.906:1. With such a transformer in place, the speaker will load the amplifier to just the right degree, drawing power at the correct voltage and current levels for the most efficient power transfer to load.

B. No Need for Matching

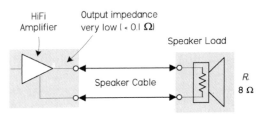

Virtually all audio power from amplifer is transferred into speaker.

(b) Note that most transistor or IC hi-fi amplifier and speaker systems have amplifiers with output impedances much lower than the speaker impedance. A typical speaker impedance is 8 Ω, for example, but most hi-fi amplifiers have an output impedance of 0.1 Ω or less. This not only ensures that most of the audio energy is transferred to the speaker, but also that the amplifier's low output impedance provides good electrical damping for the speaker's moving voice coil—giving higher fidelity.

Older valve amplifiers needed a different form of impedance matching, because output valves generally had fairly fixed and relatively high output impedance so they couldn't deliver audio energy efficiently into the low load impedance of a typical speaker. So an output transformer had to be used to produce a closer impedance match. The transformer stepped up the impedance of the speaker, so that it gave the output valve an effective load of a few thousand ohms; this was at least comparable to the valve's own output impedance, so only a small amount of energy was wasted as heat in the valve.

C. Microphone Input Transformer

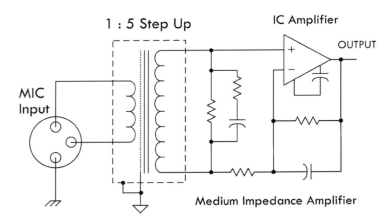

(c) The only area in audio where impedance matching (of a different kind) tends to be important is with transducers such as microphones, turntable pickups, and tape heads. Here, the transducer often needs to be provided with particular load impedance, but not in order to maximize power or signal transfer.

The diagram in (c) shows a typical matching arrangement for a microphone connected through a matching transformer to an input stage (unity gain stage) of an audio amplifier IC.

FIGURE 3.122

3.9 Fuses and Circuit Breakers

Fuses and circuit breakers are devices designed to protect circuits from excessive current flows, which are often a result of large currents that result from shorts or sudden power surges. A fuse contains a narrow strip of metal that is designed to melt when current flow exceeds its current rating, thereby interrupting power to the circuit. Once a fuse blows (wire melts), it must be replaced with a new one. A circuit breaker is a mechanical device that can be reset after it "blows." It contains a spring-loaded contact that is latched into position against another contact. When the current flow exceeds a breaker's current rating, a bimetallic strip heats up and bends. As the strip bends, it "trips" the latch, and the spring pulls the contacts apart. To reset the breaker, a button or rockerlike switch is pressed to compress the spring and reset the latch.

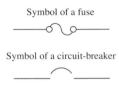

FIGURE 3.123

In homes, fuses/circuit breakers are used to prevent the wires within the walls from melting under excessive current flow (typically upwards of 15 A); they are not designed to protect devices, such as dc power supplies, oscilloscopes, and other line-powered devices, from damage. For example, if an important current-limiting component within a test instrument (powered from the main line) shorts out, or if it is connected to an extremely large test current, the circuit within the device may be flooded with, say, 10 A instead of 0.1 A. According to $P = I^2 R$, the increase in wattage will be 10,000 times larger, and as a result, the components within the circuit will fry. As the circuit is melting away, no help will come from the main 15-A breaker—the surge in current through the device may be large, but not large enough to trip the breaker. For this reason, it is essential that each individual device contain its own properly rated fuse.

Fuses come in *fast-action* (quick-blow) and *time-lag* (slow-blow) types. A fast-acting fuse will turn off with just a brief surge in current, whereas a time-lag fuse will take a while longer (a second or so). Time-lag fuses are used in circuits that have large turn-on currents, such as motor circuits, and other inductive-type circuits.

In practice, a fuse's current rating should be around 50 percent larger than the expected nominal current rating. The additional leeway allows for unexpected, slight variations in current and also provides compensation for fuses whose current ratings decrease as they age.

Fuses and breakers that are used with 120-V ac line power must be placed in the hot line (black wire in the United States) and must be placed before the device they are designed to protect. If a fuse or circuit breaker is placed in the neutral line (white wire), the full line-voltage will be present at the input, even if the fuse/circuit breaker blows. Circuit breakers that are used to protect 240-V ac appliances (e.g., stoves, clothes dryers) have fuses on all three input wires (both the hot wires and the neutral wire) (see Fig. 3.124). Power distribution and home wiring are covered in detail in App. A.

Home wiring

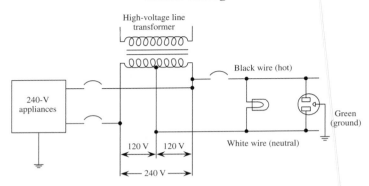

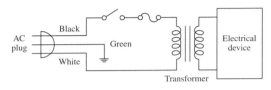

Electrical device

FIGURE 3.124

3.9.1 Types of Fuses and Circuit Breakers

Glass and Ceramic

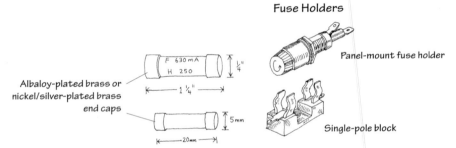

Fuse Holders

Panel-mount fuse holder

Single-pole block

These are made by encapsulating a current-sensitive wire or ceramic element within a glass cylinder. Each end of the cylinder contains a metal cap that acts as a contact lead. Fuses may be fast-acting or time-lagging. They are used in instruments, electric circuits, and small appliances. Typical cylinders come in $\frac{1}{4} \times 1\frac{1}{4}$ in or 5×20 mm sizes. Current ratings range from around $\frac{1}{4}$ to 20 A, with voltage ratings of 32, 125, and 250 V.

Blade

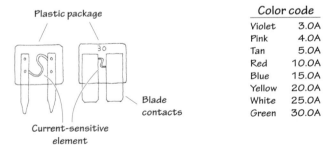

Color code	
Violet	3.0A
Pink	4.0A
Tan	5.0A
Red	10.0A
Blue	15.0A
Yellow	20.0A
White	25.0A
Green	30.0A

These are fast-acting fuses with bladelike contacts. They are easy to install and remove from their sockets. Current ratings range from 3 to 30 A, with voltage ratings of 32 and 36 V. They are color-coded according to current rating and are used primarily as automobile fuses.

FIGURE 3.125

Miscellaneous Fuses

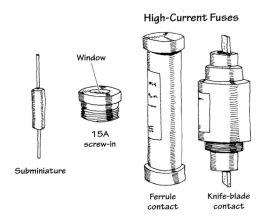

High-Current Fuses

Window

15A
screw-in

Subminiature

Ferrule
contact

Knife-blade
contact

Other types of fuses include subminiature fuses and high-current screw-in and cartridge fuses. Subminiature fuses are small devices with two wire leads that can be mounted on PC boards. Current ratings range from 0.05 to 10 A. They are used primarily in miniature circuits. Cartridge fuses are designed to handle very large currents. They are typically used as main power shutoffs and in subpanels for 240-V applications such as electric dryers and air conditioners. Cartridge fuses are wrapped in paper, like shotgun shells, and come with either ferrule or knife-blade contacts. Ferrule-contact fuses protect up to 60 A, while knife-blade contact fuses are rated at 60 A or higher.

Circuit Breakers

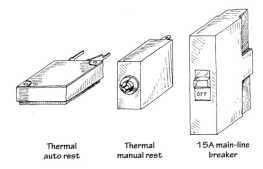

Thermal
auto rest

Thermal
manual rest

15A main-line
breaker

These come in both rocker and push-button forms. Some have manual resets, while others have thermally actuated resets (they reset themselves as they cool down). Main-line circuit breakers are rated at 15 to 20 A. Smaller circuit breakers may be rated as low as 1 A.

FIGURE 3.125   (*Continued*)

CHAPTER 4

Semiconductors

4.1 Semiconductor Technology

The most important and perhaps most exciting electrical devices used today are built from semiconductive materials. Electronic devices, such as diodes, transistors, thyristors, thermistors, photovoltaic cells, phototransistors, photoresistors, lasers, and integrated circuits, are all made from semiconductive materials, or semiconductors.

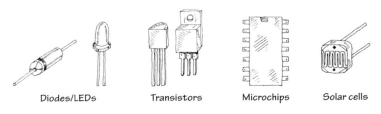

Diodes/LEDs   Transistors   Microchips   Solar cells

FIGURE 4.1

4.1.1 What Is a Semiconductor?

Materials are classified by their ability to conduct electricity. Substances that easily pass an electric current, such as silver and copper, are referred to as *conductors*. Materials that have a difficult time passing an electric current, such as rubber, glass, and Teflon, are called *insulators*. There is a third category of material whose conductivity lies between those of conductors and insulators. This third category of material is referred to as a *semiconductor*. A semiconductor has a kind of neutral conductivity when taken as a group. Technically speaking, semiconductors are defined as those materials that have a conductivity σ in the range between 10^{-7} and 10^3 mho/cm (see Fig. 4.2). Some semiconductors are pure-elemental structures (e.g., silicon, germanium), others are alloys (e.g., nichrome, brass), and still others are liquids.

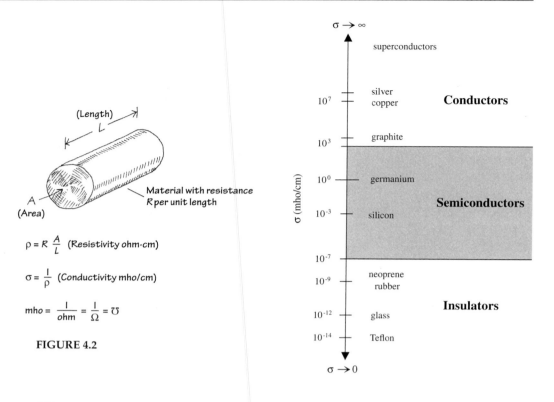

$\rho = R\,\dfrac{A}{L}$ (Resistivity ohm·cm)

$\sigma = \dfrac{1}{\rho}$ (Conductivity mho/cm)

$mho = \dfrac{1}{ohm} = \dfrac{1}{\Omega} = \mho$

FIGURE 4.2

Silicon

Silicon is the most important semiconductor used in building electrical devices. Other materials such as germanium and selenium are sometimes used, too, but they are less popular. In pure form, silicon has a unique atomic structure with very important properties useful in making electrical devices.

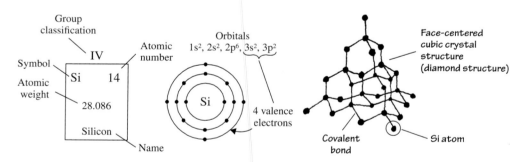

FIGURE 4.3

Silicon is ranked second in the order of elements appearing in the earth's crust, an average of 27 percent occurring in igneous rocks. It is estimated that a cubic mile of seawater contains about 15,000 tons of silicon. It is extremely rare to find silicon in its pure crystalline form in nature, and before it can be used in making electronic devices, it must be separated from its binding elements. After individuals—chemists, material scientists, etc.—perform the purification process, the silicon is melted and spun into a large "seed" crystal. This long crystal can then be cut up into slices or wafers that semiconductor-device designers use in making electrical contraptions.

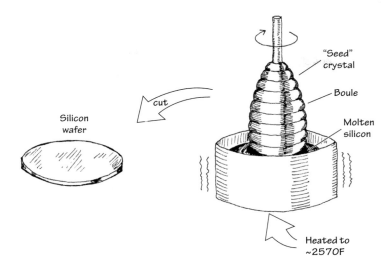

"Seed" crystal

Boule

Molten silicon

Silicon wafer

cut

Heated to ~2570F

FIGURE 4.4

For the semiconductor-device designer, a silicon wafer alone does not prove very useful. A designer would not use the silicon wafer in its pure form to build a device; it just does not have quite the right properties needed to be useful. A semiconductor-device designer is looking for a material that can be made to alter its conductive state, acting as a conductor at one moment and as an insulator at another moment. For the material to change states, it must be able to respond to some external force applied at will, such as an externally applied voltage. A silicon wafer alone is not going to do the trick. In fact, a pure silicon wafer acts more as an insulator than a conductor, and it does not have the ability to change conductive states when an external force is applied. Every designer today knows that a silicon wafer can be transformed and combined with other transformed silicon wafers to make devices that have the ability to alter their conductive states when an external force is applied. The transforming process is referred to as *doping*.

Doping

Doping refers to the process of "spicing up" or adding ingredients to a silicon wafer in such a way that it becomes useful to the semiconductor-device designer. Many ingredients can be added in the doping process, such as antimony, arsenic, aluminum, and gallium. These ingredients provide specialized characteristics such as frequency response to applied voltages, strength, and thermal integrity, to name a few. By far, however, the two most important ingredients that are of fundamental importance to the semiconductor-device designer are boron and phosphorus.

When a silicon wafer is doped with either boron or phosphorus, its electrical conductivity is altered dramatically. Normally, a pure silicon wafer contains no free electrons; all four of its valence electrons are locked up in covalent bonds with neighboring silicon atoms (see Fig. 4.5). Without any free electrons, an applied voltage will have little effect on producing an electron flow through the wafer.

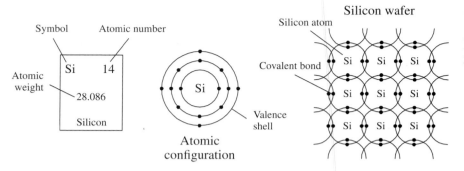

A silicon wafer in pure form doesn't contain any free charge carriers; all the electrons are locked up into covalent bonds between neighboring atoms.

FIGURE 4.5

However, if phosphorus is added to the silicon wafer, something very interesting occurs. Unlike silicon, phosphorus has five valence electrons instead of four. Four of its valence electrons will form covalent bonds with the valence electrons of four neighboring silicon atoms (see Fig. 4.6). However, the fifth valence electron will not have a "home" (binding site) and will be loosely floating about the atoms. If a voltage is applied across the silicon-phosphorus mixture, the unbound electron will migrate through the doped silicon toward the positive voltage end. By supplying more phosphorus to the mixture, a larger flow of electrons will result. Silicon that is doped with phosphorus is referred to as *n-type silicon,* or negative-charge-carrier-type silicon.

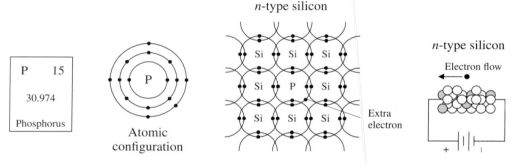

A phosphorus atom added to a silicon wafer provides an extra unbound electron that aids in conduction. Silicon doped with phosphorus is called *n*-type silicon.

FIGURE 4.6

Now, if you take pure silicon and add some boron, you will see a different kind of conduction effect. Boron, unlike silicon or phosphorus, contains only three valence electrons. When it is mixed with silicon, all three of its valence electrons will bind with neighboring silicon atoms (see Fig. 4.7). However, there will be a vacant spot—called a *hole*—within the covalent bond between one boron and one silicon atom. If a voltage is applied across the doped wafer, the hole will move toward the negative voltage end, while a neighboring electron will fill in its place. Holes are considered positive charge carriers even though they do not contain a physical charge per se. Instead, it only appears as if a hole has a positive charge because of the charge imbalance between the protons within the nucleus of the silicon atom that receives the hole and the electrons in the outer orbital. The net charge on a particular silicon atom with a hole will appear to be positive by an amount of charge equivalent to one proton (or a "negative electron"). Silicon that is doped with boron is referred to as *p-type silicon,* or positive-charge-carrier-type silicon.

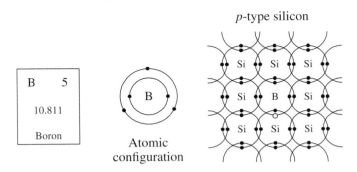

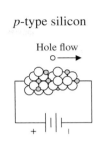

When boron is added to silicon, a hole is formed. This hole acts like a positive charge (see text) that aids in conduction. Silicon doped with boron is called *p*-type silicon.

FIGURE 4.7

As you can see, both *n*-type and *p*-type silicon have the ability to conduct electricity; one does it with extra unbound electrons (*n*-type silicon), and the other does it with holes (*p*-type silicon).

A Note to Avoid Confusion

Boron atoms have three valence electrons, not four like silicon. This means that the combined lattice structure has fewer free valence electrons as a whole. However, this does not mean that a *p*-type silicon semiconductor has an overall positive charge; the missing electrons in the structure are counterbalanced by the missing protons in the nuclei of the boron atoms. The same idea goes for *n*-type silicon, but now the extra electrons within the semiconductor are counterbalanced by the extra protons within the phosphorus nuclei.

Another Note to Avoid Confusion (Charge Carriers)

What does it mean for a hole to flow? A hole is nothing, right? How can nothing flow? Well, it is perhaps misleading, but when you hear the phrase "hole flow" or "flow of positive charge carriers in *p*-type silicon," electrons are in fact flowing. You may say, doesn't that make this just like the electron flow in *n*-type silicon? No. Think about tipping a sealed bottle of water upside down and then right side up (see Fig. 4.8). The bubble trapped in the bottle will move in the opposite direction of the water. For the bubble to proceed, water has to move out of its way. In this analogy, the water represents the electrons in the *p*-type silicon, and the holes represent the bubble. When a voltage is applied across a *p*-type silicon semiconductor, the electrons around the boron atom will be forced toward the direction of the positive terminal. Now here is where it gets tricky. A hole next to a boron atom will be pointing toward the negative terminal. This hole is just waiting for an electron from a neighboring atom to fall into it, due in part to the lower energy configuration there. Once an electron, say, from a neighboring silicon atom, falls into the hole in the boron atom's valence shell, a hole is briefly created in that silicon atom's valence shell. The electrons in the silicon atom lean toward the positive terminal, and the newly created hole leans toward the negative terminal. The next silicon atom over will let go of one of its electrons, and the electron will fall into the hole, and the hole will move over again—the process continues, and it appears as if the hole flows in a continuous motion through the *p*-type semiconductor.

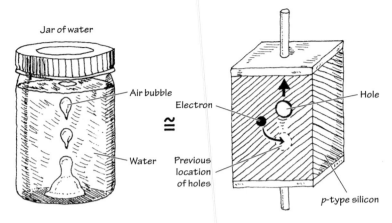

FIGURE 4.8

A Final Note to Avoid Confusion

And finally, why is a hole called a positive charge carrier? How can "nothing" carry a positive charge? Well, what's going on here is this: A hole, as it moves through the mostly silicon-based crystal, causes a brief alteration in the electrical field strength around the silicon atom in the crystal where it happens to be situated. When an electron moves out of the way, thus creating a new hole, this silicon atom as a whole will be missing an electron, and hence the positive charge from the nucleus of the silicon atom will be felt (one of the protons is not counterbalanced). The "positive charge carrier" attributed to holes comes from this effective positive nuclear charge of the protons fixed within the nucleus.

4.1.2 Applications of Silicon

You may be asking yourself, why are these two new types of silicon (*n*-type and *p*-type) so useful and interesting? What good are they for semiconductor-device designers? Why is there such a fuss over them? These doped silicon crystals are now conductors. Big deal, right? Yes, we now have two new conductors, but the two new conductors have two unique ways of passing an electric current—one does it with holes, the other with electrons. This is very important.

The manners in which *n*-type and *p*-type silicon conduct electricity (electron flow and hole flow) are very important in designing electronic devices such as diodes, transistors, and solar cells. Some clever people figured out ways to arrange slabs, chucks, strings, etc. made of *n*-type and *p*-type silicon in such a way that when an external voltage or current is applied to these structures, unique and very useful features result. These unique features are made possible by the interplay between hole flow and electron flow between the *n*-type and *p*-type semiconductors. With these new *n*-type/*p*-type contraptions, designers began building one-way gates for current flow, opening and closing channels for current flow controlled by an external electrical voltage and/or current. Folks figured out that when an *n*-type and a *p*-type semiconductor were placed together and a particular voltage was applied across the slabs, light, or photons, could be produced as the electrons jumped across the junction between the interface. It was noticed that this process could work backward as well. That is, when light was exposed at the *np* junction, electrons were made to flow, thus resulting in an electric current. A number of clever contraptions have been built using *n*-type and *p*-type semiconductor combinations. The following chapters describe some of the major devices people came up with.

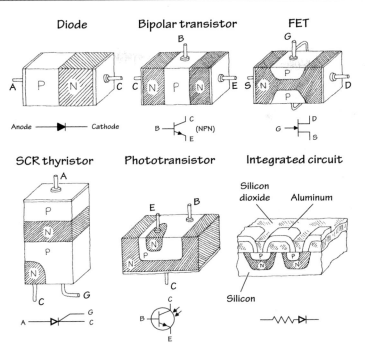

FIGURE 4.9

4.2 Diodes

A *diode* is a two-lead semiconductor device that acts as a one-way gate to electric current flow. When a diode's *anode* lead is made more positive in voltage than its *cathode* lead—a condition referred to as *forward biasing*—current is permitted to flow through the device. However, if the polarities are reversed (the anode is made more negative in voltage than the cathode)—a condition referred to as *reversed biasing*—the diode acts to block current flow.

FIGURE 4.10

Diodes are used most commonly in circuits that convert ac voltages and current into dc voltages and currents (e.g., ac/dc power supply). Diodes are also used in voltage-multiplier circuits, voltage-shifting circuits, voltage-limiting circuits, and voltage-regulator circuits.

4.2.1 How p-n Junction Diodes Work

A *p-n junction diode* (*rectifier diode*) is formed by sandwiching together *n*-type and *p*-type silicon. In practice, manufacturers grow an *n*-type silicon crystal and then abruptly change it to a *p*-type crystal. Then either a glass or plastic coating is placed around the joined crystals. The *n* side is the cathode end, and the *p* side is the anode end.

The trick to making a one-way gate from these combined pieces of silicon is getting the charge carriers in both the *n*-type and *p*-type silicon to interact in such a way that when a voltage is applied across the device, current will flow in only one direction. Both *n*-type and *p*-type silicon conducts electric current; one does it with electrons (*n*-type), and the other does it with holes (*p*-type). Now the important feature to note here, which makes a diode work (act as a one-way gate), is the manner in which the two types of charge carriers interact with each other and how they interact with an applied electrical field supplied by an external voltage across its leads. Below is an explanation describing how the charge carriers interact with each other and with the electrical field to create an electrically controlled one-way gate.

Forward-Biased ("Open Door")

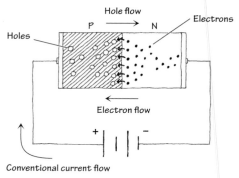

When a diode is connected to a battery, as shown here, electrons from the *n* side and holes from the *p* side are forced toward the center (*pn* interface) by the electrical field supplied by the battery. The electrons and holes combine, and current passes through the diode. When a diode is arranged in this way, it is said to be *forward-biased*.

Reverse-Biased ("Closed Door")

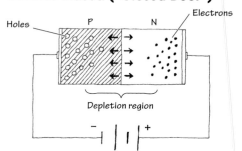

When a diode is connected to a battery, as shown here, holes in the *n* side are forced to the left, while electrons in the *p* side are forced to the right. This results in an empty zone around the *p-n* junction that is free of charge carriers, better known as the *depletion region*. This depletion region has an insulative quality that prevents current from flowing through the diode. When a diode is arranged in this way, it is said to be *reverse-biased*.

FIGURE 4.11

A diode's one-way gate feature does not work all the time. That is, it takes a minimal voltage to turn it on when it is placed in forward-biased direction. Typically for silicon diodes, an applied voltage of 0.6 V or greater is needed; otherwise, the diode will not conduct. This feature of requiring a specific voltage to turn the diode on may seem like a drawback, but in fact, this feature becomes very useful in terms of acting as a voltage-sensitive switch. Germanium diodes, unlike silicon diodes, often require a forward-biasing voltage of only 0.2 V or greater for conduction to occur. Figure 4.12 shows how the current and voltage are related for silicon and germanium diodes.

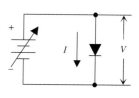
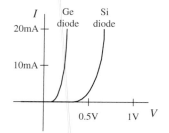

FIGURE 4.12

Another fundamental difference between silicon diodes and germanium diodes, besides the forward-biasing voltages, is their ability to dissipate heat. Silicon diodes do a better job of dissipating heat than germanium diodes. When germanium diodes get hot—temperatures exceeding 85°C—the thermal vibrations affect the physics inside the crystalline structure to a point where normal diode operation becomes unreliable. Above 85°C, germanium diodes become worthless.

4.2.2   Diode Water Analogy

A *diode* (or *rectifier*) acts as a one-way gate to current flow—see the water analogy in Fig. 4.13. Current flows in the direction of the arrow, from anode (+) to cathode (−), provided the *forward voltage* V_F across it exceeds what's called the *junction threshold voltage*. As a general rule of thumb, silicon p-n junction diodes have about a 0.6-V threshold, germanium diodes a 0.2-V threshold, and Schottky diodes a 0.4-V threshold. However, don't take this rule too seriously, because with real-life components, you'll find these thresholds may be a few tenths of a volt off. For example, it's entirely possible for a silicon p-n junction diode's threshold to be anywhere between 0.6 and 1.7 V; for germanium, 0.2 to 0.4 V; and for Schottky diodes, 0.15 to 0.9 V.

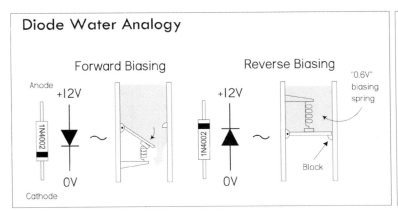

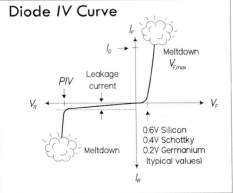

FIGURE 4.13

Note that if you actually put 12 V across a forward-biased diode as shown in Fig. 4.13, a very large current would flow, probably destroying the diode. Also, the axes of Fig. 4.13 are not to scale.

In terms of limits, avoid supplying a diode with a forward current I_F beyond its *peak current rating* $I_{0(max)}$. If you do, you'll get internal junction meltdown. Likewise, avoid applying a reverse voltage V_R any bigger than the diode's *peak inverse voltage* (PIV) *rating*. This, too, can render a diode worthless. See the graph in Fig. 4.13.

4.2.3   Kinds of Rectifiers/Diodes

There are numerous types of diodes, each specifically designed to work better in one application or another. Diodes for high-power applications (switching, power supplies, etc.) which draw lots of current or rectify high voltages typically go by the name *rectifier diodes*. On the other hand, diodes that go by names such as *signal, switching, fast recovery,* or *high speed* are designed to provide a low internal capacitance (they store less charge but usually have weaker junctions for large currents). At high speeds, these diodes will reduce RC switching time constants, which means fewer time delays and lower signal losses.

Schottky diodes have a particularly low junction capacitance and faster switching (~10 ns) when compared to silicon p-n junction diodes due to their special metal-semiconductor-junction interface. They also have a lower junction threshold voltage—as low as 0.15 V, but usually a bit more (0.4 V average). Both these characteristics enable them to detect low-voltage, high-frequency signals that ordinary p-n junction diodes would not see. (A Schottky with a 0.3-V threshold can pass signals greater than 0.3 V, but a silicon p-n junction diode with a 0.7-V threshold can only pass signals greater than 0.7 V.) For this reason, Schottky diodes are very popular in low-voltage signal rectifiers in RF circuits, signal switching in telecommunication, small dc/dc converters, small low-voltage power supplies, protection circuits, and voltage clamping arrangements. Their high-current density and low-voltage drop also make them great in power supplies, since they generate less heat, requiring smaller heat sinks to be used in design. Therefore, you'll find both rectifier and fast-switching Schottky diodes listed in the catalogs.

Germanium diodes are used mostly in RF signal detection and low-level logic design due to their small threshold voltage of about 0.2 V. You do not see them in high-current rectifying applications, since they are weaker and leak more than silicon

Common Diode/Rectifier Packages

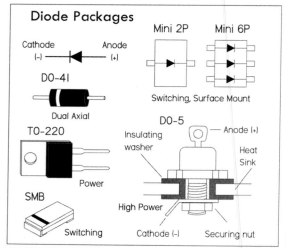

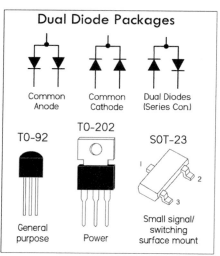

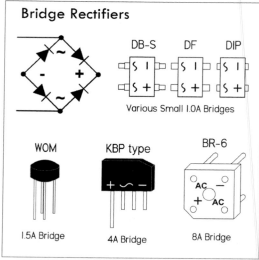

FIGURE 4.14

diodes when temperatures rise. In many applications, a good Schottky signal diode can replace a germanium diode.

4.2.4 Practical Considerations

Five major specs to consider when choosing a diode are peak inverse voltage, PIV; forward current-handling capacity, $I_{O(max)}$; response speed t_R (time for diode to switch on and off); reverse-leakage current, $I_{R(max)}$; and maximum forward-voltage drop, $V_{F(max)}$. Each of these characteristics can be manipulated during the manufacturing process to produce the various special-purpose diodes. In rectification applications (e.g., power supplies, transient protection), the most important specs are PIV and current rating. The peak negative voltages that are stopped by the diode must be smaller in magnitude than the PIV, and the peak current through the diode must be less than $I_{O(max)}$. In fast and low-voltage applications, t_R and V_F become important characteristics to consider. In the applications section that follows, you'll get a better sense of what all these specs mean.

TABLE 4.1 Selection of Popular Diodes

DEVICE	TYPE	PEAK INVERSE VOLTAGE PIV (V)	MAX. FORWARD CURRENT $I_{O\,(MAX)}$	MAX. REVERSE CURRENT $I_{R\,(MAX)}$	PEAK SURGE CURRENT I_{FSM}	MAX. VOLTAGE DROP V_F (V)
1N34A	Signal (Ge)	60	8.5 mA	15 μA		1.0
1N67A	Signal (Ge)	100	4.0 mA	5 μA		1.0
1N191	Signal (Ge)	90	5.0 mA			1.0
1N914	Fast Switch	90	75 mA	25 nA		0.8
1N4148	Signal	75	10 mA	25 nA	450 mA	1.0
1N4445	Signal	100	100 mA	50 nA		1.0
1N4001	Rectifier	50	1 A	0.03 mA	30 A	1.1
1N4002	Rectifier	100	1 A	0.03 mA	30 A	1.1
1N4003	Rectifier	200	1 A	0.03 mA	30 A	1.1
1N4004	Rectifier	400	1 A	0.03 mA	30 A	1.1
1N4007	Rectifier	1000	1 A	0.03 mA	30 A	1.1
1N5002	Rectifier	200	3 A	500 μA	200 A	
1N5006	Rectifier	600	3 A	500 μA	200 A	
1N5008	Rectifier	1000	3 A	500 μA	200 A	
1N5817	Schottky	20	1 A	1 mA	25 A	0.75
1N5818	Schottky	30	1 A		25 A	
1N5819	Schottky	40	1 A		25 A	0.90
1N5822	Schottky	40	3 A			
1N6263	Schottky	70	15 mA		50 mA	0.41
5052-2823	Schottky	8	1 mA	100 nA	10 mA	0.34

Diodes come in a variety of different packages. Some are standard two-lead deals; others are high-power packages with heat-sink attachments (e.g., TO-220, DO-5). Some come in surface-mount packages, and others contain diode arrays in IC form, used for switching applications. Dual-diode and diode-bridge rectifiers also come in a variety of package sizes and shapes for varying power levels.

4.2.5 Diode/Rectifier Applications

Voltage Dropper

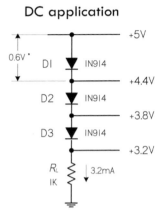

DC application

FIGURE 4.15

When current passes through a diode, there is a voltage drop across it of about 0.6 V, for a silicon p-n junction diode. (Germanium diodes have around a 0.2-V drop; Schottky, around 0.4 V—all these values vary slightly, depending on the specific diode used.) By placing a number of diodes in series, the total voltage drop across the combination is the sum of the individual voltage drops across each diode. Voltage droppers are often used in circuits where a fixed small voltage difference between two sections of a circuit is needed. Unlike resistors that can be used to lower the voltage, the diode arrangement typically doesn't waste as much power to resistive heating and will supply a stiffer regulated voltage that is less dependent on current variations. Later in this chapter, you'll see that a single zener diode can often replace a multiple series diode arrangement like the one shown here.

Voltage Regulator

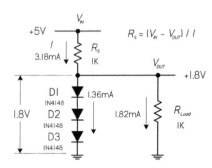

FIGURE 4.16

Here's a spin-off of the last circuit, making use of the three diodes to create a simple regulated (steady) low-voltage output equal to the sum of the threshold voltages of the diodes: 0.6 V + 0.6 V + 0.6 V = 1.8 V. The series resistor is used to set the desired output current (I) and should be less than the value calculated using the following formula, but not so low that it exceeds the power ratings of itself and the diodes:

$$R_S = \frac{V_{in} - V_{out}}{I}$$

Diodes and the series resistor must have proper power ratings for the amount of current drawn. Use $P = IV$. Note that for higher-power critical voltage sources, you'll typically use a zener diode regulator or, more commonly, a special regulator IC instead.

Reverse-Polarity Protection

Battery reversal or power polarity reversal can be fatal to portable equipment. The best design is to use a mechanical block to safeguard against reverse installation. However, even momentarily fumbling around and making contacts can lead to problems. This is especially true for one or more single-cell battery applications that use AA-alkaline, NiCad, and NiMH batteries. For these systems you must ensure that any flow of reverse current is low enough to avoid damaging the circuit or the battery.

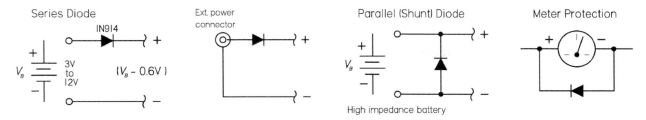

FIGURE 4.17 *Series diode:* This is the simplest battery-reversal protection. It can be used with external power connections, too—plug-and-jack. The diode allows current from a correctly installed battery to flow to load, but blocks current flow from a backward-installed battery. The drawback with a series diode is that the diode must handle the full load current. Also, the forward voltage drop of the diode shortens the equipment's operating time—cuts off about 0.6 V right away. Schottky diodes with low thresholds can do better. See Problem 1 at the end of Sec. 4.2 for another reverse-polarity protection circuit.

Parallel diode: In applications that call for alkaline or other batteries that have high output impedances, you can guard against reverse installations by using a parallel (shunt) diode, while eliminating the diode's voltage drop. This approach protects the load but draws high current from the backward-installed battery. The diode must be properly rated for current and power. Another application of the parallel diode is in meter protection, where it acts to divert large currents entering the negative terminal of the meter.

Note: In more sophisticated battery-powered designs, special ICs or transistor arrangements are used to provide essentially zero voltage drop protection, while providing a number of other special features, such as reverse polarity protection, thermal shutdown, and voltage level monitor.

Transient Suppression with Fly-Back Diodes

Transient Protection

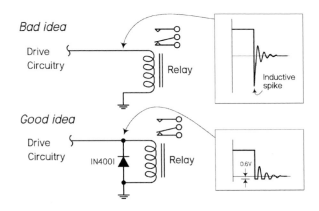

When current flowing through an inductor is suddenly switched off, the collapsing magnetic field will generate a high-voltage spike in the inductor's coils. This voltage spike or transient may have an amplitude of hundreds or even thousands of volts. This is particularly common within relay coils. A diode—referred to as a fly-back diode for this type of application—placed across the relay's coil can protect neighboring circuitry by providing a short circuit for the high-voltage spike. It also protects the relay's mechanical contacts, which often get viciously slapped shut during an inductive spike. Notice, however, that the diode is ineffective during turn-on time. Select a rectifier diode with sufficient power ratings (1N4001, 1N4002, etc.). Schottky rectifiers (e.g., 1N5818) work well, too.

Transistor Relay Driver with Protection Diodes

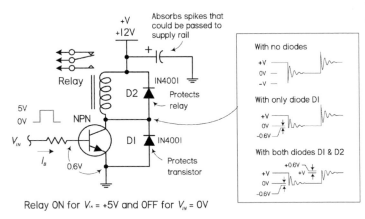

Relay ON for V_{in} = +5V and OFF for V_{IN} = 0V

FIGURE 4.18

Here's a more practical example for driving a relay that contains an extra diode placed across a transistor driver in order to protect the transistor from damage due to inductive spikes generated from the relay's coil when the transistor is turned off. This arrangement also deadens spikes during turn-on time. This dual diode arrangement is sometimes used in voltage regulator circuits, where one diode is wired from the output to the input and another is wired from ground to the output. This prevents any attached loads from sending damaging spikes back into the IC's output.

Motor Inductive Kickback Protection

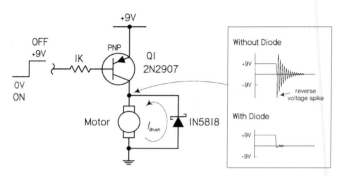

FIGURE 4.18 *(Continued)*

Here's another example of how inductive kickback from a motor that is running and then suddenly turned off can generate a voltage transient that can potentially damage connected electronics—in this case, a 2N2907 transistor. The diode reroutes or shorts the induced voltage to the opposite terminal of the motor. Here a 1N5818 Schottky diode is used—though you could use other p-n junction types, too. The Schottky diode happens to be a bit faster and will clip the transient voltage a bit lower down—at around 0.4 V.

Note: Devices such as TVs and varistors are specially designed for transient suppression. See the section on transient suppressors later in this chapter.

Diode Clamps

Diode clamps are used to clip signal levels, or they can shift an ac waveform up or down to create what's called a pulsing dc waveform—one that doesn't cross the 0-V reference.

Adjustable Waveform Clipper

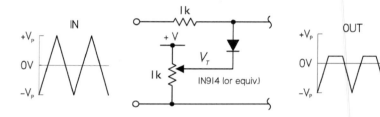

In the adjustable waveform clipper circuit, the maximum output is clipped to a level determined by the resistance of the potentiometer. The idea is to set the negative end of the diode to about 0.6 V lower than the maximum desired output level, to account for the forward voltage drop of the diode. That's what the potentiometer is intended to do. +V should be a volt or so higher than the peak input voltage.

Adjustable Attenuator

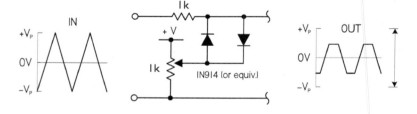

The adjustable attenuator is similar to the last circuit, but the additional opposing diode allows for clipping on both positive and negative swings. You can use separate potentiometers if you want separate positive and negative clipping level control. +V should be a volt or so higher than the peak input voltage.

Diode Voltage Clamp (DC Restoration)

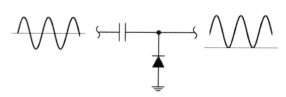

FIGURE 4.19

The diode voltage clamp provides dc restoration of a signal that has been ac-coupled (capacitively coupled). This is important for circuits whose inputs look like diodes (e.g., a transistor with grounded emitter); otherwise, an ac-couple signal would fade away.

Diode Switch

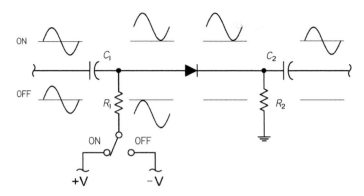

In the diode switch circuit, an input waveform is ac-coupled to the diode through C_1 at the input and C_2 at the output. R_2 provides a reference for the bias voltage. When the switch is thrown to the ON position, a positive dc voltage is added to the signal, forward-biasing the diode and allowing the signal to pass. When the switch is thrown to the OFF position, the negative dc voltage added to the signal reverse-biases the diode and the signal does not get through.

FIGURE 4.19   *(Continued)*

Half-Wave Rectifier

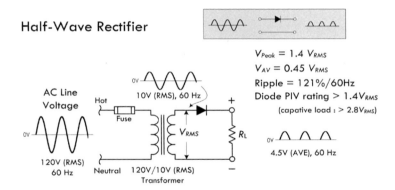

$V_{Peak} = 1.4\ V_{RMS}$
$V_{AV} = 0.45\ V_{RMS}$
Ripple = 121%/60Hz
Diode PIV rating > $1.4V_{RMS}$
(capative load : > $2.8V_{RMS}$)

Half-wave rectifier: Used to transform an ac signal into pulsing dc by blocking the negative swings. A filter is usually added (especially in low-frequency applications) to the output to smooth out the pulses and provide a higher average dc voltage. The peak inverse voltage (PIV)—the voltage that the rectifier must withstand when it isn't conducting—varies with load, and must be greater than the peak ac voltage (1.4 × V_{rms}). With a capacitor filter and a load drawing little or no current, it can rise to 2.8 × V_{rms} (capacitor voltage minus the peak negative swing of voltage from transformer secondary).

Full-Wave Center-Tap Rectifier

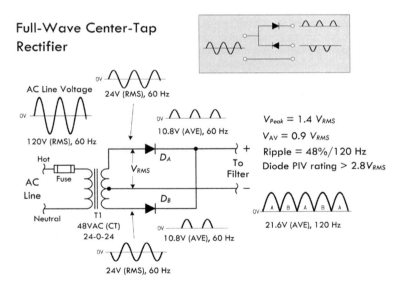

$V_{Peak} = 1.4\ V_{RMS}$
$V_{AV} = 0.9\ V_{RMS}$
Ripple = 48%/120 Hz
Diode PIV rating > $2.8V_{RMS}$

FIGURE 4.20

Full-wave center-tap rectifier: This commonly used circuit is basically two combined half-wave rectifiers that transform both halves of an ac wave into a pulsing dc signal. When designing power supplies, you need only two diodes, provided you use a center-tap transformer. The average output voltage is 0.9 V_{rms} of half the transformer secondary; this is the maximum that can be obtained with a suitable choke-input filter. The peak output voltage is 1.4 × V_{rms} of half the transformer secondary; this is the maximum voltage that can be obtained from a capacitor-input filter. The PIV impressed on each diode is independent of the type of load at the output. This is because the peak inverse voltage condition occurs when diode A conducts and diode B does not conduct. As the cathodes of diodes A and B reach a positive peak (1.4 V_{rms}), the anode of diode B is at the negative peak, also 1.4 V_{rms}, but in the opposite direction. The total peak inverse voltage is therefore 2.8 V_{rms}. The frequency of the output pulses is twice that of the half-wave rectifier, and thus comparatively less filtering is required. Since the diodes work alternately, each handles half of the load current. The current rating of each rectifier need be only half the total current drawn from the supply.

Full-Wave Bridge Rectifier

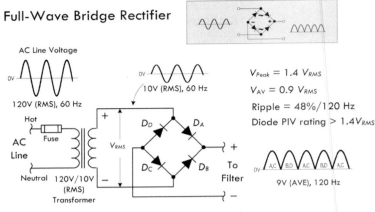

$V_{Peak} = 1.4\ V_{RMS}$
$V_{AV} = 0.9\ V_{RMS}$
Ripple = 48%/120 Hz
Diode PIV rating > $1.4 V_{RMS}$

9V (AVE), 120 Hz

FIGURE 4.20   *(Continued)*

Full-wave bridge rectifier: This rectifier produces a similar output as the last full-wave rectifier, but doesn't require a center-tap transformer. To understand how the device works, follow the current through the diode one-way gates. Note that there will be at least a 1.2-V drop from zero-to-peak input voltage to zero-to-peak output voltage (there are two 0.6-V drops across a pair of diodes during a half cycle). The average dc output voltage into a resistive load or choke-input filter is $0.9 \times V_{rms}$ of the transformer's secondary; with a capacitor filter and a light load, the maximum output voltage is $1.4 \times V_{rms}$. The inverse voltage across each diode is $1.4\ V_{rms}$; there the PIV of each diode is more than $1.4\ V_{rms}$.

See the following text for the pros and cons of the various rectifier configurations.

Voltage Multiplier Circuits

Half-Wave Voltage Doubler

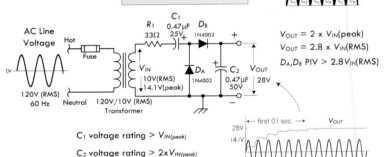

$V_{OUT} = 2 \times V_{IN}$(peak)
$V_{OUT} = 2.8 \times V_{IN}$(RMS)
D_A, D_B PIV > $2.8 V_{IN}$(RMS)

C_1 voltage rating > $V_{IN(peak)}$

C_2 voltage rating > $2 \times V_{IN(peak)}$

Half-wave voltage doubler: Takes an ac input voltage and outputs a dc voltage that is approximately equal to twice the input's peak voltage (or 2.8 times the input's RMS voltage). (The actual multiplication factor may differ slightly, depending on the capacitor, resistor, and load values.) In this circuit, we take V_{IN} to mean the secondary voltage from the transformer. During the first negative half cycle, D_A conducts, charging C_1 to the peak rectified voltage V_{IN} (peak), or $1.4\ V_{IN}$ (RMS). During the positive half cycle of the secondary voltage, D_A is cut off and D_B conducts, charging capacitor C_2. The voltage delivered to C_2 is the sum of the transformer peak secondary voltage, V_{IN} (peak) plus the voltage stored in C_1, which is the same, so the sum gives $2\ V_{IN}$ (peak), or $2.8\ V_{IN}$ (RMS). On the next negative cycle, D_B is nonconducting and C_2 will discharge into an attached load. If no load is present, the capacitors will remain charged—C_1 to $1.4\ V_{IN}$ (RMS), C_2 to $2.8\ V_{IN}$ (RMS). When a load is connected to the output, the voltage across C_2 drops during the negative half cycle and is recharged up to $2.8\ V_{IN}$ (RMS) during the positive half cycle. The output waveform across C_2 resembles that of a half-wave rectifier circuit because C_2 is pulsed once every cycle. Figure 4.21 shows levels to which the two capacitors are charged throughout the cycle. In actual operation, the capacitor will not discharge all the way to zero, as shown.

Full-Wave Voltage Doubler

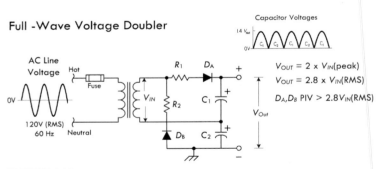

$V_{OUT} = 2 \times V_{IN}$(peak)
$V_{OUT} = 2.8 \times V_{IN}$(RMS)
D_A, D_B PIV > $2.8 V_{IN}$(RMS)

FIGURE 4.21

Full-wave doubler: During the positive half cycle of the transformer secondary voltage, D_A conducts charging C_1 to V_{IN} (peak) or $1.4\ V_{IN}$ (RMS). During the negative half cycle, D_B conducts, charging C_2 to the same value. The output voltage is the sum of the two capacitor voltages, which will be $2\ V_{IN}$ (peak) or $2.8\ V_{IN}$ (RMS) under no-load conditions. The graph shows each capacitor alternately receiving a charge once per cycle. The effective filter capacitance is that of C_1 and C_2 in series, which is less than the capacitance of either C_1 or C_2 alone. R_1 and R_2 are used to limit the surge current through the rectifiers. Their values are based on the transformer voltage and the rectifier surge current rating, since at the instant the power supply is turned on, the filter capacitors look like a short-circuited load. Provided the limiting resistors can withstand the surge current, their current-handling capacity is based on the maximum load current from the supply. The peak inverse voltage across each diode is $2.8\ V_{IN}$ (RMS).

Pros and Cons of the Rectifier Circuits

Comparing the full-wave center-tap rectifier and the full-wave bridge rectifier, you'll notice both circuits have almost the same rectifier requirements. However, the center-tap version has half the number of diodes as the bridge. These diodes will require twice the maximum inverse voltage ratings of the bridge diodes (PIV $> 2.8\ V_{rms}$, as opposed to $>1.4\ V_{rms}$). The diode current ratings are identical for the two circuits. The bridge makes better use of the transformer's secondary than the center-tap rectifier, since the transformer's full winding supplies power during both half cycles, while each half of the center-tap circuit's secondary provides power only during its positive half-cycle. This is often referred to as the *transformer utilization factor*, which is unity for the bridge configuration and 0.5 for the full-wave center-tap configuration.

The bridge rectifier is often not as popular as the full-wave center-tap rectifier in high-current, low-voltage applications. This is because the two forward-conducting series-diode voltage drops in the bridge introduce a volt or more of additional loss, and thus consume more power (heat loss) than a single diode would within a full-wave rectifier.

In regard to the half-wave configuration, it's rarely used in 60-Hz rectification for other than bias supplies. It does see considerable use, however, in high-frequency switching power supplies in what are called forward converter and fly-back converter topologies.

Voltage Tripler

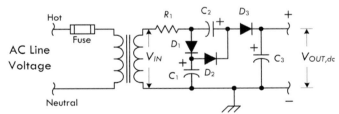

Voltage Quadrupler

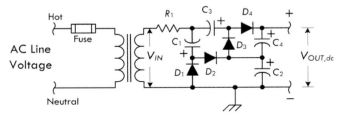

Voltage tripler: **On one half of the ac cycle, C_1 and C_3 are charged to V_{IN} (peak) through D_1, D_2, and D_3. On the opposite half of the cycle, D_2 conducts and C_2 is charged to twice V_{IN} (peak), because it sees the transformer plus the charge in C_1 as its source. (D_1 is cut off during this half cycle.) At the same time, D_3 conducts, and with the transformer and the charge in C_2 as the source, C_3 is charged to three times the transformer voltage.**

Voltage quadrupler: **Works in a similar manner as the previous one. In both these circuits, the output voltage will approach an exact multiple of the peak ac voltage when the output current drain is low and the capacitance values high.**

Capacitance values are usually 20 to 50 µF, depending on the output current drain. Capacitor dc ratings are related to VIN (peak) by:

C_1—Greater than V_{IN} (peak) or 0.7 V_{IN} (RMS)
C_2—Greater than 2 V_{IN} (peak) or 1.4 V_{IN} (RMS)
C_3—Greater than 3 V_{IN} (peak) or 2.1 V_{IN} (RMS)
C_4—Greater than 4 V_{IN} (peak) or 2.8 V_{IN} (RMS)

FIGURE 4.22

Diode Logic Gates

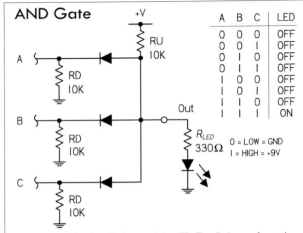

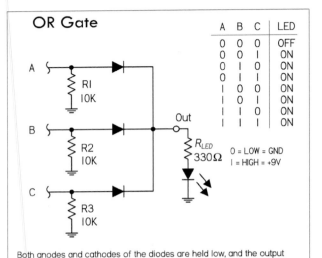

FIGURE 4.23 These simple diode logic gates are useful for learning the basics of digital logic, and can also be adapted for non-logic-level electronics (e.g., higher-voltage and power analog-like circuits)—see the following battery-backup example (Fig. 4.24). When designing high-power circuits, make sure your diodes have the proper PIV and current ratings for the job. It's also important to note that the recovery time of power diodes won't be as fast as digital logic ICs or fast-switching diodes.

Battery Backup

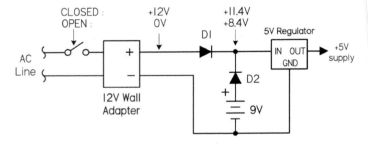

FIGURE 4.24

Devices are powered by a wall adapter with battery backup, typically diode-OR for the battery and wall-adapter connections, as shown in Fig. 4.24. Normally if the switch is closed, power is delivered to the load from the 12-V wall adapter through D_1; D_2 is reverse-biased (off), since its negative end is 2.4V more positive than its positive end. If power is interrupted (switch opened), D_1 stops conducting, and the battery kicks in, sending current through D_2 into the load; D_1 blocks current from flowing back into the wall adapter. There is a penalty for using diodes for battery backup, however, since the diode in series with the battery limits the minimum voltage at which the battery can supply power (around a 0.6-V drop for silicon p-n junction, 0.4 V for Schottky). Better battery-backup designs implement transistors or special ICs that contain an internal comparator which switches over battery power through a low-resistance transistor without the 0.6-V penalty. Check out MAXIM's website for some example ICs.

AM Detector

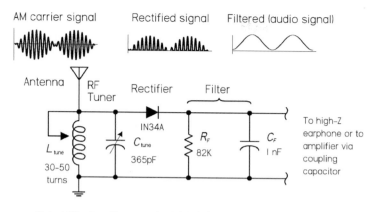

AM carrier signal   Rectified signal   Filtered (audio signal)

Antenna RF Tuner Rectifier Filter

IN34A

L_{tune}
30–50 turns

C_{tune}
365pF

R_F
82K

C_F
1 nF

To high-Z earphone or to amplifier via coupling capacitor

Output filter has time constant that is long compared to period of carrier, but short compared to period of audio signal.

FIGURE 4.25

Diodes are often used in the detection of amplitude modulated (AM) signals, as demonstrated in the simple AM radio in Fig. 4.25. Within an AM radio signal, an RF carrier signal of constant high frequency (550 to 1700 kHz) has been amplitude modulated with an audio signal (10 to 20,000 Hz). The audio information is located in duplicate in both upper and lower sidebands, or the envelope of the AM signal. Here, an antenna and LC-tuning circuit act to "resonate" in on the specific carrier frequency of interest (transform radio signal into corresponding electrical signal). A signal diode (e.g., 1N34) is then used to rectify out the negative portion of the incoming signal so it can be manipulated by the next dc stages. The rectified signal is then stripped of its high-frequency carrier by passing through a low-pass filter. The output signal is the audio signal. This signal can be used to drive a simple crystal earpiece, a modern sensitive headphone, or a telephone receiver earpiece. (Low-impedance earphones or speakers will require additional amplification via a coupling capacitor of 1 μF or so.)

Schottky Diode Termination

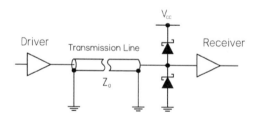

V_{cc}

Driver Transmission Line Receiver

Z_0

FIGURE 4.26

Schottky diode termination can be used to counteract the high-speed transmission line effects, which cause over/undershoots from signal reflections, reduce noise margins, and destroy timing. These types of distortions can cause false triggering in clock lines and erroneous data on address, data, and control lines, as well as contribute significantly to clock and signal jitter. For applications where transmission line impedance is variable or unknown, it's not possible to specify a termination resistance value—an alternative is needed. The Schottky diode termination has the ability to maintain signal integrity, save significant power, and permit flexible system design. A Schottky diode termination consists of a diode series combination, where one diode clamps to V_{CC}, or supply voltage, and the other to ground. The diodes at the end of the transmission line minimize the effect of reflection via a clamping operation. The top diode clamps voltages that exceed V_{CC} by the forward-bias threshold limit. This clamping will minimize overshoots caused by reflections. For falling edge signals, a clamp diode to ground affects a similar termination. This clamping function does not depend on matching the transmission line characteristic impedance, making it useful in situations where the line impedance is unknown or variable.

Read-Only Memory (ROM)

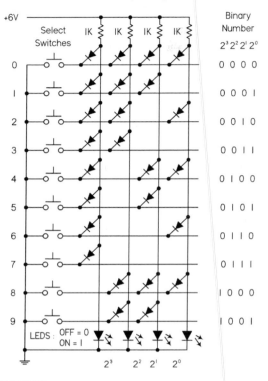

This circuit is a simple read-only memory (ROM) made with diodes. Here, the ROM acts as a decimal-to-binary encoder. With no buttons pressed, all LEDs are lit. If 1 is pressed, current from the supply is diverted away from the 2^3, 2^2, and 2^1 lines via the diodes to ground, but is allowed to pass on the 2^0 line, thus presenting 0001 on the LED readout. In reality, using a PROM such as this for encoding—or anything else, for that matter—isn't practical. Usually there is a special encoder IC you buy or you simply take care of the encoding—say, with a multiplexed keypad that's interfaced with a microcontroller—the actual encoding is taken care of at the programming level. At any rate, it's a fun circuit, and this gives you a basic idea of how read-only memory works.

FIGURE 4.27

4.2.6 Zener Diodes

A zener diode acts like a two-way gate to current flow. In the forward direction, it's easy to push open; only about 0.6 V—just like a standard diode. In the reverse direction, it's harder to push open; it requires a voltage equal to the *zener's breakdown voltage* V_Z. This breakdown voltage can be anywhere between 1.8 and 200 V, depending on the model (1N5225B = 3.0 V, 1N4733A = 5.1 V, 1N4739A = 9.1 V, etc.). Power ratings vary from around 0.25 to 50 W.

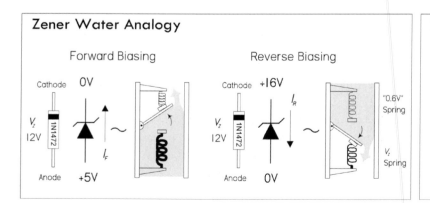

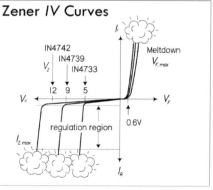

FIGURE 4.28 The reverse-bias direction is the standard configuration used in most applications, along with a series resistor. In this configuration, the zener diode acts like a pressure release value, passing as much current as necessary to keep the voltage across it constant, equal to V_Z. In other words, it can act as a voltage regulator. See application in Fig. 4.29.

Zener Voltage Regulator

These circuits act as voltage regulators, preventing any supply voltage or load current variations from pulling down the voltage supplied to the load. The following explains how the zener diode compensates for both line and load variations.

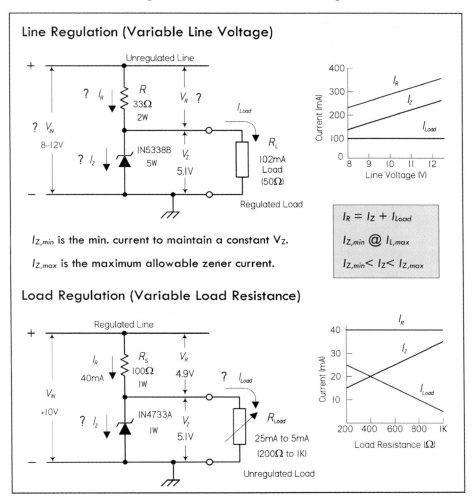

Line Regulation (Variable Line Voltage)

$I_{Z,min}$ is the min. current to maintain a constant V_Z.

$I_{Z,max}$ is the maximum allowable zener current.

$$I_R = I_Z + I_{Load}$$

$$I_{Z,min} \ @ \ I_{L,max}$$

$$I_{Z,min} < I_Z < I_{Z,max}$$

Load Regulation (Variable Load Resistance)

FIGURE 4.29   *Line regulation example:* If the line voltage increases, it will cause an increase in line current. Since the load voltage is constant (maintained by the zener), the increase in line current will result in an increase in zener current, thus maintaining a constant load current. If the line voltage decreases, less line current results, and less current is passed by the zener. See graph in Fig. 4.29, top right.

Load regulation example: If the load voltage attempts to decrease as a result of decreased load resistance (increased load current), the increase in load current is offset by the decrease in zener current. The voltage across the load will remain fairly constant. If the load voltage attempts to increase due to an increase in load resistance (decrease in load current), the decrease in load current is offset by an increase in zener current. See graph in Fig. 4.29, bottom right.

The following formulas can be used when selecting the component values:

$$R_S = \frac{V_{in,min} - V_Z}{I_{Z,min} + I_{L,max}} \ ; \quad P_R = \frac{(V_{in,max} - V_Z)^2}{R_S}$$

$$P_{Z,max} = V_Z \frac{(V_{in,max} - V_Z)}{R_S}$$

See Problem 3 at end of this section for a design example.

Note that zener regulators are somewhat temperature dependent and aren't the best choice for critical applications. A linear regulator IC, though more expensive, is less dependent on temperature variations due to an internal error amplifier. They do typically use an internal zener to supply the reference, nonetheless.

Selection of Popular Zener Diodes

Zener Diode Packages

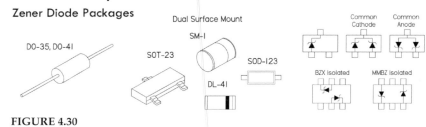

FIGURE 4.30

TABLE 4.2

ZENER VOLTS VZ VOLTS	CASE AND POWER RATING					
	AXIAL LEAD			SURFACE MOUNT		
	500 MW	1 W	5 W	200 MW	500 MW	1 W
2.4	1N5221B			BZX84C2V4, MMBZ5221B	BZT52C2V4	
2.7	1N5222B			BZX84C2V7	BZT52C2V7	
3.0	1N5225B			BZX84C3V0, MMBZ52251B	BZT52C3V0, ZMM5225B	
3.3	1N5226B	1N4728A	1N5333B	BZX84C3V3, MMBZ5226B	BZT52C3V3, ZMM5226B	ZM4728A
3.6	1N5227B	1N4729A	1N5334B	BZX84C3V6, MMBZ5227B	BZT52C3V6, ZMM5227B	
3.9	1N5228B	1N4730A	1N5335B	BZX84C3V9, MMBZ5228B	BZT52C3V9, ZMM5228B	ZM4730A
4.3	1N5229B	1N4731A	1N5336B		BZT52C4V3, ZMM5229B	ZM4731A
4.7	1N5230B	1N4732A	1N5337B	BZX84C4V7, MMBZ5230B	BZT52C4V7, ZMM5230B	ZM4732A
5.1	1N5231B	1N4733A	1N5338B	BZX84C5V1, MMBZ5231B	BZT52C5V1, ZMM5231B	SMAZ5V1, ZM4733A
5.6	1N5232B	1N4734A	1N5339B	BZX84C5V6, MMBZ5232B	BZT52C5V6, ZMM5232B	SMAZ5V6, ZM4734A
6.0	1N5233B		1N5340B		BZT52C6V0, ZMM52330B	
6.2	1N5234B	1N4735A	1N5341B	BZX84C6V2, MMBZ5234B	BZT52C6V2, ZMM5234B	SMAZ6V2, ZM4735A
6.8	1N5235B	1N4736A	1N5342B	BZX84C6V8, MMBZ5235B	BZT52C6V8, ZMM5235B	SMAZ6V8, ZM4736A
7.5	1N5236B	1N4737A	1N5343B	BZX84C7V5, MMBZ5236B	BZT52C7V5, ZMM5236B	SMAZ7V5, ZM4737A
8.2	1N5237B	1N4738A	1N5344B	BZX84C8V2, MMBZ5237B	BZT52C8V2, ZMM5237B	SMAZ8V2, ZM4738A
8.7	1N5238B		1N5345B		BZT52C8V7, ZMM5238B	
9.1	1N5239B	1N4739A	1N5346B	BZX84C9V1, MMBZ5239B	BZT52C9V1, ZMM5239B	SMAZ9V1, ZM4739A
10.0	1N5240B	1N4740A	1N5347B	BZX84C10	BZT52C10, ZMM5240B	SMAZ10, ZM4740A
11	1N5241B	1N4741A	1N5348B	BZX84C11, MMBZ5241B	BZT52C11, ZMM5241B	ZM4741A
12	1N5242B	1N4742A	1N5349B	BZX84C12, MMBZ5242B	BZT52C12, ZMM5242B	SMAZ12, ZM4742A
13	1N5243B	1N4743A	1N5350B	MMBZ5243B	BZT52C13, ZMM5243B	ZM4743A
14	1N5244B		1N5351B		BZT52C14, ZMM5244B	
15	1N5245B	1N4744A	1N5352B	BZX84C15, MMBZ5245B	BZT52C15, ZMM5245B	SMAZ15, ZM4744A
16	1N5246B	1N4745A	1N5353B	BZX84C16, MMBZ5246B	BZT52C16, ZMM5246B	SMAZ16, ZM4745A
17	1N5247B		1N5354B		ZMM5247B	
18	1N5248B	1N4746A	1N5355B	BZX84C18, MMBZ5248B	BZT52C18, ZMM5248B	SMAZ18, ZM4746A
19	1N5249B		1N5356B		ZMM5249B	
20	1N5250B	1N4747A	1N5357B	BZX84C20, MMBZ5250B	BZT52C20, ZMM5250B	SMAZ20, ZM4747A
22	1N5251B	1N4748A	1N5358B	BZX84C22, MMBZ5251B	BZT52C22, ZMM5251B	SMAZ22, ZM4748A
24	1N5252B	1N4749A	1N5359B	BZX84C24, MMBZ5252B	BZT52C24, ZMM5252B	SMAZ24, ZM4749A
25	1N5253B		1N5360B		ZMM5253B	
27	1N5254B	1N4750A	1N5361B	BZX84C27, MMBZ5254B	BZT52C27, ZMM5254B	SMAZ27, ZM4750A
28	1N5255B		1N5362B	MMBZ5255B	ZMM5255B	
30	1N5256B	1N4751A	1N5363B	BZX84C30	BZT52C30, ZMM5256B	SMAZ30, ZM4751A
33	1N5257B	1N4752A	1N5364B	BZX84C33	BZT52C33, ZMM5257B	SMAZ33, ZM4752A
36	1N5258B	1N4753A	1N5365B	BZX84C36, MMBZ5258B	BZT52C36, ZMM5258B	SMAZ36, ZM4753A
39	1N5259B	1N4754A	1N5366B	BZX84C39, MMBZ5259B	BZT52C39, ZMM5259B	SMAZ39, ZM4754A
43	1N5260B	1N4755A	1N5367B		BZT52C43, ZMM5260B	ZM4755A
47	1N5261B	1N4756A	1N5368B		BZT52C47, ZMM5261B	ZM4756A
51	1N5262B	1N4757A	1N5369B		BZT52C51, ZMM5262B	ZM4757A
56	1N5263B	1N4758A	1N5370B			ZM4758A
60	1N5264B		1N5371B			
62	1N5265B	1N4759A	1N5372B		ZMM5265B	ZM4759A
68	1N5266B	1N4760A	1N5373B		ZMM5266B	ZM4760A
75	1N5267B	1N4761A	1N5374B			ZM4761A
82	1N5268B	1N4762A	1N5375B			ZM4762A
87	1N5269B					
91	1N5270B	1N4763A	1N5377B			ZM4763A
100	1N5271B	1N4764A	1N5378B			ZM4764A

4.2.7 Zener Diode Applications

Split Supply from Single Transformer Winding

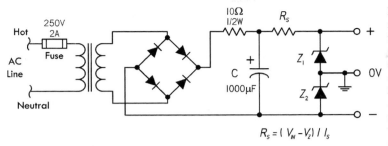

FIGURE 4.31

Here's a method for obtaining a split supply from a non-center-tapped transformer using two zener diodes. Z_1 and Z_2 are selected of equal voltage and power rating for desired split voltage and load. As with the previous example, the temperature dependency of the zener diodes makes this arrangement less accurate than a supply that uses two separate regulator ICs. However, it's a simple alternative for noncritical applications. See Chap. 11 on power supplies.

Waveform Modifier and Limiter

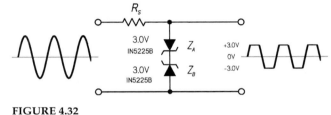

FIGURE 4.32

Two opposing zener diodes act to clip both halves of an input signal. Here a sine wave is converted to a near squarewave. Besides acting to reshape a waveform, this arrangement can also be placed across the output terminal of a dc power supply to prevent unwanted voltage transients from reaching an attached load. The breakdown voltages in that case must be greater than the supply voltage, but smaller than the maximum allowable transient voltage. A single bidirectional TVS does the same thing—see the section on transient suppressors.

Voltage Shifter

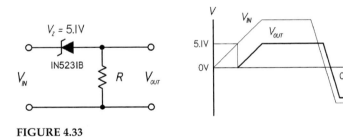

FIGURE 4.33

This circuit shifts the input voltage down by an amount equal to the breakdown voltage of the zener diode. As the input goes positive, the zener doesn't go into breakdown until it reaches 5.1 V (for the 1N5281B). After that, the output follows the input, but shifted 5.1 V below it. When the input goes negative, the output will follow the input, but shifted by 0.6 V—the forward threshold voltage drop of the zener.

Voltage Regulator Booster

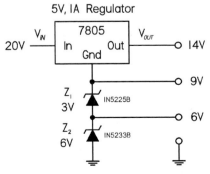

FIGURE 4.34

Zener diodes can be used to raise the level of a voltage regulator and obtain different regulated voltage outputs. Here 3-V and 6-V zener diodes are placed in series to push the reference ground of a 5 V regulator IC up 9 V to a total of 14 V. Note that in real designs, capacitors may be required at the input and output. See the section on voltage regulator ICs.

Overvoltage Protection

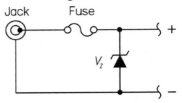

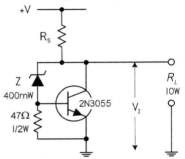

FIGURE 4.35

If excessive voltage is applied to the jack (say, via an incorrectly rated wall plug-in supply), the zener diode will conduct until the fuse is blown. The breakdown voltage of the zener should be slightly above the maximum tolerable voltage that the load can handle. Either a fast- or a slow-blow fuse can be used, depending on the sensitivity of the load. The current and voltage ratings of the fuse must be selected according to the expected limits of the application. Note that there are other, similar overvoltage protection designs that use special devices, such as TVSs and varistors. These devices are cheap and are very popular in design today.

Increasing Wattage Rating of Zener

FIGURE 4.36

Here's a simple circuit that effectively increases the wattage rating (current-handling capacity) of a zener diode by letting a power transistor take care of the majority of the regulating current. The zener itself takes a small portion of the total current and creates a base voltage/current (with the help of the base-to-ground resistor) that changes the collector-to-emitter current flow according to any variations in line or load current.

Simple LED Voltmeter

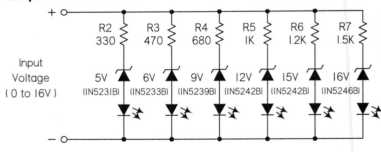

FIGURE 4.37

Here's a simple circuit voltmeter that uses the sequence of zener diodes with increasing breakdown voltages. LEDs glow in sequence as the input voltage rises. It's okay to use different zener diodes so long as the series resistors limit current through LED to a safe level. Most LEDs are happiest around 20 mA or so. You can calculate the worst-case scenario to be at the 5 V LED leg when $V_{in} = 16$ V. If you're looking for more sophistication, you can always use an analog-to-digital converter, along with a microcontroller and LCD or LED display.

4.2.8 Varactor Diodes (Variable Capacitance Diodes)

A *varactor* or variable capacitance diode (also called a *varicap*) is a diode whose junction capacitance can be altered with an applied reverse voltage. In this way, it acts as a variable capacitor. As the applied reverse voltage increases, the width of its junction increases, which decreases its capacitance. The typical capacitance range for varactors ranges from a few picofarads to over 100 pF, with a maximum reverse voltage range from a few volts to close to a hundred volts, depending on device. (Many standard diodes and zener diodes can be used as inexpensive varactor diodes, though the relationship between reverse voltage and capacitance isn't always as reliable.)

The low capacitance levels provided by a varactor usually limit its use to high-frequency RF circuits, where the applied voltage is used to change the capacitance of an oscillator circuit. The reverse voltage may be applied via a tuning potentiometer, which acts to change the overall frequency of an oscillator, or it may be applied by a modulating signal (e.g., audio signal), which acts to FM-modulate the oscillator's high-frequency carrier. See the examples that follow.

When designing with varactor diodes, the reverse-bias voltage must be absolutely free of noise, since any variation in the bias voltage will cause changes in capacitance. Unwanted frequency shifts or instability will result if the reverse-bias voltage is noisy. Filter capacitors are used to limit such noise.

Varactors come in both single and dual forms. The dual varactor configuration contains two varactors in series-opposing configuration, with common anodes and separate cathodes. In this configuration, the varactors acts as series capacitors that change capacitance levels together when a voltage is applied to the common anode lead. See Fig. 4.39.

FM Modulator

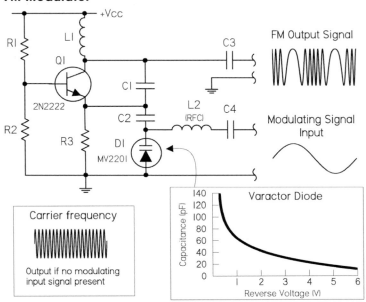

FIGURE 4.38

FM modulation: FM (frequency modulation) is produced when the frequency of a carrier is changed instantaneously according to the magnitude of an applied modulating signal. (The frequency of the carrier is usually in the megahertz, while the modulating signal is typically in the hertz to kilohertz range, e.g., audio modulating radio signal.) One way to produce FM is to use a voltage-controlled oscillator. The oscillator will have an output frequency proportional to the modulating signal's amplitude. As the amplitude of the modulating signal increases, the frequency of the carrier increases. Here a Colpitts LC oscillator uses a varactor diode in place of one of its regulator capacitors that form the tuned circuit. The modulating voltage is applied across the diode and changes the diode's capacitance in proportion. This causes the oscillator frequency to change, thus generating FM in the process. L_2 (RFC) is a radiofrequency choke that prevents high-frequency signals from feeding back into the modulating source. C_3 and C_4 are ac-coupling capacitors. The rest of the components go into making the Colpitts oscillator.

Oscillator with Pot-Controlled Varactor Tuning

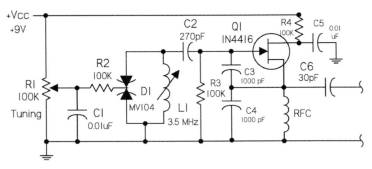

FIGURE 4.39

Unlike the preceding circuit, this circuit acts simply as a variable high-frequency oscillator, whose frequency is varied via a potentiometer (R_1). The voltage from the pot is applied to a dual varactor diode D_1 through a low-frequency filter (C_1, R_2) to ensure that the varactor bias is clean dc. This alters the effective capacitance of the D_1-L_1 tuned circuit, which changes the frequency of the entire oscillator. C_2 and C_6 are dc-blocking (ac-coupling) capacitors. Q_1 is an N-channel JFET in common drain configuration with feedback to the gate through C_3. R_3 is the gate bias resistor. R_4 is the drain voltage resistor with filter capacitor C_5.

4.2.9 PIN Diodes

PIN diodes are used as RF and microwave switches. To high-frequency signals, the PIN diode acts like a variable resistor whose value is controlled by an applied dc forward-bias current. With a high dc forward bias, the resistance is often less than an ohm. But with a small forward bias, the resistance appears very large (kiloohms) to high-frequency signals. PIN diodes are constructed with a layer of intrinsic (undoped) semiconductor placed between very highly doped *p*-type and *n*-type material, creating a PIN junction.

In terms of application, PIN diodes are used primarily as RF and microwave switches—even at high power levels. A common application is their use as transmit/receive switches in transceivers operating from 100 MHz and up. They are also used as photodetectors in fiber-optic systems. For the most part, you'll never need to use them, unless you are a graduate student in electrical engineering or physics, or are working for a high-tech firm.

RF Switching with PIN Diodes

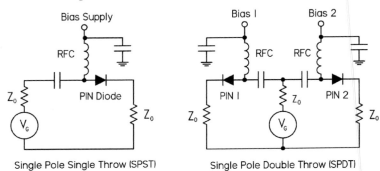

Single Pole Single Throw (SPST)

Single Pole Double Throw (SPDT)

FIGURE 4.40

At RF frequencies, switching is very finicky, requiring special design techniques to minimize signal contamination and degradation. Here are two switching circuits that make use of PIN diodes. In the SPST switch circuit, a signal from a RF generator (VG), can be allowed to pass, or can be prevented from passing to the load by applying a bias voltage to the PIN diode. The RFC is a high-frequency choke to prevent RF from entering bias supply, while the capacitor to ground is used to supply clean dc at the bias input. The SPDT switch circuit is similar to the first, but with, of course, two bias inputs.

4.2.10 Microwave Diodes (IMPATT, Gunn, Tunnel, etc.)

There are a number of diodes that you'll probably never have to use, but they are around nevertheless. These diodes are used for very special purposes at the high-frequency end—microwave and millimeter wave (>20 GHz) range, often in microwave amplifiers and oscillators. Most standard diodes and bipolar transistors usually won't cut it at such high speeds, due to the relatively slow diffusion or migration of charge carriers across semiconductor junctions. With the tunnel, Gunn, IMPATT, and other diodes, the variable effects that lead to useful alterations in, say, an amplifier's gain or an oscillator's resonant frequency involve entirely different physics—physics that allows for alterations at essentially the speed of light. The physics may be electron tunneling (through electrostatic barrier separating *p*-type and *n*-type regions, rather than being thermionically emitted over the barrier, as generally occurs in a diode)—tunnel diode. Or it may be due to a negative resistance at forward biasing because of an increase in effective mass (slowing down) of electrons due to complex conduction band symmetry—Gunn diodes. It may also be a negative resistance resulting in electrons moving to higher, less mobile bands, reducing current flow with applied forward bias—IMPATT diodes. Anyway, you get the idea—it's hairy high-frequency stuff that should probably be left to the experts. (Note: TRAPATT and Baritt diodes are also used in microwave applications.)

4.2.11 Problems

Problem 1: What does this circuit do? What's the final output voltage? What are the individual voltage drops across each diode with plug tip-positive and plug tip-negative? (Assume each diode has a 0.6 V forward voltage drop.) To prevent diode meltdown, what would be the minimum load resistance, assuming 1N4002 diodes?

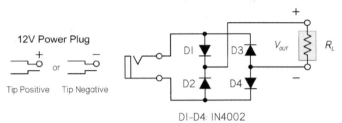

FIGURE 4.41

Answer: Polarity protection circuit that will output the same polarity regardless of the polarity applied to input. The final output voltage is 11.4 V. Tip-positive: $VD_1 = 0.6$ V, $VD_2 = 11.4$ V, $VD_3 = 11.4$ V, $VD_4 = 0.6$ V; Tip-negative: $VD_1 = 11.4$ V, $VD_2 = 0.6$ V, $VD_3 = 0.6$ V, $VD_4 = 11.4$ V. Load resistance should not drop below 11.4 Ω, assuming 1N4002 diodes, since they have a maximum current rating of 1 A. It's a good idea to keep the current to around 75 percent of the maximum value for safety, so 15 Ω would be a better limit.

Problem 2: What does the output look like for the circuit to the left in Fig. 4.42? What happens if a load of 2.2K is attached to the output?

FIGURE 4.42

Answer: Clamp circuit, where the output is shifted so that it's practically pure alternating dc, for the exception of a 0.6 V negative dip due to the diode drop. This gives a maximum peak of 27.6 V and a minimum of –0.6 V. (Recall $V_{peak} = 1.41 \times V_{rms}$.) With the load attached, the output level decreases slightly—the capacitor/load resistor acts like a high-pass filter, with a cutoff frequency of $1/(2\pi RC)$. In simulation, the output goes to 8.90 V(RMS) or 24.5 peak, –0.6 V minimum.

Problem 3: A 10- to 50-mA load requires a regulated 8.2 V. With a 12 V ± 10 percent power supply and 8.2 V zener diode. What series resistance is required? Assume from the data sheets (or experimentation) that the zener diode's minimum regulation current is 10 mA. Determine the power ratings for the resistor and zener diode.

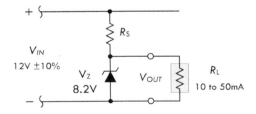

FIGURE 4.43

Answer: $V_{in,max} = 13.2$ V, $V_{in,min} = 10.8$ V: $R_S = (10.8 \text{ V} - 8.2 \text{ V})/(10 \text{ mA} + 50 \text{ mA}) = 43 \Omega$; $P_R = (13.2 \text{ V} - 8.2 \text{ V})^2/(43 \Omega) = 0.58$ W; $P_Z = 8.2 \text{ V}(13.2 \text{ V} - 8.2 \text{ V})/(43 \Omega) = 0.95$ W. See Fig. 4.29 for details.

TABLE 4.3

DIODE TYPE	SYMBOL	MODE OF OPERATION
p-n Junction	A ▶├ C	Acts as one-way gate to current-flow, from anode (A) to cathode (C). Comes in silicon and germanium types. Both require a forward-bias voltage to conduct; typically 0.6 to 1.7 V for silicon, and 0.2 to 0.4 V for germanium. Used in rectification, transient suppression, voltage multiplication, RF demodulation, analog logic, clamps, fast switches, and voltage regulation.
Schottky	A ▶[C	Similar in operation to p-n junction diode, but designed with special metal semiconductor junction instead of a p-n junction. This provides for extremely low junction capacitance that stores less charge. Results of this junction yield quicker switching times, useful in fast clamping and high-frequency applications approaching the gigahertz range. Also, generally has a lower forward-bias voltage of around 0.4 V (average)—but can be from 0.15 to 0.9 V or more. Used in similar applications as p-n junction diode, but offers better low-signal level detection, speed, and low-power loss in rectification due to low forward threshold.
Zener	A ▶├ C	Conducts from A to C like p-n junction diode, but will also conduct from C to A if the applied reverse voltage is greater than the zener's breakdown voltage rating V_z. Acts like a voltage-sensitive control valve. Comes with various breakdown voltages—1.2 V, 3.0 V, 5.1 V, 6.3 V, 9 V, 12 V, etc., and power ratings. Applications include voltage regulation, waveform clipping, voltage shifting, and transient suppression.
LED & Laser	A ▶├ C	Light-emitting diode (LED) emits a near constant wavelength of light when forward-biased (A > C) by a voltage of about 1.7 V. Comes in various wavelengths (IR through visible), sizes, power ratings, etc. Used as indicator and emitting source in IR and light-wave communications. Laser diode is similar to LED, but provides a much narrower wavelength spectrum (about 1 nm compared to around 40 nm for LED), usually in the IR region. They have very fast response times (Ins). These features provide clean signal characteristics useful in fiber-optic systems, where minimized dispersion effects, efficient coupling, and limited degradation over long distances are important. They are also used in laser pointers, CD/DVD players, bar-code readers, and in various surgical applications.
Photo	A ▶├ C	Generates a current when exposed to light, or can be used to alter current flow passing through it when the light intensity changes. Operates in reverse-bias direction (current flows from C to A) when exposed to light. Current increases with light intensity. Very fast response times (ns). Not as sensitive as phototransistors, but their linearity can make them useful in simple light meters.
Varactor (Varicap)	A ▶├├ C	Acts like a voltage-sensitive variable capacitor, whose capacitance decreases as the reverse-bias voltage on the diode increases. Designed with a junction specifically formulated to have a relatively large range of capacitance values for a modest range of reverse-bias voltages. Capacitance range in the picofarad range, so they are usually limited to RF applications, such as tuning receivers and generating FM.
PIN, IMPATT Gunn, Tunnel, etc.	A ▶├* C	Most of these are resistance devices used in RF, microwave, and millimeter wave applications (e.g., amplifiers and oscillators). Unique conduction physics yields much faster response times when compared to standard diodes that use charge carrier dispersion across a p-n junction.

4.3 Transistors

Transistors are semiconductor devices that act as either electrically controlled switches or amplifier controls. The beauty of transistors is the way they can control electric current flow in a manner similar to the way a faucet controls the flow of water. With a faucet, the flow of water is controlled by a control knob. With a transistor, a small voltage and/or current applied to a control lead acts to control a larger electric flow through its other two leads.

Transistors are used in almost every electric circuit you can imagine. For example, you find transistors in switching circuits, amplifier circuits, oscillator circuits, current-source circuits, voltage-regulator circuits, power-supply circuits, digital logic ICs, and almost any circuit that uses small control signals to control larger currents.

4.3.1 Introduction to Transistors

Transistors come in a variety of designs and come with unique control and current-flow features. Most transistors have a variable current-control feature, but a few do not. Some transistors are normally off until a voltage is applied to the base or gate, whereas others are normally on until a voltage is applied. (Here, *normally* refers to the condition when the control lead is open circuited. Also, *on* can represent a variable amount of current flow.) Some transistors require both a small current and a small voltage applied to their control lead to function, whereas others only require a voltage. Some transistors require a negative voltage and/or output current at their base lead (relative to one of their other two leads) to function, whereas others require a positive voltage and/or input current at their base.

The two major families of transistors are *bipolar transistors* and *field-effect transistors* (FETs). The major difference between these two families is that bipolar transistors require a biasing input (or output) current at their control leads, whereas FETs require only a voltage—practically no current. [Physically speaking, bipolar transistors require both positive (holes) and negative (electrons) carriers to operate, whereas FETs only require one charge carrier.] Because FETs draw little or no current, they have high input impedances ($\sim 10^{14}$ Ω). This high input impedance means that the FET's control lead will not have much influence on the current dynamics within the circuit that controls the FET. With a bipolar transistor, the control lead may draw a small amount of current from the control circuit, which then combines with the main current flowing through its other two leads, thus altering the dynamics of the control circuit.

In reality, FETs are definitely more popular in circuit design today than bipolar transistors. Besides drawing essentially zero input-output current at their control leads, they are easier to manufacture, cheaper to make (require less silicon), and can be made extremely small—making them useful elements in integrated circuits. One drawback with FETs is in amplifier circuits, where their transconductance is much lower than that of bipolar transistors at the same current. This means that the voltage gain will not be as large. For simple amplifier circuits, FETs are seldom used unless extremely high input impedances and low input currents are required.

Table 4.4 provides an overview of some of the most popular transistors. Note that the term *normally* used in this chart refers to conditions where the control lead

TABLE 4.4 Overview of Transistors

TRANSISTOR TYPE	SYMBOL	MODE OF OPERATION
Bipolar	*npn*	Normally off, but a small input current and a small positive voltage at its base (*B*)—relative to its emitter (*E*)—turns it on (permits a large collector-emitter current). Operates with $V_C > V_E$. Used in switching and amplifying applications.
	pnp	Normally off, but a small output current and negative voltage at its base (*B*)—relative to its emitter (*E*)—turns it on (permits a large emitter-collector current). Operates with $V_E > V_C$. Used in switching and amplifying applications.
Junction FET	*n-channel*	Normally on, but a small negative voltage at its gate (*G*)—relative to its source (*S*)—turns it off (stops a large drain-source current). Operates with $V_D > V_S$. Does not require a gate current. Used in switching and amplifying applications.
	p-channel	Normally on, but a small positive voltage at its gate (*G*)—relative to its source (*S*)—turns it off (stops a large source-drain current). Operates with $V_S > V_D$. Does not require a gate current. Used in switching and amplifying applications.
Metal oxide semiconductor FET (MOSFET) (depletion)	*n-channel*	Normally on, but a small negative voltage at its gate (*G*)—relative to its source (*S*)—turns it off (stops a large drain-source current). Operates with $V_D > V_S$. Does not require a gate current. Used in switching and amplifying applications.
	p-channel	Normally on, but a small positive voltage at its gate (*G*)—relative to its source (*S*)—turns it off (stops a large source-drain current). Operates with $V_S > V_D$. Does not require a gate current. Used in switching and amplifying applications.
Metal oxide semiconductor FET (MOSFET) (enhancement)	*n-channel*	Normally off, but a small positive voltage at its gate (*G*)—relative to its source (*S*)—turns it on (permits a large drain-source current). Operates with $V_D > V_S$. Does not require a gate current. Used in switching and amplifying applications.
	p-channel	Normally off, but a small negative voltage at its gate (*G*)—relative to its source (*S*)—turns it on (permits a large source-drain current). Operates with $V_S > V_D$. Does not require a gate current. Used in switching and amplifying applications.
Unijunction FET (UJT)	*UJT*	Normally a very small current flows from base 2 (*B₂*) to base 1 (*B₁*), but a positive voltage at its emitter (*E*)—relative to *B₁* or *B₂*—increases current flow. Operates with $V_{B2} > V_{B1}$. Does not require a gate current. Only acts as a switch.

(e.g., base, gate) is shorted (is at the same potential) with one of its channel leads (e.g., emitter, source). Also, the terms *on* and *off* used in this chart are not to be taken too literally; the amount of current flow through a device is usually a variable quantity, set by the magnitude of the control voltage. The transistors described in this chart will be discussed in greater detail later on in this chapter.

4.3.2 Bipolar Transistors

Bipolar transistors are three-terminal devices that act as electrically controlled switches or as amplifier controls. These devices come in either *npn* or *pnp* configurations, as shown in Fig. 4.44. An *npn* bipolar transistor uses a small input current and positive voltage at its base (relative to its emitter) to control a much larger collector-to-emitter current. Conversely, a *pnp* transistor uses a small output base current and negative base voltage (relative its emitter) to control a larger emitter-to-collector current.

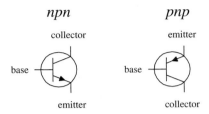

FIGURE 4.44

Bipolar transistors are incredibly useful devices. Their ability to control current flow by means of applied control signals makes them essential elements in electrically controlled switching circuits, current-regulator circuits, voltage-regulator circuits, amplifier circuits, oscillator circuits, and memory circuits.

How Bipolar Transistors Work

Here is a simple model of how an *npn* bipolar transistor works. (For a *pnp* bipolar transistor, all ingredients, polarities, and currents are reversed.)

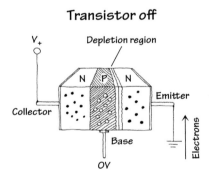

An *npn* bipolar transistor is made by sandwiching a thin slice of *p* semiconductor between two *n*-type semiconductors. When no voltage is applied at the transistor's base, electrons in the emitter are prevented from passing to the collector side because of the *p-n* junction. (Remember that for electrons to flow across a *p-n* junction, a biasing voltage is needed to give the electrons enough energy to "escape" the atomic forces holding them to the *n* side.) Notice that if a negative voltage is applied to the base, things get even worse—the *p-n* junction between the base and emitter becomes reverse-biased. As a result, a depletion region forms and prevents current flow.

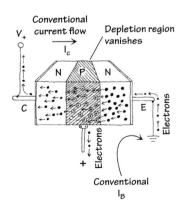

If a positive voltage (of at least 0.6 V) is applied to the base of an *npn* transistor, the *pn* junction between the base and emitter is forward-biased. During forward bias, escaping electrons are drawn to the positive base. Some electrons exit through the base, but—this is the trick—because the *p*-type base is so thin, the onslaught of electrons that leave the emitter get close enough to the collector side that they begin jumping into the collector. Increasing the base voltage increases this jumping effect and hence increases the emitter-to-collector electron flow. Remember that conventional currents are moving in the opposite direction to the electron flow. Thus, in terms of conventional currents, a positive voltage and input current applied at the base cause a "positive" current *I* to flow from the collector to the emitter.

FIGURE 4.45

Theory

Figure 4.46 shows a typical characteristic curve for a bipolar transistor. This characteristic curve describes the effects the base current I_B and the emitter-to-collector voltage V_{EC} have on the emitter/collector currents I_E and I_C. (As you will see in a second, I_C is practically equal to I_E.)

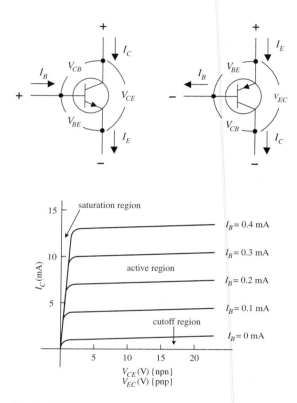

FIGURE 4.46

Some important terms used to describe a transistor's operation include saturation region, cutoff region, active mode/region, bias, and quiescent point (Q-point). *Saturation region* refers to a region of operation where maximum collector current flows and the transistor acts much like a closed switch from collector to emitter. *Cutoff region* refers to the region of operation near the voltage axis of the collector characteristics graph, where the transistor acts like an open switch—only a very small leakage current flows in this mode of operation. *Active mode/region* describes transistor operation in the region to the right of saturation and above cutoff, where a near-linear relationship exists between terminal currents (I_B, I_C, I_E). *Bias* refers to the specific dc terminal voltages and current of the transistor to set a desired point of active-mode operation, called the *quiescent point*, or *Q-point*.

The Formulas

The fundamental formula used to describe the behavior of a bipolar transistor (at least within the active region) is

$$I_C = h_{FE}I_B = \beta I_B$$

SOME IMPORTANT RULES

Rule 1   For an *npn* transistor, the voltage at the collector V_C must be greater than the voltage at the emitter V_E by at least a few tenths of a volt; otherwise, current will not flow through the collector-emitter junction, no matter what the applied voltage is at the base. For *pnp* transistors, the emitter voltage must be greater than the collector voltage by a similar amount.

Rule 2   For an *npn* transistor, there is a voltage drop from the base to the emitter of 0.6 V. For a *pnp* transistor, there is a 0.6-V rise from base to emitter. In terms of operation, this means that the base voltage V_B of an *npn* transistor must be at least 0.6 V greater than the emitter voltage V_E; otherwise, the transistor will not pass an collector-to-emitter current. For a *pnp* transistor, V_B must be at least 0.6 V less than V_E; otherwise, it will not pass a collector-to-emitter current.

where I_B is the base current, I_C is the collector current, and h_{FE} (also referred to as β) is the *current gain*. Every transistor has its own unique h_{FE}. The h_{FE} of a transistor is often taken to be a constant, typically around 10 to 500, but it may change slightly with temperature and with changes in collector-to-emitter voltage. (A transistor's h_{FE} is given in transistor spec tables.) A simple explanation of what the current-gain formula tells you is this: If you take a bipolar transistor with, say, an h_{FE} of 100 and then feed (*npn*) or sink (*pnp*) a 1-mA current into (*npn*) or out of (*pnp*) its base, a collector current of 100 mA will result. Now, it is important to note that the current-gain formula applies only if rules 1 and 2 are met, i.e., assuming the transistor is within the active region. Also, there is a limit to how much current can flow through a transistor's terminals and a limit to the size of voltage that can be applied across them. We will discuss these limits later in the chapter (Fig. 4.47).

Now, if you apply the law of conservation of current (follow the arrows in Fig. 4.47), you get the following useful expression relating the emitter, collector, and base currents:

$$I_E = I_C + I_B$$

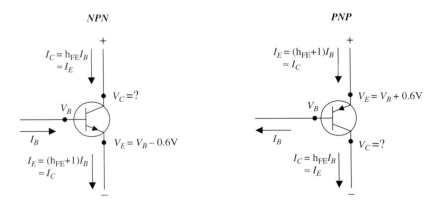

FIGURE 4.47

If you combine this equation with the current-gain equation, you can come up with an equation relating the emitter and base currents:

$$I_E = (h_{FE} + 1)I_B$$

As you can see, this equation is almost identical to the current-gain equation ($I_C = h_{FE}I_B$), with the exception of the +1 term. In practice, the +1 is insignificant as long as h_{FE} is large (which is almost always the case). This means that you can make the following approximation:

$$IE \approx IC$$

Finally, the second equation below is simply rule 2 expressed in mathematical form:

$$V_{BE} = V_B - V_E = +0.6 \text{ V } (npn)$$
$$V_{BE} = V_B - V_E = -0.6 \text{ V } (pnp)$$

Figure 4.47 shows how all the terminal currents and voltages are related. In the figure, notice that the collector voltage has a question mark next to it. As it turns out, the value of V_C cannot be determined directly by applying the formulas. Instead, V_C's value depends on the network that is connected to it. For example, if you consider the setup shown in Fig. 4.48, you must find the voltage drop across the resistor in order to find the collector voltage. By applying Ohm's law and using the current-gain relation, you can calculate V_C. The results are shown in the figure.

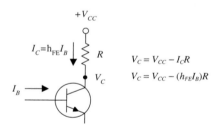

FIGURE 4.48

It is important to note that the equations here are idealistic in form. In reality, these equations may result in "unreal" answers. For instance, they tend to "screw up" when the currents and voltages are not within the bounds provided by the characteristic curves. If you apply the equations blindly, without considering the operating characteristics, you could end up with some wild results that are physically impossible.

One final note with regard to bipolar transistor theory involves what is called *transresistance* r_{tr}. Transresistance represents a small resistance that is inherently present within the emitter junction region of a transistor. Two things that determine the transresistance of a transistor are temperature and emitter current flow. The following equation provides a rough approximation of the r_{tr}:

$$r_{tr} = \frac{0.026\,\text{V}}{I_E}$$

In many cases, r_{tr} is insignificantly small (usually well below 1000 Ω) and does not pose a major threat to the overall operation of a circuit. However, in certain types of circuits, treating r_{tr} as being insignificant will not do. In fact, its presence may be the major factor determining the overall behavior of a circuit. We will take a closer look at transresistance later on in this chapter.

Here are a couple of problems that should help explain how the equations work. The first example deals with an *npn* transistor; the second deals with a *pnp* transistor.

EXAMPLE 1   Given $V_{CC} = +20$ V, $V_B = 5.6$ V, $R_1 = 4.7$ kΩ, $R_2 = 3.3$ kΩ, and $h_{FE} = 100$, find V_E, I_E, I_B, I_C, and V_C.

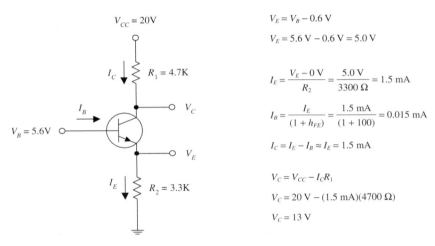

$$V_E = V_B - 0.6 \text{ V}$$

$$V_E = 5.6 \text{ V} - 0.6 \text{ V} = 5.0 \text{ V}$$

$$I_E = \frac{V_E - 0 \text{ V}}{R_2} = \frac{5.0 \text{ V}}{3300 \text{ }\Omega} = 1.5 \text{ mA}$$

$$I_B = \frac{I_E}{(1 + h_{FE})} = \frac{1.5 \text{ mA}}{(1 + 100)} = 0.015 \text{ mA}$$

$$I_C = I_E - I_B \approx I_E = 1.5 \text{ mA}$$

$$V_C = V_{CC} - I_C R_1$$

$$V_C = 20 \text{ V} - (1.5 \text{ mA})(4700 \text{ }\Omega)$$

$$V_C = 13 \text{ V}$$

FIGURE 4.49

EXAMPLE 2   Given $V_{CC} = +10$ V, $V_B = 8.2$ V, $R_1 = 560$ Ω, $R_2 = 2.8$ kΩ, and $h_{FE} = 100$, find V_E, I_E, I_B, I_C, and V_C.

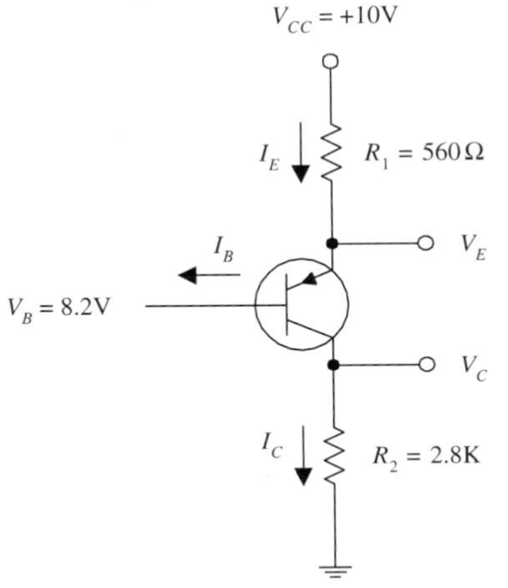

$$V_E = V_B + 0.6 \text{ V}$$

$$V_E = 8.2 \text{ V} + 0.6 \text{ V} = 8.8 \text{ V}$$

$$I_E = \frac{V_{CC} - V_E}{R_1} = \frac{10 \text{ V} - 8.8 \text{ V}}{560 \text{ }\Omega} = 2.1 \text{ mA}$$

$$I_B = \frac{I_E}{(1 + h_{FE})} = \frac{2.1 \text{ mA}}{(1 + 100)} = 0.02 \text{ mA}$$

$$I_C = I_E - I_B \approx I_E = 2.1 \text{ mA}$$

$$V_C = 0 \text{ V} + I_C R_2$$

$$V_C = 0 \text{ V} + (2.1 \text{ mA})(2800 \text{ }\Omega)$$

$$V_C = 5.9 \text{ V}$$

FIGURE 4.50

Bipolar Transistor Water Analogy

NPN WATER ANALOGY

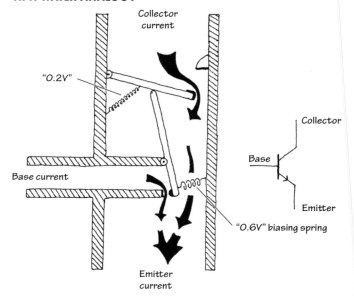

The base of the *npn* water transistor is represented by the smaller tube entering the main device from the left side. The collector is represented by the upper portion of the vertical tube, while the emitter is represented by the lower portion of the vertical tube. When no pressure or current is applied through the "base" tube (analogous to an *npn* transistor's base being open circuited), the lower lever arm remains vertical while the top of this arm holds the upper main door shut. This state is analogous to a real bipolar *npn* transistor off state. In the water analogy, when a small current and pressure are applied to the base tube, the vertical lever is pushed by the entering current and swings counterclockwise. When this lever arm swings, the upper main door is permitted to swing open a certain amount that is dependent on the amount of swing of the lever arm. In this state, water can make its way from the collector tube to the emitter tube, provided there is enough pressure to overcome the force of the spring holding the door shut. This spring force is analogous to the 0.6 V biasing voltage needed to allow current through the collector-emitter channel. Notice that in this analogy, the small base water current combines with the collector current.

PNP WATER ANALOGY

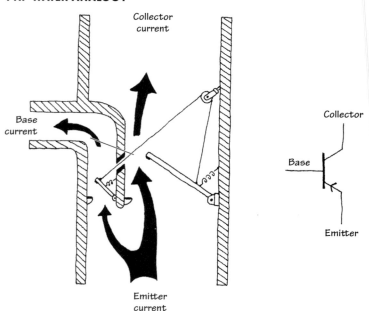

The main feature to note here is the need for a lower pressure at the base for the *pnp* water transistor to turn on. By allowing current to flow out the base tube, the lever moves, allowing the emitter-collector door to open. The degree of openness varies with the amount of swing in the lever arm, which corresponds to the amount of current escaping through the base tube. Again, note the 0.6 V biasing spring.

FIGURE 4.51

Basic Operation

TRANSISTOR SWITCH

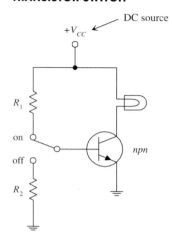

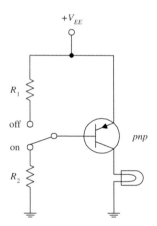

FIGURE 4.52

Here, an *npn* transistor is used to control current flow through a light bulb. When the switch is thrown to the on position, the transistor is properly biased, and the collector-to-emitter channel opens, allowing current to flow from V_{CC} through the light bulb and into ground. The amount of current entering the base is determined by

$$I_B = \frac{V_E + 0.6\,\text{V}}{R_1} = \frac{0\,\text{V} + 0.6\,\text{V}}{R_1}$$

To find the collector current, you can use the current-gain relation ($I_C = h_{FE}I_B$), provided that there is not too large a voltage drop across the light bulb (it shouldn't cause V_C to drop below 0.6 V + V_E). When the switch is thrown to the off position, the base is set to ground, and the transistor turns off, cutting current flow to the light bulb. R_2 should be large (e.g., 10 kΩ) so that very little current flows to ground.

In the *pnp* circuit, everything is reversed; current must leave the base in order for a collector current to flow.

CURRENT SOURCE

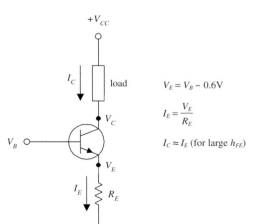

$$V_E = V_B - 0.6\,\text{V}$$

$$I_E = \frac{V_E}{R_E}$$

$$I_C \approx I_E \text{ (for large } h_{FE})$$

FIGURE 4.53

The circuit here shows how an *npn* transistor can be used to make a simple current source. By applying a small input voltage and current at the transistor's base, a larger collector/load current can be controlled. The collector/load current is related to the base voltage by

$$I_C = I_{load} = \frac{V_B - 0.6\,\text{V}}{R_E}$$

The derivation of this equation is shown with the figure.

CURRENT BIASING METHODS

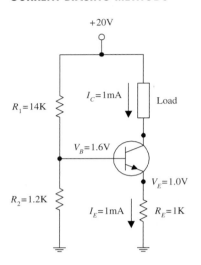

FIGURE 4.54

Two common methods for biasing a current source are to use either a voltage-divider circuit (shown in the leftmost circuit) or a zener diode regulator (shown in the rightmost circuit). In the voltage-divider circuit, the base voltage is set by R_1 and R_2 and is equal to

$$V_B = \frac{R_2}{R_1 + R_2} V_{CC}$$

In the zener diode circuit, the base voltage is set by the zener diode's breakdown voltage such that

$$V_B = V_{\text{zener}}$$

EMITTER FOLLOWER

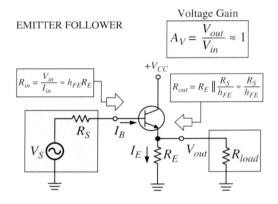

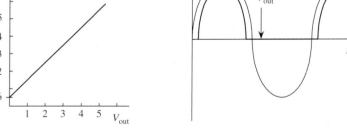

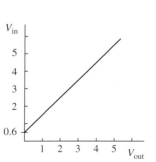

FIGURE 4.55

The network shown here is called an *emitter follower*. In this circuit, the output voltage (tapped at the emitter) is almost a mirror image of the input (output "follows" input), with the exception of a 0.6 V drop in the output relative to the input (caused by base-emitter *pn* junction). Also, whenever $V_B \leq 0.6$ V (during negative swings in input), the transistor will turn off (the *pn* junction is reversed-biased). This effect results in clipping of the output (see graph). At first glance, it may appear that the emitter follower is useless—it has no voltage gain. However, if you look at the circuit more closely, you will see that it has a much larger input impedance than an output impedance, or more precisely, it has a much larger output current (I_E) relative to an input current (I_B). In other words, the emitter follower has current gain, a feature that is just as important in applications as voltage gain. This means that this circuit requires less power from the signal source (applied to V_{in}) to drive a load than would otherwise be required if the load were to be powered directly by the source. By manipulating the transistor gain equation and using Ohm's law, the input resistance and output resistance are:

$$R_{in} = \frac{V_{in}}{I_{in}} \approx h_{FE}R_E$$

$$R_{out} = R_E \| \frac{R_S}{h_{FE}} \approx \frac{R_S}{h_{FE}}$$

$$A_V = \frac{V_{out}}{V_{in}} \approx 1 \quad \text{(Voltage Gain)}$$

EMITTER-FOLLOWER (COMMON-COLLECTOR) AMPLIFIER

The circuit shown here is called a *common-collector amplifier*, which has current gain but no voltage gain. It makes use of the emitter-follower arrangement but is modified to avoid clipping during negative input swings. The voltage divider (R_1 and R_2) is used to give the input signal (after passing through the capacitor) a positive dc level or operating point (known as the *quiescent point*). Both the input and output capacitors are included so that an ac input-output signal can be added without disturbing the dc operating point. The capacitors, as you will see, also act as filtering elements.

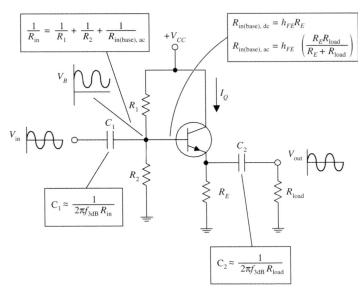

$$\frac{1}{R_{in}} = \frac{1}{R_1} + \frac{1}{R_2} + \frac{1}{R_{in(base),\,ac}}$$

$$R_{in(base),\,dc} = h_{FE}R_E$$

$$R_{in(base),\,ac} = h_{FE}\left(\frac{R_E R_{load}}{R_E + R_{load}}\right)$$

$$C_1 \approx \frac{1}{2\pi f_{3dB}\, R_{in}}$$

$$C_2 \approx \frac{1}{2\pi f_{3dB}\, R_{load}}$$

EXAMPLE

To design a common-collector amplifier used to power a 3-kΩ load, which has a supply voltage $V_{CC} = +10$ V, a transistor h_{FE} of 100, and a desired f_{3dB} point of 100 Hz, you

1. Choose a quiescent current $I_Q = I_C$. For this problem, pick $I_Q = 1$ mA.
2. Next, select $V_E = \frac{1}{2}V_{CC}$ to allow for the largest possible symmetric output swing without clipping, which in this case, is 5 V. To set $V_E = 5$ V and still get $I_Q = 1$ mA, make use of R_E, whose value you find by applying Ohm's law:

$$R_E = \frac{\frac{1}{2}V_{CC}}{I_Q} = \frac{5\text{ V}}{1\text{ mA}} = 5\text{ k}\Omega$$

3. Next, set the $V_B = V_E + 0.6$ V for quiescent conditions (to match up V_E so as to avoid clipping). To set the base voltage, use the voltage divider (R_1 and R_2). The ratio between R_1 and R_2 is determined by rearranging the voltage-divider relation and substituting into it $V_B = V_E + 0.6$ V:

$$\frac{R_2}{R_1} = \frac{V_B}{V_{CC} - V_B} = \frac{V_E + 0.6\text{ V}}{V_{CC} - (V_E + 0.6\text{V})}$$

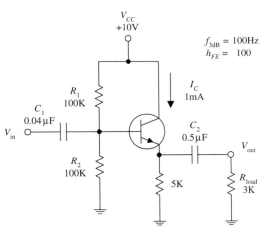

$$V_{CC}$$
$$+10\text{V}$$

$$f_{3dB} = 100\text{Hz}$$
$$h_{FE} = 100$$

R_1
100K

C_1
0.04 μF

V_{in}

I_C
1mA

C_2
0.5 μF

V_{out}

R_2
100K

5K

R_{load}
3K

FIGURE 4.56

Fortunately, you can make an approximation and simply let $R_1 = R_2$. This approximation "forgets" the 0.6-V drop but usually isn't too dramatic. The actual sizes of R_2 and R_1 should be such that their parallel resistance is less than or equal to one-tenth the dc (quiescent) input resistance at the base (this prevents the voltage divider's output voltage from lowering under loading conditions):

$$\frac{R_1 R_2}{R_1 + R_2} \le \frac{1}{10} R_{in(base),dc}$$

$$\frac{R}{2} \le \frac{1}{10} R_{in(base),dc} \quad \text{(using the approximation } R = R_1 = R_2\text{)}$$

Here, $R_{in(base),dc} = h_{FE}R_E$, or specially, $R_{in(base),dc} = 100(5\text{ k}) = 500$ k. Using the approximation above, R_1 and R_2 are calculated to be 100 k each. (Here you did not have to worry about the ac coupler load; it did not influence the voltage divider because you assumed quiescent setup conditions; C_2 acts as an open circuit, hence "eliminating" the presence of the load.)

4. Next, choose the ac coupling capacitors so as to block out dc levels and other undesired frequencies. C_1 forms a high-pass filter with R_{in} (see diagram). To find R_{in}, treat the voltage divider and $R_{in(base),ac}$ as being in parallel:

$$\frac{1}{R_{in}} = \frac{1}{R_1} + \frac{1}{R_2} + \frac{1}{R_{in(base),ac}}$$

Notice that $R_{in(base),ac}$ is used, not $R_{in(base),dc}$. This is so because you can no longer treat the load as being absent when fluctuating signals are applied to the input; the capacitor begins to pass a displacement current. This means that you must take R_E and R_{load} in parallel and multiply by h_{FE} to find $R_{in(base,ac)}$:

$$R_{in(base),ac} = h_{FE}\left(\frac{R_E R_{load}}{R_E + R_{load}}\right) = 100\left[\frac{5\text{ k}\Omega(3\text{ k}\Omega)}{5\text{ k}\Omega + 3\text{ k}\Omega}\right] = 190\text{ k}\Omega$$

Now you can find R_{in}:

$$\frac{1}{R_{in}} = \frac{1}{100\,k\Omega} + \frac{1}{100\,k\Omega} + \frac{1}{190\,k\Omega}$$

$$R_{in} = 40\ k\Omega$$

Once you have found R_{in}, choose C_1 to set the f_{3dB} point (C_1 and R_{in} form a high-pass filter.) The capacitor value C_1 is found by using the following formula:

$$C_1 = \frac{1}{2\pi f_{3dB} R_{in}} = \frac{1}{2\pi(100\,Hz)(40\,k\Omega)} = 0.04\,\mu F$$

C_2 forms a high-pass filter with the load. It is chosen by using

$$C_2 = \frac{1}{2\pi f_{3dB} R_{load}} = \frac{1}{2\pi(100\,Hz)(3\,k\Omega)} = 0.5\,\mu F$$

COMMON-EMITTER CONFIGURATION

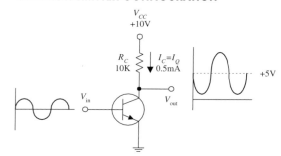

INCLUDING R_E FOR TEMPERATURE
STABILITY

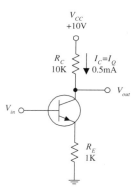

ACHIEVING HIGH-GAIN WITH
TEMPERATURE STABILITY

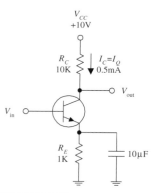

FIGURE 4.57

The transistor configuration here is referred to as the *common-emitter configuration*. Unlike the emitter follower, the common emitter has voltage gain. To figure out how this circuit works, first set $V_C = \frac{1}{2}V_{CC}$ to allow for maximum swing without clipping. Like the emitter follower, again pick a quiescent current I_Q to start with. To set $V_C = \frac{1}{2}V_{CC}$ with a desired I_Q, use R_C, which is found by Ohm's law:

$$R_C = \frac{V_{CC} - V_C}{I_C} = \frac{V_{CC} - \frac{1}{2}V_{CC}}{I_Q} = \frac{\frac{1}{2}V_{CC}}{I_Q}$$

For example, if V_{CC} is 10 V and I_Q is 0.5 mA, R_C is then 10 k. The gain of this circuit is found by realizing that $\Delta V_E = \Delta V_B$ (where Δ represents a small fluctuation). The emitter current is found using Ohm's law:

$$\Delta I_E = \frac{\Delta V_E}{R_E} = \frac{\Delta V_B}{R_E} = \Delta I_C$$

Using $V_C = V_{CC} - I_C R_C$ and the last expression, you get

$$\Delta V_C = \Delta I_C R_C = \frac{\Delta V_B}{R_E} R_C$$

Since V_C is V_{out} and V_B is V_{in}, the gain is

$$Gain = \frac{V_{out}}{V_{in}} = \frac{\Delta V_C}{\Delta V_B} = \frac{R_C}{R_E}$$

But what about R_E? According to the circuit, there's no emitter resistor. If you use the gain formula, it would appear that $R_E = 0\ \Omega$, making the gain infinite. However, as mentioned earlier, bipolar transistors have a transresistance (small internal resistance) in the emitter region, which is approximated by using

$$r_{tr} \approx \frac{0.026\ V}{I_E}$$

Applying this formula to the example, taking $I_Q = 0.5$ mA $= I_C \approx I_E$, the R_E term in the gain equation, or r_{tr} equals 52 Ω. This means the gain is actually equal to

$$Gain = -\frac{R_C}{R_E} = -\frac{R_C}{r_{tr}} = -\frac{10\,k\Omega}{52\,\Omega} = -192$$

Notice that the gain is negative (output is inverted). This results in the fact that as V_{in} increases, I_C increases, while V_C (V_{out}) decreases (Ohm's law). Now, there is one problem with this circuit. The r_{tr} term happens to be very unstable, which in effect makes the gain unstable. The instability stems from r_{tr} dependence on temperature. As the temperature rises,

V_E and I_C increase, V_{BE} decreases, but V_B remains fixed. This means that the biasing-voltage range narrows, which in effect turns the transistor's "valve" off. To eliminate this pinch, an emitter resistor is placed from emitter to ground (see second circuit). Treating R_E and r_{tr} as series resistors, the gain becomes

$$\text{Gain} = \frac{R_C}{R_E + r_{tr}}$$

By adding R_E, variations in the denominator are reduced, and therefore, the variations in gain are reduced as well. In practice, R_E should be chosen to place V_E around 1 V (for temperature stability and maximum swing in output). Using Ohm's law (and applying it to the example), choose $R_E = V_E/I_E = V_E/I_Q = 1$ V/1 mA = 1 k. One drawback that arises when R_E is added to the circuit is a reduction in gain. However, there is a trick you can use to eliminate this reduction in voltage gain and at the same time maintain the temperature stability. If you bypass R_E with a capacitor (see third circuit), you can make R_E "disappear" when high-frequency input signals are applied. (Recall that a capacitor behaves like an infinitely large resistor to dc signals but becomes less "resistive," or reactive to ac signals.) In terms of the gain equation, the R_E term goes to zero because the capacitor diverts current away from it toward ground. The only resistance left in the gain equation is r_{tr}.

COMMON-EMITTER AMPLIFIER

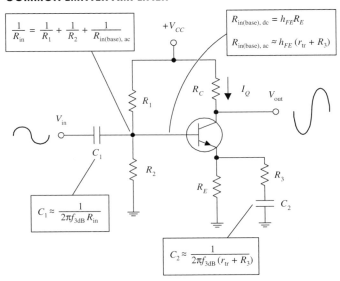

The circuit shown here is known as a *common-emitter amplifier*. Unlike the common-collector amplifier, the common-emitter amplifier provides voltage gain. This amplifier makes use of the common-emitter arrangement and is modified to allow for ac coupling. To understand how the amplifier works, let's go through the following example.

To design a common-emitter amplifier with a voltage gain of −100, an f_{3dB} point of 100 Hz, and a quiescent current $I_Q = 1$ mA, where $h_{FE} = 100$ and $V_{CC} = 20$ V:

1. Choose R_C to center V_{out} (or V_C) to $\frac{1}{2}V_{CC}$ to allow for maximum symmetrical swings in the output. In this example, this means V_C should be set to 10 V. Using Ohm's law, you find R_C:

$$R_C = \frac{V_C - V_{CC}}{I_C} = \frac{0.5V_{CC} - V_{CC}}{I_Q} = \frac{10 \text{ V}}{1 \text{ mA}} = 10 \text{k}\Omega$$

2. Next we select R_E to set $V_E = 1$ V for temperature stability. Using Ohm's law, and taking $I_Q = I_E = 1$ mA, we get $R_E = V_E/I_E = 1$ V/1 mA = 1 kΩ.

3. Now, choose R_1 and R_2 to set the voltage divider to establish the quiescent base voltage of $V_B = V_E + 0.6$ V, or 1.6 V. To find the proper ratio between R_1 and R_2, use the voltage divider (rearranged a bit):

$$\frac{R_2}{R_1} = \frac{V_B}{V_{CC} - V_B} = \frac{1.6 \text{ V}}{20 \text{ V} - 1.6 \text{ V}} = \frac{1}{11.5}$$

This means $R_1 = 11.5R_2$. The size of these resistors is found using the similar procedure you used for the common-collector amplifier; their parallel resistance should be less than or equal to $\frac{1}{10}R_{in(base),dc}$.

$$\frac{R_1 R_2}{R_1 + R_2} \leq \frac{1}{10} R_{in(base),dc}$$

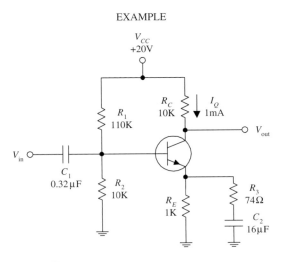

FIGURE 4.58

After plugging $R_1 = 11.5R_2$ into this expression and using $R_{in(base),dc} = h_{FE}R_E$, you find that $R_2 = 10$ kΩ, which in turn means $R_1 = 115$ kΩ (let's say, 110 kΩ is close enough for R_1).

4. Next, choose R_3 for the desired gain, where

$$\text{Gain} = -\frac{R_C}{r_{tr} + (R_E \| R_3)} = -100$$

(The double line means to take R_E and R_3 in parallel.) To find r_{tr}, use $r_{tr} = 0.026 \text{ V}/I_E = 0.026 \text{ V}/I_C = 0.026 \text{ V}/1 \text{ mA} = 26 \Omega$. Now you can simplify the gain expression by assuming R_E "disappears" when ac signals are applied. This means the gain can be simplified to

$$\text{Gain} = -\frac{R_C}{r_{tr} + R_3} = -\frac{10\text{k} \Omega}{26 \Omega + R_3} = -100$$

Solving this equation for R_3, you get $R_3 = 74 \Omega$.

5. Next, choose C_1 for filtering purposes such that $C_1 = 1/(2\pi f_{3dB} R_{in})$. Here, R_{in} is the combined parallel resistance of the voltage-divider resistors, and $R_{in(base),ac}$ looking in from the left into the voltage divider:

$$\frac{1}{R_{in}} = \frac{1}{R_1} + \frac{1}{R_2} + \frac{1}{h_{FE}(r_{tr} + R_3)} = \frac{1}{110\text{ k}\Omega} + \frac{1}{10\text{ k}\Omega} + \frac{1}{100\,(26\,\Omega + 74\Omega)}$$

Solving this equation, you get $R_{in} = 5$ kΩ. This means

$$C_1 = \frac{1}{2\pi(100 \text{ Hz})(5 \text{ k}\Omega)} = 0.32 \, \mu\text{F}$$

6. To choose C_2, treat C_2 and $r_{tr} + R_3$ as a high-pass filter (again, treat R_E as being negligible during ac conditions). C_2 is given by

$$C_2 = \frac{1}{2\pi f_{3dB}(r_{tr} + R_3)} = \frac{1}{2\pi(100\text{Hz})(26\,\Omega + 74\,\Omega)} = 16 \, \mu\text{F}$$

VOLTAGE REGULATOR

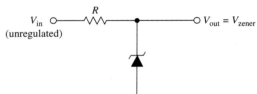

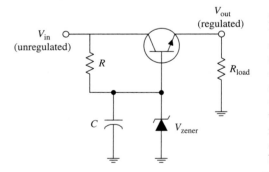

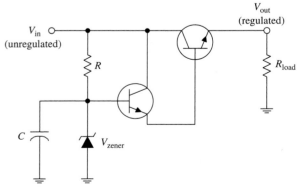

The zener diode circuit here can be used to make a simple voltage regulator. However, in many applications, the simple regulator has problems; V_{out} isn't adjustable to a precise value, and the zener diode provides only moderate protection against ripple voltages. Also, the simple zener diode regulator does not work particularly well when the load impedance varies. Accommodating large load variations requires a zener diode with a large power rating—which can be costly.

The second circuit in the figure, unlike the first circuit, does a better job of regulating; it provides regulation with load variations and provides high-current output and somewhat better stability. This circuit closely resembles the preceding circuit, except that the zener diode is connected to the base of an *npn* transistor and is used to regulate the collect-to-emitter current. The transistor is configured in the emitter-follower configuration. This means that the emitter will follow the base (except there is the 0.6-V drop). Using a zener diode to regulate the base voltage results in a regulated emitter voltage. According to the transistor rules, the current required by the base is only $1/h_{FE}$ times the emitter-to-collector current. Therefore, a low-power zener diode can regulate the base voltage of of a transistor that can pass a considerable amount of current. The capacitor is added to reduce the noise from the zener diode and also forms an *RC* filter with the resistor that is used to reduce ripple voltages.

In some instances, the preceding zener diode circuit may not be able to supply enough base current. One way to fix this problem is to add a second transistor, as shown in the third circuit. The extra transistor (the one whose base is connected to the zener diode) acts to amplify current sent to the base of the upper transistor.

FIGURE 4.59

DARLINGTON PAIR

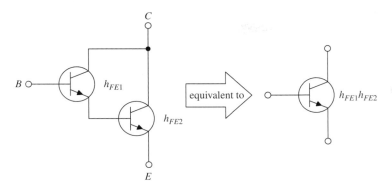

FIGURE 4.60

By attaching two transistors together as shown here, a larger current-handling, larger h_{FE} equivalent transistor circuit is formed. The combination is referred to as a *Darlington pair*. The equivalent h_{FE} for the pair is equal to the product of the individual transistor's h_{FE} values ($h_{FE} = h_{FE1}h_{FE2}$). Darlington pairs are used for large current applications and as input stages for amplifiers, where big input impedances are required. Unlike single transistors, however, Darlington pairs have slower response times (it takes longer for the top transistor to turn the lower transistor on and off) and have twice the base-to-emitter voltage drop (1.2 V instead of 0.6 V) as compared with single transistors. Darlington pairs can be purchased in single packages.

Types of Bipolar Transistors

SMALL SIGNAL

This type of transistor is used to amplify low-level signals but also can be used as a switch. Typical h_{FE} values range from 10 to 500, with maximum I_C ratings from about 80 to 600 mA. They come in both *npn* and *pnp* forms. Maximum operating frequencies range from about 1 to 300 MHz.

SMALL SWITCHING

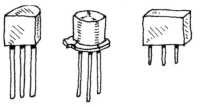

These transistors are used primarily as switches but also can be used as amplifiers. Typical h_{FE} values range from around 10 to 200, with maximum I_C ratings from around 10 to 1000 mA. They come in both *npn* and *pnp* forms. Maximum switching rates range between 10 and 2000 MHz.

HIGH FREQUENCY (RF)

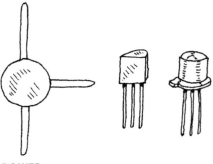

These transistors are used for small signals that run at high frequencies for high-speed switching applications. The base region is very thin, and the actual chip is very small. They are used in HF, VHF, UHF, CATV, and MATV amplifier and oscillator applications. They have a maximum frequency rating of around 2000 MHz and maximum I_C currents from 10 to 600 mA. They come in both *npn* and *pnp* forms.

POWER

FIGURE 4.61

These transistors are used in high-power amplifiers and power supplies. The collector is connected to a metal base that acts as a heat sink. Typical power ratings range from around 10 to 300 W, with frequency ratings from about 1 to 100 MHz. Maximum I_C values range between 1 to 100 A. They come in *npn, pnp,* and Darlington (*npn* or *pnp*) forms.

DARLINGTON PAIR

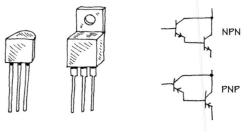

These are two transistors in one. They provide more stability at high current levels. The effective h_{FE} for the device is much larger than that of a single transistor, hence allowing for a larger current gain. They come in *npn* (*D-npn*) and *pnp* (*D-pnp*) Darlington packages.

PHOTOTRANSISTOR

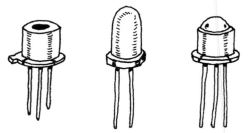

This transistor acts as a light-sensitive bipolar transistor (base is exposed to light). When light comes in contact with the base region, a base current results. Depending on the type of phototransistor, the light may act exclusively as a biasing agent (two-lead phototransistor) or may simply alter an already present base current (three-lead phototransistor). See Chap. 5 for more details.

TRANSISTOR ARRAY

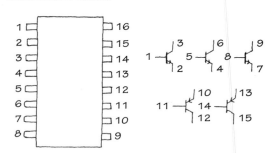

This consists of a number of transistors combined into a single integrated package. For example, the transistor array shown here is made of three *npn* transistors and two *pnp* transistors.

FIGURE 4.61 (*Continued*)

Important Things to Know about Bipolar Transistors

The current gain of a transistor (h_{FE}) is not a very good parameter to go by. It can vary from, say, 50 to 500 within same transistor group family and varies with changes in collector current, collector-to-emitter voltage, and temperature. Because h_{FE} is somewhat unpredictable, one should avoid building circuits that depend specifically on h_{FE} values.

All transistors have maximum collector-current ratings ($I_{C,max}$), maximum collector-to-base (BV_{CBO}), collector-to-emitter (BV_{CEO}), and emitter-to-base (V_{EBO}) breakdown voltages, and maximum collector power dissipation (P_D) ratings. If these ratings are exceeded, the transistor may get zapped. One method to safeguard against BV_{EB} is to place a diode from the emitter to the base, as shown in Fig. 4.62a. The diode prevents emitter-to-base conduction whenever the emitter becomes more positive than the base (e.g., input at base swings negative while emitter is grounded). To avoid exceeding BV_{CBO}, a diode placed in series with the collector (Fig. 4.62b) can be used to prevent collector-base conduction from occurring when the base voltage becomes excessively larger than the collector voltage. To prevent exceeding BV_{CEO}, which may be an issue if the collector holds an inductive load, a diode placed in parallel with the load (see Fig. 4.62c) will go into conduction before a collector-voltage spike, created by the inductive load, reaches the breakdown voltage.

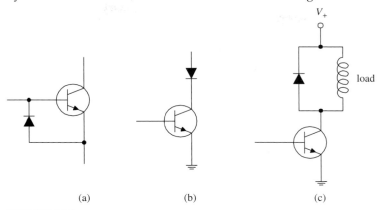

(a)   (b)   (c)

FIGURE 4.62

Pinouts for Bipolar Transistors

Bipolar transistors come in a variety of different package types. Some transistors come with plastic housings; others come with metal can-like housings. When attempting to isolate the leads that correspond to the base, emitter, and collector terminals, first check to see if the package that housed the transistor has a pinout diagram. If no pinout diagram is provided, a good cross-reference catalog (e.g., NTE Cross-Reference Catalog for Semiconductors) can be used. However, as is often the case, simple switching transistors that come in bulk cannot be "looked up"—they may not have a label. Also, these bulk suppliers often will throw together a bunch of transistors that all look alike but may have entirely different pinout designations and may include both *pnp* and *npn* polarities. If you anticipate using transistors often, it may be in your best interest to purchase a digital multimeter that comes with a transistor tester. These multimeters are relatively inexpensive and are easy to use. Such a meter comes with a number of breadboard-like slots. To test a transistor, the pins of the transistor are placed into the slots. By simply pressing a button, the multimeter then tests the transistor and displays whether the device is an *npn* or *pnp* transistor, provides you with the pinout designations (e.g., "ebc," "cbe," etc.), and will give you the transistor's h_{FE}.

Applications

RELAY DRIVER

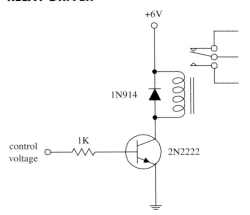

Here, an *npn* transistor is used to control a relay. When the transistor's base receives a control voltage/current, the transistor will turn on, allowing current to flow through the relay coil and causing the relay to switch states. The diode is used to eliminate voltage spikes created by the relay's coil. The relay must be chosen according to the proper voltage rating, etc.

FIGURE 4.63

DIFFERENTIAL AMPLIFIER

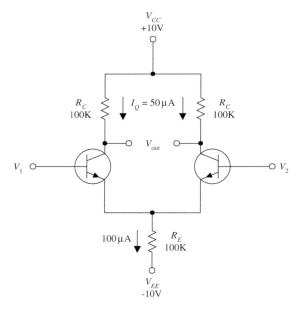

FIGURE 4.64

The differential amplifier shown here is a device that compares two separate input signals, takes the difference between them, and then amplifies this difference. To understand how the circuit works, treat both transistors as identical, and then notice that both transistors are set in the common-emitter configuration. Now, if you apply identical input signals to both V_1 and V_2, identical currents flow through each transistor. This means that (by using $V_C = V_{CC} - I_C R_C$) both transistors' collector voltages are the same. Since the output terminals are simply the left and right transistors' collector voltages, the output voltage (potential difference) is zero. Now, assume the signals applied to the inputs are different, say V_1 is larger than V_2. In this case, the current flow through the left transistor will be larger than the current flow through the right transistor. This means that the left transistor's V_C will decrease relative to the right transistor's V_C. Because the transistors are set in the common-emitter configuration, the effect is amplified. The relationship between the input and output voltages is given by

$$V_{out} \approx \frac{R_C}{r_{tr}}(V_1 - V_2)$$

Rearranging this expression, you find that the gain is equal to R_C/r_{tr}.

Understanding what resistor values to choose can be explained by examining the circuit shown here. First, choose R_C to center V_C to ½V_{CC}, or 5 V, to maximize the dynamic range. At the same time, you must choose a quiescent current (when no signals are applied), say, $I_Q = I_C = 50$ μA. By Ohm's law $R_C = (10V - 5V)/50$ μA $= 100$ kΩ. R_E is chosen to set the transistor's emitters as close to 0 V as possible. R_E is found by adding both the right and left branch's 50 μA and taking the sum to be the current flow through it, which is 100 μA. Now, apply Ohm's law: $R_E = 0$ V $- 10$ V/100 μA $= 100$ kΩ. Next, find the transresistance: $r_{tr} \approx 0.026$ V/$I_E = 0.026$ V/50 μA $= 520$ Ω. The gain then is equal to 100 kΩ/520 Ω $= 192$.

In terms of applications, differential amplifiers can be used to extract a signal that has become weak and which has picked up considerable noise during transmission through a cable (differential amplifier is placed at the receiving end). Unlike a filter circuit, which can only extract a signal from noise if the noise frequency and signal frequency are different, a differential amplifier does not require this condition. The only requirement is that the noise be common in both wires.

When dealing with differential amplifiers, the term *common-mode rejection ratio* (CMRR) is frequently used to describe the quality of the amplifier. A good differential amplifier has a high CMRR (theoretically infinite). CMRR is the ratio of the voltage that must be applied at the two inputs in parallel (V_1 and V_2) to the difference voltage ($V_1 - V_2$) for the output to have the same magnitude.

COMPLEMENTARY-SYMMETRY AMPLIFIER

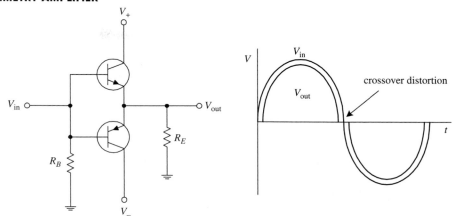

FIGURE 4.65

Recall that an *npn* emitter follower acts to clip the output during negative swings in the input (the transistor turns off when $V_B \leq V_E + 0.6$ V). Likewise, a *pnp* follower will clip the output during positive input swings. But now, if you combine an *npn* and *pnp* transistor, as shown in the circuit shown here, you get what is called a *push-pull follower*, or *complementary-symmetry amplifier*, an amplifier that provides current gain and that is capable of conducting during both positive and negative input swings. For $V_{in} = 0$ V, both transistors are biased to cutoff ($I_B = 0$). For $V_{in} > 0$ V, the upper transistor conducts and behaves like an emitter follower, while the lower transistor is cut off. For $V_{in} < 0$ V the lower transistor conducts, while the upper transistor is cut off. In addition to being useful as a dc amplifier, this circuit also conserves power because the operating point for both transistors is near $I_C = 0$. However, at $I_C = 0$, the characteristics of h_{FE} and r_{tr} are not very constant, so the circuit is not very linear for small signals or for near-zero crossing points of large signals (crossover distortion occurs).

CURRENT MIRROR

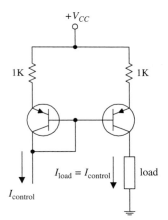

FIGURE 4.66

Here, two matched *pnp* transistors can be used to make what is called a *current mirror*. In this circuit, the load current is a "mirror image" of the control current that is sunk out of the leftmost transistor's collector. Since the same amount of biasing current leaves both transistors' bases, it follows that both transistors' collector-to-emitter currents should be the same. The control current can be set by, say, a resistor connected from the collector to a lower potential. Current mirrors can be made with *npn* transistors, too. However, you must flip this circuit upside down, replace the *pnp* transistors with *npn* transistors, reverse current directions, and swap the supply voltage with ground.

MULTIPLE CURRENT SOURCES

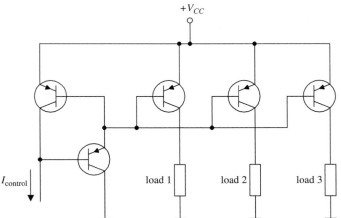

FIGURE 4.67

The circuit here is an expanded version of the previous circuit, which is used to supply a "mirror image" of control current to a number of different loads. (Again, you can design such a circuit with *npn* transistor, too, taking into consideration what was mentioned in the last example.) Note the addition of an extra transistor in the control side of the circuit. This transistor is included to help prevent one transistor that saturates (e.g., its load is removed) from stealing current from the common base line and hence reducing the other output currents.

MULTIVIBRATORS (FLIP-FLOPS)

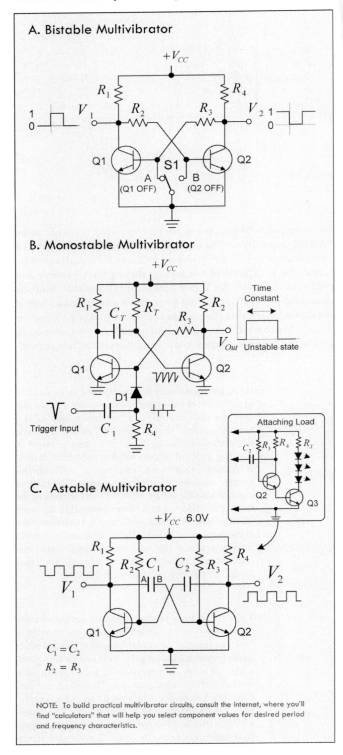

A. Bistable Multivibrator

B. Monostable Multivibrator

C. Astable Multivibrator

NOTE: To build practical multivibrator circuits, consult the internet, where you'll find "calculators" that will help you select component values for desired period and frequency characteristics.

FIGURE 4.68

A. A bistable multivibrator is a circuit that is designed to remain in either of two states indefinitely until a control signal is applied that causes it to change states. After the circuit switches states, another signal is required to switch it back to its previous state. To understand how this circuit works, initially assume that $V_1 = 0V$. This means that the transistor Q2 has no base current and hence no collector current. Therefore, current that flows through R_4 and R_3 flows into the base of transistor Q1, driving it into saturation. In the saturation state, $V_1 = 0$, as assumed initially. Now, because the circuit is symmetric, you can say it is equally stable with $V_2 = 0$ and Q1 saturated. The bistable multivibrator can be made to switch from one state to another by simply grounding either V_1 or V_2 as needed. This is accomplished with switch S1. Bistable multivibrators can be used as memory devices or as frequency dividers, since alternate pulses restore the circuit to its initial state.

B. A monostable multivibrator is a circuit that is stable in only one state, in this case $V_{out} = 0V$. It can be thrown into its unstable state ($V_{out} = V_{CC}$) by applying an external trigger signal, but it will automatically return to its stable state after a duration set by the $R_T C_T$ network. Here when a negative trigger pulse is applied at the input, the fast decaying edge of the pulse will pass through capacitor C_1 to the base of Q1 via blocking diode D1 turning Q1 ON. The collector of Q1 which was previously at V_{CC} quickly drops to below zero voltage, effectively giving C_T a reverse charge of $-0.6V$ across its plates. Transistor Q2 now has a minus base voltage at its base, holding the transistor fully OFF. This represents the circuit's unstable state ($V_{OUT} = V_{CC}$). C_T begins to discharge this $-0.6V$ through RT, attempting to charge up to V_{CC}. This negative voltage at Q2's base begins to decrease gradually at a rate set by the time constant of $R_T C_T$. As Q2's base voltage increases up to V_{CC}, it begins to conduct, causing Q1 to turn off again. The system returns to its original stable state.

C. An asatable multivibrator is a circuit that is not stable in either of two possible output states and it acts like an oscillator. It also requires no external trigger pulse, but uses positive feedback network and a RC timer network to create built-in triggering that switches the output between V_{CC} and 0V. The result is a square wave frequency generator. In the circuit to the left, Q1 and Q2 are switching transistors connected in cross-coupled feedback network, along with two time-delay capacitors. The transistors are biased for linear operation and are operated as common emitter amplifiers with 100% positive feedback. When Q1 is OFF, its collector voltage rises toward V_{CC}, while Q2 if ON. As this occurs, plate A of capacitor C_1 rises towards V_{CC}. Capacitor C_1's other plate B, which is connected to the base of Q2 is at 0.6V since Q2 is in conducting state, thus the voltage across C_1 is $6.0 - 0.6V = 5.4V$. (It's high value of charge). The instant Q1 switches ON, plate A of C_1 falls to 0.6V, causing an equal and instantaneous fall in voltage on plate B of C_1. C_1 is pulled down to -5.4 (reverse charge) and this negative voltage turns transistor Q2 hard OFF (one unstable state). C_1 now begins to charge in the opposite direction via R_3 toward the +6V supply rail, and Q2's base voltage increases toward V_{CC} with a time constant $C_1 R_3$. However, when Q2's base voltage reaches 0.6V, Q2 turns fully ON, starting the whole process over again, but now with C_2 taking the base of Q1 to –5.4V while charging up via resistor R_2 and entering the second unstable state. This process repeats over and over again as long as the supply voltage is present. The amplitude of the output is approximately V_{CC} with the time period between states determined by the RC network connected across the base terminals of the transistors. It's possible to drive low impedance loads (or current loads) such as LEDs, speakers, etc. without affecting the operation of the astable multivibrator by introducing another transistor into the circuit—as shown in figure C.

TRANSISTOR LOGIC GATES

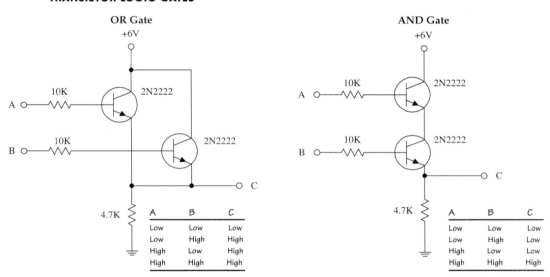

FIGURE 4.69   The two circuits here form logic gates. The OR circuit allows the output (*C*) to swing to a high voltage when either *A* or *B* or both *A* and *B* are high. In other words, as long as at least one of the transistors is biased (turned on), a high voltage will appear at the output. In the AND gate circuit, both *A* and *B* must be high in order for *C* to go high. In other words, both transistors must be biased (turned on) for a high voltage to appear at the output.

4.3.3   Junction Field-Effect Transistors

Junction field-effect transistors (JFETs) are three-lead semiconductive devices that are used as electrically controlled switches, amplifier controls, and voltage-controlled resistors. Unlike bipolar transistors, JFETs are exclusively voltage-controlled—they do not require a biasing current. Another unique trait of a JFET is that it is normally on when there is no voltage difference between its gate and source leads. However, if a voltage difference forms between these leads, the JFET becomes more resistive to current flow (less current will flow through the drain-source leads). For this reason, JFETs are referred to as *depletion devices,* unlike bipolar transistors, which are en-hancement devices (bipolar transistors become less resistive when a current/voltage is applied to their base leads).

JFETs come in either *n-channel* or *p-channel* configurations. With an *n*-channel JFET, a negative voltage applied to its gate (relative to its source lead) reduces current flow from its drain to source lead. (It operates with $V_D > V_S$.) With a *p*-channel JFET, a positive voltage applied to its gate reduces current flow from its source to drain lead. (It operates with $V_S > V_D$.) The symbols for both types of JFETs are shown at the left.

n-channel JFET

drain

gate

D

S

source

p-channel JFET

drain

gate

D

S

source

FIGURE 4.70

An important characteristic of a JFET that is useful in terms of applications is its extremely large input impedance (typically around 10^{10} Ω). This high input impedance means that the JFET draws little or no input current (lower pA range) and therefore has little or no effect on external components or circuits connected to its gate—no current is drawn away from the control circuit, and no unwanted current enters the control circuit. The ability for a JFET to control current flow while maintaining an extremely high input impedance makes it a useful device used in the construction of bidirectional analog switching circuits, input stages for amplifiers, simple two-terminal current sources, amplifier circuits, oscillators circuits, electronic gain-control logic switches, audio mixing circuits, etc.

How a JFET Works

An *n*-channel JFET is made with an *n*-type silicon channel that contains two *p*-type silicon "bumps" placed on either side. The gate lead is connected to the *p*-type bumps, while the drain and source leads are connected to either end of the *n*-type channel (see Fig. 4.71).

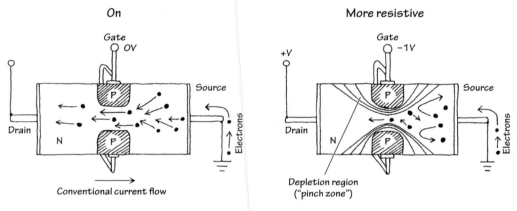

FIGURE 4.71

When no voltage is applied to the gate of an *n*-channel JFET, current flows freely through the central *n*-channel—electrons have no problem going through an *n*-channel; there are a lot of negative charger carriers already in there just waiting to help out with conduction. However, if the gate is set to a negative voltage—relative to the source—the area in between the *p*-type semiconductor bumps and the center of the *n*-channel will form two reverse-biased junctions (one about the upper bump, another about the lower bump). This reverse-biased condition forms a depletion region that extends into the channel. The more negative the gate voltage, the larger is the depletion region, and hence the harder it is for electrons to make it through the channel. For a *p*-channel JFET, everything is reversed, meaning you replace the negative gate voltage with a positive voltage, replace the *n*-channel with a *p*-channel semiconductor, replace the *p*-type semiconductor bumps with *n*-type semiconductors, and replace negative charge carriers (electrons) with positive charge carriers (holes).

JFET Water Analogies

Here are water analogies for an *n*-channel and *p*-channel JFET. Pretend water flow is conventional current flow and water pressure is voltage.

N-CHANNEL JFET WATER ANALOGY

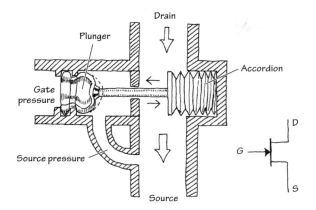

When no pressure exists between the gate and source of the *n*-channel water JFET, the device is fully on; water can flow from the drain pipe to the source pipe. To account for a real JFET's high input impedance, the JFET water analogy uses a plunger mechanism attached to a moving flood gate. (The plunger prevents current from entering the drain source channel, while at the same time it allows a pressure to control the flood gate.) When the gate of the *n*-channel JFET is made more negative in pressure relative to the source tube, the plunger is forced to the left. This in turn pulls the accordion-like flood gate across the drain-source channel, thus decreasing the current flow.

P-CHANNEL JFET WATER ANALOGY

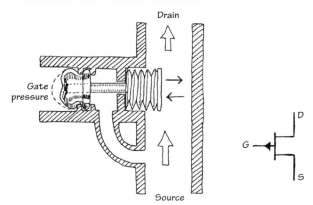

The *p*-channel water JFET is similar to the *n*-channel water JFET, except that all currents and pressures are reversed. The *p*-channel JFET is fully on until a positive pressure, relative to the source, is applied to the gate tube. The positive pressure forces the accordion across the drain-source channel, hence reducing the current flow.

FIGURE 4.72

Technical Stuff

The following graph describes how a typical *n*-channel JFET works. In particular, the graph describes how the drain current (I_D) is influenced by the gate-to-source voltage (V_{GS}) and the drain-to-source voltage (V_{DS}). The graph for a *p*-channel JFET is similar to that of the *n*-channel graph, except that I_D decreases with an increasing positive V_{GS}. In other words, V_{GS} is positive in voltage, and V_{DS} is negative in voltage.

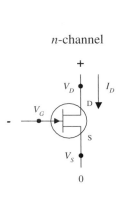

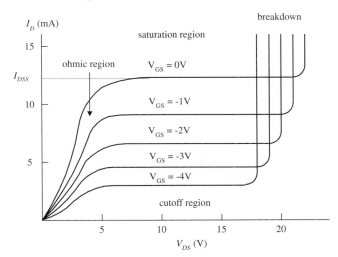

FIGURE 4.73

When the gate voltage V_G is set to the same voltage as the source ($V_{GS} = V_G - V_S = 0$ V), maximum current flows through the JFET. Technically speaking, people call this current (when $V_{GS} = 0$ V) the *drain current for zero bias*, or I_{DSS}. I_{DSS} is a constant and varies from JFET to JFET. Now notice how the I_D current depends on the drain-source voltage ($V_{DS} = V_D - V_S$). When V_{DS} is small, the drain current I_D varies nearly linearly with V_{DS} (looking at a particular curve for fixed V_{GS}). The region of the graph in which this occurs is called the *ohmic region,* or *linear region.* In this region, the JFET behaves like a voltage-controlled resistor.

Now notice the section of the graph were the curves flatten out. This region is called the *active region,* and here the drain current I_D is strongly influenced by the gate-source voltage V_{GS} but hardly at all influenced by the drain-to-source voltage V_{DS} (you have to move up and down between curves to see it).

Another thing to note is the value of V_{GS} that causes the JFET to turn off (point where practically no current flows through device). The particular V_{GS} voltage that causes the JFET to turn off is called the *cutoff voltage* (sometimes called the *pinch-off voltage V_P*), and it is expressed as $V_{GS,\text{off}}$.

Moving on with the graph analysis, you can see that when V_{DS} increases, there is a point where I_D skyrockets. At this point, the JFET loses its ability to resist current because too much voltage is applied across its drain-source terminals. In JFET lingo, this effect is referred to as *drain-source breakdown,* and the breakdown voltage is expressed as BV_{DS}.

For a typical JFET, I_{DSS} values range from about 1 mA to 1 A, $V_{GS,\text{off}}$ values range from around −0.5 to −10 V for an *n*-channel JFET (or from +0.5 to +10 V for a *p*-channel JFET), and BV_{DS} values range from about 6 to 50 V.

Like bipolar transistors, JFETs have internal resistance within their channels that varies with drain current and temperature. The reciprocal of this resistance is referred to as the *transconductance g_m.* A typical JFET transconductance is around a few thousand Ω^{-1}, where $\Omega^{-1} = 1/\Omega$ or $\mho$.

Another one of the JFET's built-in parameters is its *on resistance,* or $R_{DS,\text{on}}$. This resistance represents the internal resistance of a JFET when in its fully conducting state (when $V_{GS} = 0$). The $R_{DS,\text{on}}$ of a JFET is provided in the specification tables and typically ranges from 10 to 1000 Ω.

Useful Formulas

N-CHANNEL JFET

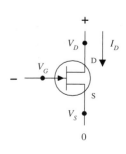

FIGURE 4.74

N-CHAN CURVES

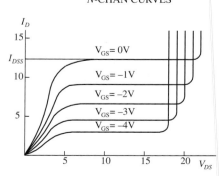

OHMIC REGION   JFET is just beginning to resist. It acts like a variable resistor.

SATURATION REGION   JFET is most strongly influenced by gate-source voltage, hardly at all influenced by the drain-source voltage.

CUTOFF VOLTAGE ($V_{GS,\text{OFF}}$)   Particular gate-source voltage where JFET acts like an open circuit (channel resistance is at its maximum).

BREAKDOWN VOLTAGE (BV_{DS})   The voltage across the drain and source that causes current to "break through" the JFET's resistive channel.

P-CHANNEL JFET

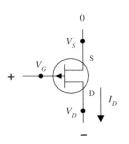

0

V_S

V_G

S

D I_D

V_D

+

−

FIGURE 4.74   (*Continued*)

P-CHAN CURVES

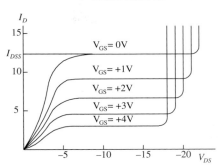

I_D

15

I_{DSS}

10

5

V_GS= 0V

V_GS= +1V

V_GS= +2V

V_GS= +3V

V_GS= +4V

−5   −10   −15   −20 V_{DS}

DRAIN-CURRENT FOR ZERO BIAS (I_{DSS}) Represents the drain current when gate-source voltage is zero volts (or gate is connected to source, $V_{GS} = 0$ V).

TRANSCONDUCTANCE (g_m) Represents the rate of change in the drain current with the gate-source voltage when the drain-to-source voltage is fixed for a particular V_{DS}. It is analogous to the transconductance ($1/R_{tr}$) for bipolar transistors.

DRAIN CURRENT (OHMIC REGION)

$$I_D = I_{DSS}\left[2\left(1-\frac{V_{GS}}{V_{GS,off}}\right)\frac{V_{DS}}{-V_{GS,off}}-\left(\frac{V_{DS}}{V_{GS,off}}\right)^2\right]$$

DRAIN CURRENT (ACTIVE REGION)

$$I_D = I_{DSS}\left(1-\frac{V_{GS}}{V_{GS,off}}\right)^2$$

DRAIN-SOURCE RESISTANCE

$$R_{DS}=\frac{V_{DS}}{i_D}\approx\frac{V_{GS,off}}{2I_{DSS}(V_{GS}-V_{GS,off})}=\frac{1}{g_m}$$

ON RESISTANCE

$$R_{DS,on}=\text{constant}$$

DRAIN-SOURCE VOLTAGE

$$V_{DS}=V_D-V_S$$

TRANSCONDUCTANCE

$$g_m=\frac{\partial I_D}{\partial V_{GS}}\bigg|_{V_{DS}}=\frac{1}{R_{DS}}=g_{m0}\left(1-\frac{V_{GS}}{V_{GS,off}}\right)=g_{m0}\sqrt{\frac{I_D}{I_{DSS}}}$$

TRANSCONDUCTANCE FOR SHORTED GATE

$$g_{m0}=\left|\frac{2I_{DSS}}{V_{GS,off}}\right|$$

> An *n*-channel JFET's $V_{GS,off}$ is negative. A *p*-channel JFET's $V_{GS,off}$ is positive.

> $V_{GS,off}$, I_{DSS} are typically the knowns (you get their values by looking them up in a data table or on the package).

> Typical JFET values:
> I_{DSS}: 1 mA to 1 A
> $V_{GS,off}$:
> −0.5 to −10 V (*n*-channel)
> 0.5 to 10 V (*p*-channel)
> $R_{DS,on}$: 10 to 1000 Ω
> BV_{DS}: 6 to 50 V
> g_m at 1 mA:
> 500 to 3000 μmho

Sample Problems

PROBLEM 1

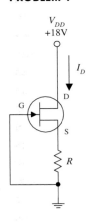

V_{DD}
+18V

I_D

G

D

S

R

FIGURE 4.75

If an *n*-channel JFET has a $I_{DSS}=8$ mA and $V_{GS,off}=-4$ V, what will be the drain current I_D if $R=1$ kΩ and $V_{DD}=+18$ V? Assume that the JFET is in the active region.

In the active region, the drain current is given by

$$I_D=I_{DSS}\left(1-\frac{V_{GS}}{V_{GS,off}}\right)^2$$

$$=8\text{mA}\left(1-\frac{V_{GS}}{-4\text{V}}\right)^2=8\text{ mA}\left(1+\frac{V_{GS}}{2}+\frac{V^2_{GS}}{16}\right)$$

Unfortunately, there is one equation and two unknowns. This means that you have to come up with another equation. Here's how you get the other equation. First, you can assume the gate voltage is 0 V because it's grounded. This means that

$$V_{GS}=V_G-V_S=0\text{ V}-V_S=-V_S$$

From this, you can come up with another equation for the drain current by using Ohm's law and treating $I_D=I_S$:

$$I_D = \frac{V_s}{R} = -\frac{V_{GS}}{R} = -\frac{V_{GS}}{1\,k\Omega}$$

This equation is then combined with the first equation to yield

$$-\frac{V_{GS}}{1\,k\Omega} = 8\ mA\left(1 + \frac{V_{GS}}{2} + \frac{V^2_{GS}}{16}\right)$$

which simplifies to

$$V_{GS}{}^2 + 10V_{GS} + 16 = 0$$

The solutions to this equation are $V_{GS} = -2$ V and $V_{GS} = -8$ V. But a V_{GS} of –8V would be below the cutoff voltage, so we can ignore it, which leaves $V_{GS} = -2$V as the correct solution. This means that $V_{GS} = -2$ V is the correct solution, so you disregard the –8-V solution. Now you substitute V_{GS} back into one of the $I_{D(active)}$ equations to get

$$I_D = -\frac{V_{GS}}{R} = -\frac{(-2V)}{1\,k\Omega} = 2mA$$

PROBLEM 2

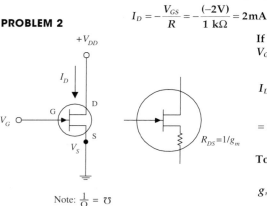

+V_{DD}

I_D

G

D

V_G

S

V_S

Note: $\frac{1}{\Omega} = \mho$

$R_{DS} = 1/g_m$

FIGURE 4.76

If $V_{GS,off} = -4$ V and $I_{DSS} = 12$ mA, find the values of I_D and g_m and R_{DS} when $V_{GS} = -2$ V and when $V_{GS} = +1$ V. Assume that the JFET is in the active region.

When $V_{GS} = -2$ V,

$$I_D = I_{DDS}\left(1 - \frac{V_{GS}}{V_{GS,off}}\right)^2$$

$$= 12\ mA\left(1 - \frac{(-2V)}{(-4V)}\right)^2 = 3.0\ mA$$

To find g_m, you first must find g_{m_0} (transconductance for shorted gate):

$$g_{m_0} = \frac{2I_{DSS}}{V_{GS,off}} = -\frac{2(12\ mA)}{(-4V)} = 0.006\mho = 6000\ \mu mhos$$

Now you can find g_m:

$$g_m = g_{m_0}\sqrt{\frac{I_D}{I_{DSS}}} = (0.006\ \mho)\sqrt{\frac{3.0\ mA}{12.0\ mA}} = 0.003\mho = 3000\ \mu mhos$$

To find the drain-source resistance (R_{DS}), use

$$R_{DS} = \frac{1}{g_m} = \frac{1}{0.003\ \mho} = 333\Omega$$

By applying the same formulas as above, you can find that when $V_{GS} = +1$ V, $I_D = 15.6$ mA, $g_m = 0.0075\ \mho = 7500\ \mu mhos$, and $R_{DS} = 133\ \Omega$.

Basic Operations

LIGHT DIMMER

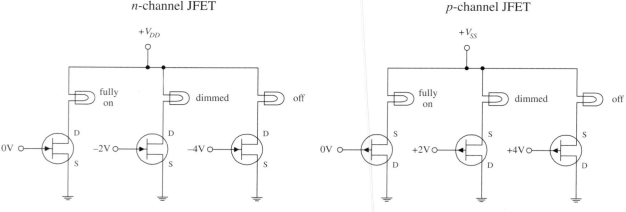

n-channel JFET

p-channel JFET

FIGURE 4.77

The two circuits here demonstrate how a JFET acts like a voltage-controlled light dimmer. In the *n*-channel circuit, a more negative gate voltage causes a larger drain-to-source resistance, hence causing the light bulb to receive less current. In the *p*-channel circuit, a more positive gate voltage causes a greater source-to-drain resistance.

BASIC CURRENT SOURCE AND BASIC AMPLIFIER

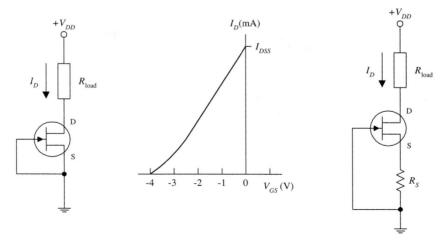

FIGURE 4.78 A simple current source can be constructed by shorting the source and gate terminals together (this is referred to as *self-biasing*), as shown in the left-most circuit. This means that $V_{GS} = V_G - V_S = 0$ V, which means the drain current is simply equal to I_{DDS}. One obvious drawback of this circuit is that the I_{DDS} for a particular JFET is unpredictable (each JFET has its own unique IDDS that is acquired during manufacturing). Also, this source is not adjustable. However, if you place a resistor between the source and ground, as shown in the right-most circuit, you can make the current source adjustable. By increasing R_S, you can decrease I_D, and vice versa (see Problem 2). Besides being adjustable, this circuit's I_D current will not vary as much as the left circuit for changes in V_{DS}. Though these simple JFET current sources are simple to construct, they are not as stable as a good bipolar or op amp current source.

SOURCE FOLLOWER

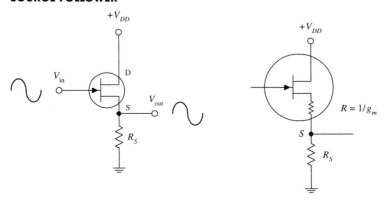

FIGURE 4.79 The JFET circuit here is called a *source follower,* which is analogous to the bipolar emitter follower; it provides current gain but not voltage gain. The amplitude of the output signal is found by applying Ohm's law: $V_S = R_S I_D$, where $I_D = g_m V_{GS} = g_m (V_G - V_S)$. Using these equations, you get

$$V_S = \frac{R_S g_m}{1 + R_S g_m} V_G$$

Since $V_S = V_{out}$ and $V_G = V_{in}$, the gain is simply $R_S g_m/(1 + R_S g_m)$. The output impedance, as you saw in Problem 2, is $1/g_m$. Unlike the emitter follower, the source follower has an extremely larger input impedance and therefore draws practically no input current. However, at the same time, the JFET's transconductance happens to be smaller than that of a bipolar transistor, meaning the output will be more attenuated. This makes sense if you treat the $1/g_m$ term as being a small internal resistance within the drain-source channel (see rightmost circuit). Also, as the drain current changes due to an applied waveform, g_m and therefore the output impedance will vary, resulting in output distortion. Another problem with this follower circuit is that V_{GS} is a poorly controlled parameter (a result of manufacturing), which gives it an unpredictable dc offset.

IMPROVED SOURCE FOLLOWER

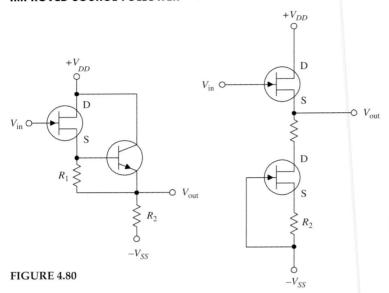

FIGURE 4.80

The source follower circuit from the preceding example had poor linearity and an unpredictable dc offset. However, you can eliminate these problems by using one of the two arrangements shown here. In the far-left circuit, you replace the source resistor with a bipolar current source. The bipolar source acts to fix V_{GS} to a constant value, which in turn eliminates the nonlinearities. To set the dc offset, you adjust R_1. (R_2 acts like R_S in the preceding circuit; it sets the gain.) The near-left circuit uses a JFET current source instead of a bipolar source. Unlike the bipolar circuit, this circuit requires no adjusting and has better temperature stability. The two JFETs used here are matched (matched JFETs can be found in pairs, assembled together within a single package). The lower transistor sinks as much current as needed to make $V_{GS} = 0$ (shorted gate). This means that both JFETs' V_{GS} values are zero, making the upper transistor a follower with zero dc offset. Also, since the lower JFET responds directly to the upper JFET, any temperature variations will be compensated. When R_1 and R_2 are set equal, $V_{out} = V_{in}$. The resistors help give the circuit better I_D linearity, allow you to set the drain current to some value other than I_{DSS}, and help to improve the linearity. In terms of applications, JFET followers are often used as input stages to amplifiers, test instruments, or other equipment that is connected to sources with high source impedance.

JFET AMPLIFIERS

SOURCE-FOLLOWER AMP

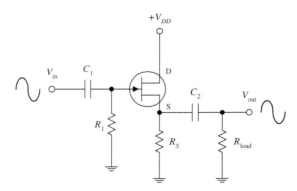

FIGURE 4.81

Recall the emitter-follower and common-emitter bipolar transistor amplifiers from the last chapter. These two amplifiers have JFET counterparts, namely, the source-follower and the common-source amplifier shown here. (The source-follower amplifier provides current gain; the common-source amplifier provides voltage gain.) If you were to set up the equations and do the math, you would find that the gain for the amplifiers would be

$$\text{Gain} = \frac{V_{out}}{V_{in}} = \frac{R_S}{R_S + 1/g_m} \quad \text{(source-follower amp.)}$$

$$\text{Gain} = \frac{V_{out}}{V_{in}} = g_m \frac{R_D R_1}{R_D + R_1} \quad \text{(common-source amp.)}$$

where the transconductance is given by

$$g_m = g_{m0} \frac{I_D}{I_{DSS}}, g_{m0} = -\frac{2I_{DSS}}{V_{GS,off}}$$

As with bipolar amplifiers, the resistors are used to set the gate voltages and set the quiescent currents, while the capacitors act as ac couplers/high-pass filters. Notice, however, that both JFET amplifiers only require one self-biasing resistor.

COMMON-SOURCE AMP

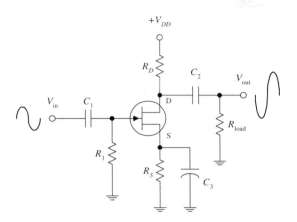

FIGURE 4.81   (*Continued*)

Now, an important question to ask at this point is, Why would you choose a JFET amplifier over a bipolar amplifier? The answer is that a JFET provides increased input impedance and low input current. However, if extremely high input impedances are not required, it is better to use a simple bipolar amplifier or op amp. In fact, bipolar amplifiers have fewer nonlinearity problems, and they tend to have higher gains when compared with JFET amplifiers. This stems from the fact that a JFET has a lower transconductance than a bipolar transistor for the same current. The difference between a bipolar's transconductance and JFET's transconductance may be as large as a factor of 100. In turn, this means that a JFET amplifier will have a significantly smaller gain.

VOLTAGE-CONTROLLED RESISTOR

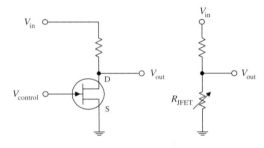

Electronic gain control

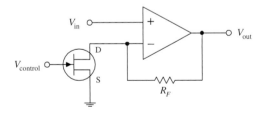

FIGURE 4.82

According to the graphs you saw earlier, if V_{DS} drops low enough, the JFET will operate within the linear (ohmic) region. In this region, the I_D versus V_{DS} curves follow approximate straight lines for V_{DS} smaller than $V_{GS} - V_{GS,off}$. This means that the JFET behaves like a voltage-controlled resistor for small signals of either polarity. For example, if you take a voltage-divider network and replace one of the resistors with a JFET, you get a voltage-controlled voltage divider (see upper left-hand figure). The range over which a JFET behaves like a traditional resistor depends on the particular JFET and is roughly proportional to the amount by which the gate voltage exceeds $V_{GS,off}$. For a JFET to be effective as a linearly responding resistor, it is important to limit V_{DS} to a value that is small compared with $V_{GS,off}$, and it is important to keep $|V_{GS}|$ below $|V_{GS,off}|$. JFETs that are used in this manner are frequently used in electronic gain-control circuits, electronic attenuators, electronically variable filters, and oscillator amplitude-control circuits. A simple electronic gain-control circuit is shown here. The voltage gain for this circuit is given by gain = $1 + R_F/R_{DS(on)}$, where R_{DS} is the drain-source channel resistance. If $R_F = 29$ kΩ and $R_{DS(on)} = 1$ kΩ, the maximum gain will be 30. As V_{GS} approaches $V_{GS,off}$, R_{DS} will increase and become very large such that $R_{DS} \gg R_F$, causing the gain to decrease to a minimum value close to unity. As you can see, the gain for this circuit can be varied over a 30:1 ratio margin.

Practical Considerations

JFETs typically are grouped into the following categories: small-signal and switching JFETs, high-frequency JFETs, and dual JFETs. Small-signal and switching JFETs are frequently used to couple a high-impedance source with an amplifier or other device such as an oscilloscope. These devices are also used as voltage-controlled switches. High-frequency JFETs are used primarily to amplify high-frequency signals (with the RF range) or are used as high-frequency switches. Dual JFETs contain two matched JFETs in one package. As you saw earlier, dual JFETs can be used to improve source-follower circuit performance.

TYPES OF JFET PACKAGES

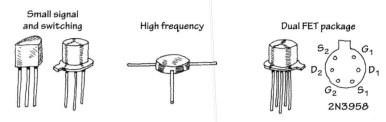

Small signal
and switching

High frequency

Dual FET package

2N3958

FIGURE 4.83

Like bipolar transistors, JFETs also can be destroyed with excess current and voltage. Make sure that you do not exceed maximum current and breakdown voltages. Table 4.5 is a sample of a JFET specification table designed to give you a feel for what to expect when you start searching for parts.

TABLE 4.5 Portion of a JFET Specification Table

TYPE	POLARITY	BV_{GS} (V)	I_{DSS} (mA) MIN (mA)	I_{DSS} (mA) MAX (mA)	$V_{GS,OFF}$ (V) MIN (V)	$V_{GS,OFF}$ (V) MAX (V)	G_M TYPICAL (μmhos)	C_{ISS} (pF)	C_{RSS} (pF)
2N5457	n-ch	25	1	5	−0.5	−6	3000	7	3
2N5460	p-ch	40	1	5	1	6	3000	7	2
2N5045	Matched-pair n-ch	50	0.5	8	−0.5	−4.5	3500	6	2

Applications

RELAY DRIVER

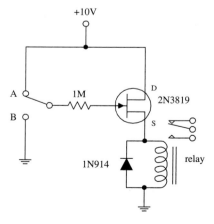

+10V

A

B

1M

2N3819

D

S

1N914

relay

FIGURE 4.84

Here, an *n*-channel JFET is used to switch a relay. When the switch is set to position *A*, the JFET is on (gate isn't properly biased for a depletion effect to occur). Current then passes through the JFET's drain-source region and through the relay's coil, causing the relay to switch states. When the switch is thrown to position *B*, a negative voltage—relative to the source—is set at the gate. This in turn causes the JFET to block current flow from reaching the relay's coil, thus forcing the relay to switch states.

AUDIO MIXER/AMPLIFIER

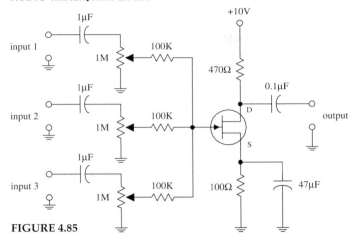

FIGURE 4.85

This circuit uses a JFET—set in the common-source arrangement—to combine (mix) signals from a number of different sources, such as microphones, preamplifiers, etc. All inputs are applied through ac coupling capacitors/filters. The source and drain resistors are used to set the overall amplification, while the 1-MΩ potentiometers are used to control the individual gains of the input signals.

ELECTRICAL FIELD METER

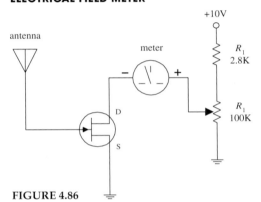

FIGURE 4.86

Here, a JFET is used to construct a simple static electricity detector. When the antenna (simple wire) is placed near a charged object, the electrons in the antenna will be drawn toward or away from the JFET's gate, depending on whether the object is positively or negatively charged. The repositioning of the electrons sets up a gate voltage that is proportional to the charge placed on the object. In turn, the JFET will either begin to resist or allow current to flow through its drain-source channel, hence resulting in ammeter needle deflection. R_1 is used to protect the ammeter, and R_2 is used to calibrate it.

4.3.4 *Metal Oxide Semiconductor Field-Effect Transistors*

Metal oxide semiconductor field-effect transistors (MOSFETs) are incredibly popular transistors that in some ways resemble JFETs. For instance, when a small voltage is applied at its gate lead, the current flow through its drain-source channel is altered. However, unlike JFETS, MOSFETs have larger gate lead input impedances ($\geq 10^{14}$ Ω, as compared with ~10^9 Ω for JFETs), which means that they draw almost no gate current whatsoever. This increased input impedance is made possible by placing a metal oxide insulator between the gate and the drain-source channel. There is a price to pay for this increased amount of input impedance, which amounts to a very low gate-to-channel capacitance (a few pF). If too much static electricity builds up on

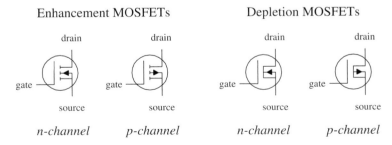

FIGURE 4.87

the gate of certain types of MOSFETs during handling, the accumulated charge may break through the gate and destroy the MOSFET. (Some MOSFETs are designed with safeguards against this breakdown—but not all.)

The two major kinds of MOSFETs are *enhancement-type MOSFETs* and *depletion-type MOSFETs* (see Fig. 4.88). A depletion-type MOSFET is normally on (maximum current flows from drain to source) when no difference in voltage exists between the gate and source terminals ($V_{GS} = V_G - V_S = 0$ V). However, if a voltage is applied to its gate lead, the drain-source channel becomes more resistive—a behavior similar to a JFET. An enhancement-type MOSFET is normally off (minimum current flows from drain to source) when $V_{GS} = 0$ V. However, if a voltage is applied to its gate lead, the drain-source channel becomes less resistive.

Both enhancement-type and depletion-type MOSFETs come in either *n*-channel or *p*-channel forms. For an *n*-channel depletion-type MOSFET, a negative gate-source voltage ($V_G < V_S$) increases the drain-source channel resistance, whereas for a *p*-channel depletion-type MOSFET, a positive gate-source voltage ($V_G > V_S$) increases the channel resistance. For an *n*-channel enhancement-type MOSFET, a positive gate-source voltage ($V_G > V_S$) decreases the drain-source channel resistance, whereas for a *p*-channel enhancement-type MOSFET, a negative gate-source voltage ($V_G < V_S$) decreases the channel resistance.

MOSFETs are perhaps the most popular transistors used today; they draw very little input current, are easy to make (require few ingredients), can be made extremely small, and consume very little power. In terms of applications, MOSFETs are used in ultrahigh input impedance amplifier circuits, voltage-controlled "resistor" circuits, switching circuits, and found with large-scale integrated digital ICs.

Like JFETs, MOSFETs have small transconductance values when compared with bipolar transistors. In terms of amplifier applications, this can lead to decreased gain values. For this reason, you will rarely see MOSFETs in simple amplifier circuits, unless there is a need for ultrahigh input impedance and low input current features.

How MOSFETs Work

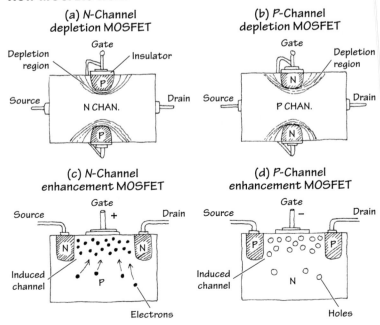

Both depletion and enhancement MOSFETs use an electrical field—produced by a gate voltage—to alter the flow of charge carriers through the semiconductive drain-source channel. With depletion-type MOSFETs, the drain-source channel is inherently conductive; charge carriers such as electrons (*n*-channel) or holes (*p*-channel) are already present within the *n*-type or *p*-type channel. If a negative gate-source voltage is applied to an *n*-channel depletion-type MOSFET, the resulting electrical field acts to "pinch off" the flow of electrons through the channel (see Fig. 4.88a). A *p*-channel depletion-type MOSFET uses a positive gate-source voltage to "pinch off" the flow of holes through its channel (see Fig. 4.88b). (The pinching-off effect results from depletion regions forming about the upper and lower gate contacts.) Enhancement MOSFETs, unlike depletion MOSFETs, have a normally resistive channel; there are few charge carriers within it. If a positive gate-source voltage is applied to an *n*-channel enhancement-type MOSFET, electrons within

FIGURE 4.88

the *p*-type semiconductor region migrate into the channel and thereby increase the conductance of the channel (see Fig. 4.88c). For a *p*-channel enhancement MOSFET, a negative gate-source voltage draws holes into the channel to increase the conductivity (see Fig. 4.88d).

Basic Operation

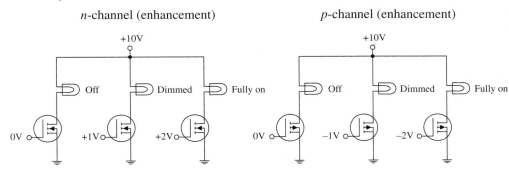

The circuits shown here demonstrate how MOSFETs can be used to control current flow through a light bulb. The desired dimming effects produced by the gate voltages may vary depending on the specific MOSFET you are working with.

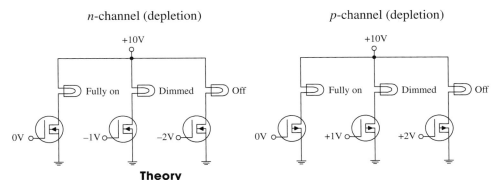

FIGURE 4.89

Theory

In terms of theory, you can treat depletion-type MOSFETs like JFETs, except you must give them larger input impedances. The following graphs, definitions, and formulas summarize the theory.

N-CHANNEL DEPLETION-TYPE MOSFET

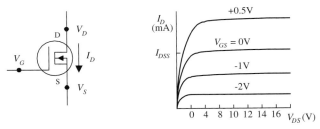

P-CHANNEL DEPLETION-TYPE MOSFET

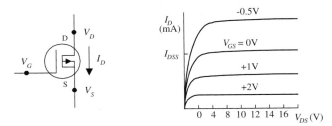

FIGURE 4.90

OHMIC REGION MOSFET is just beginning to resist. In this region, the MOSFET behaves like a variable resistor.

ACTIVE REGION MOSFET is most strongly influenced by gate-source voltage (V_{GS}) but hardly at all influenced by drain-source voltage (V_{DS}).

CUTOFF VOLTAGE ($V_{GS,OFF}$) Often referred to as the *pinch-off voltage* (V_p). Represents the particular gate-source voltage that causes the MOSFET to block almost all drain-source current flow.

BREAKDOWN VOLTAGE (BV_{DS}) The drain-source voltage (V_{DS}) that causes current to "break through" MOSFET's resistive channel.

DRAIN CURRENT FOR ZERO BIAS (I_{DSS}) Represents the drain current when gate-source voltage is zero volts (or when gate is shorted to source).

TRANSCONDUCTANCE (g_m) Represents the rate of change in the drain current with change in gate-source voltage when drain-source voltage is fixed for a particular V_{DS}. It is analogous to the transconductance I/R_{tr} for bipolar transistors.

Useful Formulas for Depletion-Type MOSFETs

DRAIN CURRENT (OHMIC REGION)

$$I_D = I_{DSS}\left[2\left(1 - \frac{V_{GS}}{V_{GS,off}}\right)\frac{V_{DS}}{-V_{GS,off}} - \left(\frac{V_{DS}}{V_{GS,off}}\right)^2\right]$$

DRAIN CURRENT (ACTIVE REGION)

$$I_D = I_{DSS}\left(1 - \frac{V_{GS}}{V_{GS,off}}\right)^2$$

DRAIN-SOURCE RESISTANCE

$$R_{DS} = \frac{V_{DS}}{I_D} \approx \frac{V_{GS,off}}{2I_{DSS}(V_{GS} - V_{GS,off})} = \frac{1}{g_m}$$

ON RESISTANCE

$$R_{DS,on} = \text{constant}$$

DRAIN-SOURCE VOLTAGE

$$V_{DS} = V_D - V_S$$

TRANSCONDUCTANCE

$$g_m = \left.\frac{\partial I_D}{\partial V_{GS}}\right|_{V_{DS}} = \frac{1}{R_{DS}} = g_{m0}\left(1 - \frac{V_{GS}}{V_{GS,off}}\right) = g_{m0}\sqrt{\frac{I_D}{I_{DSS}}}$$

TRANSCONDUCTANCE FOR SHORTED GATE

$$g_{m0} = \left|\frac{2I_{DSS}}{V_{GS,off}}\right|$$

> An n-channel MOSFET's $V_{GS,off}$ is negative.
> A p-channel MOSFET's $V_{GS,off}$ is positive.

> $V_{GS,off}$, I_{DSS} are typically the knowns (you get their values by looking them up in a data table or on the package).

> **Typical MOSFET values:**
> I_{DSS}: 1 mA to 1 A
> $V_{GS,off}$:
> −0.5 to −10 V (n-channel)
> 0.5 to 10 V (p-channel)
> $R_{DS,on}$: 10 to 1000 Ω
> BV_{DS}: 6 to 50 V
> g_m at 1 mA:
> 500 to 3000 μmho

Technical Info and Formulas for Enhancement-Type MOSFETs

Predicting how enhancement-type MOSFETs will behave requires learning some new concepts and formulas. Here's an overview of the theory.

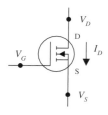

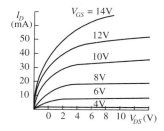

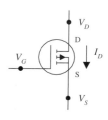

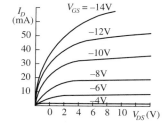

FIGURE 4.91

OHMIC REGION MOSFET is just beginning to conduct. Acts like a variable resistor.

ACTIVE REGION MOSFET is most strongly influenced by gate-source voltage V_{GS} but hardly at all influenced by drain-source voltage V_{DS}.

THRESHOLD VOLTAGE ($V_{GS,th}$) Particular gate-source voltage where MOSFET is just beginning to conduct.

BREAKDOWN VOLTAGE (BV_{DS}) The voltage across drain source (V_{DS}) that causes current to "break through" MOSFET's resistance channel.

DRAIN-CURRENT FOR GIVEN BIAS ($I_{D,on}$) Represents the amount of current I_D at a particular V_{GS}, which is given on data sheets, etc.

TRANSCONDUCTANCE (g_m) Represents the rate of change in the drain current with the change in gate-source voltage when drain-source voltage is fixed. It is analogous to the transconductance I/R_{tr} for bipolar transistors.

DRAIN CURRENT
(OHMIC REGION)

$$I_D = k[2(V_{GS} - V_{GS,th})V_{DS} - \tfrac{1}{2}V_{DS}^2]$$

The value of the construction parameter k is proportional to the width/length ratio of the transistor's channel and is dependent on temperature. It can be determined by using the construction parameter equations to the left.

DRAIN CURRENT
(ACTIVE REGION)

$$I_D = k(V_{GS} - V_{GS,th})^2$$

$V_{GS,th}$ is positive for n-channel enhancement MOSFETs.
$V_{GS,th}$ is negative for p-channel enhancement MOSFETs.

CONSTRUCTION
PARAMETER

$$k = \frac{I_D}{(V_{GS} - V_{GS,th})^2}$$

$$k = \frac{I_{D,on}}{(V_{GS,on} - V_{GS,th})^2}$$

Typical values
$I_{D,on}$: 1 mA to 1 A
$R_{DS(on)}$: 1 Ω to 10 kΩ
$V_{GS,off}$: 0.5 to 10 V
$BV_{DS(off)}$: 6 to 50 V
$BV_{GS(off)}$: 6 to 50 V

TRANSCONDUCTANCE

$$g_m = \frac{\partial I_D}{\partial V_{GS}}\bigg|V_{DS} = \frac{1}{R_{DS}}$$

$$= 2k(V_{GS} - V_{GS,th}) = 2\sqrt{kI_D}$$

$$= g_{m0}\sqrt{\frac{I_D}{I_{D,on}}}$$

$V_{GS,th}$, $I_{D,on}$, g_m at a particular I_D are typically "knowns" you can find in the data tables or on package labels.

RESISTANCE OF
DRAIN-SOURCE CHANNEL

$$R_{DS} = 1/g_m$$

$$R_{DS_2} = \frac{V_{G_1} - V_{GS,th}}{V_{G_2} - V_{GS,th}} R_{DS_1}$$

R_{DS_1} is the known resistance at a given voltage V_{G_1}. R_{DS_2} is the resistance you calculate at another gate voltage V_{G_2}.

Sample Problems

PROBLEM 1

An n-channel depletion-type MOSFET has an $I_{DDS} = 10$ mA and a $V_{GS,off} = -4$ V. Find the values of I_D, g_m, and R_{DS} when $V_{GS} = -2$ V and when $V_{GS} = +1$ V. Assume that the MOSFET is in the active region.

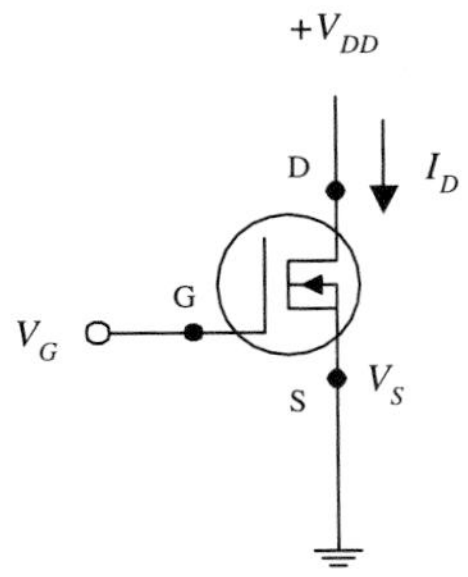

FIGURE 4.92

When considering $V_{GS} = -2$ V, use the following:

$$I_D = I_{DSS}\left(1 - \frac{V_{GS}}{V_{GS,off}}\right)^2 = (10\text{mA})\left[1 - \frac{(-2\text{V})}{(-4\text{V})}\right]^2 = 2.5 \text{ mA}$$

Before you can find g_m, you must find g_{m_0}—here's what you use:

$$g_{m_0} = \left|\frac{2I_{DSS}}{V_{GS,off}}\right| = \left|\frac{2(10 \text{ mA})}{-4\text{V}}\right| = 0.005 \text{ mhos} = 5000 \text{ μmhos}$$

Now you can substitute g_{m_0} into the following expression to find g_m:

$$g_m = g_{m_0}\left(1 - \frac{V_{GS}}{V_{GS,off}}\right) = (5000 \text{ μmhos})\left(1 - \frac{-2\text{V}}{-4\text{V}}\right) = 2500 \text{ μmhos}$$

The drain-source resistance is then found by using $R_{DS} = 1/g_m = 400$ Ω. If you do the same calculations for $V_{GS} = + 1$ V, you get $I_D = 15.6$ mA, $g_m = 6250$ μmhos, and $R_{DS} = 160$ Ω.

PROBLEM 2

An n-channel enhancement-type MOSFET has a $V_{GS,th} = +2$ V and an $I_D = 12$ mA. When $V_{GS} = +4$ V, find parameters k, g_m, and R_{DS}. Assume that the MOSFET is in the active region.

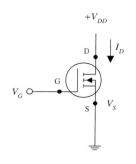

FIGURE 4.93

To find k, use the expression for the drain current in the active region:
$$I_D = k\,(V_{GS} - V_{GS,th})^2$$
Rearranging this equation and plugging in the knowns, you get

$$k = \frac{I_D}{(V_{GS} - V_{GS,th})^2} = \frac{12\ \text{mA}}{(4\ \text{V} - 2\text{V})^2} = 0.003\ \text{mhos/V} = 3000\ \mu\text{mhos/V}$$

To find g_m, use the following:

$$g_m = 2k\,(V_{GS} - V_{GS,th}) = 2\sqrt{kI_D}$$

$$= 2\sqrt{(3000\ \mu\text{mhos}/V)(12\ \text{mA})} = 0.012\ \text{mhos} = 12,000\ \mu\text{mhos}$$

The drain-source resistance is then found by using $R_{DS} = 1/g_m = 83\ \Omega$.

PROBLEM 3

In the following n-channel depletion-type MOSFET circuit, $I_{DSS} = 10$ mA, $V_{GS,off} = -4$ V, $R_D = 1$ kΩ, and $V_{DD} = +20$ V, find V_D and the gain V_{out}/V_{in}.

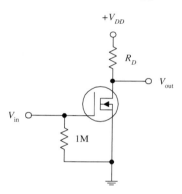

FIGURE 4.94

Applying Ohm's and Kirchhoff's laws, you can come up with the following expressions:

$$V_{DD} = V_{DS} + I_D R_D$$
$$V_{DD} = V_D + I_D R_D$$

where the last expression takes into account the grounded source terminal. (Note the 1-MΩ resistor. It is a self-biasing resistor and is used to compensate leakage currents and other parameters that can lead to MOSFET instability. The voltage drop across this resistor can be neglected because the gate leakage current is very small, typically in the nA or pA range.) Now, if you assume that there is no input voltage, you can say that $I_D = I_{DSS}$. This means that

$$V_D = V_{DD} - I_{DSS}R_D$$
$$= 20\ \text{V} - (10\ \text{mA})(1\ \text{k}\Omega) = 10\ \text{V}$$

To find the gain, use the following formula:

$$\text{Gain} = \frac{V_{out}}{V_{in}} = g_{m_0}R_D$$

where

$$g_{m_0} = \left|\frac{2I_{DSS}}{V_{GS,off}}\right| = \left|\frac{2(10\ \text{mA})}{-4V}\right| = 5000\ \mu\text{mhos}$$

Substituting g_{m_0} back into the gain formula, you get a resulting gain of 5.

PROBLEM 4

In the following n-channel enhancement-type MOSFET circuit, if $k = 1000$ μmhos/V, $V_{DD} = 20$ V, $V_{GS,th} = 2$ V, and $V_{GS} = 5$ V, what should the resistance of R_D be to center V_D to 10 V? Also, what is the gain for this circuit?

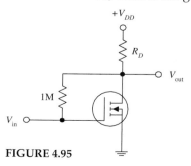

FIGURE 4.95

First, determine the drain current:

$$I_D = k\,(V_{GS} - V_{GS,th})^2$$
$$= (1000\ \mu\text{mhos/V})(5\ \text{V} - 2\ \text{V})^2 = 9\ \text{mA}$$

Next, to determine the size of R_D that is needed to set V_D to 10 V, use Ohm's law:

$$R_D = \frac{V_{DD} - V_D}{I_D} = \frac{20\text{V} - 10\text{V}}{9\ \text{mA}} = 1100\ \Omega$$

(The 1-MΩ resistor has the same role as the 1-MΩ resistor in the last example.)

To find the gain, first find the transconductance:

$g_m = 2k\,(V_{GS} - V_{GS,\text{th}}) = 2(1000\ \mu\text{mhos/V})(5\ \text{V} - 2\ \text{V}) = 6000\ \mu\text{mhos}$

Next, substitute g_m into the gain expression:

$$\text{Gain} = \frac{V_{\text{out}}}{V_{\text{in}}} = g_m R_D = 6.6$$

Important Things to Know about MOSFETs

MOSFETs may come with a fourth lead, called the *body terminal*. This terminal forms a diode junction with the drain-source channel. It must be held at a nonconducting voltage [say, to the source or to a point in a circuit that is more negative than the source (*n*-channel devices) or more positive than the source (*p*-channel devices)]. If the base is taken away from the source (for enhancement-type MOSFETs) and set to a different voltage than that of the source, the effect shifts the threshold voltage $V_{GS,\text{th}}$ by an amount equal to $\tfrac{1}{2}V_{BS}^{1/2}$ in the direction that tends to decrease drain current for a given V_{GS}. Some instances when shifting the threshold voltage becomes important are when leakage effects, capacitance effects, and signal polarities must be counterbalanced. The body terminal of a MOSFET is often used to determine the operating point of a MOSFET by applying an incremental ac signal to its gate.

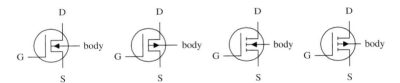

FIGURE 4.96

Damaging a MOSFET Is Easy

MOSFETs are extremely fragile. Their delicate gate-channel oxide insulators are subject to electron bombardment from statically charged objects. For example, it is possible for you to blow a hole through one of these insulators simply by walking across a carpet and then touching the gate of the MOSFET. The charge you pick up during your walk may be large enough to set yourself at a potential of a few thousand volts. Although the amount of current discharged during an interaction is not incredibly large, it does not matter; the oxide insulator is so thin (the gate-channel capacitance is so small, typically a few pF) that a small current can be fatal to a MOSFET. When installing MOSFETs, it is essential to eliminate all static electricity from your work area. In Chap. 7 you'll find guidelines for working with components subject to electrostatic discharge.

Kinds of MOSFETs

Like the other transistors, MOSFETs come in either metal can-like containers or plastic packages. High-power MOSFETs come with metal tabs that can be fastened to heat sinks. High/low MOSFET driver ICs also are available. These drivers (typically DIP in form) come with a number of independent MOSFETs built in and operate with logic signals.

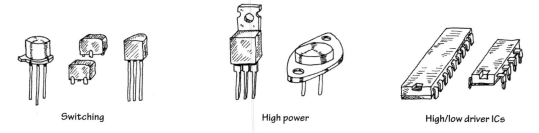

Switching   High power   High/low driver ICs

FIGURE 4.97

Things to consider when purchasing a MOSFET include breakdown voltages, $I_{D,\text{max}}$, $R_{DS(\text{on}),\text{max}}$, power dissipation, switching speed, and electrostatic discharge protection.

Applications

LIGHT DIMMER

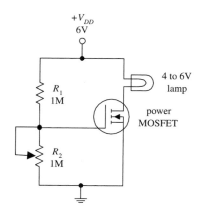

FIGURE 4.98

Here, an n-channel enhancement-type power MOSFET is used to control the current flow through a lamp. The voltage-divider resistor R_2 sets the gate voltage, which in turn sets the drain current through the lamp.

CURRENT SOURCE

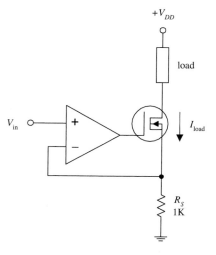

FIGURE 4.99

In the circuit shown here, an op amp is combined with an n-channel depletion-type MOSFET to make a highly reliable current source (less than 1 percent error). The MOSFET passes the load current, while the inverting input of the op amp samples the voltage across R_S and then compares it with the voltage applied to the noninverting input. If the drain current attempts to increase or decrease, the op amp will respond by decreasing or increasing its output, hence altering the MOSFETs gate voltage in the process. This in turn controls the load current. This op amp/MOSFET current source is more reliable than a simple bipolar transistor-driven source. The amount of leakage current is extremely small. The load current for this circuit is determined by applying Ohm's law (and applying the rules for op amps discussed in Chap. 8):

$$I_{\text{load}} = V_{\text{in}}/R_S$$

AMPLIFIERS

Common-source amplifier
(depletion MOSFET)

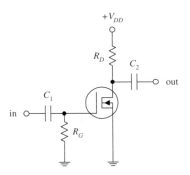

Source-follower amplifier
(depletion MOSFET)

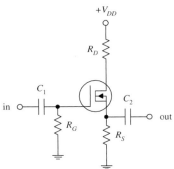

Common-source and source-follower ampli-fiers can be constructed using both deple-tion- and enhancement-type MOSFETs. The depletion-type amplifiers are similar to the JFET amplifiers discussed earlier, except that they have higher input impedances. The enhancement-type MOSFET amplifiers essentially perform the same operations as the depletion-type MOSFET amplifiers, but they require a voltage divider (as compared with a single resistor) to set the quiescent gate volt-age. Also, the output for the enhancement-type common-source MOSFET amplifier is inverted. The role of the resistors and capacitors with-in these circuits can be better understood by referring to the amplifier circuits discussed earlier.

Common-source amplifier
(enhancement MOSFET)

Source-follower amplifier
(enhancement MOSFET)

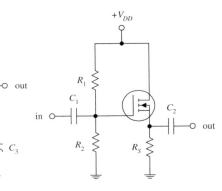

FIGURE 4.100

AUDIO AMPLIFIER

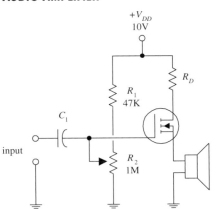

FIGURE 4.101

In this circuit, an *n*-channel enhancement-type MOSFET is used to amplify an audio signal generated by a high-impedance microphone and then uses the amplified signal to drive a speaker. C_1 acts as an ac coupling capacitor, and the R_2 voltage divider resistor acts to control the gain (the volume).

RELAY DRIVER (DIGITAL-TO-ANALOG CONVERSION)

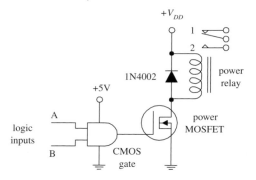

The circuit shown here uses an *n*-channel depletion-type MOSFET as an interface between a logic circuit and an analog circuit. In this example, an AND gate is used to drive a MOSFET into conduction, which in turn activates the relay. If inputs *A* and *B* are both high, the relay is switched to position 2. Any other combination (high/low, low/high, low/low) will put the relay into position 1. The MOSFET is a good choice to use as a digital-to-analog interface; its extremely high input resistance and low input current make it a good choice for powering high-voltage or high-current analog circuits without worrying about drawing current from the driving logic.

FIGURE 4.102

DIRECTION CONTROL OF A DC MOTOR

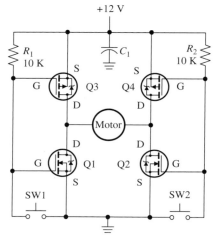

Q1, Q2, N-channel MOSFET IRF630
Q3, Q4, P-channel MOSFET IRF9630

FIGURE 4.103

A permanent magnet DC motor rotates either clockwise or counterclockwise, depending on polarity of the applied voltage across its terminals. A simple circuit that can be used to control the on/off state of a motor as well as direction of rotation is shown to the left. This circuit is an H-bridge motor-control circuit built with power MOSFETS. Transistors Q1 and Q2 are Nchannel MOSFETs, and Q3 and Q4 are P-channel MOSFETS. To turn on the motor as well as control the direction of rotation, switches SW1 and SW2 are used. Both switches are push button switches that are of the normal open variety. When SW1 is pressed, the voltage on gates Q1 and Q3 goes to zero, turning off Q1 and turning on Q3. This creates a current path from Q3, through the motor and through Q2. This causes the motor to turn clockwise. When SW1 is released, the motor turns off. When SW2 is pressed, Q2 turns off and Q4 turns on, creating a reverse current path through Q1, the motor and Q4, and the motor turns counterclockwise. To provide digital control, SW1 and SW2 can be replaced by transistors (or similar switching devices) that can be turned on and off using a microcontroller.

4.3.5 Insulated Gate Bipolar Transistors (IGBTs)

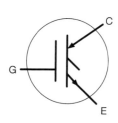

FIGURE 4.104

IGBTs are a hybrid of a MOSFET and a bipolar transistor. This is reflected in the IGBTs electronic symbol (see Fig. 4.104). This has the Gate terminal of a MOSFET and the Collector and Emitter terminals of a bipolar transistor. As you might have expect from this, the transistor is commonly used as a switch often at very high currents and voltages. The switch is voltage controlled like a MOSFET but has high current capabilities of a bipolar transistor.

IGBTs have found a niche in very high power applications such as electric vehicles, where modules are constructed from a number of IGBTs in parallel to achieve switching powers in the hundreds of amperes at high voltages. The very forgiving pulse capabilities have found application amongst hobbyists in solid-state Tesla coils.

4.3.6 Unijunction Transistors

Unijunction transistors (UJTs) are three-lead devices that act exclusively as electrically controlled switches (they are not used as amplifier controls). The basic operation of a UJT is relatively simple. When no potential difference exists between its emitter and either of

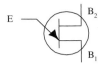

FIGURE 4.105

its base leads (B_1 or B_2), only a very small current flows from B_2 to B_1. However, if a sufficiently large positive *trigger voltage*—relative to its base leads—is applied to the emitter, a larger current flows from the emitter and combines with the small B_2-to-B_1 current, thus giving rise to a larger B_1 output current. Unlike the other transistors covered earlier—where the control leads (e.g., emitter, gate) provided little or no additional current—the UJT is just the opposite; its emitter current is the primary source of additional current.

How UJTs Work

A simple model of a UJT is shown in Fig. 4.105. It consists of a single bar of n-type semiconductor material with a p-type semiconductor "bump" in the middle. One end of the bar makes up the base 1 terminals, the other end the base 2 terminal, and the "bump" represents the emitter terminal. Below is a simple "how it works" explanation.

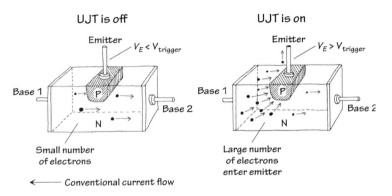

FIGURE 4.106

With no voltage applied to the emitter, only a relatively small number of electrons makes it through the n-region between base 1 and base 2. Normally, both connectors to bases 1 and 2 are resistive (each around a few thousand ohms).

When a sufficiently large voltage is applied to the emitter, the emitter-channel p-n junction is forward-biased (similar to forward-biasing a diode). This in turn allows a large number of base 1 electrons to exit through the emitter. Now, since conventional currents are defined to be flowing in the opposite direction of electron flow, you would say that a positive current flows from the emitter and combines with channel current to produce a larger base 1 output current.

Technical Info

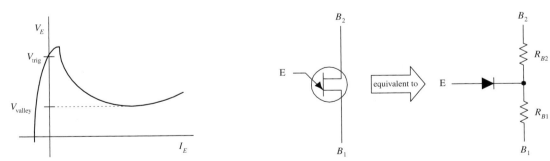

FIGURE 4.107

Figure 4.107 shows a typical V_E versus I_E graph of an UJT, as well as a UJT equivalent circuit. In terms of the UJT theory, if B_1 is grounded, a voltage applied to the emitter will have no effect (does not increase conductance from one base to another) until it exceeds a critical voltage, known as the *triggering voltage*. The triggering voltage is given by the following expression:

$$V_{trig} = \frac{R_{B1}}{R_{B1} + R_{B2}} V_{B2} = \eta \, V_{B2}$$

In this equation, R_{B1} and R_{B2} represent the inherent resistance within the region between each base terminal and the n-channel. When the emitter is open-circuited, the combined channel resistance is typically around a few thousand ohms, where R_{B1} is somewhat larger than R_{B2}. Once the trigger voltage is reached, the p-n junction is forward-biased (the diode in the equivalent circuit begins to conduct), and current then flows from the emitter into the channel. But how do we determine R_{B1} and R_{B2}? Will the manufacturers give you these resistances? They most likely will not. Instead, they typically give you a parameter called the *intrinsic standoff ratio* η. This intrinsic standoff ratio is equal to the $R_{B1}/(R_{B1} + R_{B2})$ term in the preceding expression, provided the emitter is not conducting. The value of η is between 0 to 1, but typically it hangs out at a value around 0.5.

TYPICAL APPLICATION (RELAXATION OSCILLATOR)

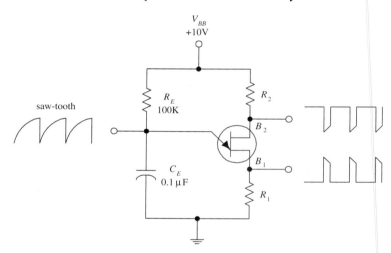

FIGURE 4.108

Most frequently, UJTs are used in oscillator circuits. Here, a UJT, along with some resistors and a capacitor, makes up a relaxation oscillator that is capable of generating three different output waveforms. During operation, at one instant in time, C_E charges through R_E until the voltage present on the emitter reaches the triggering voltage. Once the triggering voltage is exceeded, the E-to-B_1 conductivity increases sharply, which allows current to pass from the capacitor-emitter region, through the emitter-base 1 region, and then to ground. C_E suddenly loses its charges, and the emitter voltage suddenly falls below the triggering voltage. The cycle then repeats itself. The resulting waveforms generated during this process are shown in the figure. The frequency of oscillation is determined by the RC charge-discharge period and is given by

$$f = \frac{1}{R_E C_E \ln[1/(1-\eta)]}$$

For example, if $R_E = 100$ kΩ, $C_E = 0.1$ µF, and $\eta = 0.61$, then $f = 106$ Hz.

Kinds of UJTs

BASIC SWITCHING

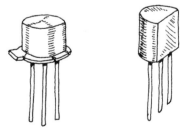

FIGURE 4.109

These UJTs are used in oscillatory, timing, and level-detecting circuits. Typical maximum ratings include 50 mA for I_E, 35 to 55 V for the inter-base voltage (V_{BB}), and 300 to 500 mW for power dissipation.

PROGRAMMABLE (PUTs)

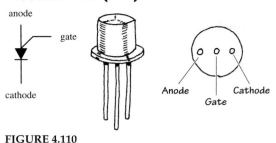

FIGURE 4.110

A PUT is similar to a UJT, except that R_{BB}, I_V (valley current level), I_P (peak current level), and η (intrinsic standoff ratio) can be programmed by means of an external voltage divider. Being able to alter these parameters is often essential in order to eliminate circuit instability. The electronic symbol for a PUT looks radically different when compared with a UJT (see figure). The lead names are also different; there is a gate lead, a cathode lead, and an anode lead. PUTs are used to construct timer circuits, high-gain phase-control circuits, and oscillator circuits. We have included a simple PUT application in the applications section that follows.

Applications

TIMER/RELAY DRIVER

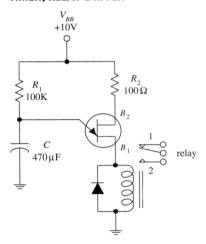

FIGURE 4.111

The circuit here causes a relay to throw its switch from one position to another in a repetitive manner. The positive supply voltage charges up the capacitor. When the voltage across the capacitor reaches the UJT's triggering voltage, the UJT goes into conduction. This causes the relay to throw its switch to position 2. When the capacitor's charge runs out, the voltage falls below the triggering voltage, and the UJT turns off. The relay then switches to position 1. R_1 controls the charging rate of the capacitor, and the size of the capacitor determines the amount of voltage used to trigger the UJT. C also determines the charge rate.

RAMP GENERATOR WITH AMPLIFIER

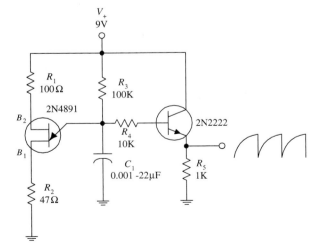

FIGURE 4.112   Here, a UJT is combined with a few resistors, a bipolar transistor, and a capacitor to make an oscillatory sawtooth generator that has controlled amplification (set by the bipolar transistor). Like the preceding oscillators, C_1 and R_3 set the frequency. The bipolar transistor samples the voltage on the capacitor and outputs a ramp or sawtooth waveform.

PUT RELAXATION OSCILLATOR

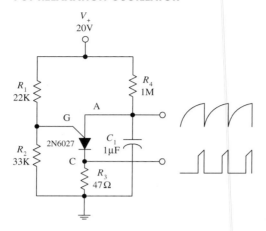

FIGURE 4.113

Here, a PUT is programmed by R_1 and R_2 to set the desired triggering voltage and anode current. These two resistors form a voltage divider that sets the gate voltage V_G (terminal used to turn PUT on or off). For the PUT to conduct, the anode voltage must exceed the gate voltage by at least 0.7 V. At a moment when the capacitor is discharged, the gate is reverse-biased, and the PUT is turned off. Over time, the capacitor begins charging through R_4, and when enough charge is collected, a large enough voltage will be present to forward-bias the gate. This then turns the PUT on (i.e., if the anode current I_A exceeds the peak current I_P). Next, the capacitor discharges through the PUT and through R_3. (*Note:* When a PUT is conducting, the voltage from the anode to the cathode is about 1 V.) As the capacitor reaches full discharge, the anode current decreases and finally stops when the gate no longer has a sufficient voltage applied to it. After that, the charging begins again, and the cycle repeats itself, over and over again. By tapping the circuit at the gate and source terminals, you can output both a spiked and sawtooth wave pattern.

CALCULATIONS
PUT begins to conduct when

$$V_A = V_G + 0.7 \text{ V}$$

where V_G is set by the voltage divider:

$$V_G = \frac{R_2}{R_2 + R_1} V_+$$

When V_A is reached, the anode current becomes:

$$I_A = \frac{V_+ - V_A}{R_1 + R_2}$$

4.4 Thyristors

4.4.1 Introduction

Thyristors are two- to four-lead semiconductor devices that act exclusively as switches—they are not used to amplify signals, like transistors. A three-lead thyristor uses a small current/voltage applied to one of its leads to control a much larger current flow through its other two leads. A two-lead thyristor, on the other hand, does not use a control lead but instead is designed to switch on when the voltage across its leads reaches a specific level, known as the *breakdown voltage*. Below this breakdown voltage, the two-lead thyristor remains off.

You may be wondering at this point, Why not simply use a transistor instead of a thyristor for switching applications? Well, you could—often transistors are indeed used as switches—but compared with thyristors, they are trickier to use because they require exacting control currents/voltages to operate properly. If the control current/voltage is not exact, the transistor may lay in between on and off states. And according to common sense, a switch that lies in between states is not a good switch. Thyristors, on the other hand, are not designed to operate in between states. For these devices, it is all or nothing—they are either on or off.

In terms of applications, thyristors are used in speed-control circuits, power-switching circuits, relay-replacement circuits, low-cost timer circuits, oscillator circuits, level-detector circuits, phase-control circuits, inverter circuits, chopper circuits, logic circuits, light-dimming circuits, motor speed-control circuits, etc.

TABLE 4.6   Major Kinds of Thyristors

TYPE	SYMBOL	MODE OF OPERATION
Silicon-controlled rectifier (SCR)		Normally off, but when a small current enters its gate (G), it turns on. Even when the gate current is removed, the SCR remains on. To turn it off, the anode-to-cathode current flow must be removed, or the anode must be set to a more negative voltage than the cathode. Current flows in only one direction, from anode (A) to cathode (C).
Silicon-controlled switch (SCS)		Similar to an SCR, but it can be made to turn off by applying a positive voltage pulse to a four-lead, called the anode gate. This device also can be made to trigger on when a negative voltage is applied to the anode-gate lead. Current flows in one direction, from anode (A) to cathode (C).
Triac		Similar to an SCR, but it can switch in both directions, meaning it can switch ac as well as dc currents. A triac remains on only when the gate is receiving current, and it turns off when the gate current is removed. Current flows in both directions, through MT1 and MT2.
Four-layer diode		It has only two leads. When placed between two points in a circuit, it acts as a voltage-sensitive switch. As long as the voltage difference across its leads is below a specific breakdown voltage, it remains off. However, when the voltage difference exceeds the breakdown point, it turns on. Conducts in one direction, from anode (A) to cathode (C).
Diac		Similar to the four-layer diode but can conduct in both directions. Designed to switch either ac or dc.

Table 4.6 provides an overview of the major kinds of thyristors. When you see the phrase *turns it on*, this means a conductive path is made between the two conducting leads [e.g., anode (A) to cathode (C), MT1 to MT2]. *Normally off* refers to the condition when no voltage is applied to the gate (the gate is open-circuited). We will present a closer look at these thyristors in the subsections that follow.

4.4.2   Silicon-Controlled Rectifiers

SCRs are three-lead semiconductor devices that act as electrically controlled switches. When a specific positive trigger voltage/current is applied to the SCR's gate lead (G), a conductive channel forms between the anode (A) and the cathode (C) leads. Current flows in only one direction through the SCR, from anode to cathode (like a diode).

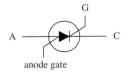

FIGURE 4.114

Another unique feature of an SCR, besides its current-controlled switching, has to do with its conduction state after the gate current is removed. After an SCR is triggered into conduction, removing the gate current has no effect. That is, the SCR will remain on even when the gate current/voltage is removed. The only way to turn the device off is to remove the anode-to-cathode current or to reverse the anode and cathode polarities.

In terms of applications, SCRs are used in switching circuits, phase-control circuits, inverting circuits, clipper circuits, and relay-control circuits, to name a few.

How SCRs Work

An SCR is essentially just an *npn* and a *pnp* bipolar transistor sandwiched together, as shown in Fig. 4.115. The bipolar transistor equivalent circuit works well in describing how the SCR works.

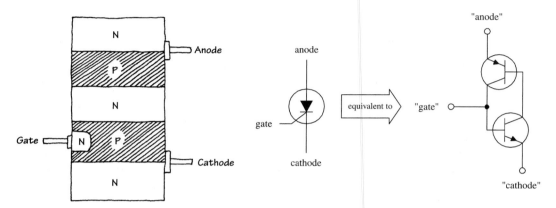

FIGURE 4.115

THE SCR IS OFF

Using the bipolar equivalent circuit, if the gate is not set to a specific positive voltage needed to turn the *npn* transistor on, the *pnp* transistor will not be able to "sink" current from its own base. This means that neither transistor will conduct, and hence current will not flow from anode to cathode.

THE SCR IS ON

If a positive voltage is applied to the gate, the *npn* transistor's base is properly biased, and it turns on. Once on, the *pnp* transistor's base can now "sink" current though the *npn* transistor's collector—which is what a *pnp* transistor needs in order to turn on. Since both transistors are on, current flows freely between anode and cathode. Notice that the SCR will remain on even after the gate current is removed. This—according to the bipolar equivalent circuit—results from the fact that both transistors are in a state of conduction when the gate current is removed. Because current is already in motion through the *pnp* transistors base, there is no reason for the transistors to turn off.

Basic SCR Applications

BASIC LATCHING SWITCH

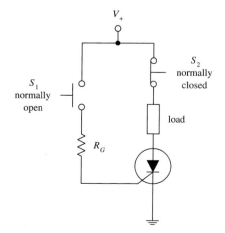

Here, an SCR is used to construct a simple latching circuit. S_1 is a momentary contact, normally open pushbutton switch, while S_2 is a momentary contact, normally closed pushbutton switch. When S_1 is pushed in and released, a small pulse of current enters the gate of the SCR, thus turning it on. Current will then flow through the load. The load will continue to receive current until the moment S_2 is pushed, at which time the SCR turns off. The gate resistor acts to set the SCR's triggering voltage/current. We'll take a closer look at the triggering specifications in a second.

FIGURE 4.116

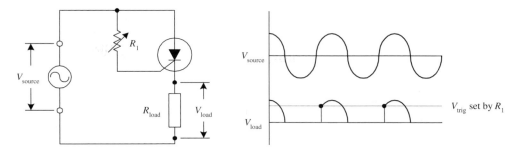

FIGURE 4.117   Here, an SCR is used to rectify a sinusoidal signal that is to be used to power a load. When a sinusoidal waveform is applied to the gate, the SCR turns on when the anode and gate receive the positive going portion of the waveform (provided the triggering voltage is exceeded). Once the SCR is on, the waveform passes through the anode and cathode, powering the load in the process. During the negative going portion of the waveform, the SCR acts like a reverse-biased diode; the SCR turns off. Increasing R_1 has the effect of lowering the current/voltage supplied to the SCR's gate. This in turn causes a lag in anode-to-cathode conduction time. As a result, the fraction of the cycle over which the device conducts can be controlled (see graph), which means that the average power dissipated by R_{load} can be adjusted. The advantage of using an SCR over a simple series variable resistor to control current flow is that essentially no power is lost to resistive heating.

DC MOTOR SPEED CONTROLLER

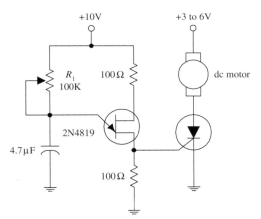

FIGURE 4.118

An SCR along with a few resistors, a capacitor, and a UJT can be connected together to make a variable-speed control circuit used to run a dc motor. The UJT, the capacitor, and the resistors make up an oscillator that supplies an ac voltage to the SCR's gate. When the voltage at the gate exceeds the SCR's triggering voltage, the SCR turns on, thus allowing current to flow through the motor. Changing the resistance of R_1 changes the frequency of the oscillator and hence determines the number of times the SCR's gate is triggered over time, which in turn controls the speed of the motor. (The motor appears to turn continuously, even though it is receiving a series of on/off pulses. The number of on cycles averaged over time determines the speed of the motor.) Using such a circuit over a simple series variable resistor to control the speed of the motor wastes less energy.

Kinds of SCRs

Some SCRs are designed specifically for phase-control applications, while others are designed for high-speed switching applications. Perhaps the most distinguishing feature of SCRs is the amount of current they can handle. Low-current SCRs typically come with maximum current/voltage ratings approximately no bigger than 1 A/100 V. Medium-current SCRs, on the other hand, come with maximum current/voltage ratings typically no bigger than 10 A/100 V. The maximum ratings for high-current SCRs may be several thousand amps at several thousand volts. Low-current SCRs come in plastic or metal can-like packages, while medium and high-current SCRs come with heat sinks built in.

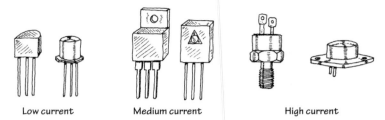

FIGURE 4.119

Low current   Medium current   High current

Technical Stuff

Here are some common terms used by the manufacturers to describe their SCRs:

V_T *On state-voltage.* The anode-to-cathode voltage present when the SCR is on.

I_{GT} *Gate trigger current.* The minimum gate current needed to switch the SCR on.

V_{GT} *Gate trigger voltage.* The minimum gate voltage required to trigger the gate trigger current.

I_H *Holding current.* The minimum current through the anode-to-cathode terminal required to maintain the SCR's on state.

P_{GM} *Peak gate power dissipation.* The maximum power that may be dissipated between the gate and the cathode region.

V_{DRM} *Repetitive peak off-state voltage.* The maximum instantaneous value of the off-state voltage that occurs across an SCR, including all repetitive transient voltages but excluding all nonrepetitive transient voltages.

I_{DRM} *Repetitive peak off-state current.* The maximum instantaneous value of the off-state current that results from the application of repetitive peak off-state voltage.

V_{RRM} *Repetitive peak reverse voltage.* The maximum instantaneous value of the reverse voltage that occurs across an SCR, including all repetitive transient voltages but excluding all nonrepetitive transient voltages.

I_{RRM} *Repetitive peak reverse current.* Maximum instantaneous value of the reverse current that results from the application of repetitive peak reverse voltage.

Here's a sample section of an SCR specifications table to give you an idea of what to expect (Table 4.7).

TABLE 4.7   **Sample Section of an SCR Specifications Table**

MNFR #	V_{DRM} (MIN) (V)	I_{DRM} (MAX) (mA)	I_{RRM} (MAX) (mA)	V_T (V)	I_{GT} (TYP/MAX) (mA)	V_{GT} (TYP/MAX) (V)	I_H (TYP/MAX) (mA)	P_{GM} (W)
2N6401	100	2.0	2.0	1.7	5.0/30	0.7/1.5	6.0/40	5

4.4.3 Silicon-Controlled Switches

A silicon-controller switch (SCS) is a device similar to an SCR, but is designed for situations requiring increased control, triggering sensitivity, and firing predictability. For example, the typical turn-off time for an SCS is from 1 to 10 microseconds as opposed to 5 to 30 microseconds for an SCR. Unlike an SCR, an SCS has lower power, current and voltage ratings, typically with a max anode current from 100 mA to 300 mA and a power dissipation from 100 to 500 mW. Unlike an SCR, a SCS can also switch OFF when a positive voltage/input current is applied to an extra anode gate lead. The SCS can also be triggered into conduction when a negative voltage/output current is applied to that same lead. The figure below shows the schematic symbol for an SCS.

Cathode Gate (G1)

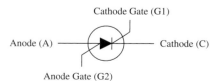

Anode (A) ——————— Cathode (C)

Anode Gate (G2)

FIGURE 4.120

SCS are used in practically any circuit that needs a switch that turns on and off through two distinct control pulses. They are found in power-switching circuits, logic circuits, lamp drivers, voltage sensors, pulse generators, etc.

How an SCS Works

Figure 4.121a shows a basic four-layer, three-junction P-N-P-N silicon model of an SCS with four electrodes, namely the cathode (C), cathode gate (G1), anode gate (G2), and anode (A). An equivalent circuit of the SCS can be modeled by the back-to-back bipolar transistor network shown in Fig. 4.121c. Using the two-transistor equivalent circuit, when a negative pulse is applied to the anode gate (G2), transistor Q1 switches ON. Q1 supplies base current to transistor Q2, and both transistors switch ON. Likewise, a positive pulse at the cathode gate G1 can switch the device on. Because the SCS uses only small currents, it can be switched off by an appropriate polarity pulse at one of the gates. At the cathode gate, a negative pulse is required to switch the device off, while at the anode gate a positive pulse is needed.

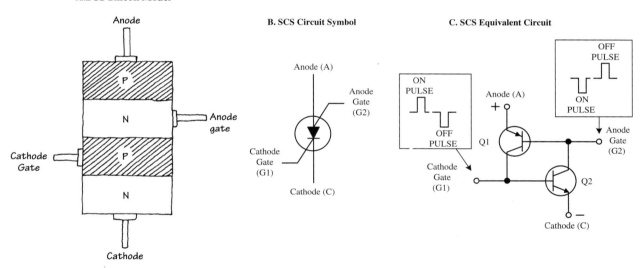

A. SCS Silicon Model

B. SCS Circuit Symbol

C. SCS Equivalent Circuit

FIGURE 4.121

Specifications

When buying an SCS, make sure to select a device that has the proper breakdown voltage, current, and power-dissipation ratings. A typical specification table will provide the following ratings: BV_{CB}, BV_{EB}, BV_{CE}, I_E, I_C, I_H (holding current), and P_D (power dissipation). Here we have assumed the alternate lead name designations.

4.4.4 Triacs

Triacs are devices similar to SCRs—they act as electrically controlled switches—but unlike SCRs, they are designed to pass current in both directions, therefore making them suitable for ac applications. Triacs come with three leads, a gate lead and two conducting leads called *MT1* and *MT2*. When no current/voltage is applied to the gate, the triac remains off. However, if a specific trigger voltage is applied to the gate, the device turns on. To turn the triac off, the gate current/voltage is removed.

FIGURE 4.122

Triacs are used in ac motor control circuits, light-dimming circuits, phase-control circuits, and other ac power-switching circuits. They are often used as substitutes for mechanical relays.

How a Triac Works

Figure 4.123 shows a simple *n*-type/*p*-type silicon model of a triac. This device resembles two SCRs placed in reverse parallel with each other. The equivalent circuit describes how the triac works.

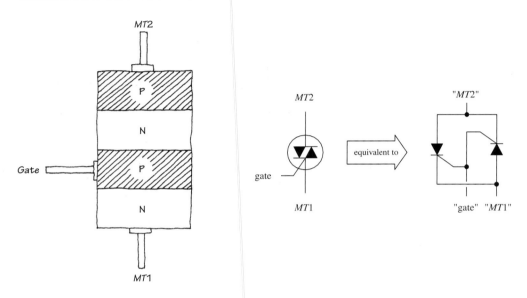

FIGURE 4.123

TRIAC IS OFF

Using the SCR equivalent circuit, when no current/voltage is applied to the gate lead, neither of the SCRs' gates receives a triggering voltage; hence current cannot flow in either direction through *MT*1 and *MT*2.

TRIAC IS ON

When a specific positive triggering current/voltage is applied to the gate, both SCRs receive sufficient voltage to trigger on. Once both SCRs are on, current can flow in either direction through *MT*1 to *MT*2 or from *MT*2 to *MT*1. If the gate voltage is removed, both SCRs will turn off when the ac waveform applied across *MT*1 and *MT*2 crosses zero volts.

Basic Applications

SIMPLE SWITCH

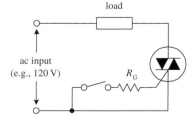

FIGURE 4.124

Here is a simple circuit showing how a triac acts to permit or prevent current from reaching a load. When the mechanical switch is open, no current enters the triac's gate; the triac remains off, and no current passes through the load. When the switch is closed, a small current slips through R_G, triggering the triac into conduction (provided the gate current and voltage exceed the triggering requirements of the triac). The alternating current can now flow through the triac and power the load. If the switch is open again, the triac turns off, and current is prevented from flowing through the load.

DUAL RECTIFIER

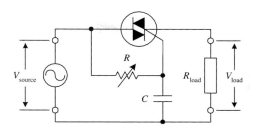 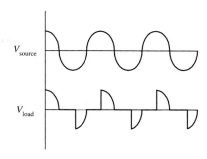

FIGURE 4.125   A triac along with a variable resistor and a capacitor can be used to construct an adjustable full-wave rectifier. The resistance R of the variable resistor sets the time at which the triac will trigger on. Increasing R causes the triac to trigger at a later time and therefore results in a larger amount of clipping (see graph). The size of C also determines the amount of clipping that will take place. (The capacitor acts to store charge until the voltage across its terminals reaches the triac's triggering voltage. At that time, the capacitor will dump its charge.) The reason why the capacitor can introduce additional clipping results from the fact that the capacitor may cause the voltage at the gate to lag the $MT2$-to-$MT1$ voltage (e.g., even if the gate receives sufficient triggering voltage, the $MT2$-to-$MT1$ voltage may be crossing zero volts). Overall, more clipping results in less power supplied to the load. Using this circuit over a simple series variable resistor connected to a load saves power. A simple series variable resistor gobbles up energy. This circuit, however, supplies energy-efficient pulses of current.

AC LIGHT DIMMER

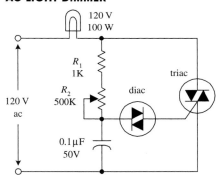

FIGURE 4.126

This circuit is used in many household dimmer switches. The diac—described in the next section—acts to ensure accurate triac triggering. (The diac acts as a switch that passes current when the voltage across its leads reaches a set breakdown value. Once the breakdown voltage is reached, the diac releases a pulse of current.) In this circuit, at one moment the diac is off. However, when enough current passes through the resistors and charges up the capacitor to a voltage that exceeds the diac's triggering voltage, the diac suddenly passes all the capacitor's charge into the triac's gate. This in turn causes the triac to turn on and thus turns the lamp on. After the capacitor is discharged to a voltage below the breakdown voltage of the diac, the diac turns off, the triac turns off, and the lamp turns off. Then the cycle repeats itself, over and over again. Now, it appears that the lamp is on (or dimmed to some degree) because the on/off cycles are occurring very quickly. The lamp's brightness is controlled by R_2.

AC MOTOR CONTROLLER

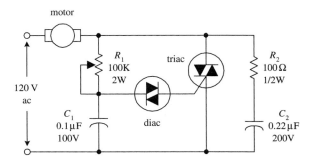

FIGURE 4.127

This circuit has the same basic structure as the light dimmer circuit, with the exception of the transient suppressor section (R_2C_2). The speed of the motor is adjusted by varying R_1.

Kinds of Triacs

Triacs come in low-current and medium-current forms. Low-current triacs typically come with maximum current/voltage ratings no bigger than 1 A/(several hundred volts). Medium-current triacs typically come with maximum current/voltage rating of up to 40 A/(few thousand volts). Triacs cannot switch as much current as high-current SCRs.

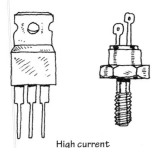

Low current   High current

FIGURE 4.128

Technical Stuff

Here are some common terms used by the manufacturers to describe their triacs:

$I_{TRMS,max}$ *RMS on-state current.* **The maximum allowable $MT1$-to-$MT2$ current**

$I_{GT,max}$ *DC gate trigger current.* **The minimum dc gate current needed to switch the triac on**

$V_{GT,max}$ *DC gate trigger voltage.* **The minimum dc gate voltage required to trigger the gate trigger current**

I_H *DC holding current.* **The minimum $MT1$-to-$MT2$ dc current needed to keep the triac in its on state**

P_{GM} *Peak gate power dissipation.* **The maximum gate-to-$MT1$ power dissipation**

I_{surge} *Surge current.* **Maximum allowable surge current**

Here's a sample section of a triac specifications table to give you an idea of what to expect (Table 4.8).

TABLE 4.8   Sample Section of a Triac Specifications Table

MNFR #	$I_{T,RMS}$ MAX. (A)	I_{GT} MAX. (mA)	V_{GT} MAX. (V)	V_{FON} (V)	I_H (mA)	I_{SURGE} (A)
NTE5600	4.0	30	2.5	2.0	30	30

4.4.5   Four-Layer Diodes and Diacs

Four-layer diodes and diacs are two-lead thyristors that switch current without the need of a gate signal. Instead, these devices turn on when the voltage across their leads reaches a particular *breakdown voltage* (or *breakover voltage*). A four-layer diode resembles an SCR without a gate lead, and it is designed to switch only dc. A diac resembles a *pnp* transistor without a base lead, and it is designed to switch only ac.

four-layer diode   diac

anode ———————— cathode

FIGURE 4.129

Four-layer diodes and diacs are used most frequently to help SCRs and triacs trigger properly. For example, by using a diac to trigger a triac's gate, as shown in Fig. 4.105a, you can avoid unreliable triac triggering caused by device instability resulting from temperature variations, etc. When the voltage across the diac reaches the breakdown voltage, the diac will suddenly release a "convincing" pulse of current into the triac's gate.

FULL-WAVE PHASE
CONTROL CIRCUIT

CIRCUIT USED TO MEASURE DIAC
CHARACTERISTICS

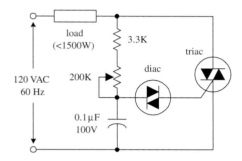
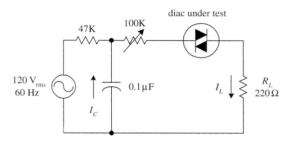

FIGURE 4.130

The circuit in Fig. 4.130 *right* is used to measure diac characteristics. The 100-kΩ variable resistor is adjusted until the diac fires once for every half-cycle.

Specifications

Here's a typical portion of a specifications table for a diac (Table 4.9).

TABLE 4.9 Sample Section of a Diac Specifications Table

MNFR #	V_{BO} (V)	I_{BO} MAX (μA)	I_{PULSE} (A)	V_{SWITCH} (V)	P_D (mW)
NTE6411	40	100	2	6	250

Here, V_{BO} is the breakover voltage, I_{BO} is the breakover current, I_{pulse} is the maximum peak pulse current, V_{switch} is the maximum switching voltage, and P_D is the maximum power dissipation.

4.5 Transient Voltage Suppressors

There are numerous devices that can be used to stomp out unwanted transients. Earlier on, we saw how a decoupling capacitor could absorb supply line fluctuations, and we also saw how diodes could clip transient spikes caused by inductive switching action. These devices work fine for such low-power applications, but there are times when transients get so large and energetic that a more robust device is required. Here we'll take a look at various transient suppressor devices, such as TVSs, varistors, multilayer varistors, Surgectors, and PolySwitches. But before we do that, here's a little lecture on transients.

4.5.1 Lecture on Transients

Transients are momentary surges or spikes in voltage or current that can wreak havoc within circuits. The peak voltage of a transient can be as small as a few millivolts or as large as several thousand volts, with a duration lasting from a few nanoseconds to more than 100 ms, depending on their origin. In some cases, the transients are repetitive, recurring in a cyclic manner, as in the case of an inductive ringing transient caused by faulty wiring of a motor.

Transients are generated both internally within a circuit and externally—where they enter the circuit via power input lines, signal input/output lines, data lines, and other wires entering and leaving the circuit's chassis. Internal transients, the predominant of the two, can result from inductive load switching, transistor/logic IC switching, arcing effect, and faulty wiring, to name a few.

In the case of inductive loads, such as motors, relay coils, solenoid coils, and transformers, the sudden switching off of these devices will cause the inductive component within the device to suddenly dump its stored energy into the supply line, creating a voltage spike—recall the inductor equation $V = LdI/dt$. In many cases, these induced voltages can exceed 1000 V, lasting anywhere from 50 ns to more than 100 ms. Any transistor or logic driver ICs as well as circuits that use the same supply line will suffer, either by getting zapped with the transient spike or suffering from erratic behavior due to propagation of the transient along the power line. (Power lines, or rails, are not perfect conductors and don't have zero output impedance.)

Switching of TTL and CMOS circuits can also result in transient current spikes of a much smaller threat, yet enough to cause erratic behavior. For example, when the output transistors of a TTL gate switch on, a sudden surge in current is drawn from the supply line. This surge is often quick enough that the supply rail or PCB trace will dip in voltage (due to the fact that a conductor has built-in impedance). All circuits connected to the rail will feel this voltage dip, and the resulting consequences lead to oscillation or some sort of instability that can cause distortion or garble digital logic levels.

Arcing is another transient generator that comes from a number of sources, such as faulty contacts in breakers, switches, and connectors, where arcs jump between the gaps. When electrons jump the gap, the voltage suddenly rises, usually resulting in an oscillatory ringing transient. Faulty connections and poor grounding can also result in transients. For example, motors with faulty windings or insulation can generate a continuous stream of transients exceeding a few hundred volts. Poor electrical wiring practices can also aggravate load-switching transients.

Transients can also attack circuits from external sources through power input lines, signal input and output lines, data lines, and any other wire coming into or going out of a chassis containing the electronics. One cause of external transients is a result of induced voltages onto lines (power, telephone, distributed computer systems, etc.) due to lightning strikes near the lines or the switching of loads, capacitor banks, and so on, at the power utility. External transients may also enter the power line to a circuit due to inductive switching that occurs within a home, such as turning on a hair dryer, microwave, or washing machine. Usually the transient is consumed by other parallel loads, so the effects aren't as pronounced. For valuable electrical equipment, such as computers, monitors, printers, fax machines, phones, and modems, it is a good idea to use a transient power surge/battery backup protector, with a phone line, too, which will handle the surges and dips in the power and signal lines.

Electrostatic discharge (ESD) is another common form of external transient that can do damage to sensitive equipment and ICs. It usually enters a system through the touch of a fingertip or handheld metal tool. Static electricity that is humidity-dependent can generate low-current transients up to 40,000 V. Systems that are interconnected with long wires, such as telephones and distributed computer systems are efficient collectors of radiated lightning energy. Close-proximity strikes can induce voltages of 300 V or more on signal lines.

Transients are to be avoided; they can cause electronics to operate erratically, perhaps locking up or producing garbled results. They can zap sensitive integrated circuits, causing them to fail immediately or sometime down the road. Today's microchips are denser than older chips and a transient voltage can literally melt, weld, pit, and burn them, causing temporary or permanent malfunctions to occur. They might also be the cause for decreased efficiency—say, a motor running at higher temperatures due to transients, which interrupt normal timing of the motor and result in microjogging. This produces motor vibration, noise, and excessive heat.

4.5.2 Devices Used to Suppress Transients

There are several devices that can be used when designing circuits to limit the harmful effects of transients. Table 4.10 provides an overview of the most popular devices.

TABLE 4.10

DEVICE TYPE	SYMBOL	APPLICATIONS	ADVANTAGES	DISADVANTAGES
Bypass capacitor Logic: 0.01–0.22 µF Power: 0.1 µF and up		Used for low-power applications, such as RC snubbers and decoupling of digital logic rails to provide clean power	Low cost, available, simple to apply, fast action, bipolar	Uneven suppression, may fail unpredictably, high capacity
Zener diodes		Diversion/clamping in low-energy circuits running at high frequencies (e.g., high-speed data lines)	Low cost, fast, calibrated clamping voltage, easy to use, standard ratings, bidirectional	Low energy handling, tend to fail open (which can hurt circuit); actually used more for regulation than transients
Transient voltage suppressor diodes (TVS)	 Unidirectional Bidirectional	Diversion/clamping in low-voltage, low-energy systems, modest frequency	Fast, calibrated low clamping voltage, available, easy to use; fails short-circuited	High capacitance limits frequency, low energy, more expensive than zeners or MOVs

TABLE 4.10 (*Continued*)

Metal oxide varistors (MOVs)		Diversion/clamping in most low to moderate frequency circuits at all voltage and current levels	Low cost, fast, available, calibrated clamping voltage, easy to use, standard ratings, and bidirectional; handles more total power than TVS; fails short-circuited	Moderate to high capacitance limits high frequency performance
Multilayer varistor (MLTV)		Diversion/clamping in low voltage (3–70 V) systems with modest frequencies	Fast, compact, high energy, low calibrated voltage bidirectional, surface mount	More expensive than zeners or MOVs, high capacitance limits frequency
Surgectors	Surgector	Diversion (crowbar) for moderate to high energy and frequency circuits and data lines	High speed/moderate energy, sharp clamp voltage, moderate cost	Cost more than other methods, exhibit follow-on current
Avalanche diode	Cathode Anode (-) (+)	Low-voltage, high-speed logic protection	Very fast (sub-nanosecond response), low shunt capacitance (50 pF)	Low surge capability
Gas discharge and spark gap TVSs		Diversion (crowbar) for very high-energy/voltage applications	Very high-energy capability—upward of 20,000 A in some cases; leakage current is almost nonexistent (within the pA range)	Cost more than other methods, slow response time
PolySwitches		Overcurrent protection for speakers, motors, power supplies, battery packs, etc.	Low cost, easy to use, overcurrent protection	Requires a cooling-down period to reset

Transient Voltage Suppressor Diodes (TVSs)

Transient voltage suppressor diodes (TVSs) are popular semiconductor devices used to instantly clamp transient voltages and currents (electrostatic discharge, inductive switching kickback, induced lightning surges, etc.) to safe levels before they can do damage to a circuit. Earlier, in the section on diodes, you saw how standard diodes and zener diodes could be used for transient suppression. Though standard diodes and zener diodes can be used for transient protection, they are actually designed for rectification and voltage regulation and are not as reliable or robust as TVSs.

TVSs come in both unipolar (unidirectional) and bipolar (bidirectional) types. The unipolar TVS breaks down in one direction (current flows in the opposite direction of the arrow—like a zener diode) when its specified breakdown voltage V_{BR} is exceeded. The bipolar TVS, unlike the unipolar TVS, can handle transients in either direction, when the applied voltage across it exceeds its breakdown voltage. See Fig. 4.131.

In terms of design, the TVS should be invisible until a transient occurs. Electrical parameters such as breakdown voltage, standby current (leakage current), and capacitance should have no effect on normal circuit performance. The TVS's V_{BR} is typically 10 percent above V_{RWM}, which approximates the circuit operating voltage to

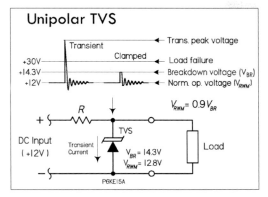

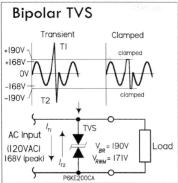

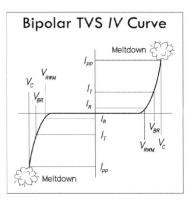

FIGURE 4.131

IMPORTANT TVS SPECIFICATIONS

Reverse stand-off voltage (V_{RWM}): Also called *working voltage*, represents the maximum rated dc operating voltage of TVS. At this point, the device will appear as a high impedance to the protected circuit. Discrete devices are available with V_{RWM}s from 2.8 to 440 V.

Maximum breakdown voltage (V_{BR}): The point where TVS begins to conduct and become a low-impedance path for a transient. Discrete devices are available with V_{BR} from 5.3 to 484 V. The breakdown voltage is measured at a test current I_T, typically 1 mA or 10 mA. V_{BR} is about 10 percent greater than V_{RWM}.

Maximum peak current (I_{PP}): The maximum permissible surge current that the device can withstand before frying.

Leakage current (I_R): The maximum leakage current measured at the working voltage.

Maximum clamping voltage (V_C): The maximum clamping voltage during the specified peak impulse current I_{PP}. Typically 35 to 40 percent higher than V_{BR} (or 60 percent higher than V_{RWM}).

Capacitance (C_J): Internal capacitance of TVS, which may become a factor in high-speed data circuits.

---

limit standby current and to allow for variations in V_{BR} caused by the temperature coefficient of the TVS. (In catalogs, they give you both—$V_{\text{BR}}/V_{\text{RWM}}$: 12.4 V/11.1 V, 15.2 V/13.6 V, 190 V/171 V, etc.) V_{RWM} should be equal to, or slightly greater than, the normal operating voltage of the protected circuit. When a transient occurs, the TVS clamps instantly to limit the spike voltage to a safe voltage level, while diverting current from the protected circuit. V_C should be, of course, less than the maximum voltage the protected circuit can handle. Note that in ac circuits, you should use the peak voltage (V_{peak}) values, not the RMS values for selecting V_{RWM} and V_{BR} ($V_{\text{peak}} = 1.4\ V_{\text{RMS}} \leq V_{\text{RWM}}$). Also, make sure to choose a TVS that can handle the maximum expected transient pulse current. Figure 4.132 shows various TVS applications.

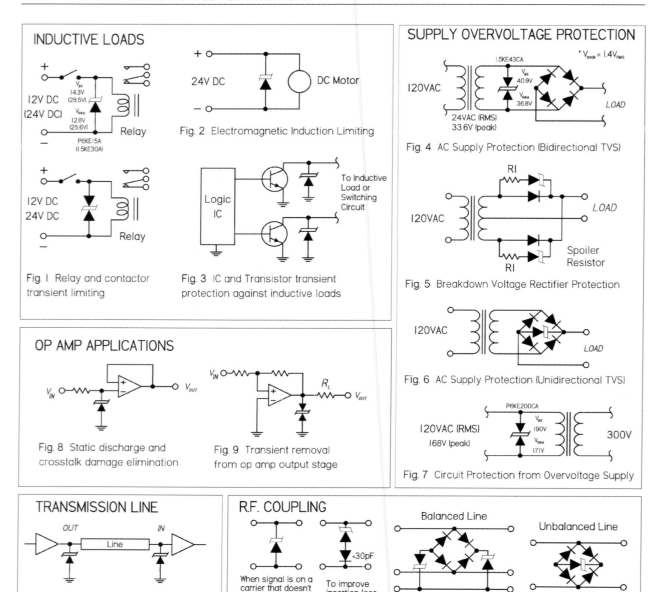

INDUCTIVE LOADS

Fig. I Relay and contactor transient limiting

Fig. 2 Electromagnetic Induction Limiting

Fig. 3 IC and Transistor transient protection against inductive loads

OP AMP APPLICATIONS

Fig. 8 Static discharge and crosstalk damage elimination

Fig. 9 Transient removal from op amp output stage

SUPPLY OVERVOLTAGE PROTECTION

Fig. 4 AC Supply Protection (Bidirectional TVS)

Fig. 5 Breakdown Voltage Rectifier Protection

Fig. 6 AC Supply Protection (Unidirectional TVS)

Fig. 7 Circuit Protection from Overvoltage Supply

TRANSMISSION LINE

Fig. 10 Absorbing transients on signal line

R.F. COUPLING

When signal is on a carrier that doesn't change polarity

To improve insertion loss

Fig. 11

FIGURE 4.132

Fig. 1–3: Very high transient voltages are generated when an inductive load is disconnected, such as motors, relay coils, and solenoids. Here, TVSs provide protection to the driving circuitry, as well as limiting damage to relay and solenoid metal contacts.

Fig. 4–7: Typical power sources employing TVS for transient protection. The TVS is chosen for the breakdown voltage that is equal to or greater than the dc output voltage. In most applications, a fuse in the line is desirable.

Fig. 8–9: Input states are vulnerable to low-current, high-voltage static discharges or crosstalk transmitted on the signal wires. Usually an op amp or other IC will have an internal clamp diode, but this provides limited protection for high currents and voltage. Here, an external TVS diode is used to provide additional protection. The second circuit has a TVS on the output of an op amp to prevent a voltage transient due to a short circuit or an inductive load from being transmitted into the output stage.

Fig. 10: Transients generated on the line can vary from a few microseconds' to several milliseconds' duration and up to 10,000 V. This threat has given rise to high-noise-immunity ICs. However, the input diodes to these devices, again, have limited internal diode protection, and IC damage is still possible, resulting in either an open circuit or slow degradation of the circuit's performance with time. Here, a TVS located on the signal line can absorb this excess energy and prevent damage.

Fig. 11: A selection of transient suppressor arrangements for RF coupling.

Metal Oxide Varistors (MOVs) and Multilayer Varistors

A *metal oxide varistor* (MOV) is a bidirectional semiconductor transient suppressor that acts like a voltage-sensitive variable resistor. Internally, it consists of a complex ceramic crystal structure with various multidirectional metal oxide p-n junction boundaries between crystal grains, all sandwiched between two electrodes. Each individual *p-n* junction is highly resistive, up until a voltage across the grain boundary in excess of around 3.6 V, where it then is bias on—has a very small resistance. The voltage at which the MOV itself switches is dependent on the average number of grains between its electrode leads. During the manufacturing process, this value can be varied to create any desired breakdown threshold. Due to the random orientation of the boundaries within the MOV, there is no directional, so the MOV acts as a bipolar device—it can be used for ac or dc applications.

In terms of applications, an MOV is usually connected across the mains input of the equipment or the circuit it's protecting, with a series filter inductor and/or fuse thrown in to protect the MOV itself. In the presence of a transient, the MOV's resistance switches from high resistance (several megaohms) to very low resistance (a few ohms), transforming itself into a high-current shunt for the transient current. MOVs are made with various clamping voltages, peak current ratings, and maximum energy ratings—reflecting the fact that an MOV can absorb a very large amount of power for a very brief time or smaller amounts over a longer time. For example, an MOV rated at 60 J can absorb 60 W for 1 s, or 600 W for 0.1 s, or 6 kW for 10 ms, or 60 kW for 1 ms, and so on.

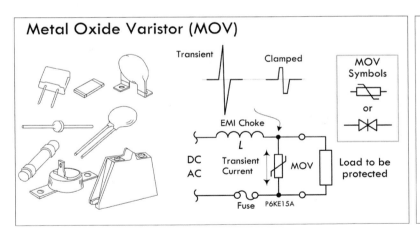

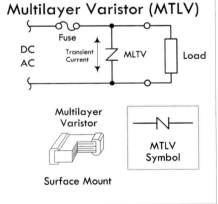

FIGURE 4.133

In many regards, MOVs resemble back-to-back zener diodes. However, unlike diodes, MOVs can handle higher-energy transients than zener diodes, since there is no single p-n junction, but rather numerous p-n junctions throughout its structure. The highly conductive ZnO grains act as heat sinks, ensuring a rapid and even distribution of thermal energy throughout the device and minimizing temperature rise. (Note that MOVs can dissipate only a relatively small amount of average power and are unsuited to applications that demand continuous power dissipation.) They are about as fast as zeners and clamp surge voltage to safe levels. Leakage is very low, which means little power is stolen from the circuit. Unlike the zener and other devices, the varistor fails shorted. A zener will also fail in an open-circuit condition that leaves equipment unprotected during a subsequent surge. This helps protect the circuit against subsequent surges; a shorted varistor across an ac or other line might fracture if the energy is high. MOVs should be fused or located where they won't effect other components should this happen.

In comparison to TVS diodes, MOVs can handle more total power/energy, while leaving a smaller footprint. TVS diodes, however, exhibit much better clamp ratios (better-quality protection) and faster response times (1 to 5 ns compared to about 5 to 200 ns for MOVs). The speed limitation of MOVs, however, is a result of parasitic inductance in the package and leads, and can be minimized with short lead design. MOVs also exhibit an inherent wear-out mechanism within their structure. As the device absorbs transient energy, the electrical characteristics (e.g., leakage, breakdown voltage) tend to drift. On the other hand, TVS diodes have no inherent wear-out mechanism. MOVs have an effective capacitance range from 75 pF for small MOVs to as high as 20,000 pF for large ones. This, combined with the lead inductance, makes practical MOVs slower than TVS diodes, but still fast—in the range of 5 to 200 ns, depending on the device. However, transients that they are designed to remove are usually much longer, so they are usually perfect for the job.

MOVs are found in power supplies of computers and other sensitive equipment, and in mains filters and stabilizers to prevent damage from mains-borne transients due to switching or lightning. They are used in telecommunication and data systems (power supply units, switching equipment, etc.), industrial equipment (control, proximity switches, transformers, motors, traffic lighting), consumer electronics (televisions and video sets, washing machines, etc.), and automotive products (all motor and electronic systems).

One variation of the MOV found in surface-mount form is the *multilayer varistor*, or MLTV. By having surface-mount contacts, lead self-inductance and series resistance are minimized, allowing for much quicker response time—less than 1 ns. A decrease in series resistance also translates into a massive increase in peak current capability per component unit volume. Even though this is the case, the energy ratings of MLTVs are rather conservative when compared to those of other varistors. One of MLTVs' strong points is their ability to survive many thousands of strikes, at full rated peak current, without degradation. MLTVs have a characteristic similar to capacitors, having an effective dielectric constant of around 800—much lower than conventional capacitors. Because of this feature, MLTVs are also used in filter circuits. MLTVs come with an operating voltage from 3.5 V to around 68 V, and they are used extensively for transient voltage protection for ICs and transistors, as well as for many ESD and I/O protection schemes.

The following are specifications for MOV and MLTVs:

Maximum continuous dc voltage ($V_{M(DC)}$): The maximum continuous dc voltage that may be applied up to the maximum operating temperature of the device. The rated dc operating voltage (working voltage) is also used as the reference point for leakage current. This voltage is always less than the breakdown voltage of the device.

Maximum continuous ac voltage ($V_{M(AC)}$): The maximum continuous sinusoidal RMS voltage that may be applied at any temperature up to the maximum operating temperature of the device. It's related to the previous dc rating by $V_{M(DC)} = 1.4 \times V_{M(AC)}$. This means that if a nonsinusoidal waveform is applied, the recurrent peak voltage should be limited to $1.4 \times V_{M(AC)}$.

Transient energy rating (W_{TM}): Energy is given in joules (watt-seconds). This represents the maximum allowable energy for a single 10/1000-µs impulse current waveform with continuous voltage applied.

Peak current rating (I_{PK}): The maximum current rating for a given maximum clamping voltage V_C.

Varistor voltage ($V_{B(DC)}$ or V_{NOM}): The voltage at which the device changes from the off state to the on state and enters its conduction mode of operation. The voltage is usually characterized at the 1-mA point and has a specified minimum and maximum voltage listed.

Clamping voltage (V_C): The clamping voltage across an MOV at a peak current I_{PK}.

Leakage at rated dc voltage (I_L): The leakage current when the device is in nonconducting mode, when a specified voltage is applied.

Capacitance (C_p): The capacitance of the device, typically specified at a frequency of 1 MHz at a bias of 1 V_{pp}. This capacitance is usually 100 pF or lower for smaller devices, and up to a few thousand for larger ones.

In terms of design, a varistor must operate under both a continuous operating (standby) mode and the predicted transient (normal) mode. Determine the necessary steady-state voltage rating (working voltage), and then establish the transient energy absorbed by the varistor. Calculate the peak transient current through the varistor and determine the power dissipation requirements. Select a model to provide the required voltage-clamping characteristics.

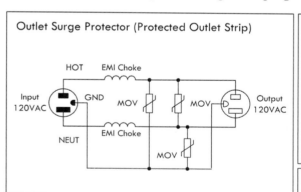

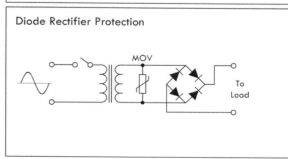

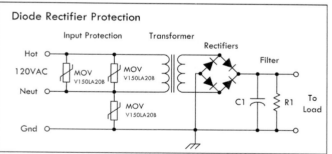

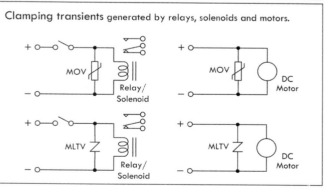

FIGURE 4.134

Surgector

There are other transient voltage suppression devices out there, such as the Surgector, gas discharge, and spark gap TVSs. The Surgector utilizes silicon thyristor technology to provide bidirectional "crowbar" clamping action for transients of either polarity. This is accomplished with a five-layer p-n junction structure. Surgectors remain in a low-leakage, reverse-bias state, presenting effectively no load to the circuit as long as the applied voltage is at or below its V_{DRM} rating. A transient voltage exceeding this value will cause the device to avalanche (breakdown), beginning the clamping action across the line to which it is connected. As the leading edge of the transient voltage attempts to rise higher, the Surgector current will increase through the circuit's source impedance until the V_{BO}, or breakover voltage mode, is reached. Thyristor action is then rapidly triggered, and the Surgector switches to its "on," or latched state. This very low impedance state crowbars the line with effectively the characteristics of a forward p-n junction, thereby short-circuiting the transient voltage.

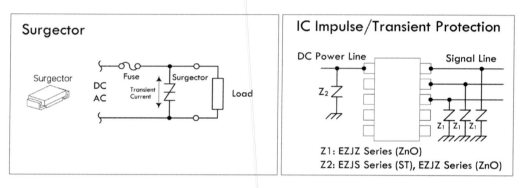

FIGURE 4.135

PolySwitch

A PolySwitch (also known as polyfuse, multiswitch, and generically resettable fuse) is a special positive temperature coefficient resistor that is constructed from a conductive polymer mix. It resembles a varistor and PTC thermistor in one. At normal temperatures, the conductive particles within the polymer form densely packed low-resistance chains, allowing current to flow easily. However, if the current flow through the PolySwitch increases to a point where its temperature rises above a critical level, the crystalline structure of the polymer suddenly changes into an expanded amorphous state. At this point, the device's resistance dramatically increases, causing a sudden drop in current flow. The point at which this occurs is referred to as the *trip current*. If the voltage level is maintained after tripping, enough holding current will generally flow, keeping the device in a tripped state. The PolySwitch will reset itself only if the voltage is reduced and the device is allowed to cool, at which point the polymer particles rapidly return to their densely packed state, and the resistance drops.

PolySwitches can be used in numerous applications wherever you need a low-cost, self-resetting solid-state circuit breaker. They are used to limit over-current in speakers, power supplies, battery packs, motors, etc. For example, Fig. 4.136

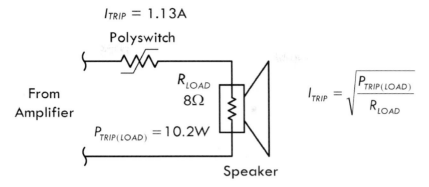

FIGURE 4.136

shows how a PolySwitch used to protect a speaker from excessive current sourced by an amplifier. The PolySwitch is rated with a trip current that is slightly higher than that rated for the power level the speaker can handle. For example, an 8-Ω, 5-W speaker has a maximum current rating determined by the generalized power law.

Avalanche Diodes

Avalanche diodes are designed to break down and conduct at a specified reverse-bias voltage. This behavior is similar to that of a zener diode, but its operation is caused by a different mechanism, called the *avalanche effect* (a reverse electric field applied across a p-n junction causes a wave of ionization, reminiscent of an avalanche, leading to a large current). However, unlike zener diodes that are rather restricted in maximum breakdown voltage, avalanche diodes are available with breakdown voltage of over 4000 V. Avalanche diodes are used in circuits to guard against damaging high-voltage transients. They are connected to a circuit so that they're reverse-biased (the cathode is set positive with respect to the anode). In this configuration, the avalanche diode is nonconducting and doesn't interfere with the circuit. However, if the voltage rises beyond a safe design limit, the diode goes into avalanche breakdown, eliminating the harmful voltage by shunting current to ground. Avalanche diodes are specified with a clamping voltage V_{BR} and a maximum-size transient that it can absorb, specified either in terms of joules of energy or as I^2t. The avalanche breakdown event is not destructive, provided the diode isn't overheated. One side effect that occurs in avalanche diodes is RF noise generation.

4.6 Integrated Circuits

An integrated circuit (IC) is a miniaturized circuit that contains a number of resistors, capacitors, diodes, and transistors stuffed together on a single chip of silicon no bigger than your fingernail. The number of resistors, capacitors, diodes, and transistors within an IC may vary from just a few to millions.

The trick to cramming everything into such a small package is to make all the components out of tiny *n*-type and *p*-type silicon structures that are embedded into the silicon chip during the production phase. To connect the little transistors, resistors, capacitors, and diodes together, aluminum plating is applied along the surface

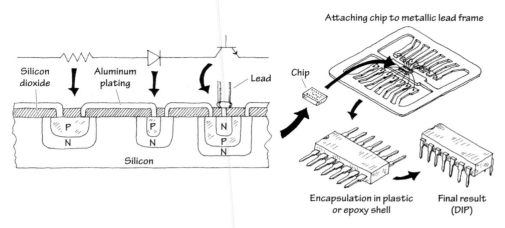

FIGURE 4.137   The structure of an IC

of the chip. Figure 4.137 shows a magnified cross-sectional view of an IC showing how the various components are embedded and linked together.

ICs come in analog, digital, or analog/digital form:

- Analog (or linear) ICs produce, amplify, or respond to varying voltages. Some common analog ICs include voltage regulators, operational amplifiers, comparators, timers, and oscillators.

- Digital (or logic) ICs respond to or produce signals having only high and low voltage states. Common digital ICs include logic gates (such as AND, OR, or NOR), microcontrollers, memories, binary counters, shift registers, multiplexers, encoders, and decoders.

- Analog/digital ICs share properties common with both analog and digital ICs. Analog/digital ICs may take a number of different forms. For example, the IC may be designed primarily as an analog timer but may contain a digital counter. Alternatively, the IC may be designed to read in digital information and then use this information to produce a linear output that can be used to drive, say, a stepper motor or LED display.

ICs are so pervasive that you are likely to use them in any project that you will undertake. You will find them used in many of the chapters that follow.

4.6.1   IC Packages

ICs come in many and various packages (see Fig. 4.138). The determining factors for the package type are the number of pins and the power dissipation. For example, a high-power voltage regulator IC may have three pins and look just like a high-power transistor.

However, the majority of ICs have many more pins and are arranged in a dual in-line (DIL) package (see Fig. 4.138) of 8, 14, 16, 20, 24, or 40 pins. There are also surface-mount versions of the DIL packages, as well as packages arranged as a square with pins on all sides. Some of the surface-mount packages have extremely small spacing between pins—sometimes as small as 0.5 mm, which is two pins every millimeter and not really intended for hand soldering.

FIGURE 4.138 IC packages

Some of the most common packages are listed in Table 4.11.

You will often find that the same ICs are available in multiple packages, making it possible to prototype in something easy to solder like DIL or SO, and then switch to a smaller package for the final product.

TABLE 4.11

PACKAGE	LONG NAME	PITCH (mm)	NOTES
DIL	Dual in-line	2.54	
SO/SOIC/SOP	Small outline IC package	1.27	
MSOP/SSOP	Mini/shrink small outline package	0.65	
SOT	Small outline transistor	0.65	
TQFP	Thin quad flat pack	0.8	Pins on four sides
TQFN	Thin quad flat no leads	0.4-0.65	No pins or pads underneath body

CHAPTER 5

Optoelectronics

Optoelectronics is a branch of electronics that deals with light-emitting and light-detecting devices. Light-emitting devices, such as lamps and light-emitting diodes (LEDs), create electromagnetic energy (e.g., light) by using an electric current to excite electrons into higher energy levels (when an electron changes energy levels, a photon is emitted). Light-detecting devices such as phototransistors and photo-resistors, on the other hand, are designed to take incoming electromagnetic energy and convert it into electric currents and voltages. This is usually accomplished using photons to liberate bound electrons within semiconductor materials. Light-emitting devices typically are used for illumination purposes or as indicator lights. Light-detecting devices are used primarily in light-sensing and communication devices, such as dark-activated switches and remote controls. This chapter examines the following optoelectronic devices: lamps, LEDs, photoresistors, photodiodes, solar cells, phototransistors, photothyristors, and optoisolators.

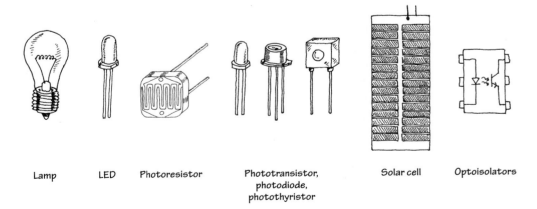

| Lamp | LED | Photoresistor | Phototransistor, photodiode, photothyristor | Solar cell | Optoisolators |

FIGURE 5.1

5.1 A Little Lecture on Photons

Photons are the elemental units of electromagnetic radiation. White light, for example, is composed of a number of different kinds of photons; some are blue photons, some are red photons, etc. It is important to note that there is no such thing as a white photon. Instead, when the combination of the various colored photons interacts with our eye, our brain perceives what we call *white light*.

Photons are not limited to visible light alone. There are also radiofrequency photons, infrared photons, microwave photons, and other kinds of photons that our eyes cannot detect.

In terms of the physics, photons are very interesting creatures. They have no rest mass, but they do carry momentum (energy). A photon also has a distant wavelike character within its electromagnetic bundle. The wavelength of a photon (horizontal distance between consecutive electrical or magnetic field peaks) depends on the medium in which it travels and on the source that produced it. It is this wavelength that determines the color of a photon. A photon's frequency is related to its wavelength by $\lambda = v/f$, where v is the speed of the photon. In free space, v is equal to the speed of light ($c = 3.0 \times 10^8$ m/s), but in other media, such as glass, v becomes smaller than the speed of light. A photon with a large wavelength (or small frequency) is less energetic than a photon with a shorter wavelength (or a higher frequency). The energy of a photon is equal to $E = hf$, where h is Planck's constant (6.63×10^{-34} J·s).

The trick to "making" a photon is to accelerate/decelerate a charged particle. For example, an electron that is made to vibrate back and forth within an antenna will produce radiofrequency photons that have very long wavelengths (low energies) when compared with light photons. Visible light, on the other hand, is produced when outer-shell electrons within atoms are forced to make transitions between energy levels, accelerating in the process. Other frequency photons may be created by vibrating or rotating molecules very quickly, while still others, specifically those with very high energy (e.g., gamma rays), can be created by the charge accelerations within the atomic nuclei.

Figure 5.2 shows the breakdown of the electromagnetic spectrum. *Radiofrequency* photons extend from a few hertz to about 10^9 Hz (wavelengths from kilometers to about 0.3 m). They are often generated by alternating currents within power lines and electric circuits such as radio and television transmitters.

Microwave photons extend from about 10^9 up to 3×10^{11} Hz (wavelengths from 30 cm to 1 mm). These photons can penetrate the earth's atmosphere and hence are used in space vehicle communications, radio astronomy, and transmitting telephone conversations to satellites. They are also used for cooking food. Microwaves are often produced by atomic transitions and by electron and nuclear spins.

Infrared photons extend from about 3×10^{11} to 4×10^{14} Hz. Infrared radiation is created by molecular oscillations and is commonly emitted from incandescent sources such as electric heaters, glowing coals, the sun, human bodies (which radiate photons in the range of 3000 to 10,000 nm), and special types of semiconductor devices.

Light photons comprise a narrow frequency band from about 3.84×10^{14} to about 7.69×10^{14} Hz and are generally produced by a rearrangement of outer electrons in atoms and molecules. For example, in the filament of an incandescent light bulb, electrons are randomly accelerated by applied voltages and undergo frequent collision. These collisions result in a wide range of electron acceleration, and as a result, a broad frequency spectrum (within the light band) results, giving rise to white light.

Ultraviolet photons extend from approximately 8×10^{14} to 3.4×10^{16} Hz and are produced when an electron in an atom makes a long jump down from a highly excited state. The frequency of ultraviolet photons—unfortunately for us—tend to react badly with human cell DNA, which in turn can lead to skin cancer. The sun

Electromagnetic Spectrum

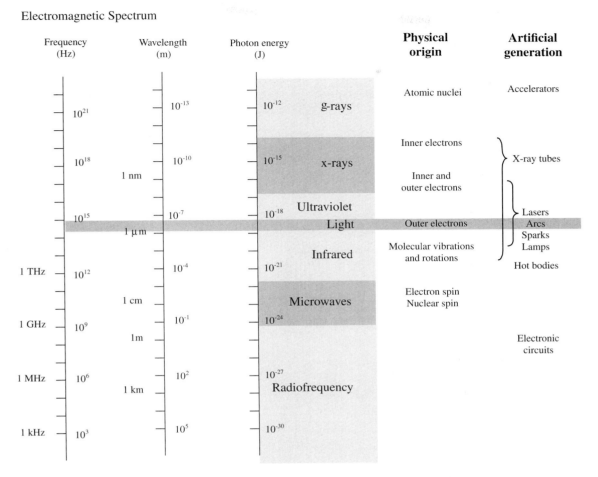

FIGURE 5.2

produces a large output of ultraviolet radiation. Fortunately for us, protective ozone molecules in the upper atmosphere can absorb most of this ultraviolet radiation by converting the photon's energy into a vibrating motion within the ozone molecule.

X-rays are highly energetic photons that extend from about 2.4×10^{16} to 5×10^{19} Hz, making their wavelengths often smaller than the diameter of an atom. One way of producing x-rays is to rapidly decelerate a high-speed charged particle. X-rays tend to act like bullets and can be used in x-ray imagery.

Gamma rays are the most energetic of the photons, whose frequency begins around 5×10^{19} Hz. These photons are produced by particles undergoing transitions within the atomic nuclei. The wavelike properties of gamma rays are extremely difficult to observe.

5.2 Lamps

Lamps are devices that convert electric current into light energy. One approach used in the conversion process is to pass a current through a special kind of wire filament. As current collides with the filament's atoms, the filament heats up, and photons are emitted. (As it turns out, this process produces a variety of different wavelength photons, so it appears that the emitted light is white in color.) Another approach used to produce light involves placing a pair of electrodes a small distance apart within a glass gas-enclosed bulb. When a voltage is set across the electrodes, the gas

Incandescent

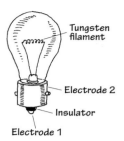

These lamps use a tungsten wire filament to produce a glowing white light when current passes through it. The filament is enclosed in an evacuated glass bulb filled with a gas such as argon, krypton, or nitrogen that helps increase the brilliance of the lamp and also helps prevent the filament from burning out (as would be the case in an oxygen-rich environment). Incandescent lamps are used in flashlights, home lighting, and as indicator lights. They come in a variety of different sizes and shapes, as well as various current, voltage, and candlelight power ratings.

Halogen

Similar to a typical incandescent lamp, these lamps provide ultrabright output light. Unlike a typical incandescent lamp, the filament is coated on the inside of a quartz bulb. Within this bulb, a halogen gas, such as bromine or iodine, is placed. These lamps are used in projector lamps, automotive headlights, strobe lights, etc.

Gas-Discharge

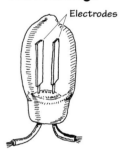

This lamp produces a dim, pale light that results from the ionization of neon gas molecules between two electrodes within the bulb. Types of gas-discharge lamps include neon, xenon flash, and mercury vapor lamps. Gas-discharge lamps have a tendency to suddenly switch on when a particular minimum operating voltage is met. For this reason, they are sometimes used in triggering and voltage-regulation applications. They are also used as indicator lights and for testing home ac power outlets.

Fluorescent

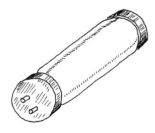 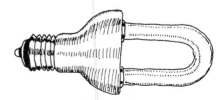

This is a lamp consisting of a mercury vapor–filled glass tube whose inner wall is coated with a material that fluoresces. At either end of the tube are cathode and anode incandescent filaments. When electrons emitted from an incandescent cathode electrode collide with the mercury atoms, ultraviolet (UV) radiation is emitted. The UV radiation then collides with the lamp's florescent coating, emitting visible light in the process. Such lamps require an auxiliary glow lamp with bimetallic contacts and a choke placed in parallel with the cathode and anode to initiate discharge within the lamp. These are highly efficient lamps that are often used in home lighting applications.

FIGURE 5.3

Xenon Flash Lamp

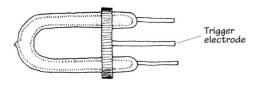

Trigger
electrode

This is a gas-discharge lamp that is filled with a xenon gas that ionizes when a particular voltage is applied across its electrodes. These lamps come with three leads: an anode, a cathode, and a trigger-voltage lead. Normally, a particular voltage is applied across the anode and cathode leads, and the lamp is off. However, when a particular voltage is applied to the trigger lead, the gas suddenly ionizes and releases an extremely bright flash. These lamps are used in photographic applications and in special-effects lighting projects.

FIGURE 5.3   (*Continued*)

ionizes (electrons are stripped from the gas atoms) emitting photons in the process. Figure 5.3 gives an overview of some of the major kinds of lamps.

Technical Stuff about Light Bulbs

A lamp's brightness is measured in what is called the *mean spherical candle power* (MSCP). Bulb manufacturers place a lamp at the center of an integrating sphere that averages the lamp's light output over its surface. The actual value of the MSCP for a lamp is a function of color temperature of the emitting surface of the lamp's filament. For a given temperature, doubling the filament's surface area doubles the MSCP. Other technical things to consider about lamps include voltage and current ratings, life expectancy, physical geometry of the bulb, and filament type. Figure 5.4 shows a number of different bulb types.

In recent years, incandescent light bulbs have become something of a rarity. In domestic lighting situations, they have almost completely been replaced by compact fluorescent designs, and indeed are not even for sale in some European countries. There is also a trend toward using arrays of LEDs for illumination in place of incandescent bulbs (see the next section). In applications that formerly would have used incandescent indicator lights, LEDs now rule supreme. LEDs have a much longer life, lower power consumption, and much greater tolerance of physical and thermal shock.

5.3   Light-Emitting Diodes

Light-emitting diodes (LEDs) are two-lead devices that are similar to *pn*-junction diodes, except that they are designed to emit visible or infrared light. When a LED's anode lead is made more positive in voltage than its cathode lead (by at least 0.6 to 2.2 V), current flows through the device and light is emitted. However, if the polarities are reversed (anode is made more negative than the cathode), the LED will not conduct, and hence it will not emit light. The symbol for an LED is shown in Fig. 5.5.

LEDs are available in a wide range of colors. Historically, red was the first LED color. Yellow and green and infrared LEDs followed. It was not until the 1990s that blue LEDs became available. These days, LEDs are available in pretty much any color, including white.

There are also high-powered LEDs that are used for illumination and organic LEDs (OLEDs) made from polymers that can be built into displays.

Often, LEDs (especially infrared LEDs) are used as transmitting elements in remote-control circuits (e.g., TV remote control). The receiving element in this case

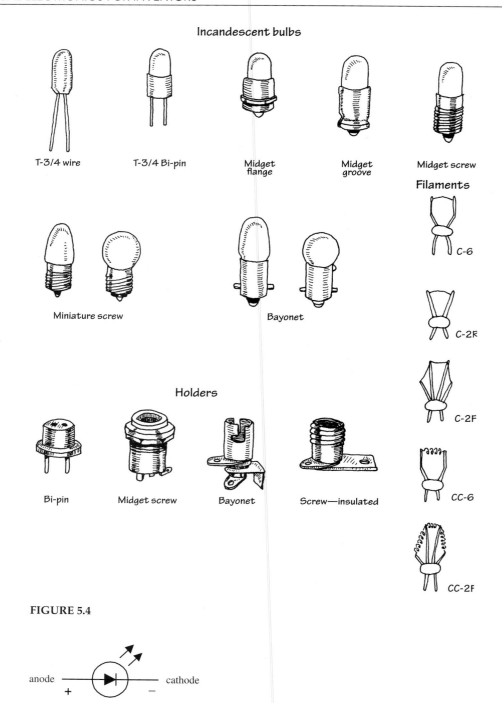

FIGURE 5.4

FIGURE 5.5

may be a phototransistor that responds to changes in LED light output by altering the current flow within the receiving circuit.

5.3.1 How an LED Works

The light-emitting section of an LED is made by joining *n*-type and *p*-type semiconductors together to form a *pn* junction. When this *pn* junction is forward-biased, electrons in the *n* side are excited across the *pn* junction and into the *p* side, where they

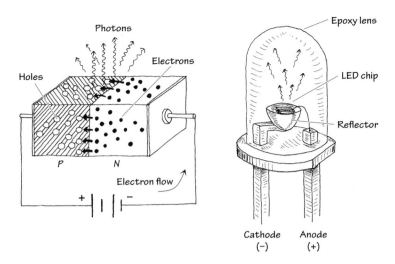

FIGURE 5.6

combine with holes. As the electrons combine with the holes, photons are emitted. Typically, the *pn*-junction section of an LED is encased in an epoxy shell that is doped with light-scattering particles to diffuse the light and make the LED appear brighter. Often a reflector placed beneath the semiconductor is used to direct light upward. The cathode and anode leads are made from a heavy-gauge conductor to help wick heat away from the semiconductor.

5.3.2 Kinds of LEDs

Visible-Light LEDs

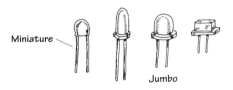

These LEDs are inexpensive and durable devices that typically are used as indicator lights. Common colors include green (~565 nm), yellow (~585 nm), orange (~615 nm), and red (~650 nm). Maximum forward voltages are about 1.8 V, with typical operating currents from 10 to 30 mA.

Infrared LEDs

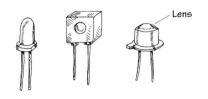

These LEDs are designed to emit infrared photons that have wavelength between approximately 880 and 940 nm. They are used in conjunction with a photosensor (e.g., photodiode, photoresistor, phototransistor) in remote-control circuits (e.g., TV remotes, intrusion alarms). They tend to have a narrower viewing angle when compared with a visible-light LED so that transmitted information can be directed efficiently. Photon output is characterized in terms of output power per specific forward current. Typical outputs range from around 0.50 mW/20 mA to 8.0 mW/50 mA. Maximum forward voltages at specific forward currents range from about 1.60 V at 20 mA to 2.0 V at 100 mA.

Blinking LEDs

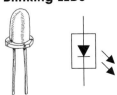

These LEDs contain a miniature integrated circuit within their package that causes the LED to flash from 1 to 6 times each second. They are used primarily as indicator flashers, but may be used as simple oscillators.

FIGURE 5.7

Multicolor LEDs

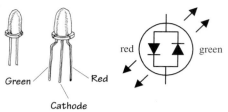

Multicolor LEDs mount a number of LEDs in a single package. The color they produce is a mixture of the colors produced by each LED. The ultimate multicolor LED is an RGB LED that has red, green, and blue LEDs within a single package. This allows any color to be mixed by controlling the brightness of each of the LEDs, with white light being produced if all the LEDs are on with equal intensity. The packages usually have a common anode or common cathode connection.

A variation on this has red and green LEDs back to back to indicate polarity (the polarity of the supply voltage will determine which LED is forward-biased and on, and which is reverse-biased and off).

LED Displays

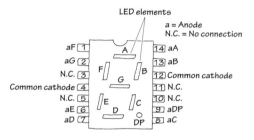

These are used for displaying numbers and other characters. In the LED display shown here, seven individual LEDs are used to make up the display. When a voltage is applied across one of the LEDs, a portion of the 8 lights up. Unlike liquid-crystal displays, LED displays tend to be more rugged, but they also consume more power.

As well as such seven-segment LEDs, LED packages containing LEDs in a matrix are also available.

FIGURE 5.7   *(Continued)*

5.3.3   *More on LEDs*

LEDs emit visible light, infrared radiation, or even ultraviolet radiation when forward-biased. Visible single-tone LEDs emit relatively narrow bands of green, yellow, orange, red, and blue light (with spectrum spread usually less than 40 nm at 90 percent peak intensity). Infrared diodes emit one of several bands beyond red.

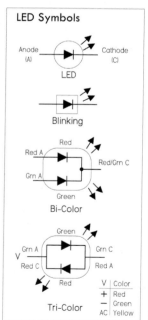

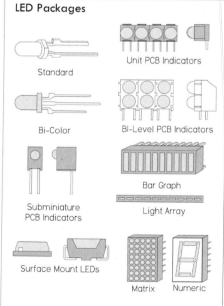

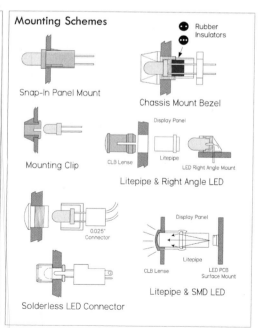

FIGURE 5.8

White LEDs provide a variety of wavelengths to mimic white light and are used in low-level lighting applications, such as backlighting, headlamps, and nightlights.

High-power LEDs (HPLEDs) are now available. These have forward currents of hundreds of mA to more than 1 A. Such LEDs are very bright but also generate a lot of heat. They must be mounted on a heat sink to prevent thermal destruction.

LEDs have very fast response times, excellent efficiency, and long lifetimes. They are current-dependent devices whose light output is directly proportional to the forward current.

To light an LED, apply a voltage greater than the LED's forward voltage V_{LED}, and limit current flow via a series resistor to a level below the LED's maximum rating, usually to I_{LED}—the manufacturer's recommended value. The following equation is used to select the series resistor:

$$R_S = \frac{V_{\text{IN}} - V_{\text{LED}}}{I_{\text{LED}}}$$

If you want brightness control, throw in a 1-K potentiometer in series, as shown in Fig. 5.10.

V_{LED} varies with LED color. Typical V_{LED} values are 1.7 V for non-high-brightness red, 1.9 V for high-brightness high-efficiency low-current red, 2 V for orange and yellow, 2.1 V for green, 3.4 to 3.6 V for bright white and most blue types, and 6 V for 430-nm blue. Given the preceding voltage drops, it's a good idea to use at least a 3-V supply voltage for lower-voltage LEDs, 4.5 V for 3.4-V types, and 6 V for 430-nm blue. If you don't know the recommended I_{LED} value of a given LED, it's usually safe to assume it will be around 20 mA. Table 5.1 shows the range of LED types and corresponding characteristic values.

TABLE 5.1 Diode Characteristics

WAVELENGTH	COLOR NAME	FWD VOLTAGE (V_F AT 20 MA)	INTENSITY (5-MM LEDS)	LED DYE MATERIAL
940	Infrared	1.5	16 mW @ 50 mA	GaAlAs/GaAs
880	Infrared	1.7	18 mW @ 50 mA	GaAlAs/GaAs
850	Infrared	1.7	26 mW @ 50 mA	GaAlAs/GaAs
660	Ultra Red	1.5–1.8	200 mcd @ 50 mA	GaAlAs/GaAs
635	High Eff. Red	2.0	200 mcd @ 20 mA	GaAsP/GaP
633	Super Red	2.2	3500 mcd @ 20 mA	InGaAlP
620	Super Orange	2.2	4500 mcd @ 20 mA	InGaAlP
612	Super Orange	2.2	6500 mcd @ 20 mA	InGaAlP

(Continued)

TABLE 5.1   Diode Characteristics (*Continued*)

WAVELENGTH	COLOR NAME	FWD VOLTAGE (V_F AT 20 MA)	INTENSITY (5-MM LEDS)	LED DYE MATERIAL
605	Orange	2.1	160 mcd @ 20 mA	GaAsP/GaP
595	Super Yellow	2.2	5500 mcd @ 20 mA	InGaAlP
592	Super Pure Yellow	2.1	7000 mcd @ 20 mA	InGaAlP
585	Yellow	2.1	100 mcd @ 20 mA	GaAsP/GaP
574	Super Lime Yellow	2.4	1000 mcd @ 20 mA	InGaAlP
570	Super Lime Green	2.0	1000 mcd @ 20 mA	InGaAlP
565	High Efficiency Green	2.1	200 mcd @ 20 mA	GaP/GaP
560	Super Pure Green	2.1	350 mcd @ 20 mA	InGaAlP
555	Pure Green	2.1	80 mcd @ 20 mA	GaP/GaP
525	Aqua Green	3.5	10,000 mcd @ 20 mA	SiC/GaN
505	Blue Green	3.5	2000 mcd @ 20 mA	SiC/GaN
470	Super Blue	3.6	3000 mcd @ 20 mA	SiC/GaN
430	Ultra Blue	3.8	100 mcd @ 20 mA	SiC/GaN
370–400	UV LED	3.9	NA	GaN
4500K	"Incandescent White"	3.6	2000 mcd @ 20 mA	SiC/GaN
6500K	Pale White	3.6	4000 mcd @ 20 mA	SiC/GaN
8000K	Cool White	3.6	6000 mcd @ 20 mA	SiC/GaN

Some other specifications to consider include power dissipation (100 mW typical), reverse voltage rating, operating temperature (–40 to +85°C typical), pulse current (100 mA typical), luminous intensity (given in millicandles, mcd), viewing angle (given in degrees), peak emission wavelength, and spectral width (20 to 40 nm typical).

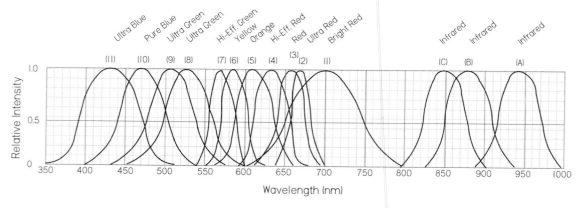

FIGURE 5.9

5.3.4 LED Applications

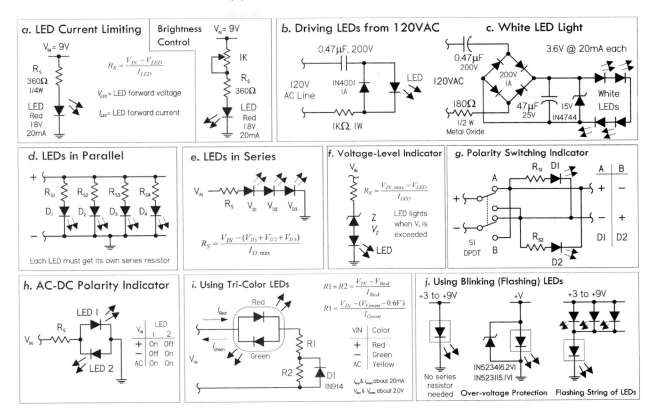

FIGURE 5.10 *(a)* It's important to keep the LED current flow below the maximum rating. Select a series current-limiting resistor according to the provided formula. A 1K pot can be added to provide variable light-intensity control. *(b)* Here's a circuit that can be used to power LEDs from an ac line. The key is the capacitor, which acts to attenuate the ac signal, and the resistor, which acts to limit the current level. Since the capacitor must pass current in both directions, a small diode is connected in parallel with the LED to provide a path for the negative half cycle and also to limit the reverse voltage across the LED. A second LED with polarity reverse may be substituted for the diode, or a tricolor LED could be used, which would appear orange with alternating current. The resistor is chosen to limit the worst-case inrush current to about 150 mA, which will drop to less than 30 mA in a millisecond, as the capacitor charges. The 0.47-μF capacitor has a reactance of 5640 Ω at 60 Hz, so the LED current is about 20 mA half wave, or 10 mA average. A larger capacitor will increase the current and a smaller one will reduce it. The capacitor must be a nonpolarized type with a voltage rating of 200 V or more. *(c)* This is also an ac-line powered arrangement used to create a nightlight using white LEDs. Like the previous circuit, the 0.47-μF input capacitor attenuates the ac voltage level, and the 180-Ω resistor acts to limit current. The bridge rectifier with attached filter capacitor creates a near constant dc voltage, while the zener diode acts to regulate the voltage level. In this case, there are four white LEDs in series, each with a 3.4-V drop, for a total voltage drop of 13.6 V; therefore, a 15-V zener is used. *(d)* Do not put LEDs in parallel with each other without using individual dropping resistors for each LED. Although this usually works, it is not reliable. LEDs become more conductive as they warm up, which may lead to unstable current distribution through paralleled LEDs. LEDs in parallel need their own individual dropping resistors, as shown in the figure. Series strings can be paralleled if each string has its own dropping resistors. *(e)* It's okay to put LEDs in series with a single common series resistor. Simply add up the voltage drops of all the diodes in series, and use that value in place of V_{LED} in the series-resistor formula. It's a good idea not to exceed 80 percent of the supply voltage in order to maintain good stability and predicable current consumption. *(f)* Here, a zener diode is used to create a simple voltage-level indicator: when the voltage exceeds the breakdown voltage of the zener, current flows, lighting the LED. *(g)* This circuit reverses the polarity at the output using a DPDT switch. D_1 goes on when the polarity is one way, and D_2 goes on when the polarity is another way. *(h)* Two LEDs in reverse-parallel make up a simple polarity indicator. Note that both LEDs appear on with ac. *(i)* Basic circuit showing how to use tricolor LEDs. *(j)* Blinking LEDs don't require a series resistor, but don't apply more than the recommended supply voltage—3 to 9 V is a safe range. A zener placed in reverse parallel with the blinking LED can be used for overvoltage protection. A number of standard LEDs can be made to blink using a blinking LED. You can also use the blinking LED to drive transistors that can be made to switch on and off as the blinking LED flashes.

In Chap. 13, we will look at how you drive LED displays from a microcontroller, including techniques such as multiplexing, Charlieplexing, and driving RGB LEDs with PWM signals to mix colors.

5.3.5   Laser Diodes

Laser diodes are light-emitting diodes with two "mirrors" on the surface of the diode to create a laser cavity. When the diode is forward-biased, charges are injected into the active area of the junction, while electrons and holes recombine in the junction, creating spontaneous emission of photons. These photons can cause other electron-hole pairs to recombine by stimulated emission. When the current is high enough, the device lases. Laser diodes are driven by low-voltage power and usually incorporate optical feedback from a monitor photodiode (commonly built into the same laser diode package) to regulate the laser diode current.

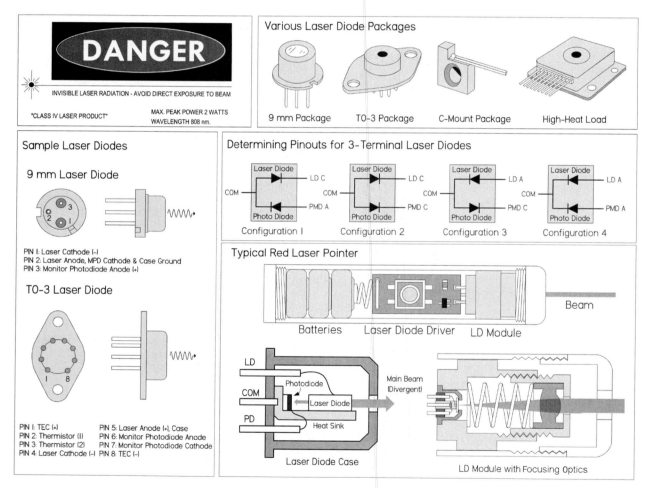

FIGURE 5.11   Figure shows a rough diagram of a laser diode of the type found in a laser pointer or CD player. The size of the package is typically 5 to 10 mm overall, but the laser diode chip is less than 1 mm in length. The maxim beam as it emerges from the laser diode is wedge-shaped and highly divergent (unlike a helium-neon laser) with a typical spread of 10 by 30°. External optics are required to produce anything approaching a parallel (collimated) beam. A simple (spherical) short-focal-length convex lens will work reasonably well for this purpose, but diode laser modules and laser pointers might use a lens that at least on the surface is aspheric (not ground to a spherical shape, as are most common lenses). The beam from the back end of the laser diode chip hits a built-in photodiode that is normally used in an optoelectronic feedback loop to regulate current and thus beam power.

Compared to LEDs, laser diodes have quicker response times and very narrow spectrum spread (around 1 nm), and can focus radiation to a spot as small as 1 μm in diameter—even for a cheap laser diode using simple optics found in a CD player. Unlike gas lasers, however, a laser diode's output beam is divergent, typically elliptical or wedge-shaped, and astigmatic, requiring refocusing.

The output wavelength of a laser diode is usually fixed to a single mode: for example red (635 nm, 670 nm), infrared (780 nm, 800 nm, 900 nm, 1,550 nm, etc.), and green, blue, violet. However, multimode laser diodes also exist where the emission spectrum consists of several individual spectral lines with a dominant line (line with greatest intensity) occurring at the nominal wavelength of the device. Multimode laser diodes are often desirable because problems with mode hops are suppressed—consequently, they generally have a better signal-to-noise ratio. Mode hops are slight changes in the wavelength caused by thermal expansion of the laser cavity.

Typical optical output for low-powered laser diodes range from around 1 mW to 5 mW, while high-power laser diodes can reach 100 W or more. The highest-power units consist of arrays of laser diodes—not a single device alone.

Laser diodes are used in CD players, CD-ROM drives, and DVD and Blu-ray players. They are also used in laser printers, laser fax machines, laser pointers, sighting and alignment scopes, measurement equipment, high-speed fiber-optic and free-space communication systems; as pump source for other lasers; in bar-code and UPC scanners; and in high-performance imagers. In those applications requiring high-speed modulation or pulse rates (into the gigahertz range), special integrated driver chips are needed to control the laser diode drive current (more on that in a bit).

A variety of small laser diodes are found in CD players, CD-ROM drives, laser printers, and bar-code scanners. The most common laser diodes around are those used in CD players and CD-ROM drives. These produce a mostly invisible beam in the near infrared spectrum at 780 nm. The optical output from the actual laser diode itself may be up to 5 mW, but once it passes through the optics leading to the CD, the power drops to 0.3 to 1 mW. Higher-power IR laser diodes found in read-write drives have more power output—up to 30 mW or so. Even higher-power blue laser diodes can be found in Blu-ray players.

Visible laser diodes have replaced the helium-neon lasers used in bar-code scanners, laser pointers, positioning devices in medicine (e.g., CT and MRI scanners), and many other applications. The first visible-light laser diode emitted with a wavelength around 670 nm in the deep red spectrum. More recently, 650- and 635-nm red laser diodes have dropped in price. 635 to 650-nm laser diodes are used in DVD technology. The shorter wavelength compared to 780 nm is one of the several improvements that enable DVDs to store about eight times the amount of information as CDs (4 to 5 GB per layer and up to two layers on each side of the disc, as compared to a typical CD that stores only around 650 MB). Like the IR laser diodes, the visible laser diodes have a typical maximum power around 3 to 5 mW, and cost about $10 to $50 for the basic laser diode device—more with optics and drive electronics. Higher-power types are also available but can cost upward of several hundred dollars for something like a 20-mW module. Very high-power diode lasers using arrays of laser diodes or laser diode bars with power output of watts or greater may cost thousands of dollars.

It's important to note that you should never look into a laser beam or any specular reflection of the laser beam—"you might poke your eye out." Also, laser diodes are extremely sensitive to electrostatic discharge (ESD), so it's important to use grounding straps and grounded equipment when working with them, as well as following manufacturer's suggested handling precautions.

Laser Diode Drive Circuits

You should never drive a laser diode without the proper drive circuitry. Without the proper drive circuitry, you can run into all sorts of problems stemming from swings in operating temperature with unstable injection current. The results can lead to a fried laser diode, or one whose life is short. For this reason, it's important to have a driver circuit that can provide stable current without the possibility of supply transients screwing things up. Two basic techniques used to achieve stable optical output from a laser diode are described here.

Automatic current control (ACC) or constant current circuit: This involves driving the laser diode without a photodiode feedback loop; the laser diode is simply driven at constant current. This is a simple method to use, but the optical output will fluctuate as the laser diode temperature changes. However, there are circuits and corresponding laser diodes that are designed to control the diode's operating temperature (without using the photodiode) that are very popular. The constant current circuit with temperature control provides faster control loop and precision current reference for accurately monitoring laser current. In addition, in many cases, the laser diode's internal photodiode may exhibit drift and have a poor noise characteristic. If performance of the internal photodiode is inferior, the laser diode's optical output is liable to be noisy and unstable. Constant current operation without temperature control, however, is generally not a good idea, except in cheap, low-power situations (cheap laser pointers, etc.). If the operating temperature of the laser diode decreases significantly, the optical power output will increase, and this could easily exceed its maximum rating.

Automatic power control (APC) circuit: This drive circuit is based on a photodiode feedback loop that monitors the optical output and provides a control signal for the laser diode to maintain a constant optical output level. Constant power control prevents the possibility of the optical power output increasing as the laser diode's temperature decreases. However, when operating in the constant power mode and without temperature control, mode hops and changes in wavelength will still occur. Also, if the diode's heat sink is inadequate and the temperature is allowed to rise, the optical power will decrease. In turn, the drive circuit will increase the injection current, attempting to maintain the optical power at a constant level. Without an absolute current limit, thermal runaway is possible and the laser may be damaged or destroyed in the process.

Regardless of the type of drive circuit used, the key is to prevent the drive current from overshooting the maximum operating level. Exceeding the maximum optical output, even for a nanosecond, will damage the mirror coatings on the laser diode end facets. Your typical laboratory power supply should not be used to directly drive a laser diode—it simply doesn't provide enough protection. Typical drive circuits incorporate slow-start circuitry, capacitive filtering, and other provisions to eliminate supply spikes, surges, and other switching transients.

Figure 5.12 shows a few do-it-yourself laser diode supplies. Though these drive circuits will work with many low-powered laser diodes that don't require modulation, it's worth your while to check out the laser data manufacturer's data sheets for recommended drive circuits. Creating your own drive circuits can be very tricky, and you'll probably end up frying a few expensive laser diodes in the process. Of course, you can also buy a laser diode drive chip, which may be the best bet if you're doing something a bit more complex than creating a laser pointer. These drive chips may support high-bit-rate modulation in addition to providing the constant current needed for optically stable power. Other types of chips can be adapted to linear and switching regulators. Some companies worth checking out include MAXIM, Linear Technology, Sharp, Toshiba, Mitsubishi, Analog Devices, and Burr-Brown. Often, these manufacturers provide free samples.

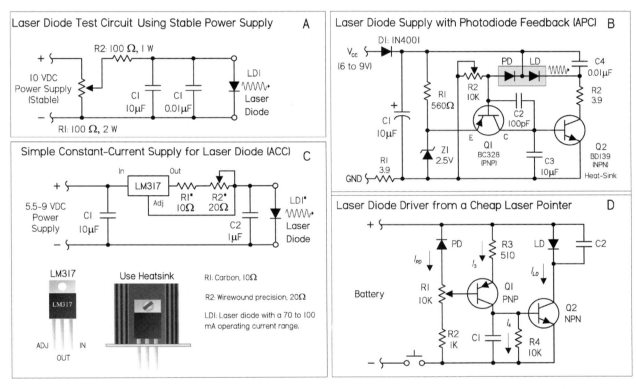

FIGURE 5.12 *(a)* This circuit can be used to identify the proper connections and polarity and then to drive the laser diode for testing purposes. Here a 0- to 10-VDC supply with current-limiting resistor is in series with the diode. If your power supply has a current limiter, set it at 20 to 25 mA to start. You can always increase it later. R_2 limits the maximum current. If you know the specs for your diode, this is a good idea (and to protect your power supply, too). You can always reduce its value if your laser diode requires more than about 85 mA (with $R_2 = 100\ \Omega$). *(b)* This runs on a (wall adapter) power supply from 6 to 9 V. There is heavy capacitive filtering in this circuit. Changes would be needed to enable this circuit to be modulated at any reasonable rate. Note that C_4 is estimated; also, an LM341 shunt regulator set up as a 2.5-V reference could replace the 2.5-V zener. *(c)* The resistor value depends on your specific laser diode current requirements. Power is provided by a 5.5- to 9-VDC battery. For the resistor, a small 10-Ω carbon resistor in series with a precision 20-Ω adjustable resistor can be used. It's a good idea to place three ordinary diodes in series instead of a laser diode, then measure the current through them and adjust the resistor until the necessary current level is reached, in this case 50 to 60 mA. You can turn up the current, never exceeding your diode's maximum limit. The dim glow will increase in intensity, but at some point a distinctive step in intensity is reached. *(d)* This circuit is from an inexpensive laser pointer. It includes some capacitive filtering, as well as a power-adjust pot R_1. Unlike the previous circuit, this one doesn't have any absolute reference, so power output will be dependent on the battery voltage to some extent. It's possible to modulate this module at a reasonable frequency by removing or greatly reducing the value of the filter capacitor C_1.

DRIVE CIRCUIT PRECAUTIONS

Even with a suitable drive circuit, watch out for intermittent or unreliable connections between the laser diode and the drive circuit. An intermittent contact in the photodiode feedback circuit will usually destroy a laser diode. Even if a power-control potentiometer's wiper breaks contact with the resistive element, there can be problems. Also, never use a switch or relay to make or break the connection between the drive circuit and the laser diode. The following are some other laser diode precautions:

Power measurement: It's not safe to assume that the optical power output of a laser diode will match what's stated in the manufacturer's minimum-maximum data—each diode will have a unique operating characteristic as a result of manufacturing tolerances. For accuracy, you must measure the output with an optical power meter or a calibrated photodiode. Remember, once the laser diode is past the threshold point, stimulated emission is achieved and the optical output increases significantly for a small increase in forward current. Therefore, a very slight increase in drive current may cause the optical output to exceed the absolute maximum. Also, make sure to include optical losses through any lenses or other components when making measurements and calculations.

Operating temperature and heat sinks: In most applications, laser diodes require heat sinks, especially when operating continuously. Without a heat sink, the laser diode junction temperature will quickly increase, causing optical output to degrade. If the laser diode temperature continues to rise, exceeding the maximum operating temperature, the diode can be catastrophically damaged or the long-term performance may degrade significantly. Generally, a lower operating temperature will help extend the diode's lifetime. Visible laser diodes with lower wavelengths (e.g., ~635 nm) are typically more sensitive to temperature than the IR laser diodes. Often, thermoelectric cooling is required to keep temperatures down. When using heat sinks, a small amount of non-silicone-type heat sink compound improves thermal conductivity between laser diode and heat sink.

Windows: Keep laser diode windows and other optics in the path clean. Dust or fingerprints will cause diffraction or interference in the laser output that can result in lower output or anomalies in the far-field pattern. The window should be cleaned using a cotton swab and ethanol.

The Easy Way Out: Laser Diode Modules and Laser Pointers

If all you really want is a visible laser to shoot around, a commercial diode laser module or some brands of laser pointers (those that include optical feedback based on laser power regulation) may be the ticket. Both the modules and laser pointers include a driver circuit capable of operating reliably on unregulated low-voltage dc input and a collimating lens matched to the laser diode. Many of the modules will permit fine adjustment of the lens position to optimize the collimation or permit focusing to a point at a particular distance. However, neither the module nor the pointer is designed to be modulated at any more than a few hertz, due to heavy internal filtering designed to protect the laser diode from power spikes. Therefore, they are generally not suited for laser communication applications. In general, it's a whole lot easier starting out with the module or pointer rather than a laser diode and homemade power supply, or even a commercial driver, if it isn't explicitly designed for your particular laser diode. There is no way to know how reliable or robust an inexpensive laser pointer will be, or if the beam quality will be acceptable. Diode laser modules are generally more expensive and of higher quality than the pointers, so they may be better for serious applications. Also, consider a helium-neon laser, since even the cheapest type is likely to generate a beam with better beam quality than the typical diode laser module or laser pointer.

Laser Diode Specifications

Lasing wavelength, λp: The wavelength of light emitted by the laser diode. For a single-mode device, this is the wavelength of the single spectral line of the laser output. For a multimode device, this is the wavelength of the spectral line with the greatest intensity.

Threshold current, I_{th}: The boundary between spontaneous emission and the stimulated emission shown on the optical power output versus forward current curve. Below the threshold current point, the output resembles the incoherent output from an LED; at or above the specified threshold current, the device begins to produce laser output. Once past the threshold point, stimulated emission is achieved and the optical output increases significantly for small increases in forward current.

Operating current, I_{op}: The amount of forward current through the laser diode necessary to produce the specified typical optical output at a specified operating temperature.

Operating voltage, V_{op}: The forward voltage across the laser diode when the device produces its specified typical optical output at a specified operating temperature.

Optical power output, P_O: Maximum allowable instantaneous optical power output in either continuous or pulse operation.

Operating temperature range: Range of case temperatures within which the device may be safely operated.

Monitor current, I_m: The current through the photodiode, at a specified reverse-bias voltage, when the laser diode is producing its typical optical power output.

Photodiode dark current, $I_{D(PD)}$: The current through the reverse-biased internal monitor photodiode when the laser is not emitting.

Reverse voltage, V_R: Maximum allowable voltage when reverse bias is applied to the laser diode or photodiode. For laser diodes with an internal monitor photodiode, the reverse voltage is specified for the laser diode as $V_{L(LD)}$ and for the photodiode as $V_{R(PD)}$.

Aspect ratio, AR: The ratio of the laser diode's divergence angles, θ⊥ (perpendicular) and θ∥ (parallel). A diode with a 30° perpendicular divergence and a 10° parallel divergence has an elliptical beam with an aspect ratio of 3:1.

Astigmatism, A_s or D_{as}: The laser beam appears to have different source points for the directions perpendicular and parallel to the junction plane. The astigmatic distance is defined as the distance between the two apparent sources. A laser diode with a large amount of astigmatism must have the astigmatism corrected (or reduced) if the laser diode output is to be accurately focused—otherwise, the resulting focused beam will be astigmatic.

Beam divergence, θ⊥ and θ∥: Also referred to as *radiation angles*. The beam divergence is measured as the full angle and at the half-maximum intensity point, known as full-width half-maximum, or FWHM. Angular specifications are provided for both the perpendicular axis and the parallel axis.

Polarization ratio: The output from a single-cavity laser diode is linearly polarized parallel to the laser junction. Spontaneous emission with a random polarization and/or with a polarization perpendicular to the laser junction is also present. The polarization ratio is defined as the parallel component divided by the perpendicular component. For a diode operating near its maximum power, the ratio is typically greater than 100:1. When operating near the threshold point, the ratio would be considerably lower as the spontaneous emission becomes more significant.

Slope efficiency, SE: Also referred to as *differential efficiency*. This is the mean value of the incremental change in optical power for an incremental change in forward current when the device is operating in the lasing region of the optical power output versus forward current curve.

Rise time: Time required for the optical output to rise from 10 percent to 90 percent of its maximum value.

Positional accuracy (D_x, D_y, D_z): Also referred to as *emission point accuracy*. These specifications define the positional accuracy of the laser diode emitter with respect to the device package. D_x and D_y are measured as the planar displacement of the chip from the physical axis of the package. D_z is measured perpendicular to the reference surface. Specifications may list both angular error expressed in degrees and the linear error in microns.

5.4 Photoresistors

FIGURE 5.13

Photoresistors are light-controlled variable resistors. These are also known as light-dependent resistors (LDRs). In terms of operation, a photoresistor is usually very resistive (in the megaohms) when placed in the dark. However, when it is illuminated, its resistance decreases significantly; it may drop as low as a few hundred ohms, depending on the light intensity. In terms of applications, photoresistors are used in light- and dark-activated switching circuits and in light-sensitive detector circuits. Figure 5.13 shows the symbol for a photoresistor.

5.4.1 How a Photoresistor Works

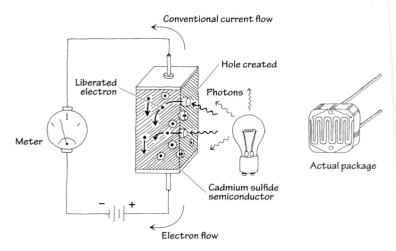

FIGURE 5.14

Photoresistors are made from a special kind of semiconductor crystal, such as cadmium sulfide (for light) or lead sulfide (for infrared). When this semiconductor is placed in the dark, electrons within its structure do not want to flow through the resistor because they are too strongly bound to the crystal's atoms. However, when illuminated, incoming photons of light collide with the bound electrons, stripping them from the binding atom, thus creating a hole in the process. These liberated electrons can now contribute to the current flowing through the device (the resistance goes down).

5.4.2 Technical Stuff

Photoresistors may require a few milliseconds or more to fully respond to changes in light intensity and may require a few seconds to return to their normal *dark resistance* once light is removed. In general, photoresistors pretty much all function in a similar manner. However, the sensitivity and resistance range of a photoresistor may vary greatly from one device to the next. Also, certain photoresistors may respond better to light that contains photons within a particular wavelength of the spectrum. For example, cadmium sulfide photoresistors respond best to light within the 400- to 800-nm range, whereas lead sulfide photoresistors respond best to infrared photons.

5.4.3 Applications

Simple Light Meter

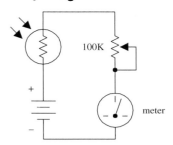

Here, a photoresistor acts as the light-sensing element in a simple light meter. When dark, the photoresistor is very resistive, and little current flows through the series loop; the meter is at its lowest deflection level. When an increasingly bright light source is shone on the photoresistor, the photoresistor's resistance begins to decrease, and more current begins to flow through the series loop; the meter starts to deflect. The potentiometer is used to adjust the sensitivity of the meter.

Light-Sensitive Voltage Divider

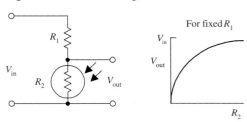

These circuits resemble the voltage-divider circuit described in Chap. 3. As before, the output voltage is given by

$$V_{out} = \frac{R_2}{R_1 + R_2} V_{in}$$

As the intensity of light increases, the resistance of the photoresistor decreases, so V_{out} in the top circuit gets smaller as more light hits it, whereas V_{out} in the lower circuit gets larger.

A voltage divider like this is usually used when using the photoresistor with a microcontroller.

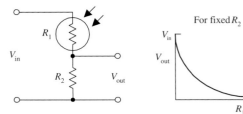

FIGURE 5.15

Dark-Activated Relay

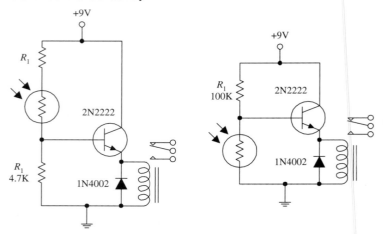

The two circuits shown here make use of light-sensitive voltage dividers to trip a relay whenever the light intensity changes. In the light-activated circuit, when light comes in contact with the photoresistor, the photoresistor's resistance decreases, causing an increase in the transistor's base current and voltage. If the base current and voltage are large enough, the transistor will allow enough current to pass from collector to emitter, triggering the relay. The dark-activated relay works in a similar but opposite manner. The value of R_1 in the light-activated circuit should be around 1 kΩ but may need some adjusting. R_1 in the dark-activated circuit (100 kΩ) also may need adjusting. A 6- to 9-V relay with a 500-Ω coil can be used in either circuit.

FIGURE 5.15   (*Continued*)

5.5   Photodiodes

Photodiodes are two-lead devices that convert light energy (photon energy) directly into electric current. If the anode and cathode leads of a photodiode are joined together by a wire and then the photodiode is placed in the dark, no current will flow through the wire. However, when the photodiode is illuminated, it suddenly becomes a small current source that pumps conventional current from the anode through the wire and into the cathode. Figure 5.16 depicts the symbol for a photo diode.

FIGURE 5.16

anode —————— cathode

Photodiodes are used most commonly to detect fast pulses of near-infrared light used in wireless communications. They are often found in light-meter circuits (e.g., camera light meters, intrusion alarms, etc.) because they have very linear light/current responses.

5.5.1   How a Photodiode Works

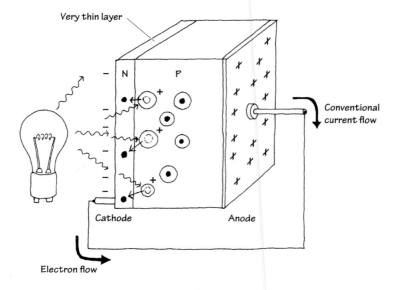

FIGURE 5.17

A photodiode is built by sandwiching a very thin *n*-type semiconductor together with a thicker *p*-type semiconductor. (The *n* side has an abundance of electrons; the *p* side has an abundance of holes.) The *n* side of the combination is considered the cathode, while the *p* side is considered the anode. Now, if you shine light on this device, a number of photons will pass through the *n*-semiconductor and into the *p*-semiconductor. Some of the photons that make it into the *p* side will then collide with bound electrons within the *p*-semiconductor, ejecting them and creating holes in the process. If these collisions are close enough to the *pn* interface, the ejected electrons will then cross the junction. What you get in the end is extra electrons on the *n* side and extra holes on the *p* side. This segregation of positive and negative charges leads to a potential difference being formed across the *p-n* junction. Now if you connect a wire from the cathode (*n* side) to the anode (*p* side), electrons will flow through the wire, from the electron-abundant cathode end to the hole-abundant anode end (or if you like, a conventional positive current will flow from the anode to cathode). A commercial photodiode typically places the *p-n* semiconductor in a plastic or metal case that contains a window. The window may contain a magnifying lens and a filter.

5.5.2   Basic Operations

Photovoltaic Current Source

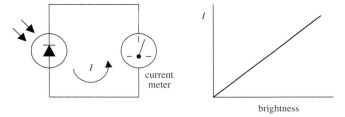

current meter

brightness

Here, a photodiode acts to convert light energy directly into electric current that can be measured with a meter. The input intensity of light (brightness) and the output current are nearly linear.

Photoconductive Operation

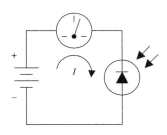

FIGURE 5.18

Individual photodiodes may not produce enough current needed to drive a particular light-sensitive circuit. Usually they are incorporated along with a voltage source. Here, a photodiode is connected in reversed-biased direction with a battery. When dark, a small current called the *dark current* (within the nA range) flows through the photodiode. When the photodiode is illuminated, a larger current flows. This circuit, unlike the preceding circuit, uses the battery for increased output current. A resistor placed in series with the diode and battery can be used to calibrate the meter. Note that if you treat the photodiode as if it were an ordinary diode, conduction will not occur; it must be pointed in the opposite direction.

5.5.3   Kinds of Photodiodes

Photodiodes come in all shapes and sizes. Some come with built-in lenses, some come with optical filters, some are designed for high-speed responses, some have large surface areas for high sensitivity, and some have small surface areas. When the surface area of a photodiode increases, the response time tends to slow down. Table 5.2 presents a sample portion of a specifications table for a photodiode.

FIGURE 5.19

TABLE 5.2   Part of a Specifications Table for a Photodiode

MNFR #	DESCRIPTION	REVERSE VOLTAGE (V) V_R	MAX. DARK CURRENT (nA) I_D	MIN. LIGHT CURRENT (μA) I_L	POWER DISSIPATION (mW) P_D	RISE TIME (ns) t_r	TYPICAL DETECTION ANGLE (°)	TYPICAL PEAK EMISSION WAVELENGTH (nm) λ_P
NTE3033	Infrared	30	50	35	100	50	65	900

5.6 Solar Cells

Solar cells are photodiodes with very large surface areas. The large surface area of a solar cell makes the device more sensitive to incoming light, as well as more powerful (larger currents and voltages) than photodiodes. For example, a single silicon solar cell may be capable of producing a 0.5-V potential that can supply up to 0.1 A when exposed to bright light.

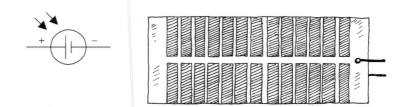

FIGURE 5.20

Solar cells can be used to power small devices such as solar-powered calculators or can be added in series to recharge nickel cadmium batteries. Often solar cells are used as light-sensitive elements in detectors of visible and near-infrared light (e.g., light meters, light-sensitive triggering mechanism for relays). Like photodiodes, solar cells have a positive and negative lead that must be connected to the more positive and more negative voltage regions within a circuit. The typical response time for a solar cell is around 20 ms.

5.6.1 Basic Operations

Power Sources

INCREASED VOLTAGE   INCREASED CURRENT

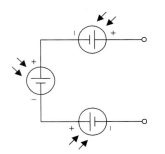
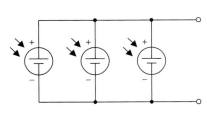

Like batteries, solar cells can be combined in series or parallel configurations. Each solar cell produces an open-circuit voltage from around 0.45 to 0.5 V and may generate as much as 0.1 A in bright light. By adding cells in series, the output voltage becomes the sum of the individual cell voltages. When cells are placed in parallel, the output current increases.

Battery Recharger

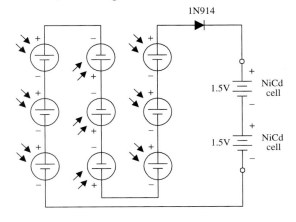

The circuit here shows how nine solar cells placed in series can be used to recharge two 1.5-V NiCd cells. (Each cell provides 0.5 V, so the total voltage is 4.5 V minus a 0.6-V drop due to the diode.) The diode is added to the circuit to prevent the NiCd cells from discharging through the solar cell during times of darkness. It is important not to exceed the safe charging rate of NiCd cells. To slow the charge rate, a resistor placed in series with the batteries can be added.

FIGURE 5.21

5.7 Phototransistors

Phototransistors are light-sensitive transistors. A common type of phototransistor resembles a bipolar transistor with its base lead removed and replaced with a light-sensitive surface area. When this surface area is kept dark, the device is off (practically no current flows through the collector-to-emitter region). However, when the light-sensitive region is exposed to light, a small base current is generated that controls a much larger collector-to-emitter current. Field-effect phototransistors (photoFETs) are light-sensitive field-effect transistors. Unlike photobipolar transistors, photoFETs use light to generate a gate voltage that is used to control a drain-source current. PhotoFETs are extremely sensitive to variations in light but are more fragile (electrically speaking) than bipolar phototransistors.

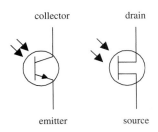

FIGURE 5.22

5.7.1   How a Phototransistor Works

Figure 5.23 shows a simple model of a two-lead bipolar phototransistor. The details of how this device works are given below.

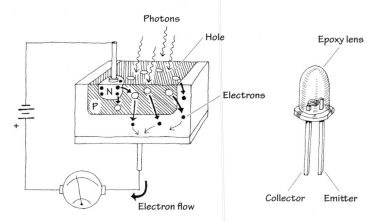

FIGURE 5.23

The bipolar phototransistor resembles a bipolar transistor (with no base lead) that has an extra large p-type semiconductor region that is open for light exposure. When photons from a light source collide with electrons within the p-type semiconductor, they gain enough energy to jump across the pn-junction energy barrier—provided the photons are of the right frequency/energy. As electrons jump from the p region into the lower n region, holes are created in the p-type semiconductor. The extra electrons injected into the lower n-type slab are drawn toward the positive terminal of the battery, while electrons from the negative terminal of the battery are drawn into the upper n-type semiconductor and across the n-p junction, where they combine with the holes. The net result is an electron current that flows from the emitter to the collector. In terms of conventional currents, everything is backward. That is, you would say that when the base region is exposed to light, a positive current I flows from the collector to the emitter. Commercial phototransistors often place the pnp semiconductor in an epoxy case that also acts as a magnifying lens. Other phototransistors use a metal container and a plastic window to encase the chip.

5.7.2   Basic Configurations

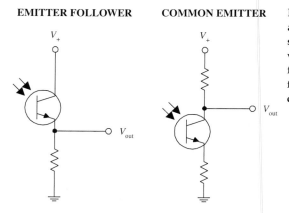

EMITTER FOLLOWER   **COMMON EMITTER**

In many ways, a phototransistor is much like an ordinary bipolar transistor. Here, you can see the emitter-follower (current gain, no voltage gain) and the common-emitter amplifier (voltage gain) configurations. The emitter-follower and common-emitter circuits are discussed in Chap. 4.

FIGURE 5.24

5.7.3 Kinds of Phototransistors

Three-Lead Phototransistor

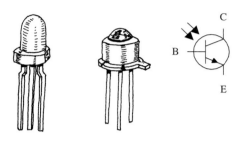

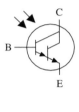

Two-lead phototransistors may not be able to inject enough electrons into the base region to promote a desired collector-to-emitter current. For this reason, a three-lead phototransistor with a base lead may be used. The extra base lead can be fed external current to help boost the number of electrons injected into the base region. In effect, the base current depends on both the light intensity and the supplied base current. With optoelectronic circuits, three-lead phototransistors are often used in place of two-lead devices, provided the base is left untouched.

Photodarlington

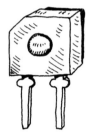

This is similar to a conventional bipolar Darlington transistor but is light-sensitive. Photodarlingtons are much more sensitive to light than ordinary phototransistors, but they tend to have slower response times. These devices may or may not come with a base lead.

FIGURE 5.25

5.7.4 Technical Stuff

Like ordinary transistors, phototransistors have maximum breakdown voltages and current and power dissipation ratings. The collector current I_C through a phototransistor depends directly on the input radiation density, the dc current gain of the device, and the external base current (for three-lead phototransistors). When a phototransistor is used to control a collector-to-emitter current, a small amount of leakage current, called the *dark current* I_D, will flow through the device even when the device is kept in the dark. This current is usually insignificant (within the nA range). Table 5.3 presents a portion of a typical data sheet for phototransistors.

TABLE 5.3 Part of a Specifications Table for Phototransistors

MNFR #	DESCRIPTION	COLLECTOR TO BASE VOLTAGE (V) BV_{CBO}	MAX. COLLECTOR CURRENT (mA) I_C	MAX. COLLECTOR DARK CURRENT (nA) I_D	MIN. LIGHT CURRENT (mA) I_L	MAX. POWER DISSIPATION (mW) P_D	TYPICAL RESPONSE TIME (µS)
NTE3031	*npn*, Si, visible and IR	30 (V_{CEO})	40	100 at 10 V V_{CE}	1	150	6
NTE3036	*npn*, Si, Darlington, visible and near IR	50	250	100	12	250	151

5.7.5 Applications

Light-Activated Relay

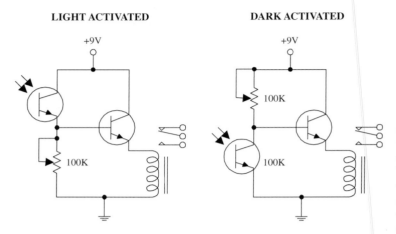

LIGHT ACTIVATED   **DARK ACTIVATED**

Here, a phototransistor is used to control the base current supplied to a power-switching transistor that is used to supply current to a relay. In the light-activated circuit, when light comes in contact with the phototransistor, the phototransistor turns on, allowing current to pass from the supply into the base of the power-switching transistor. The power-switching transistor then turns on, and current flows through the relay, triggering it to switch states. In the dark-activated circuit, an opposite effect occurs. When light is removed from the phototransistor, the phototransistor turns off, allowing more current to enter the base of the power-switching transistor. The 100-k potentiometers are used to adjust the sensitivity of both devices by controlling current flow through the phototransistor.

Receiver Circuit

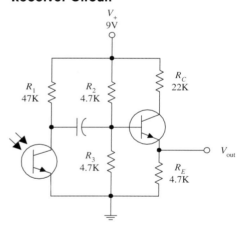

The circuit shown here illustrates how a phototransistor can be used as a modulated lightwave detector with an amplifier section (current gain amplifier). R_2 and R_3 are used to set the dc operating point of the power-switching transistor, and R_1 is used to set the sensitivity of the phototransistor. The capacitor acts to block unwanted dc signals from entering the amplifier section.

Tachometer

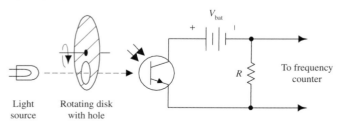

Light source   Rotating disk with hole

Here is a simple example of how a phototransistor can be used as a simple input stage for a frequency counter, or tachometer. A rotating disk with a hole (connected to a rotating shaft) will pass light through its hole once every revolution. The light that passes will then trigger the phototransistor into conduction. A frequency counter is used to count the number of electrical pulses generated.

FIGURE 5.26

5.8 **Photothyristors**

Photothyristors are light-activated thyristors. Two common photothyristors include the light-activated SCR (LASCR) and the light-activated triac. A LASCR acts like a switch that changes states whenever it is exposed to a pulse of light. Even when the light is removed, the LASCR remains on until the anode and cathode polarities are reversed or the power is removed. A light-active triac is similar to a LASCR but is designed to handle ac currents. The symbol for a LASCR is shown below.

LASCR

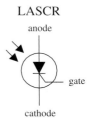

FIGURE 5.27

5.8.1 *How LASCRs Work*

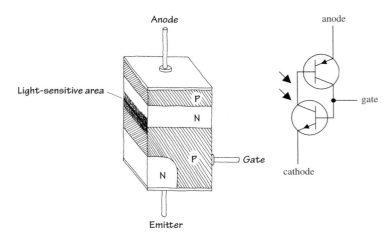

Light-sensitive area

FIGURE 5.28

The equivalent circuit shown here helps explain how a LASCR works. Again, like other *p-n* junction optoelectronic device, a photon will collide with an electron in the *p*-semiconductor side, and an electron will be ejected across the *p-n* junction into the *n* side. When a number of photons liberate a number of electrons across the junction, a large enough current at the base is generated to turn the transistors on. Even when the photons are eliminated, the LASCR will remain on until the polarities of the anode and cathode are reversed or the power is cut. (This results from the fact that the transistors' bases are continuously simulated by the main current flowing through the anode and cathode leads.)

5.8.2 *Basic Operation*

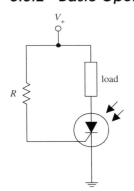

FIGURE 5.29

When no light is present, the LASCR is off; no current will flow through the load. However, when the LASCR is illuminated, it turns on, allowing current to flow through the load. The resistor in this circuit is used to set the triggering level of the LASCR.

5.9 Optoisolators

Optoisolators/optocouplers are devices that interconnect two circuits by means of an optical interface. For example, a typical optoisolator may consist of an LED and a phototransistor enclosed in a light-tight container. The LED portion of the optoisolator is connected to the source circuit, whereas the phototransistor portion is connected to the detector circuit. Whenever the LED is supplied current, it emits photons that are detected by the phototransistor. There are many other kinds of source-sensor combinations, such as LED-photodiode, LED-LASCR, and lamp-photoresistor pairs.

In terms of applications, optoisolators are used frequently to provide electrical isolation between two separate circuits. This means that one circuit can be used to control another circuit without undesirable changes in voltage and current that might occur if the two circuits were connected electrically. Isolation couplers typically are enclosed in a dark container, with both source and sensor facing each other. In such an arrangement, the optoisolator is referred to as a *closed pair* (see Fig. 5.30*a*). Besides being used for electrical isolation applications, closed pairs are also used for level conversions and solid-state relaying. A *slotted coupler/interrupter* is a device that contains an open slot between the source and sensor through which a blocker can be placed to interrupt light signals (see Fig. 5.30*b*). These devices are frequently used for object detection, bounce-free switching, and vibration detection. A *reflective pair* is another kind of optoisolator configuration that uses a source to emit light and a sensor to detect that light once it has reflected off an object. Reflective pairs are used as object detectors, reflectance monitors, tachometers, and movement detectors (see Fig. 5.30*c*).

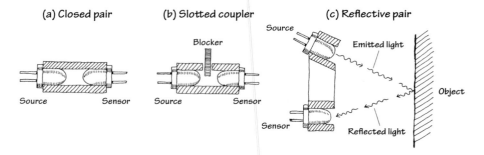

FIGURE 5.30

5.9.1 Integrated Optoisolators

Closed-pair optoisolators usually come in integrated packages. Figure 5.31 shows two sample optoisolator ICs.

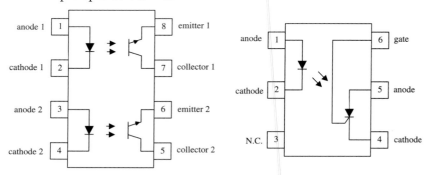

FIGURE 5.31

5.9.2 Applications

Basic Isolators/Level Shifters

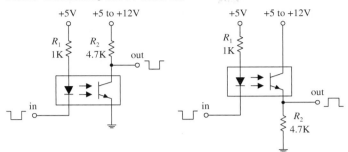

Here, a diode/phototransistor optoisolator is used to provide electrical isolation between the source circuit and the sensor circuit, as well as providing a dc level shift in the output. In the leftmost circuit, the output is noninverted, while the output in the rightmost circuit is inverted.

Optocoupler with Amplifier

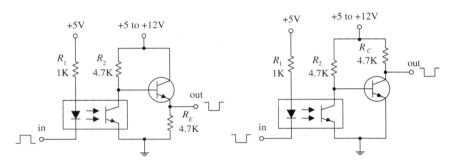

FIGURE 5.32

In optoelectronic applications, the phototransistor section of an optoisolator may not be able to provide enough power-handling capacity to switch large currents. The circuits to the right incorporate a power-switching transistor to solve this problem.

Solid-State Relay

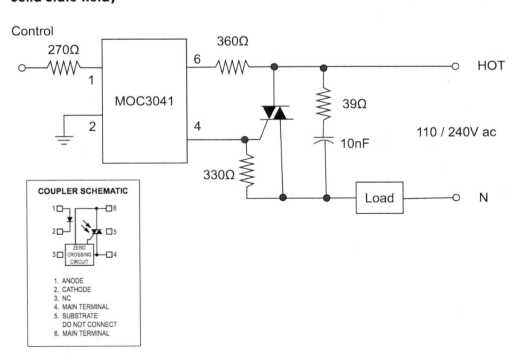

Optoisolated triacs like the MOC3041 have zero-crossing detection built in so that they turn on only at the zero-crossing point, minimizing current surges. This device can be used to control a regular triac to switch 110 V ac.

FIGURE 5.33

5.10 Optical Fiber

While strictly speaking, optical fibers are "optical" rather than "optoelectrical," they are often used with photodiodes and LEDs/laser diodes as a medium through which to carry data encoded on beams of light. Figure 5.34 shows how an optical fiber works.

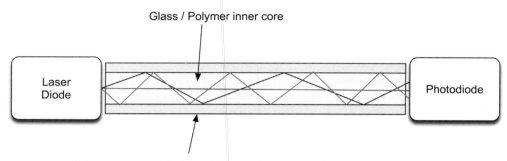

FIGURE 5.34   Optical fiber.

The boundary between the two layers, with an outer layer of higher reflective index, ensures that the fiber as a whole acts as a wave guide, with minimal losses of light as it bounces down the fiber. Therefore, optical fibers are much better than wires for transmitting data over long distances. They can also achieve staggering bandwidths, being immune to the induction and capacitive problems associated with wire. As such, they are heavily employed in the telecom industry, especially for high-bandwidth "pipes" between cities and also as submarine cables. It is becoming more and more common for fiber to replace copper in the final mile of telecom connections, actually terminating in homes to provide very high-speed Internet and cable TV.

As Fig. 5.34 implies, light will follow various paths through the fiber, and these paths will be of different lengths, limiting the effective bandwidth. This effect will reduce as the diameter of the internal core reduces. Taken to its limit, a single-mode fiber has a core of between 8 μm and 10.5 μm, and a cladding material diameter of 125 μm. This can be thought of as allowing only the straight-through path of light shown in Fig. 5.34. These fibers can achieve bandwidths in the order of 50 Gb/s over hundreds of miles.

Data are sent using a light source at the sending end (LED or laser diode) and a photodiode or transistor at the other end. Laser diodes are used in telecom systems because the coherent light that they produce travels better, but LEDs are also used in low-cost systems, such as fiber-based audio links in consumer digital audio systems.

CHAPTER 6

Sensors

Sensors are devices that measure a physical property, such as temperature, humidity, stress, and so on. As this is an electronics book, we are concerned with converting such measurements into electrical signals. We have also stretched the definition of a sensor a little to include things such as global positioning systems (GPSs) that locate your position in space.

Many sensors will provide an output signal that is simply a voltage proportional to the property being measured. Others are digital devices that provide digital data about the property. In both cases, these measurements will often be the input to a microcontroller.

6.1 General Principals

Before we look at some of the vast array of sensors that are available, we will cover some of the general principles that underpin all sensing.

6.1.1 Precision, Accuracy, and Resolution

When looking at a sensor, it is important to understand the difference between precision and accuracy, particularly as precision often gives a misleading impression of the accuracy of the sensor. A classic example of this is digital weighting scales. These may confidently tell you that you weigh 85.7 kilograms when you actually weigh 92.1 kilograms. This is only an error of 10 percent or so, and may not matter if you always use the same scales, but as an absolute measurement, it might be well off the mark.

So precision is a matter of the number of digits reported by the sensor. For an analog sensor, the readings are continuous.

When it comes to digital sensors, there will be a number of bits of resolution. That might be 8 bits (1 in 256), 12 bits (1 in 4096), or higher; that is, the reading is no longer continuous, but has been quantized.

TYPES OF SENSORS		Sensors in bold are covered in this chapter.
Category of Sensor	**What It Does**	**Example Devices**
Position Measuring Devices	 Designed to detect and respond to changes in angular position or in linear position of the device.	**Potentiometer** **Encoders:** Linear Position Sensor -**Quadrature** Hall Effect Position Sensor -Incremental Rotary Magnetoresistive Angular -**Absolute Rotary** -Optical
Proximity, Motion Sensors	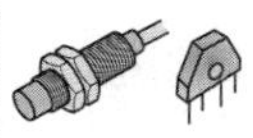 Designed to detect and respond to movement outside of the component but within the range of the sensor.	**Ultrasonic Proximity** **Inductive Proximity** **Optical Reflective** **Capacitive Proximity** **Optical Slotted** Reed Switch **PIR (Passive Infrared)** Tactile Switch
Inertial Devices	 Inertia Devices designed to changes in the physical movement of the sensor.	**Accelerometer** **Tilt Sensor** **Potentiometer** **Piezo Shock Sensor** Inclinometer LVDT/RVDT Gyroscope Vibration Sensor/Switch
Pressure/Force	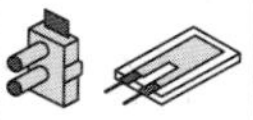 Pressure Devices designed to detect a force being exerted against it.	**IC Barometer** Piezoresistive sensor **Strain Gauge** Capacitive transducer Pressure potentiometer LVDT Silicon transducer
Optical Devices	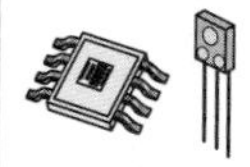 Optical Devices designed to detect the presence of light or a change in the amount of light on the sensor.	**LDR** IrDA Transceiver **Photodiodes** Solar Cells **Phototransistors** Photo interrupters **Reflective Sensors** LTV (Light to Voltage) Sensors
Image, Camera Devices	 Image, Camera Devices designed to detect and change a viewable image into a digital signal.	CMOS Image Sensor
Magnetic Devices	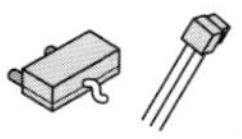 Magnetic Devices designed to detect and respond to the presence of a magnetic field.	**Hall Effect sensor** Magnetic Switch Linear Compass IC Reed Sensor
Media Devices	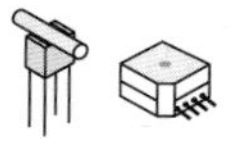 Media Devices designed to detect and respond to the presence or the amount of a physical substance on the sensor.	**Gas** Fluid Flow **Smoke** **Humidity, Moisture** Dust Float Level
Current and Voltage Devices	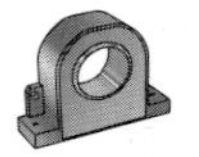 Current Devices designed to detect and respond to changes in the flow of electricity in a wire or circuit.	Hall Effect current sensor DC current sensor AC current sensor Voltage Transducer
Temperature	 Temperature Devices designed to detect the amount of heat using different techniques and in different mediums.	**Thermistor NTC** **Digital IC** **Thermistor PTC** **Analog IC** **Resistance Temp Detectors (RTD)** **Thermocouple** **Thermopile** Infrared Thermometer/Pyrometer
Specialized	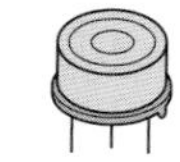 Specialized Devices designed to provide detection, measurement, or response in specialized situations, which also may include multiple functions.	**Audio Microphone** **Geiger-Müller Tube** **Chemical**

FIGURE 6.1   Sensor types.

6.1.2   The Observer Effect

The observer effect states that the act of observing a property changes it. This is the case in a car tire, for example. When you measure the pressure with a conventional tire gauge, you will let a little of the air out, thus changing the pressure that you are measuring. For most sensors, this change will have a negligible effect, but it is something worth keeping at the back of your mind when deciding whether you are getting a true reading.

6.1.3   Calibration

If your invention is a low-cost consumer item that will be mass-produced, then it will be much cheaper and easier to produce if there is no calibration to do. In fact,

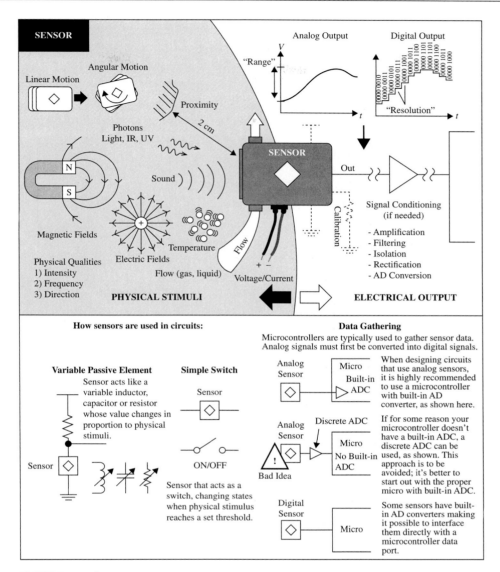

FIGURE 6.2   Sensors.

individual calibration to take differences in individual sensors into account is likely to be prohibitively expensive.

If, on the other hand, your invention is specialized, high-value equipment, then individual calibration of sensors is possible.

The principal of calibration is the same whatever property the sensor measures. The idea is that you take a number of readings from the sensor of unknown accuracy while the sensor is measuring a standard value of known accuracy. So, to calibrate a temperature sensor, your accurate standard might be boiling distilled water at 100°C, or more likely, the temperature in a highly accurate oven constructed for the purpose of calibrating the sensor. This oven will itself need to have been calibrated against an even more accurate standard.

When you have discovered the degree to which your sensor deviates from the standard, then you can compensate for it in some way. Since the sensor will almost certainly be supplying information to a microcontroller, a common way to make the calibration is to change the values in a lookup table. This table contains accurate values against raw values. For example, the raw value straight from a 12-bit analog-to-digital converter is a number between 0 and 1023. These numbers, perhaps in increments of 5, might be the left-hand side of the table, with the right-hand side of the table containing the equivalent temperature as a decimal in degrees Celsius. Interpolation can be used in the gaps between the raw readings by assuming that the small segment of the curve is linear (see Fig. 6.3).

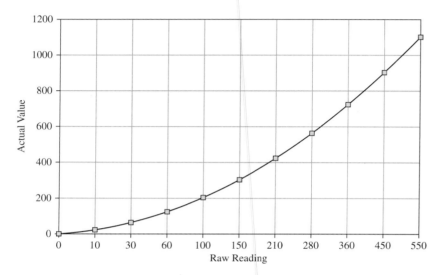

FIGURE 6.3   Actual reading against raw value for a fictional sensor.

Some IC sensors are actually calibrated individually during the manufacturing process, with the values for the lookup table being written into the read-only memory (ROM) of the sensor. This allows accurate sensing at low cost.

6.2   Temperature

Temperature sensors are one of the most common types of sensors. Computers and many types of equipment will sense their own temperature to prevent overheating. In addition, as well as electronic thermometers, there are thermostats that keep temperature constant by controlling the power to a load, usually just turning it on and off. Figure 6.4 shows a variety of temperature sensors.

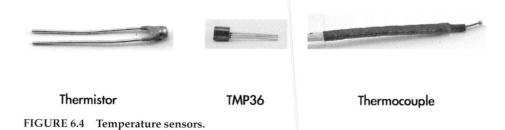

Thermistor   TMP36   Thermocouple

FIGURE 6.4   Temperature sensors.

6.2.1 Thermistors

The term "thermistor" is a combination of the words "thermal" and "resistor." A thermistor is a resistor whose resistance changes markedly with temperature. Thermistors come in two flavors: negative temperature coefficient (NTC) and positive temperature coefficient (PTC). An NTC thermistor's resistance will decrease with increasing temperature, and a PTC thermistor will behave the opposite way, with the resistance increasing as temperature increases. The NTC is the more common type.

The relationship between temperature and resistance in a thermocouple is not a linear one. Even over relatively short temperature ranges—say 0 to 100°C, a linear approximation will introduce considerable errors (see Fig. 6.5).

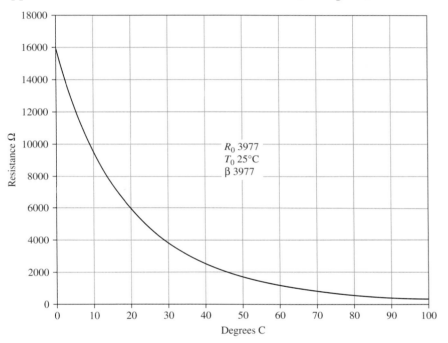

FIGURE 6.5 Resistance against temperature for an NTC thermistor.

The Steinhart-Hart equation is a third-order approximation used to determine the temperature as a function of the thermistor's resistance. It works for both NTC and PTC devices. It is normally stated as:

$$\frac{1}{T} = A + B\ln(R) + C\ln^3(R) \tag{6.1}$$

T is the temperature in Kelvin; R is the resistance of the thermistor; and A, B, and C are constants specific to that thermistor. The manufacturer of a thermistor will give all three constant values.

An alternative and more common model for this relationship uses a single parameter (*Beta*) and assumes constants of T_0 and R_0, where T_0 is usually 25°C and R_0 is the resistance of the thermistor at that temperature.

Using this equation, the temperature can be approximated to:

$$\frac{1}{T} = \frac{1}{Beta}\ln\left(\frac{R}{R_0}\right) + \frac{1}{T_0} \tag{6.2}$$

Rearranging this, we can also derive an expression for R:

$$R = R_0 e^{Beta\left(\frac{1}{T} - \frac{1}{T_0}\right)} \tag{6.3}$$

As well as providing *Beta*, T_0, and R_0, the data sheet will also specify a temperature range and accuracy.

To use a thermistor as a thermometer for input to a microcontroller, a voltage is required that can be measured by the analog-to-digital converter of the microcontroller. Figure 6.6 shows a typical arrangement using a potential divider with a fixed-value resistor of the same value as R_0.

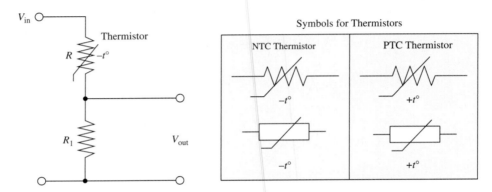

FIGURE 6.6   Using a thermistor as a thermometer.

If we place a NTC thermistor at the top of the potential divider, then as the temperature of the thermistor increases, its resistance falls and V_{out} rises.

Looking at Fig. 6.6:

$$V_{out} = \frac{R_1}{R_1 + R} V_{in} \tag{6.4}$$

Combining Formula 6.3 with Formula 6.4, we have:

$$V_{out} = \frac{R_1}{R_1 + \left(R_0 e^{Beta\left(\frac{1}{T} - \frac{1}{T_0}\right)}\right)} V_{in} \tag{6.5}$$

Example 1: Using a potential divider as shown in Fig. 6.3 with a fixed resistor of 4.7 kΩ and an NTC thermistor with a T_0 of 25°C, an R_0 of 4.7 kΩ, and a *Beta* of 3977, what is the formula for calculating V_{out}?

Answer 1: Just substituting the values into the equation, we get:

$$V_{out} = \frac{(5 \times 4700)}{4700 + \left(4700 e^{3977\left(\frac{1}{T} - \frac{1}{(25 + 273)}\right)}\right)}$$

$$V_{out} = \frac{5}{1 + \left(e^{3977\left(\frac{1}{T} - \frac{1}{(25+273)}\right)} \right)}$$

Note that 273 is added to the temperature in °C to give a temperature in °K.

Example 2: Using the same setup as Example 1, what is V_{out} at 25°C?

Answer 2: In one way, this is a trick question, since by definition the thermistor's resistance at 25°C will be 4.7 kΩ, so the voltage will be 2.5 V. However, we can apply the formula as a sanity check:

$$V_{out} = \frac{5}{1 + \left(e^{3977 \times 0} \right)}$$

$$V_{out} = \frac{5}{1+1} = 2.5 \text{ V}$$

Example 3: Using the same setup as Examples 1 and 2, what is V_{out} at 0°C?

Answer 3:

$$V_{out} = \frac{5}{1 + \left(e^{3977\left(\frac{1}{273} - \frac{1}{(25+273)}\right)} \right)}$$

$$V_{out} = \frac{5}{1 + \left(e^{3977\left(\frac{1}{273} - \frac{1}{(25+273)}\right)} \right)}$$

$$V_{out} = \frac{5}{4.34} = 1.15 \text{ V}$$

6.2.2 Thermocouples

While thermistors are good for measuring relatively small ranges in temperature—typically –40 to +125°C, for much higher temperatures and temperature ranges, thermocouples are used (see Fig. 6.7).

Any conductor subjected to a thermal gradient will experience something called the Seebeck effect; that is, it will generate a small voltage. The magnitude of this voltage is dependent on the type of metal, so if two different metals are joined, the temperature of that junction can be measured by measuring the voltage across the metal leads, at the far end from the junction. It is also necessary to measure the temperature at the other end of the thermocouple leads (often using a thermistor, since this will probably be at room temperature). This second temperature is called the "cold-junction" temperature.

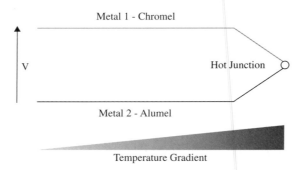

FIGURE 6.7 Thermocouples.

Lookup tables are often used to calculate the absolute temperature at the junction, based on the voltage and the cold-junction temperature, as the relationship is not completely linear and needs a fifth-order polynomial to model it accurately. Manufacturer's data sheets for thermocouples will normally contain large tables for calculating the temperature.

The most commonly used metals for a thermocouple are the alloys chromel (90 percent nickel and 10 percent chromium) and alumel (95 percent nickel, 2 percent manganese, 2 percent aluminum, and 1 percent silicon). A thermocouple made with these materials will typically be able to measure temperatures over the range –200°C to +1350°C. The sensitivity is 41 μV/°C for these metals.

6.2.3 Resistive Temperature Detectors

Resistive temperature detectors (RTDs) are perhaps the simplest temperature sensors to understand. Like thermistors, RTDs rely on their resistance changing with temperature. However, rather than use a special material that is sensitive to temperature changes (like a thermistor), they simply use a coil of wire (normally platinum) around a glass or ceramic core. The resistance of the core is often contrived to be 100 Ω at 0°C.

RTDs are much less sensitive than thermistors and can therefore be used over a much wider range of temperatures. The resistance of platinum changes in a relatively linear manner, and can be assumed to be linear for a range of 100°C or so. For the range 0 to 100°C, the resistance of a platinum RTD will vary by 0.003925 Ω/Ω/°C.

So, a 100 Ω (at 0°C) platinum RTD will, at a temperature of 100°C, have a resistance of:

$$100\ \Omega + 100°C \times 100\ \Omega \times 0.003925\ \Omega/\Omega/°C = 139.25\ \Omega$$

The first 100 Ω in the equation above is for the base resistance of the RTD at 0°C. This can be arranged in a potential divider in the same way as a thermistor.

6.2.4 Analog Output Thermometer ICs

An alternative to using a thermistor and fixed-value resistor in a potential divider arrangement is to use a special-purpose temperature measurement IC.

Devices such as the TMP36 come in a three-pin package and are used as shown in Fig. 6.8.

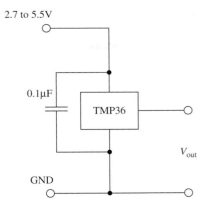

FIGURE 6.8   TMP36 temperature sensor.

Unlike a thermistor, the sensor's output voltage is almost linear at 10 mV/°C for a temperature range of –40 to +125°C. The accuracy is only ±2°C over the temperature range.

The temperature in °C is simply calculated from V_{out} using the formula:

$$T = 100V_{out} - 50$$

The constant 50 is specified in the data sheet for the TMP36.

These devices are considerably more expensive than thermistors and series resistors, but they are very easy and convenient to use.

6.2.5   Digital Thermometer ICs

An even higher technology approach to temperature measurement is to use a digital thermometer IC. These devices have a serial interface for use by microcontrollers.

A typical digital thermometer IC is the DS18B20. It uses a serial bus standard called 1-Wire that can allow multiple sensors to share the same data line (see Fig. 6.9).

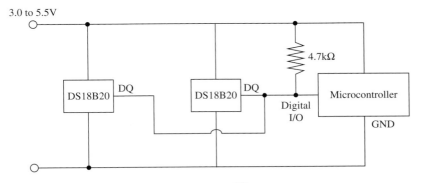

FIGURE 6.9   DS18B20 temperature sensor IC.

These devices are more accurate than the linear devices such as the TMP36. The digital thermometer IC has a stated accuracy of ±0.5°C and a temperature range of –55°C to +125°C.

Since digital thermometer ICs transmit their data digitally, they are very suitable for remote sensing, as lead length and electrical interference will have less effect than with an analog device. In fact, they will either work accurately or not at all in an electrically noisy environment.

Digital thermometer ICs can also be configured to work in "parasitic" power mode, where they harvest power from the data line, allowing them to be connected with just two wires. The GND and positive supply to the DS18B20 are both grounded, and the microcontroller must control a MOSFET transistor that pulls the data line up to the supply voltage under strict timing conditions.

Chapter 13 contains more information about using the 1-Wire interface with microcontrollers.

6.2.6 Infrared Thermometers/Pyrometers

If you have been to the doctor recently and have had your temperature taken, you will probably have seen an infrared thermometer that is placed near your ear and measures the temperature just on the inside surface of the ear without any actual contact. The ear is chosen because it is effectively a hole into the body that is free from the influence of external radiation, so that the back surface of the ear acts as a "black-body" radiator.

Pyrometers, or more specifically, broadband pyrometers, measure the radiation intensity, which has dimensions of energy per second per unit of area and use the Stefan-Boltzmann law to determine the temperature:

$$j^* = \sigma T^4$$

The radiation intensity (j^*) is proportional to the fourth power of the temperature, where σ is the Stefan-Boltzmann constant ($5.6704 \times 10^{-8} \, Js^{-1}m^{-2}K^{-4}$).

These devices use an infrared sensor such as the MLX90614 coupled with optics that focus the infrared radiation from the subject onto the sensor. When used for other measurement applications, the infrared thermometer will often have a low-power laser align with the sensor so that it can be aimed to take spot readings of temperature.

Devices such as the MLX90614 are more than just a sensor, containing a low-noise amplifier, high-precision analog-to-digital converter, and all the associated electronics to produce a digital output.

As with other IC sensors, the advantages of including all the electronics in the same package as the sensor itself are the reduction of noise and simplicity of interfacing.

Other sensors are available for high temperatures, such as those found in industrial furnaces, and the term "pyrometer," with its overtones of fire, is usually reserved for such high-temperature applications.

6.2.7 Summary

Table 6.1 shows a number of temperature sensors and their characteristics.

The figures for accuracy in Table 6.1 are somewhat arbitrary, as for most sensors, high accuracy can be obtained by individual calibration. It is also the accuracy of the overall system—including the electronics that use the sensor value—that counts. The figures assume the use of off-the-shelf components using data sheet parameters

TABLE 6.1   Temperature Sensors

SENSOR	TYPICAL TEMP RANGE (°C)	ACCURACY (±°C)	PROS	CONS	APPLICATIONS
Thermistor	−40 to 125	1	Low cost		Ambient temperature measurement
Thermocouple	−200 to 1350	3	Low cost	The lead is part of the sensor and it measures temperature difference	Industrial Domestic furnace
RTD	−260 to 800	1	Accurate Good linearity	Expensive Slow response time	
Analog IC	−40 to 125 (TMP36)	2	Simple to interface	More expensive than thermistor	Domestic thermostat Digital thermometer
Digital IC	−55 to 125 (DS18B20)	0.5	Simple to use with microcontrollers Accurate	More expensive than thermistor	Domestic thermostat Digital thermometer
Infrared thermometer/ pyrometer	−70 to 380 (MLX90614) Up to 1030 (MLX90616)	0.5	Does not need contact	More expensive than contact solutions	Medical Industrial control, especially where contact sensing is difficult

without any individual calibration and are meant as an indication of the likely accuracy of measurement that can be achieved with such a sensor.

6.3   Proximity and Touch

This section deals with detecting objects or measuring their distance from the sensor. Figure 6.10 shows a selection of such sensors.

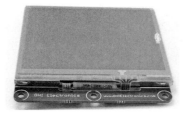  

Touch Screen   Ultrasonic Rangefinder   IR Proximity Sensor

FIGURE 6.10   Proximity and touch sensors.

6.3.1   Touch Screens

Touch screens are commonly used in cellular phones and tablet computers. There are many different technologies used in touch screens. The most commonly used are resistive touch screens.

Resistive touch screens are one of the older and more readily used touch screen technologies. They rely on a transparent sheet on top of the display. This sheet is flexible and also conductive. Figure 6.11 shows a typical arrangement for a four-wire resistive touch screen.

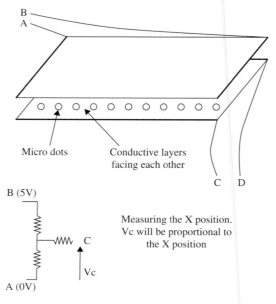

FIGURE 6.11   Resistive touch screen.

Both the top and bottom surfaces are coated with a conductive layer. Insulating dots are evenly sprayed onto the rigid bottom surface to keep the layers apart, except when they are pressed together.

To determine the X position of a touch, A is set to 0 V and B to 5 V, establishing a voltage gradient across the top surface. The voltage measured at C, or for that matter D, will be proportional to the X position. This is converted into coordinates using an analog-to-digital converter.

The conductive layer acts just like a potentiometer with C as the slider. If the device measuring the voltage at C has a very high input impedance, then the resistance of the track from the surface to the terminal C can be ignored. Most microcontrollers will have an analog-to-digital converter with a high input resistance, typically several MΩ. So, the voltage at C will be between 0 and 5 V in direct proportion to the distance from A of the touch.

When it comes to reading the Y position, some crafty footwork has to take place. Now C will be set to 0V, D to 5V, and the voltage measured at A or B.

All this processing will be carried out by either a special-purpose controller chip or a microcontroller.

6.3.2 Ultrasonic Distance

Ultrasonic distance-measuring devices are much loved by the developers of hobby robots. They are also found in commercial equivalents to the tape measure.

These devices send out a pulse of ultrasound and time how long it takes for the reflection to come back. A simple calculation involving the speed of sound will determine the distance to the object (see Fig. 6.12).

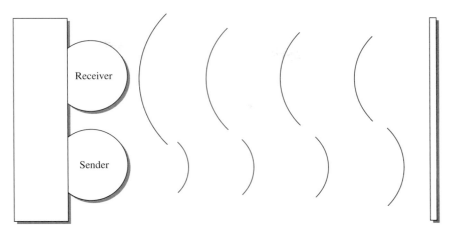

FIGURE 6.12   Ultrasonic distance measurement.

These devices have a measurement range of up to around 5 meters (15 ft), but this depends very much on the size and sound reflecting properties of the object. They also suffer limits of accuracy due to variations in the speed of sound, which is dependent on many factors, including atmospheric pressure, humidity, and temperature. For example, at 0°C, the speed of sound in dry air is 331 meter/second, but this rises to 346 meter/second at 25°C—a difference of 4.5 percent.

Sometimes separate transducers are used for sending and receiving the pulses of ultrasound, and sometimes a single transducer is used in both roles.

Some devices rely on a microcontroller to initiate the pulse and time the echo. Other, more expensive, devices will have their own digital processing that generates an output that can be read from a microcontroller as one of the following:

- An analog output voltage proportional to the distance

- Serial data

- A train of varying pulse lengths dependent on the distance

In some cases, all three outputs are provided, such as with the SparkFun device SKU: SEN-00639.

Example: Assuming that in dry air at 20°C, the speed of sound is 340 meter/second, if the time period between a pulse of ultrasound being sent is 10 milliseconds, how far away is the reflecting object?

Answer: distance = velocity × time = 340 × 0.01 = 3.4 m

However, that is the distance for the entire round-trip, so the actual distance to the object is half of that, or 1.7 meters.

6.3.3 Optical Distance

Another useful device for measuring distance at close range is the infrared optical sensor (see Fig. 6.13). This type of device uses the amount of reflection of a modulated infrared pulse to determine the distance to the object. The Sharp GP2Y0A21YK (SparkFun SKU: SEN-00242) is a typical sensor of this type.

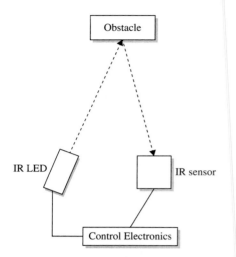

FIGURE 6.13   Optical distance measurement.

Infrared optical sensors are used for closer range measurements than the ultra-sonic devices and produce an analog output. This is not linear or particularly accurate, so these devices are most useful for simple proximity detection rather than actual measurement.

At close range, the measurements become ambiguous, as shown in the distance/output voltage plot of Fig. 6.14. Therefore, the sensors are usually mounted recessed so that no object can get closer than the ambiguous distance, which is about 5 cm for the sensor in Fig. 6.14.

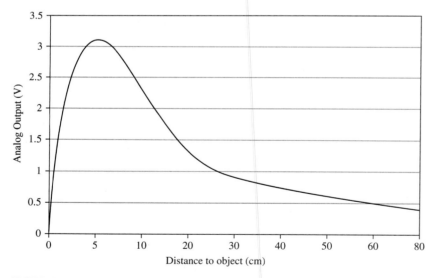

FIGURE 6.14   Output against distance for a Sharp GP2Y0A21YK.

When this type of sensor is used with a microcontroller, a lookup table is used to convert the measured voltage to a distance reading.

In some circumstances, you just need to know whether or not something is present. This can be detected with a slotted optical sensor, where the infrared source and sensor are aligned with a slot in between. When the beam is broken, the object is detected.

6.3.4 Capacitive Sensors

Capacitive sensors are frequently used as proximity or touch sensors as a replacement for mechanical push switches. They sense conductors and therefore are ideal for sensing proximity of a hand or finger.

There are various different approaches to capacitive sensing, but they all follow the basic principals shown in Fig. 6.15.

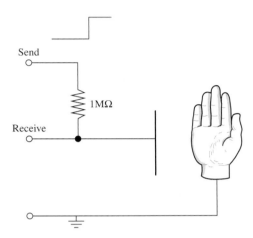

FIGURE 6.15   Capacitive sensing.

In this case, the sensing is accomplished using two general-purpose input/output (GPIO) pins of a microcontroller. The Send pin is configured as an output, and the Receive pin as an input. A single fixed resistor between the two pins forms one half of an RC arrangement, with the capacitor formed by the object being sensed and a sensor plate forming the other half of the capacitor.

When a hand moves close to the plate, the capacitance increases. This is sensed as the control software toggles the state of the Send pin and times how long it takes for the Receive pin to catch up with the changed state. In this way, it effectively measures the capacitance and hence the proximity of the object being sensed. The longer it takes for the Receive pin to change to the same state as the Send pin, the closer the object is to the plate.

This type of sensing has the advantages that it requires very little special hardware and can sense through glass, plastic, and other insulators. More advanced versions of this approach can sense in two dimensions to make a capacitative touch screen.

6.3.5 Summary

Table 6.2 compares the relative merits of the different types of distance sensors.

Another type of sensor that is used in industrial settings is the inductive sensor. This is in essence a mini-metal detector.

Proximity of a magnet can also be detected using magnetic sensors such as Hall effect sensors (discussed in Sec. 6.6.3) and even the humble read switch, which consists of a pair of contacts in a glass envelope that close when near a magnet.

TABLE 6.2 Distance Sensors

SENSOR	DISTANCE RANGE	ACCURACY	PROS	CONS	APPLICATIONS
Ultrasonic	150 mm to 6 m	25 mm	Simple to interface to	Relatively high cost	Industrial control Security Robotics
Optical reflective	100 to 800 mm	10 mm	Simple to interface to Low cost	Measurement sensitive to IR reflectivity of object	Industrial control Security Robotics
Optical slotted	N/A	N/A	Low cost	Detects only presence or absence, not distance	Industrial control
Capacitive	0 to 30 cm	N/A	Low cost	Detects conductive object with resistive link to earth only	Push switch replacement Touch screens

6.4 Movement, Force, and Pressure

Many different types of sensors can tell us about how things are moving. This has become a common occurrence in smartphones, where an accelerometer is used to detect the orientation of the device for automatic switching of the screen between landscape and portrait format, as well as for game playing. Figure 6.16 shows a variety of different movement-detecting devices.

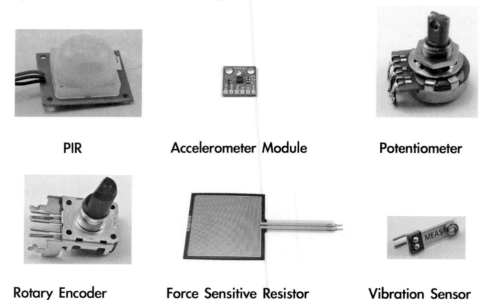

PIR   Accelerometer Module   Potentiometer

Rotary Encoder   Force Sensitive Resistor   Vibration Sensor

FIGURE 6.16 Movement sensors.

6.4.1 Passive Infrared

Passive infrared (PIR) detectors are most commonly used in intruder alarms. They detect changes in infrared heat. Because they require a plastic lens, they are usually sold as a detector module on a printed circuit board.

More advanced movement detectors will have multiple detectors that respond to a rapid change in the level of infrared heat. When this is detected, the module may activate a relay, or in some modules, switch a transistor with an open collector arrangement.

Figure 6.17 shows how to use an open collector device such as the PIR sensor from SparkFun (SKU: SEN-08630) to give a digital output.

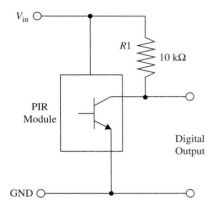

FIGURE 6.17   Using a passive infrared sensor.

6.4.2 Acceleration

The ready use of accelerometers in cellular phones has led to them becoming a low-cost and readily available component. The most common of these use a technology called microelectromechanical systems (MEMS). The basic principal for these devices is to measure the position of a mass attached to a spring that stretches under acceleration (see Fig. 6.18). The neat trick is that all of this is fabricated onto an IC.

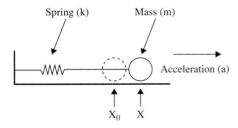

FIGURE 6.18   Mass and spring accelerometer.

This arrangement is basically the same as a spring balance of the sort favored by fishermen and travelers worried about exceeding their baggage allowance. In fact, turning it through 90 degrees, you would be measuring the gravitational constant whose units are those of an acceleration. This is how an accelerometer can determine its orientation on the vertical axis.

The force acting on the mass due to the acceleration is:

$$F = k\,(x - x_0) = m\,a$$

So:

$$a = \frac{k(x - x_0)}{m}$$

Since the manufacturer will know k and m, we just need to be able to measure the displacement of the mass m. Measuring this movement often uses the capacitive effect of the mass, in relation to one or two plates. A common arrangement is for the mass to move between two plates making a capacitive divider (see Fig. 6.19 and Sec. 2.23.14 in Chap. 2).

If the mass, shown as a rectangle in between two plates in Fig. 6.19, is halfway between the two plates A and C, then the capacitance between A and B and between B and C will be equal. An acceleration to the left of the diagram will move the weight to the right, changing the ratio of the capacitance.

Most accelerometers come in a little IC package that not only includes three accelerometers—one for each dimension—but also all the necessary electronics, analog-to-digital converters, and digital electronics to provide serial communication of the data back to a microcontroller using an I2C serial interface.

A much used low-cost accelerometer is the MMA8452Q (SparkFun SKU: COM-10953).

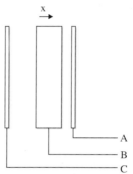

FIGURE 6.19 Measuring the offset of the mass.

6.4.3 Rotation

Measuring rotation can be as simple as using a potentiometer as a voltage divider, where the voltage at the slider is proportional to the angle of rotation (see Fig. 6.20). However, the potentiometer will not measure a full 360 degrees, and it has stops at both ends of the track that prevent it from turning all the way around.

A more flexible alternative to the potentiometer is the quadrature encoder. This type of device rotates without any stops, so you can just keep turning it. These are often used as an alternative to potentiometers in electrical appliances. They actually just measure steps to the left or steps to the right, rather than the absolute position.

A quadrature encoder is really just a pair of concentric tracks that each has a contact. They act as a pair of switches producing pulses as the shaft is rotated (see Fig. 6.21).

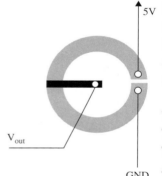

FIGURE 6.20 Measuring rotation with a potentiometer.

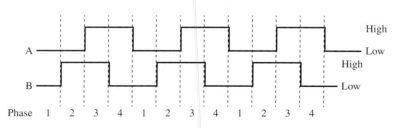

FIGURE 6.21 Outputs from a quadrature encoder.

The transitions of A and B are used to determine the direction in which the shaft is being rotated. For example, a clockwise rotation would be read as going from phase 1 to 4 and a counterclockwise from 4 down to 1.

Clockwise (AB): LH HL HH HL

Counterclockwise (AB): HL HH LH LL

As you can tell, this is a digital device that will be connected directly to a microcontroller.

Quadrature encoders come in various resolutions. Low-cost ones may have only 12 pulses per revolution. Better-quality devices can have up to 200 pulses per revolution and be designed for high-speed operation of up to 30,000 rpm. Such devices will use optical sensors rather than contacts.

A variation on the quadrature encoders that measure relative rotation are absolute rotary encoders. This type of device is essentially a rotary switch with a number of tracks that then indicate the angle by providing a binary output.

Rather than count in "normal" binary, the binary numbers for each position are arranged so that only one bit changes between one position of the switch and the next. This type of encoding is called a Gray code. For example, the 3-bit Gray code looks like this: 000, 001, 011, 010, 110, 111, 101, 100.

Table 6.3 shows some of the pros and cons of rotation sensors.

TABLE 6.3   Rotation Sensors

SENSOR	ACCURACY	PROS	CONS	APPLICATIONS
Potentiometer	2 to 5°	Cheap Measures absolute angle	End stops limited to 300°	Volume control, etc.
Quadrature encoder (low cost)	30°	Cheap	Measures changes in angle, not absolute angle	Volume control, etc.
Quadrature encoder (optical)	2°	Accurate High speed	Expensive	Industrial control
Absolute rotary encoder	22° (4-bit encoder)	Measures absolute angle	Not very accurate	Industrial control

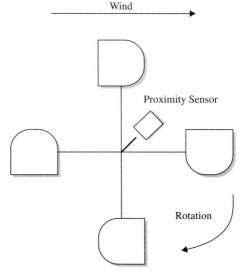

Wind

Proximity Sensor

Rotation

FIGURE 6.22   A cup anemometer.

6.4.4   Flow

The simplest sensors for measuring flow are variations on some kind of paddle wheel or turbine that is placed in the flow and spins at a speed proportional to the flow. The problem of measuring the rate of rotation remains. An anemometer that measures the wind speed is a typical example of this (see Fig. 6.22).

The wind will cause the arrangement of cups to spin at a rate of rotation proportional to the wind strength. Some fixed point on the cup assembly passes close to a proximity sensor that produces a pulse for each full rotation. In this way, you get a stream of pulses at a frequency proportional to the wind speed. Measuring the speed is then just a matter of using a microcontroller to count the number of pulses in a given period of time.

The proximity sensor may be a fixed magnet on the cup assembly and a magnetic sensor such as the Hall effect sensor (described in Sec. 6.6.3), or may be an optical sensor where a beam is interrupted.

Various types of fluid flow sensors are summarized in Table 6.4.

TABLE 6.4   Flow Sensors

SENSOR	DESCRIPTION	PROS	CONS
Hotwire	A current is passed through the wire to keep it at a constant temperature. Fluid flow draws heat away from the wire, and the current required to maintain the temperature is an indication of the rate of flow.	Simple	Hot
Ultrasonic (transit time)	Ultrasound is injected into the fluid from an upstream transducer, and the time to reach a downstream transducer is measured. The process is then reversed, and the fluid velocity can be determined from the difference in times.	No moving parts	Very dependent on geometry of the arrangement
Ultrasonic (Doppler)	A transmitter and receiver are used side by side to measure the difference in frequency between the transmission and ultrasound reflected from particles in the fluid.	No moving parts	Requires particles in the fluid
Laser (Doppler)	Similar to ultrasonic Doppler, monochromatic light from a laser is used. The difference in wavelength of the reflected light is measured.	Accurate	Requires particles in the fluid Expensive

6.4.5   Force

Force can be sensed using a force-sensitive resistor. These devices operate in a similar manner to a resistive touch screen. They are two layers, with conducting tracks and a microdot insulating layer, but the harder you press, the more of the top conductor comes into contact with the bottom conductor and the lower the resistance. These devices are not usually very accurate.

Strain gauges operate on the principal of deforming the geometry of a resistor as it is put under tension or compression (see Fig. 6.23).

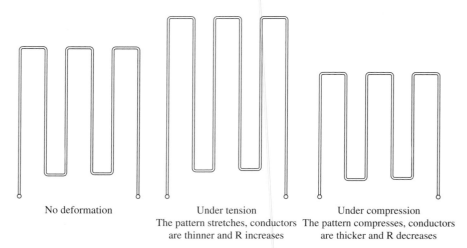

No deformation

Under tension
The pattern stretches, conductors are thinner and R increases

Under compression
The pattern compresses, conductors are thicker and R decreases

FIGURE 6.23   Strain gauge.

For a more accurate sensing of force, a load cell is used. This arranges two or more strain gauges bonded onto a deformable metal block. They will often also include temperature compensation.

6.4.6   Tilt

Although an accelerometer can be used to measure the angle of tilt, or just detect tilt from the horizontal, if all that is required is to know that the sensor has tilted from the horizontal, then a simpler sensor can be used. The simplest of these use a small metal ball in a case, which sits on contacts when horizontal but rolls off when the device is tipped.

6.4.7   Vibration and Mechanical Shock

Piezoelectric materials are good for detecting shock and vibration. The vibration sensor in Fig. 6.16 uses a small weight (just a rivet) on the end of a flexible strip of piezo-electric material. When the weight wobbles due to vibration, a voltage is generated across the terminals. The output of these voltages will drift, so for practical use, place a 10 MΩ resistor in parallel with them. This same approach will work for detecting mechanical shock.

6.4.8   Pressure

Unless you plan to work in process control or build a weather station, sensing pressure is unlikely to be something that you will need to do. Except for when you have very particular sensing requirements, the solution is nearly always to buy a digital sensor on a chip, such as the two listed in Table 6.5.

TABLE 6.5   Pressure Sensors

SENSOR	RANGE	DESCRIPTION	FEATURES	APPLICATIONS
Silicon (general purpose) (MPX2010)	0 to 10 kPa	A sensor on a chip with nozzles that measures pressure differential	Temperature compensated Good linearity Factory-calibrated Low cost	Respiratory diagnostics (medical) Air-movement control
Silicon (Atmospheric) (KP125)	40kPa to 115 kPa	A sensor on a chip top surface sensor that measures atmospheric pressure	Accurate (±1.2 kPa) Factory-calibrated	Altimeter Barometer

Other types of pressure sensors are on a larger scale, usually based on traditional pressure measurement, such as bellows or a Bourdon tube. These create a physical displacement that is then measured by one of the techniques discussed earlier, such as a potentiometer, a strain gauge, or capacitative position sensing like that used in an IC accelerometer.

6.5   Chemical

There are many sensors available for detecting chemicals of various kinds—far too many to list here. The selection shown in Fig. 6.24 illustrates a range of the sensors available, and we'll look at their principals of operation in the following sections.

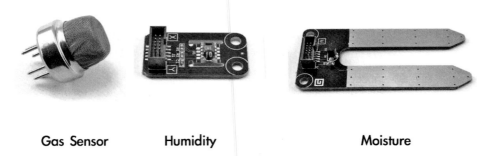

Gas Sensor   **Humidity**   **Moisture**

FIGURE 6.24   **Chemical sensors.**

6.5.1   Smoke

Domestic smoke detectors are (or should be) a feature of every home. The most common devices are actually quite sophisticated in their sensing (see Fig. 6.25).

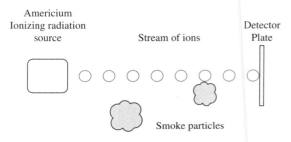

FIGURE 6.25   **An ionizing smoke detector.**

They use a radioisotope (Americium) that generates a stream of ionized particles in a chamber that is open to the air. This allows a small current to flow between the metal casing of the radiation source and a detecting plate at the far side of the chamber.

If smoke particles enter the chamber—even microscopic particles—they attach themselves to the ions, neutralizing them and reducing the current flowing. This is detected by the control electronics.

The other common type of smoke detector (optical) uses a focused infrared LED with a photodiode, off-axis from the infrared LED. When smoke enters the sensor, the smoke particles scatter light, which can be detected by the photodiode.

6.5.2   Gas

Gas detectors, such as the MQ-4 from Hanwei Electronics (SparkFun SKU: SEN-09404) shown in Fig. 6.24, contain a small heating element and a catalytic detector to detect concentrations of methane as low as 200 parts per million. Other types of sensors use different catalysts to make them sensitive to other gases.

Using such a device involves supplying a voltage (often 5 V) to the heating element (which typically draws a few tens of milliamps) and putting the sensor pins in a voltage divider arrangement with a fixed resistor to create a measurable output voltage.

6.5.3 Humidity

Although highly accurate techniques are available for measuring humidity, in many applications, an accuracy of 1 or 2 percent is acceptable and can be achieved with a low-cost capacitive sensor. These sensors are often combined with a temperature sensor, control electronics, and a serial interface for use by a microcontroller.

Capacitative sensing relies on humidity changing the dielectric constant of a polymer separating the two plates. These devices achieve better accuracy by laser calibration; that is, they are calibrated during manufacture by using a laser to write into digital memory on the device, setting a calibration parameter.

Unless extreme accuracy is required in a very specialized application, there is little reason to use anything other than an IC humidity sensor.

6.6 Light, Radiation, Magnetism, and Sound

Figure 6.26 shows a selection of various types of sensors associated with detecting radiation or magnetism.

Photoresistor (LDR)   **Photodiode**   **Geiger-Müller tube**

FIGURE 6.26   Light, radiation, and magnetic sensors.

6.6.1 Light

Light detection is used in many different ways, from optical transmission of data to proximity detection. There are many types of sensors for this purpose, including some we have already discussed, such as the photoresistor. For information about the use of photoresistors, photodiodes, and phototransistors, see Chap. 5.

6.6.2 Ionizing Radiation

To sense ionizing radiation, a Geiger–Müller tube is usually used. These tubes contain a low-pressure inert gas such as neon. A high voltage (400 to 500 V) is applied between the outer conducting cathode and a wire anode running down the middle of the tube (see Fig. 6.27).

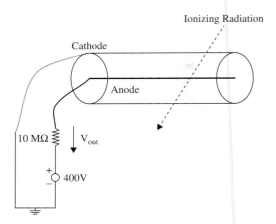

FIGURE 6.27 A Geiger–Müller tube.

When ionizing radiation passes through the tube, the gas is ionized and conducts, producing a pulse at V_{out}. These pulses are counted using a Geiger counter, traditionally producing a click sound with each event.

Different tubes are designed to be more sensitive to different types of radiation. To detect alpha particles, tubes with a very thin mica end will allow the easily stopped alpha particles to pass through and be detected.

6.6.3 Magnetic Fields

In discussing sensing rotational speed in a flow meter, we suggested detecting each rotation by sensing when a magnet passed a sensor. This sensor could just be a coil of wire, in which a current would be induced. However, the slower the magnet moves past the coil, the smaller the signal available. An alternative to this is to measure the static magnetic field and just detect the presence or absence of the magnet. A common device for doing this is called a Hall effect sensor.

The Hall effect is an electrical effect that occurs when a current passes through a conductor in the presence of a magnetic field. Referring to Fig. 6.28, when current is flowing along a bar of thickness d, in the presence of a magnetic field B, a voltage V_H will appear between the top and bottom surfaces of the conductor.

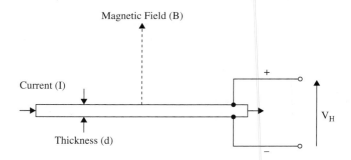

FIGURE 6.28 Hall effect sensor.

The voltage from the Hall effect is given by the equation:

$$V_H = \frac{-IB}{ned}$$

where n is the charge carrier density, e is the charge of an electron, d is the length of the conductor, I is the current flowing through it, and B is the magnetic field density.

Practical Hall effect sensors usually combine the sensor itself with high-gain amplification in a single IC. Others take this a step further and actually incorporate a serial interface to send the data to a microcontroller.

There are two types of Hall effect sensors. The most common type also acts as a switch and just provides an on-off indication of the presence of a nearby magnet. A more flexible type of Hall effect sensor (a linear sensor) provides a linear output voltage proportional to the strength of the magnetic field.

6.6.4 Sound

Sensing sound relies on using a microphone and amplifier (see Chap. 15). If you are sensing the sound level, then you will also need to use a low-pass filter and rectification to determine an indication of the level of sound (see Fig. 6.29).

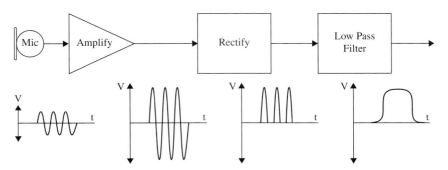

FIGURE 6.29   Sensing sound level.

6.7   GPS

A side effect of consumer devices, such as smartphones and dedicated satellite navigation systems, is that GPS modules are readily available at relatively low cost.

GPS relies on a constellation of satellites (see Fig. 6.30). Each satellite contains a highly accurate clock that is synchronized with all the other satellites in the constellation. The satellites then broadcast this time signal.

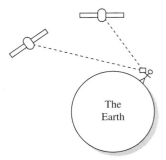

FIGURE 6.30   Global positioning system.

A GPS receiver on the ground will attempt to receive time signals from as many satellites as possible and use the differences in those times, due to the time for the signals to travel, to calculate the position of the receiver.

GPS modules are generally a single IC, but require a special antenna that is usually the largest part of the module. Figure 6.31 shows the Venus 638 FLPx GPS module from SparkFun with its antenna attached. This module and modules like it have a serial interface that transmits a stream of updates giving the position of the receiver and various other information, such as the time and number of satellites currently being received.

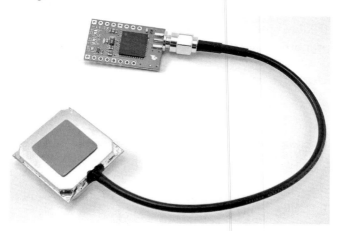

FIGURE 6.31   The SparkFun Venus638FLPx GPS module.

CHAPTER 7

Hands-on Electronics

7.1 Safety

7.1.1 Lecture on Safety

Probably the most hazardous thing in electronics is your household line voltage, around 120 V at 60 Hz in the United States (240 V at 50 Hz in many other countries). What often happens is an ungrounded metal object comes in contact with a hot wire, and then you touch the "hot object," discovering in the process that your body acts as a conducting medium for current to flow to ground. (The ground and ground wires are, in effect, at the same potential as the neutral wire—they are tied together at the utility box.) The frequency and amplitude of the line voltage tend to be at the perfect levels for inducing involuntary muscle contractions that prevent you from letting go of a hot or energized object. This freezing effect is extremely dangerous. The longer you're stuck, the more damage current will inflict upon your internal tissues (tissue heating), and the more likely you'll suffer from cardiac and/or respiratory arrest.

Voltage and frequency levels below household line levels can cause cardiac/respiratory arrest, too (especially if your internal resistance is low, such as with a wet body), but are less likely to cause the dreaded freezing effect. Frequencies above household line levels actually are less likely to cause severe muscle contractions, so the chances of fatality decrease somewhat. At very high voltages, there is an arcing effect that can occur, which can give you a nasty jolt even if you're not touching an object directly. The electric potential between you and the energized object is so great that the air becomes a conducting medium. Ironically, the initial jolt you receive from an arc will often be large enough to throw you clear of danger. However, besides being extremely painful, such jolts may induce cardiac/respiratory arrest, and depending on where you're standing, the fall may be what kills you.

In more specific terms, what kills you or what causes greatest tissue damage is the amperage. When you come in contact with a live wire or energized object, the amount of current that passes through your body to ground depends on the voltage level (assuming an ideal voltage source) and your internal resistance. A heavily callused dry palm may present a resistance of 1 MΩ, while a thin, wet palm may present only 100 Ω. Resistance tends to be lower in children, and different body tissues exhibit a range of resistances. Nerves, arteries, and muscles are low in resistance. Bone, fat, and tendon are relatively high in resistance. Across the chest of an average adult, the resistance is about 70 to 100 Ω. The low resistance across the chest and the higher resistance through the air-filled lungs provide a path of least resistance through the heart and spinal cord regions—the critical regions for life support. A 100- to 1000-mA current is sufficient to induce cardiac/respiratory arrest. Thermal burns due to I^2R losses through body parts can also be significant, resulting in loss of life or limb, even long after the initial exposure. Since the thermal heating effects increase with the square of the current, higher current levels truly cause nastier burns, whether they are internal or external.

As a rough guideline, a 10-mA, 50- to 60-Hz cycle line current, from hand to foot, will merely give you a tingly sensation. However, above 10 mA, at the same frequency, it's possible you'll freeze to the circuit you're touching. Outlets using ground fault circuit interrupters (GFCIs) will cut power if they sense sudden ground current flow in this current range—the tripping point around 5 to 10 mA. Currents from 20 to 100 mA may be fatal, but it's the 100-mA to 1-A range that's the deadliest. Above 1 A, the heart is thrown into a single contraction, and internal heating becomes significant. You may be thrown clear of the power source, but then cardiac and/or respiratory arrest may ensue.

The most lethal form of electrocution occurs via hand-to-hand current flow; current passes right across the heart, lungs, and spinal cord. Hand-to-foot electrocution is less fatal, carrying a mortality rate of around 20 percent. For this reason, when working on line voltages, make sure to keep one hand in your pocket. Another safety practice is to use the back of your hand when touching unknown entities. This prevents the "grip of death."

Consumer electronics equipment—such as TVs, computer monitors, microwave ovens, and electronic flash units—use voltage at power levels that are potentially lethal. Normally, the hazardous circuitry is safely enclosed to prevent accidental contact. However, during servicing, the cabinet or enclosure will likely be open and safety interlocks may be defeated. Depending on the type of circuitry and your general state of health, there is a wide variation of voltage, current, and total energy that can kill.

Microwave ovens are probably the most dangerous household appliance to mess around with. They operate at several thousand volts (5000 V or more), at higher current levels (more than an amp may be available momentarily). This is instantly a lethal combination.

Old-fashioned cathode-ray tube TVs and monitors may have 35 kV on the cathode-ray tube, with current lower than a couple of milliamps. However, the cathode-ray tube capacitance can hold a painful charge for quite some time. In addition, portions of the circuitry within TVs and monitors, as well as all other devices that plug into the wall socket (such as switching power supplies), are line connected. These circuits' internal grounds may be several hundred volts higher than the utility earth ground. Without isolation between the 120 V and the circuit, a real potential shock hazard exists if you come in contact with the circuit's floating ground (see the discussion of isolation transformers in Sec. 7.5.12). Switch-mode power supplies, electronic flash units, and strobe lights have large energy storage capacitors that can deliver a lethal discharge, long after the power has been removed. This even applies to disposable cameras with a flash. Even portions of apparently harmless devices like VCRs and CD players, or vacuum cleaners and toasters, can be hazardous (though the live parts may be insulated or protected, you shouldn't count on it).

The following list of safety tips will help you prolong your life.

ELECTRONICS SAFETY TIPS

1. Do not attempt repairs on line-powered circuits with the power on. Always remove the main power first.
2. Use one hand when taking measurements, and keep your other hand at your side or in your pocket. If you do get shocked, it's less likely that current will pass through your heart.
3. If you need to probe, solder, or otherwise touch circuits with the power off, discharge (across) a large power supply filter or an energy-storage capacitor with a 2-W or greater resistor of 100 to 500 Ω/V; for example, for a 200-V capacitor, use a 20- to 100-KΩ resistor. (A screwdriver's metal tip is often used to discharge a capacitor, but the sudden discharge may not be well tolerated by the capacitor.) Monitor while discharging and verify that there is no residual charge with a suitable voltmeter. Large capacitors can store a lethal amount of charge and may retain the charge for a number of days. Even capacitors rated at voltages as low as 5 or 10 V can be dangerous.
4. Perform as many tests as possible with the power off and the equipment unplugged. For example, the semiconductors in the power supply section can be tested for short circuits with an ohmmeter.
5. Avoid standing in a position that could be dangerous if you were to lose muscle control due to shock. Often the fall alone is more dangerous than the initial shock.
6. When working on high-power circuits, bring someone else along who can assist you if something goes wrong. If you see that someone cannot let go of a "hot" object, do not grab onto that person. Instead, use a stick or insulated object to push the person away from the source. It's not a bad idea to learn CPR, too.
7. All high-voltage test instruments (such as power supplies, signal generators, and oscilloscopes) that are operated on a 120 or 240 V ac supply should use three-wire line cable.
8. Use only shielded (insulated) lead probes when testing circuits. Never allow your fingers to slip down to the metal probe tip when testing a "hot" circuit. Also, make sure to remove power from a circuit when making a connection with wires and cables.
9. Connect/disconnect any test leads with the equipment unpowered or unplugged. Use clip leads or solder temporary wires to reach cramped or difficult-to-access locations. If you must probe live, put electrical tape over all but the last $\frac{1}{16}$ of the test probe to avoid the possibility of an accidental

(Continued)

short that could cause damage to various components. Clip the reference end of the meter or scope to the appropriate ground return so that you need to probe with only one hand.

10. If circuit boards need to be removed from their mounting, put insulating material between the boards and anything they may short to. Hold them in place with string or electrical tape. Prop them up with insulation sticks made of plastic or wood.

11. Set up your work area away from possible grounds that you may accidentally contact.

12. When working on ac-line circuits, wearing rubber-bottomed shoes or sneakers, or standing on a sheet of rubber or wood, can reduce possible shocks.

13. The use of a GFCI-protected outlet is a good idea, but it will not protect you from shock from many points in a line-connected device, such as a TV, monitor, or the high-voltage side of a microwave oven. (Note that a GFCI may be a nuisance by tripping at power-on or at other random times due to leakage paths, like your scope probe ground, or due to the highly capacitive or inductive input characteristics of line-powered equipment.)

14. A fuse or circuit breaker is too slow and insensitive to provide any protection for you or, in many cases, your equipment. However, these devices may save your scope probe ground wire should you accidentally connect it to a live chassis.

15. Know your equipment. TVs and monitors may use parts of the metal chassis as the ground return, yet the chassis may be electrically live with respect to the earth ground of the ac line. Microwave ovens use the chassis as the ground return for the high voltage. In addition, do not assume that the chassis is a suitable ground for your test equipment! Use an isolation transformer if there is any chance of contacting line-connected circuits. A Variac is not isolated, so you need a combination of a Variac and an isolation transformer for safety.

16. Make sure all components that are connected to ac power lines meet required power ratings.

17. When building power supplies and other instruments, make sure that all wires and components are enclosed in a metal box or insulated plastic enclosure. If you use a metal box, it is important that you ground the conductive shell (attach a wire from the inner surface of the box to the ground wire of the power cable, preferably with a screw along with solder). Grounding a metal box eliminates shock incurred as a result of a hot wire coming loose and falling onto the box, thereby making the whole box hot.

18. When making holes in metal boxes through which to insert an ac line cable, place a rubber grommet around the inner edge of the hole to eliminate the chance of fraying the cable.

19. Wear eye protection: large plastic-lens eyeglasses or safety goggles.

20. Don't wear any jewelry or other articles that could accidentally contact circuitry and conduct current.

7.1.2   Damaging Components with Electrostatic Discharge

Scuffling across a carpet while wearing sneakers on a dry day can result in a transfer of electrons from the carpet to your body. In such a case, it is entirely possible that you will assume a potential of 1000 V relative to ground. Handling a polyethylene bag can result in static voltages of 300 V or more, whereas combing your hair can result in voltages as high as 2500 V. The drier the conditions (lower the humidity), the greater the chance is for these large voltages to form. Now, the amount of electrostatic discharge (ESD) that can result from an electrostatically charged body coming in contact with a grounded object is not of much concern in terms of human standards. However, the situation is entirely different when subjecting certain types of semiconductor devices to similar discharges.

Devices that are particularly vulnerable to damage include field-effect transistors, such as MOSFETs and JFETs. For example, a MOSFET, with its delicate gate-channel oxide insulator, can be destroyed easily if an electrostatically charged individual touches its gate. The gate-channel breakdown voltage will be exceeded, and a hole will be blown through the insulator, which will destroy the transistor. Here is a rundown of the vulnerable devices out there:

- *Extremely vulnerable:* MOS transistors, MOS ICs, JFETs, laser diodes, microwave transistors, and metal film resistors

- *Moderately vulnerable:* CMOS ICs, LS TTL ICs, Schottky TTL ICs, Schottky diodes, and linear ICs

- *Somewhat vulnerable:* TTL ICs, small signal diodes and transistors, and piezoelectric crystals

- *Not vulnerable:* Capacitors, carbon-composite resistors, inductors, and many other analog devices

7.1.3   Component Handling Precautions

Devices that are highly vulnerable to damage are often marked with "Caution, components subject to damage by static electricity." If you see a label like this, use the following precautions:

- Store components in their original packages, in electrically conductive containers (such as a metal sheet or aluminum foil), or in conductive foam packages.

- Do not touch leads of ESD-sensitive components.

- Discharge the static electricity on your body before touching components by touching a grounded metal surface such as a water pipe or large appliance.

- Never allow your clothing to make contact with components.

- Ground tabletops and soldering irons (or use a battery-powered soldering iron). You also should ground yourself with a conductive wrist guard that's connected with a wire to ground.

Never install or remove an ESD-sensitive component into or from a circuit when power is applied. Once the component is installed, the chances for damage are greatly reduced.

While these risks are real, and you will no doubt occasionally lose a component to static, it is also important to keep a sense of perspective and take the necessary precautions around devices that you know are expensive or vulnerable. You certainly do not need to wear a grounded wrist strap all the time.

7.2 Constructing Circuits

This section briefly discusses the steps involved in building a working circuit: drawing a circuit schematic, building a prototype, making a permanent circuit, finding a suitable enclosure for the circuit, and applying a sequence of troubleshooting steps to fix improperly functioning circuits.

7.2.1 Drawing a Circuit Schematic

The *circuit schematic*, or *circuit diagram*, is a blueprint of a circuit. For a schematic to be effective, it must include all the information necessary so that you or anyone else reading it can figure out what parts to buy, how to assemble the parts, and possibly what kind of output behavior to expect. The following are guidelines for making an easy-to-read, unambiguous schematic:

■ The standard convention used when drawing a schematic is to place inputs on the left, outputs on the right, positive supply terminals on the top, and negative supply or ground terminals on the bottom of the drawing.

■ Keep functional groups—such as amplifiers, input stages, filters, and so on—separated within the schematic. This will make it easier to isolate problems during the testing phase.

■ Give all circuit components symbol designations (such as R_1, C_3, Q_1, and *IC4*), and provide the exact size or type of component information (such as 100 k, 0.1 μF, 2N2222, or 741). It is also important to include the power rating for certain devices, such as resistors, capacitors, relays, speakers, and so on.

■ Use abbreviations for large-valued components (for example, 100 kΩ instead of 100,000 Ω, or 100 pF instead of 100×10^{-12} F). Common unit prefixes include $p = 10^{-12}$, $n = 10^{-9}$, $\mu = 10^{-6}$, $k = 10^{3}$, and $M = 10^{6}$.

■ When labeling ICs, place lead designations (such as pin numbers) on the outside of the device symbol, and place the name of the device on the inside.

■ In certain circuits where the exact shape of a waveform is of importance (such as logic circuits and inverting circuits), it is helpful to place a picture of the expected waveforms at locations of interest on your circuit diagram. This will help isolate problems later during the testing phase.

■ Power-supply connections to op amps and digital ICs are usually assumed and typically left out of the schematic. However, if you anticipate confusion later on, include these supply voltages in your drawing.

■ To indicate joined wires, place a dot at the junction point. Unjoined wires simply cross (do not include a dot in such cases).

■ Include a title area near the bottom of the page that contains the circuit name, the designer's name, and the date. It is also useful to leave room for a list of revisions.

Figure 7.1 shows a sample schematic.

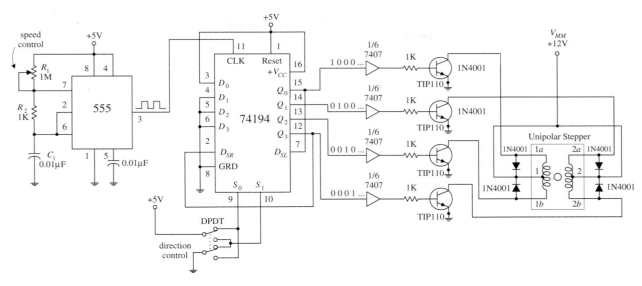

FIGURE 7.1

Once you have completed the circuit schematic, see if anything looks fishy. Are there missing connections or missing component values? Are component polarities indicated? Have you considered the power ratings of the components? Have you made connections as simple as possible? It is better to check things now; finding an error when you are soldering things together is much more annoying than erasing a few lines on a drawing.

You can draw the schematic with a pencil on paper, and often this is good enough while you are developing the project, especially if it is a simple one. However, at some stage, you will probably want to draw the circuit using an electronic computer-aided design (CAD) tool. The advantages of using a CAD tool are as follows:

- When you modify your schematic, the diagram stays clear in a way that a schematic with a lot of erasing doesn't.

- A good CAD system will apply electrical design rules to your schematic—at the very least, warning you about unconnected leads.

- A CAD system provides two views of a project: the schematic and the printed circuit board (PCB) layout. You can usually automatically route and lay out a PCB from the schematic.

- To be able to lay out a PCB, the CAD package needs to know about the geometry of the component. A catalog of components is supplied that contains information about the component package and pin assignments. Manufacturers will often provide component libraries for CAD packages.

There are many electronic CAD packages, including some that are very expensive. One of the most popular is EAGLE CAD. A free version of this product is available that restricts use to relatively small PCBs and a maximum of two layers of copper. Options for CAD packages are listed in Sec. 7.5.22. Most require a reasonable investment in time to learn to use properly.

If a full-blown CAD system is more than you need (or are prepared to learn), an alternative is to use a general-purpose drawing tool such as Microsoft Visio or OmniGraffle, both of which have libraries of electronics component symbols available.

7.2.2 A Note on Circuit Simulator Programs

Now, before you build a circuit, or even before you finish the schematic, you might consider using a circuit simulator program to test your idea to see if it works. Circuit simulator programs allow you to construct a computer model of your circuit and then test it (measure voltages, currents, wave patterns, logic states, and so on) without ever needing to touch a real component. A typical simulator program contains a library of analog and digital devices, both discrete and integrated in form. If you wish to model an oscillator circuit—one that is built from a few bipolar transistors, some resistors, a capacitor, and a dc power supply—all you do is select the parts from the library, set the values of these parts, and then arrange the parts to form an oscillator circuit.

To test the circuit, simply choose one of the simulator's test instruments, and then attach the test instrument's probes to the desired test points within the circuit. For example, if you are interested in what the output waveform of the oscillator looks like, choose the simulator's oscilloscope, and then attach the test leads and measure the output. The computer screen will then display a voltage versus time graph of the output. Some other instruments found in simulator programs include multimeters, logic analyzers, function generators, and bode plotters.

Why make a computer simulation of a circuit before building a real one? For several reasons:

- When you make a computer simulation, you do not need to worry about working with faulty components.

- There is no need to worry about destroying components with excessive current; the computer program does not "care about the numbers."

- A simulator program does all the mathematical work for you. The simulator allows you to fiddle around with component values until the circuit is working as desired.

- Using a simulator program can make learning electronics an intuitive process and saves time spent at the workbench.

Some popular simulator programs include Electronics Workbench, CircuitMaker, and MicroSim/Pspice. Electronics Workbench and CircuitMaker are relatively easy to use, while Pspice is a bit more technical.

7.2.3 Making a Prototype of Your Circuit

Once you are satisfied with the schematic, the next step is to make a prototype of your circuit. The most common tool used during the prototype phase is a solderless modular breadboard. A breadboard acts as a temporary assembly board on which electrical parts such as resistors, transistors, and ICs are placed and joined together by wires or built-in conductive pathways hidden underneath the surface of the breadboard (see Fig. 7.2).

Breadboards come with an array of small square sockets spaced about 0.100 in from center to center. When a wire or component lead is inserted into one of these sockets, a spring-like metal sleeve built into the socket acts to hold the wire or lead in place. Breadboard sockets are designed to accept 22-gauge wire but can expand to fit wire diameters between 0.015 and 0.032 in (0.38 and 0.81 mm). The upper and lower

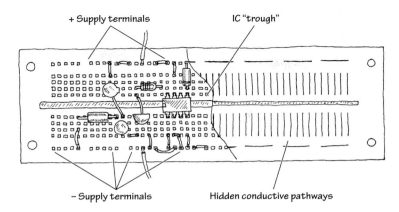

+ Supply terminals   IC "trough"

− Supply terminals   Hidden conductive pathways

FIGURE 7.2

rows of sockets of a breadboard typically are reserved for power-supply connections, while the sockets between the central gap region are reserved for DIP ICs.

See Sec. 7.5.17 for more on prototyping boards.

7.2.4 The Final Circuit

Once you are finished making a successful prototype, the next step is to construct a more permanent circuit. At this point, you must choose the type of mounting board on which to place your circuit. Your choices include a perforated board, a wire-wrap board, a pre-etched board, or a custom-etched PCB. Let's take a closer look at each of these boards.

Perforated Board

A *perforated board* is an insulated board with an array of holes drilled into it (see Fig. 7.3). To join a lead from one electrical component to another, each of the components' leads is placed through neighboring holes. The lead ends sticking out the back side of the board are then twisted together (and possibly soldered).

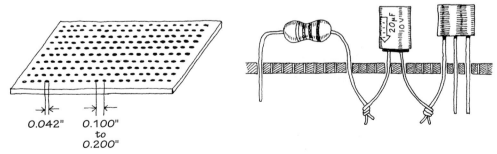

0.042"   0.100"
to
0.200"

FIGURE 7.3

Constructing a circuit on a perforated board is easy. Few supplies are needed, and making connections does not require much skill. However, what you get in the end is a fairly bulky circuit that is liable to fall apart over time and may pick up noise inadvertently (jumper wires will act like little antennas). In general, perforated boards are used for constructing simple, noncritical types of circuits.

Wire-Wrap

Using a *wire-wrap board* is another way to assemble moderately complex circuits containing ICs. Every wire-wrap board is made up of an array of sockets, each of which has a corresponding pin-like extension sticking out of the opposite side of the board (see Fig. 7.4).

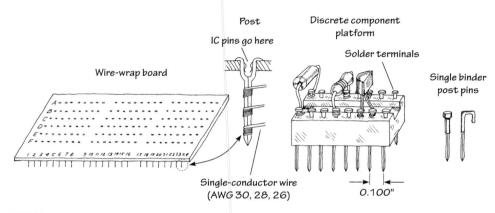

FIGURE 7.4

IC leads are inserted directly into the wire-wrap's sockets, while discrete components, such as resistors, capacitors, and transistors, must be mounted on special platforms like blocks or single post-like pins (see Fig. 7.4b). Each of these platforms contains a number of nail-like heads on which discrete component leads are attached, either by coiling the leads around the nail head or by fastening the leads to the nail head with solder. The nail-like tips of the platforms are then inserted into the board's sockets.

To connect components together, the pins on the back side of the wire-wrap board are joined together with wires (typically 30-, 28-, or 26-gauge single-conductor wires). In order to fasten the joining wires to the pins, a special wire-wrap tool is used (see Fig. 7.5). This tool wraps the joining wire around the pin by means of a bit section that rotates around a hollow core. The wire is inserted into the bit, the hollow core is placed over the pin, and then the wire-wrap tool is twisted several times (usually around seven times) and removed.

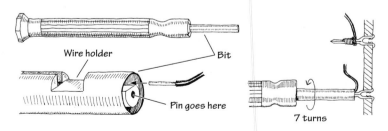

FIGURE 7.5

In practice, to save time and avoid making mistakes, it is desirable to do all the wrapping in a single pass. To follow this approach, some bookkeeping is required. Notice in Fig. 7.4a that each socket/pin is given a row/column designation.

For example, a pin located three rows down and two rows in from the left is given a C2 designation, and a pin that is located five rows down and seven rows in is given an E7 designation.

To figure out how to arrange electrical components on the board, making a simple sketch of the wire-wrap board is helpful. On the sketch, draw in all the components, fixing each component lead to a specific row/column coordinate. Once the sketch is complete, simply grab your wire-wrap tool and start making wire connections between the pins, using the sketch as a guide.

Wire-wrap boards are suitable platforms for circuits that contain a number of ICs, such as logic circuits. However, because the sockets of these boards are not designed for linear component leads (you must use platforms in such cases), it may be easier to use a pre-etched, or custom-etched PCB for building analog circuits.

Pre-etched Perforated Boards

A *pre-etched perforated board* is made of an insulating material that is coated with a pre-etched copper pattern and has a number of holes drilled into it. To join electrical components, simply place the leads of the components in the appropriate holes that are joined by a copper-etched strip, and then apply solder.

Pre-etched boards come with a variety of different etched patterns. Figure 7.6 shows samples of the kinds of patterns available.

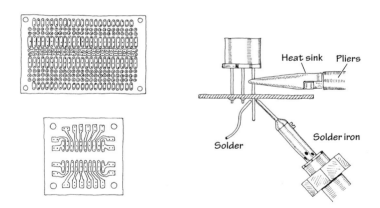

FIGURE 7.6

Custom-Etched Boards (PCBs)

If you are looking to build a circuit with a professional appearance, designing a *custom-etched circuit board*, or *PCB*, is the ticket, and the next section describes how to do this.

Designing custom-etched circuit boards takes a bit longer, but the time spent is often well worth the effort. There are times when making a custom-etched board is essential to ensure proper circuit operation, especially when dealing with circuits that contain components whose properties are greatly influenced by the length of the leads. For example, very high-speed logic circuits require the unique microstrip line geometries and precise placement of components to achieve fast rise times and at the same time avoid crosstalk among circuit elements. Sensitive low-level amplifier circuits also benefit from well-placed microstrip interconnections; the short and direct interconnections help eliminate noise pickup.

7.2.5 Making a PCB

To create the PCB layout, you can either draw the design by hand or use a CAD tool. Once you have the design, you have a number of choices for actually creating the board:

- Using an etch-resistant pen
- Printing the layout onto overhead projector (OHP) transparency film and using photosensitive board
- Using laser printer toner transfer
- Using CAD and a desktop router to cut out the pattern on the copper-clad board
- Sending the Gerber design files away to a PCB shop

Each of these approaches has pros and cons. The last option has started to become cost-effective, with prices as low as a dollar a board for a run of ten boards, and will produce professional-quality results. However, you do need to wait, sometimes for a week or two.

We will look at each of these approaches in turn.

Etch-Resistant Pen

Custom etching involves using graphic and chemical techniques to convert a copper-covered board into a custom-etched one. By doing your own custom etching, you can construct highly reliable, tightly compact circuits that require few jumper wires (see Fig. 7.7).

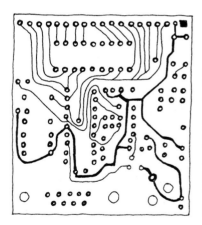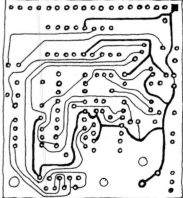

FIGURE 7.7

To design a custom-etched PCB, you first need an insulated board (usually $\frac{1}{16}$ in. thick and made from a fire-resistant epoxy-bonded fiberglass) that is completely covered on one or both sides with a very thin copper coating. Next, you must transform your circuit schematic into a PCB layout. This involves rearranging components in such ways as to make all conductive pathways short and direct. The layout should also eliminate any wire crossing, if possible.

Once you feel that your hardwired sketch is complete, the next step is to transfer it onto the copper-coated board. Afterward, the trick is to etch out all the undesired copper-coated sections while leaving the conductive pathways intact.

At this point, there are a number of different transferring/etching techniques from which you can choose. Perhaps the simplest technique involves using a PCB kit, which you can buy for a few bucks from a store such as RadioShack. A typical kit comes with a single or dual copper-coated board, a bottle of etching solvent, a permanent marker, a bottle of rubbing alcohol, and a drill bit.

To make your custom-etched board, you first transfer your hardwired sketch onto the surface of the board with, say, a pencil. Next, you drill in the appropriate holes where component leads are to go. Now, with the etch-resistant pen, you trace over the pencil sketch, making sure to encircle the drilled-out holes. After that, you place the board in a tub of etching solvent (typically ferric chloride) and wait until the copper dissolves away from the sections of the board that are not coated with magic-marker ink (the ink does not dissolve in the solvent; it acts to protect the underlying copper). After the board is removed from the solvent bath, it is washed off with water, and the magic-marker ink is then removed with a rag doused in rubbing alcohol.

Using a PCB kit is great for simple, single-run productions. PCB kits are easy to use, inexpensive, and require practically no special equipment other than what is provided in the kit itself. However, one problem with these kits is that you can construct only one circuit board at a time. Another problem with these kits is the limited precision you get by using a magic marker to create conductive pathways. If you are interested in making multiple copies of a circuit, and if you are looking for greater line precision, a more sophisticated technique that involves photochemical processes is required.

Photo-etching with OHP Transparency Film

Only the simplest of designs are really suited to hand-drawn PCB design. If you invest the time to learn a CAD system, it is unlikely that you will go back to the manual approach.

A great advantage of having your PCB design in a CAD system is that you can print it. For this technique, you print it onto transparent film, of the type that used to be used on overhead projectors. This film is available in letter and A4 sizes and comes in two types: one for laser printers and one for inkjet printers. Don't put the inkjet film into a laser printer (it is likely to melt onto the drum).

Using EAGLE PCB, as shown in Fig. 7.8, you can see the schematic (a), the PCB with all the component positions and labels (b), and, finally, just the copper (c).

The process followed here is to draw the schematic and then switch to the board view. Here, you drag the components onto the board, leaving airwires connecting the components according to the schematic. You can then load a set of design rules and click a button, and a PCB layout will be generated automatically. You will probably want to intervene, and rip up some of the wires and route them your own way, but generally, the automated layout isn't a bad start.

Once the design is printed onto the film, it is placed on top of presensitized copper-clad board. This board is PCB with a photosensitive layer that hardens when exposed to UV light. This is available from most hobby electronic stores. While short exposure to room lighting will do it no harm, you should keep it in a light-proof container.

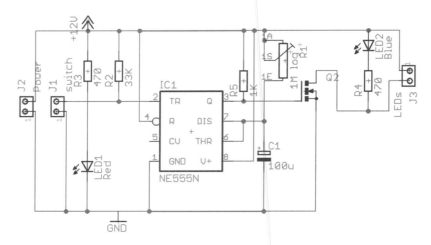

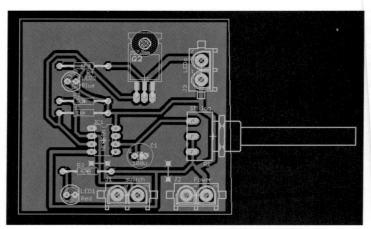

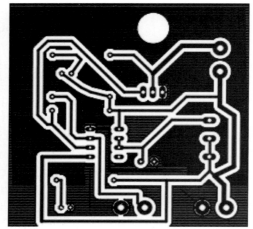

FIGURE 7.8 **EAGLE PCB, schematic and copper layer.**

FIGURE 7.9 **A photo-etching kit.**

The board-film sandwich is often held in a frame (a clip photo frame will do) while an exposure is made using a UV light box. The board is then put into a tray of developer, and the image of the PCB tracks will become visible on the board, just like an old-fashioned photograph being developed.

Next, the board is etched in a chemical that dissolves the copper except where it is protected by the photographic image of the PCB tracks. This part of the process is the same as for a pen-drawn PCB.

Figure 7.9 shows a homemade photo-etching kit that uses an array of UV LEDs inside an index card box to provide the UV exposure. Commercial kits are available, of course.

Note that the black areas will end up as copper. Quite large areas of black can be seen on the film. This reduces the quantity of copper that needs to be dissolved and hence extends the life of the chemical. The large areas of copper are generally connected to GND and are termed a *ground plane*.

Laser Printer Toner Transfer

A variation on the transparency film option is to print the PCB layout onto special paper using a laser printer, and then use a domestic iron to transfer the toner onto the copper-clad board. In this case, the image must be printed reversed.

With practice, good results can be achieved with this technique, which has the advantage of not requiring special tools. There are a number of Internet videos and tutorials describing this approach.

Creating a PCB with a Router

Low-cost desktop CNC routers offer a chemical-free method of producing PCBs by using normal copper-clad PCB, but then using a computer-controlled CNC router to cut away the unwanted copper (see Fig. 7.10).

FIGURE 7.10   A CNC router cutting a PCB.

The process is similar to the photo-etching method. Once the PCB artwork is done, the copper layer is dispatched to the router as if it were a printer.

Using a PCB Service

It used to be that designing a PCB was a very specialized skill, and in some ways, it still is. For very complex projects heading toward quantity production, the services of a specialist PCB design company are worth considering.

However, programs like EAGLE PCB will take a lot of the difficulty away, both by automating the track layout on the PCB and enforcing design rules such as minimum track widths, separations, and so on. This makes it entirely feasible to design your own PCBs and submit the finished design files to a PCB service. This service is likely to be almost entirely automated, so you will need to make sure your design is right before you send it, because these services will not check your design.

A PCB service will produce a great-looking result. All of these services generally allow a two-layer board, which provides a top and bottom copper layer. What is great is that they also do the following:

- Create the vias that link a track on one layer to one beneath it.
- Provide a silk-screen layer that can label the components and mark their positions.
- Provide a mask layer that covers all the copper not actually intended to be soldered.

Figure 7.11 shows a typical PCB produced by such a service. The cost for ten of these boards can be as low as $10 plus shipping.

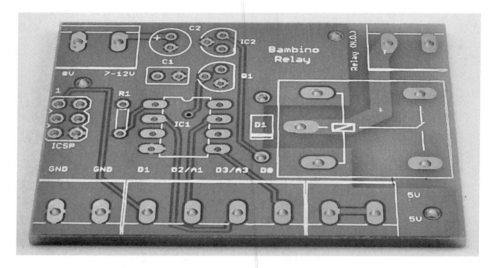

FIGURE 7.11   A PCB created by a PCB services.

The design files that you send to a PCB service are called Gerber files. Table 7.1 shows the files that you should submit. Each of the files has a different extension that indicates its contents. The computer-aided manufacturing (CAM) job feature of EAGLE CAD will produce these files for you automatically, using a CAM job file that you will find in the instructions on the PCB service's web site.

TABLE 7.1   Gerber File Types for CAM

FILE	CONTENTS
Myboard.GTL	Top layer (copper)
Myboard.GBL	Bottom layer (copper)
Myboard.GTS	Solder stop mark (top)
Myboard.GBS	Solder stop mark (bottom)
Myboard.GTO	Silk screen (top)
Myboard.GBO	Silk screen (bottom)
Myboard.TXT	Drill

All that remains to do is to pay, create a zip file containing the files, and e-mail them to the PCB service. Your boards will come back in the post days or weeks later, depending on the service level you chose.

Some PCB services will also accept EAGLE CAD .brd files and other CAD formats without requiring the generation of Gerber files.

When schematics get even a little complicated, you rapidly find that trying to route everything on one layer of copper becomes impossible, and a two-layer board becomes necessary. The top and bottom layers can either connect where a component lead naturally passes through both layers or using vias that are holes in the board just to allow a trace to jump layers.

New and better PCB deals come along all the time, so check the newsgroups to see what people are using. Two useful providers are Itead Studio (www.iteadstudio.com) and Elektor PCB Service (www.elektor.com).

A Note About Board Layout

When arranging components on a circuit board, ICs and resistors should be placed in rows and should all be pointing in the same direction. Also, make sure to leave about a 2 mm border around the circuit board to allow room for card lifters, guides, and standoffs.

Bring power supply leads or other I/O leads to the edge of the board, connecting them through an edge connector, D-connector, barrier-strip connector, or single-binder posts fixed to the edge of the board. Avoid mounting heavy components on circuit boards to prevent damage in case of a fall.

It is also a good idea to place polarity marking on the board next to devices such as diodes and electrolytic capacitors. Placing labels next to IC pins is also helpful. Consider labeling test points, trimmer functions (such as zero adjustment), inputs and outputs, indicator light functions, and power-supply terminals as well.

7.2.6 *Special Pieces of Hardware Used in Circuit Construction*

Three pieces of hardware are used frequently during the construction phase: prototyping boards with I/O interfacing gold-plated fingers, IC and transistor socket holders, and heat sinks.

Prototyping boards with gold-plated fingers typically are inserted into a card cage along with a number of other boards. Each board is inserted through a plastic guide and into an edge connector. Separate boards can be linked by means of a flat multiple-conductor cable (see Fig. 7.12a). The nice thing about these boards is that you can easily remove them from a cage to work on them, without making a mess of things in the process. When designing multiple-board devices, it is wise to use a separate board for each functional group of a circuit (for example, amplifier sections, memory-chip sections, and so on). This will make it easy to find and fix problems later on.

IC sockets are used in situations where there is a good chance that the device they house will need replacing (see Fig. 7.12b). It is tempting to use such holders everywhere within a circuit; however, placing too many IC sockets in a circuit board can lead to headaches later on. Often, the socket sections of these holders are poorly designed and may prove unreliable over time.

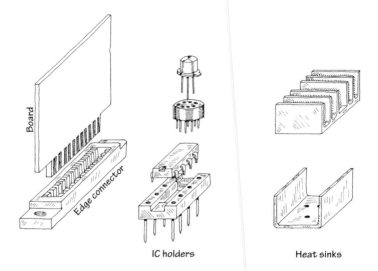

Board

Edge connector

IC holders

Heat sinks

FIGURE 7.12

Heat sinks are metal devices with large surface areas that are connected to heat-generating devices (such as power diodes and transistors) to help dissipate heat energy. A heat sink is usually connected to a component by means of a screw and washer fastener (see Fig. 7.12c). Silicon grease placed between the washer and heat sink is often used to enhance the thermal conductivity between the electrical component and heat sink.

7.2.7 Soldering

Solder is a tin-lead (these days, mostly without the lead) alloy used to join component leads together. Electrical solder often comes with a rosin flux mixed in, which is used to dissolve oxides that are present on the metal surfaces to be joined. Recent legislation in Europe (known as the Restriction of Hazardous Substances Directive, or RoHS) has outlawed the use of lead in consumer electronics. In the United States, there are tax incentives to reduce the use of lead. This has led (no pun intended) to the development of lead-free solders.

Before solder can be applied with a soldering iron, all metal surfaces must be cleansed of oils, silicones, waxes, or grease. Use a solvent, steel wool, or fine sandpaper to remove undesirable residue.

When soldering PCBs, use a low-wattage iron (25 to 40 W). To ensure good soldering connections, a thin, bright coating of molten solder should be present on the tip of the iron. With time, this coat becomes contaminated with oxides and should be renewed by wiping its surface with a sponge and then reapplying fresh solder. (Applying a fresh coat of solder to the tip of an iron is referred to as *tinning* an iron.)

The trick to making good soldering connections is to first heat the two metal pieces to be joined. Do not melt the solder first; otherwise, you will not be able to control the placement of the molten solder. Solder likes to flow toward hot spots.

Also when working on a circuit, make sure not to splatter solder on your board. If a small piece of solder lands between two separate conductive lines, you will end up shorting them. When you are finished soldering, inspect your work carefully for stray solder spatters and for a good sound joint.

To protect sensitive components from the heat of the soldering iron's tip, heat-sink components by gripping the component lead with a needle-nose pliers. Special heat-sink clips are also available for this purpose.

See Sec. 7.5.16 for more on soldering equipment.

7.2.8 Desoldering

If you make a bad connection or need to replace a component, you must melt the solder and start over again. But simply melting the solder and then attempting to yank the part while the solder is still wet is not always easy. This is especially true when it comes to dealing with ICs.

The trick to freeing a component from the solder's hold is to first melt the solder and then remove the solder with an aspiration tool, or "sucker." An aspiration tool typically resembles a turkey baster or a large syringe-like device. Another method that can be used to remove solder is to use a solder wick. This device resembles a braided copper wire and acts to draw solder away from a connection by means of capillary action.

7.2.9 Enclosing the Circuit

Circuits typically are enclosed in either an aluminum or plastic box. Aluminum enclosures are often used when designing high-voltage devices, whereas plastic containers typically are used for lower-voltage applications. If you design a high-voltage circuit housed in an aluminum box, make sure to ground the box to avoid getting shocked.

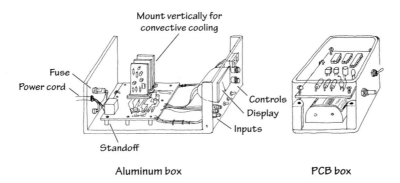

FIGURE 7.13

Circuit boards that are placed within an aluminum box should be supported off the ground with standoffs. If the circuit is ac-powered, drill a hole through the back side of the box, and insert a strain relief that grips the cable and a grommet around the edges of the hole. Place frequently used switches, dials, and indicators on the front panel, and place seldom-used switches and fuses on the back panel.

If you expect that your circuit will be generating a lot of heat (running more than 10 W or so), consider installing a blower fan. For circuits running on moderately low power, simple perforated holes placed on the top and bottom of the box will aid in conductive cooling.

Place major heat-producing components, such as power resistors and transistors, toward the back of the box, connecting them through the back panel to heat sinks. Make sure to orient heat sinks with their fins in the vertical direction. Also, if you are building a multiple-board cage device, align all boards vertically to allow for efficient ventilation.

Plastic boxes usually come with built-in standoffs on which to rest the circuit board. Typically, these boxes allow for extra room underneath the board for items such as batteries and speakers.

7.2.10 Useful Items to Keep Handy

The following items are worth having at your workbench:

- Needle-nose pliers
- Snippers
- Solder
- Soldering iron
- Solder sucker
- IC inserter
- Lead bender
- Solvent
- Clip-on heat sinks
- Circuit-board holder
- Screws (flathead and roundhead)
- Nuts
- Flat and lock washers (4-40, 6-32, 10-24)
- Binding posts
- Grommets
- Standoffs
- Cable clamps
- Line cord
- Hookup wire
- Shrink tubing in assorted sizes
- Eyelets
- Fuse holders

Sections 7.5.18 through 7.5.21 provide a more complete list.

7.2.11 Troubleshooting the Circuits You Build

If your circuit is malfunctioning, see if you have overlooked any suggestion in the troubleshooting flowchart in Fig. 7.14.

STARTING POINT

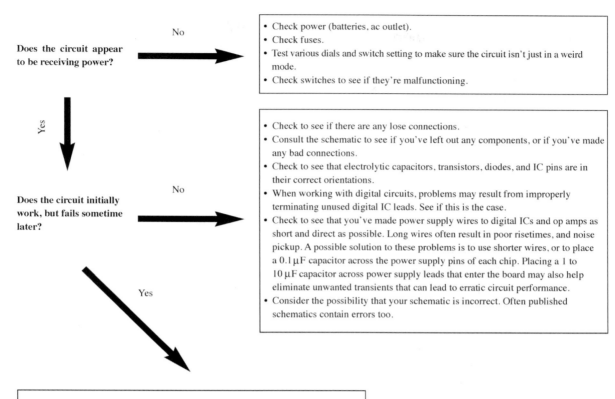

Does the circuit appear
to be receiving power?

No

• Check power (batteries, ac outlet).
• Check fuses.
• Test various dials and switch setting to make sure the circuit isn't just in a weird mode.
• Check switches to see if they're malfunctioning.

Yes

Does the circuit initially
work, but fails sometime
later?

No

• Check to see if there are any lose connections.
• Consult the schematic to see if you've left out any components, or if you've made any bad connections.
• Check to see that electrolytic capacitors, transistors, diodes, and IC pins are in their correct orientations.
• When working with digital circuits, problems may result from improperly terminating unused digital IC leads. See if this is the case.
• Check to see that you've made power supply wires to digital ICs and op amps as short and direct as possible. Long wires often result in poor risetimes, and noise pickup. A possible solution to these problems is to use shorter wires, or to place a 0.1 μF capacitor across the power supply pins of each chip. Placing a 1 to 10 μF capacitor across power supply leads that enter the board may also help eliminate unwanted transients that can lead to erratic circuit performance.
• Consider the possibility that your schematic is incorrect. Often published schematics contain errors too.

Yes

Check to see if any components are becoming excessively hot to the touch. Do you smell anything burning? If so, the problem may be that you're overheating a component. In such a case, simply replacing the component with another that has a higher power rating may be the answer. Or perhaps, a heat-sink is all that's needed.

If all else fails . . .

Attempt to isolate the area of the circuit where you think the problem lays. Perhaps, simply replacing a suspect part may be the answer. If you're still in doubt, consult literature that talks about circuits similar to yours. These sources often mention possible sources of error, and then give you a solution.

FIGURE 7.14

7.3 Multimeters

A *multimeter*, or volt-ohm-milliammeter (VOM), is an instrument used to measure current, voltage, and resistance. The two most common types of multimeters include the analog VOM and the digital VOM, as shown in Fig. 7.15.

The obvious difference between the two types of VOMs is that an analog VOM uses a moving-pointer mechanism that swings along a calibrated scale, while the digital VOM uses some complex digital circuitry to convert input measurements into a digitally displayed reading. Technically speaking, analog VOMs are somewhat less accurate than digital VOMs (they typically have a 3 percent higher error in reading

Analog multimeter

Digital multimeter

FIGURE 7.15

than a digital VOM), and multiple scales make them harder to read. Also, the resolution (displayable accuracy) for an analog VOM is roughly 1 part in 100, as compared with a 1 part in 1000 resolution for a digital VOM. Despite these limitations, analog VOMs are superior to digital VOMs when it comes to testing circuits that contain considerable electrical noise. Unlike digital VOMs, which may go blank when noise is present, analog VOMs are relatively immune to such disturbances.

Here, we'll look at how these devices work. Also see Sec. 7.5.3, which discusses which type of multimeter you may need for your electronics laboratory.

7.3.1 Basic Operation

Measuring Voltages

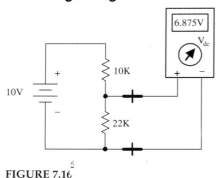

FIGURE 7.16

The trick to measuring voltages with a VOM is to turn the selector knob to the voltage setting. If you want to measure a dc voltage, the knob is turned to the appropriate dc voltage-level setting. If you wish to measure an ac voltage, the knob is turned to the ac voltage setting (V_{ac}, or V_{rms}). Note that the displayed voltage in the V_{ac} setting is the RMS voltage ($V_{rms} = 0.707\ V_{peak-to-peak}$). Once the VOM is set correctly, the voltage between two points in a circuit can be measured by touching the VOM's probes to these points (the VOM is placed in parallel). For example, Fig. 7.16 shows the procedure used to measure the voltage drop across a resistor.

Measuring Currents

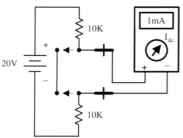

FIGURE 7.17

Measuring currents with a VOM is almost as easy as measuring voltages. The only difference (besides changing the setting) is that you must break the test circuit at the location where you wish to make a current reading. Once the circuit is open, the two probes of the VOM are placed across the break to complete the circuit (VOM is placed in series). Fig. 7.17 shows how this is done. When measuring ac currents, the VOM must be set to the RMS current setting.

Measuring Resistances

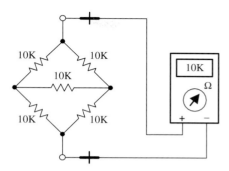

FIGURE 7.18

Measuring resistances with a VOM is simple enough: remove the power to the resistive section of interest, and then place the VOM's probes across this section. Of course, make sure to turn the VOM selector knob to the ohms setting beforehand.

7.3.2   How Analog VOMs Work

An analog VOM contains an ammeter, voltmeter, and ohmmeter all in one. In principle, understanding how each one of these meters works individually will help to explain how an analog VOM works as a whole.

Ammeter

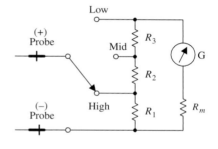

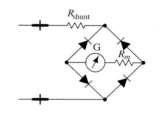

FIGURE 7.19

An ammeter uses a D-Arsonval galvanometer that consists of a current-controlled electromagnetic that imparts a torque on a spring-loaded rotatable needle. The amount of deflection of the needle is proportional to the current flow through the electromagnet. The electromagnetic coil has some inherent resistance, which means you need to throw R_m into the circuit, as shown in Fig. 7.19. (R_m is typically around 2 k or so.)

Now, a galvanometer alone could be used to measure currents directly; however, if the input current is excessively large, it will force the needle beyond the viewable scale. To avoid this effect, a number of *shunt resistors* placed in parallel with the galvanometer make up a current divider capable of diverting some of the "needle-bending" current away from the galvanometer. The current value read from the display must be read from the appropriate ruler marking on the display that corresponds to the shunt resistance chosen.

To make this device capable of measuring ac currents, a bridge rectifier can be incorporated into the design to provide a dc current to the galvanometer (see the lower circuit). The dc current will produce a needle swing that is proportional to the alternating voltage measured. A typical ammeter has about a 2-k input resistance. Ideally, an ammeter should have zero input resistance.

Voltmeter

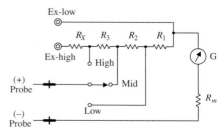

FIGURE 7.20

An analog voltmeter, like the ammeter, also uses a D-Arsonval galvanometer. Again, the galvanometer has some internal resistance (R_m). When the voltmeter's leads are placed across a voltage difference, a current will flow from the higher potential to lower potential, going through the galvanometer in the process. The current flow and the needle deflection are proportional to the voltage difference.

Again, as with the ammeter, shunt resistors are used to calibrate and control the amount of needle deflection. To make ac voltage measurements, a bridge rectifier, like the one shown in the previous example, can be incorporated into the meter's design. A typical voltmeter has an input resistance of 100 k. An ideal voltmeter should have infinite input resistance.

Ohmmeter

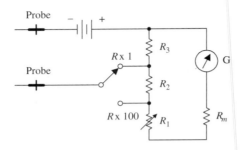

FIGURE 7.21

To measure resistance, an ohmmeter uses an internal battery to supply a current through the measured load and through a galvanometer (the load and galvanometer are in series). If the tested load is small, a large current will flow through the galvanometer, and a large deflection will occur. However, if the tested load resistance is large, the current flow and the deflection will be small. (In a VOM, the ohmmeter calibration markings are backward: 0 W is set to the right of the scale.)

The amount of current flow through the galvanometer is proportional to the load resistance. The ohmmeter is first calibrated by shorting the probe leads together and then zeroing the needle. Like the other meters, an ohmmeter uses a number of shunt resistors to control and calibrate the needle deflections. A typical ohmmeter has an input resistance of about 50 Ω. An ideal ohmmeter should have zero input resistance.

7.3.3 How Digital Multimeters Work

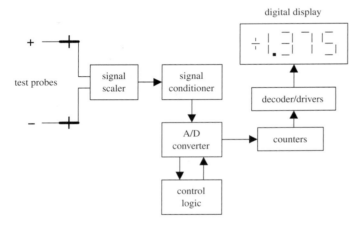

FIGURE 7.22

A digital multimeter is composed of a number of functional groups, as shown in the block diagram. The signal scaler is an attenuator amplifier that acts as a range selector. The signal conditioner converts the scaled input signal to a dc voltage within the range of the analog-to-digital converter (A/D converter). In the case of ac voltage measurements, the ac voltage is converted into a dc voltage via a precision rectifier-filter combination. The gain of the active filter is set to provide a dc level equal to the RMS value of the ac input voltage or current.

The signal conditioner also contains circuits to convert current and/or resistance into proportional dc voltages. The A/D converter converts the dc analog input voltage into a digital output voltage. The digital display provides a digital readout of the measured input. Control logic is used to synchronize the operation of the A/D converter and digital display.

7.3.4 A Note on Measurement Errors

When measuring the current through (or voltage/resistance across) a load, the reading obtained from the VOM will always be different when compared with the true value present before the meter was connected. This error comes from the internal resistance of the VOM.

For each setting (ammeter, voltmeter, and ohmmeter), there will be a different internal resistance. A real ammeter typically will have a small input resistance of around 2 k, while a voltmeter may have an internal resistance of 100 k or more. For an ohmmeter, the internal resistance is usually around 50 Ω. It is crucial to know these internal resistances in order to make accurate measurements. The following examples show how large the percentage error in readings can be for meters with corresponding input resistances.

Current-Measurement Error

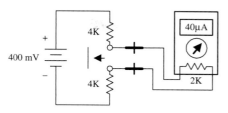

FIGURE 7.23

If an ammeter has an input resistance of 2 k, calculate the percentage error in reading for the circuit shown here.

$$I_{true} = \frac{400\,mV}{4\,k + 4\,k} = 50\,\mu A$$

$$I_{measured} = \frac{400\,mV}{4\,k + 4\,k + 2\,k} = 40\,\mu A$$

$$\% \text{ error } = \frac{50\,\mu A - 40\,\mu A}{50\,\mu A} \times 100\% = 20\%$$

Voltage-Measurement Error

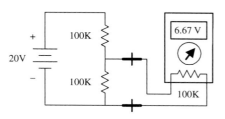

FIGURE 7.24

If a voltmeter has an input resistance of 100 k, calculate the percentage error in reading for the circuit shown here.

$$V_{true} = \frac{100\,k}{100\,k + 100\,k}(20\,V) = 10\,V$$

$$V_{measured} =$$

$$\frac{100\,k}{100\,k + (100\,k \times 100\,k)/(100\,k \times 100\,k)} = 6.67\ V$$

$$\% \text{ error} = \frac{10\,V - 6.67\,V}{10\,V} \times 100\% = 33\%$$

Resistance-Measurement Error

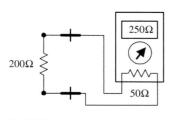

FIGURE 7.25

If an ohmmeter has an input resistance of 50 Ω, calculate the percentage error in reading for the circuit shown here.

$$R_{true} = 200\ \Omega$$

$$R_{measured} = 200\ \Omega + 50\ \Omega = 250\ \Omega$$

$$\% \text{ error} = \left| \frac{200\ \Omega - 250\ \Omega}{200\ \Omega} \right| \times 100\% = 25\%$$

To minimize the percentage error, an ammeter's input resistance should be less than the Thevenin resistance of the original circuit by 20 times or more. Conversely, a voltmeter should have an input resistance that is larger than the Thevenin resistance of the original circuit by 20 times or more. The same goes for the ohmmeter: it should have an input resistance that is at least 20 times the Thevenin resistance of the original circuit. By following these simple rules, it is possible to reduce the error to below 5 percent.

Another approach (perhaps a bit more tedious) is to look up the internal resistances of your VOM, make your measurements, and then add or subtract the internal resistances afterward. (See Appendix B for information about detailed error analysis.)

7.4 Oscilloscopes

Oscilloscopes measure voltages, not currents or resistances—just voltages. This is an important point to get straight from the start. An oscilloscope is an extremely fast xy plotter capable of plotting an input signal versus time or versus another input signal.

As a signal is supplied to the input of a scope, a luminous spot appears on the screen. As changes in the input voltage occur, the luminous spot responds by moving up or down, or left or right. In most applications, the oscilloscope's vertical-axis (y axis) input receives the voltage part of an incoming signal and then moves the spot up or down depending on the value of the voltage at a particular instant in time. The horizontal axis (x axis) is usually used as a time axis, where an internally generated linear ramp voltage is used to move the spot across the screen from left to right at a rate that can be controlled by the operator.

If the signal is repetitive, such as a sinusoidal wave, the oscilloscope can make the sinusoidal pattern appear to stand still. This makes a scope a useful tool for analyzing time-varying voltages.

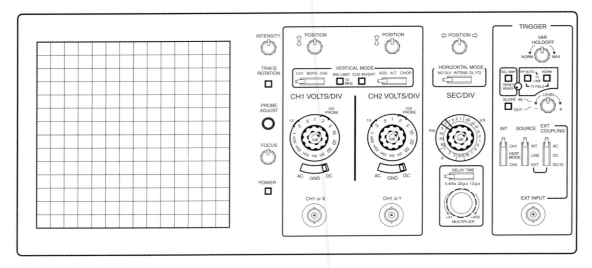

FIGURE 7.26

Even though oscilloscopes measure only voltage, it is possible to convert quantities such as current, strain, acceleration, pressure, and so on into voltages that the scope can use. To convert a current into a voltage, a resistor of known resistance is used; the current is measured indirectly by measuring the voltage drop across the resistor and then applying Ohm's law. To convert strain, movement, and so on into voltage requires the use of transducers (electromechanical devices). By applying some calibrating tricks, the magnitude of, say, a pressure applied to a pressure transducer can be measured accurately.

Here, we'll look at how oscilloscopes work. Also see Sec. 7.5.5, which discusses which type of scope you may need for your electronics laboratory.

7.4.1 How Scopes Work

While it is still possible to buy an analog oscilloscope that uses a cathode-ray tube, the majority of modern oscilloscopes are effectively a computer with fast analog-to-digital converters and an LCD or OLED display. As well as reducing the size and weight of the oscilloscope, the use of digital signal processing allows the scope to do things that an analog scope cannot do, such as generate clearer color-coded signals, provide the memory to pan left and right, allow you to export the display image to your computer, and show a frequency domain display. Digital scopes are comparable in price to their analog counterparts.

Interestingly, a veteran user of an analog oscilloscope would recognize the controls of a digital oscilloscope. The basic procedure for operating a digital oscilloscope is exactly the same as for an analog oscilloscope. It is actually easier to understand what an oscilloscope does by studying the workings of an analog oscilloscope than a digital one.

An analog oscilloscope is built around a cathode-ray tube. All the circuits inside the scope are designed to take an input signal and modify it into a set of electrical instructions that supply the tube's electron gun with aiming instructions (the location where to focus the beam). Most of the knobs and switches on the face of a scope that are connected to the interior circuitry are designed to help modify the instructions sent to the cathode-ray tube. For example, these controls set voltage scale, time scale, intensity of beam, focus of beam, channel selection, triggering, and so on.

Cathode-Ray Tube

A cathode-ray tube consists of an electron gun (filament, cathode, control grid, and anode), a second anode, vertical deflection plates, horizontal deflection plates, and a phosphor-coated screen. When current flows through the filament, the filament heats the cathode to a point where electrons are emitted. The control grid controls the amount of electrons that flow through the electron gun, and thus controls the intensity of the beam. If this grid is made more negative in voltage, more electrons will be repelled away from the grid, thus reducing the electron flow.

The electron beam is focused into a sharp point-like beam by applying a controlling voltage or focus voltage to anode 1. The second anode is supplied with a large voltage that is used to give the electrons within the beam the additional momentum needed to cause a photon emission once they collide with the phosphor screen. The beam-focusing section of the tube is referred to as the *electron gun*.

There are two sets of electrostatic deflection plates (vertical and horizontal) that are set between the second anode and the inner face of the phosphor screen. One set of plates is used to deflect the electron beam vertically; the other set is used to deflect the beam horizontally. For example, when one of the plates of a pair of plates is made more negative in charge than the other, the electron beam will bend away from the negative plate and veer toward the more positive plate. (The electrons in the beam are usually moving with sufficient forward velocity that they never actually come in contact with the plates.) When a sawtooth voltage is applied to the horizontal plates, the gradually rising potential across the plates pulls the electron beam from the negative plate to the positive plate, causing the beam to scan across the phosphor screen. The vertical plates cause the electron beam to move up and down.

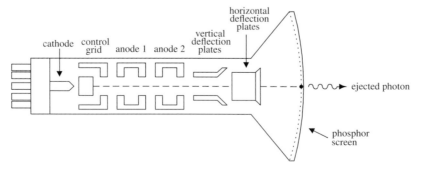

FIGURE 7.27

The next step in understanding how an oscilloscope works is knowing how an incoming signal is converted into a set of electrical signals or applied voltages that control the beam-aiming mechanisms of the cathode-ray tube. This is where the interior circuitry comes in.

7.4.2 Interior Circuitry of a Scope

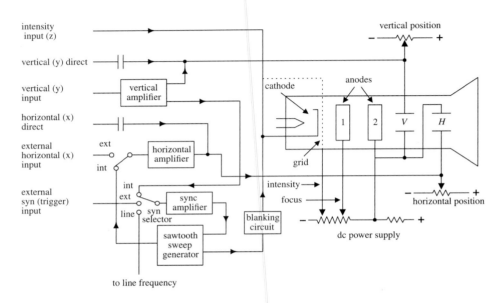

FIGURE 7.28

Let's take a sinusoidal signal and see how the interior circuits of a scope convert it into something you can see on a display. The first thing you do is apply the sinusoidal signal to the vertical input. From the vertical input, the sinusoidal signal is sent to a vertical amplifier, where it is amplified so that it can supply enough voltage to deflect the electron beam. The vertical amplifier then sends a signal to the sweep selector. When the sweep selector is switched to the internal position (the other positions are explained in Sec. 7.4.5), the signal from the vertical amplifier will enter the sync amplifier.

The sync amplifier is used to synchronize the horizontal sweep (sawtooth in this case) with the signal under test. Without the sync amplifier, the display pattern would drift across the screen in a random fashion. The sync amplifier then sends a signal to the sawtooth sweep generator, telling it to start a cycle. The sawtooth sweep generator then sends a sawtooth signal to a horizontal amplifier (when horizontal input is set to internal). At the same time, a signal is sent from the sawtooth sweep generator to the blanking circuit. The blanking circuit creates a high negative voltage on the control grid (or high positive voltage on the cathode-ray tube cathode), which turns off the beam as it snaps back to the starting point. Finally, voltages from the vertical and horizontal amplifiers (sawtooth) are sent to the vertical and horizontal plates in a synchronized fashion. The final result is a sinusoidal pattern displayed on the scope's screen.

The other features—such as vertical direct and horizontal direct inputs, external horizontal input, external trigger, line frequency, and *xy* mode—are described in Sec. 7.4.5.

It is important to note that the scope does not always use a sawtooth voltage applied to the horizontal plates. You can change the knobs and inputs and use another input signal for the horizontal axis. Controls such as intensity, focus, and horizontal and vertical position of the beam can be understood by looking at the oscilloscope circuit diagram.

7.4.3 Aiming the Beam

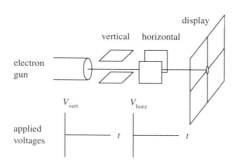

When no voltages are applied to the horizontal and vertical plates, the electron beam is focused at the center of the scope's display.

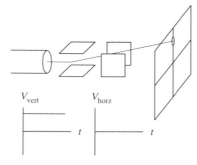

When a dc voltage is applied to the vertical plates, while no voltage is applied to the horizontal plates, the electron beam shifts up or down depending on the sign of the applied voltage.

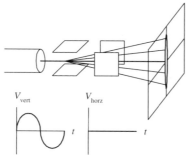

When a sinusoidal voltage is applied to the vertical plates, while no voltage is applied to the horizontal plates, a vertical line is traced on the y axis.

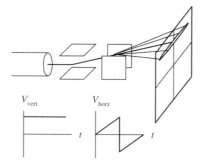

When a sawtooth voltage is applied to the horizontal plates, while no voltage is applied to the vertical plate, the electron beam traces a horizontal line from left to right. After each sawtooth, the beam jumps back to the left and repeats its left-to-right sweep.

FIGURE 7.29

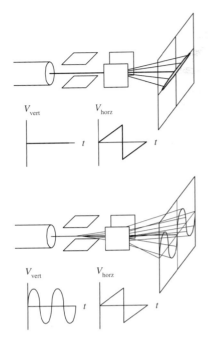

When a dc voltage is applied to the vertical plates, while a sawtooth voltage is applied to the horizontal plates, a horizontal line is created that is shifted up or down depending on the sign of the dc vertical plate voltage (+ or −).

When a sinusoidal voltage is applied to the vertical plates and a sawtooth voltage is applied to the horizontal plates, the electron beam moves up as the signal voltage increases and at the same time moves to the left as the sawtooth voltage is applied to the horizontal plates. The display gives a sinusoidal graph. If the applied sinusoidal frequency is twice that of the sawtooth frequency, two cycles appear on the display.

FIGURE 7.29   (*Continued*)

7.4.4   *Scope Usage*

DC Voltmeter

AC Voltmeter/AC Frequency Meter

T = period

f = frequency

$f = 1/T$

$$V_{rms} = \frac{1}{\sqrt{2}} V_{max}$$

Phase Relationships Between Two Signals

source 1→chan 1
source 2→chan 2

The scope can be used to compare two source signals (e.g., measure phase shifts, voltage and frequency differences, etc.).

Digital Measurements

A scope can be used to create timing diagrams for digital circuits.

xy Graphics (xy Mode)

chan 1 input→x axis
chan 2 input→y axis

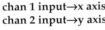

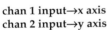

The scope no longer uses the x axis as the time axis, but uses signal voltage from another external source.

FIGURE 7.30

Measurements Using Transducers

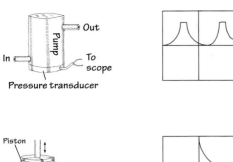

Pressure transducer

By using transducers to convert an input quantity, such as pressure, into a voltage, the scope can be transformed into a pressure meter.

y axis → pressure
x axis → time

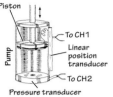

Here, a scope is used in xy mode, where
y axis → pressure
x axis → piston

FIGURE 7.30   (*Continued*)

7.4.5   *What All the Little Knobs and Switches Do*

Figure 7.31 shows a typical layout of an oscilloscope control panel. The control panel of the scope you use may look slightly different (knob positions, digital display, number of input channels, and so on), but the basic ingredients are the same. If you do not find what you need in this section, refer to the oscilloscope user manual that comes with the scope.

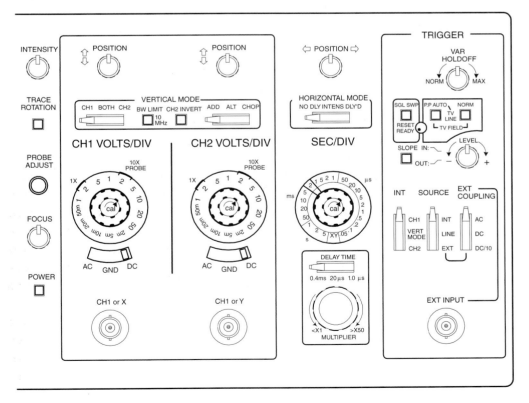

FIGURE 7.31

The control panel of an oscilloscope is divided into the following sections:

Vertical mode: This section of the scope contains all the knobs, buttons, and so on that control the vertical graphics of the scope. Most of these are associated with the voltage amplitude of an incoming signal.

Horizontal mode: This section of the scope contains all the knobs, buttons, and so on that control the horizontal portion of the graphics display. These usually are associated with the time base for the scope.

Trigger mode: This section of the scope contains all the knobs, buttons, and so on that control the way in which the scope "reads" an incoming signal. This section of the scope is probably the most technical. To understand triggering, read the upcoming "Trigger Mode" section.

Vertical Mode

The following are the vertical mode controls:

CH1 and CH2 coaxial inputs: Where input signals enter the scope.

AC, GRD, and DC switches:

- *AC*: Blocks the dc component of the signal, passing only the ac part of the signal.

- *DC*: Measures direct input of both ac and dc components of the input signal.

- *GRD*: Grounds input, causing vertical plates in cathode-ray tubes to become uncharged, thus eliminating electron beam deflection. Used to recalibrate the vertical component of the electron beam to a reference position on the display after altering the vertical position of the knob.

CH1 VOLTS/DIV and CH2 VOLTS/DIV knobs: Used to set the voltage scale on the display. For example, 5 V/div means that each division (1 cm) on the display is 5 V high.

MODE switches:

- *CH1, BOTH (DUAL), CH2 switch:* This switch allows you to pick between displaying a signal from channel 1 or channel 2, or display both channels at the same time.

NORM, INVERT: This switch lets you choose to display a signal normally or inverted.

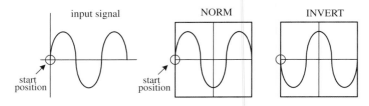

FIGURE 7.32

ADD, ALT, and CHOP:
- *ADD*: Adds signals from channel 1 and channel 2 together arithmetically.

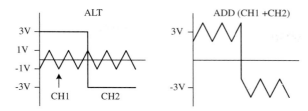

FIGURE 7.33

- *ALT:* Alternate sweep is selected regardless of sweep time, and the NORM-CHOP switch has no effect.

- *CHOP:* Operates the triggering SOURCE switch, providing automatic or manual selection of the alternate or chop method of dual-trace sweep generation.

POSITION knob: This knob allows you to move the displayed image up or down on the screen.

XY mode: When selected, the sweep rate (time base) supplied by the scope is switched off, and an external signal voltage applied to the channel 2 input replaces it.

Horizontal Mode

The following are the horizontal mode controls:

SEC/DIV knob: This knob sets the sweep speed or the scale for the horizontal time display. For example, 0.5 ms/DIV means each division (1 cm) on the display is 0.5 ms wide.

MODE switches:

- *NO DLY:* This setting takes the horizontal signal and presents it to display immediately.

- *DLY'D:* This setting delays the horizontal signal for a time you specify on the delay time section of scope. Use this to set the delay time of a signal.

SWEEP-TIME variable control: Sometimes known as the sweep frequency control, fine frequency control, or frequency vernier. Used as a fine sweep-time adjustment. In the extreme clockwise (CAL) position, the sweep time is calibrated by using the SWEEP TIME/CM switch. In the other positions, the variable control provides a continuously variable sweep rate.

POSITION knob: Moves the horizontal display left or right. This feature is useful when comparing two input signals. It allows you to align the wave patterns for comparison.

Trigger Mode

The following are the trigger mode controls:

EXT TRIG jack: Input terminal for external trigger signals.

CAL terminal: Provides a calibrated 1-kHz, 0.1-V peak-to-peak squarewave signal. The signal can be used to calibrate the vertical-amplifier attenuators and to check frequency compensation of probes used with the scope.

HOLDOFF control: Used to adjust holdoff time (ignore triggers until holdoff time has expired).

TRIGGERING mode switch:

- *SINGLE*: When a signal is nonrepetitive, or if it varies in amplitude, shape, or time, a conventional repetitive display can produce an unstable presentation. SINGLE enables the RESET switch for triggered single-sweep operation. The signal sweep can be used to photograph a nonrepetitive signal. Pushing the RESET button initiates a single sweep that begins when the next sync trigger occurs.

- *NORM*: Used for triggered sweep operation. The triggering threshold is adjustable by means of the triggering LEVEL control. No sweep is generated in the absence of the triggering signal or if the LEVEL control is set in such a way as to allow the threshold to exceed the amplitude of the triggering signal (see Fig. 7.35).

- *AUTO*: Selects automatic sweep operation, where the sweep generator free-runs and generates a sweep without a trigger signal (this is often referred to as a *recurrent sweep operation*). In AUTO mode, the sweep generator automatically switches to triggered sweep operation if an acceptable trigger signal is present. The AUTO position is useful when first setting up the scope in order to observe a waveform. It provides a sweep for waveform operation until other controls can be set properly. DC AUTO sweep must be used for dc measurements and for signals of such low amplitude that the sweep is not triggered.

- *FIX*: This is the same as the AUTO mode, except that triggering always occurs at the center of the sync trigger waveform regardless of the LEVEL control setting.

SLOPE button: Selects the point at which a scope will trigger. When positive slope is selected, the scope will begin a sweep only when the signal voltage crosses the LEVEL voltage during a positive sloping rise (see the description of the LEVEL knob). A negative slope setting initiates a sweep to occur when the signal crosses the LEVEL voltage during a negative sloping fall (see Fig. 7.34).

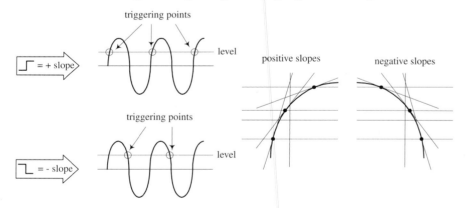

FIGURE 7.34

RESET button: When the triggering MODE switch is set to SINGLE, pushing the RESET button initiates a single sweep that begins when the next sync trigger occurs.

READY/TRIGGER indicator: In SINGLE trigger mode, an indicator light turns on when the RESET button is pushed in, indicating that a sweep is beginning. The light turns off when the sweep is completed. In the NORM, AUTO, and FIX triggering modes, the indicator turns on for the duration of the triggered sweep. The indicator also shows when the LEVEL control is set properly to obtain triggering.

LEVEL knob: Used to trigger a sweep. LEVEL sets the point when the scope will trigger based on the amplitude of the applied signal. The level can be shifted up or down. The READY/TRIGGER indicator turns on when the sweep is triggered, indicating that the triggering LEVEL control is within the proper range (see Fig. 7.35).

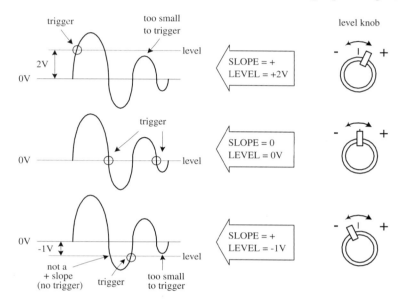

FIGURE 7.35

COUPLING switch: Used to select input coupling for the sync trigger signal.

- *AC:* This is the most commonly used position. AC position permits triggering from 10 Hz to typically over 35 MHz (depending on your scope) and blocks any dc component of the sync trigger signal.

- *LF REJ:* The dc signals are rejected, and signals below 10 kHz are attenuated; the sweep is triggered by only the higher-frequency components of the signal. This is useful for providing stable triggering when the trigger signal contains line-frequency components, such as 60-Hz hum.

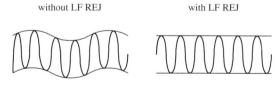

without LF REJ   with LF REJ

FIGURE 7.36

- *HF REJ:* Attenuates signals above 100 kHz. This is used to reduce high-frequency noise or to initiate a trigger from the amplitude of a modulated envelope rather than the carrier frequency.

- *VIDEO:* Used to view composite video signals.

- *DC:* Permits triggering from dc to typically over 35 MHz. The DC coupling can be used to provide stable triggering for low-frequency signals that would

without HF REJ   with HF REJ

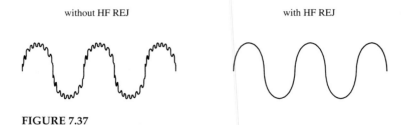

FIGURE 7.37

otherwise be attenuated if measured in the AC setting. The LEVEL control can be adjusted to provide triggering at the desired dc level on the waveform.

7.4.6 Measuring Things with Scopes

A scope's buttons and knobs must be set properly to obtain accurate measurements. If just one of these buttons or switches is set wrong, things can go haywire. You must make sure every button is set correctly.

This section covers a number of applications for oscilloscopes, such as making phase measurements between two signals. For each application, the procedure is to set the scope to the initial settings listed next, and then adjust particular buttons and knobs to put the scope in the proper configuration needed for that particular application.

Initial Scope Settings

First, start with the following settings:

- *Power switch:* Off
- *Internal recurrent sweep (TRIGGER mode switch):* Off (NORM or AUTO position)
- *Focus:* Lowest setting
- *Gain:* Lowest setting
- *Intensity:* Lowest setting
- *Sync controls* (LEVEL, HOLDOFF): Lowest settings
- *Sweep selector:* External (EXT)
- *Vertical position control:* Midpoint
- *Horizontal position control:* Midpoint
   Next, adjust the settings as follows:
- *Power switch:* On
- *Focus:* Until beam is in focus
- *Intensity:* Desired luminosity
- *Sweep selector:* Internal (use the linear internal sweep if more than one sweep is available)
- *Vertical position control:* Until beam is centered on display
- *Horizontal position control:* Until beam is centered on display
- *Internal recurrent sweep:* On; set the sweep frequency to any frequency above 100 Hz

- *Horizontal gain control:* Check that the luminous spot has expanded into a horizontal trace or line; return horizontal gain control to zero or lowest setting

- *Internal recurrent sweep:* Off

- *Vertical gain control:* To midpoint; touch the vertical input with a finger; the stray signal pickup should cause the spot to be deflected vertically into a trace or line; check that the line length is controllable by adjusting the vertical gain control; return the vertical gain control to zero or the lowest setting

- *Internal recurrent sweep:* On; advance the horizontal gain control to expand the spot into a horizontal line

Measuring a Sinusoidal Voltage Signal

1. Connect the equipment as shown in Fig. 7.38.

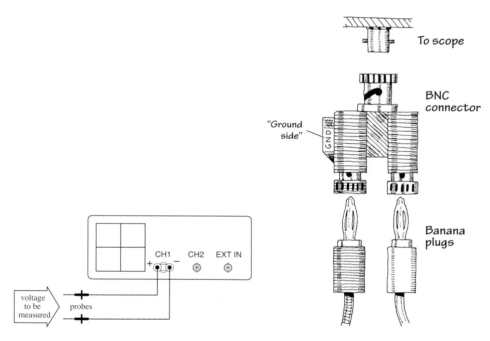

FIGURE 7.38

2. Set the scope to the initial settings listed in the previous section.
3. Fiddle with the vertical VOLT/DIV knob until the signal comes into view.
4. Set the input selector (AC/GRD/DC) to ground (GRD).
5. Switch the scope to the internal recurrent sweep. Fiddle with the SEC/DIV knob until the electron beam is tracing out a desired path of the screen.
6. Now you should have a horizontal line in view. Center this line on the x axis or some desired reference position by adjusting the vertical position knob. Make sure you do not fiddle around with the vertical position knob after it has been set to the desired reference point. If you do, your measurements will be offset. If you think you have accidentally moved the vertical position line, set the input selector to GRD and recalibrate.

7. Set the input selector switch (AC/GRD/DC) to DC. Connect the probe to the signal being measured.

8. Fiddle with the vertical and horizontal VOLT/DIV knob and SEC/DIV knob to get the signal into view.

9. Once you have an image of your signal on the screen, take a look at your VOLT/DIV and SEC/DIV knobs and record these settings. Now, visually measure the period, peak-to-peak voltage, and so on of the displayed image, using the centimeter grid lines on the scope's screen as a ruler. To find the actual voltages and times, multiply the measurement made in centimeters by the VOLT/DIV (or VOLTS/cm) and SEC/DIV (SEC/cm) recorded set values. The example in Fig. 7.39 shows how to calculate the peak-to-peak voltage, root-mean-square voltage, period, and frequency of a sinusoidal waveform.

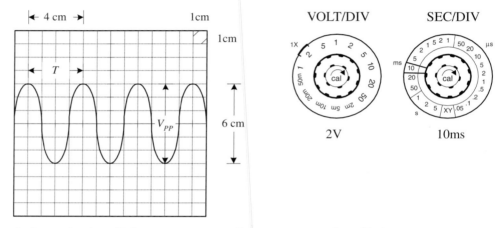

Peak-to-peak voltage (V_{pp}):

$$V_{pp} = (6 \text{ cm}) \frac{(2 \text{ V})}{(1 \text{cm})} = 12 \text{ V}$$

Period (T):

$$T = (4 \text{ cm}) \frac{(10 \text{ ms})}{(1 \text{ cm})} = 40 \text{ ms}$$

Root-mean-square voltage (V_{rms})

$$V_{rms} = \frac{1}{\sqrt{2}} V_{pp} = 8.5 \text{ V}$$

Frequency (f):

$$f = \frac{1}{T} = \frac{1}{40 \text{ ms}} = 25 \text{ Hz}$$

FIGURE 7.39

Measuring Current

As mentioned earlier, oscilloscopes can measure only voltages; they do not directly measure currents. However, with the help of a resistor and Ohm's law, you can trick the scope into making current measurements. You simply measure the voltage drop across a resistor of known resistance and let $I = V/R$ do the rest. Typically, the resistor's resistance must be small to avoid disturbing the operating conditions within the circuit that is being measured. A high-precision 1-Ω resistor is often used for such instances.

Let's now take a look at the specifics of how to measure currents with a scope.

1. Set up your equipment as shown in Fig. 7.40.

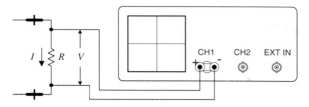

FIGURE 7.40

2. Set the scope to the initial settings listed earlier in the "Initial Scope Settings" section.

3. Apply a dc current to be measured through the resistor. Here, we'll use a 1-Ω resistor to make the calculations simple and to avoid altering the dynamics of the circuit being tested. The wattage of the resistor must be at least 2 Ω times the square of the maximum current (expressed in amps). For example, if the maximum anticipated current is 0.5 A, the minimum wattage of the resistor should be 2 $\Omega \times$ (0.5 A)^2 = 1/2 W.

4. Measure the voltage drop across the resistor using the scope. The unknown current will equal the magnitude of the voltage measured, provided that you stick with the 1 Ω resistor. Figure 7.41 shows some example measurements, two of which describe how to measure RMS and total (dc + ac) effective currents.

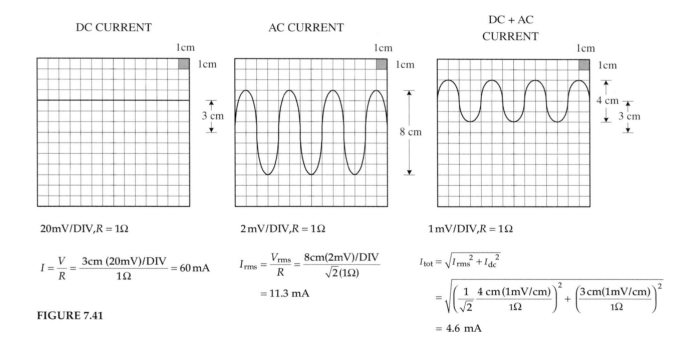

DC CURRENT

AC CURRENT

DC + AC
CURRENT

20mV/DIV,$R = 1\Omega$

2 mV/DIV,$R = 1\Omega$

1 mV/DIV,$R = 1\Omega$

$$I = \frac{V}{R} = \frac{3\text{cm (20mV)/DIV}}{1\Omega} = 60\,\text{mA}$$

$$I_{\text{rms}} = \frac{V_{\text{rms}}}{R} = \frac{8\text{cm(2mV)/DIV}}{\sqrt{2}\,(1\Omega)}$$
$$= 11.3\ \text{mA}$$

$$I_{\text{tot}} = \sqrt{I_{\text{rms}}^2 + I_{\text{dc}}^2}$$
$$= \sqrt{\left(\frac{1}{\sqrt{2}}\frac{4\,\text{cm}\,(1\text{mV/cm})}{1\Omega}\right)^2 + \left(\frac{3\,\text{cm}(1\text{mV/cm})}{1\Omega}\right)^2}$$
$$= 4.6\ \text{mA}$$

FIGURE 7.41

Phase Measurements Between Two Signals

Suppose that you wish to compare the phase relationship between two voltage signals. To do so, apply one of the signals to CH1 and the other to CH2. Then, using the DUAL setting (or BOTH setting), you can display both signals at the same time and align them side by side to compare the phased difference between them. Here's how to view both signals:

1. Step up your equipment as shown in Fig. 7.42.

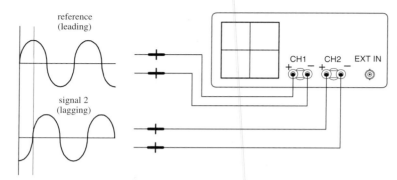

FIGURE 7.42

2. Set the scope to the initial settings listed earlier in the "Initial Scope Settings" section. Note that cables should be short, of the same length, and have similar electrical characteristics. At high frequencies, a difference in cable length or a difference in electrical characteristics between cables can introduce improper phase shifts.

3. Switch the scope's internal recurrent sweep to on.

4. Set the scope to dual trace (DUAL) mode.

5. Fiddle with the CH1 and CH2 VOLT/DIV settings until both signals are of similar amplitudes. This makes measuring phase differences easier.

6. Determine the phase factor of the reference signal. If one period (360°) of a signal is 8 cm, then 1 cm equals one-eighth of 360°, or 45°. The 45° value represents the phase factor (see Fig. 7.43).

7. Measure the horizontal distance between corresponding points (for example, corresponding peaks or troughs) of the two waveforms. Multiply this measured distance by the phase factor to get the phase difference (see Fig. 7.43). For example, if the measured difference between the two signals is 2 cm, then the phase difference is $2 \times 45°$, or 90°.

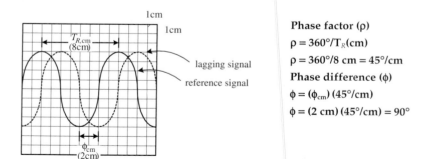

FIGURE 7.43

7.4.7 Scope Applications

The ability for an oscilloscope to "freeze" a high-frequency waveform makes it an incredibly useful instrument for testing electronic components and circuits whose response curves, transient characteristics, phase relationships, and timing relationships are of fundamental importance. For example, scopes are used to study the shape of particular waveforms (such as squarewave, sawtooth, and so on). They are used to measure static noise (current variation caused by poor connections between components), pulse delays, impedances, digital signals, and other values. The list goes on. This section presents a few example scope applications.

Checking Potentiometers for Static Noise

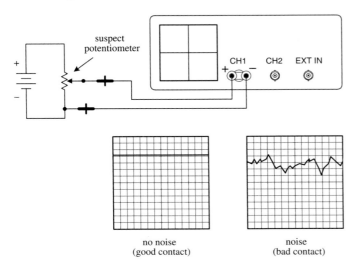

Here, a scope is used to determine if the sliding contact of a potentiometer is faulty. A good potentiometer will present a solid voltage line on the scope's screen, whereas a bad potentiometer will present a noisy pattern on the display. Before concluding that a potentiometer is bad, make sure that noise was not present beforehand. For example, the cables used in this test may have been at fault.

no noise
(good contact)

noise
(bad contact)

FIGURE 7.44

Pulse Measurements

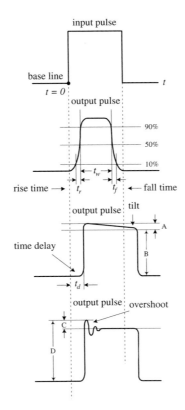

Scopes are often used to study how square pulses change as they pass through a circuit. This figure, along with the following definitions, shows some of the pulse alterations that can occur.

Rise time (t_r): The time interval during which the amplitude of the output pulse changes from 10 to 90 percent of the maximum value

Fall time (t_f): The time interval during which the amplitude of the output pulse changes from 90 to 10 percent of maximum value

Pulse width (t_w): The time interval between the two 50 percent maximum values of the output pulse

Time delay (t_d): The time interval between the beginning of the output pulse ($t = 0$) and the 10 percent maximum value of the output pulse

Tilt: A measure of the fall of the upper portion of the output pulse

$$\text{Percent tilt} = \frac{A}{B} \times 100\%$$

Overshoot: A measure of how much of the output pulse exceeds the upper portion of the input pulse

$$\text{Percent overshoot} = \frac{C}{D} \times 100\%$$

FIGURE 7.45

7.4.8 Measuring Impedances

The method of measuring impedance presented here makes use of comparing the reflected pulse with the output pulse. When the output signal travels down a transmission line, part of the signal will be reflected and sent back along the line to the source whenever the signal encounters a mismatch or difference in impedance. The line has a characteristic impedance. If the line impedance is greater than the source impedance (thing being measured), the reflected signal will be inverted. If the line impedance is lower than the source impedance, the reflected signal will not be inverted.

1. Set up the equipment as shown in Fig. 7.46.

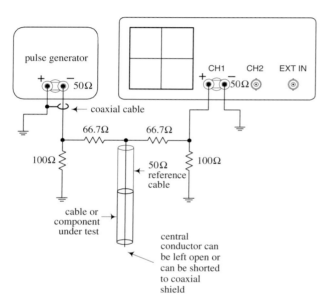

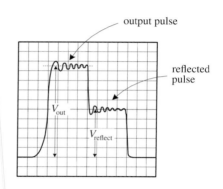

FIGURE 7.46

2. Set the knobs and switches to the initial settings listed in Sec. 7.4.6.

3. Switch the internal recurrent sweep on.

4. Set the sweep selector to INTERNAL.

5. Set the sync selector to INTERNAL.

6. Switch on the pulse generator.

7. Fiddle with the VOLT/DIV, SEC/DIV knobs until the output pulse is displayed.

8. Observe the output and reflected pulses on the scope. Measure the output voltage (V_{out}) and the reflected voltage (V_{reflect}).

9. To find the unknown impedance, use the following equation:

$$Z = \frac{50\,\Omega}{2\,V_{\text{out}}\,/\,V_{\text{reflect}}} - 1$$

The 50-Ω value represents the characteristic impedance of the coaxial reference cable.

Digital Applications

I/O RELATIONSHIPS

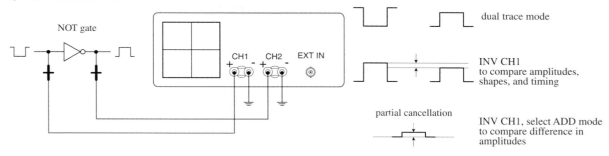

FIGURE 7.47

CLOCK TIMING RELATIONSHIPS

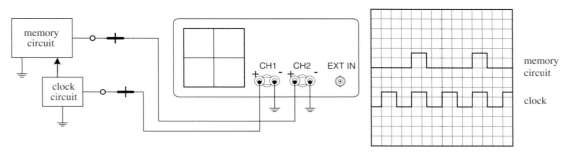

FIGURE 7.48

FREQUENCY-DIVISION RELATIONSHIPS

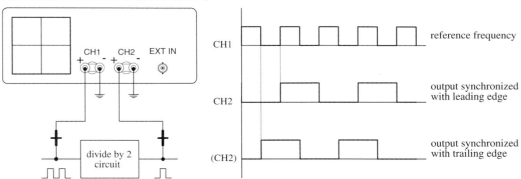

FIGURE 7.49

CHECKING PROPAGATION TIME DELAY

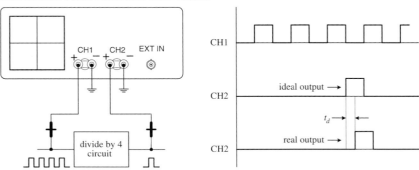

FIGURE 7.50

CHECKING LOGIC STATES

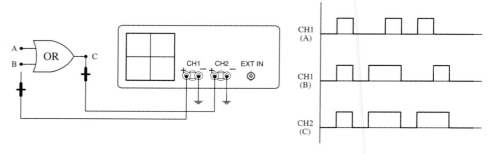

FIGURE 7.51

7.5 The Electronics Laboratory

To make electronics a safe and enjoyable experience, you need a decent electronics laboratory. In this section, you'll learn key features common to all good electronics laboratories. You'll discover how to set up a work area to limit external electromagnetic noise from coupling with your circuits, as well as how to prevent ESD from damaging sensitive ICs. You'll also learn about the various kinds of test equipment, prototyping equipment, and tools needed for troubleshooting and designing circuits. If you are serious about electronics, you will naturally end up with a setup like this. However, to get started, you do not need to have all this equipment; a desk, soldering iron, and multimeter will get you a long way.

Keep in mind that the most important tool you'll use when doing electronics is your brain. So if you lack a piece of equipment or tool listed here, see how you can work around the problem. However, be aware that you may spend more time than is necessary locating problems or constructing "flaky" circuits, simply because you lack the right gear.

Finally, if some of the information that follows seems a bit over your head, don't panic. Most of the technical information relates to test equipment specifications and troubleshooting nomenclature. The aim is to give you a good enough understanding to determine which features are important when purchasing items.

7.5.1 Work Area

Get a sturdy, large workbench so you can spread stuff out. Ideally, it should be equipped with a ground plane of metal that can be easily connected to and disconnected from earth ground (the green wire ground of an ac outlet). The ground plane helps prevent RF background radiation, 60-Hz noise, and other external electromagnetic and electrostatic disturbances from coupling with your circuit. Place an insulating sheet or cardboard between the bench and the circuit you're testing so that nothing shorts to ground.

For casual measurements that don't require high precision, you can usually get by without a ground plane. If you run into interference problems and your bench doesn't have a ground plane, simply place your circuit on a double-insulated metal sheet that's soldered to a wire connected to ground. A grounded, single-sided copper-clad board, placed copper side up, with a cardboard sheet over the top, will also work.

You can find a number of electronics workbenches online, or you can make your own. A design plan for a five-shelf, 31 × 72 × 72-in. metal-frame workbench constructed from parts from your local hardware store is provided at the end of this chapter (Sec. 7.5.23).

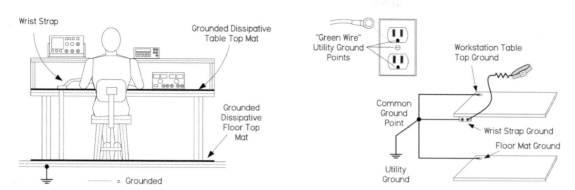

FIGURE 7.52   **A workstation that has been set up to minimize ESD by using grounded dissipative mats and a wrist strap. For reliable grounding, use separate grounding wires for each mat and wrist strap, running them directly to a common ground point.**

Damaging ESD between your body and sensitive ICs can be eliminated at your workbench by making sure your tabletop work area, your body, and the floor area are grounded to earth or a common reference. Note that you should not directly ground yourself—the risk of line voltage shock is a real potential hazard. Instead, invest in a wrist strap that houses a series 1-MΩ resistor. The resistor limits potentially lethal current flow, while providing a sufficient path for ESD.

If you're using a rubber mat on a concrete floor to limit the risk of electrical shock, make sure it's made with antistatic rubber; otherwise, you may actually be increasing the risk of ESD. For carpeted areas, you can buy a special antistatic ground mat or spray it with antistatic spray. DESCO, 3M, and a number of other manufacturers make ESD-protective work mats and complete workstation grounding kits.

Also, make sure that all equipment is properly grounded. This includes grounding your soldering iron's tip when soldering sensitive components. Though most equipment is already grounded to earth through the three-wire power cords, don't assume the ground is without fault. Test all grounds to see if they are at the same potential. Use an ohmmeter to test the resistance between earth ground and equipment ground posts, plugs, and chassis bodies. If an item is not properly grounded and you touch it, you'll become the best path to ground. Good grounding also eliminates ground loops and limits external noise from coupling with your circuit.

7.5.2   *Test Equipment*

You can obtain a vast assortment of equipment for your laboratory. The limiting factor is your budget. A great place to look for used equipment is eBay. You'll find new, top-of-the-line items all the way down to the clunkers. However, be aware that you can find new equipment that's not that much more expensive than the used stuff by checking out electronics sites such as www.web-tronics.com and others listed in Sec. 7.5.20.

Do your homework before purchasing anything. Make sure the specs (bandwidth, input impedance, accuracy, and so on) stand up to the testing you have in mind.

Don't be misled into thinking that you must have the newest and most expensive instruments to work with electronics. Unabused equipment that is 20 years old or even older can be remarkably reliable. Just be certain that the seller can verify that the equipment is in working order and has been recently calibrated. Also, ask for an inspection period before the deal is final.

7.5.3 Multimeters

If you plan to do a lot of serious bench work, get a bench digital multimeter (DMM) with at least five-digit resolution. For more digits and higher accuracy, expect to pay considerably more. Be sure the meter comes with autorange lockout so it can achieve the highest accuracy and speed. For many circuits, it's important that the meter come with a high-impedance input (>10,000 MΩ) that stays at high impedance up to 20 V or so. This prevents the meter from loading down your circuit and reduces measurement errors.

There are a number of meters with 10-MΩ inputs, which are okay if you don't see the 10 MΩ being a problem. Look for meters with four-wire resistance measurement capability—a feature that eliminates errors caused by test lead resistance. A few decent meters you can find on eBay include the HP3468A (5.5-digit), the HP34401A (6.5-digit), and the HP3485A (8.5-digit).

If you are on a budget, then look for a hand-held DMM with a pull-out stand on the back, a large display, and a backlight. These make a reasonable alternative to a high-quality multimeter.

FIGURE 7.53 *Left:* **HP 34401A bench DMM with 6.5-digit resolution and computer interface.** *Center:* **Metex M-3860M 3¾ digital DMM loaded with handy features, such as an RS232 interface, true RMS, temperature, inductance, capacitance, logic analyzer, transistor hFE tester, diode tester, and frequency counter.** *Right:* **A simple analog meter useful for monitoring trends.**

It's also nice to have a few auxiliary meters handy for taking a number of measurements at the same time. Get a hand-held DMM with true RMS and at least three to four digits of resolution. These meters will run you anywhere from $50 to $300, although you shouldn't need to pay more than $100 to $150 for a decent one. Hand-held meters with additional features—such as a capacitance tester, an inductance tester, diode and transistor testers, a frequency counter, and a thermometer—come in handy, especially when prototyping. You never know when you'll need to check polarity of an LED, check the hFE of a transistor, or check whether your guess at a capacitance label is on target.

Analog meters, though not as popular today, are still useful. They have inferior accuracy and resolution when compared with a DMM, but when you watch an ordinary analog meter, your eye can detect trends or rates of change that are hard to spot on a DMM, especially in the presence of noise or jitter. Another advantage of an analog meter is its passive nature; it won't inject noise into your circuit, as a digital meter is capable of doing. An analog meter will also tell you things about a slowly moving signal, as you don't just get an instantaneous reading but can see how fast the reading is changing. This is something that you cannot really see on a digital meter. Some digital meters get around this problem by providing a barograph-type display to accompany the numbers.

As a final option, you can check out some PC-based DMMs. These devices plug in to your computer directly through an expansion slot (such as the PCI slot) or through a box that is linked to your computer via a serial, parallel, or USB cable. Current technology has made PC-based performance much better than in the past. Check out National Instruments' NI PXI-4070 DMM, which is a 6½-digit DMM—a very impressive device. Consider these instruments if you plan to do data logging.

7.5.4 DC Power Supplies

Probably the single most used piece of equipment that you will use is a lab dc power supply to provide a steady dc voltage. Get a variable dc constant-voltage/constant-current supply that can be configured for single- and split-polarity applications, as follows:

- A single-polarity arrangement—say, a +12-V supply used for driving a dc motor—would use the positive terminal (+) as the source and the negative terminal (–) as the return.

- A split supply—say, a ±15-V one used to power an operational amplifier that swings both positive and negative in reference to a common ground—would use the positive terminal (+15 V), the negative terminal (–15 V), and the common terminal (COM), which acts as a 0-V reference.

Some supplies, like the center supply shown in Fig. 7.54, house two separate variable voltage sources. To create a split power source using such a supply, the two variable sections are connected in series using a jumper cable; the common becomes the junction point where the jumper cable is located. (You'll typically need to flip a switch on the supply to specify series mode.)

Buy a model that has built-in current and voltage displays. The current being drawn by a circuit is a great indicator of health or ill health.

FIGURE 7.54 Constant current/voltage supplies with current lockout and voltage controls. The supply to the left comes with a negative, common, and positive terminal. The center supply houses three independent dc sources, two of which are variable, while the other is fixed at 5 V. The variable sections can be used independently, in series, or in parallel. To create a split supply, the two sections are placed in series, and the common becomes the junction between the two sections. The supply to the right houses split and fixed (5-V) sections. It can save and recall three voltage/current settings.

Your supply should also come with an earth ground terminal that is internally connected to earth ground through the ac power cord's ground wire, which has a conductive path to the ground rod buried outside your home. Many circuits, and most test equipment (oscilloscopes, function generators, and so on), will use the earth ground as a reference. For example, the outer BNC sleeve of an oscilloscope's input channel or the output sleeve of a function generator's output is typically earth ground.

Also look for supplies that come with an additional fixed +5-V output useful for powering logic circuits. These outputs are typically rated for 3 A.

Your supply should come with fine and coarse adjustment controls and current-limit lockout. Avoid digital current/voltage controls, since they don't let you continuously sweep up and down while monitoring for trends.

A nice supply, available for about $500 used, is the HP3631A shown in Fig. 7.54. It comes with save and recall features that allow you to store and recall up to three power supply setups. It also comes with an RS-232 interface. If you can't afford a supply like this, you can find other supplies—minus all the bells and whistles—that will do the job. There are a number of other programmable supplies out there, but they tend to be rather expensive. Another option is to build your own supply, as discussed in Chap. 11.

Switch-mode variable power supplies can be bought for less than $100 for a single output and around $150 for dual-channel. If you are trying to keep costs down and are not going to be doing much work that requires a split power supply, then a single-channel device may be just fine for you.

Another useful troubleshooting tool is a set of batteries—alkaline, NiHM, or whatever is suitable. You can use a stack of them to create any desired voltage—for example, two 9-V batteries for +18 V or ±9 V. Batteries are a useful alternative power supply for low-noise circuits, such as in a preamplifier. If the preamp's output doesn't get quiet when you substitute your batteries in place of your ordinary supply, don't blame the power supply. You can also use batteries to power low-noise circuits, like those sealed in a metal box, without contaminating their signals with power supply noise.

7.5.5 Oscilloscope

Get a bench oscilloscope (which displays voltage waveforms) with at least two channels so that you can track two signals at once. Also, make sure it comes with a bandwidth of at least 100 MHz.

Even when working on slow amplifier circuits, a wide-bandwidth scope is important to catch high-frequency oscillations. A wide bandwidth is also important for catching higher-frequency components or harmonics of nonsinusoidal waveforms. For example, to catch the fifth harmonic of a 100-MHz square waveform, you'll need a more expensive scope (and probe) with a bandwidth of 500 MHz.

As a rule of thumb, to make accurate frequency measurements, your scope's bandwidth should be three to five times greater than the fundamental frequency waveform you're measuring. To make truly accurate amplitude measurements that are not dominated by the scope's frequency response, you'll need a bandwidth greater than ten times the frequency being measured. Signal measurements on frequencies beyond the scope's bandwidth will result in attenuation drops greater than −3 dB, and rise and fall times that can literally turn a nice, clean squarewave into a weak sine wave. Figure 7.56 shows attenuation and rise-time effects on a 50-MHz squarewave measured with 20-MHz, 100-MHz, and 500-MHz scopes.

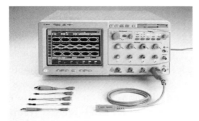

FIGURE 7.55   From vintage to high-tech, scopes will cost you accordingly. *Left:* **An Eiko TR-410 10-MHz scope for around $30 on eBay. It's an incredibly slow scope by today's standards, but it's okay for general-purpose work on low-frequency circuits.** *Center:* **Tektronix 2246 100-MHz, four-channel analog oscilloscope with voltage and time cursors with on-screen readout. For about $400, used, on eBay, it's a good investment for general-purpose testing.** *Right:* **Cutting-edge Infiniium 54850 oscilloscope with InfiniiMax 1130 probes has a 6-GHz bandwidth with a 20-Gsa/s sample rate on all four of its channels. It comes with a user interface based on Microsoft Windows XP Pro, and supports CD-RW, dual-monitor, and third-party software. Don't expect to be able to buy one for personal use, unless you plan on robbing a bank. At the other end of the cost spectrum, new LCD-based scopes can be bought for around $55 (check out http://store. nkcelectronics.com/digital-storage-oscilloscope-very-low-cost.html) and take up a lot less bench space than a vintage scope.**

When buying a scope (Fig. 7.55), you'll need to decide between an analog or a digitizing scope. In terms of recent technology, digital scopes are more powerful and more responsive. Analog scopes do have some positives, such as familiar controls, instantaneous display updating for real-time adjustments, direct dedicated controls for often-used adjustments, and a reasonable price range. However, they lack accuracy, have no pretrigger viewing capability and limited bandwidth (seldom exceeds 400 MHz), and can't store waveforms into memory for later recall.

Modern digitizing scopes, on the other hand, have display storage; high accuracy; pretrigger viewing capability; peak/glitch detection; automatic measurements; computer and printer connectivity; waveform-processing capability, including waveform math functions; display modes such as averaging and infinite persistence; and self-calibration. A good digitizing scope will cost you a considerable amount. Expect to pay $1,000 or more for a used one, and as much as $50,000 for a new, top-of-the-line model. However, the price of digital scopes is falling all the time, and a dual-channel 50 MHz or 250 M samples/second (MS/s) with more features than you could shake a stick at can be bought for less than $500. The price rises steeply with bandwidth, so check what is available and read the reviews before you buy.

If you decide on a digitizing scope, make sure it comes with a high sample rate, which is the rate at which the scope can take a "snapshot" of the incoming signals. A higher sample rate translates into more real-time bandwidth and better real-time resolution. Most manufacturers use a sample rate–to–real-time bandwidth ratio of at least 4:1 (if digital reconstruction is employed) or 10:1 (without reconstruction) to prevent aliasing. Also, make sure to consider the scope's memory depth.

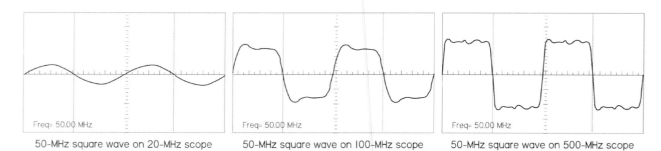

Freq= 50.00 MHz

Freq= 50.00 MHz

Freq= 50.00 MHz

50-MHz square wave on 20-MHz scope   50-MHz square wave on 100-MHz scope   50-MHz square wave on 500-MHz scope

FIGURE 7.56   **Screenshots show the same 50-MHz squarewave displayed on three different scopes with different bandwidths. The 500-MHz scope shows the best high-frequency detail and representation of rise times. As you move down in bandwidth, notice the increase in rise time and amplitude attenuation. The bandwidth of your scope should be at least three times the fundamental frequency of the fastest signal you expect to measure, and as much as ten times more for accurate amplitude measurements.**

An alternative to a stand-alone oscilloscope is a PC-based scope. These scopes use your computer and the software running on it as the testing component. Most PC-based scopes are composed of an interface card or adapter. The card adapter connects to your PC via an expansion slot (such as an ISA or a PCI slot), via a serial, parallel, or USB port, depending on the model. A test probe then connects to the interface. Software on your PC interprets the data coming through the interface and displays the results on the monitor. Low-cost (low-performance) PC-based scopes cost around $100, but the price goes up considerably with bandwidth.

Unless you're focused more on data acquisition than circuit testing and troubleshooting, we suggest a stand-alone scope over a PC-based one. Having to boot up a computer before you can use the scope becomes irritating. Although a number of high-end PC-based scopes come with decent bandwidth, the affordable ones will usually come with crummy bandwidth (20 MHz or so). Expect to pay upward of $1,000 for one with bandwidth of at least 100 MHz.

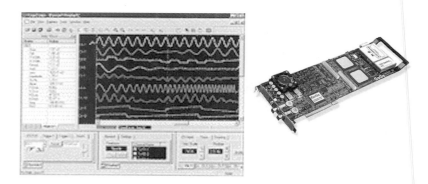

CompuScope 8500 can sample one analog input at speeds up to 500 MS/s with 8-bit resolution and data storage in on-board memory. It has two separate inputs: a 1-MΩ input and a very high-bandwidth 50-Ω input. GageScope software running on your computer captures data from the CompuScope plug-in, analyzes the data, and automatically calculates the result. CompuScope 8500 can measure rise times, fall times, frequency, pulse width, and amplitude, and it can even perform spectral analysis.

FIGURE 7.57

7.5.6 Oscilloscope Probes

Get suitable probes for your scope, preferably three: one for a trigger and two for separate channels. Though it's possible to connect a scope to your circuit with just a bare wire, that's a bad idea. The bare wire can load the scope's input amplifier with its high capacitance and inductance, but it can also cause a short.

Multimeter test leads or hookup wire rigged to a BNC connector will work for certain applications (usually low frequency), but they can pick up external radiation (such as 60-Hz power, radio, TV, and fluorescent lighting) that can corrupt your signals. Using a shielded, unterminated coaxial cable with make-shift probe wires can reduce stray pickup; however, the coaxial cable, in conjunction with a standard scope, can introduce a new set of problems. The inherent capacitance of a coaxial cable (typically 100 pF/m) has the tendency to capacitively load a circuit under test, while the cable will experience resonant effects at certain frequencies, generating signal reflections that result in signal corruption. For this reason, it's a bad idea to lengthen a dedicated probe cable by attaching a coaxial cable, unless you're dealing with special 50-Ω probes and 50-Ω scopes.

Without the appropriate probe, you can run into all sorts of problems. The main problem you'll face is circuit loading, where the thing you're using to probe your circuit will draw too much current and result in a voltage drop. Also, if the thing you're using has built-in capacitance and inductance (for example, coaxial cable), you'll run into capacitive-loading problems that will affect timing measurements, as well as inductive-loading problems that will distort signals. With a long ground probe wire, the distributed internal inductance of the wire will interact with the probe's capacitance and cause ringing—seen as a sinusoid of decaying amplitude impressed on pulses. These loading effects may also cause a working circuit to malfunction or a nonworking circuit to spring to life.

Even when using dedicated oscilloscope probes, you'll still run into these loading problems, and they will grow more severe with increasing frequency. At higher frequencies, the capacitive reactance of the probe's tip decreases, resulting in increased loading. This will limit the bandwidth and increase rise times. In terms of resistive loading, a probe with lower built-in resistance will cause greater attenuation.

The best thing to do when selecting a probe is to find your scope's manual and see what kind of probes it recommends. If a manual is nowhere to be found, check out the oscilloscope or probe manufacturers' websites for suggestions. Otherwise, consider the following:

■ Make sure that the probe's input connectors match those of your scope. Most scopes come with BNC-type input connectors; others may use SMA connectors. More sophisticated scopes may have specially designed connectors to support readout, trace, probe power (for active probes and differential probes), and other special functions.

- Choose a probe whose input resistance and capacitance match the input resistance and capacitance of your scope. Matching is critical for ensuring proper signal transfer and fidelity. The input resistance and capacitance of a probe are used to describe its loading effects. At low frequencies (<1 MHz), the probe input resistance is the key factor for loading of the circuit being tested. At higher frequencies, the probe input capacitance becomes the significant factor.

Scopes come with either a 1-MΩ input resistance or a 50-Ω input resistance (some come with both). A 1-MΩ input is most common and is used for general-purpose testing, while a 50-Ω input is used within 50-Ω environments where high speed and low loading are needed. You must use 1- MΩ probes with 1-MΩ inputs, and 50-Ω probes with 50-Ω inputs. An exception to this one-to-one resistance matching occurs when using attenuator probes. For example, a 10X probe used with a 1-MΩ input will have a 10-MΩ input resistance, while a 10X probe used with a 50-Ω input will have a 500-Ω input resistance.

The input capacitance of a scope, unlike its standard input resistance (1 MΩ or 50 MΩ) can vary, depending on the scope's bandwidth and other design features. Though a common input capacitance for many 1-MΩ scopes is usually 20 pF, it's entirely possible your scope has a different value, between 5 pF and 100 pF or so. To match the probe's capacitance with that of the scope, choose a probe whose capacitance is within the same range of your scope, and then narrow things down by adjusting the probe's compensation network, using the probe's trimmer capacitor. This is referred to as "compensating your probe". It's important to note that trimming will have little effect if the input capacitance of the scope is outside the compensation range of the probe.

Now that you have some probe theory behind you, let's take a look at some real probes.

Passive Probes

Practically speaking, the most common probe used for general-purpose testing is a passive probe built for 1-MΩ input scopes. These passive probes are constructed of wires, connectors, and, when needed, compensation or attenuation resistors and capacitors. There are no active components within these probes, such as transistors or amplifiers, and thus they are quite rugged, relatively cheap, and easy to use.

Passive voltage probes come in 1X, 10X, 100X, and 1000X forms, presenting no attenuation (1X), 10 times the attenuation (10X), 100 times the attenuation (100X), or 1000 times the attenuation (1000X). The attenuating probes act to multiply the measurement range of the scope by using an internal resistance that, when taken with the input resistance of the scope, creates a voltage divider. For example, a typical 10X probe used with a 1-MΩ scope houses an internal 9-MΩ resistor that creates a 10:1 attenuation ratio at the input channel. This means the displayed signal will be one-tenth the magnitude of the measured signal. This allows you to test a signal that might otherwise overload the scope circuits. (Table 7.2 shows a schematic of a typical 10X probe.)

A 10X probe, when compared to a 1X probe, will also cause less circuit loading, since it draws less current as it measures. A 10X probe has significantly larger bandwidth (around 60 to 300 MHz) when compared to that of a 1X probe (around 4 to

TABLE 7.2 **Overview of the Various Kinds of Oscilloscope Probes**

PASSIVE PROBES	CHARACTERISTICS AND APPLICATIONS
Typical 10X passive probe 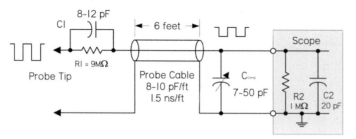	Most common probe, used to measure low to intermediate signals less than 500 V. The probe tip is usually 9 MΩ, which, when taken with the input resistance of the scope, provides a 10:1 attenuation ratio. A compensation capacitor located in the termination box at the end of probe cable is adjusted to match input capacitance of scope. It's best suited for general probing and troubleshooting. The probe's high input impedance is usually 1 MΩ or greater, and is ideal for measuring summing junctions of op amps. The trade-offs include relatively low bandwidth and high capacitive loading when compared to low-impedance divider probes.
Typical low-impedance probe—50 Ω or Z_o 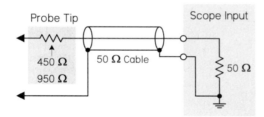	Provides the lowest possible capacitance (<1 pF) for high-frequency signals. It exhibits a frequency response that is essentially flat through their rated frequency range. The tip contains a resistor, usually either 450 or 950 Ω, to give a 10:1 or 20:1 attenuation ratio. The probe cable's characteristic impedance is 50 Ω, which is terminated with a 50-Ω scope input. Its benefits include low capacitive loading and very high bandwidth (in the gigahertz range). It also has a very low cost compared to active probes. It's used for probing low-voltage signals (less than 50 V), such as ECL circuits and 50-Ω transmission lines. One trade-off is relatively heavy resistive loading, so you must have a good understanding of the circuit being measured. The scope must also come with 50-Ω input. Applications include high-speed device characterization in microwave communications, propagation delays in logic circuits, circuit board impedance testing, and high-speed sampling.
Compensated high-resistance passive divider probe (high-voltage probe) 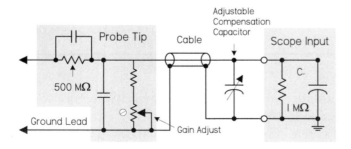	Used for safely measuring voltage signals over 500 V, where typical probes would break down. The probe to the left has a tip resistance of 500 MΩ. The probe cable used is very similar to the one that is in the standard compensated high-resistance passive divider probe. An adjustable compensation capacitor at the connection end to the scope is used to match probe and scope capacitances. The divider ratio of the probe can be selected and has a very high dynamic range. Trade-offs include a physically larger probe and a lower bandwidth compared with standard probes. Applications include high-voltage video signals, switching power supplies, and large power transmission signals.

(Continued)

TABLE 7.2 Overview of the Various Kinds of Oscilloscope Probes (Continued)

PASSIVE PROBES	CHARACTERISTICS AND APPLICATIONS
Active probe 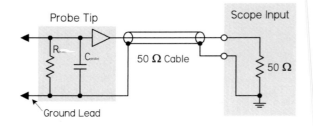	Usually comes in two varieties: FET and bipolar input probes. Both types contain an active amplifier in addition to an RC network. The active amplifier drives a 50-Ω cable that is connected to a 50-Ω input scope. (Note that there are some active probe packages designed for use with 1-MΩ scopes.) Active probes present very little resistive or capacitive loading and have a very high bandwidth (500 MHz to 4 GHz), making them much less intrusive than standard passive probes. The probe can be used to analyze many different kinds of circuits, including ECL, CMOS, and GaAs. You can also look at typical analog circuits, transmission lines, and basically any circuit that has a source resistance between 0 and 10 KW. Active probe trade-offs include high cost, limited dynamic range (±40 V typical), and ESD susceptibility.
Typical active differential probes 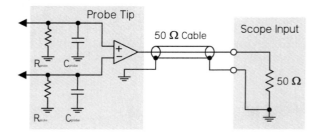	The major difference between this type and the previous active probe is that the tip has two inputs: a positive (noninverting) and a negative (inverting) input. These two inputs feed a differential amplifier, which in turn drives a 50-Ω cable that is connected to a 50-Ω scope input. (Note that there are some differential probe packages that are designed for 1-MΩ systems.) Differential probes allow you to make accurate differential measurements, measurements between two point voltages, not just one point referenced to ground. The primary features of an active probe allow for adjustment of dc offset, dc reject, and coupling. These probes also have high common-mode rejection (CMRR), e.g., 3000:1 at 1 MHz. This allows for accurate viewing of small signals in the presence of large dc offsets or other common-mode signals. These probes are used for measuring differential amplifiers, troubleshooting power supplies, and testing other differential sources. They tend to be more expensive than passive probes, have less dynamic range, and require external power and a control module.

34 MHz). The 10X probe introduces about ten times less capacitance. Bandwidth differences also apply to 100X and 1000X probes.

Look for switchable 1X/10X, 10X/100X, and 100X/1000X probes that house two probes in one. You switch between the two modes by flipping the switch on the probe's body. These probes are very convenient when jumping between signals of varying magnitudes. However, keep in mind that when you switch modes, your bandwidth will change along with the voltage scale.

Also note that typical 1X and 10X probes have a maximum voltage rating of around 400 to 500 V. The 100X and 1000X probes fare better at around 1.4 kV and 20 kV, respectively. For high-voltage work extending up to 20 kV, get a specially designed high-voltage passive probe.

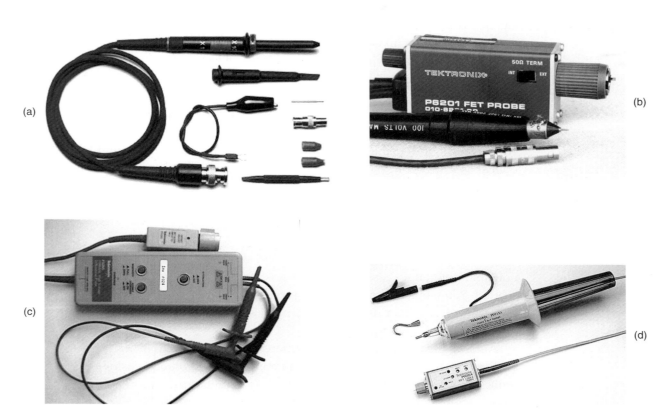

FIGURE 7.58   (a) *General-purpose 1X/10X passive probe* with assortment of probe attachments designed for use with a 1-MΩ, 20-pF scope. At 1X, there is no attenuation, but the bandwidth is limited to between 4 and 34 MHz. At 10X, signals are reduced to one-tenth their original magnitude, while the bandwidth increases from between 60 and 300 MHz, depending on the probe and scope. (b) *Active probe* used primarily for high-frequency, low-voltage work. An active probe's internal FET provides extremely low-input capacitance (>1 pF) that allows for high bandwidths (500 MHz to 6 GHz) and minimal circuit loading. The probe typically comes with a 50-Ω output impedance to drive a 50-Ω cable, making it possible to increase probe cable length. It may require a special scope with a power connector/signal connector, or an external power source. A limited voltage range of ±40 V is typical. (c) *Differential probe* houses an internal differential amplifier that allows you to make differential measurements (between any two points, not just between one point and ground). Differential probes come in 50-Ω and 1-MΩ packages, and may require a special scope with a power connector/signal connector or an external power source. (d) *High-voltage probe* used for heavy-duty measurements of voltages over 2.5 kV. The P6015A from Tektronix can measure dc voltages up to 20 kV (RMS) and pulses up to 40 kV. It comes with a bandwidth of 75 MHz. Other high-voltage probes, often termed 100X, 1000X, will provide varying maximum voltage and bandwidth limits.

Let's not forget 50-Ω *passive probes* and 50-Ω scopes. Probes used with 50-Ω scopes are often referred to as Z_O probes. The Z_O refers to the characteristic impedance of a cable, which in this case is 50 Ω for the coaxial probe cable. These probes offer much higher bandwidth than the best 1-MΩ probes, approaching the gigahertz range, with rise times of 100 picoseconds or faster. Fifty-ohm systems are used primarily for high-speed circuits.

Active Probes

For extremely fast (>500 MHz) or high-impedance circuits, passive probes (even 10X) used with 1-MΩ scopes may not cut it. They can cause severe circuit loading, as well as attenuation and timing degradation. A better probe for high-speed work is an active probe. *Active probes* house internal FETs or other active devices that present an extremely high input resistance and low input capacitance (1 pF or so) to the input signal. With external power, the active devices amplify the signal without drawing power from the circuit being tested.

Active probe bandwidths range from 500 MHz to 4 GHz. An FET probe usually has a 50-Ω output impedance and drives a 50-Ω cable, though specially designed active probes with external boxed circuitry are made for 1-MΩ scopes. This allows the distance from the probe tip to the instrument to be increased within the practical limits of the probe amplifier system and the limitations of the coaxial cable. Another benefit to using active probes is that the length of the ground lead isn't as critical as it was with passive probes due to the probe's low input capacitance that reduces ground lead effects. One shortcoming of active probes is their limited voltage range—typically from ±0.6 to ±10 V—and their maximum voltage rating of ±40 V.

Differential Probes

Another useful probe is a *differential probe*, which measures differential signals— potential differences between any two points, not just between one point and earth ground. You would use a differential probe for, say, measuring the signal developed across a collector load resistor or any other situation where signals are, in essence, "floating" above ground.

Although it's possible to make a differential measurement in an indirect manner using two standard passive probes (you measure the two points using two separate probes and scope channels, each of which is referenced to ground, then select the scope's subtract math function—channel A minus channel B—to get the differential measurement), this approach doesn't work well at high frequencies or for small signals approaching noise levels. The main problem with this approach is the two separate paths down each probe and through each scope channel. Any delay differences between these paths result in time skewing of the two signals. On high-speed signals, this skew can result in significant amplitude and timing errors in the computed difference signal.

Another problem is that the probes don't provide adequate common-mode noise rejection. A differential probe, on the other hand, uses a differential amplifier to subtract the two signals, resulting in one differential signal for measurements by one channel of the scope. This provides substantially higher common-mode rejection performance over a broader frequency range. In modern differential probes, bandwidths of 1 GHz, with common-mode rejection ratios ranging from 60 dB at 1 MHz to 30 dB at 1 GHz, are typical.

Current Probes

One last probe for oscilloscopes worth mentioning is a current probe. A current probe provides a noninvasive way to measure current flow through a conductor. Two types of current probes are available: the traditional ac-only probes and the Hall effect semiconductor probes. The ac-only probes use a transformer to convert current flux into ac voltage signals that are monitored by the scope. These probes usually have a frequency response of a few hundred hertz to a gigahertz. By combining a Hall effect device with an ac transformer, you get a probe whose response is from dc to around 50 MHz. The key reason for using current probes is their noninvasive nature; the current probe typically imposes less loading than other probe types. You would use them for measurements that would otherwise be unsafe or disruptive to the circuit's performance when gathered by other testing means.

Probe Suggestions

Here are a few suggestions for working with probes:

Compensate your probes: Most probes are designed to match the input characteristics of specific oscilloscope models. There are, however, slight input variations from scope to scope and even between different input channels of the same scope. To deal with these variations, most probes, especially attenuating probes (10X and 100X) have built-in compensation networks. If your probe does, you should adjust this network to compensate the probe for the scope channel that you are using. Do this by attaching the probe to the scope, and then attach the probe tip to the probe compensation test point on the scope's front panel, which gives you a 1- to 10-kHz squarewave. Use the adjustment tool provided with the probe or any nonmagnetic screwdriver to adjust the compensation network to obtain a calibration waveform display that has flat tops with no overshoot or rounding. If the scope has a built-in calibration routine, run this routine for increased accuracy. An uncompensated probe can lead to various measurement errors, especially when measuring pulse rise or fall times. Check compensation frequently—whenever you change scope channels or whenever you change probe tip adapters.

Use appropriate probe tip adapters: Avoid using short lengths of wire soldered to circuit points as a substitute for a probe. Even an inch of wire can cause significant impedance changes at high frequencies.

Keep ground leads short: For passive probes, longer probe ground leads introduce significant inductance that will result in ringing and signal distortion. Do not attempt to lengthen a probe ground lead.

7.5.7 General-Purpose Function Generator

Get a general-purpose function generator capable of producing sine, square, and triangle waveforms. Make sure it comes with a decent upper frequency limit (preferably upward of 5 MHz) and a decent voltage range. Some function generators also come with ramp, pulse, variable symmetry, counted bursts, gate, linear/logarithmic sweeps, AM, FM, VCO, dc offset, phase lock, and external modulation inputs—all of which are nice to have, but not entirely necessary. As with oscilloscopes, the cost of a function generator will increase with bandwidth. Figure 7.59 shows typical function generators that you can find, used, on eBay and other used equipment sites.

For a very low-cost (possibly zero-cost) function generator in the audio range, try downloading a function generator app for your smartphone. There are several free apps for both iPhone and Android. You will need a headphone plug with test leads attached. These apps will generally allow you to set a frequency of 50 Hz to 20 KHz and select

FIGURE 7.59 *Left:* Leader LFG-1300S, 2-MHz function generator with simple controls and many waveforms to select from, including amplitude modulation and linear/log sweep. Found for around $60, used, on eBay. *Middle:* HP 3312A function generator with a 0.1-Hz to 13-MHz bandwidth, providing sine, square, triangle, ramp, or variable symmetry pulses. Also has internal AM, FM, sweep, trigger, gate, or burst capability, along with external AM, FM, sweep, trigger, and gate capability. A great little instrument if you can find one for under $150. *Right:* Agilent 33120A 15-MHz function/arbitrary waveform generator, a high-performance signal source that provides sine, square, or triangular waveform signals over its 0.1- to 15-MHz frequency range. This instrument also provides a continuously variable dc offset and variable duty cycle, and the clock signal voltage levels adjusted to TTL components. It comes with both HPIB and RS-232 interfaces. Expect to pay around $500 for a used one.

the usual waveform shapes. You will need to use one channel of your scope to measure the somewhat arbitrary and frequency-dependent amplitude of the waveform. This approach is obviously quite limited, but it is good enough in some circumstances.

7.5.8 Frequency Counter

For accurate high-speed frequency measurements, such as measuring a crystal's operating frequency in a digital or an RF circuit, get a frequency counter. Oscilloscopes are simply too inaccurate (5 percent error or more) for such tasks. A counter with a bandwidth from 0 to 250 MHz isn't so expensive—around $100 to $300.

Some multimeters have a frequency counter feature, but the maximum frequency is often quite low.

FIGURE 7.60 A selection of used frequency counters. *Left:* Tektronix CFC-250, dc to 100-MHz counter found new in the box for $95. *Center:* HP 5385A, 10-Hz to 1-GHz counter with eight-digit display, which comes with two input channels—found for around $200, used, on eBay. *Right:* HP 5342A, 10-Hz to 18-GHz counter with 12-digit display—expensive.

7.5.9 Computer

It's a good idea to have a separate computer dedicated for your laboratory. An old one will usually work fine.

You'll probably use your laboratory computer most often for viewing manufacturer's component specs, circuit diagrams, electronics online catalogs, and the like.

With the Internet, you can grab a complex IC while at your bench and get a complete data sheet with spec and pin assignments from your computer as you prototype—just enter the part number into your favorite search engine. Professional component suppliers Farnell, RS Components, Mouser, Digi-Key, and Newark have links to the data sheets of most of the parts they sell, so even if you do not buy from them, this access is a very useful resource.

Of course, you'll also use your computer to program microcontrollers. Again, having Internet access is important for downloading other people's microcontroller source codes, microcontroller editor programs, and program updates.

Having circuit simulator and PCB layout programs on your computer is also handy. (See Sec. 7.5.22 for more on electronics CAD software.)

If you intend to install plug-in test equipment, make sure your computer comes with the appropriate number of expansion slots and ports. You'll need a decent computer to run sophisticated plug-in test software.

7.5.10 Miscellaneous Test Equipment

The amount of test equipment for electronics is extensive, and we've covered only the standard items so far. Other items include LCR meters, impedance analyzers, logic analyzers, spectrum analyzers, modulation analyzers, cable testers, power meters, network analyzers, and various specialized telecommunication equipment, to name a few. Fortunately, most of these items are not necessary for most hobby-type experimentation. If you're dealing with modulation analysis and spectrum analysis, chances are you're working or studying at an institution that has the necessary equipment; there's no need to shell out huge bucks for something you can borrow.

FIGURE 7.61 **(a)** BK-Precision LCR/ESR meter for measuring impedance (Z), inductance (L), capacitance (C), DCR, ESR, D, Q, and $\emptyset$, with wide range of test frequencies up to 100 kHz. Comes with four-wire testing that reduces error due to lead length. Used for spot frequency testing at various frequencies. **(b)** Agilent 4263B LCR meter capable of measuring more parameters ($|Z|$, $|Y|$, q, R, X, G, B, C, L, D, Q), with greater precision than the preceding meter. **(c)** Agilent impedance analyzer, unlike an LCR meter, can do continuous frequency sweeps and provide graphic analysis. Used to measure impedance and even capable of measuring material permitivity and permeability. Measures parameters such as $|Z|$, $|Y|$, q, R, X, G, B, L, C, D, *and* Q from a range of 40 Hz to 110 MHz. Extremely small variations in component characteristics can be precisely evaluated with sweep measurements of 0.08 percent accuracy. Good instrument for evaluation of capacitors, inductors, resonators, semiconductors, and other materials such as PCBs and toroidal cores. **(d)** Rohde & Schwartz spectrum analyzer that measures the frequency spectrum of a signal (signal amplitude versus frequency). Spectrum analyzers are used over a range of frequencies and may have different names, depending on their application. Used primarily for studying noise levels, dynamic range, frequency range, and transmitting power levels when troubleshooting radio equipment.

7.5.11 Multifunction PC Instruments

Multifunction PC instruments, both expensive and reasonably priced, are growing in popularity. One in particular that caught our eye is the TINALab II by DesignSoft. In conjunction with TINAPro circuit simulator software (also by DesignSoft), the little TINALab II box (see Fig. 7.62) acts as a multimeter (optional), oscilloscope, logic analyzer, signal analyzer, signal/function generator, and spectrum analyzer—all in one. The box is connected to your laptop or desktop using a USB or RS232 interface. In TINAPro, with a click of the mouse, you can jump between different screens representing different test instrument panels and test plots.

Perhaps the most intriguing feature of DesignSoft's package is the ability for TINAPro to act as both a simulator and a real-life tester at the same time. For example, if you're designing an amplifier circuit, using an op amp, a few capacitors, and resistors, you can first create a model of the amplifier in TINAPro and run a simulation, using the simulator's virtual instruments (virtual oscilloscope, virtual Bode plotter, and so on) to display the circuit's behavior. Next, using a prototype board and discrete components, you can build the real-life circuit, and then test it using the TINALab II interface. Results of the simulation and real-time test results measured by TINALab II can be displayed next to each other on the same screen. If something is wrong with either the simulation or the real circuit, you can alter the circuit schematic and see if the simulation works, and then go to the real circuit, make the same alteration, and see if it works. This is not only a great troubleshooting tool, but also a nice visual, hands-on way to learn electronics.

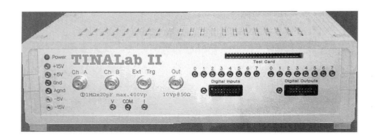 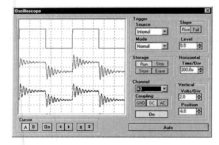

FIGURE 7.62   TINALab II from DesignSoft acts as a multifunction PC test instrument. Comes with digital oscilloscope, multimeter (optional), logic analyzer, signal analyzer, signal generator, and spectrum analyzer. Requires TINAPro circuit simulation software.

Here's a rundown of TINALab II's features:

Digital oscilloscope: Two-channel, 50-MHz bandwidth, 10/12-bit resolution, 4 GS/s equivalent sample rate on repetitive signals, and 20 MS/s for single-shot mode. Full-scale input range is ±400 V, with 5-mV/div to 100-V/div ranges.

Multimeter (optional): DC/AC from 1 mV to 400 V, and 100 μA to 2 A, with dc resistance from 1 Ω to 10 MΩ.

Function generator: Synthesized sine, square, ramp, triangle, and arbitrary waveforms from dc to 4 MHz, with logarithmic and linear sweep, and modulation up to 10 V peak to peak. Arbitrary waveforms can be programmed via the high-level language of TINAPro's interpreter.

Signal analyzer: Works in conjunction with the function generator, and measures and displays Bode amplitude and phase diagrams and Nyquist diagrams. It also works as a spectrum analyzer.

Logic generator and logic analyzer: Separate 16-channel digital inputs and outputs for generating or testing digital signals up to 40 MHz.

Other: Power supplies (±5 V, ±15 V), test card slot for plugging in experimenter modules.

TINALab II costs about $1,700, and the TINAPro simulator program is around $300 for the classic version and around $600 for the industrial version. This is a decent utility, considering all the instruments and software you get in a small package. However, be aware of certain limitations, such as bandwidth, that could lead to problems later on. For more information, check out DesignSoft's website at www.tina.com.

7.5.12   Isolation Transformers

If you plan to do work on line-power circuits such as TVs or switch-mode power supplies, or any line-power circuits that have no input isolation (no input transformer) and have a "floating ground" at a potential other than earth ground, use an isolation transformer between the line power and the circuit being tested. Attempting any repairs or simply sticking an oscilloscope's ground lead into such circuits without using an isolation transformer can lead to nasty shocks, blown circuit components, and melted test equipment probe tips.

AC Line Isolation Transformer

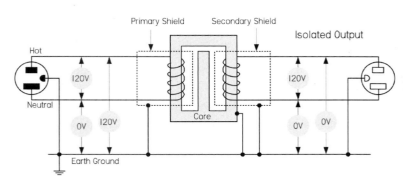

FIGURE 7.63   Basic schematic of a mains isolation transformer used to isolate the load from the source, as well as provide ground fault protection. An isolation transformer should be used whenever you work on nongrounded equipment with no input isolation, such as switch-mode power supplies. Notice that the voltage difference from secondary winding leads to earth ground is 0 V, unlike the 120-V difference in the primary side from hot to earth ground.

Figure 7.63 shows a simple mains isolation transformer. Isolation transformers are 1:1, meaning that the windings of the primary and secondary are equal in number so there is no increase or decrease in current or voltage between the primary and the secondary. The transformer is designed to isolate the load from the source and provide ground fault protection.

In your home wiring, the neutral (white) and the ground (green) connections are tied together at the main junction box, so they are basically at the same potential: 0 V, or earth ground. If you accidentally touch the hot wire while being in contact with a grounded object, current will pass through your body and give you a potentially fatal shock.

With an isolation transformer, the secondary winding leads act as a 120-V source and return, similar to the mains' hot and neutral, but with an important difference. Neither the secondary source nor the return runs are tied to earth ground! This means that if you touch the secondary source or return while being in contact with a grounded object, no current will flow through your body. Current wants to pass only between the secondary source and the return runs. Note that all transformers provide isolation, not just line isolation transformers. Therefore, equipment with input power transformers already has basic isolation protection built in. Figure 7.64 shows a simple way to construct an isolation transformer using two standard 120-V/12-V transformers, back to back.

Isolation transformers are also typically constructed with two isolated Faraday shields between the primary and the secondary windings. The use of the two shields diverts high-frequency noise, which would normally be coupled across the transformer to ground. Increasing the separation between the two Faraday shields minimizes the capacitance between the two and hence the coupling of noise between the two. Therefore, the isolation transformer acts to clean up line-power noise before being delivered to a circuit.

When do you need an isolation transformer? Well, for instance, within some TV sets, the inside cases are about 80 to 90 V above earth ground. If you open up a set and start fiddling around, you are liable to receive a nasty shock, since the test equipment and yourself are at a ground (earth ground) that is 90 V below the TV's internal ground. If you connect a scope to your TV's ground, you create a ground loop, which is bad. In an older TV, there is no provision to protect its circuit from this kind of short.

Likewise, switch-mode power supplies all require the use of isolation transformers when servicing. For example, a basic switch-mode power supply, like the one shown in Fig. 7.64, has a hot side and a cold side. There is an extreme shock hazard on the hot side of any switch-mode supply. Also, there is usually a diode bridge on the input. This means that the negative side of the filter capacitor is always one diode drop removed from the hot side of the ac power line. If you were to connect a scope ground to the "floating ground" side (point A) of the filter capacitor, you would blow out at least one of the bridge diodes, regardless of whether a fuse is incorporated into the design. The bridge diode will typically break down first, since connecting the ground lead of the scope places the full line voltage directly across the diode. In addition, you'll probably vaporize the scope probe lead, burn yourself, and receive a potentially lethal shock.

Isolation transformers are specified in terms of the amount of isolation they can provide, given as the RMS voltage, and by their power ratings, given in volt-amperes (VA). Additional specifications include efficiency and tolerance of voltage regulation. Typically, a 200-VA isolation transformer is sufficient for most modern equipment.

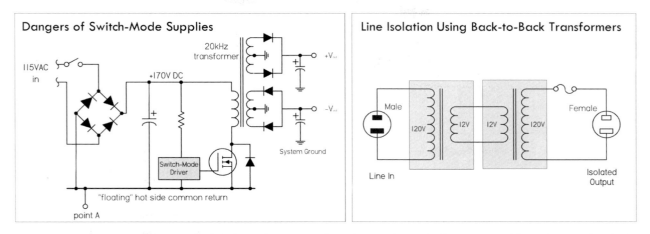

20kHz
transformer

115VAC
in

+170V DC

+V

−V

System Ground

Switch-Mode
Driver

"floating" hot side common return

point A

Line Isolation Using Back-to-Back Transformers

Male

120V

12V   12V

120V

Female

Line In

Isolated
Output

FIGURE 7.64   The illustration on the left shows how a switch-mode supply can be dangerous. Without the use of an isolation transformer, a ground lead of a scope connected to point A will blow out one of the bridge diodes, while vaporizing the scope probe tip. The illustration on the right shows how to construct a simple isolation transformer using two standard 120-V/12-V transformers, back to back.

Be extremely careful when dealing with ungrounded equipment. Always use an isolation transformer, and be extremely careful where you place your scope's ground lead. Do not attempt repairs unless you have a decent scope, total safety isolation, and accurate service information. Again, remember that any misplaced scope ground lead can instantly ruin a line-operated switch-mode supply. Always use an isolation transformer, and think before you connect and measure.

7.5.13 Variable Transformers, or Variacs

A variable transformer, or Variac, is a very useful device that acts like an adjustable ac voltage source. Its construction is that of an autotransformer, whose primary is connected to the hot and neutral of the 120-V line voltage, while the secondary leads consist of the neutral and an adjustable wiper that moves along the single core winding (see Fig. 7.65). Make sure the neutral is the common lead taken at the output. Don't use the hot lead; otherwise, the whole apparatus may be raised to live potential with respect to ground.

Being able to adjust the line voltage is a very useful trick when troubleshooting line-power equipment, where the fuse instantly blows at normal line voltage. Even without a fuse blowing, troubleshooting at around 85 V may reduce the fault current. Also, gradually increasing the voltage supplied to recently repaired equipment, such as a monitor, is a good way to ensure there are no problems; you can monitor the exact point where failure occurs.

Be aware that a Variac by itself does not provide isolation protection like a standard or isolation transformer, since the primary and secondary share a common winding. Therefore, if you plan to do work on ungrounded, "hot chassis" equipment, like that mentioned in the previous section, you must place an isolation transformer before the Variac, never after. If you don't, shock hazards await. Figure 7.65c shows a schematic of such an arrangement. It includes a switch and fuse protection, as well as

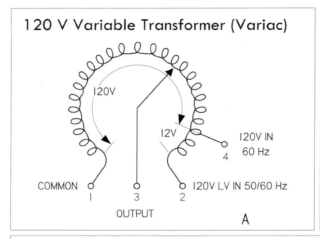

120 V Variable Transformer (Variac)

120V

12V

120V IN
60 Hz
4

COMMON 1

3

2 120V LV IN 50/60 Hz

OUTPUT

A

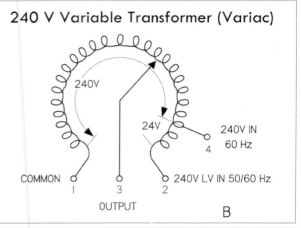

240 V Variable Transformer (Variac)

240V

24V

240V IN
60 Hz
4

COMMON 1

3

2 240V LV IN 50/60 Hz

OUTPUT

B

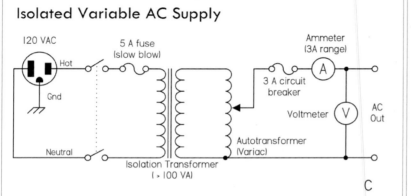

Isolated Variable AC Supply

120 VAC

Hot

Gnd

Neutral

5 A fuse
(slow blow)

Isolation Transformer
(> 100 VA)

3 A circuit
breaker

Ammeter
(3A range)

A

Voltmeter V

Autotransformer
(Variac)

AC
Out

C

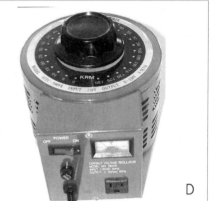

D

FIGURE 7.65 (a) Nonisolated 120-V Variac whose output voltage is varied by rotating a wiper. (b) Nonisolated 240-V Variac. (c) A homemade variable ac supply with isolation protection provided by means of an isolation transformer. (d) ac power supply that houses an isolation transformer, Variac, switch, fuse, ac outlet, and meter.

current and voltage meters, all of which create an adjustable, fully isolated ac power source. A 2-A Variac should suffice for most applications, although a 5-A or larger model isn't bad.

To avoid the hassle of cascading a Variac and isolation transformer together, simply get an ac power supply that houses both elements in one package. B+K Precision's 1653A ac power supply, for example, has a 0- to 150-V ac, 2-A Variac, an isolation transformer, and a current/voltage meter, all in one package. Similar ac power supplies can be found on eBay for around $50 (see Fig. 7.65d).

7.5.14 Substitution Boxes

Resistance, capacitance, inductance, and RC substitution boxes are nice to have around when you are trying to find the best resistance, capacitance, inductance, or RC circuit values needed within a circuit. Figure 7.66 shows a variety of commercial R, C, L, and RC-combination substitution boxes from IET Labs, Inc. (www.ietlabs .com). To select the desired resistance, capacitance, or inductance, simply dial in the value using the labeled thumbwheel switches. Voltage-divider substitution boxes are

 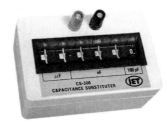  

FIGURE 7.66   Selection of resistance, capacitance, inductance, and RC boxes from IET Labs, Inc. Resistance boxes come in seven- (0 to 9,999,999 Ω, 1-Ω resolution) and nine- (0 to 99,999,999.9 Ω, 0.1-Ω resolution) decade types. Capacitance boxes come in six (0 to 99.9999 vF, 100-pF resolution) decade types. Inductance boxes come in three (999 mH, 1-mH resolution) and four (9.999 mH, 1-mH resolution) decade types. RC boxes come with mixed decade selections given in provision R and C boxes. Uses thumbwheel switches to dial in resistance, capacitance, and inductance values. Accuracy is typically 1 percent or better.

also available and can be extremely handy when determining the best voltage divider network for a given prototype. Substitution boxes aren't cheap, but the convenience and accuracy they provide is often well worth the cost.

Figure 7.67 shows a circuit diagram of a typical 0 to 9,999,999-Ω decade resistance box with 1-Ω resolution. Each wheel represents a decade, or a multiple of 10. Each resistor within a given wheel is of the same value—1, 10, 100, 1K, 10K, 100K, 1M, and so on—depending on the wheel it happens to be in. To select a desired resistance, you turn each switch to the appropriate position, thus placing resistors in series.

One-half-watt metal film resistors are a good choice for the resistors within the box. If you wish to create a decade box with a $\frac{1}{10}$-ohm range, the resistors of the $\frac{1}{10}$ place wheel should be of a wirewound variety or made of resistive wire. Be aware, however, that wirewound resistors are coiled like an inductor, so they can create inductive effects that can mess up the operation of some high-frequency circuits.

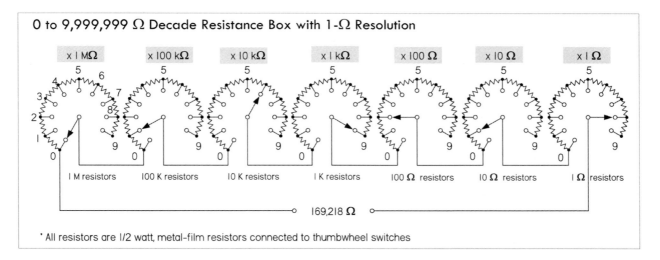

FIGURE 7.67   Resistor decade box made using thumbwheel switches.

Although making a resistance decade box is feasible, trying to put together a capacitance, inductance, or RC decade box isn't worth the effort; the low-tolerance components are too expensive, while the construction process is messy and time-consuming. It's better to simply buy a custom box from the manufacturer. However, you can still put together some simple substitution boxes, like the RC boxes in Fig. 7.68, without too much work. Though they aren't decade boxes—they can't dial in any possible component value—they can provide an adequate range or at least provide values that coincide with popular standard component values (10, 22, 33, 47, 56, 68, 82, and so on) that you find in the component catalogs.

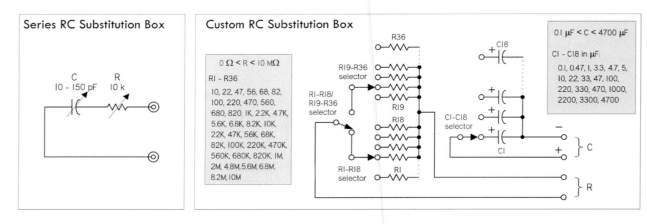

FIGURE 7.68   *Left:* A Series RC substitution box that can be constructed using plastic-film-dielectric tuning capacitor of 10 to 150 pF (or whatever is suitable) and a 10-kΩ, one-turn potentiometer. *Right:* Custom RC substitution box that can be switched to various modes: resistance-only, capacitance-only, RC-series, and RC-parallel.

For example, a simple low-capacitance series RC substitution box, shown to the left in Fig. 7.68, is made with a plastic-film-dielectric tuning capacitor of 10 to 150 pF (or whatever is suitable) and a 10-kΩ, one-turn potentiometer. Another custom RC substitution box that can be switched to various modes (resistive-only, capacitance-only, RC-series, and RC-parallel) is shown to the right in Fig. 7.68.

The selection of component values is up to you. Here, component values were chosen to coincide with popular component values produced by manufacturers. Eighteen-position rotary switches can be used to select *R* and *C* values. A typical selection of capacitor types is 100- to 900-pF mica, 0.001 to 0.009 µF in polystyrene, 0.01- to 0.9-µF polycarbonate, 1- to 9-µF polyester, and 10-µF and up tantalum or electrolytic (be careful to maintain the proper polarity across polarized capacitors). Often, for high-precision, low-capacitance values, air-dielectric tuning capacitors are used. One-half-watt, 1 percent metal film resistors provide good tolerance for all resistances above 1 Ω. Below 1 Ω, wirewound resistors or resistive wire can be used, though, again, be aware of inductive effects of wirewound resistors at high frequencies.

7.5.15   *Test Cables, Connectors, and Adapters*

Make sure you have a variety of test cables, connectors, and adapters at your disposal. This includes BNC, banana, hook, alligator, 0.100 male headers and 0.156 sockets, phone, RCA, and F. You never know what kind of connector will be needed when you start troubleshooting some new circuit or piece of equipment.

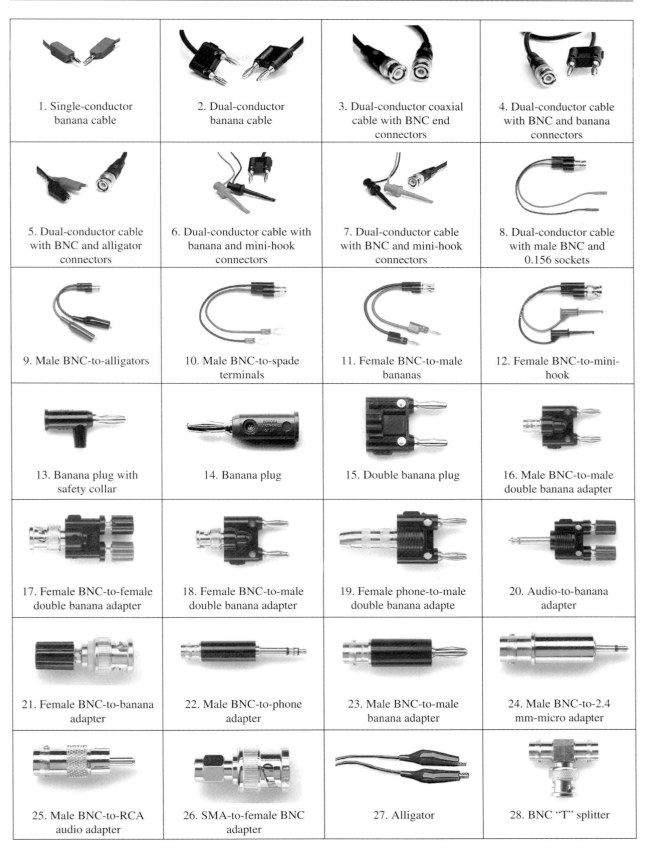

1. Single-conductor banana cable

2. Dual-conductor banana cable

3. Dual-conductor coaxial cable with BNC end connectors

4. Dual-conductor cable with BNC and banana connectors

5. Dual-conductor cable with BNC and alligator connectors

6. Dual-conductor cable with banana and mini-hook connectors

7. Dual-conductor cable with BNC and mini-hook connectors

8. Dual-conductor cable with male BNC and 0.156 sockets

9. Male BNC-to-alligators

10. Male BNC-to-spade terminals

11. Female BNC-to-male bananas

12. Female BNC-to-mini-hook

13. Banana plug with safety collar

14. Banana plug

15. Double banana plug

16. Male BNC-to-male double banana adapter

17. Female BNC-to-female double banana adapter

18. Female BNC-to-male double banana adapter

19. Female phone-to-male double banana adapte

20. Audio-to-banana adapter

21. Female BNC-to-banana adapter

22. Male BNC-to-phone adapter

23. Male BNC-to-male banana adapter

24. Male BNC-to-2.4 mm-micro adapter

25. Male BNC-to-RCA audio adapter

26. SMA-to-female BNC adapter

27. Alligator

28. BNC "T" splitter

FIGURE 7.69

7.5.16 Soldering Equipment

Soldering Iron

For most electronics work, a low-wattage, pencil-shaped soldering iron, from 25 to 40 W, will work fine. For very small components and pads, you may need to go to a 15-W iron. Very large connections may require a 50-W iron.

Adjustable temperature-controlled irons with digital displays are best, but this doesn't mean you can't get good results from $10 fixed-temperature irons with the proper tip wattage. Large irons and solder guns should not be used on any electronic assemblies or PCBs. Use them only on very large joints, such as between a large stranded wire (14-gauge or larger) and an aluminum chassis.

Figure 7.71 shows plans for building a soldering iron with low standby power, as well as plans for building a temperature-adjustable soldering iron using a fixed-temperature soldering iron.

FIGURE 7.70   Various soldering items: temperature-controlled soldering iron with digital display of tip temperature, roll of 60/40 rosin-core solder wire, solder pump, PanaVise Electronic Work Center for holding circuit boards while soldering.

In terms of tip size and shape, make sure to get an iron tip that is big enough for the job. The tip of the soldering iron should be small enough so that the joint being soldered can be seen easily, but should be large enough to quickly transfer the heat required to increase the joint temperature past the solder melting point. A chisel (spade) tip that is between 0.05 in. and 0.08 in. across the spade is ideal for general-purpose work. Smaller tips are required for small pads and surface-mount components.

Make sure your iron is powerful enough and hot enough. An iron that hasn't reached full temperature or that is set at too low a temperature can easily delaminate a trace or pad and can also fry components. This occurs due to overheating—a result of needing to apply heat for too long a duration to get the solder to melt.

Also, make sure you know whether your iron's tip is grounded (ESD-safe) or floating. In certain situations, a grounded soldering iron is required (for example, when soldering static-sensitive devices).

Solder and Fluxes

The solder types most commonly used for electrical connections are 60/40 (SN60) and 63/37 (SN63) solder. The terms 60/40 and 63/37 specify the alloy content: 60 percent tin/40 percent lead and 63 percent tin/37 percent lead, respectively. Other solders may come with different percentages and metals, such as 62/36/2 (62 percent tin/36 percent lead/2 percent silver). The melting point for 60/40 and 63/37 solder is 361°F. The 63/37 solder is excellent for small, heat-sensitive components and PCB pads; 60/40 is reserved more for general-purpose work. Both 60/40 and 63/37 solder come with either a rosin flux core or a solid (without flux) core. Rosin flux core solder is usually the preferred choice for electronics.

If you plan to use solid core solder, you'll need some flux paste or flux liquid. The flux paste or liquid must be applied to the metal surfaces to be joined before soldering takes place. When the soldering iron is placed on the flux-coated surfaces and heated, the flux will act as a chemical cleaner, removing oxidation from the metal surfaces and ensuring a good conductive solder joint. After soldering is complete, the remaining sticky flux residue should be removed with a defluxer (sometimes isopropyl alcohol is used, or even water, if the flux core is water-soluble). This prevents dirt from collecting in left-behind flux residue—something that can create low-resistance pathways between PCB pads.

When using flux core solder, the inner flux core seeps out while the outer alloy coating is melted. Generally, when using flux core solder, no additional flux is required. However, before soldering takes place, the surfaces to be soldered should not be visibly oxidized. Fine-grade steel wool or sandpaper can be used to polish the surfaces prior to soldering.

As a final note in regard to solder fluxes, never use corrosive (acid core) or conductive fluxes to solder electronic components. Use only mild fluxes such as those contained in rosin core solder or rosin flux.

Solder wire comes in a variety of wire diameters. The following list describes standard diameters and their intended uses.

0.020/0.508 mm (25 gauge) or smaller: This is excellent for soldering very small PCB pads and hand-soldering surface-mount components. However, it is too small for use as a general-purpose bench solder. It takes excessive heating time to apply sufficient solder to larger joints.

0.031/0.79 mm (21 gauge): This is excellent all-around solder for PCBs and general building and electronic repair.

0.040/1 mm (19 gauge) or larger: This is good for larger connections like tinning or connecting 14-gauge or larger wires to terminal strips, and soldering large stranded wires and other large components to aluminum chassis. It is not good for PCB soldering because excessive solder can be easily applied to pads, increasing the potential for undesired solder bridges between points.

When soldering traditional lead solder, your iron should be set to a temperature of around 330°C (625°F). Lead-free solders will require a higher temperature to melt well, typically around 400°C (750°F). If a joint is proving difficult to solder and will not flow, then a flux pen can be useful to get the solder flowing.

Solder-Removing Tools

To free components or clean up unwanted solder spillover, make sure you have a desoldering tool you're comfortable using. Here is an overview of the various tools that are available:

Desoldering pump (solder sucker): The tool is applied to the joint that has been heated to the solder's melting point. A plunger is activated, which sucks the solder into the tool's reservoir. If done properly, this method can remove solder to the point that the component leads can be lifted away from the metal to which they were joined. You may need an antistatic solder sucker when dealing with static-sensitive components; an ordinary solder sucker can generate high voltages as a result of internal friction. If you can't get everything with the sucker, try cleaning up the fine stuff using a desoldering braid.

Desoldering braid (solder wick): This consists of a copper braid impregnated with noncorrosive flux. The braid is laid on top of the solder joint to be desoldered, and the hot soldering tip is applied to the braid. As a result of capillary action, solder will flow toward the heat of the tip and away from the joint. This is used mainly if the joint doesn't have too much solder and when dealing with narrow PC traces where a sucker may cause problems, or in static-sensitive situations where a non-static sucker isn't available. It's a good choice for cleaning out solder holes in PCBs, as well as removing small solder splatters and solder bridges between IC leads.

Desoldering irons: If you anticipate the need, desoldering irons with attached vacuum suckers can really speed things up. The iron melts the solder, and then the vacuum is energized and sucks the molten solder into a reservoir for disposal. Desoldering irons are fairly expensive devices, but they are convenient.

Your soldering station should also have a damp sponge to keep your soldering tip free from excess solder and contamination. Special solder tip cleaning pastes also exist, and they do a decent job of removing oxidized material from the iron's tip. A wire brush, a file, and steel wool are nice to have around for removing built-up oxidation that may coat the solder tip. However, be sure to "tin" the tip immediately afterward by melting solder with it, then brushing off the excess with a soft cloth to form a smooth, silvery soldering surface; otherwise, the tip will oxidize.

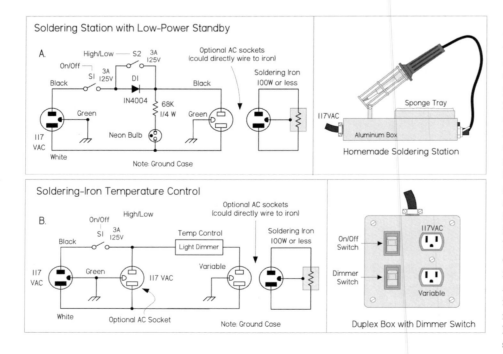

FIGURE 7.71 Soldering iron tips and heating elements will last longer if operated at reduced temperature. The circuit shown in (a) accomplishes this by half-wave rectification of the applied ac cycle. With current flowing in only one direction, only one electrode of the neon bulb glows. By closing the switch, the diode is effectively short-circuited and full power is applied to the soldering iron, igniting both bulb electrodes brightly. For safety, the circuit should be enclosed in an aluminum chassis. A 30- to 40-W soldering iron works well. A square metal tray can be used to hold a sponge. Use machine screws and nuts to secure the iron holder. A sealer must be used around holes in the sponge tray to prevent moisture from leaking into the electrical components. A $^3/_{16}$ ID grommet can be used to install the neon bulb. Heat shrink is used to insulate its leads.

A temperature control gives greater flexibility than the simple control just described. An incandescent-light dimmer can be used to control the working temperature of the tip. (b) shows a temperature control built into an electrical box. A dimmer and a duplex outlet are mounted in the box, as shown in the wiring diagram. The dimmer controls only one of two ac outlets. Normally, a jumper on the duplex outlet connects the hot terminals of both outlets together. This jumper must be removed. The hot terminal is narrower than the neutral one and typically uses a brass connecting screw. The neutral terminal remains interconnected. The dimmer can be purchased at any hardware or electrical store.

7.5.17 Prototyping Boards

Solderless Breadboards

Get a large *solderless breadboard* with plenty of rows and columns, IC channels, and power buses, along with banana terminals for external power supply connections. You'll find additional breadboards and *socket-and-bus strips* helpful as well. Breadboards will accept component lead diameters from 0.3 to 0.8 mm (20 to 30 AWG).

A note of caution: solderless breadboards are not good for building high-frequency RF circuits. The spring-loaded metal strips inside the board add too much stray capacitance to the circuit. Also, breadboards should not be used for high-current circuits above 100 mA or so.

Design Labs or Project Boards

Design labs or project boards are worth checking out, especially if you're into experimenting. Along with a large breadboard, these devices usually have a built-in power supply (fixed and variable) and function generator, as well as on-board components such as potentiometers, switches, LED indicators, and speakers. Though some individuals may shy away from such labs because of their "amateur appeal," they are, in fact, very convenient and practical—you avoid a lot of repetitive wiring, while keeping your work area in order.

FIGURE 7.72 *Left:* Typical design lab. *Center:* Solderless breadboard. *Right:* Various perforated boards.

Prototype PCBs

Following are various types of PCBs:

Blank copper-plated boards: It's nice to have both single- and double-sided clad boards. These are useful in ground-plane construction, where the circuit is built on the unetched side of the board, and wherever a component connects to ground, its lead is soldered to the copper board. Ungrounded connections between components are made point to point.

Bare perforated construction boards: These are phenolic or fiberglass circuit boards with perforations every 0.1-in., allowing common electronic devices to be mounted. These boards have no pads or buses, and are good for making "quick-and-dirty" circuits by bending the leads underneath the board and tacking them with solder. Large boards can be cut to any size desired.

Perforated boards with pads and buses: There are a variety to choose from, with varying pad and bus geometries, such as pad-per-hole, buses only, three-hole solder pads, ground plane, and volt/ground planes. A wide variety is nice to stock, but you can choose according to your needs.

PCB kit: These are designed for making your own custom PCBs. You can pick up a kit from RadioShack for $15, complete with two $3 \times 4\frac{1}{2}$-in. copper-clad circuit boards, a resist-ink pen to create circuit patterns, etching and stripping solutions, an etching tank, and $\frac{1}{16}$-in. drill bit, complete with instructions. There are more professional methods for making PCBs, but such kits are simple and good for small projects.

SURFBOARDS: These boards are an important breadboarding medium for surface-mount components. They act as adapters that convert surface-mount component footprints to a single in-line (SIP) format with 0.100-in. centers, making them compatible with breadboard sockets. They are used for constructing arrays (such as resistor, capacitor, diode, and transistor arrays) and other surface-mount sub-assemblies, and for evaluating surface-mount ICs.

7.5.18 Hand Tools

Wire Strippers and Cutters

First and foremost, get yourself a good set of *wire strippers* with built-in inner cutters, like those made by Ideal and GB Electronics. You'll probably want two: one covering

    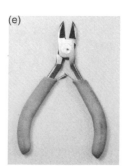

FIGURE 7.73 (a) Stripper (16–26 AWG). (b and c) Solderless crimping tools with additional features. (d) D-Sub crimping tool. (e) Diagonal cutter.

wire gauges from 10 to 18, and another for covering 16 to 26. An additional pair of 4- or 5-in. *diagonal wire cutters* or nippers is also vital for cutting wire, especially in awkward positions where the stripper's cutters won't reach.

Crimpers

It's good to have a general-purpose solderless terminal crimping tool for connecting various spade terminal and butt connectors to wires. The tool should have a separate section for insulated closure and noninsulated closure types, and a crimp gauge range from around 10 to 22.

You might also want to get a *D-Sub crimping tool* for crimping computer pin sockets, butted insulated connectors, telephone spade lugs, pin and socket contacts, and other noninsulated connectors to wires. The gauge range should be from 14 to 26.

There are many other special-purpose crimping tools, such as BNC, IDC, and modular crimping tools, which you'll need if you plan to attach connectors to BNC cables, ribbon cables, and telephone or CAT5 cables.

Other Tools

Here are some other tools that you should have:

Screwdriver set: Both Philips and standard.

Pliers set: Standard needle-nose, long-nose, and curved-nose.

Wrenches and nut driver: In a range of sizes.

Sheet metal shears, nibbling tool, and sheet metal bender: For cutting sheet metal (shears) or removing small pieces of metal (nibbling tool). With a bender, you can create your own metal enclosures.

IC remover: To get small ICs out of sockets without damaging the pins.

Tweezers: For small work, such as positioning surface-mount components.

Calipers: Used for determining component lead diameters, component sizes, board thickness, and so on.

Magnifying glass: For inspecting boards, wires, and components for cracks, flaws, hairline solder shorts, and cold-soldered joints.

X-acto knife: For general-purpose cutting.

Clip-on heat sink: Clamps on component leads to absorb or dissipate heat during soldering and desoldering.

Dremel tool: With cutting, sanding, milling, and polishing attachments.

Power drill (drill press): For drilling holes in PCBs, enclosures, and so on.

Files: For enlarging holes and slots; removing burrs; shaping metal, wood, or plastic; and cleaning metal surfaces before soldering.

Wire-wrap tools: Only if you plan to use wire-wrap construction techniques. Get a proper-sized hand-wrapping tool, along with bits for your wire-wrap wires (30 AWG is the most common).

Vise and circuit board holder: PanaVise makes some excellent vises and board holders.

Hack saw: Used for cutting bolts and sheet and beam metal, as well as PCBs.

7.5.19 Wires, Cables, Hardware, and Chemicals

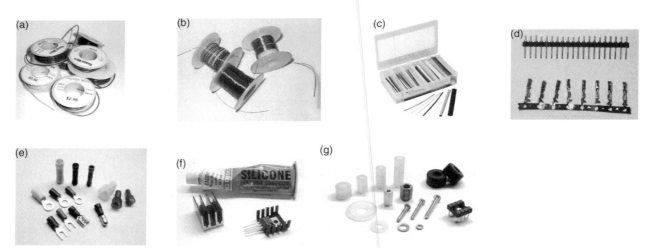

FIGURE 7.74 (a) Hookup wire. (b) Magnet wire. (c) Heat-shrinkable tubing. (d) 0.100-in. male headers and 0.156-in. female crimp sockets. (e) Butt connectors, ring, spade, crimp-on quick disconnects, wire nuts. (f) Heat sinks and heat sink compound. (g) Nylon and aluminum standoffs, washers, bolts, rubber feet, and IC socket.

Wire and Cable

Get a selection of *solid* and *stranded hookup wire* with different color coatings and gauges (16, 22, and 24 gauges are sufficient for most purposes). Jumper wire used for solderless breadboards can be made using solid 22-AWG hookup wire or ready-made male-to-male jumpers. If you desire flexible jumpers that won't break as easily as solid-wire jumpers, you can use 22- or 24-AWG stranded hookup wire with 0.156-in. female crimp pins (with optional 0.100-in. male headers soldered in). See Fig. 7.75 for construction details.

Other types of wire and cable worth keeping in stock include flat ribbon cable (28 AWG), CAT5 network cable, twisted-pair cable (24 AWG), coaxial cable (RG-59, RG-11, and so on), and household electrical wire, such as NM-B (indoor) or UF-B (outdoor). And don't forget wire-wrap wire if you intend on doing wire-wrap circuit

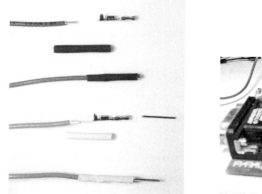
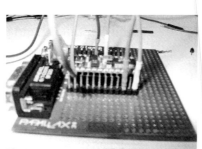

FIGURE 7.75 Making your own stranded-wire, square header jumpers, using hookup wire, female 0.156-in. crimp sockets (solder a 0.100-in. male header pin into the socket for a male jumper), and $\frac{1}{16}$- to $\frac{1}{8}$-in. shrink tubing. A crimping tool is needed to fasten sockets to wire, although you might get away with using needle-nose pliers and a bit of finesse.

board construction. The #30 wire with a Kynar jacket is perhaps the most popular type, although you might need larger wire for higher current runs.

Magnet Wire

Magnet wire (see Fig. 7.74b) is used to build custom coils and electromagnets or anything that requires a large number of loops, such as a tuning element in a radio receiver. Magnet wire is built of solid-core wire and is insulated by a varnish coating. A nice assortment of magnet wire ranging in gauge from 22 to 30 comes in handy.

Heat-Shrinkable Tubing

Heat-shrinkable tubing (see Fig. 7.74c) is a necessity for covering exposed wires and terminal connections, as well as combining a number of different wires together into a tight, single multiconductor cable. You can buy spools of tubing or purchase kits with short pieces of various color and diameter. Standard inner diameters of shrinkable tubing before heat is applied include $\frac{3}{64}$, $\frac{1}{16}$, $\frac{3}{32}$, $\frac{1}{8}$, $\frac{3}{16}$, $\frac{1}{4}$, $\frac{5}{16}$, $\frac{3}{8}$, $\frac{1}{2}$, $\frac{5}{8}$, $\frac{3}{4}$, 1, 2, 3, and 4 in. A common shrink ratio is 2:1 (50 percent), so a $\frac{1}{8}$-in. tube would shrink to $\frac{1}{16}$ in., and so forth. However, you might also find shrink ratios of 3:1. A heat gun, or often even a hair dryer, can be used to shrink the tubes.

Interconnects

For temporary wire-to-wire connections, wire nuts, butt connectors, and various pin-and-socket connectors should be at hand (see Fig. 7.74d and e). Wire nuts simply twist wires together and enclose the connection. Butt connectors join two different wire ends together within a metal-insulated crimped tube. Pin-and-socket connectors use various friction-fit mechanisms—male end on one wire, female end on the other.

For connecting wires to more permanent structures, such as PCBs and chassis, a variety of connectors can be used. For PCBs, 0.100-in. male headers (straight or right-angle) with corresponding 0.156-in. female crimp sockets are extremely useful. For larger connections, PCB-mount terminal blocks or chassis terminal blocks with corresponding solderless terminals (what gets attached to the wire) are ideal. Look for solderless terminal kits that include male and female spade terminals, quick disconnect, and ring terminals, to accommodate various wire gauges.

More Hardware

Here are some additional hardware items to keep on hand (see Fig. 7.74g):

Battery holders: AAA, AA, C, D, 9-V, and coin.

Heat sinks: TO-3, TO-92, TO-202, TO-218, TO-220, and DIP case styles. Also get heat-sink compound.

Standoffs: Aluminum and brass thread standoffs of 4-40, 6-32 thread of various lengths.

Hardware: Machine screws (4-40, 6-32), hex nuts (4-40, 6-32), and flat, split-lock washers (4, 6, and 8).

Adhesive rubber feet: For project boxes, to prevent slipping and surface scratches.

Transistor and IC sockets: Primarily 8, 14, and 16-pin DIP sockets.

Instrumentation knobs: ⅛-in. and ¼-in. shaft diameters with a screw to secure the knob to the shaft. These are used on potentiometers and the like.

Enclosures: Plastic and aluminum, with PCB mounting hardware.

Sheet metal: This is usually sold in large sheets, 4 × 8 ft or larger, and is used for making metal chassis.

Wire-wrap hardware: You will need this if you plan on using wire-wrap construction techniques. Get assorted posts, IC sockets, and so on.

Chemical-Related Items

Following are some other items to stock:

Epoxy: Two-part epoxy resin works well for joining odd-shaped items together. It creates a thick, strong bond. Try to find one that dries clear.

Silicon adhesive: This is useful for gluing components, such as Molex connectors, to circuit boards. Silicon adhesive can stand high temperatures, which is good for anything that holds a part to be soldered. It also dries to a rubbery consistency, so it can be peeled off if the component needs to be removed.

Deoxidizers and cleaners: An example is DeoxIT, which acts as a deoxidizer, cleaner, and preservative for metal electrical connections, such as switches, relay contact pads, and banana and audio plugs.

Defluxer: This is used for removing flux from soldered boards. It prevents low-resistance bridges from forming.

Antioxidant joint compound: An example is Nolox, which is good for ensuring that electrical connections don't corrode under moist conditions. For example, if you're running low-voltage landscape wires to lights or other outdoor projects using wire-nut connections, apply a dab of the compound to the wire connections before securing things together with a wire nut.

Conductive writer and circuit sealer pens: A conductive silver-based ink pen, such as the CircuitWriter by CAIG Laboratories Inc., is handy for fixing up corroded or damaged circuit board traces. The pen is used to redraw the ruined traces. A carbon-based ink pen is good for coating button contacts and membranes, where soldering is out of the question. In addition, an acrylic-based sealer pen can be used to provide conductive and oxidation protection to unprotected and newly drawn traces.

Circuit chiller: This is used to rapidly cool components and for troubleshooting intermittently faulting capacitors, resistors, semiconductors, and other defective components. It also detects cold solder joints, cracks in PCBs, and oxidized junctions.

PCB chemicals: Kits will usually already have all the necessary items, but extra resist, tape, resist pens, paint, rub-on transfer, and etchant can't hurt.

7.5.20 Electronics Catalogs

Order a variety of paperback catalogs from companies such as Digi-Key, Jameco, and Mouser Electronics, and keep them handy in your laboratory. Here are some popular sources to check out:

COMPANY	CATEGORY	WEBSITE
All Electronics	Electronics, surplus, mechanical	www.allcorp.com
Allied Electronics	Electronics	www.alliedelec.com
Alltronics	Electronics, surplus, etc.	www.alltronics.com
B.G. Micro	Electronics, surplus, kits	www.bgmicro.com
Debco Electronics	Electronics, kits, etc.	www.debcoelectronics.com
Digi-Key	Electronics	www.digikey.com
Electronic Goldmine	Electronics	www.goldmine-elec.com
Electronix Express	Electronics	www.elexp.com
Gateway Electronics	Electronics, kits, gadgets	www.gatewayelex.com
Halted Specialties	Electronics, gadgets, etc.	www.halted.com
Jameco Electronics	Electronics	www.jameco.com
JDR Microdevices	Electronics, kits, etc.	www.jdr.com
Martin P. Jones	Electronics, surplus	www.mpja.com
MECI	Electronics, surplus	www.meci.com
Mouser Electronics	Electronics	www.mouser.com
Newark Electronics	Electronics	www.newark.com
NTE Electronics	Electronics, replacement	www.nteinc.com
RadioShack	Electronics	www.radioshack.com
SparkFun	Electronics, especially modules	www.sparkfun.com
Web-tronics	Electronics, equipment, etc.	www.web-tronics.com

Many of these suppliers will ship worldwide. The following are some other international suppliers of note:

COMPANY	CATEGORY	WEBSITE
CPC	Electronics	cpc.farnell.com
Farnell	Electronics, professional supply worldwide	www.farnell.com
Maplin Electronics	Electronics, gadgets, etc., retail shops in UK	www.maplins.com
RS Components	Electronics, professional supply worldwide	www.rs-online.com

The Octopart website (www.octopart.com) is a search engine designed for finding electronic parts, and it's a great way to locate suppliers for a particular part. There are also online auction sites.

7.5.21 Recommended Electronics Parts

If you're serious about electronics, you'll want a decent stockpile of electronics components that you can resort to when unforeseen needs arise. Figs. 7.76a and 7.76b show a suggested stockpile for your laboratory. While putting together your stockpile, you can cut down on cost by investing in component kits, such as resistor, capacitor, transistor, diode, LED, digital IC, and analog IC kits. Jameco (www.jameco.com), Digi-Key (www.digikey.com), and Mouser Electronics (www.mouser.com) sell various component kits.

Decent Stockpile of Electrical Components (Part I)

Resistors

Carbon Film: 1Ω - 1MΩ, 1/8, 1/4, 1/2 watts, ± 5%

Metal Film: 1Ω - 1MΩ, 1/8, 1/4, 1/2 watts, ± 1%

Metal Oxide: 1Ω - 1MΩ, 1/2 watts, ± 5%

Power Resistors: 0.1Ω - 50kΩ, 10-100 watts, ± 5%

Standard resistor values for first two digits
10 11 **12** 13 **15** 16 18 **20 22** 24 **27** 30 **33**
36 **39** 43 **47** 51 56 **62 68** 75 **82** 91 **100**

1/2-Watt Single-Turn Potentiometers
500Ω, 1K, 2K 5K, 10K, 20K, 50K, 100K, 500K, 1M

3/4-Watt Multi-Turn Pots
500Ω, 1K, 2K 5K, 10K, 20K, 50K, 100K, 500K, 1M

1/2-Watt Single- and Multi-Turn Cermet Trimmers
100Ω, 500Ω, 1K, 2K 5K, 10K, 20K, 50K, 100K

Thick Film Resistor Networks: 2% SIP/DIP

Digital Potentiometers: e.g. DS1804-100

Capacitors

Ceramic Disc (10pF -0.47μF)
10pF, 22pF, 47pF, 100pF, 470pF, 0.001μF, 0.01μF, 0.022μF, 0.1μF, 0.47μF   (50V ± 20%)

Electrolytic (0.1μF - 4700μF)
0.1μF (50V), 1μF (50V), 1μF (100V), 2.2μF (50V), 3.3μF (50V), 4.7μF (50V), 10μF (50V), 100μF (50V), 220μF (25V), 470μF (25V), 1000μF (25V), 2200μF (25V), 3300μF (25V), 4700μF (35V)

Tantalum (0.1μF - 1000μF, ±10%)
0.1μF (35V), 0.22μF (35V), 0.47μF (35V), 1μF (25V), 1μF (35V), 2.2μF (16V), 3.3μF (35V), 4.7μF (35V), 6.8μF (35V), 10μF (16V), 10μF (35V), 15μF (25V), 22μF (16V), 33μF (25V), 47μF (25V)

Mylar (100V ± 20%)
0.001μF, 0.004μF, 0.01μF, 0.022μF, 0.033μF, 0.047μF, 0.1μF, 0.22μF, 0.47μF, 1μF

Polyester/Polypropylene Film
(0.01μF - 10μF, ±10%)
0.01μF, 0.022μF, 0.033μF, 0.047μF, 0.068μF, 0.1μF, 0.22μF, 0.47μF, 1μF, 2.2μF, 4.7μF, 10μF

Metallized Polyester (1000pF - 0.47μF, ±10%)
1000pF, 0.01μF, 0.1μF, 0.22μF, 0.33μF, 0.47μF, 1.0μF (63V); 4700pF, 0.1μF, 0.47μF (100V); 0.1μF (250V), 0.047μF, 0.1μF (400V)

Dipped Mica (1pF - 2000pF, ±5%)
1, 2, 3 (300V); 5, 10, 22, 33, 39, 47, 56, 100, 220, 270, 330, 390, 470, 560, 680, 820, 1000, 1200, 1500, 2000 (500V)

Ceramic Trimmer Capacitors
1-3 pF, 3-10pF, 5-20pF, 10-50pF, 20-70pF (200V)

Audio

Speakers: 4-Ω, 8-Ω; 70-20kHz; ferrite and piezo

Piezo Buzzers: 1.5-28VDC, 120VAC.

Microphones: Electret microphone cartridges

Choke Coils, Inductors, Ferrites

RF chokes (0.22μH to 1000μH)
0.22, 0.47, 1.0, 2.2, 3.3, 4.7, 10, 15, 22, 33, 47, 68, 100, 220, 330, 470, 680, 1000

EMI Shield Beads & Beads on Leads
43 (broadband), 61 (high freq.), 73 (low freq.)

Axial Molded Inductors (0.10μH to 6800μH)
0.10, 0.22, 0.33, 0.47, 1.0, 2.2, 3.3, 4.7, 5.6, 8.2,10, 12, 15, 22, 33, 39, 47, 56, 100, 220, 330, 470, 1000, 2200, 3300, 4700, 5600, 6800

Low Current Chokes: (0.33μH to 1000μH)

Medium Current Chokes (330μH to 33000μH):

High Choke Coils (0.7μH to 10μH)

Tunable Coils (Unshielded and Shielded)

Crystals

1.8432, 2.0, 2.4576, 3.2768, 3.579545, 3.6864, 4.0, 4.194304, 4.43361, 4.9152, 5.0, 5.0688, 6.0, 6.5536, 8.0, 10.0, 11.0592, 12.0, 16.0, 18.0, 18.432, 20.0, 24.0MHz

Transient Suppressors

Metal Oxide Varistors: AC and DC., various clamping voltages.

TVSs: unipolar and bipolar, various VBR ratings:

Fuses and Holders

1/4 x 1-1/4 and 5mm X 20mm Fast and Slow Blow

250V: 63mA, 1/16, 1/8, 3/10, 1/4, 3/8, 1/2, 3/4, 1-1/4, 1-1/2, 2, 2-1/2, 3, 5, 6, 8, 10, 15, 20A

Holders: panel mount, in-line, fuse clips and blocks

Switches

Types: Tactile, pushbutton, rotary, toggle, DIP, slide, snap-action, binary, hexadecimal, thumbwheel

Configurations: SPST, SPST-NC, SPST-NO, SPDT, DPDT, DP3T, 4PDT, 4P3T, 6PDT, (4,5,6,8,10,12-position rotary)

Relays

Low Profile Power (5A, 8A, 12A)
12VDC, 24VDC, 120VAC: SPDT, DPDT, 3PDT

Low Signal Relays (3A)
5VDC, 12VDC, 24VDC: DPDT, 4PDT

DC/AC and DC/DC Solid State Relays:

SIP/DIP Relays: 5V, 6V, 12V

Mechanical

DC Motors: 1.3V, 3V, 5V, 6V, 12V, 24V; spider shaft couplers, gears, etc.

Solenoids: 12VDC, 24VDC

DC Brushless Fans: **5VDC, 12VDC, 115VAC**

Stepper Motors: 5, 12, 24V, Bipolar, Unipolar

RC Servos: 4.8-6V, 30 to 200 oz-in.

Diodes/Zeners/Bridges

Diodes/Rectifiers

1N270	Ge, 50 PRV, 200mA
1N67A	Ge, 100 PRV, 4mA
1N914	Switch, 75 PRV, 10mA
1N3600	Switch, 50 PRV, 200 mA
1N4001	Rectifier, 50 PRV, 1A
1N4004	Rectifier, 400 PRV, 1A
1N4007	Rectifier, 1000 PRV, 1A
1N4148	Switch, 100 PRV, 25 mA
1N5404	Rectifier, 400 PRV, 3A
1N5408	Rectifier, 1000 PRV, 3A
1N5819	Schottky, 40 PRV, 1A

Zener Diodes (1-watt)

1N4730A	3.9V	1N4739A 9.1V	1N4746A 18V
1N4733A	5.1V	1N4742A 12V	1N4747A 20V
1N4735A	6.2V	1N4744A 15V	1N4749A 24V

Bridge Rectifiers: 200-600 PRV, 1-35A: DF04M, WO4G, KBP04M, BR82D, etc.

LEDs

Infrared: 940, 935, 880, 850, 800nm
Colored: Red, orange, yellow, green, blue, etc.
Others: high-output, white, blinking, tri-color
Sizes: T1, T 1 3/4, square, etc. .
Mounting Hardware: panel, litepipes, etc.

Transistors

Small-Signal General-Purpose Bipolar Transistors

2N2219A	NPN, hFE 100@150mA, TO-39
2N2222A	NPN, hFE 100@150mA, TO-92
2N2369A	NPN, 0.2A, 40-120 hFE, TO-92
2N2907A	PNP, hFE 100@150mA, TO-92
2N3904	NPN, hFE 100@10mA, TO-92
2N3906	PNP, hFE 100@10mA, TO-92

Power Bipolar Transistors

TIP31C	NPN, 100V, 3A, TO-220
TIP32C	PNP, 100V, 3A, TO-220
TIP41C	NPN, 100V, 6A, TO-220
TIP42C	PNP, 100V, 6A, TO-220
TIP48	NPN, 300V, 3A, TO-220
TIP120	NPN, 60V, 5A, TO-220
TIP121	NPN, 80V, 5A, TO-220
TIP122	NPN, 100V, 5A, TO-220
TIP125	PNP, 60V, 5A, TO-220
TIP132	NPN, 100V, 8A, TO-220
TIP140	NPN, 60V, 10A, TO-220
TIP145	PNP, 60V, 10A, TO-220
MJE2955T	PNP, 60V, 10A, TO-220
MJE3055T	NPN, 60V, 10A, TO-220

N-Channel MOSFET Transistors

IRF840	500VDC, 8A, TO-220
IRF511	200 PRV, 4A, TO-126
IRF9520	200 PRV, 10A, TO-220

Triacs

2N6071A	200 PRV, 4A, TO-126
SC146B	200 PRV, 10A, TO-220

Displays

Parallel LCD Modules: 20x1, 16x2, 16x4, 20x4

LCD Numeric Panel Meter: 3.5 and 4.5-digit

LED Numeric Displays: 7-seg., dot matrix

FIGURE 7.76a

Decent Stockpile of Electrical Components (Part II)

Voltage Regulators

Positive Voltage Regulators

78L05	5V, 0.1A, TO-92 case
7805T	5V, 1.0A, TO-220 case
7808T	8V, 1.0A, TO-220 case
78L09	9V, 0.1A, TO-92 case
78L12	12V, 0.1A, TO-92 case
7812T	12V, 1.0A, TO-220 case
7815T	15V, 1.0A, TO-220 case
78L24	24V, 0.1A, TO-92 case

Negative Voltage Regulators

7905T	-5V, 1.0A, TO-220 case
7912T	-12V, 1.0A, TO-220 case
7915T	-15V, 1.0A, TO-220 case

Adjustable Voltage Regulators

LM317T	1.2-37V, 1.5A, TO-220 case
LM317LZ	1.2-37V, 100mA, TO-92 case
LM317HVT	1.2-57V, 1.5A, TO-220 case
LM337T	-1.2-37V, 1.5A, TO-220 case

Op Amps & Audio Amps

TL082CP	JFET input op amp
TL084CP	JFET input op amp
LM301N	Precision op amp
LM308N	Precision op amp
LM324N	Low power quad op amp
LM351N	BIFET op amp
LM356N	JFET input, wide band op amp
LM358N	Low power dual op amp
LM380N	2-watt audio power amp
LM383T	9-watt audio power amp
LM384N	5-watt audio amp
LM386N-1	Low voltage audio amp, 250mW/6V
LM386N-3	Low voltage audio amp, 500mW/9V
LF411CN	Low offset drift JFET input op amp
LF412CN	Dual LF411CN op amp
LM741CN	General-purpose op amp
LM747CN	Dual 741 op amp
LM1458	Dual general purpose op amp
LM5532	Dual low noise op amp

Voltage Comparators

LM311N	Voltage comparator
LM339N	Quad, low power, low offset voltage
LM393N	Dual, low power, low offset voltage

Misc. Linear ICs

LMC555CN	Timer (MC1455P,NE555V)
XRL555	Micropower 555
LM556N	Dual 555
NE558N	Quad 555
LM564N	Hi-freq. phase locked loop
LM565N	Phase locked loop
LM3909	LED Flasher/Oscillator
LM567V	Tone Decoder
LM2907N	Frequency to Volt Converter
LM566	Voltage Controlled Oscillator
LM334Z	Adj. Current Source
LM34CZ	Temp. Sensor (-40°F to +230°F)
LM35DT	Temp. Sensor (-55°C to +150°C)
LM334Z	Adj. Current Source
ULN2003A	Hi-volt/curr. darl. transistor array
ULN2083A	Hi-volt/curr. darl. transistor array

4000 CMOS Logic ICs

4011	Quad 2-input NAND gate
4017	Decade counter/divider
4020	14-stage binary/ripple counter
4024	7-stage binary counter
4046	Micropower phase-locked loop
4049	Hex/buffer/converter (inverting)
4050	Hex/buffer/converter (non inverting)
4051	Single 8-chan. multi./demultiplexer
4066	Quad bilateral switch
4069	Hex inverter
4071	Binary up/down counter with clear
4584	Hex Schmitt trigger

Microcontrollers

Arduino Unot
Arduino Protoshield
Arduino LCD Shield
Arduino Min
ATtiny85
ATMega328
BASIC Stamps (Parallax): SB2, SB2e, SB2sx, SB2p
PICs (Microchip): PIC12Cxx, PIC12Fxx, PIC16Cxx, PIC16Fxx, PIC18xx, PIC18Fxx
Others: OOPPIC, Intel 8051, Motorola 68HC11

Optoelectronic

IR LEDs: 950, 940, 900, 880, 860 nm

NPN Photo Transistors: 910, 900, 860, 800 nm

Photo Diodes: 960, 950, 850, 820 nm

Optoisolators and Optical Switches: various outputs (photo transistor, triac, FET) and # of channels

Photo Cells: various sizes and power outputs

Photoresistors: various spectral and resistance ranges

Lamps: various types (incandescent, halogen, neon, xenon flash) and contacts (T-1, bi-pin, screw, etc.)

Wire/Connectors/Other

Wire and Cable: 16, 18, 20, 22, 24 AWG stranded and solid hook-up wire; 28 AWG flat ribbon cable; cat5 network cable; 24 AWG twisted pair cable; high-voltage wire: NM-B (indoor), UF-B (outdoor).

Shrinkable Tubing: 3/32" to 1", assorted colors.

Heat sinks: TO-3, TO-92, TO-202, TO-218, TO-220 and DIP case styles. Also get heat sink compound.

Hardware: battery holders (AAA, AA, C, D, 9V), transistor and IC socket holders, aluminum standoffs, PCB posts, strain reliefs, machine and sheet metal screws, nylon spacers, adhesive rubber feet, plastic and aluminum boxes.

AC Plugs Connectors: inlet (male), outlet (female)
AC & AC-to-DC Wall Transformers: 3 - 24V
DC Power Jacks and Plugs: 2.1mm, 2.5mm, 3.5mm
.100" Male Headers and .156 Female Crimp Pins
Terminal Blocks and Solderless Terminals: ring, spade, disconnect: 14-24 AWG
Wire-nuts: 10-18, 14-22, 16-24 AWG, etc.
Butt Splices: 10-12, 14-16, 18-24 AWG

Sensors

TMP36	Temp. Sensor
UGN3142	Hall-Effect Sensor (Allegro Microsystems)
HIH3605A	Humidity Sensor (Honeywell)
MPX2202D	Pressure Sensor (Motorola)
Sparkfun SEN-00639	Ultrasonic Range Finder
GP1S36	Tilt Sensor (Sharp)
GP2Y0A21YK IR Proximity Sensor	
MMA8452Q 3-axis accelerometer	

FIGURE 7.76b   (*Continued*)

Another great source for parts can be found within discarded consumer products, such as microwaves, stereos, printers, and breadmakers. Hack them apart, and you'll be amazed what you find: high-voltage transformers, large motors, stepper motors, laser diodes, gears, switches, relays, capacitors, wire connectors, and so on. Store what you find in boxes, and later retrieve what you need.

7.5.22 Electronic CAD Programs

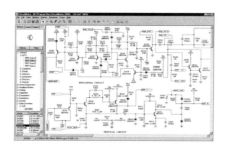 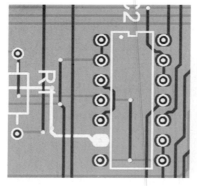 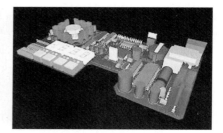

FIGURE 7.77

Electronic CAD programs are very important tools for learning, simulating, and constructing sophisticated circuits. Many software packages include schematic drawing, often simulation, and sometimes PCB design, autorouting, and even realistic 3D modeling. You can learn more about the various programs by surfing the Internet. Most of the software companies allow you to download a free demo version of their software. In addition, some of the most popular electronic CAD tools for hobby use are available for free (with restricted board size and layers) or as open source. Here are some CAD programs to check out:

CAD PROGRAM	DESCRIPTION
EAGLE CAD (www. cadsoftusa.com)	Available for Windows, Mac, and Linux systems. Free for noncommercial use for small boards, 2-layer PCB.
	Easily the most used CAD software by the hobby community, this product has a quirky user interface that takes a little getting used to. It does not have any simulation features and is centered around schematic design and PCB production. It offers auto layout of PCBs and Gerber file output, and a comprehensive scripting language. It has a large component library, as well as community-supplied popular libraries.
KiCad (kicad. sourceforge.net)	Available for Windows, Mac, and Linux systems. Similar feature set to EAGLE CAD, but it's open source. At the time of writing, the Mac version is available as a prerelease.

Here are a few commercial software packages worth checking out:

CAD PACKAGE	DESCRIPTION
Altium Designer (www.altium.com)	Formerly known as CircuitMaker 2000, the package contains a schematic design editor and run and analysis simulation. A free demo is available.
	It comes with a probe tool to examine the output waveform for any node in a circuit. The program automatically transfers component and connectivity information from a schematic to the Altium Designer 2000 PCB editor, ready for layout and routing. It includes a true mixed-mode simulator that lets you view both analog and digital output side by side. It has an extensive component library, as well as a custom symbol editor and SPICE 2 and SPICE 3 model import.
	Virtual instruments include an oscilloscope, multimeter, Bode plotter, curve tracer, data sequencer, signal generator, logic probe, and logic pulser.

CAD PACKAGE	DESCRIPTION
	After designing and testing your circuit, you simply click a button, and Altium Designer automatically generates a PCB netlist, opens the PCB editor, defines a board outline to your specifications, loads the netlist, and autoplaces the components on the board. Autorouting supports up to eight electrical layers (six signal plus power and ground planes), silk-screen overlays, and solder and paste masks. It supports through-hole and surface-mount components. It comes with a library of component footprints for PCB design, both surface-mount and through-hole devices.
ExpressPCB (www. expresspcb.com)	This is only a PCB layout program. The manufacturer anticipates you using their manufacturing service, so they give you the layout software for free. It's limited to two-sided boards, with plated through-holes.
NI MultiSim (www.ni.com/ multisim/)	This is a product from the component manufacturer National Instruments and comes with a schematic editor with a component library of over 16,000 parts. It also comes with a component editor, SPICE netlist import, and PCB footprint library. It simulates SPICE, VHDL, and Verilog models together. It comes with automatic wiring, wire drag, and exports to Ultiboard and other PCB programs. With an upgrade, you can get an RF design kit to work beyond 100 MHz, where SPICE normally becomes unreliable. Virtual instruments include oscilloscope, function generator, multimeter, Bode plotter, network analyzer, word generator, logic analyzer, spectrum analyzer, distortion analyzer, wattmeter, dc operating point, and temperature.
Ultiboard PCB Layout (www.ni.com/ ultiboard/)	A partner product of NI MultiSim, this can create any shape of board up to 2 m × 2 m. It includes DXF import, prebuilt library of standard board shapes, layer configuration, 64 signal layers, 64 mechanical layers, 2 × solder and post marks, 2 × silk screen, drill drawing and drive guide, fabrication, assembly information, glue layers, identification of violation, push-and-shove component placement, footprint library. It has built-in mechanical CAD capability, and is ideal for creating front panels, enclosures, and so on, ensuring alignment and position for attachment to the PCB. It also includes 3D viewing to visualize the complete, populated board.
TINAPro 6 (www.tina.com)	This is truly affordable, yet powerful, with a student appeal. It has a schematic editor; large component library (20,000); parameter extractor; analog, digital, and mixed-mode simulation; spectral analysis; Fourier analysis; noise analysis; network analysis; tolerance analysis; and so on. Its virtual instruments include oscilloscope, function generator, multimeter, signal analyzer/Bode plotter, network analyzer, spectrum analyzer, logic analyzer, digital signal generator, and XY recorder. The EDS3 (PCB automation tool for TINA) comes with an autorouter specially tailored to interface with TINA. It automatically generates a PCB rat's nest from schematics or netlist documents. Appropriate footprints are placed on the PCB, and you can reposition them as required. EDS 3 uses shape-based design checking to make sure the connectivity of your design is correct. All copper areas are checked for overlaps.

7.5.23 Building Your Own Workbench

After much deliberation, one of the authors decided he wanted a new workbench with a large tabletop, five shelves, a metal frame to which to attach things, a built-in backboard with power outlets, an on/off switch for a hanging fluorescent lamp, and a large wire-spool holding rod. The plans for building the workbench are shown in Fig. 7.78a and b. The whole design is based on a Pro Rack model 07200 from Do+Able Products, Inc. The rack cost only about $90, and the whole project ended up costing around $400 plus a good day's work. That author got all the material at Home Depot, where they even cut all the wood for free.

Inexpensive Do-It-Yourself Workbench

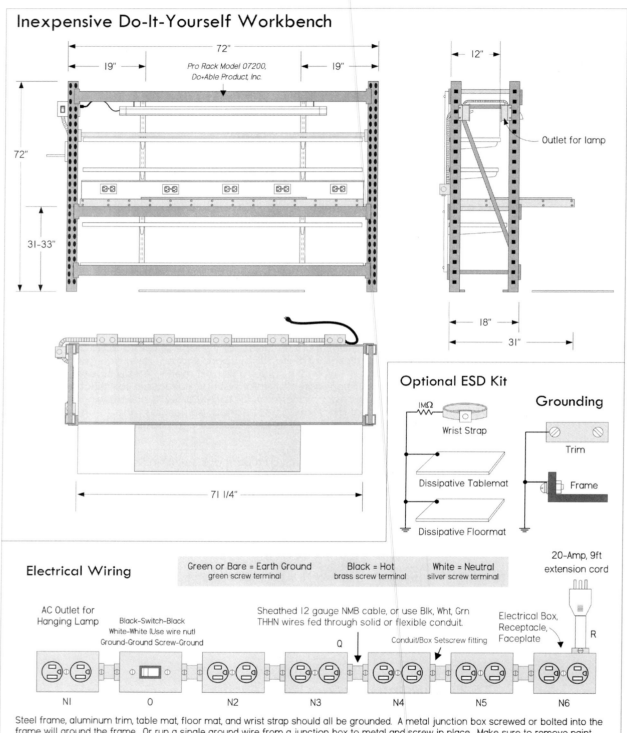

FIGURE 7.78a

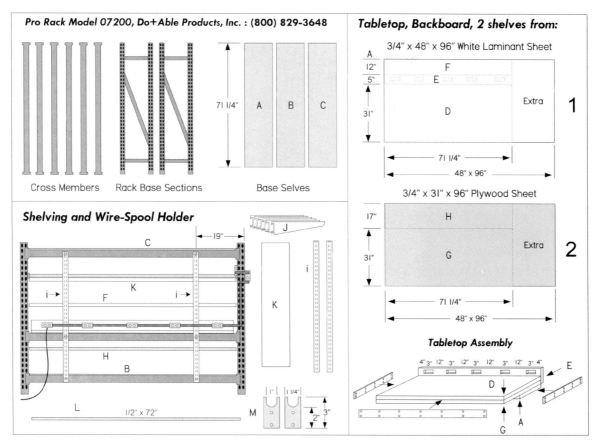

Pro Rack Model 07200, Do+Able Products, Inc. : (800) 829-3648

Cross Members   Rack Base Sections   Base Selves

Tabletop, Backboard, 2 shelves from:

3/4" x 48" x 96" White Laminant Sheet

3/4" x 31" x 96" Plywood Sheet

Shelving and Wire-Spool Holder

Tabletop Assembly

Parts List: I purchased everything at The Home Depot. They will even cut the wood for you for free.

1. Pro Rack Model 07200, Do+Able Products, Inc.: Includes metal cross members, rack bases, and three 3/4" x 16" x 71 1/4" particle board shelves, A, B, C.

2. Main Tabletop: Following boards are sandwiched together with a number of 1" drywall screws (see figure):
G. 3/4" x 71 1/4" x 31" plywood cut from sheet 2.
D. 3/4" x 71 1/4" x 31" white laminant board cut from sheet 1.
A: 3/4" x 71 1/4" x 31" particle board that comes with rack.
Trim: Select 1/16" x 1/2" edge trim, such as plastic or coated aluminum. Glue or screw into place.

3. Outlet Backboard: Cut out five 3" x 2" square electrical box-size holes in the board listed below.
E: 3/4" x 5" x 71 1/4" white laminant board cut from sheet 1.
Attach Outlet Backboard to Main Tabletop with a number of 2 1/2" drywall screws.
Place Main tabletop/backboard assembly onto middle frame shelve. Secure in place with sheetmetal screws.

4. Shelving: Made with Dorfile heavy-duty shelving uprights and brackets (double slot). Bolt uprights to frame.
i: Two 68" Dorfile uprights .
J: Six 12" Dorfile corresponding heavy-duty brackets with holes for securing screws.
F: 3/4" x 12" x 71 1/4" white laminant board cut from sheet
H: 3/4" x 17" x 71 1/4" plywood cut from sheet 2.
K: 3/4" x 12" x 71 1/4" white laminant board or 72" prefab. white shelf.
L: 1/2" x 72" solid metal conduit for wire-spool rod. Usually comes in 10-ft. lengths—cut with hacksaw.
M: Two 1/8" x 1 1/4" x 3" flat aluminum bars cut as shown in figure above. Used to support wire-spool rod.

5. Electrical: Not including dissipating pads and wristband. Follow grounding rules presented previously.
N: Six 2"x 3" electrical boxes with corresponding duplex receptacles and face plates.
O: One 2" x 3" electrical box with corner bracket to mount on rack frame, with switch and faceplate.
P: Solid or flexible conduit, conduit/box setscrew fittings, and wire nuts.
Q: Sheathed 12-gauge NMB hookup cable, or use Blk, Wht, Grn THHN wire.
R: 9-ft. extension cord (rated for 20 Amps). Snip off female end and feed into an outlet receptacle, use wire clamp.
S: Fluorescent 4-ft.work light with chain to be hung on front upper cross-member of rack frame.

6. Hardware:
1" and 2 1/2" drywall screws, box of 3/16" x 1" bolts with washers and nuts, small 3/4" sheet metal screws.

7. Tools:
Hack saw, jig saw with wood-cutting blade, power drill with 13/64" and 5/32" carbide bits, screw drivers, metal file, hammer, tape measure, wire strippers, knife, electrical tape.

FIGURE 7.78b *(Continued)*

CHAPTER 8

Operational Amplifiers

Operational amplifiers (op amps) are incredibly useful high-performance differential amplifiers that can be employed in a number of amazing ways. A typical op amp is an integrated device with a noninverting input, an inverting input, two dc power supply leads (positive and negative), an output terminal, and a few other specialized leads used for fine-tuning. The positive and negative supply leads, as well as the fine-tuning leads, are often omitted from circuit schematics. If you do not see any supply leads, assume that a dual supply is being used.

Note that we have labeled the supply voltages $+V_s$ and $-V_s$, as they are usually the same. However, they do not need to be, as you will see when we look at single-supply op amps in this chapter.

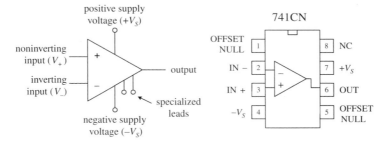

FIGURE 8.1

By itself, an op amp's operation is simple. If the voltage applied to the *inverting terminal* V_- is more positive than the voltage applied to the *noninverting terminal* V_+, the output saturates toward the *negative supply voltage* $-V_S$. Conversely, if $V_+ > V_-$, the output saturates toward the positive supply voltage $+V_S$ (see Fig. 8.2). This "maxing out" effect occurs with the slightest difference in voltage between the input terminals.

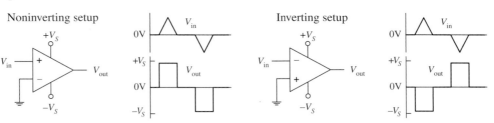

FIGURE 8.2

At first glance, it may appear that an op amp is not a very impressive device—it switches from one maximum output state to another whenever there's a voltage difference between its inputs. Big deal, right? By itself, it does indeed have limited applications. The trick to making op amps useful devices involves applying what is called *negative feedback*.

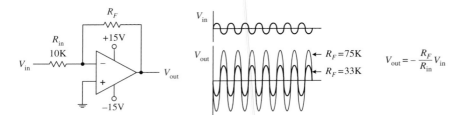

FIGURE 8.3

When voltage is "fed" back from the output terminal to the inverting terminal (this is referred to as *negative feedback*), the gain of an op amp can be controlled—the op amp's output is prevented from saturating. For example, a feedback resistor R_F placed between the output and the inverting input, as shown in Fig. 8.3, acts to convey the state of the output back to the op amp's input. This feedback information basically tells the op amp to readjust its output voltage to a value determined by the resistance of the feedback resistor. The circuit in Fig. 8.3, called an *inverting amplifier,* has an output equal to $-V_{in}(R_F/R_{in})$ (you will learn how to derive this formula later in this chapter). The negative sign means that the output is inverted relative to the input. The gain is then simply the output voltage divided by the input voltage, or $-R_F/R_{in}$ (the negative sign indicates that the output is inverted relative to the input). As you can see from this equation, if you increase the resistance of the feedback resistor, there is an increase in the voltage gain. On the other hand, if you decrease the resistance of the feedback resistor, there is a decrease in the voltage gain.

By adding other components to the negative-feedback circuit, an op amp can be made to do a number of interesting things besides pure amplification. Other interesting op amp circuits include voltage-regulator circuits, current-to-voltage converters, voltage-to-current converters, oscillator circuits, mathematical circuits (adders, subtractors, multipliers, differentiators, integrators, etc.), waveform generators, active filter circuits, active rectifiers, peak detectors, sample-and-hold circuits, etc. Most of these circuits will be covered in this chapter.

Besides negative feedback, there's positive feedback, where the output is linked through a network to the noninverting input. Positive feedback has the opposite effect as negative feedback; it drives the op amp harder toward saturation. Although positive feedback is seldom used, it finds applications in special comparator circuits that are often used in oscillator circuits. Positive feedback also will be discussed in detail in this chapter.

8.1 Operational Amplifier Water Analogy

This is the closest thing we could come up with in terms of a water analogy for an op amp. To make the analogy work, you have to pretend that water pressure is analogous to voltage and water flow is analogous to current flow.

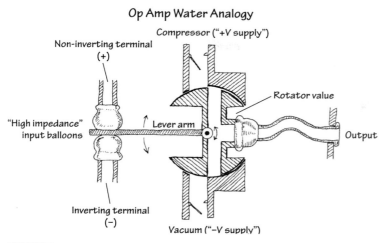

Op Amp Water Analogy

FIGURE 8.4

The inverting and noninverting terminals of the water op amp are represented by the two tubes with elastic balloon ends. When the water pressure applied to both input tubes is equal, the lever arm is centered. However, if the water pressure applied to the noninverting tube is made larger than the pressure applied to the inverting tube, the noninverting balloon expands and forces the lever arm downward. The lever arm then rotates the rotator valve counterclockwise, thus opening a canal from the compressor tube (analogous to the positive supply voltage) to the output tube. (This is analogous to an op amp saturating in the positive direction whenever the noninverting input is more positive in voltage than the inverting input.) Now, if the pressure applied at the noninverting tube becomes less than the pressure applied at the inverting tube, the lever arm is pushed upward by the inverting balloon. This causes the rotator valve to rotate clockwise, thus opening the canal from the vacuum tube (analogous to the negative supply voltage) to the output. (This is analogous to an op amp saturating in the negative direction whenever the inverting input is made more positive in voltage than the noninverting input.) See what you can do with the analogy in terms of explaining negative feedback. Also note that in the analogy there is an infinite "input water impedance" at the input tubes, while there is a zero "output water impedance" at the output tube. As you will see, ideal op amps also have similar input and output impedance. In real op amps, there are always some leakage currents.

8.2 How Op Amps Work (The "Cop-Out" Explanation)

An op amp is an integrated device that contains a large number of transistors, several resistors, and a few capacitors. Figure 8.5 shows a schematic diagram of a typical low-cost general-purpose bipolar operational amplifier.

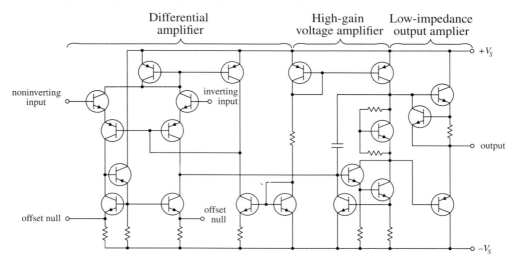

FIGURE 8.5

This op amp basically consists of three stages: a high-input-impedance differential amplifier, a high-gain voltage amplifier with a level shifter (permitting the output to swing positive and negative), and a low-impedance output amplifier. However, realizing that an op amp is composed of various stages does not help you much in terms of figuring out what will happen between the input and output leads. That is, if you attempt to figure out what the currents and voltages are doing within the complex system, you will be asking for trouble. It is just too difficult a task. What is important here is not to focus on understanding the op amp's internal circuitry but instead to focus on memorizing some rules that individuals came up with that require only working with the input and output leads. This approach seems like a "cop-out," but it works.

8.3 Theory

There is essentially only one formula you will need to know for solving op amp circuit problems. This formula is the foundation on which everything else rests. It is the expression for an op amp's output voltage as a function of its input voltages V_+ (noninverting) and V_- (inverting) and of its *open-loop voltage gain A_o*:

$$V_{out} = A_o(V_+ - V_-)$$

This expression says that an *ideal op amp* acts like an ideal voltage source that supplies an output voltage equal to $A_o(V_+ - V_-)$ (see Fig. 8.6). Things can get a little more complex when we start talking about *real op amps,* but generally, the open-loop voltage expression above pretty much remains the same, except now we have to make some slight modifications to our equivalent circuit. These modifications must take into account the nonideal features of an op amp, such as its input resistance R_{in} and output resistance R_{out}. Figure 8.6 *right* shows a more realistic equivalent circuit for an op amp.

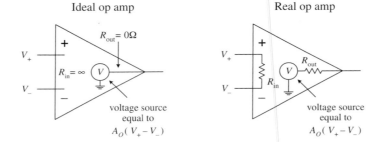

FIGURE 8.6

To give meaning to the open-loop voltage gain expression and to the ideal and real equivalent circuits, the values of A_o, R_{in}, and R_{out} are defined within the following rules:

Rule 1: For an ideal op amp, the open-loop voltage gain is infinite ($A_o = \infty$). For a real op amp, the gain is a finite value, typically between 10^4 to 10^6.

Rule 2: For an ideal op amp, the input impedance is infinite ($R_{in} = \infty$). For a real op amp, the input impedance is finite, typically between 10^6 (e.g., typical bipolar op amp) to 10^{12} Ω (e.g., typical JFET op amp). The output impedance for an ideal op amp is zero ($R_{out} = 0$). For a real op amp, R_{out} is typically between 10 to 1000 Ω.

Rule 3: The input terminals of an ideal op amp draw no current. Practically speaking, this is true for a real op amp as well—the actual amount of input current is usually (but not always) insignificantly small, typically within the picoamps (e.g., typical JFET op amp) to nanoamps (e.g., typical bipolar op amp) range.

Now that you are armed with $V_{out} = A_o(V_+ - V_-)$ and rules 1 through 3, let's apply them to a few simple example problems.

EXAMPLE 1

Solve for the gain (V_{out}/V_{in}) of the circuit below.

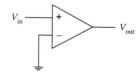

FIGURE 8.7

Since V_- is grounded (0 V) and V_+ is simply V_{in}, you can plug these values into the open-loop voltage gain expression:

$$V_{out} = A_o(V_+ - V_-)$$
$$= A_o(V_{in} - 0 \text{ V}) = A_o V_{in}$$

Rearranging this equation, you get the expression for the gain:

$$\text{Gain} = \frac{V_{out}}{V_{in}} = A_o$$

If you treat the op amp as ideal, A_o would be infinite. However, if you treat the op amp as real, A_o is finite (around 10^4 to 10^6). This circuit acts as a simple noninverting comparator that uses ground as a reference. If $V_{in} > 0$ V, the output ideally goes to $+\infty$ V; if $V_{in} < 0$V, the output ideally goes to $-\infty$ V. With a real op amp, the output is limited by the supply voltages (which are not shown in the drawing but assumed). The exact value of the output voltage is slightly below and above the positive and negative supply voltages, respectively. These maximum output voltages are called the *positive* and *negative saturation voltages*.

EXAMPLE 2

Solve for the gain (V_{out}/V_{in}) of the circuit below.

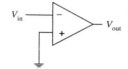

FIGURE 8.8

Since V_+ is grounded (0V) and V_- is simply V_{in}, you can substitute these values into the open-loop voltage gain expression:

$$V_{out} = A_o(V_+ - V_-)$$
$$= A_o(0 \text{ V} - V_{in}) = -A_o V_{in}$$

Rearranging this equation, you get the expression for the gain:

$$\text{Gain} = \frac{V_{out}}{V_{in}} = -A_o$$

If you treat the op amp as ideal, $-A_o$ is negatively infinite. However, if you treat the op amp as real, $-A_o$ is finite (around -10^4 to -10^6). This circuit acts as a simple inverting comparator that uses ground as a reference. If $V_{in} > 0$ V, the output ideally goes to $-\infty$ V; if $V_{in} < 0$ V, the output ideally goes to $+\infty$ V. With a real op amp, the output swings are limited to the saturation voltages.

8.4 Negative Feedback

Negative feedback is a wiring technique where some of the output voltage is sent back to the inverting terminal. This voltage can be "sent" back through a resistor, capacitor, or complex circuit or simply can be sent back through a wire. So exactly what kind of formulas do you use now? Well, that depends on the feedback circuit, but in reality, there is nothing all that new to learn. In fact, there is really only one formula you need to know for negative-feedback circuits (you still have to use the rules, however). This formula looks a lot like our old friend $V_{out} = A_o(V_+ - V_-)$. There is, however, the V_- in the formula—this you must reconsider. V_- in the formula changes because now the output voltage from the op amp is "giving" extra voltage (positive or negative) back to the inverting terminal. What this means is that you must replace V_- with fV_{out}, where f is a fraction of the voltage "sent" back from V_{out}. That's the trick!

There are two basic kinds of negative feedback, voltage feedback and operational feedback, as shown in Fig. 8.9.

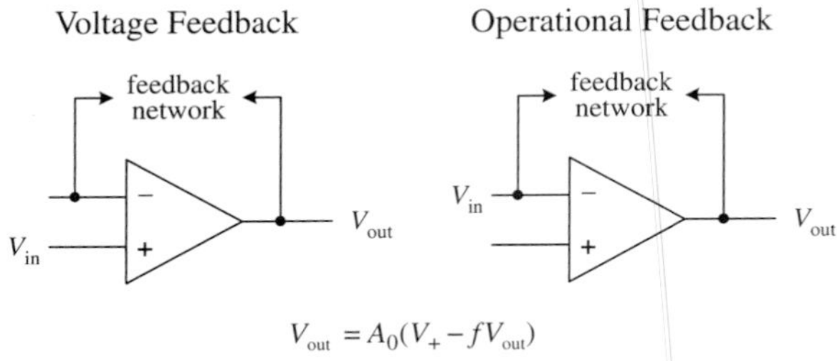

FIGURE 8.9

Voltage Feedback

Operational Feedback

feedback network

feedback network

V_{in}

V_{out}

V_{in}

V_{out}

$$V_{out} = A_0(V_+ - fV_{out})$$

Now, in practice, figuring out what the fraction f should be is not important. That is, you do not have to calculate it explicitly. The reason why we have introduced it in the open-loop voltage expression is to provide you with a bit of basic understanding as to how negative feedback works in theory. As it turns out, there is a simple trick for making op amp circuits with negative feedback easy to calculate. The trick is as follows: If you treat an op amp as an ideal device, you will notice that if you rearrange the open-loop voltage expression into $V_{out}/A_o = (V_+ - V_-)$, the left side of the equation goes to zero—A_o is infinite for an ideal op amp. What you get in the end is then simply $V_+ - V_- = 0$. This result is incredibly important in terms of simplifying op amp circuits with negative feedback—so important that the result receives its own rule (the fourth and final rule).

Rule 4: Whenever an op amp senses a voltage difference between its inverting and noninverting inputs, it responds by feeding back as much current/voltage through the feedback network as is necessary to keep this difference equal to zero ($V_+ - V_- = 0$). This rule only applies for negative feedback.

The following sample problems are designed to show you how to apply rule 4 (and the other rules) to op amp circuit problems with negative feedback.

Negative Feedback Example Problems

BUFFER (UNITY GAIN AMPLIFIER)

Solve for the gain (V_{out}/V_{in}) of the circuit below.

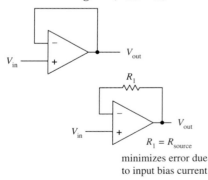

V_{in}

V_{out}

R_1

V_{in}

V_{out}

$R_1 = R_{source}$

minimizes error due to input bias current

FIGURE 8.10

Since you are dealing with negative feedback, you can apply rule 4, which says that the output will attempt to make $V_+ - V_- = 0$. By examining the simple connections, notice that $V_{in} = V_+$ and $V_- = V_{out}$. This means that $V_{in} - V_{out} = 0$. Rearranging this expression, you get the gain:

$$\text{Gain} = \frac{V_{out}}{V_{in}} = 1$$

A gain of 1 means that there is no amplification; the op amp's output follows its input. At first glance, it may appear that this circuit is useless. However, it is important to recall that an op amp's input impedance is huge, while its output impedance is extremely small (rule 2). This feature makes this circuit useful for circuit-isolation applications. In other words, the circuit acts as a buffer. With real op amps, it may be necessary to throw in a resistor in the feedback loop (lower circuit). The resistor acts to minimize voltage offset errors caused by input bias currents (leakage). The resistance of the feedback resistor should be equal to the source resistance. We will discuss input bias currents later in this chapter.

INVERTING AMPLIFIER

Solve for the gain (V_{out}/V_{in}) of the circuit below.

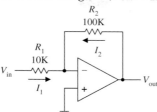

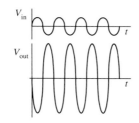

FIGURE 8.11

Because you have negative feedback, you know the output will attempt to make the difference between V_+ and V_- zero. Since V_+ is grounded (0 V), this means that V_- also will be 0 V (rule 4). To figure out the gain, you must find currents I_1 and I_2 so you can come up with an expression containing V_{out} in terms of V_{in}. Using Ohm's law, you find I_1 and I_2 to be

$$I_1 = \frac{V_{in} - V_-}{R_1} = \frac{V_{in} - 0\,V}{R_1} = \frac{V_{in}}{R_1}$$

$$I_2 = \frac{V_{out} - V_-}{R_2} = \frac{V_{out} - 0\,V}{R_2} = \frac{V_{out}}{R_2}$$

Because an ideal op amp has infinite input impedance, no current will enter its inverting terminal (rule 3). Therefore, you can apply Kirchhoff's junction rule to get $I_2 = -I_1$. Substituting the calculated values of I_1 and I_2 into this expression, you get $V_{out}/R_2 = -V_{in}/R_1$. Rearranging this expression, you find the gain:

$$\text{Gain} = \frac{V_{out}}{V_{in}} = -\frac{R_2}{R_1}$$

The negative sign tells you that the signal that enters the input will be inverted (shifted 180°). Notice that if $R_1 = R_2$, the gain is –1 (the negative sign simply means the output is inverted). In this case you get what's called a *unity-gain inverter*, or an *inverting buffer*. When using real op amps that have relatively high input bias currents (e.g., bipolar op amps), it may be necessary to place a resistor with a resistance equal to $R_1 \| R_2$ between the noninverting input and ground to minimize voltage offset errors.

NONINVERTING AMPLIFER

Solve for the gain (V_{out}/V_{in}) of the circuit below.

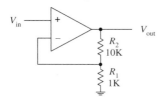

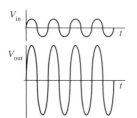

FIGURE 8.12

By inspection, you know that $V_+ = V_{in}$. By applying rule 4, you then can say that $V_- = V_+$. This means that $V_- = V_{in}$. To come up with an expression relating V_{in} and V_{out} (so that you can find the gain), the voltage divider relation is used:

$$V_- = \frac{R_1}{R_1 + R_2} V_{out} = V_{in}$$

Rearranging this equation, you find the gain:

$$\text{Gain} = \frac{V_{out}}{V_{in}} = \frac{R_1 + R_2}{R_1} = 1 + \frac{R_2}{R_1}$$

Unlike the inverting amplifier, this circuit's output is in phase with its input—the output is "noninverted." With real op amps, to minimize voltage offset errors due to input bias current, set $R_1 \| R_2 = R_{source}$.

SUMMING AMPLIFIER

Solve for V_{out} in terms of V_1 and V_2.

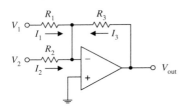

Since you know that V_+ is grounded (0 V), and since you have negative feedback in the circuit, you can say that $V_+ = V_- = 0\,\text{V}$ (rule 4). Now that you know V_-, solve for I_1, I_2, and I_3 in order to come up with an expression relating V_{out} with V_1 and V_2. The currents are found by applying Ohm's law:

$$I_1 = \frac{V_1 - V_-}{R_1} = \frac{V_1 - 0\,\text{V}}{R_1} = \frac{V_1}{R_1}$$

$$I_2 = \frac{V_2 - V_-}{R_2} = \frac{V_2 - 0\,\text{V}}{R_2} = \frac{V_2}{R_2}$$

$$I_3 = \frac{V_{\text{out}} - V_-}{R_3} = \frac{V_{\text{out}} - 0\,\text{V}}{R_3} = \frac{V_{\text{out}}}{R_3}$$

Like the last problem, assume that no current enters the op amp's inverting terminal (rule 3). This means that you can apply Kirchhoff's junction rule to combine I_1, I_2, and I_3 into one expression: $I_3 = -(I_1 + I_2) = -I_1 - I_2$. Plugging the results above into this expression gives the answer:

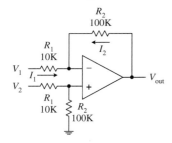

FIGURE 8.13

$$V_{\text{out}} = -\frac{R_3}{R_1}V_1 - \frac{R_3}{R_2}V_2 = -\left(\frac{R_3}{R_1}V_1 + \frac{R_3}{R_2}V_2\right)$$

If you make $R_1 = R_2 = R_3$, $V_{\text{out}} = -(V_1 + V_2)$. Notice that the sum is negative. To get a positive sum, you can add an inverting stage, as shown in the lower circuit. Here, three inputs are added together to yield the following output: $V_{\text{out}} = V_1 + V_2 + V_3$. Again, for some real op amps, an additional input-bias compensation resistor placed between the noninverting input and ground may be needed to avoid offset error caused by input bias current. Its value should be equal to the parallel resistance of all the input resistors.

DIFFERENCE AMPLIFIER

Determine V_{out}.

FIGURE 8.14

First, you determine the voltage at the non-inverting input by using the voltage divider relation (again, assume that no current enters the inputs):

$$V_+ = \frac{R_2}{R_1 + R_2}V_2$$

Next, apply Kirchhoff's current junction law to the inverting input ($I_1 = I_2$):

$$\frac{V_1 - V_-}{R_1} = \frac{V_- - V_{\text{out}}}{R_2}$$

Using rule 4 ($V_+ = V_-$), substitute the V_+ term in for V_- in the last equation to get

$$V_{\text{out}} = \frac{R_2}{R_1}(V_2 - V_1)$$

If you set $R_1 = R_2$, then $V_{\text{out}} = V_2 - V_1$.

INTEGRATOR

Solve for V_{out} in terms of V_{in}.

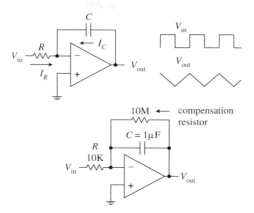

FIGURE 8.15

Because you have feedback, and because $V_+ = 0$ V, you can say that V_- is 0 V as well (rule 4). Now that you know V_-, solve for I_R and I_C so that you can come up with an expression relating V_{out} with V_{in}. Since no current enters the input of an op amp (rule 3), the displacement current I_C through the capacitor and the current I_R through the resistor must be related by $I_R + I_C = 0$. To find I_R, use Ohm's law:

$$I_R = \frac{V_{in} - V_-}{R} = \frac{V_{in} - 0\,\text{V}}{R} = \frac{V_{in}}{R}$$

I_C is found by using the displacement current relation:

$$I_C = C\frac{dV}{dt} = C\frac{d(V_{out} - V_-)}{dt} = C\frac{d(V_{out} - 0\,\text{V})}{dt} = C\frac{dV_{out}}{dt}$$

Placing these values of I_C and I_R into $I_R + I_C = 0$ and rearranging, you get the answer:

$$dV_{out} = -\frac{1}{RC}V_{in}\,dt$$

$$V_{out} = -\frac{1}{RC}V_{in}t$$

Such a circuit is called an *integrator;* the input signal is integrated at the output. Now, one problem with the first circuit is that the output tends to drift, even with the input grounded, due to nonideal characteristics of real op amps such as voltage offsets and bias current. A large resistor placed across the capacitor can provide dc feedback for stable biasing. Also, a compensation resistor may be needed between the noninverting terminal and ground to correct voltage offset errors caused by input bias currents. The size of this resistor should be equal to the parallel resistance of the input resistor and the feedback compensation resistor.

DIFFERENTIATOR

Solve for V_{out} in terms of V_{in}.

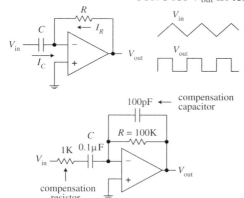

FIGURE 8.16

Since you know that V_+ is grounded (0 V), and since you have feedback in the circuit, you can say that $V_- = V_+ = 0$ V (rule 4). Now that you know V_-, solve for I_R and I_C so that you can come up with an expression relating V_{out} with V_{in}. Since no current enters the input of an op amp (rule 3), the displacement current I_C through the capacitor and the current I_R through the resistor must be related by $I_R + I_C = 0$. To find I_C, use the displacement current equation:

$$I_C = C\frac{dV}{dt} = C\frac{d(V_{in} - V_-)}{dt} = C\frac{d(V_{in} - 0\,\text{V})}{dt} = C\frac{dV_{in}}{dt}$$

The current I_R is found using Ohm's law:

$$I_R = \frac{V_{out} - V_-}{R} = \frac{V_{out} - 0\,\text{V}}{R} = \frac{V_{out}}{R}$$

Placing these values of I_C and I_R into $I_R + I_C = 0$ and rearranging, you get the answer:

$$V_{out} = -RC\frac{dV_{in}}{dt}$$

Such a circuit is called a *differentiator;* the input signal is differentiated at the output. The first differentiator circuit shown is not in practical form. It is extremely susceptible to noise due to the op amp's high ac gain. Also, the feedback network of the differentiator acts as an RC low-pass filter that contributes a 90° phase lag within the loop and may cause stability problems. A more practical differentiator is shown below the first circuit. Here, both stability and noise problems are corrected with the addition of a feedback capacitor and input resistor. The additional components provide high-frequency rolloff to reduce high-frequency noise. These components also introduce a 90° lead to cancel the 90° phase lag. The effect of the additional components, however, limits the maximum frequency of operation—at very high frequencies, the differentiator becomes an integrator. Finally, an additional input-bias compensation resistor placed between the noninverting input and ground may be needed to avoid offset error caused by input bias current. Its value should be equal to the resistance of the feedback resistor.

8.5 Positive Feedback

Positive feedback involves sending output voltage back to the noninverting input. In terms of the theory, if you look at our old friend $V_{out} = A_o(V_+ - V_-)$, the V_+ term changes to fV_{out} (f is a fraction of the voltage sent back), so you get $V_{out} = A_o(fV_{out} - V_-)$. Now, an important thing to notice about this equation (and about positive feedback in general) is that the voltage fed back to the noninverting input will act to drive the op amp "harder" in the direction the output is going (toward saturation). This makes sense in terms of the equation; fV_{out} adds to the expression. Recall that negative feedback acted in the opposite way; the fV_{out} (= V_-) term subtracted from the expression, preventing the output from "maxing out." In electronics, positive feedback is usually a bad thing, whereas negative feedback is a good thing. For most applications, it is desirable to control the gain (negative feedback), while it is undesirable to go to the extremes (positive feedback).

There is, however, an important use for positive feedback. When using an op amp to make a comparator, positive feedback can make output swings more pronounced. Also, by adjusting the size of the feedback resistor, a comparator can be made to experience what is called *hysteresis*. In effect, hysteresis gives the comparator two thresholds. The voltage between the two thresholds is called the *hysteresis voltage*. By obtaining two thresholds (instead of merely one), the comparator circuit becomes more immune to noise that can trigger unwanted output swings. To better understand hysteresis, let's take a look at the following comparator circuit that incorporates positive feedback.

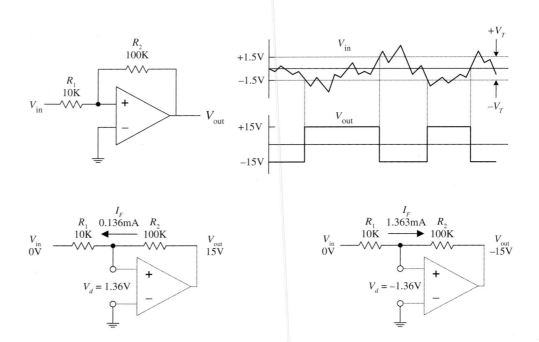

FIGURE 8.17

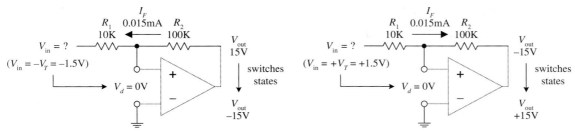

FIGURE 8.17 (*Continued*)

Assume that the op amp's output is at positive saturation, say, +15 V. If V_{in} is 0 V, the voltage difference between the inverting input and noninverting input (V_d) will be 1.36 V. You get this by using Ohm's law:

$$I_F = (V_{out} - V_{in})/(R_1 + R_2)$$
$$V_d = I_F R_1$$

This does not do anything to the output; it remains at +15 V. However, if you reduce V_{in}, there is a point when V_d goes to 0 V, at which time the output switches states. This voltage is called the *negative threshold voltage* ($-V_T$). The negative threshold voltage can be determined by using the previous two equations—the end result being $-V_T = -V_{out}/(R_2/R_1)$. In the example, $-V_T = -1.5$ V. Now, if the output is at negative saturation (−15 V) and 0 V is applied to the input, $V_d = -1.36$ V. The output remains at −15 V. However, if the input voltage is increased, there is a point where V_d goes to zero and the output switches states. This point is called the *positive threshold voltage* ($+V_T$), which is equal to $+V_{out}/(R_2/R_1)$. In the example, $+V_T = +1.5$ V. Now the difference between the two saturation voltages is the hysteresis voltage: $V_h = +V_T - (-V_T)$. In the example, $V_h = 3$ V.

8.6 Real Kinds of Op Amps

General Purpose

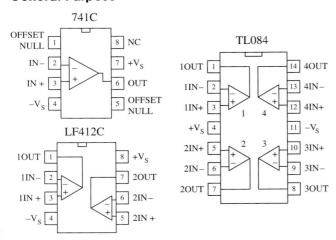

FIGURE 8.18

There is a huge selection of general-purpose and precision op amps to choose from. Precision op amps are specifically designed for high stability, low offset voltages, low bias currents, and low drift parameters. Because the selection of op amps is so incredibly large, we will leave it to you to check out the electronics catalogs to see what devices are available. When checking out these catalogs, you will find that op amps (not just general-purpose and precision) fall into one of the following categories (based on input circuitry): bipolar, JFET, MOSFET, or some hybrid thereof (e.g., BiFET). In general, bipolar op amps, like the 741 (industry standard), have higher input bias currents than either JFET or MOSFET types. This means that their input terminals have a greater tendency to "leak in" current. Input bias current results in voltage drops across resistors of feedback networks, biasing networks, or source impedances, which in turn can offset the output voltage. The amount of offset a circuit can tolerate ultimately depends on the application. Now, as we briefly mentioned earlier in this chapter, a compensation resistor placed between the noninverting terminal and ground (e.g., bipolar inverting amplifier circuit) can reduce these offset errors. (More on this in a minute.)

Precision

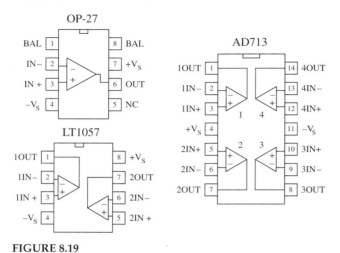

FIGURE 8.19

A simple way to avoid problems associated with input bias current is to use a FET op amp. A typical JFET op amp has a very low input bias current, typically within the lower picoamp range as compared with the nanoamp range for a typical bipolar op amp. Some MOSFET op amps come with even lower input bias currents, often as low as a few tenths of a picoamp. Though FET op amps have lower input bias current than bipolar op amps, there are other features they have that are not quite as desirable. For example, JFET op amps often experience an undesired effect called *phase inversion*. If the input common-mode voltage of the JFET approaches the negative supply too closely, the inverting and noninverting input terminals may reverse directions—negative feedback becomes positive feedback, causing the op amp to latch up. This problem can be avoided by using a bipolar op amp or by restricting the common-mode range of the signal. Here are some other general comments about bipolar and FET op amps: offset voltage (low for bipolar, medium for JFET, medium to high for MOSFET), offset drift (low for bipolar, medium for FET), bias matching (excellent for bipolar, fair for FET), bias/temperature variation (low for bipolar, fair for FET).

To avoid getting confused by the differences between the various op amp technologies, it is often easier to simply concentrate on the specifications listed in the electronics catalogs. Characteristics to look for include speed/slew rate, noise, input offset voltages and their drift, bias currents and their drift, common-mode range, gain, bandwidth, input impedance, output impedance, maximum supply voltages, supply current, power dissipation, and temperature range. Another feature to look for when purchasing an op amp is whether the op amp is internally or externally frequency compensated. An externally compensated op amp requires external components to prevent the gain from dropping too quickly at high frequencies, which can lead to phase inversions and oscillations. Internally compensated op amps take care of these problems with internal circuitry. All the terms listed in this paragraph will be explained in greater detail in a minute.

Programmable Op Amp

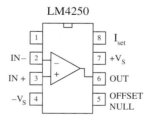

FIGURE 8.20

A programmable op amp is a versatile device that is used primarily in low-power applications (e.g., battery-powered circuits). These devices can be programmed with an external current for desired characteristics. Some of the characteristics that can be altered by applying a programming current include quiescent power dissipation, input offset and bias currents, slew rate, gain-bandwidth product, and input noise characteristics—all of which are roughly proportional to the programming current. The programming current is typically drawn from the programming pin (e.g., pin 8 of the LM4250) through a resistor and into ground. The programming current allows the op amp to be operated over a wide range of supply currents, typically from around a few microamps to a few millamps. Because a programmable op amp can be altered so as to appear as a completely different op amp for different programming currents, it is possible to use a single device for a variety of circuit functions within a system. These devices typically can operate with very low supply voltages (e.g., 1 V for the LM4250). A number of different manufacturers make programmable op amps, so check the catalogs. To learn more about how to use these devices, check out the manufacturers' literature (e.g., for National Semiconductor's LM4250, go to *www.national.com*).

Single-Supply Op Amps

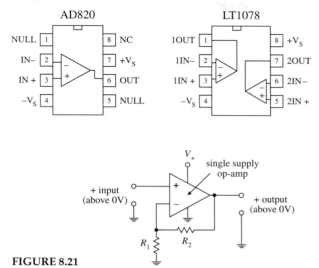

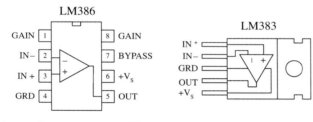

FIGURE 8.21

These op amps are designed to be operated from a single positive supply (e.g., +12 V) and allow input voltages all the way down to the negative rail (normally tied to ground). Figure 8.21 shows a simple dc amplifier that uses a single-supply op amp. It is important to note that the output of the amplifier shown cannot go negative; thus it cannot be used for, say, ac-coupled audio signals. These op amps are frequently used in battery-operated devices.

Audio Amplifiers

```
        LM386
GAIN  [1]        [8]  GAIN
IN–   [2]   -    [7]  BYPASS
IN +  [3]   +    [6]  +V_S
GRD   [4]        [5]  OUT

low-voltage power amplifier
```

```
        LM383
IN +  ⊏
IN–   ⊏
GRD   ⊏    - +
OUT   ⊏
+V_S  ⊏

8-watt power amplifier
```

FIGURE 8.22

These are closely related to conventional op amps but designed specifically to operate best (low audio band noise, crossover distortion, etc.) within the audio-frequency spectrum (20 to 20,000 Hz). These devices are used mainly in sensitive preamplifiers, audio systems, AM-FM radio receivers, servo amplifiers, and intercom and automotive circuits. There are a number of audio amplifiers to choose from. Some of these devices contain unique features that differ when compared with those of conventional op amps. For example, the popular LM386 low-voltage audio amplifier has a gain that is internally fixed at 20 but which can be increased to up to 200 with an external capacitor and resistor placed across its gain leads (pins 1 and 8). This device is also designed to drive low-impedance loads, such as an 8-Ω speaker, and runs off a single supply from +4 to +12 V—an ideal range for battery-powered applications. The LM383 is another audio amplifier designed as a power amplifier. It is a high-current device (3.5 A) designed to drive a 4-Ω load (e.g., one 4-Ω speaker or two 8-Ω speakers in parallel). This device also comes with thermal shutdown circuitry and a heat sink. We'll take a closer look at audio amplifiers in Chap. 15.

8.7 Op Amp Specifications

Common-mode rejection ratio (CMRR). The input to a difference amplifier, in general, contains two components: a common-mode and a difference-mode signal. The common-mode signal voltage is the average of the two inputs, whereas the difference-mode signal is the difference between the two inputs. Ideally, an amplifier affects the difference-mode signals only. However, the common-mode signal is also amplified to some degree. The common-mode rejection ratio (CMRR), which is defined as the ratio of the difference signal voltage gain to the common-mode signal voltage gain provides an indication of how well an op amp does at rejecting a signal applied simultaneously to both inputs. The greater the value of the CMRR, the better is the performance of the op amp.

Differential-input voltage range. Range of voltage that may be applied between input terminals without forcing the op amp to operate outside its specifications. If the inputs go beyond this range, the gain of the op amp may change drastically.

Differential input impedance. Impedance measured between the noninverting and inverting input terminals.

Input offset voltage. In theory, the output voltage of an op amp should be zero when both inputs are zero. In reality, however, a slight circuit imbalance within the internal circuitry can result in an output voltage. The input offset voltage is the amount of voltage that must be applied to one of the inputs to zero the output.

Input bias current. Theoretically, an op amp should have an infinite input imped- ance and therefore no input current. In reality, however, small currents, typically within the nanoamp to picoamp range, may be drawn by the inputs. The average of the two input currents is referred to as the *input bias current.* This current can result in a voltage drop across resistors in the feedback network, the bias network, or source impedance, which in turn can lead to error in the output voltage. Input bias currents depend on the input circuitry of an op amp. With FET op amps, input bias currents are usually small enough not to cause serious offset voltages. Bipolar op amps, on the other hand, may cause problems. With bipolar op amps, a com- pensation resistor is often required to center the output. We will discuss how this is done in a minute.

Input offset current. This represents the difference in the input currents into the two input terminals when the output is zero. What does this mean? Well, the input terminals of a real op amp tend to draw in different amounts of leakage current, even when the same voltage is applied to them. This occurs because there is always a slight difference in resistance within the input circuitry for the two terminals that originates during the manufacturing process. Therefore, if an op amp's two terminals are both connected to the same input voltage, different amounts of input current will result, causing the output to be offset. Op amps typically come with offset terminals that can be wired to a potentiometer to correct the offset current. We will discuss how this is done in a minute.

Voltage gain (A_V). A typical op amp has a voltage gain of 10^4 to 10^6 (or 80 to 120 dB; gain in dB = $20 \log_{10} A_0$[11]) at dc. However, the gain drops to 1 at a frequency called the *unity-gain frequency f_T,* typically from 1 to 10 MHz—a result of high-frequency limitations in the op amp's internal circuitry. We will talk more about high-frequency behavior in op amps in a minute.

Output voltage swing. This is the peak output voltage swing, referenced to zero, that can be obtained without clipping.

Slew rate. This represents the maximum rate of change of an op amp's output voltage with time. The limitation of output change with time results from inter- nal or external frequency compensation capacitors slowing things down, which in turn results in delayed output changes with input changes (propagation delay). At high frequencies, the magnitude of an op amp's slew rate becomes more critical. A general-purpose op amp like the 741 has a 0.5 V/μs slew rate—a relatively small value when compared with the high-speed HA2539's slew rate of 600 V/μs.

Supply current. This represents the current that is required from the power supply to operate the op amp with no load present and with an output voltage of zero.

Table 8.1 is a sample op amp specifications table.

TABLE 8.1   Sample Op Amp Specifications

TYPE	TOTAL SUPPLY VOLTAGE		SUPPLY CURRENT (mA)	OFFSET VOLTAGE		CURRENT		SLEW RATE TYPICAL (V/μS)	f_T TYPICAL (MHz)	CMRR MIN (dB)	GAIN MIN (mA)	OUTPUT CURRENT MAX (mA)
	MIN (V)	MAX (V)		TYPICAL (mV)	MAX (mV)	BIAS MAX (nA)	OFFSET MAX (nA)					
Bipolar 741C	10	36	2.8	2	6	500	200	0.5	1.2	70	86	20
MOSFET CA3420A	2	22	1	2	5	0.005	0.004	0.5	0.5	60	86	2
JFET LF411	10	36	3.4	0.8	2	0.2	0.1	15	4	70	88	30
Bipolar, precision LM10	1	45	0.4	0.3	2	20	0.7	0.12	0.1	93	102	20

8.8   Powering Op Amps

Most op amp applications require a dual-polarity power supply. A simple split ±15-V supply that uses a tapped transformer is presented in Chap. 11. If you are using batteries to power an op amp, one of the following arrangements can be used.

FIGURE 8.23

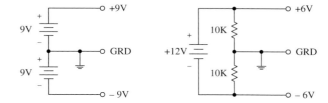

Now, it is often desirable to avoid split-supplies, especially with small battery-powered applications. One option in such a case is to use a single-supply op amp. However, as we pointed out a second ago, these devices will clip the output if the input attempts to go negative, making them unsuitable for ac-coupled applications. To avoid clipping while still using a single supply, it is possible to take a conventional op amp and apply a dc level to one of the inputs using a voltage-divider network. This, in turn, provides a dc offset level at the output. Both input and output offset levels are referenced to ground (the negative terminal of the battery). With the input offset voltage in place, when an input signal goes negative, the voltage applied to the input of the op amp will dip below the offset voltage but will not go below ground (provided you have set the bias voltage large enough, and provided the input signal is not too large; otherwise, clipping occurs). The output, in turn, will fluctuate about its offset level. To allow for input and output coupling, input and output capacitors are needed. The two circuits in Fig. 8.24 show noninverting and inverting ac-coupled amplifiers (designed for audio) that use conventional op amps that run off a single supply voltage.

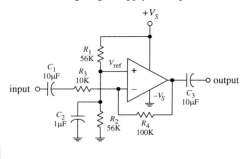

FIGURE 8.24

In the noninverting circuit, the dc offset level is set to one-half the supply voltage by R_1 and R_2 to allow for maximum symmetrical swing. C_1 (and R_2) and C_3 (and R_{load}) act as ac coupling (filtering) capacitors that block unwanted dc components and low-level frequencies. C_1 should be equal to $1/(2\pi f_{3dB}R_1)$, while C_3 should be equal to $1/(2\pi f_{3dB}R_{load})$, where f_{3dB} is the cutoff frequency (see Chaps. 9 and 15).

When using conventional op amps with single supply voltages, make sure to stay within the minimum supply voltage rating of the op amp, and also make sure to account for maximum output swing limitations and maximum common-mode input range.

8.9 Some Practical Notes

As a note of caution, never reverse an op amp's power supply leads. Doing so can result in a zapped op amp IC. One way to avoid this fate is to place a diode between the op amp's negative supply terminal and the negative supply, as shown in Fig. 8.25.

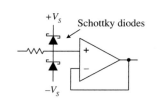

FIGURE 8.25

Also keep wires running from the power supply to the op amp's supply terminals short and direct. This helps prevent unwanted oscillations/noise from arising in the output. Disturbances also may arise from variations in supply voltage. To eliminate these effects, place bypass capacitors from the supply terminals to ground as shown in Fig. 8.25. A 0.1-μF disk capacitor or a 1.0-μF tantalum capacitor should do the trick.

Both bipolar and JFET op amps can experience a serious form of latch if the input signal becomes more positive or negative than the respective op amp power supplies. If the input terminals go more positive than $+V_S + 0.7$ V or more negative than $-V_s - 0.7$ V, current may flow in the wrong direction within the internal circuitry, short-circuiting the power supplies and destroying the device. To avoid this potentially fatal latch-up, it is important to prevent the input terminals of op amps from exceeding the power supplies. This feature has vital consequences during device turn-on; if a signal is applied to an op amp before it is powered, it may be destroyed at the moment power is applied. A "hard wire" solution to this problem involves clamping the input terminals at risk with diodes (preferably fast low-forward-voltage Schottky diodes; see Fig. 8.25). Current-limiting resistors also may be needed to prevent the diode current from becoming excessive. This protection circuitry has some problems, however. Leakage current in the diodes may increase the error. See manufacturers' literature for more information.

8.10 Voltage and Current Offset Compensation

In theory, the output voltage of an op amp should be zero when both inputs are zero. In reality, however, a slight circuit imbalance within the internal circuitry can result in an output voltage (typically within the microvolt to millivolt range). The input offset voltage is the amount of voltage that must be applied to one of the inputs to zero the output—this was discussed earlier. To zero the input offset voltage, manufacturers usually include a pair of offset null terminals. A potentiometer is placed between these two terminals, while the pot's wiper is connected to the more negative supply terminal, as shown in Fig. 8.26. To center the output, the two inputs can be shorted together and an input voltage applied. If the output saturates, the input offset needs trimming. Adjust the pot until the output approaches zero.

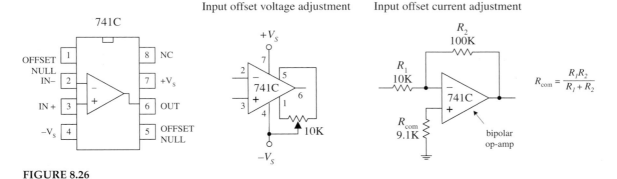

FIGURE 8.26

Notice the resistor placed between the noninverting terminal and ground within the inverting amplifier circuit shown in Fig. 8.26 *right*. What is the resistor used for? It is used to compensate for output voltage errors caused by a voltage drops across R_1 and R_2 as a result of input bias current. As discussed earlier, bipolar op amps tend to have larger input bias currents than FET op amps. With FET op amps, the input bias errors are usually so small (in the picoamp range) that the output voltage error is insignificant, and the compensation resistor is not needed. However, with bipolar op amps, this is not the case (input bias currents in the nanoamp range), and compensation is often necessary. Now, in the inverting amplifier, the bias current—assuming for now that the compensation resistor is missing—introduces a voltage drop equal to $V_{in} = I_{bias}(R_1 \| R_2)$, which is amplified by a factor of $-R_2/R_1$. In order to correct this problem, a compensation resistor with a resistance equal the $R_1 \| R_2$ is placed between the noninverting terminal and ground. This resistor makes the op amp "feel" the same input driving resistance.

8.11 Frequency Compensation

For a typical op amp, the open-loop gain is typically between 10^4 and 10^6 (80 to 120 dB). However, at a certain low frequency, called the *breakover frequency* f_B, the gain drops by 3 dB, or drops to 70.7 percent of the open-loop gain (maximum gain). As the frequency increases, the gain drops further until it reaches 1 (or 0 dB) at a frequency called the *unity-gain frequency* f_T. The unity-gain frequency for an op amp is typically around 1 MHz and is given in the manufacturer's specifications (see Fig. 8.27 *left*). The rolloff in gain as the frequency increases is caused by low-pass filter-like characteristics inherently built into the op amp's inner circuitry. If negative feedback is used, an improvement in bandwidth results; the response is flatter over a wider range of frequencies, as shown in the far-left graph in Fig. 8.27. Now op amps that exhibit an open-loop gain drop of more than 60 dB per decade at f_T are unstable due to phase shifts incurred in the filter-like regions within the interior circuitry. If these phase shifts reach 180° at some frequency at which the gain is greater than 1, negative feedback becomes positive feedback, which results in undesired oscillations (see center and right grafts in Fig. 8.27). To prevent these oscillations, frequency compensation is required. *Uncompensated* op amps can be frequency compensated by connecting an *RC* network between the op amp's frequency-compensation terminals. The network, especially the capacitor, influences the shape of the response curve. Manufacturers will supply you with the response curves, along with the component values of the compensation network for a particular desired response.

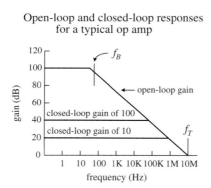

Open-loop and closed-loop responses for a typical op amp

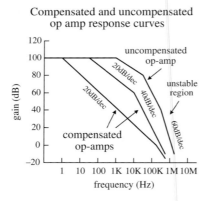

Compensated and uncompensated op amp response curves

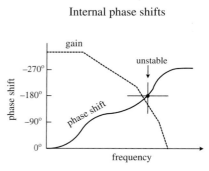

Internal phase shifts

FIGURE 8.27

Perhaps the easiest way to avoid dealing with frequency compensation is to buy an op amp that is internally compensated.

8.12 Comparators

In many situations, it is desirable to know which of two signals is bigger or to know when a signal exceeds a predetermined voltage. Simple circuits that do just this can be constructed with op amps, as shown in Fig. 8.28. In the noninverting comparator circuit, the output switches from low (0 V) to high (positive saturation) when the input voltage exceeds a reference voltage applied to the inverting input. In the inverting comparator circuit, the output switches from high to low when the input exceeds the reference voltage applied to the noninverting terminal. In the far-right circuit, a voltage divider (pot) is used to set the reference voltage.

Note that not all op amps are able to operate with a grounded negative supply, and a special-purpose comparator IC is a better approach.

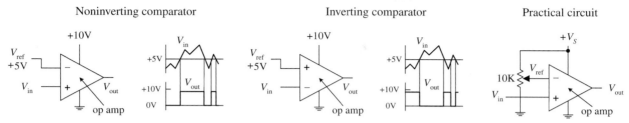

FIGURE 8.28

Another, more popular method for comparing two voltages is to use a special IC called a *comparator*. A comparator, like an op amp, has an inverting input, a noninverting input, an output, and power supply leads—its schematic looks like an op amp, too. However, unlike op amps, comparators are not frequency compensated and therefore cannot be used as linear amplifiers. In fact, comparators never use negative feedback (they often use positive feedback, as you will see). If negative feedback were used with a comparator, its output characteristics would be unstable. Comparators are specifically designed for high-speed switching—they have much larger slew rates and smaller propagation delays than op amps. Another important difference between a comparator and an op amp has to do with the output circuitry. Unlike an op amp, which typically has a push-pull output stage, a comparator uses an internal transistor whose collector is connected to the output and whose emitter is grounded. When the comparator's noninverting terminal is less positive in voltage than the inverting terminal, the output transistor turns on, grounding the output. When the noninverting terminal is made more positive than the inverting terminal, the output transistor is off. In order to give the comparator a high output state when the transistor is off ($V_- < V_+$), an external *pull-up resistor* connected from a positive voltage source to the output is used. The pull-up resistor acts like the collector resistor in a transistor amplifier. The size of the pull-up resistor should be large enough to avoid excessive power dissipation yet small enough to supply enough drive to switch whatever load circuitry is used on the comparator's output. The typical resistance for a pull-up resistor is anywhere from a few hundred to a few thousand ohms. Figure 8.29 shows a simple noninverting and inverting comparator circuit with pull-up resistors included. Both circuits have an output swing from 0 to +5 V.

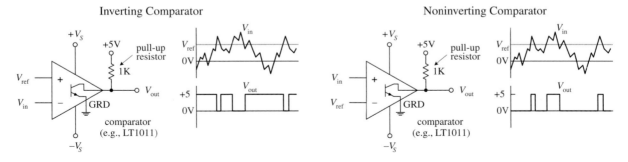

FIGURE 8.29

Comparators are commonly used in analog-to-digital conversion. A typical application might be to connect a magnetic tape sensor or photodiode to an input of a comparator (with reference voltage set at the other input) and allow the sensor to drive the comparator's output to a low or high state suitable for driving logic circuits. Analog-to-digital conversion is discussed in greater detail in Chap. 12.

8.13   Comparators with Hysteresis

Now there is a basic problem with both comparator circuits shown in Fig. 8.29. When a slowly varying signal is present that has a level near the reference voltage, the output will "jitter" or flip back and forth between high and low output states. In many situations having such a "finicky" response is undesirable. Instead, what is usually desired is a small "cushion" region that ignores small signal deviations. To provide the cushion, positive feedback can be added to the comparator to provide hysteresis, which amounts to creating two different threshold voltages or triggering points. The details of how hysteresis works within comparator circuits are given in the following two examples.

8.13.1   Inverting Comparator with Hysteresis

In the inverting comparator circuit shown in Fig. 8.30, positive feedback through R_3 provides the comparator with two threshold voltages or triggering points. The two threshold voltages result from the fact that the reference voltage applied to the noninverting terminal is different when V_{out} is high (+15 V) and when V_{out} is low (0 V)—a result of feedback current. Let's call the reference voltage when the output is high V_{ref1}, and call the reference voltage when the output is low V_{ref2}. Now let's assume that the output is high (transistor is off) and $V_{in} > V_{ref1}$. In order for the output to switch high, V_{in} must be greater than V_{ref1}. However, what is V_{ref1}? It is simply the reference voltage that pops up at the noninverting terminal when the output transistor is off and the output is high (+15 V).

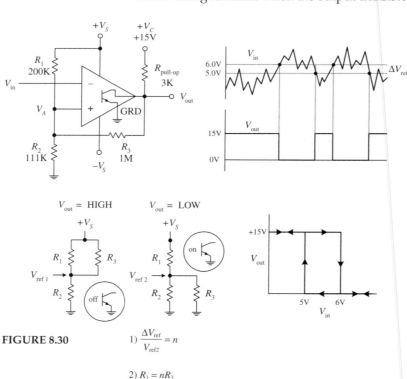

FIGURE 8.30

1) $\dfrac{\Delta V_{ref}}{V_{ref2}} = n$

2) $R_1 = nR_3$

3) $R_2 = \dfrac{R_1\|R_3}{(+V_S/V_{ref1}) - 1}$

To calculate V_{ref1}, simply use the basic resistor network shown below the main circuit and to the left:

$$V_{ref1} = \frac{+V_S R_2}{(R_1\|R_3) + R_2} = \frac{+V_S R_2 (R_1 + R_3)}{R_1 R_2 + R_1 R_3 + R_2 R_3}$$

If $V_{in} > V_{ref1}$ when the output is already high, the output suddenly goes low—transistor turns on. With the output now low, a new reference voltage V_{ref2} is in place. To calculate V_{ref2}, use the resistor network shown to the right of the first:

$$V_{ref2} = \frac{+V_S (R_2\|R_3)}{R_1 + (R_2\|R_3)} = \frac{+V_S R_2 R_3}{R_1 R_2 + R_1 R_3 + R_2 R_3}$$

When the input voltage decreases to V_{ref2} or lower, the output suddenly goes high. The difference in the reference voltages is called the *hysteresis voltage*, or ΔV_{ref}:

$$\Delta V_{ref} = V_{ref1} - V_{ref2} = \frac{+V_S R_1 R_2}{R_1 R_2 + R_1 R_3 + R_2 R_3}$$

Now, let's try the theory out on a real-life design example.

Say we want to design a comparator circuit with a $V_{ref1} = +6$ V, a $V_{ref2} = +5$ V, and $+V_C = +15$ V that drives a 100–kΩ load. The first thing to do is pick a pull-up resistor. As a rule of thumb,

$$R_{pull-up} < R_{load}$$

$$R_3 > R_{pull-up}$$

Why? Because heavier loading on $R_{pull-up}$ (smaller values of R_3 and R_{load}) reduce the maximum output voltage, thereby reducing the amount of hysteresis by lowering the value of V_{ref1}. Pick $R_{pull-up} = 3$ kΩ, and choose R_3 equal to 1 MΩ. Combining equations above gives us the practical formulas below the diagrams. With equation (1), calculate n, which is (6 V – 5 V)/5 V = 0.20. Next, using equation (2), find R_1, which is simply (0.2)(1 M) = 200 kΩ. Using equation (3), find R_3, which is 166 k/(15 V/6 V – 1) = 111 kΩ. These are the values presented in the circuit.

8.13.2 Noninverting Comparator with Hysteresis

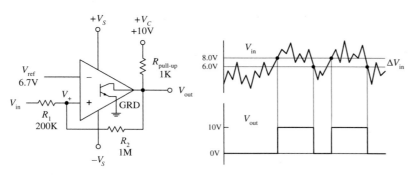

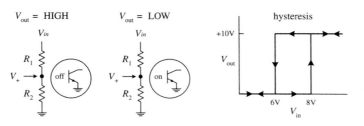

FIGURE 8.31

What's desired: $V_{in1} = 8$ V, $V_{in2} = 6$ V, given $+V_C = 10$ V and load of 100 kΩ. Question: What should V_{ref}, $R_{pull-up}$, R_2, and R_1 be?

First, choose $R_{pull-up} < R_{load}$, and $R_2 > R_{pull-up}$ to minimize the effects of loading. For $R_{pull-up}$, choose 1 kΩ, and for R_2, choose 1 M. Next, using the equations to the right, we find R_1 and V_{ref}.

$$\frac{R_1}{R_2} = \frac{\Delta V_{in}}{V_C} = \frac{10-8}{10} = 0.20$$

$$R_1 = 0.20R_2 = 0.20(1\text{ M}) = 200\text{ kΩ}$$

$$V_{ref} = \frac{V_{in1}}{1 + R_1/R_2} = \frac{8\text{ V}}{1\text{ V} - 0.20\text{ V}} = 6.7\text{ V}$$

Unlike the inverting comparator, the noninverting comparator only requires two resistors for hysteresis to occur. (Extra resistors are needed if you wish to use a voltage divider to set the reference voltage. However, these resistors do not have a direct role in developing the hysteresis voltage.) Also, the terminal to which the input signal is applied is the same location where the threshold shifting occurs—a result of positive feedback. The threshold level applied to the noninverting terminal is shifted about the reference voltage as the output changes from high $(+V_C)$ to low (0 V). For example, assume that V_{in} is at a low enough level to keep V_{out} low. For the output to switch high, V_{in} must rise to a triggering voltage, call it V_{in1}, which is found simply by using the resistor network shown to the far left:

$$V_{in1} = \frac{V_{ref}(R_1 + R_2)}{R_2}$$

As soon as V_{out} switches high, the voltage at the noninverting terminal will be shifted to a value that's greater than V_{ref} by:

$$\Delta V_+ = V_{in} + \frac{(V_{CC} - V_{in1})R_1}{R_1 + R_2}$$

To make the comparator switch back to its low state, V_{in} must go below ΔV_+. In other words, the applied input voltage must drop below what is called the lower trip point, V_{in2}:

$$V_{in2} = \frac{V_{ref}(R_1 + R_2) - V_{CC}R_1}{R_2}$$

The hysteresis is then simply the difference between V_{in1} and V_{in2}:

$$\Delta V_{in} = V_{in1} - V_{in2} = \frac{V_{CC}R_1}{R_2}$$

A practical design example is presented to the left.

8.14 Using Single-Supply Comparators

Like op amps, comparator ICs come in both dual- and single-supply forms. With single-supply comparators, the emitter and the "negative supply" are joined internally and grounded, whereas the dual-supply comparator has separate emitter (ground) and negative supply leads. A few sample comparator ICs, along with two single-supply comparator circuits are shown below.

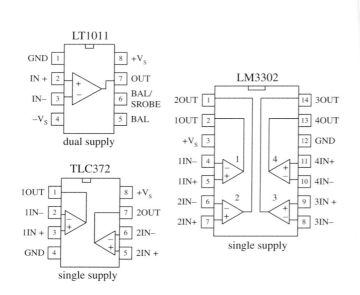

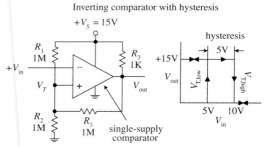

FIGURE 8.32

8.15 Window Comparator

A window comparator is a very useful circuit that changes its output state whenever the input voltage is anywhere between predetermined high and low reference voltages. The region between these two reference voltages is called the *window*. Figure 8.33 shows a simple window comparator built with two comparators (op amps also can be used). In the left-most circuit, the window is set between +3.5 V ($V_{ref,high}$) and +6.5 V($V_{ref,low}$). If V_{in} is below +3.5 V, the lower comparator's output is grounded, while the upper comparator's output floats. Only one ground is needed, however, to make $V_{out} = 0$ V. If V_{in} is above +6.5 V, the upper comparator's output is grounded, while the lower comparator's output floats—again V_{out} goes to 0 V. Only when V_{in} is between +3.5 and +6.5 V will the output go high (+5 V). The right-most circuit uses a voltage-divider network to set the reference voltages.

FIGURE 8.33

Window comparator (using comparators)

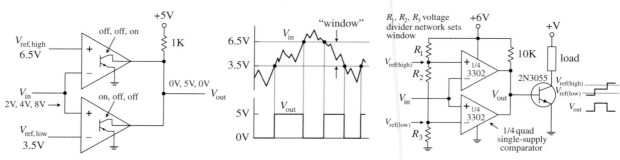

8.16 Voltage-Level Indicator

A simple way to make a voltage-level indicator is to take a number of comparators that share a common input and then supply each comparator with a different reference or triggering voltage, as shown in Fig. 8.34. In this circuit, the reference voltage applied to a comparator increases as you move up the chain of comparators (a result of the voltage-divider network). As the input voltage increases, the lower comparator's output is grounded first (diode turns on), followed in succession by the comparators (LEDs) above it. The potentiometer provides proportional control over all the reference voltages.

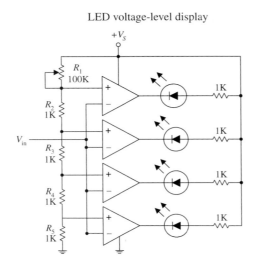

FIGURE 8.34

8.17 Instrumentation Amplifiers

Instrumentation amplifiers perform the same job as a differencing amplifier. That is they are used to amplify the difference between their positive and negative inputs. A typical practical application might be in an ECG (electrocardiogram) machine, where small differences in voltage between electrodes placed on the chest of the patient are amplified to produce the characteristic heartbeat trace so beloved of hospital dramas.

Instrumentation amplifiers have buffered inputs giving superior performance to basic differencing amplifiers and although you can construct an instrumentation amplifier from three regular op amps (see Fig. 8.35) it is more usual to use a special purpose instrumentation amplifier IC that includes the accurately matched resistors needed for precise operation. Instrumentation amplifiers have very good common mode rejection.

The resistors R_1 need to be matched. If R_g is omitted then the amplifier has unity gain. Otherwise, the gain of the amplifier is given by:

$$\frac{V_{out}}{V_{in+} - V_{in-}} = \left(1 + \frac{2R_1}{R_g}\right)\frac{R_3}{R_2}$$

The single resistor R_g is therefore able to set the gain of the whole amplifier.

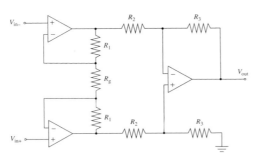

FIGURE 8.35   An instrumentation amplifier.

8.18 Applications

Op Amp Output Drivers (for Loads That Are Either On or Off)

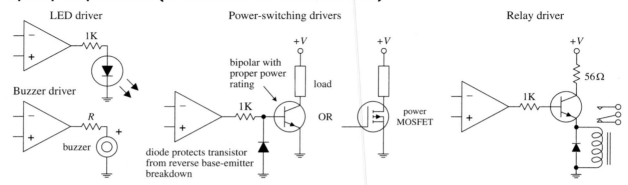

FIGURE 8.36

Comparator Output Drivers

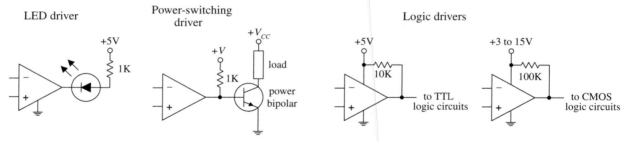

FIGURE 8.37

Op Amp Power Booster (AC Signals)

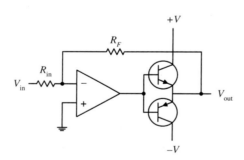

FIGURE 8.38

There are times when an op amp needs a boost in output power-handling capacity while at the same time maintaining both positive and negative output swings. A simple way to increase the output power while maintaining the swing integrity is to attach a complementary transistor push-pull circuit to the op amp's output, as shown in this circuit. At high speeds, additional biasing resistors and capacitors are needed to limit crossover distortion. At low speeds, negative feedback helps eliminate much of the crossover distortion.

Voltage-to-Current Converter

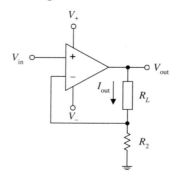

FIGURE 8.39

Here is a simple current source whose output current is determined by an input voltage applied to the noninverting terminal of the op amp. The output current and voltage are determined by the following expressions:

$$V_{\text{out}} = \frac{(R_L + R_2)}{R_2} V_{\text{in}}$$

$$I_{\text{out}} = \frac{V_{\text{out}}}{R_1 + R_2} = \frac{V_{\text{in}}}{R_2}$$

V_{in} can be set with a voltage divider.

Precision Current Source

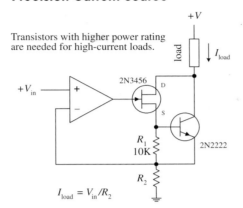

Transistors with higher power rating are needed for high-current loads.

$I_{load} = V_{in}/R_2$

FIGURE 8.40

Here, a precision current uses a JFET to drive a bipolar output transistor used to sink current through a load. Unlike the preceding current source, this circuit is less susceptible to output drift. Use of the JFET helps achieve essentially zero bias current error (a single bipolar output stage will leak base current). This circuit is accurate for output currents larger than the JFET's $I_{DS(on)}$ and provided V_{in} is greater than 0 V. For large currents, the FET-bipolar combination can be replaced with a Darlington transistor, provided its base current does not introduce significant error. The output current or load current is determined by

$$I_{load} = V_{in}/R_2$$

R_2 acts as an adjustment control. Additional compensation may be required depending on the load reactance and the transistors' parameters. Make sure to use transistors of sufficient power ratings to handle the load current in question.

Current-to-Voltage Converter

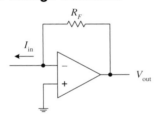

The circuit shown here transforms a current into a voltage. The feedback resistor R_F helps establish a voltage at the inverting input and controls the swing of the output. The output voltage for this circuit is given by

$$V_{out} = I_{in}R_F$$

The light-activated circuits shown below use this principle to generate an output voltage that is proportional to the amount of input current drawn through the light sensor.

photoresistor amplifier

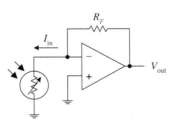

photodiode amplifier

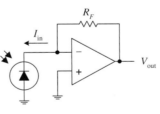

phototransistor amplifier and relay driver

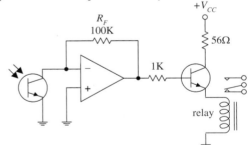

FIGURE 8.41

Overvoltage Protection (Crowbar)

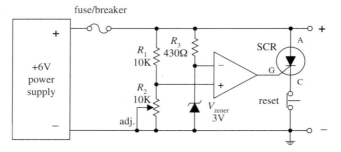

FIGURE 8.42

This circuit acts as a fast-acting overvoltage protection control used to protect sensitive loads from voltage surges generated in the power supply. Initially, let's assume that the supply is doing what it should—generating a constant +6 V. In this case, the voltage applied to the op amp's noninverting input is set to 3 V (by means of the R_1, R_2 voltage divider—the pot provides fine-tuning adjustment). At the same time, the inverting input is set to 3 V by means of the 3-V zener diode. The op amp's differential input voltage in this case is therefore zero, making the op amp's output zero (op amp acts as a comparator). With the op amp's output zero, the SCR is off, and no current will pass from anode (A) to cathode to ground. Now, let's say there's a sudden surge in the supply voltage. When this occurs, the voltage at the noninverting input increases, while the voltage at the inverting input remain at 3 V (due to the 3-V zener). This causes the op amp's output to go high, triggering the SCR on and diverting all current from the load to ground in the process. As a result, the fuse (breaker) blows and the load is saved. The switch is opened to reset the SCR.

Programmable-Gain Op Amp

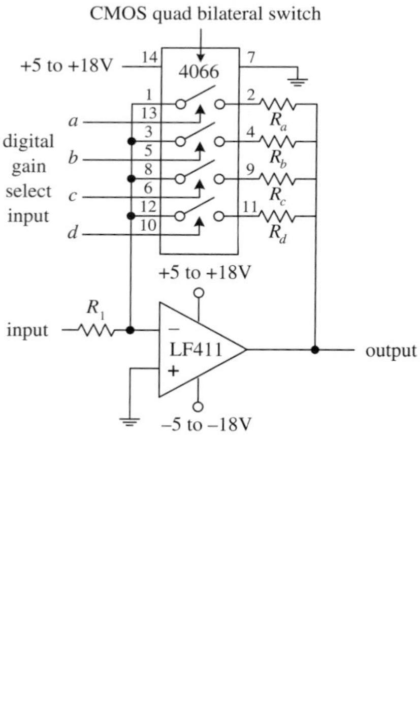

FIGURE 8.43

This circuit is simply an inverting amplifier whose feedback resistance (gain) is selected by means of a digitally controlled bilateral switch (e.g., CMOS 4066). For example, if the bilateral switch's input a is set high (+5 to +18 V) while its b through d inputs are set low (0 V), only resistor R_a will be present in the feedback loop. If you make inputs a through d high, then the effective feedback resistance is equal to the parallel resistance of resistors R_a through R_d. Bilateral switches are discussed in greater detail in Chap. 12.

Sample-and-Hold Circuits

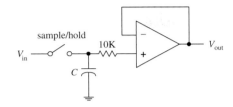

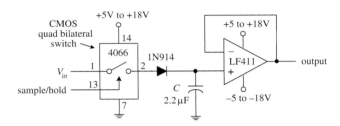

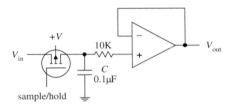

Sample-and-hold circuits are used to sample an analog signal and hold it so that it can be analyzed or converted into, say, a digital signal at one's leisure. In the first circuit, a switch acts as a sample/hold control. Sampling begins when the switch is closed and ends when the switch is opened. When the switch is opened, the input voltage present at that exact moment will be stored in C. The op amp acts as a unity-gain amplifier (buffer), relaying the capacitor's voltage to the output but preventing the capacitor from discharging (recall that ideally, no current enters the inputs of an op amp). The length of time a sample voltage can be held varies depending on how much current leaks out of the capacitor. To minimize leakage currents, use op amps with low input-bias currents (e.g., FET op amps). In the other two circuits, the sample/hold manual switch is replaced with an electrically controlled switch—the left-most circuit uses a bilateral switch, while the right-most circuit uses a MOSFET. Capacitors best suited for sample/hold applications include Teflon, polyethylene, and polycarbonate dielectric capacitors.

FIGURE 8.44

Peak Detectors

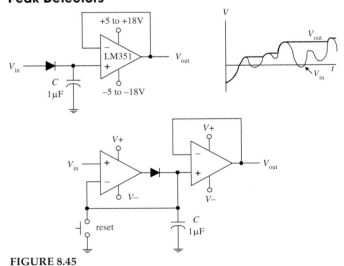

FIGURE 8.45

The circuits shown here act as peak detectors—they follow an incoming voltage signal and store its maximum voltage within C (see graph). The op amp in the upper circuit acts as a buffer—it "measures" the voltage in C, outputs that voltage, and prevents C from discharging. The diode also prevents the capacitor from discharging when the input drops below the peak voltage stored on C. The second circuit is a more practical peak detector. The additional op amp makes the detector more sensitive; it compensates for the diode voltage drop (around 0.6 V) by feeding back C's voltage to the inverting terminal. In other words, it acts as an active rectifier. Also, this circuit incorporates a switch to reset the detector. Often, peak detectors use a FET in place of the diode and use the FET's gate as a reset switch. Reducing the capacitance of C promotes faster response times for changes in V_{in}.

Noninverting Clipper Amplifier

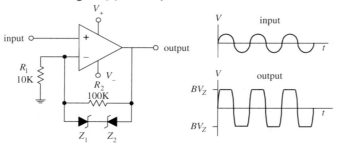

FIGURE 8.46

This simple amplifier circuit acts to clip both positive and negative going portions of the output signal. The clipping occurs in the feedback network whenever the feedback voltage exceeds a zener diode's breakdown voltage. Removing one of the zener diodes results in partial clipping (either positive or negative going, depending on which zener diode is removed). This circuit can be used to limit overloads within audio amplifiers and as a simple sinewave-to-squarewave converter.

Active Rectifiers

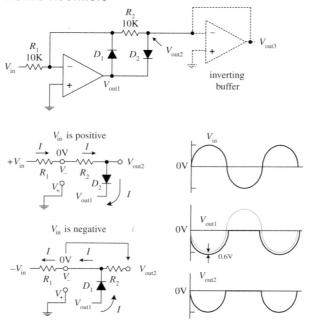

FIGURE 8.47

A single diode acts to rectify a signal—however, at the cost of a diode drop (e.g., 0.6 V). Not only does this voltage drop lower the level of the output, but it also makes it impossible to rectify low-level signals below 0.6 V. A simple solution to this problem is to construct an active rectifier like the one shown here. This circuit acts as an ideal rectifier, in so far as it rectifies signals all the way down to 0 V. To figure out how this circuit works, let's apply the rules we have learned. If V_{in} is positive, current I will flow in the direction shown in the simplified network shown below the main circuit. Since V_+ is grounded, and since we have feedback, $V_- = V_+ = 0$ V (rule 4), we can use Kirchhoff's voltage law to find V_{out1}, V_{out2}, and finally, V_{out3}:

$$0 \text{ V} - IR_2 - 0.6 \text{ V} - V_{out1} = 0$$

$$V_{out1} = 0 \text{ V} - \frac{V_{in}R_2}{R_1} - 0.6 \text{ V} = -V_{in} - 0.6 \text{ V}$$

$$V_{out2} = V_{out1} + 0.6 \text{ V} = -V_{in}$$

$$V_{out3} = V_{out2} = -V_{in}$$

Notice that there is no 0.6-V drop present at the final output; however, the output is inverted relative to the input. Now, if V_{in} is negative, the output will source current through D_1 to bring V_- to 0 V (rule 4). But since no current will pass through R_2 (due to buffer), the 0 V at V_- is present at V_{out2} and likewise present at V_{out3}. The buffer stage is used to provide low output impedance for the next stage without loading down the rectifier stage. To preserve the polarity of the input at the output, an inverting buffer (unity-gain inverter) can be attached to the output.

## CHAPTER 9

Filters

A *filter* is a circuit that is capable of passing a specific range of frequencies while blocking other frequencies. As you discovered in Chap. 2, the four major types of filters include *low-pass filters*, *high-pass filters*, *bandpass filters*, and *notch filters* (or *band-reject filters*). A low-pass filter passes low-frequency components of an input signal, while a high-pass filter passes high-frequency components. A bandpass filter passes a narrow range of frequencies centered around the filter's resonant frequency, while a notch filter passes all frequencies except those within a narrow band centered around the filter's resonant frequency.

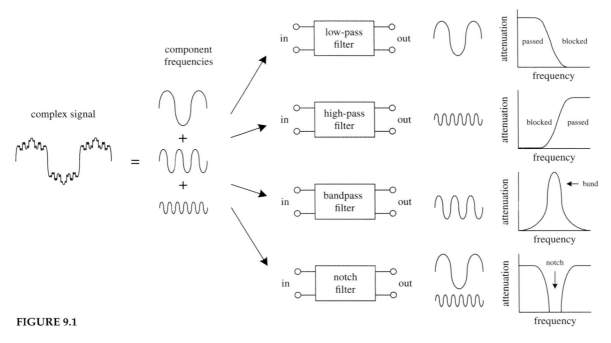

FIGURE 9.1

Filters have many practical applications in electronics. For example, within a dc power supply, filters can be used to eliminate unwanted high-frequency noise present within the ac line voltage, and they act to flatten out pulsing dc voltages generated by the supply's rectifier section. In radio communications, filters make it possible for a radio receiver to provide the listener with only the desired signal while rejecting all others. Likewise, filters allow a radio transmitter to generate only one signal while attenuating other signals that might interfere with different radio transmitters' signals. In audio electronics, filter networks called *crossover networks* are used to divert low audio signals to woofers, middle-range frequencies to midrange speakers, and high frequencies to tweeters. A high-pass filter is often used to eliminate 60 Hz mains hum from audio circuits. The list of filter applications is extensive.

There are two filter types covered in this chapter, namely, *passive filters* and *active filters*. Passive filters are designed using passive elements (e.g., resistors, capacitors, and inductors) and are most responsive to frequencies between around 100 Hz and 300 MHz. (The lower frequency limit results from the fact that at low frequencies the capacitance and inductance values become exceedingly large, meaning prohibitively large components are needed. The upper frequency limit results from the fact that at high frequencies parasitic capacitances and inductances wreak havoc.) When designing passive filters with very steep attenuation falloff responses, the number of inductor and capacitor sections increases. As more sections are added to get the desired response, greater is the chance for signal loss to occur. Also, source and load impedances must be taken into consideration when designing passive filters.

Active filters, unlike passive filters, are constructed from op amps, resistors, and capacitors—no inductors are needed. Active filters are capable of handling very low frequency signals (approaching 0 Hz), and they can provide voltage gain if needed (unlike passive filters). Active filters can be designed to offer comparable performance to *LC* filters, and they are typically easier to make, less finicky, and can be designed without the need for large-sized components. Also, with active filters, a desired input and output impedance can be provided that is independent of frequency. One major drawback with active filters is a relatively limited high-frequency range. Above around 100 kHz or so, active filters can become unreliable (a result of the op amp's bandwidth and slew-rate requirements). At radiofrequencies, it is best to use a passive filter.

9.1   Things to Know Before You Start Designing Filters

When describing how a filter behaves, a response curve is used, which is simply an attenuation (V_{out}/V_{in}) versus frequency graph (see Fig. 9.2). As you discovered in Chap. 2, attenuation is often expressed in decibels (dB), while frequency may be expressed in either angular form ω (expressed in rad/s) or conventional form f (expressed in Hz). The two forms are related by $\omega = 2\pi f$. Filter response curves may be plotted on linear-linear, log-linear, or log-log paper. In the case of log-linear graphs, the attenuation need not be specified in decibels.

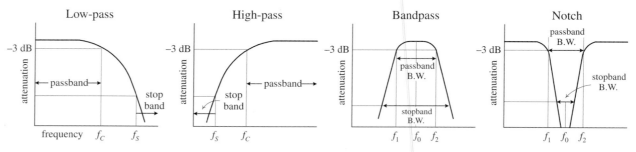

FIGURE 9.2

Here are some terms that are commonly used when describing filter response:

–3-dB Frequency (f$_{3dB}$). This represents the input frequency that causes the output signal to drop to –3 dB relative to the input signal. The –3-dB frequency is equivalent to the cutoff frequency—the point where the input-to-output power is reduced by one-half or the point where the input-to-output voltage is reduced by $1/\sqrt{2}$. For low-pass and high-pass filters, there is only one –3-dB frequency. However, for bandpass and notch filters, there are two –3-dB frequencies, typically referred to as f_1 and f_2.

Center frequency (f$_0$). On a linear-log graph, bandpass filters are geometrically symmetrical around the filter's resonant frequency or center frequency—provided the response is plotted on linear-log graph paper (the logarithmic axis representing the frequency). On linear-log paper, the central frequency is related to the –3-dB frequencies by the following expression:

$$f_0 = \sqrt{f_1 f_2}$$

For narrow-band bandpass filters, where the ratio of f_2 to f_1 is less than 1.1, the response shape approaches arithmetic symmetry. In this case, we can approximate f_0 by taking the average of –3-dB frequencies:

$$f_0 = \frac{f_1 + f_2}{2}$$

Passband. This represents those frequency signals that reach the output with no more than –3 dB worth of attenuation.

Stop-band frequency (f$_s$). This is a specific frequency where the attenuation reaches a specified value set by the designer. For low-pass and high-pass filters, the frequencies beyond the stop-band frequency are referred to as the *stop band*. For bandpass and notch filters, there are two stop-band frequencies, and the frequencies between the stop bands are also collectively called the *stop band.*

Quality factor (Q). This represents the ratio of the center frequency of a bandpass filter to the –3-dB bandwidth (distance between –3-dB points f_1 and f_2):

$$Q = \frac{f_0}{f_2 - f_1}$$

For a notch filter, use $Q = (f_2 - f_1)/f_0$, where f_0 is often referred to as the *null frequency.*

9.2   Basic Filters

In Chap. 2 you discovered that by using the reactive properties of capacitors and inductors, along with the resonant behavior of *LC* series and parallel networks, you could create simple low-pass, high-pass, bandpass, and notch filters. Here's a quick look at the basic filters covered in Chap. 2:

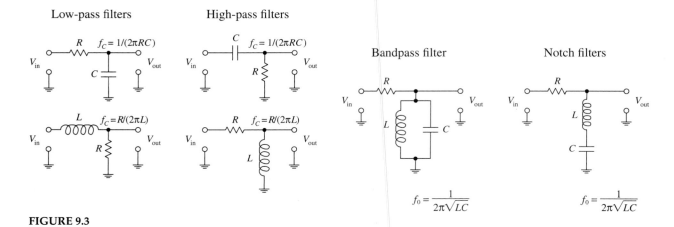

Low-pass filters High-pass filters Bandpass filter Notch filters

$f_C = 1/(2\pi RC)$ $f_C = 1/(2\pi RC)$

$f_C = R/(2\pi L)$ $f_C = R/(2\pi L)$

$f_0 = \dfrac{1}{2\pi\sqrt{LC}}$ $f_0 = \dfrac{1}{2\pi\sqrt{LC}}$

FIGURE 9.3

Now, all the filters shown in this figure have a common limiting characteristic, namely, a shallow 6-dB per octave falloff response beyond the –3-dB point(s). (You can prove this to yourself by going back to Chap. 2 and fiddling with the equations.) In certain noncritical applications, a 6-dB per octave falloff works fine, especially in cases where the signals you want to remove are set well beyond the –3-dB point. However, in situations where greater frequency selectivity is needed (e.g., steeper falloffs and flatter passbands), 6-dB per octave filters will not work. What is needed is a new way to design filters.

Making Filters with Sharper Falloff and Flatter Passband Responses

One approach used for getting a sharper falloff would be to combine a number of 6-dB per octave filters together. Each new section would act to filter the output of the preceding section. However, connecting one filter with another for the purpose of increasing the "dB per octave" slope is not as easy as it seems and in fact becomes impractical in certain instances (e.g., narrow-band bandpass filter design). For example, you have to contend with transient responses, phase-shift problems, signal degradation, winding capacitances, internal resistances, magnetic noise pickup, etc. Things can get nasty.

To keep things practical, what we will do is skip the hard-core filter theory (which can indeed get very nasty) and simply apply some design tricks that use basic response graphs and filter design tables. To truly understand the finer points of filter theory is by no means trivial. If you want in-depth coverage of filter theory, refer to a filter design handbook. (A comprehensive handbook written by Zverck covers almost everything you would want to know about filters.)

Let's begin by jumping straight into some practical filter design examples that require varying degrees of falloff response beyond 6 dB per octave. As you go through these examples, important new concepts will surface. First, we will discuss passive filters and then move on to active filters.

9.3 Passive Low-Pass Filter Design

Suppose that you want to design a low-pass filter that has a $f_{3dB} = 3000$ Hz (attenuation is –3 dB at 3000 Hz) and an attenuation of –25 dB at a frequency of 9000 Hz—which will be called the *stop frequency* f_s. Also, let's assume that both the signal-source impedance R_s and the load impedance R_L are equal to 50 Ω. How do you design the filter?

Step 1 (Normalization)

Frequency response curve

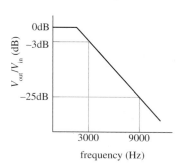

Normalized response curve

FIGURE 9.4

First, make a rough attenuation versus frequency graph to give yourself a general idea of what the response looks like (see far-left figure). Next, you must normalize the graph. This means that you set the –3-dB frequency f_{3dB} to 1 rad/s. The figure to the near left shows the normalized graph. (The reason for normalizing becomes important later on when you start applying design tricks that use normalized response curves and tables.) In order to determine the normalized stop frequency, simply use the following relation, which is also referred to as the *steepness factor*:

$$A_s = \frac{f_s}{f_{3dB}} = \frac{9000\ Hz}{3000\ Hz} = 3$$

This expression tells you that the normalized stop frequency is three times larger than the normalized –3-dB point of 1 rad/s. Therefore, the normalized stop frequency is 3 rad/s.

Step 2 (Pick Response Curve)

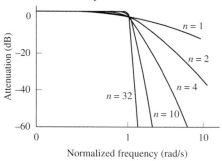

Normalized low-pass Butterworth filter response curves

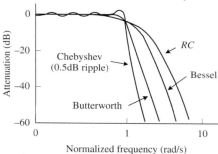

LC low-pass filter networks

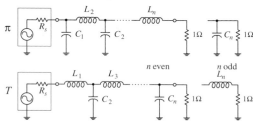

FIGURE 9.5

Next, you must pick a filter response type. Three of the major kinds to choose from include the Butterworth, Chebyshev, and Bessel. Without getting too technical here, what is going on is this: Butterworth, Chebyshev, and Bessel response curves are named after individuals who were able to model *LC* filter networks after a mathematical function called the *transfer function*, given here:

$$T(S) = \frac{V_{out}}{V_{in}} = \frac{N_m S^m + N_{m-1}S^{m-1} + \cdots + N_1 S + N_0}{D_n S^n + D_{n-1}S^{n-1} + \cdots + D_1 S + D_0}$$

The N's in the equation are the numerator's coefficients, the D's are the denominator's coefficients, and $S = j\omega$ ($j = \sqrt{-1}$, $\omega = 2\pi f$). The highest power n in the denominator is referred to as the order of the filter or the number of poles. The highest power m in the numerator is referred to as the number of zeros. Now, by manipulating this function, individuals (e.g., Butterworth, Chebyshev, and Bessel) were able to generate unique graphs of the transfer function that resembled the attenuation response curves of cascaded *LC* filter networks. What is important to know, for practical purposes, is that the number of poles within the transfer function correlates with the number of *LC* sections present within the cascaded filter network and determines the overall steepness of the response curve (the decibels per octave). As the number of poles increases (number of *LC* sections increases), the falloff response becomes steeper. The coefficients of the transfer function influence the overall shape of the response curve and correlate with the specific capacitor and inductor values found within the filter network. Butterworth, Chebyshev, and Bessel came up with their own transfer functions and figured out what values to place in the coefficients and how to influence the slope of the falloff by manipulating the order of the transfer function. Butterworth figured out a way to manipulate the function to give a maximally flat passband response at the expense of steepness in the transition region between the passband and the stop band. Chebyshev figured out a way to get a very steep transition between the passband and stop band at the expense of ripples present in the passband, while Bessel figured out a way to minimize phase shifts at the expense of both flat passbands and steep falloffs. Later we will discuss the pros and cons of Butterworth, Chebyshev, and Bessell filters. For now, however, let's concentrate on Butterworth filters.

Step 3 (Determine the Number of Poles Needed)

Attenuation curves for Butterworth low-pass filter

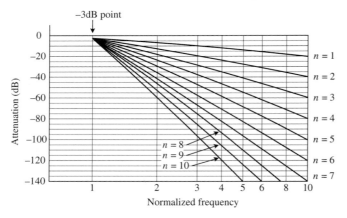

FIGURE 9.6

Continuing on with our low-pass filter problem, let's choose the Butterworth design approach, since it is one of the more popular designs used. The next step is to use a graph of attenuation versus normalized frequency curves for Butterworth low-pass filters, shown in the figure. (Response curves like this are provided in filter handbooks, along with response curves for Chebyshev and Bessel filters.) Next, pick out the single response curve from the graph that provides the desired −25 dB at 3 rad/s, as stated in the problem. If you move your finger along the curves, you will find that the $n = 3$ curve provides sufficient attenuation at 3 rad/s. Now, the filter that is needed will be a third-order low-pass filter, since there are three poles. This means that the actual filter that you will construct will have three LC sections.

Step 4 (Create a Normalized Filter)

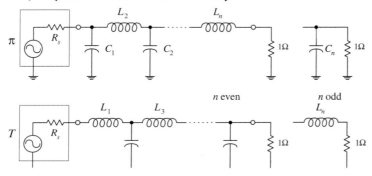

FIGURE 9.7

Now that you have determined the order of the filter, move on to the next step—creating a normalized LC filter circuit. (This circuit will not be the final filter circuit you will use—it will need to be altered.) The circuit networks that are used in this step take on either a π or the T configuration, as shown in the figure. If the source and load impedances match, either configuration can be used—though a π network is more attractive because fewer inductors are needed. However, if the load impedance is greater than the source impedance, it is better to use T configuration. If the load impedance is smaller than the source impedance, it is better to use the π configuration. Since the initial problem stated that the source and load impedances were both 50 Ω, choose the π configuration. The values of the inductors and capacitors are given in Table 9.1. (Filter handbooks will provide such tables, along with tables for Chebyshev and Bessel filters.) Since you need a third-order filter, use the values listed in the $n = 3$ row. The normalized filter circuit you get in this case is shown in Fig. 9.8.

TABLE 9.1   Butterworth Active Filter Low-Pass Values

π {T}								
	R_S	C_1	L_2	C_3	L_4	C_5	L_6	C_7
n	{$1/R_S$}	{L_1}	{C_2}	{L_3}	{C_4}	{L_5}	{C_6}	{L_7}
2	1.000	1.4142	1.4142					
3	1.000	1.0000	2.0000	1.0000				
4	1.000	0.7654	1.8478	1.8478	0.7654			
5	1.000	0.6180	1.6180	2.0000	1.6180	0.6180		
6	1.000	0.5176	1.4142	1.9319	1.9319	1.4142	0.5176	
7	1.000	0.4450	1.2470	1.8019	2.0000	1.8019	1.2470	0.4450

Note: Values of L_n and C_n are for a 1-Ω load and –3-dB frequency of 1 rad/s and have units of H and F. These values must be scaled down. See text.

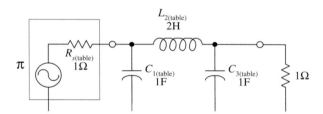

FIGURE 9.8

As mentioned a moment ago, this circuit is not the final circuit that we'll use. That is, the component values listed here will not work! This is so because the graphs and tables you used to get to this point used the normalized frequency within them. Also, you haven't considered the effects of the source and load impedances. In order to construct the final working circuit, you must frequency and impedance scale the component values listed in the circuit in Fig. 9.8. This leads us to the next step.

Step 5 (Frequency and Impedance Scaling)

$$L_{2(\text{actual})} = \frac{R_L L_{2(\text{table})}}{2\pi f_{3dB}} = \frac{(50\Omega)(2\,\text{H})}{2\pi(3000\,\text{Hz})} = 5.3\,\text{mH}$$

$$C_{1(\text{actual})} = \frac{C_{1(\text{table})}}{2\pi f_{3dB} R_L} = \frac{1\,\text{F}}{2\pi(3000\,\text{Hz})(50\,\Omega)} = 1.06\,\mu\text{F}$$

$$C_{3(\text{actual})} = \frac{C_{3(\text{table})}}{2\pi f_{3dB} R_L} = \frac{1\,\text{F}}{2\pi(3000\,\text{Hz})(50\,\Omega)} = 1.06\,\mu\text{F}$$

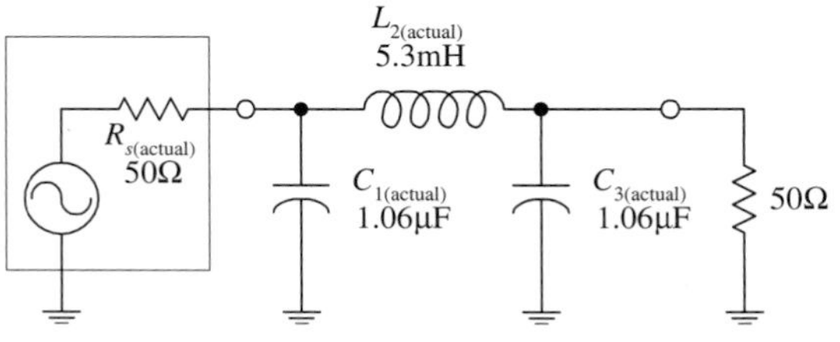

To account for impedance matching of the source and load, as well as getting rid of the normalized frequency, apply the following frequency and impedance scaling rules. To frequency scale, divide the capacitor and inductor values that you got from the table by $\omega = 2\pi f_c$. To impedance scale, multiply resistor and inductor values by the load impedance and divide the capacitor values by the load impedance. In other words, use the following two equations to get the actual component values needed:

$$L_{n(\text{actual})} = \frac{R_L L_{n(\text{table})}}{2\pi f_{3dB}}$$

$$C_{n(\text{actual})} = \frac{C_{n(\text{table})}}{2\pi f_{3dB} R_L}$$

The calculations and the final low-pass circuit are shown in the figure.

FIGURE 9.9

9.4   A Note on Filter Types

It was briefly mentioned earlier that Chebyshev and Bessel filters could be used instead of Butterworth filters. To design Chebyshev and Bessel filters, you take the same approach you used to design Butterworth filters. However, you need to use different low-pass attenuation graphs and tables to come up with the component values placed in the π and T LC networks. If you are interested in designing Chebyshev and Bessel filters, consult a filter design handbook. Now, to give you a better understanding of the differences between the various filter types, the following few paragraphs should help.

Butterworth filters are perhaps the most popular filters used. They have very flat frequency response in the middle passband region, although they have somewhat rounded bends in the region near the −3-dB point. Beyond the −3-dB point, the rate of attenuation increases and eventually reaches $n \times 6$ dB per octave (e.g., $n = 3$, attenuation = 18 dB/octave). Butterworth filters are relatively easy to construct, and the components needed tend not to require as strict tolerances as those of the other filters.

Chebyshev filters (e.g., 0.5-dB ripple, 0.1-dB ripple Chebyshev filter) provide a sharper rate of descent in attenuation beyond the −3-dB point than Butterworth and Bessel filters. However, there is a price to pay for the steep descent—the cost is a ripple voltage within the passband, referred to as the *passband ripple.* The size of the passband ripple increases with order of the filter. Also, Chebyshev filters are more sensitive to component tolerances than Butterworth filters.

Now, there is a problem with Butterworth and Chebyshev filters—they both introduce varying amounts of delay time on signals of different frequencies. In other words, if an input signal consists of a multiple-frequency waveform (e.g., a modulated signal), the output signal will become distorted because different frequencies will be displaced by different delay times. The delay-time variation over the passband is called *delay distortion,* and it increases as the order of the Butterworth and Chebyshev filters increases. To avoid this effect, a Bessel filter can be used. Bessel filters, unlike Butterworth and Chebyshev filters, provide a constant delay over the passband. However, unlike the other two filters, Bessel filters do not have as sharp an attenuation falloff. Having a sharp falloff, however, is not always as important as good signal reproduction at the output. In situations where actual signal reproduction is needed, Bessel filters are more reliable.

9.5   Passive High-Pass Filter Design

Suppose that you want to design a high-pass filter that has an $f_{3dB} = 1000$ Hz and an attenuation of at least −45 dB at 300 Hz—which we call the *stop frequency f_s.* Assume that the filter is hooked up to a source and load that both have impedances of 50 Ω and that a Butterworth response is desired. How do you design the filter? The trick, as you will see in a second, involves treating the high-pass response as an inverted low-pass response, then designing a normalized low-pass filter, applying some conversion tricks on the low-pass filter's components to get a normalized high-pass filter, and then frequency and impedance scaling the normalized high-pass filter.

Frequency Response Curve

Normalized translation to low-pass filter

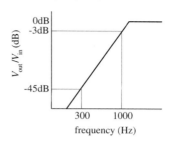

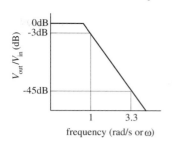

First, make a simple sketch of the response curve for the high-pass filter, as shown in the far-left graph. Next, take the high-pass curve and flip it around in the horizontal direction to get a low-pass response. Then normalize the low-pass response. (This allows you to use the low-pass design techniques. Later you will need to apply a transformation trick on the normalized component values of the low-pass filter to get the desired high-pass filter.) To find the steepness factor A_s and normalized stop-band frequency f_s, follow the same basic procedure as you used in the low-filter example, except now you must take f_{3dB} over f:

$$A_s = \frac{f_{3dB}}{f_s} = \frac{1000 \text{ Hz}}{300 \text{ Hz}} = 3.3$$

This expression tells us that the normalized stop-band frequency is 3.3 times larger than the normalized –3-dB frequency. Since the normalized graph sets f_{3dB} to 1 rad/s, f_s becomes 3.3 rad/s.

Next, take the low-pass filter response from the preceding step and determine which response curve in Fig. 9.6 provides an attenuation of at least –45 dB at 3.3 rad/s. The $n = 5$ curve does the trick, so you create a fifth-order LC network. Now, the question to ask is, Do you use the π or the T network? Initially, you might assume the π network would be best, since the load and source impedances are equal and since fewer inductors are needed. However, when you apply the transformational trick to get the low-pass filter back to a high-pass filter, you will need to interchange inductors for capacitors and capacitors for inductors. Therefore, if you choose the low-pass T network now, you will get fewer inductors in the final high-pass circuit. The fifth-order normalized low-pass filter network is shown in the figure.

To convert the low-pass into a high-pass filter, replace the inductors with capacitors that have value of $1/L$, and replace the capacitors with inductors that have values of $1/C$. In other words, do the following:

Start with a "T" low-pass filter...

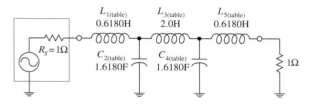

Transform low-pass filter into a high-pass filter...

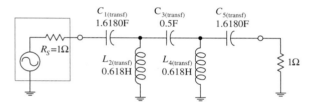

Impedance and frequency scale high-pass filter to get final circuit

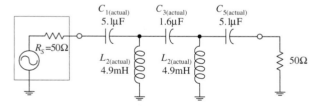

FIGURE 9.10

$$C_{1(transf)} = 1/L_{1(table)} = 1/0.6180 = 1.6180 \text{ F}$$

$$C_{3(transf)} = 1/L_{3(table)} = 1/2.0 = 0.5 \text{ F}$$

$$C_{5(transf)} = 1/L_{5(table)} = 1/0.6180 = 1.6180 \text{ F}$$

$$L_{2(transf)} = 1/C_{2(table)} = 1/1.6180 = 0.6180 \text{ H}$$

$$L_{4(transf)} = 1/C_{4(table)} = 1/1.6180 = 0.6180 \text{ H}$$

Next, frequency and impedance scale to get the actual component values:

$$C_{1(actual)} = \frac{C_{1(trans)}}{2\pi f_{3dB} R_L} = \frac{1.618 \text{ H}}{2\pi(1000 \text{ Hz})(50 \text{ }\Omega)} = 5.1 \text{ }\mu\text{F}$$

$$L_{2(actual)} = \frac{L_{2(trans)} R_L}{2\pi f_{3dB}} = \frac{(0.6180 \text{ F})(50 \text{ }\Omega)}{2\pi(1000 \text{ Hz})} = 4.9 \text{ mH}$$

$$C_{3(actual)} = \frac{C_{3(trans)}}{2\pi f_{3dB} R_L} = \frac{0.5 \text{ H}}{2\pi(1000 \text{ Hz})(50 \text{ }\Omega)} = 1.6 \text{ }\mu\text{F}$$

$$L_{4(actual)} = \frac{L_{4(trans)} R_L}{2\pi f_{3dB}} = \frac{(0.6180 \text{ F})(50 \text{ }\Omega)}{2\pi(1000 \text{ Hz})} = 4.9 \text{ mH}$$

$$C_{5(actual)} = \frac{C_{5(trans)}}{2\pi f_{3dB} R_L} = \frac{1.618 \text{ H}}{2\pi(1000 \text{ Hz})(50 \text{ }\Omega)} = 5.1 \text{ }\mu\text{F}$$

9.6   Passive Bandpass Filter Design

Bandpass filters can be broken down into narrow-band and wide-band types. The defining difference between the two is the ratio between the upper −3-dB frequency f_1 and lower −3-dB frequency f_2. If f_2/f_1 is greater than 1.5, the bandpass filter is placed in the wide-type category. Below 1.5, the bandpass filter is placed in the narrow-band category. As you will see in a moment, the procedure used to design a wide-band bandpass filter differs from that used to design a narrow-band filter.

Wide-Band Design

The basic approach used to design wide-band bandpass filters is simply to combine a low-pass and high-pass filter together. The following example will cover the details. Suppose that you want to design a bandpass filter that has −3-dB points at $f_1 = 1000$ Hz and $f_2 = 3000$ Hz and at least −45 dB at 300 Hz and more than −25 dB at 9000 Hz. Also, again assume that the source and load impedances are both 50 Ω and a Butterworth design is desired.

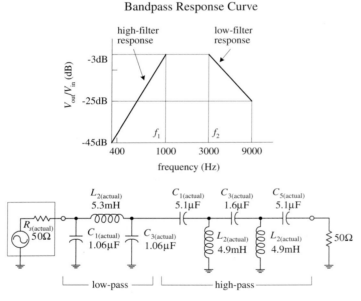

FIGURE 9.11

The basic sketch shown here points out the basic response desired. The ratio $f_2/f_1 = 3$, which is larger than 1.5, so you do indeed have a wide-band situation. Notice how the sketch resembles low-pass and high-pass response curves placed together on the same graph. If you break up the response into low- and high-phase curves, you get the following results:

Low-pass	−3 dB at 3000 Hz
	−25 dB at 9000 Hz
High-pass	−3 dB at 1000 Hz
	−45 dB at 300 Hz

Now, to design the wide-band bandpass filter, construct a low-pass and high-pass filter using the values above and the design technique used in the preceding two example problems. Once you have done this, simply cascade the low-pass and high-pass filters together. The nice thing about this problem is the low-pass and high-pass filters that are needed are simply the filters used in the preceding low-pass and high-pass examples. The final cascade network is shown at the bottom of the figure.

Narrow-Band Design

Narrow-band filters ($f_2/f_1 < 1.5$), unlike wide-band filters, cannot be made simply by cascading low-pass and high-pass filters together. Instead, you must use a new, slightly tricky procedure. This procedure involves transforming the −3-dB bandwidth ($\Delta f_{BW} = f_2 - f_1$) of a bandpass filter into the −3-dB frequency f_{3dB} of a low-pass filter. At the same time, the stop-band bandwidth of the bandpass filter is transformed into the corresponding stop-band frequency of a low-pass filter. Once this is done, a normalized low-pass filter is created. After the normalized low-pass filter is created, the filter must be frequency scaled in a special way to get the desired bandpass filter.

(The normalized circuit also must be impedance scaled, as before.) When frequency scaling the components of the normalized low-pass filter, do not divide by $\omega = 2\pi f_{3dB}$, as you would do with low-pass scaling. Instead, the normalized low-pass filter's components are divided by $2\pi(\Delta f_{BW})$. Next, the scaled circuit's branches must be resonated to the bandpass filter's center frequency f_0 by placing additional inductors in parallel with the capacitors and placing additional capacitors in series with the inductors. This creates LC resonant circuit sections. The values of the additional inductors and capacitors are determined by using the LC resonant equation (see Chap. 2 for the details):

$$f_0 = \frac{1}{2\pi\sqrt{LC}}$$

NARROW-BANDWIDTH BANDPASS FILTER EXAMPLE

Suppose that you want to design a bandpass filter with –3-dB points at $f_1 = 900$ Hz and $f_2 = 1100$ Hz and at least –20 dB worth of attenuation at 800 and 1200 Hz. Assume that both the source and load impedances are 50 Ω and that a Butterworth design is desired.

Low-pass bandpass relationship

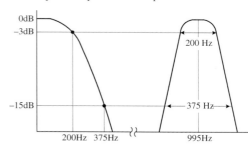

Normalized low-pass response

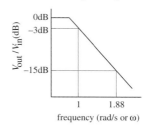

Normalized low-pass filter

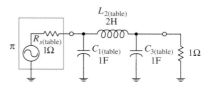

FIGURE 9.12

Since $f_2/f_1 = 1.2$, which is less than 1.5, a narrow-band filter is needed. The initial step in designing a narrow-band bandpass filter is to normalize the bandpass requirements. First, the geometric center frequency is determined:

$$f_0 = \sqrt{f_1 f_2} = \sqrt{(900\text{ Hz})(1100)} = 995\text{ Hz}$$

Next, compute the two pairs of geometrically related stop-band frequencies by using

$$f_a f_b = f_0^2$$

$$f_a = 800\text{ Hz} \qquad f_b = \frac{f_0^2}{f_a} = \frac{(995\text{ Hz})^2}{800\text{ Hz}} = 1237\text{ Hz} \qquad f_b - f_a = 437\text{ Hz}$$

$$f_b = 1200\text{ Hz} \qquad f_a = \frac{f_0^2}{f_b} = \frac{(995\text{ Hz})^2}{1200\text{ Hz}} = 825\text{ Hz} \qquad f_b - f_a = 375\text{ Hz}$$

Notice how things are a bit confusing. For each pair of stop-band frequencies, you get two new pairs—a result of making things "geometrical" with respect to f_0. Choose the pair having the least separation, which represents the more severe requirement –375 Hz.

The steepness factor for the bandpass filter is given by

$$A_s = \frac{\text{stop-band bandwidth}}{\text{3-dB bandwidth}} = \frac{375\text{ Hz}}{200\text{ Hz}} = 1.88$$

Now choose a low-pass Butterworth response that provides at least –20 dB at 1.88 rad/s. According Fig. 9.6, the $n = 3$ curve does the trick. The next step is to create a third-order normalized low-pass filter using the π configuration and Table 9.1.

Next, impedance and frequency scale the normalized low-pass filter to require an impedance level of 50 Ω and a –3-dB frequency equal to the desired bandpass filter's bandwidth ($\Delta f_{BW} = f_2 - f_1$)—which in this example equals 200 Hz. Notice the frequency-scaling trick! The results follow:

$$C_{1(\text{actual})} = \frac{C_{1(\text{table})}}{2\pi(\Delta f_{BW})R_L} = \frac{1\text{ F}}{2\pi(200\text{ Hz})(50\ \Omega)} = 15.92\ \mu\text{F}$$

$$C_{3(\text{actual})} = \frac{C_{3(\text{table})}}{2\pi(\Delta f_{BW})R_L} = \frac{1\text{ F}}{2\pi(200\text{ Hz})(50\ \Omega)} = 15.92\ \mu\text{F}$$

$$L_{2(\text{actual})} = \frac{L_{2(\text{table})}R_L}{2\pi(\Delta f_{BW})} = \frac{(2\text{ H})(50\ \Omega)}{2\pi(200\text{ Hz})} = 79.6\text{ mH}$$

Impedance and frequency scaled low-pass filter

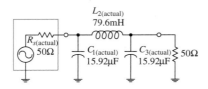

Final bandpass filter

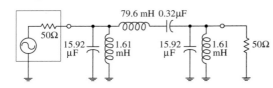

FIGURE 9.12 (*Continued*)

The important part comes now. Each circuit branch of the low-pass filter must be resonated to f_0 by adding a series capacitor to each inductor and a parallel inductor to each capacitor. The *LC* resonant equation is used to determine the additional component values:

$$L_{(\text{parallel with } C1)} = \frac{1}{(2\pi f_0)^2 C_{1(\text{actual})}} = \frac{1}{(2\pi \cdot 995 \text{ Hz})^2 (15.92 \text{ μF})} = 1.61 \text{ mH}$$

$$L_{(\text{parallel with } C3)} = \frac{1}{(2\pi f_0)^2 C_{3(\text{actual})}} = \frac{1}{(2\pi \cdot 995 \text{ Hz})^2 (15.92 \text{ μF})} = 1.61 \text{ mH}$$

$$C_{(\text{series with } L2)} = \frac{1}{(2\pi f_0)^2 L_{2(\text{actual})}} = \frac{1}{(2\pi \cdot 995 \text{ Hz})^2 (79.6 \text{ mH})} = 0.32 \text{ μF}$$

The final bandpass circuit is shown at the bottom of the figure.

9.7 Passive Notch Filter Design

To design a notch filter, you can apply a technique similar to the one you used in the narrow-band bandpass example. However, now you use a high-pass filter instead of a low-pass filter as the basic building block. The idea here is to relate the notch filter's –3-dB bandwidth ($\Delta f_{BW} = f_1 - f_2$) to the –3-dB frequency of a high-pass filter and relate the notch filter's stop-band bandwidth to the stop-band frequency of a high-pass filter. After that, a normalized high-pass filter is created. This filter is then frequency scaled in a special way—all its components are divided by $2\pi\Delta f_{BW}$. (This circuit also must be impedance scaled, as before.) As with the narrow-band bandpass filter example, the scaled high-pass filter's branches must be resonated to the notch filter's center frequency f_0 by inserting additional series capacitors with existing inductors and inserting additional parallel inductors with existing capacitors.

EXAMPLE

Suppose that you want to design a notch filter with –3-dB points at $f_1 = 800$ Hz and $f_2 = 1200$ Hz and at least –20 dB at 900 and 1100 Hz. Let's assume that both the source and load impedances are 600 Ω and that a Butterworth design is desired.

High-pass bandpass relationship

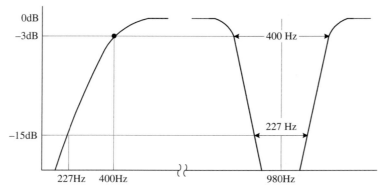

FIGURE 9.13

First, you find the geometric center frequency:

$$f_0 = \sqrt{f_1 f_2} = \sqrt{(800 \text{ Hz})(1200 \text{ Hz})} = 980 \text{ Hz}$$

Next, compute the two pairs of geometrically related stop-band frequencies:

$$f_a = 900 \text{ Hz} \qquad f_b = \frac{f_0^2}{f_a} = \frac{(980 \text{ Hz})^2}{900 \text{ Hz}} = 1067 \text{ Hz}$$

$$f_b - f_a = 1067 \text{ Hz} - 900 \text{ Hz} = 167 \text{ Hz}$$

$$f_b = 1100 \text{ Hz} \qquad f_a = \frac{f_0^2}{f_b} = \frac{(980 \text{ Hz})^2}{1100 \text{ Hz}} = 873 \text{ Hz}$$

$$f_b - f_a = 1100 \text{ Hz} - 873 \text{ Hz} = 227 \text{ Hz}$$

Choose the pair of frequencies that gives the more severe requirement—227 Hz.

Normalized low-pass filter

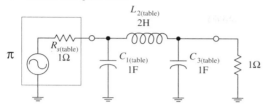

Normalized high-pass filter

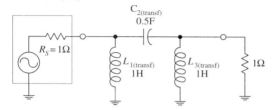

Actual high-pass filter

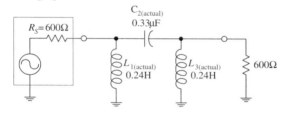

Final bandpass filter

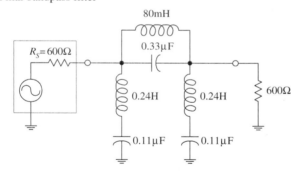

FIGURE 9.13 (*Continued*)

Next, compute the notch filter's steepness factor, which is given by

$$A_S = \frac{\text{3-dB bandwidth}}{\text{stop-band bandwidth}} = \frac{400 \text{ Hz}}{227 \text{ Hz}} = 1.7$$

To come up with the final notch filter design, start out by treating the steepness factor as the steepness factor for a high-pass filter. Next, apply the same tricks you used earlier to construct a high-pass filter. Horizontally flip the high-pass response to get a low-pass response. Then normalize the low-pass response (setting the normalized stop frequency to 1.7 rad/s) and use Fig. 9.6 ($n = 3$ provides at least −20 dB at 1.7 rad/s). Next, use Table 9.1 and the π network to come up with a normalized low-pass design. Then apply the low-pass to high-pass transformational tricks to get a normalized high-pass filter:

$$L_{1(\text{transf})} = 1/C_{1(\text{table})} = 1/1 = 1 \text{ H}$$

$$L_{3(\text{transf})} = 1/C_{3(\text{table})} = 1/1 = 1 \text{ H}$$

$$C_{2(\text{transf})} = 1/L_{2(\text{table})} = 1/1 = 0.5 \text{ F}$$

The first two circuits in the figure show the low-pass to high-pass transformational process.

Next, impedance and frequency scale the normalized high-pass filter to require an impedance level of 600 Ω and a −3-dB frequency equal to the desired notch filter's bandwidth ($\Delta f_{BW} = f_2 - f_1$)—which in the example equals 400 Hz. Notice the frequency-scaling trick! The results follow:

$$L_{1(\text{actual})} = \frac{R_L L_{1(\text{transf})}}{2\pi(\Delta f_{BW})} = \frac{(600 \ \Omega)(1 \text{ H})}{2\pi(400 \text{ Hz})} = 0.24 \text{ H}$$

$$L_{3(\text{actual})} = \frac{R_L L_{3(\text{transf})}}{2\pi(\Delta f_{BW})} = \frac{(600 \ \Omega)(1 \text{ H})}{2\pi(400 \text{ Hz})} = 0.24 \text{ H}$$

$$C_{2(\text{actual})} = \frac{C_{1(\text{transf})}}{2\pi(\Delta f_{BW})R_L} = \frac{(0.5 \text{ F})}{2\pi(400 \text{ Hz})(600 \ \Omega)} = 0.33 \ \mu\text{F}$$

And finally, the important modification—resonate each branch to the notch filter's center frequency f_0 by adding a series capacitor to each inductor and a parallel inductor to each capacitor. The values for these additional components must be

$$C_{(\text{series with } L1)} = \frac{1}{(2\pi f_0)^2 L_{1(\text{actual})}} = \frac{1}{(2\pi \cdot 400 \text{ Hz})^2 (0.24 \text{ H})} = 0.11 \ \mu\text{F}$$

$$C_{(\text{series with } L3)} = \frac{1}{(2\pi f_0)^2 L_{3(\text{actual})}} = \frac{1}{(2\pi \cdot 400 \text{ Hz})^2 (0.24 \text{ H})} = 0.11 \ \mu\text{F}$$

$$L_{(\text{parallel with } L1)} = \frac{1}{(2\pi f_0)^2 C_{2(\text{actual})}} = \frac{1}{(2\pi \cdot 400 \text{ Hz})^2 (0.33 \ \mu\text{F})} = 80 \text{ mH}$$

The final circuit is shown at the bottom of the figure.

9.8 Active Filter Design

This section covers some basic Butterworth active filter designs. We already discussed the pros and cons of active filter design earlier in this chapter. Here we will focus on the actual design techniques used to make unity-gain active filters. To begin, let's design a low-pass filter.

9.8.1 Active Low-Pass Filter Example

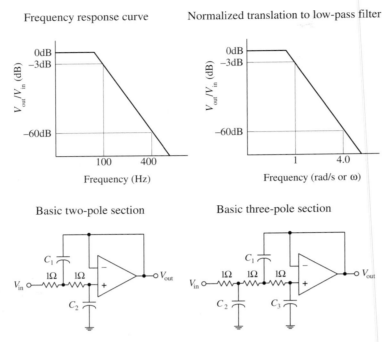

FIGURE 9.14

Suppose that you wish to design an active low-pass filter that has a 3-dB point at 100 Hz and at least 60 dB worth of attenuation at 400 Hz—which we'll call the *stop frequency* f_s.

The first step in designing the filter is to normalize low-pass requirements. The steepness factor is

$$A_s = \frac{f_s}{f_{3dB}} = \frac{400 \text{ Hz}}{100 \text{ Hz}} = 4$$

This means that the normalized position of f_s is set to 4 rad/s. See the graphs in Fig. 9.14. Next, use the Butterworth low-pass filter response curves in Fig. 9.6 to determine the order of filter you need. In this case, the $n = 5$ curve provides over −60 dB at 4 rad/s. In other words, you need a fifth-order filter.

Now, unlike passive filters design, active filter design requires the use of a different set of basic normalized filter networks and a different table to provide the components of the networks. The active filter networks are shown in Fig. 9.14—there are two of them. The one to the left is called a *two-pole section*, while the one on the right is called a *three-pole section*. To design a Butterworth low-pass normalized filter of a given order, use Table 9.2. (Filter handbooks provide Chebyshev and Bessel tables as well.) In this example, a five-pole filter is needed, so according to the table, two sections are required—a three-pole and a two-pole section. These sections are cascaded together, and the component values listed in Table 9.2 are placed next to the corresponding components within the cascaded network. The resulting normalized low-pass filter is shown in Fig. 9.15.

TABLE 9.2   Butterworth Normalized Active Low-Pass Filter Values

ORDER n	NUMBER OF SECTIONS	SECTIONS	C_1	C_2	C_3
2	1	2-pole	1.414	0.7071	
3	1	3-pole	3.546	1.392	0.2024
4	2	2-pole	1.082	0.9241	
		2-pole	2.613	0.3825	
5	2	3-pole	1.753	1.354	0.4214
		2-pole	3.235	0.3090	
6	3	2-pole	1.035	0.9660	
		2-pole	1.414	0.7071	
		2-pole	3.863	0.2588	
7	3	3-pole	1.531	1.336	0.4885
		2-pole	1.604	0.6235	
		2-pole	4.493	0.2225	
8	4	2-pole	1.020	0.9809	
		2-pole	1.202	0.8313	
		2-pole	2.000	0.5557	
		2-pole	5.758	0.1950	

Normalized low-pass filter

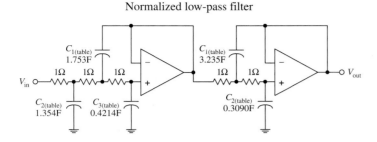

The normalized filter will provide the correct response, but the component values are impractical—they're too big. In order to bring these values down to size, the circuit must be frequency and impedance scaled. To frequency scale, simply divide the capacitor values by $2\pi f_{3dB}$ (you need not frequency scale the resistors—they aren't reactive). In terms of impedance scaling, you do not have to deal with source/load impedance matching. Instead, simply multiply the normalized filter circuit's resistors by a factor of Z and divide the capacitors by the same factor. The value of Z is chosen to scale the normalized filter components to more practical values. A typical value for Z is 10,000 Ω. In summary, the final scaling rules are expressed as follows:

$$C_{(actual)} = \frac{C_{(table)}}{Z \cdot 2\pi f_{3dB}}$$

$$R_{(actual)} = ZR_{(table)}$$

Taking Z to be 10,000, you get the final low-pass filter circuit shown at the bottom of the figure.

Final low-pass filter

FIGURE 9.15

9.8.2 Active High-Pass Filter Example

The approach used to design active high-pass filters is similar to the approach used to design passive high-pass filters. Take a normalized low-pass filter, transform it into a high-pass circuit, and then frequency and impedance scale it. For example, suppose that you want to design a high-pass filter with a −3-dB frequency of 1000 Hz and 50 dB worth of attenuation at 300 Hz. What do you do?

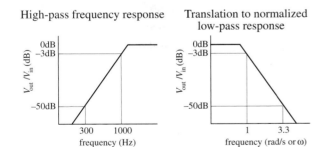

Normalized low-pass filter

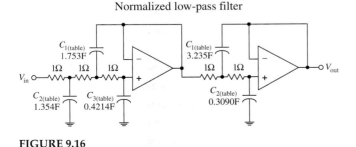

FIGURE 9.16

The first step is to convert the high-pass response into a normalized low-pass response, as shown in the figure. The steepness factor for the low-pass equivalent response is given by

$$A_s = \frac{f_{3dB}}{f_s} = \frac{1000 \text{ Hz}}{300 \text{ Hz}} = 3.3$$

This means that the stop frequency is set to 3.3 rad/s on the normalized graph. The Butterworth response curve shown in Fig. 9.6 tells you that a fifth-order ($n = 5$) filter will provide the needed attenuation response. Like the last example, a cascaded three-pole/two-pole normalized low-pass filter is required. This filter is shown in Fig. 9.16.

Next, the normalized low-pass filter must be converted into a normalized high-pass filter. To make the conversion, exchange resistors for capacitors that have values of $1/R$ F, and exchange capacitors with resistors that have values of $1/C$ Ω. The second circuit in Fig. 9.16 shows the transformation.

Normalized high-pass filter (transformed low-pass filter)

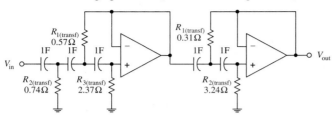

FIGURE 9.16   *(Continued)*

Like the last example problem, to construct the final circuit, the normalized high-pass filter's component values must be frequency and impedance scaled:

$$C_{(actual)} = \frac{C_{(transf)}}{Z \cdot 2\pi f_{3dB}}$$

$$R_{(actual)} = ZR_{(transf)}$$

Again, let $Z = 10{,}000$. The final circuit is shown in Fig. 9.17.

Final high-pass filter

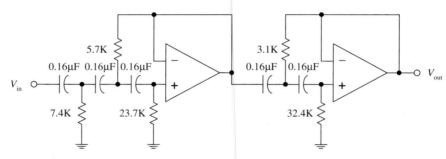

FIGURE 9.17

9.8.3   Active Bandpass Filters

To design an active bandpass filter, it is necessary to determine if a wide-band or narrow-band type is needed. If the upper 3-dB frequency divided by the lower 3-dB frequency is greater than 1.5, the bandpass filter is a wide-band type; below 1.5, it is a narrow-band type. To design a wide-band bandpass filter, simply cascade a high-pass and low-pass active filter together. To design a narrow-band bandpass filter, you have to use some special tricks.

Wide-Band Example

Suppose that you want to design a bandpass filter that has −3-dB points at $f_1 = 1000$ Hz and $f_2 = 3000$ Hz and at least −30 dB at 300 and 10,000 Hz. What do you do?

Normalized low-pass/low-pass initial setup

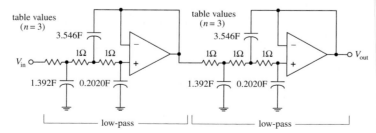

FIGURE 9.18

First, confirm that this is a wide-band situation:

$$\frac{f_2}{f_1} = \frac{3000 \text{ Hz}}{1000 \text{ Hz}} = 3$$

Yes it is—it is greater than 1.5. This means that you simply have to cascade a low-pass and high-pass filter together. Next, the response requirements for the bandpass filter are broken down into low-pass and high-pass requirements:

Low-pass: −3 dB at 3000 Hz

   −30 dB at 10,000 Hz

High-pass: −3 dB at 1000 Hz

   −30 dB at 300 Hz

Normalized and transformed bandpass filter

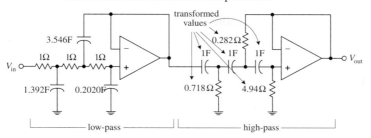

low-pass ———————————— high-pass

Final bandpass filter

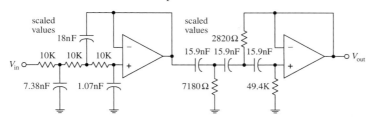

FIGURE 9.18 (*Continued*)

The steepness factor for the low-pass filter is

$$A_s = \frac{f_s}{f_{3dB}} = \frac{10,000 \text{ Hz}}{3000 \text{ Hz}} = 3.3$$

while the steepness factor for the high-pass filter is

$$A_s = \frac{f_{3dB}}{f_s} = \frac{1000 \text{ Hz}}{300 \text{ Hz}} = 3.3$$

This means that the normalized stop frequencies for both filters will be 3.3 rad/s. Next, use the response curves in Fig. 9.6 to determine the needed filter orders—$n = 3$ provides over −30 dB at 3.3 rad/s. To create the cascaded, normalized low-pass/high-pass filter, follow the steps in the last two examples. The upper two circuits in the figure show the steps involved in this process. To construct the final bandpass filter, the normalized bandpass filter must be frequency and impedance scaled.

Low-pass section:

$$C_{(actual)} = \frac{C_{table}}{Z \cdot 2\pi f_{3dB}} = \frac{C_{table}}{Z \cdot 2\pi (3000 \text{ Hz})}$$

High-pass section:

$$C_{(actual)} = \frac{C_{table}}{Z \cdot 2\pi f_{2dB}} = \frac{C_{table}}{Z \cdot 2\pi (1000 \text{ Hz})}$$

Choose $Z = 10,000 \ \Omega$ to provide convenient scaling of the components. In the normalized circuit, resistors are multiplied by a factor of Z. The final bandpass filter is shown at the bottom of the figure.

Narrow-Band Example

Suppose that you want to design a bandpass filter that has a center frequency $f_0 = 2000$ Hz and a −3-dB bandwidth $\Delta f_{BW} = f_2 - f_1 = 40$ Hz. How do you design the filter? Since $f_2/f_1 = 2040 \text{ Hz}/1960 \text{ Hz} = 1.04$, it is not possible to used the low-pass/high-pass cascading technique you used in the wide-band example. Instead, you must use a different approach. One simple approach is shown below.

Narrow-band filter circuit

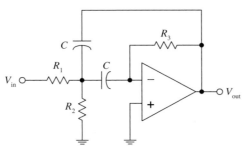

FIGURE 9.19

In this example, simply use the circuit in Fig. 9.19 and some important equations that follow. No detailed discussion will ensue.

First, find the quality factor for the desired response:

$$Q = \frac{f_0}{f_2 - f_1} = \frac{2000 \text{ Hz}}{40 \text{ Hz}} = 50$$

Next, use the following design equations:

$$R_1 = \frac{Q}{2\pi f_0 C} \qquad R_2 = \frac{R_1}{2Q^2 - 1} \qquad R_3 = 2R_1$$

Final filter circuit

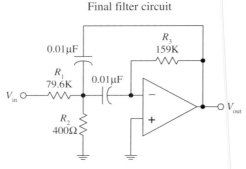

FIGURE 9.19 (*Continued*)

Picking a convenient value for C—which we'll set to 0.01 µF—the resistors' values become

$$R_1 = \frac{50}{2\pi(2000 \text{ Hz})(0.01 \text{ µF})} = 79.6 \text{ k}\Omega$$

$$R_2 = \frac{79.6 \text{ k}\Omega}{2(50)^2 - 1} = 400 \text{ }\Omega$$

$$R_3 = 2(79.6 \text{ k}\Omega) = 159 \text{ k}\Omega$$

The final circuit is shown at the bottom of the figure. R_2 can be replaced with a variable resistor to allow for tuning.

9.8.4 Active Notch Filters

Active notch filters come in narrow- and wide-band types. If the upper –3-dB frequency divided by the lower –3-dB frequency is greater than 1.5, the filter is called a *wide-band notch filter*—less than 1.5, the filter is called a *narrow-band notch filter*.

Wide-Band Notch Filter Example

To design a wide-band notch filter, simply combine a low-pass and high-pass filter together as shown in Fig. 9.20.

Basic wide-band notch filter

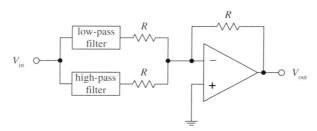

FIGURE 9.20

For example, if you need a notch filter to have –3-dB points at 500 and 5000 Hz and at least –15 dB at 1000 and 2500 Hz, simply cascade a low-pass filter with a response of

3 dB at 500 Hz

15 dB at 1000 Hz

with a high-pass filter with a response of

3 dB at 5000 Hz

15 dB at 2500 Hz

After that, go through the same low-pass and high-pass design procedures covered earlier. Once these filters are constructed, combine them as shown in the circuit in Fig. 9.20. In this circuit, $R = 10$ k is typically used.

Narrow-Band Notch Filters Example

To design a narrow-band notch filter ($f_2/f_1 < 1.5$), an RC network called the *twin-T* (see Fig. 9.21) is frequently used. A deep null can be obtained at a particular frequency with this circuit, but the circuit's Q is only ¼. (Recall that the Q for a notch filter is given as the center or null frequency divided by the –3-dB bandwidth.) To increase the Q, use the active notch filter shown in Fig. 9.22.

Twin-T passive notch filter

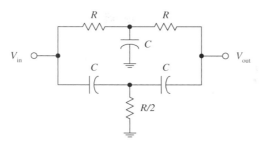

FIGURE 9.21

Improved notch filter

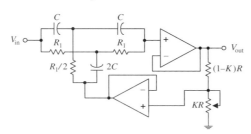

FIGURE 9.22

Like the narrow-bandpass example, let's simply go through the mechanics of how to pick the component values of the active notch filter. Here's an example.

Suppose that you want to make a "notch" at $f_0 = 2000$ Hz and desire a –3-dB bandwidth of $\Delta f_{BW} = 100$ Hz. To get this desired response, do the following. First determine the Q:

$$Q = \frac{\text{"notch" frequency}}{\text{–3-dB bandwidth}} = \frac{f_0}{\Delta f_{BW}} = \frac{2000 \text{ Hz}}{100 \text{ Hz}} = 20$$

The components of the active filter are found by using

$$R_1 = \frac{1}{2\pi f_0 C} \quad \text{and} \quad K = \frac{4Q - 1}{4Q}$$

Now arbitrarily choose R and C; say, let $R = 10$ k and $C = 0.01$ μF. Next, solve for R_1 and K:

$$R_1 = \frac{1}{2\pi f_0 C} = \frac{1}{2\pi(2000 \text{ Hz})(0.01 \text{ μF})} = 7961 \text{ }\Omega$$

$$K = \frac{4Q - 1}{4Q} = \frac{4(20) - 1}{4(20)} = 0.9875$$

Substitute these values into the circuit in Fig. 9.22. Notice the variable potentiometer—it is used to fine-tune the circuit.

9.9 Integrated Filter Circuits

A number of filter ICs are available on the market today. Two of the major categories of integrated filter circuits include the state-variable and switched-capacitor filter ICs. Both these filter ICs can be programmed to implement all the second-order functions described in the preceding sections. To design higher-order filters, a number of these ICs can be cascaded together. Typically, all that's needed to program these filter ICs is a few resistors. Using IC filters allows for great versatility, somewhat simplified design, good precision, and limited design costs. Also, in most applications, frequency and selectivity factors can be adjusted independently.

An example of a state-variable filter IC is the AF100 made by National Semiconductor. This IC can provide low-pass, high-pass, bandpass, and notch filtering capabilities (see Fig. 9.23). Unlike the preceding filters covered in this chapter, the state-variable filter also can provide voltage gain.

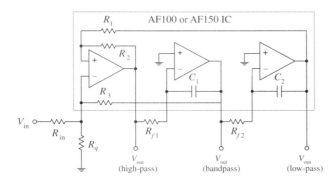

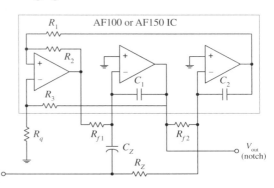

FIGURE 9.23

For the AF100, the low-pass gain is set using resistors R_1 and R_{in} (gain $= -R_1/R_{in}$). For the high-pass filter, the gain is set by resistors R_2 and R_{in} (gain $= -R_2/R_{in}$). (The negative sign indicates that the output is inverted relative to the input.) Setting the gain for the bandpass and notch functions is a bit more complex. Other parameters, such as Q, can be tweaked by using design formulas provided by the manufacturer. A good filter design handbook will discuss state-variable filters in detail and will provide the necessary design formulas. Also, check out the electronics catalogs to see what kinds of state-variable ICs exist besides the AF100.

Switched-capacitor filters are functionally similar to the other filters already discussed. However, instead of using external resistors to program the desired characteristics, switched-capacitor filters use a high-frequency capacitor-switching network technology. The capacitor-switching networks act like resistors whose values can be changed by changing the frequency of an externally applied clock voltage. The frequency of the clock signal determines which frequencies are passed and which frequencies get rejected. Typically, a digital clock signal is used to drive the filter—a useful feature if you are looking to design filters that can be altered by digital circuits. An example of a switched-capacitor IC is National Semiconductor's MF5 (see Fig. 9.24). By simply using a few external resistors, a power source, and an applied clock signal, you can program the filter for low-pass, high-pass, and bandpass functions. Again, like the state-variable ICs, manufacturers will provide you with necessary formulas needed for selecting the resistors and the frequency of the clock signal.

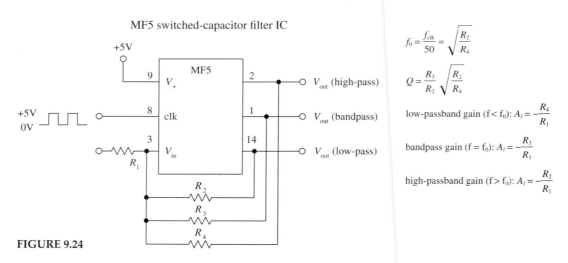

MF5 switched-capacitor filter IC

$$f_0 = \frac{f_{clk}}{50} = \sqrt{\frac{R_2}{R_4}}$$

$$Q = \frac{R_3}{R_2}\sqrt{\frac{R_2}{R_4}}$$

low-passband gain $(f < f_0)$: $A_l = -\dfrac{R_4}{R_1}$

bandpass gain $(f = f_0)$: $A_l = -\dfrac{R_3}{R_1}$

high-passband gain $(f > f_0)$: $A_l = -\dfrac{R_2}{R_1}$

FIGURE 9.24

Switched-capacitor filters come in different filter orders. For example, the MF4 is a fourth-order Butterworth low-pass filter, and the MF6 is a sixth-order low-pass Butterworth filter; both are made by National Semiconductor. These two ICs have unity passband gain and require no external components, but they do require a clock input. There are a number of different kinds of switched-capacitor filters out there, made by a number of different manufacturers. Check the catalogs.

As an important note, the periodic clock signal applied to a switched-capacitor filter can generate a significant amount of noise (around 10 to 25 mV) present in the output signal. Typically, this is not of much concern because the frequency of the noise—which is the same as the frequency of the clock—is far removed from the signal band of interest. Usually a simple *RC* filter can be used to get rid of the problem.

CHAPTER 10

Oscillators and Timers

Within practically every electronic instrument there is an oscillator of some sort. The task of the oscillator is to generate a repetitive waveform of desired shape, frequency, and amplitude that can be used to drive other circuits. Depending on the application, the driven circuit(s) may require either a pulsed, sinusoidal, square, sawtooth, or triangular waveform.

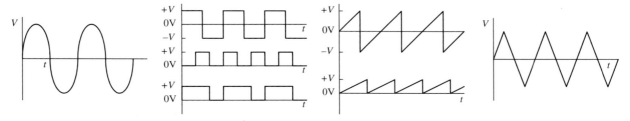

FIGURE 10.1

In digital electronics, squarewave oscillators, called *clocks*, are used to drive bits of information through logic gates and flip-flops at a rate of speed determined by the frequency of the clock. In radio circuits, high-frequency sinusoidal oscillators are used to generate carrier waves on which information can be encoded via modulation. The task of modulating the carrier also requires an oscillator. In oscilloscopes, a sawtooth generator is used to generate a horizontal electron sweep to establish the time base. Oscillators are also used in synthesizer circuits, counter and timer circuits, and LED/lamp flasher circuits. The list of applications is endless.

The art of designing good oscillator circuits can be fairly complex. There are a number of designs to choose from and a number of precision design techniques required. The various designs make use of different timing principles (e.g., *RC* charge/discharge cycle, *LC* resonant tank networks, crystals), and each is best suited for use within a specific application. Some designs are simple to construct but may have limited frequency stability. Other designs may have good stability within a certain frequency range, but poor stability outside that range. The shape of the generated waveform is obviously another factor that must be considered when designing an oscillator.

This chapter discusses the major kinds of oscillators, such as the *RC* relaxation oscillator, the Wien-bridge oscillator, the *LC* oscillator, and the crystal oscillator. The chapter also takes a look at popular oscillator ICs.

10.1   *RC* Relaxation Oscillators

Perhaps the easiest type of oscillator to design is the *RC* relaxation oscillator. Its oscillatory nature is explained by the following principle: Charge a capacitor through a resistor and then discharge it when the capacitor voltage reaches a certain threshold voltage. After that, the cycle is repeated, continuously. In order to control the charge/discharge cycle of the capacitor, an amplifier wired with positive feedback is used. The amplifier acts like a charge/discharge switch—triggered by the threshold voltage—and also provides the oscillator with gain. Figure 10.2 shows a simple op amp relaxation oscillator.

Simple Square-Wave Relaxation Oscillator

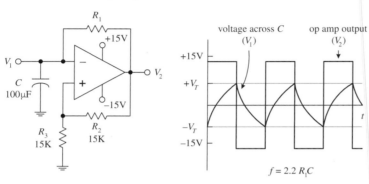

FIGURE 10.2

Assume that when power is first applied, the op amp's output goes toward positive saturation (it is equally likely that the output will go to negative saturation—see Chap. 8 for the details). The capacitor will begin to charge up toward the op amp's positive supply voltage (around +15 V) with a time constant of R_1C. When the voltage across the capacitor reaches the threshold voltage, the op amp's output suddenly switches to negative saturation (around –15 V). The threshold voltage is the voltage set at the non-inverting input, which is

$$V_T = \frac{R_3}{R_3 + R_2} = V_2 \frac{15 \text{ k}\Omega}{15 \text{ k}\Omega + 15 \text{ k}\Omega}(+15 \text{ V}) = +7.5 \text{ V}$$

The threshold voltage set by the voltage divider is now –7.5 V. The capacitor begins discharging toward negative saturation with the same R_1C time constant until it reaches –7.5 V, at which time the op amp's output switches back to the positive saturation voltage. The cycle repeats indefinitely, with a period equal to 2.2 R_1C.

Here's another relaxation oscillator that generates a sawtooth waveform (see Fig. 10.3). Unlike the preceding oscillator, this circuit resembles an op amp integrator network—with the exception of the PUT (programmable unijunction transistor) in the feedback loop. The PUT is the key ingredient that makes this circuit oscillate. Here's a rundown on how this circuit works.

Simple Sawtooth Generator

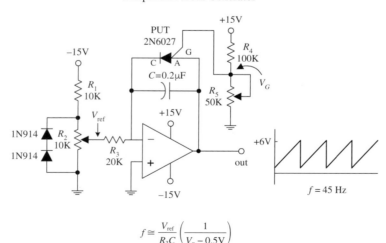

FIGURE 10.3

Let's initially pretend the circuit shown here does not contain the PUT. In this case, the circuit would resemble a simple integrator circuit; when a negative voltage is placed at the inverting input (–), the capacitor charges up at a linear rate toward the positive saturation voltage (+15 V). The output signal would simply provide a one-shot ramp voltage—it would not generate a repetitive triangular wave. In order to generate a repetitive waveform, we must now include the PUT. The PUT introduces oscillation into the circuit by acting as an active switch that turns on (anode-to-cathode conduction) when the anode-to-cathode voltage is greater by one diode drop than its gate-to-cathode voltage. The PUT will remain on until the current through it falls below the minimum holding current rating. This switching action acts to rapidly discharge the capacitor before the output saturates.

When the capacitor discharges, the PUT turns off, and the cycle repeats. The gate voltage of the PUT is set via voltage-divider resistors R_4 and R_5. The R_1 and R_2 voltage-divider resistors set the reference voltage at the inverting input, while the diodes help stabilize the voltage across R_2 when it is adjusted to vary the frequency. The output-voltage amplitude is determined by R_4, while the output frequency is approximated by the expression below the figure. (The 0.5 V represents a typical voltage drop across a PUT.)

Here's a simple dual op amp circuit that generates both triangular and square waveforms (see Fig. 10.4). This circuit combines a triangle-wave generator with a comparator.

Simple Triangle-Wave/Square-Wave Generator

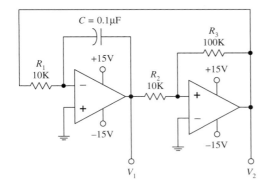

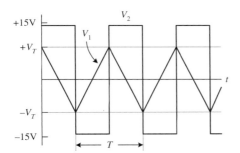

The rightmost op amp in the circuit acts as a comparator—it is wired with positive feedback. If there is a slight difference in voltage between the inputs of this op amp, the V_2 output voltage will saturate in either the positive or negative direction. For sake of argument, let's say the op amp saturates in the positive direction. It will remain in that saturated state until the voltage at the noninverting input (+) drops below zero, at which time V_2 will be driven to negative saturation. The threshold voltage is given by

$$V_T = \frac{V_{sat}}{(R_3 - R_2)}$$

where V_{sat} is a volt or so lower than the op amp's supply voltage (see Chap. 8) Now this comparator is used with a ramp generator (leftmost op amp section). The output of the ramp generator is connected to the input of the comparator, while its output is fed back to the input of the ramp generator. Each time the ramp voltage reaches the threshold voltage, the comparator changes states. This gives rise to oscillation. The period of the output waveform is determined by the R_1C time constant, the saturation voltage, and the threshold voltage:

$$T = \frac{4V_T}{V_{sat}} R_1 C$$

The frequency is $1/T$.

FIGURE 10.4

Now op amps are not the only active ingredient used to construct relaxation oscillators. Other components, such as transistors and digital logic gates, can take their place.

Unijunction oscillator

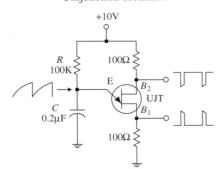

FIGURE 10.5

Here is a unijunction transistor (UJT), along with some resistors and a capacitor, that makes up a relaxation oscillator that is capable of generating three different output waveforms. During operation, at one instant in time, C charges through R until the voltage present on the emitter reaches the UJT's triggering voltage. Once the triggering voltage is exceeded, the E-to-B_1 conductivity increases sharply, which allows current to pass from the capacitor-emitter region through the emitter-base 1 region and then to ground. When this occurs, C suddenly loses its charge, and the emitter voltage suddenly falls below the triggering voltage. After that, the cycle repeats itself. The resulting waveforms generated during this process are shown in the figure. The frequency of oscillation is given by

$$f = \frac{1}{R_E C_E \ln[1/(1-\eta)]}$$

where η is the UJT's intrinsic standoff ratio, which is typically around 0.5. See Chap. 4 for more details.

Here a simple relaxation oscillator is built from a Schmitt trigger inverter IC and an RC network. (Schmitt triggers are used to transform slowly changing input waveforms into sharply defined, jitter-free output waveforms [see Chap. 12]). When power is first applied to the circuit, the voltage across C is zero, and the output of the inverter is high (+5 V). The capacitor starts charging up toward the output voltage via R. When the capacitor voltage reaches the positive-going threshold of the inverter (e.g., 1.7 V), the output of the inverter goes low (~0 V). With the output low, C discharges toward 0 V. When the capacitor voltage drops below the negative-going threshold voltage of the inverter (e.g., 0.9 V), the output of the inverter goes high. The cycle repeats. The on/off times are determined by the positive- and negative-going threshold voltages and the RC time constant.

The third example is a pair of CMOS inverters that are used to construct a simple squarewave RC relaxation oscillator. The circuit can work with voltages ranging from 4 to 18 V. The frequency of oscillation is given by

$$f = \frac{1}{4RC \ln 2} \approx \frac{1}{2.8RC}$$

R can be adjusted to vary the frequency. We will discuss CMOS inverters in Chap. 12.

Digital oscillator
(using a Schmitt trigger inverter)

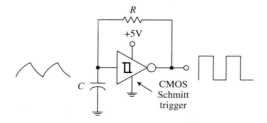

Digital oscillator (using inverters)

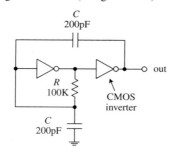

FIGURE 10.5 (*Continued*)

All the relaxation oscillators shown in this section are relatively simple to construct. Now, as it turns out, there is even an easier way to generate basic waveforms. The easy way is to use an IC especially designed for the task. An incredibly popular squarewave-generating chip that can be programmed with resistors and a capacitor is the 555 timer IC.

10.2 The 555 Timer IC

The 555 timer IC is an incredibly useful precision timer that can act as either a timer or an oscillator. In timer mode—better known as *monostable mode*—the 555 simply acts as a "one-shot" timer; when a trigger voltage is applied to its trigger lead, the chip's output goes from low to high for a duration set by an external RC circuit. In oscillator mode—better known as *astable mode*—the 555 acts as a rectangular-wave generator whose output waveform (low duration, high duration, frequency, etc.) can be adjusted by means of two external RC charge/discharge circuits.

The 555 timer IC is easy to use (requires few components and calculations) and inexpensive and can be used in an amazing number of applications. For example, with the aid of a 555, it is possible to create digital clock waveform generators, LED and lamp flasher circuits, tone-generator circuits (sirens, metronomes, etc.), one-shot timer circuits, bounce-free switches, triangular-waveform generators, frequency dividers, etc.

10.2.1 How a 555 Works (Astable Operation)

Figure 10.6 is a simplified block diagram showing what is inside a typical 555 timer IC. The overall circuit configuration shown here (with external components included) represents the astable 555 configuration.

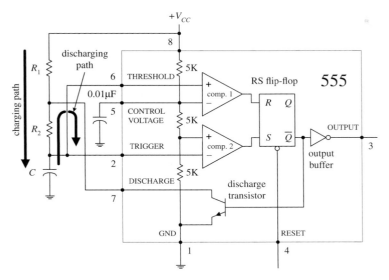

FIGURE 10.6

The 555 gets its name from the three 5-kΩ resistors shown in the block diagram. These resistors act as a three-step voltage divider between the supply voltage (V_{CC}) and ground. The top of the lower 5-kΩ resistor (+ input to comparator 2) is set to $\frac{1}{3}V_{CC}$, while the top of the middle 5-kΩ resistor (− input to comparator 1) is set to $\frac{2}{3}V_{CC}$. The two comparators output either a high or low voltage based on the analog voltages being compared at their inputs. If one of the comparator's positive inputs is more positive than its negative input, its output logic level goes high; if the positive input voltage is less than the negative input voltage, the output logic level goes low. The outputs of the comparators are sent to the inputs of an *SR* (set/reset) flip-flop. The flip-flop looks at the *R* and *S* inputs and produces either a high or a low based on the voltage states at the inputs (see Chap. 12).

Pin 1 (ground).   IC ground.

Pin 2 (trigger).   Input to comparator 2, which is used to set the flip-flop. When the voltage at pin 2 crosses from above to below $\frac{1}{3}V_{CC}$, the comparator switches to high, setting the flip-flop.

Pin 3 (output).   The output of the 555 is driven by an inverting buffer capable of sinking or sourcing around 200 mA. The output voltage levels depend on the output current but are approximately $V_{\text{out(high)}} = V_{CC} - 1.5$ V and $V_{\text{out(low)}} = 0.1$ V.

Pin 4 (reset).   Active-low reset, which forces $\overline{Q}$ high and pin 3 (output) low.

Pin 5 (control).   Used to override the $\frac{2}{3}V_{CC}$ level, if needed, but is usually grounded via a 0.01-µF bypass capacitor (the capacitor helps eliminate V_{CC} supply noise). An external voltage applied here will set a new trigger voltage level.

Pin 6 (threshold).   Input to the upper comparator, which is used to reset the flip-flop. When the voltage at pin 6 crosses from below to above $\frac{2}{3}V_{CC}$, the comparator switches to a high, resetting the flip-flop.

Pin 7 (discharge).   Connected to the open collector of the *npn* transistor. It is used to short pin 7 to ground when $\overline{Q}$ is high (pin 3 low). This causes the capacitor to discharge.

Pin 8 (Supply voltage V_{CC}).   Typically between 4.5 and 16 V for general-purpose TTL 555 timers. (For CMOS versions, the supply voltage may be as low as 1 V.)

In the astable configuration, when power is first applied to the system, the capacitor is uncharged. This means that 0 V is placed on pin 2, forcing comparator 2 high. This in turn sets the flip-flop so that $\overline{Q}$ is low and the 555's output is high (a result of the inverting buffer). With $\overline{Q}$ low, the discharge transistor is turned off, which allows the capacitor to charge toward V_{CC} through R_1 and R_2. When the capacitor voltage exceeds $\frac{1}{3}V_{CC}$, comparator 2 goes low, which has no effect on the *SR* flip-flop. However, when the capacitor voltage exceeds $\frac{2}{3}V_{CC}$, comparator 1 goes high, resetting the flip-flop and forcing $\overline{Q}$ high and the output low. At this point, the discharge transistor turns on and shorts pin 7 to ground, discharging the capacitor through R_2. When the capacitor's voltage drops below $\frac{1}{3}V_{CC}$, comparator 2's output jumps back to a high level, setting the flip-flop and making $\overline{Q}$ low and the output high. With $\overline{Q}$ low, the transistor turns off, allowing the capacitor to start charging again. The cycle repeats over and over again. The net result is a squarewave output pattern whose voltage level is approximately V_{CC} − 1.5 V and whose on/off periods are determined by the C, R_1 and R_2.

10.2.2 Basic Astable Operation

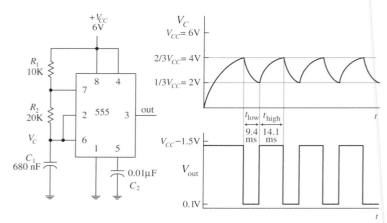

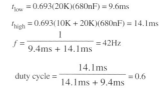

$t_{low} = 0.693(20K)(680nF) = 9.6ms$

$t_{high} = 0.693(10K + 20K)(680nF) = 14.1ms$

$f = \dfrac{1}{9.4ms + 14.1ms} = 42Hz$

$\text{duty cycle} = \dfrac{14.1ms}{14.1ms + 9.4ms} = 0.6$

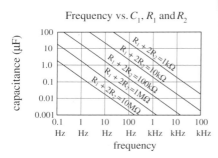

Frequency vs. C_1, R_1 and R_2

When a 555 is set up in astable mode, it has no stable states; the output jumps back and forth. The time duration V_{out} remains low (around 0.1 V) is set by the R_2C_1 time constant and the $\frac{1}{3}V_{CC}$ and $\frac{2}{3}V_{CC}$ levels; the time duration V_{out} stays high (around $V_{CC} - 1.5$ V) is determined by the $(R_1 + R_2)$ C_1 time constant and the two voltage levels (see graphs). After doing some basic calculations, the following two practical expressions arise:

$t_{low} = 0.693R_2C_1$
$t_{high} = 0.693(R_1 + R_2)C_1$

The duty cycle (the fraction of the time the output is high) is given by

$$\text{Duty cycle} = \frac{t_{high}}{t_{high} + t_{low}}$$

The frequency of the output waveform is

$$f = \frac{1}{t_{high} + t_{low}} = \frac{1.44}{(R_1 + 2R_2)C_1}$$

For reliable operation, the resistors should be between approximately 10 kΩ and 14 MΩ, and the timing capacitor should be from around 100 pF to 1000 µF. The graph will give you a general idea of how the frequency responds to the component values.

FIGURE 10.7

Low-Duty-Cycle Operation (Astable Mode)

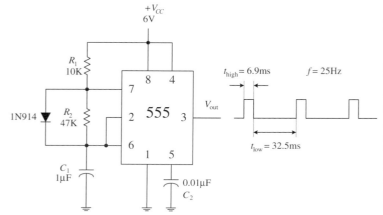

Now there is a slight problem with the last circuit—you cannot get a duty cycle that is below 0.5 (or 50 percent). In other words, you cannot make t_{high} shorter than t_{low}. For this to occur, the R_1C_1 network (used to generate t_{low}) would have to be larger the $(R_1 + R_2)C_1$ network (used to generate t_{high}). Simple arithmetic tells us that this is impossible; $(R_1 + R_2)C_1$ is always greater than R_1C_1. How do you remedy this situation? You attach a diode across R_2, as shown in the figure. With the diode in place, as the capacitor is charging (generating t_{high}), the preceding time constant $(R_1 + R_2)C_1$ is reduced to R_1C_1 because the charging current is diverted around R_2 through the diode. With the diode in place, the high and low times become

FIGURE 10.8

$t_{high} = 0.693(10K)(1\,\mu F) = 6.9\,ms$
$t_{low} = 0.693(47K)(1\,\mu F) = 32.5\,ms$

$$f = \frac{1}{6.9\text{ ms} + 32.5\text{ ms}} = 25\text{ Hz}$$

$$\text{duty cycle} = \frac{6.9\text{ ms}}{6.9\text{ ms} + 32.5\text{ ms}} = 0.18$$

$t_{high} = 0.693R_1C_1$

$t_{low} = 0.693R_2C_1$

To generate a duty cycle of less than 0.5, or 50 percent, simply make R_1 less than R_2.

10.2.3 How a 555 Works (Monostable Operation)

Figure 10.9 shows a 555 hooked up in the monostable configuration (one-shot mode). Unlike the astable mode, the monostable mode has only one stable state. This means that for the output to switch states, an externally applied signal is needed.

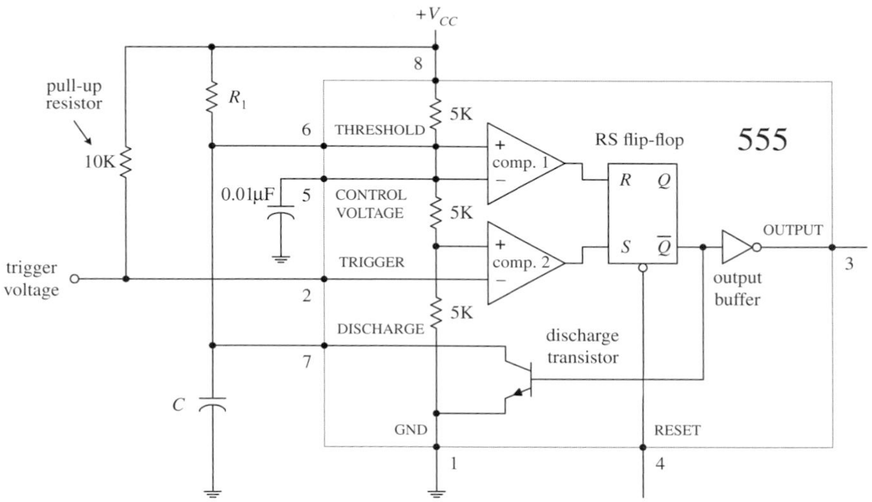

FIGURE 10.9

In the monostable configuration, initially (before a trigger pulse is applied) the 555's output is low, while the discharge transistor is on, shorting pin 7 to ground and keeping C discharged. Also, pin 2 is normally held high by the 10-k pull-up resistor. Now, when a negative-going trigger pulse (less than $\frac{1}{3}V_{CC}$) is applied to pin 2, comparator 2 is forced high, which sets the flip-flop's $\overline{Q}$ to low, making the output high (due to the inverting buffer), while turning off the discharge transistor. This allows C to charge up via R_1 from 0 V toward V_{CC}. However, when the voltage across the capacitor reaches $\frac{2}{3}V_{CC}$, comparator 1's output goes high, resetting the flip-flop and making the output low, while turning on the discharge transistor, allowing the capacitor to quickly discharge toward 0 V. The output will be held in this stable state (low) until another trigger is applied.

10.2.4 Basic Monostable Operation

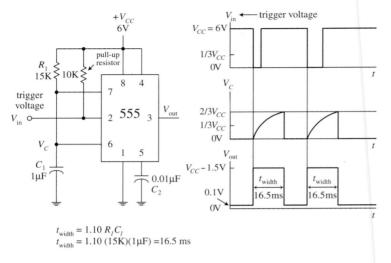

$$t_{width} = 1.10\,R_1 C_1$$
$$t_{width} = 1.10\,(15K)(1\mu F) = 16.5 \text{ ms}$$

FIGURE 10.10

The monostable circuit only has one stable state. That is, the output rests at 0 V (in reality, more like 0.1 V) until a negative-going trigger pulse is applied to the trigger lead—pin 2. (The negative-going pulse can be implemented by momentarily grounding pin 2, say, by using a pushbutton switch attached from pin 2 to ground.) After the trigger pulse is applied, the output will go high (around $V_{CC} - 1.5$ V) for the duration set by the $R_1 C_1$ network. Without going through the derivations, the width of the high output pulse is

$$t_{width} = 1.10 R_1 C_1$$

For reliable operation, the timing resistor R_1 should be between around 10 kΩ and 14 MΩ, and the timing capacitor should be from around 100 pF to 1000 μF.

10.2.5 Some Important Notes about 555 Timers

555 ICs are available in both bipolar and CMOS types. Bipolar 555s, like the ones you used in the preceding examples, use bipolar transistors inside, while CMOS 555s use MOSFET transistors instead. These two types of 555s also differ in terms of maximum output current, minimum supply voltage/current, minimum triggering current, and maximum switching speed. With the exception of maximum output current, the CMOS 555 surpasses the bipolar 555 in all regards. A CMOS 555 IC can be distinguished from a bipolar 555 by noting whether the part number contains a *C* somewhere within it (e.g., ICL7555, TLC555, LMC555, etc.). (Note that there are hybrid versions of the 555 that incorporate the best features of both the bipolar and CMOS technologies.) Table 10.1 shows specifications for a few 555 devices.

TABLE 10.1 Sample Specifications for Some 555 Devices

TYPE	SUPPLY VOLTAGE MIN. (V)	SUPPLY VOLTAGE MAX. (V)	SUPPLY CURRENT (V_{CC} = 5 V) TYP. (μA)	SUPPLY CURRENT (V_{CC} = 5 V) MAX. (μA)	TRIG. CURRENT (THRES. CURRENT) TYP. (nA)	TRIG. CURRENT (THRES. CURRENT) MAX. (nA)	TYPICAL FREQUENCY (MHZ)	$I_{out,max}$ (V_{CC} = 5 V) SOURCE (mA)	$I_{out,max}$ (V_{CC} = 5 V) SINK (mA)
SN555	4.5	18	3000	5000	100	500	0.5	200	200
ICL7555	2	18	60	300	—	10	1	4	25
TLC555	2	18	170	—	0.01	—	2.1	10	100
LMC555	1.5	15	100	250	0.01	—	3	—	—
NE555	4.5	15	—	6000	—	—	—	—	200

If you need more than one 555 timer per IC, check out the 556 (dual version) and 558 (quad version). The 556 contains two functionally independent 555 timers that share a common supply lead, while the 558 contains four slightly simplified 555 timers. In the 558, not all functions are brought out to the pins, and in fact, this device is intended to be used in monostable mode—although it can be tricked into astable mode with a few alterations (see manufacturer's literature for more information).

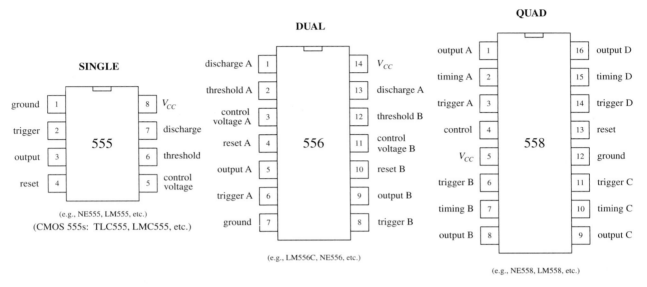

FIGURE 10.11

PRACTICAL TIP

To avoid problems associated with false triggering, connect the 555's pin 5 to ground through a 0.01-μF capacitor (we applied this trick already in this section). Also, if the power supply lead becomes long or the timer does not seem to function for some unknown reason, try attaching a 0.1-μF or larger capacitor between pins 8 and 1.

10.2.6 Simple 555 Applications

Relay Driver (Delay Timer)

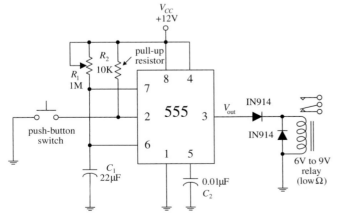

FIGURE 10.12

The monostable circuit shown here acts as a delay timer that is used to actuate a relay for a given duration. With the pushbutton switch open, the output is low (around 0.1 V), and the relay is at rest. However, when the switch is momentarily closed, the 555 begins its timing cycle; the output goes high (in this case ~10.5 V) for a duration equal to

$$t_{delay} = 1.10 R_1 C_1$$

The relay will be actuated for the same time duration. The diodes help prevent damaging current surges—generated when the relay switches states—from damaging the 555 IC, as well as the relay's switch contacts.

LED and Lamp Flasher and Metronome

LED Flasher

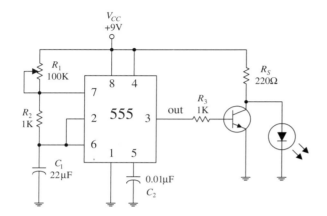

Lamp Flasher

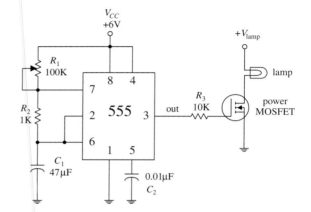

Metronome

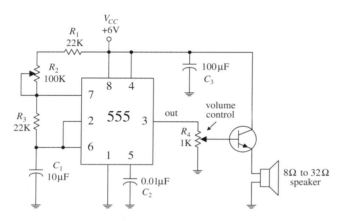

All these circuits are oscillator circuits (astable multivibrators). In the LED flasher circuit, a transistor is used to amplify the 555's output in order to provide sufficient current to drive the LED, while R_S is used to prevent excessive current from damaging the LED. In the lamp-flasher circuit, a MOSFET amplifier is used to control current flow through the lamp. A power MOSFET may be needed if the lamp draws a considerable amount of current. The metronome circuit produces a series of "clicks" at a rate determined by R_2. To control the volume of the clicks, R_4 can be adjusted.

FIGURE 10.13

10.3 Voltage-Controlled Oscillators

Besides the 555 timer IC, there are a number of other voltage-controlled oscillators (VCOs) on the market—some of which provide more than just a squarewave output. For example, the NE566 function generator is a very stable, easy-to-use triangular-wave and squarewave generator. In the 566 circuit below, R_1 and C_1 set the center frequency, while a control voltage at pin 5 varies the frequency; the control voltage is applied by means of a voltage-divider network (R_2, R_3, R_4). The output frequency of the 566 can be determined by using the formula shown in Fig. 10.14.

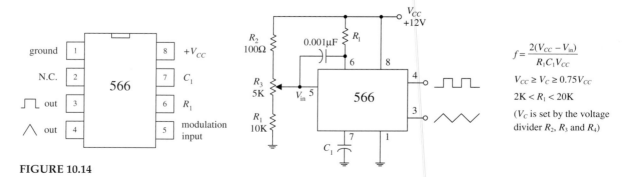

$$f = \frac{2(V_{CC} - V_{in})}{R_1 C_1 V_{CC}}$$

$$V_{CC} \geq V_C \geq 0.75 V_{CC}$$

$$2K < R_1 < 20K$$

(V_C is set by the voltage divider R_2, R_3 and R_4)

FIGURE 10.14

Other VCOs, such as the 8038 and the XR2206, can create a trio of output waveforms, including a sine wave (approximation of one, at any rate), a square wave, and triangular wave. Some VCOs are designed specifically for digital waveform generation and may use an external crystal in place of a capacitor for improved stability. To get a feel for what kinds of VCOs are out there, check the electronics catalogs.

10.4 Wien-Bridge and Twin-T Oscillators

A popular *RC*-type circuit used to generate low-distortion sinusoidal waves at low to moderate frequencies is the Wien-bridge oscillator. Unlike the oscillator circuits discussed already in this chapter, this oscillator uses a different kind of mechanism to provide oscillation, namely, a frequency-selective filter network.

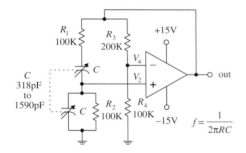

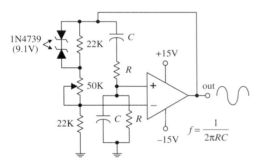

FIGURE 10.15

The heart of the Wien-bridge oscillator is its frequency-selective feedback network. The op amp's output is fed back to the inputs in phase. Part of the feedback is positive (makes its way through the frequency-selective *RC* branch to the noninverting terminal), while the other part is negative (is sent through the resistor branch to the inverting input of the op amp). At a particular frequency $f_0 = 1/(2\pi RC)$, the inverting input voltage (V_4) and the noninverting input voltage (V_2) will be equal and in phase—the positive feedback will cancel the negative feedback, and the circuit will oscillate. At any other frequency, V_2 will be too small to cancel V_4, and the circuit will not oscillate. In this circuit, the gain must be set to +3. The resistors must satisfy the condition $R3/R4 = 2$ (which gives a noninverting gain of 3). Anything less than this value will cause oscillations to cease; anything more will cause the output to saturate. With the component values listed in the figure, this oscillator can cover a frequency range of 1 to 5 kHz. The frequency can be adjusted by means of a two-ganged variable-capacitor unit.

The second circuit shown in the figure is a slight variation of the first. Unlike the first circuit, the positive feedback must be greater than the negative feedback to sustain oscillations. The potentiometer is used to adjust the amount of negative feedback, while the *RC* branch controls the amount of positive feedback based on the operating frequency. Now, since the positive feedback is larger than the negative feedback, you have to contend with the "saturation problem," as encountered in the last example. To prevent saturation, two zener diodes placed face to face (or back to back) are connected across the upper 22-kΩ resistor. When the output voltage rises above the zener's breakdown voltage, one or the other zener diode conducts, depending on the polarity of the feedback. The conducting zener diode shunts the 22-kΩ resistor, causing the resistance of the negative feedback circuit to decrease. More negative feedback is applied to the op amp, and the output voltage is controlled to a certain degree.

10.5 *LC* Oscillators (Sinusoidal Oscillators)

When it comes to generating high-frequency sinusoidal waves, commonly used in radiofrequency applications, the most common approach is to use an *LC* oscillator. The *RC* oscillators discussed so far have difficulty handling high frequencies, mainly because it is difficult to control the phase shifts of feedback signals sent to the amplifier input and because, at high frequencies, the capacitor and resistor values often become impractical to work with. *LC* oscillators, on the other hand, can use small inductances in conjunction with capacitance to create feedback oscillators that can reach frequencies up to around 500 MHz. However, it is important to note that at low frequencies (e.g., audio range), *LC* oscillators become highly unwieldy.

LC oscillators basically consist of an amplifier that incorporates positive feedback through a frequency-selective *LC* circuit (or tank). The *LC* tank acts to eliminate from the amplifier's input any frequencies significantly different from its natural resonant frequency. The positive feedback, along with the tank's resonant behavior, acts to promote sustained oscillation within the overall circuit. If this is a bit confusing, envision shock exciting a parallel *LC* tank circuit. This action will set the tank circuit into sinusoidal oscillation at the *LC*'s resonant frequency—the capacitor and inductor will "toss" the charge back and forth. However, these oscillations will die out naturally due to internal resistance and loading. To sustain the oscillation, the amplifier is used. The amplifier acts to supply additional energy to the tank circuit at just the right moment to sustain oscillations. Here is a simple example to illustrate the point.

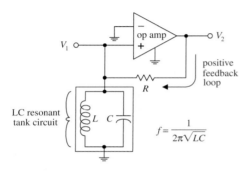

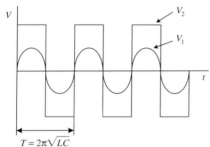

$$f = \frac{1}{2\pi\sqrt{LC}}$$

FIGURE 10.16

Here, an op amp incorporates positive feedback that is altered by an *LC* resonant filter or tank circuit. The tank eliminates from the noninverting input of the op amp any frequencies significantly different from the tank's natural resonant frequency:

$$f = \frac{1}{2\pi\sqrt{LC}}$$

(Recall from Chap. 2 that a parallel *LC* resonant circuit's impedance becomes large at the resonant frequency but falls off on either side, allowing the feedback signal to be filtered out to ground.) If a sinusoidal voltage set at the resonant frequency is present at V_1, the amplifier is alternately driven to saturation in the positive and negative directions, resulting in a square wave at the output V_2. This square wave has a strong fundamental Fourier component at the resonant frequency, part of which is fed back to the noninverting input through the resistor to keep oscillations from dying out. If the initially applied sinusoidal voltage at V_1 is removed, oscillations will continue, and the voltage at V_1 will be sinusoidal. Now in practice (considering real-life components and not theoretical models), it is not necessary to apply a sine wave to V_1 to get things going (this is a fundamentally important point to note). Instead, due to imperfections in the amplifier, the oscillator will self-start. Why? With real amplifiers, there is always some inherent noise present at the output even when the inputs of the amplifier are grounded (see Chap. 8). This noise has a Fourier component at the resonant frequency, and because of the positive feedback, it rapidly grows in amplitude (perhaps in just a few cycles) until the output amplitude saturates.

Now, in practice, *LC* oscillators usually do not incorporate op amps into their designs. At very high frequencies (e.g., RF range), op amps tend to become unreliable due to slew-rate and bandwidth limitations. When frequencies above around 100 kHz are needed, it is essential to use another kind of amplifier arrangement. For high-frequency applications, what is typically used is a transistor amplifier (e.g., bipolar or FET type). The switching speeds for transistors can be incredibly high—a 2000-MHz ceiling is not uncommon for special RF transistors. However, when using a transistor amplifier within an oscillator, there may be a slight problem to contend with—one that you did not have to deal with when you used the op amp. The problem stems from the fact that transistor-like amplifiers often take their outputs at a location where the output happens to be 180° out of phase with its input (see Chap. 4). However, for the feedback to sustain oscillations, the output must be in phase with the input. In certain *LC* oscillators this must be remedied by incorporating a special phase-shifting network between the output and input of the amplifier. Let's take a look at a few popular *LC* oscillator circuits.

Hartley *LC* Oscillator

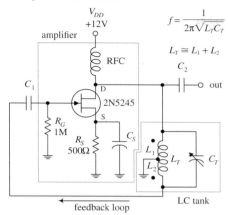

$$f = \frac{1}{2\pi\sqrt{L_T C_T}}$$

$$L_T \cong L_1 + L_2$$

A Hartley oscillator uses an inductive voltage divider to determine the feedback ratio. The Hartley oscillator can take on a number of forms (FET, bipolar, etc.)—a JFET version is shown here. This oscillator achieves a 180° phase shift needed for positive feedback by means of a tapped inductor in the tank circuit. The phase voltage at the two ends of the inductor differ by 180° with respect to the ground tap. Feedback via L_2 is coupled through C_1 to the base of the transistor amplifier. (The tapped inductor is basically an autotransformer, where L_1 is the primary and L_2 is the secondary.) The frequency of the Hartley is determined by the tank's resonant frequency:

$$f = \frac{1}{2\pi\sqrt{L_T C_T}}$$

This frequency can be adjusted by varying C_T. R_G acts as a gate-biasing resistor to set the gate voltage. R_S is the source resistor. C_S is used to improve amplifier stability, while C_1 and C_2 act as a dc-blocking capacitor that provides low impedance at the oscillator's operating frequency while preventing the transistor's dc operating point from being disturbed. The radiofrequency choke (RFC) aids in providing the amplifier with a steady dc supply while eliminating unwanted ac disturbances.

The second circuit is another form of the Hartley oscillator that uses a bipolar transistor instead of a JFET as the amplifier element. The frequency of operation is again determined by the resonant frequency of the *LC* tank. Notice that in this circuit the load is coupled to the oscillator via a transformer's secondary.

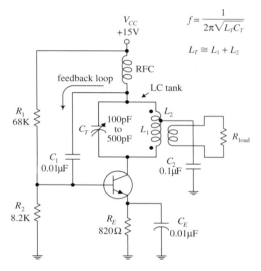

$$f = \frac{1}{2\pi\sqrt{L_T C_T}}$$

$$L_T \cong L_1 + L_2$$

FIGURE 10.17

Colpitts *LC* Oscillator

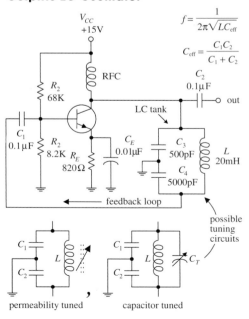

$$f = \frac{1}{2\pi\sqrt{LC_{\text{eff}}}}$$

$$C_{\text{eff}} = \frac{C_1 C_2}{C_1 + C_2}$$

The Colpitts oscillator is adaptable to a wide range of frequencies and can have better stability than the Hartley. Unlike the Hartley, feedback is obtained by means of a tap between two capacitors connected in series. The 180° phase shift required for sustained oscillation is achieved by using the fact that the two capacitors are in series; the ac circulating current in the *LC* circuit (see Chap. 2) produces voltage drops across each capacitor that are of opposite signs—relative to ground—at any instant in time. As the tank circuit oscillates, its two ends are at equal and opposite voltages, and this voltage is divided across the two capacitors. The signal voltage across C_4 is then connected to the transistor's base via coupling capacitor C_1, which is part of the signal from the collector. The collector signal is applied across C_3 as a feedback signal whose energy is coupled into the tank circuit to compensate for losses. The operating frequency of the oscillator is determined by the resonant frequency of the *LC* tank:

$$f = \frac{1}{2\pi\sqrt{LC_{\text{eff}}}}$$

where C_{eff} is the series capacitance of C_3 and C_4:

$$\frac{1}{C_{\text{eff}}} = \frac{1}{C_1} + \frac{1}{C_2}$$

FIGURE 10.18

C_1 and C_2 are dc-blocking capacitors, while R_1 and R_2 act to set the bias level of the transistor. The RFC choke is used to supply steady dc to the amplifier. This circuit's tank can be exchanged for one of the two adjustable tank networks. One tank uses permeability tuning (variable inductor), while the other uses a tuning capacitor placed across the inductor to vary the resonant frequency of the tank.

Clapp Oscillator

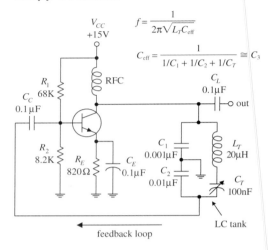

$$f = \frac{1}{2\pi\sqrt{L_T C_{\text{eff}}}}$$

$$C_{\text{eff}} = \frac{1}{1/C_1 + 1/C_2 + 1/C_T} \cong C_3$$

FIGURE 10.19

feedback loop

The Clapp oscillator has exceptional frequency stability. It is a simple variation of the Colpitts oscillator. The total tank capacitance is the series combination of C_1 and C_2. The effective inductance L of the tank is varied by changing the net reactance by adding and subtracting capacitive reactance via C_T from inductive reactance of L_T. Usually C_1 and C_2 are much larger than C_T, while L_T and C_T are series resonant at the desired frequency of operation. C_1 and C_2 determine the feedback ratio, and they are so large compared with C_T that adjusting C_T has almost no effect on feedback. The Clapp oscillator achieves its reputation for stability since stray capacitances are swamped out by C_1 and C_2, meaning that the frequency is almost entirely determined by L_T and C_T. The frequency of operation is determined by

$$f = \frac{1}{2\pi\sqrt{L_T C_{\text{eff}}}}$$

where C_{eff} is

$$C_{\text{eff}} = \frac{1}{1/C_1 + 1/C_2 + 1/C_T} \approx C_3$$

10.6 Crystal Oscillators

When stability and accuracy become critical in oscillator design—which is often the case in high-quality radio and microprocessor applications—one of the best approaches is to use a crystal oscillator. The stability of a crystal oscillator (from around 0.01 to 0.001 percent) is much greater than that of an *RC* oscillator (around 0.1 percent) or an *LC* oscillator (around 0.01 percent at best).

When a quartz crystal is cut in a specific manner and placed between two conductive plates that act as leads, the resulting two-lead device resembles an *RLC* tuned resonant tank. When the crystal is shock-excited by either a physical compression or an applied voltage, it will be set into mechanical vibration at a specific frequency and will continue to vibrate for some time, while at the same time generating an ac voltage between its plates. This behavior, better know as the *piezoelectric effect,* is similar to the damped electron oscillation of a shock-excited *LC* circuit. However, unlike an *LC* circuit, the oscillation of the crystal after the initial shock excitation will last longer—a result of the crystal's naturally high *Q* value. For a high-quality crystal, a *Q* of 100,000 is not uncommon. *LC* circuits typically have a *Q* of around a few hundred.

The *RLC* circuit shown in Fig. 10.20 is used as an equivalent circuit for a crystal. The lower branch of the equivalent circuit, consisting of R_1, C_1, and L_1 in series, is called

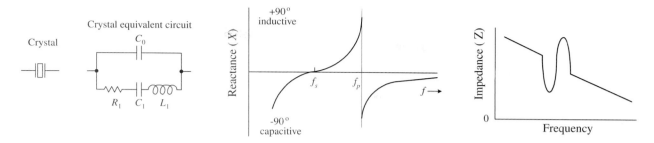

FIGURE 10.20

the *motional arm*. The motional arm represents the series mechanical resonance of the crystal. The upper branch containing C_0 accounts for the stray capacitance in the crystal holder and leads. The *motional inductance* L_1 is usually many henries in size, while the motional capacitance C_1 is very small (<<1 pF). The ratio of L_1 to C_1 for a crystal is much higher than could be achieved with real inductors and capacitors. Both the internal resistance of the crystal R_1 and the value of C_0 are both fairly small. (For a 1-MHz crystal, the typical components values within the equivalent circuit would be $L_1 = 3.5$ H, $C_1 = 0.007$ pF, $R_1 = 340$ Ω, $C_0 = 3$ pF. For a 10-MHz fundamental crystal, the typical values would be $L_1 = 9.8$ mH, $C_1 = 0.026$ pF, $R_1 = 7$ Ω, $C_0 = 6.3$ pF.)

In terms of operation, a crystal can be driven at *series resonance* or *parallel resonance*. In series resonance, when the crystal is driven at a particular frequency, called the *series resonant frequency f_s*, the crystal resembles a series-tuned resonance LC circuit; the impedance across it goes to a minimum—only R_1 remains. In parallel resonance, when the crystal is driven at what is called the *parallel resonant frequency f_p*, the crystal resembles a parallel-tuned LC tank; the impedance across it peaks to a high value (see the graphs in Fig. 10.20).

Quartz crystals come in series-mode and parallel-mode forms and may either be specified as a fundamental-type or an overtone-type crystal. Fundamental-type crystals are designed for operation at the crystal's fundamental frequency, while overtone-type crystals are designed for operation at one of the crystal's overtone frequencies. (The fundamental frequency of a crystal is accompanied by harmonics or overtone modes, which are odd multiples of the fundamental frequency. For example, a crystal with a 15-MHz fundamental also will have a 45-MHz third overtone, a 75-MHz fifth overtone, a 135-MHz ninth overtone, etc. Figure 10.21 below shows an equivalent RLC circuit for a crystal, along with a response curve, both of which take into account the overtones.) Fundamental-type crystals are available from around 10 kHz to 30 MHz, while overtone-type crystals are available up to a few hundred megahertz. Common frequencies available are 100 kHz and 1.0, 2.0, 4, 5, 8, and 10 MHz.

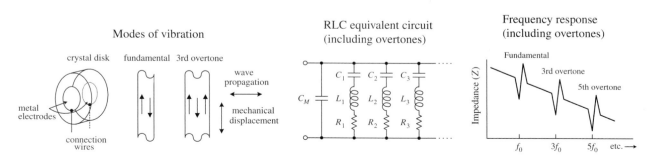

FIGURE 10.21

Designing crystal oscillator circuits is similar to designing *LC* oscillator circuits, except that now you replace the *LC* tank with a crystal. The crystal will supply positive feedback and gain at its series or parallel resonant frequency, hence leading to sustained oscillations. Here are a few basic crystal oscillator circuits to get you started.

Basic Crystal Oscillator

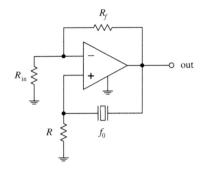

The simple op amp circuit shown here resembles the *LC* oscillator circuit in Fig. 10.16, except that it uses the series resonance of the crystal instead of the parallel resonance of an *LC* circuit to provide positive feedback at the desired frequency. Other crystal oscillators, such as the Pierce oscillator, Colpitts oscillator, and a CMOS inverter oscillator, shown below, also incorporate a crystal as a frequency-determining component. The Pierce oscillator, which uses a JFET amplifier stage, employs a crystal as a series-resonant feedback element; maximum positive feedback from drain to gate occurs only at the crystal's series-resonant frequency. The Colpitts circuit, unlike the Pierce circuit, uses a crystal in the parallel feedback arrangement; maximum base-emitter voltage signal occurs at the crystal's parallel-resonant frequency. The CMOS circuit uses a pair of CMOS inverters along with a crystal that acts as a series-resonant feedback element; maximum positive feedback occurs at the crystal's series resonant frequency.

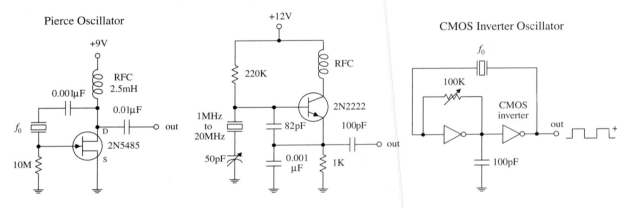

FIGURE 10.22

There are a number of ICs available that can make designing crystal oscillators a breeze. Some of these ICs, such as the 74S124 TTL VCO (squarewave generator), can be programmed by an external crystal to output a waveform whose frequency is determined by the crystal's resonant frequency. The MC12060 VCO, unlike the 74S124, outputs a pair of sine waves. Check the catalogs to see what other types of oscillator ICs are available.

Now there are also crystal oscillator modules that contain everything (crystal and all) in one single package. These modules resemble a metal-like DIP package, and they are available in many of the standard frequencies (e.g., 1, 2, 4, 5, 6, 10, 16, 24, 25, 50, and 64 MHz, etc.). Again, check out the electronics catalogs to see what is available.

10.7 Microcontroller Oscillators

In Chap. 13, we will also see how a microcontroller can be used to generate a waveform using a digital-to-analog convertor. The basic technique is to store the waveform in memory and then play it through the digital-to-analog converter.

In the case where just a squarewave is required, a simple 8-pin microcontroller with a built-in clock can be an effective alternative to a 555 timer, requiring fewer external components.

CHAPTER 11

Voltage Regulators and Power Supplies

Circuits usually require a dc power supply that can maintain a fixed voltage while supplying enough current to drive a load. Batteries make good dc supplies, but their relatively small current capacities make them impractical for driving high-current, frequently used circuits. An alternative solution is to take a 120-V ac, 60-Hz line voltage and convert it into a usable dc voltage.

There are two approaches to this. The more traditional approach is to use a step-down transformer. The other approach is to use a "switch-mode" power supply. This latter method has all but taken over from step-down transformers in recent years, and is the reason your "wall-wart" black plastic power adapters have become smaller and lighter. They also have the advantage that they will often work without modification or switch flipping on the higher line voltages found in other parts of the world. We'll discuss both methods in this chapter, beginning with the step-down transformer approach.

The trick to converting the ac line voltage into a usable (typically lower-level) dc voltage is to first use a transformer to step down the ac voltage. After that, the transformed voltage is applied through a rectifier network to get rid of the negative swings (or positive swings if you are designing a negative voltage supply). Once the negative swings are eliminated, a filter network is used to flatten out the rectified signal into a nearly flat (rippled) dc voltage pattern. Figure 11.1 shows the process in action.

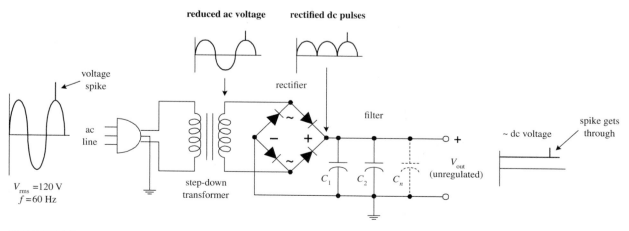

FIGURE 11.1

Now there is one problem with this supply—it is *unregulated*. This means that if there are any sudden surges within the ac input voltage (spikes, dips, etc.), these variations will be expressed at the supply's output (notice the spike that gets through in Fig. 11.1). Using an unregulated supply to run sensitive circuits (e.g., digital IC circuits) is a bad idea. The current spikes can lead to improper operating characteristics (e.g., false triggering, etc.) and may destroy the ICs in the process. An unregulated supply also has a problem maintaining a constant output voltage as the load resistance changes. If a highly resistive (low-current) load is replaced with a lower-resistance (high-current) load, the unregulated output voltage will drop (Ohm's law).

Fortunately, there is a special circuit that can be placed across the output of an unregulated supply to convert it into a regulated supply—a supply that eliminates the spikes and maintains a constant output voltage with load variations (see Fig. 11.2). This special circuit is called a *voltage regulator*.

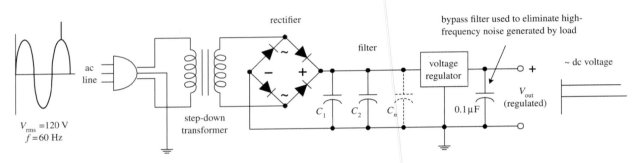

FIGURE 11.2

A voltage regulator is designed to automatically adjust the amount of current flowing through a load—so as to maintain a constant output voltage—by comparing the supply's dc output with a fixed or programmed internal reference voltage. A simple regulator consists of a sampling circuit, an error amplifier, a conduction element, and a voltage reference element (see Fig. 11.3).

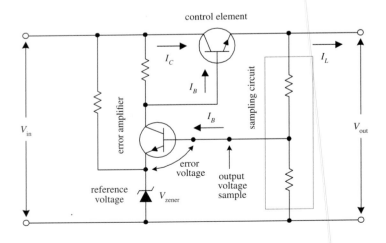

FIGURE 11.3

The regulator's sampling circuit (voltage divider) monitors the output voltage by feeding a *sample voltage* back to the error amplifier. The reference voltage element (zener diode) acts to maintain a constant *reference voltage* that is used by the error amplifier. The error amplifier compares the output sample voltage with the reference voltage and then generates an *error voltage* if there is any difference between the two. The error amplifier's output is then fed to the current-control element (transistor), which is used to control the load current.

In practice, you do not have to worry about designing voltage-regulator circuits from scratch. Instead, what you do is spend 50 cents for a voltage-regulator IC. Let's take a closer look at these integrated devices.

11.1 Voltage-Regulator ICs

There are a number of different kinds of voltage-regulator ICs on the market today. Some of these devices are designed to output a fixed positive voltage, some are designed to output a fixed negative voltage, and others are designed to be adjustable.

11.1.1 Fixed-Regulator ICs

Positive voltage regulator

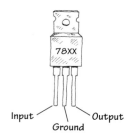

positive voltage regulator

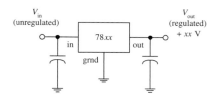

Negative voltage regulator

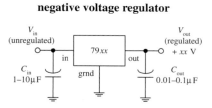

FIGURE 11.4

One popular line of regulators includes the three-terminal LM78xx series shown here. The "xx" digits represent the output voltage, e.g., 7805 (5 V), 7806 (6 V), 7808 (8 V), 7810 (10 V), 7812 (12 V), 7815 (15 V), 7818 (18 V), and 7824 (24 V). These devices can handle a maximum output current of 1.5 A if properly heat-sunk. To remove unwanted input or output spikes/noise, capacitors can be attached to the regulator's input and output terminals, as shown in the figure. A popular series of negative voltage-regulator IC is the LM79xx regulators, where "xx" represent the negative output voltage. These devices can handle a maximum output current of 1.5 A. A number of different manufacturers make their own kinds of voltage regulators. Some of the regulators can handle more current than others.

These devices are also available as SMD parts, typically in a SOT-89 package. Be sure to check the data sheets though, as these parts often have a lower maximum output current than their through-hole siblings.

Check out the catalogs to see what is available.

11.1.2 Adjustable-Regulator ICs

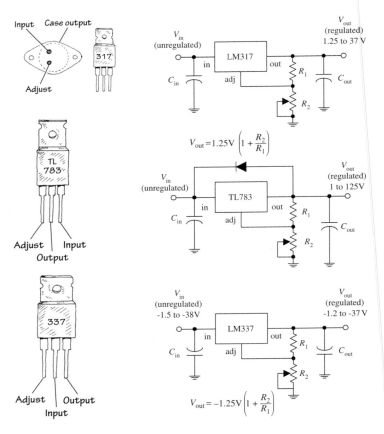

$$V_{out} = 1.25V\left(1 + \frac{R_2}{R_1}\right)$$

$$V_{out} = -1.25V\left(1 + \frac{R_2}{R_1}\right)$$

FIGURE 11.5

The LM317 regulator shown here is a popular 3-terminal adjustable positive voltage regulator. Unlike the 7800 fixed voltage-regulator series, the LM317 is a floating regulator—it sees only the input-to-output differential voltage—and it can be programmed via two external resistors to set the output voltage. In operation, the LM317 develops a nominal 1.25 V reference voltage between the output and adjust terminals. This reference voltage is impressed across program resistor R_1, and since this voltage is constant, a constant current I_1 flows through the output set resistor R_2, giving an output voltage given by the equation shown in the figure. Increasing R_2 forces the regulator's output to a higher level. The LM317 is designed to accept an unregulated input voltage of up to 37 V and can output a maximum current of 1.5 A. The TL783 is another positive adjustable regulator that can output a regulated voltage of from 1 to 125 V, with a maximum output current of 700 mA. The LM337T, unlike the previous two regulators, is an adjustable negative voltage regulator. It can output a regulated voltage of from –1.2 V to –37 V, with a maximum output current of 1.5 A. Again, check the electronics catalogs to see what other kinds of adjustable regulators are available. (C_{in} should be included if the regulator is far from the power source; it should be around 0.1 µF or so. C_{out} is used to eliminate voltage spikes at the output; it should be around 0.1 µF or larger.)

11.1.3 Regulator Specifications

The specifications tables for regulators typically will provide you with the following information: output voltage, accuracy (percent), maximum output current, power dissipation, maximum and minimum input voltage, 120-Hz ripple rejection (decibels), temperature stability ($\Delta V_{out}/\Delta T$), and output impedance (at specific frequencies). A regulator's ripple rejection feature can greatly reduce voltage variations in a power supply's output, as you will discover later in this chapter.

A voltage regulator from the 78xx or for that matter, a variable regulator like the LM317 needs the input voltage to be at least 2 V higher than the regulated output. LDO (Low Drop Out) voltage regulators reduce this 2 V to just 0.5 V in some circumstances. This has the advantage that the regulator can run from a lower input voltage and therefore run a lot cooler. For an example, LDO regulator take a look at the data sheet for an LM2940.

11.2 A Quick Look at a Few Regulator Applications

Before we take a look at how voltage regulators are used in power supplies, it is worthwhile seeing how they are used in other types of applications. Here are a few examples.

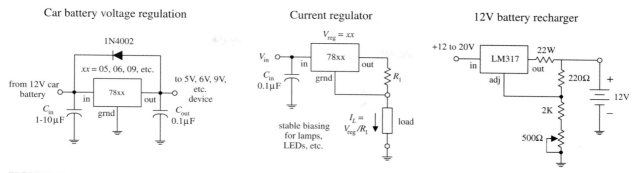

Car battery voltage regulation   Current regulator   12V battery recharger

FIGURE 11.6

The constant current regulator is often used as a power supply for LEDs, especially the higher-power devices.

11.3   The Transformer

It is important that you choose the right transformer for your power supply. The transformer's secondary voltage should not be much larger than the output voltage of the regulator; otherwise, energy will be wasted because the regulator will be forced to dissipate heat. However, at the same time, the secondary voltage must not drop below the required minimum input voltage of the regulator (typically 2 to 3 V above its output voltage).

11.4   Rectifier Packages

Three basic rectifier networks used in power supply designs include the half-wave, full-wave, and bridge rectifiers, shown in Fig. 11.7. To understand how these rectifiers work, see Chap. 4.

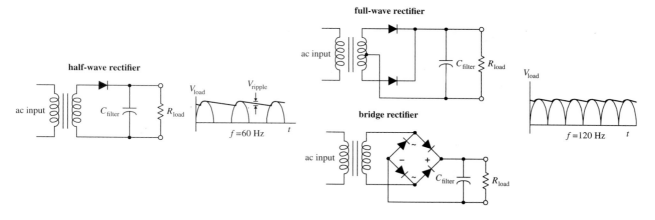

FIGURE 11.7

Half-wave, full-wave, and bridge rectifiers can be constructed entirely from individual diodes. However, both full-wave and bridge rectifiers also come in preassembled packages (see Fig. 11.8).

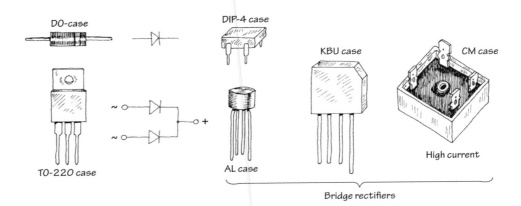

FIGURE 11.8

Bridge rectifiers

Make sure that the power supply's rectifier diodes have the proper current and peak-inverse-voltage (PIV) ratings. Typical rectifier diodes have current ratings from 1 to 25 A, PIV ratings from 50 to 1000 V, and surge-current ratings from around 30 to 400 A. Popular general-purpose rectifier diodes include the 1N4001 to 1N4007 series (rated at 1 A, 0.9-V forward voltage drop), the 1N5059 to 1N5062 series (rated at 2 A, 1.0-V forward voltage drop), the 1N5624 to 1N5627 series (rated at 5 A, 1.0-V forward voltage drop), and the 1N1183A-90A (rated at 40 A, 0.9-V forward voltage drop). For low-voltage applications, Schottky barrier rectifiers can be used; the voltage drop across these rectifiers is smaller than a typical rectifier (typically less than 0.4 V); however, their breakdown voltages are significantly smaller. Popular full-wave bridge rectifiers include the 3N246 to 3N252 series (rated at 1 A, 0.9-V forward voltage drop) and the 3N253 to 3N259 series (rated at 2 A, 0.85-V forward voltage drop).

11.5 A Few Simple Power Supplies

Regulated +5-V Supplies

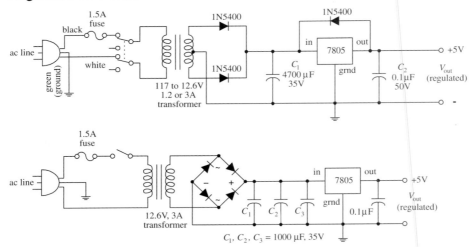

The first supply uses a center-tapped transformer rated at 12.6 V at 1.2 to 3 A. The voltage after rectification resides at an 8.9-V peak pulse. The filter capacitor (C_1) smoothes the pulses, and the 7805 outputs a regulated +5 V. C_2 is placed across the output of the regulator to bypass high-frequency noise that might be generated by the load. The diode placed across the 7805 helps protect the

FIGURE 11.9

regulator from damaging reverse-current surges generated by the load. Such surges may result when the power supply is turned off. For example, the capacitance across the output may discharge more slowly than the capacitance across the input. This would reverse-bias the regulator and could damage it in the process. The diode diverts the unwanted current away from the regulator. The second power supply is similar to the first but uses a bridge rectifier.

Dual Polarity Power Supply with Variable Output (± 1.2 to 35V)

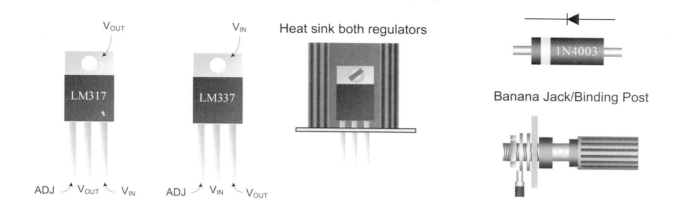

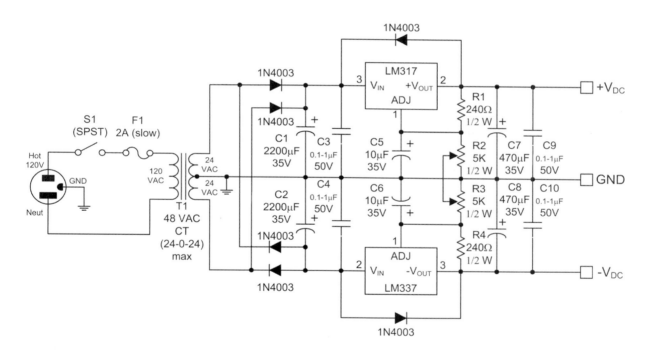

FIGURE 11.10a This dual-polarity linear power supply will provide any positive or negative voltage between 1.2 and 35 V. The complementary regulators—LM317 (+) and LM337 (−)—can deliver up to 1.5 A (output voltage dependent) if provided adequate heat sinking. This is sufficient for testing or powering a wide variety of everyday circuits. The key is the center-tap transformer with secondary tapped to ground that supplies both positive and negative voltages relative to ground. It converts the line voltage from 120 VAC to 48 VAC at the secondary, which is center-tapped and divided into 24-VAC portions. By using the center tap as a ground or common connection, it is possible to get both positive and negative outputs relative to ground. The diodes rectify the ac from the transformer output into a pulsing dc waveform. C_1 and C_2 electrolytic capacitors perform bulk filtering of the pulsing dc waveform, resulting in a raw dc voltage. C_3 and C_4 are small film capacitors bypassing the electrolytic capacitors to improve transient response and filter high-frequency line noise. The LM317 (+) and LM337 (−) are complementary adjustable voltage regulators, whose output can be programmed with two external resistors. The output voltage is given by:

$$V_{out} = 1.25\left(1 + \frac{R_2}{R_1}\right)$$

C_5 and C_6 improve the regulators' ripple rejection from 65 dB to 80 dB by preventing ripple voltage from being amplified at the output of the regulator. C_7 and C_8 electrolytic capacitors stiffen the output voltage and reduce output impedance. C_9 and C_{10} are small bypass capacitors to filter any high-frequency noise present at the output. All bypass capacitors should have low impedance (e.g., polyester, polypropylene, polystyrene, or Mylar film capacitors are okay). A 48-VAC CT (24-0-24) transformer is about the practical upper limit of commonly available models. This limit is set by the maximum input voltage of the regulators, and also by the bulk filter capacitors that are rated at 35 VDC.

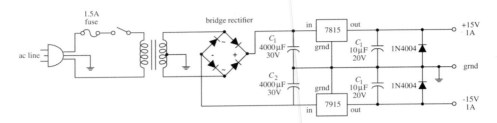

FIGURE 11.10b Here, a 7815 positive voltage regulator and a 7915 negative voltage regulator are used to construct a ±15-V supply.

Dual Polarity Power Supply (± 12 V)

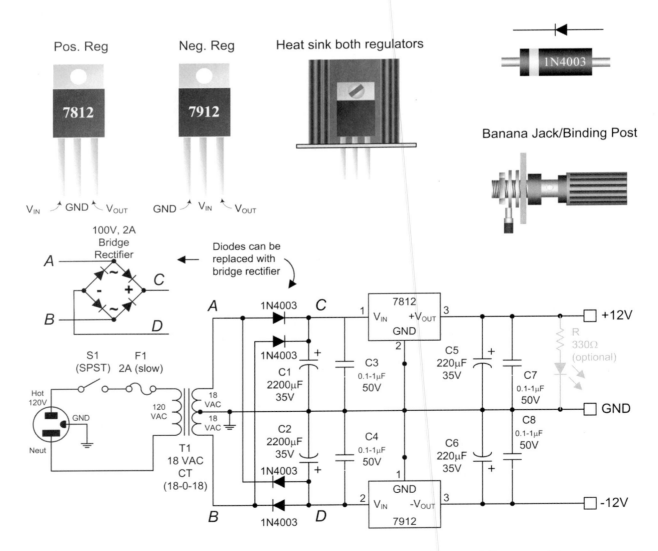

FIGURE 11.10c Here's a fixed +/−12 V made by using a center-tap 18-VAC transformer, a full-wave rectifier, and 7812 and 7912 regulators. This circuit is particularly useful when powering op amp circuits.

Adapter power engage, battery power disconnect circuit

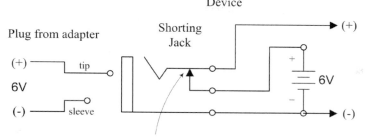

This figure shows a simple way to add an external dc power supply input to a battery-powered device. With no external power (external dc adapter plug not connected), the shorting jack acts to switch power to the battery. However, when external dc power is applied via a dc adapter plug, the shorting jack switches power from the battery to the dc adapter connects.

FIGURE 11.11

11.6 Technical Points about Ripple Reduction

When using a supply to power sensitive circuits, it is essential to keep the variation in output voltage as small as possible. For example, when driving digital circuits from a 5-V supply, the variation in output voltage should be no more than 5 percent, or 0.25 V, if not lower. In fact, digital logic circuits usually have a minimum 200-mV noise margin around critical logic levels. Small analog signal circuits can be especially finicky when it comes to output variations. For example, they may require a variation less than 1 percent to operate properly. How do you keep the output variation error low? You use filter capacitors and voltage regulators.

Filter capacitors act to reduce the fluctuations in output by storing charge during the positive-going rectifier cycle and then releasing charge through the load—at a slow enough rate to maintain a level output voltage—during the negative-going rectifier cycle. If the filter capacitor is too small, it will not be able to store enough charge to maintain the load current and output voltage during the negative-going cycle.

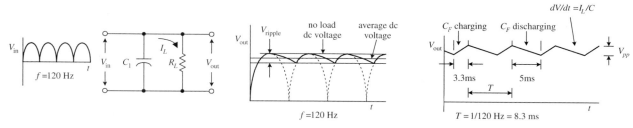

FIGURE 11.12

As it turns out, the amount of current drawn by a load influences the rate of capacitor discharge. If a low resistance, high-current load is placed across the supply's output, the capacitor will discharge relatively quickly, causing the voltage across the capacitor and therefore the voltage across the load to drop relatively quickly. On the other hand, a highly resistive, low-current load will cause the capacitor to discharge more slowly, which means the dip in output voltage will not be as significant. To calculate the drop in voltage of the capacitor during its discharge cycle, use

$$I = C\frac{dV}{dt} \approx C\frac{\Delta V}{\Delta t}$$

where I is the load current, Δt is the discharge time, and ΔV is the fluctuation in output above and below the average dc level. ΔV is also referred to as the *peak-to-peak ripple voltage* $V_{ripple(pp)}$. (Here, we cheated by substituting straight lines for exponential decays to describe the discharge cycle; see Fig. 11.12.) Δt can be approximated by dividing 1 by the rectified output voltage frequency. For a full-wave rectifier, the period is 1/120 Hz, or 8.3×10^{-3} s. In reality, the actual amount of time the capacitor spends discharging during a peak-to-peak variation is about 5 ms. The other 3.3 ms is used up during the charge. To make life easier on you, the following equation simplifies matters:

$$V_{ripple(rms)} = (0.0024 \text{ s}) \frac{I_L}{C_f}$$

Note that the ripple voltage is not given in the peak-to-peak form but is given in the rms form (recall that $V_{pp} = \sqrt{2} \, V_{rms}$). To test the equation out, let's find the ripple voltage for a 5-V supply that has a 4700-μF filter capacitor and a maximum load current of 1.0 A. (Here we are pretending that no voltage regulator is present at this point.) After plugging the numbers into the equation, you get $V_{ripple(rms)} = 510$ mV. But now, recall that a minute ago we said that the amount of variation in the output had to be ± 0.25 V% to run digital ICs—510 mV is too big. Now you could keep fiddling around with the capacitance value in the equation and come up with an even better answer, say, letting C equal infinity. In theory, this is fine, but in reality, it is not fine. It is not fine for three basic reasons. The first reason has to do with the simple fact that you cannot find a capacitor at Radio Shack that has an infinite capacitance. If an infinite-capacitance capacitor existed, the universe would not be the same, and you and we probably would not be around to talk about it. The second reason has to do with capacitor tolerances. Unfortunately, the high-capacity electrolytic capacitors used in power supplies have some of the worst tolerances among the capacitor families. It is not uncommon to see a 5 to 20 percent or even larger percentage tolerance for these devices. The mere fact that the tolerances are so bad makes being "nitpicky" about the equation a questionable thing to do. The third and perhaps most important reason for avoiding fiddling around with the equation too much has to do with the inherent ripple-rejection characteristics of voltage regulators. As you will see, the voltage regulator can save us.

Voltage regulators often come with a ripple-rejection parameter given in decibels. For example, the 7805 has a ripple-rejection characteristic of about 60 dB. Using the attenuation expression, you can find the extent of the ripple reduction:

$$-60 \text{ dB} = 20 \log_{10} \frac{V_{out}}{V_{in}}$$

$$-3 = \log_{10} \frac{V_{out}}{V_{in}}$$

$$10^{-3} = \frac{V_{out}}{V_{in}}$$

The last expression says that the output ripple is reduced by a factor of 1000. This means that if you use the regulator with the initial setup, you will only have an output ripple of 0.51 mV—a value well within the safety limits. At this point, it is important to note that the 7805 requires a minimum voltage difference between its input and output of 3 V to function properly. This means that to obtain a 5-V output, the input to the regulator must be at least 8 V. At the same time, it is important to note the voltage drop across the rectifier (typically around 1 to 2 V). The secondary voltage from the transformer therefore should be even larger than 8 V. A transformer with a secondary voltage of 12 V or so would be a suitable choice for the 5-V supply.

Now let's see how well the LM319 adjustable regulator rejects the ripple. Let's say that an LM317 is used in a power supply that has a transformer secondary rms voltage of 12.6 V. The capacitor's peak voltage during a cycle will be 17.8 V (the peak-to-peak voltage of the secondary). An LM317 has a ripple-rejection characteristic of around 65 dB, but this value can be raised to approximately 80 dB by bypassing the LM317's voltage divider with a 10-μF capacitor (see Fig. 11.13).

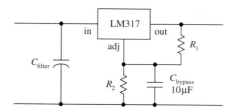

FIGURE 11.13

If you let $C = 4700 \ \mu F$ and assume a 1.5-A maximum load current, you get a ripple voltage of

$$V_{r(rms)} = 0.0024 \ s(1.5 \ A / 4700 \ \mu F) = 760 \ mV$$

Again, the ripple voltage is too much for sensitive ICs to handle. However, if you consider the LM319's ripple-rejection characteristic (assuming that you use the bypassing capacitor), you get a reduction of

$$-80 \ db = 20 \ \log_{10} \frac{V_{out}}{V_{in}}$$

$$-4 = \log_{10} \frac{V_{out}}{V_{in}}$$

$$10^{-4} = \frac{V_{out}}{V_{in}}$$

In other words, the output ripple is reduced by a factor of 10,000, so the final output ripple voltage is only 0.076 mV.

11.7 Loose Ends

Line Filter and Transients Suppressors

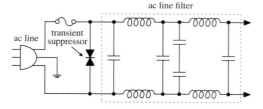

FIGURE 11.14

A line filter is an *LC* filter circuit that is inserted into a supply to filter out unwanted high-frequency interference present in the input line supply. Line filters also can help reduce voltage spikes, as well as help eliminate the emission of radiofrequency interference by the power supply. Line filters are placed before the transformer, as shown in the figure. AC line filters can be purchased in preassembled packages. See the electronics catalogs for more info.

A transient suppressor is a device that acts to short out when the terminal voltage exceeds safe limits (e.g., spikes). These devices act like bidirectional high-power zener diodes. They are inexpensive, come in diode-like packages, and come with low-voltage and peak-pulse-voltage ratings.

Overvoltage Protection

crowbar

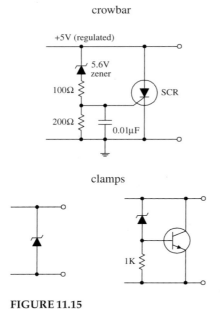

clamps

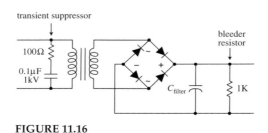

FIGURE 11.15

The crowbar and clamp circuits shown here can be placed across the output of a regulated supply to protect a load against an unregulated voltage that would be present at the output if the voltage regulator failed (shorted internally).

CROWBAR

For the crowbar circuit, when the supply voltage exceeds the zener diode's breakdown voltage by 0.6 V, the zener diode conducts, triggering the SCR into conduction. The SCR then diverts potentially harmful current to ground. The crowbar's SCR will not turn off until the power supply is turned off or the SCR's anode-to-cathode current is interrupted, say, by means of a switch.

CLAMP

A zener diode placed across the output of a supply also can be used for overvoltage protection. However, it may "fry" if the unregulated current is too large. To avoid frying the zener diode, use a high-power transistor to help divert the current. When the zener diode's breakdown voltage is exceeded, some of the current that flows through it will enter the transistor's base, allowing the excessive current to flow toward ground. Using clamps can eliminate false triggering caused by voltage spikes; the crowbar, on the other hand, would need to be reset in such a case.

Bleeder Resistors and Transient Suppressors

transient suppressor

bleeder resistor

FIGURE 11.16

When a resistor is placed across the output of the unregulated supply, it will act to discharge the high-voltage (potentially lethal) filter capacitor when the supply is turned off and the load removed. Such a resistor is referred to as a *bleeder resistor*. A 1-k, 1/2-W resistor is suitable for most applications.

An *RC* network placed across the primary coil of the transformer can prevent large, potentially damaging inductive transients from forming when the supply is turned off. The capacitor must have a high-voltage rating. A typical *RC* network consists of a 100-Ω resistor and a 0.1-μF, 1-kV capacitor. Special z-lead transient suppressor devices can also be used, as was mentioned earlier.

11.8 Switching Regulator Supplies (Switchers)

A switching power supply, or *switcher*, is a unique kind of power supply that can achieve power conversion efficiencies far exceeding those of the linear supplies covered earlier in this chapter. With linear regulated supplies, the regulator converts the input voltage that is higher than needed into a desired lower output voltage. To lower the voltage, the extra energy is dissipated as heat from the regulator's control element. The power-conversion efficiency (P_{out}/P_{in}) for these supplies is typically lower than 50 percent. This means that more than half the power is dissipated as heat.

Switchers, on the other hand, can achieve power-conversion efficiencies exceeding 85 percent, meaning that they are much more energy efficient than linearly regulated supplies. Switchers also have a wide current and voltage operating range and can be configured in either step-down (output voltage smaller than input voltage), step-up (output voltage larger than input voltage), or inverting (output is the opposite polarity of the input) configurations. Also, switchers can be designed to run directly off ac line power, without the need for a power transformer. By eliminating the hefty power transformer, the switcher can be made light and small. This makes switchers good supplies for computers and other small devices.

A switching supply resembles a linear supply in many ways. However, two unique features include an energy-storage inductor and a nonlinear regulator network. Unlike a linear supply, which provides regulation by varying the resistance of the regulator's control element, a switcher incorporates a regulation system in which the control element is switched on and off very rapidly. The on/off pulses are controlled by an oscillator/error amplifier/pulse-width modulator network (see Fig. 11.17).

Basic Switching Regulator

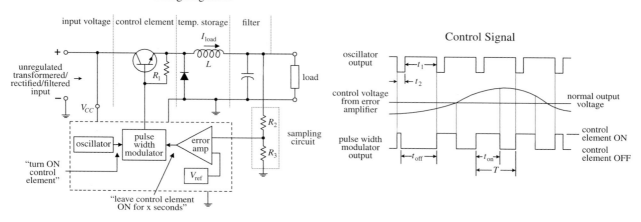

FIGURE 11.17

During the on cycle, energy is pumped into the inductor (energy is stored in the magnetic fields about the inductor's coil). When the control element is turned off, the stored energy in the inductor is directed by the diode into the filter and into the load. The sampling circuit (R_2 and R_3) takes a sample of the output voltage and feeds the sample to one of the inputs of the error amplifier. The error amplifier then compares the sample voltage with a reference voltage applied to its other input. If the sample voltage is below the reference voltage, the error amplifier increases its output control voltage. This control voltage is then sent to the pulse-width modulator. (If the sample voltage is above the reference voltage, the error amplifier will decrease the output voltage it sends to the modulator.) While this is going on, the oscillator is supplying a steady series of triggering voltage pulses to the pulse-width modulator. The modulator uses both the oscillator's pulses and the error amplifier's output to produce a modified on/off signal that is sent to the control element's base. The modified signal represents a squarewave whose on time is determined by the input error voltage. If the error voltage is low (meaning the sample voltage is higher than it should be), the modulator sends a short-duration on pulse to the control element. However, if the error voltage is high (meaning the sample voltage is lower than it should be), the pulse-width modulator sends a long-duration on pulse to the control element. (The graph in Fig. 11.17 shows how the oscillator, error amplifier, and pulse-width modulator outputs are related.) Using a series of on/off pulses that can be varied in frequency and duration gives the switching regulator its exceptional efficiency; releasing a series of short pulses of energy over time is more efficient than taking excessive supply energy and radiating it off as heat (linear supply).

Figure 11.18 shows a typical switching regulator arrangement. The 556 dual-timer IC houses both the oscillator and pulse-width modulator, while the UA723 voltage-regulator IC acts as the error amplifier. R_2 and R_3 comprise the sampling network, R_6 and R_7 set the reference voltage, and R_4 and R_5 set the final control voltage that is sent to the pulse-width modulator.

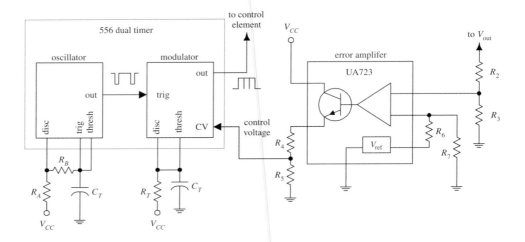

FIGURE 11.18

Step-Up, Step-Down, and Inverting Configurations

The switching regulator in Fig. 11.17 is referred to as a *step-down regulator*. It is used when the regulated output voltage is to be lower than the regulator input voltage. Now, switching regulators also come in step-up and inverting configurations. The step-up version is used when the output is to be higher than the input, whereas the inverting version is used when the output voltage is to be the opposite polarity of the input voltage. Here's an overview of the three configurations.

STEP-DOWN REGULATOR

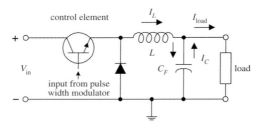

FIGURE 11.19

This is used when output voltage is to be lower than input voltage. When the control element is on, L stores energy, helps supply load current, and supplies current to the filter capacitor. When the control element is off, the energy stored in L helps supply load current but again restores charge on C_F—the charge on C_F is used to supply the load when the control element turns off and L has discharged its energy.

STEP-UP SWITCHER REGULATOR

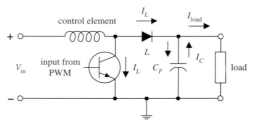

FIGURE 11.20

This is used when output is to be higher than input voltage. When the control element is on, energy is stored in the inductor. The load, isolated by the diode, is supplied by the charge stored in C_F. When the control element is off, the energy stored in L is added to the input voltage. At the same time, L supplies load current, as well as restoring the charge on C_F—the charge on C_F is used to supply the load current when the control element is off and when the energy in L is discharged.

INVERTING SWITCHER REGULATOR

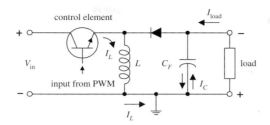

FIGURE 11.21

This is used when output voltage is to be the opposite polarity of the input voltage. When the control element is on, energy is stored in L, while the diode isolates L from load. The load current is supplied by the charge on C_F. When the control element is turned off, the energy stored in L charges C_F to a polarity such that V_{out} is negative. I_L supplies load current and restores the charge on C_F while it is discharging its energy. C_F supplies load current when the control element is off and the inductor is discharged. An inverting switcher regulator can be designed to step up or step down the inverted output.

Switching Regulator ICs

Since switching regulators have become so common, special purpose ICs have been developed to simplify circuit design. There are many of these available. Switching regulator ICs that operate at higher frequencies use smaller inductors and reduce the cost of the design.

One popular and very easy to use device is the LM2575 which is available in fixed and adjustable output versions. Just six components are needed including the regulator IC to produce a 5-V power supply. That's only three more components than an 7805. Figure 11.22 shows the schematic for a 5-V regulator using this IC.

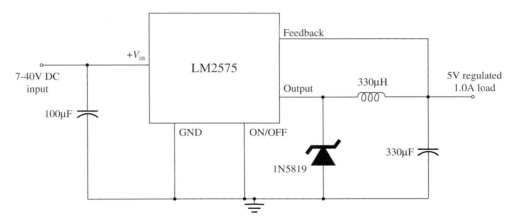

FIGURE 11.22.   A 5-V regulator using the LM2575 switching regulator IC.

11.9   Switch-Mode Power Supplies (SMPS)

The logical extension of a switching regulator is to ditch the transformer entirely and work directly from rectified line voltage.

Transformer-based designs have become a rare occurrence in consumer electronics. Transformers are relatively expensive and heavy components. Watt for watt, a switch-mode power supply is generally cheaper to produce.

By using the unique switching action of the switcher, it is possible to design a supply that does not require the hefty 60-Hz power transformer at the input stage.

In other words, you can design a switching power supply to run directly off a 120-V ac line—you still must rectify and filter the line voltage before feeding it to the regulator. However, if you remove the power transformer, you remove the protective isolation that is present between the 120-V ac line and the dc input to the supply. Without the isolation, the dc input voltage will be around 160-V. To avoid this potentially "shocking" situation, the switching regulator must be modified. One method for providing isolation involves replacing the energy-storage inductor with the secondary coil of a high-frequency transformer while using another high-frequency transformer or optoisolator to link the feedback from the error amplifier to the modulating element (see Fig. 11.23).

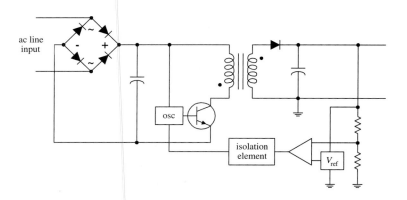

FIGURE 11.23

Now, you may be wondering how removing one transformer and adding another transformer (if not two) makes things smaller and lighter. Well, according to the laws of physics, as the frequency of an alternating signal increases, the need for a large iron core within the transformer decreases. You can use the high-frequency transformer(s) because the switcher's oscillator is beating so fast (e.g., 65 kHz). The difference in size and weight between a switching supply that uses high-frequency transformers and a supply that uses a 60-Hz power transformer is significant. For example, a 500-W switching supply takes up around 640 in^3 as compared with 1520 in^3 for a linear supply rated at the same power. Also, switching supplies run cooler than linear supplies. In terms of watts per cubic inch, a switching power supply can achieve 0.9 W/in^3, while a linear supply usually provides 0.4 W/in^3.

There is a slight problem with switchers that should be noted. As a result of the on/off pulsing action of the switching regulator, a switching supply's output will contain a small switching ripple voltage (typically in the tens of millivolts). Usually the ripple voltage does not pose too many problems (e.g., 200-mV noise margins for most digital ICs are not exceeded). However, if a circuit is not responding well to the ripple, an external high-current, low-pass filter can be added.

11.10 Kinds of Commercial Power Supply Packages

To make life easier, you can forget about designing your own supplies and buy one that has been made by the pros. These supplies come in either linear or switcher form and come in a variety of different packages. Here are some of the packages that are available.

Small Modular Units

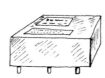 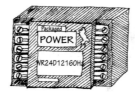

FIGURE 11.24

These are used in low-power applications (e.g., ±5, ±10, ±15 V). Supplies are housed in small modules, usually around 2.5 × 3.5 × 1 in. They often come with pinlike leads that can be mounted directly into circuit boards or come with terminal-strip screw connections along their sides. These supplies may come with single (e.g., +5 V), dual (e.g., ±15 V), or triple (e.g., +5 V, ±15 V) output terminals. Linear units have power ratings from around 1 to 10 W, while switching units have power ratings from around 10 to 25 W. You must supply the fuses, switches, and filters.

Open Frame

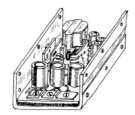 

FIGURE 11.25

This supply's circuit board, transformer, etc. are mounted on a metal platform (if it is a low-voltage supply, it may simply be mounted on a circuit board) that is inserted into an instrument. These supplies come in linear and switching types and come with a wide range of voltage, current, and power ratings (around 10 to 200 W for linear supplies, 20 to 400 W for switching supplies). You will probably have to supply the fuses, switches, and filters.

Enclosed

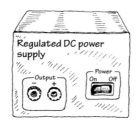 

FIGURE 11.26

These supplies are enclosed in a metal box that is especially designed to efficiently radiate off excessive heat. They come in both linear and switching forms. Power rating ranges from around 10 to 800 W for linear supplies and 20 to 1500 W for switching supplies.

Wall Plug-In

 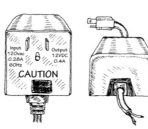

FIGURE 11.27

Wall plug-in power supplies (often called "wall-warts") get inserted directly into an ac wall socket. Some of these devices only provide ac transformation, others supply an unregulated dc voltage, and others supply a regulated dc output. Typical output voltages include +3, +5, +6, +7.5, +9, +12, and +15 V. They also come in dual-polarity form.

These devices are mostly switched mode, and are manufactured in millions, as they are not specific to a product. They are small and light, and it is possible to get a unit that can supply 2 A at 12 V at a cost of less than $10.

This approach is helped by the largely standard 6.3 mm plug for the dc voltage, which, in most consumer appliances, has a positive inner connection and a negative outer sheath connection. Note that this is often the other way around in musical equipment, such as guitar effects pedals.

11.11   Power Supply Construction

When building a power supply, the following suggestions should help:

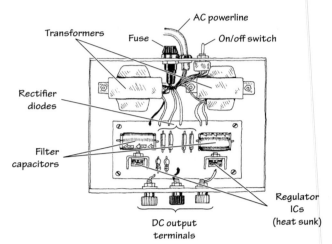

FIGURE 11.28

- Mount the transformer directly to the metal enclosure box, toward the rear.
- Install fuses, power switch, and binding posts at the rear of the box.
- Mount circuit boards on standoffs within the box.
- Place diode or rectifier modules, along with the capacitors and voltage regulators, on the circuit board.
- Make sure to heat-sink voltage regulators.
- Place supply output jacks on the front of the box.
- Drill holes in box to allow cooling.
- Ground the box.
- Place the power-line core through a hole in the rear. Use a rubber grommet for strain relief.
- To avoid shocks, make sure to insulate all exposed 120-V power connections inside the box with heat-shrink tubing.

## CHAPTER 12

Digital Electronics

Before beginning, we must warn you that there is a lot of information in this chapter, and it may be difficult to absorb all this at once. Some information is present largely for historical interest and to provide a better understanding of how complex digital systems such as microcontrollers work. Our advice is to skim to your heart's content, and pull out whatever information you find practical. The basic principles are still the same, but if you find that your design uses more than three ICs, you probably could be using a microcontroller (the subject of Chap. 13).

12.1 The Basics of Digital Electronics

Until now, we have mainly covered the analog realm of electronics—circuits that accept and respond to voltages that vary continuously over a given range. Such analog circuits included rectifiers, filters, amplifiers, simple RC timers, oscillators, simple transistor switches, and so on. Although each of these analog circuits is fundamentally important in its own right, these circuits lack an important feature: they cannot store and process bits of information needed to make complex logical decisions. To incorporate logical decision-making processes into a circuit, you need to use digital electronics.

This chapter is concerned with laying the foundations of digital electronics. The actual implementation of digital electronics these days is either handled by microcontrollers (see Chap. 13) or programmable logic devices (see Chap. 14).

Analog Signal

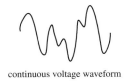

continuous voltage waveform

Digital Signal

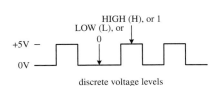

discrete voltage levels

Using a switch to demonstrate logic states

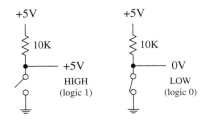

FIGURE 12.1

12.1.1 Digital Logic States

In digital electronics, there are only two voltage states present at any point within a circuit. These voltage states are either *high* or *low*. The voltage being high or low at a particular location within a circuit can signify a number of things. For example, it may represent the on or off state of a switch or saturated transistor, one bit of a number, whether an event has occurred, or whether some action should be taken.

The high and low states can be represented as true and false statements, which are used in Boolean logic. In most cases, high equals true and low equals false. However, this does not need to be the case—you could make high equal to false and low equal to true. The decision to use one convention over the other is a matter left ultimately to the designer. In digital lingo, to avoid people getting confused over which convention is in use, the term *positive true logic* is used when high equals true, while the term *negative true logic* is used when high equals false.

In Boolean logic, the symbols 1 and 0 are used to represent true and false, respectively. Now, unfortunately, 1 and 0 are also used in electronics to represent high and low voltage states, where high equals 1 and low equals 0. As you can see, things can get a bit confusing, especially if you are not sure which type of logic convention is being used: positive true or negative true logic. In Sec. 12.3, you will see some examples that deal with this confusing issue.

The exact voltages assigned to high or low voltage states depend on the specific logic IC that is used (as it turns out, digital components are IC-based). As a general rule of thumb, +5 V is considered high, while 0 V (ground) is considered low. However, as you will see in Sec. 12.4, this does not need to be the case. For example, some logic ICs may interpret a voltage from +2.4 to +5 V as high and a voltage from +0.8 to 0 V as low. Other ICs may use an entirely different range.

12.1.2 Number Codes Used in Digital Electronics

Binary

Because digital circuits work with only two voltage states, it is logical to use the binary number system to keep track of information. A binary number is composed of two binary digits, 0 and 1, which are also called *bits* (for example, 0 = low voltage and 1 = high voltage). By contrast, a decimal number such as 736 is represented by successive powers of 10:

$$736_{10} = 7 \times 10^2 + 3 \times 10^1 + 6 \times 10^0$$

Similarly, a binary number such as 11100 (28_{10}) can be expressed as successive powers of 2:

$$11100_2 = 1 \times 2^4 + 1 \times 2^3 + 1 \times 2^2 + 0 \times 2^1 + 0 \times 2^0$$

The subscript tells which number system is in use (X_{10} = decimal number and X_2 = binary number). The highest-order bit (leftmost bit) is called the *most significant bit* (MSB), while the lowest-order bit (rightmost bit) is called the *least significant bit* (LSB). Methods used to convert from decimal to binary and vice versa are shown in Fig. 12.2.

It should be noted that most digital systems deal with 4, 8, 16, or 32 bits at a time. The decimal-to-binary conversion example given here has a 7-bit answer. In an 8-bit system, you would need to put an additional 0 in front of the MSB (for example, 01101101). In a 16-bit system, nine additional 0s would need to be added (for example, 0000000001101101).

Binary-to-Decimal Conversion

Decimal-to-Binary Conversion

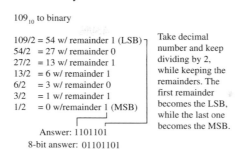

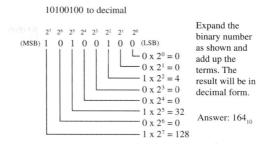

FIGURE 12.2

As a practical note, the easiest way to convert a number from one base to another is to use a calculator. For example, to convert a decimal number into a binary number, type in the decimal number (in base 10 mode) and then change to binary mode (which usually entails a second function key). The number will now be in binary (1s and 0s). To convert a binary number to a decimal number, start out in binary mode, type in the number, and then switch to decimal mode.

Octal and Hexadecimal

Two other number systems used in digital electronics include the octal and hexadecimal systems. In the octal system (base 8), there are 8 allowable digits: 0, 1, 2, 3, 4, 5, 6, and 7. In the hexadecimal system (base 16), there are 16 allowable digits: 0, 1, 2, 3, 4, 5, 6, 7, 8, 9, A, B, C, D, E, and F. Here are examples of octal and hexadecimal numbers with decimal equivalents:

$$247_8 \text{ (octal)} = 2 \times 8^2 + 4 \times 8^1 + 7 \times 8^0 = 167_{10} \text{ (decimal)}$$
$$2D5_{16} \text{ (hex)} = 2 \times 16^2 + D\ (=13_{10}) \times 16^1 + 9 \times 16^0 = 725_{10} \text{ (decimal)}$$

Of course, binary numbers are the natural choice for digital systems, but since these binary numbers can become long and difficult to interpret by our decimal-based brains (a result of our ten fingers), it is common to write them out in hexadecimal or octal form.

Unlike decimal numbers, octal and hexadecimal numbers can be translated easily to and from binary. This is because a binary number, no matter how long, can be broken up into 3-bit groupings (for octal) or 4-bit groupings (for hexadecimal). You simply add zero to the beginning of the binary number if the total numbers of bits is not divisible by 3 or 4. Figure 12.3 should paint the picture better than words.

Octal to Binary	Binary to Octal	Hex to Binary	Binary to Hex
537_8 to binary	$111\ 001\ 100_2$ to octal	$3E9_{16}$ to binary	$1001\ 1111\ 1010\ 0111_2$ to octal
5 3 7	111 001 100	3 E 9	1001 1111 1010 0111
101 011 111	7 1 4	0011 1110 1001	9 F A 7
Answer: 101011111_2	Answer: 714_8	Answer: $0011\ 1110\ 1001_2$	Answer: $9FA7_{16}$

FIGURE 12.3

A 3-digit binary number is replaced for each octal digit, and vice versa. The 3-digit terms are then grouped (or octal terms are grouped).

A 4-digit binary number is replaced for each hex digit, and vice versa. The 4-digit terms are then grouped (or hex terms are grouped).

Today, the hexadecimal system has essentially replaced the octal system. The octal system was popular at one time, when microprocessor systems used 12-bit and 36-bit words, along with a 6-bit alphanumeric code, which are all divisible by 3-bit units (1 octal digit). Today, microprocessor systems mainly work with 8-bit, 16-bit, 20-bit, 32-bit, or 64-bit words, which are all divisible by 4-bit units (1 hex digit). In other words, an 8-bit word can be broken down into 2 hex digits, a 16-bit word into 4 hex digits, a 20-bit word into 5 hex digits, and so on.

Hexadecimal representation of binary numbers pops up in many memory and microprocessor applications that use programming codes (for example, within assembly language) to address memory locations and initiate other specialized tasks that would otherwise require typing in long binary numbers. For example, a 20-bit address code used to identify one of a million memory locations can be replaced with a hexadecimal code (in the assembly program) that reduces the count to five hex digits. Note that a compiler program later converts the hex numbers within the assembly language program into binary numbers (machine code), which the microprocessor can use. Table 12.1 shows a conversion table.

TABLE 12.1 Decimal, Binary, Octal, Hex, BCD Conversion Table

DECIMAL	BINARY	OCTAL	HEXADECIMAL	BCD
00	0000 0000	00	00	0000 0000
01	0000 0001	01	01	0000 0001
02	0000 0010	02	02	0000 0010
03	0000 0011	03	03	0000 0011
04	0000 0100	04	04	0000 0100
05	0000 0101	05	05	0000 0101
06	0000 0110	06	06	0000 0110
07	0000 0111	07	07	0000 0111
08	0000 1000	10	08	0000 1000
09	0000 1001	11	09	0000 1001
10	0000 1010	12	0A	0001 0000
11	0000 1011	13	0B	0001 0001
12	0000 1100	14	0C	0001 0010
13	0000 1101	15	0D	0001 0011
14	0000 1110	16	0E	0001 0100
15	0000 1111	17	0F	0001 0101
16	0001 0000	20	10	0001 0110
17	0001 0001	21	11	0001 0111
18	0001 0010	22	12	0001 1000
19	0001 0011	23	13	0001 1001
20	0001 0100	24	14	0010 0000

Binary-Coded Decimal

Binary-coded decimal (BCD) is used to represent each digit of a decimal number as a 4-bit binary number. For example, the number 150_{10} in BCD is expressed as follows:

1 5 0 $150_{10} = 0001\ 0101\ 000_{(BCD)}$

/ | \

0001 0101 000

To convert from BCD to binary is vastly more difficult, as shown in Fig. 12.4. Of course, you could cheat by converting the BCD into decimal first and then convert to binary, but that does not show you the mechanics of how machines do things with 1s and 0s. You will rarely need to do BCD-to-binary conversion, so we will not dwell on this topic.

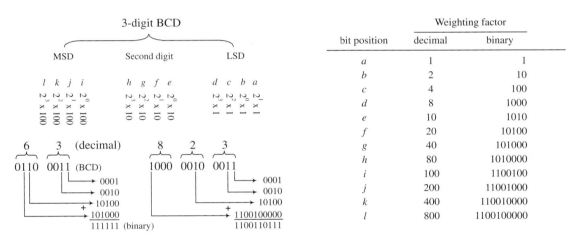

FIGURE 12.4

BCD is commonly used when outputting to decimal (0–9) displays, such as those found in digital clocks and multimeters. BCD will be discussed in Sec. 12.3.

Sign-Magnitude and 2's Complement Numbers

Up to now, we have not considered negative binary numbers. How do you represent them? A simple method is to use *sign-magnitude representation*. In this method, you simply reserve a bit, usually the MSB, to act as a sign bit. If the sign bit is 0, the number is positive; if the sign bit is 1, the number is negative (see Fig. 12.5).

Although the sign-magnitude representation is simple, it is seldom used, because adding requires a different procedure than subtracting (as you will see in the next section). Occasionally, you will see sign-magnitude numbers used in display and analog-to-digital applications, but you will hardly ever see them in circuits that perform arithmetic.

A more popular choice when dealing with negative numbers is to use *2's complement representation*. In 2's complement, the positive numbers are exactly the same as unsigned binary numbers. A negative number, however, is represented by a binary number, which when added to its corresponding positive equivalent results in zero. In this way, you can avoid two separate procedures for doing addition and subtraction. You will see how this works in the next section. A simple procedure outlining how to convert a decimal number into a binary number and then into a 2's complement number, and vice versa, is outlined in Fig. 12.5.

Decimal to 2's complement

$+41_{10}$ to 2's complement

true binary $= 0010\ 1001$
2's comp $\quad= 0010\ 1001$

If the decimal number is positive, the 2's complement number is equal to the true binary equivalent of the decimal number.

-41_{10} to 2's complement

true binary $= 0010\ 1001$
1's comp $\quad= 1101\ 0110$
Add 1 $\quad=\quad\quad +1$
2's comp $\quad= 1101\ 0111$

If the decimal number is negative, the 2's complement number is found by:
1) Complementing each bit of the true binary equivalent of the decimal (making 1's into 0's and vice versa). This is is called taking the 1's complement.
2) Adding 1 to the 1's complement number to get the magnitude bits. The sign bit will always end up being 1.

2's complement to decimal

1100 1101 (2's comp) to decimal

2's comp $\quad\quad= 1100\ 1101$
Complement $= 0011\ 0010$
Add 1 $\quad\quad=\quad\quad +1$
True binary $= 0011\ 0011$
Decimal eq. $=\quad -51_{10}$

If the 2's complement number is positive (sign bit = 0), perform a regular binary-to-decimal conversion.

If the 2's complement number is negative (sign bit = 1), the decimal sign will be negative. The decimal is found by:
1) Complementing each bit of the 2's complement number.
2) Adding 1 to get the true binary equivalent.
3) Performing a true binary-to-decimal conversion, and including negative sign.

FIGURE 12.5

Decimal, Sign-Magnitude, 2's Complement Conversion Table

DECIMAL	SIGN-MAGNITUDE	2'S COMPLEMENT
+7	0000 0111	0000 0111
+6	0000 0110	0000 0110
+5	0000 0101	0000 0101
+4	0000 0100	0000 0100
+3	0000 0011	0000 0011
+2	0000 0010	0000 0010
+1	0000 0001	0000 0001
0	0000 0000	0000 0000
−1	1000 0001	1111 1111
−2	1000 0010	1111 1110
−3	1000 0011	1111 1101
−4	1000 0100	1111 1100
−5	1000 0101	1111 1011
−6	1000 0110	1111 1010
−7	1000 0111	1111 1001
−8	1000 1000	1111 1000

Arithmetic with Binary Numbers

Adding, subtracting, multiplying, and dividing binary numbers, hexadecimal numbers, and other representations can be done with a calculator set to that particular base mode. But that's cheating, and it doesn't help you understand the mechanics of how it is done. The mechanics become important when designing the actual arithmetical circuits. Here are the basic techniques used to add and subtract binary numbers.

ADDING

FIGURE 12.6

Adding binary numbers is just like adding decimal numbers. Whenever the result of adding one column of numbers is greater than one digit, a 1 is carried over to the next column to be added.

SUBTRACTING

Subtraction done the long way

2's comp. subtraction

$+19_{10} = 0001\ 0011$
$-7_{10} = 1111\ 1001$
Sum $= 0000\ 1100$

FIGURE 12.7

Subtracting binary numbers is not as easy as it looks. It is similar to decimal subtraction but can be confusing. For example, you might think that if you were to subtract a 1 from a 0, you would borrow a 1 from the column to the left. No! You must borrow a 10 (2_{10}). It becomes a headache if you try to do this by hand. The trick to subtracting binary numbers is to use the 2's complement representation that provides the sign bit, and then just add the positive number with the negative number to get the sum. This method is often used by digital circuits because it allows both addition and subtraction, without the headache of needing to subtract the smaller number from the larger number.

ASCII

American Standard Code for Information Interchange (ASCII) is an alphanumeric code used to transmit letters, symbols, numbers, and special nonprinting characters between computers and computer peripherals (such as printers and keyboards). ASCII consists of 128 different 7-bit codes.

Codes from 000 0000 (or hex 00) to 001 1111 (or hex 1F) are reserved for nonprinting characters or special machine commands like ESC (escape), DEL (delete), CR (carriage return), and LF (line feed). Codes from 010 0000 (or hex 20) to 111 1111 (or hex 7F) are reserved for printing characters like a, A, #, &, {, @, and 3. Tables 12.2 and 12.3 show the ASCII nonprinting and printing characters.

In practice, when ASCII code is sent, an additional bit is added to make it compatible with 8-bit systems. This bit may be set to 0 and ignored, it may be used as a parity bit for error detection (Sec. 12.3.8 covers parity bits), or it may act as a special function bit used to implement an additional set of specialized characters.

TABLE 12.2 ASCII Nonprinting Characters

DEC	HEX	7-BIT CODE	CONTROL CHAR	CHAR	MEANING	DEC	HEX	7-BIT	CONTROL CHAR	CHAR	MEANING
00	00	000 0000	CTRL-@	NUL	Null	16	10	001 0000	CTRL-P	DLE	Data line escape
01	01	000 0001	CTRL-A	SOH	Start of heading	17	11	001 0001	CTRL-Q	DC1	Device control 1
02	02	000 0010	CTRL-B	STX	Start of text	18	12	001 0010	CTRL-R	DC2	Device control 2
03	03	000 0011	CTRL-C	ETX	End of text	19	13	001 0011	CTRL-S	DC3	Device control 3
04	04	000 0100	CTRL-D	EOT	End of transmit	20	14	001 0100	CTRL-T	DC4	Device control 4
05	05	000 0101	CTRL-E	ENQ	Enquiry	21	15	001 0101	CTRL-U	NAK	Neg acknowledge
06	06	000 0110	CTRL-F	ACK	Acknowledge	22	16	001 0110	CTRL-V	SYN	Synchronous idle
07	07	000 0111	CTRL-G	BEL	Bell	23	17	001 0111	CTRL-W	ETB	End of transmit block
08	08	000 1000	CTRL-H	BS	Backspace	24	18	001 1000	CTRL-X	CAN	Cancel
09	09	000 1001	CTRL-I	HT	Horizontal tab	25	19	001 1001	CTRL-Y	EM	End of medium
10	0A	000 1010	CTRL-J	LF	Line feed	26	1A	001 1010	CTRL-Z	SUB	Substitute
11	0B	000 1011	CTRL-K	VT	Vertical tab	27	1B	001 1011	CTRL-[	ESC	Escape
12	0C	000 1100	CTRL-L	FF	Form feed	28	1C	001 1100	CTRL-\	FS	File separator
13	0D	000 1101	CTRL-M	CR	Carriage return	29	1D	001 1101	CTRL-]	GS	Group separator
14	0E	000 1110	CTRL-N	SO	Shift out	30	1E	001 1110	CTRL-^	RS	Record separator
15	0F	000 1111	CTRL-O	SI	Shift in	31	1F	001 1111	CTRL-_	US	Unit separator

TABLE 12.3 ASCII Printing Characters

DEC	HEX	7-BIT CODE	CHAR	DEC	HEX	7-BIT	CHAR	DEC	HEX	7-BIT CODE	CHAR
32	20	010 0000	SP	64	40	100 0000	@	96	60	110 0000	'
33	21	010 0001	!	65	41	100 0001	A	97	61	110 0001	a
34	22	010 0010	"	66	42	100 0010	B	98	62	110 0010	b
35	23	010 0011	#	67	43	100 0011	C	99	63	110 0011	c
36	24	010 0100	$	68	44	100 0100	D	100	64	110 0100	d
37	25	010 0101	%	69	45	100 0101	E	101	65	110 0101	e
38	26	010 0110	&	70	46	100 0110	F	102	66	110 0110	f
39	27	010 0111	'	71	47	100 0111	G	103	67	110 0111	g
40	28	010 1000	(	72	48	100 1000	H	104	68	110 1000	h
41	29	010 1001	)	73	49	100 1001	I	105	69	110 1001	i
42	2A	010 1010	*	74	4A	100 1010	J	106	6A	110 1010	j
43	2B	010 1011	+	75	4B	100 1011	K	107	6B	110 1011	k
44	2C	010 1100	,	76	4C	100 1100	L	108	6C	110 1100	l
45	2D	010 1101	-	77	4D	100 1101	M	109	6D	110 1101	m
46	2E	010 1110	.	78	4E	100 1110	N	110	6E	110 1110	n
47	2F	010 1111	/	79	4F	100 1111	O	111	6F	110 1111	o
48	30	011 0000	0	80	50	101 0000	P	112	70	111 0000	p
49	31	011 0001	1	81	51	101 0001	Q	113	71	111 0001	q
50	32	011 0010	2	82	52	101 0010	R	114	72	111 0010	r
51	33	011 0011	3	83	53	101 0011	S	115	73	111 0011	s
52	34	011 0100	4	84	54	101 0100	T	116	74	111 0100	t
53	35	011 0101	5	85	55	101 0101	U	117	75	111 0101	u
54	36	011 0110	6	86	56	101 0110	V	118	76	111 0110	v
55	37	011 0111	7	87	57	101 0111	W	119	77	111 0111	w
56	38	011 1000	8	88	58	101 1000	X	120	78	111 1000	x
57	39	011 1001	9	89	59	101 1001	Y	121	79	111 1001	y
58	3A	011 1010	:	90	5A	101 1010	Z	122	7A	111 1010	z
59	3B	011 1011	;	91	5B	101 1011	[	123	7B	111 1011	{
60	3C	011 1100	<	92	5C	101 1100	\	124	7C	111 1100		
61	3D	011 1101	=	93	5D	101 1101	]	125	7D	111 1101	}
62	3E	011 1110	>	94	5E	101 1110	^	126	7E	111 1110	~
63	3F	011 1111	?	95	5F	101 1111	_	127	7F	111 1111	DEL

12.1.3 Clock Timing and Parallel versus Serial Transmission

Before moving on to the next section, let's take a brief look at three important items: clock timing, parallel transmission, and serial transmission.

Clock Timing

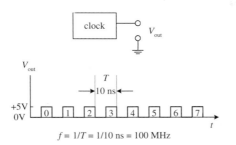

Most digital circuits require precise timing to function properly. Usually, a clock circuit that generates a series of high and low pulses at a fixed frequency is used as a reference on which to base all critical actions executed within a system. The clock is also used to push bits of data through the digital circuitry. The period of a clock pulse is related to its frequency by $T = 1/f$. So, if $T = 10$ ns, then $f = 1/(10$ ns$) = 100$ MHz.

FIGURE 12.8

$f = 1/T = 1/10$ ns $= 100$ MHz

Serial versus Parallel Representation

Binary information can be transmitted from one location to another in either a serial or parallel manner. The serial format uses a single electrical conductor (and a common ground) for data transfer. Each bit from the binary number occupies a separate clock period, with the change from one bit to another occurring at each falling or leading clock edge; the type of edge depends on the circuitry used.

Figure 12.9 shows an 8-bit (10110010) word that is transmitted from circuit A to circuit B in 8 clock pulses (0–7). In computer systems, serial communications are used to transfer data between keyboard and computer, as well as to transfer data between two computers via a telephone line.

Serial transmission of an 8-bit word (10110010)

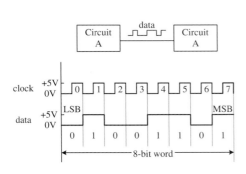

Parallel transmission of an 8-bit word (10110010)

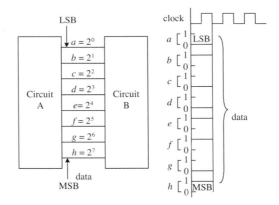

FIGURE 12.9

Parallel transmission uses separate electrical conductors for each bit (and a common ground). In Fig. 12.9, an 8-bit string (01110110) is sent from circuit A to circuit B. As you can see, unlike serial transmission, the entire word is transmitted in only one clock cycle, not eight clock cycles. In other words, it is eight times faster. Parallel communications are most frequently found within microprocessor systems that use multiline data and control buses to transmit data and control instructions from the microprocessor to other microprocessor-based devices (such as memory and output registers).

12.2 Logic Gates

Logic gates are the building blocks of digital electronics. The fundamental logic gates include the INVERT (NOT), AND, NAND, OR, NOR, exclusive OR (XOR), and exclusive NOR (XNOR) gates. Each of these gates performs a different logical operation. Figure 12.10 provides a description of what each logic gate does and gives a switch and transistor analogy for each gate.

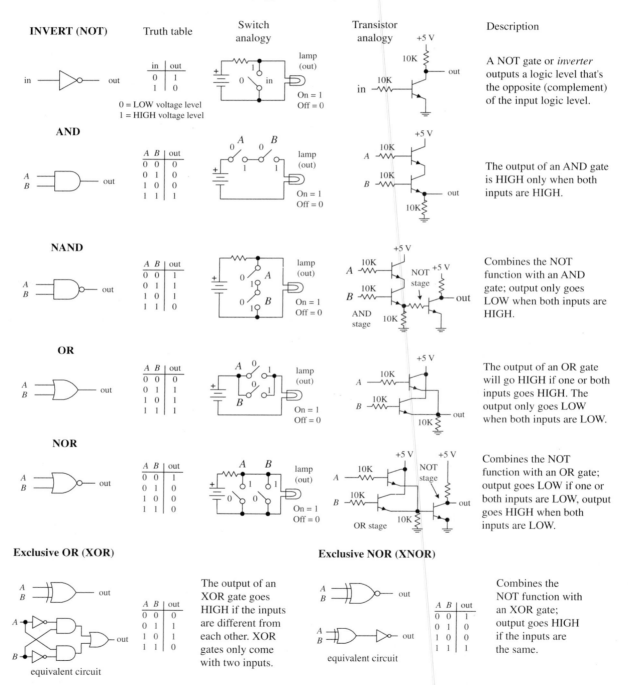

FIGURE 12.10

12.2.1 Multiple-Input Logic Gates

AND, NAND, OR, and NOR gates often come with more than two inputs (this is not the case with XOR and XNOR gates, which require two inputs only). Figure 12.11 shows a four-input AND, an eight-input AND, a three-input OR, and an eight-input OR gate. With the eight-input AND gate, all inputs must be high for the output to be high. With the eight-input OR gate, at least one of the inputs must be high for the output to go high.

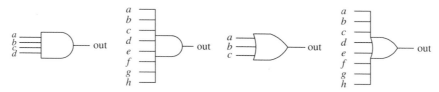

FIGURE 12.11

12.2.2 Digital Logic Gate ICs

As we mentioned in the introduction to this chapter, the use of individual logic ICs has almost completely been superseded by the use of microcontrollers, the programmable logic devices such as field-programmable gate arrays (FPGAs). However, one or two logic ICs are still often used together in simple applications.

There are a number of technologies used in the fabrication of digital logic. The most popular technology is complementary MOSFET (CMOS) logic.

A logic IC typically houses more than one logic gate (for example, a quad two-input NAND, hex inverter, and so on). Each of the gates within the IC shares a common supply voltage that is implemented via two supply pins: a positive supply pin ($+V_{CC}$ or $+V_{DD}$) and a ground pin (GND). The most popular "HC" range of digital logic gate ICs will operate from a supply of 2-6 V with 5 V being something of a standard operating voltage as a relic from the age of transistor-transistor logic (TTL).

Generally speaking, input and output voltage levels are assumed to be 0 V (low) and +5 V (high). However, the actual input voltage required and the actual output voltage provided by the gate are not set in stone. For example, the 74HCxx series will recognize a high input from 2.5 to 5 V and a low from 0 to 2.1 V. However, for the CMOS 4000B series (V_{CC} = +5 V), recognizable input voltages range from 3.3 to 5 V for high and 0 to 1.7 V for low. Guaranteed high and low output levels range from 4.9 to 5 V and 0 to 0.1 V, respectively. Again, we will discuss specifics later in Sec. 12.4. For now, let's just get acquainted with what some of these ICs look like, as shown in Figs. 12.12 and 12.13. The CMOS devices listed in the figures include 74HCxx and 4000(B). The TTL devices shown include the 74xx, 74Fxx, and 74LS.

74HC00 Series

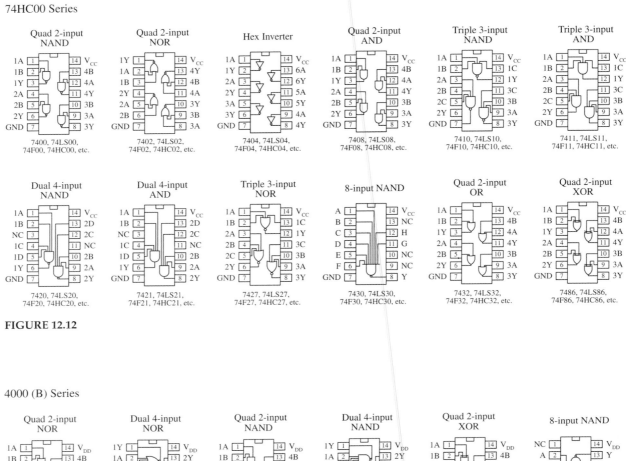

FIGURE 12.12

4000 (B) Series

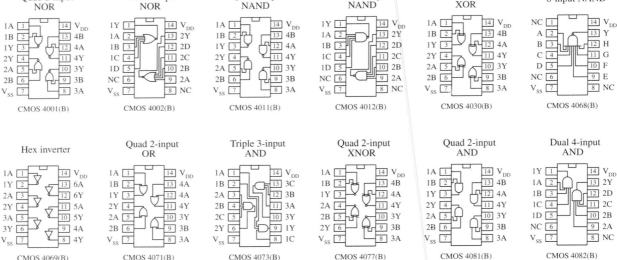

FIGURE 12.13

12.2.3 Applications for a Single Logic Gate

Before we jump into the heart of logic gate applications that involve combining logic gates to form complex decision-making circuits, let's take a look at a few simple applications that require the use of a single logic gate.

Enable/Disable Control

An enable/disable gate is a logic gate that acts to control the passage of a given waveform. The waveform—say, a clock signal—is applied to one of the gate's inputs, while the other input acts as the enable/disable control lead. Enable/disable gates are used frequently in digital systems to enable and disable control information from reaching various devices. Figure 12.14 shows two enable/disable circuits: the first uses an AND gate, and the second uses an OR gate. NAND and NOR gates are also frequently used as enable gates.

Using an AND as an enable gate

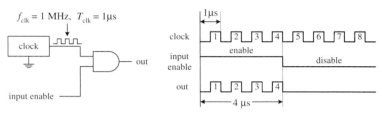

Using an OR as an enable gate

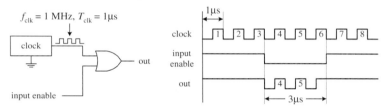

FIGURE 12.14

In the upper part of the figure, an AND gate acts as the enable gate. When the input enable lead is made high, the clock signal will pass to the output. In this example, the input enable is held high for 4 µs, allowing 4 clock pulses (where $T_{clk} = 1$ µs) to pass. When the input enable lead is low, the gate is disabled, and no clock pulses make it through to the output.

Below, an OR gate is used as the enable gate. The output is held high when the input enable lead is high, even as the clock signal is varying. However, when the enable input is low, the clock pulses are passed to the output.

Waveform Generation

By using the basic enable/disable function of a logic gate, as illustrated in the previous example, it is possible, with the help of a repetitive waveform generator circuit, to create specialized waveforms that can be used for the digital control of sequencing circuits.

An example waveform generator circuit is the Johnson counter. The Johnson counter will be discussed in Sec. 12.8. For now, let's simply focus on the outputs. In Fig. 12.15, a Johnson counter uses clock pulses to generate different output

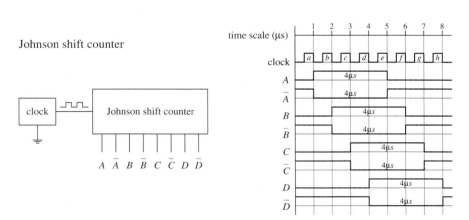

FIGURE 12.15

waveforms, as shown in the timing diagram. Outputs A, B, C, and D go high for 4 μs (four clock periods) and are offset from each other by 1 μs. Outputs $\overline{A}$, $\overline{B}$, $\overline{C}$, and $\overline{D}$ produce waveforms that are complements of outputs A, B, C, and D, respectively.

Now, there may be certain applications that require 4-μs high/low pulses applied at a given time, as the counter provides. However, what would you do if the application requires a 3-μs high waveform that begins at 2 μs and ends at 5 μs (relative to the time scale indicated in Fig. 12.15)? This is where the logic gates come in handy. For example, if you attach an AND gate's inputs to the counter's A and B outputs, you will get the desired 2- to 5-μs high waveform at the AND gate's output: from 1 to 2 μs the AND gate outputs a low ($A = 1$, $B = 0$), from 2 to 5 μs the AND gate outputs a high ($A = 1$, $B = 1$), and from 5 to 6 μs the AND gate outputs a low ($A = 0$, $B = 1$). See the leftmost area of Fig. 12.16.

Connections for 1μs to 5μs waveform

Other possible connections and waveforms

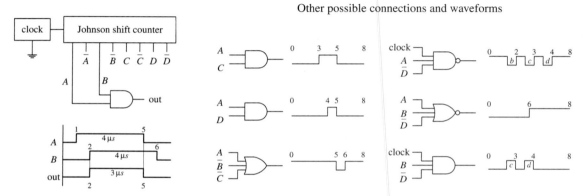

FIGURE 12.16

Various other specialized waveforms can be generated by using different logic gates and tapping different outputs of the Johnson shift counter. In Fig. 12.16, six other possibilities are shown.

12.2.4 Combinational Logic

Combinational logic involves combining logic gates together to form circuits capable of enacting more useful, complex functions. For example, let's design the logic used to instruct a janitor-type robot to recharge itself (seek out a power outlet) only when a specific set of conditions is met. The "recharge itself" condition is specified as follows:

Either its battery is low (indicated by a high output signal from a battery-monitor circuit)

OR

The workday is over (indicated by a high output signal from a timer circuit)]

OR

[When vacuuming is complete (indicated by a high voltage output from a vacuum-completion monitor circuit)

AND

When waxing is complete (indicated by a high output signal from a wax-completion monitor circuit)]

Let's also assume that the power-outlet-seeking routine circuit is activated when a high is applied to its input.

Two simple combinational circuits that perform the desired logic function for the robot are shown in Fig. 12.17. The two circuits use a different number of gates but perform the same function. Now, the question remains, how did we come up with these circuits? In either circuit, it is not hard to predict which gates are needed. You simply exchange the word *and* present within the conditional statement with an AND gate within the logic circuit, and exchange the word *or* present within the conditional statement with an OR gate within the logic circuit. Common sense takes care of the rest.

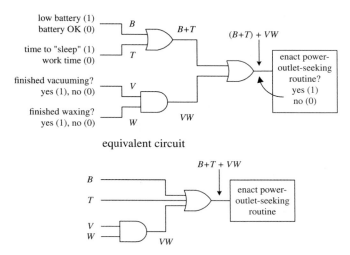

B	T	V	W	$B+T$	VW	$(B+T)+VW$
0	0	0	0	0	0	0
0	0	0	1	0	0	0
0	0	1	0	0	0	0
0	0	1	1	0	1	1
0	1	0	0	1	0	1
0	1	0	1	1	0	1
0	1	1	0	1	0	1
0	1	1	1	1	1	1
1	0	0	0	1	0	1
1	0	0	1	1	0	1
1	0	1	0	1	0	1
1	0	1	1	1	1	1
1	1	0	0	1	0	1
1	1	0	1	1	0	1
1	1	1	1	1	1	1

FIGURE 12.17

However, when you begin designing more complex circuits, using intuition to figure out what kind of logic gates to use and how to join them together becomes exceedingly difficult. To make designing combinational circuits easier, a special symbolic language called *Boolean algebra* is used, which uses only true and false variables. A Boolean expression for the robot circuit would appear as follows:

$$E = (B + T) + VW$$

This expression amounts to saying that if B (battery-check circuit's output) *or* T (timer circuit's output) is true, or V *and* W (vacuum and waxing circuit outputs) are true, then E (enact power-outlet circuit input) is true.

Note that the word or is replaced by the symbol +, and the word and is simply expressed in a way similar to multiplying two variables together (placing them side by side or using a dot between variables). Also note that the term true in Boolean algebra is expressed as a 1, and false is expressed as a 0. Here, we are assuming positive logic, where true equals high voltage. Using the Boolean expression for the robot circuit, we can come up with some of the following results (the truth table in Fig. 12.17 provides all possible results):

$$E = (B + T) + VW$$
$$E = (1 + 1) + (1 \cdot 1) = 1 + 1 = 1 \quad \text{(battery is low, time to sleep, finished with chores =}$$
go recharge)
$$E = (1 + 0) + (0 \cdot 0) = 1 + 0 = 1 \quad \text{(battery is low = go recharge)}$$
$$E = (0 + 0) + (1 \cdot 0) = 0 + 0 = 0 \quad \text{(hasn't finished waxing = don't recharge yet)}$$
$$E = (0 + 0) + (1 \cdot 1) = 0 + 1 = 1 \quad \text{(has finished all chores = go recharge)}$$
$$E = (0 + 0) + (0 \cdot 0) = 0 + 0 = 0 \quad \text{(hasn't finished vacuuming and waxing = don't}$$
recharge yet)

The robot example showed you how to express AND and OR functions in Boolean algebraic terms. But what about the negation operations (NOT, NAND, and NOR) and the exclusive operations (XOR and XNOR)? How do you express these in Boolean terms?

- For a NOT condition, place a line over the NOT'ed variable or variables.
- For a NAND expression, place a line over an AND expression.
- For a NOR expression, place a line over an OR expression.
- For exclusive operations, use the symbol $\oplus$.

Figure 12.18 shows a rundown of all the possible Boolean expressions for the various logic gates.

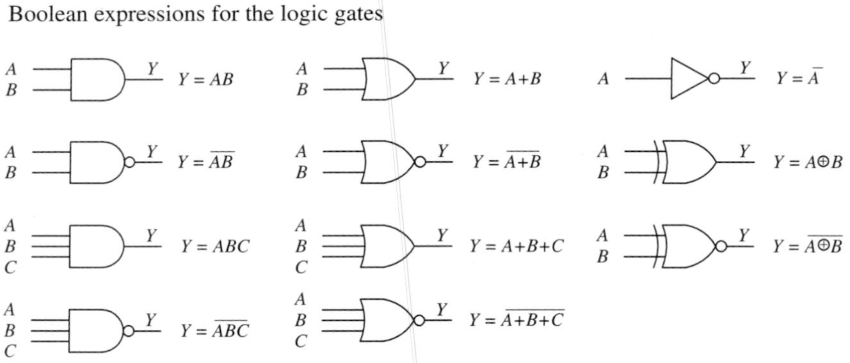

Boolean expressions for the logic gates

FIGURE 12.18

Like conventional algebra, Boolean algebra has a set of logic identities that can be used to simplify the Boolean expressions and thus make circuits more compact. These identities go by names such as the *commutative law of addition, associate law of addition,* and *distributive law*. Instead of worrying about what the various identities are called, simply make reference to the list of identities provided on the next page. Most of these identities are self-explanatory, although a few are not so obvious, as you will see in a minute. The various circuits in Fig. 12.19 show some of the identities in action.

LOGIC IDENTITIES

1) $A + B = B + A$

2) $AB = BA$

3) $A + (B + C) = (A + B) + C$

4) $A(BC) = (AB)C$

5) $A(B + C) = AB + AC$

6) $(A + B)(C + D) = AC + AD + BC + BD$

7) $\bar{1} = 0$

8) $\bar{0} = 1$

9) $A \cdot 0 = 0$

10) $A \cdot 1 = A$

11) $A + 0 = A$

12) $A + 1 = 1$

13) $A + A = A$

14) $AA = A$

15) $\bar{\bar{A}} = A$

16) $A + \bar{A} = 1$

17) $A\bar{A} = 0$

18) $\overline{A + B} = \bar{A}\bar{B}$

19) $\overline{AB} = \bar{A} + \bar{B}$

20) $A + \bar{A}B = A + B$

21) $\bar{A} + AB = \bar{A} + B$

22) $A \oplus B = \bar{A}B + A\bar{B} = (A + B)(\overline{AB})$

23) $\overline{A \oplus B} = AB + \bar{A}\bar{B}$

FIGURE 12.19

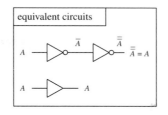

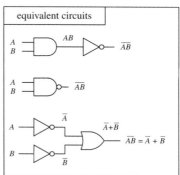

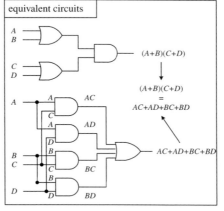

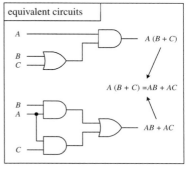

Example

Let's find the initial Boolean expression for the circuit in Fig. 12.20, and then use the logic identities to come up with a circuit that requires fewer gates.

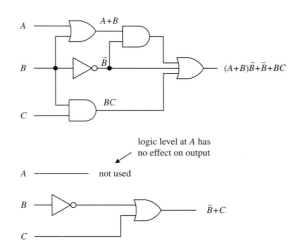

FIGURE 12.20

The circuit shown here is expressed by the following Boolean expression:

$out = (A + B)\bar{B} + \bar{B} + BC$

This expression can be simplified by using Identity 5:

$(A + B)\bar{B} = A\bar{B} + B\bar{B}$

This makes:

$out = A\bar{B} + B\bar{B} + \bar{B} + BC$

Using Identities 17 $(B\bar{B} = 0)$ and 11 $(\bar{B} + 0 = \bar{B})$, you get:

$out = A\bar{B} + 0 + \bar{B} + BC = A\bar{B} + BC + \bar{B}$

Factoring a $\bar{B}$ from the preceding term gives:

$out = \bar{B}(A + 1) + BC$

Using Identity 12, you get:

$out = \bar{B}(1) + BC = \bar{B} + BC$

Finally, using Identity 21, you get the simplified expression:

$out = \bar{B} + C$

Notice that A is now missing. This means that the logic input at A has no effect on the output and therefore can be omitted. From the reduction, you get the simplified circuit in the bottom part of the figure.

Dealing with Exclusive Gates (Identities 22 and 23)

Now let's take a look at a couple of not-so-obvious logic identities: those that involve the XOR (Identity 22) and XNOR (Identity 23) gates. The leftmost section in Fig. 12.21 shows equivalent circuits for the XOR gate. In the lower two equivalent circuits, Identity 22 is proved by Boolean reduction. Equivalent circuits for the XNOR gate are shown in the rightmost section of the figure. To prove Identity 23, you can simply invert Identity 22.

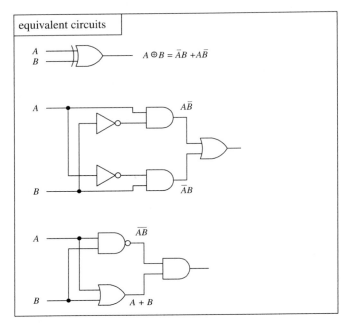

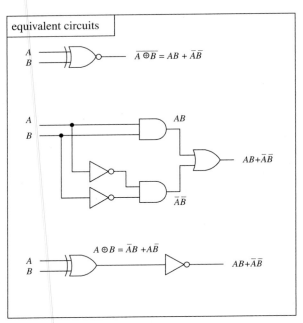

FIGURE 12.21

De Morgan's Theorem (Identities 18 and 19)

To simplify circuits containing NANDs and NORs, you can use an incredibly useful theorem known as *De Morgan's theorem*. This theorem allows you to convert an expression having an inversion bar over two or more variables into an expression having inversion bars over single variables only. De Morgan's theorem (Identities 18 and 19) is as follows:

$$\overline{A \cdot B} = \overline{A} + \overline{B} \quad \text{(2 variables)} \qquad \overline{A \cdot B \cdot C} = \overline{A} + \overline{B} + \overline{C} \quad \text{(3 or more variables)}$$

$$\overline{A + B} = \overline{A} \cdot \overline{B} \qquad\qquad\qquad \overline{A + B + C} = \overline{A} \cdot \overline{B} \cdot \overline{C}$$

The easiest way to prove that these identities are correct is to use Fig. 12.22, noting that the truth tables for the equivalent circuits are the same. Note the inversion bubbles present on the inputs of the corresponding leftmost gates. The inversion bubbles mean that before inputs A and B are applied to the base gate, they are inverted (negated). In other words, the bubbles are simplified expressions for NOT gates.

A	B	$\overline{A\cdot B}$
0	0	1
0	1	1
1	0	1
1	1	0

A	B	$\bar{A}+\bar{B}$
0	0	1
0	1	1
1	0	1
1	1	0

A	B	$\overline{A+B}$
0	0	1
0	1	0
1	0	0
1	1	0

A	B	$\bar{A}\cdot\bar{B}$
0	0	1
0	1	0
1	0	0
1	1	0

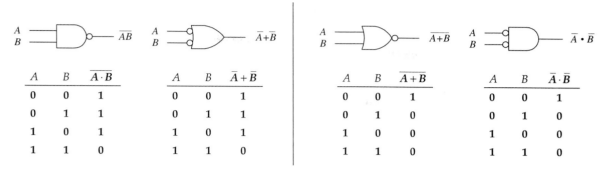

FIGURE 12.22

Why do you use the inverted-input OR gate symbol instead of a NAND gate symbol? Or why would you use the inverted-input AND gate symbol instead of a NOR gate symbol? This is left up to the designer to choose whatever symbol seems most logical to use. For example, when designing a circuit, it may be easier to think about ORing or ANDing inverted inputs than to think about NANDing or NORing inputs. Similarly, it may be easier to create truth tables or work with Boolean expressions using the inverted-input gate. It is typically easier to create truth tables and Boolean expressions that do not have variables joined under a common inversion bar. Of course, when it comes time to construct the actual working circuit, you probably will want to convert to the NAND and NOR gates because they do not require additional NOT gates at their inputs.

Bubble Pushing

A shortcut method for forming equivalent logic circuits, based on De Morgan's theorem, is to use what's called *bubble pushing*.

Bubble pushing involves the following tricks:

- Change an AND gate to an OR gate or change an OR gate to an AND gate.
- Add inversion bubbles to the inputs and outputs where there were none, while removing the original bubbles.

That's it. You can prove to yourself that this works by examining the corresponding truth tables for the original gate and the bubble-pushed gate, or you can work out the Boolean expressions using De Morgan's theorem. Figure 12.23 shows examples of bubble pushing.

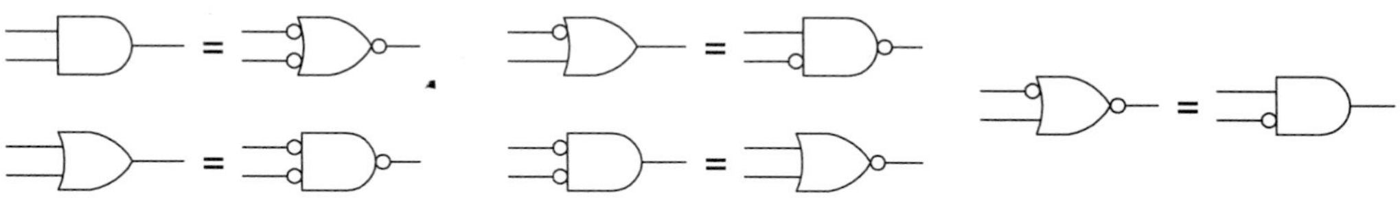

FIGURE 12.23

Universal Capability of NAND and NOR Gates

NAND and NOR gates are referred to as *universal gates* because each alone can be combined together with itself to form all other possible logic gates. The ability to create any logic gate from NAND or NOR gates is obviously a handy feature.

For example, if you do not have an XOR IC handy, you can use a single multigate NAND gate (such as 74HC00) instead. Figure 12.24 shows how to wire NAND or NOR gates together to create equivalent circuits of the various logic gates.

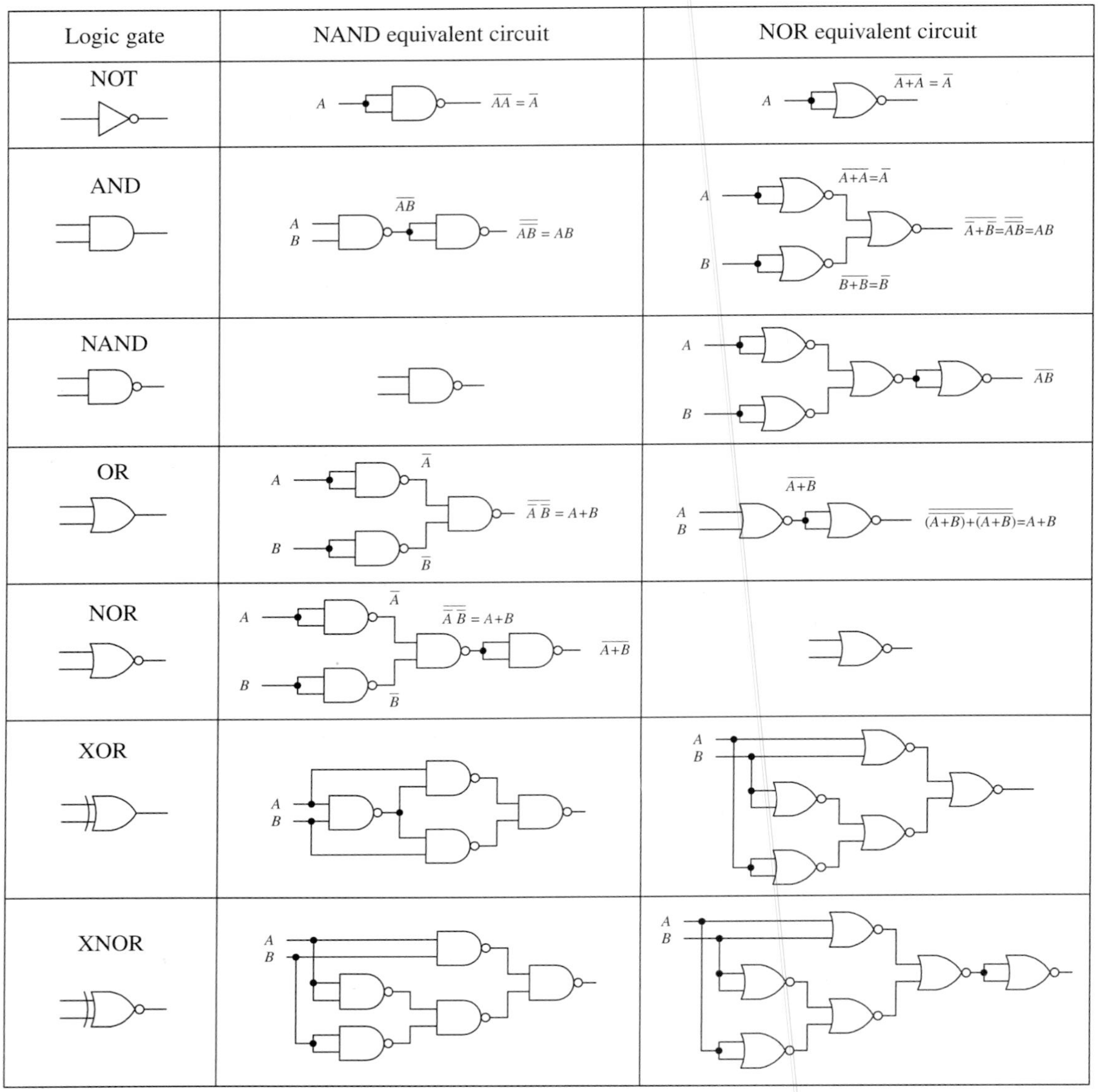

Logic gate	NAND equivalent circuit	NOR equivalent circuit
NOT		
AND		
NAND		
OR		
NOR		
XOR		
XNOR		

FIGURE 12.24

AND-OR-INVERTER Gates

When a Boolean expression is reduced, the equation that is left over typically will be of one of the following two forms: *product of sums* (POS) or *sum of products* (SOP). A POS expression appears as two or more OR'ed variables AND'ed together with two or more additional OR'ed variables. An SOP expression appears as two or more AND'ed variables OR'ed together with additional AND'ed variables. Figure 12.25 shows two circuits that provide the same logic function (they are equivalent), but the circuit to the left is designed to yield a POS expression, while the circuit to the right is designed to yield a SOP expression.

Logic circuit for POS expression

Logic circuit for SOP expression

Table made using SOP expression
(it's easier than POS)

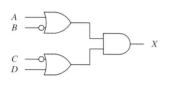

$X = (A + \bar{B})(\bar{C} + D)$

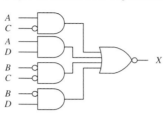

$X = A\bar{C} + AD + \bar{B}\bar{C} + \bar{B}D$

A	B	C	D	$A\bar{C}$	AD	$\bar{B}\bar{C}$	$\bar{B}D$	X
0	0	0	0	0	0	1	0	1
0	0	0	1	0	0	1	1	1
0	0	1	0	0	0	0	0	0
0	0	1	1	0	0	0	1	1
0	1	0	0	0	0	0	0	0
0	1	0	1	0	0	0	0	0
0	1	1	0	0	0	0	0	0
0	1	1	1	0	0	0	0	0
1	0	0	0	1	0	1	0	1
1	0	0	1	1	1	1	1	1
1	0	1	0	0	0	0	0	0
1	0	1	1	0	1	0	1	1
1	1	0	0	1	0	0	0	1
1	1	0	1	1	1	0	0	1
1	1	1	0	0	0	0	0	0
1	1	1	1	0	1	0	0	1

FIGURE 12.25

Which circuit is best for design: the one that implements the POS expression or the one that implements the SOP expression? The POS design shown here would appear to be the better choice because it requires fewer gates. However, the SOP design is nice because it is easy to work with the Boolean expression. For example, which Boolean expression in Fig. 12.25 (POS or SOP) would you rather use to create a truth table? The SOP expression seems the obvious choice.

A more down-to-earth reason for using an SOP design has to do with the fact that special ICs called AND-OR-INVERTER (AOI) gates are designed to handle SOP expressions. For example, the 74LS54 AOI IC shown in Fig. 12.26 creates an inverted SOP expression at its output, via two two-input AND gates and two three-input AND gates NOR'ed together. A NOT gate can be attached to the output to get rid of the inversion bar, if desired. If specific inputs are not used, they should be held high, as shown in the example circuit in Fig. 12.26. AOI ICs come in many different configurations—check out the catalogs to see what's available.

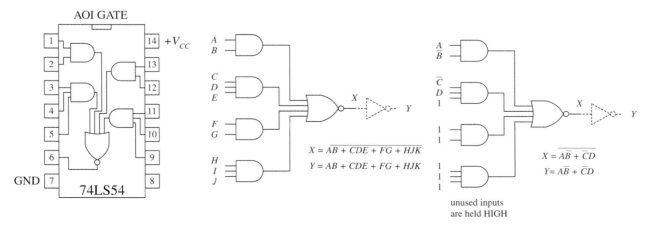

FIGURE 12.26

In Chap. 14 you will see how a combinational logic circuit like the ones described above can be programmed onto a special configurable chip called a field-programmable gate array (FPGA) without the need for large number of logic gate ICs.

12.2.5 Keeping Circuits Simple (Karnaugh Maps)

We have just covered how using the logic identities can simplify a Boolean expression. This is important because it reduces the number of gates needed to construct the logic circuit. However, as we are sure you will agree, having to work out Boolean problems in longhand is not easy. It takes time and ingenuity. A simple way to avoid the unpleasant task of using your ingenuity is to get a computer program that accepts a truth table or Boolean expression, and then provides you with the simplest expression, and perhaps even the circuit schematic.

However, let's assume that you do not have such a program to help you out. Are you stuck with the Boolean longhand approach? No. You can use a technique referred to as *Karnaugh mapping*. With this technique, you take a given truth table (or Boolean expression that can be converted into a truth table), convert it into a Karnaugh map, apply some simple graphic rules, and come up with the simplest (most of the time) possible Boolean expression for your final circuit. Karnaugh mapping works best for circuits with three to four inputs—below this, things usually do not require much thought anyway; beyond four inputs, things get quite tricky.

Here's a basic outline showing how to apply Karnaugh mapping to a three-input system:

1. Select a desired truth table. Let's choose the one shown in Fig. 12.27. (If you have a Boolean expression, transform it into an SOP expression and use the SOP expression to create the truth table.)

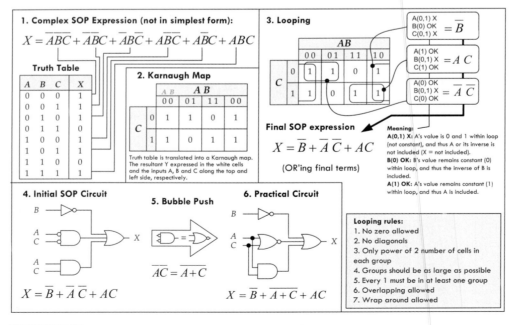

FIGURE 12.27

2. Translate the truth table into a Karnaugh map. A Karnaugh map is similar to a truth table but has its variables represented along two axes. Translating the truth table into a Karnaugh map reduces the number of 1s and 0s needed to present the information. Figure 12.27 shows how the translation is carried out.

3. After you create the Karnaugh map, you can proceed to encircle adjacent cells of 1s into groups of 2, 4, 8, ..., 2^n. The more cells you can encircle in one grouping the simpler the final equation will be. You are not allowed to encircle 0s nor allowed to make diagonal loops. Every 1 must be in at least one group. Overlapping and wrapping around map are allowed.

4. Identify the variable that remain constant within each loop, and write out an SOP equation by OR'ing these variables together. Here, constant means that a variable and its inverse are not present together within the loop. For example, in Figure 12.27 section 3, the loop that contains four 1s is reduced to $\overline{B}$ because $B = 0$ throughout the loop, while A and B aren't constant because they take on values of 0 and 1 throughout that loop. (If a variable is constant 0, the final expression is the inverse of that variable; if the variable is constant 1 throughout the loop, the final expression is the non-inverted variable.)

5. The SOP expression you end up with is the simplest possible expression. With it, you can create your logic circuit. You may need to apply some bubble pushing to make the final circuit practical, as shown in Fig. 12.27.

To apply Karnaugh mapping to four-input circuits, you apply the same basic steps used in the three-input scheme. However, now you must use a 4×4 Karnaugh map to hold all the necessary information. Figure 12.28 shows an example of how a four-input truth table (or unsimplified fourvariable SOP expression) can be mapped and converted into a simplified SOP expression that can be used to create the final logic circuit.

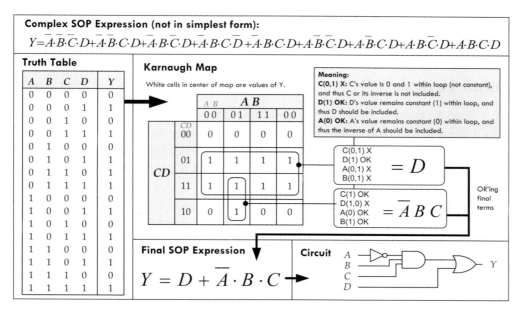

Complex SOP Expression (not in simplest form):

$$Y=\overline{A}\cdot\overline{B}\cdot\overline{C}\cdot D+\overline{A}\cdot\overline{B}\cdot C\cdot D+\overline{A}\cdot B\cdot\overline{C}\cdot D+\overline{A}\cdot B\cdot C\cdot\overline{D}+\overline{A}\cdot B\cdot C\cdot D+A\cdot\overline{B}\cdot\overline{C}\cdot D+A\cdot\overline{B}\cdot C\cdot D+A\cdot B\cdot\overline{C}\cdot D+A\cdot B\cdot C\cdot D$$

FIGURE 12.28

Figure 12.29 shows and example that uses an AOI IC to implement the final SOP expression after mapping.

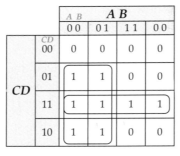

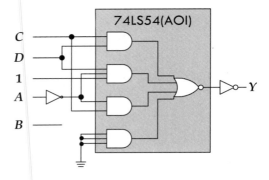

FIGURE 12.29

$$Y = CD + \overline{A} \cdot D + \overline{A} \cdot C$$

Other Looping Configurations

Figure 12.30 shows examples of other looping arrangements used with 4×4 Karnaugh maps.

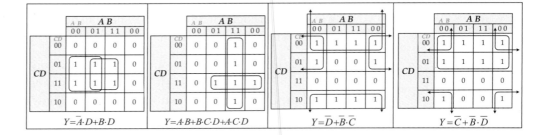

FIGURE 12.30

If you want to make your life easier, try one of the free online Karnaugh mapping calculators. Do a Google search for "Karnaugh Map Explorer 2.0". This tool will save you the hassle of having to draw maps and apply looping tricks yourself. You simply enter values into a truth table (or into the Karnaugh map) and the program calculates the final logical expression.

12.3 Combinational Devices

Now that you know a little something about how to use logic gates to enact functions represented within truth tables and Boolean expressions, it is time to take a look at some common functions that are used in the real world of digital electronics. As you will see, these functions are usually carried out by an IC that contains all the necessary logic.

As with almost everything discussed in this chapter, before using these ideas, you need to ask yourself if using a microcontroller would be more appropriate. However, many of the devices described here can be used with a microcontroller, especially when it comes to decoders. They can be a useful and low-cost solution for tasks such as driving more LEDs than there are pins on the microcontroller that you are using.

A word on IC part numbers before we begin. As with the logic gate ICs, the combinational ICs that follow will be of either the 4000 or 7400 series. It is important to note that an original TTL IC, like the 74138, is essentially the same device (usually with the same pinouts and function, but not always) as its newer counterparts, such as the 74F138, 74HC128 (CMOS), and 74LS138. The practical difference resides in the overall performance of the device (speed, power dissipation, voltage level rating, and so on). We will get into these gory details in a bit.

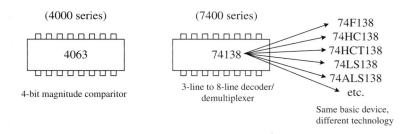

FIGURE 12.31

12.3.1 Multiplexers (Data Selectors) and Bilateral Switches

Multiplexers or data selectors act as digitally controlled switches. The term *data selector* appears to be the accepted term when the device is designed to act like an SPDT switch, while the term *multiplexer* is used when the throw count of the switch exceeds two, such as an SP8T. We will stick with this convention (although others may not).

A simple 1-of-2 data selector built from logic gates is shown in Fig. 12.32. The data select input of this circuit acts to control which input (*A* or *B*) gets passed to the output: When data select is low, input *A* passes while *B* is blocked. When data select is high, input *B* is passed while *A* is blocked. To understand how this circuit works, think of the AND gates as enable gates.

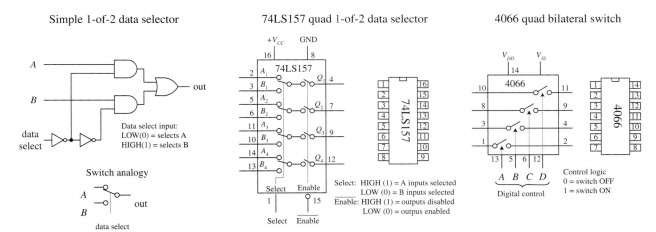

FIGURE 12.32

There are a number of different types of data selectors that come in IC form. For example, the 74LS157 quad 1-of-2 data selector IC, shown in Fig. 12.32, acts like an electrically controlled quad SPDT switch (or if you like, a 4PDT switch). When its select input is set high (1), inputs A_1, A_2, A_3, and A_4 are allowed to pass to outputs Q_1, Q_2, Q_3, and Q_4. When its select input is low (0), inputs B_1, B_2, B_3, and B_4 are allowed to pass to outputs Q_1, Q_2, Q_3, and Q_4. Either of these two conditions, however, ultimately depends on the state of the enable input.

When the enable input is low, all data-input signals are allowed to pass to the output; however, if the enable is high, the signals are not allowed to pass. This type of enable control is referred to as *active-low* enable, since the active function (passing the data to the output) occurs only with a low-level input voltage. The active-low input is denoted with a bubble (inversion bubble), and the outer label of the active-low input is represented with a line over it. Sometimes people omit the bubble and place a bar over the inner label. Both conventions are used commonly.

Figure 12.33 shows a 4-line-to-1-line multiplexer built with logic gates. This circuit resembles the 2-of-1 data selector shown in Fig. 12.32 but requires an additional select input to provide four address combinations.

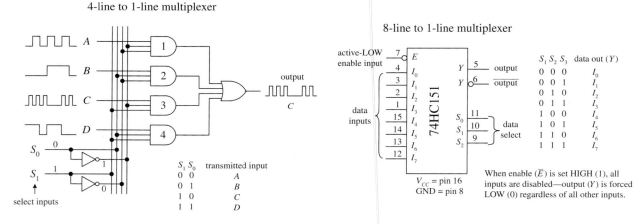

FIGURE 12.33

In terms of ICs, there are multiplexers of various input line capacities. For example, the 74151 8-line-to-1-line multiplexer uses three select inputs (S_0, S_1, S_2) to choose among one of eight possible data inputs (I_0 to I_7) to be funneled to the output. Note that this device actually has two outputs: one true (pin 5) and one inverted (pin 6). The active-low enable forces the true output low when set high, regardless of the inputs.

To create a larger multiplexer, you combine two smaller multiplexers together. For example, Fig. 12.34 shows two 8-line-to-1-line 74HC151s combined to create a 16-line-to-1-line multiplexer. Another alternative is to use a 16-line-to-1-line multiplexer IC like the 74HC150 shown in the figure. Check the catalogs to see what other kinds of multiplexers are available.

Finally, let's take a look at a very useful device called a *bilateral switch*. An example bilateral switch IC is the 4066, shown to the far right in Fig. 12.32. Unlike the multiplexer, this device merely acts as a digitally controlled quad SPST switch or quad

Combining two 8-line-to-1-line multiplexers to create a
16-line-to-1-line multiplexer

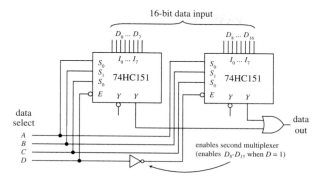

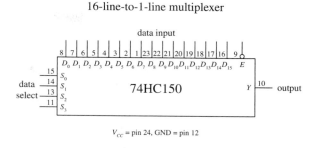

FIGURE 12.34

transmission gate. Using a digital control input, you select which switches are on
and which switches are off. To turn on a given switch, apply a high level to the corresponding switch select input; otherwise, keep the select input low.

In Sec. 12.9, we will look at analog switches and multiplexers. These devices use
digital select inputs to control analog signals. Analog switches and multiplexers
become important when you start linking the digital world to the analog world.

12.3.2 Demultiplexers (Data Distributors) and Decoders

A demultiplexer (or data distributor) is the opposite of a multiplexer. It takes a single
data input and routes it to one of several possible outputs. A simple four-line demultiplexer built from logic gates is shown on the left side of Fig. 12.35. To select the
output (A, B, C, or D) to which you want to send the input signal (applied at E), you
apply logic levels to the data select inputs (S_0, S_1), as shown in the truth table.

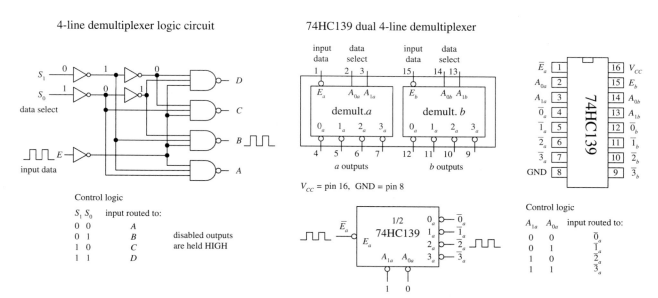

FIGURE 12.35

Notice that the unselected outputs assume a high level, while the selected output varies with the input signal. An IC that contains two functionally separate four-line demultiplexers is the 74HC139, shown on the right side of Fig. 12.35. If you need more outputs, check out the 75xx154 16-line demultiplexer. This IC uses four data select inputs to choose from 1 of 16 possible outputs. Check out the catalogs to see what other demultiplexers exist.

A decoder is somewhat like a demultiplexer, but it does not route input data to a specific output via data select inputs. Instead, it simply uses the data select inputs to choose which output (or outputs) among many are to be made high or low. The number of address inputs, the number of outputs, and the active state of the selected output vary from decoder to decoder. The variance is based on what the decoder is designed to do. For example, the 74LS138 1-of-8 decoder shown in Fig. 12.36 uses a 3-bit address input to select which of eight outputs will be made low; all other outputs are held high. Like the demultiplexer in Fig. 12.35, this decoder has active-low outputs.

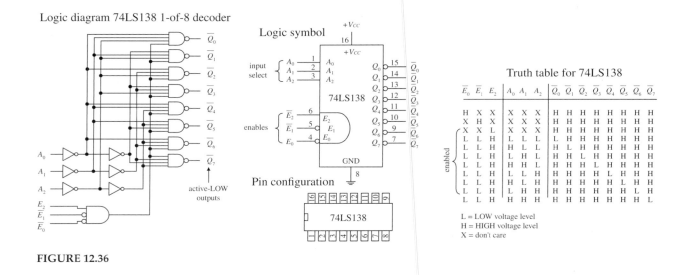

Logic diagram 74LS138 1-of-8 decoder

FIGURE 12.36

Now what exactly does it mean to say an output is an active-low output? It simply means that when an active-low output is selected, it is forced to a low logic state; otherwise, it is held high. Active-high outputs behave in the opposite manner. An active-low output is usually indicated with a bubble, although sometimes it is indicated with a barred variable within the IC logic symbol—no bubble included. Active-high outputs have no bubbles. Both active-low and active-high outputs are equally common among ICs.

By placing a load (for example, a warning LED) between $+V_{CC}$ and an active-low output, you can sink current through the load and into the active-low output when the output is selected. By placing a load between an active-high output and ground, you can source current from the active-high output and sink it through the load when the output is selected. The limits to how much current an IC can source or sink will be discussed in Sec. 12.4, and various schemes used to drive analog loads will be presented in Sec. 12.9.

Now let's get back to the 74LS138 decoder and discuss the remaining enable inputs $(\bar{E}_0, \bar{E}_1, E_2)$. For the 74LS138 to "decode," you must make the active-low inputs $\bar{E}_0$ and $\bar{E}_1$ low, while making the active-high input E_2 high. If any other set of enable inputs is applied, the decoder is disabled, making all active-low outputs high regardless of the selected inputs.

Other common decoders include the 7442 BCD-to-DEC (decimal) decoder, the 74154 1-of-16 (hex) decoder, and the 7447 BCD-to-seven-segment decoder shown in Figure 12.37. Like the preceding decoder, these devices also have active-low outputs. The 7442 uses a binary-coded decimal input to select 1 of 10 (0 through 9) possible outputs. The 74154 uses a 4-bit binary input to address 1 of 16 (0 through 15) outputs, making that output low (all others high), provided the enables are both set low.

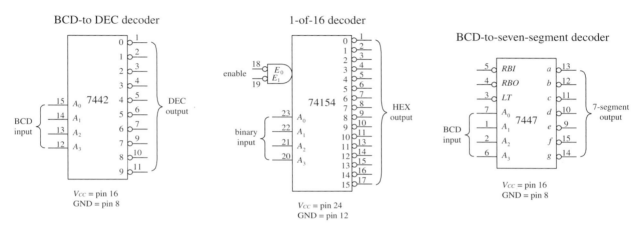

FIGURE 12.37

Now the 7447 is a bit different from the other decoders. With this device, more than one output can be driven low at a time. This is important because it allows the 7447 to drive a seven-segment LED display; to create different numbers requires driving more than one LED segment at a time. For example, in Fig. 12.38, when the BCD number for 5 (0101) is applied to the 7447's inputs, all outputs except $\bar{b}$ and $\bar{e}$ go low. This causes LED segments a, c, d, f, and g to light up—the 7447 sinks current through these LED segments, as indicated by the internal wiring of the display and the truth table.

7447 BCD-to-7-segment decoder/LED driver IC

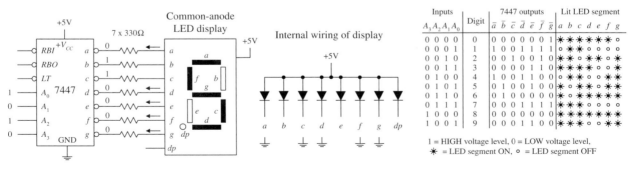

FIGURE 12.38

The 7447 also comes with a lamp test active-low input $(\overline{LT})$ that can be used to drive all LED segments at once to see if any of the segments are faulty. The ripple blanking input $(\overline{RBI})$ and ripple blanking output $(\overline{RBO})$ can be used in multi-stage display applications to suppress a leading-edge and/or trailing-edge zero in a multidigit decimal. For example, using the ripple blanking inputs and outputs, it is possible to take an eight-digit expression like 0056.020 and display 56.02, suppressing the two leading zeros and the one trailing zero. Leading-edge zero suppression is obtained by connecting the ripple blanking output of a decoder to the ripple blanking input of the next lower-stage device. The most significant decoder stage should have its ripple blanking input grounded. A similar procedure is used to provide automatic suppression of trailing zeros in the fractional part of the decimal.

12.3.3   Encoders and Code Converters

Encoders are the opposite of decoders. They are used to generate a coded output from a single active numeric input. To illustrate this in a simple manner, let's take a look at the simple decimal-to-BCD encoder circuit shown in Fig. 12.39.

Simple decimal-to-BCD encoder

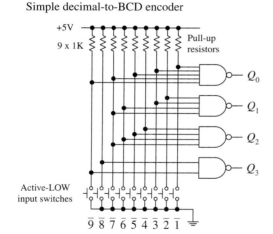

Truth table

$\overline{1}$	$\overline{2}$	$\overline{3}$	$\overline{4}$	$\overline{5}$	$\overline{6}$	$\overline{7}$	$\overline{8}$	$\overline{9}$	Q_3	Q_2	Q_1	Q_0	BCD (pos. logic)
H	H	H	H	H	H	H	H	H	L	L	L	L	$0000\ (0_{10})$
L	H	H	H	H	H	H	H	H	L	L	L	H	$0001\ (1_{10})$
H	L	H	H	H	H	H	H	H	L	L	H	L	$0010\ (2_{10})$
H	H	L	H	H	H	H	H	H	L	L	H	H	$0011\ (3_{10})$
H	H	H	L	H	H	H	H	H	L	H	L	L	$0100\ (4_{10})$
H	H	H	H	L	H	H	H	H	L	H	L	H	$0101\ (5_{10})$
H	H	H	H	H	L	H	H	H	L	H	H	L	$0110\ (6_{10})$
H	H	H	H	H	H	L	H	H	L	H	H	H	$0111\ (7_{10})$
H	H	H	H	H	H	H	L	H	H	L	L	L	$1000\ (8_{10})$
H	H	H	H	H	H	H	H	L	H	L	L	H	$1001\ (9_{10})$

H = High voltage level, L = Low voltage level

FIGURE 12.39

In this circuit, normally all lines are held high by the pullup resistors connected to +5 V. To generate a BCD output that is equivalent to a single selected decimal input, the switch corresponding to that decimal is closed. (The switch acts as an active-low input.) The truth table in Fig. 12.39 explains the rest.

Figure 12.40 shows a 74LS147 decimal-to-BCD (ten-line-to-four-line) priority encoder IC. The 74LS147 provides the same basic function as the circuit shown in Fig. 12.39, but it has active-low outputs. This means that instead of getting an LLHH output when 3 is selected, as in the previous encoder, you get HHLL. The two outputs represent the same thing (3); one is expressed in positive true logic, and the other (the 74LS147) is expressed in negative true logic. If you do not like negative true logic, you can slap inverters on the outputs of the 74LS147 to get positive true logic.

74LS147 decimal-to-4-bit BCD Priority Encoder IC

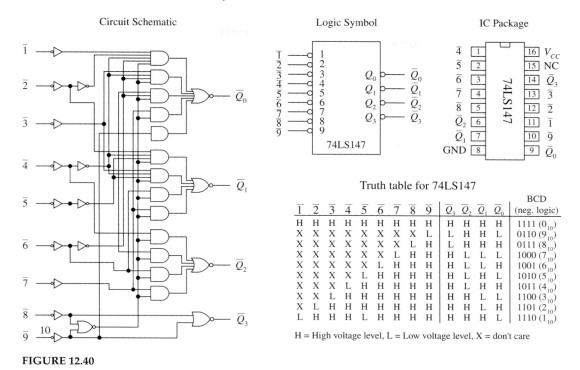

Circuit Schematic   Logic Symbol   IC Package

Truth table for 74LS147

$\bar{1}$	$\bar{2}$	$\bar{3}$	$\bar{4}$	$\bar{5}$	$\bar{6}$	$\bar{7}$	$\bar{8}$	$\bar{9}$	$\bar{Q}_3$	$\bar{Q}_2$	$\bar{Q}_1$	$\bar{Q}_0$	BCD (neg. logic)
H	H	H	H	H	H	H	H	H	H	H	H	H	1111 (0_{10})
X	X	X	X	X	X	X	X	L	L	H	H	L	0110 (9_{10})
X	X	X	X	X	X	X	L	H	L	H	H	H	0111 (8_{10})
X	X	X	X	X	X	L	H	H	H	L	L	L	1000 (7_{10})
X	X	X	X	X	L	H	H	H	H	L	L	H	1001 (6_{10})
X	X	X	X	L	H	H	H	H	H	L	H	L	1010 (5_{10})
X	X	X	L	H	H	H	H	H	H	L	H	H	1011 (4_{10})
X	X	L	H	H	H	H	H	H	H	H	L	L	1100 (3_{10})
X	L	H	H	H	H	H	H	H	H	H	L	H	1101 (2_{10})
L	H	H	H	L	H	H	H	H	H	H	H	L	1110 (1_{10})

H = High voltage level, L = Low voltage level, X = don't care

FIGURE 12.40

The choice to use positive or negative true logic really depends on what you are planning to drive. For example, negative true logic is useful when the device that you wish to drive uses active-low inputs.

Another important difference between the two encoders is the priority that is used with the 74LS147 and not used with the encoder in Fig. 12.39. The term *priority* is applied to the 74LS147 because this encoder is designed so that if two or more inputs are selected at the same time, it will select only the larger-order digit. For example, if 3, 5, and 8 are selected at the same time, only the 8 (negative true BCD LHHH or 0111) will be output. The truth table in Fig. 12.40 demonstrates this; look at the "don't care" or "X" entries. With the nonpriority encoder, if two or more inputs are applied at the same time, the output will be unpredictable.

The circuit shown in Fig. 12.41 provides a simple illustration of how an encoder and a decoder can be used together to drive an LED display via a 0 through 9 keypad. The 74LS147 encodes a keypad's input into BCD (negative logic). A set of inverters then converts the negative true BCD into positive true BCD. The transformed BCD is then fed into a 7447 seven-segment LED display decoder/driver IC.

Figure 12.42 shows a 74148 octal-to-binary priory encoder IC. It is used to transform a specified single octal input into a binary 3-bit output code. As with the 74LS147, the 74148 comes with a priority feature, so if two or more inputs are selected at the same time, only the higher-order number is selected.

A high applied to the input enable ($\overline{EI}$) forces all outputs to their inactive (high) state and allows new data to settle without producing erroneous information at the outputs. A group signal output ($\overline{GS}$) and an enable output ($\overline{EO}$) are also provided to allow for system expansion. The $\overline{GS}$ output is active level low when any input is low

A decimal-to-BCD encoder being used to convert keypad instructions into BCD instructions used to drive a LED display circuit (7447 decoder plus common-anode display)

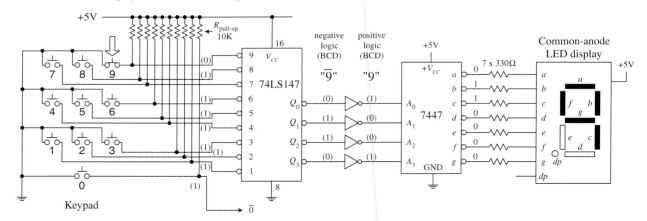

FIGURE 12.41

74148 octal-to-binary priority encoder

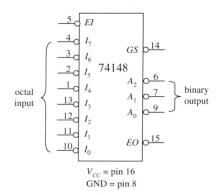

FIGURE 12.42

V_{CC} = pin 16
GND = pin 8

74148 truth table

EI	I_0	I_1	I_2	I_3	I_4	I_5	I_6	I_7	$\overline{GS}$	$\overline{A_0}$	$\overline{A_1}$	$\overline{A_2}$	$\overline{EO}$
H	X	X	X	X	X	X	X	X	H	H	H	H	H
L	H	H	H	H	H	H	H	H	H	H	H	H	L
L	X	X	X	X	X	X	X	L	L	L	L	L	H
L	X	X	X	X	X	X	L	H	L	H	L	L	H
L	X	X	X	X	X	L	H	H	L	L	H	L	H
L	X	X	X	X	L	H	H	H	L	H	H	L	H
L	X	X	X	L	H	H	H	H	L	L	L	H	H
L	X	X	L	H	H	H	H	H	L	H	L	H	H
L	X	L	H	H	H	H	H	H	L	L	H	H	H
L	L	H	H	H	H	H	H	H	L	H	H	H	H

(active). The $\overline{EO}$ output is low (active) when all inputs are high. Using the output enable along with the input enable allows priority coding of N input signals. Both $\overline{EO}$ and $\overline{GS}$ are active high when the input enable is high (device disabled).

Figure 12.43 shows a 74184 BCD-to-binary converter (encoder) IC. This device has eight active-high outputs ($Y_1 - Y_8$). Outputs Y_1 to Y_5 are outputs for regular BCD-to-binary conversion, while outputs Y_6 to Y_8 are used for a special BDC code called *nine's complement* and *ten's complement*. The active-high BCD code is applied to inputs A through E. The $\overline{G}$ input is an active-low enable input.

A sample 6-bit BCD-to-binary converter and a sample 8-bit BCD-to-binary converter that use the 74184 are shown to the right in Fig. 12.43. In the 6-bit circuit, since the LSB of the BCD input is always equal to the LSB of the binary output, the connection is made straight from input to output. The other BCD bits are applied directly to inputs A through E. The binary weighing factors for each input are $A = 2$, $B = 4$, $C = 8$, $D = 10$, and $E = 20$. Because only 2 bits are available for the MSD BCD input, the largest BCD digit in that position is 3 (binary 11). To get a complete 8-bit BCD converter, you connect two 74184s together, as shown to the far right in Fig. 12.43.

74184 BCD-to-binary converter

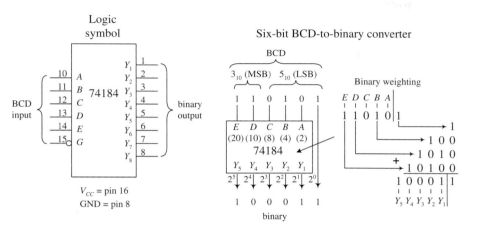

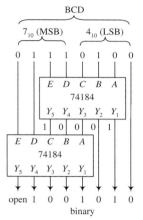

FIGURE 12.43

Figure 12.44 shows a 74185 binary-to-BCD converter (encoder). It is essentially the same as the 74184, but in reverse. The figure shows 6-bit and 8-bit binary-to-BCD converter arrangements.

74185 binary-to-BCD converter

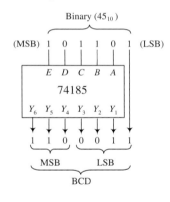

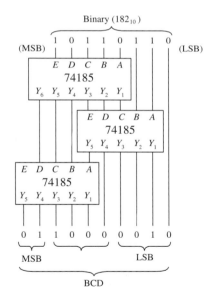

FIGURE 12.44

12.3.4 *Binary Adders*

If you find yourself needing to do arithmetic in logic, then that is a pretty sure sign that you need to use a microcontroller. However, that microcontroller will contain exactly the sort of logic that we describe here in its arithmetic logic unit (ALU), so it is instructive to see how it works under the hood.

With a few logic gates, you can create a circuit that adds binary numbers. The mechanics of adding binary numbers is basically the same as that of adding decimal numbers. When the first digit of a two-digit number is added, a 1 is carried and added to the next row whenever the count exceeds binary 2 (for example., $1 + 1 = 10$, or $= 0$ carry a 1). For numbers with more digits, you have multiple carry bits.

To demonstrate how you can use logic gates to perform basic addition, start out by considering the half-adder circuits in Fig. 12.45. Both half-adders shown are equivalent; one simply uses XOR/AND logic, while the other uses NOR/AND logic. The half-adder adds two single-bit numbers A and B and produces a 2-bit number. The LSB is represented as Σ_0, and the MSB, or carry bit, is represented as C_{out}.

The most complicated operation the half-adder can do is $1 + 1$. To perform addition on a two-digit number, you must attach a full-adder circuit (shown in Fig. 12.45) to the output of the half-adder. The full-adder has three inputs: two to input the second digits of the two binary numbers (A_1, B_1), and another that accepts the carry bit from the half-adder (the circuit that added the first digits, A_0 and B_0, of the two numbers). The two outputs of the full-adder will provide the 2d-place digit sum Σ_1 and another carry bit that acts as the third-place digit of the final sum. Now, you can keep adding more full-adders to the half-adder/full-adder combination to add larger numbers, linking the carry bit output of the first full-adder to the next full-adder, and so forth. To illustrate this point, a 4-bit adder is shown in Fig. 12.45.

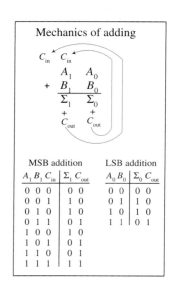

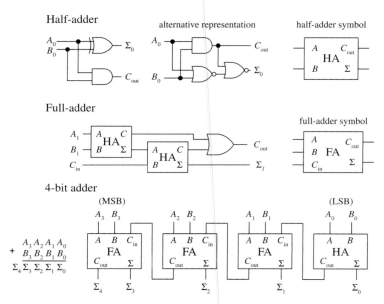

FIGURE 12.45

A number of 4-bit full-adder ICs are available, such as the 74LS283 and 4008. These devices will add two 4-bit binary numbers and provide an additional input carry bit, as well as an output carry bit, so you can stack them together to get adders that are 8-bit, 12-bit, 16-bit, and so on. For example, Fig. 12.46 shows an 8-bit adder made by cascading two 74LS283 4-bit adders.

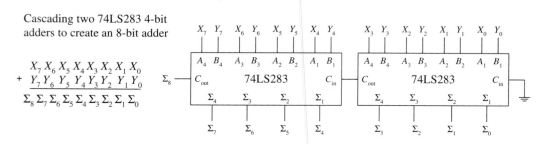

FIGURE 12.46

12.3.5   Binary Adder/Subtractor

Figure 12.47 shows how two 74LS283 4-bit adders can be combined with an XOR array to yield an 8-bit 2's complement adder/subtractor. The first number X is applied to the X_0 through X_7 inputs, while the second number Y is applied to the Y_0 through Y_7 inputs.

8-bit 2's complement adder/subtractor

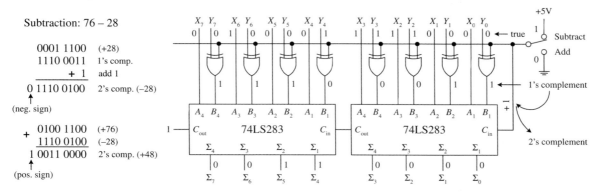

Subtraction: 76 − 28

$$
\begin{array}{ll}
0001\ 1100 & (+28)\\
1110\ 0011 & \text{1's comp.}\\
\underline{\quad +\ 1} & \text{add 1}\\
0\ 1110\ 0100 & \text{2's comp. } (-28)
\end{array}
$$
↑
(neg. sign)

$$
\begin{array}{ll}
+\ \ 0100\ 1100 & (+76)\\
\ \ \ \underline{1110\ 0100} & (-28)\\
1\ 0011\ 0000 & \text{2's comp. } (+48)
\end{array}
$$
↑
(pos. sign)

FIGURE 12.47

To add X and Y, the add/subtract switch is thrown to the add position, making one input of all XOR gates low. This has the effect of making the XOR gates appear transparent, allowing Y values to pass to the 74LS283s' B inputs (X values are passed to the A inputs). The 8-bit adder then adds the numbers and presents the result to the Σ outputs.

To subtract Y from X, you must first convert Y into 1's complement form; then you must add 1 to get Y into 2's complement form. After that you simply add X to the 2's complemented form of Y to get $X-Y$. When the add/subtract switch is thrown to the subtract position, one input to each XOR gate is set high. This causes the Y bits that are applied to the other XOR inputs to become inverted at the XOR outputs—you have just taken the 1's complement of Y. The 1's complement bits of Y are then presented to the inputs of the 8-bit adder. At the same time, C_{in} of the left 74LS283 is set high via the wire (see Fig. 12.47) so that a 1 is added to the 1's complement number to yield a 2's complement number. The 8-bit adder then adds X and the 2's complement of Y together. The final result is presented at the Σ outputs. In the figure, 76 is subtracted from 28.

12.3.6   Comparators and Magnitude Comparator ICs

A digital comparator is a circuit that accepts two binary numbers and determines whether the two numbers are equal. For example, Fig. 12.48 shows a 1-bit and a 4-bit comparator. The 1-bit comparator outputs a high (1) only when the two 1-bit numbers A and B are equal. If A is not equal to B, then the output goes low (0). The 4-bit is basically four 1-bit comparators in one. When all individual digits of each number are equal, all XOR gates output a high, which in turn enables the AND gate, making

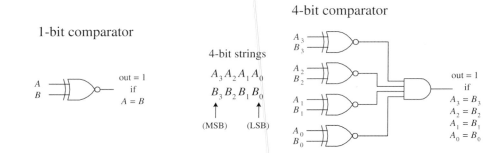

FIGURE 12.48

the output high. If any two corresponding digits of the two numbers are not equal, the output goes low.

Now, say you want to know which number, A or B, is larger. The circuits in Fig. 12.48 will not do the trick. What you need instead is a *magnitude comparator* like the 74HC85 shown in Fig. 12.49. This device not only tells you if two numbers are equal, but also which number is larger. For example, if you apply a 1001 (9_{10}) to the $A_3A_2A_1A_0$ inputs and a second number 1100 (12_{10}) to the $B_3B_2B_1B_0$ inputs, the $A < B$ output will go high (the other two outputs, $A > B$ and $A = B$, will remain low). If A and B were equal, the $A = B$ output would have gone high, and so on. If you wanted to compare larger numbers—say, two 8-bit numbers—you simply cascade two 74HC85 comparators together, as shown on the right side of Fig. 12.49. The leftmost 74HC85 compares the lower-order bits, while the rightmost 74HC85 compares the higher-order bits. To link the two devices together, you connect the output of the lower-order device to the expansion inputs of the higher-order device, as shown. The lower-order device's expansion inputs are always set low ($I_A < B$), high ($I_A = B$), and low ($I_A > B$).

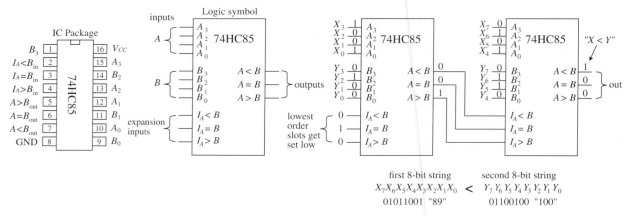

FIGURE 12.49

12.3.7 A Note on Obsolescence and the Trend Toward Microcontroller Control

We have just covered most of the combinational devices you will find discussed in textbooks and listed within electronic catalogs. Many of these devices are still used. However, some devices, such as the binary adders and code converters, are obsolete.

Today, the trend is to use software-controlled devices such as microcontrollers and FPGAs to carry out arithmetic operations and code conversions. Before you attempt to design any logic circuit, we suggest reading Chaps. 12 and 13 first.

Microcontrollers can be used to collect data, store data, and perform logical operations using the input data. They also can generate output signals that can be used to control displays, audio devices, stepper motors, servos, and so on. The specific functions a microcontroller is designed to perform depend on the program you store in its internal ROM-type memory.

Programming the microcontroller typically involves simply using a special programming unit provided by the manufacturer. The programming unit usually consists of a prototyping platform that is linked to a host computer (via a USB port) that is running a development environment. In the development environment, you typically write out a program in a high-level language such as C, or some other specialized language designed for a certain microcontroller, and then, with the press of a key, the program is converted into machine language (1s and 0s) and downloaded into the microcontroller's memory.

In many applications, a single microcontroller can replace entire logic circuits composed of numerous discrete components.

Replacing logic circuits with a microcontroller is almost certainly going to be very cost-effective as small microcontrollers often cost less than a dollar. However, replacing physical logic with a program does have performance differences. It certainly won't be as fast.

For systems where performance is important, FPGAs are probably a better approach. These are configured using either a logic gate schematic or a special purpose hardware description logic, and provide the flexibility of a microcontroller without sacrificing the raw speed of an implementation in logic gate ICs.

12.4   Logic Families

These days, pretty much all logic gates use CMOS technology.

In general, ICs made from MOSFET transistors use less space due to their simpler construction, have very high noise immunity, and consume less power than equivalent bipolar transistor ICs. Historically, the high-input impedance and input capacitance of the MOSFET transistors (due to their insulated gate leads) result in longer time constants for transistor on/off switching speeds when compared with bipolar gates, and therefore typically result in a slower device. Over years of development, however, the CMOS technology has overtaken other technologies.

Both the bipolar and MOSFET logic families can be divided into a number of subclasses and from a historical perspective, it is useful to see where we have come from and some of the more exotic technologies that exist. The major subclasses of the bipolar family include transistor-transistor logic (TTL), emitter-coupled logic (ECL), and integrated-injection logic (IIL or I^2L). The major subclasses of the MOSFET logic include P-channel MOSFET (PMOS), N-channel MOSFET (NMOS), and complementary MOSFET (CMOS). CMOS uses both NMOS and PMOS technologies. The two most popular technologies are TTL and CMOS. The other technologies are typically used in large-scale integration devices, such as microprocessors and memories. There are new technologies popping up all the time, which yield faster, more energy-efficient

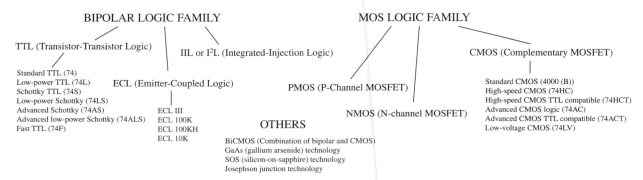

FIGURE 12.50

devices. Some examples include BiCMOS, GaAS, SOS, and Josephson junction technologies.

As you have already learned, digital logic ICs are grouped into functional categories that get placed into either the 74HC00 or 4000 or 4000 CMOS series (or the improved 4000B series).

12.4.1 CMOS Family of ICs

While the 7400 series was going through its various transformations toward CMOS technology, the CMOS series entered the picture. The original CMOS 4000 series (or the improved 4000B series) was developed to offer lower power consumption than the TTL series of devices—a feature made possible by the high input impedance characteristics of its MOSFET transistors. The 4000B series also offered a larger supply voltage range (3 to 18 V), with minimum logic high = $\frac{2}{3}V_{DD}$ and maximum logic low = $\frac{1}{3}V_{DD}$. The 4000B series, though more energy efficient than the TTL series, was significantly slower and more susceptible to damage due to electrostatic discharge. Figure 12.51 shows the internal circuitry of CMOS NAND, AND, and NOR gates. To figure out how the gates work, apply high (logic 1) or low (logic 0) levels to the inputs and see which transistor gates turn on and which transistor gates turn off.

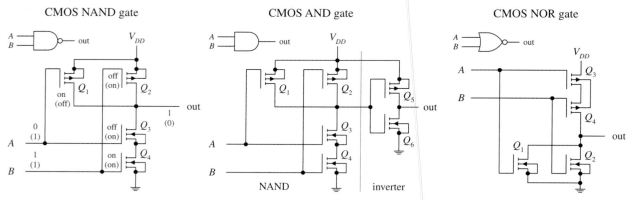

FIGURE 12.51

A further improvement in speed over the original 4000B series came with the introduction of the 40H00 series. Although this series was faster than the 4000B series, it was not quite as fast as the 74LS TTL series. The 74C CMOS series also emerged on the scene, which was designed specifically to be pin-compatible with the TTL line.

Another significant improvement in the CMOS family came with the development of the 74HC and the 74HCT series. Both these series, like the 74C series, were pin-compatible with the TTL 74 series. The 74HC (high-speed CMOS) series had the same speed as the 74LS, as well as the traditional CMOS low-power consumption. The 74HCT (high-speed CMOS TTL compatible) series was developed to be interchangeable with TTL devices (same I/O voltage level characteristics). The 74HC series is very popular today.

Still further improvements in 74HC/74HCT series led to the advanced CMOS logic (74AC/74ACT) series. The 74AC (advanced CMOS) series approached speeds comparable with the 74F TTL series, while the 74ACT (advanced CMOS TTL compatible) series was designed to be TTL compatible.

12.4.2 I/O Voltages and Noise Margins

The exact input voltage levels required for a logic IC to perceive a high (logic 1) or low (logic 0) input level differ between the various logic families. At the same time, the high and low output levels provided by a logic IC vary among the logic families. For example, Fig. 12.52 shows valid input and output voltage levels for the 74HC (CMOS) families.

In Fig. 12.52, the voltage ranges are represented as follows:

- V_{IH} represents the valid voltage range that will be interpreted as a high logic input level.

- V_{IL} represents the valid voltage range that will be interpreted as a low logic input level.

- V_{OL} represents the valid voltage range that will be guaranteed as a low logic output level.

- V_{OH} represents the valid voltage range that will be guaranteed as a high logic output level.

Valid input/output logic levels for the CMOS 74HC

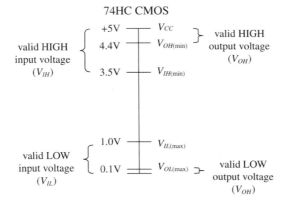

FIGURE 12.52

12.4.3   Current Ratings, Fanout, and Propagation Delays

Logic IC inputs and outputs can sink or source only a given amount of current. I_{IL} is defined as the maximum low-level input current, I_{IH} as the maximum high-level input current, I_{OH} as the maximum high-level output current, and I_{OL} as the maximum low-level output current.

The limit to how much current a device can sink or source determines the size of loads that can be attached. The term *fanout* is used to specify the total number of gates that can be driven by a single gate of the same family without exceeding the current rating of the gate. The fanout is determined by taking the smaller result of I_{OL}/I_{IL} or I_{OH}/I_{IH}. For the 7HC, it is around 50.

12.5   Powering and Testing Logic ICs

Most logic devices will work with 5 V supplies like the ones shown in Fig. 12.53.

5-V line and battery supplies for digital logic circuits

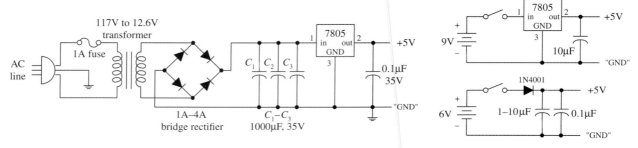

FIGURE 12.53

12.5.1   Power Supply Decoupling

When a logic device makes a low-to-high or a high-to-low level transition, there is an interval during which the conduction times in the upper and lower totem-pole output transistors overlap. During this interval, a drastic change in power supply current occurs, which results in a sharp, high-frequency current spike within the supply line. If a number of other devices are linked to the same supply, the unwanted spike can cause false triggering of these devices. The spike also can generate unwanted electromagnetic radiation.

To avoid unwanted spikes within logic systems, decoupling capacitors can be used. A decoupling capacitor, typically multilayer ceramic, from 0.01 to 0.1 μF (>5 V), is placed directly across the V_{CC}-to-ground pins of each IC in the system. The capacitors absorb the spikes and keep the V_{CC} level at each IC constant, thus reducing the likelihood of false triggering and generally electromagnetic radiation. Decoupling capacitors should be placed as close to the ICs as possible to keep current spikes local, instead of allowing them to propagate back toward the power supply. You can usually get by with using one decoupling capacitor for every five to ten gates or one for every five counter or register ICs.

12.5.2 Unused Inputs

Unused inputs that affect the logical state of a chip should not be allowed to float. Instead, they should be tied high or low, as necessary (floating inputs are liable to pick up external electrical noise, which leads to erratic output behavior). For example, a four-input NAND gate that uses only two inputs should have its two unused inputs held high to maintain proper logic operation. A three-input NOR gate that uses only two inputs should have its unused input held low to maintain proper logic operation. Likewise, the CLEAR and PRESET inputs of a flip-flop should be grounded or tied high, as appropriate.

If there are unused sections within an IC (for example, unused logic gates within a multigate package), the inputs that link to these sections may pick up unwanted charge and may reach a voltage level that causes output MOS transistors to conduct simultaneously, resulting in a large internal current spike from the supply (V_{DD}) to ground. The result can lead to excessive supply current drain and IC damage. To avoid this fate, inputs of unused sections of a CMOS IC should be grounded. Figure 12.54 illustrates what to do with unused inputs for NAND and NOR ICs.

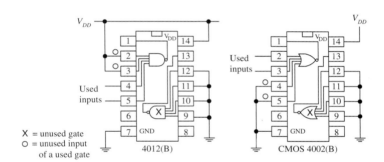

Connect unused inputs of a used NAND gate HIGH to maintain proper logic function. Connect unused inputs of a used NOR gate LOW to maintain proper logic function. Inputs of unused CMOS gates should be grounded.

FIGURE 12.54

As a last note of caution, never drive CMOS inputs when the IC's supply voltage is removed. Doing so can damage the IC's input protection diodes.

12.5.3 Logic Probes and Logic Pulsers

Two simple tools used to test logic ICs and circuits include the test probe and logic pulser, as shown in Fig. 12.55.

A typical logic probe comes in a pen-like package, with metal probe tip and power supply wires: one red and one black. Red is connected to the positive supply voltage of the digital circuit (V_{CC}), and black is connected to the ground (V_{SS}) of the circuit. To test a logic state within a circuit, the metal tip of the probe is applied. If a high voltage is detected, the probe's high LED lights up; if a low voltage is detected, the probe's low LED turns off.

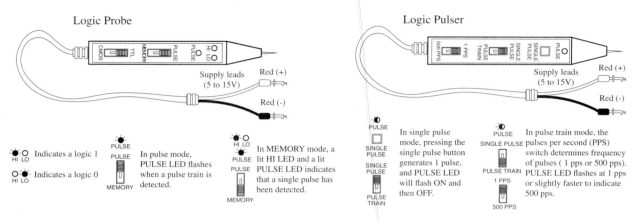

FIGURE 12.55

Along with performing simple static tests, logic probes can perform a few simple dynamic tests, such as detecting a single momentary pulse that is too fast for the human eye to detect or detecting a pulse train, such as a clock signal. To detect a single pulse, the probe's PULSE/MEMORY switch is thrown to the MEMORY position. When a single pulse is detected, the internal memory circuit remembers the single pulse and lights up both the HI LED and PULSE LED at the same time. To clear the memory to detect a new single pulse, the PULSE/MEMORY switch is toggled. To detect a pulse train, the PULSE/MEMORY switch is thrown to the PULSE position. When a pulse train is detected, the PULSE LED flashes on and off.

Logic probes usually will detect single pulses with widths as narrow as 10 ns and will detect pulse trains with frequencies around 100 MHz. Check the specifications that come with your probe to determine these minimum and maximum limits.

A logic pulser allows you to send a single logic pulse or a pulse train through an IC and circuits, where the results of the applied pulses can be monitored by a logic probe. Like a logic probe, the pulser comes with similar supply leads. To send a single pulse, the SINGLE-PULSE/PULSE-TRAIN switch is set to SINGLE-PULSE, and then the SINGLE-PULSE button is pressed. To send a pulse train, switch to PULSE-TRAIN mode. With the pulser model shown in Fig. 12.55, you get to select either 1 pulse per second (pps) or 500 pps.

12.6 Sequential Logic

The combinational circuits covered previously (encoders, decoders, multiplexers, parity generators/checkers, and so on) have the property of input-to-output immediacy. This means that when input data is applied to a combinational circuit, the output responds almost immediately. Now, combinational circuits lack a very important characteristic: they cannot store information. A digital device that cannot store information is not very interesting, practically speaking.

To provide "memory" to circuits, you must create devices that can latch onto data at a desired moment in time. The realm of digital electronics devoted to this subject is referred to as *sequential logic*. This branch of electronics is referred to as *sequential* because for data bits to be stored and retrieved, a series of steps must occur in a particular order. For example, a typical set of steps might involve first sending an enable pulse to a storage device, and then loading a group of data bits all at once (parallel load), or perhaps loading a group of data bits in a serial manner, which takes a number of individual steps. At a later time, the data bits may need to be retrieved by first applying a control pulse to the storage device. A series of other pulses might be required to force the bits out of the storage device.

To push bits through sequential circuits usually requires a clock generator. The clock generator is similar to the human heart. It generates a series of high and low voltages (analogous to a series of high and low pressures as the heart pumps blood) that can set bits into action. The clock also acts as a time base on which all sequential actions can be referenced. Clock generators will be discussed in detail in Sec. 12.6.7. Now, let's take a look at the most elementary of sequential devices: the SR flip-flop.

12.6.1 SR Flip-Flops

The most elementary data-storage circuit is the *set-reset* (SR) *flip-flop*, also referred to as a *transparent latch*. There are two basic kinds of SR flip-flops: the cross-NOR SR flip-flop and the cross-NAND SR flip-flop.

Consider the cross-NOR SR flip-flop shown in Fig. 12.56. At first, it appears that figuring out what the cross-NOR SR flip-flop does given only two input voltages is impossible, since the NOR gates' inputs depend on the outputs, and what are the outputs anyway? (For now, pretend that Q and $\overline{Q}$ are not complements but separate variables; you could call them X and Y if you like.) Well, first of all, you know that a NOR gate will output a high (logic 1) only if both inputs are low (logic 0). From this, you can deduce that if $S = 1$ and $R = 0$, Q must be 1 and $\overline{Q}$ must be 0, regardless of the outputs. This is called the *set condition*. Likewise, by similar argument, we can deduce that if $S = 0$ and $R = 1$, Q must be 0 and $\overline{Q}$ must be 1. This is called the *reset condition*.

Cross-NOR SR flip-flop

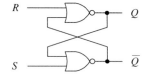
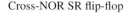

S	R	Q	$\overline{Q}$	condition
0	0	Q	$\overline{Q}$	Hold (no change)
0	1	0	1	Reset
1	0	1	0	Set
1	1	0	0	not used (race)

Cross-NAND SR flip-flop

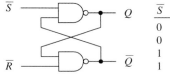

$\overline{S}$	$\overline{R}$	Q	$\overline{Q}$	condition
0	0	1	1	not used (race)
0	1	1	0	Set
1	0	0	1	Reset
1	1	Q	$\overline{Q}$	Hold (no change)

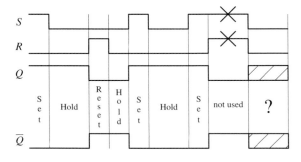

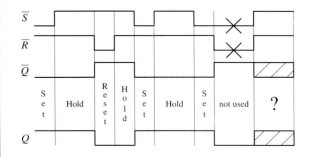

Going from $S =1$, $R =1$ back to the hold condition ($S = 0$, $R = 0$) leads to an unpredictable output. Therefore $S =1$, $R =1$ isn't used.

FIGURE 12.56

But now, what about $R = 0$ and $S = 0$? Can you predict the outputs given only input levels? No! It is impossible to predict the outputs because the outputs are essential for predicting the outputs—it is a "catch-22." However, if you know the states of the outputs beforehand, you can figure things out. For example, if you first set the flip-flop ($S = 1, R = 0, Q = 1, \bar{Q} = 0$), and then apply $S = 0, R = 0$, the flip-flop would remain set (upper gate: $S = 0, Q = 1 \rightarrow \bar{Q} = 0$; lower gate: $R = 0, \bar{Q} = 0 \rightarrow Q = 1$). Likewise, if you start out in reset mode ($S = 0, R = 1, Q = 0, \bar{Q} = 0$), and then apply $S = 0, R = 0$, the flip-flop remains in reset mode (upper gate: $S = 0, Q = 0 \rightarrow \bar{Q} = 1$; lower gate: $R = 0, \bar{Q} = 1 \rightarrow Q = 0$). In other words, the flip-flop remembers, or latches onto, the previous output state even when both inputs go low (0). This is referred to as the *hold* condition.

The last choice you have is $S = 1, R = 1$. Here, it is easy to predict what will happen because you know that as long as there is at least one high (1) applied to the input to the NOR gate, the output will always be 0. Therefore, $Q = 0$ and $\bar{Q} = 0$. Now, there are two fundamental problems with the $S = 1, R = 1$ state. First, why would you want to set and reset at the same time? Second, when you return to the hold condition from $S = 1, R = 1$, you get an unpredictable result, unless you know which input returned low last. Why? When the inputs are brought back to the hold position ($R = 0, S = 0$, $Q = 0, \bar{Q} = 0$), both NOR gates will want to be 1 (they want to be held). But let's say one of the NOR gate's outputs changes to 1 a fraction of a second before the other. In this case, the slower flip-flop will not output a 1 as planned, but will instead output 0. This is a classic example of a *race condition*, where the slower gate loses. But which flip-flop is the slower one? This unstable, unpredictable state cannot be avoided and is simply not used.

The cross-NAND SR flip-flop provides the same basic function as the NOR SR flip-flop, but there is a fundamental difference: its hold and indeterminate states are reversed. This occurs because, unlike the NOR gate, which outputs a high only when both its inputs are low, the NAND gate outputs a low only when both its inputs high. This means that the hold condition for the cross-NAND SR flip-flop is $\bar{S} = 1, \bar{R} = 1$, while the indeterminate condition is $\bar{S} = 0, \bar{R} = 0$.

Now let's look at two simple applications for SR flip-flops.

Switch Debouncer

Say you want to use the far-left switch/pullup resistor circuit (see Fig. 12.57) to drive an AND gate's input high or low (the other input is fixed high). When the switch is open, the AND gate should receive a high. When the switch is closed, the gate

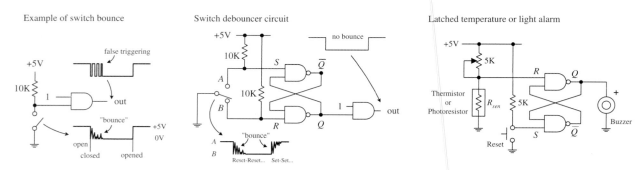

Example of switch bounce   Switch debouncer circuit   Latched temperature or light alarm

FIGURE 12.57

should receive a low. That's what should happen, but that's not what actually happens. Why? Because of switch bounce.

When a switch is closed, the metal contacts bounce a number of times before coming to rest due to inherent springlike characteristics of the contacts. Though the bouncing typically lasts no more than 50 ms, the results can lead to unwanted false triggering, as shown in the far left circuit in Fig. 12.57.

A simple way to get rid of switch bounce is to use the switch debouncer circuit, shown at center of Fig. 12.57. This circuit uses an SR flip-flop to store the initial switch contact voltage while ignoring all trailing bounces. In this circuit, when the switch is thrown from the B to A position, the flip-flop is set. As the switch bounces alternately high and low, the Q output remains high, because when the switch contact bounces away from A, the S input receives a low (R is low, too), but that's just a hold condition; the output stays the same. The same debouncing feature occurs when the switch is thrown from position A to B.

Latched Temperature or Light Alarm

The simple circuit in Fig. 12.57 uses an SR flip-flop to sound a buzzer alarm when the temperature (when using a thermistor) or the light intensity (when using a photoresistor) reaches a critical level. When the temperature or light increases, the resistance of the thermistor or photoresistor decreases, and the R input voltage goes down. When the R input voltage goes below the high threshold level of the NAND gate, the flip-flop is set, and the alarm is sounded. The alarm will continue to sound until the RESET switch is pressed and the temperature or light level has gone below the critical triggering level. The pot is used to adjust this level.

Level-Triggered SR Flip-Flop (the Beginning of Clocked Flip-Flops)

Now it would be nice to make an SR flip-flop synchronous; that is, make the S and R inputs either enabled or disabled by a control pulse, such as a clock. Only when the clock pulse arrives are the inputs sampled. Flip-flops that respond in this manner are referred to as *synchronous* or *clocked flip-flops* (as opposed to the preceding asynchronous flip-flops).

To make the preceding SR flip-flop into a synchronous or clocked device, simply attach enable gates to the inputs of the flip-flop, as shown in Fig. 12.58. The figure shows the cross-NAND arrangement, but a cross-NOR arrangement also can be used. In this setup, only when the clock is high are the S and R inputs enabled. When the clock is low, the inputs are disabled, and the flip-flop is placed in hold mode. The truth table and timing diagram in Fig. 12.58 help illustrate how this device works.

Clocked level-triggered NAND SR flip-flop

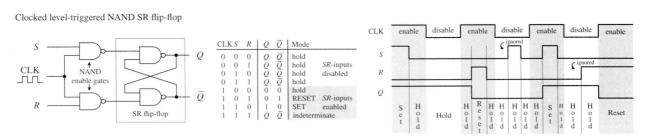

FIGURE 12.58

Edge-Triggered SR Flip-Flops

The level-triggered SR flip-flop has an annoying feature: its S and R inputs must be held at the desired input condition (set, reset, or no change) for the entire time that the clock signal is enabling the flip-flop. With a slight alteration, however, you can make the level-triggered flip-flop more flexible (in terms of timing control) by turning it into an edge-triggered flip-flop.

An edge-triggered flip-flop samples the inputs only during either a positive or negative clock edge ($\uparrow$ = positive edge, $\downarrow$ = negative edge). Any changes that occur before or after the clock edge are ignored—the flip-flop will be placed in hold mode.

To make an edge-triggered flip-flop, introduce either a positive or a negative level-triggered clock pulse generator network into the previous level-triggered flip-flop, as shown in Fig. 12.59.

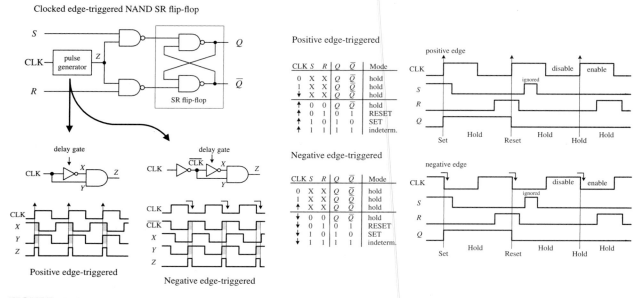

FIGURE 12.59

In a positive edge-triggered generator circuit, a NOT gate with a propagation delay is added. Since the clock signal is delayed through the inverter, the output of the AND gate will not provide a low (as would be the case without a propagation delay), but will provide a pulse that begins at the positive edge of the clock signal and lasts for a duration equal to the propagation delay of the NOT gate. It is this pulse that is used to clock the flip-flop.

Within the negative edge-triggered generator network, the clock signal is first inverted and then applied through the same NOT/AND network. The pulse begins at the negative edge of the clock and lasts for a duration equal to the propagation delay of the NOT gate. The propagation delay is typically so small (in nanoseconds) that the pulse is essentially an "edge."

Pulse-Triggered SR Flip-Flops

A pulse-triggered SR flip-flop is a level-clocked flip-flop; however, for any change in output to occur, both the high and low levels of the clock must rise and fall.

Pulse-triggered flip-flops are also called *master-slave flip-flops*; the master accepts the initial inputs and then "whips" the slave with its output when the negative clock edge arrives. Another analogy often used is to say that during the positive edge, the master gets cocked (like a gun), and during the negative clock edge, the slave gets triggered. Figure 12.60 shows a simplified pulse-triggered cross-NAND SR flip-flop.

Pulse-triggered SR flip-flop (master-slave SR flip-flop)

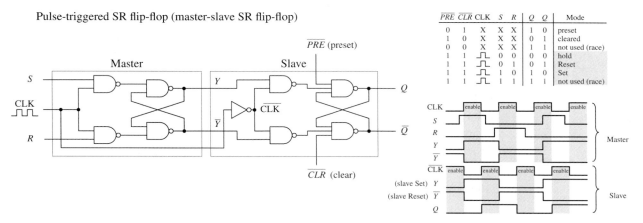

FIGURE 12.60

The master is simply a clocked SR flip-flop that is enabled during the high clock pulse and outputs Y and $\overline{Y}$ (set, reset, or no change). The slave is similar to the master, but it is enabled only during the negative clock pulse (due to the inverter). The moment the slave is enabled, it uses the Y and $\overline{Y}$ outputs of the master as inputs, and then outputs the final result.

Notice the preset $(\overline{PRE})$ and clear $(\overline{CLR})$ inputs. These are called *asynchronous inputs*. Unlike the synchronous inputs, S and R, the asynchronous inputs disregard the clock and either clear (also called *asynchronous* reset) or preset (also called *asynchronous set*) the flip-flop. When $-\overline{CLR}$ is high and $\overline{PRE}$ is low, you get asynchronous reset, $Q = 1$, $\overline{Q} = 0$, regardless of the CLK, S, and R inputs. These active-low inputs are therefore normally pulled high to make them inactive. The ability to apply asynchronous set and resets is often used to clear entire registers that consist of an array of flip-flops.

General Rules for Deciphering Flip-Flop Logic Symbols

Typically, you do not need to worry about constructing flip-flops from scratch. Instead, you buy flip-flop ICs, as discussed in the next section. Likewise, you do not need to worry about complex logic gate schematics. Instead, you use symbolic representations like the ones shown in Fig. 12.61. Although the symbols in the figure apply to SR flip-flops, the basic rules that are outlined can be applied to the D and JK flip-flops, which are discussed in the following sections.

12.6.2   SR Flip-Flop ICs

Occasionally, it can make sense to use a sequential logic IC rather than a microcontroller or FPGA, but gradually if you require more than one or two ICs, you should

Symbolic representation of level-triggered, edge-triggered, and pulse-triggered flip-flops

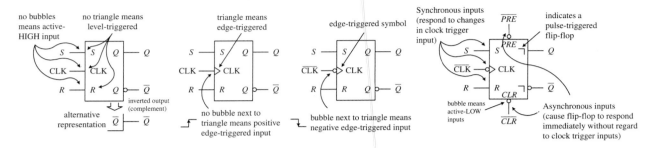

FIGURE 12.61

probably consider a microcontroller or FPGA. Figure 12.62 shows a few sample SR flip-flop (latch) ICs. Always remember that switch debouncing with hardware logic costs money to make, but software switch debouncing for a switch connected to a microcontroller is free. It's just an extra line or two of program code. The 74LS279A contains four independent SR latches (note that two of the latches have an EXTRA SET INPUT). This IC is commonly used in switch debouncers.

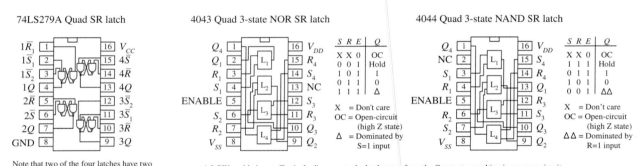

74LS279A Quad SR latch

Note that two of the four latches have two S inputs and that inputs are active-LOW

4043 Quad 3-state NOR SR latch

A LOW enable input effectively disconnects the latch states from the Q outputs, resulting in an open-circuit condition or high-impedance (Z) state at the Q outputs.

S	R	E	Q
X	X	0	OC
0	0	1	Hold
1	0	1	1
0	1	1	0
1	1	1	Δ

X = Don't care
OC = Open-circuit (high Z state)
Δ = Dominated by S=1 input

4044 Quad 3-state NAND SR latch

S	R	E	Q
X	X	0	OC
1	1	1	Hold
0	1	1	1
1	0	1	0
0	0	1	ΔΔ

X = Don't care
OC = Open-circuit (high Z state)
ΔΔ = Dominated by R=1 input

FIGURE 12.62

The 4043 contains four three-state cross-coupled NOR SR latches. Each latch has individual set and reset inputs, as well as separate Q outputs. The three-state feature is an extra bonus, which allows you to effectively disconnect all Q outputs, making it appear that the outputs are open circuits (high impedance, or high Z). This three-state feature is often used in applications where a number of devices must share a common data bus. When the output data from one latch is applied to the bus, the outputs of other latches (or other devices) are disconnected via the high-Z condition. The 4044 is similar to the 4043 but contains four three-state cross-coupled NAND RS latches.

12.6.3   D-Type Flip-Flops

A D-type flip-flop (data flip-flop) is a single input device. It is basically an SR flip-flop, where S is replaced with D and R is replaced with $\bar{D}$ (inverted D). The inverted input is tapped from the D input through an inverter to the R input, as shown in Fig. 12.63. The inverter ensures that the indeterminate condition (race, or not used state, $S = 1$, $R = 1$) never occurs. At the same time, the inverter eliminates the hold

Basic D-type flip-flop or latch

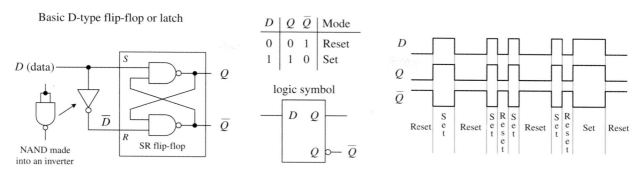

FIGURE 12.63

condition so that you are left with only set ($D = 1$) and reset ($D = 0$) conditions. The circuit in Fig. 12.63 represents a level-triggered D-type flip-flop.

To create a clocked D-type level-triggered flip-flop, first start with the clocked level-triggered SR flip-flop and throw in the inverter, as shown in Fig. 12.64.

Clocked level-triggered D flip-flop

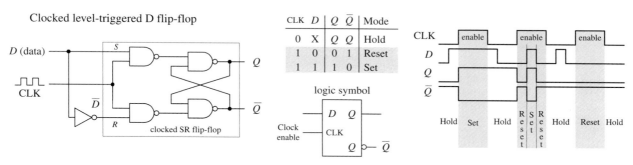

FIGURE 12.64

To create a clocked, edge-triggered, D-type flip-flop, take a clocked edge-triggered SR flip-flop and add an inverter, as shown in Fig. 12.65.

Edge-triggered D flip-flop

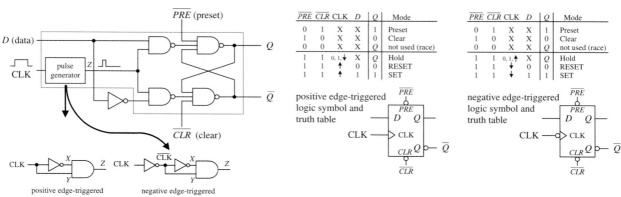

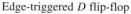

FIGURE 12.65

Figure 12.66 shows a popular edge-triggered D-type flip-flop IC, the 7474 (for example, the 74HC74). It contains two D-type positive edge-triggered flip-flops with asynchronous preset and clear inputs.

74HC74 Dual D-type positive edge-triggered flip-flop with preset and clear

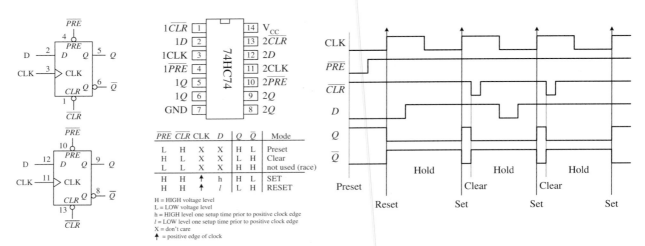

$\overline{PRE}$	$\overline{CLR}$	CLK	D	Q	$\overline{Q}$	Mode
L	H	X	X	H	L	Preset
H	L	X	X	L	H	Clear
L	L	X	X	H	H	not used (race)
H	H	↑	h	H	L	SET
H	H	↑	l	L	H	RESET

H = HIGH voltage level
L = LOW voltage level
h = HIGH level one setup time prior to positive clock edge
l = LOW level one setup time prior to positive clock edge
X = don't care
↑ = positive edge of clock

FIGURE 12.66

Note the lowercase letters l and h in the truth table in this figure. The h is similar to the H for a high voltage level, and the l is similar to the L for low voltage level; however, there is an additional condition that must be met for the flip-flop's output to do what the truth table indicates. The additional condition is that the D input must be fixed high (or low) in duration for at least one *setup time* (t_s) before the positive clock edge. This condition stems from the real-life propagation delays present in flip-flop ICs. If you try to make the flip-flop switch states too fast (do not give it time to move electrons around), you can end up with inaccurate output readings. For the 7474, the setup time is 20 ns. Therefore, when using this IC, you must not apply input pulses that are within the 20-ns limit. Other flip-flops will have different setup times, so you will need to check the manufacturer's data sheets. We will discuss setup time and some other flip-flop timing parameters in greater detail in Sec. 12.6.6.

D-type flip-flops are sometimes found in the pulse-triggered (master-slave) variety. Recall that a pulse-triggered flip-flop requires a complete clock pulse before the outputs will reflect what is applied at the input(s) (in this case, the D input). Figure 12.67 shows the basic structure of a pulse-triggered D flip-flop. It is almost exactly like the pulse-triggered SR flip-flop, except for the inverter addition to the master's input.

Now let's look at a few simple D-type flip-flop applications.

Pulse-triggered D-type flip-flop (master-slave D-type flip-flop)

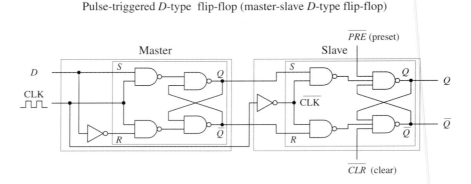

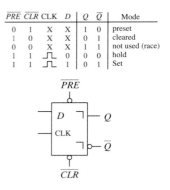

$\overline{PRE}$	$\overline{CLR}$	CLK	D	Q	$\overline{Q}$	Mode
0	1	X	X	1	0	preset
1	0	X	X	0	1	cleared
0	0	X	X	1	1	not used (race)
1	1	⊓	0	0	0	hold
1	1	⊓	1	0	1	Set

FIGURE 12.67

Stop and Go

In the stop-go indicator circuit, a simple level-triggered D-type flip-flop is used to turn on a red LED when its D input is low (reset) and turn on a green LED when the D input is high (set). Only one LED can be turned on at a time.

The divide-by-two counter uses a positive edge-triggered D-type flip-flop to divide an applied signal's frequency by two. The explanation of how this works is simple: The positive edge-triggered feature does not care about negative edges. You can figure out the rest.

External Asynchronous Control Signal

A synchronizer is used when you want to use an external asynchronous control signal (perhaps generated by a switch or other input device) to control some action within a synchronous system. The synchronizer provides a means of keeping the phase of the action generated by the control signal in synch with the phase of the synchronous system.

For example, say you want an asynchronous control signal to control the number of clock pulses that get from point A to point B within a synchronous system. You might try using a simple enable gate, as shown below the synchronizer circuit in Fig. 12.68. However, because the external control signal is not synchronous (in phase) with the clock, when you apply the external control signal, you may shorten the first or last output pulse, as shown in the lower timing diagram.

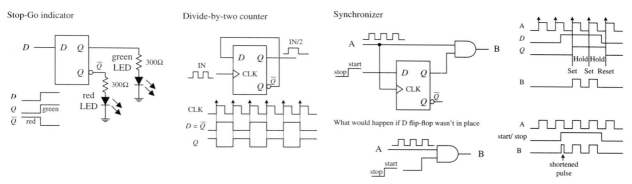

FIGURE 12.68

Certain applications do not like shortened clock pulses and will not function properly. To avoid shortened pulses, throw in an edge-triggered D-type flip-flop to create a synchronizer. The flip-flop's CLK input is tapped off the input clock line, its D input receives the external control signal, and its Q output is connected to the AND gate's enable input. With this arrangement, there will never be shortened clock pulses because the Q output of the flip-flop will not supply enable pulses to the AND gate that are out of phase with the input clock signal. This is due to the fact that after the flip-flop's CLK input receives a positive clock edge, the flip-flop ignores any input changes applied to the D input until the next positive clock edge.

12.6.4 *Quad and Octal D Flip-Flops*

Most frequently, you will find a number of D flip-flops or D latches grouped together within a single IC. For example, the 74HC75, shown in Fig. 12.69, contains four transparent D latches. Latches 0 and 1 share a common active-low enable $E_0 - E_1$, while latches 2 and 3 share a common active-low enable $E_2 - E_3$. From the function table, each Q output follows each D input as long as the corresponding enable line is high. When the enable line goes low, the Q output will become latched to the value that D was one setup time prior to the high-to-low enable transition. The 4042 is another quad D-type latch, which works as described in Fig. 12.69. D-type latches are commonly used as data registers in bus-oriented systems, as is also explained in the figure.

74HC75 quad D latch

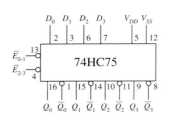

$\bar{E}$	D	Q	$\bar{Q}$	Mode
H	L	L	H	data enabled
H	H	H	L	
L	X	q	$\bar{q}$	data latched

H = High voltage level
L = Low voltage level
X = Don't care
q = Lower case letters indicate the state of referenced output one setup time prior to the High-to-Low enable transition

When $\bar{E}$ (enable) is HIGH, Q follows D.

Application: 4-bit data register

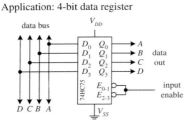

Here a 4-bit data register is created by connecting the enables together to form a single enable input. Data on bus appears at the outputs when the enable input is HIGH. When enable goes LOW, the data that was present on the bus is latched (stored) until latch input goes HIGH.

4042 quad D clocked latch

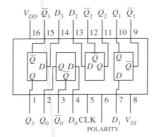

CLK	POLARITY	Q
0	0	D
$\int$	0	Latch
1	1	D
$\backslash$	1	Latch

Outputs either latch or follow the data input, depending on the clock level applied to the polarity input. When clock transistion occurs, information present at the input during the clock transistion is latched until an opposite clock transistion occurs.

Application: 4-bit data register

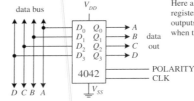

Here a 4042 is used as a 4-bit data register. Data on bus appears at outputs. Data is latched (saved) when the clock switches states.

FIGURE 12.69

D flip-flops also come in octal form—eight flip-flops per IC. These devices are frequently used as 8-bit data registers within microprocessor systems, where devices share 8-bit or $2 \times 8 = 16$-bit data or address buses. An example of an octal D-type flip-flop is the 74HCT273 shown in Fig. 12.70. All D flip-flops within the 74HCT273

74HCT273 octal edge-triggered D-type flip-flop with Clear

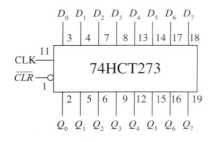

$\overline{CLR}$	CLK	D_n	Q_n	Mode
L	X	X	L	Clear
H	$\uparrow$	h	H	Set
H	$\uparrow$	l	L	Reset

V_{CC} = pin 20
GND = pin 10

H = High voltage level
L = Low voltage level
h = High voltage level one setup time prior to the low-to-high clock transition
l = Low voltage level one setup time prior to the low-to-high clock transition
X = Don't care
$\uparrow$ = Low-to-high clock transition

FIGURE 12.70

share a common positive edge-triggered clock input and a common active-low clear input. When the clock input receives a positive edge, data bits applied to D_0 through D_7 are stored in the eight flip-flops and appear at the outputs Q_0 through Q_7. To clear all flip-flops, the clear input is pulsed low.

12.6.5   JK Flip-Flops

Finally, we come to the last of the flip-flops: the JK flip-flop. A JK flip-flop resembles an SR flip-flop, where J acts like S and K acts like R. Likewise, it has a set mode ($J = 1$, $K = 0$), a reset mode ($J = 0$, $K = 1$), and a hold mode ($J = 0$, $K = 0$). However, unlike the SR flip-flop, which has an indeterminate mode when $S = 1$, $R = 1$, the JK flip-flop has a *toggle* mode when $J = 1$, $K = 1$. *Toggle* means that the Q and $\overline{Q}$ outputs switch to their opposite states at each active clock edge.

To make a JK flip-flop, modify the SR flip-flop's internal logic circuit to include two cross-coupled feedback lines between the output and input. This modification, however, means that the JK flip-flop cannot be level-triggered; it can only be edge-triggered or pulse-triggered. Figure 12.71 shows how you can create edge-triggered flip-flops based on the cross-NAND SR edge-triggered flip-flop.

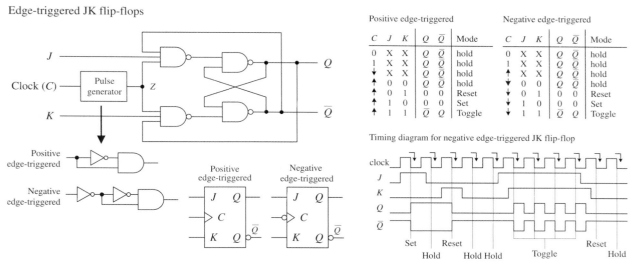

FIGURE 12.71

Edge-triggered JK flip-flops also come with preset (asynchronous set) and clear (asynchronous reset) inputs, as shown in Fig. 12.72.

There are pulse-triggered (master-slave) flip-flops, too. These devices are similar to the pulse-triggered SR flip-flops with the exception of the distinctive JK cross-coupled feedback connections from the slave's Q and $\overline{Q}$ outputs back to the master's input gates. Figure 12.73 shows a simple NAND pulse-triggered JK flip-flop.

The pulse-triggered flip-flops are not as popular as the edge-triggered JK flip-flops because of an undesired effect that can occur. Pulse-triggered JK flip-flops occasionally experience what is called *ones-catching*. In ones-catching, unwanted pulses or

Edge-triggered JK flip-flops with Preset and Clear

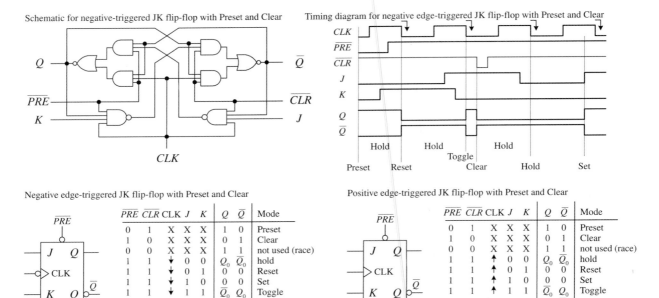

Schematic for negative-triggered JK flip-flop with Preset and Clear

Timing diagram for negative edge-triggered JK flip-flop with Preset and Clear

Negative edge-triggered JK flip-flop with Preset and Clear

$\overline{PRE}$	$\overline{CLR}$	CLK	J	K	Q	$\overline{Q}$	Mode
0	1	X	X	X	1	0	Preset
1	0	X	X	X	0	1	Clear
0	0	X	X	X	1	1	not used (race)
1	1	↓	0	0	Q_0	$\overline{Q}_0$	hold
1	1	↓	0	1	0	0	Reset
1	1	↓	1	0	0	0	Set
1	1	↓	1	1	$\overline{Q}_0$	Q_0	Toggle
1	1	↑ 0,1	1	1	Q_0	$\overline{Q}_0$	hold

Q_0 = state of Q before HIGH-to-LOW edge of clock.

Positive edge-triggered JK flip-flop with Preset and Clear

$\overline{PRE}$	$\overline{CLR}$	CLK	J	K	Q	$\overline{Q}$	Mode
0	1	X	X	X	1	0	Preset
1	0	X	X	X	0	1	Clear
0	0	X	X	X	1	1	not used (race)
1	1	↑	0	0	Q_0	$\overline{Q}_0$	hold
1	1	↑	0	1	0	0	Reset
1	1	↑	1	0	0	0	Set
1	1	↑	1	1	$\overline{Q}_0$	Q_0	Toggle
1	1	↓ 0,1	1	1	Q_0	$\overline{Q}_0$	hold

Q_0 = state of Q before LOW-to-HIGH edge of clock.

FIGURE 12.72

glitches caused by electrostatic noise appear on J and K while the clock is high. The flip-flop remembers these glitches and interprets them as true data. Ones-catching normally is not a problem when clock pulses are of short duration; it is when the pulses get long that you must watch out. To avoid ones-catching altogether, stick with edge-triggered JK flip-flops.

Two major applications for JK flip-flops are found within counter and shift register circuits. Here, we will introduce a counter application. We will discuss shift registers in Sec. 12.8 and additional counter circuits in Sec. 12.7.

Pulse-triggered JK flip-flop (master-slave JK flip-flop)

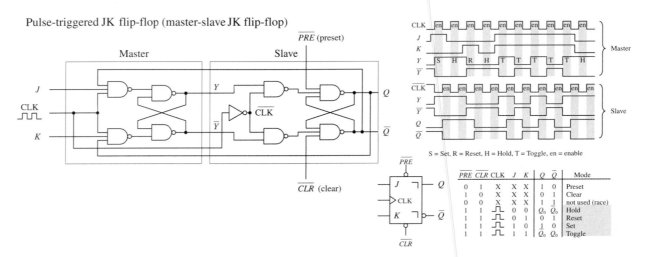

S = Set, R = Reset, H = Hold, T = Toggle, en = enable

$\overline{PRE}$	$\overline{CLR}$	CLK	J	K	Q	$\overline{Q}$	Mode
0	1	X	X	X	1	0	Preset
1	0	X	X	X	0	1	Clear
0	0	X	X	X	1	1	not used (race)
1	1	⊓	0	0	Q_0	$\overline{Q}_0$	Hold
1	1	⊓	0	1	0	1	Reset
1	1	⊓	1	0	1	0	Set
1	1	⊓	1	1	Q_0	$\overline{Q}_0$	Toggle

FIGURE 12.73

74LS76 dual negative edge-triggered JK flip-flop with Preset and Clear

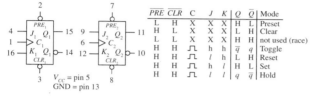

$\overline{PRE}$	$\overline{CLR}$	C	J	K	Q	$\overline{Q}$	Mode
L	H	X	X	X	H	L	Preset
H	L	X	X	X	L	H	Clear
L	L	X	X	X	H	H	not used (race)
H	H	↓	h	h	$\overline{q}$	q	Toggle
H	H	↓	l	h	L	H	Reset
H	H	↓	h	l	H	L	Set
H	H	↓	l	l	q	$\overline{q}$	Hold

V_{CC} = pin 5
GND = pin 13

74109 dual JK positive edge-triggered flip-flop with Preset and Clear

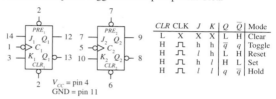

$\overline{PRE}$	$\overline{CLR}$	C_p	J	$\overline{K}$	Q	$\overline{Q}$	Mode
L	H	X	X	X	H	L	Preset
H	L	X	X	X	L	H	Clear
L	L	X	X	X	H	H	not used (race)
H	H	↑	h	l	$\overline{q}$	q	Toggle
H	H	↑	l	h	L	H	Reset
H	H	↑	h	l	H	L	Set
H	H	↑	l	h	q	$\overline{q}$	Hold

V_{CC} = pin 16
GND = pin 8

7476 dual pulse-triggered JK flip-flop with Preset and Clear

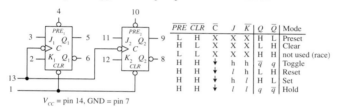

$\overline{PRE}$	$\overline{CLR}$	C	J	K	Q	$\overline{Q}$	Mode
L	H	X	X	X	H	L	Preset
H	L	X	X	X	L	H	Clear
L	L	X	X	X	H	H	not used (race)
H	H	⊓	h	h	$\overline{q}$	q	Toggle
H	H	⊓	l	h	L	H	Reset
H	H	⊓	h	l	H	L	Set
H	H	⊓	l	l	q	$\overline{q}$	Hold

V_{CC} = pin 5
GND = pin 13

74HC73 dual pulse-triggered JK flip-flop with Clear

$\overline{CLR}$	CLK	J	K	Q	$\overline{Q}$	Mode
L	X	X	X	L	H	Clear
H	⊓	h	h	$\overline{q}$	q	Toggle
H	⊓	l	h	L	H	Reset
H	⊓	h	l	H	L	Set
H	⊓	l	l	q	$\overline{q}$	Hold

V_{CC} = pin 4
GND = pin 11

74114 dual pulse-triggered JK flip-flop with common Clock

$\overline{PRE}$	$\overline{CLR}$	C	J	$\overline{K}$	Q	$\overline{Q}$	Mode
L	H	X	X	X	H	L	Preset
H	L	X	X	X	L	H	Clear
L	L	X	X	X	H	H	not used (race)
H	H	↓	h	h	$\overline{q}$	q	Toggle
H	H	↓	l	h	L	H	Reset
H	H	↓	h	l	H	L	Set
H	H	↓	l	l	q	$\overline{q}$	Hold

V_{CC} = pin 14, GND = pin 7

H = HIGH voltage level steady state
h = HIGH voltage level one setup time prior to the HIGH-to-LOW Clock transition
L = LOW voltage level steady state
l = LOW voltage level one setup time prior to the HIGH-to-LOW Clock transition
q = Lowercase letters indicate the state of the referenced output prior to the HIGH-to-LOW Clock transition
X = Don't care
⊓ = Positive Clock pulse
↓ = Negative Clock edge
↑ = Positive Clock edge

FIGURE 12.74

Ripple Counter (Asynchronous Counter)

A simple counter, called a MOD-16 *ripple counter* (or *asynchronous counter*), can be constructed by joining four JK flop-flops together, as shown in Fig. 12.75. (*MOD-16*, or *modulus 16*, means that the counter has 16 binary states.) This means that it can count from 0 to 15—the 0 is one of the counts.

The ripple counter in Fig. 12.75 also can be used as a divide-by-2, -4, -8, or -16 counter. Here, you simply replace the clock signal with any desired input signal that

MOD-16 ripple counter/divide-by-2,4,8,16 counter

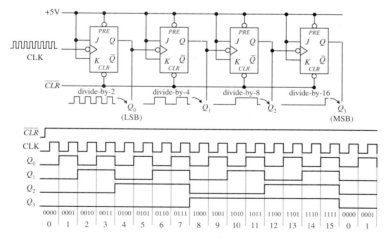

Each flip-flop in the ripple counter is fixed in toggle mode (*J* and *K* are both held high). The clock signal applied to the first flip-flop causes the flip-flop to divide the clock signal's frequency by 2 at its Q_0 output—a result of the toggle. The second flip-flop receives Q_0's output at its clock input and likewise divides by 2. The process continues down the line. What you get in the end is a binary counter with four digits. The LSB is Q_0, while the MSB is Q_3. When the count reaches 1111, the counter recycles back to 0000 and continues from there. To reset the counter at any given time, the active-low clear line is pulsed low. To make the counter count backward from 1111 to 0000, you would simply use the $\overline{Q}$ outputs.

FIGURE 12.75

you wish to divide in frequency. To get a divide-by-2 counter, you only need the first flip-flop; to get a divide-by-8 counter, you need the first three flip-flops.

Ripple counters with higher MOD values can be constructed by slapping on more flip-flops to the MOD-16 counter. But how do you create a ripple counter with a MOD value other than 2, 4, 8, 16, and so on? For example, say you want to create a MOD-10 (0 to 9) ripple counter. And what do you do if you want to stop the counter after a particular count has been reached and then trigger some device, such as an LED or buzzer? Figure 12.76 shows just such a circuit.

Ripple counter that counts from 0 to 9 then stops and activates LED

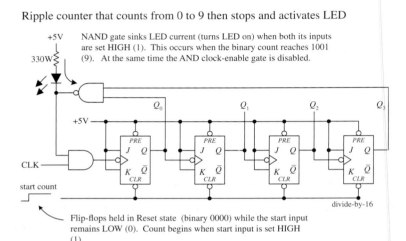

NAND gate sinks LED current (turns LED on) when both its inputs are set HIGH (1). This occurs when the binary count reaches 1001 (9). At the same time the AND clock-enable gate is disabled.

Flip-flops held in Reset state (binary 0000) while the start input remains LOW (0). Count begins when start input is set HIGH (1).

FIGURE 12.76

To make a MOD-10 counter, you simply start with the MOD-16 counter and connect the Q_0 and Q_3 outputs to a NAND gate. When the counter reaches 9 (1001), Q_0 and Q_3 will both go high, causing the NAND gate's output to go low. The NAND gate then sinks current, turning the LED on, while at the same time disabling the clock-enable gate and stopping the count. (When the NAND gate is high, there is no potential difference across the LED to light it up.) To start a new count, the active-low clear line is momentarily pulsed low. Now, to make a MOD-15 counter, you would apply the same basic approach used to the left, but you would connect Q_1, Q_2, and Q_3 to a three-input NAND gate.

Synchronous Counter

There is a problem with the ripple counter just discussed. The output stages of the flip-flops further down the line (from the first clocked flip-flop) take time to respond to changes that occur due to the initial clock signal. This is a result of the internal propagation delay that occurs within a given flip-flop. A standard TTL flip-flop may have an internal propagation delay of 30 ns. If you join four flip-flops to create a MOD-16 counter, the accumulative propagation delay at the highest-order output will be 120 ns. When used in high-precision synchronous systems, such large delays can lead to timing problems.

To avoid large delays, you can create what is called a *synchronous counter*. Synchronous counters, unlike ripple (asynchronous) counters, contain flip-flops whose clock inputs are driven at the same time by a common clock line. This means that output transitions for each flip-flop will occur at the same time.

With this approach, unlike with the ripple counter, you must use some additional logic circuitry placed between various flip-flop inputs and outputs to give the desired count waveform. For example, to create a 4-bit MOD-16 synchronous counter requires adding two additional AND gates, as shown in Fig. 12.77. The AND gates act to keep a flip-flop in hold mode (if both inputs of the gate are low) or toggle mode (if both inputs of the gate are high), as follows:

- During the 0–1 count, the first flip-flop is in toggle mode (and always is); all the rest are in hold mode.

MOD-16 synchronous counter

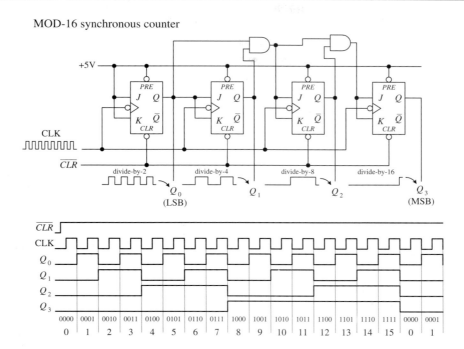

FIGURE 12.77

- When it is time for the 2–4 count, the first and second flip-flops are placed in toggle mode; the last two are held in hold mode.

- When it is time for the 4–8 count, the first AND gate is enabled, allowing the third flip-flop to toggle.

- When it is time for the 8–15 count, the second AND gate is enabled, allowing the last flip-flop to toggle.

You can work out the details for yourself by studying the circuit and timing waveforms.

The ripple (asynchronous) and synchronous counters discussed so far are simple but hardly ever used. In practice, if you need a counter—ripple or synchronous—you purchase a counter IC. These ICs are often MOD-16 or MOD-10 counters and usually come with many additional features. For example, many ICs allow you to preset the count to a desired number via parallel input lines. Others allow you to count up or to count down by means of control inputs. Counter ICs are discussed in Sec. 12.7.

12.6.6 *Practical Timing Considerations with Flip-Flops*

When working with flip-flops, it is important to avoid race conditions. For example, a typical race condition would occur if, say, you were to apply an active clock edge at the very moment you apply a high or low pulse to one of the inputs of a JK flip-flop. Since the JK flip-flop uses what is present on the inputs at the moment the clock edge arrives, having a high-to-low input change will cause problems because you cannot determine if the input is high or low at that moment—it is a straight line.

To avoid this type of race condition, you must hold the inputs of the flip-flop high or low for at least one setup time t_s before the active clock transition. If the input changes during the t_s to the clock edge region, the output levels will be unreliable.

To determine the setup time for a given flip-flop, you must look through the manufacturer's data sheets. For example, the minimum setup time for the 74LS76 JK flip-flop is 20 ns. Other timing parameters, such as hold time and propagation delay, are also given by the manufacturers. A description of what these parameters mean is given in Fig. 12.78.

Flip-Flop Timing Parameters

Clock to output delays, data setup and hold times, clock pulse width

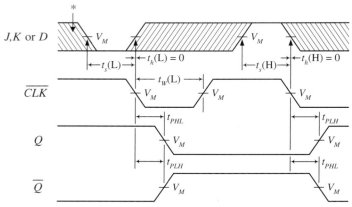

* Shaded region indicates when input is permitted to change for predictable output performance.

Preset and Clear to output delays, Preset and Clear pulse widths

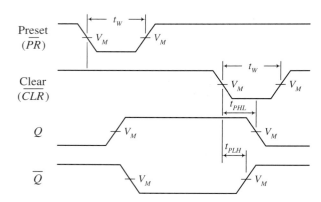

IMPORTANT TERMS

Setup time t_s: The time that the input must be held before the active clock edge for proper operation. For a typical flip-flop, t_s is around 20 ns.

Hold time t_h: The time that the input must be held after the active clock edge for proper operation. For most flip-flops, this is 0 ns—meaning inputs need not be held beyond the active clock signal.

T_{PLH}: Propagation delay from clock trigger point to the low-to-high Q output swing. A typical T_{PLH} for a flip-flop is around 20 ns.

T_{PHL}: Propagation delay from clock trigger point to the high-to-low Q output swing. A typical T_{PLH} for a flip-flop is around 20 ns.

f_{max}: Maximum frequency allowed at the clock input. Any frequency above this limit will result in unreliable performance. This can vary greatly.

$t_W(L)$: Clock pulse width (low), the minimum width (in nanoseconds) that is allowed at the clock input during the low level for reliable operation.

$t_W(H)$: Clock pulse width (high), the minimum width (in nanoseconds) that is allowed at the clock input during the high level for reliable operation.

Preset or clear pulse width: Also given by $t_W(L)$, the minimum width (in nanoseconds) of the low pulse at the preset or clear inputs.

FIGURE 12.78

12.6.7 Digital Clock Generators and Single-Pulse Generators

You have already seen the importance of clock and single-pulse control signals. Now let's take a look at some circuits that can generate these signals.

Clocks (Astable Multivibrators)

A clock is simply a squarewave oscillator. Chapter 10 discusses ways to generate squarewaves, so you can refer there to learn the theory. Here, we will simply present

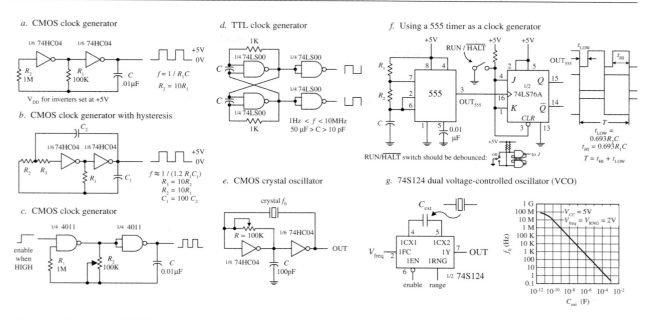

a. CMOS clock generator

b. CMOS clock generator with hysteresis

c. CMOS clock generator

d. TTL clock generator

e. CMOS crystal oscillator

f. Using a 555 timer as a clock generator

g. 74S124 dual voltage-controlled oscillator (VCO)

Figure a. Here, two CMOS inverters are connected together to form an *RC* relaxation oscillator with squarewave output. The output frequency is determined by the *RC* time constant, as shown in the figure.

Figure b. The previous oscillator has one problem: it may not oscillate if the transition regions of its two gates differ, or it may oscillate at a slightly lower frequency than the equation predicts due to the finite gain of the leftmost gate. The oscillator shown here resolves these problems by adding hysteresis via the additional *RC* network.

Figure c. This oscillator uses a pair of CMOS NAND gates and *RC* timing network along with a pot to set the frequency. A squarewave output is generated with a maximum frequency of around 2 MHz. The enable lead could be connected to the other input of the first gate, but here it is brought out to be used as a clock enable input (the clock is enabled when this lead is high).

Figure d. Here, a TTL SR flip-flop with dual feedback resistors uses an *RC* relaxation-type configuration to generate a squarewave. The frequency of the clock is determined by the *R* and *C* values, as shown in the figure. Changing the C_1-to-C_2 ratio changes the duty cycle.

Figure e. When high stability is required, a crystal oscillator is the best choice for a clock generator. Here, a pair of CMOS inverters and a feedback crystal are used (see Chap. 9 for details). The frequency of operation is determined by the crystal (such as 2 MHz or 10 MHz). Adjustment of the pot may be needed to start oscillations.

Figure f. A 555 timer in astable mode can be used to generate squarewaves. Here, we slap on a JK flip-flop that is in toggle mode to provide a means of keeping the low and high times the same, as well as providing clock-enable control. The timing diagram and the equations provided within the figure paint the rest of the picture.

Figure g. The 74S124 dual voltage-controlled oscillator (VCO) outputs squarewaves at a frequency that is dependent on the value of an external capacitor and the voltage levels applied its frequency-range input (V_{RNG}) and its frequency control input (V_{freq}). The graph in this figure shows how the frequency changes with capacitance, while V_{RNG} and V_{freq} are fixed at 2 V. This device also comes with active-low enable input. Other VCOs that are designed for clock generation include the 74LS624, 4024, and 4046 PLL (Phase Locked Loop). You will find many more listed in the catalogs.

FIGURE 12.79

some practical circuits. Digital clocks can be constructed from discrete components such as logic gates, capacitors, resistors, and crystals, or can be purchased in IC form. Figure 12.79 shows some sample clock generators.

Monostables (One-Shots)

To generate single-pulse signals of a desired width, you can use a discrete device called a *monostable multivibrator*, or *one-shot* for short. A one-shot has only one stable state, high (or low), and can be triggered into its unstable state, low (or high), for a duration of time set by an *RC* network. One-shots can be constructed from simple gates, capacitors, and resistors. These circuits, however, tend to be "finicky" and simply are not worth talking about. If you want a one-shot, you can buy a one-shot IC, which typically costs around 50 cents.

Two popular one-shots, shown in Fig. 12.80, are the 74121 nonretriggerable monostable multivibrator and the 74123 retriggerable monostable multivibrator.

Note that a 555 timer IC can also be used as a monostable and is a lower-cost device.

74121 nonretriggerable monostable multivibrator (one-shot)

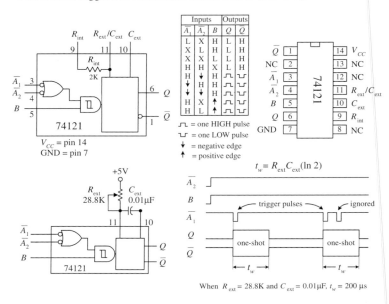

$$t_w = R_{ext} C_{ext}(\ln 2)$$

When $R_{ext} = 28.8K$ and $C_{ext} = 0.01\mu F$, $t_w = 200 \, \mu s$

The 74121 has three trigger inputs $(\bar{A}_1, \bar{A}_2, B)$, true and complemented outputs $(Q, \bar{Q})$, and timing inputs to which an RC network is attached $(R_{ext}/C_{ext}, C_{ext})$. To trigger a pulse from the 74121, you can choose between five possible trigger combinations, as shown in the truth table in the figure. Bringing the input trigger in on B, however, is attractive when dealing with slowly rising or noisy signals, since the signal is directly applied to an internal Schmitt-triggered inverter (recall hystersis). To set the desired output pulse width (t_w), a resistor/capacitor combination is connected to the R_{ext}/C_{ext} and C_{ext} inputs, as shown. (An internal 2-k resistor is provided, which can be used alone by connecting pin 9 to V_{CC} and placing the capacitor across pins 10 and 11, or which can be used in series with an external resistor attached to pin 9. Here, the internal resistor will not be used.) To determine which values to give to the external resistor and capacitor, use the formula given by the manufacturer, which is shown to the left. The maximum t_w should not exceed 28 s ($R = 40$ k, $C = 1000 \, \mu F$) for reliable operation. Also, note that with a nonretriggerable one-shot like the 74121, any trigger pulses applied when the device is already in its astable state will be ignored.

74123 retriggerable monostable multivibrator (one-shot)

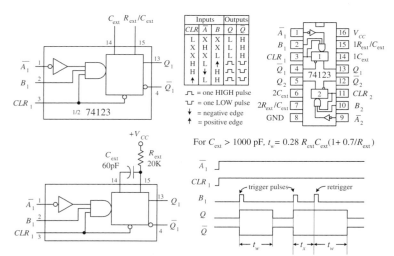

For $C_{ext} > 1000$ pF, $t_w = 0.28 \, R_{ext} C_{ext}(1 + 0.7/R_{ext})$

The 74123 is a dual, retriggerable one-shot. Unlike nonretriggerable one-shots, this device will not ignore trigger pulses that are applied during the astable state. Instead, when a new trigger pulse arrives during an astable state, the astable state will continue to be astable for a time of t_w. In other words, the device is simply retriggered. The 74123 has two trigger inputs $(\bar{A}, B)$ and a clear input (CLR). When CLR is low, the one-shot is forced back into its stable state ($Q = $ low). To determine t_w, use the formula given to the left, provided $C_{ext} > 1000$ pF. If $C_{ext} < 1000$ pF, use $t_w/C_{ext}/R_{ext}$ graphs provided by the manufacturer to find t_w.

FIGURE 12.80

Besides acting as simple pulse generators, one-shots can be combined to make time-delay generators and timing and sequencing circuits (see Fig. 12.81).

Time delay circuit   Timing and sequencing circuit

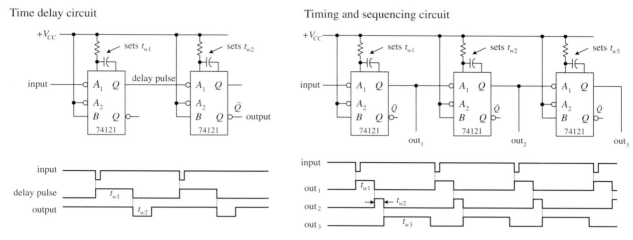

FIGURE 12.81

If you do not have a one-shot IC like the 74121, you can use a 555 timer (discussed in Chap. 9) wired in its monostable configuration, as shown in Fig. 12.82.

Using a 555 timer as a one-shot to generate unique output waveforms

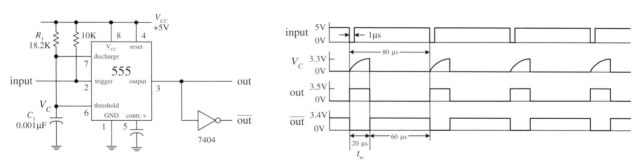

FIGURE 12.82

One-Shot/Continuous-Clock Generator

The circuit shown in Fig. 12.83 is a handy one-shot/continuous clock generator that is useful when you start experimenting with logic circuits.

12.6.8 *Automatic Power-Up Clear (Reset) Circuits*

In sequential circuits, it is usually a good idea to clear (reset) devices when power is first applied. This ensures that devices, such as flip-flops and other sequential ICs, do not start out in a weird mode (for example, counter IC does not start counting at, say, 1101 instead of 0000). Figures 12.84 and 12.85 show some techniques used to provide automatic power-up clearing.

One-shot/continuous-clock generator

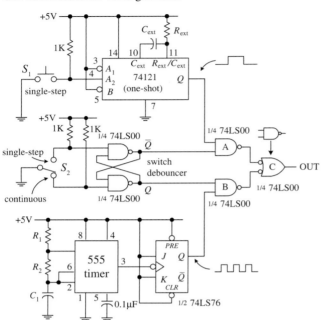

FIGURE 12.83

In this circuit, switch S_2 is used to select whether a single-step or a continuous-clock input is to be presented to the output. When S_2 is in the single-step position, the cross-NAND SR flip-flop (switch debouncer) is set ($Q = 1$, $\overline{Q} = 0$). This disables NAND gate B while enabling NAND gate A, which will allow a single pulse from the one-shot to pass through gate C to the output. To trigger the one-shot, press switch S_1. When S_2 is thrown to the continuous position, the switch debouncer is reset ($Q = 0$, $\overline{Q} = 1$). This disables NAND gate A and enables NAND gate B, allowing the clock signal generated by the 555/flip-flop to pass through gate C and to the output. (Just as a note to avoid confusion, you need gate C to prevent the output from being low and high at the same time.)

Automatic power-up CLEAR circuit

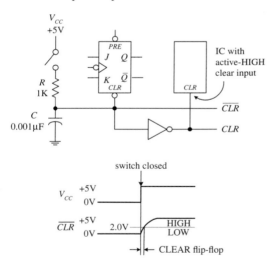

FIGURE 12.84

Let's pretend that one of the devices in a circuit has a JK flip-flop that needs clearing during power-up. In order to clear the flip-flop and then quickly return it to synchronous operations, you would like to apply a low (0) voltage to its active-low clear input; afterward, you would like the voltage to go high (at least above 2.0 V for a 74LS76 JK flip-flop). A simple way to implement this function is to use an RC network like the one shown in the figure. When the power is off (switch open), the capacitor is uncharged (0 V). This means that the $\overline{CLR}$ line is low (0 V). Once the power is turned on (switch closed), the capacitor begins charging up toward V_{CC} (+5 V). However, until the capacitor's voltage reaches 2.0 V, the $\overline{CLR}$ line is considered low to the active-low clear input. After a duration of $t = RC$, the capacitor's voltage will have reached 63 percent of V_{CC}, or 3.15 V; after a duration of $t = 5RC$, its voltage will be nearly equal to +5 V. Since the 74LS76's $\overline{CLR}$ input requires at least 2.0 V to be placed back into synchronous operations, you know that $t = RC$ is long enough. Thus, by rough estimate, if you want the $\overline{CLR}$ line to remain low for 1 μs after power-up, you must set $RC = 1$ μs. Setting $R = 1$ k and $C = 0.001$ μF does the trick.

This automatic resetting scheme can be used within circuits that contain a number of resettable ICs. If an IC requires an active-high reset (not common), simply throw in an inverter and create an active-high clear line, as shown in the figure. Depending on the device being reset, the length of time that the clear line is at a low will be about 1 μs. As more devices are placed on the clear line, the low time duration will decrease due to the additional charging paths. To prevent this from occurring, a larger capacitor can be used.

An improved automatic power-up clear circuit is shown in Fig. 12.85. Here a Schmitt-triggered inverter is used to make the clear signal switch off cleanly. With CMOS Schmitt-triggered inverters, a diode and input resistor (R_2) are necessary to protect the CMOS IC when power is removed.

Improved automatic power-up CLEAR circuit

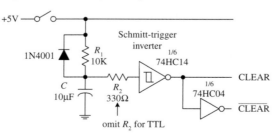

FIGURE 12.85

12.6.9 Pullup and Pulldown Resistors

As you learned when dealing with the switch debouncer circuits, a pullup resistor is used to keep an input high that would otherwise float if left unconnected. If you want to set the "pulled up" input low, you can ground the pin, say, via a switch.

It is important to get an idea of the size of pullup resistor to use. The key here is to make the resistor value small enough so that the voltage drop across it does not weigh down the input voltage below the minimum high threshold voltage ($V_{IH,min}$) of the IC. At the same time, you do not want to make it too small; otherwise, when you ground the pin, excessive current will be dissipated.

In the left diagram in Fig. 12.86, a 10-k pullup resistor is used to keep a 74LS device's input high. To make the input low, close the switch. To figure out if the resistor is large enough so as not to weigh down the input, use $V_{in} = +5\ \text{V} - RI_{IH}$, where I_{IH} is the current drawn into the IC during the high input state, when the switch is open. For a typical 74LS device, I_{IH} is around 20 µA. Thus, by applying the simple formula, you find that $V_{in} = 4.80$ V, which is well above the $V_{IH,min}$ level for a 74LS device. Now, if you close the switch to force the input low, the power dissipated through the resistor ($P_D = V^2/R$) will be $(5\ \text{V})^2/10\ \text{k} = 25$ mW. The graph shown in Fig. 12.86 provides V_{in} versus R and P_D versus R curves. As you can see, if R becomes too large, V_{in} drops below the $V_{IH,min}$ level, and the output will not go high as planned. As R gets smaller, the power dissipation skyrockets. To determine what value of R to use for a specific logic IC, you look up the $V_{IH,min}$ and $I_{IH,max}$ values within the data sheets and apply the simple formulas. In most applications, a 10-k pullup resistor will work fine.

Using a pullup resistor to keep input normally HIGH

Using a pulldown resistor to keep input normally LOW

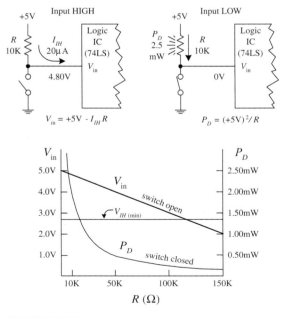

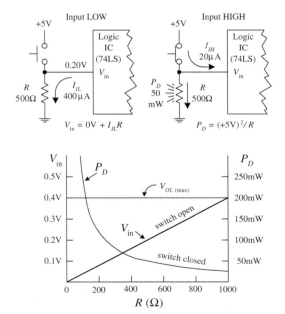

FIGURE 12.86

You will run into situations where a pulldown resistor is used to keep a floating terminal low. Unlike a pullup resistor, the pulldown resistor must be smaller because the input low current I_{IL} (sourced by IC) is usually much larger than I_{IH}. Typically, a pulldown resistor is around 100 to 1 kΩ. A lower resistance ensures that V_{in} is low enough to be interpreted as a low by the logic input. To determine if V_{in} is low enough, use $V_{in} = 0 \text{ V} + I_{IL}R$. As an example, use a 74LS device with an $I_{IL} = 400$ μA and a 500-Ω pulldown resistor. When the switch is open, the input will be 0.20 V—well below the $V_{IL,max}$ level for the 74LS (~0.8 V). When the switch is closed, the power dissipated by the resistor will be $(5 \text{ V})^2/500 \text{ Ω} = 50$ mW. The graph shown in Fig. 12.86 provides V_{in} versus R and P_D versus R curves. As you can see by the curves, if R becomes too large, V_{in} surpasses $V_{IL,max}$, and the output will not be low as planned. As R gets small, the power dissipation skyrockets. If you need to use a pulldown resistor/switch arrangement, be wary of the high power dissipation through the resistor when the switch is closed.

12.7 Counter ICs

In Sec. 12.6.5, you saw how flip-flops could be combined to make both asynchronous (ripple) and synchronous counters. In practice, using discrete flip-flops is to be avoided. Instead, use a prefabricated counter IC. These ICs cost a dollar or two and come with many additional features, like control enable inputs, parallel loading, and so on. A number of different kinds of counter ICs are available. They come in either synchronous (ripple) or asynchronous forms and are usually designed to count in binary or binary-coded decimal (BCD).

12.7.1 Asynchronous Counter (Ripple Counter) ICs

Asynchronous counters work fine for many noncritical applications, but for high-frequency applications that require precise timing, synchronous counters work better. Recall that unlike an asynchronous counter, a synchronous counter contains flip-flops that are clocked at the same time, and hence the synchronous counter does not accumulate nearly as many propagation delays as is the case with the asynchronous counter. Let's look at a few asynchronous counter ICs you will find in the electronics catalogs.

7493 4-Bit Ripple Counter with Separate MOD-2 and MOD-8 Counter Sections

The 7493's internal structure consists of four JK flip-flops connected to provide separate MOD-2 (0-to-1 counter) and MOD-8 (0-to-7 counter) sections. Both the MOD-2 and MOD-8 sections are clocked by separate clock inputs. The MOD-2 section uses C_{p0} as its clock input, while the MOD-8 section uses C_{p1} as its clock input. Likewise, the two sections have separate outputs: MOD-2's output is Q_0, while MOD-8's outputs consist of Q_1, Q_2, and Q_3. The MOD-2 section can be used as a divide-by-2 counter. The MOD-8 section can be used as a divide-by-2 counter (output tapped at Q_1), a divide-by-4 counter (output tapped at Q_2), or a divide-by-8 counter (output tapped at Q_3). If you want to create a MOD-16 counter, simply join the MOD-2 and MOD-8 sections by wiring Q_0 to C_{p1}, while using C_{p0} as the single clock input.

Logic symbol

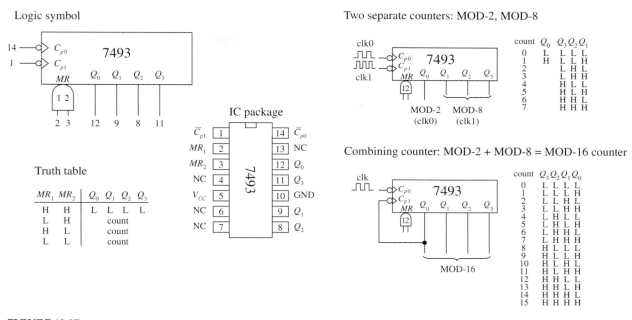

Truth table

MR_1	MR_2	Q_0	Q_1	Q_2	Q_3
H	H	L	L	L	L
L	H	count			
H	L	count			
L	L	count			

FIGURE 12.87

Logic symbol

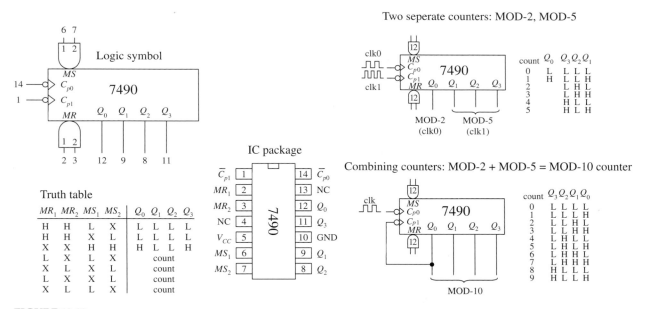

Truth table

MR_1	MR_2	MS_1	MS_2	Q_0	Q_1	Q_2	Q_3
H	H	L	X	L	L	L	L
H	H	X	L	L	L	L	L
X	X	H	H	H	L	L	H
L	X	L	X	count			
X	L	X	L	count			
L	X	X	L	count			
X	L	L	X	count			

FIGURE 12.88

The MOD-2, MOD-8, or the MOD-16 counter can be cleared by making both AND-gated master reset inputs (MR_1 and MR_2) high. To begin a count, one or both of the master reset inputs must be made low. When the negative edge of a clock pulse arrives, the count advances one step. After the maximum count is reached (1 for MOD-2, 111 for MOD-8, or 1111 for MOD-16), the outputs jump back to zero, and a new count begins.

7490 4-Bit Ripple Counter with MOD-2 and MOD-5 Counter Sections

The 7490, like the 7493, is another 4-bit ripple counter. However, its flip-flops are internally connected to provide MOD-2 (count-to-2) and MOD-5 (count-to-5) counter sections. Again, each section uses a separate clock: C_{p0} for MOD-2 and C_{p1} for MOD-5. By connecting Q_0 to C_{p1} and using C_{p0} as the single clock input, a MOD-10 counter (decade or BCD counter) can be created.

When master reset inputs MR_1 and MR_2 are set high, the counter's outputs are reset to 0—provided that master set inputs MS_1 and MS_2 are not both high (the MS inputs override the MR inputs). When MS_1 and MS_2 are high, the outputs are set to $Q_0 = 1$, $Q_1 = 0$, $Q_2 = 0$, and $Q_3 = 1$. In the MOD-10 configuration, this means that the counter is set to 9 (binary 1001). This master set feature comes in handy if you wish to start a count at 0000 after the first clock transition occurs (with master reset, the count starts out at 0001).

7492 Divide-by-12 Ripple Counter with MOD-2 and MOD-6 Counter Sections

The 7492 is another 4-bit ripple counter that is similar to the 7490. However, it has a MOD-2 and a MOD-6 section, with corresponding clock inputs C_{p0} (MOD-2) and C_{p1} (MOD-8). By joining Q_0 to C_{p1}, you get a MOD-12 counter, where C_{p0} acts as the single clock input. To clear the counter, high levels are applied to master reset inputs MR_1 and MR_2.

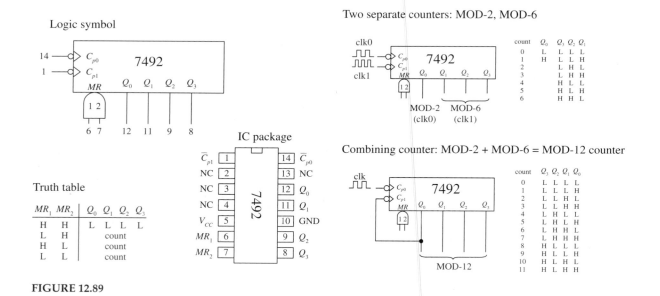

FIGURE 12.89

12.7.2 Synchronous Counter ICs

Like the asynchronous counter ICs, synchronous counter ICs come in various MOD arrangements. These devices usually come with extra goodies, such as controls for up or down counting and parallel load inputs used to preset the counter to a desired start count. Synchronous counter ICs are more popular than the asynchronous ICs, not only because of these additional features, but also because they do not have such long propagation delays as asynchronous counters. Let's take a look at a few popular IC synchronous counters.

74193 Presettable 4-Bit (MOD-16) Synchronous Up/Down Counter

The 74193 is a versatile 4-bit synchronous counter that can count up or count down and can be preset to any count desired—at least a number between 0 and 15. There are two separate clock inputs: C_{pU} is used to count up, and C_{pD} is used to count down. One of these clock inputs must be held high in order for the other input to count. The binary output count is taken from Q_0 (2^0), Q_1 (2^1), Q_2 (2^2), and Q_3 (2^3).

To preset the counter to any desired count, a corresponding binary number is applied to the parallel inputs D_0 to D_3. When the parallel load input ($\overline{PL}$) is pulsed low, the binary number is loaded into the counter, and the count, either up or down, will start from that number. The terminal count up ($\overline{TC_U}$) and terminal count down ($\overline{TC_D}$) outputs are normally high. The $\overline{TC_U}$ output is used to indicate when the maximum count has been reached and the counter is about to recycle to the minimum count (0000)—the carry condition. Specifically, this means that $\overline{TC_U}$ goes low when the count reaches 15 (1111) and the input clock (C_{pU}) goes from high to low. $\overline{TC_U}$ remains low until C_{pU} returns high. This low pulse at $\overline{TC_U}$ can be used as an input to the next high-order stage of a multistage counter. The terminal count down ($\overline{TC_D}$) output is used to indicate that the minimum count has been reached (0000) and the counter is about to recycle to the maximum count 15 (1111)—the borrow condition. Specifically, this means that $\overline{TC_D}$ goes low when the down count reaches 0000 and the input clock (C_{pD}) goes low. Figure 12.90 provides a truth table for the 74193, along with a sample load, up-count, and down-count sequence.

74193 presettable 4-bit binary up/down counter

Example load, count-up, count-down sequence

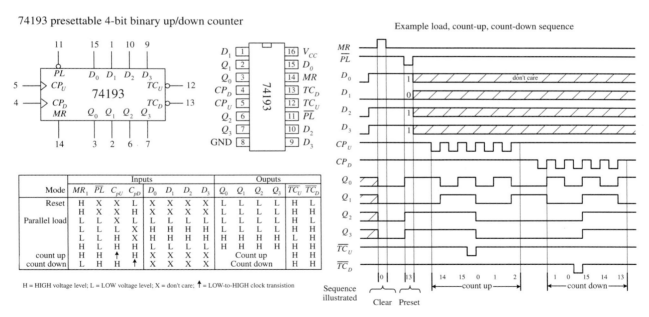

	Inputs								Ouputs					
Mode	MR_1	$\overline{PL}$	C_{pU}	C_{pD}	D_0	D_1	D_2	D_3	Q_0	Q_1	Q_2	Q_3	$\overline{TC_U}$	$\overline{TC_D}$
Reset	H	X	X	L	X	X	X	X	L	L	L	L	H	L
	H	X	X	H	X	X	X	X	L	L	L	L	H	H
Parallel load	L	L	X	L	L	L	L	L	L	L	L	L	H	L
	L	L	L	X	H	H	H	H	L	L	L	L	H	H
	L	L	H	X	H	H	H	H	H	H	H	H	L	H
	H	L	H	H	L	L	L	L	H	H	H	H	L	H
count up	H	H	↑	H	X	X	X	X	Count up				H	H
count down	L	H	H	↑	X	X	X	X	Count down				H	H

H = HIGH voltage level; L = LOW voltage level; X = don't care; ↑ = LOW-to-HIGH clock transistion

FIGURE 12.90

74192 Presettable Decade (BCD or MOD-10) Synchronous Up/Down Counter

The 74192, shown in Fig. 12.91, is essentially the same device as the 74193, except it counts up from 0 to 9 and repeats or counts down from 9 to 0 and repeats. When counting up, the terminal count up ($\overline{TC_U}$) output goes low to indicate when the maximum count is reached (9 or 1001) and the C_{pU} clock input goes from high to

74192 presettable decade (BCD) up/down counter

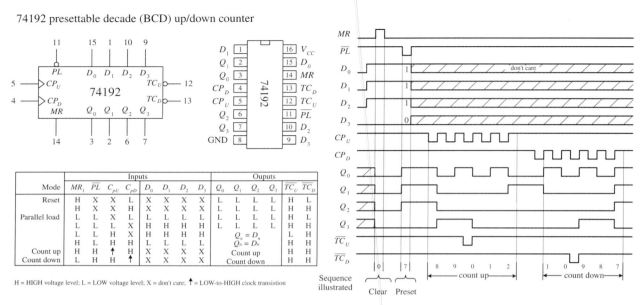

Mode	Inputs								Outputs					
	MR_1	$\overline{PL}$	C_{pU}	C_{pD}	D_0	D_1	D_2	D_3	Q_0	Q_1	Q_2	Q_3	$\overline{TC_U}$	$\overline{TC_D}$
Reset	H	X	X	L	X	X	X	X	L	L	L	L	H	L
	H	X	X	H	X	X	X	X	L	L	L	L	H	H
Parallel load	L	L	X	L	L	L	L	L	L	L	L	L	H	L
	L	L	L	X	H	H	H	H	L	L	L	L	H	H
	L	L	H	X	H	H	H	H	$Q_n = D_n$				L	H
	H	L	H	H	L	L	L	L	$Q_n = D_n$				H	H
Count up	H	H	↑	H	X	X	X	X	Count up				H	H
Count down	L	H	H	↑	X	X	X	X	Count down				H	H

H = HIGH voltage level; L = LOW voltage level; X = don't care; ↑ = LOW-to-HIGH clock transistion

Sequence illustrated

FIGURE 12.91

low. $\overline{TC_U}$ remains low until C_{pU} returns high. When counting down, the terminal count down output ($\overline{TC_D}$) goes low when the minimum count is reached (0 or 0000) and the input clock C_{pD} goes low. The truth table and example load, count-up, and count-down sequence provided in Fig. 12.91 explain how the 74192 works in greater detail.

74190 Presettable Decade (BCD or MOD-10) and 74191 Presettable 4-Bit (MOD-16) Synchronous Up/Down Counters

The 74190 and the 74191 do basically the same things as the 74192 and 74193, but the input and output pins, as well as the operating modes, are a bit different. (The 74190 and the 74191 have the same pinouts and operating modes; the only difference is the maximum count.) Like the previous synchronous counters, these counters can be preset to any count by using the parallel load ($\overline{PL}$) operation. However, unlike the previous synchronous counters, to count up or down requires using a single input: $\overline{u}/D$. When $\overline{u}/D$ is set low, the counter counts up; when $\overline{u}/D$ is high, the counter counts down.

A clock enable input ($\overline{CE}$) acts to enable or disable the counter. When $\overline{CE}$ is low, the counter is enabled. When $\overline{CE}$ is high, counting stops, and the current count is held fixed at the Q_0 to Q_3 outputs.

Unlike the previous synchronous counters, the 74190 and the 74191 use a single terminal count output (TC) to indicate when the maximum or minimum count has occurred and the counter is about to recycle. In count-down mode, TC is normally low but goes high when the counter reaches zero (for both the 74190 and 74191). In count-up mode, TC is normally low but goes high when the counter reaches 9 (for the 74190) or reaches 15 (for the 74191).

The ripple-clock output ($\overline{RC}$) follows the input clock (CP) whenever TC is high. This means, for example, that in count-down mode, when the count reaches zero, $\overline{RC}$ will go low when CP goes low. The $\overline{RC}$ output can be used as a clock input to the next

74190 presettable decade (BCD) up/down counter

74191 presettable 4-bit binary up/down counter

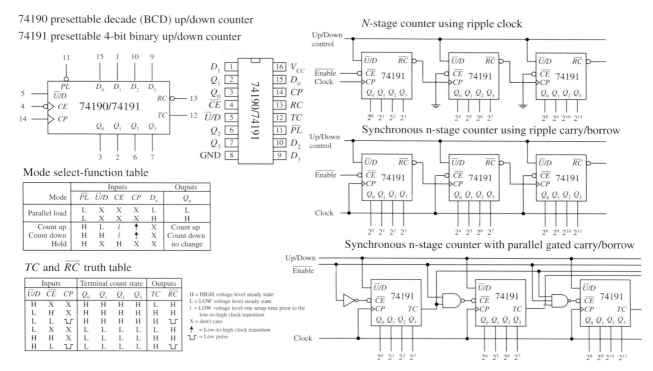

Mode select-function table

TC and $\overline{RC}$ truth table

FIGURE 12.92

higher stage of a multistage counter. This, however, leads to a multistage counter that is not truly synchronous because of the small propagation delay from CP to $\overline{RC}$ of each counter. To make a multistage counter that is truly synchronous, you must tie each IC's clock to a common clock input line. You use the TC output to inhibit each successive stage from counting until the previous stage is at its terminal count. Figure 12.92 shows various asynchronous (ripple-like) and synchronous multistage counters built from 74191 ICs.

Presettable 4-Bit (MOD-16) Synchronous Up/Down Counter

The 74160 and 74163 resemble the 74190 and 74191 but require no external gates when used in multistage counter configurations. Instead, you simply cascade counter ICs together, as shown in Fig. 12.93.

For both devices, a count can be preset by applying the desired count to the D_0 to D_3 inputs and then applying a low to the parallel enable input $(\overline{PE})$; the input number is loaded into the counter on the next low-to-high clock transition. The master reset $(\overline{MR})$ is used to force all Q output low, regardless of the other input signals. The two clock enable inputs (CEP and CET) must be high for counting to begin. The terminal count output (TC) is forced high when the maximum count is reached, but will be forced low if CET goes low. This is an important feature that makes the multistage configuration synchronous, while avoiding the need for external gating. The truth tables along with the example load, count-up, and count-down timing sequences in Figs. 12.93 and 12.94 should help you better understand how these two devices work.

74163 Synchronous 4-bit binary (MOD-16) up counter

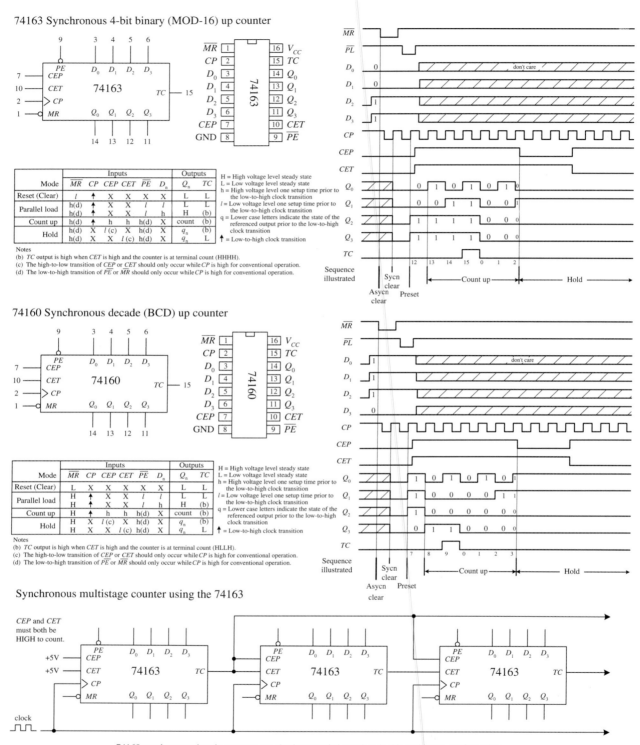

	Inputs						Outputs	
Mode	$\overline{MR}$	CP	CEP	CET	$\overline{PE}$	D_n	Q_n	TC
Reset (Clear)	l	↑	X	X	X	X	L	L
Parallel load	h(d)	↑	X	X	l	l	L	L
	h(d)	↑	X	X	l	h	H	(b)
Count up	h(d)	↑	h	h	h(d)	X	count	(b)
Hold	h(d)	X	l (c)	X	h(d)	X	q_n	(b)
	h(d)	X	X	l (c)	h(d)	X	q_n	L

H = High voltage level steady state
L = Low voltage level steady state
h = High voltage level one setup time prior to
 the low-to-high clock transition
l = Low voltage level one setup time prior to
 the low-to-high clock transition
q = Lower case letters indicate the state of the
 referenced output prior to the low-to-high
 clock transition
↑ = Low-to-high clock transition

Notes
(b) *TC* output is high when *CET* is high and the counter is at terminal count (HHHH).
(c) The high-to-low transition of *CEP* or *CET* should only occur while *CP* is high for conventional operation.
(d) The low-to-high transition of *PE* or *MR* should only occur while *CP* is high for conventional operation.

74160 Synchronous decade (BCD) up counter

	Inputs						Outputs	
Mode	$\overline{MR}$	CP	CEP	CET	$\overline{PE}$	D_n	Q_n	TC
Reset (Clear)	L	X	X	X	X	X	L	L
Parallel load	H	↑	X	X	l	l	L	L
	H	↑	X	X	l	h	H	(b)
Count up	H	↑	h	h	h(d)	X	count	(b)
Hold	H	X	l (c)	X	h(d)	X	q_n	(b)
	H	X	X	l (c)	h(d)	X	q_n	L

H = High voltage level steady state
L = Low voltage level steady state
h = High voltage level one setup time prior to
 the low-to-high clock transition
l = Low voltage level one setup time prior to
 the low-to-high clock transition
q = Lower case letters indicate the state of the
 referenced output prior to the low-to-high
 clock transition
↑ = Low-to-high clock transition

Notes
(b) *TC* output is high when *CET* is high and the counter is at terminal count (HLLH).
(c) The high-to-low transition of *CEP* or *CET* should only occur while *CP* is high for conventional operation.
(d) The low-to-high transition of *PE* or *MR* should only occur while *CP* is high for conventional operation.

Synchronous multistage counter using the 74163

74160 synchronous decade counters can also be cascaded together in this multistage configuration.

FIGURE 12.93

74LS90: Divide-by-*n* frequency counters

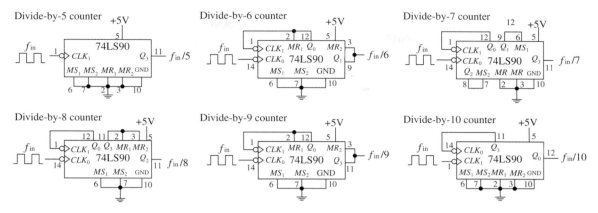

74LS90: 000 to 999 BCD counter

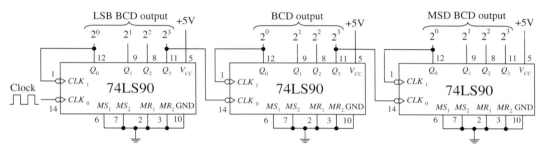

FIGURE 12.94

12.7.3 *A Note on Counters with Displays*

If you want to build a fairly sophisticated counter that can display many digits, the previous techniques are not worth pursuing, because there are simply too many discrete components to work with (for example, a separate seven-segment decoder/ driver for each digit). A common alternative approach is to use a microcontroller or FPGA that functions both as a counter and a display driver.

What microcontrollers and FPGAs can do that discrete circuits have a hard time achieving is *multiplex* a display. In a multiplexed system, corresponding segments of each digit of a multidigit display are linked together, while the common lines for each digit are brought out separately. You can see that the number of lines is significantly reduced; a nonmultiplexed 7-segment 4-digit display has 28 segment lines and 4 common lines, while the 4-digit multiplexed display has only 7 + 4, or 11, lines.

The trick to multiplexing involves flashing each digit, one after the other (and recycling), in a fast enough manner to make it appear that the display is continuously lit. In order to multiplex, the microcontroller's program must supply the correct data to the segment lines at the same time that it enables a given digit via a control signal sent to the common lead of that digit. We will talk about multiplexing displays in greater detail in Chap. 13 with microcontrollers and Chap. 14 using FPGAs.

60-Hz, 10-Hz, and 1-Hz Clock-Pulse Generator

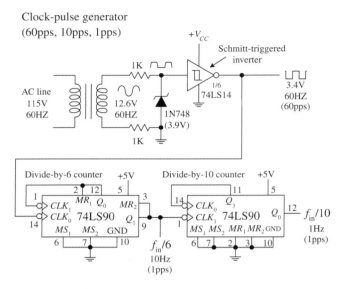

This simple clock-pulse generator provides a unique way to generate 60-, 10-, and 1-Hz clock signals that can be used in applications that require real-time counting. The basic idea is to take the characteristic 60-Hz ac line voltage (from the wall socket) and convert it into a lower-voltage squarewave of the same frequency. (Note that countries other than the United States typically use 50 Hz instead of 60 Hz. For 50 Hz operation, use an appropriate transformer and replace the divide-by-6 counter with the divide-by-5 counter shown in the upper left of Fig. 12.94.) First, the ac line voltage is stepped down to 12.6 V by the transformer. The negative-going portion of the 12.6-V ac voltage is removed by the zener diode (which acts as a half-wave rectifier). At the same time, the zener diode clips the positive-going signal to a level equal to its reverse breakdown voltage (3.9 V). This prevents the Schmitt-triggered inverter from receiving an input level that exceeds its maximum input rating. The Schmitt-triggered inverter takes the rectified/chipped sine wave and converts it into a true squarewave. The Schmitt trigger's output goes low (~0.2 V) when the input voltage exceeds its positive threshold voltage V_T^+ (~1.7 V) and goes high (~3.4 V) when its input falls below its negative threshold voltage V_T^- (~0.9 V). From the inverter's output, you get a 60-Hz squarewave (or a clock signal beating out 60 pulses per second). To get a 10-Hz clock signal, you slap on a divide-by-6 counter. To get a 1-Hz signal, you slap a divide-by-10 counter onto the output of the divide-by-6 counter.

FIGURE 12.95

Another approach used to create multidigit counters is to use a multidigit counter/display driver IC. One such IC is the ICM7217, a four-digit LED display programmable up/down counter made by Intersil. This device is typically used in hardwired applications where thumbwheel switches are used to load data and SPDT switches are used to control the chip. The ICM7217A provides multiplexed seven-segment LED display outputs that are used to drive common cathode displays.

A simple application of the ICM7217A is a four-digit unit counter shown in Fig. 12.96. If you are interested in knowing all the specifics of how this counter

ICM7217A (Intersil) 4-Digit LED Display, Programmable Up/Down Counter

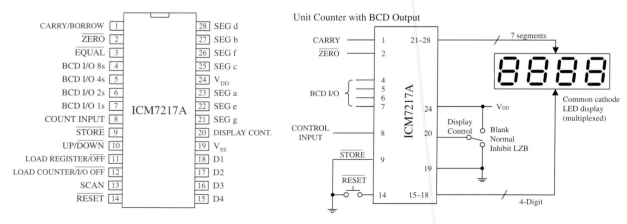

FIGURE 12.96

works, along with learning about other applications for this device, check out Maxim's data sheets at http://www.maxim-ic.com/datasheet/index.mvp/id/1501. It is better to learn from the maker in this case. Also, take a look at the other counter/display driver ICs Maxim has to offer. Other manufacturers produce similar devices, so visit their websites as well.

12.8   Shift Registers

Data words traveling through a digital system frequently must be temporarily held, copied, and bit-shifted to the left or to the right. A device that can be used for such applications is the *shift register*. A shift register is constructed from a row of flip-flops connected so that digital data can be shifted down the row either in a left or right direction. Most shift registers can handle parallel movement of data bits as well as serial movement, and also can be used to convert from parallel to serial or from serial to parallel. Figure 12.97 shows several types of shift register arrangements: serial-in/serial-out, parallel-in/serial-out, and serial-in/parallel out.

Block diagrams of the serial-in/serial-out, parallel-in/serial-out, and serial-in/parallel-out shift registers

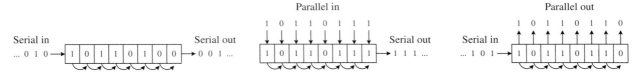

FIGURE 12.97

12.8.1   Serial-In/Serial-Out Shift Registers

Figure 12.98 shows a simple 4-bit serial-in/serial-out shift register made from D flip-flops. Serial data is applied to the D input of flip-flop 0. When the clock line receives a positive clock edge, the serial data is shifted to the right from flip-flop 0 to flip-flop 1. Whatever bits of data were present at flip-flop 2's, 3's, and 4's outputs are shifted to the right during the same clock pulse. To store a 4-bit word into this register requires four clock pulses. The rightmost circuit shows how you can rewire the flip-flops to make a shift-left register. To make larger bit-shift registers, more flip-flops are added (for example, an 8-bit shift register would require eight flip-flops cascaded together).

Simple 4-bit serial-in/serial-out shift registers

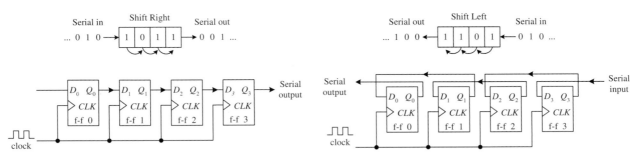

FIGURE 12.98

12.8.2 Serial-In/Parallel-Out Shift Registers

Figure 12.99 shows a 4-bit serial-in/parallel-out shift register constructed from D flip-flops. This circuit is essentially the same as the previous serial-in/serial-out shift register, except now you attach parallel output lines to the outputs of each flip-flop as shown. Note that this shift register circuit also comes with an active-low clear input ($\overline{CLR}$) and a strobe input that acts as a clock enable control. The timing diagram in the figure shows a sample serial-to-parallel shifting sequence.

4-bit serial-in/parallel-out shift register

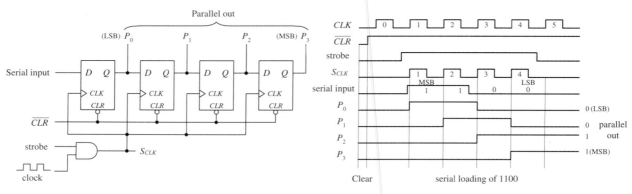

FIGURE 12.99

12.8.3 Parallel-In/Serial-Out Shift Registers

Constructing a 4-bit parallel-to-serial shift register from D flip-flops requires some additional control logic, as shown in the circuit in Fig. 12.100. Parallel data must first be loaded into the D inputs of all four flip-flops. To load data, the SHIFT/$\overline{LOAD}$ is made low. This enables the AND gates with X marks, allowing the 4-bit parallel input word to enter the D_0–D_3 inputs of the flip-flops. When strobe and CLK are both high, the 4-bit parallel word is latched simultaneously into the four flip-flops and appears at the Q_0–Q_3 outputs. To shift the latched data out through the serial output, the

Parallel-to-serial shift register

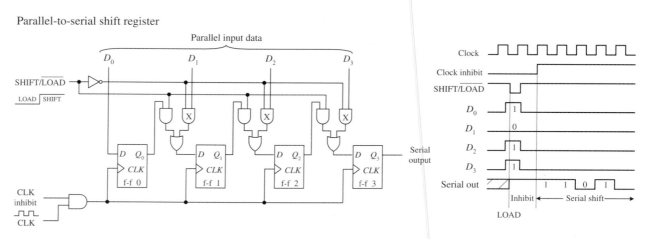

FIGURE 12.100

SHIFT/ $\overline{\text{LOAD}}$ line is made high. This enables all unmarked AND gates, allowing the latched data bit at the Q output of a flip-flop to pass (shift) to the D input of the flip-flop to the right. In this shift mode, four clock pulses are required to shift the parallel word out of the serial output.

12.8.4 Ring Counter (Shift Register Sequencer)

The ring counter (shift register sequencer) is a unique type of shift register that incorporates feedback from the output of the last flip-flop to the input of the first flip-flop. Figure 12.101 shows a 4-bit ring counter made from D-type flip-flops. In this circuit, when the $\overline{\text{START}}$ input is set low, Q_0 is forced high by the active-low preset, while Q_1, Q_2, and Q_3 are forced low (cleared) by the active-low clear. This causes the binary word 1000 to be stored within the register. When the $\overline{\text{START}}$ line is brought low, the data bits stored in the flip-flops are shifted right with each positive clock edge. The data bit from the last flip-flop is sent to the D input of the first flip-flop. The shifting cycle will continue to recirculate while the clock is applied. To start a fresh cycle, the $\overline{\text{START}}$ line is momentarily brought low.

Ring counter using positive edge-triggered D flip-flops

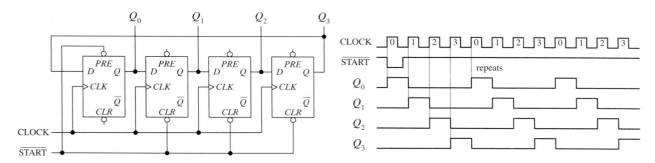

FIGURE 12.101

12.8.5 Johnson Shift Counter

The Johnson shift counter is similar to the ring counter except that its last flip-flop feeds data back to the first flip-flop from its inverted output ($\overline{Q}$). For this reason, this type is sometimes called a Moebius counter, as the bit sequence will be shifted out first "normally," then inverted, then normally, and so on. In the simple 4-bit Johnson shift counter shown in Fig. 12.102, you start out by applying a low to the $\overline{\text{START}}$ line, which sets presets Q_0 high; Q_1, Q_2, and Q_3 low; and $\overline{Q}_3$ high. In other words, you load the register with the binary word 1000, as you did with the ring counter.

Now, when you bring the $\overline{\text{START}}$ line low, data will shift through the register. However, unlike the ring counter, the first bit sent back to the D_0 input of the first flip-flop will be high because feedback is from $\overline{Q}_3$ not Q_3. At the next clock edge, another high is fed back to D_0; at the next clock edge, another high is fed back; at the next edge, another high is fed back. Only after the fourth clock edge does a low get fed back (the 1 has shifted down to the last flip-flop and $\overline{Q}_3$ goes high). At this point, the shift register is full of 1s.

Johnson counter using positive edge-triggered D flip-flops

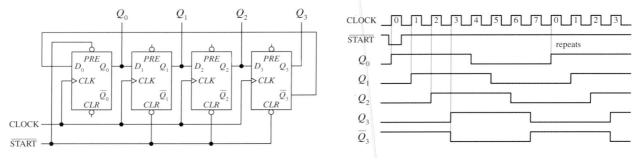

FIGURE 12.102

As more clock pulses arrive, the feedback loop supplies lows to D_0 for the next four clock pulses. After that, the Q outputs of all the flip-flops are low, while $\bar{Q}_3$ goes high. This high from $\bar{Q}_3$ is fed back to $\bar{D}_0$ during the next positive clock edge, and the cycle repeats.

As you can see, the 4-bit Johnson shift counter has eight output stages (which require eight clock pulses to recycle), not four, as is the case with the ring counter.

12.8.6 Shift Register ICs

Now that we have covered the basic theory of shift registers, let's take a look at practical shift register ICs that contain all the necessary logic circuitry inside. It is not uncommon for a serial to parallel shift register IC to be used with a microcontroller to provide it with more outputs when driving LEDs. The serial data is fed into the shift register and then the output latched to turn the LEDs on or off.

7491A 8-Bit Serial-In/Serial-Out Shift Register IC

The 7491A is an 8-bit serial-in/serial-out shift register that consists of eight internally linked SR flip-flops. This device has positive edge-triggered inputs and a pair of data inputs (A and B) that are internally ANDed together, as shown in the logic diagram in Fig. 12.103. This type of data input means that for a binary 1 to be shifted into the

7491A 8-bit serial-in/serial-out shift register IC

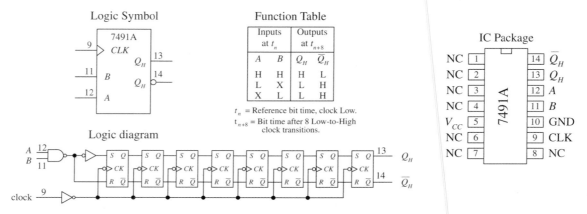

FIGURE 12.103

register, both data inputs must be high. For a binary 0 to be shifted into the register, either input can be low. Data is shifted to the right at each positive clock edge.

74164 8-Bit Serial-In/Parallel-Out Shift Register IC

The 74164 is an 8-bit serial-in/parallel-out shift register. It contains eight internally linked flip-flops and has two serial inputs, D_{sa} and D_{sb}, which are ANDed together. Like the 7491A, the unused serial input acts as an enable/disable control for the other serial input. For example, if you use D_{sa} as the serial input, you must keep D_{sb} high to allow data to enter the register, or you can keep it low to prevent data from entering the register.

Data bits are shifted one position to the right at each positive clock edge. The first data bit entered will end up at the Q_7 parallel output after the eighth clock pulse. The master reset ($\overline{MR}$) resets all internal flip-flops and forces the Q outputs low when it is pulsed low.

In the sample circuit shown in Fig. 12.104, a serial binary number 10011010 (154$_{10}$) is converted into its parallel counterpart. Note the AND gate and strobe input used in this circuit. The strobe input acts as a clock enable input; when it is set high, the clock is enabled. The timing diagram paints the rest of the picture.

The 74164 8-bit serial-in/parallel-out shift register IC

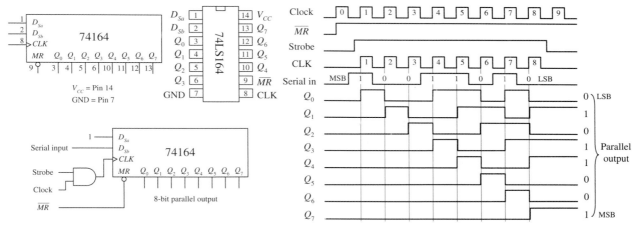

FIGURE 12.104

75165 8-Bit Serial-In or Parallel-In/Serial-Out Shift Register IC

The 75165 is a unique 8-bit device that can act as either a serial-to-serial shift register or as a parallel-to-serial shift register. When used as a parallel-to-serial shift register, parallel data is applied to the D_0–D_7 inputs and then loaded into the register when the parallel load input ($\overline{PL}$) is pulsed low. To begin shifting the loaded data out of the serial output Q_7 (or $\overline{Q}_7$ if you want inverted bits), the clock enable input ($\overline{CE}$) must be set low to allow the clock signal to reach the clock inputs of the internal D-type flip-flops. When used as a serial-to-serial shift register, serial data is applied to the serial data input DS. A sample shift, load, and inhibit timing sequence is shown in Fig. 12.105.

74165 8-Bit (serial-in or parallel-in)/serial-out shift register

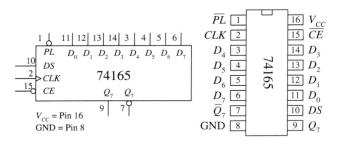

Sample shift, load, and inhibit sequence for parallel load case

Operating Modes	Inputs					Q_n Register		Outputs	
	$\overline{PL}$	$\overline{CE}$	CLK	DS	D_0-D_7	Q_0	Q_1-Q_6	Q_7	$\overline{Q}_7$
Parallel load	L	X	X	X	L	L	L-L	L	H
	L	X	X	X	H	H	H-H	H	L
Serial shift	H	L	↑	l	X	L	q_0-q_5	q_6	$\overline{q}_6$
	H	L	↑	h	X	H	q_0-q_5	q_6	$\overline{q}_6$
Hold ("do nothing")	H	H	X	X	X	q_0	q_1-q_6	q_7	$\overline{q}_7$

H = High voltage level; h = High voltage level one setup time prior to the low-to-high clock transition; L = Low voltage level; l = Low voltage level one setup time prior to the low-to-high clock transiton; q_n = Lower case letters indicate the state of the referenced output one setup time prior to the low-to-high clock transition; X = Don't care; ↑ = Low-to-high clock transition.

FIGURE 12.105

74194 Universal Shift Register IC

Figure 12.106 shows the 74194 4-bit bidirectional universal shift register. This device can accept either serial or parallel inputs, provide serial or parallel outputs, and shift left or right based on input signals applied to select controls S_0 and S_1. Serial data can be entered into either the serial shift-right input (D_{SR}) or the serial shift-left input (D_{SL}). Select controls S_0 and S_1 are used to initiate a hold (S_0 = low, S_1 = low), shift left (S_0 = low, S_1 = high), shift-right (S_0 = high, S_1 = low), or to parallel load (S_0 = high, S_1 = high) mode. A clock pulse must then be applied to shift or parallel load the data.

In parallel load mode (S_0 and S_1 are high), parallel input data is entered via the D_0 through D_3 inputs and transferred to the Q_0 to Q_3 outputs following the next low-to-high clock transition. The 74194 also has an asynchronous master reset ($\overline{MR}$) input that forces all Q outputs low when pulsed low. To make a shift-right recirculating register, the Q_3 output is wired back to the D_{SR} input, while making S_0 = high and S_1 = low. To make a shift-left recirculating register, the Q_0 output is connected back to the D_{SL} input, while making S_0 = low and S_1 = high. The timing diagram in Fig. 12.106 shows a typical parallel load and shifting sequence.

74299 8-Bit Universal Shift/Storage Register with Three-State Interface

A number of shift registers have three-state outputs—outputs that can assume a high, low, or high impedance state (open-circuit or float state). These devices are commonly used as storage registers in three-state bus interface applications.

An example 8-bit universal shift/storage register with three-state outputs is the 74299, shown in Fig. 12.107. This device has four synchronous operating modes that are selected via two select inputs, S_0 and S_1. Like the 74194 universal shift register, the 74299's select modes include shifting right, shifting left, holding, and parallel loading (see the function

74194 4-bit bidirectional universal shift register

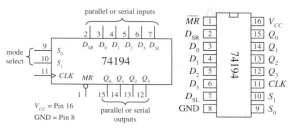

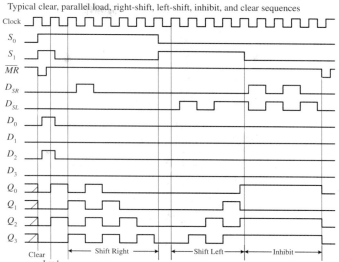

Typical clear, parallel load, right-shift, left-shift, inhibit, and clear sequences

Operating Modes	Inputs							Outputs			
	CP	$\overline{MR}$	S_1	S_0	D_{SR}	D_{SL}	D_n	Q_0	Q_1	Q_2	Q_3
Reset (clear)	X	L	X	X	X	X	X	L	L	L	L
Hold (do nothing)	X	H	l^b	l^b	X	X	X	q_0	q_1	q_2	q_3
Shift Left	↑	H	h	l^b	X	l	X	q_1	q_2	q_3	L
	↑	H	h	l^b	X	h	X	q_1	q_2	q_3	H
Shift Right	↑	H	l^b	h	l	X	X	L	q_0	q_1	q_2
	↑	H	l^b	h	h	X	X	H	q_0	q_1	q_2
Parallel Load	↑	H	h	h	X	X	d_n	d_0	d_1	d_2	d_3

H = High voltage level; h = High voltage level one setup time prior to the low-to-high clock transition; L = Low voltage level; l = Low voltage level one setup time prior to the low-to-high clock transiton; d_n (q_n) = Lower case letters indicate the state of the referenced input (or output) one setup time prior to the low-to-high clock transition; X = Don't care; ↑ = Low-to-high clock transition.

b = The high-to-low transition of S0 and S1 input should only take place while the clock is HIGH for convential operation.

FIGURE 12.106

74299 8-bit universal shift/storage register with 3-state outputs

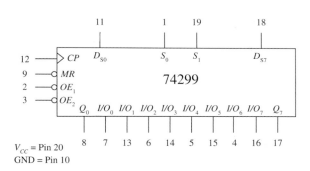

V_{CC} = Pin 20
GND = Pin 10

Operating Modes	Inputs							Outputs		
	$\overline{MR}$	CP	S_0	S_1	D_{S0}	D_{S7}	I/O_n	Q_0	Q_1 - Q_6	Q_7
Reset (clear)	L	H	X	X	X	X	X	L	L - L	L
Shift right	H	↑	h	l	l	X	X	L	q_0 - q_5	q_6
	H	↑	h	l	h	X	X	H	q_0 - q_5	q_6
Shift left	H	↑	l	h	X	l	X	q_1	q_0 - q_5	L
	H	↑	l	h	X	h	X	q_1	q_0 - q_5	H
Hold ("do nothing")	H	↑	l	l	X	X	X	q_0	q_1 - q_6	q_7
Parallel load	H	↑	h	h	X	X	l	L	L - L	L
	H	↑	h	h	X	X	h	H	H - H	H

3-state I/O port operating mode	Inputs					Inputs/Outputs
	$\overline{OE_1}$	$\overline{OE_2}$	S_0	S_1	Q_n (register)	I/O_0 -- I/O_7
Read register	L	L	L	X	L	L
	L	L	L	X	H	H
	L	L	X	L	L	L
	L	L	X	L	H	H
Load register	X	X	H	H	Q_n = I/O$_n$	I/O$_n$ = inputs
Disable I/O	H	X	X	X	X	High Z
	X	H	X	X	X	High Z

H = High voltage level; h = High voltage level one setup time prior to the low-to-high clock transition; L = Low voltage level; l = Low voltage level one setup time prior to the low-to-high clock transiton; q_n = Lowercase letters indicate the state of the referenced output one setup time prior to the low-to-high clock transition; X = Don't care; ↑ = Low-to-high clock transition.

FIGURE 12.107

table in Fig. 12.107). The mode-select inputs, serial data inputs (D_{S0} and D_{S7}), and parallel-data inputs (I/O_0 through I/O_7) are positive edge triggered. The master reset ($\overline{MR}$) input is an asynchronous active-low input that clears the register when pulsed low.

The three-state bidirectional I/O port has three modes of operation:

- The read-register mode allows data within the register to be available at the I/O outputs. This mode is selected by making both output-enable inputs ($\overline{OE}_1$ and $\overline{OE}_2$) low and making one or both select inputs low.

- The load-register mode sets up the register for a parallel load during the next low-to-high clock transition. This mode is selected by setting both select inputs high.

- The disable-I/O mode acts to disable the outputs (set to a high impedance state) when a high is applied to one or both of the output-enable inputs. This effectively isolates the register from the bus to which it is attached.

12.8.7 Simple Shift Register Applications

16-Bit Serial-to-Parallel Converter

A simple way to create a 16-bit serial-to-parallel converter is to join two 74164 8-bit serial-in/parallel-out shift registers, as shown in Fig. 12.108. To join the two ICs, simply wire the Q_7 output from the first register to one of the serial inputs of the second register. (Recall that the serial input that is not used for serial input data acts as an active-high enable control for the other serial input.)

In terms of operation, when data is shifted out of Q_7 of the first register (or data output D_7), it enters the serial input of the second (the example uses D_{Sa} as the serial input) and will be presented to the Q_0 output of the second register (or data output D_8). For an input data bit to reach the Q_7 output of the second register (or data output D_{15}), 16 clock pulses must be applied.

Using two 74164s to create a 16-bit serial-to-parallel converter

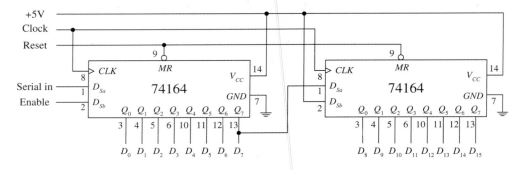

FIGURE 12.108

8-Bit Serial-to-Parallel Converter with Simultaneous Data Transfer

Figure 12.109 shows a circuit that acts as a serial-to-parallel converter that outputs the converted 8-bit word only when all 8 bits have been entered into the register. Here, a 74164 8-bit serial-in/parallel-out shift register is used, along with a 74HCT273 octal D-type flip-flop and a divide-by-8 counter. At each positive clock edge, the serial data is loaded into the 74164. After eight clock pulses, the first serial bit entered is shifted down to the 74164's Q_7 output, while the last serial bit entered resides at the 74164's Q_0 output. At the negative edge of the eighth clock pulse, the negative-edge triggered divide-by-8 circuit's output goes high. During this high transition, the data present on the inputs of the 74HCT273 (which hold the same data present at the 74164's Q outputs) is passed to the 74HCT273's outputs at the same time. (Think of the 74HCT273 as a temporary storage register that dumps its contents after every eighth clock pulse.)

8-Bit Serial-to-Parallel Data Converter

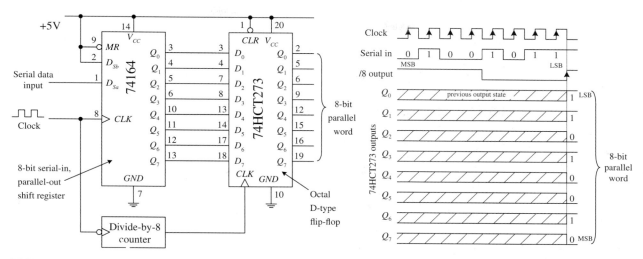

FIGURE 12.109

8-Bit Parallel-to-Serial Interface

Figure 12.110 shows a 74165 8-bit parallel-to-serial shift register used to accept a parallel ASCII word and convert it into a serial ASCII word that can be sent to a serial device. Recall that ASCII codes are only 7 bits long (for example, the binary code for & is 010 0110). How do you account for the missing bit? As it turns out, most 8-bit devices communicating via serial ASCII will use an additional eighth bit for a special purpose, perhaps to act as a parity bit or as a special function bit to enact a special set of characters. Often, the extra bit is simply set low and ignored by the serial device receiving it.

To keep things simple, let's set the extra bit low and assume that is how the serial device likes things done. This means that you will set the D_0 input of the 74165 low. The MSB of the ASCII code will be applied to the D_1 input, while the LSB of the ASCII code will be applied to the D_7 input. Now, with the parallel ASCII word applied to

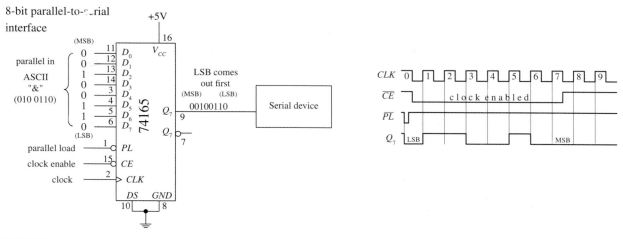

FIGURE 12.110

the inputs of the register, when you pulse the parallel load line ($\overline{PL}$) low, the ASCII word, along with the "ignored bit," is loaded into the register. Next, you must enable the clock to allow the loaded data to be shifted out serially, by setting the clock enable input ($\overline{CE}$) low for the duration it takes for the clock pulses to shift out the parallel word. After the eighth clock pulse (0 to 7), the serial device will have received all 8 serial data bits. Practically speaking, a microprocessor or microcontroller is necessary to provide the $\overline{CE}$ and $\overline{PL}$ lines with the necessary control signals to ensure that the register and serial device communicate properly.

Recirculating Memory Registers

A *recirculating memory register* is a shift register that is preloaded with a binary word that is serially recirculated through the register via a feedback connection from the output to the input. Recirculating registers can be used for a number of applications, from supplying a specific repetitive waveform used to drive IC inputs to driving output drivers used to control stepper motors.

In the leftmost circuit in Fig. 12.111, a parallel 4-bit binary word is applied to the D_0 to D_3 inputs of a 74194 universal shift register. When the S_1 select input is brought high (switch opened), the 4-bit word is loaded into the register. When the S_1 input is then brought low (switch closed), the 4-bit word is shifted in a serial fashion through the register, out Q_3, and back to Q_0 via the D_{SR} input (serial shift-right input) as positive clock edges arrive. Here, the shift register is loaded with 0111. As you begin shifting the bits through the register, a single low output will propagate down through high outputs, which in turn causes the LED attached to the corresponding low output to turn on. In other words, you have made a simple Christmas tree flasher.

The rightmost circuit in Fig. 12.111 is basically the same thing as the leftmost circuit. However, now the circuit is used to drive a stepper motor. Typically, a stepper motor has four stator coils that must be energized in sequence to make the motor turn at a given angle. For example, to make a simple stepper motor turn clockwise, you must energize its stator coils 1, 2, 3, and 4 in the following sequence: 1000, 0100, 0010, 0001, 1000, and so on. To make the motor go counterclockwise, apply the following

Simple shift register sequence generator

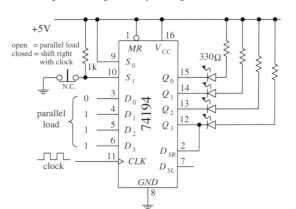

Using a universal shift register IC to control a stepper motor

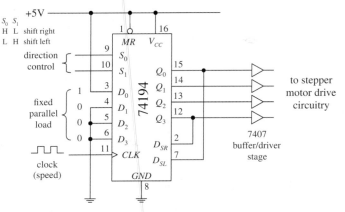

FIGURE 12.111

sequence: 1000, 0001, 0010, 0100, 1000, and so on. You can generate these simple firing sequences with the 74194 by parallel loading the D_0 to D_3 inputs with the binary word 1000. To output the clockwise firing sequence, simply shift bits to the right by setting S_0 = high and S_1 = low. As clock pulses arrive, the 1000 present at the outputs will then become 0100, then 0010, 0001, 1000, and so on.

The speed of rotation of the motor is determined by the clock frequency. To output the counterclockwise firing sequence, simply shift bits to the left by setting S_0 = low and S_1 = high. To drive steppers, it is typically necessary to use a buffer/driver interface like the 7407 shown in Fig. 12.111, as well a number of output transistors, not shown. Also, different types of stepper motors may require different firing sequences than the one shown here. Stepper motors and the various circuits used to drive them are discussed in detail in Chap. 15.

12.9   Analog/Digital Interfacing

A number of tricks are used to interface analog circuits with digital circuits. In this section, we'll take a look at two basic levels of interfacing. One level deals with simple on/off triggering. The other level deals with true analog-to-digital and digital-to-analog conversion—converting analog signals into digital numbers and converting digital numbers into analog signals. These techniques are just as applicable to connecting things to the digital input pins of a microcontroller.

12.9.1   Triggering Simple Logic Responses from Analog Signals

There are times when you need to drive logic from simple on/off signals generated by analog devices. For example, you may want to latch an alarm (via a flip-flop) when an analog voltage—say, one generated from a temperature sensor—reaches a desired threshold level. Or perhaps you simply want to count the number of times a certain analog threshold is reached. For simple on/off applications such as these, it is common to use a comparator or op amp as the interface between the analog output of the transducer and the input of the logic circuit. Often it is possible to simply use a voltage divider network composed of a transducer of variable resistance and a pullup resistor. Figure 12.112 shows some sample networks to illustrate the point.

In Fig. 12.112a, a phototransistor is used to trigger a logic response. Normally, the phototransistor is illuminated, which keeps the input of the first Schmitt inverter low. The output of the second inverter is high. When the light is briefly interrupted, the phototransistor momentarily stops conducting, causing the input to the first inverter to pulse low, while the output of the second inverter pulses high. This high pulse could be used to latch a D flip-flop, which could be used to trigger an LED or a buzzer alarm.

In Fig. 12.112b, a single-supply comparator with open-collector output is used as an analog-to-digital interface. When an analog voltage applied to V_{in} exceeds the reference voltage (V_{ref}) set at the noninverting input (+) via the pot, the output goes low (the comparator sinks current through itself to ground). When V_{in} goes below V_{ref}, the output goes high (the comparator's output floats, but the pullup resistor pulls the comparator's output high).

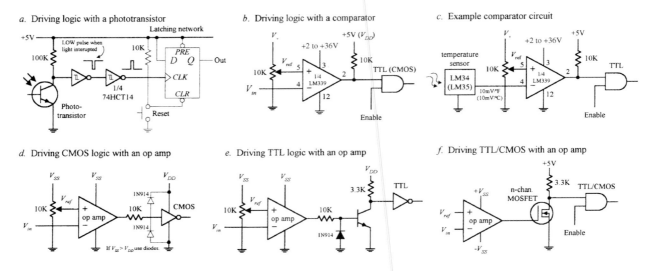

a. Driving logic with a phototransistor

b. Driving logic with a comparator

c. Example comparator circuit

d. Driving CMOS logic with an op amp

e. Driving TTL logic with an op amp

f. Driving TTL/CMOS with an op amp

FIGURE 12.112

In Fig. 12.112c, a simple application of the previous comparator interface is shown. The input voltage is generated by an LM34 or LM35 temperature sensor. The LM34 generates 10 mV/°F, while the LM35 generates 10 mV/°C. The resistance of the pot and V_+ determine the reference voltage. If we want to drive the comparator low when 75°C is reached, we set the reference voltage to 750 mV, assuming we're using the LM35.

In Fig. 12.112d, an op amp set in comparator mode can also be used as an analog-to-digital interface for simple switching applications. CMOS logic can be driven directly through a current limiting resistor, as shown. If the supply voltage of the op amp exceeds the supply voltage of the logic, protection diodes should be used (as shown in the figure).

Protection diodes were not necessary with the LM339 because that has open-collector outputs.

In Fig. 12.112e, an op amp that is used to drive TTL typically uses a transistor output stage like the one shown here. The diode acts to prevent base-to-emitter reverse breakdown. When V_{in} exceeds V_{ref}, the op amp's output goes low, the transistor turns off, and the logic input receives a high.

In Fig. 12.112f, an *n*-channel MOSFET transistor is used as an output stage to an op amp.

12.9.2   *Using Logic to Drive External Loads*

Driving simple loads such as LEDs, relays, buzzers, or any device that assumes either an on or off state is relatively simple. When driving such loads, it is important to first check the driving logic's current specifications—how much current, say, a gate can sink or source. After that, you determine how much current the device to be driven will require. If the device draws more current than the logic can source or sink, a high-power transistor typically can be used as an output switch. Figure 12.113 shows some sample circuits used to drive various loads.

In Fig. 12.113a, LEDs can be driven directly by logic through a current-limiting resistor. Current can either be sourced or sunk. If an LED requires more current than the logic can supply or sink, a transistor output stage like the one shown in Fig. 12.113f can be used.

Figure 12.113b shows a simple way to get dual-lighting action from a pair of LEDs. When the gate's output goes low, the upper green LED turns on, while the lower red LED turns off. The LEDs switch states when the output goes high.

Relays will draw considerable current. To avoid damaging the logic device, in Fig. 12.113c, a power MOSFET transistor is attached to the logic output. The diode is used to protect the circuit from current spikes generated by relay as it switches states.

A handy method for interfacing standard logic with loads is to use a gate with an open-collector output as a go-between. Recall that open-collector gates cannot source current; they can only sink current. However, they typically can sink ten times the current of a standard logic gate. In Fig. 12.113d, an open-collector gate is used to drive a relay. Check the current ratings of specific open-collector devices before using them to be sure they can handle the load current.

Figure 12.113e shows another open-collector application. In Fig. 12.113f, a bipolar transistor is used to increase the output drive current used to drive a high-current LED. Make sure the transistor is of the proper current rating.

Figure 12.113g is basically the same as the previous example, but the load can be something other than an LED.

In Fig. 12.113h, an optocoupler is used to drive a load that requires electrical isolation from the logic driving it. Electrical isolation is often used in situations where external loads use a separate ground system. The voltage level at the load side of the optical interface can be set via V_{CC}. There are many different types of optocouplers available (see Chap. 5).

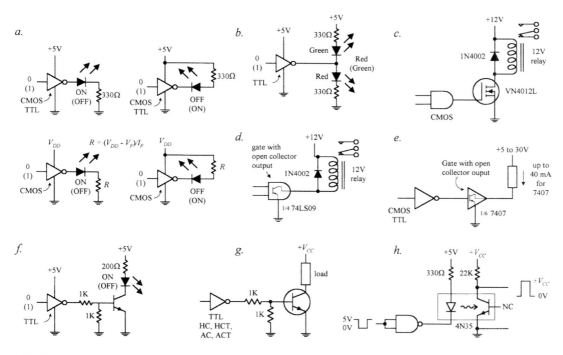

FIGURE 12.113

12.9.3 Analog Switches

Analog switches are ICs designed to switch analog signals via digital control. The internal structure of these devices typically consists of a number of logic control gates interfaced with transistor stages used to control the flow of analog signals.

Figure 12.114 shows various types of analog switches. The CMOS 4066B quad bilateral switch uses a single-supply voltage from 3 to 15 V. It can switch analog or digital signals within ±7.5 V and has a maximum power dissipation of around 700 mW. Individual switches are controlled by digital inputs A through D. The TTL-compatible AH0014D DPDT analog switch can switch analog signals of ±10 V via the A and B logic control inputs. Note that this device has separate analog and digital supplies: $V+$ and $V-$ are analog; V_{CC} and GND are digital. The DG302A dual-channel CMOS DPST analog switch can switch analog signals within the ±10-V range at switching speeds up to 15 ns.

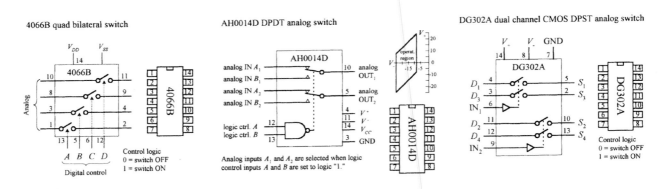

FIGURE 12.114

A number of circuits use analog switches. They are found in modulator/demodulator circuits, digitally controlled frequency circuits, analog signal-gain circuits, and analog-to-digital conversion circuits, where they often act as sample-hold switches. They can, of course, be used simply to turn a given analog device on or off.

12.9.4 Analog Multiplexer/Demultiplexer

Recall from Sec. 12.3 that a digital multiplexer acts like a data selector, while a digital demultiplexer acts like a data distributor. Analog multiplexers and demultiplexers act the same way but are capable of selecting or distributing analog signals. (They still use digital select inputs to select which pathways are open and which are closed to signal transmission.)

A popular analog multiplexer/demultiplexer IC is the 4051B, shown in Fig. 12.115. This device functions as either a multiplexer or demultiplexer, since its inputs and outputs are bidirectional (signals can flow in either direction). When used as a multiplexer, analog signals enter through I/O lines 0 through 7, while the digital code that selects which input is passed to the analog O/I line (pin 3) is applied to digital inputs A, B, and C. See the truth table in the figure. When used as a demultiplexer, the connections are reversed: The analog input comes in through the analog O/I line

4051B analog multiplexer/demultiplexer

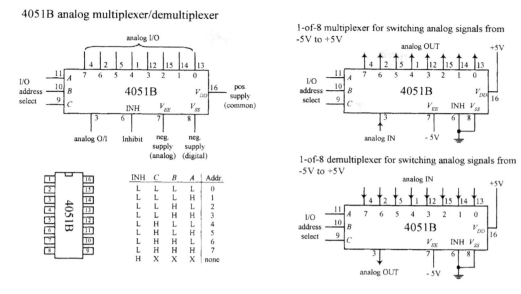

FIGURE 12.115

(pin 3) and passes out through one of the seven analog I/O lines. The specific output is again selected by the digital inputs *A*, *B*, and *C*. Note that when the inhibit line (INH) is high, none of the addresses are selected.

The I/O analog voltage levels for the 4051B are limited to a region between the positive supply voltage V_{DD} and the analog negative supply voltage V_{EE}. Note that the V_{SS} supply is grounded. If the analog signals you are planning to use are all positive, V_{EE} and V_{SS} can both be connected to a common ground. However, if you plan to use analog voltages that range from, say, −5 to +5 V, V_{EE} should be set to −5 V, while V_{DD} should be set to +5 V. The 4051B accepts digital signals from 3 to 15 V, while allowing for analog signals from −15 to +15 V.

12.9.5   Analog-to-Digital and Digital-to-Analog Conversion

In order for analog devices (temperature sensors, strain gauges, position sensors, light meters, and so on) to communicate with digital circuits in a manner that goes beyond simple threshold triggering, we use an analog-to-digital converter (ADC). An ADC converts an analog signal into a series of binary numbers, each number proportional to the analog level measured at a given moment. Typically, the digital words generated by the ADC are fed into a microprocessor or microcontroller, where they can be processed, stored, interpreted, and manipulated. Analog-to-digital conversion is used in data-acquisition systems, digital sound recording, and within simple digital display test instruments (such as light meters and thermometers).

In order for a digital circuit to communicate with the analog world, we use a digital-to-analog converter (DAC). A DAC takes a binary number and converts it to an analog voltage that is proportional to the binary number. By supplying different binary numbers, one after the other, a complete analog waveform is created. DACs are commonly used to control the gain of an op amp, which in turn can be used to create digitally controlled amplifiers and filters. They are also used in waveform generator and modulator circuits and as trimmer replacements, and are found in a number of process-control and autocalibration circuits.

Many digital consumer products such as MP3 players, DVDs, and CD players use digital signal processing ADCs and DACs often contained in a microcontroller.

ADC and DAC Basics

Figure 12.116 shows the basic idea behind analog-to-digital and digital-to-analog conversion. In the analog-to-digital figure, the ADC receives an analog input signal along with a series of digital sampling pulses. Each time a sampling pulse is received, the ADC measures the analog input voltage and outputs a 4-bit binary number that is proportional to the analog voltage measured during the specific sample. With 4 bits, we get 16 binary codes (0000 to 1111) that correspond to 16 possible analog levels (for example, 0 to 15 V).

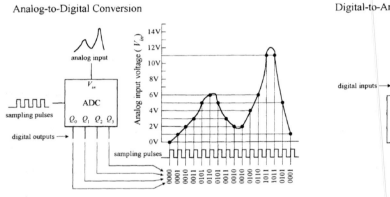

FIGURE 12.116

In the digital-to-analog conversion figure, the DAC receives a series of 4-bit binary numbers. The rate at which new binary numbers are fed into the DAC is determined by the logic that generates them. With each new binary number, a new analog voltage is generated. As with the ADC example, we have a total of 16 binary numbers to work with and 16 possible output voltages.

As you can see from the graphs, both these 4-bit converters lack the resolution needed to make the analog signal appear continuous (without steps). To make things appear more continuous, a converter with higher resolution is used. This means that instead of using 4-bit binary numbers, we use larger-bit numbers, such as 6-bit, 8-bit, 10-bit, 12-bit, 16-bit, or even 18-bit or higher numbers. If our converter has a resolution of 8 bits, we have $2^8 = 256$ binary numbers to work with, along with 256 analog steps. Now, if this 8-bit converter is set up to generate 0 V at binary 00000000 and 15 V at binary 11111111 (full scale), then each analog step is only 0.058 V high ($\frac{1}{256} \times 15$ V). With an 18-bit converter, the steps get incredibly tiny because we have $2^{18} = 262,144$ binary numbers and steps. With 0 V corresponding to binary 000000000000000000 and 15 V corresponding to 111111111111111111, the 18-bit converter yields steps that are only 0.000058 V high! As you can see in the 18-bit case, the conversion process between digital and analog appears practically continuous.

Simple Binary-Weighted DAC

Figure 12.117 shows a simple 4-bit DAC that is constructed from a digitally controlled switch (74HC4066), a set of binary-weighted resistors, and an operational amplifier. The basic idea is to create an inverting amplifier circuit whose gain is controlled by changing the input resistance R_{in}. The 74HC4066 and the resistors together act as a digitally controlled R_{in} that can take on one of 16 possible values. You can think of the 74HC4066 and resistor combination as a digitally controlled current source. Each new binary code applied to the inputs of the 74HC4066 generates a new discrete current level that is summed by R_F to provide a new discrete output voltage level.

We choose scaled resistor values of R, $R/2$, $R/4$, and $R/8$ to give R_{in} discrete values that are equally spaced. To find all possible values of Rin, we use the formula provided in Fig. 12.117. This formula looks like the old resistors-in-parallel formula, but we must exclude those resistors that are not selected by the digital input code—that's what the coefficients A through D are for (a coefficient is either 1 or 0, depending on the digital input).

To find the analog output voltage, we simply use $V_{out} = -V_{ref}(R_F/R_{in})$—the expression used for the inverting amplifier (see Chap. 8). Figure 12.117 shows what we get when we set $V_{ref} = -5$ V, $R = 100$ kΩ, and $R_F = 20$ kΩ, and take all possible input codes.

The binary-weighted DAC shown in Fig. 12.117 is limited in resolution (4-bit, 16 analog levels). To double the resolution (make an 8-bit DAC), you might consider adding another 74HC4066 and $R/16$, $R/32$, $R/64$, and $R/128$ resistors. In theory, this works; in reality, it doesn't. The problem with this approach is that when we reach the $R/128$ resistor, we must find a 0.78125-kΩ resistor, assuming $R = 100$ kΩ. Assuming we can find or construct an equivalent resistor network for $R/128$, we're still in trouble because the tolerances of these resistors will cause problems. This scaled-resistor approach becomes impractical when we deal with resolutions of more than a few bits. To increase the resolution, we scrap the scaled-resistor network and replace it with an $R/2R$ ladder network. The manufacturers of DAC ICs do this as well.

Simple binary-weighted digital-to-analog converter

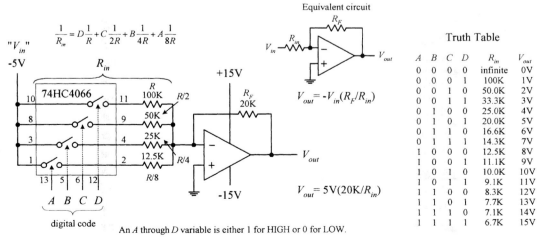

FIGURE 12.117

$R/2R$ ladder 4-bit digital-to-analog converter

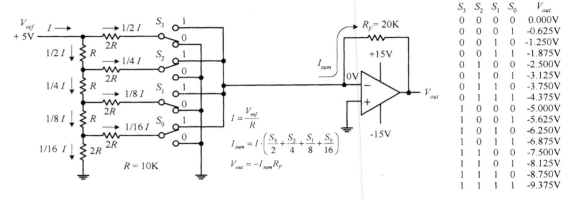

S_3	S_2	S_1	S_0	V_{out}
0	0	0	0	0.000V
0	0	0	1	-0.625V
0	0	1	0	-1.250V
0	0	1	1	-1.875V
0	1	0	0	-2.500V
0	1	0	1	-3.125V
0	1	1	0	-3.750V
0	1	1	1	-4.375V
1	0	0	0	-5.000V
1	0	0	1	-5.625V
1	0	1	0	-6.250V
1	0	1	1	-6.875V
1	1	0	0	-7.500V
1	1	0	1	-8.125V
1	1	1	0	-8.750V
1	1	1	1	-9.375V

FIGURE 12.118

R/2R Ladder DAC

An $R/2R$ DAC uses an $R/2R$ resistor ladder network instead of a scaled-resistor network, as was the case in the previous DAC. The benefit of using the $R/2R$ ladder is that we need only two resistor values: R and $2R$. Figure 12.118 shows a simple 4-bit $R/2R$ DAC. For now, assume that the switches are digitally controlled (in real DACs, they are replaced with transistors).

The trick to understanding how the $R/2R$ ladder works is realizing that the current drawn through any one switch is always the same, no matter if it is thrown up or down. If a switch is thrown down, current will flow through the switch into ground (0 V). If a switch is thrown up, current will flow toward virtual ground—located at the op amp's inverting input (recall that if the noninverting input of an op amp is set to 0 V, the op amp will make the inverting input 0 V, via negative feedback). Once you realize that the current through any given switch is always constant, you can figure that the total current (I) supplied by V_{ref} will be constant as well. Once you have that, you figure out what fractions of the total current pass through each of the branches within the $R/2R$ network using simple circuit analysis. Figure 12.118 shows that $\frac{1}{2}I$ passes through S_3 (MSB switch), $\frac{1}{4}I$ through S_2, $\frac{1}{8}I$ through S_1, and $\frac{1}{16}I$ through S_0 (LSB switch). If you're interested in how that was figured out, the circuit reduction shown in Fig. 12.119 should help.

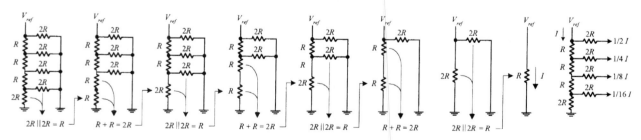

FIGURE 12.119

Now that we have a means of consistently generating fractions of $\frac{1}{2}I$, $\frac{1}{4}I$, $\frac{1}{8}I$, and $\frac{1}{16}I$, we can choose, via the digital input switches, which fractions are summed together by the amplifier. For example, if switches S_3, S_2, S_1, and S_0 are thrown to

0101 (5), $\frac{1}{4}I + \frac{1}{16}I$ combine to form I_{sum}. But what is I? Using Ohm's law, it's just $I = V_{ref} / R = +5$ V $/ 10$ k$\Omega = 500$ µA. This means that $I_{sum} = \frac{1}{4}(500$ µA$) + \frac{1}{16}(500$ µA$) = 156.25$ µA. The final output voltage is determined by $V_{out} = -I_{sum}R_F = -(156.25$ µA$)(20$ k$\Omega) = -3.125$ V. The formulas and the table in Fig. 12.118 show the other possible binary/analog combinations.

To create an $R/2R$ DAC with higher resolution, we simply add more runs and switches to the ladder.

Integrated DACs

Often, making DACs from scratch isn't worth the effort. The cost as well as the likelihood for conversion errors is great. The best thing to do is to simply buy a DAC IC. You can buy these devices from a number of different manufacturers (such as National Semiconductor, Analog Devices, and Texas Instruments). The typical resolutions for these ICs are 6, 8, 10, 12, 16, and 18 bits. DAC ICs also may come with a serial digital input, as opposed to the parallel input scheme shown in Figs 12.117 and 12.118. Before a serial-input DAC can make a conversion, the entire digital word must be clocked into an internal shift register.

Most often, DAC ICs come with an external reference input that is used to set the analog output range. There are some DACs that have fixed references, but these are becoming rare.

Often, you'll see a manufacturer list one of its DACs as being a multiplying DAC. A multiplying DAC can produce an output signal that is proportional to the product of a varying input reference level (voltage or current) times a digital code. As it turns out, most DACs, even those that are specifically designated as multiplying DAC on the data sheets, can be used for multiplying purposes simply by using the reference input as the analog input. However, many such ICs do not provide the same quality multiplying characteristics, such as a wide analog input range and fast conversion times, as those that are called *multiplying* DACs.

Multiplying is most commonly applied in systems that use ratiometeric transducers (for example, position potentiometers, strain gauges, and pressure transducers). These transducers require an external analog voltage to act as a reference level on which to base analog output responses. If this reference level is altered, say, by an unwanted supply surge, the transducer's output will change in response, and this results in conversion errors at the DAC end. However, if we use a multiplying DAC, we eliminate these errors by feeding the transducer's reference voltage to the DAC's analog input. If any supply voltage/current errors occur, the DAC will alter its output in proportion to the analog error.

DACs are capable of producing unipolar (single-polarity output) or bipolar (positive and negative) output signals. In most cases, when a DAC is used in unipolar mode, the digital code is expressed in standard binary. When used in bipolar mode, the most common code is either offset binary or 2's complement. Offset binary and 2's complement codes make it possible to express both positive and negative values. Figure 12.120 shows all three codes and their corresponding analog output levels (referenced from an external voltage source).

Note that in the figure, FS stands for full scale, which is the maximum analog level that can be reached when applying the highest binary code. It is important to realize that at full scale, the analog output for an n-bit converter is actually $(2^n - 1) / 2^n \times V_{ref}$

Common Digital Codes Used by DACs

	Unipolar Operation		Bipolar Operation				
	Binary	Analog Output	Offset Binary	Analog Output	2's Comp.	Analog Output	
FS	1111 1111	$V_{ref}\left(\frac{255}{256}\right)$	FS 1111 1111	$+V_{ref}\left(\frac{127}{128}\right)$	FS 0111 1111	$+V_{ref}\left(\frac{127}{128}\right)$	
FS-1	1111 1110	$V_{ref}\left(\frac{254}{256}\right)$	FS-1 1111 1110	$+V_{ref}\left(\frac{126}{128}\right)$	FS-1 0111 1110	$+V_{ref}\left(\frac{126}{128}\right)$	
↓			↓		↓		
$\frac{FS}{2}$	1000 0000	$V_{ref}\left(\frac{128}{256}\right)=\frac{V_{ref}}{2}$	0+1LSB 1000 0001	$+V_{ref}\left(\frac{1}{128}\right)$	0+1LSB 0000 0001	$+V_{ref}\left(\frac{1}{128}\right)$	
			0 1000 0000	$V_{ref}\left(\frac{0}{128}\right)=0$	0 0000 0000	$V_{ref}\left(\frac{0}{128}\right)=0$	
LSB	0000 0001	$V_{ref}\left(\frac{1}{256}\right)$	0−1LSB 0111 1111	$-V_{ref}\left(\frac{1}{128}\right)$	0−1LSB 1111 1111	$-V_{ref}\left(\frac{1}{128}\right)$	
LSB−1	0000 0000	$V_{ref}\left(\frac{0}{256}\right)=0$	↓		↓		
			−FS+1 0000 0001	$-V_{ref}\left(\frac{127}{128}\right)$	−FS+1 1000 0001	$-V_{ref}\left(\frac{127}{128}\right)$	
	(FS = full scale)		−FS 0000 0000	$-V_{ref}\left(\frac{128}{128}\right)$	1000 0000 −FS	$-V_{ref}\left(\frac{128}{128}\right)$	

FIGURE 12.120

not $2^n/2^n \times V_{ref}$. For example, for an 8-bit converter, the number of binary numbers is $2^8 = 256$, while the maximum analog output level is $255/256\ V_{ref}$, not $256/256\ V_{ref}$, since the highest binary number is 255 (1111 1111). The "missing count" is used up by the LSB-1 condition (0 state).

Example DAC ICs

DAC0808 8-BIT DAC

The DAC0808 (National Semiconductor) is a popular 8-bit DAC that requires an input reference current and supplies 1 of 256 analog output current levels. Figure 12.121 shows a block diagram of the DAC0808, along with its IC pin configuration and a sample application circuit.

In the application circuit, the analog output range is set by applying a reference current (I_{ref}) to pin 14 (+V_{ref}). In this example, I_{ref} is set to 2 mA via an external +10 V/5 kΩ resistor combination. Note that another 5-kΩ resistor is required between pin 15 (−V_{ref}) and ground.

DAC0808 8-bit Digital-to-Analog Converter (resolution = 256 steps)

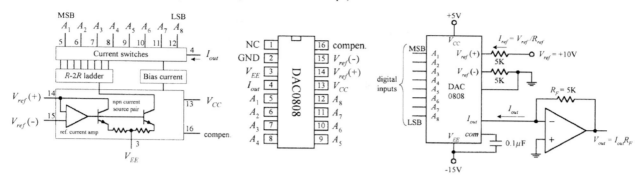

FIGURE 12.121

To determine the DAC's analog output current (I_{out}) for all possible binary inputs, we use the following formula:

$$I_{out} = I_{ref}\left(\frac{A_1}{2} + \frac{A_2}{4} + \ldots + \frac{A_8}{256}\right) = \frac{\text{decimal equivalent of input binary number}}{256}$$

At full scale (all A's high or binary 255), $I_{out} = I_{ref}(255/256) = (2\text{ mA})(0.996) = 1.99$ mA. Considering that the DAC has 256 analog output levels, we can figure that each corresponding level is spaced 1.99 mA/256 = 0.0078 mA apart.

To convert the analog output currents into analog output voltages, we attach the op amp. Using the op amp rules from Chap. 8, we find that the output voltage is $V_{out} = I_{out} \times R_f$. At full scale, $V_{out} = (1.99\text{ mA})(5\text{ k}\Omega) = 9.95$ V. Each analog output level is spaced 9.95 V/256 = 0.0389 V apart.

The DAC0808 can be configured as a multiplying DAC by applying the analog input signal to the reference input. In this case, however, the analog input current should be limited to a range from 16 μA to 4 mA to retain reasonable accuracy. See the National Semiconductor's data sheets for more details.

DAC8043A SERIAL 12-BIT INPUT MULTIPLYING DAC

The DAC8083A (Analog Devices) is a high-precision 12-bit CMOS multiplying DAC that comes with a serial digital input. Figure 12.122 shows a block diagram, pin configuration, and write cycle timing diagram for this device.

DAC8043 12-bit serial input multiplying D/A converter

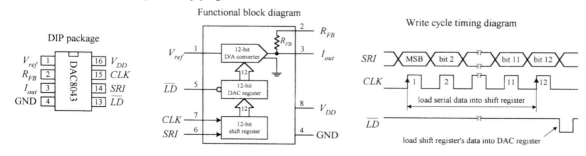

FIGURE 12.122

Before the DAC8043 can make a conversion, serial data must be clocked into the input register by supplying an external clock signal (each positive edge of the clock load one bit). Once loaded, the input register's contents are dumped off to the DAC register by applying a low pulse to the $\overline{LD}$ line. Data in the DAC register is then converted to an output current through the I_{out} terminal.

In most applications, this current is then transformed into a voltage by an op amp stage, as is the case within the two circuits shown in Fig. 12.123. In the unipolar (two-quadrant) circuit, a standard binary code is used to select from 4096 possible analog output levels. In the bipolar (four-quadrant) circuit, an offset binary code is used again to select from 4096 analog output levels, but now the range is broken up to accommodate both positive and negative polarities.

Unipolar Operation (2-Quadrant)

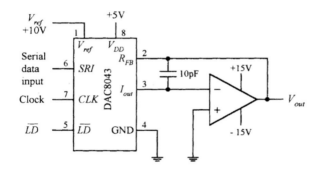

Digital Input (Binary)	Nominal Analog Output
1111 1111 1111	$-V_{ref}\left(\dfrac{4095}{4096}\right)$
1000 0000 0001	$-V_{ref}\left(\dfrac{2049}{4096}\right)$
1000 0000 0000	$-V_{ref}\left(\dfrac{2048}{4096}\right) = -\dfrac{V_{ref}}{2}$
0111 1111 1111	$-V_{ref}\left(\dfrac{2047}{4096}\right)$
0000 0000 0001	$-V_{ref}\left(\dfrac{1}{4096}\right)$
0000 0000 0000	$-V_{ref}\left(\dfrac{0}{4096}\right) = 0$

Bipolar Operation (4-Quadrant)

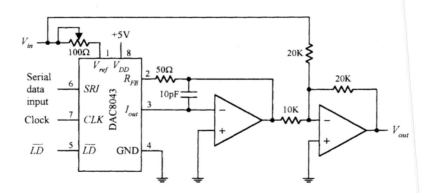

Digital Input (Offset Binary)	Nominal Analog Output
1111 1111 1111	$+V_{ref}\left(\dfrac{2047}{2048}\right)$
1000 0000 0001	$+V_{ref}\left(\dfrac{1}{2048}\right)$
1000 0000 0000	
0111 1111 1111	$-V_{ref}\left(\dfrac{1}{2048}\right)$
0000 0000 0001	$-V_{ref}\left(\dfrac{2047}{2048}\right)$
0000 0000 0000	$-V_{ref}\left(\dfrac{2048}{2048}\right)$

FIGURE 12.123

If you're interested in learning more about the DAC8043, go to Analog Device's website and check out the data sheet.

Another very similar device worth considering is the MAX522.

12.9.6 Analog-to-Digital Converters

There are a number of techniques used to convert analog signals into digital signals. The most popular techniques include successive approximation conversion and parallel-encoded conversion (or flash conversion). Other techniques include half-flash conversion, delta-sigma processing, and pulse-code modulation (PCM). In this section, we'll focus on the successive approximation and parallel-encoded conversion techniques. Most microcontrollers will have built-in ADC channels using one of the techniques described here.

Successive Approximation

Successive approximation analog-to-digital conversion is the most common approach used in integrated ADCs. In this conversion technique, each bit of the binary output is found, one bit at a time—MSB first. This technique yields fairly fast conversion times (from around 10 to 300 μs) with a limited amount of circuitry. Figure 12.124 shows a simple 8-bit successive approximation ADC, along with an example analog-to-digital conversion sequence.

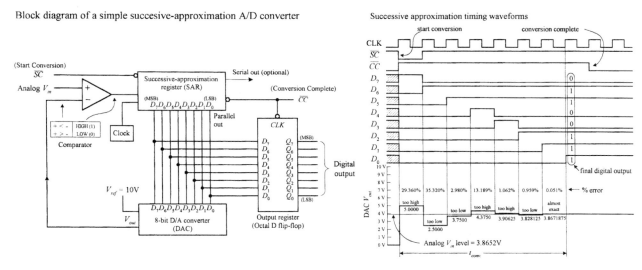

FIGURE 12.124

To begin a conversion, the $\overline{SC}$ (start conversion) input is pulsed low. This causes the successive approximation register (SAR) to first apply a high on the MSB (D_7) line of the DAC. With only D_7 high, the DAC's output is driven to one-half its full-scale level, which in this case is +5 V because the full-scale output is +10 V. The +5-V output level from the DAC is then compared with the analog input level, via the comparator. If the analog input level is greater than +5 V, the SAR keeps the D_7 line high; otherwise, the SAR returns the D_7 line low. At the next clock pulse, the next bit (D_6) is tried. Again, if the analog input level is larger than the DAC's output level, D_6 is left high; otherwise, it is returned low.

During the next six clock pulses, the rest of the bits are tried. After the last bit (LSB) is tried, the CC (conversion complete) output of the SAR goes low, indicating that a valid 8-bit conversion is complete, and the binary data is ready to be clocked into the octal flip-flop, where it can be presented to the Q_0–Q_7 outputs.

The timing diagram shows a 3.8652-V analog level being converted into an approximate digital equivalent. Note that after the first approximation (the D_7 try), the percentage error between the actual analog level and corresponding digital equivalent is 29.360 percent. However, after the final approximation, the percentage error is reduced to only 0.051 percent.

Until now, we've assumed that the analog input to our ADC was constant during the conversion. But what happens when the analog input changes during conversion time? Errors result. The more rapidly the analog input changes during the conversion time, the more pronounced the errors will become. To prevent such errors, a sample-and-hold circuit is often attached to the analog input. With an external control signal, this circuit can be made to sample the analog input voltage and hold the sample while the ADC makes the conversion.

With the exception of very high-speed ADCs, separate ADC ICs are now largely redundant and have been replaced with microcontrollers containing 12-bit or higher ADC channels.

Parallel-Encoded Analog-to-Digital Conversion (Flash Conversion)

Parallel-encoded analog-to-digital conversion, or flash conversion, is perhaps the easiest conversion process to understand. To illustrate the basics behind parallel encoding (also referred to as *simultaneous multiple comparator* or *flash converting*), let's take a look at the simple 3-bit converter in Fig. 12.125.

Simple 3-bit parallel-encoded A/D converter

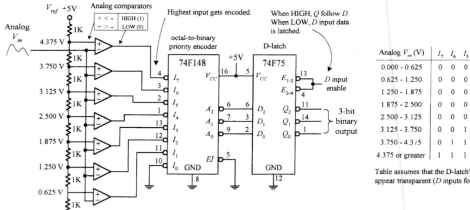

Truth Table

Analog V_{in} (V)	I_7	I_6	I_5	I_4	I_3	I_2	I_1	I_0	$\bar{A}_2$	$\bar{A}_1$	$\bar{A}_0$	$\bar{Q}_2$	$\bar{Q}_1$	$\bar{Q}_0$
0.000 - 0.625	0	0	0	0	0	0	0	0	1	1	1	0	0	0
0.625 - 1.250	0	0	0	0	0	0	1	0	1	1	0	0	0	1
1.250 - 1.875	0	0	0	0	0	1	1	0	1	0	1	0	1	0
1.875 - 2.500	0	0	0	0	1	1	1	0	1	0	0	0	1	1
2.500 - 3.125	0	0	0	1	1	1	1	0	0	1	1	1	0	0
3.125 - 3.750	0	0	1	1	1	1	1	0	0	1	0	1	0	1
3.750 - 4.375	0	1	1	1	1	1	1	0	0	0	1	1	1	0
4.375 or greater	1	1	1	1	1	1	1	0	0	0	0	1	1	1

Table assumes that the D-latch's enable input is set HIGH, making the latch appear transparent (D inputs follow Q outputs).

FIGURE 12.125

The set of comparators is the key feature to note in this circuit. Each comparator is supplied with a different reference voltage from the 1 kΩ voltage divider network. Since we've set up a +5V reference voltage, the voltage drop across each resistor within the voltage divider network is 0.625 V. From this, you can determine the specific reference voltages given to each comparator (see Fig. 12.125).

To convert an analog signal into a digital number, the analog signal is applied to all the comparators at the same time, via the common line attached to the inverting inputs of all the comparators. If the analog voltage is between, say, 2.500 and 3.125 V, only those comparators with reference voltage below 2.500 V will output a high. To create a 3-bit binary output, the eight comparator outputs are fed into an octal-to-binary priority encoder. A *D* latch also can be incorporated into the circuit to provide enable control of the binary output. The truth table should fill in the rest.

12.10 Displays

A number of displays can be interfaced with control logic to display numbers, letters, special characters, and graphics. Two popular displays that we'll consider here include the light-emitting diode (LED) display and the liquid-crystal display (LCD).

12.10.1 LED Displays

LED displays come in three basic configurations: numeric (numbers), alphanumeric (numbers and letters), and dot-matrix forms (see Fig. 12.126). Numeric displays consist of seven LED segments. Each LED segment is given a letter designation, as shown in the figure. Seven-segment LED displays are most frequently used to generate numbers (0–9), but they also can be used to display hexadecimal (0–9, A, B, C, D, E, F). The 14-segment, 16-segment, and special 4 × 7 dot matrix displays are alphanumeric. The 5 × 7 dot matrix display is both alphanumeric and graphic— you can display unique characters and simple graphics. See Chap. 5 for information about other types of LED displays.

Various types of displays

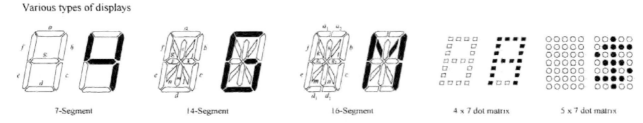

| 7-Segment | 14-Segment | 16-Segment | 4 x 7 dot matrix | 5 x 7 dot matrix |

FIGURE 12.126

Direct Drive of Numeric LED Displays

Seven-segment LED displays come in two varieties: common anode and common cathode. Figure 12.128 shows single digital eight-segment (seven digit segments + decimal point) displays of both varieties.

When driving a multidigit display, say, one with eight digits, the previous technique becomes awkward. It requires eight discrete decoder/driver ICs. One way to avoid this problem is to use a special direct-drive LED display driver IC.

For example, National Semiconductor's MM5450, shown in Fig. 12.133, is designed to drive 4- or 5-digit alphanumeric common anode LED displays. It comes with 34 TTL-compatible outputs that are used to drive desired LED segments within a display. Each of these outputs can sink up to 15 mA. In order to specify which output lines are driven high or low, serial input data are clocked into the driver's serial input. The serial data chain that is entered is 36 bits long. The first bit is a start bit (set to 1), and the remaining 35 bits are data bits. Each data bit corresponds to a given output data

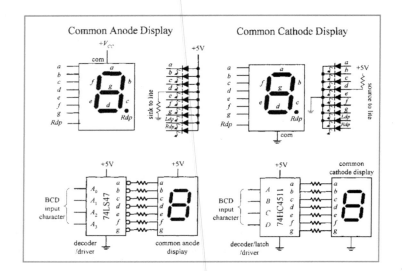

FIGURE 12.127

To drive a given segment of a common anode display, current must be sunk out through the corresponding segment's terminal. With the common cathode display, current must be sourced into the corresponding segment's terminal. A simple way to drive these displays is to use BCD to seven-segment display decoder/drivers, like the ones show in the figure. Applying a BCD input character results in a decimal digit being displayed (e.g., 0101 applied to A_0–A_3, or A–D displays a "5"). The 74LS47 active-low open-collector outputs are suited for a common anode display, while the 74HC4511's active-high outputs are suited for a common cathode display. Both ICs also come with extra terminals used for lamp testing and ripple blanking, as well as leading zero suppression (controlling the decimal point).

MM5450 (National Semiconductor) LED Display Driver

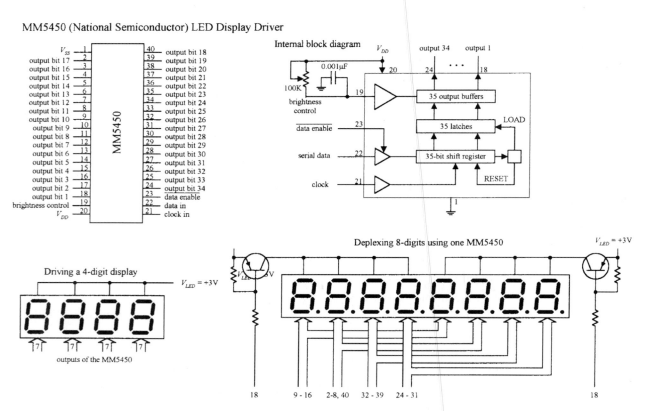

FIGURE 12.128

line that is used to drive a given LED segment within the display. At the thirty-sixth positive clock signal, a LOAD signal is generated that loads the 35 data bits into the latches (see the block diagram in Fig. 12.133). At the low state of the clock, a $\overline{\text{RESET}}$ signal is generated that clears the shift register for the next set of data. You can learn more about the MM5450 at http://www.micrel.com/_PDF/mm5450.pdf.

Multiplexed LED Displays

Another technique used to drive multidigit LED displays involves multiplexing. Multiplexing can drastically reduce the number of connections needed between display and control logic. In a multiplexed display, digits share common segment lines. Also, only one digit within the display is lighted at a time. To make it appear that a complete readout is displayed, all the digits must be flashed very rapidly in sequence, over and over again. The simple example in Fig. 12.129 shows multiplexing in action. To reduce the component count further, you can do away with the 74HC4511 and just use 7 digital outputs from the microcontroller.

Simple multiplexing scheme

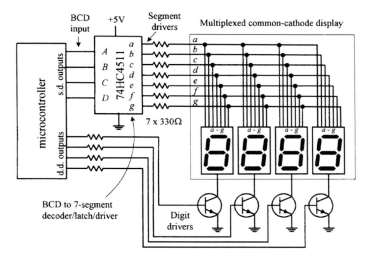

FIGURE 12.129

Here, we have a multiplexed common-cathode display—all digits share common segment lines (*a–g*). To supply a full one-digit readout, digits must be flashed rapidly, one at a time. To enable a given digit, the digit's common line is grounded via one of the digital drivers (transistors)—all other digits' common lines are left floating. In this example, the drivers are controlled by a microcontroller. To light the segments of a given digit, the microcontroller supplies the appropriate 4-bit BCD code to the seven-segment decoder/driver (74HC4511). As an example, if we wanted to display 1234, we would need to program the microcontroller (using software) to turn off all digits except the MSD (leftmost digit) and then supply the decoder/driver with the BCD code for 1. Then the next significant digit (2) would be driven, and then the next significant digit (3), and then the LSD (4). After that, the process would recycle for as long as we wanted our program to display 1234.

12.10.2 Liquid-Crystal Displays

In low-power CMOS digital systems (for example, battery- or solar-powered electronic devices), the dissipation of an LED display can consume most of a system's power requirements, which is something you want to avoid, especially since you are looking to save power when using CMOSs. LCDs, on the other hand, are ideal for low-power applications.

Unlike an LED display, an LCD is a passive device. This means that instead of using electric current to generate light, it uses light that is already externally present (such as sunlight, room lighting). For the LCD's optical effects to occur, the external light source needs to supply only a minute amount of power (within the mW/cm^2 range).

Simple Alphanumeric Display

Alphanumeric dual display (internally wired for mutliplexing)

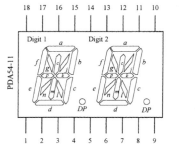

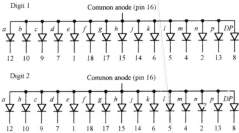

FIGURE 12.130

Figure 12.130 shows a common anode, 2-character, 14-segment (+ decimal) alphanumeric display. Notice that the segments of the two characters are internally wired together. This means that the display is designed for multiplexing. Though it is possible to use a microcontroller along with transistor drivers to control this display, the number of lines required is fairly large. Another option is to use a special driver IC, like Intersil's ICM7243B 14-segment 6-bit ASCII driver. Another alternative is simply to avoid using this kind of display and use a "smart" alphanumeric display that contains all the necessary control logic (drivers, code converters, and so on).

"Smart" Alphanumeric Display

HPDL-1414 (Hewlett Packard) 4-character smart alphanumeric display

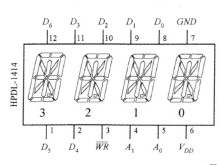

Write Truth Table

$\overline{WR}$	A_1	A_0	D_6 to D_0	DIG_3	DIG_2	DIG_1	DIG_0
0	0	0		NC	NC	NC	A
0	0	1	ASCII	NC	NC	A	NC
0	1	0	code	NC	A	NC	NC
0	1	1	"A"	A	NC	NC	NC
1	X	X	X	previously written data

Character Set

BITS	D_3 D_2 D_1 D_0	0 0 0 0	0 0 0 1	0 0 1 0	0 0 1 1	0 1 0 0	0 1 0 1	0 1 1 0	0 1 1 1	1 0 0 0	1 0 0 1	1 0 1 0	1 0 1 1	1 1 0 0	1 1 0 1	1 1 1 0	1 1 1 1
D_6 D_5 D_4	HEX	0	1	2	3	4	5	6	7	8	9	A	B	C	D	E	F
0 1 0	2	space	!	"	#	$	%	&	'	(	)	*	+	,	-	.	/
0 1 1	3	0	1	2	3	4	5	6	7	8	9	:	;	<	=	>	?
1 0 0	4	@	A	B	C	D	E	F	G	H	I	J	K	L	M	N	O
1 0 1	5	P	Q	R	S	T	U	V	W	X	Y	Z	[	\	]	^	_

FIGURE 12.131

The HPDL-1414 is a "smart," 4-character, 16-segment display. This device is complete with LEDs, on-board 4-word ASCII memory, a 64-word character generator, 17-segment drivers, 4-digit drivers, and scanning circuitry necessary to multiplex the four LED characters. It is TTL-compatible and relatively easy to use. The seven data inputs D_0 to D_6 accept a 7-bit ASCII code, while the digital select inputs A_0 and A_1 accept a 2-bit binary code that is used to specify which of the four digits is to be lighted. The WRITE ($\overline{WR}$) input is used to load new data into memory. After a character has been written to memory, the IC decodes the ASCII data, drives the display, and refreshes it without the need for external hardware or software.

One disadvantage with LCDs is their slow switching speeds (the time it takes for a new digit/character to appear). Typical switching speeds for LCDs range from around 40 to 100 ms. At low temperatures, the switching speeds get even worse. Another problem with LCDs is the requirement that external light be present. Though there are LCD displays that come with backlighting (such as an LED behind the display), obviously, this will increase power consumption.

Basic Explanation of How an LCD Works

An LCD consists of a number of layers that include a polarizer, a set of transparent electrodes, a liquid-crystal element, a transparent back electrode, a second polarizer, and a mirror (see the leftmost illustration in Fig. 12.132).

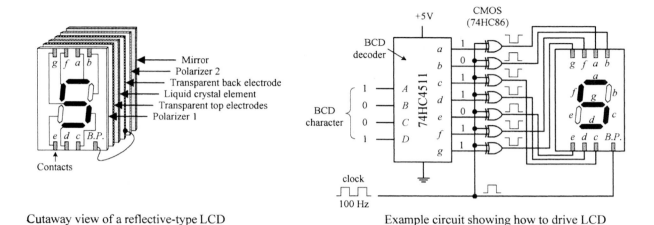

Cutaway view of a reflective-type LCD   Example circuit showing how to drive LCD

FIGURE 12.132

The transparent top electrodes are used to generate the individual segments of a digit, character, and so on, while the transparent back electrode forms a common plane, often referred to as the *back plane* (BP). The top electrode segments and the back electrode are wired to external contacts. With no potential difference between a given top electrode and the back electrode, the region where the top electrode is located appears silver in color against a silver background. However, when a potential is applied between a given top electrode and back electrode, the region where the top electrode is located appears dark against a silver background.

The circuit in Fig. 12.132 shows a basic way to drive a seven-segment LCD. It uses a 74HC4511 BCD decoder and XOR gates to generate the prior drive signals for the LCD. A very important thing to note in this circuit is the clock. As it turns out, an LCD actually requires ac drive signals (for example, squarewaves) instead of dc drive signals. If dc were used, the primary component of the display—namely, the liquid crystal—would undergo electrochemical degradation (more on the liquid crystal in a moment). The optimal frequency of the applied ac drive signal is typically from around 25 Hz to a couple hundred hertz. Now that we understand that, it is easy to see why we need the XOR gates.

As the clock delivers squarewaves to the back electrode (back plane, or BP), the XOR gates act as enable gates that pass and invert a signal and apply it to a given top electrode segment. For example, if a BCD code of 1001 (5) is applied to the decoder, the decoder's outputs a, c, d, f, and g go high, while outputs b and e go low. When a positive clock pulse arrives, XOR gates attached to the outputs that are high invert the high levels. XOR gates attached to outputs that are low pass on the low levels. During the same pulse duration, the back plane is set high. Potentials now are present between a, c, d, f, and g segments and the back plane, and therefore these segments appear dark. Segments b and e, along with the background, appear silverish because no potential exists between them and the back plane. Now, when the clock pulse goes low, the display remains the same (provided the BCD input hasn't changed), since all that has occurred is a reverse in polarity. This has no effect on the optical properties of the display.

Detailed Explanation of How an LCD Works (the Physics)

Figure 12.133 shows how an LCD generates a clear (silverish) segment. When control signals sent to the transparent top and back electrodes are in phase, no potential exists between the two electrodes. With no potential present, the cigar-shaped organic liquid crystals (nematic crystals) arrange themselves in spiral state, as shown in the figure.

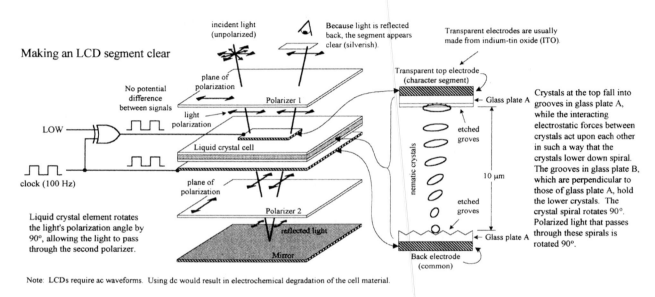

Note: LCDs require ac waveforms. Using dc would result in electrochemical degradation of the cell material.

FIGURE 12.133

The upper crystal aligns itself horizontal to the page, while the lowest crystal aligns itself perpendicular to the page. The upper crystal and the lower crystal are held in place by tiny grooves that are etched into the inner surfaces of the glass surfaces of the cell. Crystals in between the upper crystal and the lower crystal progressively spiral 90° due to electrostatic forces that exist between neighboring crystals. When polarized light passes through a region of the display that contains these spirals, the polarization angle of the light is rotated 90°.

Now, looking at the display as a whole, when incident unpolarized light passes through polarizer 1 (as shown in the figure), the light becomes polarized in the same direction of the plane of polarization of the first polarizer. The polarized light then passes through the transparent top electrode and enters the liquid-crystal cell. As it passes through the cell, its polarization angle is rotated 90°. The polarized light that exits the cell then passes through the transparent back electrode and the second polarizer without problems. (If we were to remove the liquid-crystal cell, all polarized light that passed through the first polarizer would be absorbed, since we would have crossed polarizers.) The light that passes through the second polarizer then reflects off the mirror, passes through the second polarizer, on through the liquid-crystal cell (getting rotated 90°), through the first polarizer, and finally reaches the observer's eye. This reflected light appears silver in color. Note that the background of LCDs constantly appears silver because no potential exists across the liquid-crystal cell in the background region.

Figure 12.134 shows how an LCD generates a dark segment. When control signals sent to the top and back electrodes are out of phase, a potential difference exists between the two electrodes. This causes the crystals to align themselves in a parallel manner, as shown in the figure. When the polarized light from the first polarizer passes through the cell region containing these parallel crystals, nothing happens—the polarization angle stays the same. However, when the light comes in contact with the second polarizer, it is absorbed because the angle of polarization of the light and the plane of polarization of the second polarizer are perpendicular to each other. Since light reaches the mirror, no light is reflected back to the observer's eye, and hence the segment appears dark.

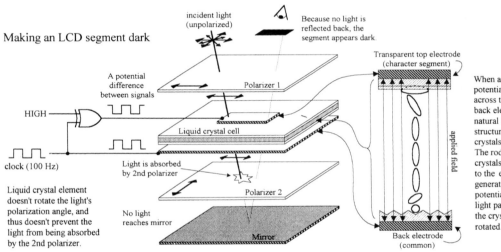

FIGURE 12.134

The LCD shown in Fig. 12.133 represents what is referred to as a *standard twisted nematic display*. Another common LCD is the *supertwist nematic display*. Unlike the standard twisted display, this display's nematic crystals rotate 270° from top to bottom. The extra 180° twist improves the contrast and viewing angle.

Driving LCDs

CD4543B CMOS BCD-TO-SEVEN-SEGMENT LATCH/DECODER/DRIVER

The CD4543B (Texas Instruments), shown in Fig. 12.135, is a BCD-to-seven-segment latch/decoder/driver that is designed for LCDs, as well as for LED displays. When used to drive LCDs, a squarewave must be applied simultaneously to the CD4543B's Phase (*Ph*) input and to the LCD's back plane. When used to drive LED displays, a high is required at the Phase input for common cathode displays, while a low is

CMOS BCD-to-Seven-Segment Latch/Decoder/Driver for LCDs

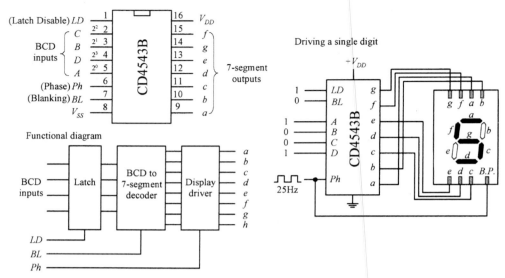

FIGURE 12.135

required for common anode displays. To blank the display (set outputs a–g low), the *BL* input is set high. The CD4543B also comes with a Latch Disable input (*LD*), which can be used to latch onto input data, preventing new input data from altering the display.

MM5453 LCD DRIVER

The MM5453 (National Semiconductor) is a 40-pin IC that can drive up to 33 segments of an LCD, which can be used to drive $4\frac{1}{2}$-digit seven-segment displays. It houses an internal oscillator section (requiring an external RC circuit) that generates the necessary squarewaves used to drive the LCD. To activate given segments within the display, a serial code is applied to the data input. The code first starts out with a start bit (high) followed by data bits that specify which outputs should be driven high or low. Figure 12.136 shows an example display circuit, along with corresponding data format required to drive a $4\frac{1}{2}$-digit display.

VI-322-DP LCD AND ICL7106 $3\frac{1}{2}$-DIGIT LCD, ADC DRIVER

There are a number of specialized LCDs that can be found in the electronics catalogs. An example is Varitronix's VI-322-DP $3\frac{1}{2}$-digit (plus ~, +, BAT, Δ) LCD, shown in Fig. 12.136. This display is configured in a static drive arrangement (each segment has a separate lead) and is found in many test instruments. To drive this display, you first check to see what kind of driver the manufacturer suggests. In this case, the manufacturer suggests using Intersil's ICL7106. This IC is a $3\frac{1}{2}$-digit LCD/LED display driver as well as an ADC. This dual-purpose feature makes it easy to interface transducers directly to the same IC that is driving the display. To learn how to use the ICL7106, check out Intersil's data sheet at http://www.intersil.com.

MM5453 (National Semiconductor) Liquid Crystal Display Driver

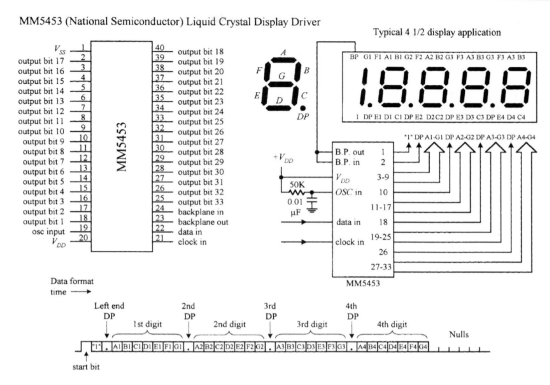

FIGURE 12.136

VI-322-DP (Varitronix) static-drive LCD

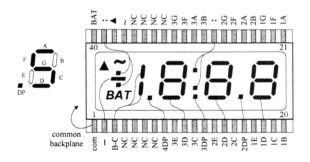

FIGURE 12.137

Multiplexed LCDs

We have just seen examples of static-drive-type LCDs, where each segment (to electrode) had its own lead, and a single common plane was used as the back electrode.
Another type of LCD is designed with multiplexing in mind and is referred to as
dynamic drive or *multiplexed display*.

As with the multiplexed LED display, multiplexed LCDs can greatly reduce the
number of external connections required between the display and driver. However,
they require increased complexity in drive circuitry (or software) to drive. In a

multiplexed LCD, appropriate segments are connected together to form groups that are sequentially addressed by means of multiple back-plane electrodes.

"Intelligent" Dot-Matrix LCD Modules

Dot-matrix LCDs are used to display alphanumeric characters and other symbols. These displays are used in cell phones, calculators, vending machines, and many other devices that provide the user with simple textual information. Dot-matrix LCDs are also used in laptop computer screens; however, these displays incorporate special filters, multicolor back lighting, and so on. For practical purposes, we'll concentrate on the simple alphanumeric LCDs.

An alphanumeric LCD screen is usually divided into a number of 5×8 pixel blocks, with vertical and horizontal spaces separating each block. Figure 12.138 shows a display with 20 columns and 4 rows of 5×8 pixel blocks. Other standard configurations come with 8, 16, 20, 24, 32, or 40 columns and 1, 2, or 4 rows. To generate a character within a given block requires that each pixel within the block be turned on or off. As you can imagine, to control so many different pixels (electrode segments) requires a great deal of sophistication. For this reason, an intelligent driver IC is required.

Alphanumeric LCD module

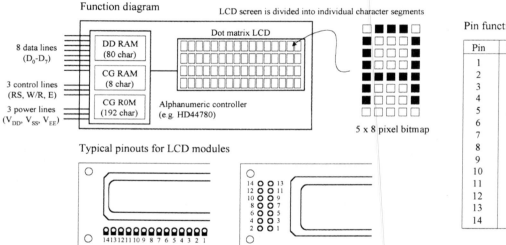

FIGURE 12.138

Almost all alphanumeric LCD modules are controlled by Hitachi's HD44780 (or equivalent) driver IC. This driver contains the following:

- A permanent memory (CG ROM) that stores 192 alphanumeric characters
- A random access memory (DD RAM) used to store the display's contents
- A second random access memory (CG RAM) used to hold custom symbols
- Input lines for data and instruction control signals
- Multiplexed outputs for driving LCD pixels
- Additional outputs for communicating with expansion chips to drive more LCD pixels

This driver is built right into the LCD module. (You could attempt to construct your own module by interfacing the driver with an LCD, but it would not be worth the effort—the numerous tiny connections would drive you nuts.) From now on, all modules described in this section are assumed to be HD44780-driven.

BASIC OVERVIEW OF THE PINS

The standard LCD module comes with a 14-pin interface: eight data lines (D_0–D_7), three control lines (RS, W/R, and E), and three power lines (V_{DD}, V_{SS}, and V_{EE}).

V_{DD} (pin 2) and V_{SS} (pin 1) are the module's positive and negative power supply leads. Usually, V_{DD} is set to +5 V, while V_{SS} is grounded. V_{EE} (pin 3) is the display's contrast control. By changing the voltage applied to this lead, the contrast of the display increases or decreases. A potentiometer placed between supply voltages, with its wiper connected to V_{EE}, allows for manual adjustment.

D_0–D_7 (pins 7–14) are the data bus lines. Data can be transferred to and from the display either as a single 8-bit byte or as two 4-bit nibbles. In the latter case, only the upper four data lines (D_4–D_7) are used.

RS (pin 4) is the Register Select line. When this line is low, data bytes transferred to the display module are interpreted as commands, and data bytes read from the display module indicate its status. When the RS line is set high, character data can be transferred to and from the display module.

R/W (pin 5) is the Read/Write control line. To write commands or character data to the module, R/W is set low. To read character data or status information from the module, R/W is set high.

E (pin 6) is the Enable control input, which is used to initiate the actual transfer of command or character data to and from the module. When writing to the display, data on the D_0–D_7 lines is transferred to the display when the enable input receives a high-to-low transition. When reading from the display, data become available to the D_0–D_7 lines shortly after a low-to-high transition occurs at the enable input and will remain available until the signal goes low again.

Figure 12.139 shows the instruction set and standard set of characters for an LCD module. Next, we'll go through some examples illustrating how to use the instructions and how to write characters to the display.

TEST CIRCUIT USED TO DEMONSTRATE HOW TO CONTROL THE LCD MODULE

Figure 12.140 shows a simple test circuit that is quite useful for learning how to send commands and character data to the LCD module. (In reality, the LCD module is connected to a microprocessor or microcontroller, as shown to the left in the figure.) In this circuit, switches connected to data inputs use pullup resistors in order to supply a high (1) when the switch is open or supply a low (0) when the switch is closed.

The enable input receives its high and low levels from a debounced toggle switch. Debouncing the enable switch prevents the likelihood of multiple enable signals being generated. Multiple enable signals tend to create unwanted effects, such as generating the same character over and over again across the display. The 5-kΩ pot is used for contrast. Note that in this circuit, we've grounded the R/W line, which means we'll only deal with writing to the display.

LCD Instruction Set

INSTRUCTION	R/S	R/W	D_7	D_6	D_5	D_4	D_3	D_2	D_1	D_0
Clear Display	0	0	0	0	0	0	0	0	0	1
Display & Cursor Home	0	0	0	0	0	0	0	0	1	X
Character Entry Mode	0	0	0	0	0	0	0	1	I/D	S
Display & Cursor On/Off	0	0	0	0	0	0	1	D	C	B
Display/Cursor Shift	0	0	0	0	0	1	D/C	R/L	X	X
Function Set	0	0	0	0	1	DL	N	F	X	X
Set CGRAM Address	0	0	0	1	A	A	A	A	A	A
Set Display Address	0	0	1	A	A	A	A	A	A	A
Poll the "Busy Flag"	0	0	BF	X	X	X	X	X	X	X
Write Character to Display [a]	1	0	D	D	D	D	D	D	D	D
Read Character on Dsplay [b]	1	1	D	D	D	D	D	D	D	D

I/D = Increment (I/D = 1)*/Decrement (I/D = 0) each byte written to display
S = Display shift on (S = 1), Display shift off (S = 0)*
D = Turn display on (D = 1), Turn display off (D = 0)*
C = Show cursor (C = 1), Hide cursor (C = 0)
B = Underline cursor (B = 0, C = 1), Blink cursor (B = 1, C = 1)
D/C = Move display (D/C = 1), Move cursor (D/C = 0)
R/L = Direction of shift: Shift right (R/L = 1), Shift left (R/L =0)
DL = Set data interface length: 8-bit interface (DL = 1)*, 4-bit interface (DL = 0)
N = Number of display lines: 2 line mode (N = 1), 1 line mode (N = 0)*
F = Character font format: 5 x 10 dot (F = 1), 5 x 7 dot (F = 0)*
BF = Poll the Busy Flag: controller not busy (BF = 0), controller busy (BF = 1)
A = CGRAM or display address bit
D = Character data bit
a = Write character to display at the current cursor position
b = Read character on display at the current cursor position
X = Don't care
* = Initialization settings

Standard LCD Character Table

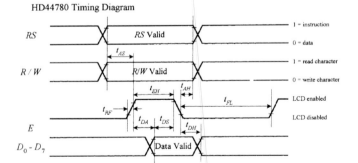

Steps used to Read and Write data to and from LCD module (HD44780 controlled)

HD44780 Timing Diagram

Steps for displaying a character
Set to Write Character to Display mode: R/W = 0, RS = 1.
Apply data bits (character code) to D_7 - D_0.
Breifly set E = 1, then set E = 0

Steps for reading data from display
Set to Read Character on Display mode: R/W = 1, RS = 1.
Set E = 1.
Read data from D_7 - D_0.
Set E = 0

t_{AS} (Address setup time) - For data inputs to be interpreted correctly, they must be held for a minimum time of t_{AS} (~140ns) prior to the Enable signal.
t_{EH} (Enable high time) - E must be held HIGH for a minimum of t_{EH} (~ 450ns) for proper operation.
t_{DS} (Data setup time) - Data inputs must be held stable for a time t_{DS} (~ 200ns) prior to the E signal for proper operation.
t_{AH} (Address hold time) - Control lines RS and R/W lines must not change for a duration of t_{AH} (~10ns) after the E line goes LOW for proper operation.
t_{DH} (Data hold time) - Data lines D_0 - D_7 must not change for a duration of t_{DH} (~20ns) after the E line goes LOW for prior operation.
t_{EL} (enable low time) - E line must not be set HIGH again (for the next command), for at least t_{EL} (~500ns) for proper operation.
t_{RF} (rise and fall time) - Rise and fall times are each ~ 25ns each.

FIGURE 12.139

Simple experimental setup that uses switches to write to LCD module

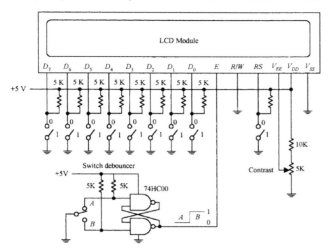

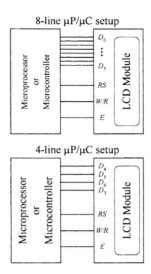

FIGURE 12.140

WHEN POWER IS FIRST APPLIED

When power is first applied to the display, the display module resets itself to its initial settings. Initial settings are indicated in the LCD instruction set with an asterisk. As indicated, the display is actually turned off during the initial setting condition. If we attempt to write character data to the display now, nothing will show up. In order to show something, we must issue a command to the module telling it to turn on its display.

According to the instruction set, the Display & Cursor On/Off instruction can be used to turn on the display. At the same time, this instruction also selects the cursor style. For example, if we apply the command code 0000 1111 to D_7–D_0, making sure to keep RS low so the module will interpret data as a command, a blinking cursor with an underline should appear at the top leftmost position on the display. But before this command can take effect, it must be sent to the module by momentarily setting the Enable (E) line low.

Another important instruction that should be implemented after power-up is the Function Set command. When a two-line display is used, this command tells the module to turn on the second line. It also tells the module what kind of data transfer is going to be used (8-bit or 4-bit), and whether a 5×10 or 5×7 pixel format will be used (5×10 is found in some one-line displays). Assuming that the display used in our example circuit is a two-line display, we can send the command 0011 1000 telling the display to turn on both lines, use an 8-bit transfer, and provide a 5×7 pixel character format. Again, to send this command, we set RS low, then supply the command data to D_7–D_0, and finally pulse E low.

Now that the module knows what format to use, we can try writing a character to the display. To do this, we set the module to character mode by setting RS high. Next, we apply one of the 8-bit codes listed in the standard LCD character set table to the data inputs D_7–D_0. For example, if we want to display the letter Q, we apply 01010001 (hex 51 or 51_H). To send the character data to the LCD module, we pulse E low. A Q should then appear on the display. To clear the screen, we use the Clear Display command 0000 0001, remembering to keep RS low and then pulsing E low.

ADDRESSING

After power-up, the module's cursor is positioned at the far-left corner of the first line of the display. This display location is assigned a hexadecimal address of 00_H. As new characters are entered, the cursor automatically moves to the right to a new address of 01_H, then 02_H, and so on. Although this automatic incrementing feature makes life easy when entering characters, there are times when it is necessary to set the cursor position to a location other than the first address location.

To set the cursor to another address location, a new starting address must be entered as a command. There are 128 different addresses to choose from, although not all these addresses have their own display location. In fact, there are only 80 display locations laid out on a single line in one-line mode or 40 display locations laid out on each line in two-line mode. Now, as it turns out, not all display locations are necessarily visible on the screen at one time. This will be made more apparent in a moment. Let's first try a simple address example with the LCD module set to two-line mode (provided that two lines are actually available).

To position the cursor to a desired location, we use the Set Address command. This command is specified with the binary code 1000 0000 + (binary value of desired hex address). For example, to send a command telling the cursor to jump to the 07_H address location, we apply (1000 0000 + 0000 0111) = 1000 0111 to the D_7–D_0 inputs, remembering to hold RS low and then pulsing E low. The cursor should now be located at the eighth position over from the left.

It is important to realize that the relationship between addresses and display locations varies from module to module. Most displays are configured with two lines of characters, with the first line starting at address 00_H and the second line at address 40_H. Figure 12.141 shows the relationship between the address and display locations for various LCD modules. Note that the four-line module is really a two-line type with the two lines split, as shown in the figure.

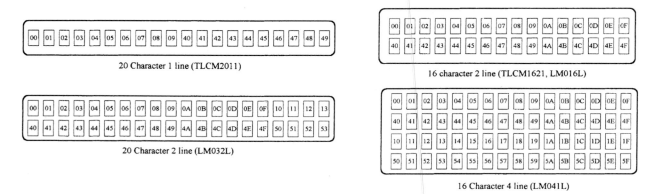

20 Character 1 line (TLCM2011)

16 character 2 line (TLCM1621, LM016L)

20 Character 2 line (LM032L)

16 Character 4 line (LM041L)

FIGURE 12.141

SHIFTING THE DISPLAY

Regardless of their size, LCD modules have 80 display locations that can be written to. With smaller displays, not all 80 locations can be displayed at once on the screen. For example, if we were to enter all the letters of the alphabet onto the first line of a 20-character display, only letters *A* through *T* would appear on the screen. Letters *S* through *Z*, along with the cursor, would be "pushed off" to the right of the screen, hidden from view.

To bring these hidden characters into view, we can apply the Cursor/Display Shift command to shift all display locations to the left. The command for shifting to the left is 0001 1000. Every time this command is issued, the characters shift one step to the left. In our example, it would take seven of these commands to bring T through Z and the cursor into view.

To shift things to the right, we apply the command 0001 1100. To bring the cursor back to address 00_H and shift the display address 00_H back to the left-hand side of the display, a Cursor Home command (0000 0010) can be issued. Another alternative is to use the Clear Display command 0000 0001. However, this command also clears all display locations.

CHARACTER ENTRY MODE

If you do not want to enter characters from left to right, you can use the Character Entry Mode command to enter characters from right to left. To do this, the cursor must first be sent to the rightmost display location on the screen. After that, the Character Entry Mode command 0000 0111 is entered into the module. This sets the entry mode to autoincrement/display shift left. Now, when characters are entered, they appear on the right-hand side, while the display shifts left for each character entered.

USER-DEFINED GRAPHICS

Commands 0100 0000 to 0111 1111 are used to program user-defined graphics. To program these graphics on-screen, the display is cleared, and the module is sent a Set Display Address command to position the cursor at address 00H. At this point, the contents of the eight user character locations can be viewed by entering binary data 0000 0000 to 0000 0111 in sequence. These characters will appear initially as garbage.

To start defining the user-defined graphics, a Set CGRAM command is sent to the module. Any value between 0100 0000 (40H) and 0111 1111 (7F) will work. Data entered from now on will be used to construct the user-defined graph, row by row. For example, to create a light bulb, the following data entries are made: 0000 1110, 0001 0001, 0001 0001, 0001 0001, 0000 1110, 0000 1010, 0000 1110, 0000 0100. Notice that the first three most significant bits are always 0 because there are only 5 pixels per row. Other user-defined graphics can be defined by entering the 8-byte sequence, and so on. Figure 12.142 shows how the CGRAM address corresponds to the individual pixels of the user-defined graphic.

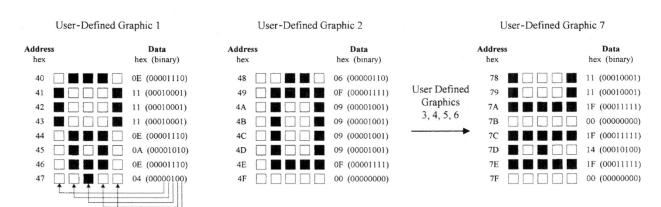

FIGURE 12.142

There are up to eight user-defined graphics that can be programmed. These then become part of the character set and can be displayed by using codes 0000 0000 to 0000 1111 or 0000 1000 to 0000 1111, both of which produce the same result.

One problem when creating user-defined graphics is they will be lost when power is removed from the module—a result of the volatile CGRAM. Typically, the user-defined graphic data is actually stored in an external nonvolatile EPROM or EEPROM, where the data is copied by a microprocessor and loaded into the display module sometime after power-up.

4-BIT DATA TRANSFER

As indicated in the Function Set command, the LCD module is capable of both 8-bit and 4-bit data transfer. In 4-bit mode, only data lines D_4–D_7 are used. The other four lines, D_0–D_3, are left either floating or tied to the power supply. To send data to the display requires sending two 4-bit chunks instead of one 8-bit word.

When power is first applied, the module is set up for 8-bit transfer. To set up 4-bit transfer, the Function Set command with binary value 0010 0000 is sent to the display. Note that since there are only four data lines in use, all 8 bits cannot be sent. However, this is not a problem, since the 8-bit/4-bit selection is on data bit D_4. From now on, 8-bit character and command bits must be sent in two halves, the first 4 most significant bits and then the remaining 4 bits. For example, to write character data 0100 1110 to the display requires setting RS high, applying 0100 to the data lines, pulsing E low, then applying 1110 to the data lines, and pulsing E low again.

The 4-bit transfer is frequently used when the LCD module is interfaced with a microcontroller that has limited I/O lines. See Fig. 12.142.

12.11   Memory Devices

Memory devices provide a means of storing data on a temporary or permanent basis for future recall. The storage medium used in a memory device may be a semiconductor-based IC (primary memory), a magnetic tape, a magnetic disk, or an optical disk (secondary memories). In most cases, the secondary memories are capable of storing more data than primary memories because their surface areas are larger. However, secondary memories take much longer to access (read or write) data because memory locations on a disk or tape must be physically positioned to the point where they can be read or written to by the read/write mechanism. Within a primary memory device, memory locations are arranged in tiny regions within a large matrix, where each memory location can be accessed quickly (matter of nanoseconds) by applying the proper address signals to the rows within the matrix.

Figure 12.143 shows an overview of primary and secondary memories. In this section, we'll discuss only the primary memories, since these devices are used more frequently in designing gadgets than secondary memories. Secondary memories are almost exclusively used for storing large amounts of computer data, audio data, or video data.

Today, the technology used in the construction of primary memory devices is almost exclusively based on MOSFET transistors. Bipolar transistors are also used within memory ICs. However, these devices are less popular because the amount of data they can store is significantly smaller than that of a memory IC built with

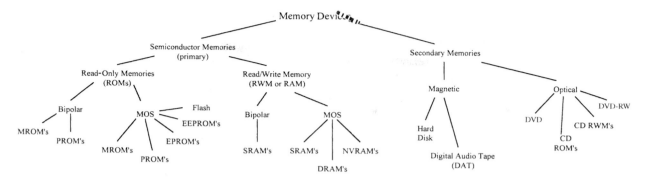

FIGURE 12.143

MOSFET transistors. At one time, bipolar memories had a significant edge in speed over MOSFET memories, but today the speed gap has almost disappeared.

Memory devices consist of two basic subfamilies: *read-only memory* (ROM) and *read/write memory* (RWM), which is more commonly referred to as *random-access memory* (RAM). Within each of these subfamilies exist more subfamilies, as shown in Fig. 12.143. Let's start out by discussing the ROM devices.

12.11.1 Read-Only Memory

ROM is used to store data on a permanent basis. These devices are capable of random access, like RAM devices, but unlike RAM devices, they do not lose stored data when power is removed from the IC.

ROM is used in nearly all computers to store boot-up instructions (such as stack allocation, port and interrupt initializations, and instructions for retrieving the operating system from disk storage) that are enacted when the computer is first turned on.

In some microcontroller applications (simple-function gadgets, appliances, toys, and so on), the entire stand-alone program resides in ROM. The microcontroller's central processing unit (CPU) retrieves the program instructions and uses volatile RAM for temporary data storage as it runs through the ROM's stored instructions.

In some instances, you find ROM within discrete digital hardware, where it is used to store lookup tables or special code-conversion routines. For example, digital data from an ADC could be used to address stored words that represent, say, a binary equivalent to a temperature reading in Celsius or Fahrenheit. This also can be used to replace a complex logic circuit, where, instead of using a large number of discrete gates to get the desired function table, you simply program the ROM to provide the designed output response when input data is applied. The last few applications mentioned, however, are becoming a bit obsolete—the microcontroller seems to be taking over everything.

ROM is generally used for read-only operations and not written to after initially programmed. However, some ROM-like devices, such as EPROM, EEPROM, and flash memory, are capable of erasing stored data and rewriting data to memory. Before we take a look at these erasable ROM-like devices, let's first cover some memory basics.

12.11.2 Simple ROM Made Using Diodes

To get a general idea of how ROM works, let's consider the simple circuit in Fig. 12.144.

In reality, today's ROM devices rarely use diode memory cells. Instead, they typically use transistor-like memory cells formed on silicon wafers. Also, a more realistic ROM device comes with three-state output buffers that can be enabled or disabled (placed in a high Z state) by applying a control signal. The three-state buffers make it possible to effectively disconnect the memory from a data bus to which it is attached. (In our simple diode memory circuit, the data is always present on the output lines.) The basic layout, with address decoder and memory cells, is pretty much the same for all memory devices. There are additional features, however, and we'll discuss these in a minute. First, let's cover some memory nomenclature.

12.11.3 Memory Size and Organization

A ROM that is organized in an $n \times m$ matrix can store n different m-bit words; in other words, it can store $n \times m$ bits of information. To access n different words requires $\log_2 n$ address lines. For example, our simple ROM in Fig. 12.144 requires $\log_2 8 = 3$ address inputs (this may look more familiar: $2^3 = 8$). Note that within multiplexed memories and memories that come with serial inputs, the actual physical number of address inputs is ether reduced or the address information is entered serially, along with data and other protocol information.

In terms of real memory ICs, the number of address inputs is typically eight or higher (for parallel input devices at any rate). Common memory sizes are indicated

Basic Diode ROM

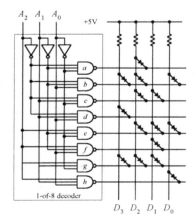

$A_2\ A_1\ A_0$	$D_3\ D_2\ D_1\ D_0$	Active gate (sink)
0 0 0	1 0 1 1	a
0 0 1	0 0 1 0	b
0 1 0	1 0 0 1	c
0 1 1	0 1 1 1	d
1 0 0	1 0 0 0	e
1 0 1	1 1 0 1	f
1 1 0	0 0 0 1	g
1 1 1	1 1 1 0	h

FIGURE 12.144

This is a simple ROM device that uses an address decoder IC to access eight different 4-bit words stored in a diode matrix. Data to be read is output via the D_3–D_0 lines. The diode matrix is broken up into rows and columns. The intersection of a row and column represents a bit location. When a given row and column are linked together with a diode, the corresponding data output line goes low (0) when the corresponding column is selected by the address decoder via the A_2–A_0 inputs. When a specific row is addressed, the NAND gate sinks current, so the current from the supply passes through the diode and into the NAND gate's output. This makes the corresponding data line low. When no diode is placed between a given column and row, the corresponding data line goes high (0) when the corresponding row is selected by the address decoder. (There is no path to ground in this case.) In this particular example, we have an 8×4 ROM (eight different 4-bit words). By increasing the width of the matrix (adding more columns), it is possible to increase the word size. By increasing the height of the matrix (adding more rows—more addresses), it is possible to store more words. In other words, we could make an $m \times n$ ROM.

TABLE 12.2   Common Memory Sizes

NO. OF ADDRESS LINES	NO. OF MEMORY LOCATIONS	NO. OF ADDRESS LINES	NO. OF MEMORY LOCATIONS	NO. OF ADDRESS LINES	NO. OF MEMORY LOCATIONS
8	256	14	16,384 (16 K)	20	1,048,576 (1 M)
9	512	15	32,768 (32 K)	21	2,097,152 (2 M)
10	1,024 (1 K)	16	65,536 (64 K)	22	4,194,304 (4 M)
11	2,048 (2 K)	17	131,072 (128 K)	23	8,388,608 (8 M)
12	4,096 (4 K)	18	262,144 (256 K)	24	16,777,216 (16 M)
13	8,192 (8 K)	19	524,288 (540 K)	25	33,554,432 (32 M)

in Table 12.2. Note that in the table, 1K is used to represent 1024 bits, not 1000 bits, as the k (kilo) would lead you to believe. By digital convention, we say that $2^1 = 2$, $2^2 = 4$, $2^3 = 8$, ... $2^8 = 256$, $2^9 = 512$, $2^{10} = 1,024$ (or 1 K), $2^{11} = 2,048$ (or 2 K), ... $2^{18} = 262,144$ (256K), $2^{19} = 524,288$ (540 K), $2^{20} = 1,048,576$ (or 1 M, for mega), $2^{21} = 2,097,152$ (2 M), ... $2^{30} = 1,073,741,824$ (or 1 G, for giga), and so on.

If this convention confuses you, it should. It is not exactly obvious, and leaves you scratching your head. Also, when a data sheet says 64 K, you need to read further to figure out what the actual organization is, say, 2048×32 (2 K × 32), 4096×16 (4 K × 16), 8192×8 (8 K × 8), $16,384 \times 4$ (16 K × 4), or another system.

In Table 12.2, watch out for terms such as kB, MB, and GB. These terms refer to bytes not bits; the *B* signifies 1 byte, or 8 bits. This means that a memory that stores 1 kB actually stores 1 K × 8 (8 K) bits of data. Likewise, memories that store 1 MB and 1 GB actually store 1 M × 8 (8 M) and 1 G × 8 (8 G) bits of data, respectively.

12.11.4 Simple Programmable ROM

Figure 12.145 shows a more accurate representation of ROM-type memory. Unlike the diode ROM, each memory cell contains a transistor and fusible link. Initially, the ROM has all programmable links in place. With every programmable link in place, every transistor is biased on, causing high voltage levels (logic 1s) to be stored throughout the array. When a programmable link is broken, the corresponding memory cell's transistor turns off, and the cell stores a low voltage level (logic 0). Note that this ROM contains three-state output buffers that keep the output floating until a low is applied to the Chip Enable ($\overline{CE}$) input. This feature allows the ROM to be interfaced with a data bus.

A basic ROM circuit schematic is shown in Fig. 12.145, along with the appropriate address and chip-enable waveforms needed to enact a read operation. To read data stored at a given address location, the Chip Enable input is set high to disable the chip (remove old data from data outputs)—see time t_0. At time t_1, a new address is placed on the 3-bit address bus (A_2, A_1, and A_0). At time t_2, the Chip Enable input is set low, which allows addressed data stored in memory to be output via D_3, D_2, D_1, and D_0.

In reality, the stored data is not output immediately but is delayed for a very short time (from t_2 to t_3) due to the propagation delay that exists between the initial chip

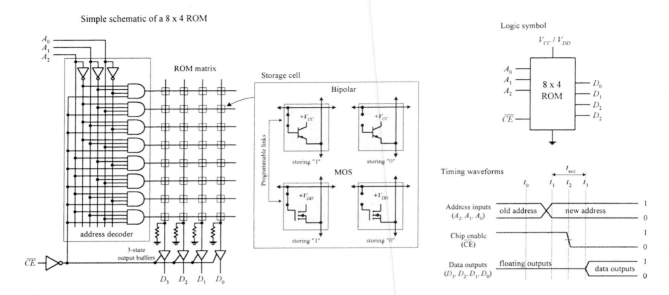

FIGURE 12.145

enable signal and the signal that reaches the enable leads of the output buffers. In memory lingo, the time from t_1 to t_4 is referred to as the *access time*, which is between around 10 ns and a couple hundred nanoseconds, depending on the specific technology used.

Now two important questions need addressing:

- How does one "break" a programmable link? In other words, how do we program the ROM?

- Is it possible to restore a broken programmable link back to its unbroken" state? In other words, is it possible to reprogram the ROM?

These lead to the next topic.

12.11.5 ROM Devices

There are basically two kinds of ROMs: those that can be programmed only once and those that can be reprogrammed any number of times. One-time programmable memories include the mask ROM (MROM) and the programmable ROM (PROM). ROMs that can be reprogrammed include the erasable programmable ROM (EPROM), electrically erasable programmable ROM (EEPROM), and flash memory.

MROM

An MROM is a custom memory device that is permanently programmed by the manufacturer simply by adding or leaving out diodes or transistors within a memory matrix. In order to create a desired memory configuration, you must supply the manufacturer with a truth table stating which data configuration is desired. Using the truth table, the manufacturer then generates a *mask* that is used to create the interconnections within the memory matrix during the fabrication process.

As you can imagine, producing a custom MROM is not exactly cheap; in fact, it is rather costly (more than $1,000). It is only worthwhile using an MROM if you plan to mass produce some device that requires the same data instructions (for example, program instructions) over and over again—no upgrades to memory needed in the future. In this case, the cost for each IC—after the initial mask is made—is relatively cheap, assuming you need more than a couple thousand chips.

MROMs are commonly found in computers, where they are used to store system operating instructions and data that is used to decode keyboard instructions.

PROM

PROMs are fusible-link programmable ROMs. Unlike the MROM, with PROM devices, data is not etched in stone. Instead, the manufacturers provide you with a memory IC whose matrix is clean (full of 1s). The number of bits and the configuration (n × m) of the matrix vary depending on specific ROM. To program the memory, each fusible link must be blown with a high-voltage pulse (such as 21 V).

The actual process of blowing individual fuses requires a PROM programming unit. This PROM programmer typically includes a hardware unit (where the actual PROM IC is attached), along with programming cable that is linked to a computer (such as via a serial or parallel port). Using software provided by the manufacturer, you enter the desired memory configuration in the program running on the computer and then press a key, which causes the software program to instruct the external programming unit to blow the appropriate links within the IC.

PROMs are relatively easy to program once you have figured out how to use the software, but as with MROMs, once the device is programmed; the memory cannot be altered. In other words, if you mess things up, you must begin afresh with a new chip. These devices were popular some years ago, but today they are considered obsolete.

The most popular ROM-type devices used today are EPROM, EEPROM, and flash memory. These devices, unlike MROM and PROM devices, can be erased and reprogrammed—a very useful feature when prototyping or designing a gadget that requires future memory alterations.

EPROM

An EPROM is a device whose memory matrix consists of a number of specialized MOSFET transistors. Unlike a conventional MOSFET transistor, the EPROM transistor has an additional floating gate that is buried beneath the control gate—insulated from both the control gate and drain-to-source channel by an oxide layer (see Fig. 12.146).

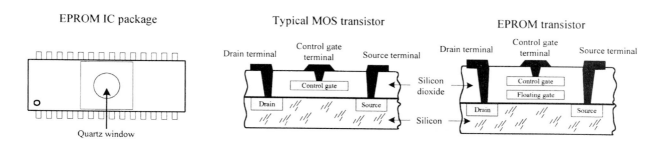

FIGURE 12.146

In its erased (unprogrammed) state, the floating gate is uncharged and does not affect the normal operation of the control gate (which when addressed results in a high voltage or logic 1 being passed through to the data lines). To program an individual transistor, a high-voltage pulse (around 12 V) is applied between the control gate and the drain terminal. This pulse, in turn, forces energetic electrons through the insulating layer and onto the floating gate (referred to as *hot electron injection*). After the high voltage is removed, a negative charge remains on the floating gate and will stay there for decades under normal operating conditions.

With the negative charge in place, the normal operation of the control gate is inhibited; when the control gate is addressed, the charge on the floating gate prevents a high voltage from reaching the data line—the addressed data appears as a low, or logic 0.

In order to reprogram (erase) an EPROM, you must first remove the device from the circuit and then remove a sticker covering its quartz window. After that, you remove all stored charges on the floating gates by shining ultraviolet (UV) light through the window onto the interior transistor matrix. The UV light liberates the stored electrons within the floating gate region by supplying them with enough energy to force them through the insulation. It usually takes 20 minutes of UV exposure for the whole memory matrix to be erased. The number of times an EPROM can be reprogrammed is typically limited to a couple hundred cycles. After that, the chip degrades considerably.

EPROM is often used as nonvolatile memory within microprocessor-based devices that require the provision for future reprogramming. They are frequently used in prototyping and then substituted with MROMs during the mass-production phase. EPROMs are also integrated within microcontroller chips where their sole purpose is to store the microcontroller's main program (more on this in Chap. 13).

EEPROM

An EEPROM device uses a technology somewhat related to the EPROM, but it does not require out-of-circuit programming or UV erasing. Instead, an EEPROM device is capable of selective memory cell erasure by means of controlled electrical pulses.

In terms of architecture, an EEPROM memory cell consists of two transistors: one transistor resembles the EPROM transistor and is used to store data, and the other transistor is used to clear charge from the first transistor's floating gate. By supplying the appropriate voltage level to the second transistor, it is possible to selectively erase individual memory cells instead of having to erase the entire memory matrix, as is the case with EPROM. The only major disadvantage with EEPROM over EPROM is size—due to the two transistors. However, today, with the introduction of new fabrication processes, size is becoming less of an issue.

In terms of applications, EEPROM is ideal for remembering configuration and calibration settings of a device when the power is turned off. For example, EEPROM is found within TV tuners, where it is used to remember the channel, volume setting of the audio amplifier, and so on when the TV is turned off. EEPROM is also found on microcontrollers, where it can be used to store the main program or to hold other nonvolatile data

Flash Memory

Flash memory is generally regarded as the next evolutionary step in ROM technology that combines the best features of EPROM and EEPROM. These devices have the advantage of both in-circuit programming (like EEPROM) and high storage density (like EPROM).

Some variants of flash memory are electrically erasable, like EEPROM, but must be erased and reprogrammed on a device-wide basis, similar to EPROM. Other devices are based on a dual transistor cell and can be erased and reprogrammed on a word-by-word basis. Flash devices are noted for their fast write and erase times, which exceed those of EEPROM devices.

Flash memories are becoming very popular as mass-storage devices. They are found in digital cameras, where a high-capacity flash memory card is inserted directly into a digital camera and can store hundreds or thousands of high-resolution images. They are also used in digital music players, cellular phones, tablets, and so on.

Microcontrollers often include flash memory to contain their program.

Serial Access Memory

So far, we have seen only memories that incorporated parallel access. These devices sit directly on the address and data buses, making it easy for processors to quickly access the memory. Serial access memory is easy to use in principle; however, since all their address lines are typically tied to an address bus within a microprocessor-based system, it is not uncommon for the data to be inadvertently destroyed when the processor runs amuck (issues an undesired write).

Another type of memory that can "hide" the memory from the processor, as well as reduce the total number of pins, uses a serial access format. To move data to and from memory and the processor, a serial link is used. This serial link imposes a strict protocol on data transfers that practically eliminates the possibility that the processor can destroy data accidentally.

Figure 12.147 shows a few serial EPROM and EEPROM devices from Microchip. The SDA pin found in the EEPROM devices acts as a bidirection data lead used to transfer address and data information into the memory IC, as well as transfer data out to the processor. The SCL pin is the serial clock input used to synchronize the data transfer from and to the device. The 24xx64 and 24LC01B/02B EEPROMs also

Sample serial EPROMs and EEPROMs from Microchip

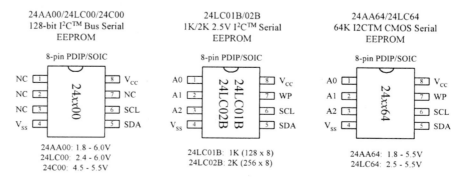

FIGURE 12.147

come with special device address inputs A_0, A_1, and A_2, which are used for multiple device operation. WP is used to enable normal memory operation (read/write entire memory) or inhibit write operations.

Controlling a serial memory device is a bit complex, due to the serial protocol and variations in protocol from IC to IC. If you want to learn more about these serial memories (and you should—they are very handy in microcontroller applications for logging data and storing programs and similar tasks), check out the various manufacturers' websites and read through their data sheets.

12.11.6 RAM

The erasable programmable ROM devices, like EEPROM, have limited read/write endurance—around 100,000 cycles—and take considerable time to write to memory. For applications that require constant and quick read and write cycles, it is necessary to use RAM. This type of memory is used for temporary storage of data and program instructions in microprocessor-based applications. Unlike ROM devices, however, RAM devices are volatile, which means they lose their data if power to the IC is interrupted.

There are two basic types of RAM:

- *Static RAM (SRAM)*: In an SRAM device, data is stored in memory cells that consist of flip-flops. A bit that is written into an SRAM memory cell stays there until overwritten or until the power is turned off.

- *Dynamic RAM (DRAM)*: In a DRAM device, a bit written to the memory cell will disappear within milliseconds if not refreshed, or supplied with periodic clocking to replenish capacitor charge lost to leakage.

In general, the major practical differences between SRAM and DRAM include overall size, power consumption, speed, and ease of use. In terms of size, DRAM devices can hold more data per unit area than SRAM devices, since a DRAM's capacitor takes up less space than an SRAM's flip-flop. In terms of power consumption, SRAMs are more energy-efficient because they do not require constant refreshing. In terms of speed and ease of use, SRAMs are superior because they do not require refresh circuitry.

In terms of applications, SRAMs are used when relatively small amounts of read/write memory are needed and are typically found within application-specific ICs that require extremely low standby power. For example, they are frequently used within portable equipment such as pocket calculators. SRAM is also integrated into all modern microprocessors, where it acts as on-chip cache memory that provides a high-speed link between the processor and memory. On the other hand, DRAM is used in applications where a large amount of read/write memory (within the megabyte range) is needed, such as within computer memory modules.

In most situations, you do not need to worry about dealing with discrete RAM memory ICs. Most of the time, RAM is already built into a microcontroller or conventionally housed on PCB memory modules that simply plug into a computer's memory banks. In both these cases, you really do not need to know how to use the memory, because you can let the existing hardware and software take care of the addressing, refreshing, and so on. For this reason, we will not discuss the finer details of the various discrete SRAM and DRAM ICs out there. Instead, we will take a look at some SRAM and DRAM block diagrams that illustrate the basics, and then we will discuss some memory packages, such as SIMMs and DIMMs, that are used within computers.

Very Simple SRAM

Figure 12.148 shows a very elementary SRAM that is set up with a 4096 (4 K) × 1-bit matrix. It uses 12 address lines to address 4096 different memory locations; each location contains a flip-flop. The memory matrix is set up as a 64 × 64 array, with A_0 to A_5 identifying the row and A_6 to A_{11} identifying the column to pinpoint the specific location to be used. The box labeled "Row Select" is a 6-to-64 decoder for identifying the appropriate 1-of-64 row. The box labeled "Column Select" is also a 6-to-64 decoder for identifying the appropriate 1-of-64 column.

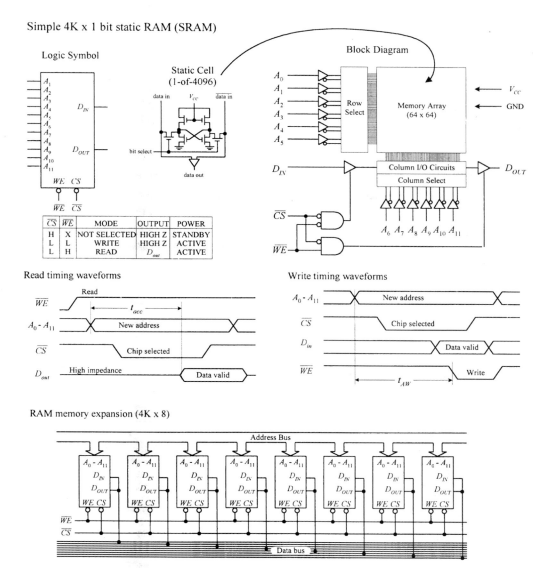

Simple 4K x 1 bit static RAM (SRAM)

FIGURE 12.148

To write a new bit of data to memory, the bit is applied to D_{IN}, the address lines are set, the Chip Select input ($\overline{CS}$) is set low (to enable the chip), and the Write Enable input ($\overline{WE}$) is set low (to enable the D_{IN} buffer). To read a bit of data from memory, the address lines are set, $\overline{CS}$ is set low, and $\overline{WE}$ is set high (to enable the D_{OUT} buffer). See the timing waveforms in Fig. 12.148.

By combining eight $4\,K \times 1$ SRAM ICs together, as shown in the lower circuit in Fig. 12.148, the memory can be expanded to form a $4\,K \times 8$ configuration, which is useful in simple 8-bit microprocessor systems. When an address is applied to the address bus, the same address locations within each memory IC are accessed at the same time. Therefore, each data bit of an 8-bit word applied to the data bus is stored in the same corresponding address locations within the memory ICs.

There are other SRAM ICs that come with configurations larger than $n \times 1$. For example, they may come in, say, an $n \times 4$ or $n \times 8$ configuration. As with the $n \times 1$ devices, these SRAMs can be expanded (two $n \times 8$ devices could be combined to form an $n \times 16$ expanded memory, four $n \times 8$ devices could be combined to form an $n \times 32$ expanded memory, and so on).

Serial SRAM with a similar interface to serial EEPROM is also available. For an example of the types of serial SRAM available see http://ww1.microchip.com/downloads/en/DeviceDoc/22127a.pdf.

Note on Nonvolatile SRAMs

In many applications, it would be ideal to have a memory device that combines both the speed and cycle endurance of an SRAM with the nonvolatile characteristics of ROM devices. To solve this problem, manufacturers have created what are called *nonvolatile SRAMs*. One such device incorporates a low-power CMOS SRAM together with a lithium battery and power-sensory circuitry. When the power is removed from the chip, the battery kicks in, providing the flip-flops with sufficient voltage to keep them set (or reset). SRAMs with battery backup, however, have limited lifetimes due to the life expectancies of the lithium batteries—around ten years.

Another nonvolatile SRAM that requires no battery backup is referred to as *non-volatile RAM* (NOVRAM). These chips incorporate a backup EEPROM memory array in parallel with an ordinary SRAM array. During normal operation, the SRAM array is written to and read from just like an ordinary SRAM. When the power supply voltage drops, an onboard circuit automatically senses the drop and performs a store operation that causes all data within the volatile SRAM array to be copied to the nonvolatile EEPROM array. When power to the chip is turned on, the NOVRAM automatically performs a recall operation that copies all the data from the EEPROM array back into the SRAM array. A NOVRAM has essentially unlimited read/write endurance, like a conventional SRAM, but has a limited number of store-to-EEPROM cycles—around 10,000.

DRAM

Figure 12.149 shows a very basic $16\,K \times 1$ DRAM. Normally, to access all 16,384 memory locations (capacitors) would require 14 address lines. However, in this DRAM (as within most large-scale DRAMs), the number of address lines is cut in half by multiplexing.

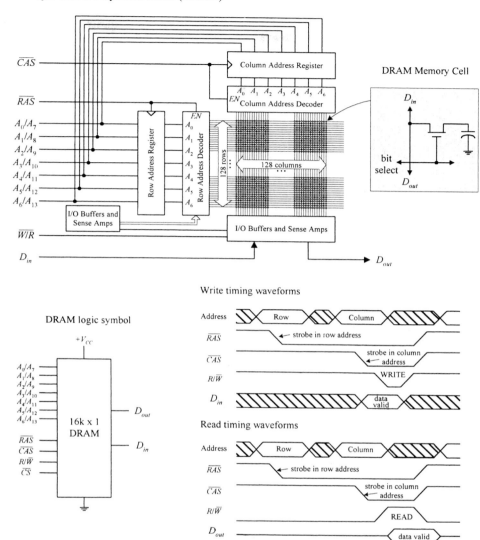

Simple 16K x 1 Dynamic RAM (DRAM)

DRAM Memory Cell

DRAM logic symbol

Write timing waveforms

Read timing waveforms

FIGURE 12.149

To address a given memory location is a two-step process. First, a 7-bit row address is applied to A_0–A_6, and then Row Address Strobe ($\overline{RAS}$) is sent low. Second, a 7-bit column address is applied to A_0–A_6, and then Column Address Strobe ($\overline{CAS}$) is sent low. At this point the memory location is latched and can now be read or written to by using the $\overline{WE}$ input. When $\overline{WE}$ is low, data is written to the RAM via D_{in}. When $\overline{WE}$ is high, data is read from the RAM via D_{out}. See the timing waveforms in Fig. 12.149.

Simple DRAM devices like this must be refreshed every 2 ms or sooner to replenish the charge on the internal capacitors. For our simple device, there are three ways to refresh the cells: use a Read cycle, use a Write cycle, or use an $\overline{RAS}$-only cycle. Unless you are reading or writing to and from all 128 rows every 2 ms, the $\overline{RAS}$-only cycle is the preferred technique. To perform this cycle, $\overline{CAS}$ is set high, A_0–A_6 are set up with the row address 000 0000, $\overline{RAS}$ is pulsed low, the row address is then incremented by 1, and the last two steps are repeated until all 128 rows have been accessed.

As you can see, needing to come up with the timing waveforms to refresh the memory is a real pain. For this reason, manufacturers produce DRAM controllers or actually incorporate automatic refreshing circuitry within the DRAM IC. In other words, today's DRAMs have all the "housekeeping" functions built in. Practically speaking, this makes the DRAM appear static to the user.

DRAM technology is changing very rapidly. Today, there are a number of DRAM-like devices that go by such names as ECC DRAM, EDO DRAM, SDRAM, SDRAM II, RDRAM, and SLDRAM (we will discuss some of these at the end of this chapter).

Computer Memory

As mentioned, you typically do not need to worry about RAM. (The only real exception would be NOVRAM that is used in many EEPROM-like applications.) RAM is usually either already integrated into a chip, such as a microcontroller, or is placed in reduced pin devices like single in-line memory modules (SIMMs) or dual in-line memory modules (DIMMs) that slide (snap) into a computer's memory bank sockets. In both cases, not much thought is needed—assuming you are not trying to design a microcontroller or computer from scratch. The main concern nowadays is figuring out what kind of RAM module to buy for your computer.

Within computers, RAM is used to hold temporary instructions and data needed to complete tasks. This enables the computer's CPU to access instructions and stored data in memory very quickly. For example, when the CPU loads an application, such as a word processor or page layout program, into memory, the CPU can quickly find what it needs, instead of needing to search for bits and pieces from, say, the hard drive or external drive. For RAM to be quick, it must be in direct communication with the computer's CPU. Early on, memory was soldered directly onto the computer's system board (motherboard). However, over time, as memory requirements increased, having fixed memory onboard became impractical. Today, computers house expansion slots arranged in memory banks. The number of memory banks and the specific configuration vary, depending of the computer's CPU and how the CPU receives information.

Historically, computers initially used either SIMM or DIMM memory modules. Both types of modules use dynamic RAM ICs as the core element. The actual SIMM or DIMM module resembles a PCB and houses a number of RAM ICs that are expanded onboard to provide the necessary bit width required by the CPU using the module. To install a SIMM or DIMM module, simply insert the module into one of the computer memory banks sockets found on the motherboard. Many computer systems use 168-pin DIMMs. Older Pentium and later 486 PCs commonly use 72-pin SIMMs, while still older 486 PCs commonly use 30-pin SIMMs.

A variation on standard sizes of DIMM memory is the SODIMM. SODIMM (Small Outline DIMM) are electrically the same as standard DIMMs but smaller in size and intended for use in laptop computers.

DIMMs

DIMMs have opposing pins electrically isolated to form two separate contacts. DIMMs are often used in computer configurations that support a 64-bit or wider memory bus.

Each new generation of memory brings with it new formats of DIMM, with strategically placed slots to prevent accidental insertion of the wrong type of memory.

DDR3 memory comes in 240-pin DIMMs. Figure 12.150 shows a somewhat simpler sample package from an old 16 M × 64-bit synchronous DRAM that comes in a 168-pin DIMM package.

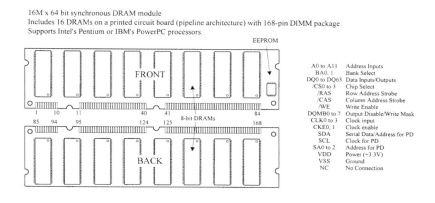

16M x 64 bit synchronous DRAM module
Includes 16 DRAMs on a printed circuit board (pipeline architecture) with 168-pin DIMM package.
Supports Intel's Pentium or IBM's PowerPC processors.

A0 to A11	Address Inputs
BA0, 1	Bank Select
DQ0 to DQ63	Data Inputs/Outputs
/CS0 to 3	Chip Select
/RAS	Row Address Strobe
/CAS	Column Address Strobe
/WE	Write Enable
DQMB0 to 7	Output Disable/Write Mask
CLK0 to 3	Clock input
CKE0, 1	Clock enable
SDA	Serial Data/Address for PD
SCL	Clock for PD
SA0 to 2	Address for PD
VDD	Power (+3.3V)
VSS	Ground
NC	No Connection

FIGURE 12.150

DRAM Technology Used in Computer Memories

A number of DRAM technologies are incorporated into computer memory modules these days. Extended data out (EDO) memory is a technology that allows the CPU (ones that support EDO) to access memory 10 to 20 percent faster than standard DRAM chips.

Another variation of DRAM is the synchronous DRAM (SDRAM), which uses a clock to synchronize signals input and output on a memory chip. The clock is coordinated with the CPU clock so the timing of the memory chips and the timing of the CPU are in synch. Synchronous DRAMs save time in executing commands and transmitting data, thereby increasing the overall performance of the computer. SDRAM allows the CPU to access memory approximately 25 percent faster than EDO memory.

Double-data-rate SRAM (DDR or SDRAMM II) is a faster version of SDRAM that is able to read data on both the rising and falling edges of the system clock, thus doubling the data rate of the memory chip. Rambus DRAM (RDRAM) is an extremely fast DRAM technology that uses a high-bandwidth "channel" to transmit data at speeds about ten times faster than a standard DRAM.

CHAPTER 13

Microcontrollers

The microcontroller is essentially a computer on a chip. It contains a processing unit, ROM, RAM, serial communications ports, ADCs, and so on. In essence, a microcontroller is a computer, but without the monitor, keyboard, and mouse. These devices are called *microcontrollers* because they are small (micro) and because they control machines, gadgets, and so on.

With one of these devices, you can build an "intelligent" machine. You write a program on a host computer; download the program into the microcontroller via the USB, parallel, or serial port of the PC; and then disconnect the programming cable and let the program run the machine. For example, in the microwave oven, a single microcontroller has all the essential ingredients to read from a keypad, write information to the display, control the heating element, and store data such as cooking time.

There are literally thousands of different kinds of microcontrollers available. Some are one-time-programmable (OTP), meaning that once a program is written into its ROM (OTP-ROM), no changes can be made to the program. OTP microcontrollers are used in devices such as microwaves, dishwashers, automobile sensor systems, and many application-specific devices that do not require changing the core program. Other microcontrollers are reprogrammable, meaning that the microcontroller's program stored in ROM (which may be EPROM, EEPROM, or flash) can be changed if desired, which is a useful feature when prototyping or designing test instruments that may require future I/O devices.

Microcontrollers are found in bicycle light flashers, data loggers, toys such as model airplanes and cars, antilock braking systems, VCRs, microwave ovens, alarm systems, fuel injectors, exercise equipment, and many other items. They also can be used to construct robots, where the microcontroller acts as the robot's brain, controlling and monitoring various input and output devices, such as light sensors, stepper and servo motors, temperature sensors, and speakers. With a bit of programming, you can make the robot avoid objects, sweep the floor, and generate various sounds to indicate that it has encountered difficulties (such as being low on power or tipped over) or has finished sweeping. The list of applications for microcontrollers is endless, and because they are so widely used, their cost is low.

13.1 Basic Structure of a Microcontroller

Figure 13.1 shows the basic ingredients found within many microcontrollers. These include a CPU, ROM (OTP-ROM, EPROM, EEPROM, of flash), RAM, I/O ports, timing circuitry/leads, interrupt control, a serial port adapter (such as UART or USART), and an ADC/DAC.

A very simplistic view of the basic components of a microcontroller

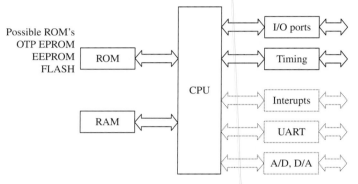

FIGURE 13.1   Possible internal architectures: RISC, SISC, CISC, Harvard, Von-Neuman

The CPU retrieves program instructions that the user programs into ROM, while using RAM to store temporary data needed during program execution. The I/O ports are used to connect external devices that send or receive instructions to or from the CPU.

The serial port adapter is used to provide serial communications between the microcontroller and a PC or between two microcontrollers. It is responsible for controlling the different rates of data flow common between devices. Example serial port adapters found within microcontrollers are the universal asynchronous receiver transmitter (UART) and universal synchronous/asynchronous receiver transmitter (USART). The UART can handle asynchronous serial communications, while the USART can handle either asynchronous or synchronous serial communications. Some microcontrollers take this a step further and include an interface for a Universal Serial Bus (USB) interface on the chip.

An interrupt system is used to interrupt a running program in order to process a special routine called the *interrupt service routine*. This boils down to the ability of a microcontroller to respond to external data that requires immediate attention, such as data conveyed by an external sensor indicating important shutdown information, say, when things get too hot or objects get too close. A timer/counter is used to "clock" the device—to provide the driving force needed to move bits around. Most microcontrollers that come with built-in ADCs and DACs can be used to interface with analog transducers, such as temperature sensors, strain gauges, and position sensors.

13.2 Example Microcontrollers

There are many different families of microcontrollers. Two of the most popular are made by the manufacturers Atmel and Microchip. In this section, we will take a close look at microcontrollers from these two manufacturers.

13.2.1 The ATtiny85 Microcontroller

The Atmel ATtiny85 microcontroller is an 8-pin IC available in both surface-mount and through-hole DIL packages. The device is designed to operate with a minimum of external components. Figure 13.2 shows what is inside one of these little packages.

ATtiny block diagram

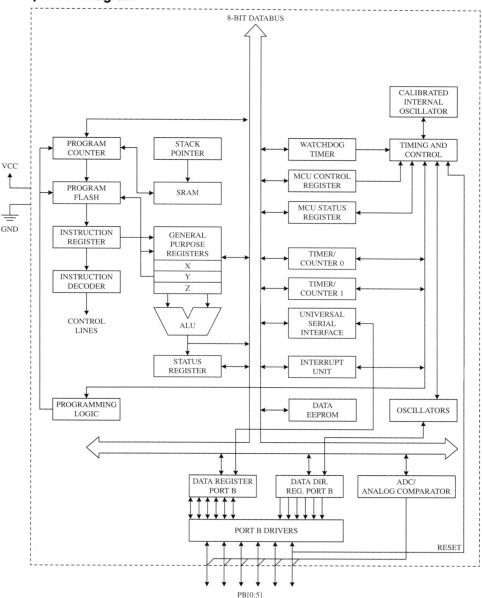

FIGURE 13.2

The ATtiny 85 has three different types of memory:

- 8 kB of flash memory in which the program instructions are stored.

- 256 bytes of SRAM, which is used to contain data during execution of instructions.

- 512 bytes of EEPROM that is used to store nonvolatile data that needs to be retained after a loss of power.

The watchdog timer allows the microcontroller to be put into a sleep mode, in which it consumes negligible power. The watchdog timer will wake up the microcontroller after a certain amount of time has passed.

The device can either use the inaccurate internal oscillator or two of the pins that would otherwise be used for inputs or outputs can be sacrificed to use an external crystal oscillator.

Two timers can be used to generate internal interrupts; that is, to trigger some code to be executed periodically. External interrupts that are triggered by a change in the level at a pin are also possible.

All of the I/O pins can also be used with the internal ADC.

The ATtiny also has a Universal Serial Interface, which can communicate with a number of different types of serial buses including USB, Inter-Integrated Circuit (I^2C), and serial. There is more about these serial protocols in Sec. 13.5.

Minimizing External Components

Figure 13.3 illustrates just how few components you need to make something with an ATtiny. The potentiometer is connected to a pin that will be used as an analog input, which could, for example, be used to control the rate at which the LED flashes.

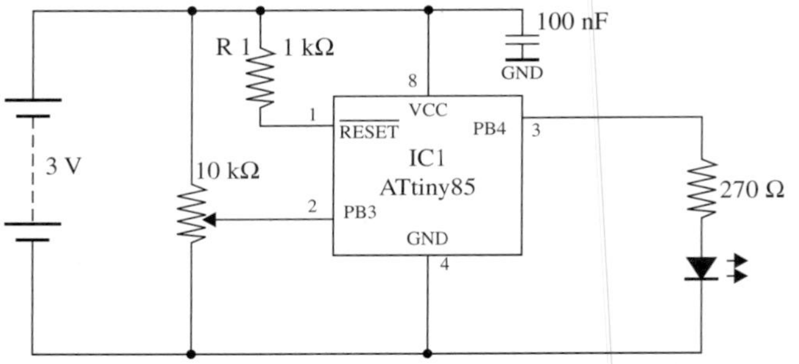

ATtiny85 LED flasher

FIGURE 13.3

The resistor R1 could be replaced by a direct connection from the RESET pin to VCC, but by using a resistor here, it becomes possible to have the RESET pin low force a reset—something that is necessary during programming.

The chip will operate from a supply voltage of between 2.7 V and 5.5 V at a clock frequency of 10 MHz or below, making it suitable for running from a 3 V lithium cell or a pair of AA batteries. The clock frequency can be set during programming, and it can also be changed from program code while the ATtiny is actually running. The main reason for controlling the clock frequency is to reduce the power consumption. At 1 MHz, the power can be reduced to just 300 µA and in power-down mode, waiting for an interrupt from the watchdog timer, power consumption is just 0.1 µA.

You may be wondering why we used a microcontroller to create something that we could have made with a 555 timer. Well, why not? The microcontroller is more expensive than a 555 timer, but only around a dollar, and also we can use slightly fewer components with a microcontroller. We also have a great deal more flexibility. Using this same hardware, we could do clever tricks such as turning the LED completely off if the potentiometer is at its most counterclockwise position. And there are three other unused I/O pins that we could do something with.

The catch is that to use a microcontroller, you need to program. But, if you are serious about developing something that will eventually become a product, chances are it will have a microcontroller in it.

Programming the ATtiny with AVR Studio

Atmel, the makers of the ATtiny, supply an integrated development environment (IDE) called AVR Studio (see Fig. 13.4) that takes away some of the pain of microprocessor programming.

AVR Studio with a Blink program

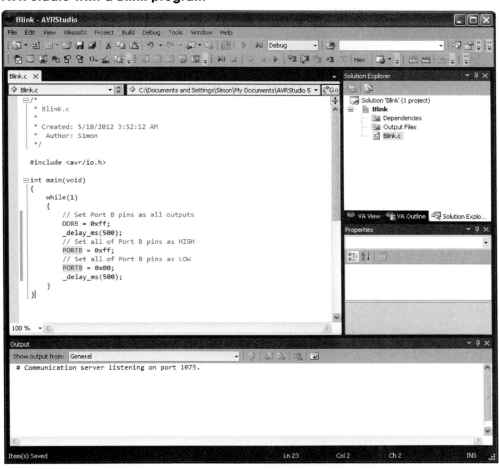

FIGURE 13.4

The ATtiny is usually programmed in C, which offers a good compromise between performance and readability. We define *readability* as the property of a program that allows it to be understood by someone other than the person who wrote it.

The standard AVR Studio way of doing things is powerful and flexible, but the C that you need to write is at a fairly low level, and not nearly as accessible as the BASIC Stamp language.

Programming the ATtiny with Arduino

Many people use the Arduino library to simplify the writing of code. This library includes all sorts of utility functions, rather like the commands found in the BASIC

Stamp language. The Arduino library was developed from a project called Wiring, which is a useful library for AVR microcontrollers. It is used mostly with the Arduino development boards (see Sec. 13.4). However, the Arduino IDE (different from AVR Studio) can also be used to write the programs for most processors in the AVR 8-bit range, including the ATtiny85.

Figure 13.5 shows the program that would control the hardware of Fig. 13.4 within the Arduino IDE.

The Arduino IDE used with ATtiny

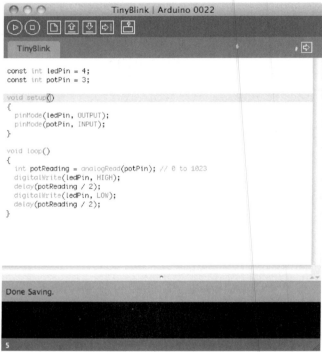

FIGURE 13.5

The Arduino IDE is open source and runs on Mac, Linux, and Windows systems. You need to install some extra configuration and library files to make the Arduino IDE work with the ATtiny microcontrollers. These instructions were developed at MIT. and details for installing and downloading can be found at http://hlt.media .mit.edu/?p=1695.

The advantages of using an ATtiny microcontroller here rather than a full Arduino board are reduced cost (an ATtiny costs $1 or less) and lower power consumption. We will meet the Arduino in Sec. 13.4.

Whether you are programming the microcontroller with AVR Studio or Arduino, you will need USB programmer hardware that is connected to a USB port of your computer. The programmers use a technique called in-circuit system programming (ICSP), which uses a six pin header connected to pins on the ATtiny. This is one of the reasons that you use a resistor to tie up the RESET pin, as it is one of the pins used by the programmer. It is quite common to design this header into the PCB for the circuit, during development, allowing changes to be easily made to the firmware.

Other ATtiny Microcontrollers

If you find that you need more I/O pins, or you can make do with less memory and save cost, you may want to consider one of the other microcontrollers in the ATtiny range. The Atmel website contains comparison tables for all its microcontrollers at http://www.atmel.com/products/microcontrollers/avr/tinyAVR.aspx?tab=parameters.

13.2.2   The PIC16Cx Microcontrollers

Having briefly explored the ATtiny85 microcontroller, we turn our attention to some rival devices from Microchip. Like the ATtiny range, these microcontrollers are 8-bit.

Figure 13.6 shows Microchip's PIC16Cx range of microcontrollers. As you can see in the internal architecture diagram, both microcontrollers house on-chip CPU, EPROM, RAM, and I/O circuitry. The architecture is based on a register file concept that uses separate buses and memories for programs and data (Harvard architecture). This allows execution to occur in parallel. As an instruction is being "pre-fetched," the current instruction is executing on the data bus.

Microchip's PIC16C57 microcontrollers

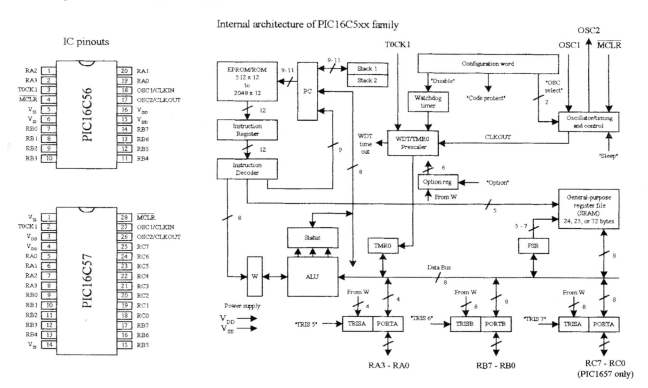

FIGURE 13.6

The PIC16C56's program memory (EPROM) has space for 1024 words, while the PIC16C57 has space for 2048 words. An 8-bit-wide ALU contains one temporary working register, and performs arithmetic and Boolean functions between data held in the working register and any file register. The ALU and register file are composed of up to 80 addressable 8-bit registers, and the I/O ports are connected via the

8-bit-wide data bus. Thirty-two bytes of RAM are directly addressable, while the access to the remaining bytes work through bank switching.

In order for bit movement to occur (clock generation), the PIC controllers require a crystal or ceramic resonator connected to pins OSC1 and OSC2. The PIC microcontrollers reach a performance of 5 million instructions per second (5 MIPS) at a clock frequency of 20 MHz. A watchdog timer is also included, which is a free-running on-chip RC oscillator that works without external components. It continues running when the clock has stopped, making it possible to generate a reset independent of whether the controller is working or sleeping.

These chips also come with a number of I/O pins that can be linked to external devices such as light sensors, speakers, LEDs, or other logic circuits. The PIC16C56 comes with 12 I/O pins that are divided into three ports: port A (RA3–RA0), port B (RB7–RB0), and port C (RC7–RC0). The PIC16C57 comes with eight more I/O pins than the PIC16C56.

Programming the PIC Microcontroller

Microcontrollers use a set of machine-code instructions (1's and 0's) to perform various tasks such as adding, comparing, sampling, and outputting data via I/O ports. These machine-code instructions are typically programmed into onboard ROM (EPROM, EEPROM, or flash) via a programming unit linked to a PC. The actual programming, however, isn't written out in machine code. It is written in a high-level language within an editor program running on the PC. The high-level language used may be a popular language such as C or a specially tailored language that the manufacturer has created to optimize all the features present within its microcontrollers. For ultimate performance, the user can revert to assembly language, which could also reduce memory usage and program size but at the expense of readability of the code.

Using a manual and software you get from the manufacturer, you learn to write statements that tell the microcontroller what to do. You type the statements in an editor program, and then use the compile/run option to check for syntax errors. Once you think the program is ready, you save it and run a compiler program to translate it into machine language. If there is an error in your program, the compiler may refuse to perform the conversion. In this case, you must return to the text editor and fix the bugs before moving on.

Once the bugs are eliminated and the program is compiled successfully, a third piece of software is used to load the program into the microcontroller. This may require physically removing the microcontroller from the circuit and placing it into a special programmer unit linked to the host PC or may use ICSP, as described in the previous section.

Another way to do things involves using an interpreter instead of a compiler. An *interpreter* is a high-level language translator that resides within the microcontroller's ROM, rather than in the host PC. This often means that an external ROM (EPROM, EEPROM, or flash) is needed to store the actual program. The interpreter receives the high-level language code from the PC and, on the spot, interprets the code and places the translated code (machine code) into the external ROM, where it can be used by the microcontroller.

The interpreter approach may seem like a waste of memory, since the interpreter consumes valuable on-chip memory space. Also, using an interpreter significantly slows things down—a result of needing to retrieve program instructions from external memory. However, using an interpreter provides a very important advantage. By having the interpreter onboard to translate on the spot, an immediate, interactive relationship between the host program and microcontroller is created. This allows you to build your program, immediately try out small pieces of code, test the code by downloading the chunks into the microcontroller, and then see if the specific chunks of code work. The host programs used to create the source code often come with debugging features that let you test to see where possible programming or hard-wiring errors may result by displaying the results (such as the logic state at a given I/O pin) on the computer screen while the program is executing within the micro-controller. This allows you to perfect specific tasks within the program, such as a sound-generation routine, a stepper motor control routine, and so forth.

A halfway house between using interpreted code (BASIC Stamp, which we'll look at next) and compiled machine code (AVR Studio) is to use a microcontroller with a boot loader installed into its EEPROM (Arduino). In this approach, a small boot loader program is installed once into the flash memory of the microcontroller. The boot loader then runs after every reset of the microcontroller and quickly checks for incoming programming commands on the serial port. If it finds them, it reads the serial data into the flash memory of the device so that it can then be run. This removes the need for special programming hardware.

Programming the PIC with BASIC Stamp

The BASIC Stamp is essentially a microcontroller with interpreter software built in. These devices also come with additional support circuitry, such as an EEPROM, volt-age regulator, ceramic oscillator, and so on. BASIC Stamps are ideal for beginners because they are easy to program, quite powerful, and relatively cheap—a whole startup package costs around $80 or so. These devices are also very popular among inventors and hobbyists, and you'll find a lot of helpful literature, application notes, and fully tested projects on the Internet.

The original Stamp was introduced in 1993 by Parallax, Inc. It got its name from the fact that it resembled a postage stamp. The early version of the BASIC Stamp was the REV D. Later improvements led to the BASIC Stamp I (BSI) and then BASIC Stamp II (BSII).

Both the BSI and BSII have a specially tailored BASIC interpreter firmware built into the microcontroller's EPROM. For both Stamps, a PIC microcontroller is used. The actual program that is to be run is stored in an onboard EEPROM. When the battery is connected, the Stamps run the BASIC program in memory. Stamps can be reprogrammed at any time by temporarily connecting them to a PC running a simple host program. The new program is typed in, a key is hit, and the program is loaded into the Stamp. I/O pins can be connected with other digital devices such as sense switches, LED displays, LCDs, servos, and stepper motors.

Here, we'll focus on the BSII. To get started with the BSII, you will need program-ming software, programming cable, the manual, the BASIC Stamp module, and an appropriate carrier board (optional). These all come in the BSII startup kit, at a lower cost than purchasing each part separately.

For more information about the BASIC Stamp family, visit http://www.parallax .com/tabid/436/Default.aspx.

Note: To fully understand all the finer details needed to program BASIC Stamps, it is necessary to read through the user's manual. However, reading the user's manual alone tends not to be the best learning strategy, as it is easy to lose your place within all the technical terms, especially if you are a beginner. A good source to learn more about BASIC Stamps is the book *Programming and Customizing the Basic Stamp Computer* by Scott Edwards (McGraw-Hill, 2001). This book is geared toward beginners and is easy reading.

BASIC STAMP II

BSII is a module that comes in a 24-pin DIL package (see Fig. 13.7). The brain of the BSII is the PIC16C57 microcontroller that is permanently programmed with a PBASIC2 instruction set within its internal one-time programmable EPROM (OTP-EPROM). When programming the BSII, you tell the PIC16C57 to store symbols, called *tokens*, in external EEPROM memory. When the program runs, the PIC16C57 retrieves tokens from memory, interprets them as PBASIC2 instructions, and carries out those instructions. The PIC16C57 can execute its internal program at a rate of 5 MIPS. However, each PBASIC2 instruction takes up many machine instructions, so the PBASIC2 executes more slowly, around 3 to 4 MIPS.

The BSII comes with 16 I/O pins (P0–P15) that are available for general use by your programs. These pins can be interfaced with all modern 5-V logic, from TTL

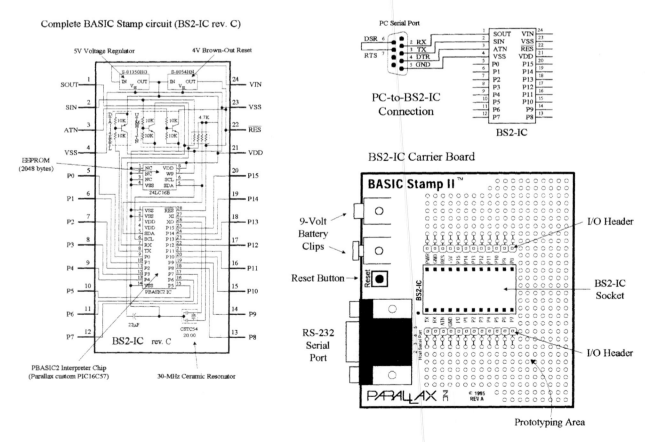

FIGURE 13.7

through CMOS (technically, they have characteristics like the 74HCT logic series). The direction of a pin—either input or output—is set during the programming phase. When a pin is set as an output pin, the BSII can send signals to other devices, like LEDs and servos. When a pin is set as an input pin, it can receive signals from external devices, such as switches and photosensors. Each I/O pin can source 20 mA and sink 25 mA. Pins P0–P7 and pins P8–P15, as groups, can each source a total of 40 and sink 50 mA per group.

2048-BYTE EEPROM

The BSII's PIC's internal OTP-EPROM is permanently programmed at the factory with Parallax's firmware, which turns this memory into a PBASIC2 interpreter chip. Because they are interpreters, the Stamp PICs have the entire PBASIC language permanently programmed into their internal program memory. This memory cannot be used to store your PBASIC2 program. Instead, the main program must be stored in the EEPROM, which retains data without power and can be reprogrammed easily. At runtime, the PBASIC2 program created on the host computer is loaded into the BSII's EEPROM starting at the highest address (2047) and working downward. Most programs do not use the entire EEPROM, which means that PBASIC2 lets you store data in the unused lower portion of the EEPROM. Since programs are stored from the top of the memory downward, data is stored in the bottom of the memory working upward. If there is an overlap, the Stamp host software will detect this problem and display an error message.

RESET CIRCUIT

The BSII comes with a reset circuit. When power is first connected to the Stamp, or if it falters due to a weak battery, the power supply voltage can fall below the required 5 V. During such brownouts, the PIC is in a voltage-deprived state and will have the tendency to behave erratically. For this reason, a reset chip is incorporated into the design, forcing the PIC to reset to the beginning of the program and hold until the supply voltage is within acceptable limits.

POWER SUPPLY

To avoid supplying the BSII with unregulated supply power, a 5-V regulator is incorporated into the BSII. This regulator accepts a voltage range from slightly over 5 V up to 15 V and regulates it to a steady 5 V. It provides up to 50 mA. The regulated 5 V is available at output V_{DD}, where it can be used to power other parts of your circuits, as long as no more than 50 mA is required.

CONNECTING BSII TO A HOST PC

To program a Stamp requires connecting it to a PC that runs host software to allow you to write, edit, download, and debug PBASIC2 programs. The PC communicates with the BSII through an RS-232 (COM port) interface consisting of pins S_{IN}, S_{OUT}, and *ATM* (serial in, serial out, and attention, respectively).

During programming, the BSII host program pulses *ATM* high to reset the PIC and then transmits a signal to the PIC through S_{IN} indicating that it wants to download a new program. PC-to-BSII connector hookup is shown in Fig. 13.7. This connection allows the PC to reset the BSII for programming, download programs, and receive debug data from the BSII. The additional pair of connections, pin 6 and 7 of the DB9 socket, lets the BSII host software identify the port to which the BSII is connected.

Usually, when programming a BSII, you use a special BSII carrier board, which comes with a prototyping area, I/O header, BSII-IC socket, 9-V battery clips, and an RS-232 serial port connector, as shown in Fig. 13.7. These boards, along with programming cable and software, can be purchased as startup packages.

THE PBASIC LANGUAGE

Even though the BASIC Stamp has BASIC in its name, it cannot be programmed in Visual BASIC. It does not have a graphical user interface, a hard drive, or a lot of RAM. The BASIC Stamp must be programmed only with Parallel's BASIC, PBASIC, which has been specifically designed to exploit all the BASIC Stamp's capabilities.

PBASIC is a hybrid form of the BASIC programming language, with which many people are familiar. PBASIC is called a *hybrid* because, while it contains some simplified forms of normal BASIC control constructs, it also has special commands to efficiently control I/O pins. PBASIC is an easy language to master and includes familiar instructions such as GOTO, FOR . . . NEXT, and IF . . . THEN. It also contains Stamp-specific instructions, such as PULSOUT, DEBUG, and BUTTON, which will be discussed shortly.

The actual program to be downloaded into the Stamp is first written using BSII editor software running on a Microsoft Windows PC, or on a Linux or Mac system running Windows using virtualization software. After you write the code for your application, you simply connect the Stamp to a serial port or USB-to-serial adapter connected to your computer, provide power to the Stamp, and download the code into the Stamp. As soon as the program has been downloaded successfully, it begins executing its new program from the first line of code.

The size of the program that can be stored in a Stamp is limited. For the BSII, 2048 bytes worth of program space are available, which is enough for around 500 to 600 lines of PBASIC code. The amount of program memory for the Stamps cannot be expanded, since the interpreter chip (PIC) expects the memory to be specific and fixed in size. However, in terms of data memory, expansion is possible. You can interface EEPROM or other memory devices to the Stamp's I/O pins to gain more data storage area. This requires that you supply the appropriate code within your PBASIC program to make communication between the Stamp and external memory device you choose possible. Additional data memory is often available with Stamp-powered applications that monitor and record data (such as from an environmental field instrument).

The PBASIC language, like other high-level computer languages, involves defining variables and constants and using address labels, mathematical and binary operators, and various instructions (including branching, looping, numerics, digital I/O, serial I/O, analog I/O, sound I/O, EEPROM access, time, power control, and so on). Here's a quick rundown of the elements of the PBASIC2 language.

Comments: Comments can be added within the program to describe what you're doing. They begin with an apostrophe (') and continue to the end of the line.

Variables: These are locations in memory that your program can use to store and recall values. These variables have limited range. Before a variable can be used in a PBASIC2 program, it must be declared. The common way used to declare variables is to use a directive VAR:

```
symbolvar  size
```

where the symbol can be any name that starts with a letter; can contain a mixture of letters, numbers, and underscore; and must not be the same as PBASIC keywords or labels used in the program. The size establishes the number of bits of storage the variable is to contain. PBASIC2 provides four sizes: bit (1 bit), nib (4 bits), byte (8 bits), and word (16 bits). Here are some examples of variable declarations:

```
'Declare variables.
sense_invar bit      'Value can be 0 or 1.
speedvar nib    'Value in range 0 to 15.
lengthvar byte    'Value in range 0 to 255.
nvar word    'Value in range 0 to 65535.
```

Constants: Constants are unchanging values that are assigned at the beginning of the program and may be used in place of the numbers they represent within the program. Defining constants can be accomplished by using the CON directive:

```
beeps        con  5         'number of beeps
```

By default, PBASIC2 assumes that numbers are in decimal (base 10). However, it is possible to use binary and hexadecimal numbers by defining them with prefixes. For example, when the prefix % is placed in front of a binary number (for example, %0111 0111), the number is treated as a binary number, not a decimal number. To define a hexadecimal number, the prefix $ is used (as in $EF). Also, PBASIC2 will automatically convert quoted text into the corresponding ASCII codes. For example, defining a constant as A will be interpreted as the ASCII code for *A* (65).

Address labels: The editor uses address labels to refer to addresses (locations) within the program. This is different from other versions of BASIC, which use line numbers. In general, an address label name can be any combination of letters, numbers, and underscores. However, the first character in the label name cannot be a number, and the label name must not be the same as a reserved word, such as a PBASIC instruction or variable. The program can be told to go to the address label and follow whatever instructions are listed after. Address labels are indicated with a terminating colon (for example, loop:).

Mathematical operators: PBASIC2 uses two types of operators: unary and binary. Unary operators take precedence over binary operators. Also, unary operations are always performed first. For example, in the expression 10 – SQR 16, the BSII first takes the square root of 16 and then subtracts it from 10. The unary operators are as follows:

ABS   Returns absolute value

SQR   Returns square root of value

DCD 2^n-power decoder

NCD   Priority encoder of a 16-bit value

SIN   Returns 2's complement sine

COS   Returns 2's complement cosine

The binary operators are as follows:

+   Addition

–   Subtraction

/	Division
/ /	Remainder of division
*	Multiplication
**	High 16 bits of multiplication
*/	Multiplies by 8-bit whole and 8-bit part
MIN	Limits a value to specified low
MAX	Limits a value to specified high
DIG	Returns specified digit of number
≪	Shifts bits left by specified amount
≫	Shifts bits right by specified amount
REV	Reverses specified number of bits
&	Bitwise AND of two values
|	Bitwise OR of two values
^	Bitwise XOR of two values

Table 13.1 shows the PBASIC instructions used by BSII.

DEBUGGING

To debug PBASIC programs, the BASIC Stamp editor comes with two handy features: syntax checking and a DEBUG command.

Syntax checking alerts you to any syntactical error and is automatically performed on your code the moment you try to download to the BASIC Stamp. Any syntax errors will cause the download process to abort and will cause the editor to display an error message, pointing out the error in the source code.

The DEBUG command, unlike syntax checking, is an instruction that is written into the program to find logical errors—ones that the Stamp does not find, but ones that the designer had not intended. DEBUG operates similar to the PRINT command in the BASIC language and can be used to print the current status of specific variables within your PBASIC program as it is executed within the BASIC Stamp. If your PBASIC code includes a DEBUG command, the editor opens a special window at the end of the download process to display the result for you.

Making a Robot Using BSII

To demonstrate how easy it is to make interesting gadgets using BSII, let's take a look at a robot application. In this application, the main objective is to prevent the robot from running into objects. The robot aimlessly moves around, and when it comes close to an object, the robot stops and then backs up and moves off in another direction. In this example, the robot is constructed as follows:

- A BSII acts as the robot's brain.
- Two servos connected to wheels act as its legs.
- A pair of infrared transmitters and sensors acts as its eyes.
- A piezoelectric speaker acts as its voice.

TABLE 13.1 PBASIC Instructions

INSTRUCTION	DESCRIPTION
Branching	
IF *condition* THEN *addressLabel*	Evaluate condition and, if true, go to the point in the program marked by *addressLabel*. (Conditions: =, <> not equal, >, <, >=, and <=)
BRANCH *offset*, [*address0, address1, . . . addressN*]	Go to the address specified by offset (if in range).
GOTO *addressLabel*	Go to the point in the program specified by *addressLabel*.
GOSUB *addressLabel*	Store the address of the next instruction after GOSUB, then go to the point in the program specified by *addressLabel*, with the intention of returning to the point at which the subroutine was called once the subroutine has returned.
RETURN	Return from subroutine.
Looping	
FOR *variable=start to end{STEP stepVal}* . . . NEXT	Create a repeating loop that executes the program lines between FOR and NEXT, incrementing or decrementing *variable* according to *stepVal* until the value of the variable passes the *end* value.
Numerics	
LOOKUP *index*, [*value0, value1, . . . valueN*], *resultVariable*	Look up the value specified by the index and store it in a variable. If the index exceeds the highest index value of the items in the list, the variable is unaffected. A maximum of 256 values can be included in the list.
LOOKDOWN *value, {comparisonOp,}* [*value0, value1, . . . valueN*], *resultVariable*	Compare a value to a list of values according to the relationship specified by the comparison operator. Store the index number of the first value that makes the comparison true in *resultVariable*. If no value in the list makes the comparison true, *resultVariable* is unaffected.
RANDOM *variable*	Generate a pseudo-random number using a byte or word variable where the bits are scrambled to produce a random number.
Digital I/O	
INPUT *pin*	Make the specified pin an input.
OUTPUT *pin*	Make the specified pin an output.
REVERSE *pin*	If the pin is an output, make it an input. If the pin is an input, make it an output.
LOW *pin*	Make the specified pin's output low.
HIGH *pin*	Make the specified pin's output high.
TOGGLE *pin*	Invert the state of a pin.
PULSIN *pin, state, resultVariable*	Measure the width of a pulse in 2-μs units.
PULSOUT *pin, time*	Output a timed pulse by inverting a pin for some time ($\times$ 2 μs).
BUTTON *pin, downstate, delay,rate,bytevariable, targetstate, address*	Debounce button input, perform auto-repeat, and branch to the address if the button is in target state. Button circuits may be active-low or active-high.
SHIFTIN *dpin, cpin, mode,* [*result{\bits}{,result{\ bits} . . . }]*	Shift data in from a synchronous serial device.

(Continued)

TABLE 13.1   PBASIC Instructions (*Continued*)

INSTRUCTION	DESCRIPTION
SHIFTOUT *dpin, cpin, mode,* *[data{\bits}{,data{* *bits} . . . }]*	Shift data out to a synchronous serial device.
COUNT *pin, period, variable*	Count the number of cycles (0-1-0 or 1-0-1) on the specified pin during *period* number of milliseconds and store that number in variable.
XOUT *mpin, zpin, [house* *keyORCommand{\cycles}* *{,house\keyOrCommand{* *cycles} . . . }]*	Generate X-10 powerline control codes.
Serial I/O	
SERIN *rpin{\fpin}, baudmode,* *{plabe}{timeout,tlabe,}[input* *Data]*	Receive asynchronous serial transmission.
SEROUT *tpin, baudmode,* *{pace,} [outputData]*	Send data serially with optional byte pacing and flow control.
Analog I/O	
PWM *pin, duty, cycles*	Output fast pulse-width modulation, then return pin to input. This can be used to output an analog voltage (0–5 V) using a capacitor and resistor (see Sec. 13.5).
RCTIME *pin, state,* *resultVariable*	Measure an RC charge/discharge time. This can be used to measure the position of a potentiometer or capacitance in capacitative sensing (see Chap. 6).
Sound	
FREQOUT *pin, duration,* *freq1{,freq2}*	Generate one or two sine-wave tones of a specified frequency for a specified duration.
DTMFOUT *pin, {ontime,offtime,}* *{,tone . . . }*	Generate dual-tone, multifrequency tones (DTMF, i.e., telephone touch tones).
EEPROM Access	
DATA	Store data in EEPROM before downloading the PBASIC program.
READ *location, variable*	Read EEPROM location and store the value in a variable.
WRITE *address,byte*	Write a byte of data to the EEPROM at the appropriate address.
Time	
PAUSE *milliseconds*	Pause the program (do nothing) for the specified number of milliseconds. Pause execution for 0–65,535 ms.
Power Control	
NAP *period*	Enter sleep mode for a short period. Power consumption is reduced to about 50 μA assuming no loads are being driven. The duration is $(2^{period}) \times$ 18 ms.
SLEEP *seconds*	Sleep from 1–65,535 seconds to reduce power consumption by ~50 μA.
END	Sleep until the power cycles or the PC connects ~50 μA.
Program Debugging	
DEBUG *outputData{,outputData . . . }*	Display variables and messages on the PC screen within the BSII host program. *outputData* consists of one or more of the following: text strings, variables, constants, expressions, formatting modifiers, and control characters.

Components and connections used to create object-avoiding robot

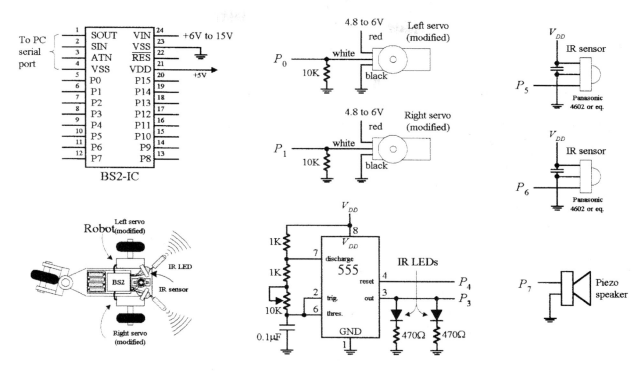

FIGURE 13.8

Figure 13.8 shows the completed robot, along with the various individual components.

THE SERVOS

The directional movement of the robot is controlled by right and left servo motors that have been modified so as to provide a full 360 degrees worth of rotation (modifying a servo is discussed in Chap. 15). To control a servo requires generating pulses ranging from 1000 to 2000 μs in width at intervals of approximately 20 ms. With one of the servos used in our example, when the pulse width sent to the servo's control line is set to 1500 μs, the servo is centered—it doesn't move. However, if the pulse width is shortened to, say, 1300 μs, the modified servo rotates clockwise. Conversely, if the pulse width is lengthened to, say, 1700 μs, the modified servo rotates counterclockwise.

The actual control pulses used to drive one of the servos in the robot are generated by the BSII using the PULSOUT *pin, time1* and the PAUSE *time2* instructions. The *pin* represents the specific BSII pin that is linked to a servo's control line, and *time1* represents how long the pin will be pulsed high. Note that for the PULSOUT instruction, the decimal placed in the *time1* slot actually represents half the time, in microseconds (s), that the pin is pulsed high. For example, PULSOUT 1, 1000 means that the BSII will pulse pin 1 high for 2000 μs, or 2 ms. For the PAUSE instruction, the decimal placed in the *time2* slot represents a pause in milliseconds. For example, PAUSE 20, represents a 20-ms pause. Figure 13.9 shows sample BSII code used to generate desired output waveforms to control a servo.

```
BS2 code          Comments
pulsout 1, 750    'pulse width of 1500us on pin 1
```

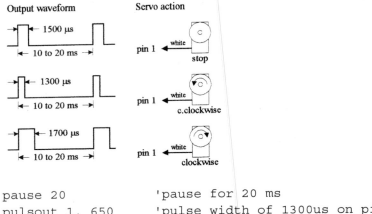

FIGURE 13.9

```
pause 20              'pause for 20 ms
pulsout 1, 650        'pulse width of 1300us on pin 1
pause 20              'pause for 20 ms
pulsout 1, 850        'pulse width of 1700us on pin 1
pause 20              'pause for 20 ms
```

The pulse widths need to be repeated for them to have a practical effect on the motors. It may be worth expanding this fragment to show loops of a second or more for each of the pulse widths.

Note that this sequence of pulses must be repeated every 20 ms or so for the servo to maintain its position.

IR TRANSMITTERS AND RECEIVERS

The robot's object-detection system consists of a right and left set of infrared (IR) LED transmitters and IR detector modules. The IR LEDs are flashed via a 555 timer at a high frequency, which in this example happens to be 38 kHz, 50 percent duty cycle. This frequency is used to avoid interference from other household sources of IR light, primarily incandescent lights, and to match the IR sensor shown in the figure. (Many types of IR LED transmitters and sensors could be used in this robot, and they may work best using a different frequency.) It is also possible to generate these pulses using the BASIC Stamp, but we have chosen to use external hardware to keep the program simple.

The IR photons emitted by the LED rebound off objects in the path of the robot and reflect back to the IR detector module. When a detector module receives photons, the I/O pin of the BSII connected to the module goes low. Note that the BSII can execute only around 4000 instructions per second, while the number of pulses generated by the detector module is 38,000. In this case, the actual number of pulses received by the BSII will be less—around 10 or 20.

PIEZOELECTRIC SPEAKER

A piezoelectric speaker is linked to one of the BSII I/O terminals and is used to generate different sounds when the robot is moving forward or backing up. To provide the piezoelectric speaker with a sinusoidal waveform to generate sound, the FREQOUT *pin, time, frequency* instruction is used. The instruction FREQOUT 7, 1000, 440 creates a 440-Hz sinusoidal frequency on pin 7 that lasts for 1000 ms.

THE PROGRAM

The following is a program used to control the robot. It is first created using the PBASIC2 host software, and then downloaded into the BSII during runtime.

```
'Program for object-avoiding robot
'Define variables and constants
'----------------------------------------------
nvar word        'n acts as a variable that changes.
right_IRvar in5      'Sets pin 5 as an input for right IR detector.
left_IRvar in6       'Sets pin 6 as an input for left IR detector.
right_servo  con 0        'Assigns 0 which will be used to identify right servo.
left_servo   con 1        'Assigns 1 to identify left servo.
IR_out       con 3        'Assigns 3 to identify IR output.
delay        con 10       'A constant that will be used in the program.
speed        con 100      'Used to set servo speed.
turn_speed   con 50       'Used to set turn speed of robot.
'Main program
'----------------------------------------------
highIR_out                              'Sets pin 3 "high"
pause 50                                 'Pauses for 50 milliseconds
sense:                                   'Label used to specify IR-sense routine.
ifleft_IR = 0 and right_IR = 0 then backup   'Object in front, jump to back_up routine.
ifleft_IR = 0 then turn_right            'Object on left side, jump to turn_right routine.
ifright_IR = 0 then turn_left            'Object on right, jump to "turn_left" routine.
'Sound Routines
'----------------------------------------------
forward_sound:                          'Label
freqout 7,1000, 440                      'Generate 1000ms, 440 Hz tone on pin 7
back_sound:                             'Label
freqout 7,1000,880                      'Generate 1000ms, 880 Hz tone on pin 7
'Motion routines
'----------------------------------------------
forward:                                'Label used to specify forward routine.
gosubforward_sound                      'Tells program to jump to forward sound subroutine.
debug "forward"                         'Tells stamp to display the word "forward" on debug window.
pause 50                                'Pause for 50ms.
for n = 1 to delay*2                     'For...Next loop that starts x = 1 and repeats until x = 20.
pulsoutleft_servo, 750-speed            'Make left servo spin to make robot move forward.
pulsoutright_servo, 750+speed           'Make right servo spin to make robot move forward.
pause 20                                'Pauses for 20ms, path of servo control.
next                                    'End of For...Next loop.
goto sense                              'Once forward routine is finished go back to sense routine.
backup:                                 'Label used to specify back-up routine.
gosubbackup_sound                       'Tells program to jump to back-up sound subroutine.
debug "backward"                        'Displays "backward" on the debug window.
pause 50                                'Pause for 50ms to ensure
for n = 1 to delay*3                     'For...Next loop that starts x = 1 and repeats until x = 60
pulsoutleft_servo, 750+speed           'Makes left servo spin to make robot move backward.
pulsoutright_servo, 750-speed          'Makes right servo spin to make robot move backward.
pause 20                                'Pauses for 20ms, part of servo control.
next                                    'End of For...Next loop.
turn_left:                              'Label used to specify turn-left routine.
debug "left"                            'Displays "left" on the debug window.
pause 50                                'Pause for 50ms.
for x = 1 to delay*1                     'For...Next loop that starts x = 1 and repeats until x = 10.
pulsoutleft_servo, 750-turn_speed      'Makes left servo spin to make robot turn left.
pulsoutright_servo, 750-turn_speed     'Makes right servo spin to make robot turn left.
pause 20                                'Pause for 20ms, part of servo control.
next                                    'End of For...Next loop.
goto sense                              'Once left-turn routine is finished, jump back to sense.
turn_right:                             'Label used to specify turn-right routine.
debug "right"                           'Displays "right" on debug window.
pause 50                                'Pause for 50 ms.
for x = 1 to delay*1                     'For...Next loop.
```

```
pulsoutleft_servo, 750+turn_speed    'Makes left servo spin to make robot turn right.
pulsoutright_servo, 750+turn_speed   'Makes right servo spin to make robot turn right.
pause 20                             'Pause for 20 ms, part of servo control.
next                                 'End of For...Next loop
goto sense                           'Once right-turn is finished, jump back to sense.
```

Note: Due to production tolerances, the motors may not stop rotating when pulsing at 1500 μs, so a small "fudge factor" may need to be added to or subtracted from the PULSOUT values.

Thinking about Mass Production

Recall that the major components of the BASIC Stamp circuit are the PIC (houses the CPU and ROM for storing PBASIC interpreter), external EEPROM (stores the program), and the resonator. In large-scale runs, it would be nice to get rid of the external memory and remove the interpreter program, and simply download a compiled PBASIC code directly into the PIC. This would save space and money. As it turns out, the BASIC Stamp editor software includes a feature to program PBASIC code directly into a PIC microcontroller using Parallax's PIC16Cxx programmer.

The major benefit of starting out with the Stamp is that you can easily fine-tune your code, test chunks, and immediately see if it works, which is important when creating prototypes. When prototyping with a PIC, checking for errors is much harder because you must compile everything at once—you can't test out chunks of code.

13.2.3 32-Bit Microcontrollers

The microcontrollers that we have explored in the previous sections use an 8-bit data bus with a clock frequency in the tens of megahertz and a few kilobytes of storage. Anyone familiar with the microprocessor-based home computers of the 1980s and 1990s would recognize the specifications. In comparison to a modern smartphone with a 32-bit processor, clock frequencies in the gigahertz, and hundreds of megabytes of RAM, these microcontrollers offer lamentable performance. However, the important thing here is what they are being used for. The adage of "not using a sledgehammer to crack a nut" was never more appropriate.

Atmel, Microchip, and most of the other microcontroller manufacturers all produce high-performance microcontrollers that use a 32-bit data bus and have more memory and processing performance than most desktop computers had ten years ago. These are useful for some high-performance applications. If you find that you need this kind of performance, then it is worth considering using one from a manufacturer whose 8-bit devices you are already familiar with, as they generally use the same or similar software tools and can be programmed in much the same way. They just cost a lot more and work a lot faster.

13.2.4 Digital Signal Processing

If you have a microcontroller that has an ADC input and a DAC output, you can digitize an audio signal (let's say music), process the data in some way, and then send it back out through the DAC. You might create a graphics equalizer or a dynamic voice changer that raises the pitch of your voice by an octave. This is called *digital signal processing* (DSP).

While you can do simple low-quality DSP with a standard 8-bit microcontroller, the ADCs are often too slow and the algorithms such as Fourier transforms that you apply to the audio signal must work in real time, and therefore benefit greatly from a fast CPU.

Microchip (among others) has variants of its standard microcontroller lines specifically designed for DSP. The dsPIC from Microchip is one such device that is frequently used in low-cost DSP applications. It has a 16-bit internal data bus, a 40 MHz clock, and 2kB of RAM.

Note: DSP is a complex area, and there are many good books devoted to this topic. *Understanding Digital Signal Processing* by Richard G. Lyons (Pearson, 1996) is one such book.

13.3   Evaluation/Development Boards

The microcontroller manufacturers are very keen to get their products into your products, so most will offer low-cost development and evaluation boards for their microcontrollers. These will often take the form of a PCB that contains the microcontroller, supporting components such as a crystal oscillator and voltage regulator, and a prototyping area where you can add your own components. They can also usually be programmed from a USB or sometimes RS-232 serial port, for which you will need to find an ancient PC, or more likely, use a USB-to-serial convertor. These boards include the use of the manufacturers' preferred software development tools, although these are sometime limited in some way unless you buy the professional version of the software.

As well as boards produced by the microcontroller manufacturers, other boards are offered by third parties. These boards can be useful during the development process, as they make prototyping much easier than starting from scratch.

Table 13.2 lists some of the most popular development boards available at the time of writing.

TABLE 13.2   Popular Microcontroller Evaluation Boards

MANUFACTURER	BOARD NAME	MICROPROCESSOR	URL	NOTES
Atmel	AVR Butterfly	ATMega169	http://www.atmel.com/tools/ AVRBUTTERFLY.aspx	Includes LCD screen Serial programmer
Freescale (Motorola)	DEMO908JL16	MC68HC08JL16 family	http://www.freescale.com/files/ microcontrollers/doc/user_guide/ DEMO908JL16UM.pdf	USB programmed
Microchip	PICkit 1 Flash Starter Kit	PIC12F675	http://www.microchip.com/ stellent/idcplg?IdcService= SS_GET_PAGE&nodeId=1406&dDo cName=en010053	Includes USB programming hardware
Microchip	MPLAB Starter Kit for dsPIC	dsPIC33FJ256GP506	http://www.microchip.com/ stellent/idcplg?IdcService= SS_GET_PAGE&nodeId=1406&dDo cName=en534506	USB Intended for audio Includes amplifier
Arduino	Arduino Uno	ATmega328	http://www.arduino.cc	See Sec. 13.4

13.4   Arduino

Arduino is an open source hardware platform for microcontroller prototyping. It encompasses both a microcontroller development board and an IDE. The IDE is simple to use and available for Mac, Linux, and Windows computers.

Arduino boards are extremely popular as a starting point for using microcontroller technologies. Their popularity is due to a number of factors, including the following:

- Low cost (around $30)
- Open source design
- Easy-to-use and cross-platform IDE
- Availability of plug-in shields (expansion hardware)

13.4.1   A Tour of Arduino

The most popular Arduino board is the Arduino Uno (see Fig. 13.10). This board is based on an Atmel microcontroller, similar to the ATtiny called ATmega328 (although the ATtiny can also be programmed using the Arduino IDE).

The ATmega328 microcontroller has 32kB of flash memory for storing programs, 2kB of RAM, and 1kB of EEPROM. It also has a hardware serial interface, or UART, as well as the usual timers and interrupt capabilities.

The microcontroller itself is the large 28-pin IC in the bottom right of the board, as shown in Fig. 13.10. Beneath this are six analog pins that can also be used as digital I/O pins, and then a block of power connections.

The Arduino can either be powered through the DC input socket (7–12 V dc) or from USB, and will switch over automatically to whichever is supplied.

An Arduino Uno board

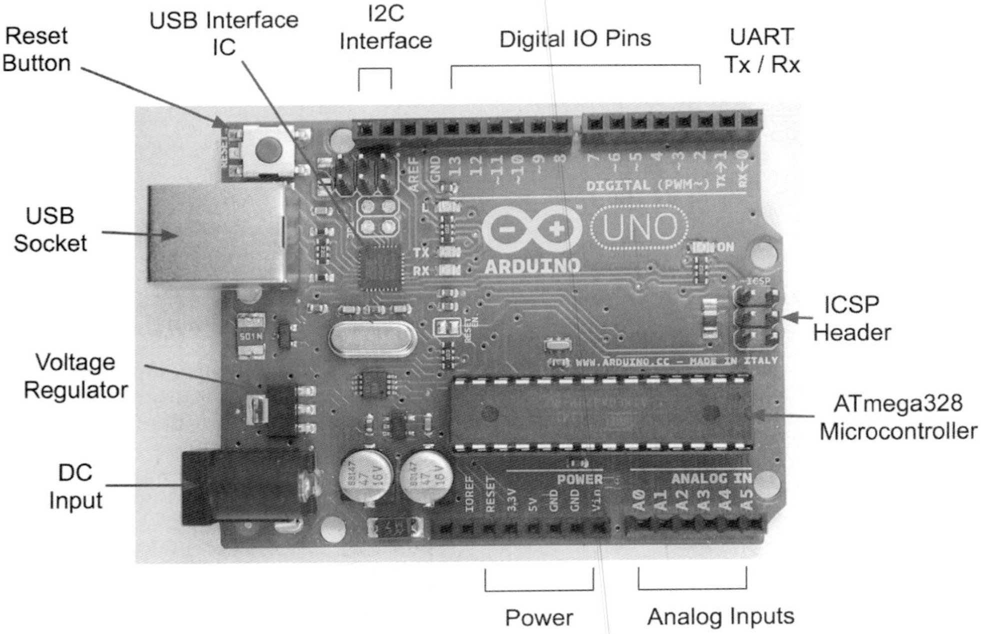

FIGURE 13.10

The connectors on the top side of the board offer an I²C interface, which actually uses two of the analog pins (A4 and A5) in the Arduino Uno, but they are repeated here for future boards that may have a separate I²C interface. There is also a row of digital I/O pins, some marked as being pulse-width modulation (PWM) capable. Two of these, D0 and D1, double as the Rx and Tx pins on the UART.

13.4.2   The Arduino IDE

The Arduino IDE provides a simple-to-use editor into which you can type your programs and upload them onto your Arduino board over USB (see Fig. 13.11). As well as the program editing area, the Arduino IDE also provides the following features:

- Color syntax highlighting
- A status area where the memory usage of your completed program is shown
- Links to the Arduino library documentation
- A serial monitor that allows two-way communication with the Arduino's USB port

The Arduino IDE

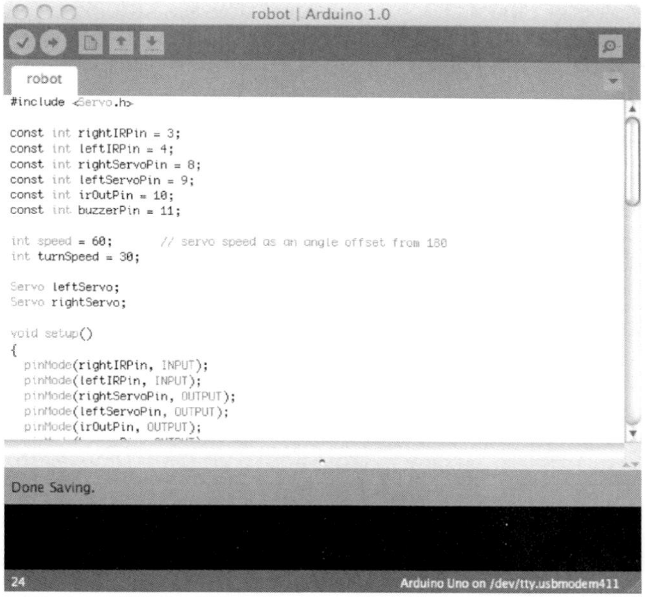

FIGURE 13.11

Most of the Arduino boards, including the Arduino Uno, have a USB connector through which they can be programmed. So, after writing your program, or "sketch" as it is called in Arduino parlance, you select the type of board and click the Upload button. Your program will be compiled and loaded into the flash memory of the microcontroller.

13.4.3   Arduino Board Models

Along with the Arduino Uno, there are many other Arduino boards to suit different uses. They are all programmed in the same way, but have different sizes, costs, and numbers of I/O pins available.

New Arduino models are released quite often. As an open source project, different manufacturers frequently take a basic model and add some different features to it.

Some of the most used official Arduino models are listed in Table 13.3.

Along with the boards listed in Table 13.3, other manufacturers offer boards that simply replicate the features of the official designs with minor differences. More interesting are the special-purpose Arduino boards. Some of these are listed in Table 13.4.

13.4.4 Shields

The success of Arduino had been in no small part due to the wide range of plug-in shields that add useful features to a basic Arduino board. A shield is designed to

TABLE 13.3 Kinds of Arduino Boards

MODEL	FEATURES	NOTES
Uno R3	At the time of writing, the latest version of Arduino Uno	Almost identical to the original Arduino Uno, but with extra header sockets for I^2C and power status
Uno	14 digital, 6 analog or digital I/O pins, 32kB of flash memory, 2kB of SRAM, 1kB of EEPROM	The most popular Arduino board and a good all-rounder to start with New USB interface means no USB drivers needed
Leonardo	Same I/O and memory spec as an Arduino Uno	Lower cost than Uno, but has a nonremovable SMD microcontroller with programmable USB features
Duemilanove	Either the same spec as Uno, or on some models using the ATMega168, half of all the memory capacities of the Uno	Predecessor to the Uno Uses an FTDI-based USB interface that requires drivers to be installed on Windows
Lilypad	14 digital, 6 analog or digital I/O pins, 16kB of flash memory, 1kB of SRAM, and 512 bytes of EEPROM running at 8 MHz	Intended to be stitched into clothing using conductive thread to connect to other Lilypad devices such as LEDs and accelerometers Requires a separate USB-to-serial converter available from SparkFun (SKU: DEV-09716)
Mega 2560	54 digital, 16 analog or digital I/O pins, 4 UARTs, 256kB of flash memory, 8kB of SRAM, 4kB of EEPROM	The device for you if you need a *lot* of I/O pins You can fit Arduino Uno style shields on it, but there are occasional compatibility problems
Mini	Same I/O and memory spec as an Arduino Uno	Much smaller than an Arduino Uno Requires a USB-to-serial converter to program it
Nano	Same I/O and memory spec as an Arduino Uno	Much smaller than an Arduino Uno, and fits directly onto the breadboard Includes a mini USB socket for programming
Fio	Similar spec to Arduino, but running at 8 MHz XBee wireless socket	Intended for mobile wireless applications Includes a LiPo battery charger IC
Ethernet	Same I/O and memory spec as an Arduino Uno	Arduino Uno with built-in Ethernet

TABLE 13.4 **Unofficial Arduino Variants**

MODEL	FEATURES	NOTES	URL
DFRobot-ShopRover	Built-in motor driver	Intended for robots	http://www.dfrobot.com/
Electric Sheep	Arduino Mega with built-in USB host connection	Often used to link to Android phones that support the Open Accessory standard	http://www.sparkfun.com/products/10745
EtherTen	Arduino Uno with built-in Ethernet connection		http://www.freetronics.com/
Lightuino	LED drivers	70 constant current LED channels	http://www.toastedcircuits.com
USBDroid	Arduino Uno with built-in USB host connection		http://www.freetronics.com/
Teensyduino	Similar spec to Leonardo	Tiny breadboard-friendly device with USB capabilities	http://www.pjrc.com/teensy/teensyduino.html

fit into the header sockets of the main Arduino board (see Fig. 13.12). Most shields will then pass through these connections in another row of header sockets, making it possible to construct stacks of shields with an Arduino at the bottom. Shields that have a display on them will not normally pass through in this way. You also need to be aware that if you stack shields in this way, you need to make sure that there are no incompatibilities, such as two of the shields using the same pin. Some shields get around this problem by providing jumpers to add some flexibility to pin assignments.

There are shields available for almost anything you could want an Arduino to do. They range from relay control to LED displays and audio file players. Most of these are designed with the Arduino Uno in mind, but are also usually compatible with the Arduino Mega.

Arduino Ethernet shield on an Arduino

FIGURE 13.12

TABLE 13.5   Common Arduino Shields

SHIELD	DESCRIPTION	URL
Motor	Ardumoto shield	http://www.sparkfun.com/products/9815
	Dual H-bridge bidirectional motor control at up to 2 A per channel	
Ethernet	Ethernet and SD card shield	http://arduino.cc/en/Main/ArduinoEthernetShield
Relay	Controls four relays	http://www.robotshop.com/seeedstudio-arduino-relay-shield.html
	Screw terminals for relay contacts	
LCD	16 × 2 character alphanumeric LCD shield with joystick	http://www.freetronics.com/products/lcd-keypad-shield

An encyclopedic list that includes useful technical details about the pin usage of these shields can be found at http://shieldlist.org/. Some of the author's favorite shields are listed in Table 13.5.

13.4.5 The Arduino C Library

You may hear people refer to the "Arduino language," but Arduino is actually just programmed in the C programming language, which has been around for many years. But Arduino provides a set of Arduino core functions that you can use in your programs, or sketches.

There are a large number of commands available in the Arduino library. A selection of the most commonly used commands are listed in Table 13.6.

The main Arduino core, including all the commands listed in Table 13.6, is automatically included in every sketch that you write. However, there are a number of other libraries that come bundled with the Arduino IDE that are added to your code only when you use them. To include them, use the include command followed by the name of the library, like this:

```
#include <Servo.h>
```

This command includes the Servo library that we will use in the example Arduino project that follows.

TABLE 13.6   Arduino Library Functions

COMMAND	EXAMPLE	DESCRIPTION
Digital I/O		
pinMode	pinMode(8, OUTPUT);	Sets pin 8 to be an output. The alternative is to set it to INPUT.
digitalWrite	digitalWrite(8, HIGH);	Sets pin 8 high. To set it low, use the constant LOW instead of HIGH.
digitalRead	inti;	Sets the value of i to HIGH or LOW depending on the voltage at the pin specified (in this case, pin 8).
	i = digitalRead(8);	
pulseIn	i = pulseIn(8, HIGH)	Returns the duration in microseconds of the next HIGH pulse on pin 8.
tone	tone(8, 440, 1000);	Makes pin 8 oscillate at 440 Hz for 1000 ms.

(Continued)

TABLE 13.6 Arduino Library Functions (*Continued*)

COMMAND	EXAMPLE	DESCRIPTION
Digital I/O		
noTone	noTone();	Cuts short the playing of any tone that was in progress.
Analog I/O		
analogRead	int r; r = analogRead(0);	Assigns a value to r of between 0 and 1023. 0 for 0 V 1023 if pin 0 is 5 V (3.3 V for a 3 V board).
analogWrite	analogWrite(9, 127);	Outputs a PWM signal (see Sec. 13.5). The duty cycle is a number between 0 and 255, 255 being 100%. This must be used by one of the pins marked as PWM on the Arduino board (3, 5, 6, 9, 10, and 11).
Time Commands		
millis	unsigned long l; l = millis();	The variable type long in Arduino is represented in 32 bits. The value returned by millis() will be the number of milliseconds since the last reset. The number will wrap around after approximately 50 days.
micros	long l; l = micros();	Like millis, except this is microseconds since the last reset. It will wrap after approximately 70 minutes.
delay	delay(1000);	Delay for 1000 ms, or 1 second.
delayMicroseconds	delayMicroseconds(100000);	Delay for 100,000 microseconds. Note the minimum delay is 3 microseconds; the maximum is around 16 ms.
Interrupts		
attachInterrupt	attachInterrupt(1, myFunction, RISING);	Associates the function myFunction with a rising transition on interrupt 1 (D3 on an Uno).
detachInterrupt	detachInterrupt(1);	Disables any interrupt on interrupt 1.

The libraries that are included in the Arduino IDE are listed in Table 13.7.

As well as the official Arduino libraries, as an open system, anyone can write a library and contribute it to the community, and many of these are extremely useful. Some of these other libraries are listed in Table 13.8.

For more information about Arduino, the official Arduino website (http://www.arduino.cc) should be your first port of call.

TABLE 13.7 Standard Arduino Libraries

LIBRARY	DESCRIPTION
EEPROM	Reading and write to EEPROM from your sketches
Ethernet	TCP/IP communications when using an Ethernet board or shield, including DNS, DHCP, HTTP, and UDP
Fermata	A protocol for turning pins on and off, reading analog values, etc., using serial commands
LiquidCrystal	Interface to the de facto standard alphanumeric LCD modules based on the HD44780 IC (most alphanumeric LCD modules)
SD	Read and write to an SD card; shields that combine an SD card socket with an Ethernet or real-time clock interface are available

(*Continued*)

TABLE 13.7 Standard Arduino Libraries (*Continued*)

LIBRARY	DESCRIPTION
Servo	Control a number of servos simultaneously (see Sec. 13.4.6)
SoftwareSerial	Use any two pins to receive and transmit data; the Arduino has one hardware serial port (UART)
SPI	Serial Peripheral Interface bus library
Stepper	Control stepper motors
Wire	I²C library

TABLE 13.8 Contributed Arduino Libraries

LIBRARY	SOURCE	DESCRIPTION
Android Accessory	http://developer.android.com/guide/topics/usb/accessory.html	For serial communications between an Android phone and an Arduino
Bounce	http://www.arduino.cc/playground/Code/Bounce	Software debouncing of switches
Dallas Temperature Control	http://milesburton.com/index.php?title=Dallas_Temperature_Control_Library	Not simply "HOT!" but a library to interface with the DS18B20 family of temperature sensors (see Chap. 6)
Handbag	http://rancidbacon.com/p/android-arduino-handbag/	An alternative mechanism for communicating with Android devices
IRRemote	https://github.com/shirriff/Arduino-IRremote	Sending and receiving IR remote commands using an IR LED sender and IR receiver
Keypad	http://arduino.cc/playground/Code/Keypad	Decodes keypresses from matrix keypads
OneWire	http://arduino.cc/playground/Learning/OneWire	1-Wire Interface library
RTC library	http://jeelabs.org/2010/02/05/new-date-time-rtc-library/	Interfaces with various real-time clock (RTC) ICs
Si4703_ Breakout	http://www.doctormonk.com/2011/09/sparkfun-si4703-fm-receiver-breakout.html	Allows easy control of the Si4703 radio receiver IC
USB Host Shield	http://www.circuitsathome.com	Allows the use of USB devices such as keyboards; also used for Android accessories
VirtualWire	http://www.open.com.au/mikem/arduino/VirtualWire.pdf	Provides serial communication between two Arduinos over a 433 MHz FM radio link
xbee	http://code.google.com/p/xbee-arduino/	Communicates with XBee data modules

13.4.6 Arduino Example Project

We are going to repeat the BASIC Stamp example project, using the same external electronics, but controlled by an Arduino rather than a BSII. The following listing shows the code needed to control the robot.

```
#include <Servo.h>

constintrightIRPin = 3;
constintleftIRPin = 4;
constintrightServoPin = 8;
constintleftServoPin = 9;
```

```
constintirOutPin = 10;
constintbuzzerPin = 11;

int speed = 60;        // servo speed as an angle offset from 180
intturnSpeed = 30;

Servo leftServo;
Servo rightServo;

void setup()
{
pinMode(rightIRPin, INPUT);
pinMode(leftIRPin, INPUT);
pinMode(irOutPin, OUTPUT);
pinMode(buzzerPin, OUTPUT);
digitalWrite(irOutPin, HIGH);
leftServo.attach(leftServoPin);
rightServo.attach(rightServoPin);
}

void loop()
{
   if (digitalRead(leftIRPin) == LOW &&digitalRead(rightIRPin) == LOW)
   {
     backup();
   }
   else if (digitalRead(leftIRPin) == LOW)
   {
     turnRight();
   }
   else if (digitalRead(rightIRPin) == LOW)
   {
     turnLeft();
   }
   else
   {
     forward();
   }
}

void forward()
{
  tone(buzzerPin, 440, 1000);   // play 440Hz for 1 second
  leftServo.write(180 - speed);
  rightServo.write(180 + speed);
}

void backup()
{
  tone(buzzerPin, 880, 1000);   // play 440Hz for 1 second
  leftServo.write(180 + speed);
  rightServo.write(180 - speed);
}

void turnLeft()
{
  leftServo.write(180 - turnSpeed);
  rightServo.write(180 - turnSpeed);
}
```

```
void turnRight()
{
  leftServo.write(180 + turnSpeed);
  rightServo.write(180 + turnSpeed);
}
```

There are many obvious similarities with the BASIC Stamp version. However, the C language used by Arduino provides a little more structure to the program, and seasoned programmers will probably feel happier using C rather than BASIC.

As noted in the previous section, a feature of the Arduino platform is the large number of libraries, both official and user-supplied, that can be included in the program. The first line includes a library for controlling servos. Strictly speaking, this is C++, the object-oriented extension to C. So, to control a servo, we first need to create an instance (well, two in this case):

```
Servo leftServo;
Servo rightServo;
```

Then we need to associate it with a particular pin using the following:

```
leftServo.attach(leftServoPin);
rightServo.attach(rightServoPin);
```

From now on, all we need to do is give the servo an angle in degrees, using the write command, like this:

```
leftServo.write(180 - speed);
```

There is no need to worry about pulse lengths; all this is handled for us automatically.

13.4.7 Taking the Arduino Offboard

For all its convenience and ease of use, an Arduino Uno is just a microcontroller with the necessary support components to provide it with a regulated voltage and allow it to be programmed over USB. When the time comes to create a product, if it is a one-off, you will probably just use the Arduino as it is, and buy yourself a replacement for the next project. If however, you will be producing a number of devices, you will probably want to lose the Arduino board and just use the programmed microcontroller in your project.

Taking the Arduino offboard means that you will probably design your own PCB for the microcontroller that includes just those features of the Uno that you need in your project, along with those extra components that may have been provided by a shield or connections to other electronics on, say, a breadboard.

The ATmega328 is not as easy to run using an internal oscillator as the ATtiny. The IDE expects an external clock (crystal or ceramic resonator). You will probably also want to use a voltage regulator IC to provide a stable voltage to the microcontroller.

The schematic shown in Fig. 13.13 was produced in Eagle CAD, along with the corresponding board shown in Fig. 13.14.

EagleCAD schematic for offboard Arduino project

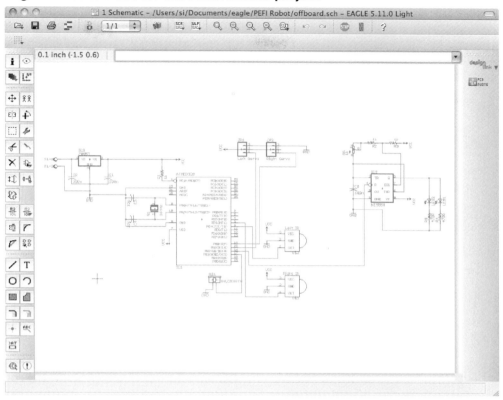

FIGURE 13.13

PCB layout for offboard Arduino project

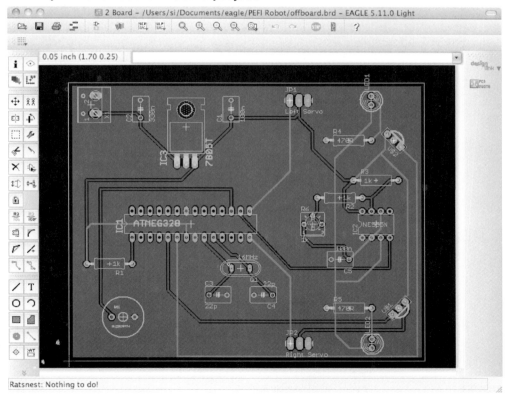

FIGURE 13.14

13.5   Interfacing with Microcontrollers

Whether you are using an ATtiny, a PIC, or an Arduino, you can be fairly certain that you will need to connect some components to it. At the very least, there will probably be a switch or two.

You can use three types of interfaces to your microcontroller:

- *Digital*: Switches as inputs, LEDs or similar as outputs
- *Analog*: Sensors of various types (see Chap. 6)
- *Serial*: A serial communications protocol of which there are four main types: TTL serial, I²C, 1-Wire, and Serial Peripheral Interface (SPI).

In the following sections, we will assume that your microcontroller has both analog and digital inputs, as well as digital and PWM outputs. It is also assumed that the microcontroller is operating at 5V. This may not always be the case, as many microcontrollers can be used at lower voltages, and 3.3 V is another common choice. If this is the case, you will need to adapt some of the schematics.

13.5.1   Switches

Single Switches

Switches are easy to connect to a digital input (see Fig. 13.15). Note the use of pullup resistors that keep the pin high until it is closed. When the switch is of the normally closed variety, then that current will be drawn continually, so you may want to use a high value of resistor there—say, 10 kΩ. However, for a normally open switch, current will flow only when the switch is pressed, so 1 kΩ is fine.

With the switch to GND, as shown in all the examples in Fig. 13.15, when the switch is closed, the digital input will go low. This means that the logic of a button being pressed is inverted, as shown in the following sample Arduino C code.

```
if (digitalRead(4) == LOW)
{
    // the key was pressed, do something
}
```

Connecting switches to a digital input

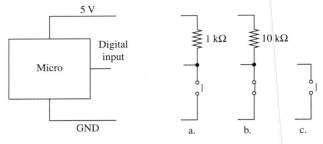

a. N.O.
b. N.C.
c. N.O. internal pullup resistor

FIGURE 13.15

It is also possible to swap over the switch and resistor, so that the resistor is now a pulldown resistor, and the switch being closed will result in a logical HIGH at the input.

The choice of pullup resistor depends on how electrically noisy your environment is and how long the leads are from the microcontroller to the switch. Essentially, it is a compromise between immunity to noise and current consumption. Given that for a normally open switch, current will flow only when the button is pressed, 5 mA using a 1 kΩ resistor is not normally a problem. In fact, some would advocate a lower value, such as 270 Ω.

Many microcontrollers include internal pullup resistors that can be turned on and off for a particular digital input. On an ATmega and ATtiny microcontroller, this resistor typically has a value of 20 k to 40 kΩ, so in a noisy environment or if the lead to the switch is long, an external pullup resistor may be better.

Multiple Switches to One Analog Input

If you have a lot of switches and do not want to tie up a load of digital inputs, then a common technique is to use an analog input and a number of resistors. The voltage at the analog input will then depend on the switches that are pressed (see Fig. 13.16).

Figure 13.16 is taken from the schematic diagram for the Freetronics Arduino LCD shield, where it is used for the five switches of a joystick type arrangement of push buttons (thanks to Freetronics for permission to use this diagram). Note how the decimal values for 10-bit A to D for each button are given as a table.

Multiple switches and an analog input

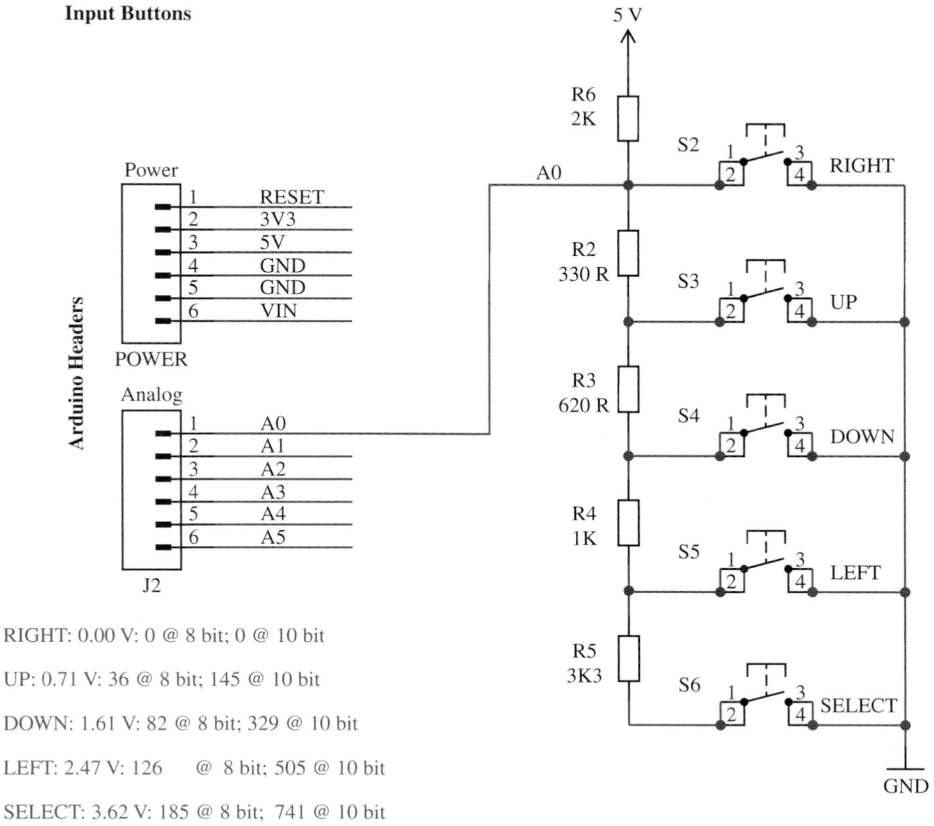

RIGHT: 0.00 V: 0 @ 8 bit; 0 @ 10 bit

UP: 0.71 V: 36 @ 8 bit; 145 @ 10 bit

DOWN: 1.61 V: 82 @ 8 bit; 329 @ 10 bit

LEFT: 2.47 V: 126   @ 8 bit; 505 @ 10 bit

SELECT: 3.62 V: 185 @ 8 bit; 741 @ 10 bit

FIGURE 13.16

The analog reading will not normally be exactly the value required due to resistor tolerances and power supply voltage changes, so in the code that interprets this, you would normally specify a band that would indicate a certain button, rather than just one value.

Using a Matrix Keypad

Keypads use switches arranged in a matrix, as shown in Fig. 13.17. The 4 × 3 keypad shown in the figure has a key at the intersection of each row and column. To determine which keys are pressed, the microcontroller will take each of the output pins Q0 to Q2 high in turn, and see what value is presented at each of the inputs I0 to I3. Note that if the microprocessor does not support internal pullup resistors, then these pullup resistors would be required on each input.

A keypad matrix

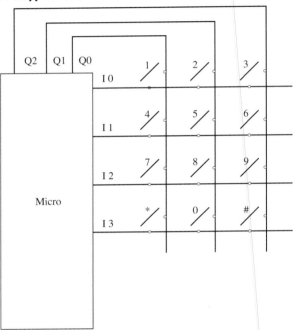

FIGURE 13.17

In practice, this is such a common component for microcontrollers that to write your own code for it would be needlessly reinventing the wheel. As an example, the following is the Arduino code for this that uses a library.

```
#include <Keypad.h>

char keys[4][3] = {
  {'1','2','3'},
  {'4','5','6'},
  {'7','8','9'},
  {'*','0','#'}
};

byte rowPins[4] = {2, 7, 6, 4};
byte colPins[3] = {3, 8, 5};

Keypad keypad = Keypad(makeKeymap(keys), rowPins, colPins, 4, 3);
```

```
void setup()
{
  Serial.begin(9600);
}

void loop()
{
  char key = keypad.getKey();
  if (key != null)
  {
    Serial.println(key);
  }
}
```

This example code will send any key that is pressed out through the Arduino serial monitor.

Debouncing

Attach an oscilloscope to the output of any of the circuits in Fig. 13.15, and you are likely to see an output something like what is shown in Fig. 13.18 when the switch is closed. This is called *bouncing*.

Switch bouncing can cause problems. Imagine the situation where pressing a button toggles an LED on and off. If there are an even number of bounces, then the LED will toggle on and then immediately off again, giving the impression that nothing happened.

It is therefore a good idea to *debounce* any switches that are connected to a microcontroller input. Although it is perfectly possible to do this with hardware—say, with a monostable that ignores subsequent pulses from the switch after triggering—it reduces the component count if you do the debouncing in software.

Switch bouncing

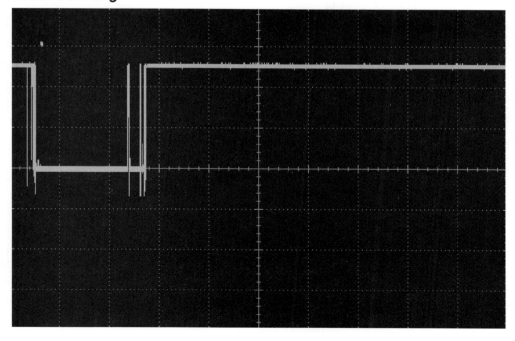

FIGURE 13.18

As with keyboard matrices, debouncing is a common problem that has been solved many times. The essence of software debouncing is the same as hardware debouncing, and that is to take action on the first transition and then ignore any subsequent transitions of the output until a safe debounce period has elapsed. Depending on what else the microcontroller has to do, this can be as simple as inserting a delay for the debounce period in the code that immediately follows the detection of the first transition. However, sometimes this is not possible, such as when the microcontroller has other responsibilities (like refreshing an LED display). In these cases, a common approach is to set a variable to the milliseconds tick after the first transition and make a condition of actioning the button press that sufficient debounce time has elapsed. The Arduino code for this is as follows.

```
constintdebouncePeriod = 100;
long lastKeyPressTime = 0;

void loop()
{
  long timeNow = millis();
  if (digitalRead(5) == LOW &&lastKeyPressTime>timeNow + debouncePeriod)
  {
    // button pressed and enough time elapsed since last press
    // do what you need to do
    lastKeyPressTime = timeNow;
  }
}
```

13.5.2 Analog Inputs

Many of the sensors described in Chap. 6 provide an analog output to indicate the property that they are reading. For example, the TMP36 temperature sensor IC would typically be connected directly to an analog input of a microcontroller, as shown in Fig. 13.19.

If you are measuring a voltage that is outside the range of the microcontroller's analog input (say 0 to 10 V), then you can just use two resistors as a voltage divider to reduce the voltage appropriately. If there is a risk that the voltage may exceed the

Reading the voltage from a TMP36 sensor

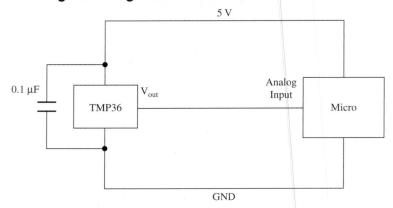

FIGURE 13.19

ADC voltage reduction and input protection

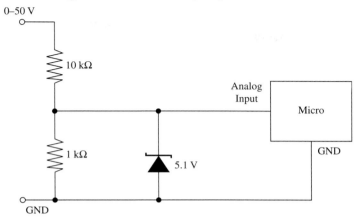

FIGURE 13.20

expected range, you can protect the microcontroller's analog input by adding a zener diode (see Fig. 13.20).

In practice, the zener diode will start to conduct before 5.1 V, to the detriment of the linearity of the readings, which is why the input range is labeled as 0 to 50 V, rather than 0 to 55 V. Remember the voltage divider is 1:11, not 1:10.

If switching behavior is required, then this should be implemented in code rather than through hardware. This approach allows more flexibility, for instance, for adding hysteresis or changing the set temperature. The following Arduino C code illustrates a simple temperature-control algorithm with 4 (±2) degrees of hysteresis.

```
const float hysteresis = 2.0;
const float setPoint = 20.0;
float temp = readTemperature();
if (temp <setPoint - hysteresis)
{
   digitalWrite(heaterControlPin, HIGH);
   }
else if (temp >setPoint + hysteresis)
{
   digitalWrite(heaterControlPin, LOW);
}
```

This example assumes that there is a user-supplied function called read Temperature.

For measuring resistance, sensors that are resistive, like LDRs and thermistors, will usually simply be used as one leg of a potential divider to produce a voltage that can be read. This is discussed in the relevant sensor sections in Chap. 6.

13.5.3 High-Power Digital Outputs

Most microcontrollers will reliably provide us with only around 20 mA of source or sink current as a direct digital output. If you want to drive a higher power load, such as a relay or a high-power LED, then you need to use a transistor.

Note that Arduino will handle up to 40 mA per pin with a maximum per chip of 200 mA. These figures should be derated by 25 percent for a production product,

Handling high-power outputs with bipolar transistors

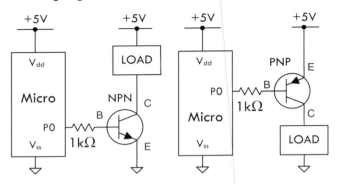

A bipolar transistor can be used to switch on and off loads. In the circuit to the left, when P0 is set HIGH, the NPN transistor is turned on—a low resistance is now present between C and E. In the circuit to the right, when P0 is set LOW, the PNP transistor is turned on. Microcontroller I/O current levels are usually large enough to provide enough base current for a bipolar transistor. Select a bipolar transistor with the appropriate current/voltage ratings for your load.

FIGURE 13.21

but Atmel says that the chip can comfortably cope with the absolute maximum current ratings, as these figures are already derated.

Figure 13.21 shows how to accomplish this with bipolar transistors, and Fig. 13.22 shows how to use MOSFETs.

MOSFETs have a number of advantages over bipolar transistors. One is that for most applications, they do not need a gate resistor. However, MOSFETs are a capacitative load, so the inrush current when the pin changes state can be very high, albeit for a very short duration. Microcontrollers will generally cope with this, but for ultimate adherence to design rules, use a gate resistor of around 1 kΩ.

Another advantage of using MOSFETs as switches is their exceptionally low drain-source on resistance and high off resistance. This makes a small MOSFET

Handling high-power outputs with MOSFETs

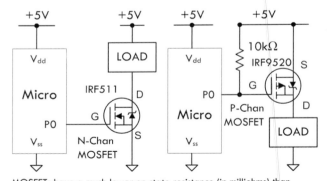

MOSFETs have a much lower on-state resistance (in milliohms) than bipolar transistors (10's to 100's of milliohms). This means that a MOSFET driver will experience a smaller voltage drop, and in general, can handle much larger currents. MOSFETs also have very high input impedances; they draw very little gate current from a microcontroller's I/O pins. Some MOSFETs are capable of handling 60A or more. In the circuit to the left, an N-Chan MOSFET is triggered with a HIGH on P0. In the right circuit, a P-Chan MOSFET is triggered with a LOW on P0. Separate load supplies can be used, and are recommended if there are inductive loads present.

FIGURE 13.22

capable of controlling quite big loads. However, you should check the gate threshold voltage to make sure that it is not above the logic level. For instance, an N-channel MOSFET with a gate threshold voltage of 6 V is not going to turn on when the gate goes to just 5 V. This is more of a problem with high-power MOSFETs. When using high-power MOSFETs, look for those described as "logic-level" MOSFETs, meaning that they have a gate threshold significantly less than 5 V.

Relays and Other Inductive Loads

With the exception of some reed relays, very few relays will switch with a current less than 50 mA, and therefore you will nearly always need to use a transistor as just described. You also need to remember to use a reverse-biased diode across the relay's coil, to prevent voltage spikes damaging the transistor during switching. Figure 13.23 shows this arrangement.

Controlling a relay or dc motor from a digital output

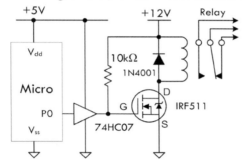

For high-current applications, say a 12V relay, a MOSFET transistor is a better choice than a bipolar transistor. Here an N-Chan MOSFET is driven by the microcontroller through a 74HC07 buffer. The diode is used to stomp out inductive spikes generated by coil. When P0 is set HIGH, MOSFET is turned on, and relay switches states.

FIGURE 13.23

Pulse-Width Modulation

The schematic shown in Fig. 13.23 is also suitable for controlling inductive loads like dc motors. If the digital output is driven as PWM, then this circuit can also be used to control the power going to the motor, and hence its speed (see Fig. 13.24).

The waveforms on the right in Fig. 13.24 show how you can control the motor speed by adjusting the duty-cycle (proportion of time the power is on). The pseudo-code on the right shows how this is accomplished. Note that some microcontrollers also have dedicated hardware support to simplify the process of generating PWM signals.

If your microcontroller has fairly robust output drivers, then there is probably little point in using the 74HC07 buffer.

Directional Motor Control

Directional control of motors can be achieved using an H-bridge, as shown in Fig. 13.25.

Most useful for controlling motor currents of less than a couple of amps are IC H-bridges such as the TB6612FNG, which combine all the transistors into

Controlling a dc motor

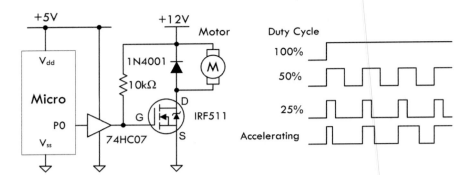

Example pseudo code segments:
Declare variables x
'100% duty cycle
Set P0 = 1   '100% duty cycle

'50% duty cycle
For x = 1 to 200
  Set P0 = 1
  Pause 5 'ON for 5 milliseconds
  Set P0 = 0 'Set pin 0 LOW
  Pause 5 'OFF for 5 milliseconds

'25% duty cycle
For x = 1 to 100
  Set P0 = 1 'Set pin0 HIGH
  Pause 5   'ON for 5 milliseconds
  Set P0 = 0 'Set pin 0 LOW
  Pause 15   'OFF for 15 milliseconds
Next

'Accelerating
For x = 100 to 1
  Set P0 = 1 'Set pin0 HIGH
  Pause 15 'ON for 15 milliseconds
  Set P0 = 0 'Set pin0 LOW
  Pause x
Next

FIGURE 13.24

Bidirectional motor control

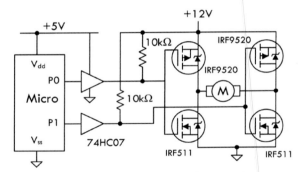

The H-bridge circuit built using MOSFETs provides forward and reverse directional control of a dc motor. The H-bridge provides built-in dynamic breaking action useful for applications requiring greater control. To make motor go in one direction, P0 is set HIGH while P1 is set LOW. To switch directions, P0 is set LOW while P1 is set HIGH.   Buffer stage (74HC07) could be replaced with an optoisolator to provide greater electrical isolation from motor section of circuit.

FIGURE 13.25

one package. They often also have features such as thermal shutdown, to protect against overloading.

Servo Motor Control

We have already touched on controlling servo motors in our robot example project. Since the servo uses a control signal, this can be provided directly from a digital output (see Fig. 13.26).

Stepper Motor Control

A stepper motor has a number of coils that must be energized in the correct sequence to move the rotor. The arrangement shown in Fig. 13.27 can be used to accomplish this.

Controlling a servo

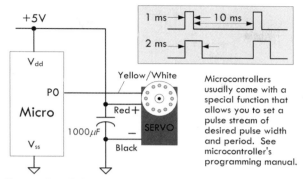

Here a relatively low-current servo can be controlled by a microcontroller. P0 sends the control signals to the servo, as shown in figure. A series of control pulses each 1ms in duration with a period of 10ms between pulses, causes servo's shaft to rotate to one extreme. Control pulses with a duration of 2ms (same period as before) sets servo's shaft to the opposite extreme direction. Anything in-between results in "in-between" positions. Without a pulse stream, the servo cannot hold its position.

FIGURE 13.26

Microcontrollers usually come with a special function that allows you to set a pulse stream of desired pulse width and period. See microcontroller's programming manual.

Stepper motor control

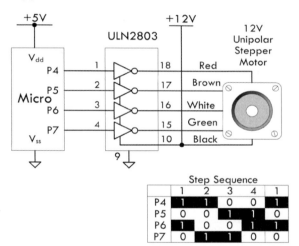

Here a 12V unipolar stepper motor is controlled using a TTL open collector driver IC connected to microcontroller. The step sequence for rotating motor is shown to the left. There are many new stepper motor drivers circuits out there that have many bells and whistles. Your best bet is to check the Internet to see what new technology exists and to study various example codes used to drive the stepper motors.

FIGURE 13.27

See Chap. 15 for more information about motors.

13.5.4 Sound Interfaces

Figure 13.28 shows a schematic for detecting sound. The second comparator stage is optional, and the output of the first stage could be fed directly into an analog input, allowing the sound to be sampled. Most microcontroller ADCs are not terribly fast, but even so, they should be able to sample at above 10 kHz, allowing some primitive digital signal processing.

When it comes to generating sound, being digital devices, few activities come more naturally to a microcontroller than generating a squarewave. All it needs to do

Detecting sound

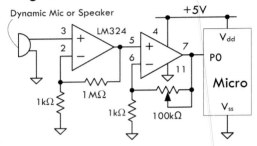

FIGURE 13.28

This circuit uses an LM324 comparator IC connected to a dynamic microphone or speaker. When a specific sound level is reached—set by pot, the output suddenly changes, supplying a HIGH to the microcontroller's input.

is set a pin high, wait, set it low, then wait again, and keep repeating those steps. As demonstrated earlier in the chapter, both the Arduino library and the BASIC Stamp provide commands to do this directly. If you are using a piezo speaker, this can be driven directly from a digital output. If you are using a electromagnetic loudspeaker, then this will be beyond the drive capabilities of an output pin, and you will need to amplify the signal. For a range of audio amplifier circuits, refer to Chap. 16. But given that a squarewave sounds pretty harsh, then high-quality amplification is unnecessary, and a circuit like the one shown in Fig. 13.21, where the load is a loudspeaker, will work just fine. Make sure you do the math to check that the transistor can cope with the collector current, as most loudspeakers are 8 Ω.

Generating a sine wave requires a bit of thought and effort. A first idea may be to use the PWM output of one of the pins to write out the waveform. However, the PWM switching frequency for most microcontrollers is at an audio frequency, so without a lot of care, the signal will sound as bad as a squarewave. A better way is to use a DAC, which has a number of digital inputs and produces an output voltage proportional to the digital input value. Fortunately, it is very easy to make a simple DAC—all you need are resistors.

Figure 13.29 shows a DAC using an R-2R resistor network. It uses resistors of a value R and twice R, so R might be 5 KΩ and $2R$ 10 KΩ. Each of the digital inputs will be connected to an Arduino digital output. The four digits represent the 4 bits of a digital number, so this gives us 16 different analog outputs. Higher-resolution DACs can be made by using more stages. Alternatively, DAC ICs, which can be more convenient to use, are available.

13.5.5 Serial Interfaces

There are a number of different standards for serial interfaces to microcontrollers, which use different numbers of pins and approaches to communication. In this section, we will explore some of them and look at how they can be used to connect things to a microcontroller.

When communicating with a peripheral, whichever serial interface it uses, there are a number of ways that the microcontroller might interact with the device. You may simply issue commands from the microcontroller, usually in the form of a 1-byte

A simple DAC

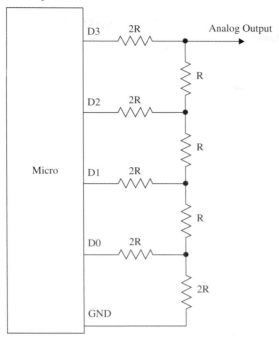

FIGURE 13.29

code that means something like "take a temperature reading," or in the case of serial EEPROM, "store this data here." The device then may respond with a result or value. Another common, but far less intuitive, approach is for the device to use registers, and some of the commands concerned with fetching and setting bits in the register that then control the electronics of the device. So, for instance, setting an I²C FM receiver IC to operate in stereo rather than mono involves setting the appropriate bit in a register using a general-purpose write-register command, rather than a command specific to setting the mode to mono or stereo.

1-Wire Bus

As the name implies, the 1-Wire serial bus uses just a single connection (apart from a common ground) to communicate. This standard was developed by Dallas Semiconductors and is used in a variety of sensors and other devices such as ADCs and EEPROM. It can operate at either 5 V or 3.3 V, so always check that a device you are connecting to your microcontroller operates at the same voltage. If it doesn't, then damage may ensue.

The DS18B20 temperature sensor uses the 1-Wire interface. This sensor was introduced in Chap. 6. In this chapter, we will look at how the sensor can be used in parasitic power mode, so that only two connections are needed from the microcontroller to the device. Furthermore, up to 255 devices can be connected to the same wire.

Figure 13.30 shows a DS18B20 attached to a microcontroller. 1-Wire devices act as either a master or slave. The microcontroller will be the master, and the peripheral devices, such as sensors, the slave. The slave devices contain a capacitor that is charged from the bus when no data is being transferred and used to power the slave device while the bus is being used for data. When the DS18B20 is being used this way, its GND and Vdd connections are tied together. The communication is two-way,

DS18B20 in parasitic power mode

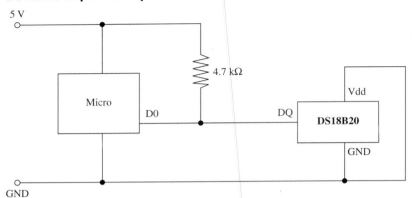

FIGURE 13.30

so the microcontroller will use the pin as both an input and an output, changing the pin's direction while the program is running. Every slave device has a unique 64-bit identifier that is programmed into ROM during manufacture.

Communication is always initiated by the master (microcontroller), which will put the data line into output mode and send a command as a sequence of pulses. The data line is pulled up to 5 V, so pulses are from 5 V to GND. A pulse of 60 µS signifies a 0, and 15 µS indicates a 1.

When the microcontroller needs to issue a command, it first sends a reset pulse of at least 480 µS, followed by the command sequence that includes the identifier of the device. The available device IDs are found by a special search protocol where the master sends a command that requests devices with a particular bit in their ID to respond. If more than one responds, then it tries another bit, and in this way, efficiently identifies all the devices.

Any microcontroller that you use with 1-Wire will have a library and example code for using the bus, so there is little point in looking at the low-level protocol. The following fragments of code illustrate how the Arduino OneWire library is used with a DS18B20.

```
#include <OneWire.h>

OneWire  ds(10);    // DS18B20 on pin 10
byte data[12];      // buffer for data
byte addr[8];       // 64 bit device address

void setup(void)
{
  Serial.begin(9600);
  if (ds.search(addr))
  {
    Serial.println("Slave Found");
  }
  else
  {
    Serial.println("Slave Not Found");
  }
}
```

The first step is to include the OneWire library and define some byte arrays to hold the data and the device ID for the DS18B20. The setup function opens a serial port, so that the temperature readings can be sent to the Arduino serial monitor, and then searches for devices on the 1-Wire bus. There should only be one, and if it is found, then a suitable message is displayed.

```
void loop(void)
{
  Serial.println(getReading());
  delay(1000);
}
```

The main loop simply calls the function getReading, sends it to the Arduino serial monitor, and then pauses for a second.

```
float getReading()
{
  ds.reset();
  ds.select(addr);
  ds.write(0x44, true);     // command: start temp conversion,
                            // true for parasitic power mode
  delay(750);

  ds.reset();
  ds.select(addr);
  ds.write(0xBE);           // command: Read Scratchpad

  for (inti = 0; i< 9; i++)
  {
    data[i] = ds.read();
  }
  return (((data[1] << 8) + data[0]) * 0.0625);
}
```

The getReading function is where most of the work goes on. It has two commands: one to start the temperature conversion and another to read the data resulting from the conversion.

Each command is preceded by a reset. Note how the slave to be communicated with is set using ds.select(). We then read the response into the byte array called data. To actually decode the temperature, we need only the first 2 bytes of the data, which are combined into a 16-bit integer and multiplied by the scaling factor (defined in the DS18B20 data sheet) as 0.0625.

Next, we have this line:

```
(((data[1] << 8) + data[0]) * 0.0625);
```

This first shifts the byte contained in data[1] left by 8 bits, and then adds in the lower 8 bits contained in data[0]. This results in a 16-bit integer that must be multiplied by 0.0625 to produce a temperature in degrees Celsius (see the data sheet for the DS18B20).

The resulting trace in the Arduino serial monitor should look something like Fig. 13.31.

Trace from DS18B20 test program

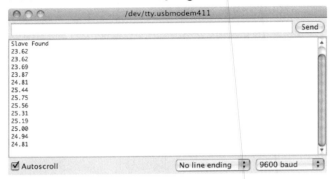

FIGURE 13.31

We have touched on only two of the DS18B20's commands. For a full list of commands and more information about the protocol, look at the DS18B20 data sheet (http://datasheets.maxim-ic.com/en/ds/DS18B20.pdf).

I²C (TWI)

On the face of it, the I²C, also sometimes known as the Two-Wire Interface (TWI), serves much the same purpose as 1-Wire, although it has two wires rather than one for data. Like 1-Wire, it is a bus and can support multiple devices connected to the same two wires. It also can run at either 5 V or 3.3 V. However, it is faster than 1-Wire, with top speeds of up to 400 kbits/s.

The two data lines of I²C are open-drain connections that operate as both inputs and outputs at the microcontroller. They must have pullup resistors in the same way as 1-Wire, but there is no equivalent to the 1-Wire parasitic mode, so remote sensors will generally require four wires in total: two for data and two for power.

Figure 13.32 shows how two microcontrollers might communicate using I²C.

I²C microcontroller-to-microcontroller communication

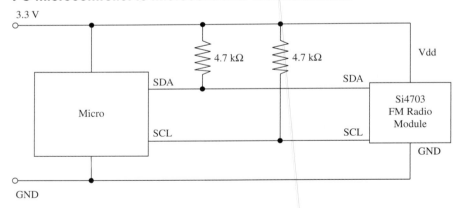

FIGURE 13.32

I²C devices are either masters or slaves, and there can be more than one master device per bus. In fact, devices are allowed to change roles, although this is not usually done. It is common for microcontrollers to have an I²C interface and use it to exchange data between microcontrollers.

The serial clock line (SCL) is a clock, and the serial data line (SDA) carries the data. The timing of these pins is shown in Fig. 13.33. The master supplies the SCL

Timing diagram for I²C

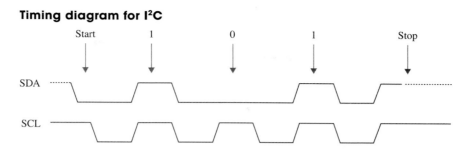

FIGURE 13.33

clock, and when there is data to be transmitted, the sender (master or slave) takes the SDA line out of tri-state and sends data as logic highs or lows in time with the clock signal. When transmission is complete, the clock can stop and the SDA pin be taken back to tri-state.

Whether using I²C or 1-Wire, from a microcontroller, the code is likely to be similar, and a library is provided to hide the low-level timing of the protocol.

The following example, in Arduino C, shows I²C in action to send data from one microcontroller to another. When using I²C to interface with a sensor or other I²C slave device, the process is similar, but the messages will generally be packed into byte arrays. For this kind of application, every device will be different, and the data sheet for the device should be studied to determine the format of the messages that it expects. These examples are adapted from the examples provided with the Arduino environment. Thanks to Nicholas Zambetti for making this code public domain.

We start with the code for the transmitting microcontroller.

```
#include <Wire.h>

void setup()
{
Wire.begin(); // join i2c bus
}

void loop()
{
  Wire.beginTransmission(4);  // transmit to device #4
  Wire.write("Hello");        // friendly greeting
  Wire.endTransmission();     // stop transmitting
  delay(1000);
}
```

The transmission is very simple. We just say which device on the bus we want to send to, and then send it the data. In this case, the data is a string, but the write method can also take a single byte or a byte array as arguments for the data to be sent.

Receiving the data is a little more complex.

```
#include <Wire.h>

void setup()
{
  Wire.begin(4);              // join i2c bus with address #4
```

```
    Wire.onReceive(receiveEvent);  // register event
    Serial.begin(9600);            // start serial for output
}

void loop()
{
}

void receiveEvent(inthowMany)
{
  while(Wire.available())
  {
    char c = Wire.read();     // read a byte as a char
    Serial.print(c);          // print the character
  }
  Serial.print('\n');         // end of line
}
```

In this case, the receiver is a slave device and must identify itself—in this case, using the number 4 as its argument to Wire.begin. It then registers a function receiveEvent, which should be invoked whenever there is incoming data for this device. This function simply loops over each byte of data in the message, displaying it on the Arduino serial monitor.

Serial Peripheral Interface

Yet another microcontroller bus standard is the SPI bus. This one uses four data lines and is faster than the previous buses that we have looked at (up to 80 Mbits/s).

Figure 13.34 shows how a number of peripherals can be connected to the bus. Note that there can only be one master device.

SPI connections

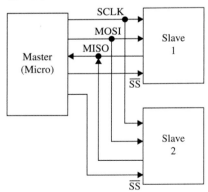

FIGURE 13.34

The slave devices are not assigned addresses. Instead, the master (usually a microcontroller) must have a dedicated Slave Select (SS) line for each of the slave devices, just selecting the one it communicates with. The other extra line is required because separate lines are used for each direction of communication. The Master Out/Slave In (MOSI) line carries the data from the master to the slave device, and the Master In/Slave Out (MISO) line does the reverse.

Many different data protocols have been layered over the physical serial interface, but the basic principal is the same as for the other buses that we have looked at. The approach to take is to find the SPI library for the microcontroller that you intend to use and read the data sheet for the device you wish to communicate with.

The SPI specification does not define the bit order for sending data, so make sure that your code agrees with the device in this respect.

SPI is also used as a means of ICSP on some microcontrollers, such as the ATmega and ATtiny families.

Serial

Many devices use yet another type of interface called just *serial*. This is a very old standard with its roots dating back to the days of teletypes. Some computers with serial ports can still be found. In the "good old days," people used to attach modems to them for communicating over phone lines with other computers.

The normal voltages used in the signals for serial ports conform to the standard RS-232 and use voltages that swing both positive and negative with respect to GND. This is not terribly convenient when using microcontrollers. For this reason, microcontrollers use the same communication protocol, but at logic levels. This is called TTL Serial, although more and more, it is being used by devices using 3.3 V rather than 5 V. See the next section for information about level conversion.

Electrically, TTL Serial uses two data pins: Tx and Rx (Transmit and Receive). It is not a bus, and the connection is point to point, so there are no problems with addressing different devices.

Another remnant from early computer history is the nomenclature around the bandwidth of serial connections. A serial connection must be set to the same baud rate at both ends of the connection. The baud rate is the number of bits per second, but that does include start, stop, and potentially parity bits, so the actual transmission of data is a little slower than the baud rate. To simplify matching up the baud rates at each end of the connection, a set of standard baud rates is used: 110, 300, 600, 1200, 2400, 4800, 9600, 14400, 19200, 38400, 57600, 115200, 128000, and 256000. Of these, 1200 is probably the slowest baud rate commonly in use, and many TTL serial devices will not go as high as 115200. 9600 is a very commonly used baud rate, and devices will often default to this rate, but be configurable to other rates.

As well as the baud rate, other parameters that define a serial connection are the number of bits per word, the type of parity bit, and the number of start and stop bits. Almost universally, these are defined as 8, none, and 1, respectively, which is often abbreviated to 8N1.

Bits are simply sent as high or low logic levels (see Fig. 13.35). As there is no separate clock signal, timing is critical, so after the start bit, the receiver will sample at the

TTL Serial

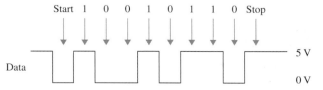

FIGURE 13.35

appropriate rate until it has read the 8 data bits and the 1 stop bit. The least significant bit of the data is sent first.

Most microcontrollers will either have dedicated hardware for TTL Serial (a UART) or manufacturer-developed software libraries for serial.

13.5.6 Level Conversion

There is a recent trend for microcontrollers and other ICs to use 3.3 V or even 1.8 V rather than 5 V. Lower-voltage devices use less current and can be more convenient to power from batteries. The same is also true of modules that the microcontrollers need to communicate with. While some 3.3 V devices can tolerate 5 V, many cannot. This means that if you are communicating with them using one of the bus and serial interfaces discussed previously, you will need to make sure that you convert voltage levels appropriately.

SPI and TTL Serial Level Conversion

Converting levels on SPI and TTL Serial is quite easy, because they have separate lines for each direction of communication. Figure 13.36 shows how resistors can be used as simple voltage dividers.

TTL Serial 5 V to 3.3 V level conversion

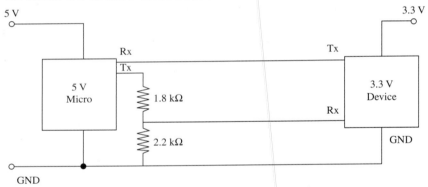

FIGURE 13.36

The Tx output of the 3.3 V device can be connected directly to the Rx input of the 5 V microprocessor, because it will see any input over about 2.5 V as a logical high anyway. The voltage divider is required when the 5 V Tx output of the microprocessor must be reduced to prevent damage to the 3.3 V device.

I²C and 1-Wire Level Conversion

The problem is more complex when pins change modes, from being an input and being an output, as they do with I²C and 1-Wire. In both these cases, the best solution is to use a custom level-shifting IC such as the TXS0102, which can convert two levels (ideal for I²C). Figure 13.37 shows the TXS0102 used to convert levels for I²C. Alternative ICs that perform the same role are the MAX3372, PCA9509, and PCA9306.

13.5.7 LED Display Interfaces

LED displays made up of a number of LEDs encapsulated in a single package can be a challenge to control. Such displays will normally be controlled using a

TXS0102 level converter used for I²C

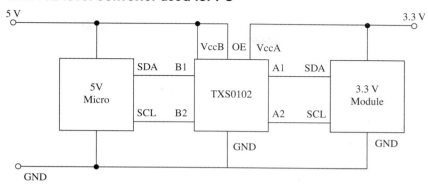

FIGURE 13.37

microcontroller, however, it is not necessary to use a microcontroller to pin each individual LED. Instead, multi-LED displays are organized as common anode or common cathode, with all the LED terminals of the anode or cathode connected together and brought out through one pin. Figure 13.38 shows how a common anode seven-segment display might be wired internally.

A common cathode seven-segment LED display

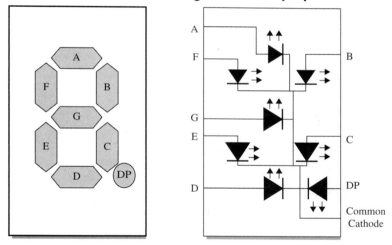

FIGURE 13.38

In a common cathode display like this, the common cathode would be connected to ground, and each segment anode driven by a microcontroller pin through a separate current-limiting resistor. Do not be tempted to use one resistor on the common pin and no resistors on the noncommon connections, as the current will be limited no matter how many LEDs are lit, and so the display would get dimmer as more LEDs were illuminated.

Multiplexing LED Displays

It is quite common for multiple displays to be contained in the same case. For example, Fig. 13.39 shows a three-digit seven-segment common cathode LED display. In this kind of display, each digit of the display is like the single-digit display

A three-digit common cathode seven-segment LED display

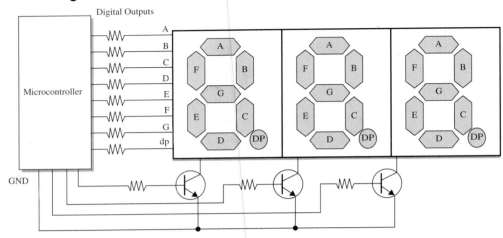

FIGURE 13.39

of Fig. 13.39 and has its own common cathode. But, in addition, all the A segment anodes are connected together, as is each segment.

The microcontroller, or LED driver IC, using the display will activate each common cathode in turn, turn on the appropriate segments for that digit, and then move on to the next digit. This refresh happens very quickly, so that the display appears to show different numbers on each digit. This is called *multiplexing*. The same approach can be used with LED matrices, where each column is activated in turn, and then the appropriate pins are set for the rows of that column.

Note the use of transistors to control the common cathodes. This is simply to handle the current of potentially eight LEDs at once, which would be too much for most microcontrollers.

Charlieplexing

When looking to minimize the number of pins used to display a matrix of LEDs, an interesting technique called Charlieplexing can be used. (The name comes from the inventor Charlie Allen at Maxim.) This technique takes advantage of the feature of modern microcontroller I/O pins that allows them to be changed from outputs to high-impedance inputs while a program is executing. Figure 13.40 shows the arrangement for controlling six LEDs with three pins.

Charlieplexing LEDs

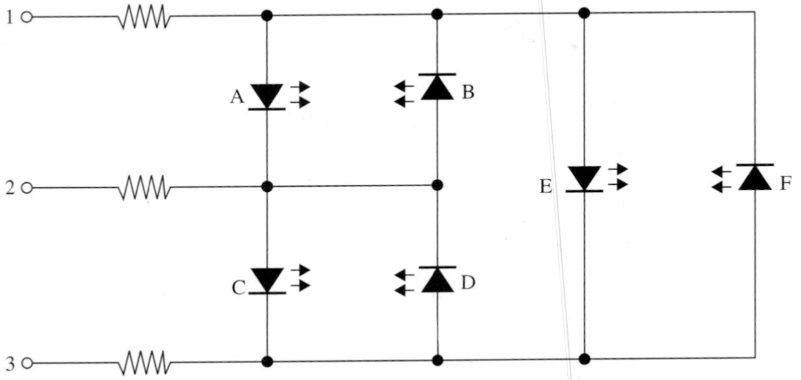

FIGURE 13.40

Charlieplexing is a dynamic activity, so rather like multiplexing, not all the LEDs that you want to be lit are lit at the same time, but they will appear lit as the display is refreshed faster than the eye can keep up. To do this, the pins will be high, low, or high impedance input, as shown in Table 13.9.

TABLE 13.9   Charlieplexing LED Addressing

LED	PIN 1	PIN 2	PIN 3
A	High	Low	Input
B	Low	High	Input
C	Input	High	Low
D	Input	Low	High
E	High	Input	Low
F	Low	Input	High

The number of LEDs that can be controlled per microcontroller pin is given by the following formula:

$$\text{LEDs} = n^2 - n$$

So, if we use 4 pins, we can have 16 – 4, or 12, LEDs, and 10 pins would give us a massive 90 LEDs. However, there are problems with scaling Charlieplexing up. One is due to the fact that the refresh rate needs to be fast enough to fool the eye, and a large number of pins will need a lot of sequence steps to energize all the LEDs that need energizing in a refresh cycle. This will also result in the LEDs becoming dim, as their duty cycle will be low. You can compensate for this to some extent by increasing the current through the LEDs, which will cope with fairly large peak currents for a small duration. This does lead to the problem that if the microcontroller freezes for some reason, the LEDs could burn out.

Controlling the Color of RGB LEDs

RGB LEDs are actually three LEDs in one package (one red, one green, and one blue). The package will often be common anode or common cathode. By controlling the power to each of the LEDs separately, it's possible to set the overall color of the LED module to any color at any intensity.

While you could change the intensity of each color channel by controlling the current to the LED in an analog fashion, it is far better to control it with a PWM signal. The duty cycle will control the brightness of the LED in a far more linear manner than controlling the current in an analog fashion.

CHAPTER 14

Programmable Logic

Designs that use the combinational and sequential logic described in Chap. 12 can be built using lots of separate ICs. Maybe you have a design that requires a 10-stage counter divider and binary-to-decimal decoder plus a NAND gate or two. While you can still buy the chips to make such a circuit, this would never be done for a commercial product. The chip count for the logic part of all but the most complex designs is rapidly heading toward one. This one chip might be a microcontroller as described in Chap. 13, but this essentially moves the design problem from hardware design to software programming. Programming is a discipline with different roots to electronics and is fraught with problems of maintaining code and managing complexity. Software solutions can also be slow, as the software effectively has to ape what the equivalent logic circuit would be doing.

When using a microcontroller, you write code to be run on the device as the device is in use. An alternative approach (programmable logic) is to use a field-programmable gate array (FPGA) or for smaller projects a complex programmable logic device (CPLD).

Using an FPGA or CPLD involves creating a combinational or sequential logic design, either by actually drawing out the gates, counter, shift registers, etc., into a CAD system, or describing the logic design using a hardware definition language (HDL). This description of the logic, either pictorial or textual, is then used to configure general-purpose logic cells in the chip to the hardware that you want. It's like creating your own custom chip containing just the logic that you need.

Programmable logic has a reputation for being difficult and inaccessible to the nonprofessional. While the technology of reconfigurable hardware takes some getting used to, the manufacturers of programmable logic devices have become less proprietary in their approach in recent years and even offer free to use design software for the inventor. In this chapter, you will learn how to get started with programmable logic, in particular using the Xilinx software tool ISE Studio and the Verilog hardware description language.

Since the best way to learn is to actually try things out, this chapter will use the Elbert 2 FPGA development board from Numato Labs (http://numato.com/elbert-v2-spartan-3a-fpga-development-board.html).

14.1   Programmable Logic

In the early days of digital computing, large numbers of logic gate ICs were combined onto printed circuit boards (PCBs) to make computer boards that would then be attached to a back-plane. You might have found hundreds of ICs on a single PCB, each of these chips containing a handful of gates or shift registers.

The arrival of large-scale integration (LSI) and the invention of the microprocessor reduced the IC count enough to use in home computing. However, open up a home computer from the early 1980s and you may still see rows of logic chips providing all the other functions that the computer needed: keyboard scanning, a cassette tape data storage interface, video output, etc.

All this clutter could easily fit on a single LSI chip, but having your own application-specific IC (ASIC) is a seriously expensive thing to do and was only possible for large production runs. Enter the programmable array logic (PAL). The PAL designed by Monolithic Memories Inc. was not the first programmable logic IC, but it was the first to really take off commercially.

The internal structure of the PAL is a sum of products arrangement of gates (see the section "AND-OR-INVERTER Gates" in Chap. 12). Figure 14.1 shows such an arrangement.

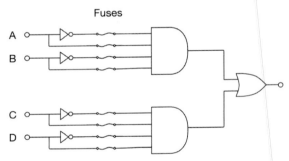

FIGURE 14.1   A sum of products term.

The inverted and noninverted inputs to the AND (sum) part of the logic are switched using "fuses" that can be blown during the manufacturing process or using special programming hardware. Some devices could only be programmed once while others were reprogrammable.

This idea of using cells of configurable logic underpins all modern types of programmable logic, although the scale and complexity has increased and the effort involved in programming them has decreased.

FPGAs generally have hundreds of thousands of logic cells. This is great if your FPGA is doing something complex, but may be excessive if you just want the equivalent of a few logic gates. To address such small needs, CPLDs are used. These are the natural successors to PALs and operate in a similar manner, using "macro cells" that implement the sum of products–type terms of Fig. 14.1. FPGAs use a different arrangement that will be explained in the next section.

The design tools that are used to "program" CPLDs and FPGAs are now so sophisticated that there is little need for the designer to think in terms of the actual logic gates on the silicon. Instead, they can simply draw their design using logic gates chosen from a palette, connect them all up, specify the inputs and outputs, and then let the tool manage the process of translating that design into configuration of the programmable logic device (see Fig. 14.2).

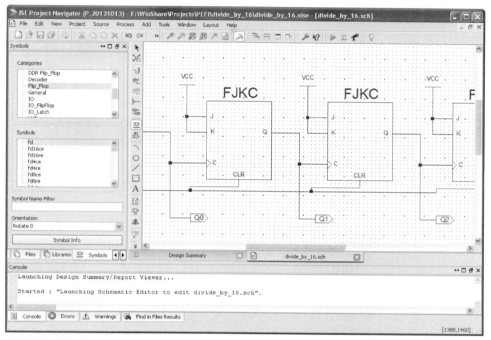

FIGURE 14.2   Designing with logic gates.

Taking this a step further, you can skip the logic diagram stage entirely and express your design in a hardware description language like Verilog or VHDL and then have your software tool convert that into configuration information for the programmable logic chip.

Getting into mind-bending territory, you can (and people do) design a microcontroller on the FPGA (along with other logic circuitry) that then runs a program.

14.2   FPGAs

The main difference between FPGAs and CPLDs is that FPGAs do not use logic cells in a sum of products arrangement, but rather the cells use a lookup table (LUT). The lookup table will have a number of inputs say six inputs and a single output. You can think of this as a 64 × 1 bit ROM, with the inputs being the address lines of the ROM and the output being the bit stored at that address. The contents of these LUTs, combined with other routing information, are what give the FPGA its logic.

LUTs are often not exactly arranged as a single six-input unit, but may comprise two five-input units, the sixth input being a select input that selects between the two LUTs. This allows extra flexibility when it comes to how the design software connects everything together.

The LUT will often be combined with extra components like a flip-flop to make an individual logic block.

Figure 14.3 shows a logical view of how this is all arranged.

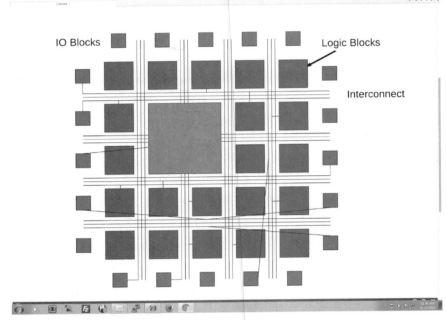

FIGURE 14.3   The logical structure of an FPGA.

General-purpose IO (GPIO) pins on the FPGA chip are connected to special-purpose IO (input/output) blocks that provide buffered microcontroller-like inputs and outputs that can typically source or sink a few tens of milliamps.

The vast bulk of the functional units in the FPGA will be logic blocks and a typical modern FPGA may have from 200,000 to several million of these blocks. There may also be a fixed RAM block for use when the FPGA is to be configured as a processor. Taking this to its extreme you find high-end SoC (systems on a chip) FPGAs that include fixed high-performance processor cores and memory on a chip that also includes configurable logic cells. FPGAs are also often used to prototype an ASIC for very large production runs.

Routing between such vast numbers of logic blocks is pretty tricky, but fortunately for us, we don't have to do it, that's what the design software is for.

The information in the LUTs and the routing matrix (that defines the interconnections) is volatile. When you lose power, all that information disappears and the FPGA reverts to its original state. To configure the FPGA the configuration is usually stored outside of the FPGA in EEPROM. The FPGA will generally have a fixed hardware loading interface built in to it that will pull in the configuration as the FPGA starts up. This typically takes less than 200 ms.

14.3   ISE and the Elbert V2

There are quite a few FPGA vendors now, but the two biggest players are Xilinx (Xilinx.com) and Altera (altera.com); between them, they have almost 90 percent of the programmable logic market. Of the two, Xilinx has the largest share.

All the FPGA manufacturers have their own design tools that work specifically with their hardware. In this chapter we will use the Xilinx Sparta 3A FPGA chip with Xilinx's design tool called ISE.

To get some practical experience of using an FPGA, there are a number of FPGA development boards that combine an FPGA with its configuration ROM and a selection of input/output devices, such as LEDs, switches, audio and video interfaces. The device that has been selected for this chapter is the Elbert V2 board, which is widely available and low cost, and includes its own USB interface for programming.

14.3.1   Installing ISE

The design tools of the FPGA manufacturers are frankly bloated monsters. The ISE design tool is a 7-GB (that's right, gigabyte!) download. In many ways getting and installing the design tool is the hardest part of getting started with FPGAs. Once you have done that everything else seems relatively simple.

The first step in obtaining ISE is to visit Xilinx.com with your web browser and find the ISE download page, which you will find by following the links: Developer Zone->ISE Design Suite->Downloads. Scroll down the "downloads" area to ISE Design Suite (we used version 14.7). Do not be tempted to download the newer Vivado Design Suite. This is only for newer Xilinx FPGAs and does not support the Spartan 3A used on the Elbert 2.

There are Windows and Linux versions of the tool. In this book we will just describe the process of getting up and running in Windows.

When you click on the "Download" button, a long complicated survey will appear that you have to complete followed by a second long and complicated registration form. Persevere and eventually the download will start and you can go and do something else for a few hours while the download completes.

After installation you have to click on the link to request a free license key for ISE Web. This will be e-mailed to you as an attachment; save the file and then from ISE open the license manager from the Manage License option on the Help menu of ISE and add the license.

14.4   The Elbert 2 Board

Figure 14.4 shows the Elbert 2 board.

These boards are available direct for Numato Labs (http://numato.com) or from Amazon.com and various other sources. The price of the entire board was just $29.95 at the time of writing. The only other thing you will need is a USB to mini-USB lead.

The board has the following features:

- Spartan XC3S50A FPGA in TQG144 package
- 16-MB SPI flash memory for configuration
- USB 2.0 interface for on-board flash programming
- Eight LEDs
- Six push buttons
- 8-way DIP switch
- VGA connector

FIGURE 14.4   The Elbert 2–Spartan 3A FPGA development board.

- Stereo jack
- Micro SD card adapter
- Three seven-segment displays
- 39 IOs for user-defined purposes
- On-board voltage regulators

14.4.1   Installing the Elbert Software

The Elbert board has a software utility for programming the board. This handles just the final step of copying the binary file generated by ISE onto the Elbert V2's flash memory. There is also a USB driver to install for Windows users. To set up your computer to use the Elbert, visit the product page for the Elbert V2 at numato.com and click on the Downloads tab.

You will need to download:

- Configuration tool: This is used to program the board
- Numato Lab USB CDC driver
- User manual

To install the USB driver on Windows, plug the Elbert V2 board into your computer and the New Hardware Wizard should start. Point the wizard at the extracted "numatocdc" folder and the driver should install and the new hardware be recognized.

After this, the Elbert V2 will be connected to one of the virtual COM ports of your PC. To find out which port, open the Windows Device Manager and you should find it listed in the Ports section as shown in Fig. 14.5.

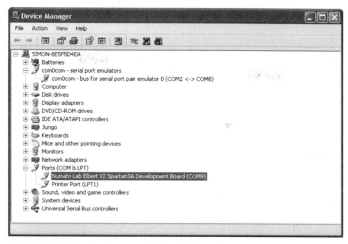

FIGURE 14.5   Finding the Elbert V2 port.

14.5   Downloads

All the code examples used in this book can be found in the GitHub repository (https://github.com/simonmonk/pefi4).

To install the examples on your PC, click on the "Download ZIP" button in the bottom right of the GitHub page and then extract the file. You will find the examples in the folder "fpga."

It is worth working through the examples below to build-up the projects and get used to using ISE, but if you get stuck and need to compare what you have done with the final working design, then these files will come in handy.

14.6   Drawing Your FPGA Logic Design

The ISE design tool gives you two ways of programming your FPGA. One is to draw a familiar logic diagram and the second is to use the Verilog hardware description language (HDL). We will start with the schematic approach, although seasoned FPGA designers nearly always use Verilog or its rival VHDL.

For this first example, we will go into quite a lot of detail on using the ISE tool to get you up and running.

14.6.1   Example 1: A Data Selector

The first example that we will make is the data selector that you first met in Chap. 12 (Fig. 12.32). The schematic for this is repeated in Fig. 14.6.

The three inputs (A, B, and "data select") for this circuit will be hooked up to three of the push buttons on the Elbert V2 and the output will be connected to one of the LEDs so that we can see the circuit actually in use.

Step 1: Create a New Project

The first step is to fire up ISE and then select File->New Project from the menu. This will open the New Project Wizard shown in Fig. 14.7.

Simple 1-of-2 data selector

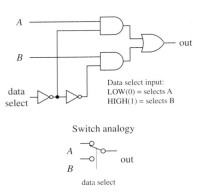

FIGURE 14.6   A simple data selector.

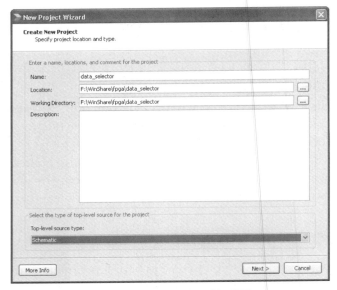

FIGURE 14.7   The New Project Wizard in ISE.

Enter "data_selector" in the Name field. In the Location field navigate to the folder where you want to keep your ISE designs. The Working Directory field will automatically update to match this directory, so you don't need to change the Working Directory field.

Change the "Top-level source type" drop-down to "Schematic" and then click "Next."

This will take you to the Project Settings shown in Fig. 14.8.

Change the settings so that they match Fig. 14.8 and then click "Next" again. The Wizard will then show you a summary of the new project and you can then click "Finish."

This will create for you the new, but empty project shown in Fig. 14.9.

The screen is divided into four main areas:

In the top left you have the Project View. This is where you can find the various files that go to make up a project. It is organized as a tree structure. Initially there are

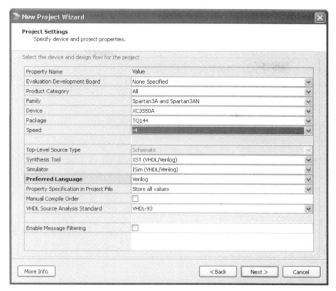

FIGURE 14.8   The New Project Wizard—Project Settings.

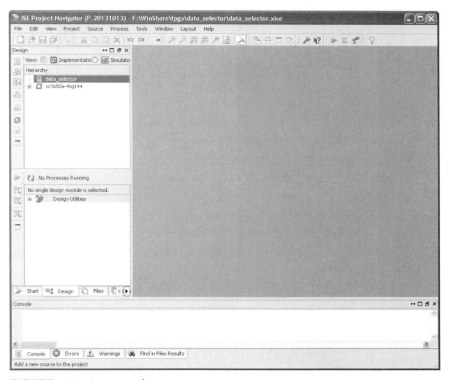

FIGURE 14.9   A new project.

two entries in this area. There is the entry that says "data_selector" and the second entry that has a seemingly random name (xc3s50a-4tq144). The latter will eventually contain two files, the schematic drawing that we are about to create and an implementation constraints file that defines how the inputs and outputs in the schematic connect to the actual switches and LEDs on the Elbert V2.

You can also double-click on "xc3s50a-4tq144" to open the project properties. So, if you made a mistake setting the project properties using the New Project Wizard, you can always correct it by double-clicking on this entry.

To the left, beneath the Project View is the Design View. This will eventually list useful actions that we can apply to our design including generating the binary file for programming the Elbert V2.

The wide area at the bottom of the window is the console. This is where error messages will appear.

The large area to the right of the window is the editor area. When it comes to drawing the schematic, this is where you will do it.

Step 2: Create a New Schematic Source

To create a new schematic, right click on data_selector in the Project View and select the option "New Source." This will open the New Source Wizard (Fig. 14.10).

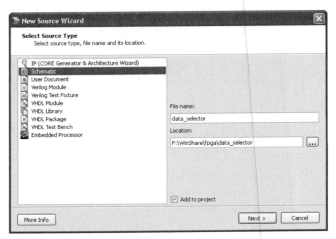

FIGURE 14.10   The New Source Wizard.

Select a Source Type of "Schematic" and enter "data_selector" in the File name field and then click "Next." A summary screen will appear, to which you can respond by clicking "Finish." This will result in a blank canvas being prepared for us in which we can draw the schematic.

This is shown in Fig. 14.11, labelling the parts that you are going to need.

The icon menu bar running vertically to the left of the editor area controls the mode of the window and also what appears on the left-hand side of the window:

- The top icon (an arrow) puts the window into select mode. You will need to click on this before you can drag circuit symbols about or change their properties.

- Click on the "Add wire" mode when you are connecting the gates and other circuit symbols together.

- IO markers are used to indicate the boundary between the schematic you are designing and the actual pins of the FPGA IC. This mode lets you add these symbols.

- Add logic symbols. This is the mode selected in Fig. 14.11. The left hand panel then divides into a top half that shows categories of circuit symbol and a bottom half that has a list of the component symbols in that category.

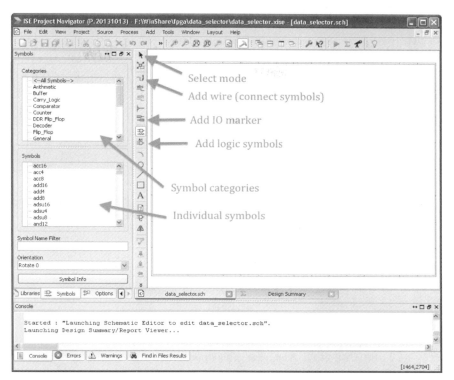

FIGURE 14.11   The schematic editor.

Step 3: Add the Logic Symbols

Put the screen into "Add logic symbols" mode. You are going to need to add two two-input AND gates, a two-input OR gate and two invertors.

Click on the category "Logic" the select "and2" (2 input AND). Then click twice in the editor area to drop the two and gates. Then select "or2" and drop an OR gate in roughly the right location to the right of the AND gates, and final add in the two invertors ("inv") below and to the left of the AND gates.

Zoom in a bit (Toolbar at the top of the window) and the editor area should look something like Fig. 14.12.

Step 4: Connect the Gates

Click on the "Add wire" icon and then connect the gates together in the arrangement of Fig. 14.6. To make a connection, click the mouse on one of the square connection points and drag out to the connection point or line that you want to connect to. The software will automatically put bends in the line for you. If you need to more things to tidy the diagram up then you can change to "Select" mode and drag the symbols and wires about.

The end result should be Fig. 14.13.

Step 5: Add the IO Markers

Click on the "Add IO Markers" icon and then add markers to all the inputs and the output by dragging the mouse out from the wire in question. Notice how the software figures out that the output is an output.

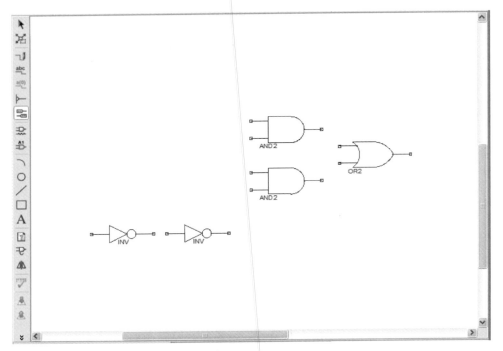

FIGURE 14.12 The logic gates in position.

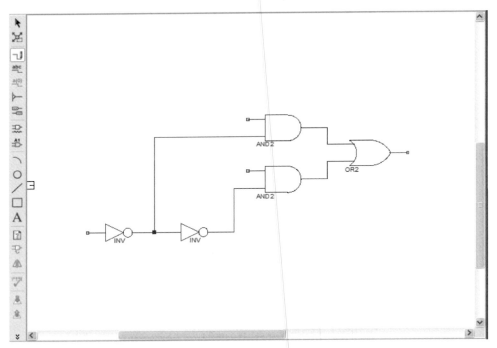

FIGURE 14.13 Connecting the symbols with wires.

Initially, the connections are all given names like XLXN_1, etc. To change these names to more meaningful names, change to "Select" mode, right-click on an IO connector and chose the menu option "Rename Port." Change the port names so that they agree with Fig. 14.14.

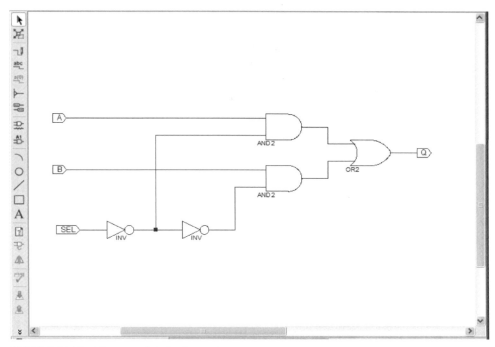

FIGURE 14.14   The completed schematic.

Notice that we have called the output Q. This is because the word "OUT" is reserved for use by ISE, so you cannot call any of your connections "OUT" or you will get an error when you try and build the project.

The schematic is now complete, and now would be a good time to do "File->Save" to save the schematic design to file.

Step 6: Create an Implementation Constrains File

Now you need to get back to the original Project View to be able to create a new source file. Click on the little "Xs" in the top right corners of the various views that will have become layered on top of the Project View. You may also have to click the "Design" tab at the bottom of the Design View.

Right-click on "data_selector" and again select "New Source" to open the New Source Wizard. This time, select "Implementation Constraints File" and enter the file name "data_selector_elbert" as shown in Fig. 14.15. Click "Next" and finish the wizard.

This will open an empty text editor window where you need to type the following text:

```
# User Constraint File for data selector
# implementation on Elbert V2

# Push buttons
NET "A" LOC = P80;   # SW1
NET "B" LOC = P79;   # SW2
NET "SEL" LOC = P78; # SW3
```

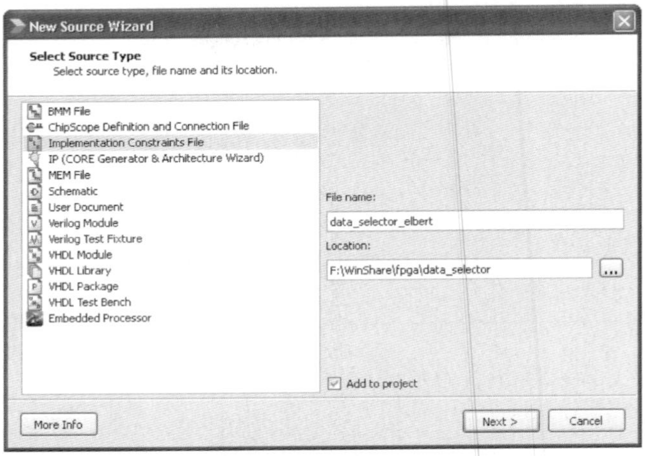

FIGURE 14.15 Creating an Implementation Constraints File.

```
# Internal pull-ups need to be enabled
NET "A" PULLUP;
NET "B" PULLUP;
NET "SEL" PULLUP;

# LED
NET "Q" LOC = P46;   # LED8
```

The lines that begin with a # are comment lines. That is, like the lines of program code starting with // in Arduino C take no part in the functioning of the program, the lines starting with # are not part of the configuration information, they are just to make it easier to see what is going on.

In the section that starts with "# Push Buttons" you can see the link between the IO Connector names on the schematic and the FPGA GPIO pins that are connected to the switches. So, SW1 is connected to P80, etc.

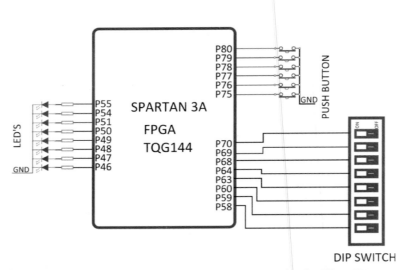

FIGURE 14.16 The switch and LED pin allocations for the Elbert V2.

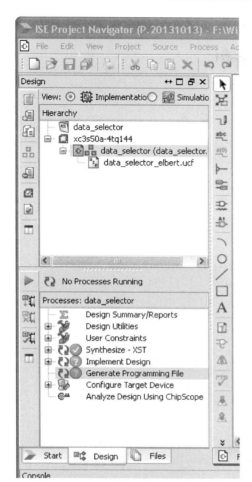

FIGURE 14.17 Generating the programming file.

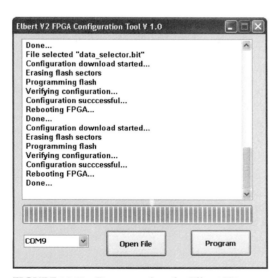

FIGURE 14.18 Programming the Elbert V2.

The connections from FPGA GPIO pins to the hardware provided on the Elbert V2 are all detailed in the Elbert user manual. Figure 14.16 taken from the manual with kind permission of Numata shows the pin allocations associated with the switches and LEDs.

The GPIO pins of the FPGA have configurable pull-up resistors and these are enabled in the implementation constraints file. The final line of the constraints file associates the output Q with pin P46.

Step 7: Generate the Programming File

You are now ready to generate the programming file to be downloaded onto the Elbert V2. So, select the data_selector entry in the hierarchy and a number of options for things to do will appear below it in the Processes section. One of those processes will be "Generate Programming File" (see Fig. 14.17). Right click on this option and select "Run."

If all is well, there will be lots of text appearing in the Console as the programming file is generated. If there are any errors this is where they will appear, so read the error message carefully and it should point to where the problem is.

Step 8: Program the Elbert V2

The end result of all this activity will be a file within the working directory of the project called data_selector.bit. It is this file that we need to transfer onto the Elbert V2 using the Elbert V2 Configuration Tool that you downloaded earlier.

Start up the program and then click on Open File. Navigate to the working directory for the project and select the file data_selector.bit. Change the COM port in the drop-down list to match the COM port allocated to the Elbert that you discovered earlier (see Fig. 14.5) and then press the Program button. After a while, you should see a reassuring message appear (Fig. 14.18).

The FPGA on the Elbert V2 is now configured to be the data selector.

Testing the Result

Initially, you should see the D8 LED on the Elbert V2 lit. If you press SW2 the LED will turn off. Release the button and LED 8 will turn on again. Pressing SW1 will have no effect. Now hold SW3 down and you will notice that SW2 no longer has any effect but SW1 does alter the LEDs state.

This is the data selector working how it should. The logic is a little confusing, because the inputs are effectively inverted as they are pulled to GND when you press the button.

Viewing the Technology Schematic and Floorplan

Although it is perfectly ok to just trust ISE to do all the work of laying out and connecting up the FPGA for us, it is fun, if not particularly useful to peek under the hood and see what exactly it did.

You can see some of this information using other tools that are accessible from the Project view.

If you expand the Synthesize XST line in the Processes, you will see an option "View Technology Schematic." If you run this, you will get to see the schematic generated by ISE that includes additional IO buffers (Fig. 14.19).

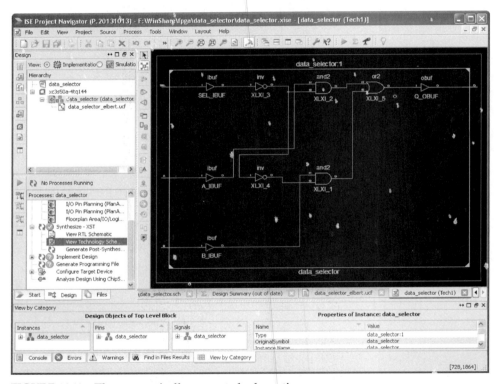

FIGURE 14.19 The automatically generated schematic.

You can actually see which parts of the FPGA silicon were used for your design if you select the menu option Tools->Planahead->Floorplan Area (Fig. 14.20).

14.6.2 Example 2: A 4-bit Ripple Counter

In the second example using the schematic approach, we move beyond simple combinational logic to a ripple counter, again taken from Chap. 12, where you will find it in Fig. 12.75. This is repeated here as Fig. 14.21 for convenience.

As you did with the selector, start by creating a new project. Give it the name "ripple_counter." You should find that when you run the New Project Wizard, this time, it remembers all the project settings from the last project.

Create a new Schematic source ("ripple_counter") and draw the schematic.

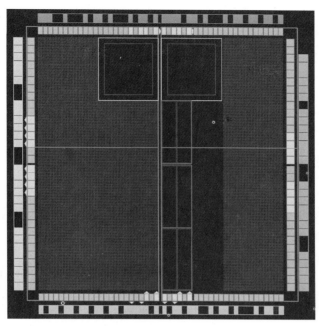

FIGURE 14.20   The FPGA floorplan.

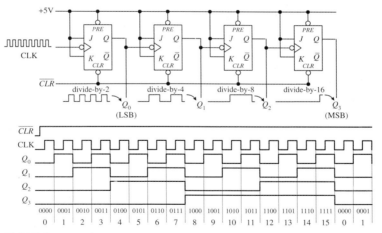

FIGURE 14.21   A 4-stage counter.

It can be hard to find the right symbols, so you will probably need to drop a selection of symbols onto the canvas before you find the right ones. Delete the ones you don't want by selecting then and then pressing the delete key. The symbol we used for each JK Flipflop was found in the category Flip_Flop and called fjkc. Add some IO Markers for CLK, CLR and Q0 to Q3. You will also need to add four VCC symbols ("General" category) to pull the J and K pins high. The end result of this should look like Fig. 14.22.

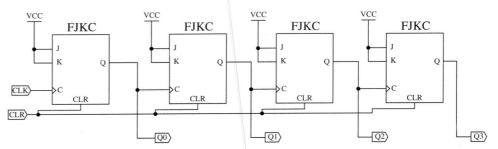

FIGURE 14.22   The ISE schematic.

You also need to create an implementation constraints file and place the following contents in it.

```
# User Constraint File for 4-bit ripple counter
# implementation on Elbert V2

NET "CLK" CLOCK_DEDICATED_ROUTE = FALSE;

# Push button switch 1 is connected to pin 80
NET "CLK" LOC = P80;    # SW1
NET "CLR" LOC = P79;    # SW2

# Internal pull-ups need to be enabled
NET "CLK" PULLUP;
NET "CLR" PULLUP;

# LEDs
NET "Q0" LOC = P55;     # LED1
NET "Q1" LOC = P54;     # LED2
NET "Q2" LOC = P51;     # LED3
NET "Q3" LOC = P50;     # LED4
```

The new first like specifies that although the CLK is a clock pin it does not need the specialized clock connectivity lines that the FPGA can provide, since we are just going to be driving the clock using the push buttons.

Generate the Programming file and then deploy it onto the Elbert V2 using the Configuration tool.

When you come to test the project using the button, you will need to keep SW2 depressed, as this is the CLR (clear button) and the input from the switch is inverted.

You will also notice that there is a fair bit of key bouncing from the push switches and the LEDs may skip past some of the binary numbers.

14.7   Verilog

Verilog is a hardware description language. Along with its rival language VHDL, it is what is most commonly used to program an FPGA.

You probably found that programming a FPGA using schematic is familiar and easy to understand, so why would you want to learn a complicated programming language to do the same thing? Well, the answer is that actually as designs become more and more complex, it can be easier to represent a design using a programming language than to draw it.

While it is feasible to use the schematic approach when designing some simple logic that could fit on a CPLD, as complexity increases, the problem of drawing all those gates and connecting them up gets impractical.

Verilog looks like a programming language and indeed you will find "if" statements, code blocks and other software-like constructions.

14.7.1   Modules

Software programmers will recognise a Verilog module as being very like a class in object-oriented programming. It defines a collection of logic with public and private properties that can be instantiated a number of times in your design.

For nonprogrammers, it is probably best to think of it as a sub-assembly of the design with defined connections to be able to wire it up to other modules.

A simple design may be all contained in a single module, but when things start to get a little complex, the design will become a load of modules that are then interconnected.

14.7.2   Wires, Registers, and Busses

What would be variables in a conventional programming language are, in Verilog, wires (connecting one thing to another) or registers (that store state and are therefore more like a programming variable). These refer to a single binary digit. Often, you don't want to work on a single bit and so you can group a number of bits into a bus and operate on the bus as a whole (sometimes also called a vector). This is rather like using a word of arbitrary length in a conventional programming language.

14.7.3   Parallel Execution

Because Verilog is describing hardware rather than software, there is an implicit parallelism in Verilog. If you have three counters in a design, all connected to different clocks, that is just fine. Each will do its own thing. It is not like using a microcontroller where there is a single thread of execution.

14.7.4   Number Format

A lot of the time, in Verilog, you will be dealing with a bus and it is convenient to assign values using numbers of any bit size in any radix. To accomplish this, Verilog uses a special number syntax. If you do not specify the number of bits and the radix, then the number is assumed to be decimal and unused bits are set to 0.

The number format starts with the number of bits, then there is an apostrophe, followed by a radix indicator (b, binary; h, hex; d, decimal) and then the number constant.

Here are some Verilog integer constants:

- 4'b1011—4-digit binary constant
- 8'hF2—8-bit hex constant

14.8   Describing Your FPGA Design in Verilog

In this section, you will work though using ISE to replicate the two earlier designs of a data selector and ripple counter in Verilog rather than using the schematic editor. You will then go on to look at some more complex designs where the modular and concise nature of using a HDL starts to pay dividends.

14.8.1   A Data Selector in Verilog

Rather than just looking at the Verilog code in isolation, let's combine it with learning how to use it in ISE.

The first step is to create a new project. This time, when the New Project Wizard appears (Fig. 14.23), give it the name "data_selector_verilog," change the drop-down list at the bottom (top-level source type) to be HDL, and click "Next" and then "Finish" at the summary screen.

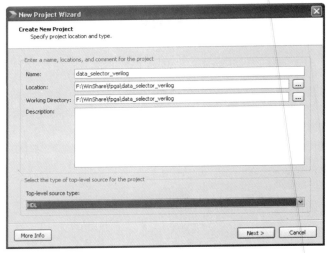

FIGURE 14.23   The New Project Wizard.

Now we need to create a new source file for the Verilog version of the data selector. So, right-click on the project and select the option "New Source." This will open the New Source Wizard (Fig. 14.24).

Select a source type of Verilog Module and give the source the name "data_selector" and then click "Next." This will allow you to define the inputs and outputs to the module (Fig. 14.25).

Use the wizard window to define three inputs (A, B, and SEL) and one output (Q). Then click "Next" and then after the summary screen click "Finish." The wizard will then generate a template file for your Verilog module using the information you entered in the wizard (Fig. 14.26).

At present, this module does not actually do anything. We will add that Verilog code to it shortly.

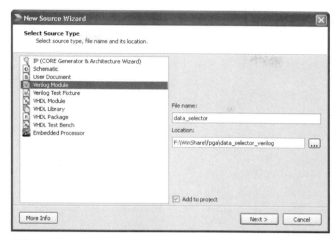

FIGURE 14.24   Creating a new Verilog source file.

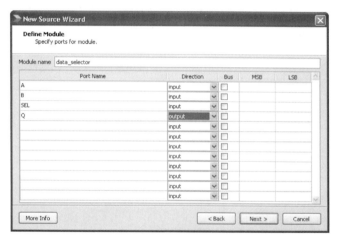

FIGURE 14.25   Defining inputs and outputs for the new Verilog source.

Let's analyze what has been generated by the Wizard. Here is the code that was generated.

```
module data_selector(
    input A,
    input B,
    input SEL,
    outputQ
    );

endmodule
```

The module starts with the "module" keyword and is followed by the name of the module. Inside the parentheses, the inputs and outputs to the module. The word "endmodule" marks the end of the module definition.

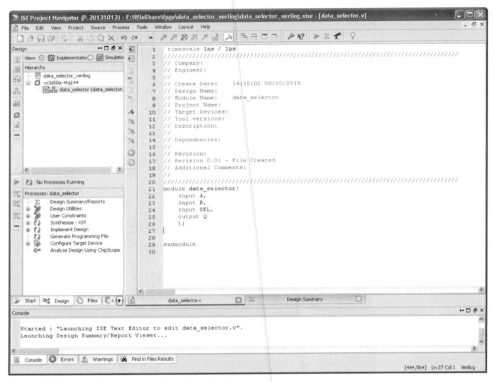

FIGURE 14.26   The generated module code.

Modify the text so that it appears as below. Note that the additions are marked in bold.

```
module data_selector(
    input A,
    input B,
    input SEL,
    output reg Q
    );

always @(A or B or SEL)
begin
  if (SEL)
        Q = A;
  else
        Q = B;
end

endmodule
```

The first change is the addition of the word "reg" to the output definition for Q. This indicates that Q is a register and can therefore be modified.

The other addition is the "always" block. Immediately after "always" is the "sensitivity" list that follows "@." This specifies the signals (separated by the word "or") to which the "always" block is sensitive. That is the code between "begin" and "end" comes into play. It is very easy to think of this code as if it were a programming language rather than a hardware definition language.

If SEL is 1 then Q will be assigned to whatever the state of A is. Otherwise Q will be set to the value at the B input. This is exactly what the selector should do.

That's all there is to the Verilog, however you still need an implementation constraints file if you want to try out the example on the Elbert V2. The one that you created for the schematic version of this project will work just fine. You can copy the implementation constraints file from the other project by right-clicking on the project name and clicking "Add Copy of Source." This will allow you to take a copy of "data_selector_elbert.ucf."

Build the project and then install it on the Elbert in the same way as you did for the schematic project. The project should work in exactly the same way.

14.8.2 A Ripple Counter in Verilog

The ripple counter schematic project can also be implemented in Verilog. This time, when you create the new project (you could call it "ripple_counter_verilog,"""), add the inputs and outputs as shown in Fig. 14.27.

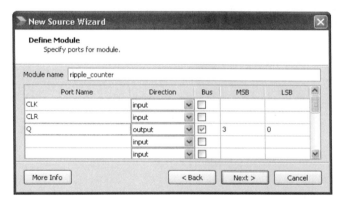

FIGURE 14.27   **Defining the inputs and outputs for the ripple counter.**

The output Q is defined as being a bus by checking the Bus checkbox. The MSB column indicates its most significant bit number (in this case 3) and the LSB of 0 is entered in the LSB column.

Finish of the wizard and the generated code will start like this:

```
module ripple_counter(
    input CLK,
    input CLR,
    output [3:0] Q
    );
```

You now need to add the counting logic for the counter, so edit the code to be:

```
module ripple_counter(
    input CLK,
    input CLR,
    output reg [3:0] Q
    );
```

```
always @(posedge CLK, posedge CLR)
begin
  if (CLR)
       Q <= 0;
  elseif (CLK)
       Q <= Q + 1;
end

endmodule
```

The added code is shown in bold. The sensitivity list in the always block includes positive edges of either the CLK or CLR signals. If either of these happen, then the code between begin and end comes into play. This simply says that if CLR goes high, then the count Q gets set back to 0 and if CLK is high then 1 is added to Q.

Note that in this case, since 0 and 1 are the same in any radix, we have not specified the radix or number of bits in the number constants.

You now need to add an implementation constraints file for the project that looks like this:

```
# User Constraint File for 4-bit ripple counter implementation
on Elbert V2
NET "CLK" CLOCK_DEDICATED_ROUTE = FALSE;

# Push button switch 1 is connected to pin 80
NET "CLK" LOC = P80;    # SW1
NET "CLR" LOC = P79;    # SW2

# Internal pull-ups need to be enabled
NET "CLK" PULLUP;
NET "CLR" PULLUP;

# LEDs
NET "Q[0]" LOC = P55;    # LED1
NET "Q[1]" LOC = P54;    # LED2
NET "Q[2]" LOC = P51;    # LED3
NET "Q[3]" LOC = P50;    # LED4
```

This is very similar to the one for the schematic-based counter, but in this case the separate bits of the Q bus are linked to the LEDs using a square bracket notation to indicate the bit linked to a particular LED.

Generate the binary file and install it on your Elbert V2 board and you should have something that behaves just like the schematic version.

14.9 Modular Design

When designing a complex system for a FPGA there is nothing to stop you putting all your Verilog code into one module. However, by splitting things up, it firstly makes it easier for others to understand what you have done as they can assume that the component modules perform the role they are supposed to and therefore see a bigger picture of how all the modules work together before getting into the nitty-gritty of how each one works.

Breaking things up into a number of modules also makes it a log easier to take a module that you used in one project and use it in another, or to share it with someone else to use in their project.

When you create a project with more than one module, you will always have a top-level module. This is the module that brings all the sub-modules together and also the module that will have an implementation constraints file associated with it to map the IO pins of the FPGA to the signals in the design.

14.9.1 Counter/Decoder Example

In this example, you will build on the Verilog version of the counter module and add a 7-segment LED decoder module, so that it can count in decimal on one of the 7-segment LEDS on the Elbert 2 board.

See Sec. 13.3.2 for information on 7-segment decoding. The basic idea is that a four-digit binary input will be decoded into 7 segment bits for a segment display for the decimal values 0 to 9.

Figure 14.28 shows the relationship between the three modules that will be defined for this project.

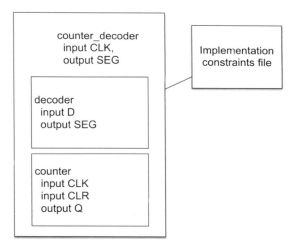

FIGURE 14.28   Counter/decoder modules.

In this case, the top-level module will use one counter and one decoder module and combine them into a new module (counter_decoder).

Let's start with the counter module, because we have already made this module once in the project ripple_counter_verilog. So, from the file menu use the option "Add copy of source" and go and find "ripple_counter.v" in the other project and add it to this project. It should look like this:

```
module ripple_counter(
    input CLK,
    input CLR,
    output reg [3:0] Q
    );
```

```
always @(posedge CLK, posedge CLR)
begin
  if (CLR)
        Q <= 0;
  elseif (CLK)
        Q <= Q + 1;
end

endmodule
```

Next, you are going to create a new Verilog source called "decoder_7_seg.v." It should have a 4-bit bus input called D (the numeric digit 0 to 9 in 4 bits) and an 8-bit bus output SEG for each of the 7 segments of the display (plus 1 for the decimal point). The module it contains should look like this:

```
module decoder_7_seg(
    input [3:0] D,
    output reg [7:0] SEG
    );

always @(D)
begin
    case(D)
        0: SEG <= 8'b00000011;
        1: SEG <= 8'b10011111;
        2: SEG <= 8'b00100101;
        3: SEG <= 8'b00001101;
        4: SEG <= 8'b10011001;
        5: SEG <= 8'b01001001;
        6: SEG <= 8'b01000001;
        7: SEG <= 8'b00011111;
        8: SEG <= 8'b00000001;
        9: SEG <= 8'b00001001;
        default: SEG <= 8'b11111111;
    endcase
end

endmodule
```

This time the always block just has a sensitivity list of the input data (D). The "case" statement will be familiar to C and Java programmers as being a "switch" statement. It is a short-hand way of chaining together a whole load of "if" statements. The "case" command takes a parameter (in this case D) and if value of D is 0, it sets the bit pattern for SEG to 8'b00000011. If D is 1 then SEG is set to the second bit pattern down, and so on. Note that the segment bits are inverted. A 0 means that that segment will be lit—that's just how the Elbert V2 is wired up.

Both the decoder and counter modules are both going to be used by the top-level module that we will call "counter_decoder." Create a new Verilog source file called counter_decoder.v with the single input CLK and the 8-bit bus output SEG that will be connected to the segment LEDs. Edit the generated code so that the module appears as below:

```
module counter_decoder(
    input CLK,
    output [7:0] SEG
);

wire [3:0] data;
wire clear = 0;

decoder_7_seg decoder(.D (data), .SEG (SEG));
ripple_counter counter(.Q (data), .CLK (CLK),
 .CLR (clear));

endmodule
```

This is a pretty sparse module, as most of the work is taking place in the two modules that it uses.

To link the data output from the counter (D) to the data input to the 7-segment decoder (a different D) a wire bus is defined called "data." Although the counter has a CLR (clear) input, this is not going to be used in the counter_decoder module and so a second wire ("clear") is defined and its value set to 0.

Next comes the part where the two sub-modules are "instantiated" and their outputs and inputs coupled.

One way to think of a module is as the name for a logic gate (perhaps an AND gate). So instantiating an AND gate would mean adding one to a schematic design. You might add (instantiate) several AND gates onto the schematic. In this case, we are instantiating first a decoder_7_seg module and then a ripple_counter module.

Looking at the line starting "decoder_7_seg": the syntax for instantiating a module is to first specify the module name ("decoder_7_seg" and then the name of the instance ("decoder"). So if your design needed more than one "decoder_7_seg" then you could call them "decoder_1", "decoder_2" and so on.

After the name of the instance comes the bit that allows you to associate the inputs and outputs of the instance with signals inside the containing module (in this case "counter_decoder"). This is contained in parentheses and is a mechanism that software programmers would consider to be like using named parameters. So, "D (data) means that the D input to the decoder should be linked to the wire bus called "data" in the module counter_decoder. Similarly, "SEG (SEG)" links the SEG output of the decoder to the SEG output of "counter_decoder" that will in turn be linked to the segment LEDs on the Elbert V2.

A "ripple_counter" instance is created in the same way, linking its Q output to "data," passing through the CLK signal and setting the CLR input to "ripple_counter" to be "clear."

Before you can build this example, you need to tell ISE which module is the top-level module. The top-level module is marked by an icon that looks like a triangle of little squares (Fig. 14.29).

You can set a module to be the top-level module by right clicking on it in the Hierarchy View and selecting the option "Set as Top Module."

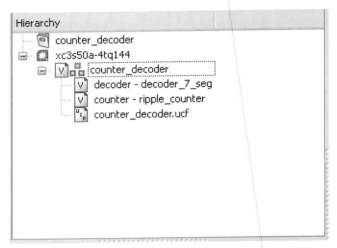

FIGURE 14.29 **The Hierarchy View indicating the top-level module.**

You also need an implementation constraints file that associates SW1 on the Elbert V2 with CLK and the segments of the LED with the SEG bus.

The pinouts associated with the 7-segment display are shown in Fig. 14.30 and the constraints file that you will need to create is listed below.

```
NET "CLK" CLOCK_DEDICATED_ROUTE = FALSE;

# Push buttons
NET "CLK" LOC = P80;     # SW1
NET "CLK" PULLUP;

# 7-segments
NET "SEG[7]" LOC = P117;
NET "SEG[6]" LOC = P116;
NET "SEG[5]" LOC = P115;
NET "SEG[4]" LOC = P113;
NET "SEG[3]" LOC = P112;
NET "SEG[2]" LOC = P111;
NET "SEG[1]" LOC = P110;
NET "SEG[0]" LOC = P114;
```

Build the project and install it on your Elbert and you should see all three digits counting as one when you press SW1. All three digits are counting, because we are not controlling the common anodes of the display digits. This is something you will remedy in the next example.

14.9.2 Multiplexed 7-Segment Counter Example

This example makes the three-digit 7-segment display of the Elbert V2 count upward from 0 to 999 incrementing the three digit number displayed once per second.

The three-digit 7-segment LED display on the Elbert V2 is multiplexed. Referring back to Fig. 14.30, you can see that three PNP transistors are used to enable the three

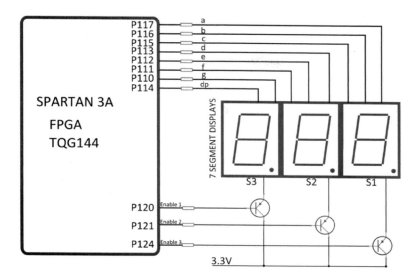

FIGURE 14.30　The Elbert V2 7-segment display.

anodes of the display and the cathode connections to the segments for the three displays are controlled by eight FPGA output pins (one for the decimal point). See the section "Multiplexed LED Displays" in Chap. 12 for background information on multiplexing LED displays.

To display a different number on each digit, it is necessary to fool the eye by turning one digit on (and the others off), setting the segment pattern to the number you want for that digit, then turning that digit off and enabling the next digit, resetting the segment pattern and so on.

In this example you will reuse the "segment_decoder" module that you created earlier and also create two new modules. The modules used in the project, which is called "seconds_counter," are:

- second_counter.v: The top-level module
- multiplexed_7_seg_display.v: The multiplexed display driver logic
- decoder_7_segment.v: The decimal digit to 7 segment decoder

Starting at the bottom of the modules, the "decoder_7_segment" module is exactly as described in Sec. 14.9.1.

The "multiplexed_7_seg_display" module makes use of the decoder_7_segment module.

```
module multiplexed_7_seg_display(
    input CLK,
    input [3:0] units, tens, hundreds,
    output [7:0] SEG,
    output reg [2:0] DIGIT
    );

reg [3:0] digit_data;
reg [2:0] digit_posn;
reg [23:0] prescaler;
```

```
decoder_7_seg decoder(.SEG      (SEG), .D (digit_data));

always @(posedge CLK)
begin
  prescaler<= prescaler + 1;
  if (prescaler == 12000) // 1 kHz
  begin
        prescaler<= 0;
        digit_posn<= digit_posn + 1;
        if (digit_posn == 0)
        begin
              digit_data<= units;
              DIGIT <= 3'b110;
        end
        if (digit_posn == 2'd1)
        begin
              digit_data<= tens;
              DIGIT <= 3'b101;
        end
        if (digit_posn == 2'd2)
        begin
              digit_data<= hundreds;
              DIGIT <= 3'b011;
        end
        if (digit_posn == 2'd3)
        begin
              digit_posn<= 0;
              DIGIT <= 3'b111;
        end
  end
end

endmodule
```

The module has a clock input (CLK) that will be used to control the switching between one digit and another. CLK will be connected to the 12-MHz clock that the Elbert V2 provides on pin P129.

There are also three 4-bit inputs—"units," "tens," and "hundreds"—that will contain the three digits to be displayed. Note how these three inputs can just follow the initial input declaration without the need to repeat the bus size for each of the inputs.

The two outputs of this module are the segment driver pins (SEG) and the digit driver pins (DIGIT).

Three registers are needed for the mechanics of the multiplexing:

- digit_data: Provides the link to the data input to the 7-segment decoder.

- digit_posn: Digit position; this cycles round from 0 to 2 to indicate which of the three digits is active during the refresh cycle.

- Prescaler: This counter is used to divide the 12-MHz clock down to a 100-Hz refresh clock signal.

An instance of the 7-segment decoder is created with the line:

```
decoder_7_seg decoder(.SEG (SEG), .D (digit_data));
```

This passes through the segment pins SEG from this module to the 7-segment decoder and links the digit_data register to the D input of the 7-segment decoder.

The "always" block is sensitive to the positive edge of the clock and pre-scales the 12-MHz clock by using the "if" to only actually do something when the prescaler gets to 12000 (decimal). When this happens, the prescaler is reset and then the business of refreshing the next digit of the display takes place.

This involves first incrementing the digit_posn register and then using a series of "if" statements to set the digit_data to either "units," "tens," or "hundreds" depending on the digit position. The DIGIT control pins are also set for the digit currently being displayed. Note that the digit control is active low.

The top-level module (seconds_counter) has one input (CLK) that uses the same 12-MHz clock as the multiplexed_7_seg_display module. The two outputs SEG and DIGIT will be linked to the FPGA pins that drive the segments and digits on the Elbert V2 multiplexed display.

```verilog
module second_counter(
    input CLK,
    output [7:0] SEG,
    output [2:0] DIGIT
    );

reg [3:0] units, tens, hundreds;
reg [23:0] prescaler;

multiplexed_7_seg_display display(.CLK (CLK),
        .units (units), .tens (tens), .hundreds (hundreds),
        .SEG (SEG), .DIGIT (DIGIT));

always @(posedge CLK)
begin
  prescaler<= prescaler + 1;
  if (prescaler == 24'd12000000)
  begin
        prescaler<= 0;
        units<= units + 1;
        if (units == 9)
        begin
              units<= 0;
              tens<= tens + 1;
        end
        if (tens == 9)
        begin
              tens<= 0;
              hundreds<= hundreds + 1;
        end
        if (hundreds == 9)
        begin
              hundreds<= 0;
        end
  end
end

endmodule
```

This module uses three registers to contain the units, tens and hundreds digits as well as a prescaler to divide the 12-MHz clock signal down to 1 Hz.

The "always" block is sensitive to the positive edge of CLK and divides the clock by 12,000,000 (decimal) to tick over the units, tens and hundreds registers. Once the units register has reached 9, it increments then tens register and then resets to 0.

If you want to try out this example and install it on the Elbert V2, load the project "seconds_counter" from the book downloads, build it and deploy it.

14.9.3 Parameterized Modules

Some modules like the ripple_counter module, that you created earlier, would benefit from parameterization. That is instead of always being the same size, it would be able to specify their size when instantiating them.

As an example of using parameters you can modify the ripple_counter module that you created earlier. The modified code is shown below:

```
module ripple_counter#(parameter SIZE=4) (
    input CLK,
    input CLR,
    output reg [SIZE-1:0] Q
    );

always @(posedge CLK, posedge CLR)
begin
  if (CLR)
        Q <= 0;
  else if (CLK)
        Q <= Q + 1;
end

endmodule
```

The addition of the size parameter is shown in bold. After the parameter name (SIZE) a default value is specified, so that if the parameter is not specified when the module is instantiated, then it will still have a size.

The parameter can then be used anywhere within the module, so in this case, the MSB of the Q output register is set to be SIZE-1.

When it comes to instantiating a parameterized module, you can specify the parameter like this:

```
ripple_counter#(4) counter(.Q (data), .CLK (CLK), .CLR (clear));
```

14.10   Simulation

A board like the Elbert 2 allows you to test out your designs using its switches and LEDs to exercise the FPGA and see how it is behaving. You could also use the Elbert's GPIO pins and test the design using a logic analyzer.

However, ISE includes a simulator that allows you to write a Verilog test fixture and then use it to exercise a module and make sure that its doing what it should, before going anywhere near real hardware.

As an example you can add a text fixture to the Verilog ripple counter project that you created earlier.

To do this, re-open the project ripple_counter_project and then add a new source to it. But, this time, when the New Source wizard opens, select the source type of Verilog Test Fixture and name it "counter_tester" (Fig. 14.31).

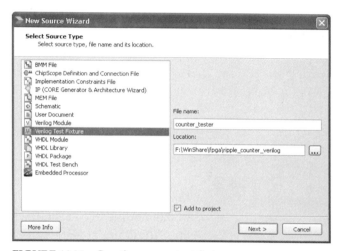

FIGURE 14.31   Creating a new test fixture.

Change the contents of counter_tester.v so that it appears as below:

```
module counter_tester;

// Inputs
reg CLK;
reg CLR;

// Outputs
wire [3:0] Q;

// Instantiate the Unit Under Test (UUT)
ripple_counteruut (
    .CLK(CLK),
    .CLR(CLR),
    .Q(Q)
);


initial
begin
  // Initialize Inputs
  CLK = 0;
  CLR = 0;
```

```
// Wait 100 ns for global reset to finish
#100;
end


always
begin
  #10
  CLK =!CLK;
end


always
begin
  #320
  CLR = 1;
  #1
  CLR = 0;
end


endmodule
```

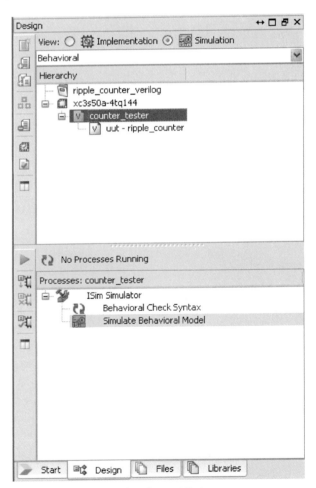

FIGURE 14.32   Running the test fixture.

The new source wizard creates a lot of the code for us. This includes some code to create an instance of the module to be tested, which is given the name "uut" (unit under test).

The new parts are highlighted in bold.

The "initial" block is run just once as the FPGA resets and just sets both CLK and CLR low. The line #100 instructs the test code to delay for 100 ns, to allow everything to stabilize.

There are then two "always" blocks. The first drives the CLK pin by first delaying for 10 ns then inverting CLK.

The second block drives the CLR pin on a much slower clock. It delays for 320 ns then provides a 1-ns pulse on the CLR pin.

To run the test fixture, click on the Simulation radio button in the Hierarchy View (Fig. 14.32).

Notice that when you select "counter_tester" in the Hierarchy View, the process "Simulate Behavioral Model" appears in the Processes area of the window. Right-click on "Simulate Behavioral Model" and run it. This will result in the ISim module opening and displaying the results of the simulation (Fig. 14.33).

To see the detail, you need to click on Q[3:0] in the name column so that the individual lines of the bus are expanded. You will also need to click on the zoom-out icon about 10 times to get to the right time

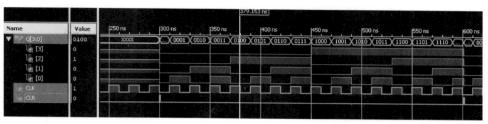

FIGURE 14.33 Simulation results for the counter.

scale and then use the horizontal scrollbar at the bottom of the simulation area to pan over to an interesting area of the simulation results.

14.11 VHDL

Verilog is not the only HDL. The FPGA community is almost evenly split between Verilog and VHDL. In fact, ISE can use both Verilog and VHDL source code and you can even mix them in a project.

VHDL tends to be more popular in some industry sectors such as aerospace and defence, as it has what programmers call "strong typing" that makes the code more rigorous and allows you to catch more potential problems at the time the project is being built rather than during simulation.

Most software programmers will tell you that once you have programmed in one language, learning a second is much easier to learn as you can just map things you have learnt in the new language to construct in the language you already know. The basic concepts are the same in both.

CHAPTER 15

Motors

Perhaps one of the most entertaining things to do with electronics is make some mechanical device move. Three very popular devices used to "make things move" include dc motors, RC servos, and stepper motors.

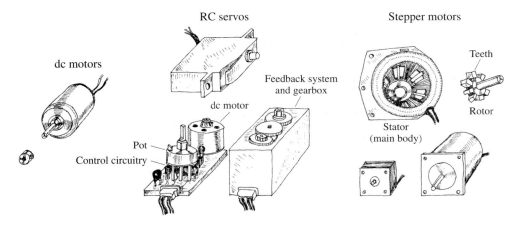

FIGURE 15.1

15.1 DC Continuous Motors

A *dc motor* is a simple two-lead, electrically controlled device that comes with a rotary shaft on which wheels, gears, propellers, etc., can be mounted. A dc motor generates a considerable amount of revolutions per minute (rpm) for its size and can be made to rotate clockwise or counterclockwise by reversing the polarity applied to the leads. At low speeds, dc motors provide little torque and minimal position control, making them impractical for pointlike position-control applications.

Generally, dc motors are available in many different shapes and sizes. Most dc motors provide rotational speeds anywhere between 3000 and 8000 rpm at a specific operating voltage typically set between 1.5 and 24 V. The operating voltage provided by the manufacturer tells you at what voltage the motor runs most efficiently. Now, the actual voltage applied to a motor can be made slightly lower to make the motor slower or can be elevated to make the motor faster. However, when the applied voltage drops to below around 50 percent of the specified operating voltage, the motor usually will cease to rotate. Conversely, if the applied voltage exceeds the operating voltage by around 30 percent, there is a chance that the motor will overheat and become damaged. In practice, as you will see in a second, the speed of a dc motor is most efficiently controlled by means of pulse-width modulation, whereby the motor is rapidly turned on and off. The width of the applied pulse, as well as the period between pulses, controls the speed of the motor. Also, it is worth noting that a freely running motor (no load) may draw little current (power). However, if a load is applied, the amount of current drawn by the motor's inner coils goes up immensely (up to 1000 percent or more). Manufacturers usually will provide what is called a *stall current* rating for their motors. This rating specifies the amount of current drawn at the moment the motor stalls. If your motor's stall current rating is not listed, it is possible to determine it by using an ammeter; slowly apply a force to the motor's shaft, and note the current level at the point when the motor stalls. Another specification given to dc motors is a torque rating. This rating represents the amount of force the motor can exert on a load. A motor with a high torque rating will exert a larger force on a load placed at a tangent to its rotational arm than a motor with a lower torque rating. The torque rating of a motor is usually given in lb/ft, g/cm, or oz/in.

15.2   Speed Control of DC Motors

Bad Designs

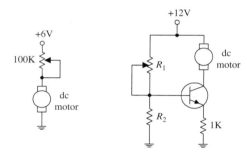

FIGURE 15.2

Better Designs

UJT/SCR Control Circuit

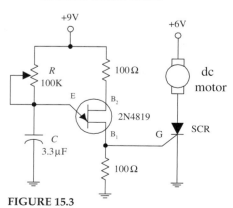

FIGURE 15.3

A seemingly obvious approach to control the speed of a dc motor would be simply to limit the current flow by using a potentiometer, as shown in the circuit to the left in the figure. According to Ohm's law, as the resistance of the pot increases, the current decreases, and the motor will slow down. However, using a pot to control the current flow is inefficient. As the pot's resistance increases, the amount of current energy that must be converted into heat increases. Producing heat in order to slow a motor down is not good—it consumes supply power and may lead to potentiometer meltdown. Another seemingly good but inefficient approach to control the speed of a motor is to use a transistor amplifier arrangement like the one shown to the right in the figure. However, again, there is a problem. As the collector-to-emitter resistance increases with varying base voltage/current, the transistor must dissipate a considerable amount of heat. This can lead to transistor meltdown.

In order to conserve energy and prevent component meltdown, an approach similar to what was used in switching power supplies is used to control the speed of the motor. This approach involves sending the motor short pulses of current. By varying the width and frequency of the applied pulses, the speed of the motor can be controlled. Controlling a motor's speed in this manner prevents any components from experiencing continuous current stress. Figure 13.3 shows three simple circuits used to provide the desired motor-control pulses.

In the first circuit, a UJT relaxation oscillator generates a series of pulses that drives an SCR on and off. To vary the speed of the motor, the UJT's oscillatory frequency is adjusted by changing the RC time constant.

CMOS/MOSFET Control Circuit

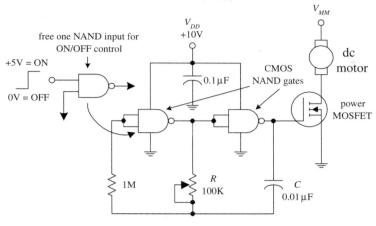

In the second circuit, a pair of NAND gates make up the relaxation oscillator section, while an enhancement-type power MOSFET is used to drive the motor. Like the preceding circuit, the speed of the motor is controlled by the oscillator's *RC* time constant. Notice that if one of the input leads of the left NAND gate is pulled out, it is possible to create an extra terminal that can be used to provide on/off controls that can be interfaced with CMOS logic circuits.

555 Timer/MOSFET Control Circuit

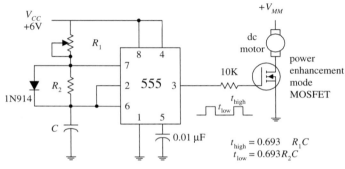

$$t_{high} = 0.693 \; R_1 C$$
$$t_{low} = 0.693 R_2 C$$

The third circuit is a 555 timer that is used to generate pulses that drive a power MOSFET. By inserting a diode between pins 7 and 6, as shown, the 555 is placed into low-duty cycle operation. R_1, R_2, and C set the frequency and on/off duration of the output pulses. The formulas accompanying the diagram provide the details.

A microcontroller-based dc control circuit with speed control is found in Chap. 13.

In many applications the 555 timer in this final circuit can be replaced by a microcontroller with a PWM output driving the MOSFET.

FIGURE 15.3   (*Continued*)

15.3   Directional Control of DC Motors

To control the direction of a motor, the polarity applied to the motor's leads must be reversed. A simple manual-control approach is to use a DPDT switch (see leftmost circuit in Fig. 15.4). Alternately, a transistor-driven DPDT relay can be used (see middle circuit). If you do not like relays, you can use a push-pull transistor circuit (see leftmost circuit). This circuit uses a complementary pair of transistors (similar betas and power rating)—one is an *npn* power Darlington, and the other is a *pnp* power Darlington. When a high voltage (e.g., +5 V) is applied to the input, the upper transistor (*npn*) conducts, allowing current to pass from the positive supply through the motor and into ground. If a low voltage (0 V) is applied to the input, the lower transistor (*pnp*) conducts, allowing current to pass through the motor from ground into the negative supply terminal.

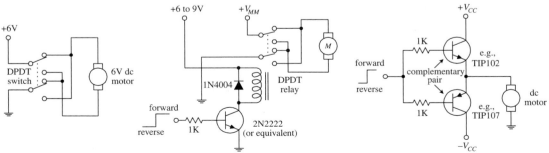

FIGURE 15.4

Another very popular circuit used to control the direction of a motor (as well as the speed) is the H-bridge. Figure 15.5 shows two simple versions of the H-bridge circuit. The left H-bridge circuit is constructed with bipolar transistors, whereas the right H-bridge circuit is constructed from MOSFETs. To make the motor rotate in the forward direction, a high (+5-V) signal is applied to the forward input, while no signal is applied to the reverse input (applying a voltage to both inputs at the same time is not allowed). The speed of the motor is controlled by pulse-width modulating the input signal. Here is a description of how the bipolar H-bridge works: When a high voltage is applied to Q_3's base, Q_3 conducts, which in turn allows the *pnp* transistor Q_2 to conduct. Current then flows from the positive supply terminal through the motor in the right-to-left direction (call it the *forward direction* if you like). To reverse the motor's direction, the high voltage signal is removed from Q_3's base and placed on Q_4's base. This sets Q_4 and Q_1 into conduction, allowing current to pass through the motor in the opposite direction. The MOSFET H-bridge works in a similar manner.

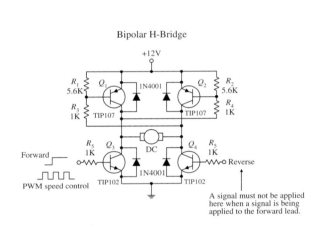

Bipolar H-Bridge

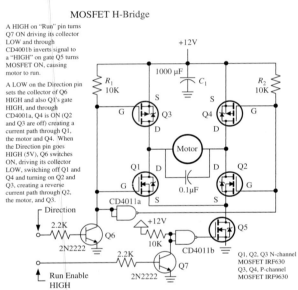

MOSFET H-Bridge

A HIGH on "Run" pin turns Q7 ON driving its collector LOW and through CD4001b inverts signal to a "HIGH" on gate Q5 turns MOSFET ON, causing motor to run.

A LOW on the Direction pin sets the collector of Q6 HIGH and also Q1's gate HIGH, and through CD4001a, Q4 is ON (Q2 and Q3 are off) creating a current path through Q1, the motor and Q4. When the Direction pin goes HIGH (5V), Q6 switches ON, driving its collector LOW, switching off Q1 and Q4 and turning on Q2 and Q3, creating a reverse current path through Q2, the motor, and Q3.

FIGURE 15.5

Now, it is possible to construct these H-bridge circuits from scratch, but it is far easier and usually cheaper to buy a motor-driven IC. For example, National Semiconductor's LMD18200 motor-driver IC is a high-current, easy-to-use H-bridge chip that has a rating of 3 A and 12 to 55 V. This chip is TTL and CMOS compatible and includes clamping diodes, shorted load protection, and a thermal warning interrupt output lead. The L293D (Unitrode) is another popular motor-driver IC. This chip is very easy to use and is cheaper than the LMD18200, but it cannot handle as much current and does not provide as many additional features. There are many other motor-driver ICs out there, as well as a number of prefab motor-diver boards that are capable of driving a number of motors. Check the electronics catalogs and Internet to see what is available.

15.4 RC Servos

Remote control (RC) servos, unlike dc motors, are motorlike devices designed specifically for pointerlike position-control applications. An RC servo uses an external pulse-width-modulated (PWM) signal to control the position of its shaft to within a small fraction of its maximum range of rotation. To alter the position of the shaft, the pulse width of the modulated signal is varied. The amount of angular rotation of an RC servo's shaft is limited to around 180 or 210° depending on the specific brand of servo. These devices can provide a significant amount of low-speed torque (due to an internal gearing system) and provide moderate full-swing displacement switching speeds. RC servos frequently are used to control steering in model cars, boats, and airplanes. They are also used commonly in robotics as well as in many sensor-positioning applications.

The standard RC servo looks like a simple box with a drive shaft and three wires coming out of it. The three wires consist of a power supply wire (usually black), a ground wire (usually red), and the shaft-positioning control wire (color varies based on manufacturer). Within the box there is a dc motor, a feedback device, and a control circuit. The feedback device usually consists of a potentiometer whose control dial is mechanically linked to the motor through a series of gears. When the motor is rotated, the potentiometer's control dial is rotated. The shaft of the motor is usually limited to a rotation of 180° (or 210°)—a result of the pot not being able to rotate indefinitely. The potentiometer acts as a position-monitoring device that tells the control circuit (by means of its resistance) exactly how far the shaft has been rotated. The control circuit uses this resistance, along with a pulse-width-modulated input control signal, to drive the motor a specific number of degrees and then hold. (The amount of holding torque varies from servo to servo.) The width of the input signal determines how far the servo's shaft will be rotated.

A typical servo-control signal and shaft-position response

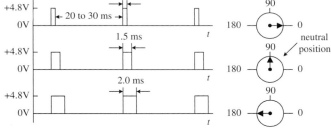

Simple servo driver

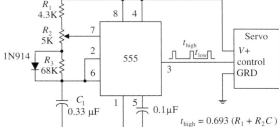

FIGURE 15.6

By convention, when the pulse width is set to 1.5 ms, the servo rotates its shaft to neutral position (e.g., 90° if the servo is constrained within a 0 to 180° range). To rotate the shaft a certain number of degrees from neutral position, the pulse width of the control signal is varied. To make the shaft go counterclockwise from neutral, a pulse wider than 1.5 ms is applied to the control input. Conversely, to make the shaft go clockwise from neutral, a pulse narrower than 1.5 ms is applied (see Fig. 15.6). Knowing exactly how much wider or narrower to make the pulse to achieve exact angular displacements depends largely on what brand of servo you are using. For example, one brand of servo may provide maximum counterclockwise rotation at 1 ms and maximum clockwise rotation at 2 ms, whereas another brand of servo may provide maximum counterclockwise rotation at 1.25 ms and maximum clockwise rotation at 1.75 ms. The supply voltage used to power servos is commonly 4.8 V but may be 6.0 V or so depending on the specific brand of servo. Unlike the supply voltage, the supply current drawn by a servo varies greatly, depending on servo's power output.

A simple 555 timer circuit like the one shown in Fig. 15.6 can be used to generate the servo control signal. In this circuit, R_2 acts as the pulse-width control. Servos also can be controlled by a microprocessor or microcontroller. See Chap. 13 for two microcontroller-based servo control circuits.

Now, when controlling servos within model airplanes, an initial control signal (generated by varying position-control potentiometers) is first sent to a radiowave modulator circuit that encodes the control signal within a carrier wave. This carrier wave is then radiated off as a radiowave by an antenna. The radiowave, in turn, is then transmitted to the model's receiver circuit. The receiver circuit recovers the initial control signal by demodulating the carrier. After that, the control signal is sent to the designated servo within the model. If there is more than one servo per model, more channels are required. For example, most RC airplanes require a four-channel radio set; one channel is used to control the ailerons, another channel controls the elevator, another controls the rudder, and another controls the throttle. More complex models may use five or six channels to control additional features such as flaps and retractable landing gear. The FCC sets aside 50 frequencies in the 72-MHz band (channels 11–60) dedicated to aircraft use only. No license is needed to operate these radios. However, with an amateur (ham) radio operator's license, it is possible to use a radio within the 50-MHz band. Also, there are frequencies set aside within the 27-MHz band that are legal for any kind of model use (surface or air). If you are interested in radio-controlled RC servos, a good starting point is to check out an RC model hobby shop. These shops carry a number of transmitter and receiver sets, along with the servos.

As a final note, with a bit of rewiring, a servo can be converted into a drive motor with unconstrained rotation. A simple way to modify the servo is to break the feedback loop. This involves removing the three-lead potentiometer (and unlinking the gear system so that it can rotate 360°) and replacing it with a pair of voltage-divider resistors (the output of the voltage divider replaces the variable terminal of the potentiometer). The voltage divider is used to convince the servo control circuit that the servo is in neutral position. The exact values of the resistors needed to set the servo in neutral position can be determined by using the old potentiometer and an ohmmeter. Now, to turn the motor clockwise, a pulse wider than 1.5 ms is applied to the control input. As long as the control signal is in place, the motor will keep turning and not stop—you have removed the feedback system. To turn the motor counterclockwise, a pulse narrower than 1.5 ms is applied to the control input.

15.5 Stepper Motors

Stepper motors, or *steppers,* are digitally controlled brushless motors that rotate a specific number of degrees (a step) every time a clock pulse is applied to a special translator circuit that is used to control the stepper. The number of degrees per step (resolution) for a given stepper motor can be as small as 0.72° per step or as large as 90° per step. Common general-purpose stepper resolutions are 15 and 30° per step. Unlike RC servos, steppers can rotate a full 360° and can be made to rotate in a continuous manner like a dc motor (but with a lower maximum speed) with the help of proper digital control circuitry. Unlike dc motors, steppers provide a large amount of torque at low speeds, making them suitable in applications where low-speed and high-precision position control is needed. For example, they are used in printers to control paper feed and are used to help a telescope track stars. Steppers are also found in plotter- and sensor-positioning applications. The list goes on. To give you a basic idea of how a stepper works, take a look at Fig. 15.7.

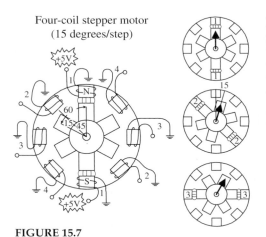

Four-coil stepper motor
(15 degrees/step)

FIGURE 15.7

Here is a simple model depicting a 15° per step variable-reluctance stepper. The stationary section of the motor, called the *stator*, has eight poles that are spaced 45° apart. The moving section of the motor, called the *rotor*, is made from a ferromagnetic material (a material that is attracted to magnetic fields) that has six teeth spaced 60° apart. To make the rotor turn one step, current is applied, at the same time, through two opposing pole pairs, or coil pairs. The applied current causes the opposing pair of poles to become magnetized. This in turn causes the rotor's teeth to align with the poles, as shown in the figure. To make the rotor rotate 15° clockwise from this position, the current through coil pair 1 is removed and sent through coil pair 2. To make the rotor rotate another 15° clockwise from this position, the current is removed from coil pair 2 and sent through coil pair 3. The process continues in this way. To make the rotor spin counterclockwise, the coil-pair firing sequence is reversed.

15.6 Kinds of Stepper Motors

The model used in the last example was based on a variable-reluctance stepper. As it turns out, this model is incomplete—it does not show how a real variable-reluctance stepper is wired internally. Also, the model does not apply to a class of steppers referred to as *permanent-magnet steppers*. To make things more realistic, let's take a look at some real-life steppers.

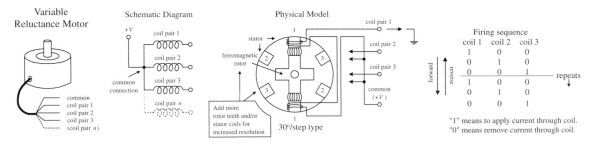

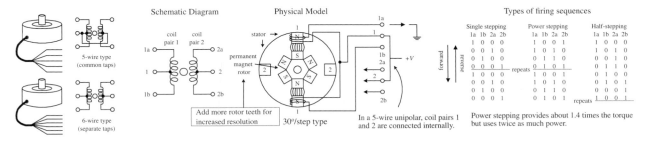

FIGURE 15.8

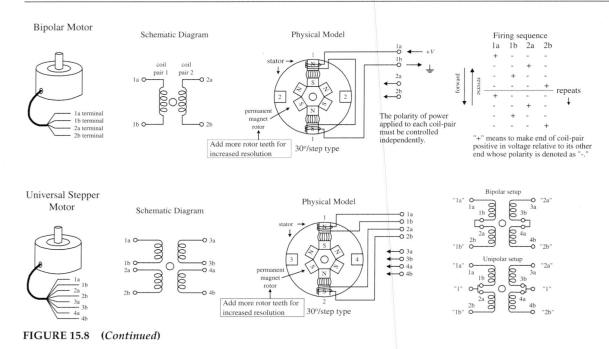

FIGURE 15.8 (*Continued*)

Variable-Reluctance Steppers

Figure 15.8 shows a physical model and schematic diagram of a 30° per step variable-reluctance stepper. This stepper consists of a six-pole (or three-coil pair) stator and a four-toothed ferromagnetic rotor. Variable-reluctance steppers with higher angular resolutions are constructed with more coil pairs and/or more rotor teeth. Notice that in both the physical model and the schematic, the ends of all the coil pairs are joined together at a common point. (This joining of the coil ends occurs internally within the motor's case.) The common and the coil pair free ends are brought out as wires from the motor's case. These wires are referred to as the *phase wires*. The common wire is connected to the supply voltage, whereas the phase wires are grounded in sequence according to the table shown in Fig. 15.8.

Permanent-Magnet Steppers (Unipolar, Bipolar, Universal)

UNIPOLAR STEPPERS

These steppers have a similar stator arrangement as the variable-reluctance steppers, but they use a permanent-magnet rotor and different internal wiring arrangements. Figure 15.8 shows a 30° per step unipolar stepper. It consists of a four-pole (or two-coil pair) stator with center taps between coil pairs and a six-toothed permanent-magnetic rotor. The center taps may be wired internally and brought out as one wire or may be brought out separately as two wires. The center taps typically are wired to the positive supply voltage, whereas the two free ends of a coil pair are alternately grounded to reverse the direction of the field provided by that winding. As shown in the figure, when current flows from the center tap of winding 1 out terminal 1a, the top stator pole "goes north," while the bottom stator pole "goes south." This causes the rotor to snap into position. If the current through winding 1 is removed, sent through winding 2, and out terminal 2a, the horizontal poles will become energized, causing the rotor to turn 30°, or one step. In Fig. 15.8, three firing sequences are shown. The first sequence provides full stepping action (what we just discussed). The second sequence, referred to as the *power stepping sequence*, provides full stepping action with 1.4 times the torque but twice the power consumption. The third sequence provides half stepping (e.g., 15° instead of the rated 30°). Half stepping is made possible by energizing adjacent poles at the same time. This pulls the rotor in-between the poles, thus resulting in one-half the stepping angle. As a final note, unipolar steppers with higher angular resolutions are constructed with more rotor teeth. Also, unipolars come in either five- or six-wire types. The five-wire type has the center taps joined internally, while the six-wire type does not.

BIPOLAR STEPPERS

These steppers resemble unipolar steppers, but their coil pairs do not have center taps. This means that instead of simply supplying a fixed supply voltage to a lead, as was the case in unipolar steppers (supply voltage was fixed to center taps), the supply voltage must be alternately applied to different coil ends. At the same time, the opposite end of a coil pair must be set to the opposite polarity (ground). For example, in Fig. 15.8, a 30° per step bipolar stepper is made to rotate by applying the polarities shown in the firing sequence table to the leads of the stepper. Notice that the firing sequence uses the same basic drive pattern as the unipolar stepper, but the "0" and "1" signals are replaced with "+" and "−" symbols to show that the polarity matters. As you will see in the next section, the circuitry used to drive a bipolar stepper requires an H-bridge network for every coil pair. Bipolar steppers are more difficult to control than both unipolar steppers and variable-reluctance steppers, but their unique polarity-shifting feature gives them a better size-to-torque ratio. As a final note, bipolar steppers with higher angular resolutions are constructed with more rotor teeth.

UNIVERSAL STEPPERS

These steppers represent a type of unipolar-bipolar hybrid. A universal stepper comes with four independent windings and eight leads. By connecting the coil windings in parallel, as shown in Fig. 15.8, the universal stepper can be converted into a unipolar stepper. If the coil windings are connected in series, the stepper can be converted into a bipolar stepper.

15.7   Driving Stepper Motors

Every stepper motor needs a driver circuit that can control the current flow sent through the coils within the stepper's stator. The driver, in turn, must be controlled by a logic circuit referred to as a *translator*. We will discuss translator circuits after we have covered the driver circuits.

Figure 15.9 shows driver networks for a variable-reluctance stepper and for a unipolar stepper. Both drivers use transistors to control current flow through the motor's individual windings. In both driver networks, input buffer stages are added to protect the translator circuit from the motor's supply voltage in the event of transistor collector-to-base breakdown. Diodes are added to both drivers to protect the transistors and power supply from inductive kickback generated by the motor's coils. (Notice that the unipolar driver uses extra diodes because inductive kickback can leak out on either side of the center tap. As you will see in a moment, a pair of diodes within this driver can be replaced with a single diode, keeping the diode count to four.) The single driver section shown in Fig. 15.9 provides a general idea of what kinds of components can be used within the driver networks. This circuit uses a high-power Darlington transistor, a TTL buffer, and a reasonably fast protection diode (the extra diode should be included in the unipolar circuit). If you do not want to bother with discrete components, transistor-array ICs, such as the ULN200x series by Allegro Microsystems or the DS200x series by National Semiconductor, can be used to construct the driver section. The ULN2003, shown in Fig. 15.9, is a TTL-compatible chip that contains seven Darlington transistors with protection diodes included. The 7407 buffer IC can be used with the ULN2003 to construct a full-stepper driver. Other ICs, such as Motorola's MC1414 Darlington array IC, can drive multiple motor winding directly from logic inputs.

Single Driver Section

Transistor and Buffer Arrays

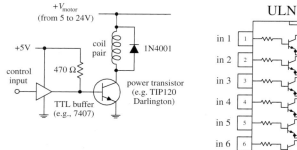

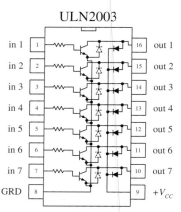

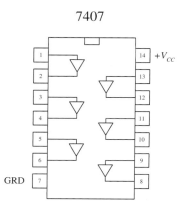

Variable-Reluctance Driver Network

Unipolar Driver Network

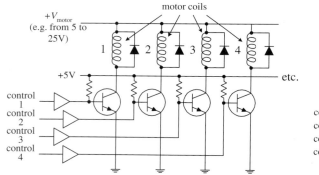

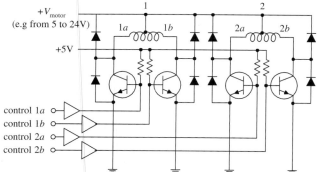

FIGURE 15.9

The circuitry used to drive a bipolar stepper requires the use of an H-bridge circuit. The H-bridge circuit acts to reverse the polarity applied across a given coil pair within the stepper. (Refer back to the section on dc direction control for details on how H-bridges work.) For each coil pair within a stepper, a separate H-bridge is needed. The H-bridge circuit shown in Fig. 15.10 uses four power Darlington transistors that are protected from the coil's inductive kickback by diodes. An XOR logic circuit is added to the input to prevent two high (1's) signals from being applied to the inputs at the same time. [If two high signals are placed at both inputs (assuming that there is no logic circuit present), the supply will short to ground. This is not good for the supply.] The table in Fig. 15.10 provides the proper firing sequence needed to create the desired polarities.

H-bridge used with Bipolar Stepper

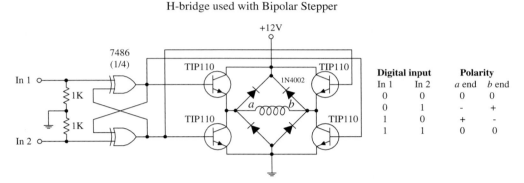

FIGURE 15.10

As mentioned in the dc motor section of this chapter, H-bridges can be purchased in IC form. SGS Thompson's L293 dual H-bridge IC is a popular choice for driving small bipolar steppers drawing up to 1 A per motor winding at up to 36 V. The L298 dual H-bridge is similar to the L293 but can handle up to 2 A per winding. National Semiconductor's LMD18200 H-bridge IC can handle up to 3 A, and unlike the L293 and L298, it has protection diodes built in. More H-bridge ICs are available, so check the catalogs.

15.8   Controlling the Driver with a Translator

A translator is a circuit that enacts the sequencing pulses used to drive a driver. In some instances, the translator may simply be a computer or programmable interface controller, with software directly generating the outputs needed to control the driver leads. In most cases, the translator is a special IC that is designed to provide the proper firing sequences from its output leads when a clock signal is applied to one of its input leads; another input signal may control the direction of the firing sequence (the direction of the motor). There are a number of stepper translator ICs available that are easy to use and fairly inexpensive. Let's take a look at one of these devices in a second. First, let's take a look at some simple translator circuits that can be built from simple digital components.

A simple way to generate a four-phase drive pattern is to use a CMOS 4017 decade counter/divider IC (or a 74194 TTL version). This device sequentially makes 1 of 10 possible outputs high (others stay low) in response to clock pulses. Tying the fifth output (Q_4) to ground makes the decade counter into a quad counter. To enact the drive sequence, a clock signal is applied to the clock input (see Fig. 15.11). Another four-phase translator circuit that provides power stepping control as well as direction control can be constructed with a CMOS 4027 dual JK flip-flop IC (or a 7476 TTL version). The CMOS 4070 XOR logic (or 7486 TTL XOR logic) is used to set up directional control.

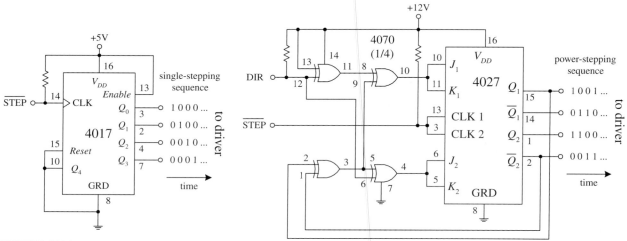

FIGURE 15.11

Figure 15.12 shows a circuit that contains the translator, driver, and stepper all in one. The motor, in this case, is a unipolar stepper, while the translator is a TTL 74194 shift counter. The 555 timer provides clock signals to the 74194, while the DPDT switch acts to control the direction of the motor. The speed of the motor is dependent on the frequency of the clock, which in turn is dependent on R_1's resistance. The translator in this circuit also can be used to control a variable reluctance stepper. Simply use the variable-reluctance driver from Fig. 15.9 and the firing sequence shown in Fig. 15.8 as your guides.

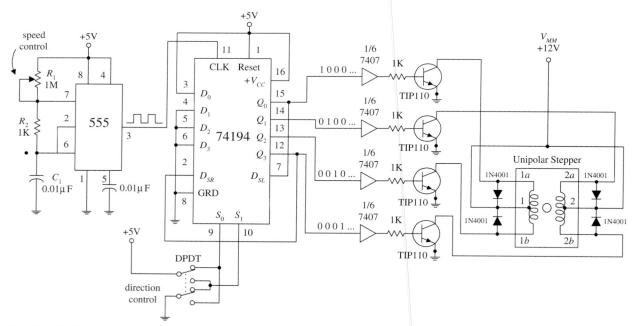

FIGURE 15.12

Perhaps the best translator circuits you can hope for come in integrated packages. A number of manufacturers produce stepper motor controller ICs that house both the translator and driver sections. These chips are fairly simple to use and inexpensive. A classic stepper controller chip is the Philips SAA1027. The SAA1027 is a bipolar IC that is designed to drive four-phase steppers. It consists of a bidirectional four-state counter and a code converter that are used to drive four outputs in sequence. This chip has high-noise-immunity inputs, clockwise and counterclockwise capability, a reset control input, high output current, and output voltage protection. Its supply voltage runs from 9.5 to 18 V, and it accepts input voltages of 7.5 V minimum for high (1) and 4.5 V maximum for low (0). It has a maximum output current of 500 mA. Figure 15.13 will paint the rest of the picture.

As mentioned, the SAA1027 is a classic chip (old chip). Newer, better stepper control ICs are available from a number of manufacturers. If you are interested in learning more about these chips, try searching the Internet. You will find some useful websites that discuss stepper controller ICs in detail. Also, these websites often will provide links to manufacturers and distributors of stepper motors and controller ICs.

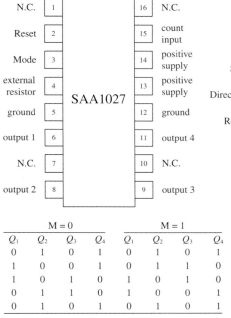

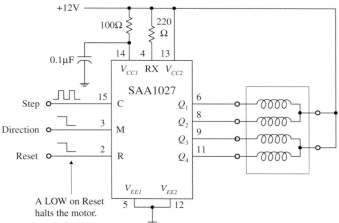

	M = 0				M = 1		
Q_1	Q_2	Q_3	Q_4	Q_1	Q_2	Q_3	Q_4
0	1	0	1	0	1	0	1
1	0	0	1	0	1	1	0
1	0	1	0	1	0	1	0
0	1	1	0	1	0	0	1
0	1	0	1	0	1	0	1

FIGURE 15.13

Count input C (pin 15)—A low-to-high transition at this pin causes the outputs to change states.

Mode input M (pin 3)—Controls the direction of the motor. See table to the left.

Reset input R (pin 2)—A low (0) at the R input resets the counter to zero. The outputs take on the levels shown in the upper and lower line of the table to the left.

External resistor RX (pin 4)—An external resistor connected to the RX terminal sets the base current of the transistor drivers. Its value is based on the required output current.

Outputs Q_1 through Q_4 (pins 6, 8, 9, 11)—Output terminals that are connected to the stepper motor.

An alternative to using a hardware translator is to use a microcontroller that generates the signals for the coil drivers. Microcontrollers and microcontroller boards such as the Arduino will have libraries of code that provide all the sequencing necessary to drive a stepper motor.

15.9   A Final Word on Identifying Stepper Motors

When it comes to identifying the characteristics of an unknown stepper, the following suggestions should help. The vast majority of the steppers on the market today are unipolar, bipolar, or universal types. Based on this, you can guess that if your stepper has four leads, it is most likely a bipolar stepper. If the stepper has five leads, then the motor is most likely a unipolar with common center taps. If the stepper has six leads, it is probably a unipolar with separate center taps. A motor with eight leads would most likely be a universal stepper. (If you think your motor might be a variable-reluctance stepper, try spinning the shaft. If the shaft spins freely, the motor is most likely a variable-reluctance stepper. A coglike resistance indicates that the stepper is a permanent-magnet type.)

Once you have determined what kind of stepper you have, the next step is to determine which leads are which. A simple way to figure this out is to test the resistance between various leads with an ohmmeter.

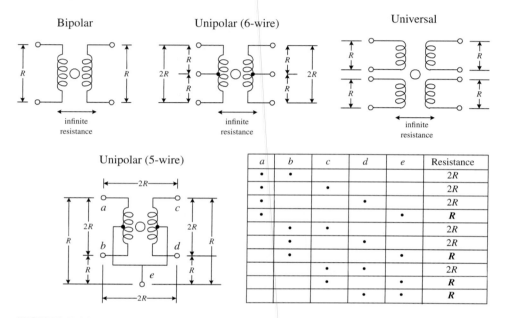

a	b	c	d	e	Resistance
•	•				2R
•		•			2R
•			•		2R
•				•	R
	•	•			2R
	•		•		2R
	•			•	R
		•	•		2R
		•		•	R
			•	•	R

FIGURE 15.14

Decoding the leads of a bipolar stepper is easy. Simply use an ohmmeter to determine which wire pair yields a low resistance value. A low resistance indicates that the two wires are ends of the same winding. If the two wires are not part of the same winding, the resistance will be infinite. A universal stepper can be decoded using a similar approach. Decoding a six-wire unipolar stepper requires isolating two three-wire pairs. From there, you figure out which wire is the common center tap by noticing which measured pair among the isolated three wires gives a unit R worth of resistance and which pair gives a unit of $2R$ worth of resistance (see Fig. 15.14). Now, decoding a five-wire unipolar (with common center tap) is a bit more tricky than the others because of the common, but hidden, center tap. To help decode this stepper, you can use the diagram and table shown in Fig. 15.14. (The dots within the table represent where the ohmmeter's two probes are placed within the diagram.) With the table you isolate e (common tap wire) by noting when the ohmmeter gives a resistance of R units. Next, you determine which of the two wires in your hand is actually e by testing one of the two with the rest of the wires. If you always get R, then you are holding e, but if you get $2R$, you are not holding e. Once the e wire is determined, any more ohmmeter deducing does not work—at least in theory—because you will always get $2R$. The best bet now is to connect the motor to the driver circuitry and see if the stepper steps. If it does not step, fiddle around with the wires until it does.

CHAPTER 16

Audio Electronics

Audio electronics, in part, deals with converting sound signals into electrical signals. This conversion process typically is accomplished by means of a microphone. Once the sound is converted, what is done with the corresponding electrical signal is up to you. For example, you can amplify the signal, filter out certain frequencies from the signal, combine (mix) the signal with other signals, transform the signal into a digitally encoded signal that can be stored in memory, modulate the signal for the purpose of radiowave transmission, use the signal to trigger a switch (e.g., transistor or relay), etc.

Another aspect of audio electronics deals with generating sound signals from electrical signals. To convert electrical signals into sound signals, you can use a speaker. (If you are not interested in retaining frequency response—say, you are only interested in making a warning alarm—you can use an audible sound device such as a dc buzzer or compression washer.) The electrical signals used to drive a speaker may be sound-generated in origin or may be artificially generated by special oscillator circuits.

16.1 A Little Lecture on Sound

Before you start dealing with audio-related circuits, it is worthwhile reviewing some of the basic concepts of sound. Sound consists of three basic elements: *frequency, intensity (loudness)*, and *timbre (overtones)*.

The frequency of a sound corresponds to the vibrating frequency of the object that produced the sound. In terms of human physiology, the human ear can perceive frequencies from around 20 to 20,000 Hz; however, the ear is most sensitive to frequencies between 1000 and 2000 Hz.

The intensity of a sound corresponds to the amount of sound energy transported across a unit area per second (or W/m^2) and depends on the amplitude of oscillation of the vibrating object. As you move further away from the vibrating object, the intensity drops in proportion to one over the distance squared. The human ear can perceive an incredible range of intensities, from 10^{-12} to $1\ W/m^2$. Because this range

947

is so extensive, it is usually more convenient to use a logarithmic scale to describe intensity. For this purpose, decibels are used. When using decibels, sound intensity is defined as $dB = 10 \log_{10} (I/I_0)$, where I is the measured intensity in watts per meter, and $I_0 = 10^{-12}$ W/m² is defined as the smallest intensity that is perceived as sound by humans. In terms of decibels, the audio intensity range for humans is between 0 and 120 dB. Figure 16.1 shows a number of sounds, along with their frequency and intensity ranges.

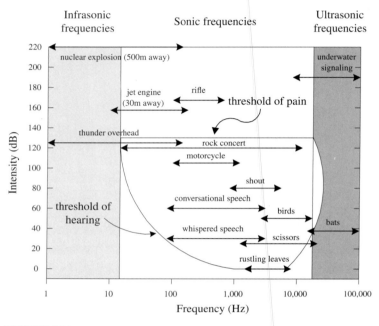

FIGURE 16.1

Tonal quality, or timbre, represents the complex wave pattern that is generated when the overtones of an instrument, voice, etc., are present along with the fundamental frequency. To demonstrate what overtones mean, consider a simple tuning fork that has a resonant frequency of 261.6 Hz (middle C). If you treat the fork as an ideal vibrator, when it is hit, it will vibrate off soundwaves with a frequency of 261.6 Hz. In this case, you have no overtones—you only get one frequency. But now, if you play middle C on a violin, you get an intensely sounding 261.1 Hz, along with a number of other higher, typically less intense frequencies called *overtones* (or *harmonics*). The most intense frequency sounded is typically referred to as the *fundamental frequency*. The overtones of importance have frequencies that are integer multiples of the fundamental frequency (e.g., 2×261.1 Hz is the first harmonic, 3×261.1 Hz is the second harmonic, and $n \times 261.1$ Hz is the nth harmonic). It is the specific intensity of each overtone within the harmonic spectrum of an instrument, voice, etc., that is largely responsible for giving the instrument, voice, etc., its unique tonal quality. (The reason for an instrument's unique set of overtones depends on the construction of the instrument.) Figure 16.2a shows a harmonic spectrum (spectral plot) for an oboe that is tuned to middle C—the fundamental frequency.

In theory, you can create the sound from any type of instrument (e.g., violin, tuba, banjo, etc.) by examining the harmonic spectrum of that instrument. To illustrate how

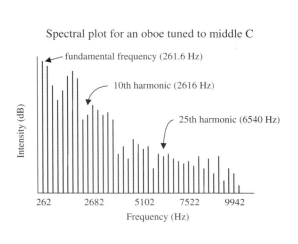

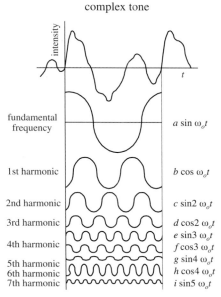

FIGURE 16.2

this can be done, pretend that you have a number of ideal tuning forks. One fork represents the fundamental frequency; the other forks represent the various overtone frequencies. Using the harmonic spectrum of an instrument as a guide, you can mimic the sound of the instrument by varying the intensity of each "overtone fork." (In reality, to accurately mimic an instrument, you also must consider rise and decay times of certain overtones—controlling the intensities of the overtones is not enough.) Mathematically, you can express a complex sound as the sum of all its overtones:

$$\text{Signal} = a \sin \omega_0 t + b \cos \omega_0 t + c \sin 2\omega_0 t + d \cos 2\omega_0 t + e \sin 3\omega_0 t + f \cos \omega_0 t + \cdots$$

The coefficients a, b, c, d, etc., are the intensities of the overtones, and the fundamental frequency $f_0 = \omega_0/2\pi$. This expression is referred to as a *Fourier series*. The coefficients must be calculated from the given waveform or the data from which it is plotted—although an instrument called a *harmonic analyzer* can automatically compute the coefficients. Figure 16.2*b* shows a complex sound made up of seven of its harmonics.

The art of synthesizing sounds via electric circuits is fairly complex business. To accurately mimic an instrumental sound, train whistle, bird chirp, etc., you must design circuits that can generate complex waveforms that contain all the overtones and decay and rise time information. For this purpose, special oscillator and modulator circuits are needed.

16.2 Microphones

A microphone converts variations in sound pressure into corresponding variations in electric current. The amplitude of the ac voltage generated by a microphone is proportional to the intensity of the sound, while the frequency of the ac voltage corresponds to the frequency of the sound. (Note that if overtones are present within the sound signal, these overtones also will be present in the electrical signal.) Three commonly used microphones are listed next.

Dynamic

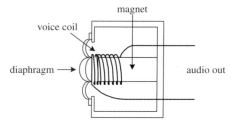

FIGURE 16.3

This type of microphone consists of a plastic diaphragm, voice coil, and a permanent magnet. The diaphragm is connected to one end of the voice coil, while the other end of the coil is loosely supported around (or within) the magnet. When an alternating pressure is applied to the diaphragm, the voice coil alternates in response. Since the voice coil is accelerating through the magnet's magnetic field, an induced voltage is set up across the leads of the voice coil. You can use this voltage to power a very small load, or you can use an amplifier to increase the strength of the signal so as to drive a larger load. Dynamic microphones are extremely rugged, provide smooth and extended frequency response, do not require an external dc source to drive them, perform well over a wide range of temperatures, and have a low impedance output. Some dynamic microphones house internal transformers within their bodies, which give them the ability to have either a high- or low-impedance output—a switch is used to select between the two. Dynamic microphones are widely used in public address, hi-fi, and recording applications.

Condenser

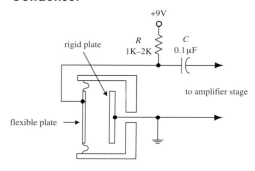

FIGURE 16.4

This type of microphone consists of a pair of charged plates that can be forced closer or further apart by variations in air pressure. In effect, the plates act like a sound-sensitive capacitor. One plate is made of a rigid metal that is fixed in place and grounded. The other plate is made of a flexible metal or metal-coiled plastic that is positively charged by means of an external voltage source. A very low-noise, high-impedance amplifier is required to operate this type of microphone and to provide low output impedance. Condenser microphones offer crisp, low-noise sound and are used for high-quality sound recording.

Electret

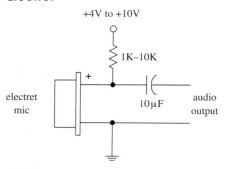

FIGURE 16.5

An electret microphone is a variation of the condenser microphone. Instead of requiring an external voltage source to charge the diaphragm, it uses a permanently charged plastic element (electret) placed in parallel with a conductive metal backplate. Most electret microphones have a small FET amplifier built into their cases. This amplifier requires power to operate—typically a voltage between +1.5 and +10 V is needed. This voltage is fed into the microphone through a resistor (1–10 K) (see figure). Electret microphones used to suffer from poor performance, but modern designs can achieve results comparable to those of a condenser microphone.

16.3 Microphone Specifications

A microphone's *sensitivity* represents the ratio of electrical output (voltage) to the intensity of sound input. Sensitivity is often expressed in decibels with respect to a reference sound pressure of 1 dyn/cm^2.

The *frequency response* of a microphone is a measure of the microphone's ability to convert different acoustical frequencies into ac voltages. For speech, the frequency response of a microphone need only cover a range from around 100 to 3000 Hz. However, for hi-fi applications, the frequency response of the microphone must cover a wider range, from around 20 to 20,000 Hz.

The *directivity characteristic* of a microphone refers to how well the microphone responds to sound coming from different directions. *Omnidirectional* microphones respond equally well in all directions, whereas *directional* microphones respond well only in specific directions.

The *impedance* of a microphone represents how much the microphone resists the flow of an ac signal. Low-impedance microphones are classified as having an impedance of less than 600 Ω. Medium-impedance microphones range from 600 to 10,000 Ω, whereas high-impedance microphones extend above 10,000 Ω. In modern audio systems, it is desirable to connect a lower-impedance microphone to a higher-impedance input device (e.g., 50-Ω microphone to 600-Ω mixer), but it is undesirable to connect a high-impedance microphone to a low-impedance input. In the first case, not much signal loss will occur, whereas in the second case, a significant amount of signal loss may occur. The standard rule of thumb is to allow the load impedance to be 10 times the source impedance. We'll take a closer look at impedance matching later on in this chapter.

16.4   Audio Amplifiers

Electrical signals within audio circuits often require amplification to effectively drive other circuit elements or devices. Perhaps the easiest and most efficient way to amplify a signal is to use an op amp. General-purpose op amps such as the 741 will work fine for many noncritical audio applications, but they may cause distortion and other undesirable effects when audio signals get complex. A better choice for audio applications is to use an audio op amp especially designed to handle audio signals. Audio amplifiers have high slew rates, high gain-bandwidth products, high input impedances, low distortion, high voltage/power operation, and very low input noise. There are a number of good op amps produced by a number of different manufacturers. Some high-quality op amps worth mentioning include the AD842, AD847, AD845, AD797, NE5532, NE5534, NE5535, OP-27, LT1115, LM833, OPA2604, OP249, HA5112, LM4562, OPA134, OPA2134, and LT1057.

16.4.1   Inverting Amplifier

The following two circuits act as inverting amplifiers. The gain for both circuits is determined by $-R_2/R_1$ (see Chap. 8 for the theory), while the input impedance is approximately equal to R_1. The first op amp circuit uses a dual power supply, while the second op amp circuit uses a single power supply.

Inverting amplifier (dual power supply)

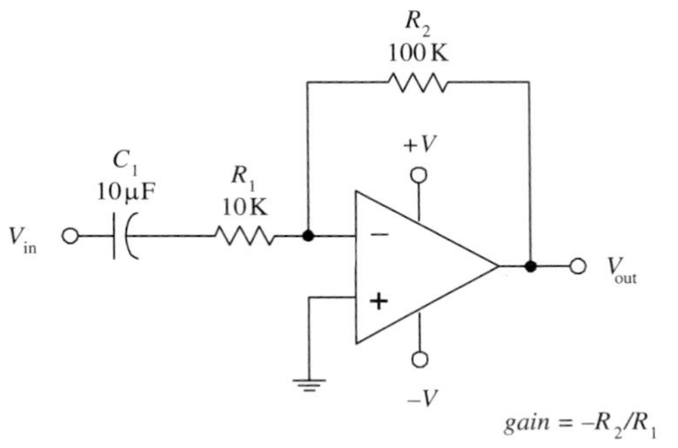

In both amplifier circuits, C_1 acts as an ac coupling capacitor—it acts to pass ac signals while preventing unwanted dc signals from passing from the previous stage. Without C_1, dc levels would be present at the op amp's output, which in turn could lead to amplifier saturation and distortion as the ac portion of the input signal is amplified. C_1 also helps prevent low-frequency noise from reaching the amplifier's input.

$$gain = -R_2/R_1$$

FIGURE 16.6

Inverting amplifier (single power supply)

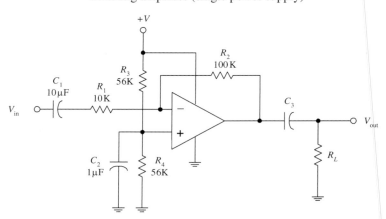

FIGURE 16.6   (*Continued*)

In the single-power-supply circuit, biasing resistors R_3 and R_4 are needed to prevent the amplifier from clipping during negative swings in the audio input signals. They act to give the op amp's output a dc level on which the ac signal can safely fluctuate. Setting $R_3 = R_4$ sets the dc level of the op amp's output to 1/2 (+V). For reliable results, the biasing resistors should have resistance values between 10 and 100 k. Now, to prevent passing the dc level onto the next stage, C_3 (ac coupling capacitor) must be included. Its value should be equal to $1/(2\pi f_C R_L)$, where R_L is the load resistance, and f_C is the cutoff frequency. R_C acts as a filtering capacitor used to eliminate power-supply noise from reaching the op amp's noninverting input.

Notably, many audio op amps are especially designed for single-supply operation—they do not require biasing resistors.

16.4.2 Noninverting Amplifier

The preceding inverting amplifier works fine for many applications, but its input impedance is not incredibly large. To achieve a larger input impedance (useful when bridging a high-impedance source to the input of an amplifier), you can use one of the following noninverting amplifiers in Fig. 16.7. The left amplifier circuit uses a dual power supply, whereas the right amplifier circuit uses a single power supply. The gain for both circuits is equal to $R_2/R_1 + 1$.

16.4.3 Digital Amplifiers

Digital power amplifiers are also known as class-D amplifiers or PWM amplifiers. They are extremely efficient and so run quite cool. This makes them ideal for very

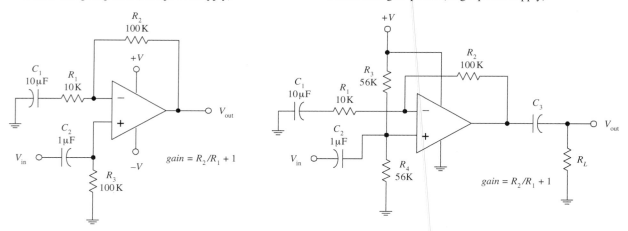

FIGURE 16.7   Components R_1, C_1, R_2, and the biasing resistors serve the same function as was seen in the inverting amplifier circuits. The noninverting input offers an exceptionally high input impedance and can be matched to the source impedance more readily by adjusting C_2 and R_3 (dual-supply circuit) or R_4 (single-supply circuit). The input impedance is approximately equal to R_3 (dual-supply circuit) or R_4 (single-supply circuit).

Digital amplifier

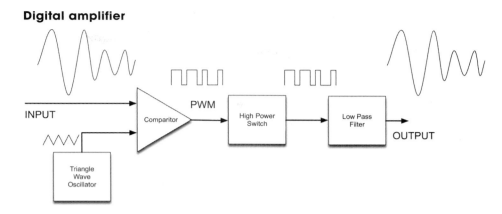

FIGURE 16.8

high-power amplifiers of powers in the hundreds of watts or even the kilowatts range.

Figure 16.8 shows the block diagram for a class-D amplifier. The input signal is converted into a PWM signal by comparing it with a triangle wave. This is a very neat trick. As the triangle wave rises, at some point, it will become greater than the signal voltage. If the signal voltage is high, this will take longer; if the voltage is low, it will happen sooner. In this way, the pulse length produced will be proportional to the instantaneous input voltage.

Figure 16.9 shows the result of simulating the comparator with a triangle waveform of 10 kHz sampling a sine wave input signal at 1 kHz. The squarewave is the PWM output.

PWM encoding using a triangle waveform and comparator

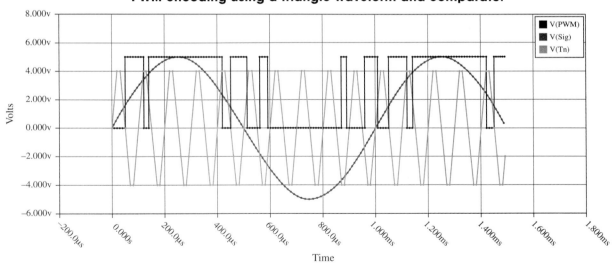

FIGURE 16.9

We have digitized our audio signal. It is now just on or off. So we can use that to switch as high a current as we like, using big MOSFETs in a complementary arrangement or some other switching transistor. This high-power PWM signal will be conveying the right energy at the right time, but it is being carried in the high-frequency switching squarewave. This needs low-pass filtering to remove this carrier, leaving just the original signal in amplified form.

The quality of digital amplifiers can be very variable, but their efficiency makes them an attractive proposition. There are a number of ICs available that either encapsulate the whole amplifier in one device or provide outputs suitable for driving a complementary pair of MOSFETs. A couple devices to look at are the NCP2704 and LX1720.

16.4.4 Reducing Hum in Audio Amplifiers

If your project includes an audio amplifier that you are designing yourself, then you need to consider *hum*. Domestic power lines and devices will easily induce an annoying 60-Hz hum in an audio amplifier. Some of this signal can arrive through the amplifier's power supply. Hence, design of power supplies for hi-fi amplifiers is almost as important as the designs for the amplifiers themselves. The main thing is to add as much smoothing capacitance to the power supply as possible.

Other 60-Hz signals will arrive in your audio amplifier circuit by mutual induction with wires and tracks on your PCB. You should make sure that you keep all PCB tracks and wires as short as possible. Wires should also be screened with the screening layer grounded.

16.5 Preamplifiers

In most audio applications, the term *preamplifier* refers to a control amplifier that is used to control features such as input selection, level control, gain, and impedance levels. Here are a few simple microphone preamplifier circuits to get you started. (Note that "high Z" refers to a microphone with a high input impedance—one that is greater than ~600 Ω.)

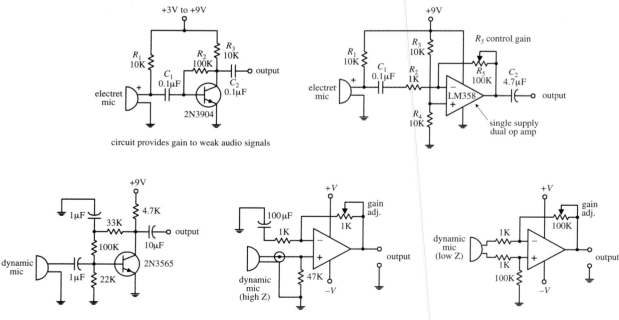

FIGURE 16.10

16.6 Mixer Circuits

Audio mixers are basically summing amplifiers—they add a number of different input signals together to form a single superimposed output signal. The two circuits below are simple audio mixer circuits. The left circuit uses a common-emitter amplifier as the summing element, while the right circuit uses an op amp. The potentiometers are used as independent input volume controls.

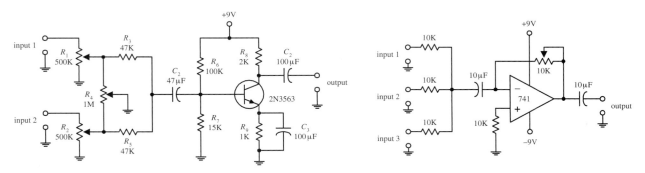

FIGURE 16.11

16.7 A Note on Impedance Matching

Is matching impedances between audio devices necessary? Not any more, at least when it comes to connecting a low-impedance source to a high-impedance load. In the era when vacuum-tube amplifiers were the standard, it was important to match impedances to achieve maximum power transfer between two devices. Impedance matching reduced the number of vacuum-tube amplifiers needed in circuit design (e.g., the number of vacuum-tube amplifiers needed along a telephone transmission line). However, with the advent of the transistor, more efficient amplifiers were created. For these new amplifiers, what was important—and still is important—was maximum voltage transfer, not maximum power transfer. (Think of an op amp with its extremely high input impedance and low output impedance. To initiate a large-output-current response from an op amp, practically no input current is required into its input leads.) For maximum voltage transfer to occur, it was found that the destination device (called the *load*) should have an impedance of at least 10 times that of the sending device (called the *source*). This condition is referred to as *bridging*. (Without applying the bridging rule, if two audio devices with the same impedances are joined, you would see around 6 dB worth of attenuation loss in the transmitted signal.) Bridging is the most common circuit configuration used when connecting modern audio devices. It is also applied to most other electronic source-load connections, with the exception of certain radiofrequency circuits where matching impedance is usually desired and in cases where the signal being transmitted is a current rather than a voltage. If the transmitted signal is a current, the source impedance should be larger than the load impedance.

Now, if you consider a high-impedance source connected to a low-impedance load (e.g., a high-impedance microphone connected to a low-impedance mixer), voltage transfer can result in significant signal loss. The amount of signal loss in this case would be equal to

$$dB = 20 \ \log_{10} \frac{R_{load}}{R_{load} + R_{source}}$$

As a rule of thumb, a loss of 6 dB or less is acceptable for most applications.

16.8 Speakers

Speakers convert electrical signals into audible signals. The most popular speaker used today is the dynamic speaker. The dynamic speaker operates on the same basic principle as a dynamic microphone. When a fluctuating current is applied through a moving coil (voice coil) that surrounds a magnet (or that is surrounded by a magnet), the coil is forced back and forth. A large paper cone attached to the coil responds to the back-and-forth motion by "drumming off" sound waves.

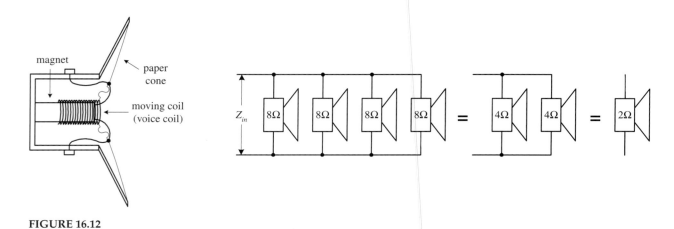

FIGURE 16.12

Every speaker is given a *nominal impedance* Z that represents the average impedance across its leads. (In reality, a speaker's impedance varies slightly with frequency, above and below the nominal level.) In terms of applications, you can treat the speaker like a simple resistive load of impedance Z. For example, if you attach an 8-Ω speaker to an amplifier's output, the amplifier will treat the speaker as an 8-Ω load. The amount of current drawn from the amplifier will be $I = V_{out}/Z_{speaker}$. However, if you replace the 8-Ω speaker with a 4-Ω speaker, the current drawn from the amplifier will double.

Driving two 8-Ω speakers in parallel is equivalent to driving a 4-Ω speaker. Driving two 4-Ω speakers in parallel is equivalent to driving a 2-Ω speaker. By using high-power resistors, it is possible to change the overall impedance sensed by the amplifier. For example, by placing a 4-Ω resistor in series with a 4-Ω speaker, you can create a load impedance of 8 Ω. However, using a series resistor to increase the impedance may hurt the sound quality. There are speaker-matching transformers that can change from 4 to 8 Ω, but a high-quality transformer like this can cost as much as a new speaker and can add slight frequency response and dynamic range errors.

Another important characteristic of a speaker is its frequency response. The *frequency response* represents the range over which a speaker can effectively vibrate off audio signals. Speakers that are designed to respond to low frequencies (typically less than 200 Hz) are referred to as *woofers*. *Midrange speakers* are designed to handle frequencies typically between 500 and around 3000 Hz. A *tweeter* is a special type of speaker (typically dome or horn type) that can handle frequencies above midrange. Some speakers are designed as full-range units that are capable of reproducing frequencies from around 100 to 15,000 Hz. A full-range speaker's sound quality tends to be inferior to a speaker system that incorporates a woofer, midrange, and tweeter speaker all together.

16.9   Crossover Networks

To design a decent speaker system, it would be best to incorporate a woofer, midrange speaker, and tweeter together so that you get good sound response over the entire audio spectrum (20 to 20,000 Hz). However, simply connecting these speakers in parallel will not work because each speaker will be receiving frequencies outside its natural frequency-response range. What you need is a filter network that can divert high-frequency signals to the tweeter, low-frequency signals to the woofer, and midrange-frequency signals to the midrange speaker. The filter network that is used for this sort of application is called a *crossover network*.

There are two types of crossovers: passive or active. Passive crossover networks consist of passive filter elements (e.g., capacitors, resistors, inductors) that are placed between the power amplifier and the speaker—they are placed inside the speaker cabinet. Passive crossover networks are cheap to make, easy to make foolproof, and can be tailored for a specific speaker. However, they are nonadjustable and always use up some amplifier power. Active crossover networks consist of a set of active filters (op amp filters) that are placed before the amplifier section. The fact that active crossover networks come before the power amplifier section makes it easier to manipulate the signal because the signal is still tiny (not amplified). Also, a single active crossover network can be used to control a number of different amplifier-speaker combinations at the same time. Since active crossover networks use active filters, the audio signal will not suffer as much attenuation loss as it would if it were applied through a passive crossover network.

Figure 16.13 shows a simple passive crossover network used to drive a three-speaker system. The graph shows typical frequency-response curves for each speaker. To produce an overall flat response from the system, you use a low-pass, bandpass, and high-pass filter. C_1 and R_t form the low-pass filter, L_1, C_1, and R_m form the band-pass filter, and L_2 and R_w form the low-pass filter (R_t, R_m, and R_w are the nominal impedances of the tweeter, midrange, and woofer speakers).

To determine the component values needed to get the desired response, use the following: $C_1 = 1/(2\pi f_2 R_t)$, $L_1 = R_m/(2\pi f_2)$, $C_2 = 1/(2\pi f_1 R_m)$, and $L_2 = R_w/2\pi f_1$, where f_1 and f_2 represent the 3-dB points shown in the graph. Usually, passive crossover networks are a bit more sophisticated than the one presented here. They often incorporate higher-order filters along with additional elements, such as an impedance compensation network, attenuation network, series notch filter, etc., all of which are used to achieve a flatter overall response.

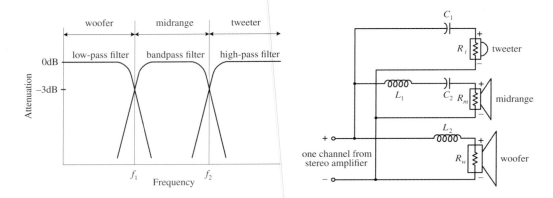

FIGURE 16.13

Here is a more practical passive crossover network (crossover frequency 1.8 kHz) used to drive a two-speaker system consisting of an 8-Ω tweeter and an 8-Ω woofer. An $18 \times 12 \times 8$ in fiberboard box acts as a good resonant cavity for this system.

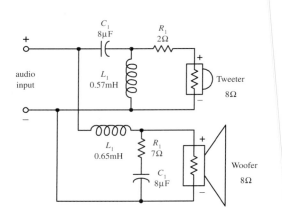

FIGURE 16.14

The following is an active crossover network that is used to drive a two-speaker system that has a crossover frequency (3-dB point) around 600 Hz and 18 dB per octave response. A LF356 high-performance op amp is used as the active element. Remember that for active filters the output signals must be amplified before they are applied to the speaker inputs.

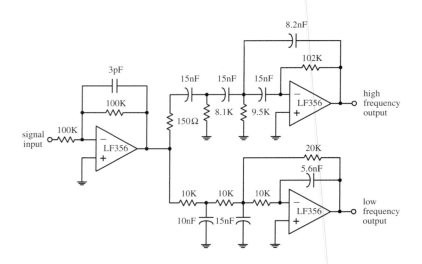

FIGURE 16.15

16.10 Simple ICs Used to Drive Speakers

Audio Amplifier (LM386)

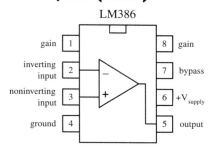

LM386

gain	1		8	gain
inverting input	2		7	bypass
noninverting input	3		6	+V$_{supply}$
ground	4		5	output

The LM386 audio amplifier is designed primarily for low-power applications. A +4- to +15-V supply voltage is used to power the IC. Unlike a traditional op amp, such as the 741, the 386's gain is internally fixed to 20. However, it is possible to increase the gain to 200 by connecting a resistor-capacitor network between pins 1 and 8. The 386's inputs are referenced to ground, while internal circuitry automatically biases the output signal to one-half the supply voltage. This audio amplifier is designed to drive an 8-Ω speaker.

Audio Amplifier (gain of 20)   ### Audio Amplifier (gain of 200)

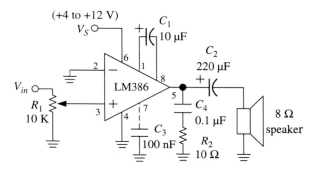

FIGURE 16.16

Audio Amplifier (LM383)

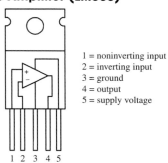

1 = noninverting input
2 = inverting input
3 = ground
4 = output
5 = supply voltage

The LM383 is a power amplifier designed to drive a 4-Ω speaker or two 8-Ω speakers in parallel. It contains thermal shutdown circuitry to protect itself from excessive loading. A heat sink is required to avoid shutdown at levels below the maximum power output.

8-watt amplifier   #### 16-watt amplifier

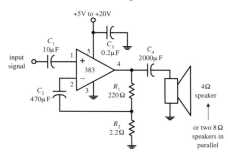

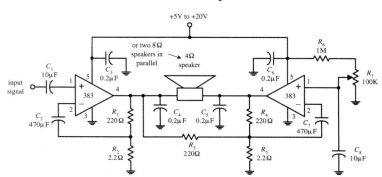

FIGURE 16.17

16.11 Audible-Signal Devices

There are a number of unique audible-signal devices that are used as simple warning signal indicators. Some of these devices produce a continuous tone, others produce intermittent tones, and still others are capable of generating a number of different frequency tones, along with various periodic on/off cycling characteristics. Audible-signal devices come in both dc and ac types and in various shapes and sizes. Some of these devices are extremely small—no bigger than a dime. A good electronics catalog will provide a listing of audio-signal devices, along with their size, sound type, dB ratings, voltage ratings, and current-drain specifications.

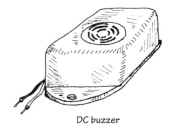

FIGURE 16.18   Sonalert® audible sound device   Compression washer   DC buzzer

16.12 Miscellaneous Audio Circuits

Simple Tone-Related Circuits

Tone generator

+9V to +12V

R_1 1K
R_3 100Ω
R_2 100K
E
B_2
UJT transistor
B_1
C 0.1μF
8Ω speaker

Vary R_2 to change the speaker's audio output frequency.

Metronome

+6V to +10V

R_1 10K
2N2907
R_2 100K
2N2222
C 22μF
8Ω speaker

R_2 controls the number of "clicks" per second that are emitted from the speaker.

Tone generator

+9V

R_1 5K
2N3638
R_2 100K
2N2222
8Ω speaker
C_1 0.01μF to 0.1μF

R_2 controls the audio output frequency of the speaker.

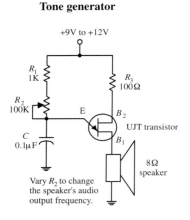

Tone generator

+5V

8   4
R_1 1M
7
R_2 1K
6
555 timer IC
2
3
8Ω speaker
C_1 0.1μF
C_2 0.01μF
1

Adjust R_1 to vary speaker tone. To boost the signal, place an amplifier between the 555's output and the speaker.

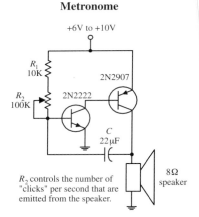

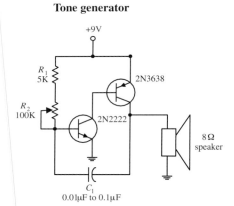

Warbler siren

+5V to +12V

8   4
R_1 10K
7
R_2 1M
6
555 timer IC
2
C_1 0.22μF
1
R_3 47K
3
R_4 10K
R_5 1K
C_2 0.1μF
8   4
7
3
6
555 timer IC
2
1
8Ω speaker

To change the pitch and speed of the warble, alter R_2 and R_5. To boost the signal, place an amplifier between the 555's output and the speaker.

FIGURE 16.19

Simple Buzzer Circuits

Buzzer volume control

Digitally actuated buzzers

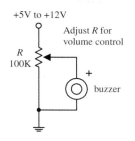

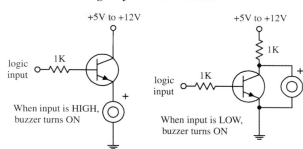

FIGURE 16.20

Megaphone

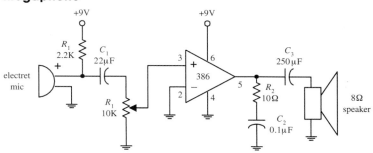

FIGURE 16.21

Sound-Activated Switch

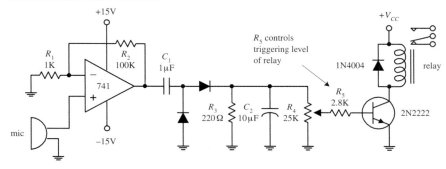

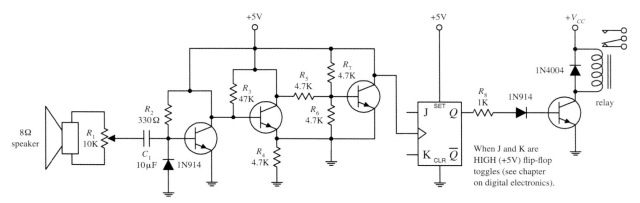

FIGURE 16.22

## CHAPTER 17

Modular Electronics

Electronics has changed over the past few years, and more and more people have an invention, but don't necessarily want to learn degree-level electronics to make it a reality. Suppliers such as SparkFun, Seeed Studio, Pololu, and others support this by supplying modules and breakout boards that simplify the process of using complex devices. In addition, there are ICs for most of what you might want to do in electronics, and these will greatly simplify your project build. Also, there are whole systems, such as Arduino (see Chap. 13), .NET Gadgeteer, and Netduino, that provide plug-together modules for pretty much anything that you might want to build that has a microcontroller at its heart.

17.1   There's an IC for It

Although plenty of useful, general-purpose ICs are available, there are also some very specialized devices. Before designing anything complex from discrete devices, you should always check that you are not reinventing the wheel. There may be an IC that you can use that will reduce the component count and cost of your project.

Table 17.1 lists some of the ICs that you may find useful for your projects. Some of them are general purpose; others fill a very narrow niche. This is not intended to be any exhaustive list, but rather to provide inspiration. We have not listed part URLs. You can find places to buy components using the Octopart parts search engine (http://www.octopart.com), where you will also be able to track down data sheets.

17.2   Breakout Boards and Modules

You do not always need to start your design from scratch. Many ready-made modules and breakout boards are available to inventors to incorporate into their designs.

The difference between a breakout board and a module is often a little blurred. The intention of a breakout board is that it simply "breaks out" the difficult-to-access pins of a surface-mounted device (SMD) IC into 0.1-in connections that are much easier to use. However, breakout boards often add a few extra components like a decoupling capacitor, voltage regulator, or level conversion. So the point at which a breakout board becomes a module is a bit arbitrary.

TABLE 17.1 ICs for Electronics Projects

IC	DESCRIPTION
Audio	
HT9200	DTMF code generator, for use in applications such as phone auto-dialers; easy connection to microcontroller
ICL7611	Low-voltage, single power supply op amp
LM358	Low-cost dual op amp
RTS0072	Voice changer IC that distorts or transposes audio signals
ISD1932	Voice recorder IC
SAE800	Door chime generator
TDA7052	1-W audio power amp
TDA2003	10-W audio power amp
DG201B	Quad analog switch
Power Control	
L298	Two-channel 2-A H-bridge motor controller
S202T01F	Solid-state relay, 2-A, 600-V
MAX1551	Lithium polymer battery charger
L297	Stepper motor controller
LED Drivers	
MAX6958	4 × 9 segment LED driver with I²C interface
LM3914	Bar graph LED display driver (10 LED outputs, analog input)
LM3404	1-A constant current LED driver
Miscellaneous	
NE555	Timer IC
DS1302	Real-time clock, I²C interface
24C1024	128 kbit × 8 bit I2C EEPROM
SST25VF010A	1-Mbit SPI flash

Whatever you chose to call them, breakout boards and modules can be very useful in prototyping when you want to get something working to prove a concept using prebuilt modules, and then advance to a final design without the module.

There are many sensor modules (see Table 17.2), and some of these are covered in more detail in Chap. 6. Table 17.2 lists some of the more interesting modules that we have come across.

17.2.1 Radio Frequency Modules

Radio frequency (RF) electronics is a specialized part of electronics that is a discipline in its own right. At high frequencies, PCB layout becomes critical, and design is not as

TABLE 17.2   Some Modules and Breakout Boards

MODULE	USAGE	SOURCES
RF Modules		
I²C FM radio receiver	FM radio	SparkFun: BOB-10344
433/315-MHz transmitters/ receivers	Data link between microcontrollers, sensors, and actuators	SparkFun: WRL-10533, WRL-10535
		Seeed Studio: WLS105B5B
Bluetooth	Microcontroller/cellular phone/computer link	SparkFun: WRL-10269, WRL-10253
		Seeed Studio: WLS123A1M
WiFi	Wireless network to microcontroller	SparkFun: WRL-10004
		Seeed Studio: WLS48188P
XRF modules	Data link between microcontrollers, sensors, and actuators	http://shop.ciseco.co.uk/wireless/
XBee	Data link between microcontrollers, sensors, and actuators; medium range	SparkFun: WRL-10414
XBee Pro	Long-range data link	SparkFun: WRL-09085
RFID tag reader	Security	SparkFun: SEN-08419
		Seeed Studio: RFR101A1M
GSM modem	GPS tracking, telemetry	SparkFun: CEL-09533
Audio Modules		
Audio power amps	Driving speakers	SparkFun: BOB-11044
MP3 encoder/decoder/ player	Sound file playing	SparkFun: DEV-10628
Microphone with pre-amp	Sound detection; sound recording or DSP	SparkFun: BOB-09964
MIDI decoder	Musical instruments	SparkFun: BOB-08953
MP3 player	Playing sound samples	SparkFun: BOB-10608
Power Control		
H-bridge	Bidirectional motor control	SparkFun: ROB-09457, DEV-10182
Relay (wireless)	Wireless power control, home automation	Seeed Studio: WLS120B5B
Display Modules		
LCD alphanumeric	2- and 4-line by 16- or 20-character displays	SparkFun: LCD-00255
		Seeed Studio: LCD108B6B
LCD graphical	128 × 64 up to 320 × 240 pixel displays, color and grayscale	SparkFun: LCD-00569, LCD-10089
		Seed Studio: LCD101B6B, LCD105B6B
OLED displays	Bright color displays	Spark Fun: LCD-09678
		Seeed Studio: OLE42178P
LED matrix displays (serial interface)	Large color displays and signs	SparkFun: COM-00760

(Continued)

TABLE 17.2 Some Modules and Breakout Boards (*Continued*)

MODULE	USAGE	SOURCES
Sensor Modules		
Humidity sensor	Weather stations, humidity control	SparkFun: SEN-10239
		Seeed Studio: SEN111A2B
Compass	Direction finding	SparkFun: SEN-07915
		Seeed Studio: SEN101D1P
Magnetometer	Measurement of magnetic field strength	SparkFun: SEN-00244
Color sensor	Measurement of light color, for example in industrial control	SparkFun: SEN-10904
		Seeed Studio: SEN60256P
Temperature sensor	Digital thermometers and thermostats	SparkFun: SEN-09418
		Seeed Studio: SEN01041P

straightforward as with more typical analog and digital design. For this reason, you will often find ready-to-use RF modules that can either be soldered onto your own PCB as part of a bigger design or plugged into socket headers.

Figure 17.1 shows a variety of RF modules:

- A 433-MHz receiver and transmitter pair (A)
- A TEA5767 FM receiver board (B)
- A Bluetooth modem, which is actually one module on top of another, with the larger one providing level conversion (C)
- An XRF wireless serial module that fits a standard XBee socket (D)

These represent just a small subset of the wide range of RF modules available to the inventor (see Table 17.2 for a more complete list).

433-MHz and 315-MHz Modules

The 433-MHz and 315-MHz modules (Fig. 17.1A) are very low cost. They find their way into all sorts of consumer electronics requiring remote control, such as wireless door bells, remote car unlocking, smart meters, and remote-control toys.

RF modules

FIGURE 17.1

The data rates are generally low (8 kb/s is usually the top end, and 2 kb/s is quite common), but power consumption is also low. Their maximum range is normally about 100 yards, but can be much less indoors.

These modules are usually separate transmitters and receivers, rather than a combined transceiver, and so they are generally used in situations where data is flowing in only one direction.

Bluetooth Modules

Bluetooth modules provide a great means of interfacing your electronics to a mobile phone. The module in Fig. 17.1C is actually a basic 3.3-V Bluetooth module designed to be surface mounted onto another board. In this case, it is mounted on a level changing board that allows the module to operate at 5-V TTL serial levels, but 3.3-V modules like this often find themselves incorporated into products. Again, economies of scale mean that these modules can be bought for just a few dollars if you look around.

XBee Modules

XBee is a proprietary standard of Digi International, but the sockets have been adopted for a range of radio link modules from many manufacturers. These modules, such as the XRF module from Ciseco Plc (see Fig. 17.2), act as a transparent serial interface between two devices. For example, one of the devices might be an Arduino with an XBee socket shield with an XRF module installed communicating with a low-power remote sensor with an XBee socket and another XRF module, as shown on the left in Fig. 17.2.

XRF radio module and wireless temperature sensor module

FIGURE 17.2

GSM/GPRS Modem Modules

The GSM/GPRS modem modules are essentially cellular phone modules that provide most of the features of a cellular phone, including sending Short Message Service (SMS) text messages and Global System for Mobile Communications (GSM)/General Packet Radio Service (GPRS) data. Serial commands are sent to the module from a microcontroller to control its operation.

17.2.2 Audio Modules

Although audio electronics is not as difficult as RF electronics, you still need to be careful to avoid earth loops and the dreaded 60-Hz hum that all too easily finds its way on to the signal path given half a chance.

Audio power amp module

FIGURE 17.3

You will find a good selection of audio amplifier modules such as the class D amplifier based on the XMA2012, which is a 2 × 3 W power amp IC (see Fig. 17.3).

This kind of module is very convenient, as it comes with screw terminals for power and speakers and a socket for audio input. Economies of scale make the cost of such modules low.

As well as power amp modules, you will also find audio modules for pre-amps, MP3 players, MIDI interfaces, and tone controls.

17.3 Plug-and-Play Prototyping

When taken to its extreme, the modular approach can result in a system such as .NET Gadgeteer that allows complete plug-and-play development of a prototype (see Fig. 17.4). This system is based on attaching a wide range of sensors and other modules to a microcontroller main board using plug-in cables.

The .NET Gadgeteer system

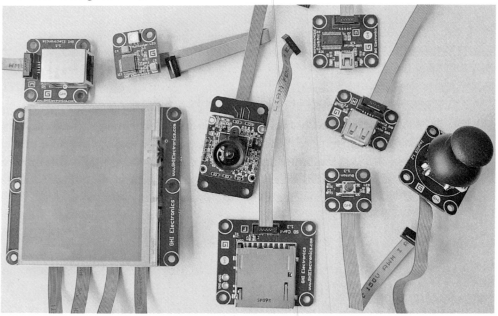

FIGURE 17.4

Connecting modules in Visual Studio

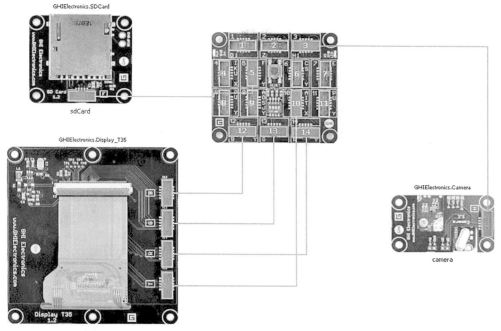

FIGURE 17.5

.NET Gadgeteer is programmed using Microsoft Visual Studio. When used with .NET Gadgeteer, this IDE includes a graphical editor that generates much of the boilerplate code, simply by connecting modules in the design window (see Fig. 17.5).

New modules for the .NET Gadgeteer are being developed all the time. The following are most of the modules available at the time of writing:

- Accelerometer
- Barometer
- Bluetooth
- Buttons
- Camera
- CAN (vehicle engine management unit)
- Cellular radio/GSM/GPRS modem
- Compass
- Current measurement (for smart meters and energy monitors)
- Display, touch screen LCD
- GPS
- Gyroscope
- Joystick
- LED, multicolor
- Light sensor
- Moisture sensor

- Motor driver
- MP3 music player
- OLED display
- Potentiometer
- Pulse oximeter (heart pulse rate measurement)
- Relay
- SD card
- USB interface
- Video output
- Wi-Fi
- XBee

The module manufacturers also supply main boards of different sizes and specifications. For up-to-date lists, see the websites of the main suppliers of .NET Gadgeteer hardware: GHI Electronics, Seeed Studio, Sytech Designs, and DFRobot.

If you want to learn more about using .NET Gadgeteer, the book *Getting Started with .NET Gadgeteer* by Simon Monk (Make, 2012) is a good place to start.

17.4   Open Source Hardware

Open source software has been with us for many years now, but open source hardware is a relatively new concept. The term *open source* really applies to the design files for the hardware, and means that the schematic and PCB design are made publicly available (usually in EAGLE CAD format). The implications of this are that anyone is free to take those design files and produce their own boards using the design. Often, the originator of a new piece of open source hardware will also manufacture the boards and sell them directly or through distributors. The originator will generally be known in the community, and these will be considered to be the "original" boards and therefore the most valuable. Copies and clones may appear if the design is successful, but the original boards are likely to remain the most used.

Perhaps the most successful open source hardware design is the Arduino (see Chap. 13). All the Arduino boards and most of the shields available for the Arduino are released under an open source or creative commons license. Other well-known open source hardware projects include the following:

- **BeagleBoard**   A single-board computer
- **MIDIbox**   MIDI music hardware
- **Monome**   A button and LED grid for controlling virtual synthesizers
- **Ultimaker, RepRap, and MakerBot**   3D printer designs
- **Chumby**   An embedded computer
- **Open EEG**   A medical device

There are also many small modules that are released as open source designs.

So, why would anyone want to give away their ideas for free for the world to use? Well, for one thing, many creative people make things for the fun of making them and not to get rich. There is a world of difference between having a good idea and building a business. If you want people to see and use what you have done, then why not release it into the wild? As with open source software, there are also opportunities for businesses to develop with the so called "halo effect," providing consultation and support for the hardware.

Power Distribution and Home Wiring

A.1 Power Distribution

Figure A.1 shows a typical power-distribution system found in the United States (region in California). The voltages listed are sinusoidal and are represented by their rms values. Note that the system in your area may look a bit different from the system shown here. Contact your local utility to learn about how things are set up in your region.

As a note, ac is used in electrical distribution instead of dc because ac can be stepped up or down easily by using a transformer. Also, over long distances, it is more efficient to send electricity via high-voltage/low-current transmission lines. By reducing the current, less power is lost to resistive heating during transmission (lowering I within $P = I^2R$ lowers P). Once electricity reaches a substation, it must be stepped down to a safe level before entering homes and businesses.

Notice that industry typically uses three-phase electricity. The natural sequencing of the three phases is particularly useful for devices that perform rhythmic tasks. For example, three-phase electric motors (found in grinders, lathes, welders, air conditioners, and other high-power devices) often turn in near synchrony with the rising and falling voltages of the phases. Also, with three-phase electric power, there is never a time when all three phases are at the same voltage. With single-phase power, whenever the two phases have the same voltage, there is temporarily no electric power available. This is why single-phase electric devices must store energy to carry them over these dry spells. In three-phase power, a device can always obtain power from at least one pair of phases.

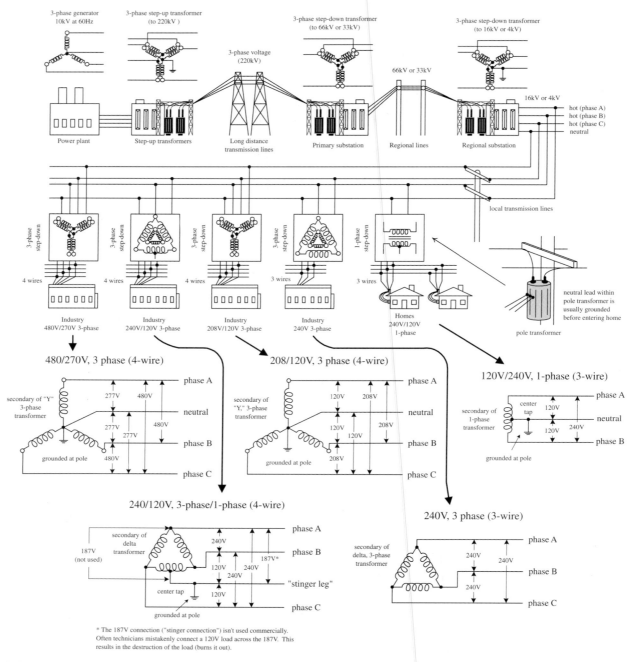

FIGURE A.1

A.2   A Closer Look at Three-Phase Electricity

The simple generator shown in Fig. A.2 can be used to generate single-phase voltage. As the magnet is rotated by a mechanical force, a voltage is induced within the two coils (spaced 180° apart) that yields a single sinusoidal voltage. The output voltage is usually expressed as an rms voltage ($V_{rms} = 1/\sqrt{2}\ V_0$).

In the three-phase generator, three separate voltages are generated by using three different coils spaced 120° apart. As the magnet rotates, a voltage is induced across each of the generator's coils. All the coil voltages are equal in magnitude but are 120°

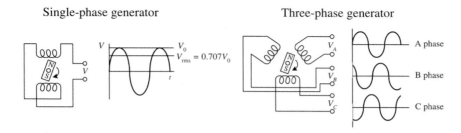

Single-phase generator

Three-phase generator

FIGURE A.2

out of phase with each other. With this generator, you could power three separate loads of equal resistance, or you could drive a three-phase motor with a similar coil configuration; however, this requires using six separate wires. To reduce the number of wires, there are two tricks that can be used. The first trick involves rearranging the three-phase coil connections into what is called a *delta connection*—which yields three wires. The second trick involves rearranging the coil connection into what is called a *Y connection*—which yields four wires. Here are the details.

Y Connection

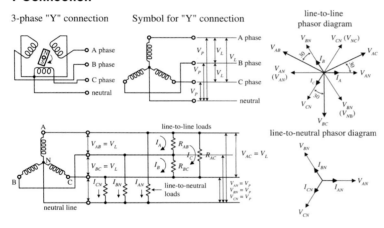

3-phase "Y" connection   Symbol for "Y" connection

line-to-line phasor diagram

line-to-line loads

line-to-neutral loads

line-to-neutral phasor diagram

neutral line

FIGURE A.3

A Y configuration is made by connecting one end of each of the generator's coils together to form what is called the *neutral* lead. The remaining three coil ends are brought out separately and are considered the "hot" leads. The voltage between the neutral and any one of the hot lines is called the *phase voltage V_p*. The total voltage, or *line voltage V_L* is the voltage across any two hot leads. The line voltage is the vector sum of the individual phase voltages. In a Y-loaded circuit, each load has two phases in series. This means that the current and voltage through and across a load must be determined by superimposing the phase currents and phase voltages. One way to do this is to make a phasor diagram, as shown in the figure. For practical purposes, what is important to note is the line voltage is about $\sqrt{3}$ times the phase voltage. Also, the line currents are 30° out of phase with the line voltages. When a neutral is used, the line currents and line voltages are equal to the individual phases from the generator. No current flows in the neutral when the loads are equal across each phase (balanced load).

Delta Connection

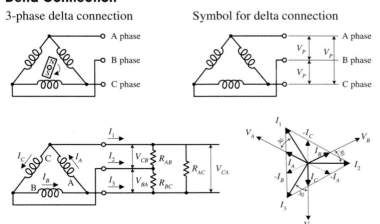

3-phase delta connection   Symbol for delta connection

FIGURE A.4

A delta connection is made by placing the three-phase generator's coils end to end, as shown in the figure. Because the delta configuration has no neutral lead, the phase voltages are equal to the line voltages. Again, like the Y configuration, all line voltages are 120° out of phase with each other. However, unlike the Y configuration, the line currents (I_1, I_2, I_3) are equal to the vector sum of the phase currents (I_A, I_B, I_C). When each of the phases are equally loaded, the line currents are all equal but 120° apart. The line currents are 30° out of phase with the line voltages, and they are $\sqrt{3}$ times the phase currents.

A.3 Home Wiring

In the United States, three wires run from the pole transformer (or "green box" transformer) to the main service panel at one's home. One wire is the A-phase wire (black in color), another is the B-phase wire (black in color), and the third is the neutral wire (white in color). (Figure A.5 shows where these three wires originate within the pole/green box transformer.) The voltage between the A-phase and B-phase wires, or the "hot to hot" voltage, is 240 V, while the voltage between the neutral wire and either A-phase or B-phase wire, or the "neutral to hot" voltage, is 120 V. (These voltages are nominal and may vary from region to region.)

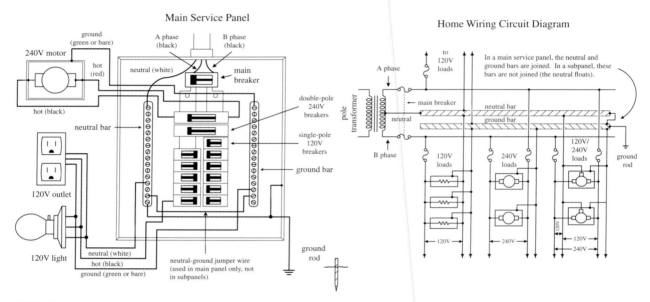

FIGURE A.5

At the home, the three wires from the pole/green box transformer are connected through a wattmeter and then enter a main service panel that is grounded to a long copper rod driven into the ground or to the steel in a home's foundation. The A-phase and B-phase wires that enter the main panel are connected through a main disconnect breaker, while the neutral wire is connected to a terminal referred to as the *neutral bar* or *neutral bus*. A *ground bar* also may be present within the main service panel. The ground bar is connected to the grounding rod or to the foundation's steel supports.

Within main service panels, the neutral bar and the ground bar are connected together (they act as one). However, within subpanels (service panels that get their power from the main service panel but which are located some distance from the main service panel), the neutral and ground bars are not joined together. Instead, the subpanel's ground bar receives a ground wire from the main service panel. Often the metal conduit that is used to transport the wires from the main service panel to the subpanel is used as the "ground wire." However, for certain critical applications (e.g., computer and life-support systems), the ground wire probably will be included within the conduit. Also, if a subpanel is not located in the same building as the main panel, a new ground rod typically is used to ground the subpanel. Note that different regions within the United States may use different wiring protocols. Therefore, do not assume that what we are telling you is standard practice where you live. Contact your local electrical inspector.

Within the main service panel, there are typically two bus bars into which circuit breaker modules are inserted. One of these bus bars is connected to the A-phase wire; the other bus bar is connected to the B-phase wire. To power a group of 120-V loads (e.g., upstairs lights and 120-V outlets), you throw the main breaker to the off position and then insert a single-pole breaker into one of the bus bars. (You can choose either the A-phase bus bar or the B-phase bus bar. The choice of which bus bar you use only becomes important when it comes to balancing the overall load—more on that in a second.) Next, you take a 120-V three-wire cable and connect the cable's black (hot) wire to the breaker, connect the cable's white (neutral) wire to the neutral bar, and connect the cable's ground wire (green or bare) to the ground bar. You then run the cable to where the 120-V loads are located, connect the hot and neutral wires across the load, and fasten the ground wire to the case of the load (typically a ground screw is supplied on an outlet mounting or light figure for this purpose). To power other 120-V loads that use their own breakers, you basically do the same thing you did in the last setup. However, to maximize the capacity of the main panel (or subpanel) to supply as much current as possible without overloading the main circuit breaker in the process, it is important to balance the number of loads connected to the A-phase breakers with the number of loads connected to the B-phase breakers. This is referred to as *balancing the load*.

Now, if you want to supply power to 240-V appliances (e.g., ovens, washers, etc.), you insert a double-pole breaker between the A-phase and B-phase bus bars in the main panel (or subpanel). Next, you take a 240-V three-wire cable and attach one of its hot wires to the A-phase terminal of the breaker and attach its other hot wire to the B-phase terminal of the breaker. The ground wire (green or bare) is connected to the ground bar. You then run the cable to where the 240-V loads are located and attach the wires to the corresponding terminals of the load (typically within a 240-V outlet). Also, 120-V/240-V appliances are wired in a similar manner, except you use a fourwire cable that contains an additional neutral (white) wire that is joined at the neutral bar within the main panel (or subpanel). (As a practical note, you could use a fourwire 120-V/240-V cable instead of a 240-V three-wire cable for 240-V applications—you would just leave the neutral wire alone in this case.)

As a note of caution, do not attempt home wiring unless you are sure of your abilities. If you feel that you are capable, just make sure to flip the main breaker off before you start work within the main service panel. When working on light fixtures, switches, and outlets that are connected to an individual breaker, tag that breaker with tape so that you do not mistakenly flip the wrong breaker when you go back to test your connections.

A.4   Electricity in Other Countries

In the United States, homes receive a 60-Hz, 120-V single-phase voltage, whereas industry typically receives a 60-Hz, 208-V/120-V three-phase voltage. Most other countries, on the other hand, work with a 50-Hz, 230-V single-phase voltage and a 415-V three-phase voltage. Now, if you were to take a U.S.-built 120-V, 60-Hz device over to Norway—where 230 V is used—and plug that device directly into the outlet (you would need an adapter to do this—their outlets look different), you run a risk of damaging the device. Some devices may not "care" about the voltage and frequency differences, but others will. You could use a converter (transformer plug-in device) to step down the voltage from the outlet, but you would still be stuck with 50 Hz. The 10-Hz difference will not affect most devices, but other devices, such as TVs and VCRs, may not function properly.

Here's a listing of single-phase voltages found in some countries. Note the plug types.

Country	Voltage V	Frequency Hz	Plug Type
Australia	240	50	I
Belgium	230	50	C, E
Brazil	110/220	60	A, B, C, D, G
Canada	120	60	A, B
Chile	220	50	C, L
China	220	50	I
Congo	230	50	C, E
Costa Rica	120	60	A, B
Egypt	220	60	C
France	230	50	C, E, F
Germany	230	50	F
Hong Kong	230	50	D, G
India	230	50	C, D
Iraq	220	50	C, D, G
Italy	127/220	50	F, L
Japan	100	50/60	A, B
Korea	110/220	60	A, B, D, G, I, K
Mexico	127	60	A
Netherlands	230	50	C, E
Norway	230	50	C, F
Philippines	110/220	60	A, B, C, E, F, I
Russia & former Soviet Republics	220	50	C, F
Spain	127/220	50	C, E
Switzerland	220	50	C, E, J
Taiwan	110	60	A, B, I
US	120	60	A, B
United Kingdom	230	50	G

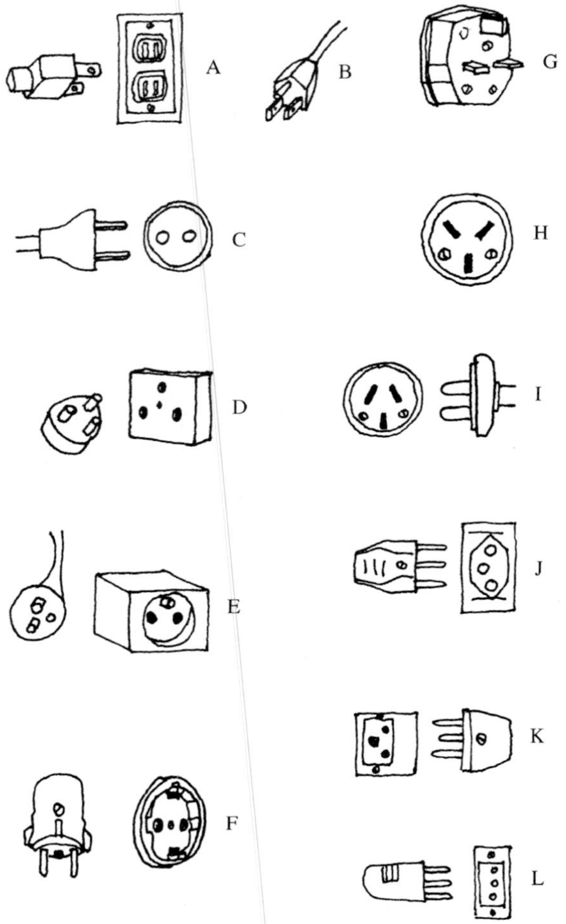

FIGURE A.6

APPENDIX B

Error Analysis

Reliability estimates of measurements greatly enhance their value. For example, saying the resistance of a resistor is $1000\ \Omega \pm 50\ \Omega$ tells you much more than simply stating the resistance is $1000\ \Omega$.

The term *error* is basically interchangeable with *uncertainty*, but does not have the same meaning as *mistake*. Mistakes, such as errors in calculations, should be corrected before estimating the experimental error. In estimating the reliability of a single quantity (such as resistance), you should recognize three different kinds of sources of error:

1. Actual variations of the quantity being measured, for example, resistance changes due to temperature variations. In electronics, for accurate measurements, you should consider the temperature ratings and the corresponding errors specified in the data sheets.

2. Test equipment in error. The errors introduced by test equipment. Make sure that all equipment is calibrated, and be certain to take into account input impedance characteristics, such as the input impedance of a multimeter and oscilloscope.

3. Human error. With digital equipment displays this is less of a problem. A scope with a graphical display will not give you high accuracy (only around 5 percent).

B.1 Absolute Error, Relative Error, and Percent Error

If Δx is the *absolute error* (or *uncertainty* with a $\pm$ in front) in a measurement whose value is x, then $\Delta x / x$ is called the *relative error* (or *fractional uncertainty*). If we multiply the relative error by 100 percent, we get the *percent error*, or $100\% \cdot (\Delta x / x)$. The term *tolerance* is interchangeable with both absolute error and percent error. For example, tolerances in length measurements are typically given in terms of absolute error values, while tolerances in resistances are typically given in percent error.

Example 1: What's the relative error and percent error of 0.125 A ± 0.01 A?

Answer:

$$\text{Relative error} = \frac{\Lambda x}{x} = \frac{0.01 \text{ A}}{0.125 \text{ A}} = 0.08$$

$$\text{Percent error} = 100\% \frac{\Lambda x}{x} = 100\% \times \text{Relative error} = 100 \times 0.08 = 8\%$$

Example 2: What are the relative error and the absolute error or uncertainty of a 3300-Ω resistor with a tolerance of 5 percent? What's the guaranteed range of resistance?

Answer: Here, tolerance represents percent error, so

$$\text{Relative error} = \frac{\text{Percent error}}{100\%} = \frac{\text{Tollerance}}{100\%} = \frac{5\%}{100\%} = 0.05$$

The absolute error or uncertainty is:

$$\Delta x = x \text{ (relative error)} = (3300 \text{ } \Omega)(0.05) = \pm 165 \text{ } \Omega$$

The resistor is guaranteed for 3300 Ω ± 165 Ω or guaranteed to be between 3135 Ω and 3465 Ω.

B.2 Uncertainty Estimates

When dealing with equations with many independent variables, like the following RC charge response equation

$$I = \frac{V_C}{R} e^{-t/RC}$$

the uncertainty or error in the final result (e.g., current) will depend on the individual uncertainties (e.g., uncertainties in resistance, capacitance, voltage, and time). The propagation of errors can be explained by first analyzing simple arithmetic cases:

1. If the desired result is the *sum* or *difference* of two measurements, the absolute uncertainties add:

   Let Δx and Δy be the errors in x and y, respectively. For the *sum*, we have:

   $$z = x + \Delta x + y + \Delta y$$

   and a relative error of

   $$(\Delta x + \Delta y)/(x + y)$$

Since the signs of Δx and Δy can be opposite, adding the absolute values gives a pessimistic estimate of the uncertainty. If errors have a Gaussian distribution and are independent, they combine in quadrature (square root of the sum of the squares):

$$\Delta z = \sqrt{\Delta x^2 + \Delta y^2}$$

For the *difference* of two measurements, we obtain a relative error of:

$$(\Delta x + \Delta y)/(x - y)$$

which becomes very large if x is nearly equal to y. This is an important point to note; you must avoid designing experiments where two large quantities are measured and their difference obtained.

2. If the desired result involves *multiplying* or *dividing* measured quantities, then the relative uncertainty of the result is the sum of the relative errors in each of the measured quantities. The most pessimistic case corresponds to adding the absolute value of each term, since the Δx_i and Δx_i can be of either sign:

$$\Delta z = z \times \left[\sum_i \left| \frac{\Delta x_i}{x_i} \right| + \sum_i \left| \frac{\Delta y_i}{y_i} \right| \right]$$

Again, if the measurement errors are independent and have a Gaussian distribution, the relative errors will add in quadrature:

$$\Delta z = z \times \sqrt{ \sum_i \left(\left| \frac{\Delta x_i}{x_i} \right| \right)^2 + \sum_i \left(\left| \frac{\Delta y_i}{y_i} \right| \right)^2 }$$

3. If the desired result is a *power* of the measured quantity, the relative error in the result is the relative error in the measured quantity multiplied by the power. For example, the uncertainty in

$$z = x^n$$

is:

$$\Delta z = z \times n \left(\frac{\Delta x}{x} \right)$$

4. For anything more complex, the following equation gives the general method for finding uncertainty in measurements. For example, if $R = f(x, y, z)$ is the functional relationship between three measurements x, y, z, then

$$dR = \frac{\partial f}{\partial x} dx + \frac{\partial f}{\partial y} dy + \frac{\partial f}{\partial z} dz$$

gives the uncertainty in R when the uncertainties dx, dy, and dz are known.

Usually, you don't have to go to this extreme—you can usually get by with the rules for adding, subtracting, multiplying, dividing, and rising to the power. To make life easy on you, a cheat sheet is provided here.

FORMULAS TO CALCULATE UNCORRELATED ERRORS

If $A = \bar{A} \pm a$, $B = \bar{B} \pm b$, and $C = \bar{C} \pm c$, where $\bar{A}$, $\bar{B}$, $\bar{C}$ are the measured values of the quantities A, B, C, and a, b, c are the respective errors, then the calculated values with error (assuming independent variables and Gaussian distribution) give:

1. $A + B = \bar{A} + \bar{B} \pm \sqrt{a^2 + b^2}$
2. $A + B + C = \bar{A} + \bar{B} + \bar{C} \pm \sqrt{a^2 + b^2 + c^2}$
3. $A - B = \bar{A} - \bar{B} \pm \sqrt{a^2 + b^2}$
4. $A \times B = \bar{A} \times \bar{B} \pm (\bar{A} \times \bar{B}) \sqrt{(a/\bar{A})^2 + (b/\bar{B})^2}$
5. $A \times B \times C = \bar{A} \times \bar{B} \times \bar{C} \pm (\bar{A} \times \bar{B} \times \bar{C}) \sqrt{(a/\bar{A})^2 + (b/\bar{B})^2 + (c/\bar{C})^2}$
6. $A/B = (\bar{A}/\bar{B}) \pm (\bar{A}/\bar{B}) \sqrt{(a/\bar{A})^2 + (b/\bar{B})^2}$
7. $A^B = (\bar{A})^{\bar{B}} \pm (\bar{A})^{\bar{B}} \times \bar{B}(b/\bar{B})$

Example 1: The voltages across two series resistors are measured using two different voltmeters. One digital meter reads 6.24 V ± 0.01 V across the first resistor; a less precise analog meter measures 14.3 V ± 0.2 V. What is the total voltage across the pair, including uncertainty in the result?

Answer: We simply add voltages and use Eq. 1 to determine the resultant uncertainty:

$$V_1 + V_2 = 6.24 \text{ V} + 14.3 \text{ V} \pm \sqrt{(0.01 \text{ V})^2 + (0.2 \text{ V})^2} = 20.5 \text{ V} \pm 0.2 \text{ V}$$

Example 2: The current through a 180-Ω, 5 percent resistor is 1.256 A ± 0.005 A. Determine the voltage across the resistor with uncertainty included.

Answer: First convert the tolerance to absolute error (or uncertainty):

$$\Delta R = \frac{\text{Tolerance}}{100\%} R = \frac{5\%}{100\%} (180 \text{ } \Omega) = \pm 9 \Omega$$

Since the equation for voltage is Ohm's law, $V = I \times R$, we use Eq. 4:

$$V = I \times R = 1.256 \text{ A} \times 180 \text{ } \Omega \pm (1.256 \text{ A} \times 180 \text{ } \Omega) \sqrt{\left(\frac{0.005 \text{ A}}{1.256 \text{ A}}\right)^2 + \left(\frac{9 \text{ } \Omega}{180 \text{ } \Omega}\right)^2} = 226 \pm 11 \text{ V}$$

APPENDIX C

Useful Facts and Formulas

C.1 Greek Alphabet

Alpha	A	α	Eta	E	η	Nu	N	ν	Tau	T	τ
Beta	B	β	Theta	Θ	θ	Xi	Ξ	ξ	Upsilon	Y	υ
Gamma	Γ	γ	Iota	I	ι	Omicron	O	o	Phi	Φ	ϕ
Delta	Δ	δ	Kappa	K	κ	Pi	Π	π	Chi	X	χ
Epsilon	E	ε	Lambda	Λ	λ	Rho	P	ρ	Psi	Ψ	ψ
Zeta	Z	ζ	Mu	M	μ	Sigma	Σ	σ	Omega	Ω	ω

C.2 Powers of 10 Unit Prefixes

PREFIX	SYMBOL	MULTIPLYING FACTOR
tera	T	$\times 10^{12}$
giga	G	$\times 10^{9}$
mega	M	$\times 10^{6}$
kilo	k	$\times 10^{3}$
centi	c	$\times 10^{-2}$
milli	m	$\times 10^{-3}$
micro	μ	$\times 10^{-6}$
nano	n	$\times 10^{-9}$
pico	p	$\times 10^{-12}$

C.3 Linear Functions ($y = mx + b$)

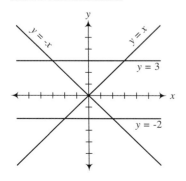

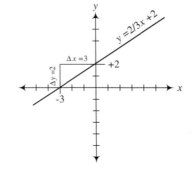

The equation $y = mx + b$ represents the equation of a line. The slope of the line ($\Delta y / \Delta x$) is equal to m, while the vertical shift, or point where the line crosses the y axis, is equal to b.

FIGURE C.1

C.4 Quadratic Equation ($y = ax^2 + bx + c$)

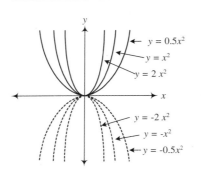

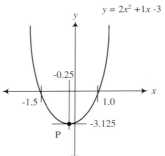

$y = 2x^2 + 1x - 3$

The equation $y = ax^2 + bx + c$ traces out a parabola in the xy plane. The narrowness of the parabola is influenced by a, the horizontal shift is given by $-b/2a$, and the vertical shift is given by $-b^2/a + c$. To determine the roots of the equation (points where the parabola crosses the x axis), use

$$x = \frac{-b \pm \sqrt{b^2 - 4ac}}{2a}$$

FIGURE C.2

C.5 Exponents and Logarithms

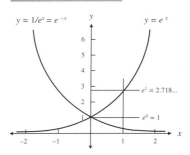

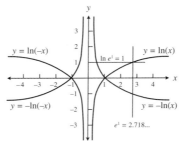

$x^0 = 1$ — Base 10: if $10^n = x$, then $\log_{10} x = n$

$1/x^n = x^{-n}$ — Base e: if $e^m = y$, then $\ln y = m$

$x^{1/n} = \sqrt[n]{x}$ — ($\log_{10} 100 = 2$, since $10^2 = 100$,

$x^m \cdot x^n = x^{m+n}$ — $\ln e = 1$, since $e^1 = e = 2.718\ldots$)

$(xy)^n = x^n \cdot y^n$ — Properties of any logarithm to the base b:

$(x^n)^m = x^{n \cdot m}$ — $\log_b 1 = 0$

$\log_b b = 1$

$\log_b 0 = \begin{cases} +\infty & b < 1 \\ -\infty & b > 1 \end{cases}$

$\log_b (x \cdot y) = \log_b x + \log_b y$

$\log_b (x/y) = \log_b x - \log_b y$

$\log_b (x^y) = y \log_b x$

FIGURE C.3

C.6 Trigonometry

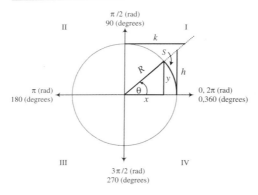

FIGURE C.4

The angle θ subtended by an arc S of a circle of radius R is equal to the ratio $\theta = S/R$, where θ is in radians. 1 radian = $180°/\pi = 57.296°$, while $1° = \pi/180° = 0.17453$ radian. If R is rotated counterclockwise from the positive x axis, θ is positive in sign. If R is rotated clockwise from the positive x axis, θ is negative in sign. The trigonometric functions of the angle θ are defined as specific ratios between the sides of the triangles shown in the figure and are expressed as

$$\sin\theta = \frac{y}{R} \quad \text{if } R = 1 \rightarrow y = \sin\theta$$

$$\cos\theta = \frac{x}{R} \quad \text{if } R = 1 \rightarrow x = \cos\theta$$

$$\tan\theta = \frac{y}{x} \quad \text{if } R = 1 \rightarrow h = \tan\theta$$

$$\cot\theta = \frac{x}{y} = \frac{1}{\tan\theta} \quad \text{if } R = 1 \rightarrow k = \cot\theta$$

$$\sec\theta = \frac{R}{x} = \frac{1}{\cos\theta} \quad \text{if } R = 1 \rightarrow \frac{1}{x} = \sec\theta$$

$$\csc\theta = \frac{R}{y} = \frac{1}{\sin\theta} \quad \text{if } R = 1 \rightarrow \frac{1}{y} = \csc\theta$$

Sine and Cosine Functions

$y = A \sin x$

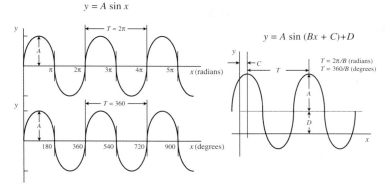

$y = A \cos(x)$

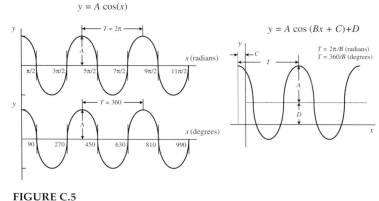

FIGURE C.5

The graph of $y = A \sin \theta$ is shown to the far left. To alter the vertical, horizontal, period, and phase of this function, alter the equation to the form $y = A \sin (Bx + C) + D$. A is amplitude, $2\pi/B$ is the period (T), C is the phase shift, and D is the vertical shift. In electronics, a voltage can be expressed as

$$V(t) = V_0 \sin (\omega t + \phi) + V_{dc}$$

V_0 is the peak voltage, V_{dc} is the dc offset, ϕ is the phase shift, and ω is the angular frequency (rad/s), which is related to the conventional frequency (cycles/s) by

$$f = \frac{1}{T} = \frac{\omega}{2\pi}$$

The graph of $y = A \cos x$ is shifted in phase by $\pi/2$ radians (or 90°) with respect to the graph $y = A \sin x$. The following relations show how the sine and cosine functions are related:

$$\sin\left(\frac{\pi}{2} \pm x\right) = +\cos x \quad \text{or} \quad \sin (90° \pm x) = +\cos x$$

$$\sin\left(\frac{3\pi}{2} \pm x\right) = -\cos x \quad \text{or} \quad \sin (270° \pm x) = +\cos x$$

$$\cos\left(\frac{\pi}{2} \pm x\right) = \pm\sin x \quad \text{or} \quad \cos (90° \pm x) = \pm\sin x$$

$$\cos\left(\frac{3\pi}{2} \pm x\right) = \pm\sin x \quad \text{or} \quad \cos (270° \pm x) = \pm\sin x$$

C.7   Complex Numbers

Complex numbers are covered in detail in Chap. 2.

C.8   Differential Calculus

Say you have a function $f(x)$. This function may represent a line, parabola, exponential curve, trigonometric curve, etc. Now pretend that you take a point and move it along the curve of $f(x)$. At the same time, you envision a tangent line touching the curve at the point. As the point moves along the curve, the slope of the tangent line changes ("teeter-totters")—provided the curve is not a line. Now, the slope of the tangent line has great significance in real-life situations. For example, if you graph a curve of the position of an object versus time, the slope at a particular time along the curve represents the instantaneous velocity of the object. Likewise, if you have an electrical charge versus time graph, the slope at time t represents the instantaneous current-flow. Now, the trick to finding the slope of a tangent line for any point along a curve involves using differential calculus. What differential calculus does is this: If you have a function, say, $y = x^2$—through the tricks of differential calculus—you can find another function, which is called the *derivative of y* (usually expressed as y' or dy/dx), that tells you the slope at every point along the curve of y. For $y = x^2$, the derivative is $dy/dx = 2x$. If you are interested in the slope of y at $x = 2$, you plug 2 into dy/dx to give you a slope of 4. But how do you find the derivative of $y = x^2$? Better yet, how do you find the derivative of any given function? The following provides the basic theory.

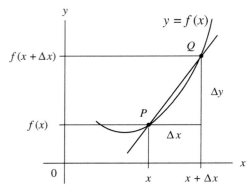

FIGURE C.6

To find the derivative of a function $y = f(x)$, you let $P(x,y)$ be a point of the graph of $y = f(x)$ and let $Q(x + \Delta x, y + \Delta y)$ be another point of the graph. The slope of the line between P and Q is then simply

$$\frac{f(x + \Delta x) - f(x)}{\Delta x}$$

Now you substitute the function into the preceding equation. For example, if $f(x) = x^2$, then $f(x + \Delta x) = (x + \Delta x)^2$, and the whole expression equals $[(x + \Delta x)^2 - x^2]/\Delta x$. Next, you hold x fixed and let Δx approach zero. If the slope approaches a value that depends only on x, you call this the slope of the curve at point P. The slope of the curve at P is itself a function of x, defined at every value of x at which the limit exists. You denote the slope by $f'(x)$ ("f prime"), or dy/dx ("dydx"), or df/dx ("dfdx"), and call any one of these terms (your choice) the derivative of $f(x)$:

$$f'(x) = \frac{dy}{dx} = \lim_{\Delta x \to 0} \frac{f(x + \Delta x) - f(x)}{\Delta x}$$

For the function $f(x) = x^2$, after carrying out the limit, you would get $f'(x) = dy/dx = 2x$ as the derivative.

Now, in practice, if you need to find the derivative of a function, you do not bother using the preceding equation. To do so would be very time-consuming and might require a number of nasty mathematical tricks, especially if you are trying to find the derivative of a complex function such as $2e^x \sin(3x + 2)$. Instead, what you do is memorize a few simple rules and memorize a few simple derivatives. The table below shows some of the rules and simple derivatives that will come in handy for many applications. In the table, a and n are constants, while u and v are functions.

DERIVATIVE	EXAMPLES
$\dfrac{d}{dx} a = 0$	$\dfrac{d}{dx} 4 = 0$
$\dfrac{d}{dx} x^n = nx^{n-1}$ (note: $\dfrac{1}{x^n} = x^{-n}$)	$\dfrac{d}{dx} x = 1, \dfrac{d}{dx} x^2 = 2x, \dfrac{d}{dx} x^5 = 5x^4, \dfrac{d}{dx} x^{-1/2} = -\tfrac{1}{2} x^{-3/2}$
$\dfrac{d}{dx} e^x = e^x$	
$\dfrac{d}{dx} \ln x = \dfrac{1}{x}$	
$\dfrac{d}{dx} \sin x = \cos x$	
$\dfrac{d}{dx} \cos x = -\sin x$	
$\dfrac{d}{dx} au(x) = a\dfrac{d}{dx} u(x)$	$\dfrac{d}{dx} 3x^2 = 3\dfrac{d}{dx} x^2 = 6x, \dfrac{d}{dx} 3e^x = 3e^x, \dfrac{d}{dx} 7\sin x = 7\cos x$
$\dfrac{d}{dx} (u + v) = \dfrac{du}{dx} + \dfrac{dv}{dx}$	$\dfrac{d}{dx} (2x + x^2) = \dfrac{d}{dx} (2x) + \dfrac{d}{dx} (x^2) = 2 + 2x$
$\dfrac{d}{dx} \left(\dfrac{u}{v}\right) = \dfrac{v \, du/dx - u \, dv/dx}{v^2}$	$\dfrac{d}{dx} \left(\dfrac{x^2 + 1}{x^2 - 1}\right) = \dfrac{(x^2 - 1) \cdot 2x - (x^2 + 1) \cdot 2x}{(x^2 - 1)^2} = \dfrac{-4x}{(x^2 - 1)^2}$
Chain rule: If u is a function of v, and v is in turn a function of x, then	
$\dfrac{d}{dx} (u[v(x)]) = \dfrac{du}{dv} \cdot \dfrac{dv}{dx}$	$\dfrac{d}{dx} \sin(ax) = a\cos(ax), \dfrac{d}{dx} e^{2x} = 2e^{2x}$

C.9 Integral Calculus

In differential calculus, our goal was to find the derivative of a function. In integral calculus, our goal is to find the *functions of a derivative*. In reality, calling one thing a derivative and another thing a function is somewhat confusing because a derivative is usually a function in itself. To avoid possible complications, we simply call anything that's written as $y = f(x)$ a function, call anything that's written as $dy/dx = df(x)/dx$ a derivative, and call anything that's written as $\int dy = \int f(x)dx$ an antiderivative or integral. In the last case, the "$\int$" is called the integral sign, the function $f(x)$ is called the integrand, and dx is called the variable of integration. To integrate a function is to find all the functions that have it as a derivative—to find all of the given function's antiderivatives. But, the term *integration* has a second, less technical meaning—to give the sum total of. In this meaning, integration represents a mathematical process that enables us to calculate the area of a region that has curved boundaries (see graph in Fig. C.7).

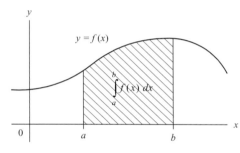

FIGURE C.7

If we're given an equation $dy/dx = f(x)$ and wish to find y, we must integrate (antidifferentiate). In this case, we first rearrange the equation into the form $dy = f(x)\,dx$. Next, we integrate both sides:

$$\int dy = \int (x)\,dx$$

$$y \pm C = \int f(x)\,dx$$

$$y = \int f(x)\,dx + C$$

Again, $\int$ is an integral sign, the function $f(x)$ is the integrand, and C is the constant of integration. We guessed that $\int dy = y \pm C$, since if we were to take the derivative of any function of the form $y \pm C$ ($y + 2$, $y - 54$, etc.) we'd get y as the answer—we worked backward. Since C could be positive or negative, it's arbitrary to give it a sign. This way, we simply add C to the left. In this form, we call the integral an indefinite integral.

Example: Given $dy/dx = 2$, find y.

$$dy = 2dx$$

$$\int dy = \int 2\,dx$$

$$y = 2x + C$$

In real-life situations, it's typically undesirable to have a constant in our answers, since we're looking for a definite result. To get rid of the constant, we apply boundary conditions. For example, if we take our last example $dy/dx = 2$, we say we're only interested in values of dy/dx from, say, 1 to 5—whatever range is appropriate for the situation. In that case, we take the definite integral

$$y = \int_a^b f(x)\,dx = F(x)\,|_a^b = F(b) - F(a)$$

Without getting too technical, F represents the definite integral without the constant sign, while the (a) and (b) terms mean to place the boundary points into the x term of F. If we consider our example $dy/dx = 2$ with boundary condition 1 to 5, we'd get:

$$y = \int_1^5 2\,dx = 2x\,|_1^5 = 2(5) - 2(1) = 8$$

We call this a definite integral. At this point, it's worth introducing a graphical approach to visualizing integration. If we consider $dy/dx = 2$, and graph dy/dx versus x, we get a straight horizontal line at $dy/dx = 2$ for all values of x. By summing up the area under the curve, we get the integral. So incorporating the boundary conditions the area is $(5 - 1) \times 2 = 8$.

Now, if we attempt to integrate more complex functions, we'll have a hard time guessing what the functions of the derivative are. Instead, we must apply some special tricks along the way. These tricks stem from a basic idea called the *fundamental theorem of calculus*, and they get fairly involved. Presenting the theory in such a short space isn't going to do. However, in practice, you typically never use the fundamental theorem of calculus to find integrals of functions. Instead, you memorize a few fundamental integrals and learn a few tricks. The following list highlights some of the most common integrals and solutions. A mathematical handbook will provide a more extensive list. In the list, a *u* represents a function of *x*, while *v* represents another function of *x*.

$$\int dx = x + C$$

$$\int a\, f(x)\,dx = a\int f(x)\,dx$$

$$\int (du(x) \pm dv(x)) = \int du(x) \pm \int dv(x)$$

$$\int u\,dv = uv - \int v\,du \quad \text{(integration by parts)}$$

$$\int u^n\, du = \frac{u^{n+1}}{n+1} + C \qquad (n \neq 1)$$

$$\int \frac{1}{u}\, du = \ln u + C$$

$$\int e^u\, dx = e^u + C$$

$$\int \ln x\, dx = x\ln x - x + C$$

$$\int \sin x\, dx = -\cos x + C$$

$$\int \cos x\, dx = \sin x + C$$

$$\int x\sin x\, dx = \sin x - x\cos x + C$$

$$\int x\cos x\, dx = \cos x + x\sin x + C$$

$$\int \sin x \cos x = \tfrac{1}{2}\sin^2 x + C$$

INDEX